The Twenty Amino Acids with Abbreviations and Messenger RNA Code Designations

Amino acid	One letter symbol	Three letter symbol	mRNA code designation
alanine	A	ala	GCU, GCC, GCA, GCG
arginine	R	arg	CGU, CGC, CGA, CGG, AGA, AGG
asparagine	P	asn	AAU, AAC
aspartic acid	D	asp	GAU, GAC
cysteine	C	cys	UGU, UGC
glutamic acid	E	glu	GAA, GAG
glutamine	Q	gln	CAA, CAG
glycine	G	gly	GGU, GGC, GGA, GGG
histidine	H	his	CAU, CAC
isoleucine	I	ile	AUU, AUC, AUA
leucine	L	leu	UUA, UUG, CUU, CUC, CUA, CUG
lysine	K	lys	AAA, AAG
methionine	M	met	AUG
phenylalanine	F	phe	UUU, UUC
proline	P	pro	CCU, CCC, CCA, CCG
serine	S	ser	UCU, UCC, UCA, UCG, AGU, AGC
threonine	T	thr	ACU, ACC, ACA, ACG
tryptophan	W	trp	UGG
tyrosine	Y	tyr	UAU, UAC
valine	V	val	GUU, GUC, GUA, GUG

Base Pairing (Hydrogen Bonding) in Deoxyribonucleic Acid

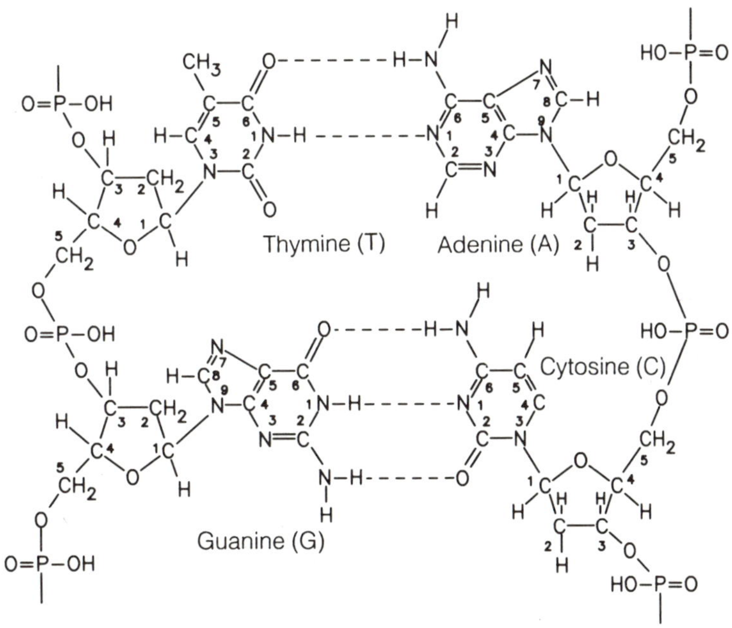

Encyclopedia of Molecular Biology and Molecular Medicine

VCH

Editorial Board

Encyclopedia of Molecular Biology and Molecular Medicine

Volume 6

Tandemly Repeated Noncoding DNA Sequences to
Zinc Finger DNA Binding Motifs

Edited by
Robert A. Meyers

VCH

Weinheim • New York • Basel • Cambridge • Tokyo

Robert A. Meyers
3715 Gleneagles Drive
Tarzana, CA 91356, USA

Management Supervised by:
Chernow Editorial Services, Inc.,
1133 Broadway, Suite 721, New York, NY, USA

Cover art courtesy of Dr. Zuzana Hostomska and Dr. Zdenek Hostomsky from Figure 6 of their article, "AIDS HIV Enzymes, Three-Dimensional Structures of."
Art prepared by David A. Matthews.

Library of Congress Cataloging-in-Publication Data
Encyclopedia of molecular biology and molecular medicine / edited by
Robert A. Meyers.
p. cm.
Contents: v. 1. Achilles' cleavage to cytoskeleton-plasma membrane interactions
Includes bibliographical references and index.
ISBN 3-527-28478-8 (set). — ISBN 3-527-28471-0 (v. 1 : alk. paper)
1. Molecular biology—Encyclopedias. 2. Molecular genetics—Encyclopedias. 3. Biotechnology—Encyclopedias. I. Meyers, Robert A. (Robert Allen), 1936–
QH506.E534 1996
611′.01816—dc20 95-44980
CIP

Updates to Library of Congress CIP Data for Volume 6:
Contents: v. 6. Tandemly repeated noncoding DNA sequences to zinc finger DNA-binding motifs
ISBN 3-527-28476-1 (Volume 6)

Die Deutsche Bibliothek – CIP – Einheitsaufnahme

Encyclopedia of molecular biology and molecular medicine / [editorial board Werner Arber . . .]. - Weinheim ; New York ; Basel ; Cambridge ; Tokyo ; VCH.
NE: Arber, Werner [Hrsg.]

Vol. 6. Tandemly repeated noncoding DNA sequences to zinc finger DNA-binding motifs / ed. by Robert A. Meyers. - 1997
ISBN 3-527-28478-8
NE: Meyers, Robert A. [Hrsg.]

Printed in the United States of America

ISBN 3-527-28476-1 (Volume 6)
ISBN 3-527-28478-8 (Set)

Printing History:
10 9 8 7 6 5 4 3 2 1

Contents of Volume 6

Preface vii

Acknowledgments ix

Contributors xi

Tandemly Repeated Noncoding DNA Sequences 1
Wolfgang Stephan

Taxol® and Taxane Production by Cell Culture 10
Arthur G. Fett-Neto and Frank DiCosmo

Telomerase 17
Silvia Bacchetti and Calvin B. Harley

Telomeres, Mammalian 27
Andrew H. Palin and Christine J. Farr

Theoretical Molecular Biology 37
Andrzej K. Konopka

Tomatoes, Gene Alterations of 53
Belinda Martineau and William R. Hiatt

Trace Element Micronutrients 58
Robert J. Cousins

Transgenic Animal Modeling 63
Carl A. Pinkert, Michael H. Irwin, and R. Jeffrey Moffatt

Transgenic Animal Patents 74
W. Lesser

Transgenic Fish 84
Thomas T. Chen, Jenn-Kan Lu, Mike J. Shamblott, Clara M. Cheng, Chun-Mean Lin, Jane C. Burns, Renate Reimschuessel, N. Chatakondi, and Rex A. Dunham

Transgenic Livestock as Bioreactors 99
Michael A. Dalrymple and Ian Garner

Translation of RNA to Protein 108
R. A. Cox and H. R. V. Arnstein

Transport Proteins 125
Milton H. Saier, Jr.

Transposons in the Human Genome 133
Abram Gabriel

Tumor Suppressor Genes 138
Prem Mohini Sharma and S. R. Dev Sharma

Ultraviolet Radiation Damage to DNA 149
David L. Mitchell

Vaccine Biotechnology 165
Tilahun D. Yilma

Varicella-Zoster Virus and Cytomegalovirus Vaccines 172
Lawrence A. Hunt

Viral Envelope Assembly and Budding 183
Milton J. Schlesinger

Virology, Molecular 192
Andrew P. Rice

Viruses, DNA Packaging of 200
Philip Serwer

Viruses, Photodynamic Inactivation of 207
James L. Matthews and Millard M. Judy

Viruses, RNA Packaging of 217
Dennis H. Bamford

Virus-Receptor Interactions 223
Richard L. Crowell and Ruliang Xu

Vitamin Receptors 235
Krishnamurti Dakshinamurti

Vitamins, Structure and Function of 244
Donald B. McCormick

Whole Chromosome Complementary Probe Fluorescence Staining 253
Heinz-Ulrich G. Weier, Daniel Pinkel, and Joe W. Gray

Xenobiotic Metabolism 259
Bernard Testa

X-Ray Diffraction of Biomolecules 275
K. Ravi Acharya and Anthony R. Rees

Yeast, DNA Repair in 283
Errol C. Friedberg

Yeast Artificial Chromosome (YAC) Construction and Applications 291
Christine Loehrlein and Ronald W. Davis

Yeast Genetics 302
Fred Sherman

Zinc Finger DNA-Binding Motifs 327
Louise Fairall and John W. R. Schwabe

Glossary of Basic Terms 337
Cumulative Contents 339
Cumulative Contributors 347
Index 369

Preface

This is a six-volume encyclopedia of molecular genetics and the molecular basis of life, with a focus on molecular medicine, including genetic screening, gene therapy, molecular medicine, structural biology, and the technology and findings of the Human Genome Project. More than 90 percent of the encyclopedia articles discuss topics in human genetics or molecular medicine. Molecular biology is treated in the balance of the encyclopedia by articles covering each of the major animal and plant genomes, and by articles covering key biotechnology advances and basics. The molecular structure, interactions, and analyses of proteins, carbohydrates, lipids, and other biological components complete the coverage of the molecular basis of life. This is the most comprehensive, detailed treatment of molecular biology and molecular medicine to date.

The most active research areas are presented. Some examples of topics include PCR technology, evolution of genetic intelligence, genomic imprinting, oligonucleotides, genetic mutations in aging, transposons in the human genome, human repetitive elements, molecular chaperones, protein aggregation, protein folding, protein designs for the specific recognition of DNA, signal transduction by heterotrimeric G proteins, transgenic livestock as bioreactors, motor proteins, molecular biology of *Caenorhabditis elegans*, colon cancer, lung cancer, oncogenes, tumor suppressor genes, cystic fibrosis, molecular basis of AIDS and AIDS therapy (four articles), Huntington's disease, infectious disease testing by ligase chain reaction, immuno-PCR, biosensors, cytochrome P450, and gene transfer techniques for use in genetic vaccination.

A comprehensive treatment of the Human Genome Project is included. It consists of core articles that present overviews and perspectives (for example, nucleic acid sequencing techniques, gene distribution in the human genome, human linkage maps, physical maps of human chromosomes, cloned DNA markers, human disease gene mapping, and human genetic predisposition to disease). These core articles are supported by detailed treatments of the various mapping methods (for example, chromosome microdissection and microcloning, high speed DNA sequencing in ultrathin gels, gene mapping by fluorescence *in situ* hybridization, and robotics and automation in genome reseach). Articles on the genetic and physical mapping status of several of the human chromosomes are presented to give the reader an introduction to the results of the Human Genome Project, which will be a complete genetic and physical map of all 23 chromosomes. This is, of course, an early view, as only a small percentage of the human genome has been mapped at this time.

The overall objective of this publication is to provide a single source for the molecular basis of life, disease diagnosis, and therapy. In addition, the theory and techniques for understanding, modifying, manipulating, expressing, and synthesizing biological molecules are presented. The result is an encyclopedic reference work with coverage of the following subject areas:

- Nucleic acids,
- Structure determination techniques (DNA, RNA, and proteins),
- Purification and processing of DNA, RNA, and proteins,
- The Human Genome Project,
- Proteins, polypeptides, and amino acids,
- Lipids,
- Immunology and biomolecular interactions,
- Molecular biology of specific organisms and organ systems,
- Molecular biology of specific diseases,
- Biochemistry and pharmacology, and
- Biotechnology.

Our purpose is to provide a multivolume reference for the expanding number of scientists who will contribute to this field, many of whom, as in the past, will enter molecular biology research from majors or careers in animal, plant, and cell biology and medicine, as well as physics, chemistry, mathematics, and engineering. It is a useful source of information for preclinical and clinical research in industry and in hospitals and for physicians who will apply the advances in genetic medicine in their practices. Teachers and professors in schools and universities will utilize this six-volume set for course preparation, and students will find a comprehensive introduction to molecular biology and genetic medicine.

The nearly 300 articles that comprise the set are designed as *self-contained treatments*. Each article begins with a *key word section*, including definitions, to assist the scientist or student who is unfamiliar with the specific subject area. The *Encyclopedia* includes more than 1,900 key words, each of which is defined within the context of the particular scientific field covered by the article. Because the same word may have different meanings in different fields, some key words have more than one definition.

In addition to the key word definitions, the *glossary of basic terms* found at the back of each volume defines the most commonly used terms in molecular biology. These definitions, along with the reference materials printed on the inside front and back covers of each volume (the genetic code, the common amino acids, and the structures of the deoxyribonucleotides), should allow most readers to understand articles in the encyclopedia without referring to a dictionary, textbook, or other reference work.

Each article begins with a concise definition of the subject and its importance, followed by the body of the article and references for further reading. Each subject is presented on a first-principles basis, including detailed figures, tables, and drawings to elucidate atomic or structural features. Because of the self-contained nature of each article, some overlap among articles on related topics exists. Cross-references to related articles are provided to help the reader expand his or her range of inquiry.

The detailed *cumulative subject index* as well as the *complete list of articles* and a *compilation of key words* at the end of volume 6 provide the reader with additional tools for locating topics of particular interest.

The articles in the encyclopedia may be categorized as follows: (1) *core articles* that give perspective on the major disciplines (such as, DNA structure, DNA replication and transcription, translation of RNA to protein, genetics, human genetic predisposition to disease, molecular genetic medicine, peptides, energetics and regulation in biological catalysis of enzymes, bioprocess engineering, immunology, medicinal chemistry, and cancer); (2) *satellite core articles* that give perspectives on particularly active areas of research and importance (for example, the core article on cancer is supported by satellite core articles on oncogenes and tumor suppressor genes); and (3) *specific subject articles* on such topics as liver cancer, lung cancer, breast cancer, colon cancer, and retinoblastoma. Cross-references are given to related articles, so that articles on cancer would reference DNA in neoplastic-disease diagnosis; steroids, antitumor; and cancer chemotherapy, theoretical foundations of. Because the subjects are advancing so rapidly, sometimes a specific field is covered within core or satellite core articles, rather than in an individual subject article.

It must be admitted that in some cases readers at the beginning of the educational hierarchy, that is, university students with a year of college chemistry, physics, and biology, may need to refer to a dictionary of scientific terms to access a few of the articles. These users will find the extra effort well worthwhile. Also, readers at the top of the hierarchy—molecular biologists and bioengineers—who are experts in one or more discipline within molecular biology, might find some of the articles in their field of research to be less detailed than desired. However, these scientists will find the articles in related disciplines to be useful in planning experimentation and interpreting results within their lines of research. The hope is that the reader will find as complete and authoritative a coverage of the molecular basis of life as can be provided in a reference work.

Robert A. Meyers

Acknowledgments

PUBLISHER'S ACKNOWLEDGMENTS

The publisher gratefully acknowledges the contributions of the following people to the development of this reference work.

Abul K. Abbas
Harvard Medical School

David Abramson
Cornell Medical Center

Thomas O. Baldwin
Texas A&M University

M. J. Bishop
University of Cambridge School of Medicine

Pamela Bjorkman
California Institute of Technology

Kristina Bostrom
University of California School of Medicine

Lisa Brannon-Peppas
Eli Lilly and Company

Arnold L. Demain
Massachusetts Institute of Technology

Thomas Devlin
Hahnemann University School of Medicine

H. Dintzis
Johns Hopkins School of Medicine

Tim Donlon
Kapiolani Medical Center

William A. Eaton
National Institutes of Health

Gadi Fibich
University of California

Ian Garner
Pharmaceutical Proteins, Ltd.

James I. Garrels
Cold Springs Harbor Laboratory

Alison Goate
University of Washington School of Medicine

Paul Goodfellow
Washington University

Jay A. Gottfried
New York University

Barney Graham
Vanderbilt University Hospital

Michael D. Griswold
Washington State University

George Guilbault
University of New Orleans

Sarah Gurr
University of Oxford

Barry Hall
University of Rochester

Patrick Hess
National Health Labs

Veronica van Heyningen
Western General Hospital

John Hildebrandt
Medical University of South Carolina

George Alan Jeffrey
University of Pittsburgh

Dick Junghans
Deaconess Hospital

David Kaplan
Natick Research, Development, and Engineering Center

Brian Kay
University of North Carolina

Roger Keynes
University of Cambridge

Richard A. King
University of Minnesota

Claude B. Klee
National Cancer Institute

Ted Klein
E.I. du Pont de Nemours & Co.

Joseph R. Lakowitz
University of Maryland School of Medicine

Charles Langley
University of California

Jenifer Levini
Cornell University

Tim Lohman
Washington University School of Medicine

Mike McKeown
Salk Institute

L. McLaughlin
Boston College

Peter B. Moore
Yale University

Krishna Murthy
Temple University Medical School

R. H. Pain
Jožf Stefan Institute

Carl A. Pinkert
University of Alabama at Birmingham

Richard Porter
State University of New York, Stony Brook

Clive Pullinger
University of California

Thomas Quinn
Johns Hopkins University

Francesco Ramirez
Mount Sinai Hospital

Andrew Rice
Baylor College of Medicine

David Schlessinger
Washington University School of Medicine

Dennis Shields
Albert Einstein College of Medicine

Barton Slatko
New England Biolabs

Jonathan Smith
Rockefeller University

Gary Spedding
Butler University

Robert F. Steiner
University of Maryland, Baltimore County

Lorne B. Taichman
State University of New York, Stony Brook

M. Thompson
University of Toronto

Cary Thrall
Synchrom, Inc.

Terrence Tiersch
Louisiana State University

Alan Tomkinson
University of Texas Health Center at San Antonio

Jan Vilcek
New York University Medical Center

Stephen S. Wachtel
University of Tennessee College of Medicine

Steve Warren
Emory University School of Medicine

Jean Weissenbach
Institute Pasteur

Michael J. Welsh
University of Iowa College of Medicine

Christine A. White
Scripps Memorial Hospital

Timothy Wilcosky
Center for Epidemiologic and Medical Studies

Steven A. Williams
Smith College and University of Massachusetss

Emily S. Winn-Dean
Applied Biosystems Inc.

June Hsieh Wu
Chang-Gung Medical College

EDITOR'S ACKNOWLEDGMENTS

I would like to acknowledge Robert A. Meyers, Jr., Tamara H. Cochrane, and Kenneth Cochrane, who maintained the encyclopedia article and topical outline database that I used to monitor author manuscript scheduling and to correlate the contents to assure completeness of coverage. I would also like to acknowledge my wife, Ilene Meyers, who assisted in this effort in the areas of project planning and scheduling. And finally, I thank Madeleine Lewis, who so ably assisted me with author recruitment during the concluding year of manuscript preparation.

VCH Verlagsgesellschaft mbH, Weinheim, Germany

Hans-Joachim Kraus, Ph.D., *Executive Editor, Life Sciences*

VCH Publishers, Inc., New York, NY

Camille Pecoul, *Director, Production and Manufacturing*

George A. Vallasi and Michael P. Beaubien, *Managing Editors, Chernow Editorial Services, New York, NY*

Brenda Griffing, *Copy Editor*
Virginia Hobbs, *Indexer*

Contributors

K. Ravi Acharya
School of Biology and Biochemistry
University of Bath
Bath, BA2 7AY, United Kingdom
X-Ray Diffraction of Biomolecules

H. R. V. Arnstein
King's College, London
London WC2R 2LS, United Kingdom
Translation of RNA to Protein

Silvia Bacchetti
Cancer Research Group
and
Institute for Molecular Biology and Biotechnology
McMaster University
Hamilton, Ontario L8N 3Z5, Canada
Telomerase

Dennis H. Bamford
Institute of Biotechnology
and
Department of Biosciences Division Genetics
University of Helsinki
FIN-00014 Helsinki, Finland
Viruses, RNA Packaging of

Jane C. Burns
Department of Pediatrics
School of Medicine
University of California
San Diego, CA 92093, USA
Transgenic Fish

N. Chatakondi
Department of Fisheries and Allied Aquacultures
Auburn University
Auburn, AL 36849, USA
Transgenic Fish

Thomas T. Chen
Biotechnology Center
University of Connecticut
Storrs, CT 06269, USA
Transgenic Fish

Clara M. Cheng
Center for Marine Biotechnology
Maryland Biotechnology Institute
Baltimore, MD 21202, USA
Transgenic Fish

Robert J. Cousins
Food Science and Human Nutrition Department
Center for Nutritional Sciences
University of Florida
Gainesville, FL 32611, USA
Trace Element Micronutrients

R. A. Cox
National Institute for Medical Research
London NW7 1AA, United Kingdom
Translation of RNA to Protein

Richard L. Crowell
Department of Microbiology and Immunology
Allegheny University of the Health Sciences
Philadelphia, PA 19129, USA
Virus-Receptor Interactions

Krishnamurti Dakshinamurti
Department of Biochemistry and Molecular Biology
Faculty of Medicine
University of Manitoba
Winnipeg, Manitoba R3E OW3
Canada
Vitamin Receptors

Michael A. Dalrymple
PPL Therapeutics Ltd.
Roslin, Edinburgh
Scotland EH25 9PP, United Kingdom
Transgenic Livestock as Bioreactors

Ronald W. Davis
Department of Biochemistry
Beckman Center
Stanford University School of Medicine
Stanford, CA 94305, USA
Yeast Artificial Chromosome (YAC) Construction and Applications

Frank DiCosmo
Department of Botany
Institute of Biomedical Engineering
and
Department of Surgery
University of Toronto
Toronto, Ontario M5S 3B2, Canada
Taxol® and Taxane Production by Cell Culture

Rex A. Dunham
Department of Fisheries and Allied Aquacultures
Auburn University
Auburn, AL 36849, USA
Transgenic Fish

Louise Fairall
Medical Research Council
Laboratory of Molecular Biology
Cambridge CB2 2QH, United Kingdom
Zinc Finger DNA-Binding Motifs

Christine J. Farr
Department of Genetics
University of Cambridge
Cambridge CB2 3EH, United Kingdom
Telomeres, Mammalian

Arthur G. Fett-Neto
Department of Botany
University of Toronto
Toronto, Ontario M5S 3B2, Canada
Taxol® and Taxane Production by Cell Culture

Errol C. Friedberg
Laboratory of Molecular Biology
Department of Pathology
University of Texas Southwestern Medical Center
Dallas, TX 75235, USA
Yeast, DNA Repair in

Abram Gabriel
Department of Molecular Biology and Biochemistry
Rutgers University
Piscataway, NJ 08854, USA
Transposons in the Human Genome

Ian Garner
PPL Therapeutics Ltd.
Edinburgh EH25 9PP, United Kingdom
Transgenic Livestock as Bioreactors

Joe W. Gray
Cancer Genetics Program
Cancer Center
University of California
San Francisco, CA 94143, USA
Whole Chromosome Complementary Probe Fluorescence Staining

Calvin B. Harley
Geron Corporation
Menlo Park, CA 94025, USA
Telomerase

William R. Hiatt
Calgene, Inc.
Davis, CA 95616, USA
Tomatoes, Gene Alterations of

Lawrence A. Hunt
Department of Microbiology and Immunology
University of Louisville School of Medicine
Louisville, KY 40292, USA
Varicella-Zoster Virus and Cytomegalovirus Vaccines

Michael H. Irwin
Department of Comparative Medicine
University of Alabama at Birmingham
Birmingham, AL 35294, USA
Transgenic Animal Modeling

Millard M. Judy
Baylor Research Institute
Baylor University Medical Center
Dallas, TX 75226, USA
Viruses, Photodynamic Inactivation of

Andrzej K. Konopka
BioLingua Research, Inc.
Gaithersburg, MD 20879, USA
Theoretical Molecular Biology

W. Lesser
Department of Agricultural Economics
New York State College of Agriculture and Life Science
Ithaca, NY 14853, USA
Transgenic Animal Patents

Chun-Mean Lin
Biotechnology Center
University of Connecticut
Storrs, CT 06269, USA
Transgenic Fish

Christine Loehrlein
Department of Biochemistry
Beckman Center
Stanford School of Medicine
Stanford, CA 94305, USA
Yeast Artificial Chromosome (YAC) Construction and Applications

Jenn-Kan Lu
Biotechnology Center
University of Connecticut
Storrs, CT 06269, USA
Transgenic Fish

Belinda Martineau
Calgene, Inc.
Davis, CA 95616, USA
Tomatoes, Gene Alterations of

James L. Matthews
Baylor Research Institute
Baylor University Medical Center
Dallas, TX 75226, USA
Viruses, Photodynamic Inactivation of

Donald B. McCormick
Department of Biochemistry
Emory University School of Medicine
Atlanta, GA 30322, USA
Vitamins, Structure and Function of

David L. Mitchell
M.D. Anderson Cancer Center
Science Park, Research Division
University of Texas
Smithville, TX 78957, USA
Ultraviolet Radiation Damage to DNA

R. Jeffrey Moffatt
Department of Comparative Medicine and the Transgenic Animal/ES Cell Resource
University of Alabama at Birmingham
Birmingham, AL 35294, USA
Transgenic Animal Modeling

Andrew H. Palin
Department of Genetics
University of Cambridge
Cambridge CB2 3EH, United Kingdom
Telomeres, Mammalian

Daniel Pinkel
Department of Laboratory Medicine
Cancer Genetics Program
Cancer Center
University of California
San Francisco, CA 94143, USA
Whole Chromosome Complementary Probe Fluorescence Staining

Carl A. Pinkert
Department of Comparative Medicine and the Transgenic Animal/ES Cell Resource
University of Alabama at Birmingham
Birmingham, AL 35294, USA
Transgenic Animal Modeling

Anthony R. Rees
School of Biology and Biochemistry
University of Bath
Bath, BA2 7AY, United Kingdom
X-Ray Diffraction of Biomolecules

Renate Reimschuessel
Department of Pathology
Aquatic Pathobiology Center
University of Maryland at Baltimore
Baltimore, MD 21201, USA
Transgenic Fish

Andrew P. Rice
Division of Molecular Virology
Baylor College of Medicine
Houston, TX 77030, USA
Virology, Molecular

Milton H. Saier, Jr.
Department of Biology
University of California at San Diego
La Jolla, CA 92093, USA
Transport Proteins

Milton J. Schlesinger
Department of Molecular Microbiology
Washington University School of Medicine
St. Louis, MO 63110, USA
Viral Envelope Assembly and Budding

John W. R. Schwabe
Medical Research Council
Laboratory of Molecular Biology
Cambridge CB2 2QH, United Kingdom
Zinc Finger DNA-Binding Motifs

Philip Serwer
Department of Biochemistry
University of Texas Health Science Center
San Antonio, TX 78284, USA
Viruses, DNA Packaging of

Mike J. Shamblott
The Johns Hopkins University School of Medicine
Baltimore, MD 21287, USA
Transgenic Fish

Prem Mohini Sharma
Department of Medicine
University of California at San Diego
La Jolla, CA 92093, USA
Tumor Suppressor Genes

S. R. Dev Sharma
PrimeX General Imaging
Carlsbad, CA 92008, USA
Tumor Suppressor Genes

Fred Sherman
Department of Biochemistry and Biophysics
University of Rochester Medical Center
Rochester, NY 14642, USA
Yeast Genetics

Wolfgang Stephan
Department of Zoology
University of Maryland
College Park, MD 20742, USA
Tandemly Repeated Noncoding DNA Sequences

Bernard Testa
Institut de Chimie Thérapeutique
Section de Pharmacie
Université de Lausanne
CH-1015 Lausanne, Switzerland
Xenobiotic Metabolism

Heinz-Ulrich G. Weier
Division of Life Sciences
Lawrence Berkeley National Laboratory
Berkeley, CA 94720, USA
Whole Chromosome Complementary Probe Fluorescence Staining

Ruliang Xu
Department of Microbiology
School of Dental Medicine
University of Pennsylvania
Philadelphia, PA 19104, USA
Virus-Receptor Interactions

Tilahun D. Yilma
Department of Veterinary Pathology, Microbiology, and Immunology
International Laboratory of Molecular Biology for Tropical Disease Agents
University of California
Davis, CA 95616, USA
Vaccine Biotechnology

Encyclopedia of Molecular Biology and Molecular Medicine

T

Tandemly Repeated Noncoding DNA Sequences

Wolfgang Stephan

1 **The C-Value Paradox**

2 **Properties of Tandem Arrays**
- 2.1 Microsatellite Sequences
- 2.2 Minisatellite Sequences
- 2.3 Satellite Sequences

3 **Genetic Mechanisms of Copy Number Change**
- 3.1 DNA Turnover Processes
- 3.2 Amplification of Disease-Related Microsatellites

4 **The Selfish DNA Hypothesis**

5 **Dynamics of Tandem Arrays**
- 5.1 The General Model
- 5.2 Copy Number Dynamics of Tandem Arrays
- 5.3 Sequence Homogenization

6 **Data Analysis**
- 6.1 Microsatellites
- 6.2 Minisatellites
- 6.3 Satellites

7 **Population Processes**

Key Words

Copy Number Polymorphism A hallmark of tandemly repeated, noncoding DNA sequences referring to the observation that the copy number of an array is characteristically variable within a population or species.

Mechanisms of Copy Number Variation DNA turnover mechanisms such as gene amplification and deletion, and evolutionary processes such as genetic drift and natural selection on array size.

Selfish DNA DNA sequences such as tandemly repeated sequences that may be responsible for significant fitness losses in the organism but are maintained by their ability to replicate quickly within the genome.

Tandemly Repeated Noncoding DNA Localized arrays of nucleotide repeats found in eukaryote genomes and usually classified as microsatellites (small tandem arrays of very short repeat units), minisatellites (larger arrays of longer repeat units), and satellite DNA (very large arrays of typically long, complex repeat units).

Tandemly repeated, noncoding DNA sequences form a substantial fraction of the genomes of many eukaryotes. Much of the variation in genome size between different eukaryotic species, which bears little relation to differences in organismal complexity or the numbers of protein-coding genes (the "C-value paradox"), is due to these nongenic sequences. This class of DNA includes satellite DNA (very highly repetitive sequences), minisatellite DNA (moderately repetitive sequences), and microsatellite DNA (short tandem arrays). In many cases, tandem arrays of noncoding DNA sequences seem to be maintained solely by their ability to replicate quickly within the genome (the "selfish DNA" hypothesis). Their behavior can result in mutations that cause human genetic diseases (e.g., fragile X syndrome). Features of the organization of tandemly repeated DNA sequences in eukaryotic genomes, and their distribution in natural populations, reflect the mechanisms acting on selfish DNA.

1 THE C-VALUE PARADOX

The amount of DNA in an unreplicated haploid genome, such as that of the sperm nucleus, is called the genome size or, alternatively, C-value, because it is usually constant in any one species. In contrast to this relative constancy of genome size within species, great variation in C-values has been found among eukaryotic species. C-values in eukaryotes range from 8.8×10^6 base pairs (bp) to 6.9×10^{11} bp. The most dramatic differences are observed among unicellular organisms. For example, between *Saccharomyces cerevisiae* (yeast) and *Amoeba dubia,* variation in genome size is roughly 80,000-fold. Within a given taxonomic group, unicellular protists show the greatest range in genome size (20,000-fold). In contrast, the three amniote classes (mammals, birds, and reptiles) vary much less (1.3- to 4-fold). Most extensively studied eukaryote groups, whether plant, animal, or microorganism, show much greater variation in C-values than the amniotes. Protozoa exhibit a 1000-fold and unicellular algae a 3000-fold range, whereas most classes of multicellular animals and plants show variation of the order of 10- to 100-fold. Great variation in C-value is, therefore, the general rule in most eukaryote groups; the relative uniformity found in mammals, birds, and reptiles appears to be an exception.

Given the relative constancy of genome size within species, it is puzzling that eukaryotes exhibit huge interspecific differences in C-values that bear little relation to differences in organismal complexity, ploidy level, or the percentage of coding DNA (Table 1). For example, several unicellular protozoans possess more DNA than mammals; and the newt *Triturus cristatus* has around six times as much DNA as humans. This lack of correspondence between genome size and genetic information content has become known in the literature as the C-value paradox. The C-value paradox is particularly apparent in comparisons of closely related species, which, by definition, show similar levels of organismal complexity. Among protists, bony fishes, amphibia, and flowering plants, many sibling

Table 1 Haploid Genome Size and Coding DNA

Species	Haploid Genome Size (base pairs × 10^9)	Coding DNA (%)
Bacterium (*E. coli*)	0.004	100.0
Yeast (*Saccharomyces*)	0.009	69.0
Nematode (*Caenorhabditis*)	0.09	25.0
Fruit fly (*Drosophila*)	0.18	33.0
Human	3.4	27.0
Newt (*Triturus*)	19.0	4.5
Lungfish (*Protopterus*)	140.0	1.2
Flowering plant		
Arabidopsis	0.2	31.0
Fritillaria	127.4	0.02

species differ greatly in their C-values. For example, the ciliate *Paramecium caudatum* has about 45 times as much DNA as *P. aurelia.*

Since a species does not contain less DNA than the amount required for specifying its vital functions in inheritance, the observed variations in genome size must be due to DNA that is in excess to this requirement. Eukaryotes show approximately 50-fold variation in the number of protein-coding genes, ranging from about 3000 in yeast to approximately 150,000 in mammals. This 50-fold variation is much too small to account for the 80,000-fold differences in nuclear genome size. Likewise, the C-value paradox cannot be explained by the interspecific variation in gene length because the severalfold variation in the average size of protein-coding genes (largely caused by variation in intron length) is also far too small to account for the C-value differences between species. On the other hand, the (repetition) number of RNA-specifying genes (e.g., ribosomal RNA genes) and the haploid DNA content are positively correlated; and a positive correlation exists also between the number of regulatory sequences (e.g., for replication) and C-value. Since, however, RNA-specifying genes and regulatory elements constitute only a minute fraction of the eukaryotic genome, they cannot explain the great variation in genome size. Based on these observations it has been concluded that the interspecific differences in C-values are largely due to nongenic DNA. It has been estimated that the amount of nongenic DNA per haploid genome varies in eukaryotes from approximately 3.0×10^6 bp to more than 1.0×10^{11} bp. Nongenic DNA per genome makes up anything from less than 30% to almost 100% of the genome.

Classical studies of the kinetics of DNA reassociation showed that the eukaryote genome can be divided roughly into four fractions: foldback DNA (i.e., palindromic DNA sequences), highly repetitive DNA, middle-repetitive DNA, and single-copy DNA. The proportion of the genome consisting of repetitive sequences varies widely between taxa. In yeast, the amount of repetitive DNA is small (20%); in mammals, up to 60% of the DNA is repetitive. In plants, the amount of repetitive DNA can exceed 80%, and higher values than that have been registered. A large fraction of the repetitive DNA is noncoding and organized in tandem arrays. It has been found that much of the variation in C-value is due to these noncoding, tandemly repeated DNA sequences.

2 PROPERTIES OF TANDEM ARRAYS

Variation in the numbers and lengths of tandemly repeated units occurs on many different scales. According to these characteristics, three major classes of tandemly repeated, noncoding DNA sequences can be distinguished: micro-, mini-, and satellite DNAs.

2.1 Microsatellite Sequences

Arrays of short (2–5 bp) nucleotide repeats, found in vertebrate, insect, and plant genomes, are called microsatellite loci; at least 30,000 are present in the human genome, located in the euchromatin. Copy numbers are characteristically variable within a population, typically with mean array sizes on the order of 100 but with multiple array size classes distributed around the mean. The distribution of microsatellites across the genomes of extensively studied organisms (e.g., man, mouse, and rat) appears to be relatively uniform. However, a significant deficit of *polymorphic* $(CA)_n$ arrays (i.e., arrays in which the dinucleotide is repeated *n* times) has been found on the X chromosomes of these species. Seemingly contrasting evidence has been reported from several *Drosophila* species. By in situ hybridization to polytene chromosomes of *D. melanogaster,* more $(CA)_n$ loci (irrespective of their degree of polymorphism) have been detected on the X chromosome than on the autosomes. All autosomes show approximately equal levels of hybridization except chromosome 4, which has no detectable stretches of $(CA)_n$. Another striking feature is the lack of hybridization of the heterochromatic regions around the centromere. The same overall chromosomal pattern of hybridization was seen for other *Drosophila* species, including *D. simulans, D. hydei,* and *D. virilis.* The abundance of different repeat types varies between species. For example, the dinucleotide repeats $(CA)_n$ are very frequent in mammals, but rare in plants; and the trinucleotide repeat array $(CTG)_n$ is overrepresented in *Drosophila,* but less frequent in other species.

2.2 Minisatellite Sequences

Minisatellite sequences are tandem arrays of longer (> 15 bp) repeats, generally involving mean array lengths of 0.5–30 kilobase pairs (kb). They are found in euchromatic regions of the genome of vertebrates, fungi, and plants, and many minisatellites are also highly variable in array sizes. In contrast to hypervariable microsatellites, which are relatively uniformly dispersed throughout mammalian genomes (with the exception of X chromosome–autosome differences), the distribution of minisatellite loci across the genomes of some extensively studied species appears to be uneven. Hypervariable minisatellites have usually been found in chromosomal regions of high recombination rates. In humans, cattle, and pigs, most hypervariable loci have been localized to subtelomeric regions which are recombination proficient. In mice, the distribution of variable minisatellite loci seems to be more uniform, but the hypervariable minisatellites tend to be located also in recombination prone regions. The genomic distribution of all minisatellites (including the less variable loci) is currently not known, not in even in humans.

2.3 Satellite Sequences

Repeat units for satellites can be similar in length (2–30 bp) to micro- and minisatellites, or much larger (> 100 bp). They are typically organized as large (up to 10^8 bp) clusters in the heterochromatic regions of chromosomes, near centromeres and telomeres, or on the Y chromosome. In plants, blocks of satellite sequences are also found interstitially (i.e., within euchromatic regions). Pulsed field gel electrophoresis has shown that satellites are apparently not as variable in array size within populations as micro- and minisatellites (on a per-nucleotide basis). However, substantial differences in array size have frequently been observed between isolated populations and closely related species of both plants and animals. Satellite DNA families that are overabundant in one species may be absent or in very low amounts in a sibling species (hence detectable only by the polymerase chain reaction (PCR) but not by conventional Southern blot analysis).

Another characteristic difference between satellite and minisatellite DNAs is the greater tendency for multimeric repeats to be found in satellites. The formation of higher order repeats is particularly often seen in larger arrays (> 200 kb). The best studied example is the alpha satellite DNA family in primates. This consists of a basic repeat unit of 171 bp that is combined into various higher order structures, ranging from dimers to (at least) 16-mers. The very highly repetitive satellite sequences can account for a large proportion of the DNA in the genome. Copy numbers of satellite DNA clusters range between 10^3 and 10^9. Arrays with shorter repeat units tend to have higher copy numbers. On average, the proportion of satellite DNA in the eukaryote genome is around 10%. Satellite DNA is virtually absent in yeast. But other species (including mammals) can possess more than 50% satellite DNA. For example, in the kangaroo rat *Dipodomys ordii,* 50–60% of the genome consists of three repeated sequences: AAG (repeated 2.4×10^9 times), TTAGGG (2.2×10^9 times), and ACACAGCGGG (1.2×10^9 times). The repeats within a tandem array are not completely identical. Interrepeat variability typically amounts to 5–10%. Another distinctive feature is related to their propensity to form higher order structures: satellite repeats are often composed of subunits of simple sequences such as homopolymeric runs or short runs of di- or trinucleotide repeats. This is illustrated by the third satellite of the kangaroo rat, which contains both a homopolymeric run and a short stretch of dinucleotide repeats. Because of this simple nucleotide composition, satellite DNAs usually form one or more ("satellite") bands in buoyant density gradients of genomic DNA which are clearly distinguishable from the main band. This property led to the discovery of satellite DNA and to the name of this class of DNA. Satellite DNAs that cannot be separated from the main band by density gradients are known as "cryptic."

Micro-, mini-, and satellite DNAs comprise only a fraction of repeated, localized DNA sequences in eukaryotes. Interspersed DNA sequences, such as transposable elements, may also form clusters of repetitive DNA in certain parts of the genome after they have lost their ability to move. For instance, these nonfunctional transposable elements may be found as scrambled, clustered arrays in regions around the centromere at the boundary between heterochromatin and euchromatin. The individual sequence elements of these numerous arrays of moderately repeated sequences are 300–1000 bp in length. A cluster, which is several kilobase pairs long, usually contains many different types of sequence, some of which may be repeated either directly or in inverted orientation. Different clusters within the same genome may share some, though not all, sequences, but the arrangement of shared sequences is not conserved between clusters. In this way, genetic and evolutionary mechanisms have created a considerable diversity of different types of repetitive DNA clusters that go beyond the three main tandem array classes discussed here.

3 GENETIC MECHANISMS OF COPY NUMBER CHANGE

Several genetic mechanisms can affect the dynamics of tandem arrays: gene amplification by processes such as replication slippage and rolling circle amplification, unequal exchange, intrastrand exchange, and mutations by base substitutions (Figures 1–3). However, the relative importance of these DNA turnover processes for micro-, mini-, and satellite DNAs, and the values of the parameters that describe them, are still uncertain.

3.1 DNA Turnover Processes

The molecular mechanisms affecting simple DNA sequences (microsatellites) are best understood. In vitro studies suggest that strand slippage during DNA replication is the major cause of the observed length polymorphism of microsatellites between individuals within populations (Figure 1). This process results in length changes of a few bases at a time. Dinucleotide repeats show a higher in vitro slippage rate than trinucleotide motifs; and those trinucleotides with the highest GC content have the lowest rates. In general, the in vitro data suggest that microsatellite arrays should be very highly variable in natural populations. The frequencies with which changes in array sizes occur at microsatellite loci are as high as a few percent per generation. These rates are much higher than normal mutation rates (due to base substitutions), but lower than expected from the in vitro

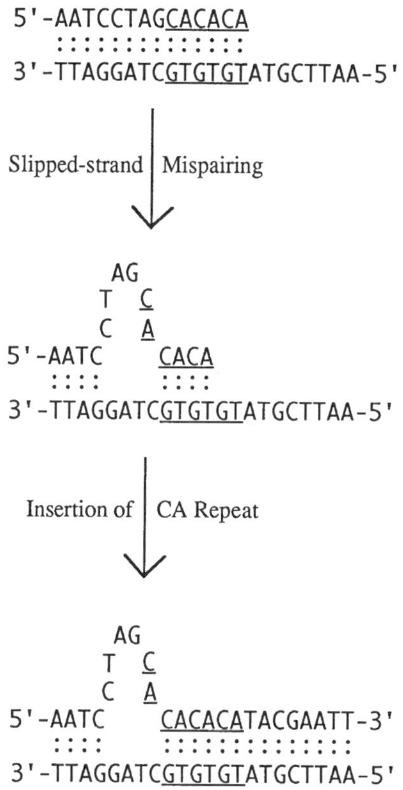

Figure 1. Replication slippage: the newly synthesized strand (upper strand) is slipping in the $3' \rightarrow 5'$ direction in the tandem array $(CA)_3$ (underlined). This leads to the insertion of an additional CA repeat as replication proceeds. Similarly, if the template strand (lower strand) slips in the $5' \rightarrow 3'$ direction, a deletion of one or a few repeat units may occur.

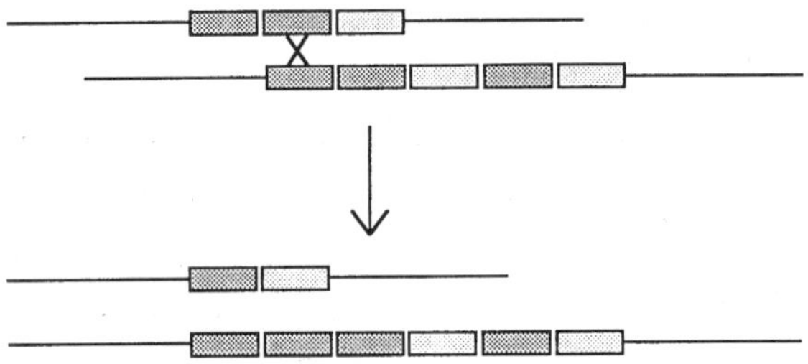

Figure 2. Unequal exchange: unequal crossing over occurring between tandem arrays on homologous chromosomes. The boxes represent repeat units; heavy and light stippling indicates that the members of a tandem array may differ. Recombination is most likely to occur between repeats with the highest degree of similarity. Unequal exchange leads to the expansion of one array (lower one) and to the contraction of the other one. Furthermore, it results in more homogeneous arrays, a process called "concerted evolution."

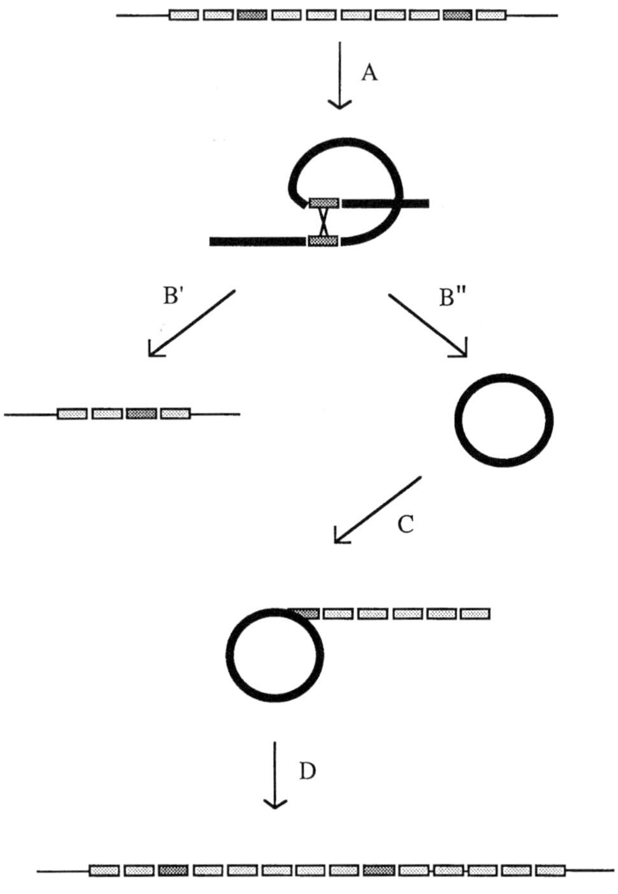

Figure 3. Intrastrand deletion and rolling circle replication: pathway (A, B′), which includes the steps A and B′, results in a deletion of repeat units, whereas pathway (A, B″, C, D) leads to an amplification of repeats. In step A, pairing occurs between the two heavily stippled copies of the tandem array shown at the top (the other repeats are not shown). In step B′, this leads to the deletion of the repeats that are located between the two pairing repeat units (except one). In step B″, an extrachromosomal circular plasmid is created that carries the deleted repeats. Most of these plasmids are lost. However, some of them may undergo "rolling circle" replication. Step C results in the replication of all copies of this circle. In step D, these amplified repeats are reintegrated into the tandem array by recombination.

results. Furthermore, there appears to be no relation between the occurrence of certain repeat types in eukaryotic genomes and their in vitro slippage properties. For example, $(GCC)_n$ repeats occur frequently in genomic DNA, but their in vitro slippage propensity is poorest among all trinucleotides. These observations suggest that mechanisms other than slippage must contribute to the dynamics of simple repeats in organisms.

There is a relation between replication slippage of microsatellite sequences and defective DNA repair, since mutations in DNA repair genes can cause tract instability in yeast when errors resulting from slippage replication remain unrepaired. Similarly, a mutation in a gene of the same mismatch repair system in humans can lead to expansion of microsatellite tracts associated with colorectal cancers. These observations may suggest that replication slippage is the primary force involved in the expansion and contraction of microsatellite arrays, but that most of the length changes due to slippage are corrected by the DNA repair system. In addition, unequal exchanges could occur in microsatellite arrays that are sufficiently large (Figure 2). Total array size, however, appears to be an important constraint for the occurrence of unequal exchange. In contrast to replication slippage, unequal exchange is an interhelical event, involving DNA molecules from two different sister chromatids or homologues. This may place some constraints on the length of the sequences undergoing unequal exchange. Large changes in copy number, which are occasionally observed at microsatellite loci in population surveys (see Section 5.2), may result from unequal exchanges.

Several processes may cause the instability of arrays of minisatellites. At some well-studied human minisatellite loci, a bias toward gains of a few repeats has been found, with gains occurring preferentially at one end of a tandem array. New repeats on a given chromosome can originate either from the same chromosome or from its homologue. A gene conversion–like mechanism causing this net increase in copy number has been proposed. In addition, unequal exchange could occur between tandem arrays on sister chromatids or between arrays on homologous chromosomes. At present, however, there is no direct evidence for unequal exchanges at minisatellite loci. The frequencies with which changes in array sizes occur at minisatellite loci are similar to those of microsatellites and can be as high as a few percent per generation. At extremely variable loci, mutations to new length alleles occur at a rate high enough to measure it by direct observation in pedigree analysis. The apparent relative stability in the population of small minisatellite alleles provides indirect evidence that the expansion and/or contraction of minisatellite arrays may occur at a frequency dependent on array size, such that longer arrays exhibit a higher rate of copy number change.

The forces governing the dynamics of satellite DNA are less well understood, mainly because the large sizes of satellite DNA clusters preclude direct experimental analysis, and also hinder the collection of meaningful population data. Even the most advanced techniques, such as pulsed field gel electrophoresis, are thus far rather inefficient in revealing the complexity of satellite and large minisatellite DNA arrays. Because of these technical constraints, most of our knowledge of the mechanisms operating on extremely large clusters is inferred from data on smaller arrays or from data on multigene families (e.g., ribosomal RNA gene clusters). Since replication slippage involves only a short stretch of DNA at a time, it probably plays only a minor role in the amplification of the longer satellite repeat units ($>$ 100 bp). Slippage may, however, be important in the initial formation of satellite repeats which are generally composed of smaller subunits (see Section 2.3).

The observation of extrachromosomal circular satellite DNA suggests the possibility that long satellite repeats may primarily be amplified by extrachromosomal rolling circle replication, followed by reinsertion into the genome (Figure 3). The occurrence of unequal exchange has been inferred from sequence analysis of satellites containing higher order repeat units. The question of whether unequal exchange occurs predominantly between sister chromatids or between homologues is still open, although data on the tandem arrays of the *PSR* supernumerary chromosome in the parasitoid wasp *Nasonia vitripennis* (which is confined to the haploid males) can be explained only by mitotic sister chromatid exchange or mitotic intrastrand exchange (intrastrand deletion). If recombination occurs either between arrays on sister strands or between different homologues, it is also likely to occur between members within a strand. This process, called intrastrand exchange or intrastrand deletion, is expected to result in deletion of members on that array (Figure 3). In fact, the finding of extrachromosomal DNA circles containing satellite repeats is the best available evidence that intrastrand exchange occurs in large tandem arrays. Most of the extrachromosomal plasmids carrying copies of the array may be lost. However, some of these plasmids may undergo rolling circle replication, which amplifies a small section of the array before integration into the array.

3.2 Amplification of Disease-Related Microsatellites

The discovery of the causative role of the expansion of microsatellites in several important human diseases (e.g., the fragile X syndrome of mental retardation) has revealed an unexpected practical significance to the studies of tandemly repeated noncoding DNA. The mutations causing neurological diseases such as the fragile X syndrome and myotonic dystrophy are preceded by "premutations" that lead to arrays of large size within the normal range of variation in the general population. It has been suggested that replication slippage plays a role in the generation of permutation alleles by increasing array sizes by a few repeats at a time.

Since, however, the major expansion, leading to the full disease stage, appears to occur in one or a few generations, it is not caused by a stepwise amplification mechanism that involves only a few repeats. The starting point for this amplification consists of large arrays within the normal range of variation, and therefore it is possible that mechanisms that usually do not work on microsatellites (perhaps because of steric hindrance due to small array size) come into play and lead to sudden expansions. These mechanisms may include processes such as those described earlier (e.g., unequal exchange) that are thought to operate on larger tandem arrays. However, it is not likely that unequal exchange is a force that causes the dramatic expansion of disease-related microsatellites in the premutation stage. Unequal exchange may involve a large number of repeats, but it cannot expand an array to more than twice its size in one cycle. Thus, unequal exchange may be involved in the most common changes in size of premutation alleles that are relatively small, but it does not allow the unstable array to expand manyfold (up to 10-fold) in length beyond a threshold value in a single generation. This and other considerations indicate that the sudden increases in the lengths of trinucleotide repeat arrays that cause human genetic diseases seem to involve other, unknown, mechanisms.

4 THE SELFISH DNA HYPOTHESIS

It has often been proposed that tandem-repetitive DNA sequences are functionally important for the host organism. The *Responder* satellite sequences of the *Drosophila melanogaster* segregation distorter system seem to affect the fitness of their carriers. The *Responder* locus in natural populations of *D. melanogaster* consists of 20–2500 copies of an 120 bp repeat unit. In a competition experiment involving a mixed population of flies with 700 copies and flies with 20 copies, it was observed that the frequency of the flies containing 20 copies decreases with time. This may suggest that flies with 700 copies have a higher fitness than flies with 20 copies. However, there is a potential problem with this experiment in that the chromosomes carrying 20 copies contain a large deletion that includes a segment of DNA flanking the *Responder* satellite.

There is also evidence for biological effects of heterochromatin (which is rich in satellite sequences), such as the binding of heterochromatin by certain proteins, and its role in achiasmatic chromosome segregation in *Drosophila.* Partial deletion of some centromeric satellites (e.g., human alphoid DNA) has been shown to be associated with centromere inactivation. Similarly, there appears to be an effect of minisatellites on the expression of genes that are located nearby. For example, the 14 bp minisatellite in the diabetes susceptibility locus *IDDM2* contains high affinity binding sites for the transcription factor, Pur-1 and may stimulate insulin transcription. It has therefore been suggested that this result demonstrates a possible function of the insulin minisatellite.

But these properties of minisatellite and centromeric satellite DNA may not have a role in the origin and maintenance of tandemly repeated sequences: they could simply be a response to their presence in the genome, that is, consequences rather than causes. That the insulin repeat is found in much lower copy numbers in other primates (e.g., chimpanzees) and is absent in nonprimates seems to support this hypothesis (although it does not completely rule out the possibility that this repetitive structure is the result of a recent adaptive process). Comparative patterns, such as the inverse correlation between development rate and amount of highly repeated DNA in groups such as flowering plants and salamanders, are compatible with nonadaptive mechanisms for maintaining satellite repeat arrays. Selection against large amounts of repetitive DNA may vary in intensity with life history; for example, low rates of cell division caused by larger genome size are probably less harmful to organisms that develop relatively slowly. More slowly developing species could therefore accumulate larger quantities of repetitive sequences generated by array size expansion. The comparative data seem to fit this interpretation better than adaptive alternatives that have been proposed. The large differences in amounts of satellite DNA sequences between closely related species (see Section 2.3), and the absence of measurable fitness effects of large deletions or duplications of heterochromatin in *Drosophila,* also suggest strongly that selection does not actively maintain the high copy numbers of many of these sequences.

It has therefore been proposed that in most cases, tandemly repeated noncoding DNA sequences are maintained solely by their ability to replicate quickly within the genome (the "selfish DNA" hypothesis). Far from conferring benefits, their behavior can result in a fitness loss to the host. Some human genetic diseases are known to be caused in this way, including mutations due to the amplification of microsatellite sequences (see Section 3.2).

5 DYNAMICS OF TANDEM ARRAYS

Because of an obvious lack of functional significance and technical constraints, many aspects of the biology of tandemly repeated, noncoding DNA sequences cannot be explored by direct experimentation. In this section, I therefore introduce some basic theoretical concepts that have been used to understand the properties of this class of DNA. I will show that many features of the organization of tandemly repeated DNA sequences in eukaryotic genomes, including their distributions in natural populations, may be explained based on the selfish DNA hypothesis. The theoretical models will establish relations between the properties of tandemly repeated DNA (discussed in Section 2) that at first seem to be paradoxical.

5.1 The General Model

The state of a haploid genome with respect to a given class of tandemly repeated DNA sequences at a particular chromosomal location is characterized by *i*, the number of repeats in the array. The state of the population is described by the distribution of values of *i* among haploid sets, such that x_i is the frequency of chromosomes with array size *i*. The mean array size is $\bar{\imath}$. The following factors affect the change of this distribution over time. Experimental evidence for these mechanisms is discussed in Section 3.

Amplification A sudden increase in array size by one or more repeat units. Depending on array type (see Section 3), it may involve replication slippage (Figure 1), gene conversion–like processes, rolling circle replication of extrachromosomal repeat units followed by reinsertion (Figure 3), or other, unknown, mechanisms (e.g., in disease-related microsatellites).

Deletion A sudden decrease in array length by one or more units, caused by replication slippage or intrastrand exchange (Figure 3).

These two processes can be included in a model in which the probability that an array of length *j* generates an array of length *i* is $\alpha_j P_{ij}$, where α_j is the probability that a change of an array of length *j* occurs per generation, and P_{ij} is the conditional probability of the change from *j* to *i*. Although there is evidence that array size affects the rate of copy number change, I assume here for simplicity that α_j is independent of *j*; that is, $\alpha \equiv \alpha_j$.

Unequal exchange This can occur either between homologous chromosomes or between sister chromatids. I will consider only the first category here (Figure 2). Let γ_{jk} be the probability of an exchange between an array of *j* and an array of *k* copies, and Q_{ijk} be the conditional probability that an exchange produces a daughter chromosome with array size *i* from parental chromosomes with copy numbers *j* and *k* ($0 < i < j + k$). There is an equal chance of production of a daughter with $j + k - i$ copies. The probability of the production of a daughter chromosome with *i* (or $j + k - i$) copies is then $\gamma_{jk} Q_{ijk}$. Again, for simplicity, I assume that all γ_{jk} are identical; that is, $\gamma \equiv \gamma_{jk}$.

Natural selection According to the selfish DNA hypothesis, selection is usually assumed to be nonexistent or very weak (i.e., fitness declines very slightly) until array sizes reach a relatively high value. Above a certain threshold value Ω, however, fitness is assumed to drop rapidly to zero with increasing numbers of repeats for individuals carrying more than Ω copies. The threshold effects observed in the dynamics of disease-related microsatellite loci (see Section 3.2) seem to suggest that such abrupt changes in fitness may occur. Base substitutions in the sequence are generally considered to be selectively neutral.

Genetic drift As a result of finite effective population size N_e, the frequencies of haploid genomes with different array sizes are subject to multinomial sampling during reproduction in the same way as a conventional multiple-allele system with allele frequencies x_i.

Mutation Mutational differences in sequence (occurring at rate μ) among members of an array may affect their probability of undergoing strand pairing during recombinational processes. Recombinational steps are believed to be important in several DNA turnover processes that operate on tandem arrays (e.g., unequal exchange, rolling circle replication, gene conversion–like amplification). Mutational differences among the repeat units also have a strong impact on the rate of strand slippage during replication because these differences, which appear as mismatches in strands that are reannealed out of register, can be more easily detected by the DNA repair system when they are more frequent.

5.2 Copy Number Dynamics of Tandem Arrays

The qualitative behavior of this model can best be studied by considering the persistence times of tandem arrays. In this approach it is assumed that a certain distribution of array sizes (with mean copy number $\bar{\imath}_0$) exists at a given initial time in the population, generated by a sudden amplification event. In the absence of further amplifications, these tandem arrays are lost from the population owing to the joint action of DNA turnover processes, genetic drift, and selection. I turn now to a discussion of unequal exchange as a typical DNA turnover process that can affect copy number; but other processes with similar properties can be substituted or added, as will be shown. Natural selection is modeled such that fitness of an individual in the population decreases slowly with increasing copy number and declines rapidly to zero when copy number is larger than a threshold value Ω. This selection scheme is consistent with the selfish DNA hypothesis. Unequal exchange can both expand and contract a tandem array with equal probability and by equal amounts; hence, on average, mean copy number $\bar{\imath}$ does not change from generation to generation. However, unequal exchange, in conjunction with genetic drift and selection, leads to a decrease in mean copy number over time until the population has reached a state in which it is fixed for chromosomes with array size 1, and unequal exchanges are no longer possible. The average time to loss of a tandem array is inversely proportional to γ, the frequency of unequal exchanges per generation, if γ is sufficiently small ($4N_e\gamma < 1$); furthermore, it increases with Ω, and decreases with increasing effective population size N_e. However, if $4N_e\gamma$ exceeds 1, the expected time to loss of a tandem array from the population increases with increasing population size. This behavior of the model is summarized in Figure 4.

Four predictions that can be compared with the data (see Section 6) follow immediately from these results:

1. Longer tandem arrays persist in regions of low rates of unequal exchanges.
2. More variable arrays are less conserved over evolutionary time.

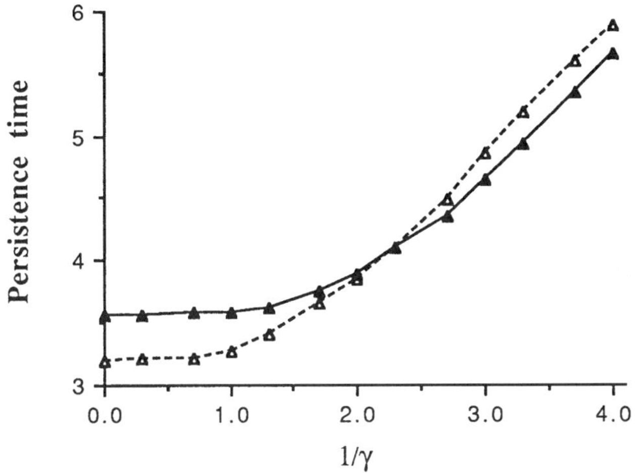

Figure 4. Persistence time of tandem arrays versus inverse recombination rate. The mean time to loss of tandemly repeated DNA sequences (in generations) under the joint action of unequal exchange, genetic drift, and selection on array size is plotted against the inverse recombination rate in a double-logarithmic fashion. It is assumed that amplification processes are not operating and that the mean copy number of the population is initially 50. The results have been generated by computer simulations for two population sizes, $N_e = 50$ (open triangles) and $N_e = 200$ (solid triangles), and a maximum array size of $\Omega = 10{,}000$. Note the dramatic increase in persistence time as the recombination rate decreases. Furthermore, note that the effect of N_e on persistence time is reversed for small and large values of γ. Because of the logarithmic scale of the *y* axis, the persistence times between the two populations differ for both small and large values of γ roughly by a factor 2. That means effective population size, in addition to recombination rate, can have a dramatic effect on the persistence of tandem arrays in populations. In contrast, the effects of initial copy number and selection (Ω) are less pronounced.

3. Tandem arrays are larger and persist longer when selective constraints on array size are weak (i.e., Ω larger).
4. Tandem arrays are larger and persist longer in populations whose effective size N_e is small.

Predictions 1 and 4 may not be intuitively obvious and need additional explanation. In regions of low rates of crossing over, arrays with high copy numbers are not frequently generated by unequal exchange; thus, selection is less effective in eliminating from the population individuals whose copy numbers are too high. As a result, mean array size decreases slowly, which explains prediction 1. Indeed, a detailed analysis of the model indicates that the effect of unequal exchange on the persistence of arrays is particularly pronounced in regions of very low rates of crossing over, suggesting that array sizes of tandemly repeated sequences should be much larger in regions of reduced rates of exchange than in those of intermediate or high rates of crossing over. Prediction 4 is a consequence of the interaction of genetic drift and selection on array length. In populations with small effective sizes, random fluctuations of the frequencies of alleles containing tandem arrays due to genetic drift are more dramatic than in populations with large effective sizes. Therefore, in small populations drift is stronger and can overcome the effects of selection pressure on array size. This reduces the efficiency with which alleles whose copy numbers are too high are removed from the population by selection, and increases the persistence times of tandem arrays.

The foregoing results are not limited to unequal exchange, which is thought to act primarily on larger tandem arrays such as minisatellites and satellite DNAs. Any other mechanism that is symmetrical with respect to copy number change can be substituted for unequal exchange such that the results are essentially unchanged. For example, replication slippage (operating on microsatellites) and unequal exchange would have qualitatively similar effects on copy number dynamics.

When recurrent amplification processes are included in the model, an equilibrium array size will result as a balance between amplification (leading to a net increase in copy number) and unequal exchange, genetic drift, and selection (causing a decrease). A particularly interesting prediction of this model that may pertain to satellite DNA is the following: as a result of amplification in combination with selection against high copy numbers, unequal exchange, and genetic drift, mean array size is high only when $4N_e\gamma$ is less than 1 and is highest when selection against arrays is weakest, as might be the case for organisms with long developmental times. Thus, satellite sequences should accumulate in chromosomal regions of very low rates of unequal exchanges, and more slowly developing species should possess more repetitive DNA.

5.3 Sequence Homogenization

Computer simulations of the general model for tandem arrays have explored the relation between copy number change and sequence variability between members of a tandem array. In these models, mutation was allowed to alter the sequence of repeat copies by base substitution. This has consequences for the probability that copies can recombine in unequal exchanges, and strands reanneal during replication slippage. In other words, the exchange rate γ and amplification rate α were assumed to be dependent on sequence similarity. As suggested by experimental evidence from several different in vivo and in vitro systems, it was assumed that a short stretch of perfect sequence identity rather than long segments of overall homology is required for a successful turnover process to occur. The outcomes of these simulations are remarkable:

1. Tandem arrays of repetitive sequences can be generated de novo, starting from a random DNA sequence.
2. Short simple repeat sequences emerged when the sequence-dependent DNA turnover processes occurred frequently; in contrast, arrays with longer repeat units (including higher order structures such as multimeric repeats) were generated when the rates of amplification and exchange were low. A related phenomenon is that the basic repeat units often exhibited an internal substructure featuring runs of homopolymeric nucleotides and/or short stretches of di- or trinucleotide repeats.
3. A positive correlation between the rate of unequal exchange γ and sequence homogeneity between repeat units within a tandem array was observed. Remarkably, however, sequence homogeneity remained relatively high and did not drop below a certain threshold level until γ was reduced dramatically relative to the mutation rate μ.
4. Weak selective constraints (i.e., large Ω values) led on average to larger repeat units and increased the tendency to form higher order structures.

6 DATA ANALYSIS

This section compares the theoretical results from Section 5 with the properties of tandem arrays.

6.1 Microsatellites

Microsatellites have very short and homogeneous repeats and a very small total array size. They are distributed in euchromatic regions of the genome in a relatively uniform manner and have been found near and within gene regions. The theoretical and experimental analyses suggest that these properties are consistent with high rates of DNA turnover mechanisms (in particular, replication slippage) and strong selective constraints on total array size (small values of Ω). As a consequence of these high turnover rates, most microsatellite arrays are highly variable within populations. Furthermore, as expected, there appears to be an inverse correlation between the degree of intraspecific variability of copy number and interspecific conservation of microsatellite loci, such that the more variable loci exist only over a narrower phylogenetic range. For example, using PCR with primers in the flanking regions, it has been demonstrated for crocodilians that more variable microsatellite loci are less conserved among species. However, there is a caveat because part of this correlation may be due to sequence divergence of the flanking regions in which the PCR primers bind. Some simple repeat loci do not follow the predictions of the proposed model in that they show fewer interspecific differences in copy number than expected based on the degree of intraspecific variation. In these cases, functional explanations may have to be invoked to describe their evolutionary dynamics.

Seemingly conflicting evidence has been reported from a number of well-studied species with respect to the abundance of microsatellite loci on X chromosomes versus autosomes (see Section 2). There is a significant lack of highly polymorphic $(CA)_n$ loci on the X chromosomes in several species such as man, mouse, and rat; on the other hand, in *Drosophila* the total number of $(CA)_n$ microsatellites (poly- and monomorphic ones) on the X chromosome is higher. This uneven distribution in *Drosophila* is consistent with the theoretical model discussed earlier. Since the rate of copy number change for *Drosophila* microsatellites is generally very low and the effective population size of the X chromosome in *Drosophila* is smaller than that of the autosomes, one should expect that the forces for removal of repetitive DNA are weaker on the X chromosome and that therefore more loci persist. A smaller effective population size increases the effect of genetic drift relative to selection and thus alleviates the constraints imposed by selection on total array size. As a consequence, the removal of repetitive DNA is less efficient. Thus, it is not necessary to invoke hypotheses about differences in the genetic mechanisms between the X chromosome and the autosomes or functional explanations to understand the uneven distribution of the number of microsatellite loci. The deficit of highly polymorphic microsatellites on the X chromosome in the three mammalian species is also compatible with this model. As shown in Figure 4, the expected time to loss of a tandem array increases with effective population size, if the DNA turnover rate is high. Thus, highly polymorphic microsatellites should be lost more quickly on the X chromosome than on the autosomes. Alternatively, the deficit of hypervariable microsatellites may be explained by a lower rate of replication slippage, owing to the fact that the X chromosome is present less often than the autosomes in the male germ line, which is thought to be more mutation prone in mammals.

6.2 Minisatellites

The salient properties of minisatellites may also be explained by the proposed model. Minisatellite loci are found in the euchromatin. Many of them are highly polymorphic, though not, on a per nucleotide basis, as polymorphic as microsatellites. Their repeat units are longer, and their total array sizes are much larger than those of microsatellites. These properties are consistent with the predictions of the model if smaller rates of DNA turnover are assumed. Indeed, the per-base-pair turnover rates that have been estimated by direct observation in pedigree analysis are substantially lower than those of microsatellites. The relatively large array sizes of minisatellites indicate that selective constraints on total array size are weaker than for microsatellites, possibly because minisatellites do not appear to be located as close to gene regions as microsatellites.

It is unknown at present why highly polymorphic minisatellite loci tend to cluster in subtelomeric regions of the human genome. However, since the genetic map is extended in these portions of chromosomes, recombinational steps may generally occur at a higher rate in these regions. Recombinational steps are involved in processes such as gene conversion–like amplification, unequal exchange, and intrastrand exchange, which are thought to operate on minisatellites (see Section 2). According to the proposed model, increased rates of DNA turnover should lead to a high degree of copy number variation and may therefore explain the overabundance of hypervariable minisatellites in these regions.

Because of a lack of exchange of flanking markers, unequal exchange between homologues is probably not the predominant mechanism of copy number change at minisatellite loci. However, there is indirect evidence for the occurrence of unequal exchange between sister chromatids. In accordance with the predictions of the model, it has been found for a sample of human minisatellites with different degrees of allelic variability in repeat copy number that a significantly negative correlation exists between copy number variability and interrepeat sequence variability (Figure 5). In other words, the most highly polymorphic loci (with respect to copy num-

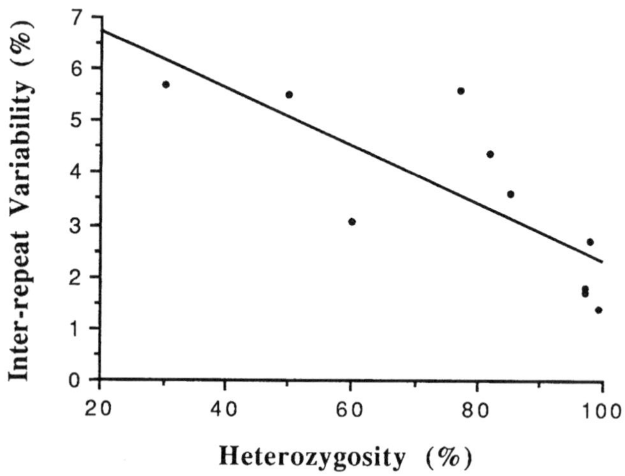

Figure 5. Interrepeat variability versus allelic heterozygosity for human minisatellite loci: allelic heterozygosity due to copy number variation versus sequence variation between members of a tandem array. The negative correlation between allelic heterozygosity and sequence heterogeneity is highly significant ($P < 0.01$). [Reproduced from *Genetics,* 136:333–341 (1994), with permission.]

ber variation) show the highest degree of sequence homogeneity between repeat units. This homogeneity in tandem arrays is difficult to explain by events that are restricted to certain parts of arrays, such as the gene conversion–like amplification mechanism mentioned in Section 3.1. Instead, long-range forces such as unequal exchange that act over the entire array must be invoked. Furthermore, as expected from the theoretical analysis, minisatellites that are highly polymorphic within species are evolutionarily unstable. For example, comparisons between humans, great apes, and monkeys have shown that minisatellite arrays that are highly polymorphic in humans (e.g., minisatellite locus *MS32*) are sometimes very short and relatively monomorphic in great apes and monkeys (3–4 repeats for *MS32*).

6.3 Satellites

The interpretation of the properties of satellite DNA is the greatest challenge for the proposed model because this data set is relatively complex and, despite a long history of research, not as comprehensive as those of micro- and minisatellites. Satellite DNA can form short repeat units that are similar in length to micro- and minisatellite repeats and also very large ones (> 100 bp). In any case, their total array sizes are much larger than those of micro- and minisatellites. In contrast to the two latter classes of DNAs, satellites are located in heterochromatic regions of chromosomes. These properties of satellite DNA are clearly compatible with weak selective constraints on array size, probably because satellites are located in heterochromatin that is devoid of functional genes (for reasons given shortly). Furthermore, the features of satellite DNAs are compatible with low rates of DNA turnover mechanisms such as unequal exchange and rolling circle amplification that produce copy number variation and sequence homogeneity in these tandem arrays. The short repeat length of some satellites may raise a caveat to this interpretation, but computer simulations of the general model have demonstrated that repeat unit lengths of the simulated arrays may vary greatly when the rate of DNA turnover is low relative to the nucleotide substitution rate μ. This interpretation is also consistent with the absence of microsatellites from heterochromatin (see Section 2.1). Tandem arrays of very short repeat units that are located in heterochromatin, often juxtaposed to satellites with longer repeat units (e.g., in humans and *Drosophila*), are much longer than microsatellites (Section 2.3).

Direct evidence for low rates of DNA turnover processes in heterochromatic regions near centromeres and telomeres has been found for meiotic recombination. Contraction of the genetic map suggests that meiotic recombination is strongly suppressed in regions near the centromeric and/or telomeric heterochromatin in several well-studied species, including human, *Drosophila,* and yeast. The causes for this reduction are unknown, although it may be selectively advantageous for recombination to be suppressed near centromeres and telomeres. It has often been hypothesized that heterochromatin causes the suppression of recombination. However, a recent report from *Drosophila* suggests that heterochromatin is the consequence of reduced recombination rather than the cause: when array size exceeds a certain threshold value, tandem arrays introduced into euchromatin may lead locally to the formation of heterochromatin and to silencing of functional genes in this region. Thus, it appears that the accumulation of tandem arrays such as satellite sequences in regions of suppressed recombination predicted by our model provides a signal for local heterochromatization. It may also explain why genes are lacking in heterochromatin and why selective constraints on array size are weaker than in euchromatin. Since recombinational steps are involved in essentially all DNA turnover mechanisms that alter copy number of satellite arrays, it is possible that there is a general reduction of the rates at which these processes occur in heterochromatin. As a consequence, unequal exchange between homologues as well as between sister chromatids, intrastrand exchange, and rolling circle amplification may be less frequent than in euchromatin. It therefore seems likely that suppression of recombination (or, in general, of recombinational processes) causes the accumulation of satellite sequences and leads to heterochromatin formation.

This hypothesis is consistent with the observation that satellite sequences show a greater tendency to higher order structure (multimeric repeats) than minisatellites. Computer simulations of the general model suggest that higher order structures are preferentially found in regions of low rates of unequal exchange and weak selective constraints on total array size. Additional evidence that rates of DNA turnover in satellites are low comes from the observation that satellites, in sharp contrast to micro- and minisatellites, show little variation in copy number within populations, relative to their large array sizes. However, substantial differences in array size are often observed between closely related species of animals and plants, indicating that processes causing expansion or contraction of satellite arrays must occur at a significant rate on an evolutionary time scale. Evidence that a minimal level of DNA turnover occurs in satellite DNAs comes also from the observation of a high degree of sequence homogeneity within satellite arrays. The simulations indicate that unequal exchange is a strong long-range homogenization force that may operate over the entire array length. However, rates of exchange cannot be too low relative to mutation; otherwise, it would not be possible to maintain the observed degree of sequence homogeneity in satellites (90–95%), which is comparable to that of minisatellites.

Besides DNA turnover processes and selective constraints that appear to be the major determinants of the chromosomal location, structural properties, and copy number variation of satellite DNA, genetic drift may have a role in the accumulation of highly repeated DNA. It has been reported that cultivated species (in particular plant cultivars) tend to have a higher portion of highly repetitive DNA than the wild progenitor species. For example, the genome of *Cucurbita pepo* (squash), a highly cultivated species, consists of around 30% satellite DNA, whereas the wild species *C. lundellina* has only 5% satellites. A similar tendency has been observed in the genus *Cucumis* which contains *C. sativus* (cucumber), a species with a relatively large amount of satellite DNA, and in the genus *Solanum* with regard to *S. tuberosum* (potato). Based on our model, this increase in highly repetitive DNA may be explained by a severe reduction in effective population size that occurred when the wild progenitor species was taken into cultivation. A smaller effective population size increases the effect of genetic drift relative to selection and thus alleviates the constraints imposed by selection on total array size. As a consequence, more repetitive DNA can accumulate.

In sum, the great majority of tandem arrays of noncoding DNA studied so far follows the predictions of the proposed model that was based on the selfish DNA hypothesis and included a suite of DNA turnover processes. A purely neutral model in which fitness of an individual is not affected by the amount of tandemly repeated DNA would have accounted for most of the properties of micro, mini, and satellite DNA. However, some of the observations, such as the distinctly different chromosomal locations of these classes of DNA in euchromatin versus heterochromatin, the effects of population size on the higher abundance of microsatellites on X chromosomes versus autosomes, and of satellites in cultivated versus progenitor species, cannot readily be explained without invoking the idea of selection acting against high copy number. Thus, it appears that tandem arrays are subject to selective constraints on array size, as predicted by the selfish DNA hypothesis.

7 POPULATION PROCESSES

Tandemly repeated noncoding DNA provides population biologists with the great opportunity to study processes such as migration and population differentiation, population structure, and genetic relatedness at the molecular level. Most prominent is the use of minisatellite loci as evidence in legal proceedings, usually in the determination of paternity or individual identification in forensic medicine. Population genetic models of tandem arrays generally assume that they are nonfunctional, that is, neutral or slightly deleterious. Base substitutions in the sequence are considered to be selectively neutral, but there may be natural selection against increased array size. In other words, these models are similar to those described in Section 5.

The frequency distributions of length alleles at more than 100 microsatellite loci in human populations fit a "stepwise mutation" model quite well. In such a model, arrays increase or decrease in repeat number by one or a few repeats, with occasional increases by larger steps. The fit of the population data to this model suggests that mutational steps usually involve only very few repeat units, as with slippage replication. This confirms the results of the in vitro studies of microsatellites reported in Section 2.

In contrast to microsatellites, the length distributions of minisatellite alleles do not fit a stepwise mutation model. This suggests that events that alter copy number occasionally involve many repeat units, as can happen with unequal exchange. On the other hand, the population data are not consistent with a model in which every new array that appears in the population is unique, as would be expected if events such as unequal exchanges with unconstrained misalignment were frequent. For satellite DNAs, only a very limited amount of population data is available at present; therefore no efforts have been made thus far to analyze the length distributions of satellite alleles within populations.

The nature of the forces controlling array sizes of tandemly repeated DNA sequences requires further research. Progress will probably come from two directions: from increased knowledge of the molecular basis of array size expansion and contraction, and from the comparison of patterns of within-population variation in array size with models based on this knowledge. Promising starts in these research areas have been made.

See also DNA Fingerprint Analysis; Fragile Sites; Fragile X-Linked Mental Retardation; Gene Distribution in the Human Genome; Genetic Analysis of Populations; Human Linkage Maps.

Bibliography

Armour, J. A. L., and Jeffreys, A. J. (1992) Biology and applications of human minisatellite loci. *Curr. Opin. Genet. Dev.* 2:850–856.

Cavalier-Smith, T., Ed. (1985) *The Evolution of Genome Size*. Wiley, Chichester.

Charlesworth, B., Sniegowski, P., and Stephan, W. (1994) The evolutionary dynamics of repetitive DNA in eukaryotes. *Nature,* 371:215–220.

John, B., and Miklos, G. L. G. (1988) *The Eukaryote Genome in Development and Evolution.* Allen & Unwin, London.

Kuhl, D. P. A., and Caskey, C. T. (1993) Trinucleotide repeats and genome variation. *Curr. Opin. Genet. Dev.* 3:404–407.

Levinson, G., and Gutman, G. A. (1987) Slipped-strand mispairing: A major mechanism for DNA sequence evolution. *Mol. Biol. Evol.* 4:203–221.

Tautz, D., and Schlötterer, C. (1994) Simple sequences. *Curr. Opin. Genet. Dev.* 4:832–837.

Tyler-Smith, C., and Willard, H. F. (1993) Mammalian chromosome structure. *Curr. Opin. Genet. Dev.* 3:390–397.

Targeting and Delivery of Drugs: *see* Drug Targeting and Delivery, Molecular Principles of.

Taxol® and Taxane Production by Cell Culture

Arthur G. Fett-Neto and Frank DiCosmo

1 Plant Material Selection and Preparation
- 1.1 Type and Source of Explant
- 1.2 Sterilization Procedure
- 1.3 Explant Size and Final Preparation

2 Establishment of Callus Cultures
- 2.1 General Procedure
- 2.2 Medium Optimization

3 Establishment of Cell Cultures

4 Analyses of Taxol® and Taxane Production
- 4.1 Extraction Procedure
- 4.2 Separation, Quantification, and Identification

5 Taxol® Yield and Possibilities of Manipulation
- 5.1 Characteristics of Taxol® Yield and Growth in *Taxus* In Vitro Cultures
- 5.2 Precursor Feeding
- 5.3 Cell Line Selection

6 Future Perspectives

Key Words

Callus A mass of disorganized plant tissue, generally with low degree of cell differentiation, usually cultured on solid media.

Cell Suspension Cells derived from callus or explant tissue and cultivated in liquid media, usually proliferating cell aggregates or single cells.

Explant The portion of the intact plant used to initiate callus cultures.

Taxane A compound related to taxol that can be described as a diterpene composed of three pentadecene rings.

Taxol® A diterpene amide with antineoplastic properties.

Taxus A genus of gymnosperm trees belonging to the family Taxaceae and commonly known as yew.

Taxol® is an anticancer drug increasingly used for the approved treatment of ovarian and breast cancers. The current supply of this compound is limited because the only commercial source is the bark of *Taxus brevifolia* (pre 1995), although other *Taxus* species also produce it. Yew trees grow relatively slowly, contain low and variable amounts of Taxol®, and in many cases are located in protected, environmentally sensitive areas. Cell culture of yew species is a viable alternate source of Taxol® and related taxanes, which can be used for production and/or semisynthesis of the drug. Cell cultures are readily renewable and yield relatively homogeneous, readily manipulated systems for biosynthetic studies on Taxol®, with potential for large-scale commercial production.

1 PLANT MATERIAL SELECTION AND PREPARATION

1.1 Type and Source of Explant

The procedures described here were established primarily from results obtained with *Taxus cuspidata* (Japanese yew), although some studies were performed using *Taxus canadensis* and *Taxus brevifolia.* The response of yew tissue to in vitro culture is dependent on the type of explant used to initiate the culture, on the physiological status of the donor plant, and on its genetic makeup. Explants that contain meristematic or parenchymal tissue are preferable. These are the types of tissue most likely to display cell proliferation in response to the growth regulators in the culture medium, and this tendency is attributed to these tissues' low degree of differentiation, which seems to be associated with a certain developmental flexibility in the presence of the appropriate stimuli. Relatively more differentiated tissues, such as sclerenchyma or xylem, are less responsive to exogenous plant growth regulators, possibly because of the genetic commitment to a specialized developmental pathway.

Young stem segments (flexible, but bearing dark green needles) are ideal for rapid initiation of callus to be used in developing cell cultures. Seed contents (megagametophyte and embryo) are better suited for establishing long-term maintenance callus cultures of *Taxus brevifolia.* Needles are also a suitable explant type for callus initiation, although initial growth rates tend to decrease more rapidly than those of young stem segments. Arils (fleshy basal covering of the seed) display different responses to in vitro culture depending on their stage of development. Green arils grow relatively fast for the first 2 weeks, but from then on growth seems to stop. Calli derived from green arils are relatively hard and not friable. Mature red arils, as expected, fail to grow in culture, possibly as result of the advanced degree of cell lysis at this later stage of their development (Table 1).

The physiological condition of the donor plant must be stable and favorable to growth. Outdoor-grown plants display higher contamination rates and tissue pigmentation when cultures are started in the winter months in Canada (November to March) relative to those started in other seasons. Genetic differences between plants of the same species are also a factor to be considered; it may be necessary to test different donor plants to develop in vitro cultures. Donor plants of the same species, although grown under similar environmental conditions, may display significant differences in the amount of growth in vitro as a result of genetic variability. It may also be beneficial to begin in vitro culture using plants shown to produce high yields of Taxol®.

Table 1 Effect of Different Explant Sources of *Taxus cuspidata* Cultivated in B5 Medium (containing 1 mg/L 2,4-D) on Callus Formation and Growth

Explant	Growth[a]	Friability
Green aril	+	No
Red aril	−	—
Seed content	++	Yes
Young stem[b]	+++	Yes
Needle[b]	++	Yes

[a] +, directly proportional to growth; −, poor growth.
[b] Also tested with *Taxus canadensis.*

1.2 Sterilization Procedure

Sterilization is accomplished using standard plant tissue culture techniques. The tissue is washed with distilled water to remove dust and solid particles, immersed in 70% ethanol for one minute, immersed with a few drops of Tween 20 (surfactant agent) in NaClO (1.5% v/v) for 30 minutes, and rinsed three times with sterile distilled and deionized water. During the sterilization period, the tissue must be kept fully immersed and under agitation. Sterilization times may have to be reduced for more sensitive and tender tissues. Our experience has yielded typical contamination rates using outdoor-grown donor plants that are less than 3% of the total number of explants used. In the winter, however, the incidence of contamination can be 10 times the regular rate. In certain cases, a particular plant may be extensively infested with endophytes that are not destroyed by surface sterilization procedures; in this situation, contamination can be very high and a change of donor plant is advisable.

1.3 Explant Size and Final Preparation

Following sterilization, the exposed cut portions of the tissue that were damaged by the ethanol and NaClO solutions are excised, and the healthy tissue is appropriately trimmed and placed in the growth medium. Excision of explants should be made smoothly, with a relatively new blade, to minimize the production of phenolics in the cut areas. In the case of stem segments, approximately 1 cm long segments give good results. Stem segments can be cultured at an approximate angle of 45° to the agar surface, with one of the transversely sectioned ends immersed in the medium and the other partly exposed to the air of the culture vessel. Alternatively, longitudinally cut stem segments can be cultured with their cut portions facing the surface of the medium. The contents of longitudinally halved seeds can be placed on the medium after separation from the seed coat with a spatula. Yew tissues seem to be very sensitive to heat, relative to tissues of other woody plants. All tools used in explant preparation and cell transfer should be sterilized in ethanol, flamed briefly to remove ethanol residue, and cooled to room temperature before touching on plant material.

2 ESTABLISHMENT OF CALLUS CULTURES

2.1 General Procedure

Cultures of yew are relatively difficult to establish because growth rates are slow and the explants produce and release phenolics; in turn, the oxidation of the phenolics leads to darkening, growth inhibition, and eventual loss of viability of the tissue. The inclusion in the medium of phenolic-binding compounds, such as polyvinylpolypyrrolidone (PVP), improves the growth of the explants. The inclusion of activated charcoal in the media can also reduce darkening, although it may also diminish explant growth as a result of binding of medium components. In addition, transfers to fresh medium (or rotation of the explants to new positions in the culture vessel containing portions of unused medium) at approximately 15-day intervals help prevent growth inhibition due to phenolic oxidation. More frequent transfers may be required in the early period of cultivation or when the donor plant is more prone to darkening upon in vitro culture.

Following an initial screening and evaluation of some standard medium formulations, the basic medium B5bPVP was established for callus initiation in yew; it consists of B5 medium salts and organic supplements, 3.0% (w/v) sucrose, 1.5% (w/v) insoluble PVP, 0.8% (w/v) agar, and 1mg/L of 2,4-dichlorophenoxyacetic acid (2,4-D). The media are prepared with distilled and deionized water, plant cell culture or reagent-grade chemicals, adjusted to pH 5.8 with NaOH and HCl (after the addition of PVP and prior to the addition of agar) and autoclaved (1.1 kg/cm^2 at 121°C for 20 mins). Preferably, the glassware utilized for plant cell culture should be washed with neutral detergent, abundantly rinsed with tap water, and then rinsed with distilled water. The cultures can be grown in plastic petri dishes (9 cm in diameter) containing 30 mL of medium (3–4 explants per plate) and sealed with Parafilm, or in 30 mL glass vials with polypropylene lids containing 10 mL of medium each (1 explant per vial). The cultures are incubated at 25°C in darkness.

2.2 Medium Optimization

Additional improvements on the basic medium formulation for yew tissue culture can be accomplished by investigating several medium components (e.g., plant growth regulator balance, salt and organic supplement composition) that promote the growth of related species or seem likely to have a stimulatory effect on the putative biosynthetic pathway to Taxol®. The effect of the medium components when included in various amounts in the basic formulation, or when replacing some of its original constituents, can be analyzed relative to growth and Taxol® yield in developing yew calli. Medium factors with a positive effect on growth and Taxol® yield are combined in new formulations to investigate the possibility of synergistic or antagonistic interactions.

A great deal of work is entailed in the high number of treatments and respective replicates needed for accurate statistical analysis, but these comprehensive experiments allow the identification of limiting factors for growth and Taxol® production. The overall analysis of the results also shows the relationship between growth and Taxol® production. From such a series of experiments, two improved media formulations were devised: B5bPhe (B5bPVP with 0.1 mM of phenylalanine) and B5Phe8/0GA (B5bPhe salts and organic supplements with 8.0 mg/L of 2,4-D and 0.5 mg/L of gibberellic acid). After autoclaving, gibberellic acid (GA_3) is added to the culture medium by filtration through 0.22 μm sterile filters. All other medium components are added prior to autoclaving. The overall analysis of the results showed a trend toward an inverse relationship between Taxol® production and growth in the developing callus cultures.

3 ESTABLISHMENT OF CELL CULTURES

Cell suspension cultures can be initiated from approximately 2-month-old callus cultures developed from stem segments. Depending on the availability of callus biomass for inoculum and on incubator space, cell cultures can be grown in 125, 250, 500, and 1000 mL Erlenmeyer flasks containing one-fifth of their total volume of liquid medium. Basic incubation conditions are 22–25°C in darkness on gyratory shakers at 110 rpm. Experiments done in callus and some additional optimization for cell suspension culture medium yielded the following basic formulations:

- B5 medium salts, threefold the regular amount of B5 organic supplements, 3.0% sucrose (w/v), 1.5% soluble PVP (MW $\approx$ 40 $\times$ 10^3), and 1 mg/L of 2,4-D (B5C1)
- the same composition of B5C1, but with a modified balance of plant growth regulators [i.e., 4 mg/L of 2,4-D and 1mg/L of kinetin (B5C2)]

Medium preparation, pH adjustment, and sterilization procedure are the same as in callus culture medium. Insoluble PVP is replaced with soluble PVP in liquid cell cultures to facilitate visual and microscopic observation of the cells and cell harvesting.

To initially establish cell suspension cultures, inoculum weights are preferably high (e.g., one-third of the volume of the liquid medium). After 1–2 weeks in culture on a gyratory shaker at 110 rpm, the cells released from the calli are collected and transferred into fresh liquid medium (1:1 or 2:1 proportion of old to fresh medium). The calli with attendant stem debris can be discarded or replated in solid medium to regenerate more callus tissue. During the first months of culture, the media can be replaced at 2-week intervals following 10 minutes of centrifugation at 600 $\times$ g at 20°C under sterile conditions to pellet the cells; in this way, the maximum amount of cell biomass is saved until the density is great enough to start suspension culture propagation in new flasks. Once suspensions have been stabilized (generally after several cycles of propagation), inoculum amounts can be reduced to one-sixth (w/v) of the liquid medium for propagation of new cell suspension flasks. Transfers to fresh medium in stock cultures can be done every 20 days. Viability checks of newly developing cell lines can be performed via dye exclusion of cells in the presence of Evans Blue (cells staining blue under the microscope are considered not viable) or other cytochemical method, and/or by analyzing the occurrence of growth in microcalli rescued from the liquid culture and plated in solid B5bPVP. Microscopic examination of cell viability and culture sterility should be performed regularly, particularly during the first few months of culture in liquid medium.

Cell cultures can also be immobilized after the first months of culture to facilitate medium replacement and to aid in the establishment of recalcitrant cell lines; this approach has been particularly useful in our efforts in establishing *Taxus cuspidata* cell lines (Figure 1). Immobilization is performed by spontaneous adhesion to 5 cm $\times$ 5 cm, 10–15 mm thick glass fiber mats with a uniform diameter of 10 μm and fiber density of approximately 61 mg/cm^2. Within the

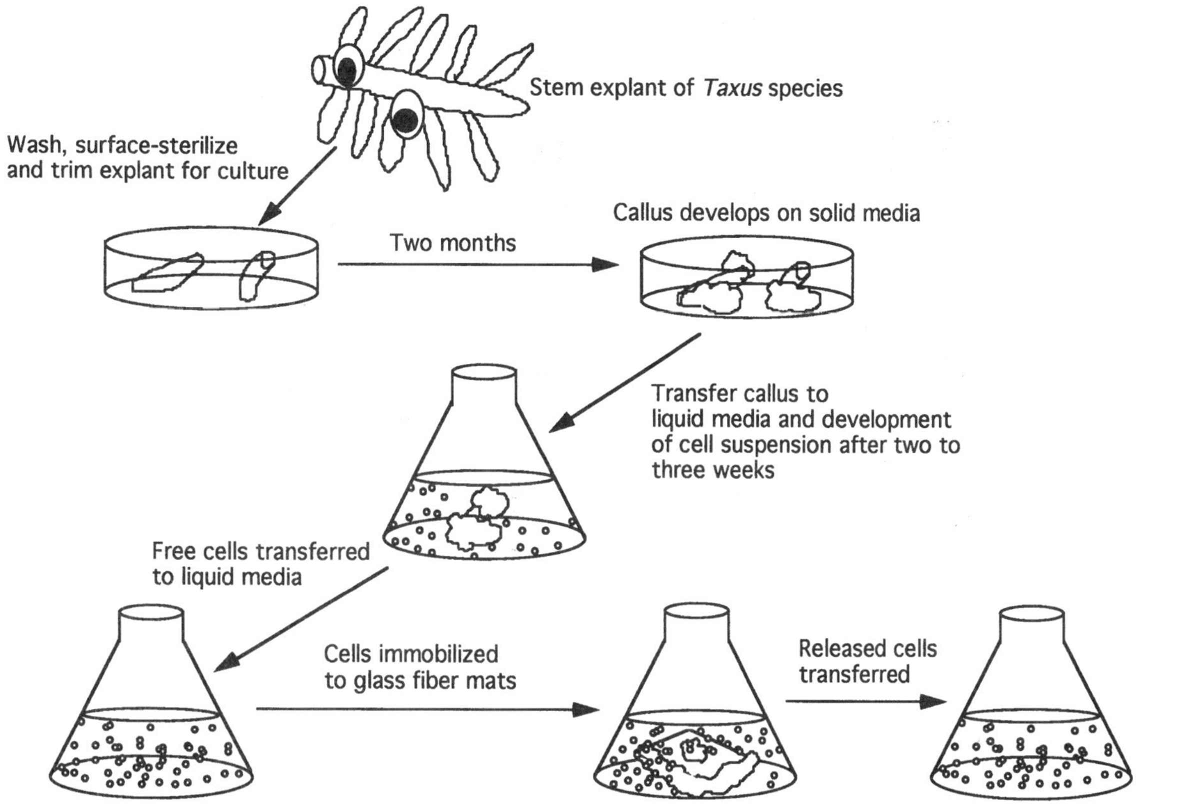

Figure 1. Establishment of immobilized cell cultures and cell suspension cultures of *Taxus cuspidata.*

first days of contact with the substrate, the cells are spontaneously immobilized. The cells grow on top of and within the mats, forming several layers of cells; after the cells have saturated the immobilization substrate, they start to be released into the liquid medium. These cells can be used to establish new suspension cultures or new immobilized cultures. Alternatively, cell cultures can be propagated by carefully removing external cell layers with a sterile spatula, or by cutting the mats in two or four portions and placing these on fresh medium flasks containing a new glass fiber mat (Figure 1).

Two-year-old cell suspensions of *Taxus cuspidata* have various degrees of cell aggregation. Moreover, the shape of the cells is variable, from round to considerably long, cylindric cells. In general, most cell suspension cultures display a relatively high number of cell aggregates of various sizes, with relatively fewer cells occurring individually. Cell aggregates may constitute a problem in large-scale cell cultures, which can be more susceptible to shear stress and may create difficulties in the mixing of cells, gases, and medium components. It is possible to select for smaller aggregate sizes by allowing larger aggregates to settle for a short period of time and then transferring the smaller aggregates and single cells from the upper portion of the flask into a new culture flask. The selection for the single-cell or small-aggregate type of suspension, however, may have implications for Taxol® productivity. Aggregated cell suspensions tend to be more productive than suspensions of the single-cell type, possibly because of their relatively higher degree of differentiation.

In our experience with cell lines of *Taxus cuspidata,* the stabilization of cell lines is a relatively lengthy process, normally requiring 6–12 months for completion. This is probably the case for all yew species, as a result of their relatively slow growth and recalcitrant behavior in vitro.

4 ANALYSES OF TAXOL® AND TAXANE PRODUCTION

4.1 Extraction Procedure

At least 0.5 g of fresh callus tissue or 1 g of fresh cells from suspensions is collected in glass test tubes (previously weighed and identified), frozen (−20°C), and homogenized with a glass rod in 1–2 mL of hexane, followed by 1 hour of sonication at room temperature. After 12 hours in hexane at 25°C, the samples are centrifuged at 2500 × *g* for 20 minutes to pellet the cells, and the supernatants are discarded. The pellets are then extracted in 1–2 mL of methanol/dichloromethane (1:1) with glass rod rehomogenization, followed by sonication for 2 hours at room temperature. After centrifugation at 2500 × *g* for 30 minutes, the supernatants containing the taxanes are collected; the pellets are used for determining the extracted dry weights at 60°C until constant weight.

The crude extracts are dried under vacuum or nitrogen at 25°C, redissolved in 1 mL of dichloromethane, partitioned with 1 mL of water, and centrifuged at 2500 × *g* for 20 minutes. The dichloro-

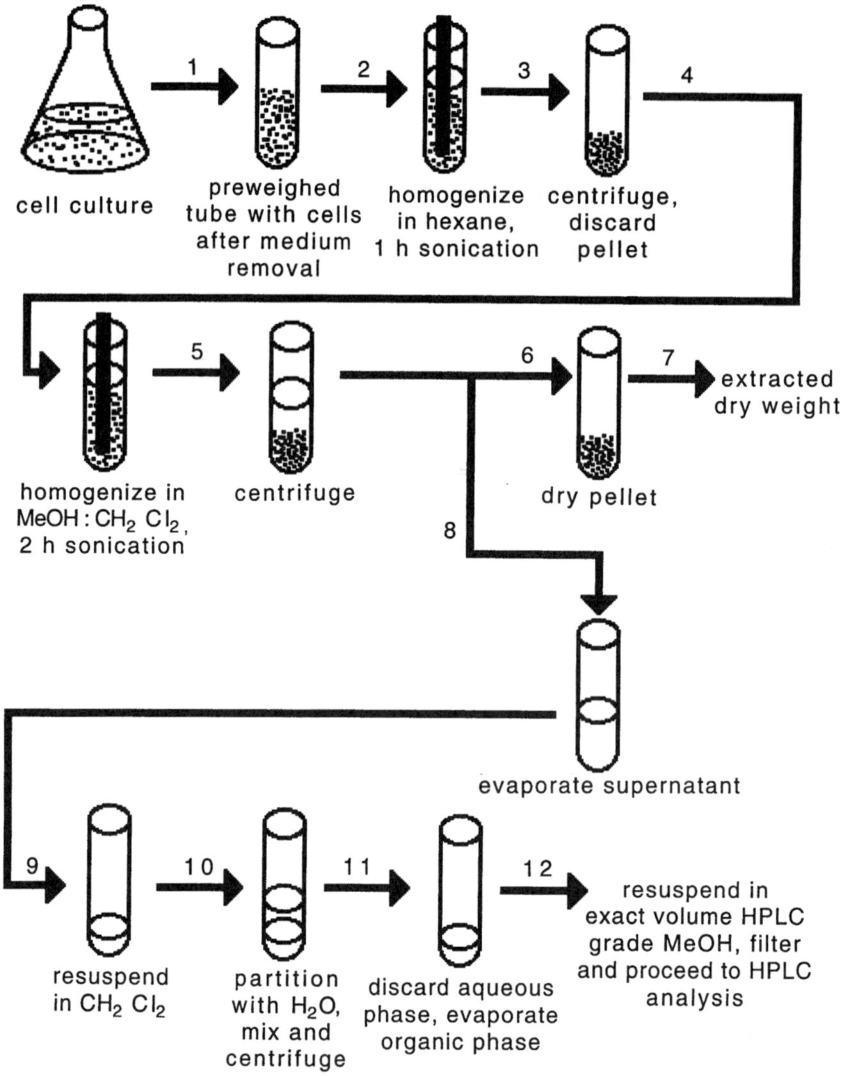

Figure 2. Extraction of Taxol® and taxanes from plant material.

methane fraction is collected, dried under vacuum or nitrogen at 25°C, redissolved in an exact volume of high performance liquid chromatography (HPLC) grade methanol (100–500 μL), and filtered through 0.45 μm filters for HPLC analysis (Figure 2).

Alternatively, the tissue can be extracted with 100% methanol. The hexane step can be avoided for cleaner extracts (e.g., cell suspension extracts) or if the column for liquid chromatography to be used in extract analysis is equipped with a precolumn, which prevents irreversible binding of impurities to the column. The Taxol® extraction step (methanol/dichloromethane, 1:1) can be repeated two or three times with the pellet of the first Taxol® extraction to achieve more complete Taxol® recoveries. In experiments involving several treatment comparisons and large number of samples with respective replicates, however, a single methanol/dichloromethane extraction step is more practical; this procedure allows for Taxol® recovery of 80–85%.

Media from cell cultures can be extracted for Taxol® with dichloromethane (1:1 v/v ratio of medium to solvent). For small volumes, the mixture contained in glass tubes is vigorously agitated on a tube mixer for 2–5 minutes, followed by sonication for an hour at room temperature. The tubes are briefly mixed again and centrifuged for 30 minutes at 2000 × *g* at 20°C. For large volumes, the mixture is agitated overnight in flask shakers and partitioned for an hour in separation funnels. The upper aqueous layer is discarded and the organic fraction is dried under vacuum at 25°C. The preparation for HPLC analysis is similar to the one just described, although the filtering of media extracts tends to be more difficult than in the case of plant material extracts.

This difficulty in filtering media extracts prior to HPLC analysis is probably a result of the presence of relatively large amounts of soluble PVP in the media. It may be necessary to use larger filters (keeping the pore size at 0.45 μm) and larger volumes of HPLC-grade methanol to filter and resuspend extracts of large volumes of medium. Once filtered, these extracts can be further concentrated prior to HPLC analysis. Other alternatives to facilitate medium sample preparation are possible: for example, the removal of PVP by pore size exclusion prior to medium extraction or the substitution of the last filtration step by centrifugation prior to HPLC analysis.

4.2 Separation, Quantification, and Identification

The taxane analyses are done by HPLC. The samples are passed through a reversed phase microbore column (e.g., Hewlett-Packard ODS Hypersil, 5 μm, 100 mm × 2.1 mm), using a mobile phase containing methanol, water, and acetonitrile; the linear gradient elution profile starts with 20:67:13 (solvent A), ends with 20:27:53 (solvent B) within 50 minutes, and includes a 5-minute wash in both solvents to reestablish the initial condition. The flow rate is 0.2 mL/min, and chromatograms are plotted at 227 nm, using an HPLC system equipped with a photodiode array detector (e.g., Hewlett-Packard 1090A HPLC system). Injection volumes are 10 μL, and solvents are all HPLC grade, filtered through 0.45 μm filters prior to use.

Microbore columns have the advantages of high sensitivity, low sample volume, and low solvent consumption. For cell suspension extracts analyzed with the foregoing solvent system, however, a better resolution of early eluting taxanes, such as baccatin III and 10-deacetyl-baccatin III, can be achieved using reversed phase analytical columns; in this case, the flow rate is increased to 1 mL/min, whereas all other chromatographic and detection parameters are not changed.

Quantification is done by the method of external standards, that is, a standard curve is generated with authentic standard (at least five different concentrations covering the concentration range of the extract samples) and run under the same chromatographic conditions as the samples. Taxol® and taxane identification is done by retention time, UV spectra and, cochromatography with standard. Peak purity checks can be performed as additional evidence for positive identification if the HPLC equipment has a diode array detector.

Further proof of identification can be obtained by collecting the corresponding HPLC peak and performing nuclear magnetic resonance (NMR) analyses. Over a series of injections, a concentrated sample is run through the HPLC system, the corresponding Taxol® peak in the sample extract is collected and pooled, until approximately 50 μg of Taxol® has been collected. The collected material is dried under vacuum at 25°C and resuspended in 150 μL of HPLC-grade methanol; then a small portion of it is injected again for purity evaluation and final quantification. If a single peak with the same retention time and UV spectrum as Taxol® is present in the amount required, the sample is dried under vacuum as before and freeze-dried for 24–48 hours to remove solvent residue. Taxol® identification, for example, may be obtained by ^{1}H NMR analysis in deuterated chloroform (≈ 50 μg of HPLC-purified sample Taxol® in 0.75 mL of deuterated chloroform with 0.05% tetramethylsilane as internal reference) at 500 MHz and 25°C on a Varian UNITY 500 spectrometer equipped with an inverse detection 5 mm probe. Taxol® and related taxane standards may be requested for nonhuman research use from the Drug Synthesis and Chemistry Branch, Developmental Therapeutics Program, Division of Cancer Treatment of the National Cancer Institute (U.S.A.). Taxol® is also available from several chemical supply companies.

5 TAXOL® YIELD AND POSSIBILITIES OF MANIPULATION

5.1 Characteristics of Taxol® Yield and Growth in *Taxus* In Vitro Cultures

Experiments with callus and cell cultures of *Taxus cuspidata* have shown that growth and Taxol® yield are generally inversely proportional. The reason for this phenomenon, also observed with the production of other secondary metabolites by a number of plant species, is not well understood. Possible explanations include the lack of expression of key enzyme genes in cells with low differentiation, diversion of substrates for secondary metabolism into growth processes, unregulated catabolism, and lack of end product transport systems and storage sites.

The analysis of Taxol® yields in different in vitro cultures of *Taxus cuspidata* supports the observation that faster growth rates and/or lower degrees of differentiation are normally associated with lower Taxol® amounts (Figure 3). Stem-derived explants of *Taxus*

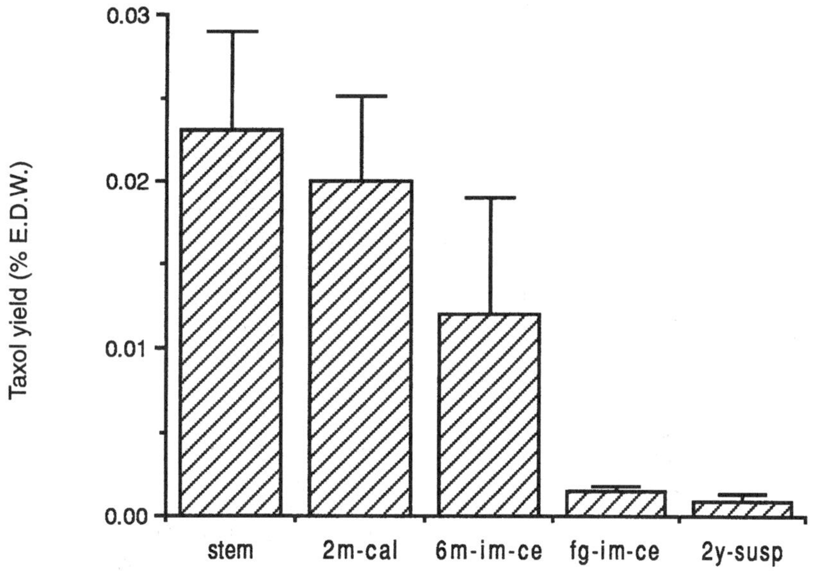

Figure 3. Taxol® yield as a percentage of the extracted dry weight from *Taxus cuspidata* sources of decreasing degree of differentiation: stem, intact plant stem; 2m-cal, 2-month-old callus, 6m-im-ce, 6-month-old immobilized cell cultures; fg-im-ce, fast-growth-selected, 6-month-old, immobilized cell cultures; 2y-susp, 2-year-old cell suspensions.

Table 2 Growth Parameters of *Taxus cuspidata* Cell Suspensions in Comparison to Some Other Plant Species (shake-flask experiments)

Species	Doubling Time (days)	Specific Growth Rate (day^{-1})
Taxus cuspidata	20.6	0.034
Brucea javanica	32.0	
Callitris drummondi	40.0	
Catharanthus roseus	3.0	0.32

cuspidata after 2 months of culture in B5bPVP display an average eight-fold increase in fresh weight and Taxol® yields of 0.020 ±0.005% of the extracted dry weight (equivalent to the yield of stems of the intact plant). Slow-growing, 6-month-old immobilized cell cultures of *T. cuspidata* grown in B5C2 have variable amounts of Taxol®, up to 0.012 ±0.007% of the dry weight (equivalent on average to the bark of the intact plant), whereas 6-month-old immobilized cell cultures visually selected for faster growth yield 0.0014 ±0.0003% of Taxol® on a dry weight basis. One-year-old cell suspension lines of *T. cuspidata* grown in B5C2 (doubling time, ≈ 15 days) have yields of Taxol® ranging from 0.0001 to 0.0010% on a dry weight basis; cell suspensions of the same age grown in B5C1 have Taxol® yields of 0.0004 ± 0.0002% of the dry weight. A 2-year-old cell suspension line of *T. cuspidata* grown in B5C2 (dry weight doubling time of 17 days) yielded 0.0014 ± 0.0005% of Taxol® (dry weight basis) when cultured as 300 mL suspensions in 1 L Erlenmeyer flasks. These yields may be altered by precursor feeding (see Section 5.2).

Studies on the kinetics of growth and Taxol® accumulation in cell suspensions of *Taxus cuspidata* (cell line FCL1F) showed a relatively slow growth rate and confirmed the relationship between growth and Taxol® yield. The doubling time was approximately 17 days for fresh weight and 21 days for dry weight. Although relatively long in contrast to data from herbaceous species, these doubling times are comparable to or even shorter than those of other woody species that grow slowly (Table 2). The maximum specific growth rates were 0.040 g^{-1} (fresh weight) and 0.034 g^{-1} (dry weight). The accumulation of Taxol® in the cells was not growth-linked. The highest taxol accumulation values occurred in cells in stationary phase, and Taxol® amounts in the cells decreased with the increase in cell biomass (Table 3).

Although the yields of other taxanes in yew callus and cell cultures have not yet been investigated in detail, HPLC evidence indicates the presence of baccatin III, 10-deacetyl-baccatin III, 1-dehydroxy-baccatin V, 7-epitaxol, and cephalomannine in both types of culture of *Taxus cuspidata.* Baccatin III yields in 2-month-old callus of this species grown in B5bPVP are 0.043 ± 0.016% of the extracted dry weight. A 2-year-old cell line suspension of *Taxus cuspidata* (FCL1) has been estimated to yield 0.018 ± 0.010% of baccatin III (extracted dry weight basis). Taxanes such as 10-deacetyl-baccatin III and baccatin III can be used to produce Taxol® through semisynthesis. These taxanes can also be used to produce Taxotere®, a semisynthetic Taxol® analogue with higher antineoplastic properties and higher water solubility than Taxol®; Taxotere® was first developed by Dr. P. Potier and associates and Rhône-Poulenc Rorer in France and is currently undergoing clinical trials for the treatment of ovarian cancer.

Table 3 Growth and Taxol® Accumulation in Cells and in Medium of Suspensions of *Taxus cuspidata*[a]

Time (days)	Fresh Weight (g/mL)	Taxol® Accumulated[b] In Cells (μg/g EDW)	In Medium (μg/mL)
0	0.15	4.0	0.013
6	0.20	nd	0.356
12	0.24	2.0	0.060
18	0.32	nd	bd
25	0.35	0.4	bd
32	0.39	1.2	bd
38	0.28	2.8	0.100
44	0.27	3.2	0.050

[a]Values are the mean of two to three determinations; inoculi were 1/6 w/v obtained from a single stationary phase donor suspension.
[b]EDW, estimated dry weight; nd, not determined; bd, below limit of detection.

Media collected from 6-month-old immobilized cell suspensions of *T. cuspidata* and extracted with dichloromethane may contain trace amounts of Taxol® and baccatin III as indicated by HPLC analysis. Estimated daily rates of Taxol® accumulation in media were approximately 4 μg/L both B5C2 and B5C1. These low amounts suggest that Taxol® in the media results from cell turnover rather than active secretion, although this estimation was done with medium collected at a single point in time and assuming no Taxol® degradation in the medium. In differentiated physiologically active cells, Taxol® may be stored inside the cells, possibly in the vacuole or other membranous compartment.

However, in a more detailed time course study using the FCL1F cell suspension line of *Taxus cuspidata,* the analysis of the volumetric total production of Taxol® at stationary phase indicated that 66% of the total Taxol® was in the medium. In contrast, at a given time, Taxol® accumulation in the medium on a per-milliliter basis was generally a small portion of that contained per gram of extracted dry weight basis in the cells (Table 3). The occurrence of a larger extracellular component of the total volumetric Taxol® at stationary phase (productive phase) may be the result of excretion by the cells that reach highest amounts of Taxol® accumulation in this phase. Studies on the production of various secondary products by plant cell cultures show that the proportion of excreted to endogenously accumulated product can be highly variable depending on species and cell line.

Taxol® accumulation in the medium of cell suspensions of *T. cuspidata* can vary considerably with the time of the growth cycle. In the kinetic study just mentioned, the accumulation of Taxol® in the medium had two peaks (Table 3). The first peak of Taxol® in the medium (≈ 0.4 μg/mL) occurred in the early portion of the exponential growth phase, soon after inoculation in fresh medium. This early release of Taxol® in the medium could be a response to the transfer into fresh medium ("transfer effect") and/or to the increase of the internal pH of the cells due to the absorption of ammonium (exchanged with H^+), which takes place in this portion of the growth cycle. Following the first peak, Taxol® was degraded, possibly by enzymes released into the medium. A second peak of Taxol® in the medium (0.1 μg/mL) occurred at stationary phase and may have resulted from excretion by the cells that reach their highest level of Taxol® at this stage and/or from a small degree of cell lysis, common at later stages of the growth cycle. Baccatin III accumulation in the medium peaked after approximately 9 days of culture at 6.2 μg/mL in the same cell line suspension. The occurrence of product excretion into the medium is useful for industrial application of plant cell culture technology, since it facilitates downstream processing (especially when using immobilized cells).

5.2 Precursor Feeding

The problem of overall low Taxol® yields, particularly at higher growth rates, can be partly overcome by feeding Taxol® precursors to the cell cultures. The appropriate combination of type, amount, and feeding time of precursors able to enter the cells can significantly increase the yield of secondary metabolites in plant cell cultures, provided the biosynthesis of the metabolite of interest is at least partly limited by poor activity of enzymes involved in precursor formation or appropriate supply of precursor to the biosynthetic reaction. In 2-month-old callus cultures of *T. cuspidata,* phenylalanine supplementation at 0.1 mM in B5bPVP can double the Taxol® yield in relation to regular B5bPVP without affecting growth significantly; similar results can be obtained with cell suspensions after a 25-day growth period in medium supplemented with the same amount of this amino acid. Phenylalanine acts as a precursor for the side chain of Taxol.®

Other potential precursors for the Taxol® side chain were investigated in feeding experiments to both callus and cell suspension cultures of *Taxus cuspidata.* Compounds that significantly enhanced Taxol® yields in cultures were phenylalanine, benzoic acid, hippuric acid, serine, and glycine. Impacts on Taxol® yields were concentration dependent. Taxol® yields were increased by two to four times when the putative precursors were provided at concentrations of 0.2 mM for callus cultures and 0.05 mM for cell suspension cultures; all cultures were fed at the onset of the stationary phase of growth. The best results were obtained with phenylalanine or benzoic acid feeding. Results suggest that precursor feeding is a promising approach to increase Taxol® yields in yew cell cultures without detrimental effects on biomass generation.

5.3 Cell Line Selection

Screening different cell lines for high yields of Taxol® and/or taxanes may lead to the establishment and propagation of elite cell lines that can be used for large-scale production purposes. The technique has been successfully applied, for instance, to the production of ajmalicine by *Catharanthus roseus,* shikonin by *Lithospermum erythrorhizon,* and berberine by *Coptis japonica.* Although laborious and time-consuming, the approach may turn out to be advantageous for long-term potential industrial applications of yew cell cultures as a source of Taxol® and taxanes.

The presence of a visual feature associated with high productivity is highly desirable for selecting high yielding cell lines (e.g., colored product, such as anthocyanin). In the case of *Taxus cuspidata,* additional studies are required to identify a reliable visual feature associated with taxane productivity. When such a facilitating screening parameter is not present at all, however, as often happens, the use of standard phytochemical extraction and separation procedures is required or, if the product of interest can be made antigenic, immunoassays can be used in the screening procedure. Antitaxol and several other antitaxane antibodies are commercially available from the Hawaii Biotechnology Company.

6 FUTURE PERSPECTIVES

Data suggest that cultured cells of *Taxus* spp., although in early development, represent a source of Taxol® and related taxanes, which can be used in the production and semisynthesis of this antineoplastic agent and its analogues. Scaled-up cell cultures studies need to be developed with emphasis on nutrient consumption rates, gas composition, cell shearing sensitivity, and precursor feeding schedules. The studies, in combination with the biochemical and growth data accumulated and ongoing with callus and cell cultures, may provide further insights on the factors that limit growth, and the metabolic pathways leading to Taxol® and related taxanes in yew cells.

See also Bioprocess Engineering; Cancer; Drug Targeting and Delivery, Molecular Principles of; Medicinal Chemistry.

Bibliography

Blechert, S., and Guénard, D. (1990) *Taxus* alkaloids. In *The Alkaloids. Chemistry and Pharmacology,* Vol. 39, A. Brossi, Ed., pp. 195–238. Academic Press, San Diego, CA.

Charlwood, B. V., and Rhodes, M. J. C., Eds. (1990) *Secondary Products from Plant Tissue Culture,* Vol. 30 in *Proceedings of the Phytochemical Society of Europe.* Clarendon Press, Oxford.

Evans, D. A., Sharp, W. R., Ammirato, P. V., and Yamada, Y., Eds. (1983–1984) *Handbook of Plant Cell Culture,* Vols. 1–6. Macmillan, New York, London.

Facchini, P. J., DiCosmo, F., Radvanyi, L. G., and Neumann, A. W. (1990) Immobilization of cells by spontaneous adhesion. In *Methods in Molecular Biology,* Vol. 6, *Plant Cell and Tissue Culture,* J. W. Pollard and J. M. Walker, Eds. Humana Press, Clifton, NJ.

Fett-Neto, A. G., DiCosmo, F., Reynolds, W. F., and Sakata, K. (1992) Cell culture of *Taxus* as a source of taxol and related taxanes. *Bio/Technology,* 10:1572–1575.

———, Melanson, S. J., Sakata, K., and DiCosmo, F. (1993) Improved growth and taxol yield in developing calli of *Taxus cuspidata* by media composition modification. *Bio/Technology,* 11:731–734.

Nijkamp, H. J. J., Van der Plas, L. H. W., and Van Aartrijk, Eds. (1990) *Progress in Plant Cellular and Molecular Biology: Proceedings of the Seventh International Congress on Plant Tissue and Cell Culture, Amsterdam, June 1990.* Kluwer Academic Publishers, Dordrecht, Boston, London.

Telomerase

Silvia Bacchetti and Calvin B. Harley

1 Telomerase and Telomere Maintenance

2 Telomerase Biochemistry
- 2.1 Telomerase: A Novel RNA-Dependent DNA Polymerase
- 2.2 Functional Sites and Catalytic Activity
- 2.3 Telomerase Components

3 Telomerase Assays
- 3.1 The Standard Assay
- 3.2 TRAP: Telomere Repeat Amplification Protocol

4 Telomeres, Telomerase, and the Replicative Capacity of Eukaryotic Cells

5 Perspectives
- 5.1 Telomerase and Human Disease
- 5.2 Telomerase Biogenesis, Structure, and Function

Key Words

dNTP Deoxyribonucleoside triphosphate.

rNTP Ribonucleoside triphosphate.

Telomeres Essential genetic elements at the termini of linear chromosomes.

Telomerase The ribonucleoprotein enzyme that synthesizes telomeric DNA.

Telomerase Primer An oligonucleotide, typically a telomeric DNA sequence, capable of being extended by telomerase.

Terminal Restriction Fragment (TRF) The telomere-containing fragment generated by digestion of genomic DNA with restriction endonucleases.

Immortal cells transmit essentially a complete genome from one division to the next. To ensure the full replication of their DNA, most immortal eukaryotic cells possess a unique enzyme, telomerase, which apparently evolved with the ends of linear chromosomes, specifically for the de novo synthesis of telomeric DNA. Telomere maintenance is the only known biochemical function of telomerase, and in the absence of telomerase, telomeric DNA is gradually lost with each round of cell division. Normal human somatic cells are mortal in the sense that they have a finite capacity for division. The observation that these cells lack telomerase, while most tumor cell populations express telomerase, suggests an important biological function for this enzyme in controlling a key difference between mortal and immortal cells.

1 TELOMERASE AND TELOMERE MAINTENANCE

Telomeres are nucleoprotein structures that cap the ends of eukaryotic chromosomes. They are essential for chromosome function and stability because they prevent nucleolytic degradation and fusion of chromosome ends, contribute to chromosome movement and localization within the nucleus through their attachment to the nuclear matrix, and can affect gene expression through position effects. Telomeres have a distinctive chromatin organization and are composed of specific proteins, of which only a few have been characterized to date, primarily from lower eukaryotes. It is however apparent that telomeric proteins are crucial to telomere structure and function. The second component of telomeres is noncoding DNA consisting of tandem reiterations of simple G-rich repeats oriented 5′-to-3′ toward the chromosome termini and in some cases overhanging the C-rich strand.

The evolutionary conservation of telomeric DNA sequence and organization suggests that both these features are important for telomere function. Moreover, telomeric DNA contributes to the complete replication of the chromosome. Since eukaryotic DNA polymerases catalyze synthesis only in the $5' \rightarrow 3'$ direction and utilize an RNA primer for initiation, linear DNA molecules cannot be replicated in their entirety (Figure 1). Telomeric DNA shoulders the loss of terminal sequences masked by the 5′-most primer, thus acting as a buffer against loss of internal and essential genes. In addi-

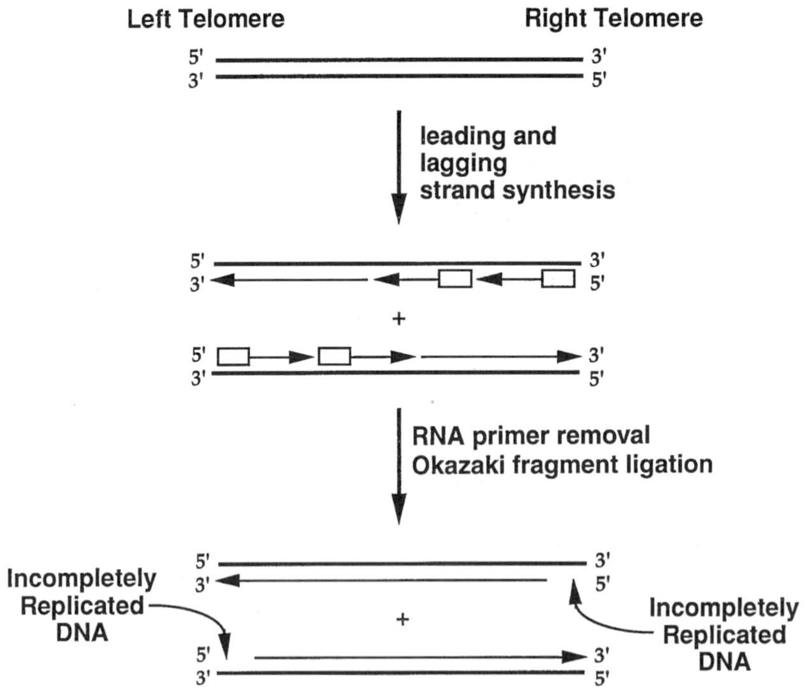

Figure 1. The end replication problem: DNA polymerases synthesize DNA in the $5' \rightarrow 3'$ direction and require an RNA primer (open boxes) for initiation of synthesis. Synthesis of the lagging strand is discontinuous and generates Okazaki fragments that are ligated to one another following removal of primers and new DNA synthesis. Removal of the 5′-most primer results in an irreparable gap at the 5′ end of the molecule. Thus, the daughter strand is shorter than the parental strand at this point. The extent of sequence loss reflects the size and position of the 5′-most primer; since the latter may vary, daughter molecules of different length may be generated.

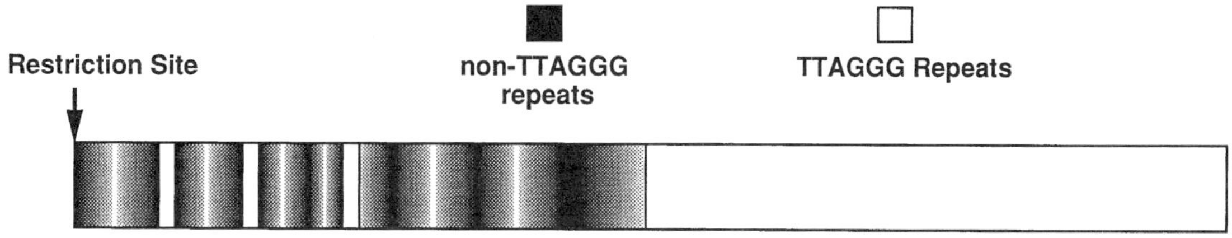

Figure 2. Structure of terminal restriction fragments: digestion of genomic DNA with restriction enzymes releases terminal restriction fragments (TRFs) comprising a telomeric region of perfect TTAGGG repeats, and a subtelomeric region containing degenerate TTAGGG repeats, possibly interspersed with perfect repeats and unrelated simple sequences.

tion, being subject to elongation by telomerase, telomeric DNA provides a means for compensating for the loss of sequences stemming from the end replication problem.

Telomere length is extremely variable among species, possibly as a result of differences in minimal functional sizes that may reflect number of telomeric proteins or protein-binding sites. In most cases, however, telomere length is unknown because current methodology does not allow direct and precise measurement of telomeric DNA tracts on individual chromosomes. True telomeric DNA does not contain sites for known restriction enzymes and is therefore cleaved from genomic DNA as part of terminal restriction fragments (TRFs), which also comprise a subtelomeric region (Figure 2). Because of the variable position of restriction enzyme sites and the unequal loss of sequences at each chromosome end (Figure 1), TRFs are extremely heterogeneous in size. At present, telomeric DNA length is usually equated to the average TRF size in a population, as determined by Southern hybridization.

2 TELOMERASE BIOCHEMISTRY

2.1 Telomerase: A Novel RNA-Dependent DNA Polymerase

Telomerase activity was first discovered in the ciliate *Tetrahymena* based on its novel property of extending a telomere-like oligonucleotide with de novo synthesized telomeric DNA. The enzyme is a ribonucleoprotein complex containing proteins and an RNA component. A short region of the RNA component, called the template domain, is complementary to the species-specific telomere sequence and directs the synthesis of the proper telomeric DNA (Figure 3). Thus, telomerase functionally is a reverse transcriptase: an RNA-dependent DNA polymerase. However, unlike viral reverse transcriptases, which can fully copy varied "exogenous" RNAs, telomerase utilizes only a small template domain of its specific integral RNA. In addition, telomerase is capable of reiteratively copying the template domain without dissociating from the extended DNA chain. Thus, after binding to a DNA primer (typically a telomere sequence), telomerase cycles between an elongation phase in which deoxyribonucleotides (dNTPs) are sequentially added to the 3′ end of the DNA, and a "translocation" event, in which telomerase advances one repeat relative to the telomere (Figure 3). The reiterative copying of the template domain accounts for the simple repeated structure of telomeric DNA at chromosome ends from most eukaryotic species.

2.2 Functional Sites and Catalytic Activity

The template region of the integral RNA component of telomerase typically contains more than one repeat of the complementary telomere sequence. The additional bases at the 3′ side of the template domain are utilized in part as an "alignment domain" to facilitate the binding of DNA to telomerase. This binding and alignment is especially important following the translocation event when the 3′ end of the DNA slips back such that it is entirely outside the template domain (Figure 4). Thus, the template region contains two domains: the alignment domain and the template domain. However, RNA:DNA hybridization is not sufficient to account for binding of the DNA primer to telomerase. An "anchor site" involving protein–DNA interactions also functions in initial primer binding as well as in maintaining the correct position of the DNA during the elongation/translocation events.

The anchor site of telomerase is somewhat promiscuous in that it is capable of utilizing various DNA sequences as primers for telomere extension. Telomerase can also utilize an RNA primer under certain circumstances and even incorporate ribonucleoside triphosphates (rNTPs) when they are present a high concentrations in the absence of the correct deoxyribonucleoside triphosphates (dNTPs). However, only dNTPs complementary to the template base are efficiently added to the growing DNA primer. Nontelomeric primers that function in a telomerase assay are generally relatively G-rich single-stranded DNAs and typically have a 3′ end with some similarity to the natural telomere sequence. This suggests that the anchor site has some affinity for exposed guanine bases and that hybridization of the 3′ end of the DNA primer to the alignment region is important in forming a proper primer:template configuration. With the 3′-OH of the primer positioned in the catalytic site of the enzyme, formation of the phosphodiester bond catalyzed by all nucleic acid polymerases can occur between the terminal 3′-OH and the α-phosphate of the incoming dNTP, which is selected based on its complementarity to the templating ribonucleotide in the catalytic site at that time.

No additional high energy bond cleavage other than that of the dNTPs appears to be necessary for elongation and translocation. It is possible that translocation is facilitated by two additional properties of telomerase: a helicase-like activity that maintains a small number of basepairs between the growing DNA chain and the template, and strain, which builds during the elongation steps when the relative position of the RNA and the catalytic site of the proteins must gradually change.

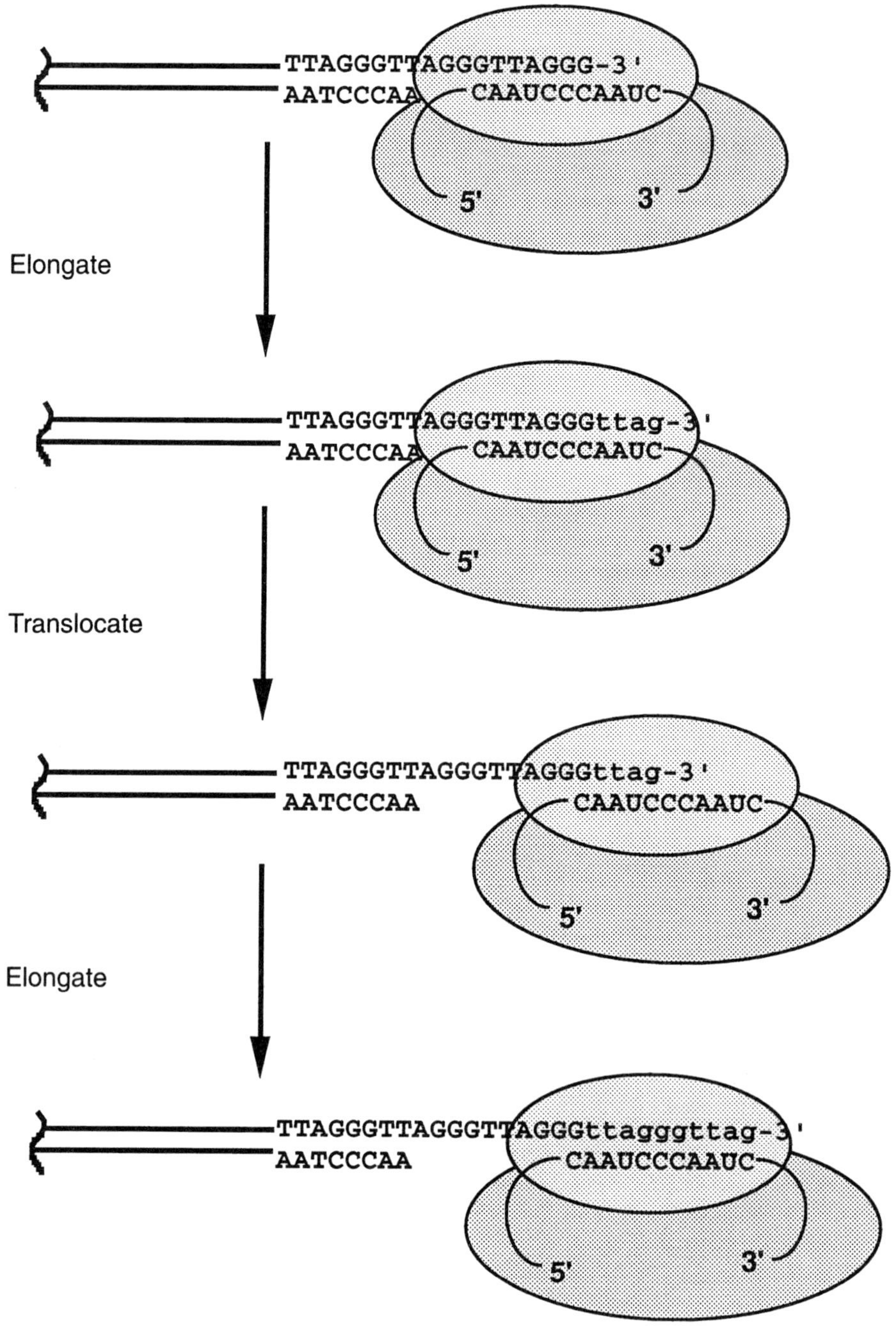

Figure 3. Telomerase mechanism of action: the spatial relationships between telomerase protein subunits (shaded ovals), telomerase RNA (curved strand showing the human template region, CAAUCCCAAUC), and the end of a human chromosome during the cyclical process of elongation and translocation.

2.3 Telomerase Components

The RNA component of telomerase has been cloned from a number of ciliates, yeasts, and mammalian species, including humans (Table 1). In every case, there is a template region comprising the putative alignment and template domains. The template domain has been changed by site-directed mutagenesis of the gene for the RNA component from several species. Cells expressing the mutated gene have been shown to generate telomerase which synthesizes a mutated telomere sequence, thus confirming that the correct gene was cloned. In addition, site-directed mutagenesis of the template and alignment domains clearly demonstrates the functional role of these sites.

Comparison of the genes for the RNA component of telomerase across different species indicates a high degree of sequence divergence. However, the RNA of ciliate telomerases has a conserved secondary, if not tertiary structure, which places the template region in an exposed loop. The positioning of the template region accounts

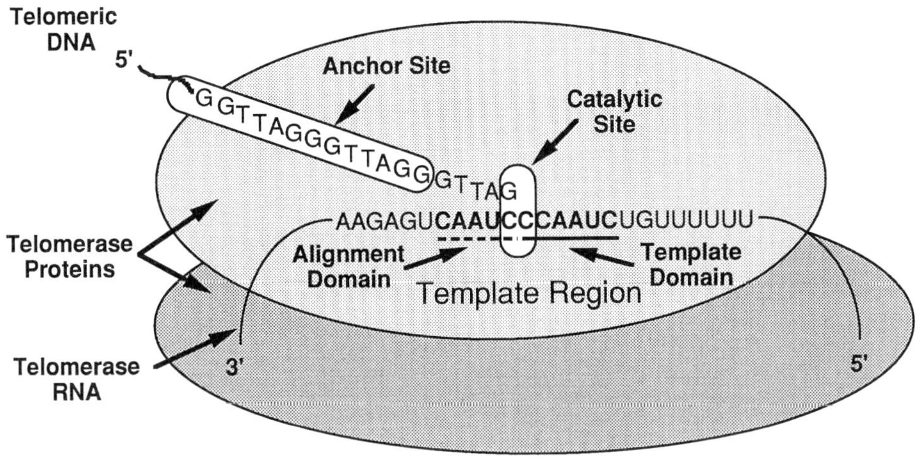

Figure 4. The template region and primer-binding site of human telomerase: the presumed primer–template configuration at the catalytic site of human telomerase is schematically illustrated, along with functional sites within the protein components (shaded ovals) and the RNA template region. This picture captures the relative position of telomeric DNA and the template region just after a translocation event: the 3′ end of the DNA is positioned opposite the alignment domain of the RNA template region, with the catalytic site ready to add dNTPs according to the sequence dictated by the template domain. The number of protein subunits in human telomerase is unknown.

for the ability of primers and antisense oligonucleotides to gain access to this part of the RNA. Other regions of the RNA appear to be well protected. The secondary structure of the ciliate telomerase RNA was determined in part from comparison of the diverged primary sequences and the discovery of "compensatory mutations" that maintain base pairing on opposite sides of predicted stem structures. The deduced structure was found to be consistent with chemical probing of exposed bases within the native holoenzyme. However, details of the tertiary structure of the RNA or of the complex as a whole are essentially unknown. Structure predictions for the yeast and mammalian telomerase RNAs should be published and confirmed in the near future.

The telomerase proteins have at least three functions: binding the integral RNA, binding DNA, and catalyzing the elongation/translocation events. Recent evidence also indicates that telomerase has a cleavage activity capable of removing the terminal nucleotide of the DNA extension product at the pause site (i.e., when the extension product is at the 5′ end of the template domain). The function of the cleavage activity is unclear, however; nor is it certain where this activity resides.

The cloning of protein components of telomerase has been reported for only one species, *Tetrahymena thermophila.* The *Tetrahymena* enzyme has two protein components, sized 80 and 95 kDa. Studies with antibodies directed against the separate subunits suggest that the smaller subunit is intimately involved in binding to the RNA. In contrast, cross-linking studies suggest that the anchor site involved in DNA primer binding resides with the 95 kDa protein. The primary sequence is unique for both subunits, clearly pointing to the novel properties of telomerase compared to other nucleic acid polymerases. However, limited sequence homology with RNA and DNA-dependent RNA polymerases and associated factors suggest that telomerase may be most closely related to RNA polymerases.

3 TELOMERASE ASSAYS

3.1 The Standard Assay

The standard telomerase reaction uses a telomeric oligonucleotide, such as $d(TTAGGG)_3$ for the human enzyme, as the primer for extension by telomerase in a cell-free extract mixed with appropriate

Table 1 Some Known Telomerase RNAs

Species[a]	Telomere Repeat	RNA Size (nucleotide)	Template Region (3′ → 5′)[b]
Human	GGTTAG	≈450	CAAUCCCAAUC
Mouse	GGTTAG	≈450	AUCCCAAUC
T. thermophila	GGGTTG	159	AACCCCAAC
O. nova	GGGTTTTG	190	CCAAAACCCCAAAAC
E. crassus	GGGTTTTG	189	CCAAAACCCCAAAAC
S. cerevisiae	$(TG)_{1-3}TG_{2-3}$	≈1300	CACCACACCCACACAC

[a]Complete cDNAs for the RNA components have been published for human, mouse, *Tetrahymena thermophila, Oxytricha nova, Euplotes crassus, Saccharomyces cerevisiae,* and a number of other ciliates and yeast species.

[b]The putative template domain of the template region is underlined. The degenerate telomere sequence of *S. cerevisiae* can be explained by variable alignment of the telomeric DNA 3′ end on the template region and/or variable sites of termination during elongation.

Table 2 Human Telomerase Assays

A. Standard Reaction

1 volume CHAPS cell extract + 1 volume 2× reaction buffer
Incubate for 30–60 minutes at 30° C
Extract DNA and resolve bands on a sequencing gel.

CHAPS Cell Extract

10^4–10^6 washed cells in 20 μL of
- 10 mM Tris-HCl (pH 7.5)
- 1 mM $MgCl_2$
- 1 mM EGTA
- 0.1 mM phenylmethylsulfonyl fluoride
- 5 mM β-mercaptoethanol
- 0.5% CHAPS detergent
- 10% glycerol

2× Reaction Buffer

100 mM Tris-HCl (pH 8)
100 mM $KO_2C_2H_3$
- 6 mM $MgCl_2$
- 5 mM EGTA
- 2 mM spermidine
- 0.2 mM spermine
- 2 mM β-mercaptoethanol
- 2 μ $(TTAGGG)_3$
- 5 mM dATP
- 5 mM dTTP
- 1.25 μM [α-^{32}P]dGTP (3000 Ci/mmol)

B. TRAP Reaction (Telomere Repeat Amplification Protocol)

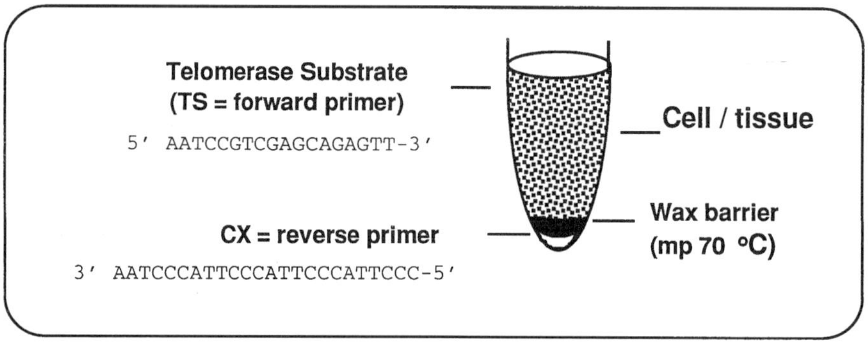

Lyophilize 0.1 μg CX primer (5′-CCCTTACCCTTACCCTTACCCTAA-3′) onto bottom of tube.
Seal with 7–10 μg molten wax (Ampliwax, Perkin-Elmer).
Add 50 μL of TRAP reaction solution above the wax barrier.
Incubate 10 minutes at room temperature, allowing telomerase, if present, to extend the TS primer.
Transfer tubes to a thermal cycler for 27 rounds at 94°C for 30 seconds, 50°C for 30 seconds, and 72°C for 90 seconds.
Resolve bands on a sequencing gel.

TRAP Reaction Solution[a]

20 mM Tris-HCl (pH 8.3)
1.5 mM $MgCl_2$
63 mM KCl
0.05% Tween-20
1 mM EGTA
50 μM each dATP, dTTP, dCTP, dGTP
0.1 μg TS substrate
1 μg of T4g32 protein (Boehringer Mannheim)
0.1 mg/mL bovine serum albumin
2 U *Taq* DNA polymerase (Boehringer Mannheim)
1–2 μL CHAPS cell extract

[a]For radioactive detection of bands, include 0.2–0.4 μGi [α^{32}P]dGTP (3000 Ci/mmol).

buffers and deoxyribonucleotides (Table 2A). Most of the telomerase activity from mammalian cells resides in the supernatant (i.e., "cytoplasmic fraction") from a 100,000 × *g* centrifugation (the "S100" fraction) of either mechanically ruptured or detergent-treated human cells. Since telomeres reside in the nucleus of a cell, it is assumed that most telomerase in vivo would be found in the nucleus and that the enzyme may leak from this compartment during preparation of the extract.

The products of the telomerase reaction in a cell-free extract can be labeled by the incorporation of [$\alpha-^{32}P$] dGTP during elongation of the primer and then detected by autoradiography after resolution of the products on a polyacrylamide gel. Because telomerase is processive and reiteratively copies the six-nucleotide template in a cyclical fashion, the products of the many simultaneous extension reactions occurring in the test tube generate a "ladder" of bands spaced at a six-nucleotide repeat distance (Figure 5). Moreover, the generation of the characteristic six-nucleotide ladder is sensitive to pretreatment of the extract with RNase, since telomerase utilizes its integral RNA as the template for telomere synthesis. Thus, a "+ RNase" reaction is often included as a negative control in telomerase assays (Figure 5). For both human and *Tetrahymena* telomerase, the major band detected at the six-nucleotide phasing of the reaction products corresponds to the translocation event, suggesting that translocation is slower than elongation.

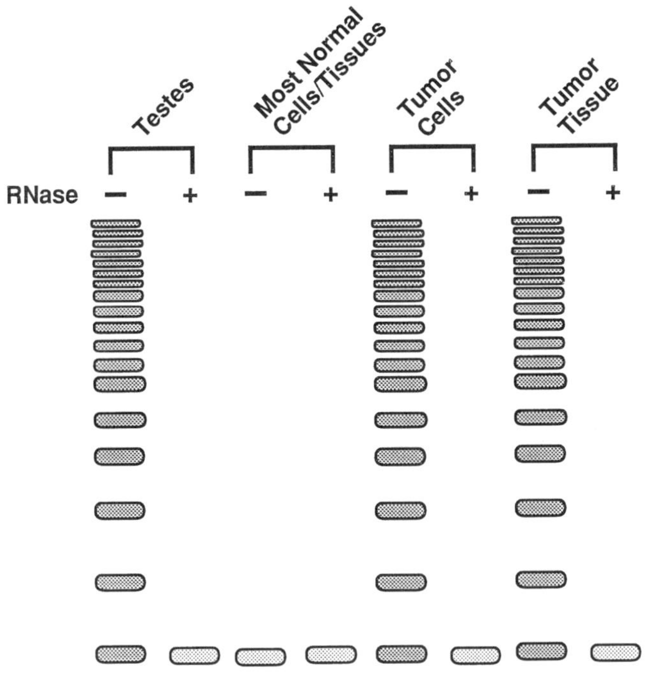

Figure 5. A standard telomerase "ladder": gel electrophoresis of telomerase products resolves the DNA oligonucleotides according to size. Since the translocation event is slow relative to elongation, there is a greater frequency of partial products at six-nucleotide intervals corresponding to the extension products that reach the 5′ end of the template domain. In telomerase-positive tissues such as testes, tumor cells in culture, or tumor biopsies, a distinct ladder of products is seen. This pattern is not seen when the extract is pretreated with RNase, since RNase destroys telomerase activity.

3.2 TRAP: Telomere Repeat Amplification Protocol

A highly sensitive and reproducible assay for telomerase utilizes a polymerase chain reaction (PCR) step to amplify the telomerase products prior to detection (Table 2B). Modifications to the reaction conditions for the telomerase portion of this "telomere repeat amplification protocol," or TRAP, streamline the overall assay and reduce the chance of PCR artifacts. A critical modification is the use of a non-TTAGGG oligonucleotide (TS), which acts as a good primer for telomerase but reduces the probability of primer/dimer artifacts during the PCR amplification step. The C-rich complementary primer for the PCR reaction is also changed from the perfect CCCTAA repeat to reduce the possibility of false positive products in the PCR phase of the assay, which may be generated by primer/dimer formation. TRAP is more reliable and sensitive than the standard assay and is capable of detecting telomerase activity in as few as 10–100 telomerase-positive cells among 10^5–10^6 telomerase-negative cells. Thus, very small tissue samples can be analyzed for the presence of potentially malignant cells. Detection systems can use either radioactivity or stains incorporated or intercalated into the double-stranded DNA created by the PCR reaction. Fresh or frozen biopsy, autopsy, and surgical specimens, including frozen histological sections, can be analyzed for telomerase activity with TRAP. TRAP, however, is an activity assay and thus cannot be used on fixed material or on poorly stored or necrotic material in which enzyme activity is compromised. To avoid false negatives, tissues should also be tested for the presence of inhibitors of telomerase or *Taq* polymerase by mixing or titration experiments.

4 TELOMERES, TELOMERASE, AND THE REPLICATIVE CAPACITY OF EUKARYOTIC CELLS

In multicellular eukaryotes, reproductive cells have unlimited proliferative potential (immortality), whereas somatic cells are capable of only a limited number of divisions. The finite replicative capacity of normal somatic cells has been termed "cellular senescence" and is believed to play a significant role in age-related disease. Conversely, the escape from cell senescence and ultimately the immortalization of somatic cells are believed to be important aspects of tumorigenesis. Thus, cell senescence can be viewed as a tumor suppressor mechanism.

It has been proposed that the proliferative capacity of somatic cells is at least partly controlled by the status of telomeres and telomerase (Figure 6). In the absence of the enzyme or of any mechanism for telomere maintenance, the loss of telomeric DNA that occurs with each round of replication is thought to act as a "mitotic clock" that registers the number of times a cell has divided and signals cessation of division (senescence), presumably when telomeres have attained a functionally suboptimal size. Somatic telomere length therefore may be both an indicator of the cell proliferative history and a predictor of its replicative potential. In accord with this model, sperm and ova express telomerase activity and maintain telomere length from one generation to the next. At least in humans, somatic cells usually lack the enzyme (Table 3) and contain telomeres that are progressively shorter with donor age or cell division in culture. This pattern of telomerase expression is consistent with somatic repression of the enzyme, a process that may curb cancer

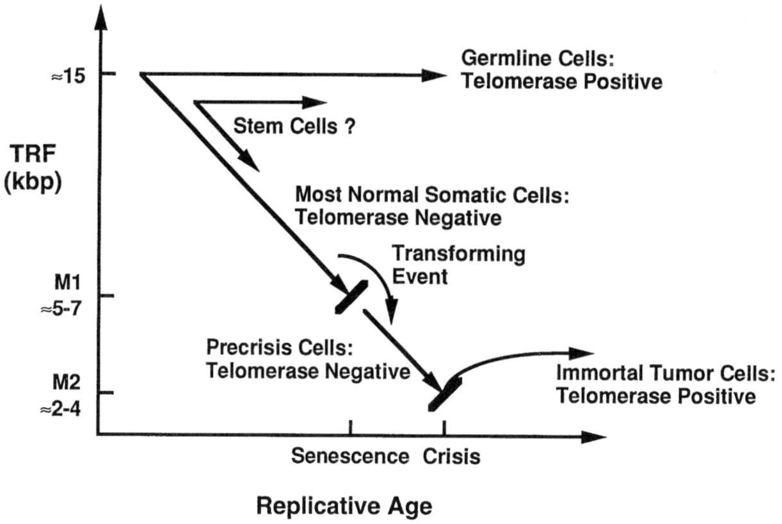

Figure 6. The telomere hypothesis: reproductive cells, and perhaps stem cells, express telomerase activity and maintain stable telomeres. During development, telomerase is repressed in most somatic cells and, as a consequence, telomeres shorten with cell division. Shortening of telomeres is thought to monitor the proliferative history of somatic cells and to signal growth arrest (senescence) when telomeres have reached a functionally suboptimal size. Transformation extends the proliferative capacity of cells but does not reactivate telomerase, and transformed cells continue to lose telomeric DNA till the end of their extended life span (crisis). At crisis, telomeres on average are extremely short and some may be absent, and chromosomes become unstable. The loss of cell viability occurring at this stage may be due to nonfunctional telomeres and/or lethal chromosome aberrations resulting from terminal fusion of unprotected chromosomes. The rare cells that reacquire telomerase activity have stable telomeres and are immortal.

Table 3 Telomerase Activity in Human Tissues and Cell Strains

Tissue of Origin	Telomerase Activity (number positive/number tested)
Normal Tissues[a]	
Adrenal gland (fetal)	2/5
Bone marrow[b]	10/13
Breast	0/8
Colon/rectum	0/11
Cord blood[b]	3/3
Liver	0/5
Myometrium	0/10
Ovary (fetal)	2/2
Ovarian follicle	1/1
Peripheral blood[b,c]	12/12
Prostate	0/8
Testis	
Fetal	2/2
Adult	1/1
Vascular intima	0/>6
Cell Strains	
Bladder	0/1
Breast	0/8
Cervix	0/1
Joint	0/1
Kidney (fetal)	0/1
Lung	0/3
Ovary	0/2
Prostate	0/2
Retina (fetal)	0/1
Skin	0/6
Uterus	0/1

[a]Assayed by the TRAP method, unless otherwise indicated.
[b]Telomerase activity in these three normal tissues typically is much less than that observed in testes, ovaries, and cancer.
[c]Purified leukocytes.

incidence by limiting somatic cell life span through its effect on telomere dynamics. This hypothesis is supported by numerous data indicating that telomerase is reactivated upon cell immortalization in vitro and is present in the vast majority of tumor-derived cell lines and tumor tissues (Table 4). Indeed, telomerase rates as the most prevalent tumor marker described to date, and it is assumed that enzymatic activity signifies the existence of at least a subset of immortal cells in the malignant tissues.

Telomerase repression in humans appears to be part of an early developmental program, since activity has been detected in certain fetal tissues whereas no activity is detectable in the corresponding tissue from young postnatal or adult donors. The stringency of telomerase repression is greater in humans than in rodents because telomerase can be easily detected in several adult mouse tissues. This difference in telomerase regulation may in part account for the much higher frequency of spontaneous cell immortalization and cancer in rodents compared to humans.

5 PERSPECTIVES

5.1 Telomerase and Human Disease

The apparent lack of telomerase in most somatic tissues and its widespread presence in malignant tissues has raised the hope that modulation of enzymatic activity may be exploited in the management of human disease (Figure 7). Reactivation of the enzyme in a transient or regulated manner could theoretically be used to extend the replicative life span of somatic cells, particularly in the case of premature aging syndromes or in aging tissues of otherwise healthy individuals. There are a number of tissues in which replicative cell senescence may play a role in the pathology of age-related diseases. Examples include senescence of dermal fibroblasts in skin aging (decreased wound repair and susceptibility to ulcers), senescence of endothelial cells in vascular aging (cardiovascular diseases),

Table 4 Telomere Dynamics and Telomerase Activity in Human Tumors[a]

Tissue	Cancer Grade and Stage[b]	TRF Length[c]	Telomerase[d,e]
Bone	GCL	S, L	3–5/5
Bone marrow/ blood	ALL	NA	18/25
	AML	S, E	8/12
	APL	NA	2/2
	CLL		
	Early	S, E	2/14
	Late	S	8/9
	NS	NA	1/2
	CML: All stages	NA	12/12
	MDS		
	Early	S, L	3/3
	Late	S, L	1/3
	Myeloma	NA	1/1
Breast	Carcinoma		
	Grade I	S, L	15/22
	Grade II	S, L	51/52
	Grade III	S	44/46
	Grade IV	S, L	7/7
Colon/rectum	Adenoma/polyp	S	0/23
	Carcinoma	S, L	22/23
Head/Neck	Squamous cell carcinoma	NA	14/16
Kidney	Renal carcinoma	S	41/55
	Wilm's tumor	S, L	6/6
Liver	Carcinoma and blastoma	S, L	1/1
Lung	NSCLC		
	Primary	S, L	68/91
	Metastasis	S, L	6/10
	SCLC		
	Primary	S, E	11/11
	Metastasis	S, E	3/3
Lymph nodes	Lymphoma		
	Low grade	NA	2/3
	High grade	NA	9/9
Muscle	Leiomyoma (fibroids)	NA	0/11
	Leiomyosarcoma	NA	3/3
	Rhabdomyosarcoma	NA	1/1
Neural	Ganglioneuroma	NA	0/4
	Neuroblastoma		
	Stages I–II	S, L	40/40
	Stages III–IV	S, L	13/36
	Stages IVs	S, E	5/8
	Glioblastoma	S, L	13/51
	Glioma	S, E	
	Retinoblastoma	NA	16/33
Ovary	Carcinoma		
	End stage	S, stable	7/7
	NS	S, E	6/7
Prostate	BPH	NA	4/35
	PIN3	NA	2/2
	Carcinoma	L	24/30
Skin	Melanoma	NA	7/7
Stomach	Adenocarcinoma	E, S, L	56/66

[a]Only tumors that have been assayed for telomerase are included.

[b]GCL, giant cell tumor; ALL, acute myelocytic leukemia; AML, acute myeloid leukemia; APL, acute promyelocytic leukemia; CLL, chronic lymphocytic leukemia; CML, chronic myelocytic leukemia; MDS, myelodysplastic syndrome; NSCLC, non-small-cell lung cancer; SCLC, small-cell lung cancer; BPH, benign prostate hyperplasia; PIN3, prostate intraepithelial neoplasia type 3.

[c]Relative to control tissue: E, equal; L, longer; S, shorter; NA, not determined or no control tissue data available.

[d]Number of telomerase-positive specimens over total tested.

[e]Telomerase activity in all or some of the samples was assayed by the conventional method rather than by TRAP.

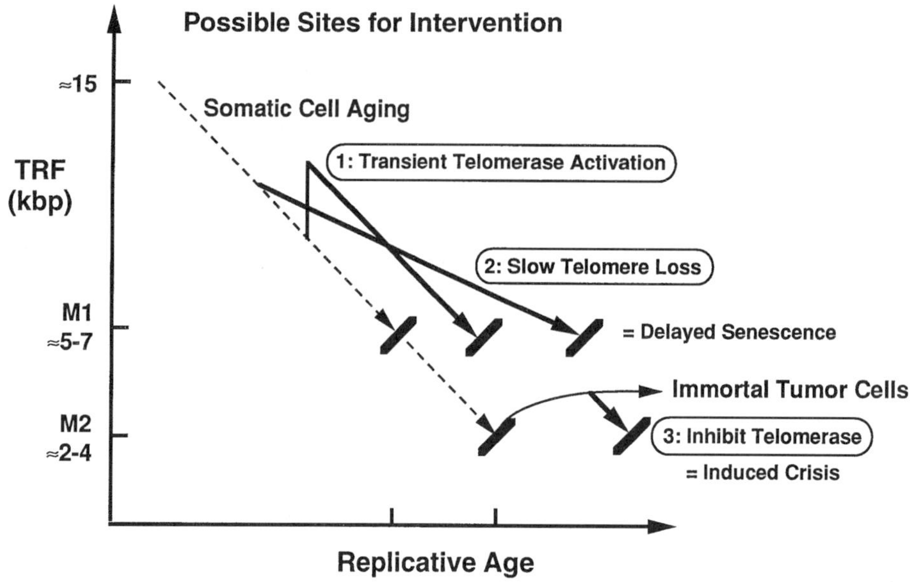

Figure 7. Telomere length modulation for age-related diseases and cancer: the proliferative life span of somatic cells may be extended by counteracting erosion of telomeres. (1) Transient activation of telomerase may lead to temporary telomere stabilization or even elongation, and thus to "rejuvenation" of telomeres and consequently of the cells. (2) Interference with loss of telomeric DNA would not reset the "mitotic clock" but may augment the number of divisions that cells can undergo prior to a telomere-dependent growth arrest signal. (3) Conversely, growth of telomerase-positive tumor cells may be arrested by inhibition of the enzyme and consequent erosion of telomeres to a nonfunctional size.

senescence of osteoblasts in bone (osteoporosis), and senescence of lymphocytes in peripheral blood (susceptibility to infection). An alternative to telomerase reactivation for increasing the life span of these cells is intervention through any mechanism that slows the rate of telomere loss in telomerase-negative tissues. Recent evidence suggests that oxidative stress may influence the rate of telomere loss.

Conceptually easier than telomerase activation is the possibility of an anticancer therapy based on inhibition of enzymatic activity. Such inhibition should lead to loss of cell viability through resumption of telomere shortening. Experimental data from lower eukaryotes and, more recently, from human cells support this expectation: deleting the telomerase gene or inhibiting telomerase biogenesis by the antisense transcript of the template RNA or other mechanisms can restore a mortal phenotype to otherwise immortal cells by causing telomere loss and ultimately senescence or crisis of the cell population. There are several issues that remain to be addressed before it may be possible to fully validate this approach for treatment of human cancer. These range from the development of reagents that inhibit telomerase efficiently and specifically to the investigation of the biological outcome of this event on normal tissues and tumors in vivo.

Last, the nearly ubiquitous presence of telomerase in tumors suggests that the enzyme should also serve as a diagnostic or prognostic marker for cancer. Telomerase activation in cancer does not appear to be associated directly with deregulation of growth or metastatic capacity. Thus, in the multihit process of tumorigenesis, there should be no direct selective advantage for telomerase activation until the tumor cells have reached crisis with critically short telomeres. The early events in tumorigenesis likely involve activation of proto-oncogenes and inactivation of tumor suppressor genes that confer various forms of growth advantage during successive rounds of clonal expansion and selection in the tumor population. Telomerase activation is therefore expected to be a late event in most cancers and would correlate with relatively aggressive forms of the disease.

5.2 Telomerase Biogenesis, Structure, and Function

Telomerase is clearly one of the most interesting new enzymes discovered in the past 20 years of modern molecular biology. Not only are its biological properties of great significance, but from a purely biochemical perspective, this enzyme offers many years of stimulating research on assembly, structure, function, and regulation of the ribonucleoprotein complex. Some of the questions that are unanswered or just beginning to be addressed are as follows:

1. What are the structures and functions of the individual components?
2. How are the subunits regulated during development and differentiation of the soma and germ line?
3. What are the spatial and temporal features of complex assembly, and what other molecules are involved in facilitating the proper protein–protein and protein–RNA interactions?
4. How does telomerase find its primary site of action at telomeres in the cell nucleus?
5. What is the nature of interaction between telomerase and telomere-binding proteins or other components of chromatin or the nuclear matrix, including telomeric DNA?
6. What regulates the initiation event (i.e., telomere binding and the formation of a functional complex)?

7. What regulates the extent of telomere addition and its timing in the cell cycle?
8. How do the spatial relationships between the proteins, RNA, and DNA change during elongation and translocation, and what role do these changes play in enzyme kinetics?
9. What role does the cleavage reaction have in telomerase function?
10. What is the error frequency of telomerase?
11. What happens to telomerase when it is not adding telomeric DNA to the ends of chromosomes?
12. What is the half-life of telomerase, and how is its turnover regulated?
13. Does telomerase have other functions besides telomere maintenance?

These, and many other questions, will keep molecular biologists active for years to come. Moreover, the knowledge acquired through investigations into the detailed structure and function of this enzyme and its mechanism of action will provide new insights into how telomerase might be manipulated for the benefit of human health.

Bibliography

Bacchetti, S., and Counter C. M. (1995) Telomeres and telomerase in human cancer. *Int. J. Oncol.* 7:423–432.

Blackburn, E. H., and Greider, C. W., Eds. (1995) *Telomeres.* Cold Spring Harbor Laboratory Press, Cold Spring Harbor, NY.

———, and Szostak, J. W. (1984) The molecular structure of centromeres and telomeres. *Annu. Rev. Biochem.* 53:163–194.

Collins, K., Kobayashi, R., and Greider, C. W. (1995) Purification of *Tetrahymena* telomerase and cloning of genes encoding the two protein components of the enzyme. *Cell,* 81:677–686.

Feng, J., Funk, W. D., Wang, S.-S., Weinrich, S. L. O., Avilion, A. A., Chiu, C.-P., Adams, R. R., Chang, E., Allsopp, R. C., Yu, J., Le, S., West, M. D., Harley, C. B., Andrews, W. H., Greider, C. W., and Villeponteau, B. (1995) The RNA component of human telomerase, *Science,* 269:1236–1241.

Harley, C. B., Kim, N. W., Prowse, K. R., Weinrich, S. L., Hirsch, K. S., West, M. D., Bacchetti, S., Hirte, H. W., Counter, C. M., Greider, C. W., Wright, W. E., and Shay, J. W. (1994) Telomerase, cell immortality, and cancer. *Cold Spring Harbor Symp. Quant. Biol.* 59:307–315.

Kipling, D. (1995) *The Telomere.* Oxford University Press, Oxford.

Shay, J. W., and Wright, W. E. (1996) Telomerase activity in human cancer. *Curr. Opin. Oncol.* 8:66–71.

Telomeres, Mammalian

Andrew H. Palin and Christine J. Farr

1 **Defining Telomeres**

2 **Cloning Telomeres**

3 **Telomeric DNA**
- 3.1 General Features
- 3.2 Possible Reasons for the G-Rich Strand Asymmetry at Telomeres
- 3.3 Structural Organization of Mammalian Telomeric DNA

4 **Telomere-Binding Proteins**

5 **Chromatin Structure at Mammalian Telomeres**

6 **G4 DNA**

7 **Organization of the Mitotic Nucleus**

8 **Telomere Behavior in Meiosis**

9 **(TTAGGG)$_n$ DNA, Chiasmata, and Recombination**

10 **Telomeric Position Effects**

11 **The End Replication Problem**

12 **Telomerase**
- 12.1 General Features
- 12.2 The RNA Component
- 12.3 The Protein Components
- 12.4 Model for Sequence Addition by Telomerase
- 12.5 Mammalian Telomerase

13 **Senescence and Immortalization**
- 13.1 Telomere Length in Normal Cells
- 13.2 Telomere Length in Cancer Cells
- 13.3 The Telomere Hypothesis of Cellular Aging
- 13.4 The Evidence for a Role of Telomere Shortening and Telomerase Reactivation in Senescence and Immortalization
- 13.5 Targeting Telomerase as a Cancer Treatment
- 13.6 Telomerase-Independent Telomere Elongation

14 **Telomeres and Rearrangements of the Mammalian Genome**
- 14.1 Cryptic Translocations and Deletion Events Involving Telomeres
- 14.2 Spontaneous Chromosome Healing by Telomerase
- 14.3 Interstitial (TTAGGG)$_n$ in the Vertebrate Genome

15 **Experimentally Induced Chromosome Fragmentation and Telomere Seeding in Mammalian Cell Lines**

16 **Application of Telomere-Associated Chromosome Fragmentation in Genome Analysis and Manipulation**
- 16.1 Chromosome Fragmentation and Genome Mapping
- 16.2 A Strategy for Generating a Mammalian Artificial Chromosome

17 **Concluding Remarks**

Key Words

Chiasmata Sites at which two homologous chromosomes appear to have exchanged material during meiosis.

Fluorescence In Situ Hybridization (FISH) Chromosomal DNA is denatured in situ on a microscope slide and hybridized with a denatured DNA probe. By using biotinylated probes, the hybridization signal can be visualized by fluorescence microscopy (using an avidin-conjugated fluorescent tag).

Immortalization The acquisition by a eukaryotic cell of the ability to escape senescence and grow indefinitely in culture.

Matrix Attachment Site (MAR) A region of DNA that attaches to the nuclear matrix. Also called scaffold attachment site (SAR).

Meiosis Involves two successive cell divisions that reduce the starting number of $4n$ chromosomes to $1n$ in each of the four products (these may mature to germ cells).

Mitosis The division of a eukaryotic somatic cell.

Nucleosome The basic structural subunit of chromatin, consisting of approximately 200 base pairs of DNA and an octomer of histone proteins.

Nuclear Envelope A layer of two membranes surrounding the nucleus. It is penetrated by nuclear pores and bounded on the interior by the nuclear laminin.

Nuclear Matrix A network of fibers surrounding and penetrating the nucleus.

Okazaki Fragments The short stretches of 1000–2000 bases produced by discontinuous replication of the lagging strand.

Processive Enzyme An enzyme that continues to act on the particular substrate (i.e., does not dissociate between repetitions of the catalytic event).

Senescence Normal diploid mammalian somatic cells undergo a finite number of population doublings in culture (generally 50–100). Cells then cease dividing (the culture enters crisis) and are referred to as senescent.

Somatic Cell Hybrid A cell line containing chromosomes derived from different parental cells, generated by fusing the cells to form a heterokaryon in which the nuclei subsequently fused (can be either inter- or intraspecific).

Transfection The introduction of exogenous DNA into eukaryotic cells.

Terminal Restriction Fragment (TRF) The $(TTAGGG)_n$ array and a portion of the subtelomeric DNA; detected by digesting genomic DNA with a frequently cutting restriction enzyme and probing the Southern blot filter with labeled oligonucleotides complementary to the telomere repeat.

Much of what we know about the biology of the telomere has emerged from studies on the ciliated protozoa and yeasts. There have been many excellent reviews of this work. This article focuses on mammalian telomeres a field that has expanded dramatically since the cloning of vertebrate telomeric DNA in the late 1980s.

1 DEFINING TELOMERES

Telomeres were first described as a theoretical element in 1938 by Müller, who had observed that irradiated *Drosophila* chromosomes always retained the distal portions of the chromosome arm irrespective of the rearrangements that had occurred. In the same year, McClintock observed in parallel that a maize chromosome that had lost the most distal DNA sequences would fuse at high frequency at the breakpoint with its newly replicated sister chromatid and form dicentric chromosomes. During the subsequent anaphase, the chromosomes were pulled apart and once again fused with the new sister chromatid (referred to as a breakage–fusion–bridge cycle). The identification of this sequence led to the concept of the telomere as a specialized structure that "caps" chromosomes, preventing deleterious rearrangements, and ensuring that the integrity of the host genome is maintained.

2 CLONING TELOMERES

Telomeres were first studied at the molecular level in ciliates. These unusual unicellular eukaryotes can have 40,000–100,000 telomeres per cell, making analysis of the terminal DNA sequences easier in ciliates than in other eukaryotes (which usually have fewer than 100 telomeres per cell). The first nonciliate telomere was isolated by Jack Szostak and Elizabeth Blackburn in 1982. These investigators showed that *Tetrahymena* telomeres would allow the replication of a plasmid as a linear molecule in yeast. They then used this linear plasmid in a functional complementation assay to clone a yeast telomere. More recently the same approach was used to isolate authentic telomeric DNA from both the human and mouse genomes.

3 TELOMERIC DNA

3.1 General Features

Telomeric DNA sequence and structure have been found to exhibit similar organizations among a wide variety of eukaryotes. Most consist of tandem, simple repeats, which are G-rich (at least at the 3′ end of the repeat) on the strand that provides the 3′ end of the chromosome. Most eukaryotic repeats conform to the loose consensus of $(T/A_{1-4})(G_{1-8})$ (Table 1). In all vertebrates studied to date (including the mammals, reptiles, amphibia, birds, and fish) the

Table 1 Telomere Repeat Sequences Found in a Variety of Eukaryotes

Organism	DNA Sequence[a]
Tetrahymena, Glaucoma	TTGGGG
Paramecium	TT(T/G)GGG
Stylonychia, Oxytricha, Euplotes	TTTTGGGG
Physarum polycephalum	TTAGGG
Dictyostelium discoideum	AG_{1-8}
Giardia lamblia	TAGGG
Plasmodium	TT(T/C)AGGG
Trypanosoma brucei	TTAGGG
Ascaris lumbricoides	TTAGGC
Parascaris univalens	TTGCA
Saccharomyces cerevisiae	$(TG)_{1-6}TG_{2-3}$
Schizosaccharomyces pombe	$TTACA_{0-1}C_{0-1}G_{1-8}$
Candida albicans	ACGGATGTCTAACTTCTTGGTGT
Cryptococcus neoformans	TTAGGGGG
Podospora anserina	TTAGGG
Fusarium oxysporum	TTAGGG
Neurospora crassa	TTAGGG
Didymium iridis, Physarum polycephalum	TTAGGG
Chlamydomonas reinhardtii	TTTTAGGG
Arabidopsis thaliana	TTTAGGG
Lycopersicon esculentum	TT(T/A)AGGG
Drosophila melanogaster	HeT-A, TART (retroposons)
Bombyx mori	TTAGG
Humans, mice	TTAGGG

[a] The DNA sequence reads 5′→3′ toward the chromosome end.

telomeric DNA consists of long tandem arrays of the hexanucleotide $(TTAGGG)_n$.

In the ciliates and in the yeast *Saccharomyces cerevisiae,* the G-rich strand extends past the C-rich strand, generating a short overhang. It is not known whether such an overhang exists in vivo in higher eukaryotes. In some species (*Tetrahymena* spp., *Physarum* spp., *Trypanosoma* spp., *S. cerevisiae*) there is evidence to suggest the presence of nicks in the terminal repeat arrays, involving both the C- and G-rich strands.

3.2 Possible Reasons for the G-Rich Strand Asymmetry at Telomeres

About half of the known sequences of eukaryotic telomeres are based on a repeated oligoguanine block that is preceded at the 5′ end by an adenine base (the AGGG repeat). From a structural viewpoint it has been argued that the DNA sequence found at natural telomeres is dictated by its ability to adopt tetraplex geometry (see Section 6). Approaching this matter from an evolutionary perspective, others have suggested that the G-rich nature of the 3′ end of telomere repeat DNA may reflect the adaptation of a much older RNA replication mechanism to a role in maintaining the chromosome end. Both telomerase (the ribonucleoprotein that maintains the chromosome end, see Sections 11 and 12) and the enzyme that adds CCA to tRNA precursors (nucleotidyl-transferase) are primer-dependent, template-independent terminal transferases that add nucleotide triphosphates one by one to the 3′ end of a nucleic acid.

3.3 Structural Organization of Mammalian Telomeric DNA

The termini of human chromosomes are composed of 2–20 kb arrays of $(TTAGGG)_n$; at the proximal end of the terminal repeat array are clusters of degenerate repeats such as $(TTGGGG)_n$ or $(TGAGGG)_n$. The subtelomeric regions of human chromosomes contain a complex mixture of repetitive sequences, which are present in varying copy number at some telomeres and entirely absent from others. Furthermore the subtelomeric repeats are conserved only within primates, and even within the primates the sequences adjacent to telomeres can be organized very differently. These observations suggest that the subtelomeric DNA sequences are not absolutely required for telomere function, and as yet no clear role has been ascribed to them. The clustering of hypervariable minisatellites in the subtelomeric regions of human chromosomes has also been detected.

Mouse terminal restriction fragment (TRF) lengths can be up to several hundred kilobase pairs (in *Mus domesticus, Mus musculus, Apodemus sylvaticus*), although the telomeres of some species (*Mus caroli* and *Mus spretus*) are very much smaller, ranging in size from 20 to 30 and 5 to 20 kb, respectively. One striking feature of the ultralong $(TTAGGG)_n$ arrays in *Mus musculus* is the highly polymorphic banding pattern within inbred strains (such as DBA/2), with new variants arising at high frequency.

4 TELOMERE-BINDING PROTEINS

Telomeres are thought to function as a nucleoprotein complex consisting of telomeric repeats and telomere-binding factors. This protective cap functions both to protect the chromosome end (from exonucleolytic degradation and end-to-end fusions) and to ensure that natural chromosome ends are masked from the cell's DNA damage detection systems. Research into these interactions has been performed predominantly in unicellular eukaryotes, but over the last 4–5 years several groups have started the search for equivalent proteins in mammalian systems. Although telomerase is present in ciliates and mammals, conservation of telomere-binding proteins (as predicted by the highly conserved nature of the G-rich telomeric repeat sequence) has not been demonstrated, and progress in identifying proteins that interact with mammalian telomeres has been limited.

Telomere-binding proteins fall into two types, based on their single-stranded or double-stranded telomeric DNA-binding properties (Table 2). Some attempts to identify components that bind specifically to the presumed vertebrate telomere termini (a 3′ overhang of TTAGGG repeats) have been reported. A number of nuclear proteins that bind to single-stranded telomeric DNA have been purified using cell extracts in gel mobility shift assays with $(TTAGGG)_n$ oligonucleotides. However, careful analysis revealed that the proteins purified bind RNA more tightly than DNA and bind specifically to an oligonucleotide having the pre-mRNA 3′ splice site sequence. It seems therefore that the abundance of proteins with a role in pre-mRNA metabolism and the similarity of their consensus sequence to the vertebrate telomere repeat has confounded many attempts to isolate genuine single-stranded, telomere-binding proteins from mammalian systems. So far the best candidate for an authentic telomere-capping protein in higher eukaryotes is the *Xenopus* telomere end factor (XTEF). This protein binds in vitro to terminal TTAGGG repeats in a 3′ overhang, does not bind to RNA $(UUAGGG)_n$ repeats, and is much more abundant in *Xenopus* eggs and ovaries than in somatic cell nuclei. Single-stranded TTAGGG repeats at internal positions, or at a 5′ protruding end, or TTAGGG repeats in double-stranded DNA, are not recognized. The binding to single-stranded 3′ extensions is very stable in high salt and is highly sequence specific. However, changes in the double-stranded sequence 5′ to the TTAGGG repeat tail are tolerated.

Recently a DNA-binding activity that interacts with double-stranded TTAGGG repeats in linear and circular DNAs has been identified. This factor, called TRF (for TTAGGG repeat factor) is ubiquitously expressed in primate, rodent, and bird cell nuclei at all stages of the cell cycle. For optimal binding, TRF requires tandemly repeated stretches of TTAGGG longer than 36 bp and does not bind heterologous telomere repeats such as TAGGG, TTTAGGG, TTTTAGGG, TTGGGG, and TTAGGC. It is postulated that TRF may bind along the length of mammalian telomeres. However, direct evidence that any of the proteins just described associate with telomeres in vivo has still to be presented.

5 CHROMATIN STRUCTURE AT MAMMALIAN TELOMERES

The study of the chromatin structure at chromosome ends in ciliates and in *S. cerevisiae* has led to the suggestion that interactions at the telomere may profoundly alter the chromatin structure and result in telomeres influencing neighboring genetic elements. Lower eukaryotes (*Tetrahymena, Oxytricha, S. cerevisiae*) show no evidence of nucleosomal structure at telomeres. The chromatin structure at

Table 2 Telomeric DNA-Binding Proteins

Protein	Organism	Comments[a]
Single-Stranded DNA-Binding Proteins (bind to single-strand G-strand DNA)		
Telomerase	*Tetrahymena* and other ciliated protozoa, *S. cerevisiae,* mouse, human	A ribonucleoprotein complex.
α and β (subunits dimerize at telomere)	*Oxytricha*	α subunit, M_r, 56 and β subunit, M_r 41 kDa; the β subunit induces G4 DNA formation in vitro.
α and β	*Stylonychia*	α subunit has weak homology with human vimentin.
α	*Euplotes*	A single subunit of 51 kDa.
TEP (telomere end-binding protein)	*Tetrahymena*	A duplex structure adjacent to the 3′ overhang is required for binding. Does not bind G4 DNA; low affinity for RNA substrates; M_r 65 kDa.
GBP (G-stand-binding protein)	*Chlamydomonas*	The predicted protein contains two RNA recognition motifs, but it does not bind r(UUUUAGGG); M_r 34 kDa.
Nsr1p	*S. cerevisiae*	Uncertain whether this protein interacts with telomeres in vivo.
Gbp2p	*S. cerevisiae*	Uncertain whether this protein interacts with telomeres in vivo.
MF3	Birds	May recognize a hairpin DNA structure.
XTEF (*Xenopus* telomere end factor)	*Xenopus*	Binds $(TTAGGG)_2$ when present in a single-stranded 3′ overhang; sequence specific.
Several heteronuclear RNPs (hnRNP A1, A2/B1, D, E)	Mouse and human	Bind the RNA equivalent of TTAGGG with high affinity. Uncertain whether these proteins interact with telomeres in vivo.
Double-Stranded DNA-Binding Proteins		
RAP1	*S. cerevisiae*	A multifunctional, essential protein, which binds telomeres in vivo. Affects both telomere length and the expression of genes in the proximity.
TBFα (TBF1)	*S. cerevisiae*	Binds the junction between the subtelomeric DNA and the poly $(C_{1-3}A)$ sequence; M_r 63 kDa.
PPT	*Physarum* (slime mold)	M_r 10 kDa
ST-1	*Trypanosoma*	M_r 39 kDa; will bind both double-stranded telomere and subtelomere repeats. Also exhibits affinity for the single-stranded, C-rich strand.
TRF (TTAGGG repeat factor)	Human, mouse, and monkey	Binds double-stranded (TTAGGG)≥6 repeats in linear and circular DNAs in vitro; $M_r \approx 50$ kDa.

[a]Relative molecular mass M_r in kilodaltons.

vertebrate telomeres remains unclear. It appears that at least some of the terminal repeat array at the ends of vertebrate chromosomes is packaged into nucleosomes. Studies on the nucleoprotein structure of rat telomeres have revealed very regular arrays of nucleosomes spaced by 157 ±2 bp on the telomeres, representing the shortest nucleosomes found in animals and plants (40 bp less than nucleosomes of bulk chromatin). This work has been extended to reptiles, birds, amphibians, fish, and the invertebrate class Echinodea, further supporting the hypothesis that animal cells have highly conserved telomere DNA sequences organized largely into short nucleosomes, with the repeat length varying in the different animals.

Some MNase digestion data have been presented which may indicate the presence of nonnucleosomal DNA at the most distal ends of telomeres. However, whether a specialized chromatin structure actually exists near the vertebrate telomere terminus remains to be determined.

6 G4 DNA

It has been well documented that the G-strand DNAs, such as those found at telomeres, can adopt non-Watson–Crick base pairing (such as Höögsteen pairing) in vitro to form quadruplex (or G4-DNA) structures. Thus the G-strand overhang at telomeres may not be single-stranded in vivo. Potential roles for the tetraplex conformation are in the protection of the chromosome end and in mediating telomere–telomere interactions. However, the biological relevance of G4-DNA is still under debate. It has been shown that the intramolecularly folded G-quartet structure is not a substrate for telomere elongation by *Oxytricha* telomerase and that the *Oxytricha* telomere-binding protein (TBP) and the *Xenopus* XTEF factor bind poorly to folded G-quartet structures. Moreover, although extremely stable, G4-DNA is very slow to form under physiological conditions in vitro. Recently, however, two proteins known to function at the ends of chromosomes, β-TBP in *Oxytricha* and RAP1 in *S. cerevisiae*, have been shown to accelerate the rate of G4-DNA formation in vitro, lending support to the idea that such unusual structures do actually exist in cells. Furthermore, a G4-DNA-dependent nuclease activity, encoded by the *KEM1/SEP1* gene, has been identified in yeast telomere extracts.

Recently it has been shown that the cytosines of the C-rich strand also have the potential to form four-stranded structures, rather than being entirely in duplex DNA. At present, however, there is no conclusive evidence demonstrating the presence in biological systems of either four-stranded guanine or four-stranded cytosine structures.

7 ORGANIZATION OF THE MITOTIC NUCLEUS

It has been suggested that telomere–telomere interactions, together with attachments to the nuclear matrix, could contribute to chromosome organization in the mitotic nucleus. However, there does not appear to be any association between telomeres and the nuclear membrane in interphase mammalian cells. In situ hybridization to detect $(TTAGGG)_n$ has revealed that mouse and human telomeres are not associated with the nuclear envelope, but are distributed throughout the volume of the interphase nucleus. Indirect evidence for the association of telomeric DNA with a nuclear matrix protein in mammalian cells has been inferred from the isolation of nuclear matrix with telomeric DNA attached specifically by the (TTAGGG) repeats in nuclei extracted by means of lithium-diiodosalicylate. It remains unclear, however, whether there is any specific interaction between telomeres and the nuclear envelope in mitotic cells.

8 TELOMERE BEHAVIOR IN MEIOSIS

Telomeres may have a specialized role in meiotic chromosome segregation by causing end-to-end associations. Attachment of meiotic telomeres to the nuclear envelope is an almost universal feature. The peripheral clustering of telomeres in mammalian spermatozoa has been revealed by FISH, and it has been suggested that this may help to facilitate chromosome pairing. The initial telomeric alignments of human chromosomes are often incorrect, which would be consistent with the high level of sequence variability of the subtelomeric regions and the observation that some nonhomologous chromosomes share common telomere-associated repeats.

9 $(TTAGGG)_n$ DNA, CHIASMATA, AND RECOMBINATION

The report of a hot spot for male-specific meiotic recombination near the telomere of murine chromosome 8 and the existence of another hot spot of male recombination coincident with an interstitial telomeric sequence in the Armenian hamster has led to the suggestion that telomere-promoted recombination may represent a second pathway of recombination, mechanistically different from the more evenly dispersed process of meiotic exchange. However, although the most distal regions of many chromosomes in human males show a high frequency of chiasmata, preliminary data from the human and mouse genetic maps do not reveal massive expansions of the terminal regions of chromosomes. It is possible that these terminal chiasmata may in fact represent distinct telomere–telomere associations, rather than crossover events.

10 TELOMERIC POSITION EFFECTS

It is well documented that both centromeres and telomeres may down-modulate gene expression and replication origins in *Drosophila* and yeasts. The situation in mammalian cells is less clear. It is unknown whether the shortened nucleosome repeat characteristic of the vertebrate $(TTAGGG)_n$ DNA extends into the subtelomeric region. As the telomere repeat length shortens, the expression of terminally located genes could potentially be influenced by the spread of altered chromatin. As yet, however, there is no evidence for telomere-associated position effects in mammalian cells; whether this is because they do not exist, or simply have not yet been detected, is unclear. The extensive subtelomeric regions at mammalian telomeres may act to prevent the repression of terminally located genes and replication origins.

11 THE END REPLICATION PROBLEM

DNA synthesis proceeds in a $5' \rightarrow 3'$ direction and is primed by short RNA molecules. These primers are subsequently removed, and the gaps are filled in by DNA polymerase. However, at the 5′ end of the lagging strand a region minimally the size of the primer and perhaps as large as the Okazaki fragment is incompletely replicated. As a result, in the absence of some other mechanism, multiple rounds of replication will cause the chromosome end to recede (Figure 1). To circumvent this progressive loss, the G-rich strand of the telomeric DNA is synthesized by a ribonucleoprotein called telomerase. Telomerase is a reverse transcriptase containing its own RNA template as an integral part of the enzyme. This enzyme adds telomere repeats, one nucleotide at a time, onto the 3′ end of the G-rich strand, with the other strand presumably being replicated by conventional lagging strand DNA synthesis. Treatment of telomerase extracts with either proteases or RNase A abolishes activity. Since its original discovery in *Tetrahymena thermophila* extracts, telomerase activity has been identified in several other ciliates, as well as in extracts of *Xenopus*, human, and mouse cells. It seems likely therefore that telomerase is responsible for the maintenance of the G-rich strand of the telomeric DNA in a diverse range of eukaryotes.

12 TELOMERASE

12.1 General Features

The only component of telomerase studied in detail as yet is the RNA moiety. The gene encoding the RNA component of telomerase has been isolated from *Tetrahymena thermophila*, from several other hypotrichous ciliates, from the yeasts *Saccharomyces cerevisiae* and *Kluveromyces lactis*, and from human and mouse cells. Less progress has been made in identifying the DNA encoding the

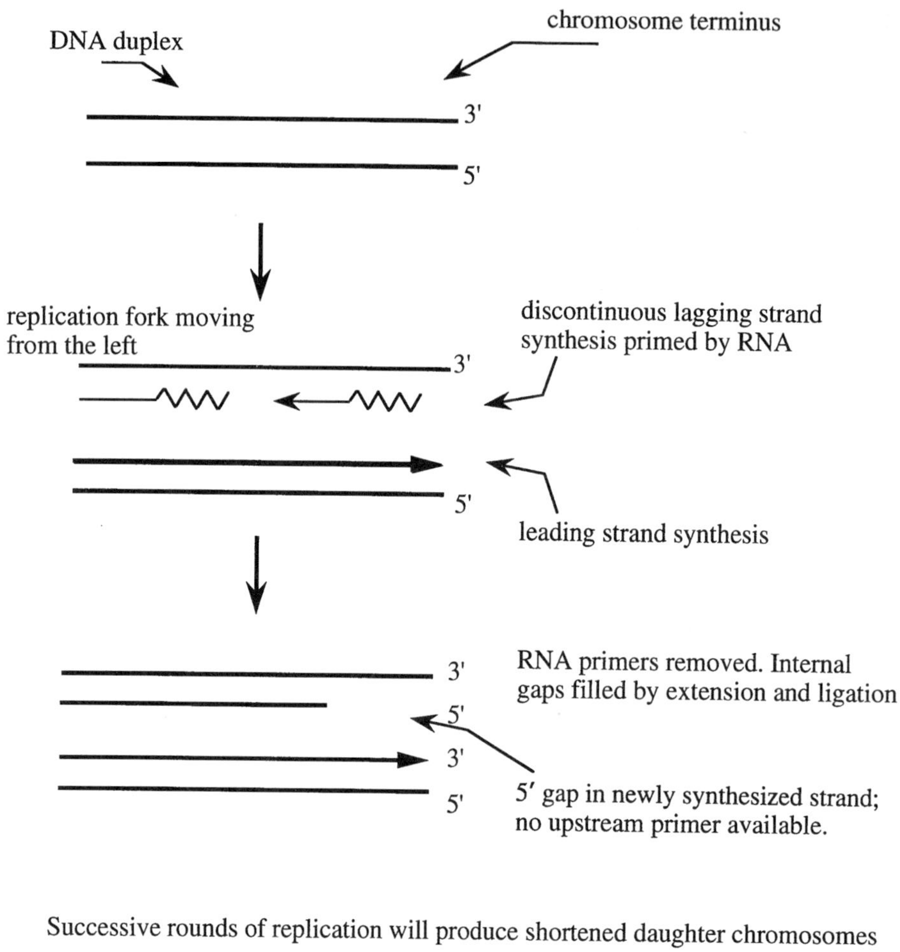

Figure 1. The end replication problem.

telomerase proteins, although the genes encoding two *Tetrahymena* proteins that copurify with telomerase activity and telomerase RNA have recently been cloned and sequenced, and a good candidate for a component of the *Saccharomyces* telomere replication, or maintenance, apparatus is the protein encoded by the *EST1* gene. Since in mammals telomerase may be present at only 100–200 molecules per cell, cloning candidates for the mammalian telomerase protein is proving to be a daunting challenge.

12.2 The RNA Component

Structural analysis of the ciliate RNA template has led to an understanding of how telomere replication is performed. The RNA component in *Thermophila* telomerase is 159 nucleotides long, appears to be transcribed by RNA polymerase III, and contains the complementary telomeric sequence 5′-CAACCCCAA-3′. Upstream of this sequence lies a conserved motif 5′-GUCAUU-3′ present in all ciliate telomerases currently described. In vitro reconstitution of *Tetrahymena* telomerase activity using wild-type and mutant telomerase RNA added to nuclease-treated extracts has shown that the template region of the RNA can be dissected into an alignment region and a true template region. Although DNA sequence across the gene coding for the RNA moiety in several ciliated protozoa has diverged by up to 50%, highly homologous structural motifs are maintained, presumably through functional selection.

The recently cloned RNA component of human telomerase is about 450 nucleotides long. It has no homology to any of the lower eukaryotic telomerase RNAs and is only 65% identical to the mouse telomerase RNA gene. In the human telomerase RNA gene, a template region has been identified that encompasses 11 nucleotides (5′-CUAACCCUAAC-3′) complementary to the vertebrate telomere sequence $(TTAGGG)_n$, while in the mouse, rat, and Chinese hamster telomerase RNAs, only 8 or 9 residues are potential templates. Preliminary data suggest that the human telomerase RNA might be an RNA polymerase II transcript.

12.3 The Protein Components

Recently the first complete telomerase complex was isolated, from *Tetrahymena thermophila.* It comprises an 80 kDa protein binding to the RNA moiety and a 95 kDa protein, which binds to the repeat array and subsequently complexes with the p80-RNA subunit. p95 contains a motif found in class II tRNA synthetases, which positions a deoxyribonucleoside triphosphate (dNTP) at the synthetase reactive site, and the p80 subunit demonstrates a putative zinc finger, which bears similarity to RNA polymerase transcription factors.

12.4 Model for Sequence Addition by Telomerase

Elongation by telomerase is thought to be initiated by the recruitment of subunit proteins. As mentioned earlier, certain subunits may recruit the RNA moiety prior to binding with a second subunit already complexed with the DNA repeat. The RNA template is then presented to the 3′ end of the G-rich strand, which serves as a primer, and they base-pair over a short number of nucleotides. The unbound RNA moiety then serves as a template on which dNTPs polymerize, extending the 3′ end by a single repeat unit. Once this addition has been made, telomerase translocates, perhaps by destabilizing the RNA moiety, and repositions itself with the newly synthesized sequence acting as the 3′ primer, as in the initial reaction. This process is predicted to repeat until the occurrence of dissociation between telomerase and the DNA strand (Figure 2). In *Tetrahymena,* in the yeasts *S. cerevisiae* and *Kluveromyces lactis,* and in human and mouse cells, the introduction of mutated versions of the DNA encoding the RNA template generates corresponding changes in the telomeric sequence in vivo. Thus it is demonstrated directly that the DNA sequence at the telomere is determined by the templating activity of telomerase, rather than being inherent in the

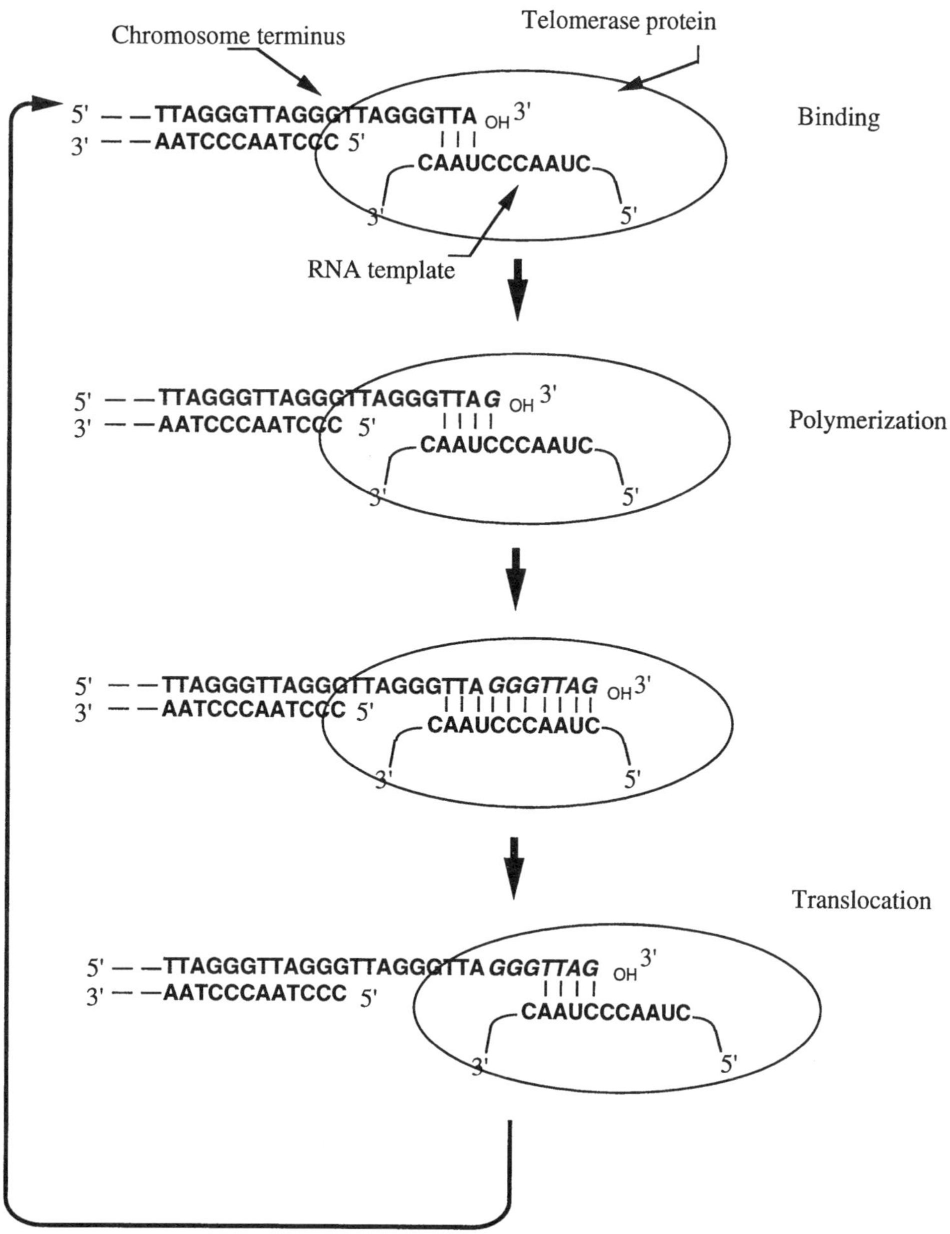

Figure 2. Model for the synthesis of human telomeric DNA by the ribonucleoprotein telomerase.

telomere DNA itself. It has yet to be established whether telomerase acts before or after semiconservative DNA replication.

12.5 Mammalian Telomerase

Mammalian telomerase was first identified in crude cell extracts of the human tumor-derived HeLa cell line. It can catalyze the addition of multiple (up to 65–70) TTAGGG repeats onto the 3′ ends of oligonucleotides containing human or other telomeric repeat sequences in vitro. The reaction is processive in vitro as judged by an oligonucleotide challenge assay. Many heterologous repeat oligonucleotides are used as efficiently as $(TTAGGG)_n$, indicating that in vitro telomerase displays minimal sequence specificity, although a variety of nontelomeric repeats and C-strand oligonucleotides do not function as primers. Mouse telomerase, like that from *Xenopus* ooctyes, appears to behave in a nonprocessive fashion in vitro, adding only one or two TTAGGG repeats onto an oligonucleotide. This reduced processivity may be related to the shorter template region in the rodent RNAs (and the consequent increase in the probability of dissociation of a growing primer from mouse telomerase compared with the human enzyme).

13 SENESCENCE AND IMMORTALIZATION

13.1 Telomere Length in Normal Cells

One function of telomeres is to stabilize the genome. As cells divide, loss of telomeric DNA due to incomplete replication (in the absence of a compensatory mechanism, like telomerase) and exonucleolytic enzyme attack, will eventually shorten the telomere. It has been known for some years that telomeres in human germ line cells are longer than those in somatic tissues. It was subsequently observed that human somatic cell telomeres shorten with each cell division in vitro (by approximately 48 bp per generation until the population enters senescence) and that telomere length correlates with the life span of fibroblasts in culture. Shortening of telomeres during aging in vivo has been observed with skin dermal and epidermal cells, peripheral blood leukocytes and colon epithelia, and candidate hemopoietic stem cells. There is also a correlation, at all ages, between replicative capacity and initial telomere length. In contrast, telomeres from sperm DNA do not decrease in length with the age of the donor, indicating that telomeres are actively maintained in the germ line. More evidence for the association of telomere length with aging has come from individuals with Down syndrome and donors with premature aging syndromes (e.g., Werner syndrome, Hutchinson–Gilford syndrome). Somatic cells from these individuals have shorter telomeres than age-matched controls and a reduced replicative capacity in vitro.

Comparable studies in the mouse have been difficult to undertake owing to the ultralong telomeres present in inbred strains. However, the finding that *Mus musculus* telomeres appear to be unusually long, with the length of telomeric DNA in *Mus spretus* and *Mus caroli* being comparable to that in human cells, has allowed researchers to explore whether changes in telomere length also occur in the mouse. A similar inverse correlation has been found between telomere length and increased cell doublings of primary fibroblasts in culture. In vivo telomere length is similar among the different tissues of a newborn mouse, whereas telomere length differs among different tissues in an adult mouse.

13.2 Telomere Length in Cancer Cells

Over the last 5 years reduction in telomere length when comparing tumor material with normal tissue has been reported from a whole range of human cancers (e.g., endometrial adenocarcinoma, leukemia, lung cancer, renal tumors, neuroblastoma, ovarian cancer). Shortened telomeres in transformed cells suggest that decreases in telomeric length may contribute to genomic instability and cellular transformation. However, telomere length does not appear to be a good marker of tumor progression, or of the rate of chromosome fusion in tumor cells. Yet since such studies look at the average length of telomeres on multiple chromosomes, it may be that they mask considerable variation at individual telomeres.

13.3 The Telomere Hypothesis of Cellular Aging

It was proposed some years ago that uncontrolled telomere shortening would lead to senescence and cell death, and the relationship between telomere length and senescence has recently been the object of intense discussion. A model has been proposed to explain the behavior of telomeres in cellular senescence and in cancer. In somatic cells, telomeres shorten with each round of cell division as a result of a lack of telomerase activity, whereas in the germ line, and in immortalized somatic cells, telomerase ensures that the telomeric DNA is stably maintained. In normal somatic cells telomere shortening could provide a signal to the cell to undergo senescence.

There is no clear understanding of how shortened telomeres might regulate the onset of senescence. Double-stranded DNA breaks cause cell-cycle arrest in a number of organisms (checkpoint control), and a nontelomeric chromosome end may induce this signal and cause exit from the cell cycle. Genes at the ends of chromosomes may be deleted or inactivated or, alternatively, silent senescence genes could become activated as a result of the removal of terminal heterochromatic regions.

13.4 The Evidence for a Role of Telomere Shortening and Telomerase Reactivation in Senescence and Immortalization

Over the past 2 years, several research groups have begun testing the hypothesis that telomere shortening and telomerase reactivation play a role in senescence and immortalization, using both tissue culture systems and human tumors. It has been proposed that telomerase activation is an obligatory step in the immortalization of human cells. Simian virus 40 (SV-40) large T antigen and the adenovirus five *E1A* and *E1B* oncogenes can extend the life span of human embryonic kidney cells without directly activating telomerase. During the extended life span phase of these cultures (an additional 20 population doublings), telomeres continue to shorten until the occurrence of a massive cell death ("crisis" point), when most cells die (termed M2 for mortality stage 2). Telomerase was detected only in the relatively rare immortal clones that survived crisis, and in these cells telomere length was stabilized.

Similar results have been obtained in human papilloma virus (HPV) transformed human epithelial cells and in human B lymphocytes immortalized by Epstein–Barr virus. Occasionally lengthening of telomeres has been observed following immortalization, suggesting that regulation of telomere length in cells that have overcome crisis is not properly controlled. From this type of study it has been proposed that cells transformed in vitro lose telomeric DNA until a critically short telomere length is reached. At that stage, only cells that can maintain functional chromosome ends are capable of continued proliferation. Although as yet unproven, it is hypothesized that crisis corresponds to the functional loss of a telomere from at least one chromosome, resulting in breakage–fusion–bridge cycles and genome damage. Only rare cells, which spontaneously activate telomerase and negate telomere sequence loss, avoid M2 crisis and continue to grow.

Reactivation of telomerase activity during immortalization has also been assayed without the introduction of exogenous agents. Human breast epithelial cells from a Li–Fraumeni patient containing a germ line mutation in p53, which were initially telomerase negative, spontaneously immortalized in cell culture and expressed telomerase activity. Rodent cells also spontaneously immortalize in culture. In *M. spretus,* which has telomere restriction fragment lengths similar to those in human cells, telomere length and telomerase activity have been assayed during spontaneous immortalization. In a mouse primary culture where telomeres shortened, telomerase activity was not detectable. After an apparent crisis, and telomere length stabilization, telomerase activity was detectable.

If telomerase is required for immortalization in vitro, it may also be necessary during tumorigenesis, although it is not known whether all cancer cells are immortal in vivo. Until recently, telomerase activity in vitro was estimated using a primer extension assay. In extracts of human cells and tissues, telomerase activity was identified in nine immortal cell lines and in metastatic ovarian carcinoma, but it was not detected in four normal somatic cell cultures or in normal tissues adjacent to the carcinoma. This method has now been modified to incorporate a polymerase chain reaction (PCR) step, considerably increasing its sensitivity. This approach (designated TRAP, for telomeric repeat amplification protocol) has allowed telomerase activity to be investigated in a much wider range of cells and tissues.

In the first reported application of this PCR-based assay, telomerase was found to be undetectable in 22 normal somatic cell cultures and 50 normal or benign tissues, while activity was detected in human oocytes and sperm and a broad range of tumor cell lines (in 98 out of 100) and primary tumors (in 90 out of 101). The data, together with the observations on telomere shortening in normal somatic tissues, support the suggestion that telomerase activity is directly involved in telomere maintenance and is a necessary step in tumorigenesis. It should be noted, however, that telomerase activity was not detected in all tumor biopsies, and it is not yet known whether all cells in a PCR-positive tumor biopsy express telomerase, or only a minority.

It has been suggested that in long-lived species, such as humans, telomerase is repressed in somatic tissues to reduce the risk of cancer. Interestingly telomerase activity is present in many mouse tissues (kidney, liver, and spleen, as well as testes). This difference in telomerase expression between human and mouse tissues may account for the occurrence of spontaneous immortalization of mouse cells in culture at least 10^6 times more frequently than in human cells.

Direct evidence for a causal link between telomere loss and cellular senescence on the one hand and telomerase reactivation and cancer on the other has long been awaited. Recent isolation of the gene for the RNA component of human telomerase allows experiments that directly manipulate the enzyme to be undertaken. Transfection of HeLa cells with an antisense human telomerase RNA expression construct led to cell crisis after 23–26 population doublings, as well as a shorter mean terminal restriction fragment length. In many of the antisense clones telomerase activity was detectable, but low. The induction of cell crisis in HeLa cells expressing antisense human telomerase RNA argues that telomere loss leads to crisis and cell death once telomeres have been shortened to a critical length.

13.5 Targeting Telomerase as a Cancer Treatment

Assuming that telomerase is necessary for the formation of at least some tumors, telomerase inhibitors could have potential as antitumor agents. Side effects from such a therapeutic approach should be limited, since telomerase is believed to be absent or minimally expressed in human somatic tissues. The main problem may be the predicted phenotypic lag, especially in tumors with long telomeres. Such drugs may therefore have only limited use against a subset of late stage tumor cells with very short telomeres. Nevertheless the antisense experiment described in Section 13.4 demonstrates the potential of telomerase inhibition as a therapy for treating human cancer.

13.6 Telomerase-Independent Telomere Elongation

Recently use of the TRAP assay has identified several telomerase-negative, in vitro immortalized mammalian cell lines. These cell lines all have chromosomes with very long telomeres, suggesting that the cells have been able to maintain and lengthen their telomeres through a novel, telomerase-independent, mechanism. Some other organisms also exhibit telomerase-independent means of lengthening their telomeres (*S. cerevisiae* and *Drosophila*). As yet, however, the significance of a telomerase-independent mechanism of telomere elongation for mammalian cells in vivo remains unclear.

14 TELOMERES AND REARRANGEMENTS OF THE MAMMALIAN GENOME

14.1 Cryptic Translocation and Deletion Events Involving Telomeres

The potential clinical significance of translocations and terminal deletions is very great. Cytogenetically invisible unbalanced translocations of chromosome ends have recently been shown to be responsible for some cases of the α-thalassemia/mental retardation syndrome (ATR-16), Wolf–Hirschhorn syndrome, Miller–Dieker syndrome, and cri-du-chat syndrome. The involvement of subtelomeric translocations in melanoma has also been identified.

14.2 Spontaneous Chromosome Healing by Telomerase

In a second type of terminal rearrangement, a chromosome is truncated and stabilized by the de novo addition of telomeric repeat DNA, a process that probably accounts for many terminal deletions. Evidence for this mechanism of chromosome stabilization has emerged from the study of a number of unrelated individuals with α-thalassemia, resulting from terminal truncations of the short arm of chromosome 16, involving the α-globin genes. In the majority of cases translocations are involved, but occasionally such deletions are the result of a chromosome breakage and healing process. In each of six cases described to date there is an abrupt transition on the broken chromosome from the sequence at the α-globin locus to a stretch of telomeric repeats. Comparison with the corresponding structure of the normal chromosome shows that there are no $(TTAGGG)_n$ repeats preexisting at the break sites. With the exception of one de novo event, the truncated chromosomes appear to be stably inherited.

The structure of these broken chromosome ends suggests that they may have been healed by a process in which telomerase directly adds telomeric repeats to unique nontelomeric DNA. Using in vitro assays, it has been shown that an oligonucleotide corresponding to one of these breakpoint sequences can act as a substrate for both *Tetrahymena* and human telomerases. More recently it has been demonstrated that in five of the six healed chromosomes, the three or four nucleotides preceding the point of TTAGGG addition are complementary to, and in register with, the RNA template of human telomerase. Telomerase activity in the germ line therefore seems to be a plausible mechanism for healing these chromosomal breakage events. An important biological conclusion to come out of the study of both spontaneously arising and experimentally induced (see Section 14.3) chromosome fragmentation/healing events is that an array of the simple hexanucleotide repeat $(TTAGGG)_n$ is sufficient to form a functional telomere.

The proximity of the α-globin genes to the 16p telomere, together with an easily recognizable phenotype associated with hemizygous deletions, has provided a means of identifying terminal chromosome deletions and analyzing the mechanisms involved. Since there appears to be a high density of genes close to telomeres, it seems likely that telomerase-mediated healing of cryptic chromosome truncations may be involved in a number of human genetic diseases, which have been localized to the terminal regions of chromosomes.

It must be emphasized, however, that in most species the stabilization of broken chromosomes by de novo terminal repeat addition is a relatively rare event. Most cells appear to restrict telomere sequence addition to telomere-like repeat arrays, presumably because promiscuous healing of breaks in germ line cells would lead to karyotype instability. The apparent inconsistency between this need to maintain a stable genome and the minimal sequence specificity of telomerase in vitro is probably accounted for by more stringent DNA-binding requirements of other essential telomere-associated factors.

14.3 Interstitial $(TTAGGG)_n$ in the Vertebrate Genome

The genomes of many eukaryotes have accumulated sites of interstitial telomere repeats. Fluorescence in situ hybridization has shown that in several classes of Mammalia, the most common nontelomeric location of the repeat sequence is around the centromere. This is particularly notable on Chinese hamster centromeres (Figure 3; See color plate 1). In the human genome the bulk of $(TTAGGG)_n$ arrays are confined to the ends of chromosomes or clustered in the subtelomeric regions (Figure 3). However, small blocks of tandem repeats also occur at distinct internal loci (e.g., at 2q13, 8p21, and 10q22-q24). Such interspersed $(TTAGGG)_n$ repeat blocks may have arisen through illegitimate recombination events or, alternatively, they may be remnants of karyotypic changes during evolution. Recently genomic clones from human chromosome 2q13 have been isolated and shown to contain extensive tracts of the vertebrate telomeric repeat in a head-to-head arrangement, 5′-$(TTAGGG)_n$ - $(CCCTAA)_m$-3′. Flanking sequences are characteristic of subtelomeric DNA, providing support for the hypothesis that human chromosome 2 originated from a telomere–telomere fusion event of two ancestral ape chromosomes.

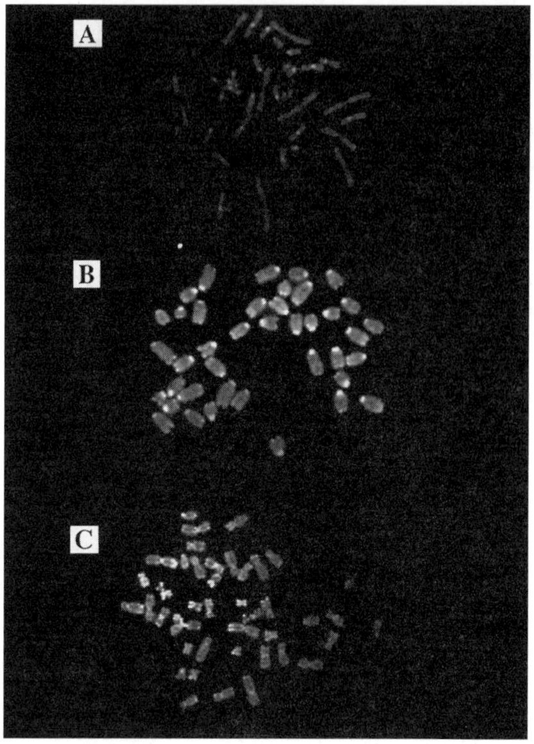

Figure 3. Identifying $(TTAGGG)_n$ DNA sequences on chromosomes by fluorescence in situ hybridization: (A) a Chinese hamster fibroblast cell line, (B) a mouse embryonic stem cell line, and (C) human peripheral blood lymphocytes. The yellow fluorescence identifies $(TTAGGG)_n$ arrays. The chromosomes are counterstained with propidium iodide and appear red. (Courtesy of R. Critcher, Department of Genetics, Cambridge University.) (See color plate 1.)

Several groups have hypothesized that interstitial arrays of $(TTAGGG)_n$ may be inherently unstable and prone to breakage. The accumulated data, however, argues against tandem $(TTAGGG)_n$ repeats being a direct cause of chromosome breakage. In fact short interstitial tracts of $(TTAGGG)_n$ appear to be quite stable. However, breakage within, or close to, interstitial $(TTAGGG)_n$ arrays could serve to stabilize spontaneous, or induced, chromosome breaks by providing a substrate for telomerase.

15 EXPERIMENTALLY INDUCED CHROMOSOME FRAGMENTATION AND TELOMERE SEEDING IN MAMMALIAN CELL LINES

It has been demonstrated that when a vertebrate telomere repeat array is introduced into a mammalian cell line, it may give rise to a new chromosome end at a previously interstitial site in the recipient genome. In the first reported experiments of this type, a 2 kb telomeric fragment containing 500 bp of $(TTAGGG)_n$ and related repeats was cloned into a construct carrying a gene that confers resistance to histidinol on mammalian cells. The construct was linearized and transfected into an established mammalian cell line. In

20% of stable transfectants, the introduced DNA was located at chromosome ends on the basis of both molecular and cytogenetic analyses. In control experiments, in which the introduced construct lacked a $(TTAGGG)_n$ array, the frequency of telomere seeding fell to undetectable levels.

The data are consistent with the integration of the telomere construct into small gaps in the DNA of mammalian cell lines through end ligation (an extremely efficient process in human cells). While the non-$(TTAGGG)_n$ end of the telomere–seed plasmid may become ligated into a double-stranded break in the mammalian genome, the telomeric repeat, which provides a substrate for further $(TTAGGG)_n$ addition by telomerase, is presumably capped by various telomere-associated factors. These may prevent joining with the other chromosomal fragment, thereby converting a gap into a chromosome break. If the newly seeded chromosome end is associated with a centromere, a stably truncated chromosome will be the end result, and the displaced acentric fragment will be lost from the cell. Telomerase by itself is only very rarely able to establish a new telomere. Efficient seeding requires the presence of a correctly oriented array of the vertebrate terminal repeat $(TTAGGG)_n$ close to, but not necessarily at, the end of the linear molecule.

16 APPLICATION OF TELOMERE-ASSOCIATED CHROMOSOME FRAGMENTATION IN GENOME ANALYSIS AND MANIPULATION

16.1 Chromosome Fragmentation and Genome Mapping

One of the first practical applications of this technology has been in the generation of a panel of somatic cell hybrids in which the long arm of the human X chromosome is truncated to varying extents. Selection against the human *HPRT* gene at Xq26 was used to isolate cells in which a hygromycin-based telomere construct had replaced the long arm telomere. Forty-two cell lines out of 85 generated using this approach displayed apparently random terminal truncations of Xq, extending from qter to the α-satellite array at the centromere. This type of somatic cell hybrid panel is a powerful and rapid way to order markers based on their presence or absence in the various derivative cell lines.

16.2 A Strategy for Generating a Mammalian Artificial Chromosome

Manipulation of the mammalian genome by telomere-associated chromosome fragmentation may provide a way of defining the structural elements required for a functional chromosome. For example, telomeric DNA has recently been used to dissect the centromere of a human Y chromosome retained in a Chinese hamster ovary cell line, and the truncated derivatives were analyzed for their ability to segregate accurately at mitosis.

A natural extension of this approach is in experiments to generate an artificial mammalian chromosome, which could be used to explore chromosome structure and function and would provide an alternative strategy for somatic gene therapy. Repeated fragmentation of mammalian chromosomes, to eliminate both chromosome arms, should provide a way of producing a minimal human chromosome, on which such a vector could be based.

17 CONCLUDING REMARKS

The present interest in mammalian telomeres has been driven largely by recent research linking telomeres to cellular senescence and immortalization. The recent cloning of the genes for the RNA component of human and mouse telomerases will allow the direct testing of the idea that telomeres and telomerase have a direct role to play in these fundamental aspects of cell biology. Excitement in the telomere field has also come from the application of cloned $(TTAGGG)_n$ repeat arrays in the manipulation of the mammalian genome and the potential of this technology for generating an artificial mammalian chromosome vector.

See also Chromatin Formation and Structure; Histones; Telomerase; Yeast Genetics.

Bibliography

Beissman, H., and Mason, J. M. (1994) Telomeric repeat sequences. *Chromosoma,* 103:154–161.

Blackburn, E. H. (1991) Structure and function of telomeres. *Nature,* 350:569–573.

———. (1992) Telomerases. *Annu. Rev. Biochem.* 61:113–129.

de Lange, T. (1994) Activation of telomerase in a human tumor. *Proc. Natl. Acad. Sci. U.S.A.* 91:2882–2885.

Farr, C. J. (1996) Mammalian telomeres and chromosome fragmentation. *Semin. Cell Dev. Biol.* 7:41–48.

Greider, C. W. (1994) Mammalian telomere dynamics: Healing, shortening and stabilisation. *Curr. Opin. Genet. Dev.* 4:203–211.

Harley, C. B., and Villeponteau, B. (1995) Telomeres and telomerase in aging and cancer. *Curr. Opin. Genet. Dev.* 5:249–255.

Kipling, D. (1995) *The Telomere.* Oxford University Press, London, New York.

Tyler-Smith, C., and Willard, H. F. (1993) Mammalian chromosome structure. *Curr. Biol.* 3:390–397.

Wagner, R. P., Maguire, M. P., and Stallings, R. L. (1993) *Chromosomes: A Synthesis.* Wiley-Liss, New York.

Theoretical Molecular Biology

Andrzej K. Konopka

1 Preliminaries
- 1.1 Scope of Theoretical Molecular Biology
- 1.2 Codes and (Biological) Function: A Logical Trap
- 1.3 Classification Codes
- 1.4 Pragmatic Inference
- 1.5 Determining Classification Code Words

2 Statistical Modeling and Frequency Counts
- 2.1 Nonoverlapping and Overlapping Frequency Counts
- 2.2 Ad hoc (Simple) Frequency Counts
- 2.3 Ranked Oligonucleotide Frequencies; Yule Distribution

3 Measuring Deviations from Discrete Uniform Distribution (DUD)
- 3.1 Entropy and Redundancy of Nucleotide Sequences
- 3.2 Coincidence Indices and Their Use in Sequence Analysis

4 Noncontiguous Patterns and Distance Charts

5 Local Compositional Complexity: A Maximum Entropy Principle
5.1 Local Compositional Complexity
5.2 Surprisal
5.3 Log–Odds Complexity Distribution
5.4 Patchy Complexity

6 Concluding Remarks
6.1 Semantic and Pragmatic Theories of Information
6.2 Logic of Heuristic Reasoning
6.3 Question of Measurement in Biology

Key Words

Code A mapping from one symbolic representation (input representation) of a system into another representation (output representation). If the input and output representations can be expressed in the form of elementary symbols chosen from finite alphabets, a code can be seen as a relation between input and output alphabets.

Distance Chart We select a particular *k*-gram to be a motif and then consider oligonucleotides separating the instances of this *k*-gram to be punctuations. The frequency of occurrence of punctuations plotted against their length is a distance chart. We refer to a distance chart in which the motif is not allowed to be a substring of any punctuation as to the shortest distance chart.

***k*-Gram (*k*-tuple)** A string of *k* symbols chosen from a given elementary alphabet. In the case of nucleotide sequences, *k*-grams correspond to oligonucleotides of length *k*. The most frequently used nucleic acid elementary alphabet is {A, C, G, T (or U)}, where the letters stand for adenine, cytosine, guanine, and thymine (or uracil) nucleotides, respectively. Other alphabets used in nucleic acid studies include {K, M}, where K is either guanine or thymine (uracil in RNA) and M is either adenine or cytosine; {R, Y}, where R is a purine (adenine or guanine) and Y is pyrimidine (either cytosine or thymine); and {S, W}, where S is either cytosine or guanine and W is either adenine or thymine (or uracil in RNA). If the elementary alphabet contains n symbols and k is fixed, we refer to a set of all n^k *k*-grams as a *k*-gram (nonelementary) alphabet.

Motifs and Punctuations Given a *k*-gram alphabet, we select *k*-grams that we call motifs and *k*-grams that we call punctuations. Heuristic principles for such choice are complex because they involve knowledge of "extrasequential" facts.

Pattern There is no general definition; a pattern can be any logical, geometrical, or (broadly) factual connection between elements of a model that attracts our attention. "Pattern" is usually understood ostensively by experienced explorers of the same model.

Sequence Pattern A string of motifs and punctuations that begins and ends with a motif. A contiguous pattern is a sequence pattern that does not contain punctuations, whereas a noncontiguous pattern is a sequence pattern that does contain punctuations.

In a narrow sense theoretical molecular biology aims toward extended interpretation of both laboratory and computational experiments in molecular biology. The "extended interpretation" consists of several activities, which often draw on methods that themselves are distinct research areas with their own theoretical foundations. Some of the methods pertain to physics (e.g., molecular mechanics) and some to chemistry (reaction kinetics); yet others are parts of applied mathematics (combinatorics) or computer science (theory of algorithms). Theoretical molecular biology differs from chemistry, physics, and even biochemistry by systematically considering symbolic interpretations (besides the mechanistic ones) of the biologic role(s) of nucleic acid and protein sequences.

In a larger sense theoretical molecular biology is concerned with cognitive systematization of concepts, theories, and phenomena and aims toward an explanation of the phenomenon of life itself. In this (larger) sense the conceptual principles of modeling biological systems appear to be more important than design and interpretations of experiments.

This article is primarily devoted to the aspects of theoretical molecular biology that pertain to sequence analysis and genomics and thereby are of interest to molecular biologists proper. However, to satisfy readers who would like to predict future research directions in the field, topics belonging to the larger aspect of theoretical molecular biology are discussed briefly.

1 PRELIMINARIES

1.1 Scope of Theoretical Molecular Biology

General tasks involved in extended interpretation of experiments involve:

1. Relating the mechanistic and the informational (i.e., symbolic) aspects of biological phenomena through appropriate generalization and modeling. EXAMPLE: Postulating the triplet nature of the genetic code before it had been determined experimentally.
2. Devising principles for analyzing inexact data (quantitative representations of qualitative descriptions included). EXAMPLES: Database design, annotation, and searches; logic of heuristic reasoning.
3. Relating principles and laws inferred from studies at the molecular level with those derived from studies of organisms, populations, and ecosystems. EXAMPLES: Molecular evolution; molecular evolutionary genetics.
4. Deriving properties of biological systems from principles of physics and chemistry. EXAMPLES: Biological thermodynamics and kinetics; theories of origin of life; molecular mechanics.
5. Modeling (usually mathematical) properties of laboratory tools and designing experiments. EXAMPLE: Analysis and interpretation of molecular spectra, X-ray data, and so on.
6. Modeling properties of computational tools and designing computational experiments. EXAMPLE: Sequence alignment.

In addition, there are several tasks pertaining to the larger aspect of theoretical biology. These include:

7. Developing a general system theory and research on specific classes of general systems. EXAMPLE: Anticipatory systems and their use for the description of (biochemical) reaction networks.

8. Developing foundations of logic and mathematic such that the resulting mathematical objects would be appropriate for modeling biological systems. EXAMPLE: Theory of fuzzy sets as a language for formulating models and theories.
9. Developing foundations of general information theory and information science. EXAMPLE: Work on semantic and pragmatic theories of information.

Activities 7–9 sometimes draw on methods from other research programs such as system theory, cognitive science, logic, and epistemology but do not belong to the current paradigms of these fields.

1.2 Codes and (Biological) Function: A Logical Trap

The focus on symbolic (or informational) aspects of biological phenomena makes "code" the central, unifying metaphor in theoretical molecular biology. From the early period in molecular biology, biopolymer sequences and structures were thought of as texts that carry encoded messages. The three best-known "biological codes" are the Watson–Crick–Chargaff base pairing rules, the translation code for protein biosynthesis (the genetic code), and the cleavage sites for restriction enzymes. Many biologists assume that more *functional codes* of this kind exist. For instance, we could think of replication code, homologous and illegitimate recombination codes, mutation hot spots code, protein or RNA folding codes, and so on. However, the adequate description of biological function is a formidable task. It calls for knowledge of a great many mechanistic details that constitute the function, and this requirement is a major methodological obstacle in specifying a functional code. In most cases in which we know or surmise that a code exists, we are unable to appropriately describe the biological function concerned (the genetic code is a notable exception to this common situation). Moreover, even if a given sequence contains "encoded" functions, the corresponding "messages" are often superimposed. Thereby, the assumption "one sequence, one function" is not valid and needs to be replaced by "one sequence, many functions" and "many sequences, many functions."

1.3 Classification Codes

The above-mentioned methodological problems can be avoided when we replace functional codes by an alternative concept of *classification codes.* In principle, these are many-to-one mappings from a set of patterns found in (or resulting from) a class of sequences into descriptors (such as a name or a list of features) of this class. We specify a classification code by just listing patterns from its domain. We call each individual pattern from the list a *classification code word* or just a *code word.* Not every pattern found in a given collection of sequences is a classification code word. To qualify as a code word, the pattern must be *significant* according to predetermined criteria.

1.4 Pragmatic Inference

The criteria of significance usually involve a complex heuristics called *pragmatic inference.* (The name indicates a focus on the "rules" of usage for patterns.) Pragmatic inference explores the entire body of available biological knowledge in addition to statistical analyses. The resulting measure of significance often does not correspond to statistical significance because the rules of pragmatic inference are different from the rules of statistical inference. However, statistical modeling is a vital part of those steps of pragmatic inference in which selection of motifs and punctuations is made. Figure 1 shows a conceptual scheme of pragmatic inference for sequence research as seen by most theoretical and computational biologists.

1.5 Determining Classification Code Words

Methods of determining code words draw on several assumptions concerning the patterns of characters (from an elementary alphabet) that are correct code words. A nonexhaustive list of assumptions includes the following:

- A1. Relative frequency of a given code word should be the same (with the appropriate accuracy) in all sufficiently long sequences from a given functional class. [For sequences of infinite length, we expect that the relative frequencies will converge to definite, invariant values (probabilities).]
- A2. Probabilities of at least some individual code words should markedly differ from one another in all sequences from a given functional class.
- A3. Probability distribution of code words should be the same (with appropriate accuracy) in all sequences from a given functional class.
- A4. Selected code words should occupy distinct relative positions in every sequence from a given functional class. (This assumption is known in linguistics as a postulate of existence of distributional structures.)

Several methods of determining classification code words from corpora of sequences explore assumptions A1–A3. The most important are perhaps those based on coincidence indices (string length test and string repetition test) or on Markov analyses (including hidden Markov chains and nonhomogeneous Markov chains). A couple of powerful methods explore distributional structures (assumption A4) in the form of distance charts or Fourier analysis.

Knowing classification code words has practical consequences. It allows us to develop algorithms for the prediction of putative functional domains in unannotated sequences. These algorithms (and the corresponding software) are of paramount importance for all genome sequencing projects, but we shall not discuss them here. The relevant discussion can be found in specialized publications on computational biology and genomics listed in the bibliography.

2 STATISTICAL MODELING AND FREQUENCY COUNTS

To determine how "unusual" the occurrence of a given k-gram in a long sequence is, we compare the occurrence frequency in the data with the frequency expected from a model of chance.

It is in general a laborious task to find the best model of chance (i.e., prior probability distribution) for a given situation. A model often used in sequence research is a *Bernoulli text* in which all 1-grams occur independently of each other and with equal probability. The corresponding probability distribution is known as *discrete uniform distribution (DUD).* (The reverse distribution assigns a k-gram to a given probability.)

Let it be determined from data that the frequency of the ith k-gram be $F(i)$. Let the expected frequency of this k-gram be $F_0(i)$ and the variance (determined from a model of chance) $V(i)$. The "degree of

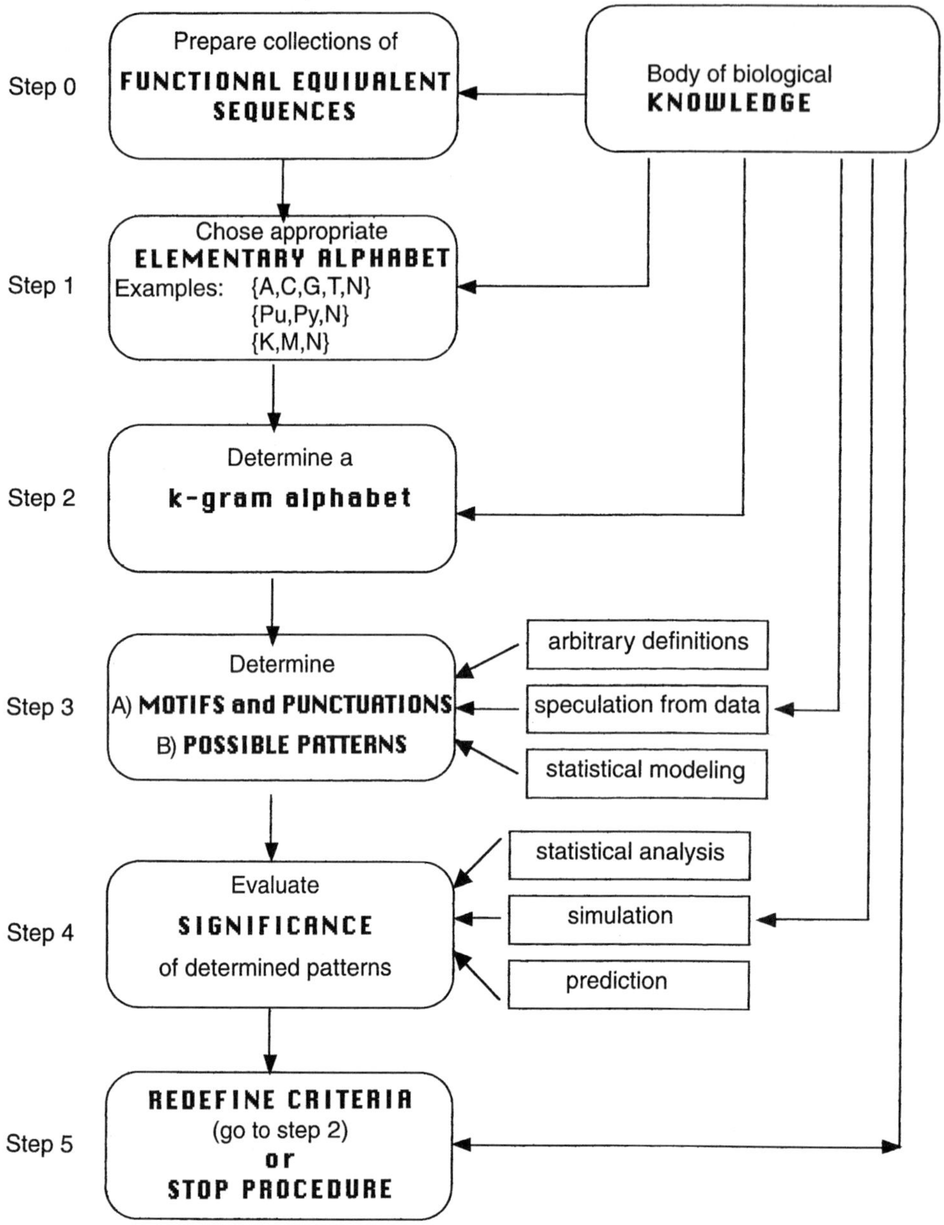

Figure 1. General protocol of pragmatic inference pertaining to computer-assisted sequence analysis. Vaguely defined "biological knowledge" is taken into account at every step of inference. [Based on A.K. Konopka (1994) Sequences and codes: Fundamentals of biomolecular cryptology. In *Biocomputing: Informatics and Genome Projects,* D. Smith, Ed., pp. 119–174. Academic Press, San Diego, CA.

unusuality" of the occurrence of ith k-gram in the data is a function of $F(i)$, $F_0(i)$, and $V(i)$. The most common such function is a *z-score* defined as follows.

$$z(i) = \frac{F(i) - F_0(i)}{[V(i)]^{1/2}} \tag{1}$$

After having calculated z-score $z(i)$, we compare it with a threshold value z_0.

If $|z(i)| > z_0$, we consider the frequency of ith k-gram unusually high for $z(i) - z_0 > 0$ (overrepresentation) or unusually low for $z(i) - z_0 < 0$ (underrepresentation).

2.1 Nonoverlapping and Overlapping Frequency Counts

There are two general ways of calculating k-gram frequencies in a given long sequence (of length L): nonoverlapping count and overlapping counts. In the case of nonoverlapping count we take into account k-grams in positions $1, 1 + k, 1 + 2k, \ldots, 1 + mk < L$. The number $(m + 1)$ of k-grams counted in this way equals the integer part of L/k. Cases of overlapping counts of k-grams differ from each other by "counting steps" of 1 through $k - 1$ (counting step k is nonoverlapping count). In the overlapping count of step 1 we take into account k-grams in positions $1, 2, 3, \ldots, L - k + 1$, and there are exactly $L - k + 1$ k-grams in the sequence of total length L.

(This counting is equivalent to shifting the window of length k by one nucleotide at a time.) If we chose a counting step of 2 and k is greater than 2, we take into account k-grams in every other position (i.e., 1, 3, 5, . . . , n). Counting step of 3 (if k is greater than 3) means taking into account every third position (i.e., 1, 4, 7, . . . , n). Most molecular biologists understand the overlapping count as a count of step 1, and we will use this convention in this article (i.e., we will not discuss cases of counting steps other than 1 and k).

When the elementary alphabet has n letters (1-grams) and when k is fixed, the k-gram alphabet contains $N = n^k$ elements. In a Bernoulli text of length L, the expected frequency of each k-gram is therefore $F_0 = L/n^k$. For the nonoverlapping count, variance is a simple function of n, k, and L given by the equation:

$$V = \frac{L}{n^k}\left(1 - \frac{2k-1}{n^k}\right) \tag{2}$$

For the overlapping count, variance is no longer a simple function of n, k, and L and must be calculated for each individual k-gram separately. The reason for this complication is the variance's dependence on the *self-overlap capacity* of individual k-grams.

The self-overlap capacity of a given k-gram S is determined from a polynomial:

$$K_S = \sum_{i=0}^{k-1} a_i x^i \tag{3}$$

Coefficients a_i are 1 if the first and last $(k - i)$-grams in S are identical and 0 otherwise. For example, the polynomial for trinucleotide GGG is $K_{GGG} = 1 + x + x^2$. For trinucleotides GAG and GAT we have $K_{GAG} = 1 + x^2$ and $K_{GAT} = 1$ (no self-overlap), respectively.

The variance of frequency distribution for a k-gram S in a Bernoulli text of length L over an elementary alphabet containing n letters is a function of $K_S(1/n)$ given by the formula:

$$V_s = \frac{L}{n^k}\left[2K_S\left(\frac{1}{n}\right) - 1 - \frac{2k-1}{n^k}\right] \tag{4}$$

Failure to consider self-overlap capacity in overlapping counts leads to huge errors (hundreds of percent) in variance calculation, particularly in cases of small-size elementary alphabets and large values of k. [Variance for nonoverlapping count, equation (2) can be derived from equation (4) by putting $K_S(1/n) = 1$ (no self-overlap).]

2.2 Ad Hoc (Simple) Frequency Counts

Examples of the overlapping 2-gram frequency counts in seven large collections of DNA sequences are shown in Figure 2. Although the "folklore" of describing and interpreting k-gram frequency counts resembles the descriptive biology of the nineteenth century (see caption to Figure 2), it plays an invaluable role in selecting elementary alphabets and k-gram candidates for motifs.

2.3 Ranked Oligonucleotide Frequencies; Yule Distribution

Ranks of word frequency in printed alphabetic languages appear to follow the so-called Zipf distribution, which is given by the equation:

$$F = aR^k, \qquad k < 0 \tag{5}$$

where F is frequency of occurrence, R is rank from most frequent to least frequent, and a and k are parameters.

The same distribution also characterizes a host of nonlinguistic phenomena such as the number of species per genus, the number of papers per author in a given field of science, the distribution of income, and the populations of cities in a nation.

Attempts to find a fit of ranked oligonucleotide frequencies to Zipf distribution failed soon after the first nucleotide sequences had been identified. It is clear now that short oligonucleotide frequencies from all known genomes do not follow Zipf distribution but a more general three-parameter density called the Yule distribution, which is given by the equation:

$$F = aR^k b^R, \qquad k < 0 \tag{6}$$

where F is frequency and R is the rank; a, b, and k are parameters.

Zipf distribution [equation (5)] is a special case of the Yule distribution [equation (6)] in which parameter $b = 1$. Even if b is very close to 1 but not exactly 1 (e.g., $b = 0.9995$), the Yule distribution differs remarkably from Zipf distribution.

Parameters of Yule distribution for hexanucleotide frequencies in large collections of intron and exon sequences are shown in Table 1. It appears that both intron and exon data fit the Yule distribution extremely well (and about equally well; see note b to Table I). Example plots (in log-rank versus log-frequency coordinates) of the Yule distribution are shown in Figure 3.

3 MEASURING DEVIATIONS FROM DISCRETE UNIFORM DISTRIBUTION (DUD)

3.1 Entropy and Redundancy of Nucleotide Sequences

The mathematical apparatus of Shannon's theory of communication ("information" theory) has been applied to biology quite against Shannon's preferences. In 1954 he excluded psychology and biology from the list of fields of science that could benefit from applying his theory because he believed that semantic aspects of communication are more important in these fields than syntactic aspects. Of course he was unaware of the important role statistical studies of nucleotide and protein sequences would play in biology. Today we know that indeed communication theory does not explain much of biology (nor did anyone sane ever expect it to do so) but, on the other hand, it is an extraordinary source of statistical methods for studying sequences.

Shannon's studies of k-grams of letters in printed alphabetic languages have been a spectacular application of his theory of communication. The central concept in these studies is an abstract device (called source) that generates sequences of symbols (messages) chosen from a finite alphabet by means of a random mechanism. Particularly important are ergodic sources in which the random mechanism leads to "typical" messages with high probability (close to 1) and to "atypical" sequences of symbols with negligible probability. (We can say, in other words, that sequences produced by an ergodic source are statistically homogeneous.)

It has long been known that frequency distributions of letters are roughly the same in all texts written in the same language. It has then been assumed (but not proved) that any alphabetic writing system can be modeled by an ergodic source. (In particular it has been suggested that a writing system can be regarded as a limit of a suc-

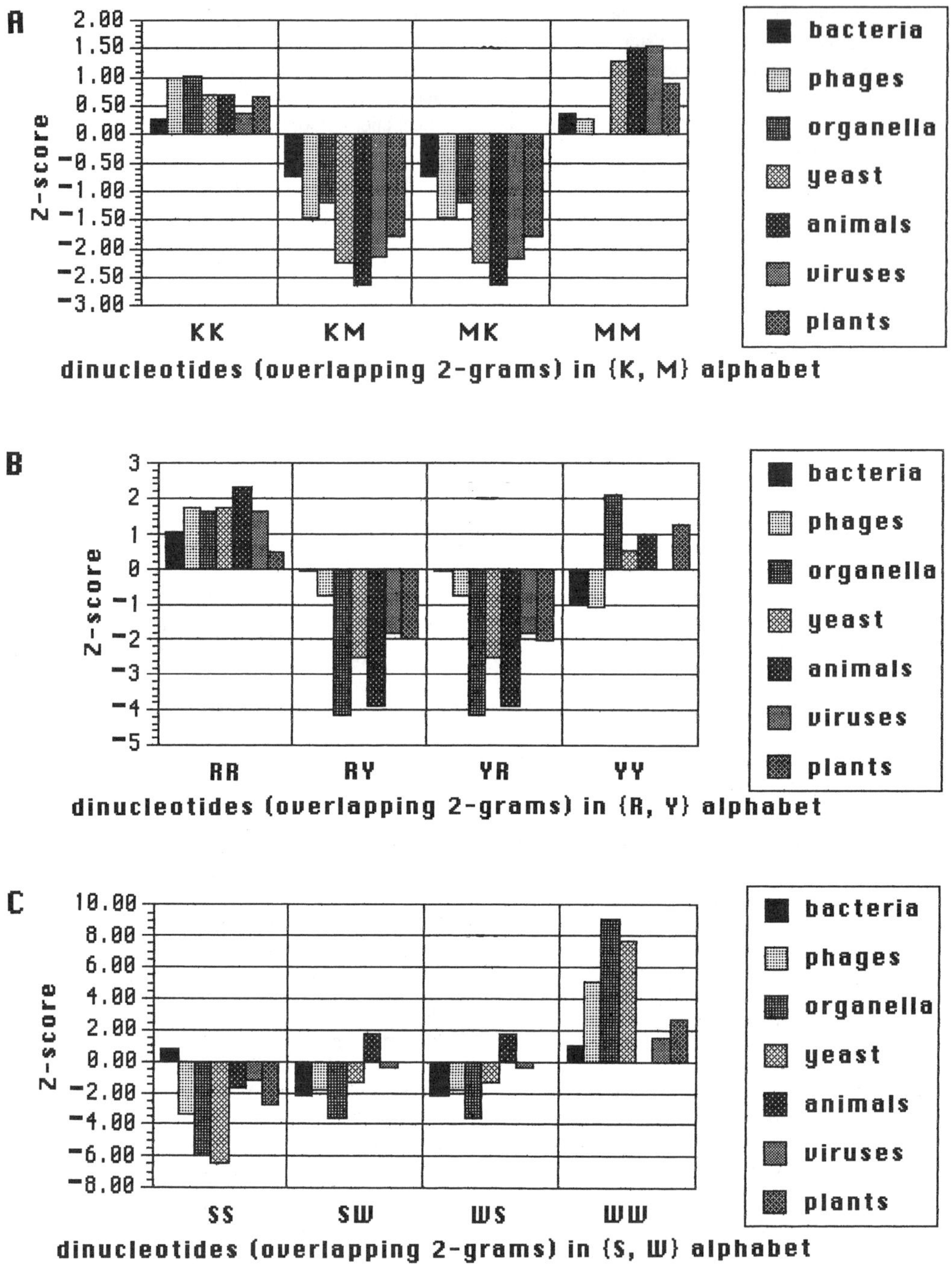

Figure 2. Examples of dinucleotide frequency counts in representative (large) samples of DNA sequences from seven classes of genomes. Three different elementary alphabets were taken into account. (A) {K, M} alphabet, where K is either guanine or thymine and M is either adenine or cytosine. Assuming that the threshold z-score values for unusually high frequency is $+3$ and for unusually low frequency is -3, none of the dinucleotides over this alphabet occur significantly more often or significantly less often than others in all sequence samples studied. However, "complex" dinucleotides KM and MK tend to be slightly but systematically underrepresented. (B) {R, Y} alphabet, where R stands for purine (adenine or guanine) and Y stands for pyrimidine (cytosine or thymine). "Complex" dinucleotides RY and YR appear to be unusually underrepresented in organella and animal (i.e., primate, rodent, other mammalian, other vertebrate and invertebrate) genomes. This underrepresentation might be due in part to the well-known avoidance of long alternating purine–pyrimidine tracks (sequences with a potential for B-DNA–Z-DNA tertiary structure transition). Note that in other genomes (bacteria, phages, yeast, viruses, and plants) slight underrepresentation of RY and YR takes place as well. (C) {S, W} alphabet, where S stands for either guanine or cytosine and W stands for either adenine or thymine. It can be seen that dinucleotide SS is underrepresented in phage, organella (mitochondria and chloroplast), and yeast genomes. In contrast, dinucleotide WW is overrepresented in these genomes. Notably, dinucleotides SW and WS are significantly underrepresented in organella ($z < -3$) and tend to be underrepresented in other genomes as well.

Table 1 Fit of The Yule Equation to Ranked Hexanucleotide Frequencies in Several Large Collections of Intron and Exon Sequences[a]

	Base Pairs	Yule Distribution[b]			
		a	*k*	*b*	ρ^2
Primate introns	346,953	.0016	−.15	.9994	.96
Primate exons	500,899	.0017	−.14	.9993	.97
Vertebrate introns	232,272	.0017	−.16	.9995	.96
Vertebrate exons	370,327	.0021	−.18	.9994	.97
Invertebrate introns	181,184	.0034	−.28	.9995	.99
Invertebrate exons	499,601	.0016	−.15	.9994	.95
Organelle introns	105,189	.0045	−.33	.9996	.99
Organelle exons	531,943	.0020	−.17	.9994	.97
Plant introns	255,535	.0027	−.24	.9995	.98
Plant exons	588,172	.0015	−.13	.9994	.95
Yeast introns	18,821	.0089	−.43	.9999	.97
Yeast exons	333,789	.0025	−.19	.9993	.99
Viral introns	75,359	.0015	−.15	.9995	.99
Viral exons	733,231	.0011	−.09	.9995	.89
Bacterial genes	876,257	.0015	−.12	.9993	.95
Phage genes	353,104	.0022	−.19	.9994	.98

[a]When plotted on log–log coordinates, all the data exhibit the convex form shown in Figure 3.

[b]The mean values of parameters *a, b,* and *k* in introns and exons are the same within error limits (*a*-intron = .0034 ±.0026; *a*-exon = .0018 ± .0005; *b*-intron = .9981 ± .0036; *b*-exon = .9994 ± .0001; *k*-intron = 0.25 ± 0.11; *k*-exon = 0.15 ± 0.03). This implies that Yule parameters should not be used to discriminate between coding and noncoding regions in nucleotide sequences (the sampling errors in noncoding regions are too high). However, high ρ^2 values show that fit to the Yule distribution is excellent for both coding and noncoding regions regardless of the species.

Source: Based on C. Martindale and A. K. Konopka, *Computers Chem.* 20(1): 35–38 (1996).

cession of ergodic Markov sources that in turn is itself an ergodic source.)

The probability distribution of letters from an alphabet of size n in a given ensemble of texts can be characterized by a function called entropy (of the distribution):

$$H1 = -\sum_{i=1}^{n} p(i) * \log p(i) \tag{7}$$

where $p(i)$ are probabilities of letters, the log is taken to the base 2 (unless otherwise specified), and a convention $0 \log(0) = 0$ is assumed.

If textual elements (letters) from a finite alphabet of size n were combined in a text by chance alone, it would have to be a Bernoulli text. Therefore the letters should occur independently of each other and with probabilities following a DUD (i.e., equal to $1/n$). The entropy of DUD obviously equals $\log(n)$, that is,

$$H0 = -n*\left(\frac{1}{n}\right)*\log\left(\frac{1}{n}\right) = \log(n) \tag{8}$$

The *entropy of strings of length k* composed of characters from an (elementary) alphabet of size n is defined as follows:

$$H^{(k)} = -\sum_{i1=1}^{n}\sum_{i2=1}^{n}\ldots\sum_{ik=1}^{n} p(i1, i2, \ldots, ik) * \log p(i1, i2, \ldots, ik) \tag{9}$$

The *k-gram entropy* is defined as follows:

$$Hk = H^{(k)} - H^{(k-1)}, \qquad \text{for } k > 1 \tag{10}$$

with $H^{(1)} = H1$ given by equation (7) and $H0$ given by equation (8).

Note that entropy has a dimension that is determined by the basis of log function and by the alphabet taken into account. In the general case when the basis of log function is r, the k-gram entropy Hk is given in r-ary units per letter, and the entropy of strings of length k is given in r-ary units per k-gram. The most frequently used value of r is 2. In this case $H^{(k)}$ is in bits per k-gram and Hk in bits per letter ("bit" is a name of 2-ary unit).

The *entropy of a source S* exists and equals $H(S)$ if

$$\lim_{k\to\infty} \frac{H^{(k)}}{k} \tag{11}$$

exists and equals $H(S)$.

The ergodic sources always have entropy. A given writing system (sometimes referred to as a language) L is assumed to be an ergodic source (a limit of a succession of ergodic Markov sources). Hence, the writing system's entropy $H(L)$ exists and is bounded by the following (obvious) chain of inequalities:

$$H(L) < \cdots < Hk < \cdots < H3 < H2 < H1 < H0 \tag{12}$$

The ratio of entropy of source to the maximum value it could have while still restricted to the same symbols is called *relative entropy* (of source). Since it follows from equation (12) that $H0$ is the maximum value of entropy, we have:

$$H'(L) = \frac{H(L)}{H0} \tag{13}$$

Similarly, we can define *relative k-gram entropy:*

$$Hk' = \frac{Hk}{H0} \tag{14}$$

Note that relative entropy is a dimensionless real number. Its minimum value is 0 and maximum value is 1.

A writing system's (language) *redundancy* is defined as follows:

$$R(L) = 1 - H'(L) \tag{15}$$

Similarly, we can define k-gram redundancy as follows:

$$Rk = 1 - Hk' \tag{16}$$

It follows from equation (12) that $R(L)$ is higher than any k-gram redundancy; that is,

$$0 < R1 < R2 < \cdots < Rk < \cdots < R(L) \tag{17}$$

Because of high sampling errors of k-gram entropies in sequences of finite length, it is not obvious that the ensembles of functionally equivalent nucleotide sequences have entropy. Dedicated computational studies will be needed in the future to determine how well k-gram entropies converge to the source entropy values. For the time being we assume that functionally equivalent sequences do have entropy and that "functionality" can be thought of as an ergodic source.

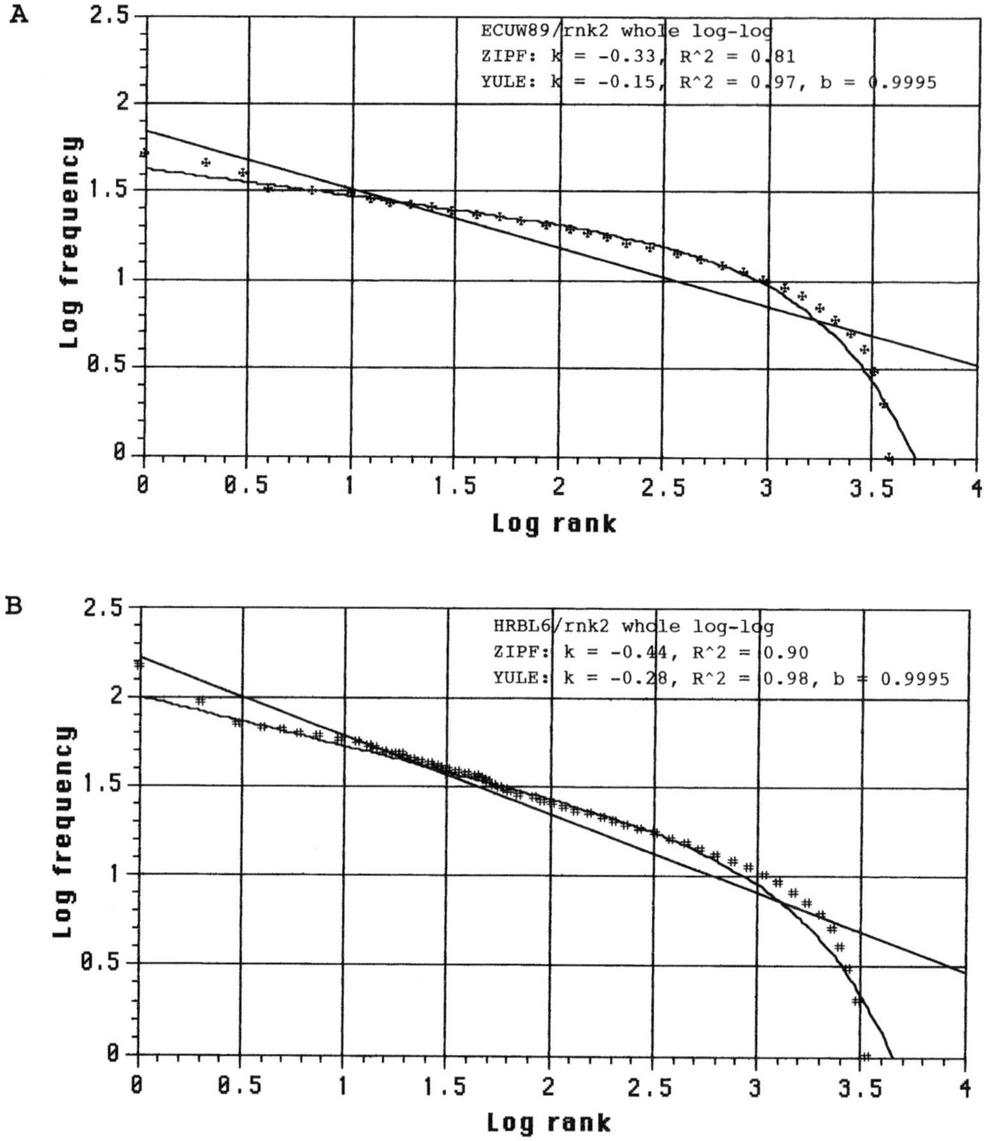

Figure 3. Ranked hexanucleotide frequencies fit the Yule distribution [convex lines on (A) and (B)] much better than the Zipf distribution [straight lines on (A) and (B)]. (A) *E. coli* chromosomal region labeled ECOUW89 is a long sequence (176,195 nt) in which 82.1% of nucleotides (nt) constitute protein-coding regions. (B) HUMRETBLAS (human retinoblastoma susceptibility gene) is a long sequence (180,388 nt) in which only 1.5% of nucleotides belong to exons. [Based on C. Martindale and A. K. Konopka, *Computers Chem.* 20(1): 35–38 (1996).]

Table 2 shows the results of redundancy calculation for several large collections of functionally equivalent nucleotide sequences. As expected, the frequent occurrence of mononucleotide repeats has resulted in very high redundancy ($\approx 20\%$) in the sequences of eukaryotic illegitimate recombination regions. Interestingly, the lowest redundancy can be seen for protein-coding genes and the highest for intergenic spacers and flanking regions of genes. However, because of very high error values, redundancy in exons and the corresponding introns appears to be the same even at the 33% level of confidence. This precludes using the "function-associated" redundancy as a criterion for determining exon and intron location in unannotated sequences. It also means that redundancy is unlikely to be a basis for a classification code.

3.2 Coincidence Indices and Their Use in Sequence Analysis

For a discrete probability distribution $[p(1), p(2), \ldots, p(n)]$ over an ensemble of n events, the *index of coincidence* can be defined as follows:

$$I = n \sum_{i=1}^{n} p^2(i) \tag{18}$$

For the text of finite length L, index of coincidence can be estimated as follows:

Table 2 Estimated Redundancy in Various Large Collections of Functionally Equivalent Nucleotide Sequences[a]

	Redundancy
Human exons (in mature mRNAs)	0.070 ±.009
Human introns	0.082 ±.017
Human 3′UTRs	0.081 ± .020
Human 5′UTRs	0.083 ±.022
Human 3′ flanks of protein-coding genes	0.160 ±.034
Human 5′ flanks of protein-coding genes	0.122 ±.025
Eukaryotic exons (in mature mRNAs)	0.076 ±.021
Eukaryotic introns	0.087 ±.022
Eukaryotic intergenic spacers (non-RNA, nonhistone)	0.112 ±.007
Eukaryotic intergenic spacers (histone)	0.142 ±.027
Eukaryotic illegitimate recombination regions	0.202 ±.032
Bacterial protein-coding genes	0.075 ±.023
Bacterial spacers (RNA genes)	0.090 ±.023
18s-rRNA genes	0.095 ±.040
Small nuclear RNAs	0.127 ±.032
16s-rRNA genes	0.118 ±.038
5s-rRNAs	0.092 ±.028
tRNAs	0.155 ±.041
23s-RNAs	0.127 ±.041
28s-RNAs	0.169 ±.024

[a]Estimates of $k \to \infty$ were obtained from regression analyses of the redundancies of 1- through 4-grams or 1- through 5-grams (depending on the length of sequences in a given collection). The error values listed are standard errors of estmate for hyperbolic regression.

Source: Based on A. K. Konopka, Sequences and codes: Fundamentals of biomolecular cryptology. In *Biocomputing: Informatics and Genome Projects,* D. Smith, Ed., pp. 119–174. Academic Press, San Diego, CA, 1994.

$$I = \frac{n \sum_{i=1}^{n} f(i)\,[f(i) - 1]}{M\,(M - 1)} \qquad (19)$$

where $f(i)$ is a frequency of ith element of the alphabet n is a size of the alphabet and M is the number of elements of the alphabet present in the text (for single letters, $M = L$; for strings of two letters, $M = L - 1$; for strings of k letters, $M = L - k + 1$).

The variance of I has been found to be:

$$\mathrm{Var}\,(I) = \frac{2i\,(n - I)}{M(M - 1)} \qquad (20)$$

In statistical linguistics and cryptanalysis tests based on coincidence, indices are reliable tools for determining substitution block encodings. They work particularly well when each string of the code alphabet has the same length as other strings. It also helps if the size of (encoded) "space bars" is negligible comparing to the size of (also encoded) "message units."

Classification codes for different ensembles of functionally equivalent sequences can be determined by the two tests based on coincidence indices. Both these measures, the *string repetition test* and the *length test,* apply to the following situation: given a sequence of length L represented in an elementary alphabet $E = \{A, C, G, T\}$, consider m code alphabets containing $N(k) = 4^k$ strings each ($k = 1, 2, \ldots, m$ and $m \ll L$); choose "the best" code alphabet (to represent a given sequence) from the m alphabets given.

In the string repetition test we pick the kth alphabet and calculate the coincidence index [equation (19)] along with its variance [equation (20)] in our sequence of length L (i.e., containing $M = L - k + 1$ "overlapping" or $M' = \mathrm{Int}\{L/k\}$ "nonoverlapping" strings belonging to the alphabet of $N(k)$ elements). For a Bernoulli text, the k-gram probability distribution is a DUD and therefore $I_{\text{expected}} = 1$. The z-score of I in a given alphabet is:

$$z(i) = \frac{I - 1}{[\mathrm{Var}(I)]^{1/2}} \qquad (21)$$

After calculating $z(I)$ for each alphabet, we consider the alphabets that maximize equation (21) to be "the best."

In the string length test, mean coincidence indices (per nucleotide) in k-grams are compared with the coincidence index over mononucleotides (1-grams) in an entire sequence. We pick the kth alphabet and calculate the coincidence index for all nucleotides at all positions $j = 1, 2, \ldots, k$ within each k-gram in a given sequence. That is:

$$I(j) = \frac{4}{M(M - 1)} \sum_{i=1}^{4} f(i, j)\,[f(i, j) - 1] \qquad (22)$$

where $f(i, j)$ is the frequency of ith nucleotide at the jth position of a k-gram.

The preceding calculations provide us with a vector $J = [I(1), I(2), \ldots, I(k)]$ from which we can determine the mean coincidence index $<I>$, that is,

$$<I> = \frac{1}{k} \sum_{j=1}^{k} I(j) \qquad (23)$$

The expected value of $<I>$ is obtained from equation (19) with $M = L$ and $n = 4$, that is,

$$E(<I>) = \frac{4 \sum_{i=1}^{4} f(i)[f(i) - 1]}{L(L - 1)} \qquad (24)$$

The variance is obtained from equation (20) by again putting $M = L$ and $n = 4$ (if we use {A, C, G, T/U} alphabet; for binary alphabets such as {R, Y} we would need to put $n = 2$), that is,

$$\mathrm{Var}(<I>) = \frac{2I(4 - I)}{L(L - 1)} \qquad (25)$$

The z-score for this test is then:

$$z(<I>) = \frac{<I> - E(<I>)}{[\mathrm{Var}(<I>)]^{1/2}} \qquad (26)$$

After calculating $z(<I>)$ for each alphabet, we consider the alphabet that maximizes equation (26) to be "the best."

Both tests indicate unambiguously (see Table 3) that the trinucleotide code is the best to represent protein-coding regions in nu-

Table 3 Examples of z-Scores Calculated for String Lengths 2–9 from the String Length Test [equation (26)] for Sequences Belonging to Different Functional Classes[a]

	Length of k-Strings (i.e., k)							
	2	3	4	5	6	7	8	9
Typical Two-Base Periodic IVS (≈ 20% of all IVS)								
>DL;FOSMS	2.1	−0.4	3.5	−0.2	7.2	2.4	4.9	−3.4
>DL;TVMSP1	2.7	−1.6	2.9	−3.1	6.1	−1.5	0.2	−5.3
>DL;HVMS3	7.5	0.1	7.0	−1.8	7.5	−1.0	7.1	−4.3
>DL;HVMS44	3.4	1.8	4.8	−3.6	5.0	−0.7	3.3	5.7
>DL;IGCMD	3.7	1.6	2.3	0.6	4.7	−0.2	4.2	−0.2
>DL;G2MSA	4.1	0.3	5.7	−2.5	5.4	−0.6	4.3	5.8
>DL;HLMSE1	2.0	0.0	6.1	0.1	1.5	−0.2	3.7	−3.0
>DL;CELDEB	2.7	−0.7	4.0	−3.0	0.6	0.6	7.3	−0.5
>DL;DROGLD	3.9	1.8	3.6	1.1	8.1	3.9	2.7	−1.0
>DL;DROLGL	4.2	−1.0	6.3	−0.5	2.3	−0.4	5.3	−1.9
>DL;DROMHC	3.9	0.1	3.6	−0.7	4.1	−2.6	3.2	−2.3
Typical Translated Regions of Genes ("true" coding regions)								
>DL;HPHU2R	−0.2	29.9	−0.5	1.4	29.1	2.5	2.8	32.5
>DL;KFHUG	−1.0	26.8	−0.3	0.3	23.8	1.2	−1.7	26.3
>DL;KXHU	1.6	71.0	4.4	3.8	73.9	−3.9	5.1	71.4
>DL;UBHUB	1.2	95.5	−0.3	−2.5	98.2	−0.6	−1.6	91.1
>DL;LPHU4A	−0.9	106.6	0.8	−1.3	106.8	−3.6	0.5	106.1
>DL;FGHUGG	1.4	17.6	0.6	−1.1	20.1	2.1	−3.6	19.9
>DL;APOCHO	4.5	151.0	2.9	−2.9	176.1	−1.1	1.0	229.9
>DL;BBO58M	0.0	51.2	0.4	−0.5	52.3	−1.4	0.2	47.3
>DL;BBOMER	−0.6	43.9	−1.5	−2.5	44.4	−1.6	−0.1	43.9
>DL;DROCID	−1.0	59.8	−1.3	0.0	59.0	−0.8	−3.4	61.5
>DL;DROCOP	0.1	175.9	−0.6	−2.9	179.8	−2.7	3.1	175.7
>DL;DROCOP	−1.1	62.5	1.2	9.2	62.7	2.7	0.6	63.6
Typical Viral Translated Regions of Genes								
>DL;ADLEIB	−0.6	22.3	−1.9	−2.0	20.6	4.6	−0.7	28.0
>DL;ADTA_1	−0.1	10.0	−0.8	0.4	9.2	0.8	−0.8	10.9
>DL;ADV40E	0.7	7.2	0.3	−1.6	6.8	−2.1	−1.5	3.5
>DL;ADVFIB	0.2	26.1	0.0	−0.1	24.9	2.4	0.8	25.9
>DL;ADX1AB	0.2	12.6	−1.1	−2.7	11.2	−0.5	−2.1	13.7
Typical Viral IVS								
>DL;ADLEIB	0.0	19.1	−1.9	−0.2	18.1	4.4	0.1	22.9
>DL;ADX1AB	−0.1	10.3	−1.1	−2.4	9.4	−1.7	−3.7	9.8
>DL;ALRCG_	−1.0	10.0	−1.8	−0.9	9.3	−2.3	−1.9	10.0
>DL;HIVBH1	−0.8	20.6	−0.9	−2.7	20.0	−3.0	0.6	24.0
>DL;PPH47C	0.0	12.6	−1.2	1.2	13.9	−1.5	−0.3	11.2

Source: A. K. Konopka (1994) Sequences and codes: Fundamentals of biomolecular cryptology. In *Biocomputing: Informatics and Genome Projects*, D. Smith, Ed., pp. 119–174. Academic Press, San Diego, CA.

cleotide sequences. This result is remarkable because it leads to a rediscovery of codons of the genetic code solely on the basis of the statistical structure of sequences. This means that knowledge of the protein-coding function of genes is not required to find genes among other sequences. Needless to say, methodology of this kind can be used for practical purposes of finding exons in unannotated sequences.

In fact, about 98.5% of known translated regions of genes (exons and bacterial genes) display excessively high, positive values of $z(<I>)$ for the alphabets of 3-grams, 6-grams, 9-grams, and generally, $3m$-grams ($m = 1, 2, 3, \ldots, 20$). In addition, the large fraction of remaining 1.5% are the exons that are subject to alternative splicing (i.e., are also parts of introns).

As far as sequences that are not translated are concerned, the results of string length test are less clear. In a large sample of nuclear eukaryotic introns, for example, about 20% of the sequences test positively for a code word of length 2 [i.e., $2m$-grams ($m = 1, 2, 3, \ldots$) are unambiguously indicated by the test given in equation (26); see Table 3]. Another 3% of intron sequences display clear preference for strings of length 4. Both these results are consistent with the 2-base periodicities in large collections of intron sequences (see Section 5). However, the test [equation (26)] also indicates that about 7% of intron sequences display a clear preference for strings of length 5, and there exist introns (only about 1.3% in the sample collection) that "favor" heptanucleotides.

There are at least three plausible interpretations of the foregoing findings. One possibility is that there are introns of many different kinds playing different biological roles before being spliced out of heteronuclear RNAs. Another possibility is that the "intronic" code alphabet contains strings of variable length. In such a case, the string length test would detect the mean length of a code word but not the actual lengths of individual code words. Finally, it may well be that only fragments of introns are involved in some (unknown) biological processes. In such a case, long stretches of sequence within introns would not code for any function, and therefore there would be no reason to "employ" a statistically distinguishable k-gram alphabet. This last situation would be similar to a text in which long strings of space bars separated individual words (or even letters) and many different symbols were allowed to stand for a space bar.

Viral introns display conspicuously high z-scores for string lengths that are multiples of 3 (examples are listed in Table 3), as do many mitochondrial introns (data not shown). This "exonic" characteristic is quite unexpected in introns but can be easily explained by noting that viral genomes contain overlapping genes. A given viral intron can be an intron in one gene and a part of exon in another (overlapping) gene. Therefore the "exonic" characteristics (in the form of preference for strings of lengths that are multiples of 3) of viral introns can be detected. Similarly, mitochondrial introns that display "exonic" characteristics (code word lengths 3, 6, 9, . . .) overlap with genes that encode the endonuclease responsible for their splicing.

4 NONCONTIGUOUS PATTERNS AND DISTANCE CHARTS

Distributional structures in nucleotide sequences from the same functional class (as introns, exons, tRNAs, 5sRNA, illegitimate recombination sites, etc.) can be seen on k-gram distance charts. If there were no distributional structure in sequences studied, there would be no particular reason for any punctuation length to occur more frequently than other lengths. In other words, with no distributional structures, punctuation lengths should follow a discrete, uniform frequency distribution. When the distributional structures are present, some punctuation lengths should be much more frequent and some others much more rare than expected from discrete, uniform distribution.

Similar reasoning applies to the shortest distance charts. In the absence of distributional structures, the frequency of punctuation lengths in the shortest distance charts would follow geometric distribution. Again, when distributional structures are present, some punctuation lengths will occur more frequently and some others less frequently than would be expected from geometric distribution.

Distance charts (z-scores with regard to discrete, uniform distribution) for motif AC in large collections of nuclear introns and exons from animal genomes (primates, rodents, other mammals, vertebrates, and invertebrates) are shown in Figure 4A. The charts for an internally mirror-symmetric motif ACA appears in Figure 4B. The corresponding shortest distance charts (z-scores with regard to the appropriate geometric distribution) are shown in Figures 4C and 4D. It is clearly seen from Figure 4 that exon and intron sequences display k-gram distributional structures and that these structures are transparently different for exons and introns. The former display a strong overrepresentation of punctuation lengths 0, 3, 6, . . . and generally $3k$ ($k = 0, 1, 2, 3, \ldots$) for trinucleotide ACA, whereas the latter favor distances 0, 2, 4, . . . and generally $2k$ ($k = 0, 1, 2, \ldots$) for dinucleotide AC. Inspection of shortest distances (Figures 4C and 4D) shows that preferred nearest distances between ACs are 0 and 2 in introns. However, exon-preferred shortest distances between ACAs (Figure 4D) are the same as distances from Figure 4B (i.e., 0, 3, 6, 9, . . . and generally $3k$). These findings suggest that AC is involved in "true" 2-base periodic patterns in introns, whereas ACA is involved in strongly pronounced 3-base, quasi-periodic patterns in exons.

Other than AC, nonhomopolymeric dinucleotides and the vast majority of nonhomopolymeric (including mirror-symmetric) trinucleotides display distance charts almost identical to those shown in Figure 4 (data not shown). This suggests that predominant patterns in exons and introns are 3-base, quasi-periodic arrangements of trinucleotides and 2-base periodic repeats of dinucleotides, respectively.

Notably, distance charts from the example just shown allowed us to rediscover the triplet structure of the genetic code from exon sequences alone (i.e., without taking into account our knowledge of the genetic code). In the light of this rediscovery, 2-base periodicities observed in introns must reflect meaningful, intron-specific distributional structures.

5 LOCAL COMPOSITIONAL COMPLEXITY: A MAXIMUM ENTROPY PRINCIPLE

5.1 Local Compositional Complexity

Local compositional complexity can be defined by two different measures, both of which are based on a concept of *repetition vector* (sometimes called *complexity state vector*). For a string of length N, the repetition vector is:

$$\tilde{n} = (n_1, n_2, \ldots, n_L) \tag{27}$$

where L is the number of letters in the alphabet, and n_i is the number of occurrences of letter i in the string (i.e., $n_1 + n_2 + \cdots + n_L = N$). The elements of the repetition vector are ordered according to size, so $n_1 \geq n_2 \geq n_3 \cdots \geq n_L$. For example, the string ACTGAAGT has the repetition vector (3,2,2,1). The repetition vector is not dependent on the order of the letters; thus any permutation of a given string will have the same repetition vector [e.g., AAATTGGC also has the repetition vector (3,2,2,1)]. Similarly, any permutation of the alphabet yields a string with the same repetition vector [e.g., TGACTTCA has the repetition vector (3,2,2,1)].

One measure of compositional complexity is:

$$K_1 = \left(\frac{1}{N}\right) \log_r \left(\frac{N!}{\prod_{i=1}^{L} n_i!} \right) \tag{28}$$

where L, N, and n_i are defined as in equation (27) and $r = L$. For a DNA fragment, for example, one alphabet would be the possible nucleotide bases in the DNA, that is, adenine, cytosine, guanine, and thymine, or A, C, G, T. In this case, L and r would be 4, and the n_is would be the numbers of As, Cs, Gs, and Ts in the fragment.

Another measure of complexity is based on Shannon entropy:

$$K_2 = -\sum_{i=1}^{L} \left[\frac{n_i}{N} \log_r \left(\frac{n_i}{N} \right) \right] \tag{29}$$

where n_i, N, L, and r are the same as before.

Although K_1 and K_2 converge as N approaches infinity, for short oligonucleotides they differ significantly.

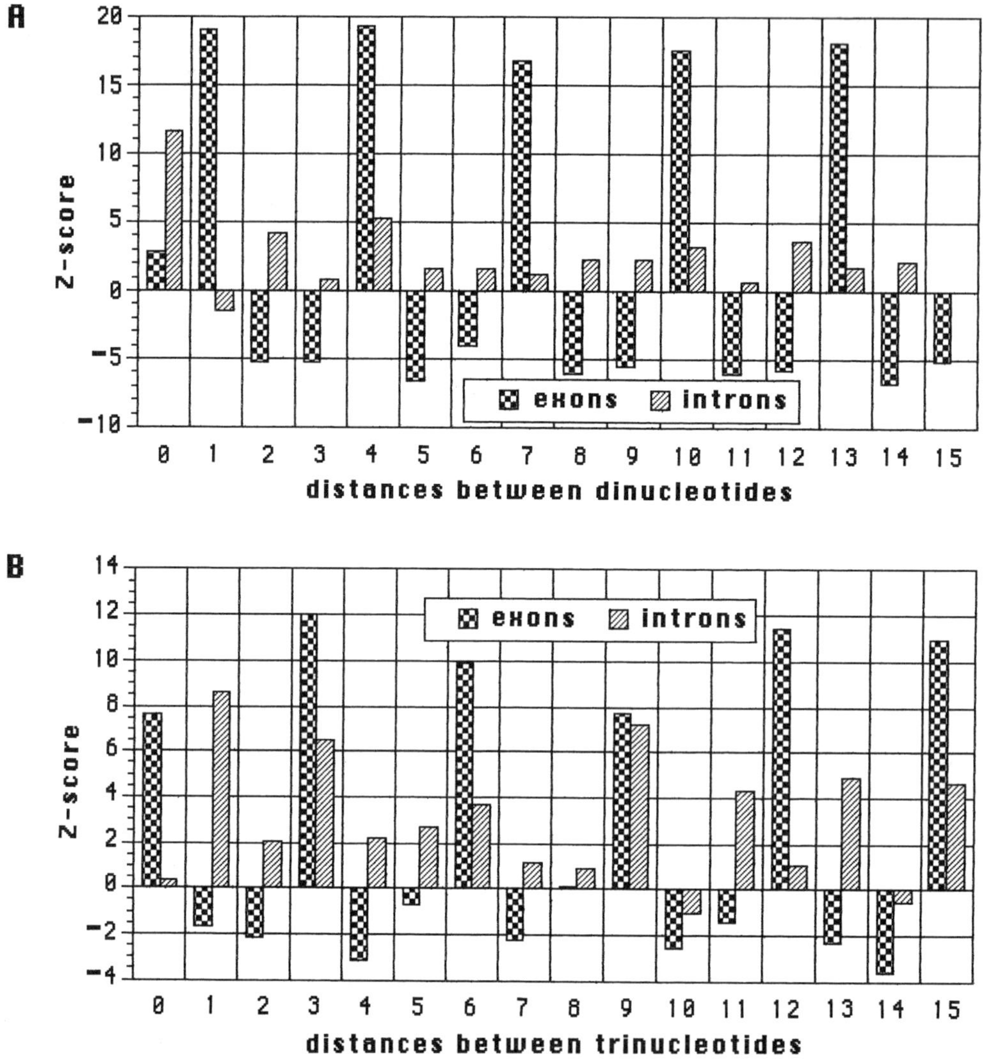

Figure 4. Examples of distance charts for nonhomopolymeric dinucleotides and the corresponding mirror-symmetric trinucleotides in large corpora of animal, nuclear exons, and introns. (A) Dinucleotides tend to be separated by gaps of lengths divisible by 2 in introns. In exons we can see a tendency of those gaps lengths to follow an arithmetic progression with increment 3. This leads to predominant distances of 1, 4, 7, 10, . . . and generally $1 + 3k$ ($k = 0, 1, 2, \ldots$). (B) Trinucleotides (including the mirror-symmetric ones) tend to be separated by gaps of lengths divisible by 3 in exons.

5.2 Surprisal

We can define a *complexity class* as a set of strings that have the same repetition vector, and therefore the same complexity. For a string of length $S \gg N$ we can calculate the *surprisal,* or log–odds ratio, for each such class. Surprisal (log–odds ratio) is defined as follows:

$$S(\tilde{n}) = \log_r \left[\frac{P_{\text{observed}}(\tilde{n})}{P_{\text{expected}}(\tilde{n})} \right] \tag{30}$$

where $\tilde{n}$ is the repetition vector $(n_1, n_2, \ldots, n_L)$, P_{observed} is the observed probability of the repetition vector n in the string, and P_{expected} is the a priori probability of the repetition vector $\tilde{n}$ in the string. The expected probability for a given repetition vector is the sum of probabilities of each of the distinct strings that share that repetition vector. Assuming that the letters are equally probable, then all these strings have equal probability, and the probability for any repetition vector class is simply the number of different strings that share that repetition vector divided by the number of possible strings. For example, for a DNA fragment and the alphabet ACGT, the total number of possible strings is L^N, the number of permutations of a given string is

$$w(\tilde{n}) = \frac{N!}{\prod_{i=1}^{L} n_i!} \tag{31}$$

and the number of distinct permutations of the alphabet for a given repetition vector is

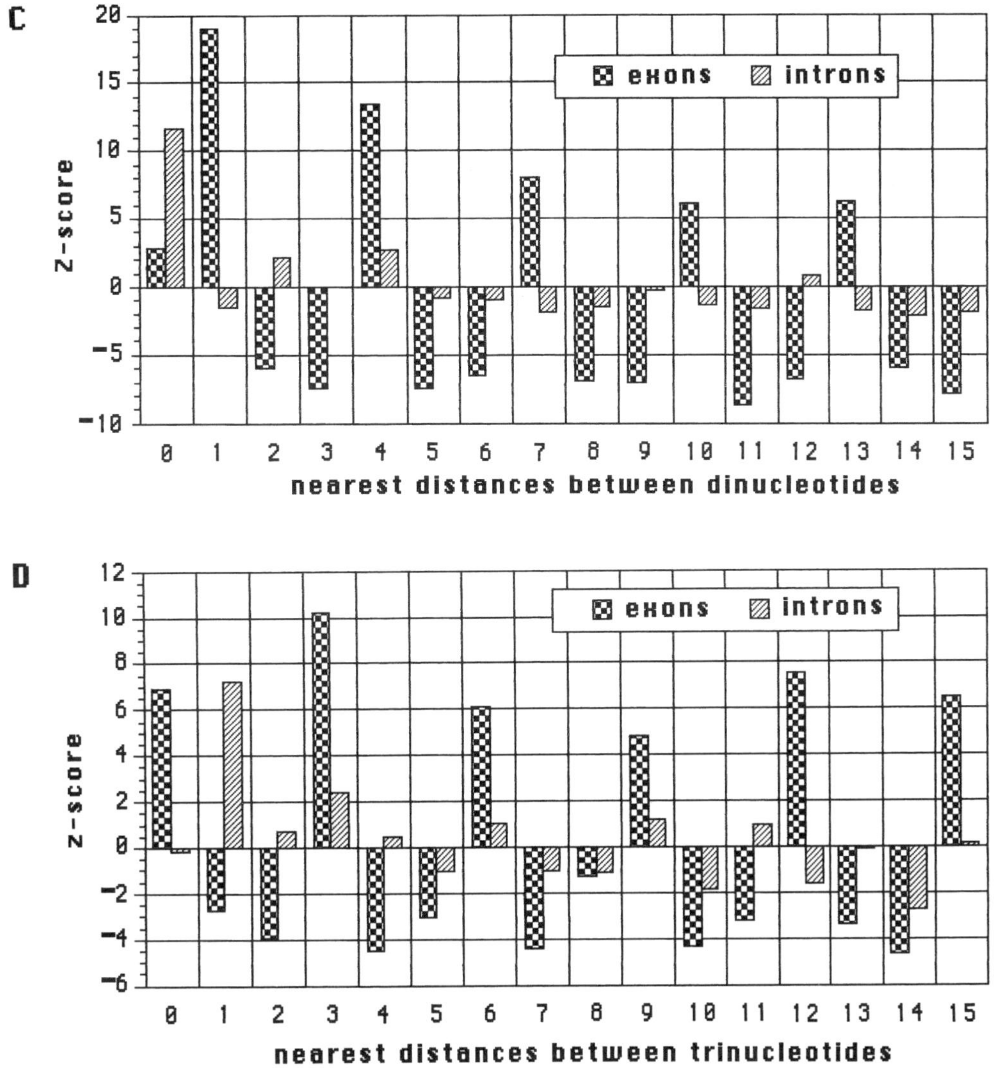

Figure 4 (continued). (C) Shortest distance chart displays the same predominant dinucleotide distances in exons as the distance chart (Figure 4A). However, the predominant shortest distance for dinucleotides in introns is 0 nucleotides. This means that 2-base periodicity implied by the distance chart is a "true" periodicity (ie., is a result of long repeats of dinucleotides). (D) Shortest distance chart for trinucleotides in exons displays the same (divisible by 3) predominant distances as the distance chart from Figure 4B. The 3-base periodicity implied by Figure 4B is not a "true" periodicity but a 3-base quasi-periodicity (i.e., is not a consequence of long repeats of trinucleotides). [Based on and modified from A. K. Konopka and G. W. Smythers, *Comput. Appl. Biosci.* 3:193–201 (1987).]

$$F(\tilde{n}) = \frac{L!}{\prod_{i=1}^{j} h_i!} \qquad (32)$$

where j is the number of distinct values among the elements of the repetition vector n, and h_i is the number of elements in $\tilde{n}$ that have the jth value. For example, for $\tilde{n} = (3,2,2,1)$, $j = 3$, $h_1 = 1$ because there is one element with the value 3, $h_2 = 2$ because there are two elements with the value 2, and $h_3 = 1$ because there is one element with the value 1. Note that $L!$ is the number of permutations of the alphabet, and it is divided to eliminate redundancy. Thus, the a priori probability for $\tilde{n}$ is:

$$P_{\text{expected}}(\tilde{n}) = \frac{F(\tilde{n})\, w(\tilde{n})}{L^N} \qquad (33)$$

The probability distribution generated from the assumption of equal probabilities for each of the letters in the alphabet we call P_0.

5.3 Log–Odds Complexity Distribution

All collections of functionally equivalent nucleotide sequences studied thus far display a linear tendency in surprisal-versus-complexity distributions (see two examples in Figure 5). The slope of this linear relation is negative in all cases studied. This is interpreted as a *maximum entropy relationship* constrained by mean complexity.

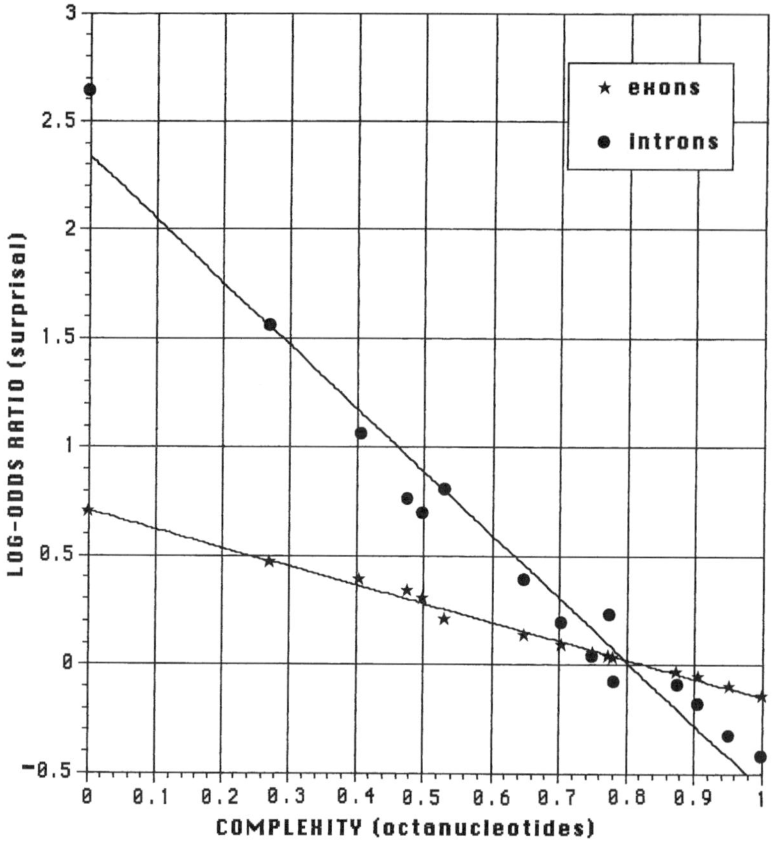

Figure 5. Example of the maximum entropy relationship between complexity and surprisal (see the text) for octanucleotides in invertebrate introns and exons. All naturally occurring nucleotide sequences studied thus far (more than 30 large corpora of sequences) display a linear trend (with negative slope) for tetra- through octanucleotides. [Based on P. Salamon and A. K. Konopka, *Computers Chem.* 16:117–124 (1992).]

It is likely that this maximum entropy relation is a general law governing all natural nucleotide sequences. (Preliminary evidence exists that protein sequences "obey" this principle as well.) On the other hand, sequences belonging to different putative functional domains display systematically different slope values. Discriminative power of the slopes is much stronger than the criteria based on sequence periodicity or uneven distribution of mononucleotides between three positions in codons. This suggests that slope values can be used as reliable classification code words.

5.4 Patchy Complexity

Strings of symbols used to determine repetition vector $\tilde{n}$ can be generated by selecting noncontiguous nucleotides from a given (long) sequence. We shall discuss complexity of strings of L symbols obtained by taking every kth nucleotide, $k = 1, 2, \ldots, 20$, from original nucleotide sequences. *Patchy oligonucleotides* defined in this way are characterized by patchiness k and block length L. For instance, "normal" (i.e., nonpatchy) hexanucleotides are characterized by $k = 0$ and $L = 6$, whereas hexanucleotides obtained by concatenating (six times) every other nucleotide in a given sequence are described by $k = 1$ and $L = 6$. Similarly, octanucleotides generated by taking 8 times every third nucleotide from the original sequence are characterized by $k = 2$ and $L = 8$, but "normal" octanucleotide are of patchiness $k = 0$.

We refer to complexity calculated over repetition vectors of patchy oligonucleotides as to *patchy complexity.* Of course the "normal" (i.e., nonpatchy) local compositional complexity is a special case of patchy complexity.

All standard results obtained for local compositional complexity of nucleotide sequences hold for patchy complexities. In particular:

1. For every value of patchiness k, the linear relation between complexity of patchy L nucleotides and surprisal is observed. The slope values are negative in majority of cases studied.
2. The larger the block length L, the more negative the slope of the surprisal-versus-complexity relationship (for a given k).

Examples of the relationship between slope values and patchiness for the block lengths $3 \leq L \leq 13$ in $\{A, C, G, T/U\}$ alphabet are shown in Figure 6. It can be seen in Figure 6A that the relationship is clearly a modulo-3 periodic function of k for exons (minimum slope values for $k = 2$, 5, and 8; a clear period of 3), but this is not the case for introns (clear minimum at $k = 1$ for $L > 4$ as well as weakly pronounced minima at $k = 1$, 3, 5, and 7, which correspond to the well-known weak period of 2; Figure 6B). Slopes for the 3′-untranslated region (3′UTR: Figure 6C) are an increasing function of k with a very weakly pronounced minima at $k = 5$ and $k = 7$ for large L. The 5′UTRs (Figure 6D) display a weakly pronounced period of 3 but generally lower slope values than exons. The differ-

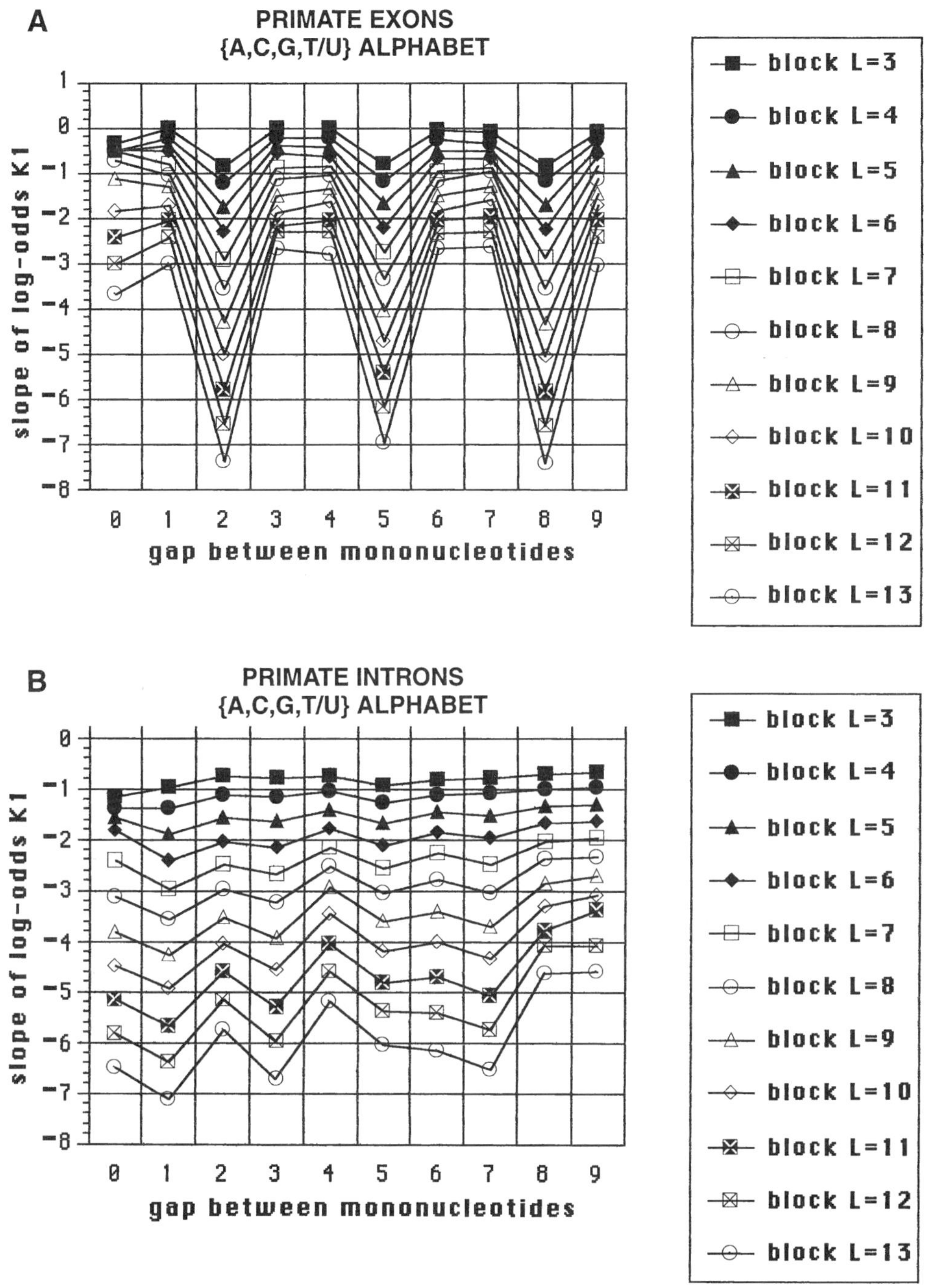

Figure 6. Slopes of log–odds patchy complexity distributions in (A) primate exons and (B) primate introns. (*continued*)

ence between slope values for different block lengths L appears to be much more dramatic than in the case of exons.

6 CONCLUDING REMARKS

With the passage of time, most of today's molecular biology seems to be transformed into biotechnology. However, the scientific remainder of molecular biology desperately needs reconciliation with the rest of biology and with other fields of science. Theoretical research toward this reconciliation involves conceptual foundations of biology on the one hand and modeling biological phenomena on the other. In the final portion of this article I briefly discuss three general topics that seem to be representative of future research in theoretical molecular biology.

6.1 Semantic and Pragmatic Theories of Information

The existing (Hartley–Shannon) theory of communication was created for telecommunication systems and digital computers. Despite

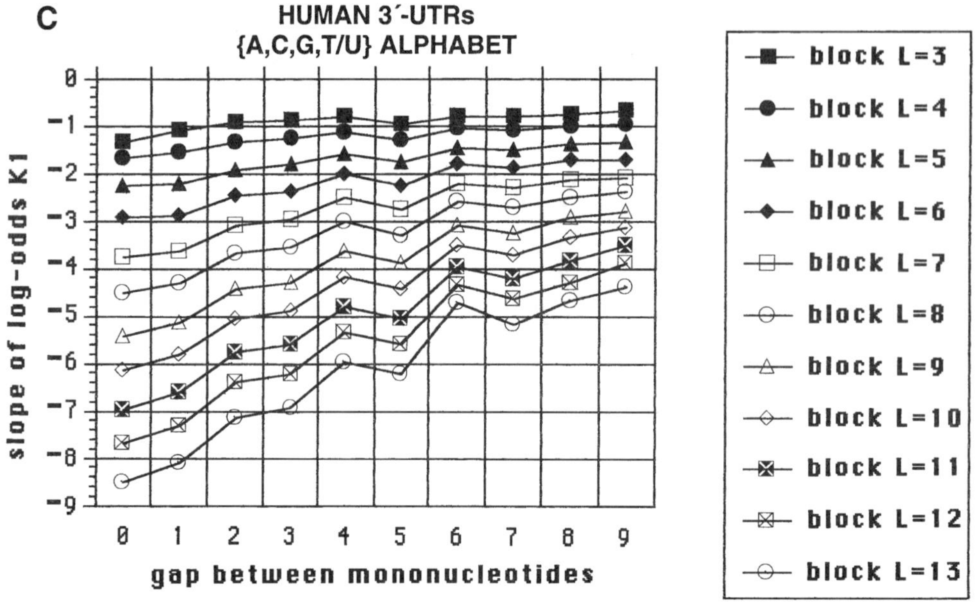

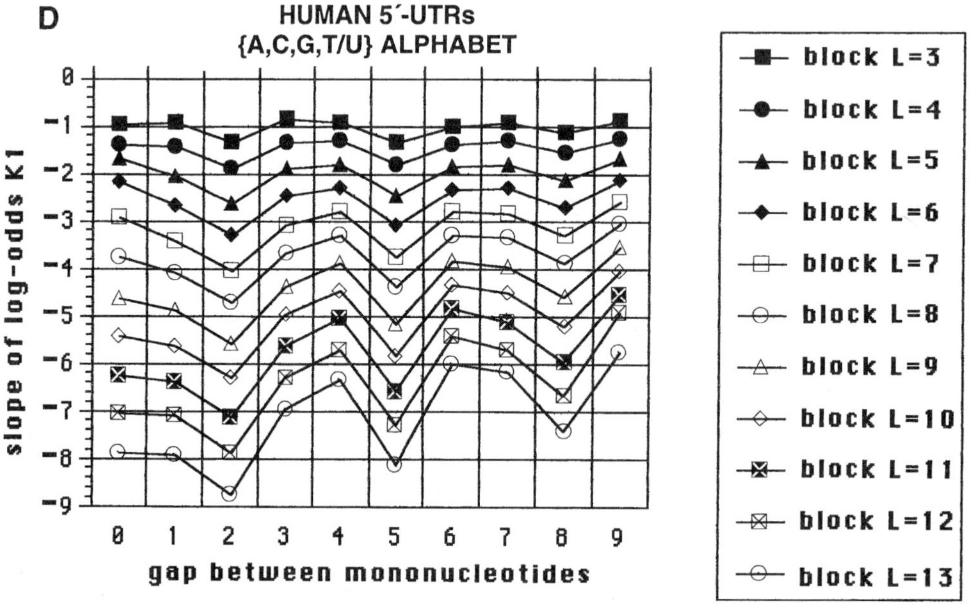

Figure 6 (continued). Slopes of log–odds patchy complexity distributions in (C) human 3′-untranslated regions of protein-coding genes (3′UTRs) and (D) human 5′-untranslated regions of protein-coding genes (5′UTRs). (Based on unpublished work by A. K. Konopka, B. E. Felts, and C. Hand, Extended Abstracts—Open Problems of Computational Molecular Biology (4), Telluride, CO, 1994, and Notes: Gene-Finding and Gene Structure Prediction workshop, Philadelphia, 1995.)

numerous proposals stating otherwise, it does not seem to provide much new insight into molecular biology (it can be a useful source of statistical methods, though). On the other hand, concepts of "information," "meaning," "function," and "organization" seem to be integral parts of biology. If we are to comprehend the phenomenon of life despite our language limitations, we need to develop a theory that can handle these concepts. The need for a theory of information pertinent to biology has been already postulated in neurobiology and in developmental biology. Although such a need has not troubled molecular biologists quite yet, it begins to emerge as evidence accumulates for the role of "epigenetic information" in controlling gene expression and regulation.

6.2 Logic of Heuristic Reasoning

As most of us know, heuristic reasoning can be conclusive. Yet biologists face the problem of not being able to specify all the rules applied to derive conclusions. Nor are they able to list all assumptions on which those rules ought to operate. It seems that these inabilities are a reflection of the complexity of biological systems themselves. Biological phenomena are often represented by models that are still too complex to be described in a communicable manner. Further and further modeling is required until our observations can be communicated in a linguistically comprehensive way. The cascade of models gives us the advantage of creating "communicable reality" but does not help us to judge the evidence pertinent to "real" (i.e., not necessarily communicable) reality. To the contrary, the more advanced a model in a cascade is, the further is its "distance" (in terms of number of modeling steps) from the modeled system.

From a logical point of view there are two problems here. First, we have no formal system of inference to judge *formal correctness* of observation sentences that have a variable true value (credibility) and that use ill-defined terms. Second, we have no formal system to judge the *material adequacy* of sentences derived from a "distant" model to the properties of modeled phenomena.

The problem of formal correctness seems to be solvable in principle. Progress in dealing with it can be noticed already in the fields of artificial intelligence (AI), pattern recognition (especially development of fuzzy mathematical techniques), and situation logic. In the near future we can expect to have formal tools to derive plausible conclusions from imprecise premises. Or, at least, we will have the option of relying on a machine (i.e., reliable AI software) that will perform plausible reasoning for us.

As far as formal judgment of material adequacy is concerned, it is unlikely that the problem is "addressable" in its generality. At least within mathematics the celebrated Gödel theorem precludes such a possibility. However, informal (more or less educated commonsense) judgments of material adequacy not only are possible but are explored by all fields of science.

6.3 Question of Measurement in Biology

The notion that only objects and phenomena that are measurable belong to reality is a persistent legacy of neopositivism. For physicists it means that "real" things need to be viewed as systems of states, with measurable observables assigned to those states. However, interpretation of biological activities in terms of microphysical phenomena will almost certainly lead to observables not known to physics so far. Moreover, preliminary evidence suggests that these new observables will be irreducible to the traditional (i.e., known) ones. How theoretical biologists will handle this measurement problem is not clear yet. Some authors suggest that perhaps it will be necessary for biologists to abandon the concept of states.

See also Human Chromosomes, Physical Maps of; Human Repetitive Elements; Nucleic Acid Sequencing Techniques; Sequence Analysis; Sequence Divergence Estimation.

Bibliography

Bell, G. I., and Marr, T., Eds. (1990) *Computers and DNA*. Addison-Wesley Longman, Reading, MA.

Calladine, C. R., and Drew, H. R. (1992) *Understanding DNA*. Academic Press, San Diego, CA.

Doolittle, R. F., Ed. (1991) *Computer Analysis of Protein and Nucleic Acid Sequences* Vol. 183, *Methods of Enzymology*. Academic Press, San Diego, CA.

Li, W.-H., and Graur, D. (1991) *Fundamentals of Molecular Evolution*. Sinauer Associates, Sunderland, MA.

Open Problems of Computational Molecular Biology (special issue). (1992) *Computers Chem.*, 16(2).

Open Problems of Computational Molecular Biology (2) (special issue). (1993) *Computers Chem.* 17(2).

Open Problems of Computational Molecular Biology (3) (special issue). (1994) *Computers Chem.* 18(3).

Open Problems of Computational Molecular Biology (4) (special issue). (1996) *Computers Chem.* 20(1).

Rosen, R. (1991) *Life Itself*. Columbia University Press, New York.

Shulman, M. J., Steinberg, C. M., and Westmoreland, N. (1981) The coding function of nucleotide sequences can be discerned by statistical analysis. *J. Theor. Biol.* 88:409–420.

Smith, D. W., Ed. (1994) *Biocomputing: Informatics and Genome Projects*. Academic Press, San Diego, CA.

Waterman, M. S., Ed. (1989) *Mathematical Methods for DNA Sequences*. CRC Press, Boca Raton, FL.

Wyman, J., and Gill, S. J. (1990) *Binding and Linkage*. University Science Books, Mill Valley, CA.

Tomatoes, Gene Alterations of

Belinda Martineau and William R. Hiatt

1 Principles
1.1 Tomato Transformation and Regeneration
1.2 Altering Gene Function Through Gene Addition

2 Techniques
2.1 Gene Isolation and Chimeric Gene Construction
2.2 *Agrobacterium* Transformation Vectors
2.3 Transformation and Regeneration of Tomato Plants
2.4 Analysis of Transformed Plants

3 Applications
3.1 Agronomic Practices
3.2 Fruit Quality

4 Perspectives

Key Words

Agrobacterium tumefaciens A soil bacterium carrying a tumor-inducing (Ti) plasmid that allows the microorganism to infect and transform plants; products of the Ti plasmid virulence (*vir*) region genes provide for transfer to and insertion of the Ti plasmid T-DNA region into the plant genome.

Antisense RNA RNA that is complementary to the mRNA transcribed from an endogenous gene.

Chimeric Gene A recombinant gene consisting of the protein-coding region from one gene and the promoter and/or other regulatory regions from another gene(s).

Promoter A segment of DNA located upstream from the coding region of a gene that regulates spatial and temporal expression.

Transform To add exogenous DNA to the genome of an organism (e.g., a bacterium or a plant).

Alteration of gene function in tomato is currently achieved using a transformation technology mediated by a harmless version of the plant bacterial pathogen *Agrobacterium tumefaciens.* A cloned gene, modified to effect changes in its spatial and/or temporal pattern of expression or to eliminate its expression, is spliced into a region of *A. tumefaciens* DNA that is inserted into the tomato genome upon infection of the plant. *Agrobacterium*-mediated gene addition experiments in tomato have increased our understanding of tomato gene expression and gene product function. The technology is now being applied to improve fruit quality and agronomic practices.

1 PRINCIPLES

1.1 Tomato Transformation and Regeneration

1.1.1 Transformation

Lycopersicon esculentum, the cultivated tomato, is most commonly transformed by utilizing the virulence mechanism of the soil bacterium *Agrobacterium tumefaciens.* This bacterium infects tomato and many other dicotyledonous plant species by inserting a segment of *Agrobacterium* DNA into a susceptible individual. Expression of the inserted DNA segment in the infected plant provides the substrates necessary for bacterial sustenance and the "crown gall" of undifferentiated cellular growth by which the disease inflicted by *A. tumefaciens* gets its name.

Genetic engineers have taken advantage of this parasitic relationship between bacterium and plant to introduce their own selected segment of DNA into tomatoes. The bacterial genes responsible for gall formation and other bacterial functions have been eliminated from the DNA that is transferred from the bacterium to the plant (T-DNA), thus "disarming" *Agrobacterium* of pathogenicity. Various genes of interest to the genetic engineer are then placed into the bacterial T-DNA and are consequently inserted into the genome of a tomato plant.

1.1.2 Regeneration

Individually or in small groups, cells from tomato leaves, stems, or other plant parts can be regenerated into an entire, fertile plant. The genetic engineer provides for tomato and bacteria cells a conducive laboratory environment in which to carry out both the DNA transfer process and tomato plant regeneration. The two organisms are incubated together under specific conditions of light, heat, and supporting media, including plant growth regulators in proper proportions. Tomato cells that have permanently acquired T-DNA (i.e., have been stably transformed) are then usually identified by virtue of a marker gene included in the T-DNA. For most gene alterations of tomato to date, this marker gene has encoded resistance to an antibiotic, kanamycin. The inclusion of kanamycin in the regeneration medium means that transformed tomato cells expressing the marker gene will be regenerated into whole, fertile tomato plants while the growth of nontransformed cells will be inhibited by the antibiotic.

1.2 Altering Gene Function Through Gene Addition

The gene alterations possible using *Agrobacterium*-mediated transformation result from the introduction of the T-DNA from the bacterium into the tomato genome. Therefore, tomato gene alterations produced using this system are gene additions. It is the composition of the added genes, customized by the genetic engineer, that dictates the alteration of the *function* of tomato genes.

2 TECHNIQUES

2.1 Gene Isolation and Chimeric Gene Construction

2.1.1 Gene Protein-Coding Regions

Any organism may serve as the source of a gene for addition to tomato. The selectable marker gene that confers resistance in transformed tomato cells to kanamycin, the *kan*^{r} gene, is a good example of a foreign gene added to tomato. This gene, encoding the enzyme aminoglycoside 3′-phosphotransferase II [APH(3′)II], was originally identified in a bacterial species. Because of the universality of the genetic code, the specific DNA sequence providing the code for the APH(3′)II protein can be deciphered by the protein translational machinery of a tomato plant as well as by bacteria. Similarly, gene protein-coding regions from any other organism can also be utilized for tomato gene alteration experiments.

2.1.2 Gene Promoters

The DNA sequences that regulate a gene's spatial and temporal function in one organism, however, are not always recognized by other organisms. Therefore, genetic engineers customize a foreign gene to have regulatory DNA sequences that are readily recognized by the tomato plant, in conjunction with the protein-coding region of the foreign gene. The resulting recombinant gene is a molecular chimera, or "chimeric gene," consisting of gene regions of diverse genetic derivation. (See depictions of sample chimeric genes in Figure 1.) The *kan*^{r} gene, for example, contains regulatory sequences that provide for expression of the bacteria-derived APH(3′)II pro-

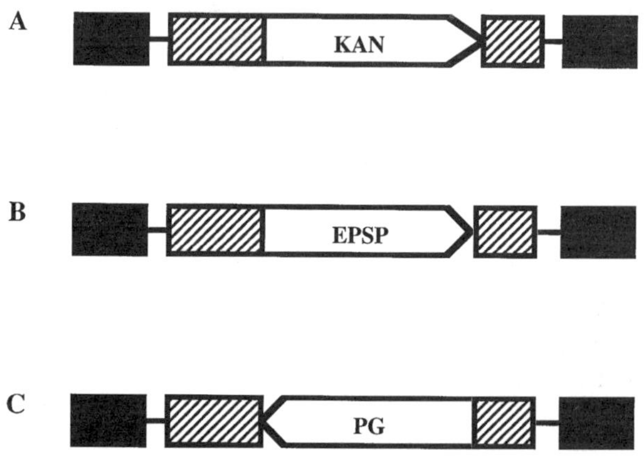

Figure 1. Schematic representations of protein-coding regions from chimeric genes: (A) a non-tomato gene, (B) a tomato (or another plant) gene, and (C) a tomato gene in the antisense orientation. Solid boxes, borders of the *Agrobacterium* T-DNA; hatched boxes, regulatory DNA sequences required for gene expression in plants; arrows, protein-coding regions; KAN, the *kan*^{r} gene protein-coding region; EPSP, the protein-coding region of an EPSP synthetase gene; PG, the protein-coding region for the polygalac turonase gene.

tein-encoding gene in tomato cells during the tomato regeneration process and throughout the life cycle of the regenerated plant. Regulatory sequences commonly utilized to provide this constitutive expression of the *kan*^{r} gene were isolated from the cauliflower mosaic virus (CaMV). Although this DNA virus primarily infects members of the plant family Cruciferae, its 35S promoter has been shown to be highly active in driving foreign gene expression in tomato (as well as many other plant species). The regulatory sequences associated with the genes located in the T-DNA region of wild-type *A. tumefaciens* are also commonly used to provide constitutive expression of chimeric genes in transgenic tomato. The bacteria derived T-DNA genes and the viral 35S promoter obviously contain expression signals that are recognized by factors in plants necessary for gene transcription.

2.1.3 Altered Regulation of Endogenous Genes

Similarly, construction of chimeric genes can also involve protein-coding regions from genes that are normally present in tomato (or other higher plants). The "normal" expression of such an endogenous tomato gene can be altered by eliminating the original regulatory DNA sequences of the cloned gene and replacing them with sequences that determine the gene expression patterns of another (usually a tomato) gene. Since the protein-coding sequences of the gene serving as the source of the regulatory sequences are eliminated in the process, this gene can be of known or unknown function. Typically, gene regulatory sequences, or promoters, are identified by virtue of the desirable level and specificity of their expression in particular plant tissues or organs. This is usually accomplished by means of a technique referred to as "differential screening" of a cDNA library. The protein-coding region from the cDNA clone with the desired specificity of expression, defined by a start ATG codon and a stop codon, is then eliminated and replaced with that of the endogenous gene. Such constructed chimeric genes, when inserted into the tomato genome via *Agrobacterium*-mediated methods, can augment the spatial and temporal expression of the original endogenous tomato gene.

2.1.4 Antisense Genes

Another method of altering tomato gene function is through expression, in transformed tomato plants, of antisense genes. Expression of antisense genes curtails the expression of an endogenous gene. In an antisense gene construction the protein-coding region (or a portion thereof) from the endogenous gene is placed in reverse orientation with regard to a promoter region, with the result that the complement of the normal mRNA (i.e., antisense RNA) is produced during transcription. The presence of antisense RNA interferes with the function of the normal mRNA, and expression of chimeric antisense genes in transformed tomato plants is, therefore, a means of creating specific mutations in selected tomato genes. The production of specific antisense mutants has been used to identify the in planta function of genes otherwise known only by virtue of their fruit-specific expression patterns. The genes encoding ACC oxidase (see Section 3.2.2), an important enzyme in the production of the plant growth regulator ethylene, and phytoene synthetase, a key enzyme in the synthesis of lycopene (the major red fruit pigment), are examples of tomato genes identified using this antisense strategy.

2.2 *Agrobacterium* Transformation Vectors

Once constructed, a chimeric gene destined for incorporation into the tomato genome must be placed in the T-DNA of *A. tumefaciens*. The T-DNA region is present on the tumor-inducing (Ti) plasmid, a plasmid so large that conventional cloning techniques are cumbersome. Therefore, the biology of the bacterium has been manipulated to make the procedure more feasible.

While it is the T-DNA of the Ti plasmid that is transferred to a plant cell during *A. tumefaciens* infection, the proteins responsible for directing that transfer are encoded elsewhere on the Ti plasmid. In fact, this "*vir*" region of the Ti plasmid can function in directing T-DNA transfer without being present on the same piece of DNA. As long as the *vir* and the T-DNA regions are present in the same *A. tumefaciens* cell, T-DNA transfer to plant cells can occur. So-called binary vector systems, in which the *vir* and T-DNA regions are located on two separate DNA plasmids, are based on this finding that *vir* gene functions can act in trans. Especially convenient binary vectors have been developed which allow manipulation of the T-DNA region in *E. coli* as well as in *A. tumefaciens*. As a result of this feature, chimeric genes can be introduced into the T-DNA using standard recombinant DNA techniques. Introduction of the recombinant T-DNA-containing binary vector into an *A. tumefaciens* strain carrying the *vir* region containing vector is accomplished via standard bacterial conjugation or transformation procedures. Each bacterium, when given the opportunity, will unwittingly insert the engineered gene within its T-DNA region into the DNA of a tomato cell. An overview of this process is shown in Figure 2.

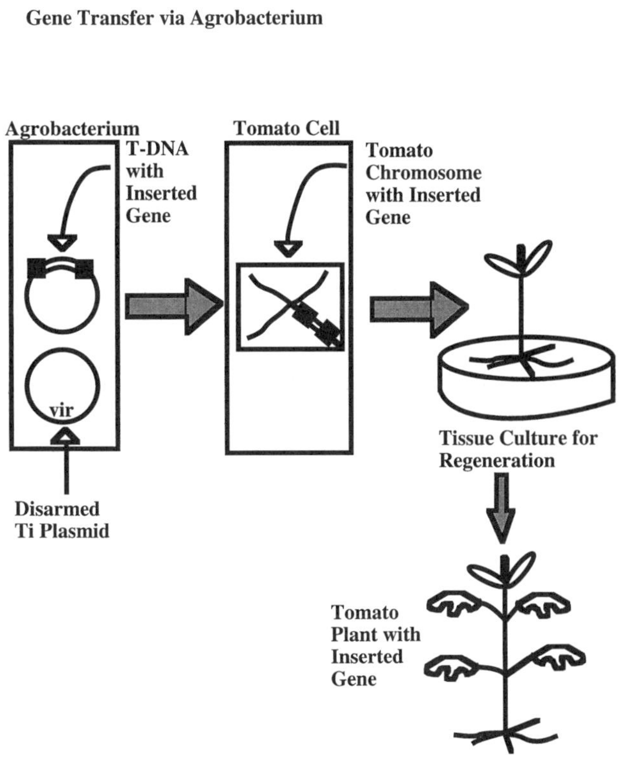

Figure 2. Schematic representation of the *Agrobacterium*-mediated transformation and regeneration process for inserting genes into tomato.

2.3 Transformation and Regeneration of Tomato Plants

The opportunity required by *A. tumefaciens* to infect tomato and other plants is wounding. Healthy, intact tomato plants are immune to infection, hence to the transforming ability of the bacterium. The genetic engineer, therefore, provides the required opportunity for bacterial infection by injuring the plant tissue. The injury can take a number of forms. In one form, specific enzymes are used to eliminate walls from the plant's cells. The wall-less cells resulting are referred to as protoplasts. Simple cuts inflicted on whole plant organs (e.g., leaves, stems, cotyledons) with implements ranging from razor blades to paper punches also fulfill the wounding requirement.

The cells on the cut surfaces of the resulting organ pieces (explants) serve not only as sites for *Agrobacterium* entry but also as sources of factors involved in the induction of the *vir* region genes required for T-DNA transfer from bacterium to plant. Various compounds in plant cell exudates that induce *vir* gene expression have been identified. The most significant *vir* gene inducer in tomato is (α-hydroxy)-acetosyringone.

Explants or protoplasts are then cultivated together with an *Agrobacterium* strain carrying the chimeric genes of interest. This cocultivation is conducted under specific conditions designed to optimize the efficiency of both the transformation (i.e., the T-DNA transfer) and the plant regeneration processes (as described in Section 1.1.2). The specific conditions depend on many variables, including organ source and age of explants, concentration of bacteria, and length of the cocultivation period.

After cocultivation, the bacteria are eliminated through application of a bactericide. A phase of pure plant regeneration is then entered. For tomato explants such as those derived from cotyledon (primary leaf) tissue, this phase—from wound infliction to young plants with roots—lasts approximately 6–8 weeks. The regeneration phase is somewhat longer for plants derived from protoplasts.

2.4 Analysis of Transformed Plants

The first indication that a tomato plant is transformed by an *Agrobacterium* strain carrying a selectable marker gene such as *kan*^{r}, as well as a specific chimeric gene of interest, is the plant's growth on media containing kanamycin. Small regenerated plant shoots, comprising the aerial portions of tomato plants, are commonly placed on growth media designed to induce root formation, for example. By including kanamycin in the root induction media, this last step in the plant regeneration process also serves as an additional check that a particular shoot actually carries a functional copy of the *kan*^{r} marker gene. Regenerated shoots lacking a functional *kan*^{r} gene will not develop roots on media containing kanamycin.

Seeds produced by a transformed plant that carries the *kan*^{r} gene will also germinate normally in liquid media containing kanamycin. Seeds from the same plant that lack the introduced gene as a result of meiotic segregation will exhibit stressful, stunted growth in the same environment. This germination test, therefore, can be used to determine the number of genetic loci into which the genes of interest carried on the T-DNA have been inserted into the tomato genome. A segregation ratio of three resistant germinated seedlings for every one that is susceptible would be indicative of a T-DNA region(s) inserted at one genetic locus, for example. A ratio of 15 resistant seedlings to one susceptible would indicate T-DNA insertion had occurred at two distinct genetic loci, and so on in typical Mendelian fashion for a dominant gene. Topical application of kanamycin to young, healthy progeny seedlings produced by transgenic tomatoes can also be utilized to carry out the same kind of segregation analysis.

Among the transformed plants there can be variability not only in the number of sites in the tomato genome into which the T-DNA has been integrated but also in the number of copies of the T-DNA region inserted at each site. Hybridization analysis of genomic DNA isolated from a transformed tomato plant can be used to determine the number of T-DNA regions inserted into that plant. This analysis is based on the availability of DNA sequence (and therefore restriction enzyme recognition site) information for the T-DNA region. Specific hybridization probes, designed utilizing the T-DNA sequence information, combined with the predictable T-DNA restriction map, allow for precise characterization of the number and structure of the introduced genes. As a result of this kind of DNA hybridization analysis, it is known that multiple copies of the T-DNA region are often inserted into the tomato genome at one genetic locus. Such multiple copies most often comprise concatameric structures consisting of T-DNA regions lined up in head-to-tail/tail-to-head arrangements.

The functioning of the genes within the T-DNA can also vary from one transformed plant to another. That is, the same gene can be expressed at a higher level in some transgenic tomato plants than in others. Local chromosomal influences at the site of insertion are believed to account for much of this variability. The number of transgenes inserted into a particular transgenic plant may also contribute to these expression effects. Various standard molecular techniques, designed to identify the number of inserted genes, the products of those genes, and the levels of any biochemical constituents altered through the functioning of those gene products, are therefore used to characterize transformed tomatoes.

Selected transformants are ultimately field-tested. Field trials are usually conducted when it is intended that the introduced gene(s) alter biochemical constituents that affect field-measured traits. Fruit-related qualities such as total yield, color, and various processing characteristics (e.g., viscosity and consistency of paste, yield of tomato solids) are examples of such traits. A field trial situation is also a good way to ensure that a transformed plant exhibits no unexpected morphological abnormalities that may have been introduced as a result of the regeneration process.

3 APPLICATIONS

There are two major areas to which *Agrobacterium*-mediated transformation methods for altering gene function in tomato have been applied. One addresses the agronomic practices of pesticide and herbicide use during tomato production. The other deals with fruit quality issues.

3.1 Agronomic Practices

3.1.1 Resistance to Insects and Pathogens

In one strategy designed to induce crop plants like tomato to produce their own insecticide, genes encoding the so-called Bt toxin proteins have been inserted into plant genomes. These proteins are toxic only to specific lepidopteran and coleopteran insect species (they produce no ill effects in other organisms) and have been used topically for decades to biologically control these major pests. The genes encoding Bt toxins were isolated from the soil bacterium (Bacillus thuringiensis and used to produce (via *Agrobacterium*-mediated methods) tomato as well as other crop plants that are unpalatable, even poisonous, for tomato hornworms among other pests.

Tomato plants that are resistant to various viral plant pathogens have also been developed using *Agrobacterium*-mediated techniques. The molecular strategy utilized most successfully to confer viral resistance involves addition of a foreign gene isolated from the targeted virus itself and encoding its protein capsid or "coat protein." Coat protein genes have been isolated from alfalfa mosaic virus (AlMV), tobacco mosaic virus (TMV) and tomato mosaic virus (ToMV), customized for efficient expression in plants, and inserted into tomato plants.

Following inoculation with the particular virus, these plants either escape infection altogether or develop disease symptoms significantly later than do control plants. Plants expressing the TMV coat protein are resistant to infection by both TMV and, to a lesser extent, ToMV. Similarly, plants that express ToMV coat protein provide some resistance to TMV in addition to resistance to ToMV. The coat proteins of these related tobamoviruses are 80–85% homologous at the amino acid sequence level, and the "cross-resistance" phenomenon may relate to this degree of homology. Plants that express both TMV and ToMV coat protein genes are equally well protected against TMV and ToMV. Although the mechanism through which resistance to virus infection and/or disease symptoms is attained is unknown, this "coat protein mediated resistance" has proven quite effective in field trial tests of artificial virus infections.

Genetic engineering strategies designed to protect tomatoes and other agriculturally significant crops from fungi, nematodes, and other viruses are also being investigated.

3.1.2 Tolerance for Environmentally Safe Herbicides

Affecting crop plant tolerances to specific herbicides is another way in which genetic engineers are applying this technology to crop production practices. The herbicide glyphosate serves as a case in point. Glyphosate kills plant cells by inhibiting 5-enolpyruvylshikimate 3-phosphate synthetase; (EPSP synthetase), an important enzyme involved in the biosynthesis of aromatic compounds. Tomato and other crop plants have been made more tolerant to glyphosate through the addition of EPSP synthetase genes that have been altered either of two different ways. The plant version of the EPSP synthetase gene has been altered so that significantly more of the enzyme is produced and the plant expressing the altered gene can therefore tolerate higher levels of the herbicide compound. A bacterial gene encoding a glyphosate-resistant version of EPSP synthetase has also been customized for and expressed in tomato. Plants altered in this fashion enable the use of this safe, biodegradable herbicide in place of other more hazardous or persistent chemicals.

Another broadleaf herbicide with very short half-life in the environment is bromoxynil. Bromoxynil appears to act by binding a component of the quinone-binding protein complex of photosystem II (PSII), thereby inhibiting electron transfer during photosynthesis. One of the reasons for the ready degradability of this herbicide is that various microbes in the environment are capable of detoxifying it. One such microbe, *Klebsiella ozaenae,* degrades bromoxynil by converting it, with the release of ammonia, to 3,5-dibromo-4-hydroxybenzoic acid, a nontoxic plant metabolite (see Figure 3). The enzyme responsible for this conversion, a bromoxynil-specific nitrilase, is encoded by the *bxn* gene. Field tests of tomato plants transformed with a chimeric *bxn* gene, designed to be expressed in plants, have been conducted. Commercially relevant resistance levels have been achieved, even in plants expressing quite low levels of the nitrilase enzyme.

3.2 Fruit Quality

3.2.1 Alteration of Cell Wall Hydrolases

The quality of a fresh tomato is judged primarily on the basis of its color, firmness, and general lack of visible damage. To date the greatest strides in improving tomato fruit quality have been attained by slowing down the process of fruit softening by specifically attenuating the function of the tomato gene encoding polygalacturonase (PG). As tomato fruit soften during the course of ripening, intercellular pectin is depolymerized and solubilized, and this process contributes to a loss of cell-to-cell cohesion and fruit integrity. PG, which breaks down pectin, had long been implicated in the fruit-softening process. Altering PG gene function by transforming tomato plants with chimeric antisense PG genes results in fruit that soften and rot less quickly.

The pectin component of tomato fruit cell walls may become

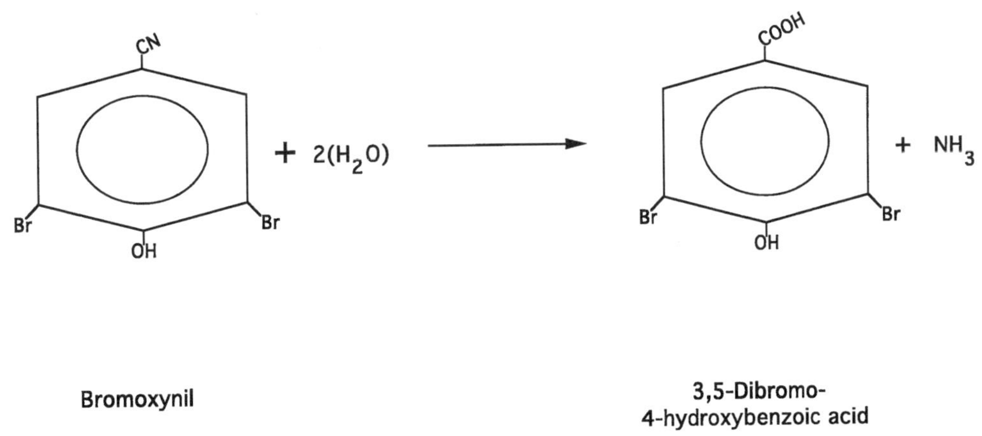

Figure 3. Depiction of the biochemical reaction in which the herbicide bromoxynil is detoxified via the bromoxynil-specific nitrilase.

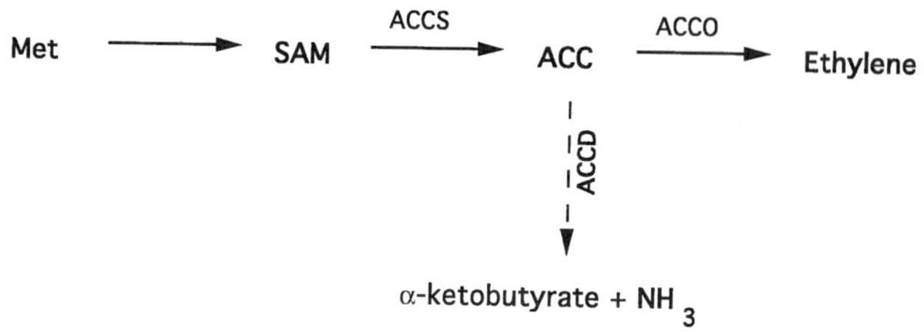

Figure 4. Schematic representation of the biosynthetic pathway for ethylene production in higher plants: solid lines, reactions known to occur in nontransformed plants; dashed line, a reaction that takes place in transformed plants expressing an introduced ACC deaminase gene originally isolated from a *Pseudomonas* bacterium; Met, methionine; SAM, *S*-adenosylmethionine; ACC, 1-aminocyclopropane-1-carboxylic acid; ACCS, ACC synthetase; ACCO, ACC oxidase; ACCD, ACC deaminase.

more susceptible to PG-induced endohydrolysis during ripening as a result of the action of another tomato fruit cell wall hydrolase, pectinmethylesterase (PME). This enzyme catalyzes the demethylation of methyl esters on pectin. PG has been shown to be more active in vitro on demethylated pectin than on the methylated form. Field trials of transgenic tomato plants containing a chimeric antisense PME, as well as plants containing chimeric genes that direct the synthesis of both antisense PME and antisense PG RNA, have been undertaken to assess the possibility that this strategy could improve tomato fruit firmness and/or processing characteristics.

3.2.2 Alteration of Ethylene Production

Ethylene influences many aspects of tomato development, including leaf and flower senescence and fruit ripening. Ethylene is biosynthesized in higher plants from methionine via *S*-adenosylmethionine (SAM) and 1-aminocyclopropane-1-carboxylic acid (ACC). Induction is required of both the enzyme responsible for production of ACC (i.e., ACC synthetase) and the enzyme that catalyzes the conversion of ACC to ethylene (ACC oxidase) for ethylene formation during fruit ripening (see Figure 4).

In an effort to control ethylene production, tomato plants have been genetically altered, via *Agrobacterium*-mediated transformation, to express genes for chimeric antisense ACC synthetase or antisense ACC oxidase. An additional strategy in which a gene foreign to tomato has been introduced into the tomato genome has also been employed to alter ethylene production. The foreign gene, encoding ACC deaminase and isolated from bacteria of the genus *Pseudomonas,* utilizes ACC in a reaction that produces α-ketobutyrate and ammonia, compounds that are common metabolites in plants. ACC deaminase provides a mechanism by which to shunt ACC away from the ethylene biosynthetic pathway and thereby limit ethylene production (see Figure 4). Tomato plants expressing any of these "antiethylene" genes at high levels show marked delays in the time required for the fruit to ripen. Thus, relative to nonengineered fruit, which ripen and soften very quickly, the genetically engineered fruit can survive a longer trip from field to market. Controlling the production of the fruit-ripening hormone, ethylene, may therefore be used to improve the quality characteristics of fresh tomato fruit available commercially.

4 PERSPECTIVES

The introduction and expression of foreign and endogenous genes in tomato has advanced our understanding of the functioning of plant genes and the products they encode. The resulting knowledge, particularly in areas relating to tomato fruit quality, is now being applied to commercial goals. As a convenient model for gene alteration experiments, tomato will continue to play an integral role in deciphering the many obscurities in our understanding of plant molecular biology and physiology. Because of tomato's agricultural relevance, future insights gained can also have wide consequences for the consumer.

See also PESTICIDE-PRODUCING BACTERIA; PLANT CELLS, GENETIC MANIPULATION OF; PLANT CELL TRANSFORMATION, PHYSICAL METHODS FOR.

Bibliography

Gasser, C. S. (1992) Transgenic crops. *Sci. Am.* 266:62–69.

Gray, J., Picton, S., Shabbeer, J., Schuch, W., and Grierson, D. (1992) Molecular biology of fruit ripening and its manipulation with antisense genes. *Plant Mol. Biol.* 19:69–87.

Hofte, H., and Whiteley, H. R. (1989) Insecticidal crystal proteins of *Bacillus thuringiensis. Microbiol. Rev.* 53:245–255.

Hooykaas, P. J., and Schilperoot, R. A. (1992) *Agrobacterium* and plant genetic engineering. *Plant Mol. Biol.* 19:15–38.

Mazur, B. J. (1989) The development of herbicide resistant crops. *Annu. Rev. Plant Physiol. Plant Mol. Biol.* 40:441–470.

Nevins, D. J., and Jones, R. A., Eds. (1987) *Tomato Biotechnology.* Liss, New York.

Redenbaugh, K., Hiatt, W., Martineau, B., Kramer, M., Sheehy, R., Sanders, R., Houck, C., and Emlay, D. (1992) *Safety Assessment of Genetically Engineered Fruits and Vegetables: A Case Study of the FLAVR SAVR™ Tomato.* CRC Press, Boca Raton, FL.

TRACE ELEMENT MICRONUTRIENTS

Robert J. Cousins

1 Scope of Trace Element Micronutrients in Molecular Biology
- 1.1 Chemical Considerations
- 1.2 Classes of Involvement
- 1.3 Cellular Distributions

2 Structural/Catalytic/Regulatory Roles
- 2.1 Metalloenzymes and Metal-Binding Domains
- 2.2 Metalloregulatory Proteins

3 Regulation of Gene Expression by Trace Element Micronutrients
3.1 Transcriptional Regulation
3.2 Translational Regulation
4 Applications and Perspectives

Key Words

Ligand A negatively charged organic molecule that donates electrons to form coordinate bonds with a metal ion.

Metal Response Element (MRE) A specific sequence of a gene's promoter region that enhances transcription through a specific transcription factor (MRE-binding protein) that may require metal occupancy for DNA binding.

Metalloenzyme An enzyme that requires one or more metal atoms for catalytic activity and/or for the structure needed for activity.

RNA-Responsive Element Binding Protein A protein that binds in response to metal occupancy to a stem loop of nucleotides called a regulatory element in the 5′-untranslated region of a specific mRNA to influence translation.

Trace Element Micronutrient An essential nutrient metal found in the diet in small (micro or milligram) amounts.

Zinc Finger Motif A zinc-binding domain of a protein having an —X-Cys-X_2-Cys-X_n-Cys-X_{2-5}-Cys-X— (or His substituted for Cys) or similar sequence that provides the opportunity for Zn(II) binding.

"Trace element micronutrients" is a term that refers to the metals (inorganic elements) in the diet that are essential for specific cellular processes. Required dietary trace elements (e.g., Cu, Fe, Mn, Se, Zn) can perform catalytic, regulatory, and structural roles related to gene expression. Zinc in particular has a major role in maintaining conformations (particularly zinc fingers) necessary for DNA binding by transcription factors and for protein–protein interactions required for regulation of specific genes. Zinc can also influence transcription through metal-binding proteins that, during metal occupancy, either bind directly to a specific DNA sequence (metal response element) in the promoter/enhancer region of the gene or allow another previously inhibited molecule the opportunity for DNA binding.

Regulation of expression by trace element micronutrients can also occur through metal-requiring, RNA-binding proteins, as has been demonstrated for iron. These metal-binding proteins influence translation of specific mRNAs via nucleotide sequences in the untranslated regions.

Trace element micronutrients have potential applications in medicine and biotechnology, either directly as signaling molecules for regulation of chimeric genes, or indirectly as therapeutic agents (by targeting specific metal-binding sites in regulatory factors).

1 SCOPE OF TRACE ELEMENT MICRONUTRIENTS IN MOLECULAR BIOLOGY

1.1 Chemical Considerations

"Micronutrients" as a general term describes both essential organic molecules (vitamins) and metals (trace elements). All living organisms require metals to sustain cellular processes. Required trace elements include copper, iron, manganese, nickel, selenium, and zinc. In humans these substances are required in the diet in microgram to milligram amounts per day to maintain balance with endogenous losses. For metal binding, cellular constituents, particularly proteins, provide ligands (e.g., amine, thiolate, carboxyl) that follow approximately the Irving–Williams order of divalent ions: Cu $>$ Zn $\geq$ Ni $>$ Co $>$ Fe $>$ Mn $>$ Mg $>$ Ca. If ligand concentrations and metal concentrations were not controlled within cells, copper binding would predominate for most ligands. Nevertheless, biochemical evidence shows that of the trace element micronutrients, zinc has the most presently recognized collective influence in molecular biology.

Cellular transport systems regulate intracellular trace element concentrations through homeostatic control of uptake and efflux. These processes limit the influence of natural abundance and thermodynamic constraints in determining which elements participate in specific cellular functions. Only trace elements that are established nutrients in mammalian systems are emphasized in this article.

1.2 Classes of Involvement

There are three main classes of involvement of trace element micronutrients (metals) in the province of molecular biology. Metalloenzymes, where metals provide a catalytic function, comprise the first class. Another class is structural, with metals facilitating interaction among various binding groups of specific motifs that provide altered conformations necessary to achieve unique opportunities for specific interaction. The third class of involvement is regulatory, as exemplified by the metal-binding, trans-acting proteins that provide signaling to initiate (or potentially inhibit) transcription of genes through interaction with specific DNA sequences, the metal response elements (MREs). Metal-responsive, RNA-binding proteins are another example of regulatory involvement. Collectively, these classes of biological function of trace element micronutrients provide a vital link between the organism and its external nutrient supply.

1.3 Cellular Distributions

Concentrations of trace elements in cells vary greatly but usually are no higher than the microgram-per-gram range. Cells differ with respect to the organelles to which trace elements are most abundantly distributed. It is metal occupancy at specific binding sites, however, rather than concentration, that determines physiologic effects. Zinc is the predominant trace element in cells and is the most important from the standpoint of molecular biology. It is most abundant in the nuclear and cytosolic fractions of cells.

Cells vary widely in responses to or dependence on the trace element content of the environment. This is particularly evident in unicellular organisms. For example, metal resistance genes with elaborate control mechanisms have evolved to deal with elevated levels of that metal present in the environment. In contrast, some cells are very dependent on metals for cell replication and metalloenzyme activity. Similarly, many mammalian cells in situ and in vitro are resistant to metal deprivation, but others are not. Therefore, the tissue metal concentrations in animals that consume diets deficient in a given essential trace element or receive an excess of that element are frequently within normal limits. Functional deficiency may occur only after depletion of a critical pool or compartment of the trace element (e.g., zinc in the nucleus), which affects specific binding sites. Our understanding of the cell biology of trace elements, however, lags behind our understanding of roles of these micronutrients in molecular biology.

2 STRUCTURAL/CATALYTIC/ REGULATORY ROLES

2.1 Metalloenzymes and Metal-Binding Domains

Characterization of individual metal-binding molecules has shown that stoichiometry is usually 1–4 metal atoms per binding molecule. There are exceptions—for example, ferritin (4500 atoms Fe/molecule) and metallothionein (7 atoms Zn or 12 atoms Cu/molecule). In addition, abundance of metal-binding molecules ranges from a few to many molecules per cell. The composite cellular and extracellular need for each trace element to satisfy the occupancy requirements of these binding sites as balanced against turnover and endogenous losses comprises the dietary requirement for the micronutrient.

Trace elements are essential to the enzymology of molecular biology. For example, all the RNA nucleotidyl transferases (RNA polymerases) are zinc-requiring metalloenzymes. Those metalloenzyme systems are not uniformly susceptible to dietary zinc restriction. However, many manipulations in molecular biology use metal chelation (e.g., with EDTA) as a method of controlling activity of metal-requiring enzymes.

2.2 Metalloregulatory Proteins

Metalloproteins that act as DNA-binding transcription factors, trans-acting MRE-binding proteins, proteins binding single-stranded nucleic acid, and proteins exhibiting metal-dependent interactions with other proteins usually bind metals through thiolate groups from cysteine (Cys) and/or nitrogen from histidine (His). Very little is known about the required dietary abundance, the transfer mechanisms across plasma membranes or among organelles, or the ligand exchange reactions that are necessary for metals to fulfill these regulatory roles. Primary amino acid sequences of proteins dictate the conformational changes induced by metal binding and the resulting alteration of transcriptional and translational rates. The sequence —X-Cys(His)-X_2-Cys(His)-X_n-Cys(His)-X_{2-5}-Cys(His)-X— comprises a classical zinc finger motif that can be found in a plethora of metalloproteins with different functions. Most proteins with zinc finger domains have been shown to have regulatory functions. Genes for zinc finger proteins may account for up to 1% of the human genome.

3 REGULATION OF GENE EXPRESSION BY TRACE ELEMENT MICRONUTRIENTS

Trace elements provide signals to systems that influence rates of either transcription or translation. They do not act directly. Rather, through recognition that requires specific coordination chemistry, metals bind to metalloregulatory proteins (sensors or receptors) of low abundance and influence signaling pathways. This sequence ultimately augments production of a specific protein. In the classical sense, these micronutrients (metals) act as inducers of the system. Since the dietary supply provides the source of the inducer, fluctuations in intake and/or physiologic factors influencing utilization or tissue distribution of the micronutrient are determinants of how effective the trace element will be in activating the metalloregulatory system. Homeostatic regulators of metal metabolism help buffer fluctuations from the dietary supply.

3.1 Transcriptional Regulation

Examples of transcriptional regulation by trace elements are found in microbial and animal cells, but detailed descriptions of the cellular apparatus involved have emerged more rapidly from microbial systems. Regulation of mercury resistance factors serves as an example. Through binding of Hg(II) to the metalloregulatory protein MerR (the receptor) and interaction of this complex with upstream elements of the *MerR* promoter, a change in transcription rate occurs that is directly dependent on extracellular concentration of the inducer, Hg(II). Selectivity is provided by lower binding affinity of other transition elements. Other bacterial genes that provide metal regulation are *DtxR, AnsR, CzeR, fliC,* and *SmtB.* Similarly, the metalloregulatory protein HAP1 of yeast senses heme iron and binds to the apocytochrome *c* gene promoter to increase transcription. In contrast, the microbial gene *Fur* (ferric iron uptake regulation) produces a Fur protein (metalloregulatory) which, through Fe(II)-binding properties, represses transcription of iron uptake genes via interaction with the promoters when iron levels are elevated. Evidence for metal regulation of microbial genes is increasing dramatically. These systems are providing avenues to explore trace element micronutrient regulation (both positive and negative) of mammalian genes.

Metallothionein gene expression has been the most widely studied system that is transcriptionally regulated by trace elements. Research was stimulated by metallothionein-related copper resistance in both *Saccharomyces* and *Candida* organisms, and with animals responding to changes in dietary trace element intake, exposure to toxic metals, and/or physiologic stimuli. Mammalian metallothionein expression is tissue specific, with high abundance in liver, intestine, bone marrow, and kidney, but it can be detected in most tissues and cultures of mammalian cells.

Elevation of dietary zinc intake level provides a positive stimulus for metallothionein expression, but changes in the copper intake level over a wide range have little effect. This difference between copper and zinc illustrates how mammalian systems differ from responses with unicellular organisms, where copper is an important inducer of metallothionein. In mammalian cells, copper metabolism may be regulated in a way that precludes interaction with proteins capable of activating the MRE system, except perhaps at toxic levels of this metal. Studies with intact animals where the metals are injected can show regulation by multiple inducer metals. This could be a direct action of the trace element on the sensor protein or an indirect action via transcriptional regulation signaled by metal-stimulated release of stress-related mediators or hormones.

The exact mechanism by which the trace element (metal) enters the cell, traverses the intracellular space and enters the nucleus for interaction with the metalloregulatory protein is an emerging area of interest. It has been shown that dietary zinc (when tagged with radioactive zinc) can be traced to the nucleus in proportion to the intake level.

Underlying questions for genes exhibiting putative regulation by a trace element include the direct role of the metal and which metal is the natural inducer of expression. A summary of eukaryotic genes where metal regulation has been demonstrated or proposed is shown in Table 1. Many genes are more responsive in vivo to activation by Cd(II), not a micronutrient, than to Zn(II). Much of this difference

Table 1 Transcriptional Regulation by Trace Element Micronutrients: All Eukaryotes

Trace Element	Metal-Binding Regulatory Factor[a]	Gene/Protein Expressed
Cobalt	ND	Heme oxygenase
Cobalt	ND	Plastocyanin
Copper	ACE1	Metallothionein
Copper/zinc	MTF[b]	α_1-Acid glycoprotein
Copper/zinc	MTF[b]	C-Reactive protein
Heme iron	HAP1	Cytochrome *c*
Iron	ND	Ferritin (L subunit only)
Iron	ND	Tartrate-resistant acid phosphatase
Manganese	ND	Mn superoxide dismutase
Zinc	MTF	Metallothionein
Zinc	MTF[b]	Apolipoprotein A_1
Zinc	ND	Creatine kinase

[a]ND, not determined; ACE1, yeast copper transcription factor; HAP1, heme sensor protein.
[b]Probable regulatory protein.

may relate as much to physiology of absorption and cellular uptake as to sensitivity of the sensor protein(s) for transcription initiation. Evidence is accumulating that a variety of genes have MRE elements and that these can influence transcriptional regulation in response to the trace element micronutrient status of the organism. Membrane metal transporters that regulate uptake and efflux are being identified. These may be regulated through metal responsive components.

A minimum of three components is, at present, considered to be necessary for transcriptional regulation by a trace element (Figure 1). First, the responsive gene must have a metal response element at one or more sites in the promoter, usually upstream from the transcription initiation site. All DNA motifs responsible for metal regulation thus far identified have a consensus 13–15 bp imperfect sequence. The consensus MRE sequence for the mammalian metallothionein promoter is CTCTGCRCNCGGCCC (the core of this consensus sequence is underlined). MREs are usually found in multiple copies of either orientation within the first few hundred base pairs of the promoter. MREs are frequently located near other regulatory elements, perhaps for synergistic or differential regulatory purposes.

The second component needed for transcriptional regulation by trace element micronutrients is a trans-acting metalloregulatory protein(s). Such metal binding/sensing proteins (depicted in Figure 1) act as transcription factors, presumably to initiate or enhance transcription of an MRE-regulated gene. They are called MRE-binding transcription factors (MTFs). Metals bind to these proteins through unique coordination chemistry [e.g., most likely trigonal for Cu(II), tetrahedral for Zn(II)]. Each MRE regulated by a specific metal may have a unique metalloregulatory protein, but this is unlikely. Multiple combinations of metal-binding proteins could form complexes that produce negative as well as positive effects on transcription of a specific MRE-regulated gene. Furthermore, tissue specificity of metal regulation requires that genes for the MTF protein be expressed to differing extents requisite to cell function and stage of development. MTF expression may respond to dietary or hormonal stimuli. Data are lacking on these points.

The third component for this metal-responsive unit is the trace element (metal), that is, the inducer. Intracellular concentrations of

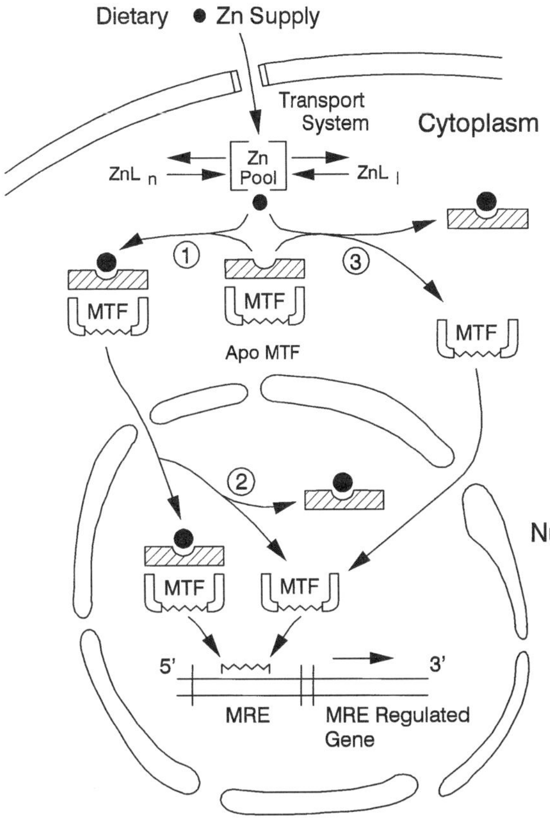

Figure 1. Hypothetical modes of transcriptional regulation of gene expression by trace element (metal) micronutrients. Metallothionein gene expression by Zn provides an example of this mode of regulation. For this system, the regulated gene(s) must have a metal response element (MRE) sequence(s) in the promoter/enhancer region. Following transport into the cell, the metal (solid circles) interacts rapidly with a variety of ligands (L). One of these is the apo (metal-free) form of an MRE-binding protein termed an MRE-binding transcription factor (MTF). Subsequent activation modes could differ: (1) The metal enters the nucleus with the MTF, and metal occupancy allows MRE binding. (2) After the metal has entered the nucleus, the metal-binding domain (inhibitor) of MTF is removed, allowing MTF to bind to the MRE. (3) Apo-MTF is viewed as an inhibited complex, in which metal occupancy removes a metal-binding inhibitor domain of MTF and allows MTF to enter the nucleus and bind to MRE sequences. In each model, the rate of transcription is proportional to the dietary metal supply, the intracellular metal pool, and the extent of MRE binding. It is possible that some MRE-binding transcription factors act as negative rather than positive regulators of transcription.

metals vary with physiologic state and dietary intake. This concentration will influence metal availability for binding to the MTF protein. The source (a low molecular weight chelate or a protein of lower binding affinity) and the cellular location (nuclear, cytoplasmic, or both) of metals destined for binding to MTF proteins remain to be defined (Figure 1). It is likely that some membrane transport proteins for metals are regulated by intracellular concentrations through MTF-like control mechanisms.

3.2 Translational Regulation

The initial concept of trace element control of translation was developed using as a model the acute induction of ferritin synthesis by

Fe(II), a process that does not require transcription. Iron-containing transferrin binds to transferrin receptor located on the cell surface, undergoes endocytosis, and contributes iron to a low molecular weight iron pool. When this cellular pool is large, ferritin synthesis is stimulated via increases in translatable ferritin mRNA. In contrast, when the cellular iron pool is low (e.g., during dietary iron deficiency), translation of transferrin receptor mRNA increases. This translational regulation is controlled by an iron-responsive element binding protein (IRE-BP), which binds iron. The stoichiometry of iron in the binding domain is three or four Fe(II) atoms per molecule. Iron binding is correlated with the *c*-aconitase activity of IRE-BP. This enzyme activity may alter cellular citrate levels and, through iron-binding properties of citrate, influence the low molecular weight iron pool. A different IRE-BP2 does not have *c*-aconitase activity. With low iron availability, three atoms are bound to IRE-BP. In this configuration, IRE-BP exhibits high binding affinity for both transferrin receptor mRNA and ferritin mRNA. This property promotes cellular iron uptake through transferrin receptor mRNA stabilization, causing increased receptor synthesis. Simultaneously, decreased (repressed) ferritin synthesis occurs because IRE-BP binding inhibits translation of ferritin mRNA. When the iron supply is sufficient, IRE-BP binds four iron atoms, and mRNA affinity is low. Consequently, transferrin receptor mRNA degradation increases and more ferritin mRNA is translated (derepressed). The result is a decrease in iron uptake and an increase in iron storage. IRE-BP may be controlled by physiologic agents such as nitric oxide. The mechanism as currently understood is shown in Figure 2.

Iron metabolism makes further use of the IRE-BP through control of heme synthesis via regulation of δ-aminolevulinate synthetase mRNA. This mRNA is repressed by IRE-BP binding analo-

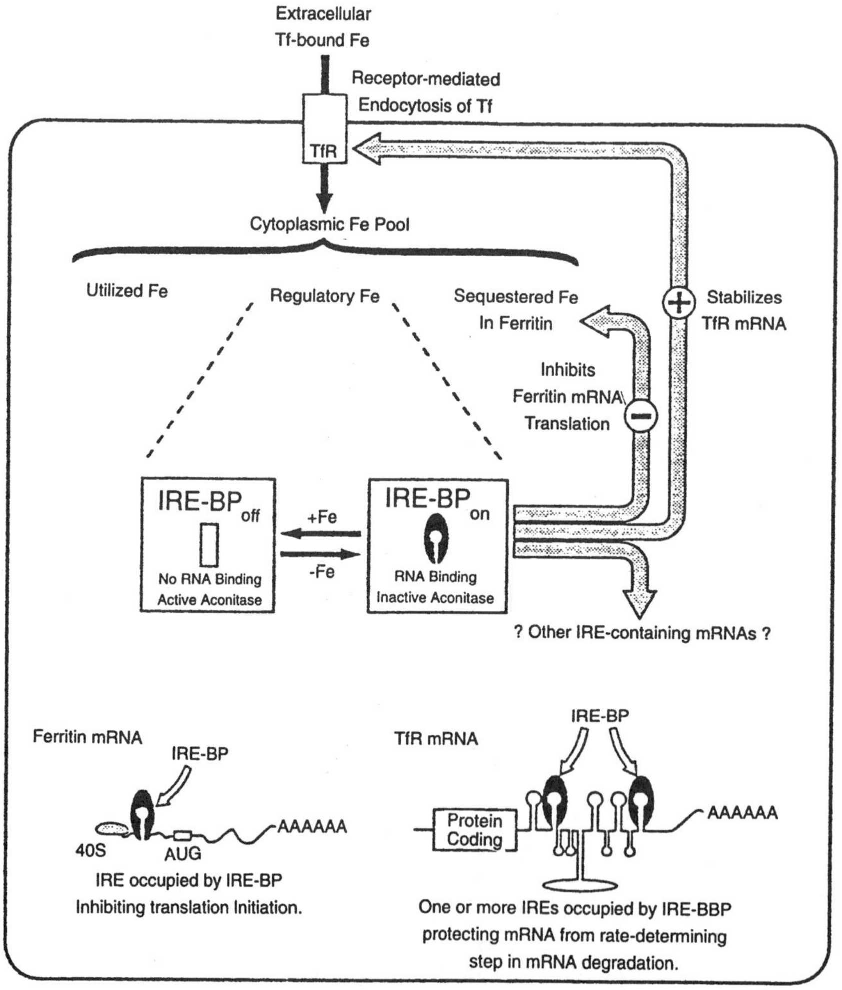

Figure 2. Translational control by iron of transferrin receptor mRNA (TfR mRNA) and ferritin mRNA. The intracellular iron pool is reflective of dietary iron availability. Cellular iron is monitored by an iron-responsive element binding protein (IRE-BP). This protein has *c*-aconitase activity, which may help regulate the low molecular weight iron pool size. When cellular iron is low, less Fe(II) is bound to IRE-BP, favoring the binding of IRE-BP to specific mRNAs. Binding protects TfR mRNA from degradation and increases translation; thus transferrin receptor synthesis increases, and more iron enters the cell. Simultaneously, low iron favors IRE-BP binding to ferritin mRNA, which inhibits translation of this mRNA. When cellular iron levels are high, Fe(II) binding to IRE-BP increases. This lowers the mRNA-binding affinity of IRE-BP, thus increasing ferritin mRNA translation and TfR mRNA degradation. As a result, ferritin synthesis increases and TfR synthesis decreases. [From Klausner (1993). Reproduced with permission of copyright holder, Cell Press.]

Table 2 Translational Regulation by Trace Element Micronutrients

Trace Element	Regulatory Factor	mRNA Regulated
Heme iron	Iron-responsive element binding protein	δ-Aminolevulinate synthetase
Iron	Iron-responsive element binding protein	Ferritin (both heavy and light subunits)
Iron	Iron-responsive element binding protein	Transferrin receptor
Selenium	Not determined	Cellular glutathione peroxidase

gous to ferritin mRNA. There may be other mRNAs controlled by IRE-BP. Likely candidates would be mRNAs for iron-sensitive factors such as those related to low iron status and iron transport proteins for regulated iron uptake.

There may be mRNA-binding proteins analogous to IRE-BP that have functional binding domains for other metals. Trace element dietary status could, in this way, influence many metabolic pathways by translational mechanisms. Knowledge of trace element mediated translational control is limited, as shown in Table 2.

4 APPLICATIONS AND PERSPECTIVES

The application of trace element micronutrients for control of biotechnological systems has been demonstrated. The dramatic growth of transgenic animals producing excess growth hormone through activation of metallothionein promoter–growth hormone transgene constructs in response to zinc in the drinking water illustrated the potential of this approach over a decade ago.

Chimeric constructs have been used to regulate plant genes by trace elements. The copper-responsive transcription factor from yeast (ACE1) has been used to stimulate expression of a gene coding for a commercially important enzyme, via copper treatment of the plant. A DNA construct with the ACE1-binding site (MRE) was inserted into the promoter of the gene coding for the enzyme. Through copper treatment of the transgenic plant, production of the enzyme is increased. Similar experiments in animals and humans will likely occur in the future.

A key to using trace elements for technological purposes is to select those that are relatively nontoxic (e.g., zinc) and will not damage the organism, but are sufficiently bioavailable to promote maximum expression of the transgenes. The unique physiology, tissue-specific sequestration, and nutritional intake of each trace element need to be considered in these strategies. In addition, there is concern over the possibility that the metal resistance genes of microbial systems may lead to overgrowth of endogenous microbial populations and induction of metal-inducible proteins, potentially providing gratuitous changes in the metabolic fates of xenobiotics.

The development of pharmaceuticals with actions related to metal chelation or metal donation is of major current interest. Many of these novel compounds will use trace element micronutrients as a vehicle for altering gene expression. Alternatively, naturally occurring cellular components (e.g., a zinc finger transcription factor) could be targets of various therapeutic or sequestering agents, each containing a specific trace element.

See also DNA Replication and Transcription; Gene Expression, Regulation of; Zinc Finger DNA Binding Motifs.

Bibliography

Chesters, J. K. (1992) Trace element–gene interactions. *Nutr. Rev.* 50:217–223.

Cousins, R. J. (1994) Metal elements and gene expression. In *Annual Review of Nutrition,* Vol. 14, R. E. Olson, Ed., pp. 449–469. Annual Reviews, Palo Alto, CA.

Cousins, R. J. (1996) Chapter 29. In *Present Knowledge in Nutrition,* 7th ed; L. J. Filer and E. E. Ziegler, Eds., pp. 293–306. International Life Sciences Institute-Nutrition Foundation, Washington, DC.

Heuchel, R., Radtke, F., Georgiev, O., Stark, G., Aguet, M., and Schaffner, W. (1994) The transcription factor MTF-1 is essential for basal and heavy metal-induced metallothionein gene expression. *EMBO J.* 13:2870–2875.

Klausner, R. D., Rouault, T. A., and Harford, J. B. (1993) Regulating the fate of mRNA: The control of cellular iron metabolism. *Cell,* 72:19–28.

Klug, A., and Schwabe, J. W. R. (1995) Zinc fingers. *FASEB J.* 9:597–604.

Leibold, E. A., and Guo, B. (1992) Iron-dependent regulation of ferritin and transferrin receptor expression by the iron-responsive element binding protein. In *Annual Review of Nutrition,* Vol. 12, R. E. Olson, Ed., pp. 345–368. Annual Reviews, Palo Alto, CA.

O'Halloran, T. V. (1993) Transition metals in control of gene expression. *Science,* 261:715–725.

Palmiter, R. D. (1994) Regulation of metallothionein genes by heavy metals appears to be mediated by a zinc-sensitive inhibitor that interacts with a constitutively active transcription factor, MTF-1. *Proc. Natl. Acad. Sci. U.S.A.* 91:1219–1223.

Steele, N. C., and Pursel, V. G. (1990) Nutrient partitioning by transgenic animals. In *Annual Review of Nutrition,* Vol. 10, R. E. Olson, Ed., pp. 213–232. Annual Reviews, Palo Alto, CA.

Thiele, D. J. (1992) Metal-regulated transcription in eukaryotes. *Nucleic Acids Res.* 20:1183–1191.

Transcription: *see* DNA Replication and Transcription.

Transgenic Animal Modeling

Carl A. Pinkert, Michael H. Irwin, and R. Jeffrey Moffatt

1 Introduction
- 1.1 Brief History and Definition of Transgenic Animals
- 1.2 Applications of Transgenic Animals
- 1.3 Societal and Ethical Considerations

2 Production of Transgenic Laboratory Animals
- 2.1 Choice of Animals
- 2.2 DNA Microinjection
- 2.3 Retroviral Introduction of Transgenes
- 2.4 Embryonic Stem Cell Technology
- 2.5 Other Laboratory Animal Models

3 Production of Transgenic Domestic Animals
- 3.1 Overview
- 3.2 Traits Affecting Domestic Animal Productivity
- 3.3 Domestic Animals as Bioreactors
- 3.4 Examples from Domestic and Miniature Pigs

4 Analysis of Transgene Integration and Expression
- 4.1 Transgene Integration
- 4.2 Transgene-Encoded mRNA Expression
- 4.3 Protein Expression

5 Conclusions and Future Directions
- 5.1 Gene Transfer Today
- 5.2 Future Directions

Key Words

DNA Microinjection A gene transfer technique utilizing micromanipulation to inject DNA constructs (transgenes) directly into the pronuclei or nuclei of cells. The foreign DNA integrates into the host cell genome at a random location and usually in concatemers. DNA microinjection is the most commonly used gene transfer technique for creating transgenic mammalian species.

Embryonic Stem (ES) Cell Transfer The transfer of genetically manipulable pluripotent embryonic stem cells into a developing embryo; achieved by microinjection of stem cells into blastocysts or coculture with earlier stage ova. The resultant transgenic animal possesses a proportion of cells descended from the ES cell lineage. In mammalian species, this gene transfer technique has been successful only in producing germ line transmission of ES cell lineages in mice.

Gene Transfer A set of techniques directed toward manipulating biological function via the introduction of foreign DNA sequences (genes) into living cells.

Transgenic Animal An animal either integrating foreign DNA segments into its genome following gene transfer, or resulting from the molecular manipulation of endogenous genomic DNA.

Transgenic Founder The first-generation transgenic organism directly produced as a result of gene transfer/germ plasm manipulation procedures. The term does not imply that the transferred gene (transgene) is heritable to future generations.

Transgenic Line A direct familial lineage of organisms derived from one or more transgenic founders, characterized by the passing of the transgene(s) to successive generations as a stable genetic element. The line includes the founder and any subsequent offspring inheriting the specific germ line manipulation.

Transgenic animals and our ability to manipulate the genome of whole animals in vivo have influenced the life sciences in dramatic fashion. In less than 15 years, practical aspects of germ line manipulation of the genetic composition of animals have allowed researchers to address fundamental questions in fields ranging from medicine and biology to biotechnology and production agriculture. Studies emanating from basic efforts in gene regulation have led the way to *in vivo* gene transfer studies focused on the expression of foreign genes and the ablation of endogenous genes. Our appreciation of basic biological development in animals and the potential ramifications of novel phenotypes resulting from genetic engineering efforts have been markedly influenced by pioneering efforts in transgenic animal technology.

1 INTRODUCTION

1.1 Brief History and Definition of Transgenic Animals

The ability to introduce functional genes into animals provides a very powerful tool for dissecting complex biological processes and systems. Hence, transgenic animals represent unique models that are custom-tailored to address specific biological questions. Furthermore, classical genetic monitoring cannot engineer a specific genetic trait in a directed fashion. For identification of interesting new models, the genetic screening and characterization of chance mutations remains a long and arduous task. Therefore, gene transfer is also of great value, particularly for animal species in which long life cycles reduce the value of classical breeding practices for genetic modification.

Although the entire procedure for microinjection into living cells was described in the late 1960s, it took 15 years before transgenic animals were created. In 1980 and 1981, six pioneering laboratories reported successful gene transfer and the production of transgenic mice. Their success would not have been possible, however, were it not for the refinement of recombinant DNA technologies needed to develop viable protocols and to document results. In gene transfer, animals receiving new genes (foreign DNA sequences integrated into their genome) are referred to as "transgenic," a term first coined by Gordon and Ruddle in 1981. As such, transgenic animals are recognized as specific species variants or strains, which result from the introduction and integration of new gene(s), or "transgenes," into the genome. More recently, the term "transgenic" has been extended to chimeric or "knockout" mice in which researchers have selectively removed or modified a gene or genes within the host genome.

For years, many studies were envisioned—if only an animal genome could be readily modified. The realization of the many technologies at hand today has opened numerous avenues of research promise. Production of transgenic mice has marked the convergence of earlier advances in the areas of recombinant DNA technology and the manipulation and culture of animal germ plasm (Figure 1). Transgenic mice provide powerful models to explore the regulation of gene expression as well as the regulation of cellular and physiological processes. The use of transgenic animals in biomedical, agricultural, and biotechnological fields requires the ability to target gene incorporation and to control the timing and level of transgene expression. Experimental designs have taken advantage of the ability to direct specific (e.g., cell, tissue, organ specificity) as well as ubiquitous (whole-body) expression in vivo. Furthermore, from embryology to virology, transgenic technology provides unique animal models for studies in various disciplines that would otherwise be all but impossible to develop spontaneously.

1.2 Applications of Transgenic Animals

A number of methods exist for gene transfer in mammalian species, including DNA microinjection, embryonic stem (ES) cell transfer, retroviral infection, blastomere/embryo aggregation, teratocarcinoma cell transfer, electrofusion, nuclear transplantation, spermatogonial cell transfer, and perhaps spermatozoa-mediated transfer (although the latter application is quite controversial).

Transgenic technology has been extended to a variety of animal species in addition to the mouse, including rats, rabbits, swine, ruminants (sheep, goats, and cattle), poultry, and fish. Transgenic am-

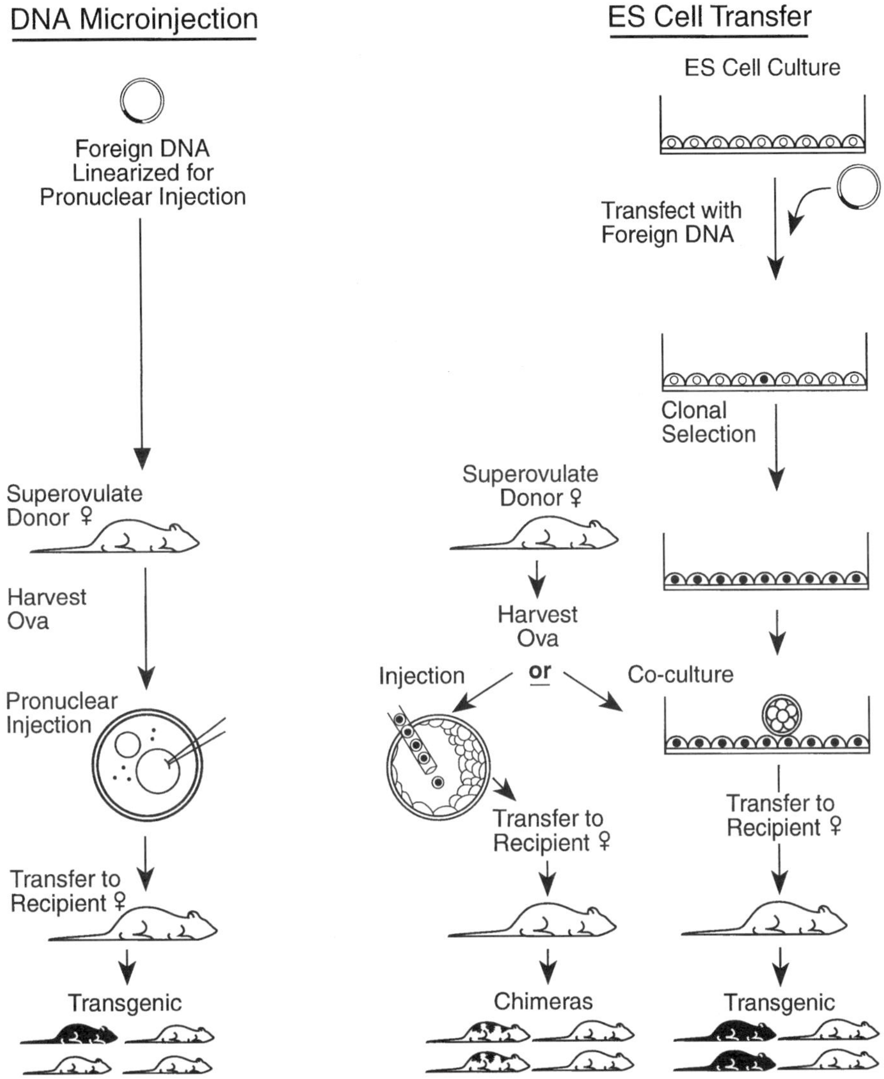

Figure 1. DNA microinjection and embryonic stem (ES) cell transfer are the two most practiced methods for producing transgenic mice. Note that an in vitro culture step is not necessary for DNA microinjection (left); DNA is injected directly into the male pronucleus of fertilized one-cell ova (pronuclear zygotes). In general, if transgenic mice (represented by black mouse) are derived by DNA microinjection, all their cells will contain the new transgene(s). On the right, after clonal selection of transfected ES cells, one of two techniques is used for ES cell transfer. Ova are harvested between the eight-cell and blastocyst stages. ES cells are either injected directly into a host blastocyst ("injection") or cocultured with eight-cell to morula stage ova, so that transfected ES cells are preferentially incorporated into the inner cell mass of the developing embryo ("coculture"). With blastocyst injection, transgenic offspring are termed "chimeric" because some of their cells are derived from the host blastocyst and some from transfected ES cells (denoted by white mice with black patches). Using coculture and tetraploid embryos, one can obtain founder mice derived completely from the transfected ES cells (denoted as all-black mice).

phibians, insects, nematodes, and members of the plant kingdom have also been produced. With advances in the understanding of promoter/enhancer elements and transcription/regulatory proteins involved in the control of gene expression, the technology continues to evolve using different model systems (Table 1). In the systems explored to date, gene transfer technology is a proven asset in science as a means of dissecting gene regulation and expression in vivo. As such, the primary questions that are addressed concern the roles of individual genes in development or in particular developmental pathways. With this caveat, considerations include the ramifications of gene activity from intracellular to inter- and extracellular events within a given tissue or cell-type milieu.

Table 1 Applications of Transgenic Mice

Transgenic mice have provided models in biomedical, veterinary, agricultural, and biotechnological disciplines in the study of gene expression and developmental biology, as well as for:

- The fields of embryology, endocrinology, genetics, immunology, neurology, oncology, pathology, physiology, toxicology, and virology
- The study of genetic bases of disease, and the design and testing of therapeutic strategies
- Gene therapy
- Disease resistance in humans and animals
- Drug and product efficacy testing/screening
- Novel product development through "molecular farming"

Normally, gene function is influenced by cis-acting elements and trans-acting factors. For transferred genes, the cis- and trans-activators, in conjunction with the gene integration/insertion event within the host genome, influence regulation of both endogenous and transferred genes. Using genes that code for reporter proteins (e.g., oncogene or *lacZ*/βgal constructs), analysis of transgenic animals has revealed the importance of these factors in determining developmental timing, tissue distribution, and relative efficiency of gene expression. Transgenic animals have also proven quite useful in determining in vivo artifacts of other model systems or techniques.

There are a number of strategies in the development of transgenic mouse models, including systems designed to study dominant gene expression, homologous recombination/gene targeting and the use of ES cells, efficiency of transformation of eggs or cells, disruption of gene expression by antisense transgene constructs, gene ablation or knockout models, reporter genes, and marking genes for identification of developmental lineages.

1.3 Societal and Ethical Considerations

While gene transfer technology continues to open exciting new biological frontiers, it also raises questions concerning regulation and commercialization. It is not within the scope of this presentation to fully address these issues; yet they do exist and will continue to influence the genetically engineered systems under development.

Contrary to the early prospects related to potential commercialization of this technology, there are societal challenges regarding perceived risks still ahead (Table 2). The risks at hand can be defined not only by scientific evidence but also in relation to public concern. Therefore, the central questions will revolve around the proper safeguards to employ and the development of coherent and unified regulatory guidelines. Can new animal reservoirs of fatal human diseases be created? Can more virulent pathogens be artificially created? What is the environmental impact of the "release" of genetically engineered animals? Do the advantages of a bioengineered product outweigh potential consequences of its use? These are but a few of the questions that we cannot ignore. Yet, the regulatory hurdles that exist today will challenge not only scientists and policy makers, but academicians, bioethicists, and legal scholars as well.

The strategies associated with the production of transgenic animals are quite variable, even between laboratories that utilize similar systems. However, most of the techniques for extensions to different systems are unique and timely for new investigators. The overall efficiency of many procedures will vary, as will the cost–benefit ratios. however, one must not let the formidable challenges of gene transfer experimentation outweigh the primary reason for instituting such studies—the desire to develop a biological model with specific genetic characteristics in a whole-animal environment.

Table 2 Societal Concerns Regarding Genetically Engineered Animals

Environmental impact following "release" or commercialization of transgenic animals
Public perceptions of perceived risk
Ethical considerations of transgenic animal exploitation
Legislative oversight of technology
Safety of transgenic products, by-products, and foodstuffs
Patent aspects and product uniformity/economics of production

2 PRODUCTION OF TRANSGENIC LABORATORY ANIMALS

2.1 Choice of Animals

The relative importance of using particular strains or breeds of animals in gene transfer experimentation will vary dramatically according to the species under consideration. Probably the most complex system is encountered in the production of transgenic mice, simply because so much work has been done with murine species. Here, well-documented differences in reproductive productivity, behavior, related husbandry requirements, and responses to various experimental procedures play major roles in determining the efficiency and degree of effort associated with production of transgenic founder animals. A general discussion of these factors therefore serves as an appropriate starting point for understanding the many processes and procedures that must be evaluated and monitored when considering production of transgenic animals.

DNA microinjection, the most direct and reproducible method for producing transgenic animals, necessitates that several "pools" of animals be maintained for specific purposes. Most commonly, donor females are induced to superovulate by means of a regimen of injections of pregnant mare serum gonadotropin (PMSG, to stimulate follicular growth and development) followed by human chorionic gonadotropin (hCG, to induce ovulation of mature eggs). These donor females are then mated to fertile males, and large numbers of fertilized one-cell eggs ("zygotes") are obtained surgically. Alternatively, ova may be collected from donor females and then subjected to in vitro fertilization (IVF) to obtain zygotes. In either case, the DNA construct, in a buffer solution, is microinjected into the male pronuclei of the fertilized eggs. Ova that survive microinjection are surgically transferred to the reproductive tracts of hormonally synchronous female recipients, which carry the embryos to term. In mice, DNA from offspring is isolated from tissue samples (e.g., tail biopsies) obtained at weaning and analyzed by Southern blot analysis and/or methods based on the polymerase chain reaction (PCR) to determine which individuals carry the transgene.

Certainly, the relative costs of different strains of mice that are chosen to serve as gamete donors or recipients is a factor. This must be balanced with the background genetics of the various strains and the effects these may potentially exert on expression of the integrated, foreign DNA construct (the transgene) and the resulting phenotype. Strains used for donor females are chosen for (1) numbers of eggs ovulated in response to hormonal superovulation regimen, (2) the relative uniformity of this response within the strain, (3) the viability of superovulated eggs both before and after microinjection, (4) the ease with which fertilized eggs are visualized and recovered from the reproductive tract, and (5) the morphology and/or size of male pronuclei (which affects the ease and speed with which eggs can be microinjected).

The strain chosen for embryo transfer recipient use is evaluated for (1) parenting ability, (2) docile behavior (which allows for inspection of pups or cross-fostering after birth), (3) ease of embryo transfer procedures, and (4) the regularity with which females exhibit readily detectable vaginal plugs and undergo a pseudopregnant response when mated with vasectomized males.

Obviously, the genetic background of the strain used for vasectomized males (and also recipient females) is less important, since these animals do not contribute to the genetic composition of the transgenic offspring. Strains are favorably considered in which vasectomized males exhibit libido and are able to form visible copulatory plugs in the recipient females both frequently (two to three matings/week) and with regularity over a period of about 12 months.

Intact males are chosen for their (1) libido, (2) absence of nonmating aggressiveness when placed with families, (3) fertility and capacity to produce visible copulatory plugs, and, if serving as IVF donors, (4) sperm concentration and quality. Finally, it is desirable to maintain a pool of synchronous, pregnant females that can serve as foster mothers in the event that recipient females begin to cannibalize pups or fail to adequately nurse their offspring.

For most other nonmurine species, it will remain important to use donor females that respond well to hormonal synchronization and superovulation, and to have recipients that are able to carry fetuses to term and care for neonates appropriately. However, differences in methods used for hormonal synchronization, selection of proestrus females, and evaluation of other forms of reproductive behavior may differ significantly from those identified for mice.

2.2 DNA Microinjection

The microinjection method, referred to earlier, involves the use of micromanipulators and an air- or oil-driven microinjection apparatus to physically inject the DNA construct solution into the male pronuclei or nuclei of fertilized one- or two-cell ova. As such, the method is also sometimes referred to as "pronuclear microinjection."

Virtually any cloned DNA construct can be used, albeit some caveats apply. Linearized (as opposed to closed, circular) DNA constructs are more readily integrated into the host genome. The presence of extraneous plasmid- or vector-related DNA sequences associated with the construct can adversely affect expression of the integrated construct/transgene. Also, greater care must be taken when handling larger DNA constructs, such as those derived from yeast artificial chromosomes (YACs), to avoid damage by shearing prior to and during microinjection.

With few exceptions, microinjected constructs integrate randomly throughout the host's genome, but usually only in single locations. This fact can be exploited to obtain functional linkage of independent transgenes by simultaneous coinjection of more than one DNA construct. In such a case, both constructs tend to integrate in the same randomly located site and may exhibit coexpression. Because integration is spatially random, transgenes may be integrated into either a somatic or a sex chromosome.

The integration process itself is poorly understood, but it apparently does not involve homologous recombination. During integration, single or multiple copies (1–200) of a transgene are incorporated into the genomic DNA, predominantly as head-to-tail concatemers. Regulatory elements in the host DNA near the site of integration, and the general availability of this region for transcription, appear to play a major role in affecting the level of transgene expression. This "positional effect" is presumed to explain why the levels of expression of the same transgene may vary dramatically between individual founder animals as well as their offspring (or "lines"). It is therefore prudent to examine transgene expression in offspring from at least three or four founder animals to determine what might be a result of position or incorporation site influences. Additionally, the presence of tissue- or cell-type-specific nuclear regulatory factors that act on host genes near the site of integration in trans may restrict transgene expression to only a subpopulation of cells even though the transgene is present in virtually all cells of a founder.

Host DNA near the site of integration frequently undergoes various forms of sequence duplication, deletion, or rearrangement as a result of transgene incorporation. Such alterations, if sufficiently drastic, may disrupt the function of normally active host genes at the integration site, an activity that constitutes "insertional mutagenesis," wherein an aberrant phenotype may result. Such events cannot be purposefully designed, but they have led to the serendipitous discovery of previously unsuspected genes and gene functions.

Because gene transfer is accomplished at the one-to two-cell stage, very early in embryonic development, the transgene is incorporated into essentially every cell that contributes to development of the embryo. Thus, transgenic animals (or "founders") produced by this method are usually considered to be "nonmosaic" or "nonchimeric," in the sense that the transgene is physically present in the nuclei of all cells of the body. Incorporation of the transgene into cells that will eventually contribute to development of germ cells (sperm or ova) is a common occurrence with this method, and it makes possible heritability of the transgene by offspring of founder animals. In such cases, the transgene has been said to "have gone germ line." However, integration of the microinjected DNA construct into the host's genome occasionally may be inexplicably delayed. In such a case, if cells of the early embryo ("blastomeres") undergo mitosis before integration occurs, some but not all of the cells will contain the transgene, and the founder animal, although still considered to be transgenic, will be classified as a mosaic or chimera.

Advantages of the microinjection method are as follows: (1) the relatively high frequency of generating transgenic animals from viable microinjected embryos transferred to recipient females (20–30%), (2) the high probability of germ line transmission of the transgene, (3) the relative lack of constraints on the size or type of DNA construct used, (4) the relative stability of the transgene as it is transmitted from generation to generation, and (5) the low frequency of mosaicism or double integrations (combined estimate of 10–30% of founders).

Disadvantages of this method include the following: (1) the random and potentially significant influence of the site of integration on transgene expression (positional effects), (2) the potential for undesired insertional mutagenesis, (3) the occasional production of mosaic founders, (4) the occasional lack of germ line incorporation, and (5) the time and expense required to obtain the skills necessary for micromanipulation and microinjection.

2.3 Retroviral Introduction of Transgenes

Transfer of foreign genes into animal genomes has also been accomplished using retroviruses. Although embryos can be infected with retroviruses up to midgestation, early eggs, usually at the 4- to 16-cell stages, are used for infection with one or more recombinant retroviruses containing a foreign gene. Immediately following infection, the retrovirus produces a DNA copy of its RNA genome using the viral enzyme reverse transcriptase. Completion of this process requires the host cell to undergo the S phase of the cell cycle. Therefore, retroviruses effectively transduce only mitotically active cells. Incomplete infections, in which not all embryonic cells acquire the retrovirus, occur more frequently when using embryos after the 4-cell stage, with resultant chimeric embryos. Modifications to the retrovirus frequently consist of removal of structural genes, such as *gag, pol,* and *env,* which support viral particle formation. Additionally, most retroviruses and complementary lines are ecotropic in that they infect only rodents (e.g., rats, mice) and rodent cell lines rather than humans.

The DNA copy of the viral genome, or provirus, integrates randomly into the host cell genome, usually without deletions or rearrangements. However, as is the case for gene transfer by microinjection, because integration is not by way of homologous recombination, this method is not used effectively for site-directed mutagenesis.

Very high rates of gene transfer are achieved with the use of retroviruses. However, disadvantages include: (1) low copy number integration, (2) the additional steps required to produce retroviruses, (3) a general limitation on the size of the foreign DNA insert (usually 9–15 kb) that can be accommodated by individual viral strains, (4) a potential for undesired genetic recombination that may alter the replicative characteristics of the retrovirus, (5) a high frequency of mosaicism (which necessitates outbreeding of founder animals to generate pure lines), and (6) possible interference by retroviral sequences on transgene expression.

2.4 Embryonic Stem Cell Technology

Gene transfer has been used to produce both random and targeted insertion or ablation of discrete DNA fragments into the mouse genome. For targeted insertions, where the integration of foreign genes is based on a recombinational gene insertion with a specific homology to cellular sequences (termed homologous recombination), the efficiency of DNA microinjection is extremely low. In contrast, the use of ES cell transfer into mouse embryos has been quite effective in allowing an investigator to preselect a specific genetic modification, via homologous recombination, at a precise chromosomal position. This preselection has led to the production of mice that (1) incorporate a novel foreign gene into their genome, (2) carry a modified endogenous gene, or (3) lack a specific endogenous gene following gene deletion or "knockout" procedures.

Technologies involving ES cells, and more recently primordial germ cells, have been used to produce a host of mouse models. Pluripotential ES cells are derived from early preimplantation embryos and maintained in culture for a sufficient period for one to perform various in vitro manipulations. The cells may be injected directly into the blastocoel of a host blastocyst or incubated in association with a zona-free morula. The host embryos are then transferred into intermediate hosts or surrogate females for continued development. Currently, the efficiency of chimeric mouse production results in about 30% of the live-born animals containing tissue derived from the injected stem cells. This ability to produce "chimeric" animals using ES cells has given researchers another tool in their armamentarium for the production of transgenic animals. In this set of techniques, the power of gene transfer technology has been catapulted forward because such processes allow for targeted insertions into the genome. Such targeting is extremely important, particularly in areas of gene therapy and correction, wherein previous technologies allowed only for random integration events.

The genome of ES cells can be manipulated *in vitro* by introducing foreign genes or foreign DNA sequences by techniques including electroporation, microinjection, precipitation reactions, transfection, or retroviral insertion. The use of ES cells to produce transgenic mice faced a number of procedural obstacles before it became competitive with DNA microinjection as a standard technique in mouse modeling.

The first procedural obstacle in creating ES cell derived transgenic animals was the extreme difficulty associated with the production and maintenance of ES cell lines. This problem lies in a general inability of embryos of most mammalian species, other than mice, to survive in vitro. A standard technique for producing murine ES cells involves culture of embryos *in vitro,* up to and beyond the point at which they would normally implant in the uterus. In mouse embryo culture, this period is mimicked through coculture on fibroblast monolayers or by the inclusion of various growth factors, (leukemia inhibitory factor [LIF], ciliary neurotropic factor [CNTF], etc.) in the medium. In the case of coculture, fibroblasts serve as a substrate for the attachment of developing embryos, enabling them to develop to a stage at which ES cells may be harvested. Additionally, although not completely understood, culture conditions and media/sera, as well as the handling of ES cells before and during transfer procedures, play critical roles in experimental efficiencies.

Following identification of true ES cells and subsequent gene transfer procedures, it is necessary to accurately and efficiently identify ES cells that have integrated foreign DNA. In this regard, the use of marker-assisted selection schemes has greatly simplified mouse experimentation. Some genes were easily selectable in vitro and therefore made screening less difficult, but most were not. In these latter cases, PCR technology has been used effectively to screen for homologous recombination events. Mutated copies of normal endogenous gene sequences have been inserted in place of normal cellular sequences, and PCR techniques were used to screen for the novel mutants, with the capability of detecting even a single base pair mutation. Hence, PCR techniques have proven to be valuable tools in selecting altered ES cell clones for use in transgenic animal production.

ES cell chimeric animals may indeed harbor cells of both host embryo and ES cell derivation, extending the study of germ line transmission of desired transgenes one or more additional generations compared to most microinjection-derived transgenic animals. Concerns associated with the difficulty of line maintenance for DNA microinjection derived transgenic animals are minimized, however, because the ES cells have the potential to give rise to multiple founder animals harboring the same transgene at the same chromosomal integration site. More recently, with the addition of coculture techniques involving tetraploid host embryos (eight-cell stage to morula), founders can be derived completely from the cocultured ES cells. Hence, the founders are no longer chimeras, having become "better engineered" models.

Still for most modeling employed today, germ line transmission of transferred genes in ES cell derived mice is influenced to a great extent by the background genetics of host and donor strains of mice. Interestingly, germ line transmission is routinely problematic in some strains but not in others. The underlying biological mechanisms responsible for such differences are not known.

While ES cell lines have been identified for species other than the mouse, production of germ line ES cell derived/chimeric animals has not been reported. If ES cells are identified that could give rise to germ line competent animals, it is probable that gene transfer into such cells would be successful, based on the wealth of available gene transfer literature. Thus, the most obvious question to pose today is: Why has it not been possible to procure and utilize embryonic-derived pluripotential stem cells (ES cells) in mammalian species other than the mouse? Is it that present-day difficulties reflect an evolutionary mutation in the mouse genome, but not other species, allowing immortalization of pluripotent cells? Or perhaps, might the idealized conditions just be beyond our immediate grasp?

2.5 Other Laboratory Animal Models

Transgenic animal protocols developed in mice have been modified to accommodate production of other transgenic species. In particular, the larger body sizes of rats and rabbits, their short estrous cycles and gestation lengths, their relatively rapid generation times, and their litter-bearing ability have helped make these animals preferred models for several areas of research employing gene transfer. As with mice, DNA microinjection has been the method used most frequently for introduction of foreign DNA. Differences between these species and mice in the size of ova, physical response to microinjection, requirements for in vitro ova culture, and numbers of fertilized eggs needed for embryo transfer (to maintain pregnancy), as well as differences in general husbandry practices are well documented and used to optimize production of transgenic rats and rabbits.

In some cases, specific physiological characteristics unique to a species have made it a more appropriate or successful model for a particular disease; such was the case for lines of transgenic rats exhibiting hypertension-induced symptoms and pathology resulting from overexpression of specific transgenes (viz., a mouse *Ren-2* renin transgene or a human angiotensinogen gene construct) that were not similarly expressed in transgenic mice.

Both transgenic rats and rabbits have provided models for the study of specific cancers: transgenic rats expressing an SV-40 T-antigen construct controlled by an albumin promoter develop hepatocellular carcinomas. Transgenic rabbits in which expression of a c-*myc* transgene construct is driven by a rabbit immunoglobulin enhancer exhibit malignant lymphocytic leukemia. Particular strains of rabbits have historically served as models for the study of lipid and lipoprotein metabolism. In this tradition, lines of transgenic rabbits bearing human apolipoprotein A-I antisense transgenes have been generated for the study of the biosynthesis of high density lipoproteins.

Finally, several studies have proven rabbits to be acceptable models for "gene farming." In these experiments, foreign proteins are expressed in transgenic animals, preferentially in mammary tissue, and are secreted into milk, from which they can be readily purified. Rabbits are considered to be suitable models for such use for three reasons: lactating females produce relatively large volumes of milk daily, rabbit milk possesses a high protein concentration, and does tend to nurse their offspring infrequently. To date, transgenic rabbits have been separately produced in which human proteins (e.g., α_1-antitrypsin, interleukin 2, growth hormone, tissue plasminogen activator) and bovine α-lactalbumin have been expressed in mammary tissue and harvested from milk.

3 PRODUCTION OF TRANSGENIC DOMESTIC ANIMALS

3.1 Overview

The success of transgenic mouse experiments led a number of research groups to study the transfer of similar gene constructs into the germ line of domestic animal species. With one exception, these efforts have been directed primarily toward either of two general goals: improving the productivity traits of domestic food animal species, and development of transgenic lines for use as "bioreactors" (i.e., producers of recoverable quantities of medically or biologically important proteins). These studies have proven revealing in identifying basic biological mechanisms as well as a need for precise regulation of gene expression. Since 1985, transgenic farm animals harboring growth-related gene constructs have been created, although ideal growth phenotypes have not been achieved because of an inability to coordinately regulate gene expression and the ensuing cascade of endocrine events.

Presently, DNA microinjection is the only method used to successfully produce transgenic livestock. Although complex and at times quite tedious, the steps in the development of transgenic models are relatively straightforward. Once a specific gene sequence has been cloned and characterized, sufficient quantities are isolated, purified, and tested in cell culture. If in vitro mRNA expression of the gene is identified, the appropriate fragment is linearized, purified, and readied for preliminary mammalian gene transfer experiments. The DNA microinjection experiments are first performed in the mouse model. While the transgenic mouse model will not identify likely phenotypic expression patterns in domestic animals, we have not observed a single construct that would function in a pig when there was no evidence of transgene expression in mice. Therefore, preliminary experimentation has been a crucial component of any gene transfer experiment in domestic animals.

Following confirmation of transgene expression in mice, it then becomes cost-effective to initiate DNA microinjection experiments in other species. In mouse experiments, less than 2 months is required from the time the purified construct is ready for microinjection through weaning of founder pups. In contrast, for pig experiments, one month to a year is required for a sufficient number of DNA injections and recipient transfers to ensure the likelihood of success. Experimental efficiencies coupled with a long generational interval (114-day gestation period, 21–28 day lactation, onset of puberty at 6–9 months of age) reflect the efforts necessary to identify and characterize transgenic pigs and illustrate the extended time lines associated with the production of any transgenic livestock model. In addition, the time frame from birth of pigs or cattle to the establishment of lines can be 1–2 years to 4–5 years, respectively (while also dependent on the sex of founders). Hence, there is an obvious advantage to characterizing transgenic mouse models to expedite what will ultimately be a lengthy undertaking.

More recently, protocols have been developed to permit removal of individual blastomeres from microinjected preimplantation embryos maintained in culture prior to transfer to recipient females. Sensitive methods involving the use of PCR allow analysis of DNA purified from individual blastomeres and subsequent identification of the embryos that bear the transgene of interest. Use of such methods offers the potential to greatly increase the efficiency associated with production of transgenic farm animals and to thereby significantly reduce the associated costs.

3.2 Traits Affecting Domestic Animal Productivity

Interest in modifying traits that determine productivity of domestic animals was greatly stimulated by early experiments conducted by Ralph Brinster, Richard Palmiter, and their colleagues, in which body size and growth rates were dramatically affected in transgenic mice expressing growth hormone transgenes driven by a metallothionein (MT) enhancer/promoter. From that starting point, similar attempts followed in swine and sheep studies to enhance growth by introduction of various growth hormone (GH) gene constructs under control of a number of different regulatory promoters. Use of these constructs was intended to allow for tight regulation of individual transgene expression by dietary supplementation. However, although resulting phenotypes included altered fat composition, feed efficiency and rate of gain, and lean/fat body composition, they were accompanied by undesirable side effects (e.g., joint pathology, skeletal abnormalities, increased metabolic rate, gastric ulcers, infertility). Such problems were attributed to chronic overexpression or aberrant expression of the growth-related transgenes and could be mimicked, in several cases, in normal animals by long-term treatment with elevated doses of GH.

Subsequent efforts to genetically alter growth rates and patterns have included production of transgenic swine and cattle expressing a foreign c-*ski* oncogene, which targets skeletal muscle, and studies of growth in lines of mice and sheep that separately express transgenes encoding growth hormone releasing factor (GRF) or insulin-like growth factor I (IGF-I). Cumulatively, it has become apparent from these studies that greater knowledge of the biology of muscle growth and development will be required before lines of domestic animals with these desired characteristics can be genetically engineered.

Other productivity traits that are major targets for genetic engineering include altering the properties or proportions of caseins, lactose, or butterfat in milk of transgenic cattle and goats, more efficient wool production, and enhanced resistance to viral and bacterial diseases (including development of "constitutive immunity" or germ line transmission of specific, rearranged antibody genes).

3.3 Domestic Animals as Bioreactors

The second general area of interest has been the development of lines of transgenic domestic animals for use as bioreactors. One of the main targets of these so-called gene farming efforts has involved attempts to direct expression of transgenes encoding biologically active human proteins. In such a strategy, the goal is to recover from serum or from the milk of lactating females large quantities of functional proteins that have therapeutic value. To date, expression of foreign genes encoding α_1-antitrypsin, tissue plasminogen activator, clotting factor IX, and protein C were successfully targeted to the mammary glands of goats, sheep, cattle, and/or swine.

Targeting expression of the cystic fibrosis transmembrane conductance regulator (CFTR) gene to the mammary epithelium, where the protein becomes associated with the apical membranes of these cells, has resulted in apocrine secretion of the CFTR protein as a constituent of the milk fat globule membrane. The demonstrated utility of such a targeting strategy has led to proposals for large-scale production and harvesting from milk of other classes of important membrane-associated proteins, such as hormone receptors, channel proteins, and transport proteins.

Similarly, lines of transgenic swine and mice have been created that produce human hemoglobin or specific circulating immunoglobulins. The ultimate goal of these efforts is to harvest proteins, from the serum of transgenic animals, for use as important constituents of blood transfusion substitutes, or for use in diagnostic testing.

3.4 Examples from Domestic and Miniature Pigs

In contrast to gene transfer in mice, the efficiency associated with the production of transgenic livestock, including swine, is quite low. However, two advantages offered by swine over other domestic species include a favorable response to hormonal superovulation protocols (20–30 ova can be collected on average) and a significant uterine capacity, in that swine are a litter-bearing species. The highest transgenic success rate in domestic animals has been with outbred domestic pigs, while use of inbred strains was reported with lower overall efficiency. Since the description in 1985 of the first transgenic livestock species, DNA microinjection has been *the only successful method* identified to produce germ line competent transgenic livestock.

An initial problem encountered during the creation of transgenic farm animal species concerned the visualization of the pronuclei or nuclei within the ova. In lipid-dense swine ova, for example, the cytoplasm is opaque and the nuclear structures are not discernible without some type of manipulation. This was a critical problem to overcome because transgenic pigs were not produced following injection of DNA into the cytoplasm of one- and two-cell ova. Fortunately, it was found that centrifugation of pig ova resulted in stratification of the cytoplasm, rendering pronuclei and nuclei visible under differential interference contrast or Nomarski/Smith microscopy. Even though the situation was greatly improved, centrifugation failed to reveal the nuclear structures in 15–33% of fertilized eggs. However, when using conditions initially optimized for transgenic mouse production, DNA microinjection into pronuclei/nuclei resulted in nearly a 10% yield of transgenic offspring (vs. total live-born/viable offspring).

The proportion of transgenic swine that develop from microinjected ova, which varies from zero to about 12%, is generally much lower than that observed in mice. The survival of microinjected pig embryos is related to several factors, including the developmental stage of ova injected, the duration of in vitro culture, synchrony of ova donors and recipients, the number of ova transferred, and donor age. Other factors that have been shown to influence the developmental efficiency of transgenic mouse production—including technician proficiency and embryo handling/transfer, pipette dimensions, DNA preparation, and the viability of microinjected ova—readily influence transgenic production efficiency for all other species.

If a foreign gene fragment is integrated into a domestic animal's genome, the gene copy number and orientation (head-to-head vs. head-to-tail arrays) vary considerably, as might be anticipated from the mouse studies. In addition, mosaic animals are sometimes produced, making analysis of transgene integration more difficult. Recently, the results of PCR analyses of founder animals were useful in detecting extreme mosaicism, where a given transgene is found in only a proportion of the animal's cells. While the identification of transgenic founders (both mosaic and nonmosaic) can be enhanced using PCR analyses, confirmatory DNA slot-blot or Southern blot analyses are necessary to avoid extreme and costly breeding efforts to identify germ line transmission of the transgene.

In one set of experiments, approximately 70% of pigs harboring a growth hormone gene construct (as determined by Southern blot analysis) expressed the transgene-encoded growth hormone. Failure of the remaining 30% of transgenic swine to express the foreign gene was attributed to integration of the transgene into an inactive chromosomal locus or alteration of the transgene sequence during its integration. Of the pigs that did express the transgene, the level of expression appeared to vary greatly among individuals. While cis- and trans-acting factors are obviously important influences affecting transgene expression, the integration site of the transgene is equally critical. Transgene transcription rates are likely influenced by the level of activity present at the locus, or the site of integration, and the properties of enhancer sequences located in genes flanking the transgene.

While it is not unreasonable for an investigator to suggest that 40 transgenic mice can be produced in a week's worth of DNA microinjections, there are significant differences in undertaking pig experiments. In the pioneering collaboration of the University of Pennsylvania, the University of Washington, and the U.S. Department of Agriculture, which first reported on the production of transgenic pigs, more than 4 years went into amassing the experimental results that culminated in the production of 40 founder transgenic pigs. The 55% of founder animals that expressed transgene-encoded mRNA was very much in line with results reported by other groups.

Currently, estimates for the cost of production for each transgenic pig run from $25,000 to $125,000 (U.S. $), with the broad range reflecting the low overall experimental efficiency. The costs are compounded (doubled or even tripled) if functional transgenes and pig survival are factored into the cost estimates. To address the efficiency of the gene transfer experiments, a modified scheduling can be employed to decrease some of the fixed and variable costs, particularly in the use and scheduling of research animals. By overlapping days of experimentation rather than synchronizing one group of donor animals per week, actual experimental costs can be reduced by more than 25%. Superovulation can induce an average of 25–30 embryos per sow. While cost effectiveness dictates the specific regimen, all pigs do not respond similarly, and only 60% respond within a specific window of opportunity for embryo collection. This low response rate may appear unacceptable, but it should be noted that the actual yield of suitable embryos is still higher than might be available using naturally synchronized females or other superovulatory schemes.

Surgical transfer of microinjected embryos usually follows the microinjection efforts. The mechanical insult to the embryos is severe, however, and only 15–25% of them will still be viable 5 days after transfer. Therefore, 30–50 microinjected embryos are routinely transferred per recipient sow, with the expectation that 50% of the recipient females will maintain pregnancy. While the number of embryos transferred may seem excessive, the basis is derived from classical studies that establish a requirement of four viable embryos at the time of implantation for a sow to initiate and maintain a successful pregnancy.

For transgene expression studies targeting pigs, the use of outbred domestic pigs is the most practical way to produce and evaluate potential models. However, miniature or laboratory swine are now used with increasing frequency in biomedical research, where desirable, well-characterized background genetics make these animals more suitable for human modeling studies (e.g., xenotransplantation research). While miniature swine are commercially available, research in developing methodologies to explore or enhance reproductive performance has been limited. Reproductive efficiency in miniature swine is low compared to commercial swine and is characterized by a low ovulation rate, low birth weight, and small litter size. In addition, extensive management is required to maintain neonates. Average litter size is between 4 and 7 pigs at birth, with each breeding sow producing 12 to 18 pigs per year. Estrous cycles and gestation length are similar to standard commercial swine; however sexual maturity in males and females occurs between 4 and 6 months of age in some breeds, which is sooner than the 6–9 month norm of commercial swine.

It is important to evaluate the potential breed differences and ultimate production efficiencies using a superovulatory protocol on groups of donor animals, to be able to assess and possibly enhance the reproductive performance prior to embarking on a gene transfer program. Trials were conducted using Yucatan miniature pigs to ascertain the feasibility of a superovulation regimen, artificial insemination, and embryo transfer to standard commercial crossbred females. Yucatan semen was collected and used to successfully fertilize ova obtained by hormonal stimulation of donor females. As it happened, only 25% of the potential eggs were recoverable and usable (fertilized one- and two-cell ova). However, fertilized ova were obtained with enhanced efficiency compared to spontaneous ovulations in both young and aged miniature pigs.

Breeding requirements for transgenic pigs are critical, and successful management calls for expertise in fields ranging from record keeping to animal husbandry. With a gestation length of 114 days, and an interval from birth to puberty of 6–12 months, the generational interval seems extraordinarily long, particularly when one is accustomed to the experimental progress observed in mouse studies. For xenotransplantation studies, the desire to breed animals to homozygosity for the transgene poses additional challenges. Breeding schemes have been reported for production efforts that, even under ideal circumstances, lead to a 3–5 year timetable (after founders have been produced and identified), assuming no lethality associated with transgene integration and insertional mutations. There are a number of additional questions that may arise in the breeding efforts. If transgenes interact, for example, their effects may not be cumulative; indeed, it might happen that such effects mask one another, perhaps in competition for trans-activators. Other issues involve stability of the genome as well as limiting the gene pool in intricate breeding schemes. Additionally, from a technical viewpoint, evaluation of homozygosity of low copy number or single-copy transgenic animals can be very difficult, with an additional breeding/generational cycle needed to ensure the homozygous status of a transgene in any particular animal.

In future experiments with domestic animals, the ability to utilize embryonic stem cells could represent a tremendous improvement over the microinjection protocols currently in use. The ability to "knock out" or replace endogenous genes has been an extremely valuable tool in mouse genetics and modeling. Yet, characterization and use of embryonic stem cells for gene transfer has not worked to date in any species other than the mouse. The need for such technology in domestic animal species is great, and the perseverance by the groups focused on this work is exemplary.

4 ANALYSIS OF TRANSGENE INTEGRATION AND EXPRESSION

4.1 Transgene Integration

When founder mice are 6–7 weeks of age, biopsies are readily performed to obtain representative tissues for analysis of transgene integration and expression. For integration analysis in mice, typically tail cuts are performed and the tissue is digested with proteinase K followed by DNA extraction using standard protocols. Preliminary determination of transgene integration by the PCR technique can be very useful when the target sequence, the transgene, possesses unique sequences (not endogenous to the genome of the animal). However, given the extreme sensitivity of PCR and the potential for mispriming and false positives, other more specific techniques should always be used to confirm the presence of the transgene in PCR-positive samples. These include DNA slot-blotting and Southern blot hybridization, the latter being not only more informative but also less likely to present false positive signals. Additionally, the transgene can be constructed to include a molecular "tag," a unique sequence that is easily detected and has minimal similarity to any endogenous sequence. This strategy is especially useful when the transgene itself is similar or nearly identical to an endogenous sequence. In such cases, another strategy that may be helpful is the introduction of new restriction sites into the transgene, without perturbing function, so that restriction fragment length polymorphism (RFLP) analysis can be used to distinguish the transgene from its endogenous counterpart. Finally, in some instances phenotypic screening is possible if the expression of the transgene leads to an identifiable change in the appearance of the animal. Phenotypic screening by coinjection of a second transgene that gives rise to a specific phenotype can also be useful, since cointegration is much more likely than integration of the phenotypic marker gene alone. Later, however, the integration of the transgene of interest must be confirmed by other means.

Other concerns related to the analysis of transgene integration include determinations of copy number of the transgene per cell, orientation of tandemly arranged copies, and the presence of multiple integration sites. These questions can be addressed by Southern blot hybridization following digestion of the genomic DNA with appropriate restriction enzymes. In addition, the functionality of the transgene is an important consideration in the analysis of transgene integration. However, this aspect is usually dealt with in the analysis of transgene expression (detection of a functional mRNA transcript or protein product). Finally, the critical question of whether the transgene is integrated into the germ line of the founder animal and is heritable as a stable genetic element must be answered.

4.2 Transgene-Encoded mRNA Expression

The analysis of expression of the transgene is, of course, absolutely essential in determining the utility of the transgenic animals produced. As with integration analysis, the presence or absence of similar or identical endogenous counterparts will determine, to a degree, the strategies that may be most useful. For transgenes that are unique (no endogenous counterpart) or contain some unique sequences, the strategies that can be used are more straightforward. The presence of a novel mRNA transcript or a unique protein (or enzyme activity) is more easily determined than in situations involving transcripts or protein products that are very similar to endogenous transcripts or proteins. As with integration analysis, molecular "tags" are also sometimes useful in that the transcripts will contain some unique identifying sequence that can be readily and unequivocally determined. Once again, care must be taken to ensure that any additions or deletions designed to aid in analysis do not interfere with expression.

Technically, the most critical step in analyzing transgene expression is the isolation of RNA. Great care must be taken to avoid contamination of RNA preparations with ribonucleases (enzymes that degrade RNA). Ribonucleases are present in significant amounts in a variety of locations from which they may gain access to samples of purified RNA (e.g., the skin surface, saliva aerosols from coughing or sneezing, airborne dust). Once a high quality RNA preparation has been obtained from the appropriate tissue, the presence of mRNA transcripts of the transgene can be determined by RNA slot-blot hybridization. A more informative technique, Northern blot hybridization, confirms not only the presence but also the size of the transcript of interest. The Northern blotting technique consists of electrophoresis into an agarose gel to separate the RNA molecules on the basis of size, transfer (blotting) of the RNA to a nylon or nitrocellulose membrane, and detection of the mRNA of interest by hybridizing to a specific, radiolabeled probe, followed by autoradiography.

Additional techniques exist for determining the presence or relative levels of mRNA transcripts from transgenic animals. In the nuclease protection assay, a radiolabeled probe is allowed to hybridize to the RNA in solution, followed by nuclease digestion of unhybridized RNA. The sample is then resolved on a polyacrylamide gel, followed by autoradiography to detect the probe. This technique is very useful in determining the steady state levels of RNA in a given tissue and offers an inherently high signal-to-noise ratio. In the reverse transcription-PCR (RT-PCR) assay, the RNA of interest is transcribed into a cDNA molecule by means of a specific primer and the enzyme reverse transcriptase. A second primer is added, and a standard PCR amplification is performed.

The advantage of RT-PCR lies in its extreme sensitivity—theoretically, one mRNA molecule can be amplified to a quantity sufficient for visualization on an agarose gel after ethidium bromide staining. The drawbacks include the possibility of genomic DNA contamination, in addition to the concerns associated with standard PCR reactions (e.g., false positives, mispriming, contamination, etc.).

In the in situ hybridization technique, a labeled probe is hybridized to a target mRNA transcript in sections of tissue to permit the identification of individual cells containing the transcript. This technique is particularly useful in identifying subsets of positive cells within a given tissue in which the relative levels of specific transcript for the tissue as a whole might be too low to be detectable by other means.

4.3 Protein Expression

In addition to the foregoing assays for transcription of the transgene, various other techniques to determine the translation of the transcript into a protein product are often employed. Most of these depend on the availability of specific antibodies to the protein of interest. The immunoblotting technique, in which proteins are resolved on a polyacrylamide gel, transferred to a membrane, and detected by means of a labeled antibody, is useful in verifying the appropriate molecular weight of the protein product of interest. Although rarely used, the radioimmunoprecipitation assay can also serve in the determination of the size of the protein product and to allow for various other manipulations, including peptide mapping and isoelectric focusing, to verify proper modification of the protein. To identify which cells within a tissue contain the protein product of the transgene, immunohistochemical "staining" of tissue sections with a labeled antibody is often employed.

Some experiments involving transgenic animal production focus on measuring the influence of various cis-regulating DNA sequences; the transcript and ultimate protein product are for the most part irrelevant. In these cases, "reporter genes" are often employed because they can simplify determination of expression levels by producing a transcript or protein that is easily and unequivocally detectable. Particularly useful in these applications are reporter genes that code for novel enzymes in that the activities of the enzymes can be assayed, thus greatly increasing the sensitivity of detection.

In analyzing the expression of any transgene, it is always important to evaluate at least two separate lines of transgenic animals, to show that the expression patterns are consistent and reproducible. It is not uncommon for the site of integration within a given line of transgenic animals to have profound influences on transgene expression, independent of any transcriptional control sequences in the transgene itself.

5 CONCLUSIONS AND FUTURE DIRECTIONS

5.1 Gene Transfer Today

The efforts and costs for gene transfer experimentation in animal biology are challenging and limiting. Innovative technologies to enhance experimental gene transfer efficiency in different species are desperately needed. Such enabling techniques would not only bring the cost of individual projects into a reasonable realm but would also increase the likelihood of breakthrough studies in many disciplines. As stated earlier, aside from mouse modeling, DNA microinjection into preimplantation embryos has been the only reproducible means for heritable gene transfer in the preponderance of mammalian species, other than multigene-targeted (nonspecific) blastomere aggregation and nuclear transplantation studies. To date, no other laboratory has been able to repeat the sperm-mediated gene transfer efforts in mice and pigs reported in 1989.

The outlined efficiencies for the production of transgenic animals represent what is considered to be current "state of the art" technology. It is envisioned that procedures will be modified and enhanced as new research efforts are reported. Under the circumstances, there are a number of specific achievements that would significantly enhance experimental productivity. Enabling technologies would center on the following areas:

- development of alternate DNA delivery systems (e.g., embryonic stem cell transfer for stable gene transfer, liposome-mediated gene transfer, or perhaps targeted somatic cell techniques)
- the identification of optimal experimental conditions for a given form of gene transfer and the identification of animal strains best suited to the individual technologies
- complete animal genome mapping and identification of homologies to human genes
- the establishment of routine embryo and germ plasm culture systems
- the development of a means to reduce the number of animals and embryos required [e.g., use of PCR amplification of blastomere DNA, or fluorescence-activated cell sorter (FACS) analyses and gating of transfected germ plasm]
- the elucidation of mechanisms by which the 129 mouse strain is more suited to the maintenance of germ line competent embryonic stem cells and extension to other strains and species.

Primary considerations for genetic manipulations and gene transfer are not limited to donor and recipient requirements, surgical manipulations, embryo handling, and gene transfer skills, but also include expertise in tissue/cell culture and molecular biological techniques. For any animal experimentation, ethical considerations, regulatory requirements, and experimental design are also important considerations that may limit research efforts. For most species, husbandry considerations are extensive in comparison to mouse requirements, running the gamut from animal and record-keeping management to female- (e.g., controlled estrus cycles/synchronization) and male-specific (e.g., sperm analysis/quantitation) requirements. For most nonrodent species, procurement of embryos and the trauma associated with most manipulations are significant factors. The refinement of nonsurgical techniques for embryo collection and transfer would bypass a huge financial and labor requirement for many large domestic animal species. More importantly, experimental animals would no longer be necessarily subjected to significant surgical manipulations. Perhaps recent advances, involving embryo aggregation and the derivation of transgenic mice completely from early passage ES cells, may shed new light on novel and efficient techniques.

5.2 Future Directions

In looking to the future, the use of embryonic stem cells for mammalian gene transfer would be most advantageous. The ability to "knock out" or replace endogenous genes has been an extremely valuable tool in mouse genetics and modeling. Yet, characterization and use of ES cells for gene transfer has not worked to date in any species other than the mouse. Numerous laboratories with appropriate tissue and embryo culture background were successful in genetic engineering efforts with mouse models, and ES cell or "ES cell–like" lines were identified for other species. Yet, with hamster, rat, rabbit, and pig stem cell experiments ongoing now for at least 5–7 years, the first bona fide transfected stem cells and chimeric founder animals have yet to be documented or reported. Regardless of whether a genetic alteration or deficiency in the mouse genome is responsible for such a dramatic difference in comparison to other species, the need for tissue/culture technology in additional species is great.

For many species, there are technical drawbacks to both ES cell and gene transfer technologies that are manifest in the lack of significant gene characterization and mapping efforts. With the complexity of the mammalian genome, this information is critical for the determination of appropriate genes to engineer and transfer, and in providing preliminary indications of the potential biological consequences of genetic engineering experimentation. What are the appropriate genetic targets in different species and how will gene targeting manipulations influence mammalian development? If genetic polymorphisms (indicative of related genes that code for different isoform proteins) or pseudogenes are present, targeting efforts can be long and arduous. For well-characterized genomes, the identification of specific genetic variations or deficiencies may provide only a glimpse of the effort necessary to develop in other species. Yet, it is hoped that with current efforts focused on the human genome, accumulating data will prove quite informative and adaptable to other species, inasmuch as conserved linkage groups are highly probable and would be exploitable.

While current experimental efficiencies for transgenic animal production are relatively poor, the mechanical aspects of producing transgenic animals should not overshadow our ability to create and evaluate biologically important animal models. Much has been learned about early development and effects of altering the growth cascade as well as other physiological processes in initial transgenic animal models. These models have had far-reaching effects and have redefined what is possible in the biological, biomedical, and agricultural sectors. While studies of transgene expression and overexpression in mammalian models may not always correlate exactly, nor reveal all the possible effects and consequences when extended to human biology, the utility of transgenic animal models to scientific discovery cannot be overestimated. The use of ES cell or related methodologies to provide efficient and targeted in vivo genetic manipulations offers the prospects of creating profoundly useful animal models for biomedical, biological, and agricultural applications. However, the road to such success will continue to be most challenging.

See also Genetic Immunization; Mammalian Genome; Mouse Genome; Transgenic Animal Patents; Transgenic Fish; Transgenic Livestock as Bioreactors.

Bibliography

Brinster, R. L. (1993) Stem cells and transgenic mice in the study of development. *Int. J. Dev. Biol.* 37:89–99.

Hogan, B., Costantini, F., and Lacy, E. (1994) *Manipulating the Mouse Embryo: A Laboratory Manual.* Cold Spring Harbor Laboratory, Cold Spring Harbor, NY.

Pinkert, C. A. (1994) *Transgenic Animal Technology: A Laboratory Handbook.* Academic Press, San Diego, CA.

Wood, S. A., Allen, N. D., Rossant, J., Auerbach, A., and Nagy, A. (1993) Noninjection methods for the production of embryonic stem cell–embryo chimaeras. *Nature,* 365:87–89.

Transgenic Animal Patents

W. Lesser

1 **Introduction**

2 **Legal Considerations**
 2.1 Legal Status: Patentable Subject Matter
 2.2 Patentability Requirements

3 **Scientific Developments**

4 **Economic Factors**
 4.1 Economic Incentives
 4.2 Impacts of Patents on Prices and Industry Structure
 4.3 Royalty Payments
 4.4 Implications for Farm Structure
 4.5 Genetic Diversity
 4.6 Alternatives to Patents

5 **Ethical Issues**
 5.1 Public Attitudes Toward Animal Biotechnology and Patents
 5.2 Animal Welfare
 5.3 Animal Rights
 5.4 Devalued Humanness

Key Words

Animal Rights A belief that animals have inherent rights beyond protection from cruelty and unnecessary suffering.

Animal Welfare A requirement, legal and/or ethical, that the suffering of animals be minimized and, when unavoidable, be justified by significant human need.

Disclosure The patent law requirement that an invention be described sufficiently to be repeatable. For living organisms, a sample deposit is sometimes required. In U.S. law (35 USC Section 112), "The specification shall contain a written description of the invention, and of the manner and process of making and using it, in such full, clear, concise, and exact terms as to enable any person skilled in the art . . . to make and use the same."

Infringement The unauthorized use of a patented invention.

Intellectual Property Rights Refers to a group of laws (patent, copyright, trademark, industrial secrets) prohibiting the unauthorized use of the protected creation.

Nonobviousness (Inventive Step) The patent law requirement that an invention must not be a trivial extension of what exists.

Patent (Utility) A legal document describing an invention and granting the owner the right to prohibit use of the invention. Patents may apply to products, processes, or products-by-process.

Royalty A fee charged for the use rights to a patent.

Scope The range of similar products, processes, or uses implicitly covered by a patent.

The term "animal patents" refers to the extension of utility patent protection to higher animals. Patents are intended to provide incentives for private investment in research and development. Such legal protection can be especially important for animals, many of which are self-regenerating. To date only a few animals, mostly laboratory specimens, have been patented in the United States, Europe, and Australia. Multiple pending applications are largely in the areas of animal disease models, livestock, and animal producers of pharmaceuticals. While the practical effects to date are limited, the field has major potential impacts, raising broad-based but varied concerns about animal welfare and rights, ethics, and economic considerations, especially regarding small farms, and the collection of royalties on progeny. There are no general agreements on how best to proceed, inasmuch as concerns tend to be individual, but a majority of the public across several countries appears to be generally supportive of animal biotech and patents. At present, only the United States expressly permits animal patents. Ten countries tacitly allow them, and 65 have exclusionary language, but interpretation of statutes (e.g., those in Europe) can be complex.

1 INTRODUCTION

The U.S. Patent and Trademark Office (PTO) created a flap when, on April 7, 1987, the commissioner declared higher animals to be patentable subject matter. The news release said, in part, "The Patent and Trademark Office now considers non-naturally occurring non-human multicellular living organisms, including animals, to be patentable subject matter." The announcement was all the more surprising because it did not result in the granting of a patent; the distinction of the first ever animal patent went on April 12, 1988, to a laboratory mouse with a heightened susceptibility to cancer.

Several additional animal patents, similarly for medical purposes, were granted subsequently. These events brought an outcry from groups including animal rights advocates and "family farm" representatives; in the Congress, there were eight bills calling for amendments and/or a moratorium on patenting animals. No legislative changes resulted, and the issue has for the present, at least, largely slipped from public attention, even though an animal patent was subsequently granted in Australia.

This article describes the legal status of animal patenting worldwide and explores the likely consequences in the areas of scientific developments, economic issues (including the payment of royalties for progeny), and ethical matters (incorporating public policy and animal welfare/rights).

2 LEGAL CONSIDERATIONS

2.1 Legal Status: Patentable Subject Matter

Patentability of higher animals is not clarified in many countries worldwide. Nations that, as of 1996, have granted a patent for an animal are the United States and Australia (1 grant). The first multinational patent is for the same animal, the "Onco-mouse" (alternatively known as the Leder or Harvard mouse). The patent* is for:

> A transgenic non-human eukaryotic animal whose germ cells and somatic cells contain an activated oncogene sequence introduced into the animal, or an ancestor of the animal, at an embryonic stage.

This genetically engineered animal is useful for medical research.

In addition to these three areas, the patent laws of six larger/wealthier countries (Japan, Australia, New Zealand, Brazil, Greece, and Turkey) do not expressly prohibit animal patents. Several former Eastern Block countries (Czechoslovakia, Hungary, Bulgaria, and Rumania) permit protection for animal varieties, but the forms of protection are more akin to plant breeders' rights than patents (see Section 4.1.2). The current status in Russia is unclear. In contrast, the laws circa 1990 of some 65 countries expressly exclude the patenting of animals using terminology similar to the following:

> plant or animal varieties or essentially biological processes for the development of plants or animals

This is Article 53(b) from the European Patent Convention (EPC), which has been incorporated into national laws of the member states, or entered via the 1979 draft of the World Intellectual Property Organization (WIPO) model law for developing countries. Additionally, the EPC excludes from patentability those inventions which are "contrary to 'ordre public' or morality" [Article 53(a)]. The treatment of that article remains in active debate following the March 1995 provisional revision of the directive of the European Union by the European Parliament. Finally, signatory countries to GATT under the Trade-Related Aspects of Intellectual Property Rights (TRIPS) section (GATT MTN/FA II-A1C) have committed to enhancing patent protection, but may exclude "plants and animals." Thus, in most countries, the status of animal patents will remain limited.

Article 53(b) notwithstanding, in 1991 a patent was granted for the Onco-mouse in Europe. Technically the issue involves "patentable subject matter." All countries explicitly disallow some forms of invention. The United States, with few exclusions, prohibits patents detrimental to national security, while many countries, as noted, exclude plants and animals in some forms along with pharmaceuticals, surgical procedures and, less commonly, food products.

In the United States, the Onco-mouse application was initially rejected as unpatentable subject matter on the grounds of being "controlled by laws of nature and not a manufacture by man that is patentable." U.S. patent law requires that a patentable product invention be a machine, manufacture, or composition of matter. Upon appeal, the internal Patent and Trademark Office Board of Patent Appeals and Interferences (BPAI) in *Ex parte Allen* referred to the famous 1980 U.S. Supreme Court decision in *Diamond v. Chakrabarty* in deciding that the Congress intended patent protection to be broadly applicable under the terminology "anything under the sun that is made by man." The Chakrabarty decision itself applied to microorganisms, but in *Ex parte Hibberd,* the BPAI had in 1985 extended protection to higher plants.

The patentable subject matter issue in Europe is complicated by the existence of Article 53(b) and, indeed, the initial application was rejected on those grounds. Moreover, the Examining Division concluded that the legislators intentionally had excluded animals as such, rendering them unpatentable subject matter.

On appeal in 1990, the Technical Board of Appeals reached four wide-ranging conclusions and referred the matter back to the Examining Division:

1. Animals were not intended to be excluded as a class, only animal varieties. One factor contributing to this decision was the absence in Europe of other forms of intellectual property protection for animals. [Plant varieties are protectable under plant breeders' rights or, technically, the Convention for the Protection of New Varieties of Plants (UPOV, to use the French acronym). UPOV gives patentlike protection for whole plants but contains several additional exemptions, or provisions for exemptions, which reduce the level of protection compared to utility patents.]
2. The pertinent issue is whether the subject matter of the application is an "animal variety."
3. That the descendants are likely to be the outcome of a biological process is not a sufficient reason to reject the application, for the reproductive process could be other than sexual.
4. The contention that patent law was not the proper legislative tool for regulating ethical problems was rejected because "precisely in cases of this kind there are compelling reasons to consider the implications of Article 53(a)." The decision "would seem to depend mainly on a careful weighing up of the suffering of animals and possible risks to the environment on the one hand, and the invention's usefulness to mankind on the other." Such arguments were used to reject an application for an animal model used for screening hair growth stimulants, although a systematic delineation for "undue" has yet to emerge.

A more recent decision regarding plants treats all transgenic plants as "varieties" which are excludable from patent protection. How this decision would apply to animals (for which there is no varietal protection) and whether it will stand at all is unclear at this time.

The legal position of animals as patentable subject matter is not

* U.S. Patent 4,736,866, April 12, 1988.

clarified in other countries where applications have yet to be acted upon. In general, where the law does not expressly prohibit animal patents (e.g., as in Japan), granting seems likely in the foreseeable future. The interpretation of ethical issues under the patent acts, however, remains unclear. Where the legislation expressly prohibits patents for animal varieties and biological processes, the outcome is far less evident. An interpretative process similar to that used by the EPO may be employed. On the other hand, many of these countries, which are predominantly developing, have a far more skeptical view of the domestic benefits of intellectual property protection. These concerns are heavily directed to agricultural and medical applications, which happen to coincide with the major uses for patented animals (see Section 3). In such an environment, a broad extension of protection to animals seems unlikely.

2.2 Patentability Requirements

2.2.1 Novelty, Utility, and Nonobviousness

The delineation of patentable subject matter merely demarcates what classes of inventions are protectable. Applicants must meet the patentability requirements of novelty, utility, and nonobviousness ("inventive step" in the EPC terminology).

Novelty is a mandate that the invention be new—not previously known or used. (Different degrees of novelty exist in different countries, ranging from the EPC requirement of absolute novelty to the one-year grace period allowed in the United States. Japan falls between these extremes. These distinctions have ramifications for how openly research can be conducted and reported but are not fundamental to the issue of animal patenting.) Society will not allow a reward for what is already known. The novelty search is the first substantive step in an application review and usually incorporates, but is not limited to, published materials including other patents.

Utility requires, in essence, that the invention function in some specified way for a particular use—in short, it must work. Utility is not a mandate that a patented invention be efficient or economically viable, or better in some practical way. Patent offices, for good reason, do not attempt to judge practicality or sales potential. That is done by the market. At another level of specificity, the concept of utility is more complex than this. Through the nineteenth century, when agricultural applications were generally not patentable, it was a component in defining patentable subject matter. On a broader scale, utility helps to distinguish (patentable) inventions from (unpatentable) discoveries; Straus has discussed this area with respect to WIPO.

Nonobviousness is the stipulation that an invention go beyond what is presently known, be innovative, and not "be obvious at the time the invention was made to a person having ordinary skill in the art." The nonobviousness requirement, in conjunction with the novelty search, is the key determinant of the scope of an invention. The invention's scope defines how different another product must be to avoid infringing on the patent. The term "doctrine of equivalents" is also used to describe scope.

2.2.2 Scope

In examination systems (which include those of the United States and the European Union) the patent examiner, working within past practice and court decisions, delineates the scope of an invention. The scope of protection can vary considerably across inventions, being quite narrow for older, specific inventions like pipe fittings and broad for new areas of endeavor, such as patented animals. The process begins with the development by the applicant of claims setting out what the invention is. Typically an application will have multiple claims, the Onco-mouse having a total of 12. Often these claims begin with the general and narrow to the particular; the first claim of the Onco-mouse application reads to "transgenic nonhuman mammal," while Claim 11 refers only to rodents and Claim 12 to mice. According to patent office practice in many countries, the examiner must document why a claim should be disallowed. Since, however, new areas like transgenic animals offer little documentation for use by the examiners, pioneering inventions tend to be broader in scope. In the case of the Onco-mouse, where the product was shown only in mice, disallowing other species would have required documentation that the claimed approach would not work or would be unreliable in other species. In the absence of such documentation, the claim to all nonhuman mammals was allowed. Subsequently, the PTO has viewed genetic engineering on animals as unpredictable, with the likely result that claims will be restricted to species on which transformations have been achieved.

2.2.3 Disclosure

A final requirement for patenting is that the invention be disclosed (described) with sufficient detail that it can be recreated. The underlying concept is for patents to be a source of technical information allowing further advances. Moreover, this technical teaching allows the invention to be copied more readily as soon as the patent lapses. To satisfy these requirements, the invention must be repeatable, within a reasonable time frame and at reasonable cost, on the basis of the description. The use of written descriptions only has proven problematic for living organisms because their use alone often requires costly and time-consuming "undue experimentation" in recreating the invention, an issue that has reached the courts in Germany and the United States. It became common for microorganisms and some seeds to require the deposit of a sample as a complement to the written disclosure. Deposits are sensitive because with them the invention can be copied readily; the sample is the invention.

For animals, deposits are doubly controversial because of the technical complexity involved. No deposit was required for the Onco-mouse patent, but this may change for future awards. In 1989 Foote concluded that the use of frozen embryos had proven practical with rodents and was likely extendible to other mammalian species.

2.2.4 Infringement

The use of a patented invention without permission is infringement. A suit and burden of proof is the responsibility of the patent holder. Two aspects of infringement are of particular relevance to patented animals: the research exemption and the legality of producing offspring.

The research exemption refers to the legal use of a patented article for subsequent research and possible improvement. Since much patent legislation lacks language establishing a research exemption, this is a matter of interpretation by the courts. The principal U.S. case applies to the use of a patented pharmaceutical product for "reverse engineering" to permit production upon the expiration of the patent. In 1984 a federal court found that use impermissible, while allowing experimentation for nonpractical reasons (*Roche Products v. Bolar Pharmaceutical Co.*). At issue now is whether that decision prohibits applied/practically oriented research on patented articles as well. Bent argues that the Roche decision was not dictum, meaning that a limitation of research to the nonpractical is not binding on the grounds of having exceeded the scope of the decision. However, this interpretation is not universally accepted, so the matter is far from clear.

Of broader importance is the production and use of offpsring. In general, it is an infringement if a patented product is used to recreate itself. The legal principle is fair use; a patented invention comes with an implied license to use the invention for its intended purpose. With animals this issue becomes complex, for there are different use classes, especially among agricultural users. The purchaser of a meat animal would not normally be expected to breed it, while someone paying tens of thousands of dollars for a prize brood bull would have every expectation of breeding use. The implied license would cover these cases. More complex is the case of the dairy cow, which must be bred annually (with her daughters used for replacements) to fulfill her purpose of producing milk. These are uncertain areas of the law and, while calls have been made for clarification, a pragmatic approach is also needed (see Section 4.3).

2.2.5 Proposed Amendments

Following the *Ex parte Allen* decision in 1987, eight bills were proposed in the U.S. Congress to limit animal patents. Four called for moratoriums (one year to permanent), but none passed. Subsequently the PTO agreed to a voluntary 8-month moratorium. The House in 1988 passed the Transgenic Animal Patent Reform Act (H.R. 4970) but it failed to emerge from the Joint Committee and did not become law. It was reintroduced in 1989 (H.R. 1556), with the same result. H.R. 4970 called for three amendments. One allows the PTO to accept biological deposits for disclosure purposes (see Section 2.2.3), while another prohibits the patenting of human beings. These two clauses largely confirm current practice. The third amendment would have allowed "farmers" to produce and sell offspring without paying royalties (see Section 4.3). At this time, further legislative action in the United States seems unlikely. However, groups like the (U.S.) Animal Legal Defense Fund are attempting to use the courts to impose an animal patent moratorium.

3 SCIENTIFIC DEVELOPMENTS

Any detailed discussion of current developments potentially leading to patentable animals would be quickly dated. Nonetheless, a brief overview of the major directions of current research will serve as an indication of the types of patented animal likely to be seen in the future. A knowledge of the species and uses of patented animals will, further, be helpful in projecting the socioeconomic consequences of animal patenting (Section 4) and, in part, the ethical issues involved (Section 5).

Here attention is placed solely on transgenic techniques of developing patentable animals. In the United States, at least, there is no inherent reason for not, at some point, patenting unique crosses created through traditional breeding practices, artificial insemination, embryo implants, and embryo splitting, among other techniques. Patentability of traditionally bred animals in Europe is uncertain, as indicated in Section 2.1. In any event, conventional measures are unlikely to meet the inventive step requirements for patents (Section 2.2).

In the early 1980s, a groundbreaking study on animal biotechnology by Palmiter and co-workers was the successful introduction of growth hormone genes into mice. This study has particular relevance because it included two dominant aspects of research continuing to the present: the search for successful growth enhancers and the frequent use of mice and other rodents. Mice, which are convenient for reasons of cost and because they reproduce rapidly, are used extensively in the development of animal disease models. Animal models serve as experimental animals for medical and related research. Indeed, the bulk of patented animals to date (Section 2.1) are such models.

Animal models have several attributes beneficial to researchers. They have a proclivity to specific diseases, which reduces the number of specimens needed. Equally important, they are often inbred (especially mice); hence their genetic makeup is known and stable, a helpful feature when investigators need to distinguish between results attributable to a treatment and those specific to an individual. With these benefits, not surprisingly, many new or improved models are under study for a variety of diseases. Particularly sought after are those for several forms of cancer. Hereditary diseases like sickle-cell anemia and forms of arthritis, as well as debilitating illnesses such as Lesch–Nyhan disease, are a second focus for developing animal models. The most intense search is for a nonprimate model for AIDS. Many of these developments will also apply to curing illness in animals.

Growth enhancements are useful particularly for agricultural livestock, to speed growth with concomitant reductions in feed usage and duration. To date, most attention has focused on the incorporation of genes and promoters, producing additional amounts of growth hormones (somatotrophin). Injected somatotrophin has been found to increase milk production up to 25% and to reduce feed usage and fat content of cattle and hogs. The genetic incorporation of growth hormone has been accomplished on an experimental basis in fish and poultry (one such superchicken, the Macro-Chicken, was the object of a European patent application), and less successfully in cattle and pigs. At an earlier level of development are genetic manipulations conferring resistance to diseases of animals and changes in product traits. As examples, efforts (not practical to date) are under way to increase wool production and to change the structure of milk sugars to make them digestible by lactose-intolerant individuals. Cold-tolerant fish have already been created. For the more distant future, numerous other genetic changes have been mentioned and seem within the realm of possibility, as reported in a volume edited by Fesseden MacDonald. The recent use of a baboon's liver as a temporary transplant organ, for example, raises the possibility of genetically engineering animals (e.g., pigs) to produce human donor organs.

The third major use of genetically altered animals is as producers of pharmaceutical and related products. An early effort involved the excretion in sheep's milk of human factor IX, a compound whose deficiency causes hepatitis B in males. Goats have also been used effectively, and one of the patented mice produces human β-interferon, a natural defense against viral infections. Such efforts represent a search for lower cost production methods. The relatively small quantities of product needed means that the number of animals involved will be small compared to any food use.

The congressional Office of Technology Assessment (OTA) found in 1989 that developments to date involved relatively few gene constructs, a very small percentage of the 50,000–100,000 genes making up higher animals. Hence, each individual genetic manipulation makes only limited changes to the sum of appearance and behavioral characteristics that define an animal species.

4 ECONOMIC FACTORS

Two broad justifications exist for the granting of patents and other forms of intellectual property rights. These are inalienable rights, in vogue through the eighteenth century, and economic incentives. Under the natural law concept, an idea belongs to its creator because it is an evident aspect of the creator's personality or self. The incentive concept is more pragmatic, allowing the private inventor to benefit financially from the invention as an incentive to investing in research and development (R&D).

Thompson (1992) reviews the philosophical bases of property, classifying them into three theories: the natural law, the instrumental theory, and the labor theory. In the seventeenth-century writings of natural law philosopher John Locke, the three theories meld into a principle of social utility. Under natural law, according to Locke's First Treatise, God confers the right of property to mankind in common. That, however, raises the necessity to distinguish between property held by individuals and that held jointly. Here Locke refers to labor, reasoning that self-ownership implies ownership of one's own labor. Locke said, in addition, that labor can be sold directly, or indirectly through the creation of a salable commodity, including an invention: "[whereby the individual] hath mixed his labor with and joined to it something that is his own, and thereby makes it his property." Disallowing the right to control the use of property created through labor is equivalent to losing control over the labor itself, to not being free.

What, though, about the appropriation of the public good for private use, such as the fencing of public land for private use? Locke took a utilitarian position: "for the provisions serving the support of human life, produced by one acre of enclosed and cultivated land are . . . ten times more than those which are yielded by an acre of land of an equal richness lying waste in common." This social utility position underlies the economic concept that society benefits overall when exclusive ownership is granted to intellectual creations. However, like the fencing of public land, the allowance of exclusive private rights also entails social cost.

4.1 Economic Incentives

The fundamental economic concept of patents, which was described by Machlup in 1958, is a simple one: the allowance of a limited, temporary monopoly right to the inventor, during which period he or she may attempt to commercialize the invention free from direct copying. The intent is quite straightforward, although its implementation becomes complex. The inventor, it is reasoned, bears the cost of discovery and improvement. Once the item or device in question has been discovered and proven, the cost of copying is a fraction of the amount already spent; thus the inventor cannot compete against the imitator and, in the absence of legal protection from direct copying, has no incentive to invest.

What patents allow is protection from direct copying for a temporary period during which the inventor may profit. These are solely negative rights—the right to prohibit others from using or selling the invention. The patent itself gives nothing and is worth nothing; all remuneration must come from the market through sale price or royalty fees.

In addition, patents are intended to serve several other objectives:

- to recognize the inventor
- to serve as a storehouse of technical information
- to provide an incentive for the full and early disclosure of inventions

The "technical teaching" aspect is the direct result of the disclosure requirement (see Section 2.2.3).

4.1.1 Theoretical Treatment

Economists have attempted to determine the optimal patent duration and scope both theoretically and empirically. Not surprisingly, the optimum varies by the type of technology, or, in more economic terms, by the demand elasticity and significance in cost reduction. Thus, little practical guidance is available. Economists have also examined the "winner take all" aspect of patents, leading to what is referred to as the patent race. The race arises because there is no benefit to being second, and "losing" is feared to increase costs and waste effort. On the other hand, more entrants and distinct approaches increase the likelihood of a quick, successful outcome. Again, economic theory does not identify a general resolution. But perhaps the matter is mooted by Dasgupta's observation a decade ago that technological competition is an ongoing process from which firms are absent at their peril. This is in contrast to the discrete decisions implicit in patent races.

Economists have subdivided of the inventive process into three stages:

- invention
- innovation
- diffusion (commercialization)

Studies indicate that the basic creative process (invention) is driven as much by the creative urge as by financial incentives. Patents are important in the long, tedious, and costly process of making the initial invention commercially viable. Thus, patents apply to the "99% perspiration" phase Edison referred to when describing the inventive process.

4.1.2 Empirical Measures

What is known about the fundamental question, How effective are patent laws in enhancing private investments in R&D? Here there is partial empirical evidence. Typically, "before and after" comparisons are used which assume that the only difference in the "after" period is the imposition of intellectual property rights. Before-and-after analysis can be used for plant breeders' rights (PBR) because they are recent, dating to only 1970 in the United States. That approach is obviously invalid for patents, however, which go back two centuries. Indeed, the first U.S. patent act, dated 1790, is one of the earliest modern laws.

PBR is the informal term used to describe the Plant Variety Protection Act of 1970, the U.S. adaptation of the International Convention for the Protection of New Varieties of Plants (UPOV, introduced in Section 2.1). PBR apply only to plants, and two additional exemptions, known as farmers' privilege and breeders' privilege, can be considered to lessen the scope of protection compared to patents. Moreover, at least in the United States, the scope of protection is narrower than for patents.* Taken together, these differences indicate that the incentive effects of PBR are weaker than for patents.

The impact of PBR in the United States has been analyzed several times. Butler and Marion examined the impacts of PBR on private firm investment in plant breeding. Plant breeding of self-pollinating varieties had previously been a public domain activity. Self-pollinating varieties, which include the staple crops of wheat, cotton, and soybeans, replicate themselves from saved seed. As a result, any farmer or propagator becomes a competitor, which means that the original breeder cannot recover the investment.

Butler and Marion conclude that private R&D expenditures accelerated rapidly in 1967–1970 in apparent anticipation of the passage of the Plant Variety Protection Act. Subsequently, expenditures kept pace with the increase in seed sales. Numbers of private firm breeders employed and new or enlarged breeding programs expanded in a similar fashion. Butler and Marion found that the sharp increases in R&D expenditures between 1967 and 1970 were largely concentrated in research facilities and nonpersonnel costs. Between 1970 and 1975, however, the number of plant breeders per firm increased significantly. Yet growth was uneven across crops, with soybeans receiving the majority and minor crops little or none. An explanation of the intercrop distribution in terms of economic and agronomic factors was published in 1983.

Subsequent studies confirm these findings. Brim reported in 1987 that Ph.D. soybean breeders employed by private firms increased from 6 in 1970 to 70 in 1987, while Foster and Perrin determined that private breeding programs had increased from one in 1970 to 34 in 1988. Questions about the productive merit of these programs were answered in part by the analysis of intercrop distribution just alluded to, which showed private protected varieties to be marginally more productive than public varieties. More recently, private varieties have dominated acres planted for several major crops, a further suggestion of productive merit. Recent evidence from Latin America concurs with these earlier findings.

Thus there is support for the theoretical prediction that the legal protection provided by PBR does indeed lead to increases in private R&D. Recognizing that self-pollinating plants and animals share the ability to recreate themselves,† by extension it can be predicted that animal patents will increase private investments in that sector. The value of patents, however, should not be overestimated. Most executives rate patents, overall, as a noncritical factor in making R&D decisions, although patents for animals are likely to be more important than average. In an empirical study published in 1991, Schankerman found that patents generate about a quarter of the private returns to inventive activity—not inconsequential, but not dominant either.

* In U.S. law, the farmers' privilege explicitly allows farmers to save and replant seeds, while the breeders' privilege allows competitors to use protected plants for breeding new varieties. Patented inventions appear to be usable in research also, but the right is not a statutory one (see Section 2.2.4).

† This is a simplification, inasmuch as the inheritability of traits is determined by Mendelian principles and cannot be predicted with complete confidence except for the simple case of a single gene carried by both parents. However see Section 4.3 for a discussion of the practical need to have a high inheritability rate for many commercial products.

4.2 Impacts of Patents on Prices and Industry Structure

Patents are a legal granting of partial monopoly rights. Market economies abhor monopoly because of the potential for high prices and low efficiency. Moreover, if large firms grew disproportionally, the problems would be enhanced. Given the very limited experience with patented animals, no real evidence for or against these positions is available. However, very similar concerns have been raised regarding plant breeders' rights. Butler and Marion, who also examined seed prices and sector concentration in the aftermath of PBR, made the following conclusion regarding prices: "PVPA has resulted in slight to substantial increases in the seed prices of protected varieties" (1985, p. 60). The pertinent question, however, is not the absolute level of increase (for the intent is to raise returns), but price relative to the productivity. In that regard, Butler and Marion found that farmers are very conservative, typically willing to pay only 25–50 cents for a dollar of expected profit. The retained profit can be considered to be a "risk premium" to cover weather and other uncontrollables. It is likely that livestock farmers using patented animals will behave similarly, indicating that the opportunity to raise prices is likewise limited. The situation for laboratory animals and other product classes is less well clarified.

As Lesser and Masson (1985) emphasize, the price issue is one of competition, essentially the availability of unpatented close substitutes. For patented animals (lab specimens, agricultural livestock, pets, racehorses, etc), the competitive situation varies. For pets and racehorses, every owner is a potential competitor. Livestock breeder numbers vary by species and class from tens of thousands for beef cattle to a half-dozen artificial breeders for dairy cattle to a dozen poultry breeders to perhaps three breeders of "synthetic" hogs. Yet even when the number of firms is small, sales are by production merit (e.g., as in agricultural breeding stock, which is typically sold with performance measures such as milk yield and gain rate calculated on a statistically valid basis, to provide a clear basis for selecting and valuing purchases). This practice, which has its analogue in seed sales, makes it difficult for patent holders to charge "too much." Nonetheless the availability of substitutes remains critical. The public sector has long served this function for agriculture, and Butler and Marion urge a continuing role for public plant breeding. The same recommendation seems prudent for livestock.

Structural issues regarding patenting revolve around the question of whether large firms have certain inherent advantages over smaller ones (e.g., capital formation, spreading risk over several projects, attracting the most talented researchers with higher salaries, size economies in research). Economists have evaluated these issues with no conclusive results. Theoretically, a strong case can be made for and against a bias toward large firms. Empirically, the results are middling; the literature has been reviewed by Scherer. However, the results vary greatly by industry, hence say little about individual sectors.

Experiences with seed breeders indicate some concentration of certificates of plant variety protection. Concern regarding firm concentrations is likely to apply to animal patents as well. General experiences with biotechnology indicate that large firms, while more bureaucratic and restrictive at the research level, have the funding and expertise often necessary to meet regulatory approval and other commercialization requirements. Thus, a degree of specialization can also be anticipated with patented animals.

The preceding discussion is particularly applicable to changes in animals that affect the efficiency of production but leave the basic product unchanged. Another class of changes would affect the product characteristics directly and could lead to very different marketing arrangements and industry structure. A hypothetical example is polyunsaturated animal fat. The patent holder may wish to market such a revolutionary new product under a brand name, promoting the product heavily to consumers. Such a strategy could be expected to increase the concentration in meat marketing, since, as in other food sectors, it is the large firms that do the bulk of advertising.

The meat consumer need not purchase the new, higher priced product, however, for the traditional one will certainly remain available. In this way, the consumer of traditional products is no worse off, while those who choose the new kinds have a choice not otherwise available. Such a brief overview is simplistic, admittedly, since it assumes that the branding of fresh meat products will continue to be largely unsuccessful and that the meat packing industry, while moderately concentrated, will remain a low margin, highly efficient operation. At minimum, however, it indicates there is no obvious reason to believe that patenting will be unambiguously deleterious for producers and consumers (see also Section 4.4).

4.3 Royalty Payments

One of the most contentious issues, especially for farmers, is the payment of royalties. The basic legal issues are of "fair use" and "implied license" (see Section 2.2.4), but they are encumbered in the case of self-reproducible organisms. When a farmer's cow produces a calf, is that legal and is a royalty payment due? Concerns have led some farm groups to request a disallowance of royalty payments. Other users of patented animals would be affected as well, but their position on this matter, if any, is not recorded.

A particular complexity with agriculture is the sheer size of the sector. According to the U.S. Department of Agriculture, there are in the United States alone some 1.2 million farms with some livestock and in 1995 a total of 110 million head of red meat cattle and more than a billion chickens on farms. Clearly, any royalty payment system must be simple to ensure that it is functional. One possibility is to make animals sterile, or hybrids, so that progeny are absent or have less value. That is feasible with high-reproducing chickens ($\approx$ 240 chicks/hen/year) and even hogs (up to 20 pigs/sow/year), but not for cattle, which reproduce at the rate of one calf a year. Since the success of genetic manipulation is very low—no more than 2%—a strategy of creating transgenic livestock in the laboratory would be wildly impractical. Moreover the delineation between production and breeding stock is not clear within the red meat species, since the brood stock varies with price in accordance with livestock cycles. Finally, the likely existence of multiple (but unknown) patented traits within the same animal is a further complication for royalty collection.

All this argues for a system of royalty collection at a concentration point (e.g., slaughterhouses), as discussed by Lesser (1994). To be workable—that is, to have the system not cost more than the value of the innovations—some averaging of value will be needed when several patents are involved. Royalties may be dispersed according to the sample-based percentage of animals they are found in. And farmers, for planning purposes, must know the fees due prior to production. From a patent law perspective, this system would be close to a compulsory license system under which applicants have the right to the invention contingent on payment of the royalty. Because patents work by exclusion, compulsory licenses are anathema to the system, but the sheer volume and intricacy of livestock enterprises, combined with the low unit profit, rule out more individualized payment systems. Hudson, for his part, calls for new patent law language to apply to this problem. It is noted parenthetically that the average net profit for feeding cattle in the United States is on the order of $30. In general, agriculture is a low net profit industry, dependent on efficient management and handling systems.

4.4 Implications for Farm Structure

"Farm structure" refers to the number and size distribution of farming operations. The industrialized countries became predominantly nonagricultural prior to World War II, and the farm segment has continued to decline as a result of intensification and specialization. Presently in the United States, where farm production is not controlled, less than 5% of the population are farmers. Total farm numbers remain relatively high (>1 million), but the bulk of production comes from the 15% of very large operations. This decline in farming has raised concerns from proponents of "family farms," while removing the economic base for many rural areas. Against this background, it is not surprising that patented animals are associated with added concerns about farm structure and rural life. Here are two short excerpts from the 1987 hearings before the U.S. House of Representatives Committee on the Judiciary, regarding patents, the Constitution, and transgenic animals:

> "[T]he patenting of animals will almost certainly lead to an increased concentration in livestock breeding and production."
>
> "[I]ndividual farmers will merely become wards of Wall Street and the biotechnology establishment."

What can be said about the validity of these statements?

Much of the consolidation of farming, beginning in the 1930s, is attributable to size economies made possible by mechanization. With tractors widely available, farmers could manage more acreage, and average farm size grew while farm numbers declined. Ongoing consolidation appears to be more related to the low unit profit of farming, necessitating that many units be produced for the farmer to obtain a living wage. Patented animals provide yet another case to consider.

Patented animals can be added without making fundamental changes in operations, as was required when horses were replaced by tractors. This suggests that limited size economies are involved. However, while operations need not be changed, management must be. Patented animals are expected to be both more costly and more productive. For these reasons, a higher level of management is required, and in agriculture, as in all enterprises, management talent is limited. Management is particularly critical when several innovations, including transgenics and computerized records, are combined. In brief, well-managed farms grow, and the less competently administered ones disappear, leading indirectly to somewhat greater concentration. Overall, the effects are likely to be a slight accentuation of existing trends rather than a new departure.

The preceding discussion is applicable to changes in the efficiency of production—that is, changes that will not lead to patented products. Patented products, such as a hypothetical "no cholesterol" meat, likely to be sold under brand names, would lead to different structural adjustments in production. The Flavr Savr™ tomato is an example in the produce area. Marketers will need a steady supply of quality meats, while producers must be assured of a market. Traditional livestock exchange systems, which rely on markets for coordination, have been found to be inadequate when these additional needs are imposed. Other commodities, including broiler chickens and processing vegetables, changed years ago to elaborate contract arrangements. In all likelihood, patented livestock, too, would be produced under contract. Contract production changes the nature and particularly the risks of production but does not directly affect structure. Presently, under 5% of U.S. beef cattle were produced under contract (custom feeding).

When discussing the implications of animal patenting on farm structure, it is important to emphasize that the effect comes principally from technological improvements, not from the existence of patents per se. That is, animal biotechnology would affect farm structure even if patenting were not an issue. In fact, Milligan and Lesser argue that concentration of production could be even greater in the absence of patents than with them. Their reasoning recognizes that profitable inventions are likely to be utilized, but control of production would have to be different. If inventors had to use contracts rather than patents (see also Section 4.6), they would, for ease of monitoring, choose large operators over multiple small ones. Small farmers would have unequal access. And, if the production benefits of the new varieties were sufficiently great, the small farms would eventually go out of business. Thus, patents, by facilitating broader access, might limit farm structure change over what could be expected in the absence of patents.

4.5 Genetic Diversity

Selective breeding is an important aspect of the industrialization of agriculture. Scientific breeding has evolved into a process based on careful record keeping and known hereditary rates for important traits. However, this approach leads to the continued improvement of a limited number of elite lines, causing, over time, a restricted genetic base. (Reference here is to agricultural livestock and other pedigreed animals including dogs. For laboratory animals, genetic uniformity within lines is one of the key requirements for experimental use.) The issue is best documented for plants, but these considerations apply also to livestock. With artificial insemination, for example, high producing bulls can sire thousands of calves, and embryo transplants greatly magnify the offspring from genetically superior dams. How does uniformity relate to patents?

Certainly, members of the public make that association, as was done with plants. Indeed, at least two witnesses before the 1987 House Judiciary Committee hearings on transgenic animals cited a causal relationship between patents and monopolization by a few firms, leading to loss of variability within livestock breeds. While this is possible, an equally strong argument can be made for the opposite effect. Since genetic engineering greatly accelerates the process of introducing new genes, diversity may instead increase rather than decrease.

Altogether, the breeding of agricultural livestock is a complex task. For the relatively long-lived and slowly reproducing red meat species, the gene pool evolves slowly in the standing herd of tens of millions of head. Within such a system, concerns about additional genetic uniformity induced by patenting have not been substantiated at this time. Certainly direct comparisons with the short-lived and rapidly reproducing plant species are misplaced.

4.6 Alternatives to Patents

Patents are but one means of extending private control over inventions as an inducement to private investment. When an invention can be kept secret or is difficult to copy, the inventor has the options of using secrecy and/or contracting. Many patented animals cannot be kept secret, and, indeed, must reproduce themselves to be commercially viable (see Section 4.3), and thus great attention is not directed here to those options. However, animals producing pharmaceutical products are likely to be needed only in small numbers, hence are suited to alternatives to patents.

Secrets are protected under "trade secret" statutes which, like patents, are a form of intellectual property. Trade secrets allow prosecution when secrets are improperly revealed. Trade secrets have the advantage of being permanent, as with the Coca-Cola syrup formula, but there is no protection against those who would decipher the secret (by applying "reverse engineering"). Trade secret legislation is also less standardized—in the United States it is based on state law—hence is difficult to use. For the public, trade secrets are less desirable because the expertise is never shared.

Contracts are a flexible means of separating ownership from control and can substitute or supplement patents in a number of ways. For example, a firm might license a patented production process and use a contract to specify the market area. Or, in the absence of a patent, the inventor may use a license to specify the total agreement. In this regard, patent law can be described as a form of standardized contract law. Contracts, however, bind only the signatory parties. Third parties must be addressed under other laws. Thus, contracts are only partial substitutes for patents.

Economists have proposed broader alternatives to patent laws and have evaluated the ramifications under limited circumstances. One alternative receiving greater attention is the use of research prizes. The concept is to have the government (or other party) give a prize for a successful invention. From a social perspective this approach might be less costly overall, although it has the equity complication of having all taxpayers subsidize inventions potentially useful to only a portion of the public. Firms benefit because uncertainty is reduced to technical matters; market forces and risks would not apply. This latter matter is particularly significant for revolutionary developments, which are traditionally slow to be accepted. It remains questionable, however, how well a government agency could anticipate and recognize these developments. A prize system seems better suited to small innovations of obvious value, but the need for an alternative to patent laws in such cases is also reduced. Overall, their conceptual and practical limitations notwithstanding, no system superior to patents has been devised.

5 ETHICAL ISSUES

The patenting of animals has proved to be highly controversial. Part of the controversy was evident in the debate over bovine somatotropin (bST), a growth and milk production stimulant. Yet most of those concerns relate to farm structure matters and the cost of surplus milk production, issues that can be classified as basically, though not exclusively, economic. In many industrialized countries, the government supports milk prices by purchasing surplus cheese and other products; U.S. costs in some years exceed $1 billion.

On the noneconomic side, Consumers Union has raised the issues of herd health and safety of the milk supply.

Animal patenting has raised ethical, welfare, and moral issues as

Table 1 Support for Genetic Engineering on Animals by Country

Support (%)	Country	Date	Source
42	United States	1992	Hoban and Kendall (1992), Figure 6
54	Japan	1991	Macer (1992), Table 4-2
58	New Zealand	1990	Macer (192), Figure 4-1

well. In a 1992 review article, Hudson listed the ethical issues in five interrelated areas: (1) the proper realm of government activity, (2) animal welfare/rights, (3) speciesism, (4) devaluing human life, and (5) "species integrity." The latter four are addressed here.

5.1 Public Attitudes Toward Animal Biotechnology and Patents

Those expressing concern (or more) with the practice of genetically manipulating animals represent substantial numbers across several countries, and a majority in the United States. Available opinion surveys were conducted at different times using different methodologies, with the result that the results are not strictly comparable, but the message of limited support is nonetheless clear (Table 1).

The major reasons given for opposition were "unnatural, interfering with nature" (22% Japan) and "unethical" (35% New Zealand, 7% Japan). As is often found, the respondents who are better educated and younger (the two categories overlap) are the more accepting. Among the public, 59% in New Zealand and 49% in Japan are supporters of animal patents.

5.2 Animal Welfare

Animal welfare is defined here as the minimization of suffering and is a subset of animal rights (see Section 5.3). Some observers contend that biotechnology, as a newer and potentially speedier technology, enhances the potential for animal suffering. The example of some malformed pigs resulting from early experiments was presented in the 1987 congressional hearings. But, this example aside, Adler notes, "No scientific rationale, however, is offered for the belief that genetic manipulation is necessarily harmful to an animal's welfare" (1988, p. 37). Moreover, he emphasizes that conventional breeding practices can be less specific, hence more harmful to individual animals, than genetic engineering. Clearly, excessive inbreeding has reduced the vigor of a number of dog breeds.

The Animal Welfare Act of 1985 has been cited to distinguish the roles of the patent office and other legislation. Some suggest that the role of applicable federal and state legislation is to prevent undue suffering and that those interests are better served by legislation than through the Patent Office, which, in any regard, is involved in but a few sources of possible suffering. The same argument can be applied to the environmental issues raised by the existence of genetically engineered animals.

The terms "undue" and "unnecessary" applied with respect to animal suffering suggest that some level of suffering is socially acceptable, but how much and for what purposes? Certainly animals are routinely used to enhance human well-being, whether it be in the form of food, sport, companionship, labor, or testing purposes. The degree to which these forms of use are considered ethical is classified by Van de Veer in five levels. Of these the middle, referred to as "interest-sensitive speciesism," is adhered to by the "majority of people." Under this standard, animal welfare morally may be subordinated to promote basic interests of humans, but inflicting suffering on animals in pursuit of peripheral or frivolous goals is held to be immoral. Thus it is conceivable that a majority of the U.S. public would condone some animal suffering in the development of a cure for AIDS but would decry use of animals for testing cosmetics. Complicating matters is the differential treatment of animals at stages of complexity ("intelligence"); for example, what is acceptable for mice might be condemned for primates. Nor is this ambivalence recent; Brody traces the diversity and evolution of attitudes back to the sixteenth century. Donnelley, McCarthy, and Singleton emphasize that ambivalence on this subject exists within as well as among individuals.

Singer contends that the choice between human welfare and animal suffering cannot be defended in nonreligious ethical terms but observes that in considering public policy in a pluralistic society, no specific religious outlook should be selected as the basis for public laws. While Singer's position necessarily leads to a utilitarian standard, Macer argues that the cultural and religious views of many different peoples are more consistent with human beings having a higher moral status than animals. In a detailed treatment of animal experimentation in European-based societies, Orlans reaches a related position advocating compromise.

This all says that animal welfare is a complex and personal issue, lying largely outside the competency and control of patent offices. Moreover, it is important to reemphasize that patents do not give permission to do anything; their sole function is to allow the owner to prohibit use of his or her invention. Use often depends on other requirements such as, in this case, meeting applicable animal welfare regulations. Within such a circumscribed context, it is difficult to depict the granting of patents as governmental condoning of certain activities, yet Hettinger records that animal patents are sometimes opposed on that basis.

5.3 Animal Rights

Animal rights are an extension of animal welfare to the position that animals have inalienable rights beyond the minimization of suffering. As Hettinger describes it, "living beings come into the world with a claim to use their physical characteristics for their own benefit" and "The satisfaction of . . . biological functions constitutes the organism's own good" (1992, 16, 18). At its extreme, known as species egalitarianism, the antithesis of speciesism, we have a position that champions equal rights for animals and humans. As Brody notes, this is a radical position that implies major restructuring of life and food patterns, although the consequences are not discussed by proponents. No evidence is put forth that species egalitarianism is a widely held position.

Another position is "species integrity," which a witness at the 1987 hearings on transgenic animals described as follows: "The crossing of species borders and the incorporation of genetic traits from one species into the germ line of another represents a fundamental assault on the principle of *species integrity and violates the right of every species to exist as a separate, identifiable creature*" (emphasis added).

The implications of this position are not detailed. For example, is the creation of mules, the natural intermixing of germ lines of horses and donkeys, a violation of the integrity of those species? Mules, being sterile, do not constitute a species, so would genetic engineering of those animals, or any others that are nonfertile, be acceptable under this principle? The term "separate, identifiable creature" is also difficult to operationalize. It is possible, for example, to refer to Holsteins, a common breed of dairy cattle, as having integrity because they fulfill their biological function of producing milk; but the manipulation of that same breed to secrete pharmaceuticals into the milk might be said to somehow violate species integrity. Yet the appearance and behavior of the cows will be unaffected by the insertion of a few genes, and their milk can still be consumed.

For scientists, the identification of particular "species" is a complex undertaking. Classification involves the use of the entirety of characteristics, including normal behavioral patterns. Within such a system, the definition of "separate, identifiable creatures" is difficult and defies knowledge of evolution. Biologists refer to species as "reproductive communities," which are fluid over time and space (microevolution). Within this structure a species can be identified only statistically; certainly the alteration of a few (20 of 50,000–100,000) genes would be insufficient to change any genetic definition of a species. Moreover, it has been argued that the right of a species to exist as a separate, identifiable creature has no biological basis precisely because species are dynamic populations, not separate individuals with fixed attributes.

But this is not what the witness who introduced the notion of "species integrity" to the 1987 congressional hearings was talking about. Rather, this witness was referring to pigs as each of us knows pigs, and to fundamental beliefs about the essential nature of living beings. At present no body exists to mediate such basic differences in beliefs. Certainly the issue exceeds the scope and competence of patent offices, just as the basic issue is genetic engineering of animals, not their patentability status. One witness at the 1987 hearings summarized the debate by noting that "compared with the ethical issues involved in our breeding, buying, selling, confining, eating, and performing research on animals, the ethical questions surrounding animal patents seem relatively less important." Nonetheless, recent discussion using the public morality clause to reject patent applications places at least some patent offices (EPO, possibly Australia) in the position of having to mediate the contentious matter.

5.4 Devalued Humanness

The position that transgenic animals somehow devalue humanness has two facets, one being the use of human genes in other species (or animal genes in humans). Human genes are likely to be widely used, for they are better described under the Human Genome Project and other projects. Indeed, in one of the earlier transgenic animal experiments, researchers introduced human growth hormone genes into pigs. The transmigration of human genes raises questions about the definition of what it is to be human and how many human components may be required before another species becomes "human." But while these are important and interesting ethical and philosophical problems, they are not practical issues for the foreseeable future, if for no other reason than the inability of today's science to make such far-reaching changes. Rather than suggesting a specific action, a group of religious leaders who testified at the 1987 hearings called for the examination of the "unique moral, ethical and theological questions, such as the sanctity of human life" raised by the combination of human genetic traits with animals.

As for more limited gene transfer, there seems to be significant public support, especially in the medical area. The use of a baboon liver in a human organ transplant recipient, while unrelated to the patent issues considered here, does suggest public support for use of animal parts in humans.

The second and broader issue is the diminution in the value of life, including human life, as a result of bioengineering and patenting animals. Reducing animals to mere factories has been described as forcing a reductionist and materialistic concept of life on people alive today, and their descendants. More generally, the subjugation of the animal kingdom to the will of mankind is seen as unnatural and arrogant, a subversion of God's will. While the significant alteration of animal life to serve better the needs of people has gone on for millennia, patents have been associated because of the use of the term "composition of matter." According to arguments by Kass, reducing animal life to compositions of matter reduces the sanctity of human life as well.

The terminology "composition of matter" is drawn from U.S. patent law as one type of patentable subject matter. While many believe animals to be more than compositions of matter—since they are animate and sense pain and pleasure, Brody argues that they are at least compositions of matter, and it is in that measure that they are patented. A cow can be a patented producer of pharmaceutical products, but it is in other respects a cow. Hence in Brody's judgment there is no inherent relationship between patenting and strict materialism.

A fundamental and unresolvable societal issue relates to the degree to which factors related to life can be partitioned. If one lives in a world in which responsibility can be divided across agencies and an animate object can serve multiple functions while retaining some internal integrity, then the patenting of animals is a less compelling moral issue. However, if the viewpoint is holistic, such that any invasion of the functioning of a creature debases it, then animal patents are a part of the problem. Such major differences in beliefs and perception are virtually unresolvable in a pluralistic society. Yet, in a democratic system the majority appears to support a more utilitarian view of animals, the outlook identified as interest-sensitive speciesism. Animal patents are compatible with that attitude.

See also Transgenic Animal Modeling; Transgenic Livestock as Bioreactors; Vaccine Biotechnology.

Bibliography

Adler, R. G. (1988) Controlling the applications of biotechnology: A critical analysis of the proposed moratorium on animal patenting. *Harvard Journal of Law and Technology*, 1:1–61.

Bent, S. A., Schwaab, R. L., Conlin, D. G., and Jeffrey, D. D. (1987) *Intellectual Property Rights in Biotechnology Worldwide.* Stockton Press, New York.

Butler, L. J., and Marion, B. W. (1985) *The Impacts of Patent Protection on the U.S. Seed Industry and Public Plant Breeding.* University of Wisconsin, Agriculture Experiment Station, NC Project 117, Monograph 16.

Connor, J. M., Rogers, R. T., Marion, B. W., and Mueller, W. F. (1985) *The Food Manufacturing Industries: Structure, Strategies, Performance and Policies.* Lexington Books, Lexington, MA.

Fessenden MacDonald, J., Ed. (1992) *Animal Biotechnology: Opportunities and Challenges,* NABC Report 4. Natural Agricultural Biotechnology Council, Ithaca, NY.

Hettinger, N. (1992) Owning varieties of life: Historical, conceptual and ethical dimensions. In *Ethics and Patenting of Transgenic Organism,* NABC Occasional Papers 1. National Agricultural Biotech Council, Ithaca, NY.

Hoban, T. J. IV, and Kendall, P. A. (1992) Consumer Attitudes about the Use of Biotechnology. In *Agriculture and Food Production.* North Carolina State University, July.

House of Representatives, Committee on the Judiciary. (1987) Hearings, "Patents and the constitution: Transgenic animals," Serial No. 23. Government Printing Office, Washington, DC.

Jewkes, J., Sauers, D., and Stillerman, R. (1969) *The Sources of Invention,* 2nd ed. Norton, New York.

Lesser, W. (1989) Implications for breeders. In *Animal Patents: The Legal, Economic and Social Issues,* W. Lesser, Ed. Stockton Press, New York.

———, (1994) Royalty collection for patented livestock. *Eur. Intellect. Prop. Rev.* 16: 441–44.

———, and Masson, R. T. (1985) *An Economic Analysis of the Plant Variety Protection Act.* American Seed Trade Association, Washington, DC.

Macer, D. (1992) *Attitudes to Genetic Engineering.* Eubios Ethics Institute, Christchurch, New Zealand.

Machlup, F. (1958) An Economic Review of the Patent System. Study of the Subcommittee on Patents, Trademarks and Copyright, Committee on the Judiciary, U.S. Senate, Study No. 15.

Office of Technology Assessment, U.S. Congress. (1989) *New development in Biotechnology: Patenting Life—Special Report No. 5,* OTA-BA-370. Government Printing Office, Washington, DC.

———. (1992) *A New Technological Era for American Agriculture,* OTA-F-474. Government Printing Office.

Scherer, F. M. (1980) *Industrial Market Structure and Economic Performance.* Rand McNally, Skokie, IL.

Straus, J. (1985) Industrial Property Protection of Biotechnological Inventions—Analysis of Certain Basic Issues, BIG/281. WIPO, Geneva.

Siebeck, W. F., Ed. (1990) *Strengthening Protection of Intellectual Property in Developing Countries.* The World Bank, Washington, DC.

Thompson, D. B. (1992) Concepts of property and the biotechnology debate. In Natural Agricultural Biotechnology Council, *Ethics and Patenting of Transgenic Organisms.* NABC Occasional Papers #1, Sept.

U.S. Department of Agriculture. Packers and Stockyards Administration (1991). Statistical Report—1989 Reporting Year. P&SA Statistical Report No. 91-1.

Transgenic Fish

Thomas T. Chen, Jenn-Kan Lu, Mike J. Shamblott, Clara M. Cheng, Chun-Mean Lin, Jane C. Burns, Renate Reimschuessel, N. Chatakondi, and Rex A. Dunham

1 Introduction

2 Selection of Model Fish Species
- 2.1 Small-Size Fish Models
- 2.2 Medium-Size Fish Models
- 2.3 Large-Size Fish Models

3 Preparation of Gene Constructs
- 3.1 Gain-of-Function Transgenes
- 3.2 Reporter Function Transgenes
- 3.3 Loss-of-Function Transgenes

4 Methodology of Gene Transfer
- 4.1 Microinjection of DNA into Embryos or Unfertilized Eggs
- 4.2 Electroporation of DNA into Embryos or Sperm
- 4.3 Transfer of Foreign Genes by Infection with Pantropic Defective Virus

5 Characterization of Transgenic Fish
- 5.1 Identification of Transgenic Fish
- 5.2 Patterns of Transgene Integration
- 5.3 Inheritance of Transgenes
- 5.4 Expression of Transgenes

6 Examples of Transgenic Fish
- 6.1 Growth Hormone Transgenic Fish
- 6.2 Insulin-like Growth Factor I Transgenic Fish
- 6.3 Other Examples of Transgenic Fish

7 Applications and Perspectives

Key Words

Electroporation A process utilizing a series of short electrical pulses to permeate cell membranes, thereby permitting the entry of DNA molecules into animal or plant cells.

F_1 Transgenic Fish Transgenic progeny derived from a cross between a P_1 transgenic individual and a nontransgenic individual or between two P_1 transgenic individuals.

P_1 Transgenic Fish Transgenic fish developed from embryos that received foreign gene transfer.

Polymerase Chain Reaction (PCR) An in vitro method for the enzymatic synthesis of a specific DNA sequence using two oligonucleotide primers hybridizing to opposite strands and flanking the region of interest in the target DNA.

Transgene In the present context: a gene construct that is introduced into fish for the production of transgenic fish.

Transgenic Fish Fish into which a foreign gene (DNA) has been introduced and stably integrated into the genome.

Since 1985, a wide range of transgenic fish has been produced by microinjecting or electroporating homologous or heterologous transgenes into newly fertilized or unfertilized eggs. This article describes the usual steps in producing a transgenic fish meeting desired specifications. It also explores the applications of transgenic fish technology in basic and applied research. Finally, it reviews the progress in transgenic fish technology and discusses the potential application of this technology in basic research and biotechnological application, using as examples results generated in the authors' laboratories.

1 INTRODUCTION

Animals or plants into whose genomes heterologous DNA has been artificially introduced and integrated are called *transgenic.* Transgenic plants, nematodes, fruit flies, sea urchins, frogs, laboratory mice, and farm mammals such as cows, pigs, and sheep have been successfully produced since the early 1980s. Transgenic plants are produced via the introduction of foreign DNA into cells by infection with *Agrobacterium tumefaciens* or by direct physical means. In animal systems, foreign DNA is injected into the pronuclei of fertilized eggs and the injected embryos are incubated in vitro or implanted into the uterus of a pseudopregnant female for subsequent development. The results of such studies have shown that multiple copies of transgenes are integrated at random locations in the genome of the transgenic individuals. If the transgenes are linked with functional promoters, expression of transgenes as well as a dis-

Table 1 Summary of Transgenic Fish Studies, 1985–1995

Gene Constructs Used[a]
Structural Genes
GH, IGF, AFP, β-*gal*, *neo*^{r}, CAT, *Lux*, δ-crystalline, hygromycin, integrase, etc.
Promoters
RSV, CMV, MT, TK, β-actin, LTR, X47, AFP, hsp
Fish species
Trout, salmon, tilapia, catfish, carp, loach, medaka, zebrafish, goldfish, northern pike

[a] GH, growth hormone; IGF, insulin-like growth factor; AFP, antifreeze protein; β-*gal*, β-galactosidase; *neo*^{r}, neomycin resistance; CAT, chloraphenicol acetyltransferase; *Lux*, luciferase; RSV, Rous sarcoma virus; CMV, cytomegalovirus; MT, metallothionein; TK, thymidine kinase; LTR, long terminal repeat sequence; hsp, heat shock protein.

play of change in phenotype is expected in some of the transgenic individuals. Furthermore, the transgenes in many transgenic individuals are also transmitted through the germ line to subsequent generations. These transgenic animals play important roles in basic research as well as in applied biotechnology. In basic research, transgenic animals provide excellent models for studying the molecular genetics of early vertebrate development, the actions of oncogenes, and the biological functions of hormones at different stages of development. In applied biotechnology, transgenic animals offer unique opportunities for producing animal models for biomedical research, improving the genetic background of broodstock for animal husbandry or aquaculture, and designing bioreactors for producing valuable proteins for pharmaceutical or industrial purposes.

There are several important steps in the production of transgenic fish species by microinjecting or electroporating homologous or heterologous transgenes into eggs. First, an appropriate fish species must be chosen. The identity of the most suitable fish species depends on the nature of the studies and the availability of the fish-holding facility. Second, a specific gene construct must be prepared. The gene construct contains the structural gene encoding a gene product of interest and the regulatory elements that regulate the expressions of the gene in a temporal, spatial, and developmental manner. Third, the gene construct must be introduced into the developing embryos, to allow the transgene to be integrated stably into the genome of every cell. Fourth, since not all instances of gene transfer are efficient, a screening method must be adopted for identifying transgenic individuals.

Although remarkable progress has been made in producing transgenic fish by gene transfer technology, a critical review of the published results shows that the majority of the research effort has been devoted to confirming the phenomenon of foreign gene transfer into various fish species (Table 1). Very few attempts have been made to explore the application of transgenic fish technology in basic as well as applied research. We have devoted a substantial amount of our research effort to this problem, and the results are very promising.

2 SELECTION OF MODEL FISH SPECIES

Gene transfer studies have been conducted in several different fish species such as salmon, rainbow trout, common carp, goldfish, loach, channel catfish, tilapia, northern pike, walleye, zebrafish, and Japanese medaka. Depending on the purpose of the transgenic fish studies, the embryos of some fish species prove to be superior than the others for serving as hosts. In selecting a fish species for gene transfer studies, a series of parameters are considered: (1) length of the life cycle, (2) year-round supply of eggs and sperm, (3) culture conditions, (4) mature size of the adult, and (5) availability of background information on the genetics, physiology, and endocrinology of the species selected.

2.1 Small-Size Fish Models

Japanese medaka *(Oryzias latipes)* and zebrafish *(Barchydanio rerio)* are regarded as ideal fish species for conducting gene transfer studies: both have short life cycles (3 months from eggs to mature adults), produce hundreds of eggs on a regular basis without exhibiting a seasonal breeding cycle, and can be maintained easily in the laboratory for 2–3 years. Their relatively large eggs (diameter of 1.0–1.5 mm), with very thin and semitransparent chorions, allow easy microinjection of DNA if appropriate glass needles are used.

Furthermore, inbred lines and various morphological mutants of both fish species are available. These fish species are suitable candidates for producing transgenic fish for studying developmental regulation of gene expression and gene action, identifying regulatory elements that regulate the expression of a gene, measuring the activities of promoters, and producing transgenic models for environmental toxicology. However, a major drawback of these two fish species is that their small body size makes them unsuitable for endocrinological or biochemical analysis.

2.2 Medium-Size Fish Models

The handicap of using medaka, zebrafish, or other species of small body size in gene transfer studies can be easily overcome. Fish species such as loach, killifish *(Fundulus)*, goldfish, and tilapia have body sizes big enough to permit the isolation of sufficient amounts of tissues from individual fish for biochemical and endocrinological studies. Another important attribute of these fish species is their shorter maturation time compared to rainbow trout or salmon. In particular, tilapia is a very desirable medium-size fish species for gene transfer studies because its entire life cycle is completed in about 3 months. It is possible to produce three generations of transgenic tilapia in a year. The lack of a well-defined genetic background and asynchronous reproductive behavior of these fish species present some problems for conducting gene transfer studies, however.

2.3 Large-Size Fish Models

Rainbow trout, salmon, channel catfish, and common carp are commonly used large body size model fish species for transgenic fish studies. Since our knowledge of the endocrinology, reproductive biology, and physiology of these species has been well worked out, the fish are well suited for studies on comparative endocrinology and as well as aquaculture applications. Generally speaking, these fish species take a very long time to mature: 2–3 years for rainbow trout and salmon, about 1–2 years for common carp or channel catfish. Thus the time factor will hamper inheritance studies. In addition, these fish species have a single spawning cycle per year, which restricts the number of gene transfer attempts that can be conducted each year.

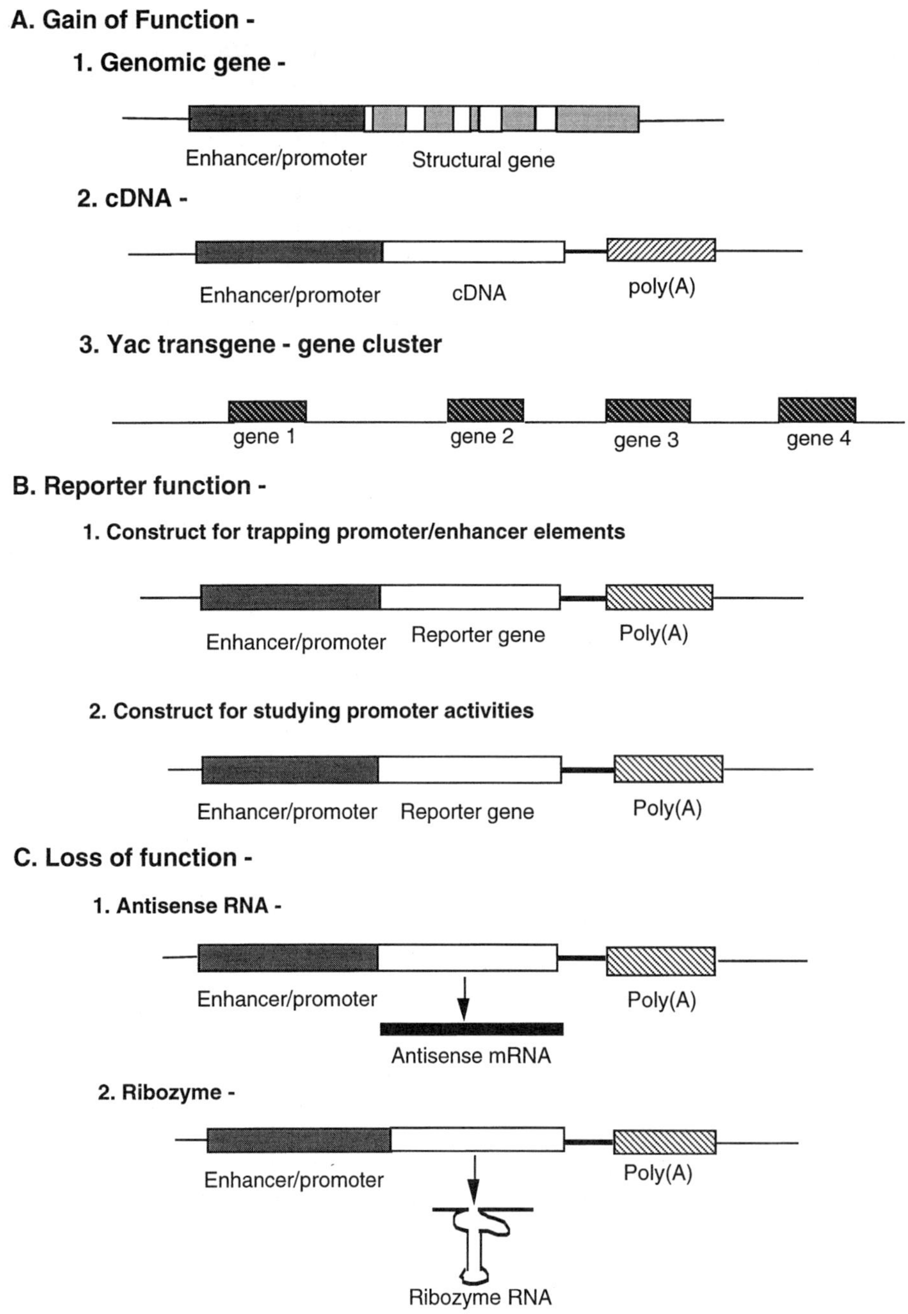

Figure 1. Prototypes of transgenes.

3 PREPARATION OF GENE CONSTRUCTS

Transgenes used in producing transgenic fish for basic research and biotechnological applications are recombinant gene constructs that will produce gene products at appropriate levels in the desired tissue(s) at the desired time(s). Transgenes are usually constructed in plasmids to contain appropriate promoter/enhancer elements and structural genes. Depending on the purpose of the gene transfer studies, transgenes can be grouped into three major types: gain-of-function, reporter function, and loss-of-function transgenes.

3.1 Gain-of-Function Transgenes

Gain-of-function transgenes are designed to add new functions to the transgenic individuals or to facilitate their identification if the genes are expressed properly in these transgenic individuals. The coding regions of the transgenes are usually homologous or heterologous genomic or cDNA sequences that encode polypeptide products (Figure 1A). The expression of the transgenes genes usually is driven by promoter/enhancer sequences of homologous or heterologous sources. If the transgenes contain functional signal peptide sequences, the gene products will be secreted out of cells once they have been synthesized. Transgenes containing the structural gene of mammalian or fish growth hormone (GH) or the respective cDNA fused to a functional promoter/enhancer of chicken or fish β-actin gene are examples of gain-of-function transgene constructs. Expression of such transgenes in transgenic individuals will result in growth enhancement. Winter flounder antifreeze protein gene and chicken δ-crystalline gene, with their own promoter/enhancer sequences, are other *gain-of-function* transgenes commonly used in various gene transfer studies.

Another type of commonly used *gain-of-function* transgene comprises those that contain the structural gene sequences of bacterial chloramphenicol acetyltransferase (CAT), β-galactosidase (β-gal), neomycine phosphotransferase II (*neo*^{r}), and luciferase *(Lux)* driven by known functional promoter/enhancer sequences such as the long terminal repeat (LTR) of Rous sarcoma virus (RSV) or mouse metallothionein gene. Although the expression of these genes does not influence fish physiology, it serves as a convenient marker for identifying the success of gene transfer studies, since the gene products of these genes can be easily monitored by color reaction or other simple methods.

3.2 Reporter Function Transgenes

Since the activities of bacterial CAT, β-galactosidase, and luciferase can be measured quantitatively, these genes are routinely adopted as reporter genes for identifying the sequence of a promoter/enhancer in question or for measuring the relative activity of a given promoter/enhancer element. Following gene transfer, the expression of the reporter gene activity is used to determine the transcriptional regulatory sequence of a gene or the strength of a promoter. The prototypes of a *reporter function* transgene are shown in Figure 1B. Since the activity of β-galactosidase can also be measured by following the conversion of a colorless lactose analogue (5-bromo-4-chloro-3-indolyl-*b*-D-galactopyranoside) into a blue indolyl dye cytochemically on tissue slices or whole-animal samples, this reporter gene is most useful for determining the tissue-specific and developmental-stage-specific expression of a promoter/enhancer sequence.

3.3 Loss-of-Function Transgenes

The "*loss-of-function*" transgenes are constructed for interfering with the expression of host genes. These genes might encode antisense RNAs to interfere with the posttranscriptional process or translation of endogenous mRNAs. Alternatively, these genes might encode catalytic RNA (ribozymes) that can cleave specific mRNAs and thereby cancel the production of the normal gene products. The prototype of "loss-of-function" transgenes are shown in Figure 1C. Although these genes have not yet been introduced into a fish model, they could potentially be employed to produce disease-resistant transgenic broodstocks for aquaculture or transgenic model fish defective in a particular gene product for basic research.

4 METHODOLOGY OF GENE TRANSFER

Techniques such as calcium phosphate precipitation, direct microinjection, retrovirus infection, electroporation, and particle gun bombardment have been widely used to introduce foreign DNA into animal cells, plant cells, and the gem lines of mammals and other vertebrates. Direct microinjection and electroporation of DNA into newly fertilized eggs have been proven to be the most reliable methods of gene transfer in fish systems.

4.1 Microinjection of DNA into Embryos or Unfertilized Eggs

Microinjection of foreign DNA into newly fertilized eggs was first developed for the production of transgenic mice in the early 1980s. Since 1985, the technique of microinjection has also been adopted for introducing transgenes into Atlantic salmon, common carp, catfish, goldfish, loach, medaka, rainbow trout, tilapia, and zebrafish. The gene constructs that were used in these studies include human or rat growth hormone (GH) gene, rainbow trout or salmon GH cDNA, chicken δ-crystalline protein gene, winter flounder antifreeze protein gene, *E. coli* β-galactosidase gene, and *E. coli* hygromycin resistance gene.

In general, gene transfer in fish by direct microinjection is conducted as follows, observing the parameters for microinjection summarized in Table 2. Eggs and sperm are collected in separate, dry containers. Fertilization is initiated by adding water and sperm to the eggs, with gentle stirring to enhance fertilization. Eggs are microinjected within the first few hours after fertilization. The injection apparatus consists of a dissecting stereomicroscope and two micromanipulators, one with a glass microneedle for injection and the other with a micropipette for holding fish embryos in place. Routinely, about 10^6–10^8 molecules of a linearized transgene in about 20 nL is injected into the egg cytoplasm. Following injection, the embryos are incubated in water until hatching. Since natural spawning in zebrafish or medaka can be induced by adjusting photoperiod and water temperature, precisely staged, newly fertilized eggs can be collected from the aquaria for gene transfer. If the medaka eggs are maintained at 4°C immediately after fertilization, the micropyle on the fertilized eggs will remain visible for 2 hours. The DNA solution can be easily delivered into the embryos by injecting through this opening.

Depending on the species, the survival rate of injected fish embryos ranges from 35 to 80% while the rate of DNA integration ranges from 10 to 70% in the survivors (Table 3). The tough chorions of the fertilized eggs in some fish species (e.g., rainbow trout, Atlantic salmon) can make insertion of glass needles difficult. This difficulty can be overcome by any one of the following methods: inserting the injection needles through the micropyle, making an

Table 2 Parameters of Gene Transfer in Fish by Microinjection and Electroporation

Parameters	Gene Transfer Method[a]	
	Microinjection	Electroporation
Developmental stage	1–2 cells	1–2 cells
DNA size	<10 kb	<10 kb
DNA concentration	10^{6-7} molecules/embryo	100 μg/mL
DNA topology	Linear	Linear
Chorion barrier	Dechorionated/micropyle	Intact chorion
Electrical field strength	N/A	500–3,000V
Pulse shape	N/A	Exponential/square
Pulse duration	N/A	Milliseconds–seconds
Medium	PBS/saline	PBS/saline

[a] N/A, not applicable; PBS, phosphate buffer solution.

Table 3 Transfer of Foreign DNA into Medaka Embryos by Different Gene Transfer Methods

		Electroporation		Pantropic Retrovireal Vector	
	Microinjection[a]	I[b]	II[c]	Electroporation[d]	Incubation[e]
Viability (at hatching)	50%	70%	90%	50%	70%
Integration rate[f]	20%	15%	25%	50%	70%
Transgene expression	Yes	Yes	Yes	Yes	Yes
Efficiency (eggs/min)	1–2	200	200	200	200

[a] Injecting is carried out via micropyle prior to blastodic formation.
[b] Exponential decay impulse mode.
[c] Square wave impulse mode.
[d] Electroporation with square wave mode at 3.5 kV.
[e] Fertilized eggs are exposed to a mixture of medaka hatching enzyme and pancreatin for 2 hours. The dechorionated embryos are incubated with the pantropic pseudotyped retrovirus overnight at room temperature.
[f] Integration rate calculated from the surviving embryos after gene transfer.

opening on the egg chorions by microsurgery, removing the chorion by mechanical or enzymatic means, preventing chorion hardening by initiating fertilization in a solution containing 1 mM glutathione, or injecting the unfertilized eggs directly.

4.2 Electroporation of DNA into Embryos or Sperm

Electroporation, a successful method for transferring foreign DNA into bacteria, yeast, and plant and animal cells in culture, recently has become popular for transferring foreign genes into fish embryos. Electroporation utilizes a series of short electrical pulses to permeate cell membranes, thereby permitting the entry of DNA molecules into embryos. The electrical pulse patterns can be emitted in a single pulse of exponential decay form (i.e., exponential decay generator) or high frequency multiple peaks of square waves (i.e., square wave generator); for the basic parameters, see Table 2. Studies conducted in our laboratory and those of others have shown that the rate of DNA integration in electroporated embryos is on the order of 20% or higher in the survivors (Table 3). Although the overall rate of DNA integration in transgenic fish produced by electroporation was equal to or slightly higher than that of microinjection, the actual amount of time required for handling a large number of embryos by electroporation is orders of magnitude less than that required for microinjection. Recently, several reports have also appeared in the literature describing successful transfer of transgenes into fish by electroporating sperm instead of embryos. Electroporation is therefore considered to be an efficient and versatile massive gene transfer technology.

4.3 Transfer of Foreign Genes by Infection with Pantropic Defective Virus

Although transgenes can be introduced into many fish species studied to date readily and with great efficiency by means of microinjection or electroporation, the resulting P_1 transgenic individuals are mosaics as a result of delayed transgene integration. Furthermore, these methods are not effective or not successful in transferring foreign DNA into embryos of some marine fish and invertebrates. Recently a new gene transfer vector, a highly defective pantropic retroviral vector, has been developed. This vector contains the long terminal repeat (LTR) sequences of Moloney murine leukemia virus (MoMLV) and transgenes in a viral envelop with the G protein of vesicular stomatitis virus (VSV). The strategy for producing this type of defective retroviral particle is summarized in Figure 2. Since entry of VSV into cells is mediated by interaction of the VSV-G protein with a phospholipid component of the cell, this pseudotyped retroviral vector has a very broad host range and is able to transfer transgene into many different cell types. Using the pantropic pseudotyped defective retrovirus as a gene transfer vector, transgenes containing *neo* or β-galactosidase have been transferred into zebrafish and medaka (Table 3). Recently, the feasibility of using pantropic pseudotyped retroviral vectors for introducing genes into marine invertebrates was tested in dwarf surf clams, and the results showed that transgenes can be readily transferred into clams by a brief electroporation with the pantropic retrovirus at about 500 V (Lu, Burns, Allen, and Chen, unpublished results).

5 CHARACTERIZATION OF TRANSGENIC FISH

5.1 Identification of Transgenic Fish

The most time-consuming step in producing transgenic fish is the identification of transgenic individuals. Traditionally, dot-blot and Southern blot hybridizations of genomic DNA were common methods of determining the presence of transgenes in the presumed transgenic individuals. These methods involve isolation of genomic DNA from tissues of putative transgenic individuals, digestion of DNA samples with restriction enzymes, and Southern blot hybridization of the digested DNA products. Although this method is expensive, laborious, and insensitive, it offers a definitive answer as to whether a transgene has been integrated into the host genome. Furthermore, it reveals the pattern of transgene integration if appropriate restriction enzymes are employed in the Southern blot hybridization analysis. To handle a large number of animals efficiently and economically, an assay based on the polymerase chain reaction (PCR) has been adopted. The strategy of the assay (outlined in Figure 3) involves the isolation of genomic DNA from a very small piece of fin tissue, PCR amplification of the transgene sequence, and Southern blot analysis of the amplified products. Although this method does not reveal whether the transgene is integrated in the host genome or is staying as an extrachromosomal unit, it serves as

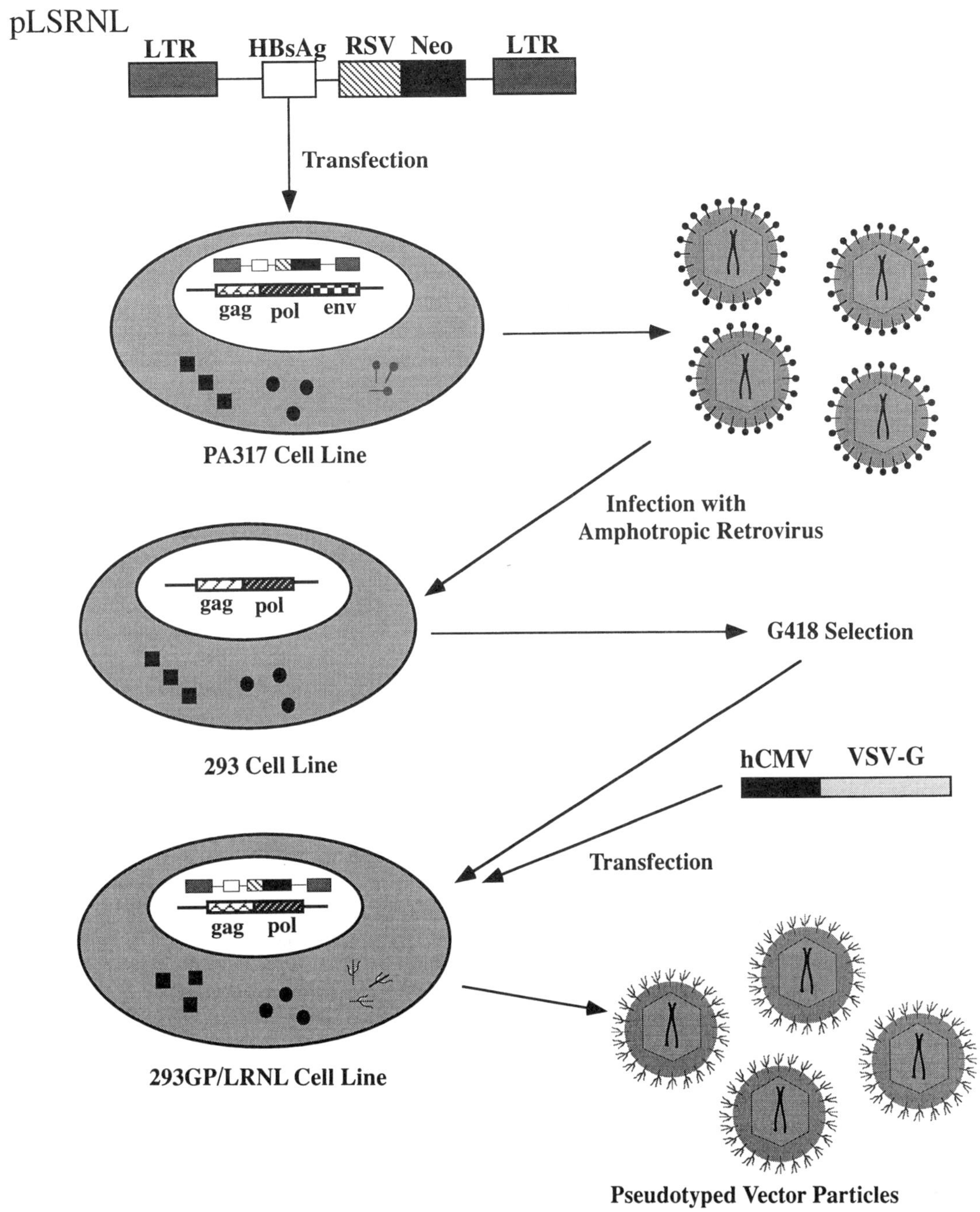

Figure 2. Strategy of producing VSV-G pantropic defective retroviral vector.

a rapid and sensitive screening method for identifying individuals that contain the transgene at the time of analysis. This method has been used in our laboratory as a routine preliminary screen for the presence of transgenes in thousands of presumed transgenic fish.

5.2 Patterns of Transgene Integration

Studies conducted in many fish species have shown that following injection of linear or circular transgene constructs into fish embryos, the transgenes are maintained as an extrachromosomal unit through many rounds of DNA replication in the early phase of the embryonic development. At a later stage of embryonic development, some of the transgenes are randomly integrated into the host genome while others are degraded, resulting in the production of mosaic transgenic fish. To determine the pattern of transgene integration, genomic DNA from PCR-positive fish will be digested with a series of restriction endonucleases and the resulting products resolved in agarose gels for Southern blot analysis. In many fish species studied to date, multiple copies of transgenes are found to be integrated in a head-to-head, head-to-tail, or tail-to-tail form, except in trans-

(A)

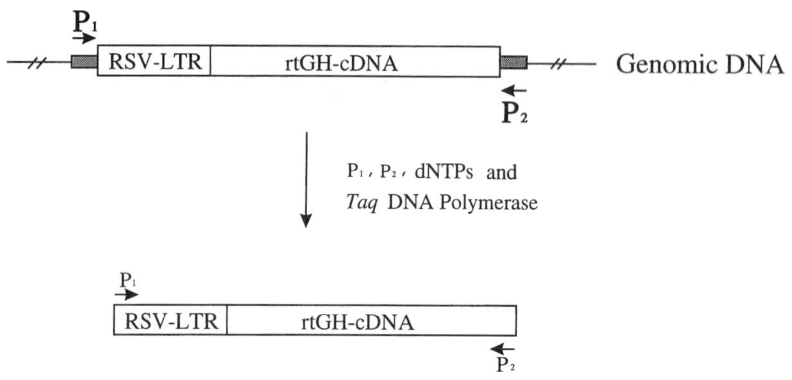

(B)

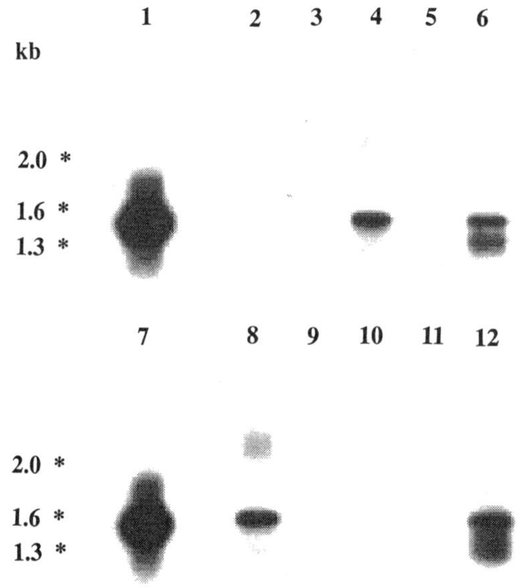

Figure 3. Strategy for identifying the presence of transgenes in the presumed transgenic fish by PCR and Southern blot hybridization. DNA samples were isolated from pectoral fin tissues of presumed transgenic fish and subjected to PCR amplification. The amplified products were analyzed by electrophoresis on agarose gels and Southern blot hybridization. (A) Strategy of PCR amplification. (B) Southern blot analysis of PCR-amplified products: lanes 2–6 and 8–12, DNA samples from presumed transgenic fish; lanes 1 and 7, transgene construct (RSV-LTR-rtGH1 cDNA). [From T. T. Chen, D. A. Powers, C.-M. Lin, K. Kight, M. Hayat, N. Chatakondi, A. C. Ramboux, P. L. Duncan, and R. A. Dunham, Expression and inheritance of RSV-LTR-rtGH1 cDNA in common carp, *Cyprinus carpio*. *Mol. Mar. Biol. Technol.* 2:88–95 (1993), with permission.]

genic common carp and channel catfish, where a single copy of transgene is integrated at multiple sites on the host chromosomes.

5.3 Inheritance of Transgenes

Stable integration of transgenes is an absolute requirement for continuous vertical transmission to subsequent generations and establishment of a transgenic fish line. To determine whether the transgene is transmitted to subsequent generations, P_1 transgenic individuals are mated to nontransgenic individuals and the progeny are assayed for the presence of transgene by the PCR assay method described earlier. Although it has been shown that the transgene may persist into the F_1 generation of transgenic zebrafish as the extrachromosomal DNA, detailed analysis of the rate of transmission of transgenes to F_1 and F_2 generations in many transgenic fish species indicates true and stable incorporation of gene constructs into the host genome. If the entire germ line of the P_1 transgenic fish is transformed with at least one copy of the transgene per haploid genome, at least 50% of the F_1 transgenic progeny will be expected in a backcross involving a P_1 transgenic with a nontransgenic control. In many such crosses, only about 20% of the progeny are transgenic. When the F_1 transgenic individual is backcrossed with a nontransgenic control, however, at least 50% of the F_2 progeny are transgenics. These results clearly suggest that the germ lines of the P_1 transgenic fish are mosaic as a result of delayed transgene integration during embryonic development.

5.4 Expression of Transgenes

An important aspect of gene transfer studies is the detection of transgene expression. Depending on the levels of transgene products in the transgenic individuals, the following methods are commonly used to detect transgene expression: RNA Northern or dot-blot hybridization, RNase protection assay, reverse transcription/poly-

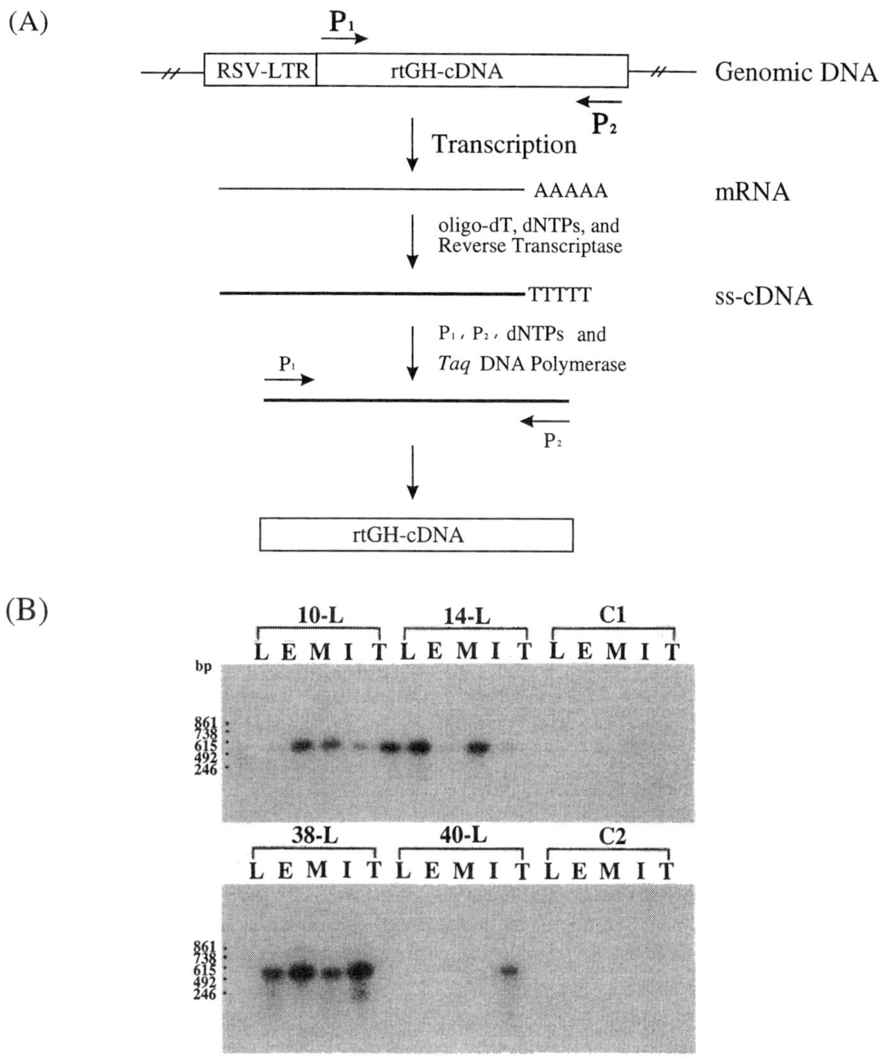

Figure 4. Strategy of detecting rtGH transgene expression by reverse transcription RT-PCR assay. (A) Strategy of RT-PCR. (B) Detection of rtGH transgene expression in transgenic carp by RT-PCR. Total RNA was isolated from liver, muscle, eyes, gut, and testes (L, E, M, I, T, respectively), of F_1 transgenic carp and controls following the acid guanidinium thiocyanate/phenol/chloroform method. Single-stranded cDNA was prepared by reverse transcription from each total RNA and used as a template for PCR amplification of rtGH using synthetic oligonucleotides as amplification primers. The resulting products were analyzed by Southern blot analysis using radio labeled rtGH cDNA as a hybridization probe. [From T. T. Chen, D. A. Powers, C.-M. Lin, K. Kight, M. Hayat, N. Chatakondi, A. C. Ramboux, P. L. Duncan, and R. A. Dunham, Expression and inheritance of RSV-LTR-rtGH1 cDNA in common carp, *Cyprinus carpio*. *Mol. Mar. Biol. Technol.* 2:88–95 (1993), with permission.]

merase chain reaction (RT-PCR), immunoblotting assay, and other biochemical assays for determining the presence of the transgene protein products.

Among these assays, RT-PCR is the most sensitive method and requires only a small amount of sample. The strategy of this assay is summarized in Figure 4. Briefly, it involves the isolation of total RNA from a small piece of tissue, synthesis of single-stranded cDNA by reverse transcription, and PCR amplification of the transgene cDNA by employing a pair of oligonucleotides specific to the transgene product as amplification primers. The resulting products are resolved on agarose gels and analyzed by Southern blot hybridization using a radio labeled transgene as a hybridization probe. Transgene expression can also be quantified by a quantitative RT-PCR method.

6 EXAMPLES OF TRANSGENIC FISH

6.1 Growth Hormone Transgenic Fish

The initial drive for transgenic fish research came from attempts to increase production of economically important fish for human consumption. The worldwide harvest of fishery products traditionally depends on natural populations of finfish, shellfish, and crustaceans in fresh and marine waters. In recent years, however, the level of total worldwide annual harvest of fish products has approached or even surpassed the maximal potential level of about 150 million metric tons (as calculated by the U.S. Department of Commerce and the U.S. National Oceanic and Atmospheric Administration). To cope with world demand for fish products and escalating prices, many countries have turned to aquaculture as a means of increasing production. In 1985 the world production of finfish, shellfish, and macroalgae by aquaculture/mariculture reached 10.6 million metric tons, or approximately 12.3% of the worldwide catch generated by international fishery efforts. Although aquaculture clearly has the potential for increasing worldwide fish production, innovative strategies are needed to improve efficiency. What can transgenic technology offer?

Success in aquaculture depends on six factors:

- complete control of the reproductive cycle of the fish species in culture
- excellent genetic background of the broodstock
- efficient prevention and detection of disease infection
- thorough understanding of the optimal physiological, environmental, and nutritional conditions for growth and development
- sufficient supply of excellent quality water
- application of innovative management techniques

By improving these factors, the aquaculture industry has grown to a remarkable extent during the last decade. To sustain this growth, however, newly developed technologies in molecular biology and transgenesis will have to be increasingly applied by the aquaculture industry. These technologies can be employed to enhance growth rates, control reproductive cycles, improve feed compositions, produce new vaccines, and develop disease-resistant and hardier genetic stocks.

In recent years, cDNAs and their genomic DNA of growth hormone have been isolated and characterized for several fish species. Expression of rainbow trout or striped bass GH cDNA in *E. coli* cells results in production of a large quantity of recombinant GH polypeptide. Since the GH polypeptide is highly hydrophobic and contains four cysteine residues, the newly synthesized recombinant GH polypeptide forms insoluble inclusion bodies in *E. coli* cells, rendering the hormone inactive. In an attempt to regain the biological activity of the recombinant hormone, Cheng et al. developed a procedure for renaturing the protein in which the insoluble recombinant hormone is dissolved in a buffer containing 8M urea and the polypeptide is renatured by slowly removing the urea from the protein solution. The biological activity of the renatured protein was then assessed by an *in vitro* sulfation assay.

In a series of in vivo studies, Agellon *et al.* showed that application of the recombinant trout GH to yearling rainbow trout resulted in a significant growth enhancement. After treatment of yearling rainbow trout with the recombinant GH for 4 weeks at a dose of 1 μg per gram of body weight per week, the weight gain among the individuals of the hormone-treated group was twice that of the controls. Significant length gain was also evident in hormone-treated animals. When the same recombinant hormone was administered to rainbow trout fry or small juveniles by immersing the fish in a GH-containing solution, the same growth-promoting effect was also observed. These results are in agreement with those reported by Sekine et al., Gill et al., and many others. However, it is important to mention that the growth enhancement effect of the biosynthetic hormone was markedly reduced when the amount of hormone applied to the test animals exceeded 2 μg/g body weight. Recently Paynter and Chen observed that administration of recombinant trout GH polypeptide to spats of juvenile oysters *(Crassostrea virginica)* by the "dipping" method also resulted in significant increases in shell height, shell weight, wet weight, and dry weight. The results summarized earlier clearly suggest that exogenous application of recombinant fish growth hormone can enhance the somatic growth of finfish and shellfish.

Although exogenous application of biosynthetic GH results in a significant growth enhancement in fish, it may not be cost effective for the following reasons: (1) producing purified biosynthetic GH is costly; (2) treating individual fish with the hormone is labor intensive; (3) the optimal hormone dosage for each fish species is difficult to identify; and (4) GH uptake into fish from an exogenous source is inefficient. If new strains of fish producing elevated but optimal levels of GH were produced, it would be possible to bypass all the problems associated with exogenous GH treatment. Moreover, once these fish strains had been generated, they would be far more cost-effective than their ordinary counterparts because these treated fish would have their own means of producing and delivering the hormone and could transmit their enhanced growth characteristics to their offspring.

Three aspects of fish growth characteristics that could be improved for aquaculture are initial growth rate (so that juveniles reach maturation earlier), enhanced somatic growth rate as adults (to provide larger body size for market), and fish with improved feed conversion efficiencies. Among these three, the achievement of enhanced somatic growth rates via manipulation of the GH gene shows considerable promise. Zhu et al. reported the first successful transfer of a human GH gene fused to a mouse metallothionein (MT) gene promoter into goldfish and loach. According to Zhu (personal communication), the F_1 offspring of these transgenic fish grew twice as large as their nontransgenic siblings. The transgenic fish studies of Zhu and his colleagues did not present compelling evidence for integration and expression of the foreign genes. Recently, however, many laboratories throughout the world have successfully confirmed Zhu's work by demonstrating that human or fish GH and many other genes can be readily transferred into embryos of a number of fish species and integrated into the genome of the host fish. While a few groups have demonstrated expression of foreign genes in transgenic fish, only Zhang *et al.*, Du *et al.*, and Lu *et al.* have documented that a foreign GH gene could be transferred to the target fish species, integrated into the fish genome, and genetically transmitted to the subsequent generations. Furthermore, these investigators showed that the expression of the foreign GH gene may result in enhancement of growth rates of both P_1 and F_1 generations of transgenic fish.

In gene transfer studies conducted in common carp and channel catfish, about 10^6 molecules of a linearized recombinant plasmid containing the long terminal repeat (LTR) sequence of avian Rous sarcoma virus (RSV) and the rainbow trout GH cDNAs were injected into the cytoplasm of one-cell, two-cell, and four-cell embryos. Genomic DNA samples extracted from the pectoral fins of presumed transgenic fish were analyzed for the presence of RSV-LTR-rtGH1-cDNA by PCR amplification and Southern blot hybridization of the amplified DNA products using radiolabeled LTR of RSV and/or trout GH1 cDNA as hybridization probes. In the case of transgenic carp studies, about 35% of the injected embryos survived at hatching, of which about 10% of the survivors had stably integrated the RSV-LTR-rtGH1-cDNA sequence. A similar percentage of transgenic fish was also obtained when RSV-LTR-csGH-cDNA construct was injected into catfish embryos. Southern blot analysis of genomic DNA samples of several transgenic carp revealed that a single copy of the RSV-LTR-rtGH1-cDNA sequence was integrated at multiple chromosomal sites.

Table 4 Percent of F_1 Progeny Inheriting RSV-LTR-rtGH1 cDNA Sequence in Transgenic Carp

Family	Mating	*N*	Observed Inheritance (%)	Expected Inheritance (%)[a]
1	$P_1 \times$ control	17	0	50
2	$P_1 \times$ control	96	32	50
3	$P_1 \times$ control	26	42[b]	50
4	$P_1 \times$ control	4	100[c]	50
5	$P_1 \times P_1$	28	21	75
6	$P_1 \times P_1$	99	21	75
7	$P_1 \times P_1$	312	31	75
8	$P_1 \times P_1$	93	30	75
9	$P_1 \times F_1$	65	23	75

[a]All observed values are less ($P < 0.05$) than expected except as indicated in chi-square test.
[b]Not significantly different ($P > 0.05$).
[c]Greater than expected ($P < 0.05$).
Source: T. T. Chen, D. A. Powers, C.-M. Lin, K. Kight, M. Hayat, N. Chatakondi, A. C. Ramboux, P. L. Duncan, and R. A. Dunham, Expression and inheritance of KSVLTR-rt GH1 cDNA in common carp, *Cyprinus carpio. Mol. Mar. Biol. Biotechnol.* 2:88–95 (1993), with permission.

The patterns of inheritance of RSV-LTR-rtGH1 cDNA in the transgenic common carp were studied by fertilizing eggs collected from nontransgenic females or P_1 transgenic females with sperm samples collected from several sexually mature P_1 male transgenic fish. DNA samples extracted from the resulting F_1 progeny were assayed for the presence of RSV-LTR-rtGH1-cDNA sequence by PCR amplification and dot-blot hybridization. The percentages of the transgenic progeny resulting from nine matings were 0, 32, 42, 100 (4 progeny only), 21, 21, 31, 30, and 23%, respectively (Table 4). If each of the transgenic parents of these nine matings carries at least one copy of the transgene in the gonad cell, about 50–75% transgenic progeny would have been expected in each pairing. Out of these nine matings, two siblots, both control $\times$ P_1, gave transgenic progeny numbers as large or larger than expected ($P < 0.05$), and the remaining had lower than expected numbers of transgenic progeny. These results indicate that though most of these P_1 transgenic fish have RSV-LTR-rtGH1 cDNA in their germ line, they may be mosaics. Similar patterns of mosaicism in the germ line of P_1 transgenic fish have been observed in many fish species studied to date.

If the transgene carries a functional promoter, some of the transgenic individuals are expected to express the transgene activity. According to Zhang et al. and Chen *et al.*, many of the P_1 and F_1 transgenic common carp produced rtGH, and the levels of rtGH produced by the transgenic individuals varied about 10-fold. Chen *et al.* confirmed these results by detecting the presence of rtGH mRNA in the F_1 transgenic carp using an assay involving RT-PCR amplification. These workers have found that different levels of rtGH mRNA were detected in liver, eyes, gonads, intestine, and muscle of the F_1 transgenic individuals.

Since the site of transgene integration differs among the individuals in any population of P_1 transgenic fish, the fish should be considered to be totally different transgenic individuals, hence inappropriate for direct comparison of the growth performance among these animals. Instead, the growth performance studies should be conducted in F_1 transgenic and nontransgenic siblings derived from the same family. Recently Chen *et al.* conducted studies to evaluate the growth performance of F_1 transgenic carp in seven families. In these experiments, when transgenic and nontransgenic full siblings were spawned, hatched, and reared communally under the same environment, growth response by families of F_1 transgenic individuals in response to the presence of rtGH1 cDNA varied widely. In comparison to the nontransgenic full siblings, the results of four out of seven growth trials showed 21, 40, 59, and 22% increase in growth, respectively (Table 5).

In three of the four families where F_1 transgenics grew faster than their nontransgenic full siblings, the maximum and minimum body

Table 5 Mean, Standard Deviation (SD), Coefficient of Variation, and Percent Difference in Body Weight of Transgenic Common Carp, *Cyprinus Carpio*, and Their Nontransgenic Full Siblings

Family	Mating	Genotype[a]	*N*	Mean Body Weight (SD)	Coefficient of Variation	Difference (%)	Range in Body Weight (g)
1	$P_1 \times$ control	T	31	120.6 (17.4)	14.4	20.8	95–173
		NT	65	99.3 (14.7)	14.8		65–129
2	$P_1 \times$ control	T	11	206.0 (45.2)	21.9	40.1	115–283
		NT	15	147.0 (48)	32.6		67–228
3	$P_1 \times P_1$	T	7	5.8 (3.4)	58.6	−26.6	1.8–11.3
		NT	21	7.9 (3.1)	39.2		3.3–17.9
4	$P_1 \times P_1$	T	28	66.1 (36.9)	55.8	58.5	18.5–338
		NT	65	41.7 (27.8)	66.6		8.3–141
5	$P_1 \times P_1$	T	17	14.7 (6.8)	46.3	21.5	6.5–30.4
		NT	82	12.1 (8.4)	69.4		3.9–56.1
6	$P_1 \times P_1$	T	97	114.2 (81.6)	71.5	−14.5	18.3–565.1
		NT	215	133.6 (83.6)	62.5		20.9–416.2
7	$P_1 \times P_1$	T	15	72.2 (58.0)	80.3	−1.5	7.1–214.4
		NT	48	73.3 (47.6)	64.5		8.7–203.3

[a]T, transgenic; NT, nontransgenic.
[b]Number of fish.
Source: T. T. Chen, D. A. Powers, C.-M. Lin, K. Kight, M. Hayat, N. Chatakondi, A. C. Ramboux, P. L. Duncan, and R. A. Dunham, Expression and inheritance of KSVLTR-rt GH1 cDNA in common carp, *Cyprinus carpio. Mol. Mar. Biol. Biotechnol.* 2:88–95 (1993), with permission.

weights of the transgenics were larger than those of the nontransgenics. In the fourth family, the minimum, but not the maximum, body weight of the transgenics was larger than that of the nontransgenics. In two of those three transgenic families in which transgenics did not grow faster than their nontransgenic full siblings, the maximum and minimum body weights of the transgenics were smaller than those of the nontransgenics. In the third family, however, one of the F_1 transgenics was the largest fish in the family. The same extent of growth enhancement was also observed in F_2 offspring derived from crossing F_1 transgenics with nontransgenic controls (Figure 5).

Since the response of the transgenic fish to the insertion of the

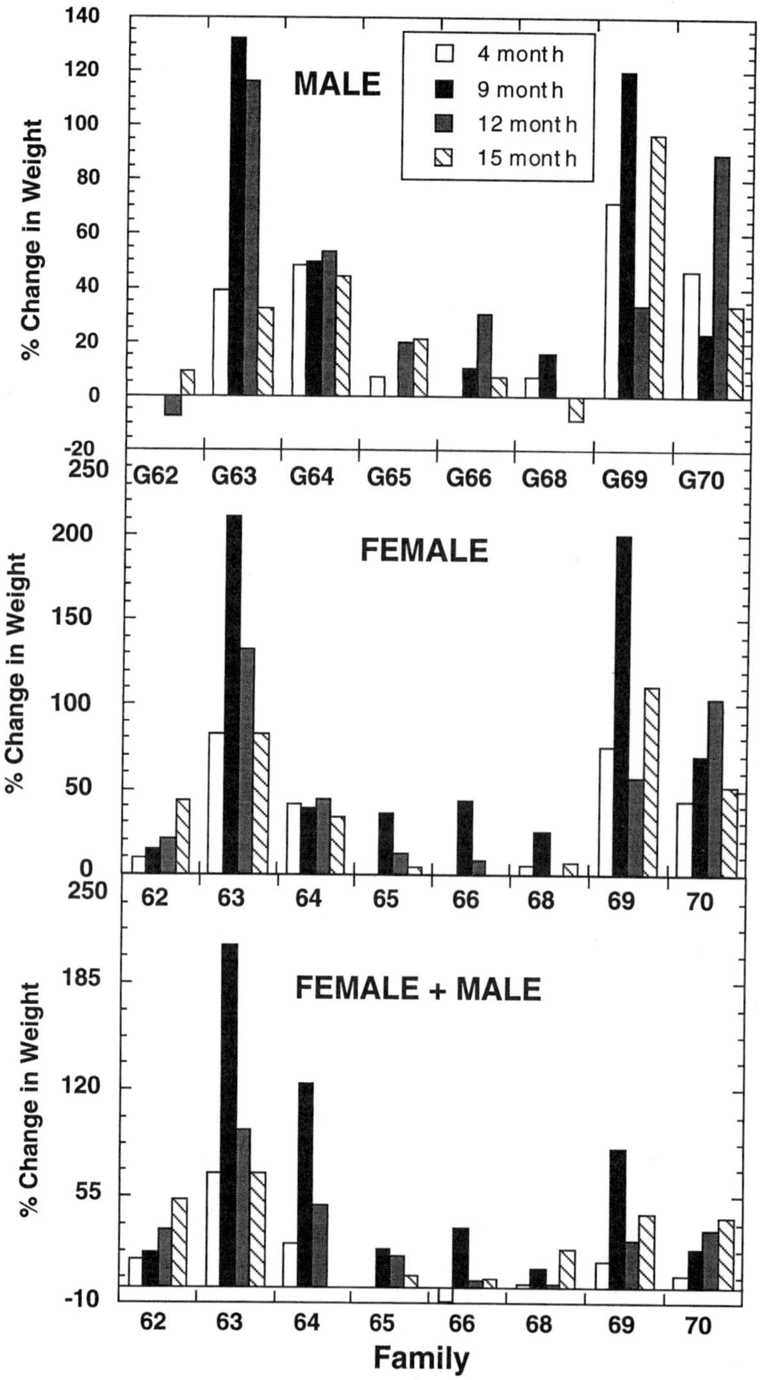

Figure 5. Body weight changes in F_2 transgenic and nontransgenic common carp carrying RSV-LTR-rtGH transgene. Each family of fish is derived from a cross of F_1 transgenic and nontransgenic fish. All eight F_1 transgenics are derived from the same P_1 transgenic individual. Percent weight change in each F_2 individual is determined by comparison to the nontransgenic siblings.

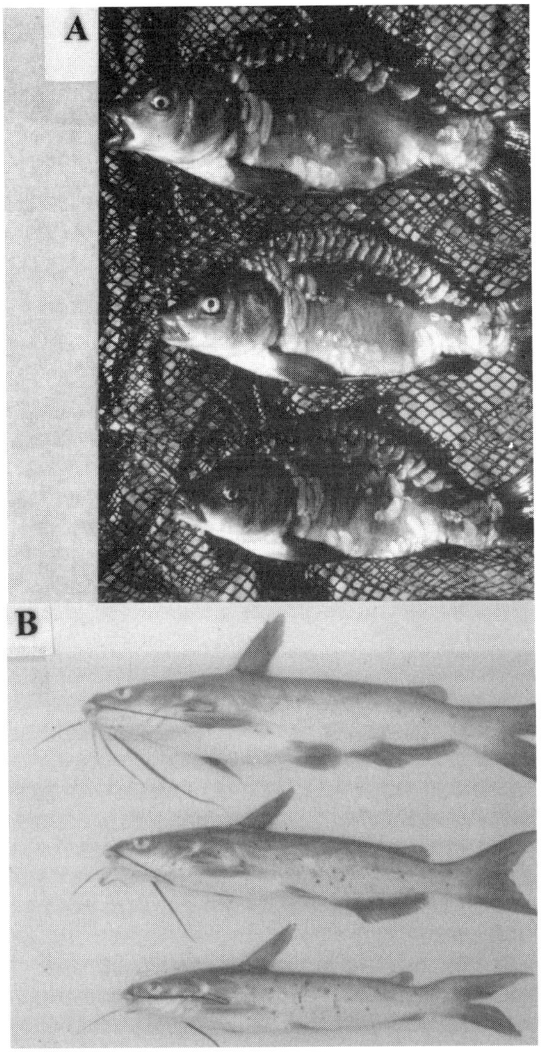

Figure 6. Transgenic common carp and catfish: (A) transgenic common carp carrying trout GH cDNA and (B) transgenic channel catfish carrying trout GH cDNA. In each case: top, F_1 transgenic individual; middle, P_1 transgenic individual; bottom, nontransgenic control.

RSV-LTR-rtGH1 cDNA appears to be variable, as a result of random integration of the transgene, the fastest growing genotype will likely be developed by utilizing a combination of family selection and mass selection of transgenic individuals following the insertion of the foreign gene. More dramatic growth enhancement in transgenic fish was obtained by introducing chinook salmon GH cDNA driven by the promoter of ocean pout antifreeze protein gene into Atlantic salmon embryos. Some of these transgenic animals grew several times faster than their controls. In the studies of transgenic medaka carrying chicken β-actin gene promoter/human GH gene construct, the F_1 transgenic individuals also grew significantly faster than the nontransgenic siblings. Figure 6 shows transgenic common carp and channel catfish harboring trout GH cDNA.

6.2 Insulin-like Growth Factor I Transgenic Fish

In higher vertebrates, growth is primarily modulated by the availability of growth hormone and insulin-like growth factors (IGFs) to their respective receptors. The secretion of GH from the pituitary gland and the binding of GH to its receptor signal the production of IGF I mRNA and the corresponding polypeptide by the liver (endocrine production) and other tissues (autocrine/paracrine function). Although the influence of GH on IGF induction and the molecular mechanism that underlies GH-controlled IGF I gene expression have been under intensive investigation in higher vertebrates for many years, very little is known about the process in lower vertebrates such as fish. Using rainbow trout as experimental animals, we are studying the mechanism by which GH and IGF control growth in lower vertebrates. Extensive analysis of PCR products and screening of a liver cDNA library have resulted in the isolation and characterization of two unique cDNA sequences. On the basis of a 98.7% nucleotide sequence homology to coho salmon IGF I, one of the cDNA sequences was identified as rainbow trout IGF I. The second cDNA sequence, which shared 43.3% identity with trout IGF I at the predicted amino acid level and 53.6% identity with human IGF II, was identified as trout IGF II. This was the first time an IGF II was identified in a fish system. Further studies revealed the presence in the rainbow trout of four size forms of IGF I mRNA with increasing nucleotide length in the E domain and one size form of IGF II mRNA. These four forms of IGF mRNA are designated as IGF IEa-1, -2, -3, and -4, in accordance with suggested revisions of IGF I nomenclature.

Both in vivo and in vitro studies were conducted to determine the dependency of IGF mRNA accumulation on GH treatment. Results of these studies showed that the production of IGF I and II mRNA in the rainbow trout liver is regulated by growth hormone.

Although continuing investigation through molecular biological approaches will shed light on the biological actions of IGFs, construction by means of gene transfer technologies of transgenic fish producing elevated levels of these polypeptides may generate alternative models for determining the effect of IGFs on fish growth and development. Toward this end, we have recently produced transgenic medaka by electroporating rtIGF I cDNA fused to a functional carp β-actin gene promoter (kindly donated by P. Hackett). In preliminary results, more than 20% of the surviving embryos integrated the rtIGF I transgene in their respective genomes. Although the number of P_1 transgenic individuals is still small, the fish are significantly larger than their nontransgenic controls ($P < 0.01$). Furthermore, it is interesting to note that both P_1 and F_1 IGF I transgenic fish hatched, on the average, 2 days earlier than their nontransgenic controls (Figure 7). Unlike the cases of studies conducted with mice and rats, our results in medaka suggest that in addition to regulating postembryonic somatic growth, IGF I may play an important role in embryonic development.

As shown in Figure 8A, a good number of P_1 and F_1 transgenic female medaka exhibit unusual enlarged abdomens, and these animals failed to spawn even after 6 months. Necropsy of these animals and histological examination have shown that the ovaries are filled with a gelatinous fluid; no mature eggs were observed (Figure 8B).

A normal medaka ovary is surrounded by a tin tunica albuginea that is covered by mesothelium (Figure 8C). Primary oocytes are small and intensely basophilic, and the nucleus is prominent. As oocytes mature, they become larger and more eosinophilic, and yolk granules are present. In the IGF I transgenic fish, several lesions in the ovary are noted (Figure 8D–F). The tunica albuginea is thickened, and numerous degenerating follicles are present in these ovaries. In some specimens, an inflammatory infiltrate of primarily macrophages surrounds necrotic follicles and extends into the tu-

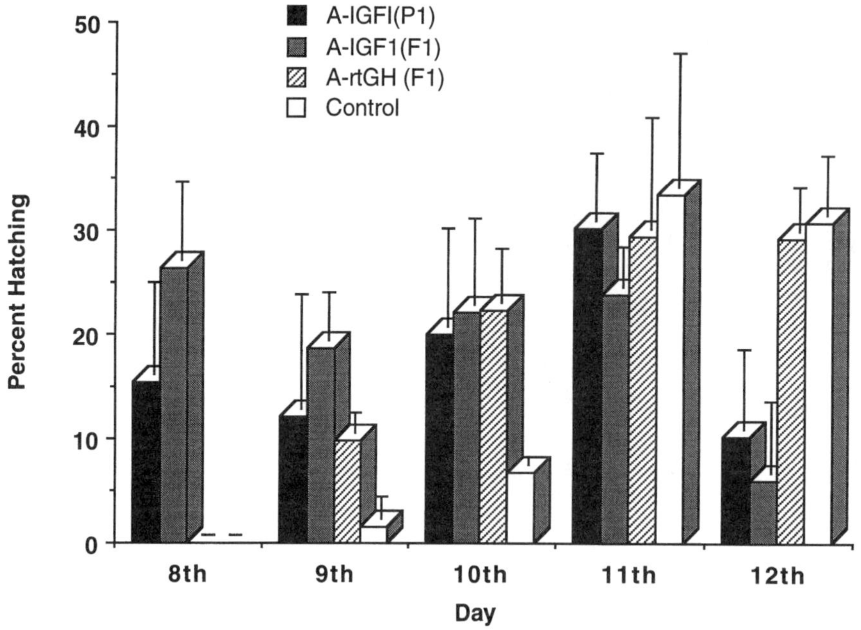

Figure 7. Effect of IGF I transgene on medaka embryonic development. Embryos were incubated at 25°C and the hatched fry were collected each day.

nica albuginea. Interestingly, a similar ovary abnormality has been observed in some of the GH transgenic female medaka (Figure 8G). These histological findings are compatible with disruptions in normal ovarian function that may have been caused by overexpression of GH or IGF genes in the transgenic individuals.

One IGF I P_1 transgenic female further developed an adenocarcinoma on the maxilla. This female was initially tumorless and spawned several batches of eggs. When the tumor appeared, spawning ceased. Histological studies have shown that the lesion was composed of epithelial cells forming small cysts (Figure 9). These cysts contained a pale eosinophilic material and some necrotic cells. The lesion was quite aggressive and replaced facial cartilage and bone, although it had not invaded the eye.

Although the observations just described are still preliminary, and more detailed studies are required, it is clear that IGF transgenic fish can serve as models for studying the involvement of IGFs on normal growth and development, reproduction, and tumor development.

6.3 Other Examples of Transgenic Fish

In addition to growth hormone and IGF gene transfer studies, a number of other gene constructs such as antifreeze protein gene, hygromycin resistance gene, neomycin resistance gene, α-globin gene, β-galactosidase gene, chicken δ-crystalline protein gene, and bacterial or insect luciferase gene have been introduced into fertilized eggs of several fish species, and these genes are driven by enhancer/promoter sequences such as mammalian or fish metallothionein promoters, simian virus 40 promoter, RSV-LTR sequence, fish β-actin, and fish antifreeze protein. In most such studies, successful gene transfer, expression, and inheritance have been reported.

Fish species inhabiting cold water, such as the winter flounder (wf), ocean pout (op), and sea raven (sr), have developed an adaptive mechanism that allows them to survive in water at subfreezing temperatures. These fish species contain antifreeze protein (AFP) genes, expression of which permits them surviving in freezing seawater, −1.8°C. Transfer of AFP genes into other fish species may also permit these fish to survive in freezing temperatures and thus make them hardier for aquaculture in cold waters. These considerations led the Fletcher, Davies, and Hew groups in Canada to collaborate on the isolation and characterization of AFP genes from winter flounder, ocean pout, and sea raven, and the production of transgenic salmon with AFP genes. Transgenic Atlantic salmon with wfAFP transgene construct have been produced, and 3% of the microinjected embryos showed expression of the transgene. When two P_1 wfAFP transgenic fish were backcrossed with nontransgenic controls, about 40% of the progeny carried the transgene as detected by PCR amplification. Although expression of the AFP transgenes were detected in some of the F_1 transgenic individuals, none of the transgenic fish showed a measurable increase of cold tolerance.

7 APPLICATIONS AND PERSPECTIVES

As reviewed by Powers, fish is an excellent model system for biological research. In the last few years, increasing numbers of scientists have turned their attention to using zebrafish and medaka as model animals for examining problems of development and transgenesis, since these fish species have a very short life cycle (3 months from embryos to reproductive mature adults) and can be made homozygous by chromosome manipulation.

Production of transgenic medaka or zebrafish may provide a powerful technology for dissecting the molecular genetics of vertebrate embryonic development. First, by introducing reporter genes linked to enhancer/promoter sequences of other genes, the temporal and spatial expression patterns of these genes can be determined. Westerfield et al. recently used the bacterial β-*gal* reporter gene to determine the activity of the mammalian homeobox *Hox* gene promoters in zebrafish. If a reporter gene is placed behind a minimal promoter sequence and the resulting transgenic offspring are exam-

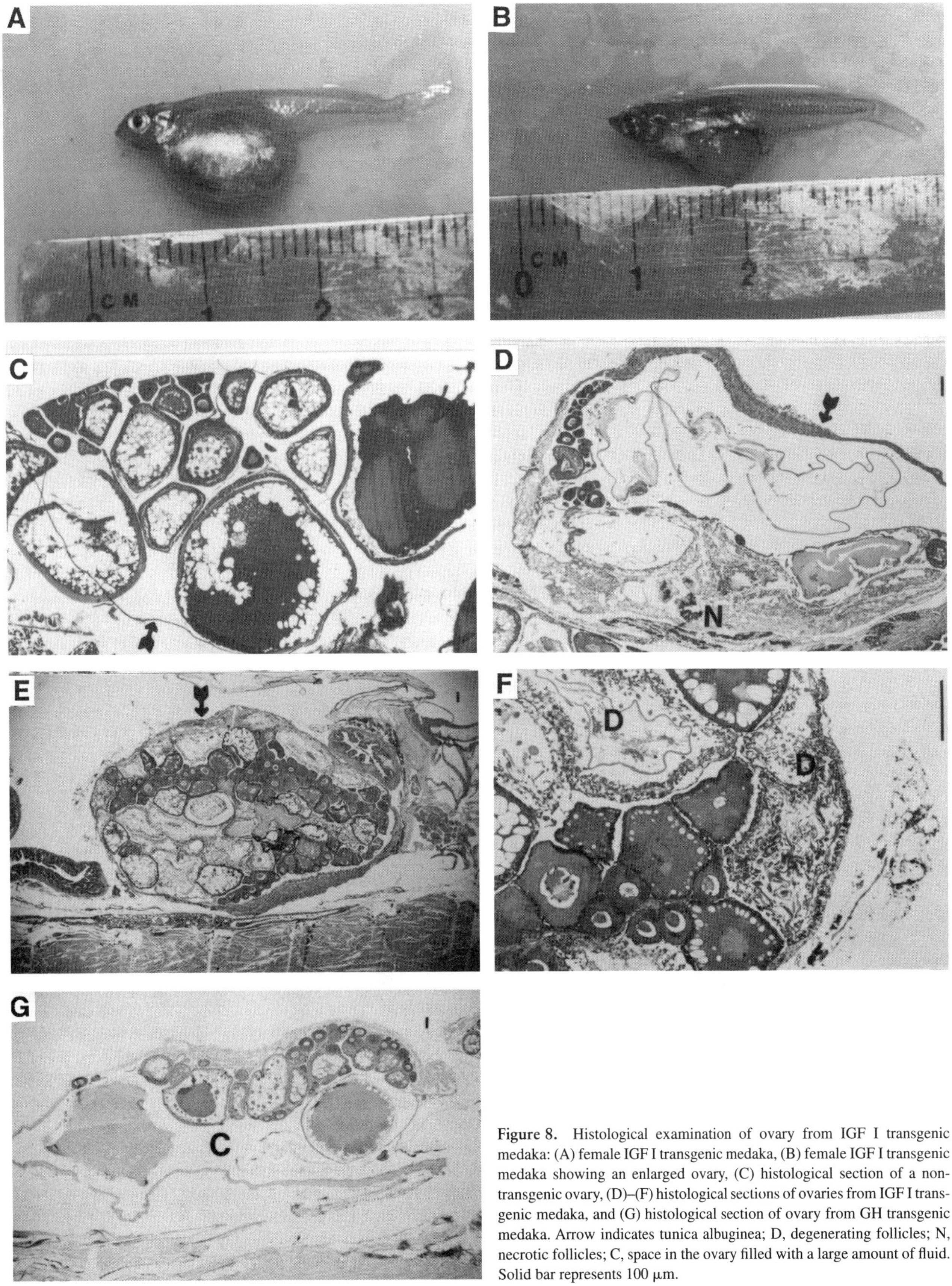

Figure 8. Histological examination of ovary from IGF I transgenic medaka: (A) female IGF I transgenic medaka, (B) female IGF I transgenic medaka showing an enlarged ovary, (C) histological section of a non-transgenic ovary, (D)–(F) histological sections of ovaries from IGF I transgenic medaka, and (G) histological section of ovary from GH transgenic medaka. Arrow indicates tunica albuginea; D, degenerating follicles; N, necrotic follicles; C, space in the ovary filled with a large amount of fluid. Solid bar represents 100 μm.

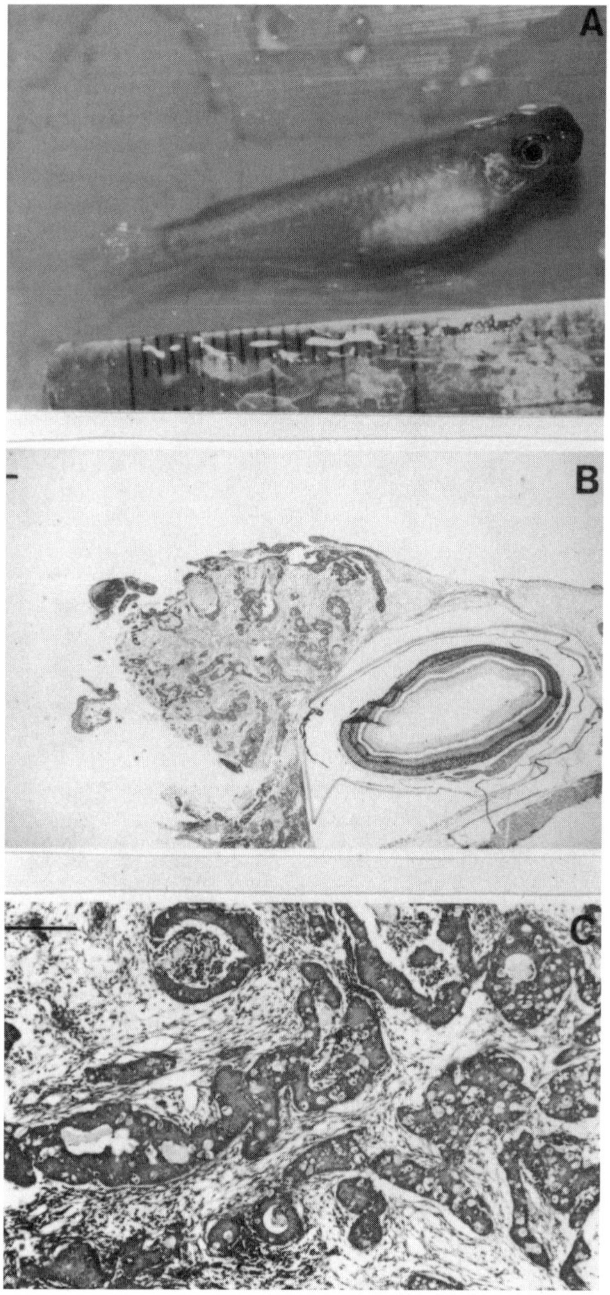

Figure 9. Histological examination of adenocarcinoma in IGF I transgenic medaka: (A) gross morphology medaka bearing an adenocarcinoma; (B) and (C) histological sections of adenocarnoma. Solid bar represents 100 μm.

ined for tissue-specific or developmental-stage-specific gene expression, specific activation of the reporter gene will indicate the interaction of the neighborhood of a specific enhancer/promoter sequence with the reporter gene. As in other systems, such a process is said to center around a "promoter or enhancer trap" and is invaluable for mapping the enhancer/promoter function of a gene in vertebrate development.

If a transgenic fish is made with a gain-of-function gene construct containing a normal fish gene or its homologue under the control of a constitutive or strong promoter/enhancer, the function of the gene product can be determined. For instance, by overexpressing a c-*myc* oncogene in a transgenic fish, the biological effect of overexpression of oncogenes can be determined. Furthermore, if a *loss-of-function* gene construct is introduced to inactivate a normal gene in a transgenic fish, the biological function of this gene product can also be determined. For example, transgenic fish carrying an inducible ribozyme transgene for interrupting the production of GH or IGF can be produced. Turning on the ribozyme transgene with an appropriate inducer at any developmental stage will serve to stop production of GH or IGF in the transgenic fish, thus allowing investigations of the effect of these hormones on fish growth and development.

Transgenic fish technology can also play an important role in biotechnological applications. Desirable genetic traits can be introduced into finfish or shellfish to produce superior transgenic strains for aquaculture. These traits may include elevated growth enhancement, improved food conversion efficiency, resistance to some known diseases, tolerance to low oxygen concentrations, and tolerance to subzero temperatures.

Recent progress in our laboratory as well as in those of others showed that stable GH transgenic fish can be produced in several fish species. The transgene not only can be transmitted into subsequent generations but also can be expressed in different tissues. The production of fast-growing transgenic fish for aquaculture calls for several important scientific breakthroughs, namely:

- developing more efficient mass gene transfer technologies
- identifying genes controlling somatic growth
- developing targeted gene transfer technology (e.g., embryonic stem cell gene transfer method, ribozyme gene inactivation method)
- identifying suitable promoters to direct the expression of transgenes at optimal levels during the desired developmental stages
- determining physiological, nutritional, immunological, and environmental factors that will maximize the performance of the transgenic individuals
- assessing safety and environmental impacts of transgenic fish

Once these problems have been resolved, the commercial application of the transgenic fish technology will be readily attained.

See also PCR TECHNOLOGY; TRANSGENIC ANIMAL PATENTS.

Bibliography

Agellon, L. B., Emery, C. J., Jones, J. M., Davies, S. L., Dingle, A. D., and Chen, T. T. (1988) Growth hormone enhancement by genetically engineered rainbow trout growth hormone. *Can. J. Fish. Aqua. Sci.* 45:146–151.

Chen, T. T., and Powers, D. A. (1990) Transgenic fish. *Trends Biotechnol.* 8:209–215.

———, Marsh, A., Shamblott, M. J., Chan, K. M., Tang, Y. L., Cheng, C. M., and Yang, B. Y. (1994) Structure and evolution of fish growth hormones and insulin-like growth factor genes. In *Fish Physiology,* C. L. Hew and N. Sherwood, Eds., pp. 179–209. Academic Press, San Diego, CA.

Du, S. J., Gong, G. L., Fletcher, G. L., Shears, M. A., King, M. J., Idler, D. R., and Hew, C. L. (1992) Growth enhancement in transgenic Atlantic salmon by the use of an "all fish" chimeric growth hormone gene construct. *Bio/Technology,* 10:176–181.

Dunham, R. A., Ramboux, A. C., Duncan, P. L., Hayat, M., Chen, T. T., Lin, C.-M., Kight, K., Gonzalez-Villasenor, L. I., and Powers, D. A. (1992) Transfer, expression and inheritance of salmonid growth hormone genes in channel catfish, *Ictalarus punctatus,* and effects on performance traits. *Mol. Mar. Biol. Biotechnol.* 1:380–389.

Fletcher, G. L., and Davis, P. L. (1991) Transgenic fish for aquaculture. In *Genetic Engineering,* Vol. 13, J. K. Seetlow, Ed., pp. 331–370. Plenum Press, New York.

Gordon, J. W. (1989) Transgenic animals. *Int. Rev. Cytol.* 155:171–229.

Hackett, P. B. (1993) The molecular biology of transgenic fish. In *Biochemistry and Molecular Biology of Fish,* Vol. 2, P. Hochachka and T. Mommsen, Eds., pp. 207–240. Elsevier Science Publishers, Amsterdam.

Jaenisch, R. (1990) Transgenic animals. *Science,* 240:1468–1477.

Lu, J. K., Chrisman, C. L., Andrisani, O. M., Dixon, J. E., and Chen, T. T. (1992) Integration expression and germ-line transmission of foreign growth hormone genes in medaka, *Oryzias latipes. Mol. Mar. Biol. Biotechnol.* 1:366–375.

———, Chen, T. T., Allen, S. K., Matsubara, T., and Burns, J. C. (1996) Production of transgenic dwarf surfclams, *Mulinis lateralis,* with pantropic retroviral vectors. *Proc. Natl. Acad. Sci. U.S.A.* 93:3482–3486.

Powers, D. A., Hereford, L., Cole, T., Creech, K., Chen, T. T., Lin, C. M., Kight, K., and Dunham, R. A. (1992) Electroporation: A method for transferring genes into gametes of zebrafish *(Brachydanio rerio)* channel catfish (*Ictalurus punctatus*), and common carp *(Cyrinus carpio). Mol. Mar. Biol. Biotechnol.* 1:301–308.

Shamblott, M. J., and Chen, T. T. (1992) Identification of a second insulin-like growth factor in a fish species. *Proc. Natl. Acad. Sci. U.S.A.* 89:8913–8917.

———, Cheng, C., Bolt, D., and Chen, T. T. (1995) Insulin-like growth factor (IGF) mRNA is accumulated in the liver and pyloric caeca of a teleost in a growth hormone-dependent manner. *Proc. Natl. Acad. Sci. U.S.A.* 92:6943–6946.

Zhang, P., Hayat, M., Joyce, C., Gonzales-Villasenor, L. I., Lin, C.-M., Dunham, R., Chen, T. T., and Powers, D. A. (1990) Gene transfer, expression and inheritance of pRSV-rainbow trout-GH-cDNA in the carp, *Cyprinus carpio* (Linnaeus). *Mol. Reprod. Dev.* 25:3–13.

Transgenic Livestock as Bioreactors

Michael A. Dalrymple and Ian Garner

1 Introduction
- 1.1 Transgenic Animals
- 1.2 Production of Transgenic Farm Animals

2 Gene Expression in Transgenic Animals
- 2.1 The Position Effect
- 2.2 The Problem with cDNAs

3 Milk and the Mammary Gland
- 3.1 Milk Proteins
- 3.2 Caseins
- 3.3 Whey Acidic Protein
- 3.4 β-Lactoglobulin
- 3.5 α-Lactalbumin

4 Expression of Heterologous Proteins in the Milk of Farm Animals
- 4.1 Examples
- 4.2 Human α_1-Antitrypsin as a Case Study

5 Problems and Future Directions
- 5.1 Problems or Opportunities?
- 5.2 The Future

Key Words

Construct An in vitro generated fusion of DNA segments consisting of regulatory and coding regions designed to spatially or temporally express the protein of interest in a novel way.

DNA Microinjection A technique for introducing nucleic acids into cells such as the oocytes of *Xenopus,* mammalian cells in culture, or fertilized mammalian eggs via a small bore glass needle using a micromanipulator.

Embryonic Stem (ES) Cells Pluripotent or totipotent cells of normal karyotype derived from the developing mammalian embryo. When introduced into other embryos of the same species, ES cells will contribute to the developing embryo in such a way that the resulting animal will be a chimera of two cell types. If the germ cells consist of two populations of cells, whole animals derived from the original ES cells can be obtained by conventional breeding.

Founder A transgenic animal resulting directly from the manipulation of a mammalian egg involving gene transfer. Such animals may be mosaics.

IVM/IVF In vitro maturation of mammalian oocytes and subsequent in vitro fertilization of eggs.

Lactation The period during which milk is produced following parturition. This varies from species to species and can be influenced by such factors as weaning times and machine milking. For example, the latter can extend lactation times dramatically. Typical lengths of lactation for the mouse, rabbit, pig, sheep, goat, and cow are 3, 4, 3, 8, 52, and 52 weeks, respectively.

Line A family of animals derived from a founder animal and possessing the same germ line modification.

Mosaic A founder transgenic animal that possesses the desired germ line modification in only a subset of its cells. Such animals are thought to arise following integration of the transgene after the first cell division of the embryo and may be incapable of transmitting the transgene to their progeny by virtue of an absence of the modification in their germ cells.

Transgenic The term applied to a plant or animal into which a gene from another species (the transgene) has been introduced. The term is also used more broadly to describe any organism whose genome has been altered by *in vitro* manipulation to induce, for example, specific gene knockout.

The ability to induce stable genetic alteration of the chromosomal material of mammals has opened up exciting new possibilities for the manipulation of farm animals. One such application is to modify the mammary gland in such a way that additional, commercially valuable proteins are secreted into milk during lactation. Such so-called, transgenic livestock effectively become walking "bioreactors," producing large quantities of (usually human) recombinant proteins in their milk at relatively low cost. Furthermore, not only has the gland an amazing capacity for synthesis, but it appears to be capable of performing posttranslational modifications that are frequently essential for full biological activity. Such animals represent an easily maintained and infinitely renewable source of disease-curing recombinant proteins.

1 INTRODUCTION

Recombinant pharmaceuticals are invariably proteins that are obtained by cloning the relevant genetic material, placing its expression under the control of a very different DNA sequence, and reintroducing this hybrid "gene" into a host organism within which the new construct will produce mRNA and protein. The choice of host has generally reflected the molecular biologists' state of knowledge and so, historically, the emphasis has shifted through the familiar laboratory "tools." A myriad of expression systems have been developed to enable very high expression of foreign genetic material in the bacterium *Escherichia coli.* Unfortunately, both the biological activity and stability of most proteins having therapeutic value are intimately connected with key posttranslational modifications, and prokaryotes are generally unable to perform the relevant enzymatic reactions. This has also been found to be the case with the lower eukaryote *Saccharomyces cerevisiae,* baker's yeast, another favorite of the genetic engineer. Mammalian cells in culture are a more suitable environment for the expression of human therapeutic proteins. However, the growth requirements of *in vitro* cultured cells are frequently demanding and expensive to maintain. Achieving a balance between high cellular expression and high cell density is problematic.

1.1 Transgenic Animals

Soon after it became possible to stably introduce "foreign" DNA into cultured mammalian cells, Ruddle's group demonstrated that by microinjection of DNA into the pronucleus of fertilized eggs, it is possible to achieve stable integration of this DNA in tissues of mice. These workers and others went on to show that the newly introduced DNA could enter the germ line and be passed to progeny in Mendelian fashion. It was also shown that expression of foreign DNA could be detected in such "transgenic" animals and, furthermore, directed to a specific tissue or organ of choice. As the work with mice gathered pace, it became clear that larger animals, including domestic farm animals, could be altered in the same way.

The power that this technology opens up is that of being able to genetically modify domestic animals in a more rapid fashion than can be achieved by conventional breeding methods. Furthermore, changes that were previously thought not possible are now a realistic goal. One such modification, perhaps the only one to truly reflect the term "bioreactor," is to direct the expression of a potentially valuable protein to a body fluid, which can then be harvested. Of preference, the garnering of this material should be noninvasive, and a renewable body secretion should be targeted. The ideal overall process would be somewhat symbiotic, benefiting both the producing animal and the collector. Several body fluids have been targeted—for example, blood, saliva, and milk. Of these, only the latter approaches the ideal just alluded to. From the first demonstration that foreign, or heterologous, proteins can be expressed in the milk of transgenic mice, this capacity has been extended to sheep, goats, pigs, rats, cows, and rabbits. Hopefully, it will soon be demonstrated in a transgenic cow. Thus, the expression of recombinant therapeutic proteins, initially performed in the bacterial cell, now returns to the whole mammal.

This article concentrates on the production of heterologous proteins in the milk of transgenic mammals and addresses several aspects of this new technology. We will survey control of gene expression in the mammary gland, the systems used to achieve heterologous gene expression in mice and other animals, and the important differences between expression in tissue culture and expression in the whole animal.

1.2 Production of Transgenic Farm Animals

Despite the efforts of many workers over several years, the development of a reliable cell culture system able to faithfully mimic the properties of the mammary gland has proved elusive. Primary cultures of mammary cells have been the most successful, but by their nature these lack a degree of consistency. If such cells were available, they would allow for a rapid and cost-effective validation of constructs designed to express heterologous proteins in the lactating mammary gland of livestock. In their absence, the mouse is used as a model system for mammary expression studies to enable the testing of transgene constructs for high expression and the ability of the mammary gland to process the product correctly prior to the production of larger transgenic animals.

Several methods can be used to introduce exogenous DNA into the developing mammalian embryo in such a way that it may ultimately become stably integrated into a chromosome of the resulting animal. *Microinjection,* which we discuss in this section, introduces DNA into one of the pronuclei of a single-cell, fertilized egg by physically injecting it through a glass needle having a microscopically small bore. The approach is labor-intensive and technically demanding, but it appears to have few limitations with respect to the type or size of DNA that can be introduced. *Retroviruses* can be genetically modified to act as vectors, allowing the infection of embryos with exogenous DNA. Superficially, this is an attractive option inasmuch as the process is relatively efficient. However, only limited sizes of exogenous DNA can be incorporated into available vectors; in addition, a high degree of mosaicism is often obtained, necessitating extensive breeding to isolate true lines; and in some cases, the transgene may be inactivated during this process. *Sperm* too, have been claimed to be efficient mediators of exogenous gene transfer into mammalian embryos by one group. Despite the obvious attractions of such an approach, this work has been dogged by the striking inability of many other groups to repeat the successes reported by the originators.

The final method of note is *DNA recombination, homologous or nonspecific, in embryonic stem (ES) cells.* Derived from a mammalian embryo and maintainable in culture, ES cells are pluripotent or totipotent cells that can contribute to a developing embryo in such a way that the resulting animal possesses a number of cells derived from these ES cells and not from the original embryo. ES cells can also be genetically manipulated while in culture. DNA can be introduced into them by a variety of techniques (e.g., electroporation, lipofection, microinjection, retroviral infection, calcium phosphate precipitation), and investigators can select cells that have incorporated exogenous DNA in a specific or nonspecific way. This, too, has obvious attractions, the most desirable of which may be that one can direct the exogenous DNA to a specific site on a given chromosome. However, to date this method has been successful only in the mouse. ES cells have been isolated from murine embryos only, the search continues for ES cells of other species.

Hitherto, the favored means of stably introducing exogenous DNA into somatic and germ line cells of the whole animal has been microinjection of DNA into the pronuclei of fertilized eggs. The procedures by which this technique is used to generate transgenic mice have been extensively described elsewhere. Other methods have been successfully used to generate transgenic mice and livestock (e.g., cytoplasmic injection of exogenous DNA). In this article we will confine our discussions to the former route and the peculiarities of the generation of transgenic livestock.

Several difficulties arise when the mouse technology is transferred to domestic animals, and in particular farm animals. The first of these is the generation time of the chosen species. Gestation time and the attainment of sexual maturity greatly influence the time at which milk, hence product, may be attainable. Indeed, realistic commercial products will be available only at the birth of at least G_2 animals. Table 1 summarizes the salient features of the more popular species. Because of the rapidity of gestation and maturation of the mouse, this animal offers the earliest source of milk. However, milk volume in this species is extremely limited, although yields are sufficient to justify its use as a test system. The next most rapid is the rabbit, which will produce about 8–10 L of milk per year over progressive lactations, yielding first product after about 7 months. The pig can produce more than 100 L of milk from a single lactation twice a year, although it is somewhat more difficult to harvest this volume than milk from the more established dairy animals. The sheep and the goat produce similar quantities of milk (≥100 L/lactation) in a time frame similar to that of the pig but are much more amenable to the milking process. Depending on the breed of sheep or goat, several hundred liters of milk can be obtained per animal over a 12-month period. The ultimate source to be considered is the cow, which may produce around 10,000 L of milk during a yearly lactation, but only after an extensive gestation and maturation phase.

Several of the foregoing species are amenable to hormonal induction of lactation, which can shorten the time to first milk; but volumes of product obtained are generally much below that obtained from a natural lactation. Thus hormonally induced lactation does not provide a satisfactory source of material for production purposes but may yield material with which early characterizations of the product can be performed and purification processes can be developed. All these factors, together with the estimated volume of recombinant protein required, influence the amount of and the speed with which a given volume of product is attainable and thus must be taken into consideration when choosing the species with which to work.

Having chosen the target livestock species, one faces the potentially formidable task of preparing hundreds of fertilized eggs at the one- or two-cell stage, followed by synchronizing suitable recipient animals for reimplantation. There is also cost to consider, both in terms of labor and with respect to the purchase and housing of the animals. Since all the main farm animals and rabbits can be superovulated with appropriate hormonal regimes, it is possible to get large numbers of embryos for microinjection. However, the process is much less efficient in livestock than in the mouse. Rabbits are the closest to the mouse, yielding similar numbers of embryos. Sheep generally produce only around 6–10 eggs per ewe, compared to about 30 per female mouse. In this respect, cows and goats are similar to sheep, although one advantage in the use of pigs is that they are closer to the mouse. Pigs are polytocous or litter bearing, thus, the number of recipient animals is also reduced. Sheep, goats, and cows do not generally bear more than twins, and indeed the life of the mother and offspring can be compromised by multiple births. While sheep, goats, and pigs are relatively cheap, good dairy cattle are very expensive. For this reason, and to avoid the need for surgical intervention, most people now use slaughterhouse-derived ovaries, which are manipulated in vitro to produce eggs that are then fertilized in vitro (so-called IVM/IVF).

Microinjection of pronuclei in farm animals is also problematic. Fertilized eggs from pigs and goats contain granular inclusions that obscure the target. In sheep, the pronucleus is very close to the cytoplasm in refractive index and thus difficult to see by all but the expert eye. For cattle, the problem is that the pronucleus is frequently loosely associated with the cytoskeleton, and very sharp needles are needed to penetrate the pronuclear membrane without moving the target! Some of these difficulties can be alleviated by centrifugation of the eggs to reveal the pronucleus prior to microinjection.

Finally, for reasons that are completely mysterious, the frequency of transgenesis in larger animals, as measured by the percentage of live births that are transgenic, is considerably reduced. It is not clear why this should be. Perhaps integration of foreign DNA in these eggs happens less frequently; or possibly eggs that integrate foreign DNA are at a developmental disadvantage and preferentially die. The end result is that the frequency of transgenic animals in livestock is usually much closer to 1% than the 10% or more associated with the mouse.

Table 1 Salient Features of Different Species

Species	Eggs/ Super- ovulation	Times (months) G_0 Birth	G_0 Adult	G_1 Birth (first milk)[a]	Potential Milk Volume (l)	Times (months) G_1 Adult	G_2 Birth	G_3 Birth (G_2 Herd)
Sheep/goat	6–10	5	13	18 (9)	>100	26	31	4
Cow	—[b]	9	23	32 (12)	>10,000	46	55	78
Pig	30–50	4	11	15	>100	22	26	37
Rabbit	≈30	1	6	7	≈1			
Mouse	≈30	0.75	2.5	3.25	≈0.001			

[a]Numbers in parentheses refer to the time of first milk using premature induction of lactation.

[b]Not applicable; IVM/IVF generally used.

2 GENE EXPRESSION IN TRANSGENIC ANIMALS

There are two important areas within which gene expression in the whole animal differs considerably from that seen in other systems, namely, the influence of local chromatin environment and the effect of introns. The first has been called the "position effect" and the latter may be termed the "cDNA problem."

2.1 The Position Effect

When transgenes integrate at single sites to form different transgenic lines, there are basically two physical features that appear to play a role in the ultimate expression level from this new locus: transgene copy number and the actual site of integration. The site of integration would appear to be the most important feature. A generally accepted interpretation of this position effect is that local chromatin structure bears global controls that mediate permissive and nonpermissive environments for subsequent activation of transcription and maintenance of expression levels. In support of this interpretation it has been shown that if poorly and strongly expressing transgenes are reisolated from transgenic mice and reintroduced into eggs to generate new lines, these lines exhibit both characteristic low and high levels of expression irrespective of the original transgenic line. This result excludes the possibility that the original expression level was a function of some drastic transgene modification at the site of integration.

The hypothesis also predicts that there are trans-acting elements that mediate control of chromatin structure in such a way as to determine the state of permissiveness for transcription. Several DNA elements have indeed been described that can mediate such controls. The most extensively studied is the so-called locus control region (LCR) from the human globin locus. This element was discovered following the identification of DNAse-hypersensitive sites in chromatin flanking the human globin gene locus, extensive molecular analysis of the locus, and studies on various naturally occurring thalassemias. Upon cloning, this region was shown to confer on globin and heterologous transgenes the property of high level, tissue-specific expression in mice. Furthermore, the individual transgenic lines exhibited remarkably similar levels of expression, which appeared to be simply a function of transgene copy number. Thus, the LCR would also appear to confer position-independent expression on transgenes to which it is linked. Similar elements have been found to be associated with other genes. Thus, the classic definition of an LCR is a DNA element that confers position-independent, copy-number-dependent, high level, and tissue-specific gene expression on a linked transgene.

Other DNA elements have been described that impact on gene expression, presumably through effects at the chromatin level. Matrix attachment regions (MARs) are pieces of DNA that are found preferentially bound to the nuclear matrix. These DNAs are often AT rich and have been shown to have insulating properties, protecting transgenes from their chromatin environment. Other, non-MAR, insulators have also been reported.

2.2 The Problem with cDNAs

We, and others, have consistently found that transgenes with no introns are expressed very poorly or below the level of detection. Engineering artificial introns into such genes or replacing one or more of the natural introns can partially restore expression but rarely brings it back to wild-type levels. This, of course, can be clearly seen using milk protein gene elements, given the high levels of expression in nature. Compared to a wild-type gene (i.e., the coding sequences with all the natural introns in place), cDNAs or partial genes almost always fall some way short in their expression levels. This is also true of heterologous genes expressed from milk protein gene promoters.

It is not clear why cDNA transgenes express so badly. It is not simply the requirement to enter the splicing pathway, because putting back one or more introns does not restore the expression levels of the natural gene. At the moment we can only speculate, but it could be that introns and intron position play a role in determining mRNA secondary structure and have an impact on nuclear–cytoplasmic transport, protection from intracellular degradation, and translation. A further possibility is that introns influence chromatin structure in some way, allowing the expression of integrated transgenes.

This poor expression of cDNA is an important consideration when considering transgenic animals as bioreactors, primarily because cDNA transgenes often are the only ones available, but also because some naturally occurring genes are very large (e.g., human factor VIII, ≈ 180 kb), and the manipulation and microinjection of large DNA sequences pose additional problems.

For the production of transgenics expressing proteins in the milk, a method has been developed by which reasonable levels of gene expression can be "rescued" from otherwise inactive cDNA transgenes. This "rescue" technology comprises a coinjection of the entire ovine β-lactoglobulin gene together with the transgene of interest. Absolute levels of mRNA and protein expression still vary considerably (indicating that the milk protein gene does not function as an LCR), but expression can approach 10–40% of the level attained with equivalent genomic constructs. It is not clear why this should be. The suggestion is that the β-lactoglobulin gene acts to make the local chromatin environment at least partially permissible, perhaps through the strength of its promoter sequences.

3 MILK AND THE MAMMARY GLAND

The mammary gland has many attractions for the biologist. For the developmental biologist, a major appeal is the morphogenesis that occurs almost entirely in the preadult or adult animal; cancer specialists have been fascinated by specific mammary tumors, in particular those caused by mouse mammary tumor virus (MMTV), while others have been drawn by the complex hormonal control of gene expression. Without doubt, the mammary gland has an impressive capacity for macromolecular synthesis, producing a renewable secretion that can be harvested by well-developed, noninvasive methods. In domestic farm animals, such as cows, goats, and sheep, hundreds to thousands of liters of protein-rich milk can be garnered per animal per year (see Table 1). It is this capability that has attracted the attention of the genetic engineer, who can manipulate farm animals and target the expression of human and other new proteins to the mammary gland.

3.1 Milk Proteins

The major milk proteins are the caseins (e.g., α_{s1}, α_{s2}, β, and κ), which form large spherical complexes called micelles, and the whey proteins (e.g., β-lactoglobulin, whey acidic protein, α-lactalbumin).

Table 2 Protein Composition of Various Milks

Protein	Concentration (g/L)[a]				
	Cow	Sheep	Mouse	Pig	Human
Caseins					
α_s1	10.0	12.0			
α_s2	3.4	3.8			
β	10.0	16.0	7.0	30.0	3.0
Whey					
α-Lactalbumin	1.0	0.8	0.1		1.6
β-Lactoglobulin	3.0	2.8		20.0	
Whey acid protein			2.0		

[a] Blanks, not present or not known.

Apart from their obvious nutritional function, these proteins have roles in maintaining the integrity of the milk itself. Although the exact combination of proteins differs across species (Table 2), between them, the two classes constitute greater than 90% of the protein in milk. Consequently, the genes for these proteins have been the main vehicles for transgenic studies.

Caseins are almost always the most abundant milk proteins. They usually have a molecular weight of around 20,000, are relatively hydrophobic and, under the ionic conditions in milk, associate into colloidal micelles. There are five casein genes known in the mouse and four major casein genes in cows and sheep. In all these animals, the genes are tightly linked on a single chromosome and fall within 400 kb of DNA. The proportion of casein differs considerably between humans (45%) and cows (80%), and this may be reflected in the fact that only two caseins are known in humans, β and κ.

Of the whey proteins, α-lactalbumin (α-lac) has been detected in all milks that contain lactose, presumably as a consequence of this protein's role in lactose synthesis. α-Lactalbumin is a subunit of lactose synthetase and has molecular mass of around 14,000 Da. It has a key role in the regulation of milk lactose content, contributing to milk osmotic pressure and volume. A deficiency in α-lac severely disrupts the ability of a mammal to lactate. β-Lactoglobulin (β-LG) is a whey protein of 162 amino acids and is the most common whey protein in the milk of ruminants. Homology to retinoic acid binding proteins has led to the suggestion that β-lactoglobulin may function in vitamin A transport. Whey acidic protein (WAP) is the major whey protein in rodent milk, constituting some 10–15% of the total poly(A+) mRNA of mouse mammary gland. It is a member of the "disulfide core" family of proteins. Initially thought to behave like a nutritional source, WAP may have a wider role, as suggested by its structural similarity to mucous protease inhibitors, proteins that participate in tissue modeling.

Given the huge synthetic capacity of the mammary gland, it was immediately obvious to some that if a heterologous gene could be driven by a milk gene promoter, it should be possible to secrete high levels of this product in milk. Assuming that high levels of expression can be achieved, there are additional advantages: the dairy industry, especially for cows but also for sheep and goats, is well established and scientifically advanced, providing the necessary infrastructure for production and processing of transgenic milk; production of milk is easy, cheap, renewable, and noninvasive; milk has a very positive image in the mind of the consumer.

Genes for the major milk proteins have been the vehicles for both the study of milk proteins in transgenic animals and the elaboration of vector systems for expression of heterologous proteins in the mammary gland. Because of the extended time scales and costs involved in work with livestock, as alluded to earlier, most initial studies are performed using the mouse as a model system. A collection of published results is included in Table 3. In the following two sections, we concentrate on some of the highlights of this work.

3.2 Caseins

Rosen and co-workers generated lines of transgenic mice containing either the entire rat β-casein gene or the rat β-casein promoter, driving expression of chloramphenicol acetyltransferase (CAT). The levels of transgene mRNA expression were only 0.01–1% of the endogenous mouse β-casein. CAT activity was specific to the mammary gland, but levels were low. Removal of the first exon and most of the first intron from the gene completely abolished expression, suggesting a role for intragenic sequences in control of expression. The authors draw a parallel between the casein family of genes and the globin situation. The whey protein genes, WAP, β-LG, and α-lactalbumin appear to be unlinked in the genome; however, the casein genes are arranged on a single chromosome and are tightly linked. This is consistent with these genes being under coordinated, long-range control. By analogy with the globins, perhaps we should be looking for an LCR, which will confer on casein genes high level, locus-independent, mammary-specific expression.

Mammary explant cultures, derived from these transgenics, have been used to examine hormonal control of β-casein. Endogenous and transgenic β-casein genes were appropriately stimulated by prolactin; however, β-casein/CAT constructs were not. Similar effects were seen with insulin and glucocorticoids, implying that intragenic sequences mediate hormonal regulation of β-casein. This type of control, however, is not entirely due to DNA downstream of the promoter, for some stimulation can be detected. Work with hormone-responsive cell lines has indicated that induction of the rat β-casein gene may be by relief of transcription repression. An analysis of the mouse β-casein gene in pregnant and lactating mice identified a pregnancy-specific mammary nuclear factor that appears to serve as a repressor mediating the inhibitory action of progesterone on β-casein. Thus, the design of efficient constructs to drive heterologous protein expression to the transgenic mammary gland is not an easy task.

Experiments designed to express heterologous proteins from casein promoters have been both successful and innovative. Interleukin 2 (IL-2) has been expressed in rabbit milk from 2 kb of rabbit β-casein promoter. Of four transgenic females analyzed, the highest level of expression was 450 ng/mL. A massive 21 kb of bovine α_{s1}-casein promoter was utilized by Meade and co-workers to drive expression of the human urokinase gene in transgenic mice. Three founder animals were obtained, but only one transmitted the transgene. The female offspring of this mouse secreted active urokinase at concentrations of 1–2 mg/mL in the milk. In both these cases, the expressed heterologous gene included some or all of its natural introns, a feature that seems to be necessary, but not the sole determining factor, for high expression.

Rosen's group have addressed the possibility of secreting both subunits of a heterodimeric protein into the milk of transgenic mice. Separate constructs placed cDNAs for both the α and β subunits of follicle-stimulating hormone (FSH) under the control of the rat β-casein promoter. Mice harboring both transgenes were made either by single injection and mating or by coinjection. Bigenic mice expressed functional, heterodimeric FSH at levels up to 15 μg/mL. Interestingly, mRNA for the α subunit accumulated to 7–17-fold

Table 3 Published Expression Levels of Foreign Proteins in Mouse Milk

Milk Gene	Protein	Gene Configuration	Level
Bovine			
α-Lac	α-Lactalbumin	Genomic	0.8 mg/mL
α-Lac	α-Lactalbumin	Genomic	1.2–3.7 mg/mL
α-lac	Ovine trophoblast interferon	cDNA	1 μg/mL
α_{s1}-Casein	h Urokinase	Genomic	1–2 mg/mL
α_{s1}-Casein	h Lactoferrin	Genomic	10.6 mg/mL
α_{s1}-Casein	h Tissue plasminogen activator	Minigene	0.5 μg/mL
α_{s1}-Casein	h Lactoferrin	cDNA	36 μg/mL
Goat			
β-Casein	β-Casein	Genomic	21–24 mg/mL
β-Casein	h Antithrombin III	Minigene	10 mg/mL
β-Casein	h Longer acting tissue plasminogen activator	cDNA	5 mg/mL
Mouse			
WAP	h Tissue plasminogen activator	cDNA	100 μg/mL
WAP	h Growth hormone	Genomic	3.5 mg/mL
WAP	h Breast cancer protein PS2	cDNA	1.5 μg/mL
WAP	h Tissue plasminogen activator	cDNA	0.25 mg/mL
WAP	h Protein C	cDNA	3 μg/mL
WAP	h Fibrinogen	cDNA	60 μg/mL
WAP	h Extracellular superoxide dismutase	cDNA	0.7 mg/mL
WAP	h Protein C	Genomic	0.7 mg/mL
Rabbit			
WAP	h α_1-antitrypsin	Genomic	10 mg/mL
WAP	h Growth hormone	Genomic	22 mg/mL
Sheep			
β-LG	h Factor IX	cDNA	100 μg/mL
β-LG	h Tissue plasminogen activator	cDNA	0.6 μg/mL
β-LG	h Growth hormone	Genomic	1.7 mg/mL
β-LG	h Protein C	Genomic	750 μg/mL
β-LG	h Protein C	cDNA	187 μg/mL
β-LG	h Fibrinogen	Genomic	≥1 mg/mL
β-LG	h γ-Interferon	cDNA	20 ng/mL
β-LG	h Monoclonal antibody	cDNA LC minigene HC	0.3 mg/mL
β-LG	h Serum albumin	Minigene	10 mg/mL
β-LG	h α_1-antitrypsin	cDNA	800 μg/mL
β-LG	h α_1-antitrypsin	Genomic	21.3 mg/mL

LC = light chain; HC = heavy chain.

higher levels than that for the β subunit. One construct placed 408 bp of MMTV LTR [carrying four glucocorticoid response elements (GREs)] upstream of the β-casein promoter, and this transgene gave more consistent expression, and expression at higher levels, than the non-MMTV counterpart.

The caprine β-casein promoter has been used to express in mouse milk a normally membrane-associated protein—cystic fibrosis transmembrane conductance regulator (CFTR). Specifically, 50 μL of milk contained the equivalent of 3×10^6 Chinese hamster ovary (CHO) cells worth of CFTR, and the protein appeared to be associated with milk lipid globules. This opens up the possibility of secreting other membrane-associated receptors in milk, although yields would appear to be limited.

3.3 Whey Acidic Protein

WAP expression has been particularly difficult to detect in mammary cell lines, leading to the conclusion that cell interactions and shape play an important role in its expression. The mouse and rat WAP genes have been cloned and reintroduced into mice by microinjection. Rosen and co-workers introduced a 4.3 kb genomic rat WAP fragment into mice. WAP expression levels ranged from 1 to 95% (average 27%) of the endogenous mouse gene, and the expression was mammary specific. However, the temporal expression of the transgene appeared to be slightly different, being switched on several days before the endogenous gene at around 7 days of pregnancy. When a different construct was used, having fewer 3′ flanking sequences, expression became uniformly high but still initiated earlier than the mouse gene. Hennighausen's group also noted this phenomenon. These workers reintroduced a 7.2 kb fragment of the mouse WAP gene, modified in a way that allowed its discrimination from the endogenous mouse gene. Six mice from 13 lines expressed the new WAP mRNA, during lactation and specifically in the mammary gland, at levels from 3 to 54% of normal. Surprisingly, these transgenic mice also exhibited a peculiar phenotype, *milchlos,* or female mice that failed to lactate. This failure was associated with an arrest in mammary development and apparently was due to the early onset of WAP expression in these mice. It is possible then that WAP plays a regulatory role in the terminal differentiation of mammary tissue. The *milchlos* phenotype was also seen in transgenic pigs expressing murine WAP.

A variety of heterologous genes have been hooked onto the WAP promoter and expressed in transgenic mice: tissue plasminogen activator (tPA), protein C, growth hormone, and breast cancer protein PS2. In general, levels of expression have been variable in mice (100 μg/mL tPA, 1.5 μg/mL PS2, 700 μg/mL protein C), probably reflecting the properties of the various proteins and constructs used.

3.4 β-Lactoglobulin

The gene for sheep β-LG was the first to be introduced into transgenic mice and, so far, has given far and away the highest reported levels of heterologous gene expression in either mice or livestock. Since rodents do not possess a β-LG gene, detection of β-LG expression in transgenic mice is easy.

Simons and co-workers took 16.2 or 10.5 kb genomic fragments of the ovine β-LG gene and introduced them into transgenic mice. These constructs differed in the amount of 3′ flanking sequences, each containing 4.2 kb of 5′ flank and 4.9 kb of intron/exon sequence. From nine G_0 mice selected for further study, expression was generally very high: one line expressed β-LG protein at around 23 mg/mL as judged by gel electrophoresis.

Developmental studies on these mice showed that expression of ovine β-LG in mice, which have no endogenous β-LG gene, roughly paralleled that found in sheep. RNA analyses on a variety of tissues showed that expression of the transgene was restricted to the mammary gland. Temporal expression of β-LG was compared with the endogenous β-casein gene. Both genes were expressed at low levels in virgin gland. From days 1 to 10 of pregnancy, there was a small accumulation of transcript; the period of 10 days to parturition saw a rise to between 65 and 80% of the final level. Following parturition, a further 1.5–2-fold increase occurred. Differences between sheep and mice were explained in terms of the relative state of differentiation of the gland in the two animals. Indices of differentiation show that, by parturition, mouse mammary cells are more differentiated than their ruminant counterparts.

The "minimal" β-LG promoter has been determined. Only 406 bp of sequence 5′ to the transcription start site is necessary for both high level and tissue-specific expression of the β-LG gene in transgenic mice.

Several heterologous genes have been expressed from the β-LG promoter. Archibald et al. expressed human α_1-antitrypsin (AAT) in mice. In this case, the AAT gene used was a full genomic fragment, minus the first intron but including all other intron sequences. This β-LG-AAT transgene (AATB) gave levels of fully functional AAT protein in the milk ranging up to 7 mg/mL. We have since attained 21.3 mg/mL with the same construct. We have also expressed human growth hormone, fibrinogen, and protein C in transgenic mice at very high levels. The best expressing β-LG-GH mice gave 1.7 mg/mL, the best β-LG-protein C mice secreted 750 μg/mL, and β-LG-fibrinogen mice secrete in excess of 1 mg/mL. In all cases, genomic gene fragments were used, introns and all. Expression data are summarized in Table 3.

Such results have reinforced the idea that introns are required to attain high levels of expression of transgenes. For example, if an AAT cDNA is used instead of a genomic fragment, expression levels of AAT in mammary gland fall to low or undetectable levels. Analysis of expression from a completely intronless FIX transgene showed no detectable RNA in mammary tissue. Our own results with human protein C show almost undetectable levels of protein in milk where there are no introns in the transgene. If the highest level attainable with full genomic protein C DNA constructs is assigned as 100%, the maximum expression we have achieved with a cDNA has been about 25%, in a transgene containing the last β-LG intron only. These observations are consistent with results obtained with other milk protein systems and, as noted earlier, indicate a fundamental feature of the control of gene expression in animals.

3.5 α-Lactalbumin

The α-lactalbumin system has also been studied in transgenics and utilized as a vector for expression of heterologous proteins. In one experiment, the entire bovine α-lactalbumin gene was introduced into mice and expression of protein and mRNA monitored. Two mice from six transgenics gave levels approaching those of the endogenous α-lactalbumin gene. The gene also shows developmental expression patterns similar to those of its endogenous counterpart. Very high expression levels of bovine α-lactalbumin have been reported in transgenic rats, reaching 2.4 mg/mL. There is one example of heterologous protein expression in the literature, and this involved the expression of ovine trophoblast interferon. Levels of approximately 1 μg/mL of ovine trophoblastin were demonstrated.

In general, the promoter sequences derived from milk protein genes direct expression quite specifically to mammary tissue in transgenic mice. There are, however, cases in which apparently aberrant expression has been detected. A β-LG-AAT transgene showed human AAT mRNA in the salivary glands of some transgenic lines. There is also one report that a guinea pig α-lactalbumin gene introduced into transgenic mice showed expression in the basal layer of the sebaceous glands in the skin. It would appear that such examples are rare, which serves to confirm the value of expressing heterologous proteins in the mammary gland as a system for producing recombinant pharmaceuticals.

4 EXPRESSION OF HETEROLOGOUS PROTEINS IN THE MILK OF FARM ANIMALS

4.1 Examples

Several papers have described the successful generation of transgenic livestock that express foreign proteins in their milk. A summary of results obtained to date is presented in Table 4.

In the early stages of the development of this technology, Clarke and colleagues generated transgenic sheep with both human AAT and factor IX transgenes driven by the ovine β-LG promoter. In both cases, a cDNA encoding either human AAT or factor IX was inserted into the 5′-untranslated region of the ovine β-LG gene. In general, constructs of these types have performed poorly over the years. Reflecting this, initial results were disappointing and were probably related to the type of construct used. Levels of factor IX and AAT obtained were of the order of 25 ng/mL and 5 μg/mL, respectively. This work did, however, demonstrate that recombinant proteins can be expressed by the transgenic mammary gland and that lines of animals can be derived from the original founder animals. This latter point, which may have seemed relatively insignificant at the time, is an essential requirement for the technology to develop into a true production process. Around the same time, Genzyme Transgenic Corporation demonstrated that the murine WAP gene could be used to direct the expression of human tPA to the milk of lactating transgenic mice. These researchers later extended their work to transgenic goats using a fusion of 2.6 kb of the murine WAP promoter region to a cDNA encoding a variant of tPA with a longer systemic half-life than the normal protein (LAtPA). Initial levels of expression attained were only 3 μg/mL, but this yield was improved to 7.5 mg/mL by switching to the caprine β-casein gene promoter. These authors have utilized the same promoter to express human antithrombin III (ATIII) at levels of about 1.3 mg/mL from an ATIII cDNA and 7 mg/mL from ATIII genomic fragment in goat milk. Both proteins exhibit biological activity.

Velander and co-workers have generated transgenic pigs expressing human protein C in their milk at levels up to 1 mg/mL. The construct used consisted of an insertion of a cDNA encoding protein C into the first exon of the murine WAP gene. However, the product was incompletely processed. We have also expressed protein C in the milk of transgenic sheep at levels of 0.1 mg/mL. In this case, the construct used was cDNA based, being driven by the β-LG promoter. The sheep-derived product is correctly processed and fully active. Perhaps these results reflect an inability of the mammary gland to process this very complex protein at high expression levels [≥ 1 g/L], or perhaps they represent a characteristic of the pig. Time will tell.

4.2 Human α_1-Antitrypsin as a Case Study

We have generated sheep containing a transgene encoding human α_1-antitrypsin (AAT), which gave up to 21.3 mg/mL in transgenic mice. Five founder animals were produced, of which 4 were female. Upon mating and subsequent milking, the females all expressed human AAT in their milk at levels greater than 1 g/L. One animal, founder 60, produced as much as 65 g/L initially, which stabilized at approximately 35 g/L and represented nearly 50% of the total milk protein throughout the lactation period.

After initial purification, the transgenic protein was shown to be more than 95% pure and of a similar size and immunosensitivity to protein derived from human plasma. Bioactivity and glycosylation comparisons were also very favorable. This examination was extended to more detailed analyses of glycosylation state, amino-terminal sequencing, isoelectric point, and molecular weight determined by mass spectrometry. The transgenic protein was found to be extremely similar to material derived from human plasma, except for a slight difference observed in the transgenic protein when analyzed by isoelectric focusing. This result could be explained by minor differences in terminal sialylation. The similari-

Table 4 Expression Levels of Foreign Proteins in Livestock Milk

Milk Gene	Protein	Gene Configuration	Levels	Animal
Bovine α_{s1}-casein	h Lactoferrin	cDNA	No results yet	Cow
Goat β-Casein	h Longer acting tissue plasminogen activator	cDNA	7.5 mg/mL	Goat
Goat β-Casein	h Antithrombin III	cDNA	2.0 mg/mL	Goat
Mouse WAP	h Longer acting tissue plasminogen activator	cDNA	3 μg/mL	Goat
Mouse WAP	h Protein C	cDNA	1 mg/mL	Pig
Mouse WAP	Whey acidic protein	Genomic	1–2 mg/mL	Pig
Sheep β-LG	h Factor IX	cDNA	25 ng/mL	Sheep
Sheep β-LG	α_1-antitrypsin	cDNA	5 μg/mL	Sheep
Sheep β-LG	h Protein C	cDNA	0.1 mg/mL	Sheep
Sheep β-LG	α_1-antitrypsin	Genomic	35 mg/mL	Sheep

ties between the transgenic and human plasma derived proteins merited the development of a production process for recombinant human AAT from these original founder sheep.

The five founder animals (G_0s) have been bred on for three more generations, G_1–G_3. This allowed us to investigate the transmission of the transgene, and expression level, from generation to generation and during sequential lactations. We have been unable to demonstrate transmission from female flounder 65 (< 1 copy AATB). This animal has produced four lambs to date, all nontransgenic, which suggests that she may be mosaic for the transgene and unable to transmit it. However, she has expressed human AAT in her milk over three lactations at levels of 1.5, 1.4, and 1.4 g/L.

Founder 60 (12–16 copies AATB) is a very interesting animal. She has produced the highest levels of expression of the human protein and has maintained high levels through three lactations: 33, 27, and 30.6 g/L. Surprisingly, her progeny have been shown to have three different genotypes: 2–4, 8–10, and 12–16 copies of AATB. Pulse field gel electrophoresis (PFGE) analyses indicate that the observed copy number differences are the result of genetic rearrangement, hence cannot be attributed to segregation of two or more transgene integration sites. Expression levels of recombinant AAT in these animals varied with copy number. Although unsuitable as a founder animal for a production process because of the observed genetic instability, founder 60 did demonstrate the enormous potential of the technology to produce large quantities of recombinant protein.

Founders 76, 77, and 78 all seem to transmit their transgene faithfully. Females 76 and 77 expressed human AAT at an average of 3.4 g/L over two lactations and 3.1 g/L over three lactations, respectively. Offspring (G_1) of these females and G_2 animals have copy numbers identical to those of their mothers, and the female progeny studied express the transgene at levels very similar to those recorded for their mothers.

The male founder 78 (3–4 copies AATB) enabled us to assess whether it would be possible to use a male animal as the founder of a transgenic production flock. After analysis of 78 and some of his progeny by Southern blotting, we believe that this animal is mosaic for the transgene. However, the mosaicism was bred out in G_1 and subsequent generations, and line 78 appears to be stable. Only 25% of this founder's offspring were transgenic, but these all had 7–8 copies of AATB. Of 10 G_2 animals and recent G_3 arrivals, all had 7–8 copies. A homozygote animal was also bred with 12–16 copies.

The expression levels from line 78 were particularly interesting. Seven transgenic G_1 ewes expressed human AAT at 13–16 g/L in their first lactations and are expressing very similar levels during their second lactations. Two G_2 animals are also expressing human AAT at similar levels, while a probable G_2 homozygote is expressing AAT at 37.0 g/L in her milk. This last result confirms that there is no insertional mutation associated with this line. The protein produced by all animals appears to be identical.

These very exciting results show for the first time that stable transmission and stable expression levels can be obtained from a transgenic livestock animal. It also is demonstrated that a transgenic production flock can be derived from one founder male, or several half-brothers containing the same integrant. This is essential for livestock animals, where the rapid expansion of a flock or herd would be greatly facilitated by using a male founder, rather than employing modern breeding techniques such as in vitro culture, embryo splitting, and transfer. We are progressing this expansion with a view to entering transgenically derived human AAT into clinical trials at the earliest opportunity.

5 PROBLEMS AND FUTURE DIRECTIONS

5.1 Problems or Opportunities?

One of the main challenges this technology has encountered since the beginning has been that of public perception. More established production routes have been available for some time, such as bacterial and yeast fermentation. Most of the proteins of interest to the biotechnology/pharmaceutical sector are relatively complex in nature, requiring correct folding and substantial posttranslational modifications for biological activity. Prokaryotic systems are incapable of performing most of these modifications. Yeast-based processes are more accomplished but still fall short of the mark for many modifications. The mainstay of more complex recombinant protein production to date has been mammalian cell culture, which has produced a regular stream of effective, safe products. Whichever route has been taken, the costs and yields involved have proved to be prohibitive to the production of some proteins. The production of proteins via the transgenic mammary gland is clearly superior to the alternatives in terms of both cost and yield. The resistance of the establishment to the use of this new production route is gradually being eroded by advances in the field. We envisage that in the future, the two approaches will live side by side, giving the industry more arrows for its bow.

One area for concern has been the ability of the mammary gland to correctly glycosylate products when required. Other recombinant systems used to produce therapeutic proteins of interest do not produce glycosylation patterns identical to those found on proteins extracted from human sources. Indeed, changing the cultured mammalian cells used to produce a protein, or even changing the conditions under which the same cells are grown, can substantially alter the glycosylation pattern of the product. It is to be expected that transgenic livestock, too, will produce patterns of glycosylation not identical to the human pattern. There may even be species-to-species variation; this remains to be seen.

What is important is that the modifications allow the product to be safe and efficacious. An example of this has been the recent production of recombinant ATIII in goat milk. Following purification, the transgenic product exhibited poor pharmacokinetics compared to human plasma derived material in a rabbit model. In vitro modeling of the carbohydrates present on therapeutic proteins from other sources is already an accepted practice. When this approach is applied to the transgenic ATIII, a dramatic improvement in the half-life of the protein is observed: specifically, it was nearly identical to that exhibited by the plasma-derived material.

All areas of product safety (including glycosylation) and efficacy will be extensively controlled by regulatory authorities before any product is allowed into the marketplace. The regulatory authorities of Europe and the United States produced guidelines on this technology in 1995, concentrating in both cases on the production in transgenic animals of products for parental use. We and others in the field will comply with these and any future regulations to ensure product safety.

One area of concern relates to animal health matters. The first major issue here is prion infection. These agents promote in animals neuropathies such as scrapie and bovine spongiform encephalopathy. We have adopted a two-pronged approach to combat this risk.

First, the purification process will be extensively validated for the removal of these particles. Second, we have taken the precaution of using sheep of New Zealand origin and cows from the United States, since these countries are considered free of the relevant ovine and bovine disease, respectively. It is interesting to note that no prion removal validation is performed in the human blood fractionation industry despite the risk of contamination of pooled plasma with Creutzfeld–Jakob disease.

A second area of concern is viral infection of producer animals. We have identified viruses of specific concern, will monitor our flocks/herds for infections, and will ensure that the purification processes used will remove or inactivate these viruses even if infections are subclinical.

5.2 The Future

The small production flock of human AAT transgenic animals is now being expanded, and preclinical trials have begun. Manufacturing for production of clinical trials material is expected in time to allow for a full validation of the purification process. Subsequent products would be expected to reach this stage within 3–4 years of the initial sheep microinjections, for example.

On a broader front, several areas will need to be improved if transgenic technology is to achieve its full potential. The first is to establish a fast and reliable preimplantation screening method. Such a screen would allow for the use of fewer animals in many cases and would permit an increased confidence level in experiments in relation to the number of transgenics to be obtained. This is more complicated than it sounds. For an effective screen, the injected embryos must be cultured in vitro, a biopsy sample of each embryo must be taken, and this material must then be screened for the presence of an integrated transgene quickly and reliably enough to allow the selection of transgenic embryos and their transfer to recipient mothers. While this series of steps is being undertaken, the remaining portions of the embryos must survive a further period in culture.

Generally, the culturing side of the process is not a problem. The major problem encountered is that DNA introduced into the developing embryo persists up to and perhaps beyond the blastocyst stage. This necessitates that one is able to distinguish between integrated and nonintegrated DNAs. To date this has proved to be impossible, and no one has developed a reliable assay. The search continues.

The second area of transgenic technology that could benefit from additional research efforts comprises investigations into mammary gland locus control regions and matrix attachment regions. Such studies are desirable because a control element of this type would confer independence from positional effects on DNA constructs. This would increase efficiencies and reduce costs since nearly all the transgenic animals produced would give high levels of expression.

Third, apart from a few exceptions mentioned earlier, cDNAs tend not to give high levels of expression, particularly in transgenic mice. Many genes are not available as genomic DNAs, and an additional problem is that some of these gDNAs are too big to allow easy manipulation and microinjection. One technique that has been mentioned, "Rescue," does alleviate the problem to some extent, but a greater understanding of introns and positional effects on expression levels is needed.

Fourth, the search for embryonic stem cells from the main livestock breeds must continue. Such cells may well help to overcome some of the remaining problems already mentioned, and they would offer a huge potential in many areas of transgenic animal production and animal breeding.

Finally, the development of a truly reliable sperm-mediated gene transfer method may improve the efficiency of the process. It would certainly make it less technically demanding!

See also Electroporation and Electrofusion; Hemoglobon, Genetic Engineering of; Mammalian Genome; Mouse Genome; Transgenic Animal Modeling; Vaccine Biotechnology.

Bibliography

Carver, A. S., Dalrymple, M. A., Wright, G., Cottom, D. S., Reeves, D. B., Gibson, Y. H., Keenan, J. K., Barrass, D., Scott, A. R., Colman, A., and Garner, I. (1993) Transgenic livestock as bioreactors: Stable expression of human alpha-1-antitrypsin by a flock of sheep. *Bio/Technology,* 11:1263–1270.

Ebert, K. M., Selgrath, J. P., Ditullio, P., Denman, J., Smith, T. E., Memon, M. A., Schindler, J. E., Monastersky, G. M., Vitale, J. A., and Gordon, K. (1991) Transgenic production of a variant of human tissue-type plasminogen activator in goat milk: Generation of transgenic goats and analysis of expression. *Bio/Technology,* 9:835–838.

Gordon, J. W., and Ruddle, F. H. (1981) Integration and stable germline transmission of genes injected into mouse pronuclei. *Science,* 214:1244–1246.

——— and ———. (1983) Gene transfer into mouse embryos: Production of transgenic mice by pronuclear injection. *Methods Enzymol.* 101:411–433.

——— and ———. (1985) DNA-mediated genetic transformation of mouse embryos and bone marrow—A review. *Gene,* 33:121–136.

Hogan, B., Constantini, F., and Lacy, E. (1994) *Manipulating the mouse embryo: A laboratory manual.* Cold Spring Harbor Laboratory Press, Plainview, NY.

Houdebine, L.-M. (1994) Production of pharmaceutical proteins from transgenic animals. *J. Biotechnol.* 34:269–287.

Velander, W. H., Johnson, J. L., Page, R. L., Russell, A., Subramanian, A., Wilkins, T. D., Gwazdauskas, F. C., Pittius, C., and Drohan, W. N. (1992) High-level expression of heterologous protein in the milk of transgenic swine using the cDNA encoding human protein C. *Proc. Natl. Acad. Sci. U.S.A.w* 89:12003–12007.

Wright, G., Carver, A., Cottom, D., Reeves, D., Scott, A., Simons, P., Wilmut, I., Garner, I., and Colman, A. (1991) High level expression of active human alpha-1-antitrypsin in the milk of transgenic sheep. *Bio/Technology,* 9:830–834.

Translation of RNA to Protein

R. A. Cox and H. R. V. Arnstein

1 Introduction

2 mRNA Structure and the Genetic Code
- 2.1 Structure
- 2.2 Prokaryotic mRNA
- 2.3 Eukaryotic mRNA
- 2.4 The Genetic Code

3 Transfer RNA

4 **Ribosome Structure and Function in Translation**
4.1 Ribosome Structure
4.2 Function of the Ribosome Cycle in Translation
5 **Translational Control of Gene Expression**
5.1 mRNA Stability
5.2 Control by Interaction of Proteins with mRNA
5.3 Control by mRNA Structure
5.4 Control Through Modification of Translation Factor Activity
5.5 Effects of Antisense Polynucleotides
5.6 Availability of Amino Acids, tRNA Abundance, and Codon Usage
5.7 Modulation of Ribosome Activity
6 **Concluding Remarks**

Key Words

Anticodon Three consecutive bases in tRNA that bind to a specific mRNA codon by complementary antiparallel base pairing.

Antiparallel Base Pairing Pairing through specific hydrogen bonds between base residues of two polynucleotide chains or two segments of a single chain with phosphodiester bonds running in the $5' \rightarrow 3'$ direction in one chain or segment and in the $3' \rightarrow 5'$ direction in the other. In DNA and RNA, the hydrogen bonds are usually formed between complementary base pairs (A with either T or U, and G with C).

Elongation The stepwise addition of amino acids to the carboxyl terminus of a growing polypeptide chain.

Initiation A multistep reaction between ribosomal subunits, charged initiator transfer RNA, and messenger RNA that results in apposition of the ribosome-bound initiator Met-tRNA [methionine-charged initiator transfer RNA] with an AUG initiation codon in mRNA. In this position the ribosome is poised to form the first peptide bond.

Polarity The asymmetry of a polymer such as a polynucleotide or polypeptide. In DNA the two strands have opposite polarity; that is, they run in opposite directions ($5' \rightarrow 3'$ and $3' \rightarrow 5'$). The polarity of a polypeptide is defined as running from the N-terminus to the C-terminus.

Reading Frame One of three possible ways of translating groups of three nucleotides in mRNA. The appropriate reading frame is determined by the initiation codon.

Template Strand The strand of the DNA double helix, also known as the coding strand, that is used as a template for transcription of RNA. It has a base sequence complementary to the RNA transcript.

Termination The end of polypeptide synthesis, which is signaled by a codon for which there is no corresponding aminoacyl-tRNA. When the ribosome reaches a termination codon in the mRNA, the polypeptide is released and the ribosome–mRNA–tRNA complex dissociates.

Translation The stepwise synthesis of a polypeptide with an amino acid sequence that is determined by the nucleotide sequence of the mRNA coding region. The genetic code relates each amino acid to a codon, a group of three consecutive nucleotides. Decoding of mRNA takes place in the $5' \rightarrow 3'$ direction, and the polypeptide is synthesized from the amino terminus to the carboxyl terminus.

Translocation The stepwise advance of a ribosome along mRNA, one codon at a time, with simultaneous transfer of peptidyl-tRNA from the A site to the P site of the ribosome.

Proteins are essential to the structure and function of living cells. The assembly of polypeptide chains from amino acids and their subsequent modifications, leading to the final three-dimensional protein structure, are exceptionally complex processes; many components are involved, and much of the cell's energy is utilized. The linear amino acid sequence of a protein is encoded within the gene as a linear deoxyribonucleotide sequence. Early steps in the biosynthesis of a protein include transcription of the gene and appropriate processing of the transcript, leading to the production of mature messenger RNA (mRNA). We describe the mechanisms involved in translating mRNA to produce a polypeptide chain having the amino acid sequence specified by the gene.

1 INTRODUCTION

A gene or cistron is defined as the region of DNA that is transcribed into a functional RNA. The transcript functions either as such (e.g., tRNA, rRNA, snRNA) or as a messenger (mRNA) that codes for a single polypeptide chain in the translation process. A polynucleotide such as RNA is an asymmetrical polymer that is assembled from nucleoside triphosphates by a stepwise mechanism linking the $3'$ position of one nucleotide by a phosphate bridge to the $5'$ position of the adjacent nucleotide. In the finished polynucleotide chain, the first nucleotide residue has a $5'$ position that is not linked to another nucleotide, whereas the last nucleotide has an unlinked $3'$ position. Thus, polynucleotide synthesis proceeds from the $5'$ to the $3'$ terminus, and the polymer is said to have a $5'$-to-$3'$ ($5' \rightarrow 3'$) polarity. Usually, linear RNA sequences are written with the $5'$ terminus on the left and the $3'$ terminus on the right (Figure 1A). Within the RNA chain, some bases may form antiparallel base pairs (Figure 1B).

The polypeptide chains of proteins are also asymmetrical polymers in which the amino acid residues are linked by peptide bonds between their α-amino and carboxyl groups (Figure 2), leaving a free α-amino group at one end (the amino terminus) of the polymer and a free α-carboxyl group at the opposite end (the carboxyl terminus). The significance of the polarity of RNA and proteins will become evident when the process of protein biosynthesis is explained.

The genetic information stored in DNA is not usable directly for making proteins. Rather, it first must be copied into mRNA by an enzymatic transcription of segments of DNA containing the genes. Messenger RNA serves as the template for protein synthesis; that is, the linear nucleotide sequence of the mRNA dictates the amino acid sequence of the polypeptide encoded originally by the gene. Conventionally, gene and mRNA nucleotide sequences are written in the $5' \rightarrow 3'$ direction, which corresponds to the direction in which mRNA is decoded during polypeptide synthesis: that is, the mRNA is read in the $5' \rightarrow 3'$ direction, and the polypeptide is synthesized from the amino terminus toward the carboxyl terminus.

The mechanism whereby RNA is translated into protein is com-

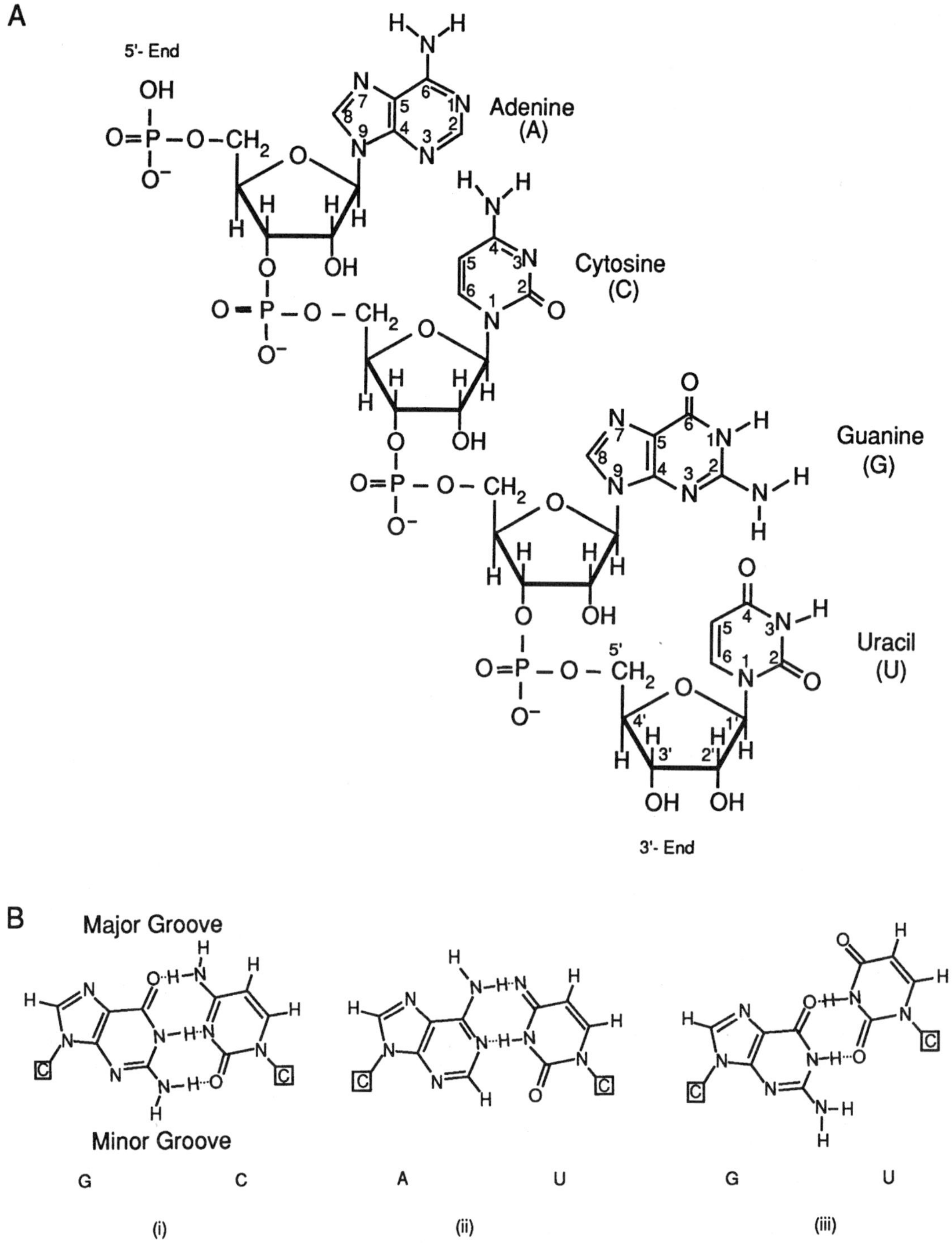

Figure 1. Structural elements of ribonucleic acids. (A) Primary structure indicating the numbering system for purines, pyrimidines, and ribose. (B) Base-pairing interactions commonly found in RNA. The G · C and A · U base pairs (i and ii) are Watson–Crick base pairs; G · U (iii) is a special type of base pairing found in intramolecular bihelical regions. The minor groove is on the side of the base pair with the glycosidic bond, of which the carbon atom C1′ of ribose is boxed. Note that DNA contains base pairs of types (i) and (ii) only but with thymine (5-methyluracil) in place of uracil. [Reproduced from Arnstein and Cox (1992) with permission of the Oxford University Press.]

plex, and the cell devotes considerable resources to the translational machinery. The components include 20 different amino acids, transfer RNAs, aminoacyl–tRNA synthetases, ribosomes, and a number of protein factors that cycle on and off the ribosomes and facilitate various steps in initiation of translation, elongation of the nascent polypeptide chain, and termination of synthesis with release of the completed polypeptide from the ribosome. The process depends on a supply of energy, which is provided by ATP and GTP. The rate of

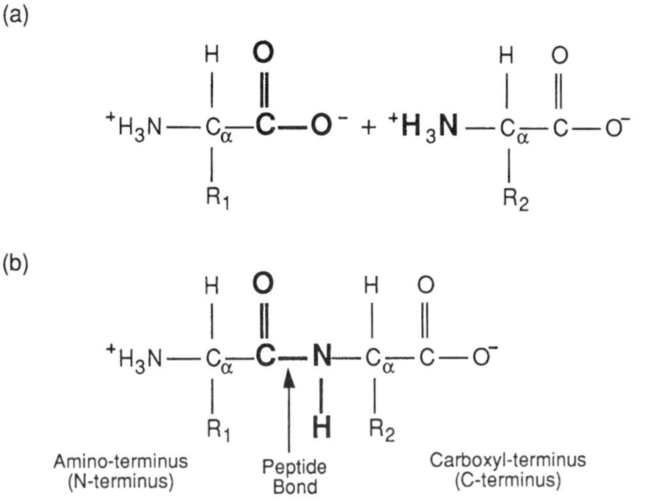

Figure 2. Structure of the peptide bond: (a) two L-amino acids with different side chains, R_1 and R_2, and (b) dipeptide formed from the two amino acids shown in (a).

protein synthesis is approximately 0.15 second per peptide bond at 37°C.

2 mRNA STRUCTURE AND THE GENETIC CODE

2.1 STRUCTURE

The sequence information of a gene is copied (transcribed) into the nucleotide sequence of RNA using one strand of DNA (called the coding strand) as the template. The primary transcript is a single strand of RNA, which is a faithful copy of the *other* strand of DNA with substitution of U residues in place of T residues found in DNA. Sometimes, the primary transcript is altered, as described later, before it functions as mRNA; in these cases the original unmodified transcript is the precursor or pre-mRNA. Usually, mRNAs have nontranslated sequences at the 5′ and 3′ ends in addition to the coding domain. These noncoding sequences sometimes affect the efficiency of translation and the stability of mRNA. The structures of two typical mRNAs are shown in Figure 3. The decoding process involves base pairing between three bases (designated a codon) in the mRNA and the three-base anticodon of a transfer RNA (tRNA). In a separate reaction, each tRNA is first linked to a particular amino acid, and thus the pairing of mRNA with tRNA determines the sequence of amino acids in the resulting protein.

2.2 PROKARYOTIC mRNA

In organisms that do not have a nucleus (prokaryotes), pre-mRNA usually undergoes little or no modification, with the result that pre-mRNA and mRNA are very similar if not identical. Since pre-mRNA is collinear with DNA, DNA and proteins are usually collinear in these organisms. Gene expression in prokaryotes usually involves the cotranscription of several adjacent genes, and translation of mRNA sequences into polypeptides may begin at the 5′ end of mRNA while transcription is still in progress at the 3′ end.

2.3 EUKARYOTIC mRNA

In a cell having a nucleus (a eukaryote), the genetic information is stored mainly in the nucleus and to a minor degree in some organelles (mitochondria and chloroplasts). The description that follows pertains only to nuclear genes. Eukaryotic genes are more complicated than prokaryotic genes because in the former the coding region is often discontinuous: the coding sequences or exons are interrupted by intervening sequences (introns). Thus, genes and proteins are usually *not* collinear in eukaryotes. In the nucleus a complicated set of splicing reactions removes all the introns from pre-mRNA and fuses the exons into a continuous coding sequence. Other processing steps involve adding a "cap" to the 5′ end of the mRNA and a poly(A) tail to the 3′ end. After completion of these nuclear maturation steps, the mRNA is transported to the cytoplasm, where it is translated. As with prokaryotic mRNA, the coding region is flanked by 5′- and 3′-untranslated sequences.

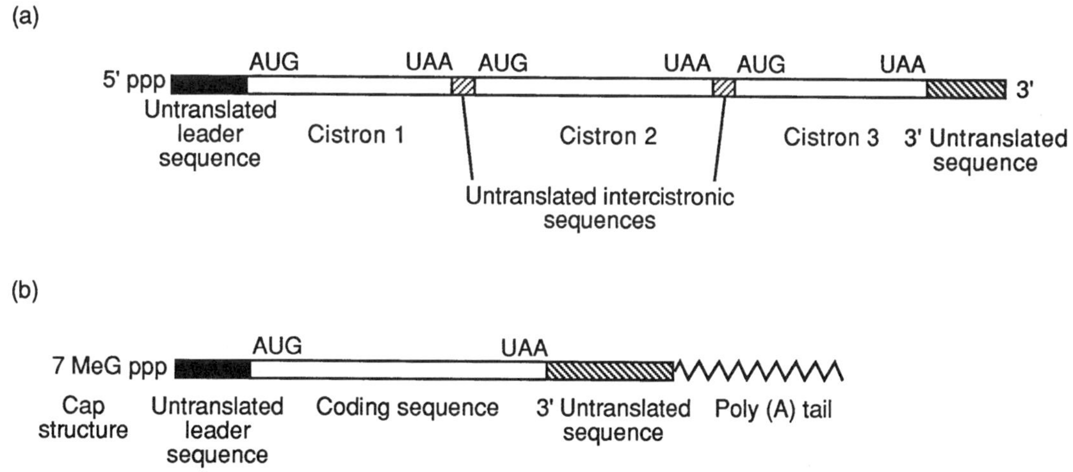

Figure 3. Structure of typical mRNAs: (a) prokaryotic mRNA and (b) eukaryotic mRNA. [Reproduced from Arnstein and Cox (1992) with permission of the Oxford University Press.]

(a) 5' |AUG|GUA|UUC|AG |3' . . .
fMet Val Phe –

(b) 5'A|UGG|UAU|UCA|G . |3' . .
Trp Tyr Ser –

(c) 5'AU|GGU|AUU|CAG|3' . . . .
Gly Ile Gln

Figure 4. Translation of a polynucleotide sequence into three alternative polypeptides using different reading frames.

2.4 The Genetic Code

With a few minor exceptions (e.g., in mitochondria), the genetic code is universal. It is triplet, commaless, and nonoverlapping, and, as a consequence, a nucleotide sequence has three possible reading frames (Figure 4). Since mRNA is normally translated into a unique polypeptide, an essential step in the translating process is the selection of the appropriate reading frame. This is achieved by starting translation at the initiation codon, usually AUG or less frequently GUG, which ensures that the codons that follow are read in phase within the required reading frame. Of the 64 theoretically possible triplets in the genetic code, 61 sense codons correspond to the 20 genetically encoded amino acids found in all, or nearly all, proteins. When GUG is used as the initiation codon, it codes for methionine by interacting with the anticodon of the initiator Met-tRNA^{Met}, whereas elsewhere it codes for valine. All other codons specify only one amino acid, but many amino acids are specified by two or more (up to six) codons; that is, the code is unambiguous but degenerate. The remaining three codons, termed nonsense codons, usually signify termination of synthesis and release of the finished polypeptide chain.

One of the nonsense codons, UGA, has an additional function in the synthesis of selenoproteins. The process involves the initial synthesis of selenocysteyl-tRNA from a novel seryl-tRNA and selenium. This tRNA contains an anticodon that is able to decode UGA and insert selenocysteine residues into the growing polypeptide chain, but only at UGA codons in a particular context of neighboring nucleotides. The insertion of selenocysteine residues into the polypeptide also requires a specific elongation factor T that differs from the factor used for the incorporation of other aminoacyl-tRNAs.

2.4.1 Deviations from the Standard Genetic Code

Mitochondria and chloroplasts, as well as certain organisms such as mycoplasma and ciliated protozoa, use genetic codes that differ in a number of minor respects from the standard code. For example, methionine is usually coded for by AUG, but in human mitochondria this codon is replaced by AUA, which usually codes for isoleucine. Variations in the genetic code are thought to have arisen as a result of the loss of some tRNA genes and mutational pressure on DNA, giving rise to a predominance of either AT- or GC-rich codons.

3 TRANSFER RNA*

By relating individual codons of mRNA to the cognate amino acids, tRNA functions as a key bilingual intermediate in the translation of the genetic code. All transfer RNAs are single-stranded molecules about 80 nucleotides long, with a common 3′-terminal CCA sequence. The secondary structure of tRNA is usually presented in two dimensions as a cloverleaf, to highlight the regions of base pairing (Figure 5a). X-ray crystallography reveals that additional hydrogen bonds give rise to an L-shaped tertiary structure (Figure 5b). The CCA sequence carrying the amino acid is located distal to the anticodon.

The decoding process involves antiparallel base pairing between the three bases of mRNA codons and the complementary anticodons of transfer RNA (tRNA) during peptide bond formation (Figure 6). The first and middle bases of the codon form conventional base pairs with the third and middle bases of the anticodon, respectively; but the third base of the codon pairs with the first base of the anticodon by a less stringent interaction. This so-called wobble considerably reduces the number of tRNA species required to decode the 61 sense codons. Thus, the protein synthesis system in the cytosol of eukaryotes contains only a few more than 40 different tRNAs, and in mitochondrial protein synthesis it is sufficient to have 22–24 tRNA species.

The attachment of amino acids to tRNA involves the formation of an ester bond between the α-carboxyl group of the amino acid and the 3′-hydroxyl group of the terminal adenosine of tRNA. It requires specific enzymes, the aminoacyl–tRNA synthetases. There are 20 different synthetases, each specific for one of the 20 amino acids, and each enzyme recognizes something unique in the structure of its cognate tRNA. The structural determinants that ensure accuracy of this charging reaction vary for different tRNAs. The anticodon may play a part, but sometimes even a single base elsewhere is sufficient to determine the specificity of the tRNA–synthetase interaction.

The accuracy of the synthetase reaction in attaching an amino acid to its cognate tRNA is critically important to the fidelity of the translation process. Once the aminoacyl–tRNA has been formed, the subsequent incorporation of the amino acid residue into a polypeptide does not depend on the amino acid itself but only on the interaction between the anticodon of the aminoacyl–tRNA with the codon of mRNA. Thus, an error in the synthetase reaction would lead to the incorporation of an inappropriate amino acid into the polypeptide.

Synthesis of aminoacyl–tRNA (**III**) from amino acids (**I**) requires

* Transfer RNA nomenclature: the amino acid linked to a charged tRNA is indicated by a prefix, and the specificity of the transfer RNA (tRNA) in the aminoacylation reaction is shown as a superscript on the right. Thus Phe-tRNA^{Phe} indicates phenylalanine-specific tRNA charged with phenylalanine. The anticodon may be indicated as a right subscript or, alternatively, in the superscript after the amino acid (e.g., tRNA$_{UGC}$ or tRNA$^{Ala/UGC}$). The right-hand subscript position is sometimes used to indicate the organism from which the tRNA is derived (e.g., tRNA$^{Val}_{yeast}$). The initiator tRNA, which is specific for methionine, is termed tRNA$^{Met}_{f}$ or tRNA$^{Met}_{i}$. In the cytosol of eukaryotes the charged initiator tRNA is termed Met-tRNA$^{Met}_{f}$ or Met-tRNA$^{Met}_{i}$. Often the superscript ^{Met} is omitted. In the mitochondria of eukaryotes and in prokaryotes Met-tRNA$^{Met}_{f}$ is formylated by a transformylase using N^{10}-formyltetrafolate as the donor of the formyl group. Commonly, the charged tRNA is termed fMet-tRNA$_{f}$. The elongator methionine-specific RNA, which inserts methionine into internal positions of the growing peptide chain, is termed tRNA$^{Met}_{m}$ (or tRNA$_{m}$) when uncharged and Met-tRNA$^{Met}_{m}$ (or Met-tRNA$_{m}$) when charged.

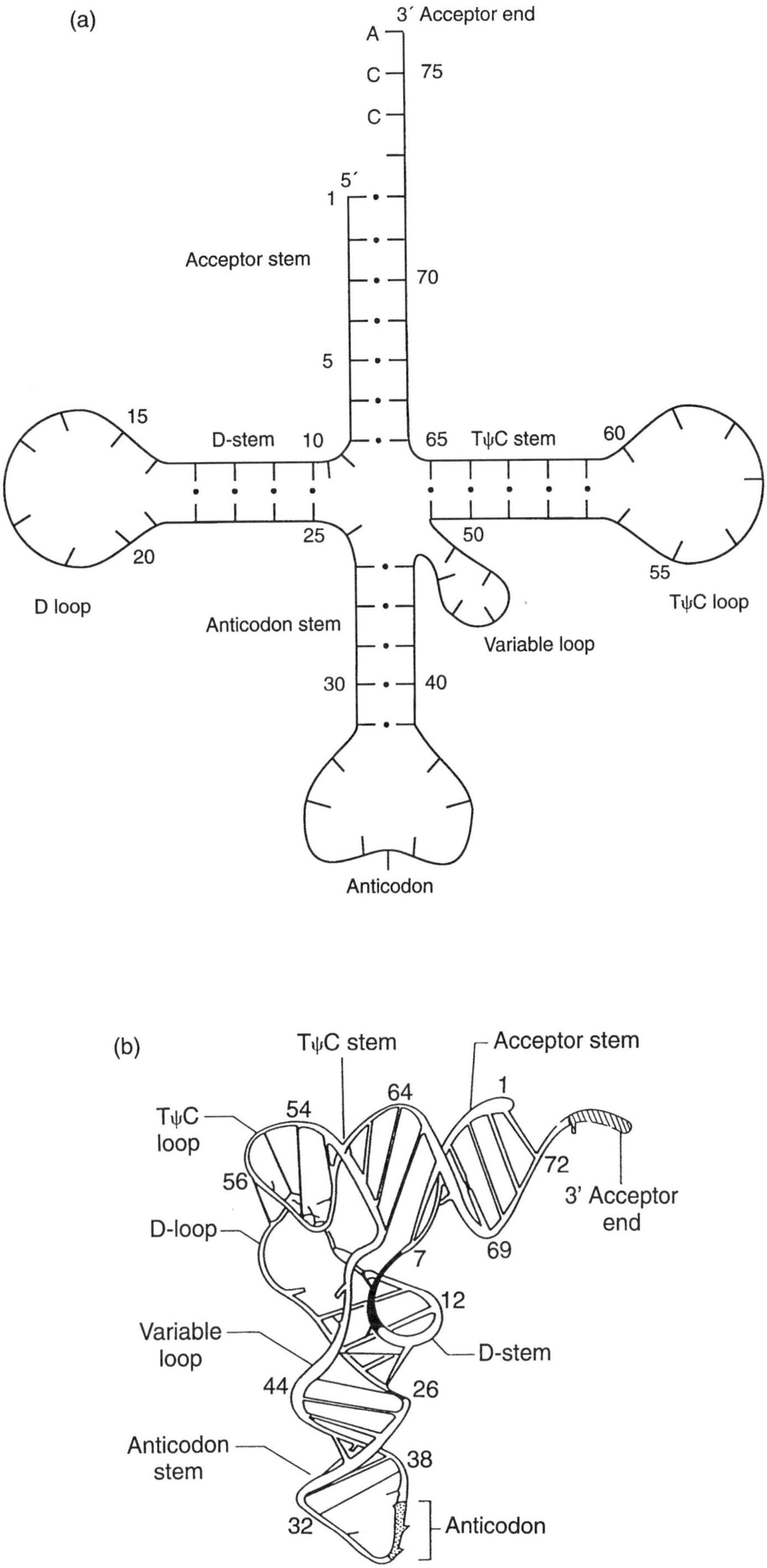

Figure 5. Structure of phenylalanyl–transfer RNA: (a) secondary cloverleaf structure and (b) tertiary structure. [Reproduced from Arnstein and Cox (1992) by permission of the Oxford University Press.]

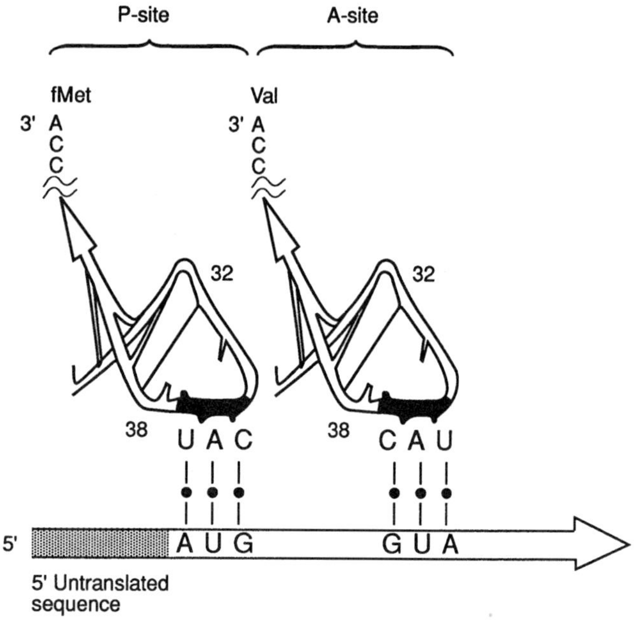

Figure 6. Schematic illustration of base pairing between a codon and its anticodon, showing the interaction at the P site of the ribosome (see Figure 9) between the initiation codon AUG of mRNA and the anticodon CAU of fMet-$tRNA_f^{Met}$ and the interaction at the A site between the codon GUA of mRNA and the anticodon UAC of Val-$tRNA^{Val}$. The polarity of each RNA species is revealed by the arrowhead, which points from the 5′ end to the 3′ end. The fragment of tRNA is representative of the general structure (see Figure 5b) placed in the appropriate orientation; the numbers refer to the nucleotide positions measured from the 5′ end. The interaction between the codon and the anticodon is antiparallel, and the three base pairs have a bihelical conformation.

the activation of the amino acid carboxyl group with formation of an intermediate enzyme-bound aminoacyladenylate (**II**).

$$\mathrm{E} + \underset{\substack{|\\ \mathrm{NH_3^+}\\[1em] \mathbf{I}}}{\mathrm{R\cdot CH}}\cdot \mathrm{CO_2^-} + \mathrm{ATP} \longrightarrow \mathrm{E}\cdot\underset{\substack{|\\ \mathrm{NH_3^+}\\[1em] \mathbf{II}}}{\mathrm{R\cdot CH}}\cdot\mathrm{CO\cdot AMP} + \underset{\substack{\downarrow\\ 2\mathrm{P_i}}}{\mathrm{PP_i}} \rightarrow$$

$$\rightarrow \underset{\substack{|\\ \mathrm{NH_3^+}\\[1em] \mathbf{III}}}{\mathrm{R\cdot CH}}\cdot\mathrm{CO\cdot tRNA} + \mathrm{AMP} + \mathrm{E}$$

where E = aminoacyl–tRNA synthetase.

The energy for the reaction is provided by ATP and stored in the ester bond of the aminoacyl–tRNA to be used subsequently for peptide bond synthesis.

4 RIBOSOME STRUCTURE AND FUNCTION IN TRANSLATION

4.1 Ribosome Structure

Ribosomes are high molecular weight complexes of RNA (rRNA) and proteins (Table 1), and the electron-dense particles are easily visualized by electron microscopy. Ribosomes from various sources (prokaryotes, eukaryotic cytoplasm, mitochondria, and chloroplasts) vary in size from 20 nm to 30 nm in diameter, but all are composed of a large and a small subparticle or subunit and perform similar functions in protein synthesis. The principal functional domains of the ribosome and associated components are given schematically in Figure 7. More detailed resolution of the ribosome structure has allowed the placement of mRNA, aminoacyl-tRNA, peptidyl-tRNA, and the nascent polypeptide chain (Figure 8; see color plate 2).

The small subunit comprises a single rRNA having a molecular mass of 0.6–1.7 × 10^6 Da (relative molecular weight, M_r of 0.6–1.7 × 10^6 Da) and single copies of 20–30 unique proteins. It has a major function in binding initiator tRNA and mRNA in the initiation of protein synthesis and in decoding the genetic message.

The large subunit comprises a high molecular mass rRNA (0.6–1.7 × 10^6 Da) and often one or two smaller rRNAs (0.03–0.05 × 10^6 Da) and 30–50 different proteins, which are present, with one exception, as single copies. The large subunit binds aminoacyl-tRNA at the A site, peptidyl-tRNA at the P site, and uncharged tRNA at the E site. The large subunit contains the peptidyl transferase, and this enzyme activity may, unusually, reside in the rRNA molecule itself rather than in the associated ribosomal proteins. This subunit is also involved in binding elongation factor G (EF-G), which is required for translocation (Figure 7).

4.2 Function of the Ribosome Cycle in Translation

Polypeptide synthesis can be divided into three stages—initiation, elongation, and termination—in a process involving binding of a ribosome to mRNA, movement of the ribosome relative to the mRNA in the 5′ → 3′ direction, and release of both mRNA and the finished polypeptide from the ribosome (Figure 9).

The ribosome cycle starts with the stepwise formation of an initiation complex from mRNA, charged initiator tRNA, and ribosomal subunits. A number of preinitiation complexes are formed as intermediates, and the process is facilitated by initiation factors. In outline, prokaryotic and eukaryotic systems are similar, but there are a few differences, particularly with regard to the complexity of the initiation factors and details of the mechanisms.

4.2.1 Formation of Preinitiation Complexes

Prokaryotic Systems Three proteins, initiation factors IF-1, IF-2, and IF-3 (see Table 2), are required for the initiation of protein biosynthesis (see Figure 10a). Ribosomal subunits are released by dissociation of ribosomes following translation of the mRNA. Dissociation is facilitated by the combined action of the initiation factors IF-1 and IF-3; IF-1 increases the rate of dissociation, and IF-3 acts as an antiassociation factor when bound to the 30S ribosomal subunit, thereby displacing the equilibrium in favor of subunit formation. Initiation factor IF-2 is also able to bind to the 30S subunit, and this association is stabilized by IF-1 and GTP, the latter acting as a steric effector without being hydrolyzed at this stage. IF-2 plays a central role in binding fMet-$tRNA_f$ to the 30S preinitiation complex by specific recognition of the *N*-formylmethionine residue attached to the tRNA, thus restricting this interaction to charged initiator tRNA. All three factors bind to the 30S ribosomal subunit near the 3′ end of the 16S ribosomal RNA at adjacent sites that are located at the interface between the small and large ribosomal subunits.

In the next step, the initiator tRNA and mRNA associate with the 30S · IF-1 · IF-2 · IF-3 complex with release of IF-3. There is evi-

Table 1 Properties of Ribosomes and Ribosomal Subunits

				rRNA			
Species	$S_{20,w}$	Size (nm) (largest dimension)[a]	Mass (MDa)	RNA: Protein (w/w)	Axial Ratio[a]	Mass (MDa)	$S_{20,w}$ (nominal)
Ribosomes							
Procaryotes: *E. coli*	70S	22.5 ± 2.5	2.6–2.9	2:1			
Eukaryotes							
Cytoplasmic	80S	28.0 ± 2.8	3.4–4.5	1:1			
Chloroplast	70S	22.5 ± 2.5	2.5–3.3	1:1			
Mitochondrial	55–77S	—	3.2–4.5	1.1:1.3			
Small Ribosomal Subunits							
Prokaryotes: *E. coli*	30S	22.0 ± 2.2	0.95	2:1	2:1	0.6	16S
Eukaryotes							
Cytoplasmic	36–41S	25.0 ± 2.5	1.4	1:1	2:1	0.7	18S
Chloroplast	28–35S	22.0 ± 2.2	1.2	1:1	2:1	0.6	16S
Mitochondrial	28–40S	—	1.1–1.7	1:3	—	0.3–0.7	12–14S
Large Ribosomal Subunits							
Prokaryotes: *E. coli*	50S	22.5 ± 2.2	1.75	2:1	1:1	1.17	23S
						0.04	5S
Eukaryotes							
Cytoplasmic	60S	28.0 ± 2.8	2.1–3.1	1:1	1:1	1.2–1.75	25–28S
						0.05	5.8S
						0.4	5S
Chloroplast	46–54S	22.5 ± 2.2	2.4	1:1	1:1	1.1	23S
						0.4	5S
						0.03[b]	4.5S[b]
Mitochondrial	39–60S	—	1.65–2.5	1:3	—	0.5–1.5	16–25S
						0.04[c]	5S[c]

[a]—, data not available.
[b]4.5S rRNA corresponding to the 100 nucleotides at the 3′ end of 23S rRNA in eubacteria occurs as a separate species in chloroplasts of higher plants.
[c]Species present only in plant mitochondria; absent from all other mitochondrial ribosomes.
Source: From Arnstein and Cox (1992), reprinted by permission of Oxford University Press.

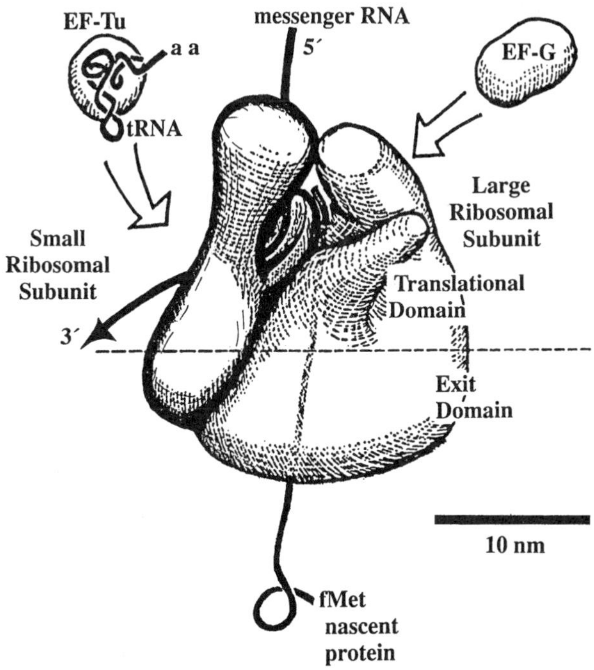

Figure 7. Model of the *E. coli* ribosome, based on low resolution electron microscopy, showing the relative orientation of the large and small subunits, as well as other functional components involved in polypeptide chain synthesis: EF-Tu, elongation factor required for binding aminoacyl-tRNA to the ribosomal A site; EF-G, elongation factor required for translocation of the peptidyl-tRNA from the A to the P site; dashed line, boundary between the translational and exit domains, which are involved in peptide bond formation and extrusion of the nascent polypeptide chain, respectively. [Based on Figure 2 of M. I. Oakes, A. Scheinman, T. Atha, G. Shankweiler, and J. A. Lake, in Hill et al. (1990), pp. 180–193.]

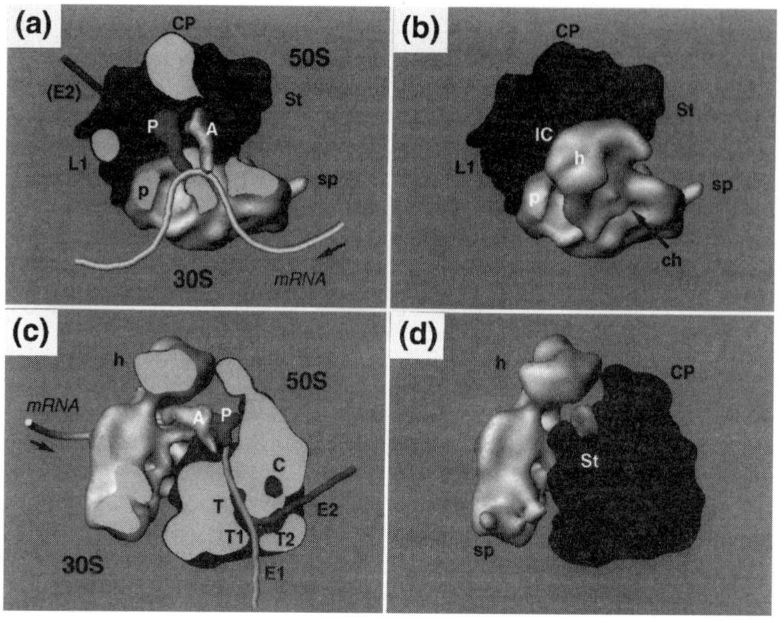

Figure 8. A model of elements involved in translating mRNA to protein based on 2.5 nm reconstructions of images of the *E. coli* ribosome obtained by cryo-electron microscopy, according to Frank et al. in Matheson et al. (1996) and in *Nature* [376, 441–444 (1995)]. The small subunit (yellow) has prominent features such as the head (h), platform (p), a "spur" (sp) at the base, and a channel (ch) that penetrates the "neck." The main features of the large subunit (blue) include the L1 arm (L1), the central protuberance (CP), the stalk (St), and the interface canyon (IC). The subunits are joined by five bridges across the subunit–subunit interface and enclose a space that accommodates mRNA, aminoacyl-tRNA (A) and peptidyl-tRNA (P). Sections through two views of the ribosome are shown in (a) and (c) to reveal the proposed locations of mRNA and tRNAs. The corresponding views of the whole ribosome are shown in (b) and (d). (a) A section through the top view of the ribosome (b) revealing the channel and the cleft into which it leads. The mRNA (orange) is shown after it has entered from the cytosol through the channel, encircling the back portion of the neck. The arrow points away from the 3′ end of mRNA and indicates the direction of movement relative to the ribosome. The complex of A (pink) and P (green) is placed with the anticodons facing the cleft region immediately next to the channel. (c) A section through a side view of the ribosome (d) revealing the bifurcating tunnel (T), its first branch (T1) and second branch (T2), and the compartment (C) contiguous with T2. E1 and E2 are alternative exit sides for the nascent polypeptide chain shown in yellow (E1) and yellow then olive green (E2). The tRNA complex is placed with the aminoacyl (or peptidyl) ends at the entry to the tunnel, indicating the location of the peptidyl transferase center. The location of this center is not known precisely but ranges from near the aminoacyl ends of the AP complex, as shown to the limit where the aminoacyl ends of the AP complex are placed close to the roof formed by the head of the small subunit. (Reproduced by permission from an original photograph by Dr. J. Frank, Wadsworth Center for Laboratories and Research, New York State Health Department, Albany, NY 12201, U.S.A.) (See color plate 2.)

dence from in vitro experiments that the binding of mRNA precedes that of the initiator tRNA.

Messenger RNA binds to the small ribosomal subunit immediately before formation of the final initiation complex with the initiation codon correctly positioned (see Figure 9a). In the case of bacterial and bacteriophage messengers, the molecular recognition mechanism proposed by Shine and Dalgarno involves base pairing between short nucleotide sequences, most often CUCC, near the 3′ end of the 16S ribosomal RNA and a complementary region, usually consisting of three to nine bases on the 5′ side of the mRNA initiation codon, which has been found to be present in nearly all of more than 150 bacterial and bacteriophage messengers. Studies with mutants and mRNA fragments indicate that in addition to the Shine–Dalgarno interaction, outlying upstream sequences in the leader region may provide recognition signals between mRNAs and ribosomes, possibly by ensuring that the Shine–Dalgarno sequence is in an appropriate conformation.

The Shine–Dalgarno mechanism is also found in chloroplast protein synthesis, as judged from limited sequence analysis of the 16S rRNA and mRNAs, but apparently not in the mammalian mitochondrial system (where the initiator codon occurs either directly at, or only a few nucleotides downstream from, the 5′ end of mRNA, which excludes the possibility of mRNA–rRNA base pairing in this region).

Figure 9. Schematic view of the biosynthesis of a polypeptide by translation of prokaryotic mRNA. (a) Formation of the initiation complex from mRNA, ribosomal subunits, initiation factors (IF), GTP and fMet-tRNA^{Met} (for details see Figure 10). The binding sites for aminoacyl-tRNA, peptidyl-tRNA, and uncharged tRNA are designated A, P, and E, respectively. (b) Decoding of the codon GUA with Val-tRNA^{Val}. (c) Formation of the peptide bond by transfer of fMet to Val-tRNA^{Val}, forming fMet-Val-tRNA^{Val}. (d) Translocation of tRNA$_f^{Met}$ to the E site and of fMet-Val-tRNA^{Val} to the P site with codon UUC aligned with the A-site. (e) Ejection of tRNA$_f^{Met}$ from the ribosome. (f) Decoding of codon UUC with Phe-tRNA^{Phe}. (g) Decoding of codon UAA with release factor. Steps (b)–(f) constitute the elongation–translocation cycle. The position shown is reached by repeating the cycle $n + 2$ times after formation of fMet-Val-tRNA. (h) Release of the completed polypeptide, fMet-Val-Phe-(aa)$_n$-Ser, ribosomal subunits, release factor, mRNA and tRNA^{Ser}. The cycle may then be repeated starting at (a).

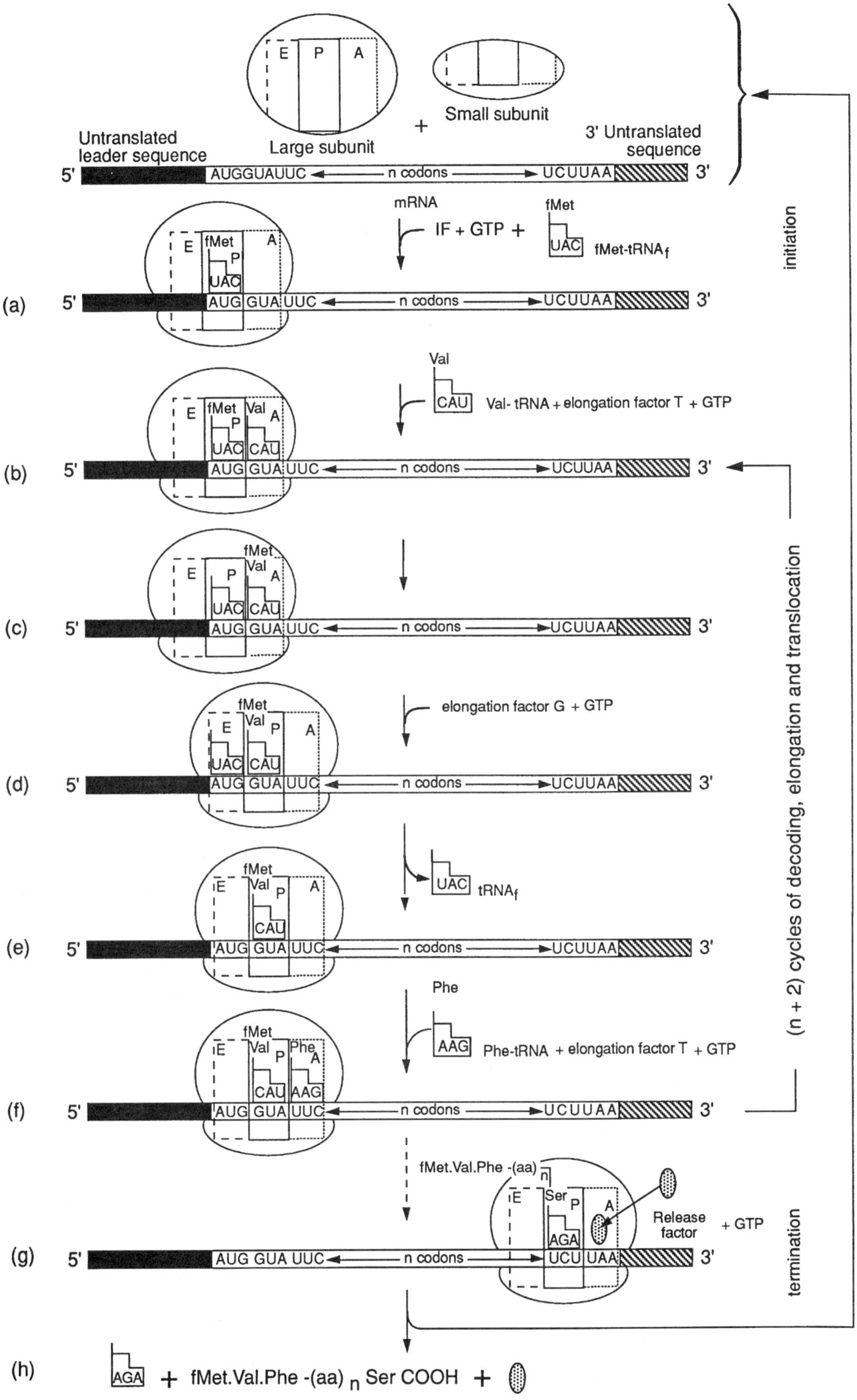

E P A
Large subunit
+
Small subunit
Untranslated leader sequence
3' Untranslated sequence
5' AUGGUAUUC ← n codons → UCUUAA 3'
initiation
mRNA
IF + GTP +
fMet
UAC fMet-tRNA$_f$
(a)
Val
CAU Val- tRNA + elongation factor T + GTP
(b)
(c)
elongation factor G + GTP
(d)
UAC tRNA$_f$
(e)
Phe
AAG Phe-tRNA + elongation factor T + GTP
(f)
(n + 2) cycles of decoding, elongation and translocation
fMet.Val.Phe -(aa)$_n$
Ser
AGA
UCU UAA
Release factor
+ GTP
(g)
termination
(h)
AGA + fMet.Val.Phe -(aa)$_n$ Ser COOH +

Table 2 Properties of Prokaryotic Initiation Factors from E. coli

Factor	Molecular Mass (kDa)	Properties and Function
IF-1	9	Stimulates activity of IF-2. Accelerates dissociation of unprogrammed ribosomes to subunits
IF-2	100	Binds fMet-tRNA$_f$ to the ribosomal P site by a GTP-requiring reaction
IF-3	22	Binds natural mRNAs to the small ribosomal subunit, probably by facilitating base pairing between the untranslated leader sequence and the 3′ end of 16S rRNA Prevents ribosomal subunit association when bound to the small subunit

Source: From Arnstein and Cox (1992), reprinted by permission of Oxford University Press.

Eukaryotic Systems At least 12 proteins, the eukaryotic initiation factors (eIFs) (see Table 3), are needed for the initiation of protein biosynthesis (Figure 10b). The dissociation of cytosolic 80S ribosomes is facilitated by a complex initiation factor, eIF-3 (≈ 500,000–700,000 Da), consisting of 9–11 polypeptide chains, which binds to the small ribosomal subunit (40S) and prevents its reassociation to 80S ribosomes. Thus, this factor has antiassociation activity. Low molecular weight proteins with similar activity have also been reported, however, and a protein, eIF-4C (20,000 Da) seems to function as an accessory factor to eIF-3 in the formation of a 43S ribosomal preinitiation complex. Also, another protein factor, eIF-6 (24,000 Da), prevents reassociation by binding to the large (60S) ribosomal subunit.

Initiation factor eIF-2 gives a stable binary complex with GTP, which binds the initiator tRNA, Met-tRNA$_f$, forming a ternary complex. Interaction of this ternary complex with the 40S ribosomal subunit containing bound initiation factors eIF-3 and eIF-4C gives rise to the 43S preinitiation complex, which is competent to bind messenger RNA in the presence of three further initiation factors (eIF-4A, eIF-4B, and eIF-4F), together with ATP.

The binding of cytosolic eukaryotic messenger RNAs to the small ribosomal subunit probably does not involve base pairing with the 18S rRNA, since no uninterrupted sequences of the Shine–Dalgarno type have been found. Instead, a "scanning model" has been proposed in which the preinitiation complex, composed of the 40S ribosomal subunit, Met-tRNA$_f^{Met}$ and associated initiation factors, binds at or near the 5′ cap of the mRNA and slides along the messenger until it encounters the first AUG triplet, at which point the 60S ribosomal subunit joins to give rise to the 80S initiation complex. Recognition of the cap is facilitated by cap-binding proteins, which mediate an ATP-dependent melting of the mRNA secondary structure at the 5′-terminal region to allow the mRNA to thread through a channel in the neck of the 40S subunit. The cap structure is required for efficient binding and translation even when the initiating AUG codon occurs hundreds of nucleotides downstream. As a rule, scanning by the 40S subunit stalls at the first AUG codon, which is recognized mainly by interaction with the anticodon of the Met-tRNA$_f^{Met}$. However. this recognition also depends in some way on eIF-2 and may be modulated as a result of deviation of the mRNA structure from the consensus sequence GCCGCC$^{A}_{G}$CC<u>AUG</u>G. Sequence context may also account for the rare cases of the occurrence of initiation downstream of the 5′-proximal AUG codon. Where this sequence context is unfavorable, initiation becomes inefficient, and thus most 40S subunits will tend to initiate further along the mRNA at another AUG triplet in a more favorable context. This model also explains rare cases in which initiation is not restricted to one particular AUG codon and translation of a single mRNA gives rise to two proteins.

4.2.2 Initiation Complex Formation: Joining of the Large Ribosomal Subunit

The last event in the initiation of protein synthesis involves the joining of the large ribosomal subunit to the preinitiation complex (Figure 9a). In the prokaryotic system, association of the 50S subunit with the 30S preinitiation complex takes place with hydrolysis of GTP by the GTPase activity of IF-2 and release of IF-1, IF-2, GDP, and inorganic phosphate (P$_i$). GTP hydrolysis is essential for the release of IF-2 from the initiation complex, which is a prerequisite for allowing the fMet-tRNA$_f$ to engage in the formation of the first peptide bond. In eukaryotic protein synthesis, the 80S initiation complex is formed by joining the 60S ribosomal subunit to the 48S preinitiation complex consisting of the 40S ribosomal subunit, eIF-2, eIF-3, GTP, Met-tRNA$_f$, mRNA, and possibly eIF-4C. This coupling reaction requires an additional factor, eIF-5, which mediates the hydrolysis of GTP to GDP with release of eIF-2 · GDP, P$_i$ and eIF-3 from the 48S preinitiation complex.

By this stage, all initiation factors have been released and are available for recycling, although the exact steps at which factors are released from intermediate complexes are not known in every case. There is thus an initiation factor cycle within the ribosome cycle, and regulation of the activity of factors, particularly eIF-2, is an important control mechanism in translation (see Section 5.4.1).

In the initiation complex, location of the charged initiator tRNA in the P site of the ribosome allows transfer of the methionine residue to the amino group of another aminoacyl-tRNA in the A site (Figure 9b) by peptidyltransferase to form dipeptidyl-tRNA (see Figure 9c). Functional insertion of Met-tRNA$_f$ directly into the P site can be demonstrated using the trinucleotide AUG as a synthetic mRNA and another trinucleotide (e.g., UUU) to bind an acceptor aminoacyl-tRNA (in this case Phe-tRNA).

It is possible to measure the peptidyl-transferase activity of the large subunit in the absence of mRNA by using the antibiotic puromycin, which resembles the 3′-terminal region of Phe-tRNA in structure, as an artificial acceptor to form methionyl puromycin from Met-tRNA$_f$.

4.2.3 Polypeptide Chain Synthesis: The Elongation–Translocation Cycle

The elongation–translocation cycle is outlined in Figure 9b–f.

Elongation The first peptide bond is formed when the aminoacyl-tRNA in the ribosomal A site is converted into the corresponding methionyl·aminoacyl-tRNA by transfer of the methionyl (or *N*-formylmethionyl) residue from the charged initiator tRNA in the P site. In artificial cell-free systems, any N-substituted aminoacyl-tRNA, such as peptidyl-tRNA or *N*-acetylaminoacyl-tRNA, can function in peptide bond synthesis as a donor in the P site in place of the charged initiator tRNA. The reaction is catalyzed by the peptidyltransferase activity of the large ribosomal subunit. No soluble cofactors appear to be involved, but monovalent cations (K^+) at a

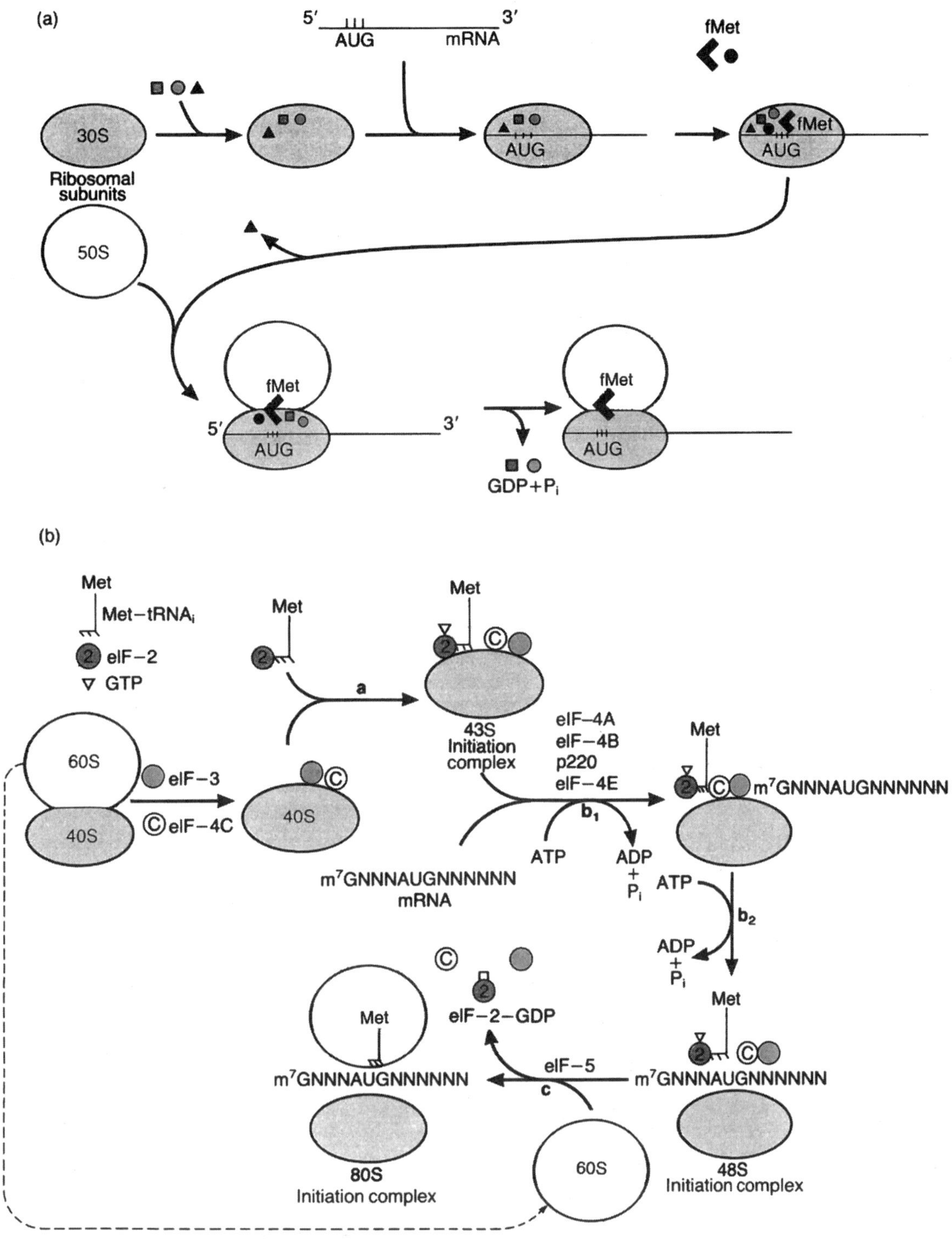

Figure 10. Schematic diagram illustrating the formation of initiation complexes. (a) Prokaryotic initiation. The following symbols are used for initiation factors and other components involved in the formation of the 70S initiation complex: □, IF-1; ○, IF-2; △, IF-3; <, fMet-tRNA$_f$; ●, GTP. (b) Eukaryotic initiation complexes in mammalian protein biosynthesis. The process may be divided into three stages: **a,** formation of a 43S preinitiation complex; **b**, binding of mRNA with formation of a 48S preinitiation complex; and **c,** synthesis of the 80S initiation complex containing the initiator tRNA in the correct position for peptide bond formation. Symbols are defined in the diagram. [Reproduced from Arnstein and Cox (1992) by permission of the Oxford University Press.]

Table 3 Eukaryotic Initiation Factors

Initiation Factor	Synonym[a]	Subunits and Molecular Mass (kDa)	Properties and Function
eIF-1		15	Stabilizes initiation complexes
eIF-2		α 38[b] β 35–50[b] γ 55	GTP-dependent binding of Met-tRNA$_f$ to small ribosomal subunit
eIF-2B	GEF	27, 37, 52, 67, 85[b]	Conversion of eIF-2 · GDP into eIF-2 · GTP
eIF-3		9–11 subunits 24–170[b]	Associates with 40S subunit to maintain dissociation; binds mRNA to 43S preinitiation complex
eIF-4A	50 kDa component of CBP-II	50	ATP-dependent unwinding of the secondary structure of the mRNA 5′ region; stimulates translation of exogenous mRNA in cell-free systems
eIF-4B		80[b]	mRNA binding; stimulates cell-free translation; ATPase activity of eIF-4A and eIF-4F; AUG recognition and recycling of eIF-4F
eIF-4C		17	Ribosome dissociation; 60S subunit joining
eIF-4D		17	Formation of first peptide bond
eIF-4E	CBP-I; 24 kDa CBP	24–28[b]	Binds mRNA cap structure
eIF-4F	CBP-II: complex	24 (CBP-I)[b] 50 (eIF-4A) 200[b]	ATPase; unwinds mRNA secondary structure; stimulates cell-free translation
eIF-5		60[b]	GTPase; release of eIF-2 and eIF-3 from preinitiation complex to allow joining of 60S subunit
eIF-6		24	Antiassociation activity; binds to 60S ribosomal subunit

[a]GEF, guanine nucleotide exchange factor; CPB, cap-binding protein.
[b]Subunit can be phosphorylated in vivo.
Source: From Arnstein and Cox (1992), reprinted by permission of Oxford University Press.

concentration of 100 mM or more and divalent cations (Mg^{2+}) below 2 mM are required.

Efficient entry of aminoacyl-tRNA into the ribosomal A site requires the participation of an elongation factor, termed EF-Tu in prokaryotes (see Table 4) and EF-1α (EF-1L) in eukaryotes (see Table 5), and GTP. This elongation factor forms a ternary complex with GTP and all aminoacyl-tRNAs except initiator tRNA, but not with uncharged tRNA, thus ensuring that only appropriately charged tRNAs are efficiently bound in the A site. A special elongation factor showing extensive homology with both EF-Tu and IF-2 is involved in the synthesis of selenoproteins (see Section 2.4) from selenocysteyl-tRNA^{UCA} in *E. coli.* In the aminoacyl-tRNA binding reaction mediated by EF-Tu, there is no hydrolysis of GTP, and the elongation factor remains associated with the charged tRNA until peptide bond synthesis has taken place, when GTP hydrolysis occurs and the elongation factor is released from the ribosome as a complex with GDP. The operation of this elongation factor is thus similar to that of initiation factor 2, which binds charged initiator tRNA to the small ribosomal subunit.

Following dissociation from the ribosome, the EF-Tu · GDP complex interacts with another elongation factor, EF-Ts, with formation of an EF-Tu · EF-Ts heterodimer and release of GDP. Reaction of the heterodimer with GTP regenerates the EF-Tu · GTP complex required for binding aminoacyl-tRNA. The sequence of events is similar in eukaryotes, with EF-1α (50,000 Da) corresponding to EF-Tu and EF-1β (30,000 Da) to EF-Ts.

Selection of the specific aminoacyl-tRNA to be bound at the ribosomal A site is by base pairing between the relevant mRNA codon and the tRNA anticodon. Since this interaction involves only a triplet of bases, hence a maximum of nine hydrogen bonds (see Fig-

Table 4 Properties of Prokaryotic (*E. coli*) Elongation and Termination Factors

Name	Molecular Mass (kDa)	Properties and Function
Elongation Factors		
EF-Tu	43	N-terminal acetylserine; heat labile; binds aminoacyl-tRNA to the ribosomal A site
EF-Ts	30	Regeneration of EFTU · GTP complex; heat stable
EF-G	77	GTP-dependent translocation of peptidyl-tRNA and its mRNA codon from the A site to the P site of the ribosome
Termination or Release Factors		
RF1	36	Requires UAA or UAG codons for hydrolysis of peptidyl-tRNA
RF2	38	Requires UAA or UGA codons for hydrolysis of peptidyl-tRNA
RF3	46	Enhances RF1 and RF2 activity

Source: From Arnstein and Cox (1992), reprinted by permission of Oxford University Press.

Table 5 Eukaryotic Elongation and Termination Factors

Name	Molecular Mass (kDa)	Properties and Function
Elongation Factors from Yeast, Animal, and Plant Cells		
EF-1α (EF-1_L or eEF-Tu)	50–60	Analogous to EF-Tu
EF-1β (eEF-Ts)	30	Analogous to EF-Ts
EF-2	105	GTP-dependent translocation analogous to EF-G; contains essential SH groups and one residue of a posttranslationally modified histidine residue
EF-3	125	GTPase and ATPase activity; function not fully defined
Termination or Release Factor (RF) from Reticulocytes		
RF	110	Two 55 kDa subunits; binds to the ribosome A site by a GTP- and termination-codon-dependent reaction; hydrolyzes peptidyl-tRNA in the P site

Source: From Arnstein and Cox (1992), reprinted by permission of Oxford University Press.

ure 1B), it is intrinsically unstable at physiological temperatures and is probably stabilized by components of the ribosome to allow sufficient time for peptide bond synthesis to occur. Also, the codon–anticodon pairing must be monitored for fidelity, to minimize errors in translation. In *E. coli* there is genetic and biochemical evidence that one of the proteins of the small ribosomal subunit, S12, is involved in ensuring the fidelity of normal translation and in causing the mistranslation that occurs in the presence of the antibiotic streptomycin as a result of incorrect codon–anticodon interactions.

Translocation Immediately after synthesis of the first peptide bond, the ribosomal A site contains dipeptidyl-tRNA, while uncharged initiator tRNA remains in the P site. Thus, both these sites are occupied, and to allow the next aminoacyl-tRNA to enter the A site, it is necessary to eject the uncharged tRNA and shift the dipeptidyl-tRNA from the A into the P site. This translocation takes place as a concerted process involving movement of messenger RNA and dipeptidyl RNA together into the P site, leaving the A site occupied by the next mRNA codon and free to accept the cognate aminoacyl-tRNA (see Figure 9d). According to the two-site model of the ribosome, ejection of the deacylated initiator tRNA occurs at the same time, but there is increasing evidence for a three-site model in which the deacylated tRNA moves first to an E (exit) site and then, when the next aminoacyl-tRNA enters the A site, is ejected.

Translocation requires the participation of another elongation factor, termed EF-G in prokaryotes (Table 4) and EF-2 in eukaryotes (Table 5), and GTP. It seems that when EF-G and GTP bind to the ribosome, translocation occurs, but GTP hydrolysis is required only afterward, to release EF-G and GDP. The location on the ribosome of the site for EF-G binding overlaps with that for EF-Tu, and thus EF-G must be released before the EF-Tu · aminoacyl-tRNA · GTP complex can enter the A site. Analogous reactions occur in eukaryotic systems.

There is little information about the details of the translocation mechanism. A continuous polyribonucleotide chain is not essential, since translocation can occur with individual trinucleotides. It seems likely that movement of the mRNA is dependent on and tightly coupled to that of the tRNA, with the binding sites for the tRNA providing the precision for movement by exactly one codon. Presumably, binding of EF-G and GTP after release of EF-Tu·GDP following peptide bond synthesis induces in the ribosome a conformational change that leads to translocation.

After translocation, the ribosomal P site is occupied by dipeptidyl-tRNA and the vacant A site contains the third mRNA codon. Entry into the A site of the next aminoacyl-tRNA, selected as before by the codon–anticodon interaction, enables peptide bond synthesis to continue, and repeated operation of the elongation–translocation cycle gives rise to a stepwise elongation of the nascent polypeptide chain, each complete cycle elongating the chain by one amino acid residue and moving the mRNA by one codon in the $5' \rightarrow 3'$ direction. When the end of the coding sequence is reached and one of the termination (or stop) codons has entered the A site, translation stops and the completed polypeptide chain is released.

Termination (see Figure 9g) The presence in the A site of one of the three termination codons, UAA, UAG, or UGA, results in the binding of a release factor (Table 4) instead of an aminoacyl-tRNA to the ribosome. In prokaryotes, two release factors have been identified: one (RF1) recognizing UAA and UAG, the other (RF2) functioning with UGA. Ribosomal binding and release of RF1 and RF2 are stimulated by a third factor, S, which interacts with GTP and GDP. In eukaryotic cells, such as reticulocytes, one release factor (RF) has been found to function with all three termination codons, and the binding of this factor to ribosomes is stimulated by GTP but not GDP. Although the details are not entirely clear, GTP hydrolysis appears to be required for the release of the finished polypeptide chain by cleavage of the peptidyl-tRNA bond and completion of the termination process, leading to dissociation of the release factor from the ribosome.

Thus, at the end of the ribosome cycle the coding sequence of messenger RNA has been translated to produce a particular polypeptide chain, and all the components involved become available for reuse in another round of the cycle (Figure 9h). Usually several ribosomes are attached to one mRNA molecule at the same time, giving rise to polyribosomes (also called polysomes) (Figure 11).

5 TRANSLATIONAL CONTROL OF GENE EXPRESSION

Cells need to synthesize specific proteins in the required amounts at particular times and to deliver them to the correct locations. These processes depend on a great many interactions, and thus there are numerous potential mechanisms for the translational control of gene expression. Examples of the ways used by cells to control the translation of mRNA are presented in the sections that follow. This material is intended to be illustrative rather than comprehensive, and it is to be expected that novel control mechanisms will continue to be discovered.

5.1 mRNA Stability

Provided all other components of the translational systems are present in optimum amounts, control of translation may be achieved through the availability of the relevant messenger RNAs at the site of protein synthesis. The steady state level of an mRNA is determined by the rates of its synthesis and degradation. Control of transcription is of major importance for synthesis in both prokaryotes

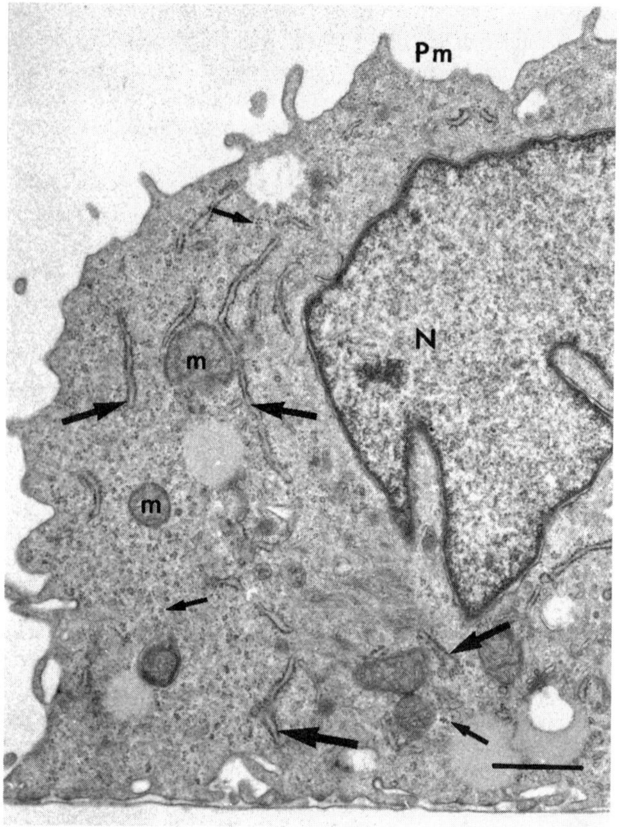

Figure 11. Electron micrograph of a thin section through part of an epithelial cell showing plasma membrane (Pm), nucleus (N), mitochondria (m), and rough endoplasmic reticulum (large arrows) to which are attached numerous ribosomes seen as electron-dense particles (≈ 28 nm diameter). Clusters of free ribosomes (small arrows) occur in the cytoplasm. Bar represents 1000 nm. (Micrograph by I. D. J. Burdett, National Institute for Medical Research, London, U.K.)

and eukaryotes. The stability of mRNA in the cell depends on its primary and secondary structure as well as on the presence of factors such as stabilizing proteins and nucleases. For example, during the cell cycle histone mRNA is destabilized after completion of DNA replication, resulting in a 30–50-fold decrease. This change appears to be due to an increase in the level of free histones, which form a complex with histone mRNA by interaction with a stem–loop structure at the extreme 3′ terminus. Formation of this histone–histone mRNA complex is thought to activate a ribosome-associated 3′–5′-exonuclease, which degrades the histone mRNA. During S phase, newly synthesized DNA binds free histones to form nucleosomes, which prevents the degradation of histone mRNA at this stage of the cell cycle.

5.2 Control by Interaction of Proteins with mRNA

Throughout the ribosome cycle, dynamic protein–mRNA interactions are functionally important in the initiation, elongation, and termination of polypeptide synthesis. In addition, more stable associations between proteins and mRNAs have been observed, particularly in eukaryotic cells. These messenger ribonucleoprotein complexes (mRNPs) occur both in polyribosomes and free in the cytosol, some of the latter being either temporarily or permanently unavailable for translation. Thus, protein–mRNA interactions contribute to the efficiency with which mRNAs are translated.

Some proteins, such as the poly(A)-binding protein (p78), are present in most if not all mRNPs, whereas others appear to be cell specific and mRNA selective. In unfertilized sea urchin eggs and *Xenopus* oocytes, for example, untranslated messenger is sequestered by association with proteins that prevent translation until later stages of development. Duck reticulocytes contain globin mRNP, which is untranslatable in vitro, whereas the mRNA obtained by deproteinizing the complex can be translated, showing that in this case translation is prevented by the mRNP proteins.

Formation of a site-specific mRNA–protein complex is involved in the translational control of the biosynthesis of ferritin, an iron storage protein, which is stimulated in response to the presence of iron. In this instance, a cytoplasmic repressor protein of 85 kDa binds to a highly conserved 28 nucleotide stem–loop structure in the 5′-untranslated region of ferritin mRNAs in the absence of iron. In the presence of iron, the protein dissociates from the mRNA, which is then available for translation. A similar loop motif occurs in the 3′-untranslated region of transferrin receptor mRNA, which is also subject to translational control by an iron-responsive repressor.

Specific regulation of gene expression at the level of translation also exists in prokaryotes. For example, the synthesis of *E. coli* threonyl–tRNA synthetase is negatively autoregulated by an interaction of the tRNA-like leader sequence of its mRNA with the synthetase, which inhibits translation by preventing the binding of ribosomes. The synthetase is displaced from the mRNA by $tRNA^{Thr}$, which thus acts as a translational antirepressor. This regulatory mechanism allows the cell to maintain a balance between the tRNA synthetase and its cognate tRNA.

Similarly, the synthesis of some ribosomal proteins is under negative translational control. In this case, selective binding of one ribosomal protein to the initiation regions of polycistronic mRNA for a number of ribosomal proteins regulates its own synthesis as well as that of the other ribosomal proteins. The mRNA secondary structure involved in binding the ribosomal protein S4, which acts as a translational repressor of the synthesis of four ribosomal proteins (S4, S11, S13, L17), appears to include an unusual "pseudoknot" linking a hairpin upstream of the ribosome-binding site with sequences 2–10 codons downstream of the initiation codon. (A pseudoknot structure contains intramolecular base pairs between base residues in the loop of a stem–loop structure and distal complementary regions of the RNA.) Stabilization of this structure by S4 would prevent the binding of ribosomes, and this control mechanism may contribute to the coordinated synthesis of the different ribosomal proteins required for ribosome assembly.

5.3 Control by mRNA Structure

The secondary structure of some eukaryotic mRNAs regulates translation by a mechanism involving a ribosomal frameshift, which gives rise to a directed change of the translational reading frame to allow the synthesis of a single protein from two or more overlapping genes by suppression of an intervening termination codon. Several retroviruses use this mechanism to move from one reading frame to another in the expression of the viral RNA-dependent DNA polymerase. Other examples of the operation of such a frameshift include the synthesis of the reverse transcriptase enzymes of avian coronavirus and several retrotransposons, such as the yeast Ty1. The mechanism of changing the reading frame involves "slippery" sequences and the complex folding of the mRNA into the type of structure described earlier as a pseudoknot.

In *E. coli,* translational control of the three cistrons of Q_β, f2, and related bacteriophage RNAs (Figure 12) accounts for the synthesis of coat protein/replicase/A protein in the approximate ratio of 20:5:1. These quantitative differences are due to the differential and independent initiation of translation at each cistron as a result of differences in the secondary structure of the mRNA initiation sites. Furthermore, in vivo there is a delay in the synthesis of coat protein, and this temporal control involves translational repression of the cistron by ribosomal protein S1. In addition, S1 functions as one of the subunits of the f2 RNA replicase, and therefore association of the newly synthesized translation product of the f2 replicase cistron with S1 will favor its dissociation from the phage RNA, thus allowing translation of the coat protein cistron to start.

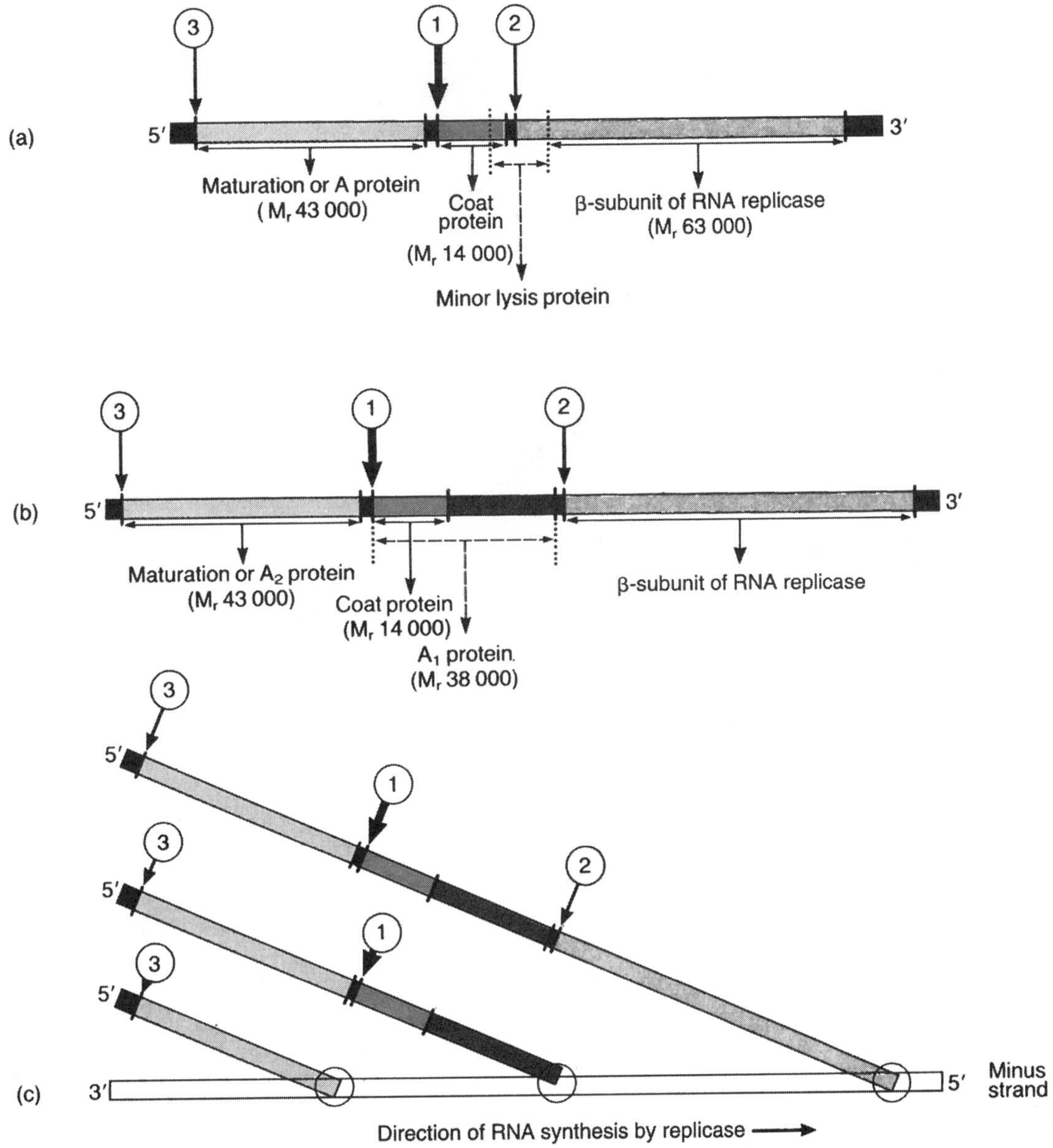

Figure 12. Translational control of bacteriophage synthesis: (a) arrangements of MS2 and f2 bacteriophage cistrons, (b) arrangement of Q_β bacteriophage cistrons, and (c) replicative intermediate synthesizing new plus strands on the complementary minus-strand copy of the original bacteriophage RNA. The ribosome-binding sites (RBS) are indicated by the numbered arrows. The major RBS (1) binds ribosomes efficiently but can be blocked by ribosomal protein S1. The secondary RBS (2) becomes available only after translation of at least part of the coat protein cistron by ribosomes. RBS 3 is masked by the secondary structure of native bacteriophage RNA but is accessible in nascent RNA (c) or in vitro when the secondary structure is destroyed. Noncoding regions are shown in black. [Reproduced from Arnstein and Cox (1992) by permission of the Oxford University Press.]

5.4 Control Through Modification of Translation Factor Activity

5.4.1 Initiation Factors

In eukaryotes, initiation of protein synthesis is inhibited by phosphorylation of the initiation factor eIF-2. Phosphorylation is stimulated by lack of heme or the presence of double-stranded RNA. Two different protein kinases capable of phosphorylating the α subunit of eIF-2 have been characterized. One enzyme, called the heme-controlled repressor (HCR) or heme-regulated inhibitor (HRI), is a cytoplasmic protein (95,000 Da) that becomes activated by phosphorylation. The other kinase (67,000 Da) is activated by phosphorylation in the presence of double-stranded RNA. Thus, a cascade of protein phosphorylation is involved. Phosphorylated eIF-2 is unable to exchange GDP for GTP, which prevents it from functioning in the binding of initiator tRNA to ribosomes.

Conditions other than lack of heme (e.g., heat shock, serum deprivation, or the presence of oxidized glutathione, which are known to inhibit protein synthesis) also give rise to the phosphorylation of eIF-2α. Conversely, the activity of eIF-4F is decreased by dephosphorylation of the 24 kDa subunit (see Table 3) rather than by phosphorylation. Thus, a number of different kinases and phosphatases are involved in modulating the activities of different factors.

Small RNAs may also be involved in regulating the translation of mRNA in eukaryotic cells. Among the stimulatory RNAs, the best characterized is a small RNA of about 160 nucleotides, which accumulates in cells after infection with adenovirus. This virus-associated RNA, VA-RNA_1, which is required to maintain general protein synthesis, acts by inhibiting the phosphorylation of the α subunit of initiation factor eIF-2.

5.4.2 Other Translation Factors

A Ca^{2+}/calmodulin-dependent protein kinase phosphorylates eEF-2. The phosphorylated factor appears to be inactive and moreover also inhibits the activity of the nonphosphorylated EF. Dephosphorylation of the factor by phosphatase restores its activity.

5.5 Effects of Antisense Polynucleotides

Antisense RNAs, which are polynucleotides with base sequences complementary to messenger RNAs, have been found in both prokaryotes and eukaryotes. It has been demonstrated that they can function as inhibitors of messenger RNA translation. In prokaryotes, the most effective inhibitors appear to have a base sequence complementary to the 5′ leader region including the Shine–Dalgarno sequence, which is involved in the binding of mRNA to the small ribosomal subunit. In eukaryotes, translation of mRNA is inhibited by polyribonucleotides complementary to the 5′ untranslated region of mRNA, indicating a direct effect on initiation similar to that in prokaryotes. Polynucleotides complementary to the 3′-untranslated region of mRNA also inhibit translation in some cells, and this effect may be due to destabilization of mRNA by ribonucleases specific for double-stranded RNA. It should be noted that the effect of antisense polynucleotides is not restricted to translation of mRNAs, but transcription and the processing of transcripts may also be inhibited.

5.6 Availability of Amino Acids, tRNA Abundance, and Codon Usage

5.6.1 Amino Acids

Polypeptide synthesis depends on an adequate supply of tRNAs charged with the 20 protein amino acids and appropriate interactions between their anticodons and the codons of mRNA. Peptide chain elongation is decreased or inhibited by lack of amino acids or by other conditions giving rise to an imbalance or deficiency in aminoacyl-tRNAs.

5.6.2 Abundance of tRNAs and Codon Usage

In the elongation cycle of protein synthesis, the rate-limiting step is the search by the ternary aminoacyl-tRNA · EF-Tu · GTP complex for the specific mRNA codon in the ribosomal A site; the next two steps, peptide bond formation and translocation, are more rapid. Since different tRNAs are present in the cytosol in unequal amounts, elongation rates are slower at codons corresponding to rare tRNA species.

The existence of synonymous codons raises the question of preferential use of some codons and its possible significance in relation to translational efficiency and control. In some bacteria [e.g., *Pseudomonas aeruginosa,* which has a high content of G+C (67.2%) in DNA], the most common codons are those with the strongest predicted codon–anticodon interaction—that is, G · C base pairs—but this preference is not universal and, for example, does not apply to *E. coli,* which has a lower proportion of G+C (50%).

Codon degeneracy allows the mRNA for a specific polypeptide to have any one of many alternative nucleotide sequences that will give rise to alternative secondary structures. The secondary structure of mRNA is important in relation to the efficient initiation of translation, whereas codon usage may play a part in determining elongation rates. However, codon usage is probably not of general importance in the translational control of protein synthesis.

5.7 Modulation of Ribosome Activity

Specific ribosomal components have an important function in relation to the fidelity of protein synthesis. Thus, in *E. coli* ribosomal protein S12 determines the accuracy of codon–anticodon interactions and modulates the translational error frequency in the presence of the antibiotic streptomycin.

Whether and to what extent reversible modification of ribosomal constituents are involved in translational control of protein synthesis is uncertain. Although the phosphorylation of ribosomal protein S6 increases with cell proliferation, it is not known whether this change is directly related to the accompanying increase in protein synthesis by an effect on the translation rate.

6 CONCLUDING REMARKS

The control of protein synthesis, either by regulation of the amount of mRNA available for translation or by the efficiency with which it is translated, is important in cell growth and development as a factor determining the level of cellular and extracellular proteins. Subversion of this control occurs in cells infected by viruses when the viral nucleic acid uses the protein-synthesizing machinery of the host cell and thereby changes normal cell metabolism in favor of the synthesis of viral proteins needed for the production of virus progeny.

The polypeptide chains of all proteins are synthesized by the process described here. This mechanism gives rise to primary polypeptide chains, which are often further modified—for example, by cleavage into smaller peptides, by structural modification of selected amino acid residues, by splicing of the polypeptide chain, or by the formation of covalent bonds between polypeptide chains. Some of these secondary modifications are related to the correct folding of polypeptide chains and to the production of active enzymes or peptide hormones from inactive precursors (e.g., insulin from proinsulin). Also, the transport of proteins within the cell or the secretion of extracellular proteins is often linked to structural changes in polypeptide chains either during or after completion of synthesis.

See also Ribosome Preparation and Protein Synthesis Techniques.

Bibliography

Arnstein, H. R. V., and Cox, R. A. (1992) *Protein Biosynthesis.* Oxford University Press, Oxford, New York.

Bock, A., Forchhammer, K., Heider, J., and Baron, C. (1991) Selenoprotein synthesis: An expansion of the genetic code. *Trends Biochem. Sci.* 16:464–467.

Hatfield, D., and Oroszlan, S. (1990) The where, what and how of ribosomal frameshifting in retroviral protein synthesis. *Trends Biochem. Sci.* 15:186–190.

———, Lee, B. J., and Pirtle, R. M., Eds. (1992) *Transfer RNA in Protein Synthesis.* CRC Press, Boca Raton, FL.

Helene, C., and Toulme, J.-J. (1990) Specific regulation of gene expression by antisense, sense and antigene nucleic acids. *Biochim. Biophys. Acta,* 1049:99–125.

Hill, W. E., Dahlberg, A., Garrett, R. A., Moore, P. B., Schlessinger, D., and Warner, J. R. (1990) *The Ribosome: Structure, Function and Evolution.* American Society for Microbiology, Washington, DC.

Hinnebusch, A. G., and Liebman, S. W. (1991) Protein synthesis and translational control in *Saccharomyces cerevisiae,* in Vol. I, *The Molecular and Cellular Biology of the Yeast Saccharomyces: Genome Dynamic, Protein Synthesis and Energetics,* E. W. Jones and J. R. Pringle, Eds. Cold Spring Harbor Laboratory Press, Plainview, NY.

Jackson, R. J., Howell, M. T., and Kaminski, A. (1990) The novel mechanism of initiation of picornavirus RNA translation. *Trends Biochem. Sci.* 15:477–483.

Klausner, R. D., Rouault, T. A., and Harford, J. B. (1993) Regulating the fate of mRNA: The control of cellular iron metabolism. *Cell,* 72:19–28.

Kozak, M. (1983) Comparison of initiation of protein synthesis in procaryotes, eucaryotes and organelles. *Microbiol. Rev.* 47:1–45.

Lewin, B. (1990) *Genes IV.* Oxford University Press, Oxford, New York.

Matheson, A. T., Davies, J., Hill, W., and Dennis, P. (1996) *Frontiers in Translation.* National Research Council of Canada, Ottawa.

Mechanism and Control of Translation. (1990). A collection of papers presented by many different authors at a meeting in Noordwijkerhout, The Netherlands, May 12–17, 1990. *Biochim. Biophys. Acta,* 1050:1–360.

Merrick, W. C. (1992) Mechanism and regulation of eukaryotic protein synthesis. *Microbiol. Rev.* 55:291–315.

Nierhaus, K. H., Subramanian, A. R., Erdmann, V. A., Franceschi, F., and Wittmann-Liebold, B., Eds. (1994) *Translational Apparatus: Structure, Function, Regulation and Evolution.* Plenum Press, New York.

Noller, H. F. (1991) Ribosomal RNA and translation. *Annu Rev. Biochem.* 60:191–227.

Rijs, B., Rattan, S. I. S., Clark, B. F. C., and Merrick, W. C. (1990) Eukaryotic protein elongation factors. *Trends Biochem. Sci.* 15:420–424.

Shub, D. A., and Goodrich-Blair, H. (1992) Protein introns: A new home for endonucleases. *Cell,* 71:183–186.

Transport Proteins

Milton H. Saier, Jr.

1 **Introduction**
2 **Types of Transport Process**
3 **Classes of Integral Membrane Transport Proteins**
4 **Common Structural Motif of Many Transport Protein Families**
5 **Mobile Carriers and Pores**
6 **Channel-Type Proteins**
 6.1 Porin Proteins of Bacteria and Eukaryotic Organelles
 6.2 Ligand-Gated Ion Channels: The Acetylcholine Receptor
 6.3 Voltage-Sensitive Ion Channels
 6.4 The Major Intrinsic Protein Family
7 **Mobile Carriers**
 7.1 Solute Uni-, Sym-, and Antiporters of Similar Topology
 7.2 The Mitochondrial Carrier Family
 7.3 Proteins of the Bacterial Phosphotransferase System
 7.4 A Class of ATP-Dependent (ABC-type) Active Transport Systems
8 **A Light-Driven Ion Pump: Bacteriorhodopsin**
9 **Electron-Flow-Driven Ion Pumps**
10 **Ion-Translocating ATPases**
 10.1 Phosphoryl Intermediate (P-type) ATPases
 10.2 F_1/F_0 (F-type) ATPases
 10.3 Vacuolar (V-type) ATPases
 10.4 Oxyanion-Specific (A-type) ATPases
11 **Chemical Versus Chemiosmotic Energization of Transport**
12 **Conclusions on an Evolutionary Theme**

Key Words

Antiport (Countertransport; Exchange Transport) Transmembrane transport of two species in opposite directions in a carrier-mediated process.

Carrier Protein A transmembrane transport protein that selectively binds a solute and translocates that solute with the binding site across the membrane.

Channel Protein A transmembrane transport protein that forms a pore through the membrane, allowing passage of solutes from one side to the other.

Group Translocation Transport coupled to chemical modification of the solute transported (e.g., sugar transport and phosphorylation catalyzed by the bacterial phosphotransferase system).

Homology (Homologous, adj.) The property of sharing a common evolutionary ancestry.

Symport (Cotransport) Transmembrane transport of two species, usually a solute and a cation, together in a tightly coupled, carrier-mediated process.

Transporter (Porter; Permease; Pump) A transmembrane solute transport system.

Uniport Transport of a single species across a membrane by a carrier-type mechanism.

Transport proteins translocate solutes across biological membranes. Different classes of transport systems are known which function by different mechanisms, couple different forms of energy to transmembrane solute translocation, and comprise distinct families of homologous proteins. Some of these permease classes, many of which extend across the prokaryotic–eukaryotic boundary, share structural and functional characteristics. While some of them probably have a common evolutionary origin, others clearly evolved independently of each other.

1 INTRODUCTION

All living organisms probably descended from a common ancestor. They diverged to give rise to two kingdoms, the prokaryotes (which lack nuclei) and the eukaryotes (which are typified by eukaryotic organelles such as nuclei, mitochondria, and chloroplasts). Extensive evidence suggests that eukaryotes may have emerged some 1.8 billion years ago, although the prokaryotes may have existed more than 2 billion years before that time. Over this enormous time interval of about 4 billion years, proteins, including transmembrane solute transport proteins, evolved to their present state of complexity.

2 TYPES OF TRANSPORT PROCESS

Transport processes have been classified on the basis of their energy-coupling mechanisms. Four classifications are presented in Table 1. Primary active transport refers to a process whereby a *primary* source of energy (chemical energy: e.g., ATP, light, electron flow) is used to drive the active accumulation of a solute into or extrusion of a solute from a cell. Two examples of transport proteins that use each of these energy sources are provided in Table 1.

Group translocation is a process whereby the solute transported across the membrane is modified during transport. A chemical agent drives uptake of the solute and simultaneously derivatizes the solute. An example of group translocation is the sugar-transporting phosphoenolpyruvate/sugar phosphotransferase system that takes free sugar from the medium and releases sugar phosphate into the cytoplasm of bacteria. In this process, phosphoenolpyruvate is converted to pyruvate.

Secondary active transport utilizes a *gradient* of a substance across the membrane, resulting from a primary active transport process, to drive accumulation of another solute. Thus, a proton electrochemical gradient, created by the efflux of protons from *E. coli* by electron carriers, can be used to drive the uptake of a sugar such as lactose because uptake of lactose is coupled to the influx of a proton. In intestinal epithelial cells, a sodium electrochemical gradient is used to drive uptake of the sugar glucose. Sodium ion efflux is driven by a transport enzyme, the Na^+/K^+-adenosine triphosphatase (ATPase). This enzyme is the primary active transport system in the intestine. The coupled flux of glucose together with a sodium ion allows accumulation of the sugar against a concentration gradient because the sodium ion is flowing down its electrochemical gradient in an energetically favorable process when it enters the cell together with glucose.

Finally, facilitated diffusion is an energy-independent process that merely allows the diffusion of a solute from a higher concentration to a lower concentration, across a membrane. In the case of channel proteins (see Section 5), the solute merely passes through a pore or hole in the membrane. In the case of a carrier protein, the solute binds to a solute recognition site on the carrier, and the solute moves to the other side of the membrane together with its combining site. The solute is then released on the other side of the membrane.

3 CLASSES OF INTEGRAL MEMBRANE TRANSPORT PROTEINS

Recent studies have established that most eukaryotic integral transmembrane solute–transport proteins possess homologous prokaryotic counterparts. For example, the three major cation-translocating ATPases, types P, F, and V, are found in both bacteria and eukaryotes (Table 2). Similarly, several mitochondrial electron carrier proteins capable of transmembrane proton translocation possess homologous counterparts in bacteria.

Gram-negative bacteria possess an inner membrane and an outer membrane. Outer membrane porin proteins, which transport many

Table 1 Types of Transport Process

Type of Transport	Energy Source	Example(s)
Primary active transport	Chemical (ATP)	P-type ATPases ABC-type ATPases
	Light ($h\nu$)	Photosynthetic reaction center Bacteriorhodopsin
	Electrical (e^- flow)	NADH dehydrogenase Cytochrome oxidase
Group translocation	Chemical (PEP)	Phosphonenolpyruvate phosphotransferase system
Secondary active transport	Proton electrochemical gradient	Solute/H^+ symporters
	Sodium electrochemical gradient	Solute/Na^+ symporters
Facilitated diffusion	None	Ion channel proteins Sugar uniporters

Table 2 Occurrence of Various Homologous Transport Protein Families Across the Prokaryotic/Eukaryotic Boundary

Protein Family	Prokaryotic	Eukaryotic
Outer membrane porins	+	+
Ligand-gated ion channels	−	+
Voltage-gated ion channels	+	+
MIP family	+	+
P-type ATPases	+	+
F-type ATPases	+	+
V-type ATPases	+	+
A-type ATPases	+	+
Bacteriorhodopsins	+	−
H^+-translocating electron carriers	+	+
Uni-, sym-, and antiporters	+	+
ABC-type, ATP-driven solute pumps	+	+
Phosphotransferase system	+	−

solutes nonselectively, have a structurally similar counterpart in the anion-selective channels of the outer mitochondrial membranes of eukaryotes. Other pore-type transporters include the ion channels of nerve and muscle cells, some of which possess known homologues in prokaryotes, and the presumed pore-type major intrinsic protein (MIP) family of proteins, members of which are found ubiquitously throughout the living world (Table 2).

In contrast to these channel-type transporters, there exist several classes of carrier-type transporters including several ubiquitous families of uni-, sym-, and antiporters. The transport mechanisms of some ion-translocating proteins such as the various ATPases, bacteriorhodopsins, and certain electron carriers may possess a combination of channel and carrier characteristics.

While many of the families of transport proteins extend across the eukaryotic–prokaryotic division, some are restricted to just one of these kingdoms. The ABC-type, ATP-driven solute pumps of bacteria have eukaryotic counterparts, but the permeases of the sugar-transporting bacterial phosphotransferase system, which couple sugar uptake to sugar phosphorylation, do not appear to have a functional equivalent in eukaryotes (Table 2). A family of porters found in eukaryotic mitochondria (the mitochondrial carrier family) appears to be restricted exclusively to eukaryotic organelles. The sections that follow describe some of these families of transport proteins.

4 COMMON STRUCTURAL MOTIF OF MANY TRANSPORT PROTEIN FAMILIES

Many but not all transmembrane solute transport proteins exhibit a common structural motif. The proteins in these families consist of units or domains that pass through the membrane six times, each time as an α-helix. The families of transport proteins which appear to exhibit this motif are listed in Table 3.

Both the amino terminus (N-terminus) and the carboxy terminus (C-terminus) of the domains of these proteins are localized to the cytoplasmic surface of the membrane. Sometimes a transport protein consists of only one such unit or domain, but in other cases, two, four, or even eight such units are present in a single polypeptide chain. An odd number of such units apparently never occurs. Frequently there are more than one polypeptide chains or subunits present in a single permease protein complex. The number of such subunits is again 1, 2, or 4 in all cases thus far documented.

Table 3 summarizes properties of many structurally related transport proteins. The voltage-sensitive Na^+ and Ca^{2+} channels of nerve and muscle cells consist of four units of six transmembrane helices (spanners), connected to each other by a cytoplasmic loop. These proteins consist of a single polypeptide chain. By contrast, the homologous K^+ channels of nerves and muscles, which are similar in overall makeup to the Ca^{2+} and Na^+ channels, consist of four such structural units, but they are present as four subunits, each of the same composition and orientation in the membrane. Members of the major intrinsic protein family of presumed channel proteins usually resemble the K^SPL channels in this regard.

Facilitative diffusion carriers (uniporters) and many (but not all) secondary active transport proteins (symporters and antiporters) of several distinct families exhibit a common structural feature, namely, two such units are linked together in a single polypeptide chain (Table 3). Mitochondrial carriers comprise a distinct family of uni-, sym-, and antiporters, and the proteins of this family usually consist of two subunits, each containing six transmembrane helices or spanners.

Table 3 Structural Features of Several Families of Transport Proteins[a]

Protein Family	Transmembrane Segments per Unit*	Number of Units per Permease[b]	Number of Transmembrane Polypeptide Chains
Voltage-sensitive Na^+ and Ca^{2+} channels	6	4	1
Voltage-sensitive K^+ channels	6	4	4
Major intrinsic protein (MIP) family	6	4 (2)[c]	4 (2)[c]
Uniporters, symporters, antiporters (MFS, SSF, SNF, and APC families)	6	2 (8)[d]	1 (4)[d]
Mitochondrial carriers	6	2	2
Phosphotransferase system permeases	6 or more	2	2
ATP-binding cassette (ABC)-type systems	6	2	1 or 2

[a] In all the permease classes tabulated, the amino- and carboxy-termini of the proteins as well as the loop regions between the six transmembrane segment domains are believed to be on the cytoplasmic side of the membrane.

[b] Unit corresponds to the common six-transmembrane structural feature, presumed to correspond to a protein domain (see text for explanatory remarks).

[c] The one reported exception to the generalization that these channel proteins are tetrameric is the glycerol channel of E. coli, which appears to be dimeric.

[d] The one reported exception to the generalization that these permeases are monomeric and span the membrane 12 times is the intestinal Na^+/glucose symporter, which appears to be tetrameric.

Finally, primary active transporters of the ABC type and the group-translocating permease proteins of the bacterial phosphotransferase system (PTS) appear to have six or more transmembrane spanners per unit (Table 3). The former proteins use ATP to drive active transport (uptake or extrusion) of the solute, while the phosphotransferase system permeases take up sugars utilizing the energy of phosphoenolpyruvate.

A common structural feature for many transport proteins has led to the suggestion that they share a common evolutionary origin. This, however, proves not to be the case for several distinct families of these proteins. The evidence is discussed briefly in Section 7.2.

5 MOBILE CARRIERS AND PORES

At least two conceptual models explain the transport of substances through a biological membrane. In the *mobile carrier* mechanism, the transporter or a solute-binding moiety of a permease is assumed to physically change its orientation in the membrane during translocation by shuttling back and forth across the bilayer. On the other hand, the *pore* model assumes that the protein is more or less fixed in its intramembrane orientation and forms a hydrophilic pore that is more or less stereospecific for its transport substrate. This pore may be gated in the sense that it opens only transiently in response to appropriate stimuli.

As model systems for studying biological transport, the *ion-translocating antibiotics* have provided evidence for both mechanisms. Valinomycin, an antibiotic isolated from a species of *Streptomyces,* is a cyclic peptide containing the sequence D-valine, L-lactate, L-valine, and D-hydroxyisovalerate repeated three times. It is selective for K^+ and makes both artificial bilayers and biological membranes permeable to this cation. The valinomycin–K^+ complex is soluble in lipid bilayers, while free K^+ is highly soluble in water. The complex of K^+ and valinomycin (or valinomycin alone) can shuttle across a membrane at a great rate like a "carrier."

A second well-studied antibiotic is gramicidin A, also isolated from a bacterium (*Bacillus brevis*). It is a 15 amino acid linear polypeptide that is a cation-specific ionophore with much less specificity than valinomycin for any particular cation. Gramicidin A forms a hole or pore through the membranes so that K^+ can freely flow down its concentration gradient, from one side of the membrane to the other. These prototypic antibiotics serve as models for our concepts of transporter carriers and channels.

6 CHANNEL-TYPE PROTEINS

6.1 Porin Proteins of Bacteria and Eukaryotic Organelles

Studies on the topographies of integral membrane proteins suggest a more or less static and asymmetric disposition of these polypeptides across biological membranes. Several classes of proteins that carry out transport do so by forming transmembrane pores or channels. These structures may be static or gated. Nonspecific transport, where solute recognition is minimal, is mediated by relatively static pore structures.

In a few instances, transmembrane pores formed by integral membrane proteins have been visualized by electron microscopy and X-ray diffraction. The outer membrane of *E. coli* is permeable to molecules with molecular weights below about 600. This permeability is conferred by a class of proteins called *porins,* which are integral constituents of the outer membranes of all Gram-negative bacteria. They are also found in the outer membranes of mitochondria and chloroplasts.

In *E. coli,* porin molecules are arranged in the outer membrane in a hexagonal lattice of trimers. The three-dimensional structures of porins have been solved by X-ray crystallography. Through the center of each porin monomer there is a pore with a diameter of about 1 nm, consistent with the molecular weight exclusion limit mentioned earlier. Unlike most transmembrane proteins studied to date, the membrane-spanning regions of porins consist almost entirely of β-type structures (β-sheets arranged in a β-barrel).

Nonspecific transmembrane pores also have been demonstrated in the plasma membranes of eukaryotic cells. Well-studied examples of these pores are the so-called *gap junctions,* which connect neighboring cells of similar types in tissues such as liver, intestine, kidney, brain, and cardiac muscle. In the region of the gap junction, the plasma membranes of two different cells are closely apposed and appear to have channels connecting them. The protein hexamers in each membrane are thought to span the bilayer, and apposition of two hexamers forms an aqueous channel between the cells.

6.2 Ligand-Gated Ion Channels: The Acetylcholine Receptor

Neurotransmitters and hormones of higher organisms are chemicals that sometimes interact with specific channel proteins in the membranes of nerve and muscle cells to regulate ion fluxes across the membranes. They thereby regulate the electrical behavior of the cell and generate, or block generation of, action potentials that are responsible for the propagation of nerve impulses. Many such neurotransmitters are known. Examples are acetylcholine, noradrenaline, dopamine, γ-amino butyric acid (GABA), and a variety of amino acids such as glutamate, glycine, and proline. These ligand-gated ion channels generally are multisubunit proteins, which do not exhibit the six-spanner units of many other transport proteins. The prototype of this family (or superfamily) of channels is the acetylcholine receptor.

The structure and properties of the acetylcholine receptor are by far the best understood among all ion-transporting neurotransmitter receptors. Release of acetylcholine by the presynaptic nerve endings depolarizes the postsynaptic membrane by virtue of the binding of this neurotransmitter to its receptor in the latter membrane. In this way, each cell, when stimulated, can attain a rapid change in membrane potential because the influx of Na^+ when the channels open produces depolarization.

Acetylcholine receptors are rosettes, apparently consisting of five subunits arranged around a central axis. Closed membrane vesicles derived from membrane fragments can be made permeable to Na^+ by adding acetylcholine, an effect that is blocked by antagonists such as snake venom toxins. The acetylcholine receptor is a glycoprotein consisting of four subunits (40, 50, 60, and 65 $\times 10^3$ Da, named α, β, γ, and δ) in a molar ratio of 2:1:1:1. The structure for the acetylcholine receptor is therefore a pentamer ($\alpha_2\beta\gamma\delta$) with a mass of about 255,000 Da. Interestingly, the four different subunits of this protein exhibit amino acid sequence similarity near their NH_2 terminals, indicating that they all arose by duplication and divergence of a single ancestral gene. Because all five subunits span the postsynaptic membrane, it is possible that the channel corresponds to the central "hole" in the rosette seen by electron microscopy of postsynaptic membranes.

Considerable evidence points to a gated-pore-type mechanism for ion permeability conferred by the acetylcholine receptor, as is the case for the Na^+ channel (see Section 6.3). A pore model, allowing more or less free diffusion of ions through the channel in its open state, rather than a slower, carrier-mediated mechanism, is therefore favored for the acetylcholine receptor.

6.3 Voltage-Sensitive Ion Channels

Nerve and muscle cells of higher eukaryotes possess homologous proteins that form voltage-sensitive ion channels. Voltage-sensitive channels selective for Na^+, Ca^{2+}, and K^+ in nerve and muscle cell membranes "sense" the membrane potential and react by opening or closing during various stages of action–potential propagation. Thus, these ion channels can apparently assume at least two conformations, one of which is more permeable to the ion in question than the other.

K^+, Ca^{2+}, and Na^+ movements through nerve membranes are mediated by separate channels that act as gated pores. During a typical action potential, Na^+ ions pass rapidly through a Na^+-specific channel. Passive Na^+ diffusion through the transmembrane Na^+ channel explains these rapid Na^+ fluxes. These fluxes are responsive to voltage changes and generate further changes in transmembrane voltage.

Proteins that comprise the voltage-sensitive channels that facilitate Na^+ or Ca^{2+} transport consist of single polypeptide chains of about 2000 amino acid residues. Each one of these channel proteins contains four homologous internal repeats that are similar in sequence and consist of about 300 residues. Each of the four repeated domains contains six hydrophobic presumed α-helical transmembrane segments or spanners. These domains are connected by hydrophilic loops. The voltage-sensitive K^+ channels are similar in sequence but differ from the Na^+ and Ca^{2+} channels in being homotetramers in which each subunit corresponds to one of the four internal repeats of the Na^+ or Ca^{2+} channels. Although their cation specificities differ, all these channels probably function by similar mechanisms. As bacterial homologues of these channel proteins have recently been identified, they evidently comprise a very ancient family. This last fact suggests that transmembrane cation channels were important in early life-forms on earth.

6.4 The Major Intrinsic Protein Family

Recent sequence analyses have revealed a family of intrinsic membrane proteins from plants, animals, yeast, and bacteria. These proteins include the major intrinsic protein from the lens of the mammalian eye and the glycerol facilitator from the bacterium *E. coli.* The former protein is probably a channel protein that allows transport of water and salt across the lens fiber cell membrane in mammals, while the latter is a channel protein that allows facilitation of glycerol across the bacterial membrane. These proteins are similar in size and have six putative membrane-spanning domains per subunit. Sequence comparisons surprisingly revealed that the six transmembrane helical segments in each of these proteins arose by intragenic duplication of a primordial gene encoding just three transmembrane helical segments.

The best characterized of these proteins, MIP, is known to exist in the membrane as a homotetramer in which both the amino- and carboxy-termini of the subunits face the cytoplasm. This arrangement is the same as that established for the voltage-responsive K^+ channels of nerves and muscle cells discussed in the preceding section. By contrast, the glycerol facilitator of *E. coli* is believed to be a dimer, similar in this regard to the phylogenetically unrelated mitochondrial carriers discussed in Section 7.2.

7 MOBILE CARRIERS

7.1 Solute Uni-, Sym-, and Antiporters of Similar Topology

It has long been believed that carrier-type systems that allow the transmembrane passage of solutes by facilitative diffusion (uniport), countertransport (antiport), or ion cotransport (symport) function by similar mechanisms. Recent sequence analyses have shown that sugar uniporters of eukaryotes are homologous to symporters and antiporters of bacteria. In fact, this large family of proteins extends from enteric bacteria and cyanobacteria to the lower eukaryotes such as yeast, algae, and protozoans, and on up to higher plants and animals including man. They are found in both plasma membranes and intracellular vesicles of animals. These porters are specific for sugars, oligosaccharides, organic acids, inorganic phosphate, organophosphate esters, bioactive amines, and drugs. They consist of protein subunits with twelve (6 + 6) transmembrane helical segments, all of which may have been derived from a common ancestral protein of half this size. This family of facilitator proteins, called the major facilitator superfamily (MFS), is a very old family that arose in prokaryotes and probably dates back more than 3.5 billion years.

Other families of facilitators may well share a common ancestry with the MFS. Thus, a large family of transport proteins specific for amino acids, polyamines, and choline (the APC family) has been described. Members of this family exhibit the same putative topology documented for the MFS proteins. These porters, like members of the MFS, catalyze solute/proton symport and/or solute/solute antiport. They occur in both prokaryotes and eukaryotes.

Another family of facilitators that exhibits structural and mechanistic similarities to the MFS and APC families, but is too distant from them in sequence to allow establishment of homology without three-dimensional data, consists of a diverse group of solute/Na^+symporters. These proteins individually exhibit specificity toward amino acids (proline in *E. coli*) and a vitamin (pantothenate in *E. coli*), as well as hexoses, myoinositol, and nucleosides (all in mammals). It is probable that members of this family, termed the sodium/solute symporter family (SSF), will also be found to be ubiquitous.

A distinct family of sodium/neurotransmitter symporters (the SNF) has been found to exhibit the characteristic 6 + 6 putative transmembrane spanner topology also observed for the other facilitator families. All symporters of the SNF thus far sequenced are from eukaryotic sources, and most are present in neural tissues of animals. None of these porters exhibits demonstrable internal repeat sequences. It is possible, but not proven, that this family of proteins arose in eukaryotes, independently of the major facilitator superfamily and other recognized families. Alternatively, a member of one of the ubiquitous families may have diverged beyond recognition and subsequently duplicated repeatedly in eukaryotes, to give rise to the members of this new family.

7.2 The Mitochondrial Carrier Family

In mitochondria, there exist a number of strict transport carriers that include an ATP/ADP exchanger, a phosphate porter, an α-ketoglutarate/malate exchanger, a tricarboxylate carrier, and the uncoupling protein of mammalian brown adipose tissue. Sequence comparisons have shown that these porters as well as several functionally uncharacterized proteins are homologous (i.e., derived from a common ancestral protein). Each subunit of such transport systems apparently spans the membrane six times, and the dimeric forms of these proteins are presumed to be the structural and functional equivalents of the monomeric forms of carriers that span the membrane 12 times. They nevertheless form a distinct family, the mitochondrial carrier family (MCF), which arose by triplication of a small primordial gene that encoded a protein having only two transmembrane helical segments.

The degree of similarity between the three internal repeats in MCF proteins is much greater than that between the two 3-spanner internal repeats in the MIP family or the very distantly related 6-spanner internal repeats in the MFS, suggesting that this triplication event occurred relatively recently in evolutionary time, possibly about 1.5 billion years ago, after the development of the eukaryotic cell. Members of the MCF have been found only in eukaryotic organelles, but are exclusively nuclear encoded, which is a point in favor of the more recent date just suggested.

7.3 Proteins of the Bacterial Phosphotransferase System

The phosphotransferase system (PTS) is responsible for the phosphorylation-coupled transport of many sugars into bacterial cells. It is energized by phosphorylation at the expense of phosphoenolpyruvate via two energy-coupling proteins, which are themselves phosphorylated. Each sugar permease of the PTS consists of at least three structurally distinct domains that either can be fused together in a single polypeptide chain or can exist as two or three interactive polypeptide chains. Two of these three proteins or protein domains (IIA and IIB) exhibit the properties of globular water-soluble constituents, and they are sequentially phosphorylated in the phosphoryl transfer reactions of the PTS. The integral membrane domain or protein (IIC) binds and transports the sugar and catalyzes phosphoryl transfer from IIB-phosphate to sugar. IIC is the only protein constituent of the PTS that is not itself phosphorylated.

The sugar transport complex must be phosphorylated on the IIB domain before the sugar substrate can be transported at an appreciable rate via the transmembrane IIC channel. In this process, called group translocation, sugar phosphorylation is tightly coupled to transmembrane transport. It is believed that group translocation arose by the superimposition of sugar phosphorylation onto the carrier-mediated transport process. Some evidence suggests that the integral transport proteins or protein domains, the enzymes IIC, may be distantly related to, and thus structurally and functionally equivalent to, simple facilitators such as the insulin-responsive glucose transporter of mammals. One of these permeases, the mannitol permease of *E. coli,* has been reported to have six transmembrane spanners with the N- and C-termini facing the cytoplasm, as do many permease classes. Another one, the glucose permease of *E. coli,* is believed to have eight transmembrane spanners, again with the N- and C-termini facing the cytoplasm.

7.4 A Class of ATP-Dependent (ABC-type) Active Transport Systems

Numerous structurally distinctive bacterial permeases as well as several homologous eukaryotic transport systems share a common organization. Two hydrophilic domains or proteins function to couple ATP hydrolysis in the cytoplasm to active substrate uptake or efflux, and two hydrophobic domains or proteins function as the transmembrane substrate channels. These proteins or protein domains constitute what is referred to as the ABC (ATP-binding cassette) superfamily. Either the two hydrophilic domains or proteins or the two hydrophobic domains or proteins (or both) may exist either as heterodimers or homodimers. If, as in most bacterial systems, each of these constituents is a distinct protein, then either two, three, or four genes will code for them, depending on whether both are homodimers, one is a homodimer and one is a heterodimer, or both are heterodimers, respectively. The best characterized of the eukaryotic proteins included in this family are the multidrug-resistance transporter and the cystic fibrosis related chloride channel of mammalian cells.

ABC-type transport systems of bacteria and eukaryotes exhibit an exceptionally wide variety of substrate specificities, each transporting one or a few structurally related sugars, oligosaccharides, amino acids, peptides, anions, intermediary metabolites, drugs, vitamins, pigments, proteins, or complex carbohydrates. Those that catalyze substrate uptake possess extracytoplasmic substrate-binding receptors that interact with the transmembrane permease proteins on the external surface of the cytoplasmic membrane and are believed to mediate substrate recognition. In Gram-negative bacteria these receptor proteins are found in the periplasm, while in Gram-positive bacteria they are present as lipoproteins, anchored to the external surface of the cytoplasmic membrane. The ligand-binding domains of the glutamate-sensitive ion channels in nerve cells of animals are homologous to some of these bacterial receptor proteins. No ABC-type export system in bacteria possesses an extracytoplasmic solute-binding receptor. As almost all currently characterized ABC-transport systems of eukaryotes catalyze efflux, no such external receptor has yet been identified in a eukaryote.

Although some controversy exists regarding the topology of these proteins, most studies suggest that the transmembrane domains or proteins that catalyze transport consist of six helical membrane-spanning segments or spanners with both the amino- and carboxy-terminal residues localized to the cytoplasmic side of the membrane. As for the PTS permeases, the various domains either exist as distinct polypeptide chains or are covalently linked to each other in various arrangements. While the individual domains in bacteria frequently consist of distinct polypeptide chains, the eukaryotic transport systems are almost always single, large, multidomain proteins. Regardless of this variability in domain linkage pattern, all ABC-type transport systems appear to conform to the $(6)_{2n}$ [n = 1–4] spanner structural pattern noted earlier for voltage-sensitive channel proteins, MIP family proteins, the various families of facilitators, and possibly the PTS permeases. For the ABC-type permeases, $n = 1$, as is true for most well-characterized facilitators.

8 A LIGHT-DRIVEN ION PUMP: BACTERIORHODOPSIN

In halophilic bacteria, such as *Halobacterium halobium,* a remarkable light-driven proton pump, bacteriorhodopsin, has evolved in part to allow the organisms to synthesize ATP under conditions of oxygen limitation. The structure of bacteriorhodopsin, the polypeptide of which spans the membrane seven times, allows it to actively transport protons from inside the cell to the outside in response to light absorption. The purple color of bacteriorhodopsin is due to the covalent association of one molecule of all-*trans* retinal, a derivative of vitamin A, to lysine 216 of the protein by means of a Schiff base linkage. It is the retinal moiety that absorbs photons and is responsible, along with at least two aspartyl residues in the protein, for the conversion of this form of energy into a proton gradient that can be used for ATP synthesis. The chromophore and amino acyl residues probably function as a "proton wire" to conduct protons through the membrane, from the cytoplasm to the external medium.

Although the photochemical reactions of bacteriorhodopsin are extremely complex, it is known that light absorption by retinal leads to deprotonation of the Schiff base to one of several intermediate forms. Associated with this process is the isomerization of the all-*trans* retinal to the 13-*cis* isomer. Reassociation of a proton with the retinal moiety then takes place more slowly, and the bacteriorhodopsin protein cycles back to the original all-*trans* form through several other intermediate conformations that can be detected spectrophotometrically.

9 ELECTRON-FLOW-DRIVEN ION PUMPS

Substrates such as NADH and succinate can be oxidized in bacteria and mitochondria with the concomitant reduction of molecular oxygen to water. The process is catalyzed by the integral membrane protein complexes of the electron transport chains of these organisms and organelles. Some of these protein complexes pump protons (H^+) across the membranes, from the inside to the outside of the cell or organelle. Additionally, transhydrogenases of many organisms, and photosynthetic reaction centers and electron transfer chains of bacteria and chloroplasts, pump protons. The proton electrochemical gradients thus generated can be used for various purposes such as ATP synthesis and the driving of secondary active transport processes.

The mechanisms by which electron flow is coupled to proton efflux are poorly understood, and in no case can a detailed picture be drawn. An analogy with light-driven, bacteriorhodopsin-dependent pumping seems likely, but other possibilities exist. Future studies will be required before the mechanistic details can be visualized.

10 ION-TRANSLOCATING ATPASES

Transmembrane, ATP-hydrolyzing enzymes use the energy in the terminal γ-phosphate anhydride bond of ATP to drive the active uptake or extrusion of an ion (or ions) into or out of a cell or organelle. Four types of these enzymes are now recognized: the P-type, plasma-membrane-localized, cation-translocating ATPases; the F-type (F_1/F_0), H^+- or Na^+-translocating ATPases; the V-type, vacuolar, H^+- or Na^+-translocating ATPases; and the A-type, oxyanion-translocating ATPases.

The P-type ATPases are sensitive to vanadate (a phosphate analogue) and form an acyl phosphate intermediate during ATP hydrolysis. They are usually composed of monomeric or multimeric complexes in which a single polypeptide chain of about 100 kDa serves as the catalytic unit.

The F-type ATPases are inhibited by dicyclohexylcarbodiimide (DCCD) and azide and do not form a phosphorylated enzyme derivative. They consist of multisubunit F_1 and F_0 moieties in which the F_1 moiety functions in ATP hydrolysis/synthesis, and the F_0 moiety functions in ion translocation.

The V-type ATPases are inhibited by DCCD at high concentrations, and also by nitrate and the antibiotic bafilomycin A_1. Like the F-type ATPases, they are large and are composed of several major polypeptide components. They are structurally and evolutionarily related to the F-type ATPases.

A-type ATPases, the fourth class of ion-translocating ATPase, are oxyanion-specific ATPases. They pump toxic arsenic derivatives out of bacterial cells. It is possible, but not proven, that they are related to the ABC-type transport systems. Homologues of these oxyanion-specific transport proteins have recently been identified in eukaryotes.

10.1 Phosphoryl Intermediate (P-type) ATPases

One of the first biological transport systems to be studied in detail was that responsible for maintaining the normally high concentrations of K^+ and low levels of Na^+ inside animal cells. For this to occur, both cations must be transported essentially unidirectionally against their concentration gradients. The sodium/potassium ATPase pumps Na^+ out of the cell and K^+ into the cell, accompanied by ATP hydrolysis. The reaction catalyzed is:

$$3Na^+_{(in)} + 2K^+_{(out)} + ATP_{(in)} \rightarrow 3Na^+_{(out)} + 2K^+_{(in)} + ADP_{(in)} + P_{i(in)}$$

This coupling of chemical energy to concentrative transport is an example of primary active transport, as well as of antiport. Thus, a single protein catalyzes both ion translocations as well as the hydrolysis of ATP. The energy released by ATP hydrolysis is "translated" by the Na^+/K^+ ATPase into potential energy in the form of Na^+ and K^+ electrochemical gradients across the plasma membrane. The Na^+/K^+ ATPase is an example of a P-type ATPase.

The Na^+/K^+ ATPase exists in the membrane as a heterodimer consisting of a 110 kDa α subunit and a 55 kDa β subunit. In the process of active Na^+ and K^+ transport, the cations become "trapped" or "occluded" in the structure of the protein. Occlusion in a cavity inside the enzyme constitutes an essential step required for active Na^+ and K^+ transport.

Transport of Na^+ out of the cell and K^+ into the cytoplasm is thought to be mediated by conformational changes in the enzyme triggered by phosphorylation, dephosphorylation, and the binding of ligands at both surfaces. Little is known about how conformational changes in the protein lead to unidirectional transport of ions through the membrane.

Muscle cells contain a specialized intracellular endoplasmic reticular membrane system. This organelle acts as a repository for Ca^{2+} in resting muscle, and electrical stimulation of muscle cells causes rapid release of this ion into the cytoplasm where it triggers contraction of the myofibrils. Reuptake of Ca^{2+} into the reticular lumen is effected by a calcium ATPase that catalyzes the following reaction:

$$2Ca^{2+}_{(out)} + ATP_{(out)} \rightarrow 2Ca^{2+}_{(in)} + ADP_{(out)}P_{i(out)}$$

In this reaction sequence, the subscript (out) refers to the outside of the endoplasmic reticulum (the cytoplasm) and (in) corresponds to the lumen of the endoplasmic reticulum.

Like the Na^+/K^+ ATPase, the Ca^{2+} ATPase spans the membrane asymmetrically. In the presence of Ca^{2+} and ATP, formation of a covalent phosphoryl enzyme can be demonstrated. Thus, many of the aspects of Ca^{2+} transport parallel those of Na^+ and K^+ translocation. These enzymes are homologous, and they function by very similar mechanisms.

A third P-type ATPase is the H^+-translocating enzyme found in the plasma membranes of epithelial cells lining the stomach. This enzyme is responsible for acidification of the lumen and is most closely related to the Na^+/K^+ ATPases. Each of these enzymes has a large catalytic polypeptide chain of about 100 kDa that is transiently phosphorylated during transport. Other P-type ATPases catalyze uptake or extrusion of a variety of ions such as H^+, Cu^{2+}, and Cd^{2+}, particularly in bacteria and other microorganisms, but also in higher organisms including man.

10.2 F_1/F_0 (F-type) ATPases

The F_1/F_0 (F-type) ATPases (or ATP synthetases) of mitochondria, chloroplasts, and bacteria appear to serve related functions. H^+or Na^+ gradients, formed as a result of electron transport, are used to synthesize ATP from ADP and P_i as a consequence of the flow of H^+ or Na^+ *down* its electrochemical gradient through the F_1/F_0 ATPase. It is also possible, however, for the bacterial F_1/F_0 ATPase to work in the reverse direction, that is, to hydrolyze ATP and translocate protons against an electrochemical gradient. Thus, the F_1/F_0 ATPase can be classified as a protein complex carrying out primary active transport. Unlike the P-type ATPases, however, there is no evidence that the F_1/F_0 ATPases are covalently phosphorylated during proton translocation.

The F_1 portion of these ATPases possesses ATPase activity and consists of five different subunits (named α through ϵ). The F_0 portion, which penetrates the membrane, has been shown to be a proton (or sodium ion) channel. In the simplest systems it consists of three different polypeptides (named *a, b,* and *c*). Removal of F_1 from membranes results in an increased permeability of the bilayer to protons, suggesting that F_1 acts as a "cap" that couples proton translocation to ATP synthesis or hydrolysis. The F_0 portion, which is hydrophobic, has been shown to catalyze proton conductivity when incorporated into phospholipid vesicles.

10.3 Vacuolar (V-type) ATPases

V-type ATPases were relatively recently discovered and are less well characterized than either the P-type or F-type enzymes. They are present in the dynamic intracellular membrane networks of animal cells. They are also found in plant tonoplasts, the vacuoles of yeast and fungi, and both archaebacteria and eubacteria. In all examples studied, these ATPases utilize cytoplasmic ATP to pump protons into the vesicular or cellular compartment. The vacuolar ATPases are distantly related to the F_1/F_0 ATPases. This third class of ATPases and the F-type ATPases have thus evolved from a common ancestor. It is possible that vacuolar ATPases of eukaryotes arose by the internalization of V-type ATPases from the cytoplasmic membrane of an ancestral prokaryotic cell.

10.4 Oxyanion-Specific (A-type) ATPases

A-type ATPases are often plasmid-encoded enzyme complexes that actively extrude toxic oxyanions such as arsenite, antimonite, and tellurite from the cytoplasm of bacteria. Because the genes for the synthesis of this enzyme are encoded on small, transferable plasmids, properties associated with the possession of this enzyme can be easily transferred from one population of bacteria to another. Since the plasmids confer on their host bacteria resistance to a variety of bacteriocidal agents, their service to the organism can be described as providing antibiotic resistance.

The arsenical resistance pump of *E. coli* consists of two polypeptides, ArsA and ArsB. ArsB is an integral membrane protein that serves as the oxyanion channel. ArsA is a peripheral membrane protein that is associated with ArsB and possesses ATPase activity. ArsA therefore energizes transport via ArsB. By contrast, in *Staphylococcus aureus,* ArsB alone serves as the transporter and consequently oxyanion transport is not coupled to ATP hydrolysis (see the next section).

11 Chemical versus Chemiosmotic Energization of Transport

As noted earlier, some transport systems drive active solute uptake or efflux by utilizing the energy in a high energy phosphorylated compound such as adenosine triphosphate or phosphoenolpyruvate, while others utilize energy available in preformed ion electrochemical gradients to secondarily drive solute accumulation or extrusion. The presence of common structural features of transport proteins utilizing dissimilar energy-coupling mechanisms suggests the possibility that a single transport system might be able to utilize either one or the other energy source, depending on conditions. Evidence for just such a possibility has resulted from studies of the arsenical resistance (A-type) pumps of Gram-negative and Gram-positive bacteria.

As noted in the preceding section, arsenical resistance transport systems are encoded by various bacterial plasmids that render bacteria resistant to the toxic effects of antimicrobial oxyanions. The topology of the integral membrane ArsB protein of *E. coli* resembles that of the iron hydroxamate ABC-type permease of *E. coli* and the MFS permeases in having 12 spanners. Although insufficient sequence similarity is observed to allow us to postulate homology, there is no question that ArsB is the actual oxyanion transport protein. The ArsA subunit of the *E. coli* arsenical resistance pump serves as the ATP-binding constituent of the permease that couples ATP hydrolysis to oxyanion efflux. ArsA superficially resembles the ATP-hydrolyzing domain of an ABC-type permease, although a degree of sequence similarity sufficient to establish homology is lacking.

In two sequenced *ars* operons present on staphylococcal plasmids, *arsB* genes are found that encode proteins exhibiting about 58% identity to the *E. coli* ArsB protein. Surprisingly, these staphylococcal operons do not encode the ArsA proteins, which would normally be expected to provide the ATP-dependent energy coupling function. A recent analysis has suggested that efflux of arsenite via the staphylococcal ArsB-dependent, ArsA-independent transport system is driven by the membrane potential. In this respect, the system resembles the MFS permeases. However, when the *E. coli arsA* gene is coexpressed with the staphylococcal *arsB* gene, an increased degree of resistance to arsenite is observed. This increased resistance is apparently due to efficient ATP-driven arsenite efflux as observed for the *E. coli* arsenical resistance pump. The system thus functionally resembles a typical ABC-type export permease when ArsA is available but more closely resembles an MFS permease when ArsA is absent. These exciting new results suggest a functional (and thus possibly an evolutionary) connection between the integral membrane constituents of the arsenical resistance pump or the structurally similar, ABC-type, ATP-driven permeases and the MFS-type protonmotive-force-driven permeases.

12 CONCLUSIONS ON AN EVOLUTIONARY THEME

Transport proteins translocate their solutes either by a channel-type mechanism or by a carrier-type mechanism, and a close relationship between these two kinetically distinguishable transport mechanisms has been postulated. While several ancient transport protein families such as those of the chemiosmotically driven major facilitators, the ATP-driven ABC-type systems, and the phosphoenolpyruvate-driven PTS group translocators may be related by common descent, it is clear that others are not. Thus, voltage-sensitive ion channels of nerve and muscle cells, the mitochondrial carrier family (MCF), the MIP family, and the MFS evidently arose independently, at different times in evolutionary history, in different cell types, using different routes. Voltage-sensitive Ca^{2+} and Na^+ channels of nerves arose by quadruplication (two duplications) of a 6-spanner-encoding primordial gene resembling those currently encoding K^+ channels. The MCF proteins arose by triplication of a 2-spanner-encoding primordial gene; the MIP family arose by duplication of a 3-spanner encoding primordial gene, and the MFS protein family probably arose by duplication of a 6-spanner-encoding primordial gene.

The unit hydrophobic domains of the constituent members of these families of transport proteins apparently exhibit a common pattern of six tightly clustered transmembrane helical segments with both the amino and carboxy termini localized to the cytoplasmic membrane surface. It may be that a three-dimensional transmembrane structure consisting of six nearly parallel α-helices is particularly well suited to transmembrane channel formation. Such constraints could account both for the retention of this structural motif during evolutionary divergence of homologous transport proteins and for convergence of evolutionarily unrelated transport proteins to give proteins possessing similar topologies. It is interesting to note that all well-characterized transporters of this type consist of two, four, or eight units of six spanners and never just one or an odd number of such units (see Table 3). This fact must have functional significance, although cooperativity with respect to substrate binding is not a general property of porters. Elucidation of the detailed three-dimensional structures of these transmembrane transport proteins and thus their common and divergent structural features will be required to fully define the structure–function relationships among transport proteins.

See also BIOINORGANIC CHEMISTRY; ELECTRON TRANSFER, BIOLOGICAL; MEMBRANE FUSION, MOLECULAR MECHANISM OF; PROTEIN TARGETING.

Bibliography

Fagan, M. J., and Saier, M. H., Jr. (1994) P-type ATPases of eukaryotes and bacteria: Sequence comparisons and construction of phylogenetic trees. *J. Mol. Evol.* 38:57–99.

Higgins, C. F. (1992) ABC transporters: From microorganisms to man. *Annu. Rev. Cell Biol.* 8:67–113.

Hille, B. (1992) *Ionic Channels of Excitable Membranes,* 2nd ed. Sinauer Associates, Sunderland, MA.

Kuan, J., and Saier, M. H., Jr. (1993) The mitochondrial carrier family of transport proteins: Structural, functional and evolutionary relationships. *Crit. Rev. Biochem. Mol. Biol.* 28:209–233.

Marger, M. D., and Saier, M. H., Jr. (1993) A major superfamily of transmembrane facilitators that catalyse uniport, symport and antiport. *Trends Biochem. Sci.* 18:13–20.

Mortlock, R. P., Ed. (1992) *The Evolution of Metabolic Function.* CRC Press, Boca Raton, FL.

Nikaido, H., and Saier, M. H., Jr. (1992) Transport proteins in bacteria: Common themes in their design. *Science,* 258:936–942.

Reizer, J., Reizer, A., and Saier, M. H., Jr. (1993) The MIP family of integral membrane channel proteins: Sequence comparisons, evolutionary relationships, reconstructed pathway of evolution, and proposed functional differentiation of the two repeated halves of the proteins. *Crit. Rev. Biochem. Mol. Biol.* 28:235–257.

Saier, M. H., Jr. (1985) *Mechanisms and Regulation of Carbohydrate Transport in Bacteria.* Academic Press, Orlando, FL.

———, Ed. (1990) Sixth Forum in Microbiology. Coupling of energy to transmembrane solute translocation in bacteria. *Res. Microbiol.* 141:281–395.

———, and Reizer, J. (1992) Proposed uniform nomenclature for the proteins and protein domains of the bacterial phosphoenolpyruvate:sugar phosphotransferase system. *J. Bacteriol.* 174:1433–1438.

Schloss, P., Mayser, W., and Betz, H. (1992) Neurotransmitter transporters. A novel family of integral plasma membrane proteins. *FEBS Lett.* 307:76–80.

Silver, S., Ji, G., Bröer, S., Dey, S., Dou, D., and Rosen, B. P. (1993) Orphan enzyme or patriarch of a new tribe: The arsenic resistance ATPase of bacterial plasmids. *Mol. Microbiol.* 8:637–642.

Weiss, M. S., Kreusch, A., Schiltz, E., Nestel, U., Welte, W., Weckesser, J., and Schulz, G. E. (1991) The structure of porin from *Rhodobacter capsulatus* at 1.8 Å resolution. *FEBS Lett.* 280:379–382.

Woese, C. R. (1987) Bacterial evolution. *Microbiol. Rev.* 51:221–271.

Yeagle, P., Ed. (1992) *The Structure of Biological Membranes.* CRC Press, Boca Raton, FL.

TRANSPOSONS IN THE HUMAN GENOME

Abram Gabriel

1 LINEs
2 SINEs
3 Endogenous Retroviruses
4 Future Prospects

Key Words

Alu Sequences The major class of SINEs in the human genome, named for the presence of a distinctive *Alu* I restriction site.

Endogenous Retroviruses Retroviruses that have inserted into the germ line and are inherited in a Mendelian fashion. All known endogenous retroviruses in humans are noninfectious.

LINEs The abundant family of long (6–7 kilobase) interspersed nuclear repeats found in all mammalian genomes.

Orthologous Genes Genes that are structurally and functionally homologous in different species.

Retrotransposons Intracellular mobile genetic elements whose replication requires the reverse transcription of an RNA intermediate.

SINEs Highly repeated families of short (100–500 bases) interspersed nuclear sequences found in all mammalian genomes. All SINEs appear to be derived from genes transcribed by RNA polymerase III.

The human genome is home to several classes of multicopy dispersed genetic elements, which in aggregate comprise more than 10% of the entire genome. These elements have many of the hallmarks of a group of mobile genes known as retrotransposons or retroposons, whose replication (i.e., transposition) is thought to involve the following steps: transcription of a genomic copy, reverse transcription of the RNA, and reinsertion of the complementary DNA into a new genomic location, with the creation of flanking target site duplications. Human endogenous retroviruses (HERVs) are found in relatively low numbers and are structurally related to infectious retroviruses, with characteristic long terminal repeats, primer-binding sites, and multiple open reading frames (ORFs). In contrast, long interspersed nuclear sequences (LINEs) and short interspersed nuclear sequences (SINEs) have accumulated to remarkably high levels within the genome and are distinct entities lacking long terminal repeats. While phylogenetic analysis suggests that human transposons are stable ancient residents that entered the genome early in primate evolution, recent evidence indicates that subclasses of LINEs and SINEs are still capable of inserting copies into new genomic loci.

The presence of transposable elements in the human genome has many significant implications.

1. Transposons are inherently mutagenic. Insertion of an element into or near a functional gene can alter its expression by causing loss of gene function, by changing the gene's tissue-specific or temporal expression, or by inducing the gene's inappropriate overexpression. Although these changes potentially provide an adaptive advantage, they may also be adaptively neutral or may even result in genetic disease or neoplastic transformation.
2. Recombination between elements can result in deletion, inversion, or duplication of the intervening sequences and can lead to either gene loss or gene duplication. These recombinational events may have contributed to the evolution of the human genome.
3. Transposons can serve as experimental tools. The presence of related elements in other mammalian species has been exploited for phylogenetic studies of mammalian evolution. The polymorphic nature of particular genomic loci resulting from the presence or absence of specific elements has been used as a tool for population genetic studies. The multiple classes of high-copy elements within the genome are useful landmarks for the construction of genetic and physical maps of the human genome.

1 LINEs

The human genome contains 50,000 to 100,000 copies of a specific LINE sequences, referred to as L1Hs (L1 of *Homo sapiens*). Full-length, 6 kilobase elements represent less than 10% of the total, with the remainder being variably truncated at their 5′ ends. The vast majority of L1Hs elements have accumulated missense mutations, nonsense mutations, deletions, inversions, and/or frameshifts and therefore lack uninterrupted ORFs. A consensus L1Hs sequence, however, predicts the presence of two long ORFs followed by a poly(dA)-rich 3′ end, all flanked by variable length target site duplications (see Figure 1). The deduced amino acid sequence of the L1 ORF2 has a limited homology to retroviral reverse transcriptase (RT). The sequence of the first full-length, intact L1Hs element (the recently isolated L1.2) confirms the predicted structure.

LINEs have been found dispersed through the genomes of all mammals, suggesting that the "ancestral" LINE may have become associated with the genome very early in mammalian evolution. Sequence data, however, indicate that intraspecies variation among LINEs is much more limited than interspecies variation, a phenomenon known as concerted evolution. These findings have led to the hypothesis that within a given species, either LINEs are homogenized by gene conversion of a few "master" copies, or existing LINEs have resulted from the relatively recent amplification of a small number of transpositionally active, full-length "molecular drivers."

Although the mechanism by which LINEs transpose is unknown, the key intermediate in the retrotransposon replication model is a full-length RNA, which serves a dual function—it is translated into proteins used in the replication process (e.g., reverse transcriptase) and also is used as the template for reverse transcription. Evidence in support of this model for L1Hs transposition comes from a variety of sources. Discrete, homogeneous, full-length poly-adenylated plus-strand LINE RNA is present in the cytoplasm of undifferentiated human teratocarcinoma cell lines, indicating that at least some full-length copies of L1Hs are specifically transcribed and processed in this setting. In more differentiated cell lines, L1Hs RNA is transcribed from both strands, is heterogeneous in size, and is localized to the nucleus. Sequences within approximately the first 100 bases at the 5′ end of L1 can function as promoters, with specificity for teratocarcinoma cells. The observed cell specificity of L1Hs expression is consistent with the conclusion that L1Hs transposition occurs via transcription of a small number of transpositionally competent L1Hs elements early in embryonic development, either within germ cells or in cells destined to become germ cells.

This conclusion is also supported by protein expression studies. Antiserum raised against bacterially expressed L1Hs ORF1 protein recognizes an appropriately sized 38 kDa endogenous polypeptide on Western blots of extracts from teratocarcinoma and choriocarcinoma cell lines. Using the same antiserum, a subset of formalin-fixed, human testicular germ cell tumors can be immunohistochemically stained, implying that these germ cell tumors also express the L1Hs ORF1 gene product. In mouse teratocarcinoma cells, the equivalent ORF1 protein cofractionates with full-length L1 RNA, suggesting the existence of an L1 ribonucleoprotein (RNP) particle. Although a function for the ORF1 gene product has not been identified, by analogy to retroviruses it may serve as a capsidlike shell in which reverse transcription occurs. The L1Hs ORF2 from the full-length L1.2 has been cloned into a yeast expression vector and shown to encode a distinct RT activity. Thus L1 encodes

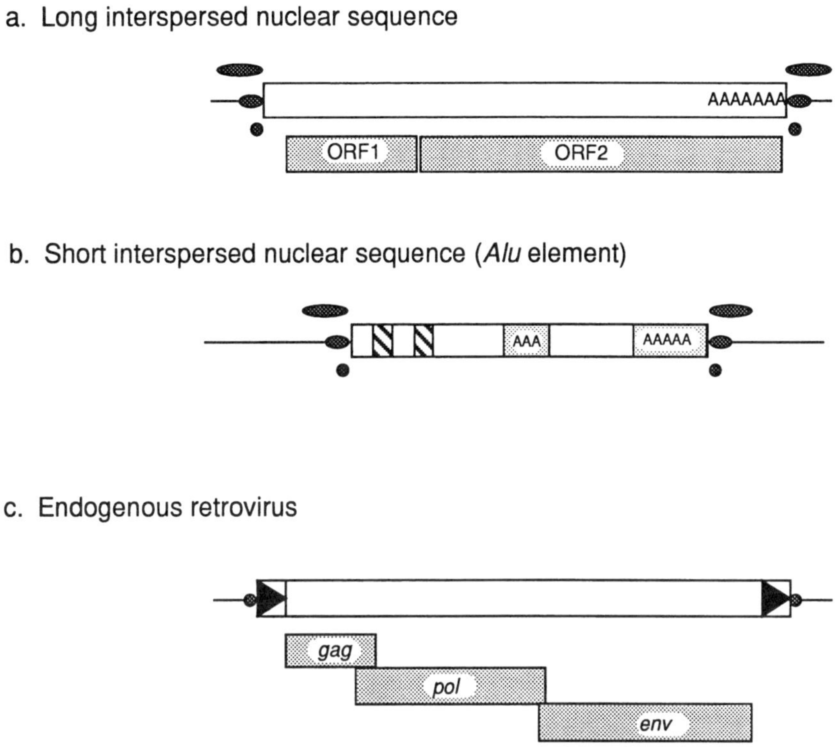

Figure 1. General structure of three classes of human transposons. (a) LINEs are 6 to 7 kilobases long and contain two ORFs. Elements are flanked by variable length target site duplications (ovals) and contain a 3′-poly(dA) stretch. (b) *Alu* elements are the most common SINEs in human DNA. They are 300 bases long and form an imperfect dimer, with poly(dA) stretches 3′ to each monomer. Two *pol* III promoter regions are present in the left monomer (striped boxes) and the elements are flanked by variable length target site duplications (ovals). (c) Endogenous retroviruses are 5 to 9 kb long and contain multiple ORFs, corresponding to *gag, pol,* and *env.* Long terminal repeat sequences (solid triangles) flank the coding regions, and characteristic length target site duplications (circles) flank the entire element.

a functional enzyme thought to be necessary for its replication. Furthermore, an RT activity identified in human teratocarcinoma cells cofractionates with L1 messenger RNA. While consistent with L1 RNPs, this activity may be due to endogenous retroviruses (see Section 3) and the cofractionation with L1 mRNA fortuitous.

Originally LINEs were thought to be evolutionary relics of a period in mammalian evolution during which these elements were widely amplified and dispersed throughout the genome. According to this model, current LINEs are pseudogenes, independently accumulating random mutations and diverging from one another. While this may be generally true, recent reports of new L1Hs transposition events lend powerful support to the idea that full-length, active elements still reside in the human genome. The best evidence comes from two patients with the X-linked disorder hemophilia A. In each case an insertion of truncated L1 DNA, accompanied by a 3′-poly(dA) tail and perfect flanking target site duplications, was found in exon 14 of the factor VIII gene. In neither case was there a family history of the disease, and in the one case studied in detail, the patient's mother was shown not to have an L1 insertion in either of her factor VIII alleles, indicating that this was a de novo event. Whether the transposition event occurred during maternal gametogenesis or during the early embryonic development of the affected offspring has yet to be determined. The truncated L1 was used as a probe to search the human genome for its progenitor, and L1.2, the first full-length L1Hs element (discussed earlier) was identified. This potentially active element is located on chromosome 22, where it has existed since before the divergence of man, chimpanzee, and gorilla. In addition to these instances of heritable L1 insertions, somatic insertions have been noted within the tumor tissue of two patients with carcinomas. Both an L1 insertion into the proto-oncogene *myc* in a patient with breast cancer as well as an insertion into the tumor suppressor *APC* gene in a patient with colon cancer suggests that transposition events may have played a role in the pathogenesis of these particular neoplasias.

2 SINEs

An estimated 500,000 to 1,000,000 copies of various families of short interspersed nuclear sequences are present in the human genome, spaced every 5 to 10 kilobases. These elements are often found inserted within intron sequences and untranslated regions of genes, where they normally do not affect the protein-coding capacity of the gene. As with LINEs, the variable length target site duplications at sites of insertion and the presence of 3′-poly(dA) stretches (see Figure 1) are hallmarks of retrotransposition. However, SINEs are quite distinct from LINEs. They average only 300 bases in length and have no protein-coding capacity. Therefore they must rely on a non-element-encoded source of RT for their dispersal, with the leading candidates being either the L1 RT or an RT derived from an endogenous retrovirus.

The *Alu* family is the most widespread class of human SINEs and bears a striking resemblance to the genes encoding the extremely abundant small cytoplasmic 7SL RNA. 7SL RNA is a component of the mammalian signal recognition particle, a structure required for transporting proteins across the membrane of the endoplasmic reticulum. While *Alu* sequences are clearly related to 7SL RNA sequences (~90% identity), the relationship between the two elements is complex. *Alu* elements are arranged as imperfect dimeric structures consisting of two internally truncated tandem copies of 7SL RNA, with each monomer flanked at its 3′ end by adenine-rich stretches that are not present in 7SL RNA (see Figure 1). Two RNA polymerase III promoter regions present in the left half are absent in the right half, yet the right half is 31 bases longer than the left. Thus, while *Alu* sequences are not simple 7SL RNA pseudogenes, they may have resulted from reverse transcription of an aberrantly processed form of this RNA. Interestingly, other families of SINEs in both rodents and primates have strong but independent sequence homology to 7SL RNA or to specific tRNAs, which are also transcribed by RNA polymerase III. These findings have led to the hypothesis that propagation of SINEs requires a member of a multicopy precursor gene family with an internal promoter (e.g., *pol* III promoter) to be altered so as to become both highly expressed in germ cells and a preferred target sequence for a reverse transcriptase.

Whereas LINEs are present in all mammals examined, *Alu* sequences are specific to primates, suggesting that these elements are of relatively recent origin. The emergence and evolution of the *Alu* family can be inferred from comparisons of their genomic arrangement and sequence in humans and in related primate species. For example, the position of the seven *Alu* sequences within the human α-globin gene cluster is identical to that found in chimpanzees, which diverged from humans approximately 5 million years ago, but is completely different from that found in the prosimian galago, which diverged about 55 million years ago. Individual elements within a species vary by roughly 14% from a derived consensus sequence, as do the orthologous human and chimpanzee sequences examined in the α-globin gene cluster. While these findings are consistent with neutral drift of nonfunctional DNA sequences, *Alu* subfamilies with much higher degrees of sequence identity have been found. This phenomenon is best documented for an *Alu* subfamily (termed HS or PV) that is present in 500 to 2000 copies per haploid genome but is not found at orthologous positions in other primates. Individual members share more than 98% sequence identity. Two de novo *Alu* insertions resulting in genetic disease are members of this subfamily (see later). All these findings indicate recent and ongoing *Alu* amplification and suggest that the members of a small subset of *Alu* elements, present in the genome before the human-ape divergence, are capable of serving as "master" copies for transposition.

The retrotransposition model of replication requires that full-length *Alu* sequences be transcribed in appropriate tissues at appropriate times in development. While in vitro transcription experiments using RNA polymerase III demonstrate correct initiation of *Alu* transcripts, in vivo results are less clear-cut. A brain-specific monkey cDNA has been shown to be a 200-nucleotide polyadenylated *Alu* transcript that is also expressed in the human brain. Other short *Alu*-subfamily-specific transcripts have been recovered from HeLa cells, as well as osteosarcoma and thyroid cancer cell lines. A problem that has complicated interpretation of the in vivo transcription data is the ubiquity of *Alu* sequences—because of their presence within introns and in noncoding regions, *Alu* sequences are nonspecifically transcribed in all tissues and cell lines.

Alu sequences can cause human genetic disease in a variety of ways. In the most common scenario for *Alu*-mediated gene inactivation, unequal meiotic crossing over between two closely linked elements results in the deletion or duplication of intervening DNA. This genetic mechanism underlies many of the mutant alleles causing familial hypercholesteremia, thalassemia, hereditary persistence of fetal hemoglobin, severe combined immunodeficiency, and hereditary angioedema. As with other mobile elements, a germ line insertion into a functional gene can cause heritable inactivation of that gene. Recently, two case reports of de novo *Alu* insertions have been published. In the first instance, a patient with neurofibromatosis was shown to have an *Alu* insertion into his paternal *NF1* allele. Although the insertion was into an intron, it resulted in an aberrant splicing event, presumably as a result of disruption of a normal branch point. In the second case, a person with acholinesterasemia was found to be homozygous for an insertion of an *Alu* sequence within exon 2 of his cholinesterase gene, resulting in a translational frameshift. In both cases, the *Alu* elements had more than 93% sequence identity to the HS family consensus sequence. A more unusual scenario has been proposed for a case of gyrate atrophy. A guanine cytosine transversion in a preexisting *Alu* element located within intron 3 of the ornithine δ-aminotransferase *(OAT)* gene created a new donor splice site. Aberrant splicing then resulted in insertion of a fragment of the *Alu* element into the mature *OAT* mRNA.

The ubiquity of *Alu* elements throughout the genome has led to speculations that *Alu* sequences may have broader evolutionary significance in shaping the structure and function of the human genome. Gene duplication by unequal crossing over between *Alu* elements may be involved in generating families of genes with similar structures that serve different functions. Nonhomologous chromosomal rearrangements may be generated by *Alu* recombination, as has been shown for a human XX male with a sex chromosome rearrangement. This process could lead to exon shuffling and the evolution of new genes. On a larger scale, chromosomal translocations might occur between *Alu* elements. The specific arrangement of repeated sequences such as *Alu* sequences on a chromosome may even serve as a cue for homologue pairing during the prophase of meiosis 1, the stage at which meiotic recombination occurs.

3 ENDOGENOUS RETROVIRUSES

Retroviruses are infectious agents that behave like transposable elements in that they insert proviral copies of themselves into the host cell's genome. While this is generally a somatic event, endogenous retroviruses are thought to arise by insertion of retroviral proviruses into the host's germ line and their subsequent vertical inheritance. The human genome contains multiple classes of endogenous retroviruses (HERVs), which vary from single-copy elements (HERV-R, S71) to families containing hundreds or thousands of members (HERV-H, THE-1). Most of the families bear strong homologies either to known exogenous retroviruses or to endogenous retroviruses found in other species. In particular, sequences similar to elements RTVL-1 and ERV1 are found at similar loci in closely related primate species, suggesting that germ line insertion of these retroviruses occurred early in primate evolution.

Sequence data from cloned HERVs reveal that these elements retain the general structure of exogenous retroviruses, with long terminal repeats and three ORFs corresponding to *gag, pol,* and *env* (see Figure 1). In all cases so far examined, however, the endogenous elements have accumulated stop codons, deletions, and/or other rearrangements that render them defective. Whether fully functional HERVs exist remains an open question. In mouse, an extensively studied class of endogenous retroviral elements termed intracisternal A-particles (IAPs) are capable of autonomously transposing to new loci within their host genome. IAPs encode a functional *gag* and *pol* gene and are assembled into viruslike particles (VLPs) within the cell, but they lack an *env* gene and are therefore strictly intracellular.

In humans the evidence for functional elements is much more circumstantial and is based on HERV expression studies. Retroviral LTRs contain strong and complex promoters and many HERVs (e.g., HERV-K, RTVL-H) are transcribed in a cell-specific manner from these promoters. Expression of some HERVs (e.g., RRHERV-I, ERV-9, HERV-K) can be induced or repressed by treatment of cells with steroid hormones or retinoic acid. Antigens in some leukemic and choriocarcinoma cell lines, as well as in placenta, cross-react with antibodies raised against various retroviral proteins. Both VLPs and/or reverse transcriptase activity have been reported in placenta, various tumor cell lines, and other human tissues, although their presence has not been definitively linked to specific HERVs. To date, there is no evidence that any HERV can form an infectious viral particle, nor have human mutations caused by de novo HERV germ line insertions been reported.

Instead HERVs have been postulated to cause human disease by more subtle means. For example, HERVs have been implicated in the origin of autoimmune disease based on the finding that sera from some patients with mixed connective tissue disorders cross-react with a 70 kDa U1 small nuclear ribonucleoprotein associated protein that shares antigenic determinants with a p30-gag protein from a human endogenous retrovirus. According to this hypothesis, fortuitous expression of the retroviral antigen might initiate an autoantibody response. Furthermore, it has been speculated that somatic transposition of HERVs into tumor suppressor genes or nearby proto-oncogenes could lead to neoplastic transformation, although there is no direct evidence of these occurrences in humans.

The potential for interaction between HERVs and exogenous retroviruses exists at many levels. Recombination between an infectious retroviral provirus and an endogenous sequence could result in the formation of a new retrovirus with altered tissue tropisms or host ranges, a phenomenon clearly observed in mice and resulting in infectious murine leukemia viruses. Recombination need not be limited to DNA. Copackaging of HERV RNA with an exogenous retroviral RNA genome could lead to a chimeric retrovirus via template jumping during reverse transcription. Interestingly, polymerase chain reaction screens of human and other primate genomes with HIV-1 related probes have revealed the existence of endogenous sequences with significant homology to HIV *env.* Exogenous retroviruses could contribute gene products (e.g., reverse transcriptase, *env* glycoprotein) *in trans* that enable a defective endogenous retrovirus to transpose, or possibly even to bud from the cell. Moreover, trans complementation of endogenous elements has been demonstrated experimentally using an endogenous mouse virus. Intracellular transposition events were observed when HeLa cells were cotransfected with a genetically marked defective virus and a helper construct, which supplied the homologous *gag* and *pol* gene products.

Evolutionary biologists have suggested that HERVs may be involved in such processes as the evolution of gene function and the regulation of tissue-specific and developmentally regulated gene expression. The best documented example of HERV involvement in human gene regulation is the case of the amylase gene cluster. In the human genome, five linked genes encode the differentially expressed enzyme amylase. Two of the genes are expressed only in the pancreas, while the other three are expressed only in the parotid gland. Analysis of the sequences flanking each gene revealed that an identical HERV-E sequence is inserted, in opposite orientation, just upstream of each gene expressed in the parotid gland. Furthermore, when the first 670 bases of the HERV were placed upstream of a reporter gene in transgenic mice, the reporter was expressed specifically in the parotid gland. Thus particular HERV sequences can act as enhancers and confer tissue specificity to nonretroviral host genes. An interesting sidelight to these findings is that rodents (the only other family of mammals that expresses amylase in both the parotid gland and the pancreas) have a completely different arrangement of amylase genes and no nearby endogenous retroviral sequences. This then is an example of convergent evolution.

While most HERVs are clearly related to other retroviruses, the high-copy, transposon-like human element (THE-1) family is an exception. Full-length copies are 2.3 kb long and are flanked by 350 bp LTRs. Large numbers of solo LTRs, thought to arise by homologous LTR–LTR recombination, are present throughout the genome. Circular forms, which may be also be products of LTR–LTR recombination, have been observed in HeLa cells. However, the internal sequences of the few copies so far examined contain no long open reading frames and bear no similarity to known retroviruses. Instead, two single-copy genes in the prosimian galago bear strong sequence similarity to THE-1, although they lack LTRs. It has been proposed that these cellular genes were recruited into a transposable element via a retrovirally mediated process. Although no de novo insertions of THE-1 have been reported, two cases of Duchenne muscular dystrophy have been shown to involve deletions in the dystrophin gene, with preexisting THE-1 sequences representing the proximal break point. Thus, THE-1 sequences may serve as sites for illegitimate recombination.

4 FUTURE PROSPECTS

While much has been learned about the structure and organization of transposons in the human genome, and there are now specific examples of interactions between these elements and their host, many questions remain unanswered. The mechanism by which LINEs and SINEs transpose is poorly understood. In particular, nothing is known about the process leading to integration into new loci. Likewise, it remains a mystery how these elements became so highly amplified within the genome and whether this amplification is an ongoing process. Do these elements serve a generally useful function or are they simply "selfish DNA"? Is somatic transposition an important component of neoplastic transformation, and do transposons significantly contribute to the generation of mutations leading to human genetic disease? The recent advances in the understanding of human transposons and their potential role in both evolution and disease pathogenesis should make further research in this area an exciting prospect.

See also GENETIC ANALYSIS OF POPULATIONS; GENETICS; MAMMALIAN GENOME; RECOMBINATION, MOLECULAR BIOLOGY OF.

Bibliography

Berg, D. E., and Howe, M. M. (1989) *Mobile DNA.* American Society for Microbiology, Washington, DC.

The definitive reference on transposable elements. Particularly relevant chapters are by Varmus and Brown, Hutchinson et al., and Deininger (Chapters 3, 26, and 27, respectively).
Additional reviews on specific topics relevant to human transposons include:
Deninger, P. L., et. al. (1992) *Trends Genet.* 8:307–311. Fanning, T. G., and Singer, M. F. (1987) *Biochim. Biophys. Acta,* 910:203–212. Larsson, E., Kato, N., and Cohen, M. (1989) *Curr. Top. Microbiol. Immunol.* 148:115–131.
Discussions of the phylogeny of retrotransposable elements can be found in:
Doolittle, R. F., Feng, D. F., Johnson, M. S., and McClure, M. A. (1989) *Q. Rev. Biol.* 64:1–30. Xiong, Y., and Eickbush, T. H. (1990). *EMBO J.* 9:3353–3362.

Tumor Suppressor Genes

Prem Mohini Sharma and S. R. Dev Sharma

1 **Definitions: Oncogenes, Tumor Suppressor Genes**
2 **History**
3 **The Retinoblastoma Gene**
4 **The *p53* Gene**
5 ***BRCA1* and *BRCA2* Genes and Breast Cancer**
6 **Wilms' Tumor Gene, WT1**
7 **The Neurofibromatosis Type 1 and 2 (NF1 and NF2) Genes**
8 **The von Hippel–Lindau (VHL) Syndrome and Renal Cell Carcinoma**
9 **Familial Adenomatous Polyposis and Colon Cancer**
10 ***KAI1* Gene and Prostate Cancer**
11 **Search for New Antioncogenes**
12 **Transgenic Animal Models**
13 **Gene Therapy**
14 **Concluding Comments**

Key Words

Cellular Oncogene An oncogene formed by mutation or rearrangement of a proto-oncogene in a tumor.

Chromosome Translocation Exchange of segments between nonhomologous chromosomes.

Familial Adenomatous Polyposis A rare inherited form of colon cancer in which affected individuals develop multiple colon adenomas (polyps).

G_0 A quiescent state in which cells are not proliferating.

G_1 The stage of the cell cycle between the end of mitosis and the beginning of DNA synthesis.

G_2 The stage of the cell cycle between the end of DNA synthesis and the beginning of mitosis.

Gene Amplification Increased number of gene copies in a cell, resulting from repeated DNA replication.

Genome The complete genetic information of a species.

KAI1 (for *kang ai:* Chinese for anticancer) A metastasis suppressor gene for prostate cancer.

Kinase An enzyme that transfers phosphate groups, usually from ATP, to another molecule.

Li–Fraumeni Cancer Family Syndrome A rare hereditary cancer susceptibility, leading to the development of multiple kinds of tumor, which result from inherited mutations of the *p53* tumor suppressor gene.

M Phase The stage of the cell cycle during which cell division takes place.

MCC A tumor suppressor gene frequently inactivated in colon and rectum carcinomas.

NF1 A tumor suppressor gene responsible for inheritance of type 1 (von Recklinghausen's) neurofibromatosis.

p53 A tumor suppressor gene that is lost or inactivated in a variety of neoplasms, including breast, colon, and lung carcinomas, sarcomas, and leukemias.

Point Mutation Alteration of a single nucleotide of DNA.

RB A tumor suppressor gene identified by genetic studies of retinoblastoma, and also frequently inactivated in osteosarcomas, rhabdomyosarcomas, and bladder, breast, lung, and prostate carcinomas.

Transcription Factor A protein that regulates transcription.

Transformation The conversion of a normal cell in culture to a cell displaying one or more of the characteristics of a tumor cell.

Tumor Suppressor Gene A gene that inhibits tumor development.

Wilms' Tumor A childhood kidney cancer.

WT1 A tumor suppressor gene that is inactivated in Wilms' tumor.

The oncogenes are one of two distinct classes of genes that contribute to the development of human cancer. They arise from proto-oncogenes as a result of genetic alterations that either increase gene expression or lead to uncontrolled function of the oncogene-encoded proteins. The products of most proto-oncogenes act to stimulate normal cell proliferation, and the unregulated action of the corresponding oncogene proteins leads to the abnormal proliferation that is characteristic of cancer cells. The other class of genes that are important in carcinogenesis, the tumor suppressor genes, represents the opposite side of the coin of cellular growth control. Whereas oncogenes stimulate cell proliferation and tumor development, the tumor suppressor genes act to inhibit these processes. If oncogenes are viewed as gas pedals, acting to accelerate the growth of cancer cells, tumor suppressor genes can be viewed as the brakes, acting to retard cell growth. In many neoplasms, genetic alterations lead to the loss or inactivation of tumor suppressor genes, thereby eliminating inhibitors of cell proliferation. Activation of oncogenes and inactivation of tumor suppressor genes thus represent complementary events in the development of cancer, both contributing to increased cell proliferation and loss of normal growth control.

The action of tumor suppressor genes is thus opposite to that of oncogenes. Tumor suppressor genes are inactivated in cancer cells, and the introduction of functional tumor suppressor genes from normal cells into cancer cells results in loss of tumorigenicity. In contrast, oncogenes are activated in cancer cells, and their introduction into normal cells leads to tumor formation. Not only do oncogenes and tumor suppressor genes both contribute to development of human cancers; the products of some tumor suppressor genes and oncogenes interact directly to regulate each other's expression of function within the cell.

1 DEFINITIONS: ONCOGENES, TUMOR SUPPRESSOR GENES

Two classes of genes are implicated in causing or helping to cause cancer. Although greater emphasis has been placed on the study of tumor-promoting oncogenes, the existence of tumor-suppressing genes has long been suspected. Dominant *oncogenes* are altered genes whose presence leads to neoplasia. *Tumor suppressor* genes are genes whose absence allows the malignancy to become manifest. Every gene has a matching copy on the paired homologous chromosome. The gene and its copy are each called an *allele*. Oncogenes function in a genetically dominant manner: the presence of an alteration in just one allele is tumorigenic. In the recessively acting tumor suppressor genes, on the other hand, both alleles must be inactivated to cause loss of growth control.

The past several years have brought important discoveries regarding the genetic predisposition to cancer. Antioncogenes/tumor suppressor genes do exist and in fact do play a major role in human carcinogenesis. The current cloned and characterized human tumor suppressor genes include *RB, p53, WT1, APC, VHL, NF1,* and *NF2* (Table 1). Several putative genes of this class have been mapped to specific chromosomal bands, and many more are forthcoming. These discoveries have illuminated our understanding of human cancer considerably and have revealed a group of genes that is important in cellular and developmental biology, making this a propitious time to review progress on the subject and to take note of some future prospects for extending this understanding.

2 HISTORY

Evidence for the existence of tumor suppressor genes has converged from several distinct lines of work. Early evidence came from somatic cell hybridization studies. Geneticists took cultures of two cell lines, one a tumor cell line and the other normal, and fused them to form cell hybrids that retained the genomes of both parents. If the tumorigenic phenotype is dominant, as oncogene researchers expected, the hybrids should all be tumorigenic because they carry the oncogenes from the tumorigenic parent. But in fact, the hybrids were all normal. Therefore, the normal cells must possess genes that are capable of suppressing the neoplastic phenotype of their tumor cell partners.

Evidence that there were specific tumor suppressor genes in normal cells came from examining revertants of these hybrid cells that regained the tumorigenic phenotype. These hybrid cells had unstable karyotypes and frequently shed chromosomes. By examining hybrids that spontaneously regained the transformed phenotype, investigators could identify the normal chromosome that was always lost from such cells. In human cells, this was commonly chromosome 11 or 13. These studies led to the concept that many tumors arise through loss of genetic material and provided the first clue that cancer cells often lose critical growth-regulating information during their progression toward full malignancy.

Human genetics provided a second clue that suggested the existence of tumor suppressor genes. The hereditary and sporadic forms of retinoblastoma led Knudson in 1971 to develop the two-hit hypothesis, in which he postulates that the development of retinoblastoma requires two rare mutations. In the sporadic form of the tumor, both mutations would have to occur within a single retinoblast, an exceedingly infrequent circumstance. In the inherited form, however, he suggests that one of the mutations is already present (inherited) in all retinal cells. Therefore, only a single additional mutation is required for tumorigenesis.

A decade later, analysis of the chromosomes in tumor cells and normal tissues of retinoblastoma patients resoundingly confirmed Knudson's hypothesis. Many of these patients carried deletions in chromosome 13q14, and the mutated gene inherited by afflicted children, termed *RB,* was mapped to this same chromosome. Thus, development of retinoblastoma appeared to require that both copies of the *RB* gene be mutated. The *RB* gene must therefore be a tumor suppressor gene that normally functions to arrest the growth of retinal precursor cells. Even one copy is sufficient to keep growth in check. Loss of both copies of *RB* eliminates the block, and a tumor develops.

A third clue for discovering tumor suppressor genes was suggested by the genetic mechanisms used by evolving tumor cells to eliminate both copies of genes such as *RB.* The first copy of a suppressor gene is inactivated by a somatic (or a germ line) mutation. The chromosomal region carrying the surviving wild-type allele may then be replaced by a duplicated copy of the homologous chromosome region that carries the mutant allele. Most tumors that lack functional copies of a suppressor gene (such as *RB*) display two identically mutated (homozygous) alleles, while the unaffected tissues can be shown to carry one mutant *RB* allele and one normal one. Therefore, the loss of heterozygosity (LOH) at particular chromosome regions in tumor cells (compared with somatic cells from the same individual) is generally regarded as evidence for the un-

Table 1 Cloned Human Tumor Suppressor Genes: Their Chromosomal Location Protein Products and Characteristic Neoplasm Associated with Germ Line Mutations

Gene	Chromosomal Location	Protein	Neoplasm
RB1	13q14	110 kDa transcription modulator of $G_1 \rightarrow S$ transition	Sarcoma, retinoblastoma
WT1	11p13	45 kDa transcription factor, zinc finger protein	Wilms' tumor
p53	17p13	53 kDa transcription factor, conditional regulator $G_1 \rightarrow S$	Sarcoma, glioma, breast carcinoma
NF1	17q11	327 kDa activator of *Ras* GTPase activity	Sarcoma, glioma
NF2	22q12	66 kDa protein at membrane–cytoskeleton interface	Schwannoma
VHL	3p25	34 kDa cell membrane protein	Pheochromocytoma, kidney carcinoma
APC	5q21	310 kDa cytoplasmic protein	Colon carcinoma
DCC		153 kDa cell adhesion molecule	

masking of mutations in tumor suppressor genes located in these regions.

The finding of a loss of genetic information at the *RB* locus in association with retinoblastoma formation spurred investigators to examine other malignancies for LOH at *RB* and other chromosomal locations. With polymorphic DNA markers used to survey tumor cell genomes systemically, numerous malignancies show LOH; these include urological neoplasms such as Wilms' tumor, von Hippel–Lindau disease, renal cell carcinoma, breast carcinoma, colon carcinoma, and bladder cancer.

3 THE RETINOBLASTOMA GENE

The human retinoblastoma gene *(RB)* is one of the best characterized tumor suppressor genes and serves as a prototype for genes in this category. The locus governs the hereditary susceptibility to many human cancers, most notable of which is retinoblastoma, a malignant tumor arising in the eyes of young children. In 1993, a 180,388 bp contig, or set of overlapping clones, encompassing the human *RB* gene was sequenced in its entirety. It is one of the largest contigs of sequenced human DNA reported to date. The gene consists of 27 exons spanning more than 180 kb of genomic DNA within human chromosome 13q14.12 or 13q14.2. The locus produces a 4.7 kb transcript that encodes a nuclear phosphoprotein consisting of 928 amino acids (Figure 1). The *RB* gene is conserved among vertebrates. Human and mouse *RB* genes share a 91% homology at the amino acid level. This homology extends into the promoter region, which contains potential sites for the transcription factors ATF, Sp1, and E2F.

The RB protein ($p110^{RB}$) is a nuclear protein of 110 kDa and is phosphorylated on both serine and threonine residues. The number of ^{32}P-labeled tryptic peptides indicates that there are at least 10 phosphorylated amino acids. These modifications are thought to be important regulators of $p110^{RB}$ and have been shown to occur in a cell-cycle-dependent manner. Cells in early G_1 phase of the cycle contain exclusively unphosphorylated or underphosphorylated p110. At an as-yet undefined point later in G_1, the protein is hyperphosphorylated ($pp110^{RB}$) and exists in this state until M phase, although the level of phorphorylation does fluctuate slightly. Several studies have implicated the cyclin-dependent kinase (cdk) family of kinases as enzymes responsible for this phosphorylation.

Inactivation of both alleles of this gene has been found in all retinoblastomas examined, and additional studies have implicated *RB* mutations in the development of a variety of tumors (e.g., in sarcomas, in small-cell carcinoma of the lung, in carcinoma of the breast, bladder, and prostate). Introduction of a wild-type copy of this gene into RB^- tumor cells suppresses their tumorigenicity in nude mice, providing further evidence that the cloned gene has properties consistent with those predicted for a tumor suppressor. The mechanism by which $p110^{RB}$ suppresses tumorigenicity remains unknown. However, a possible explanation emerged with the discovery that $p110^{RB}$ physically associates with oncoproteins of many DNA tumor viruses, namely, the adenovirus E1A protein, the SV-40-T antigen, and the E7 protein of human papilloma 16 (HPV-16), including the transcription factor E2F. The $p110^{RB}$ sequences required for these interactions have been mapped to two large domains in the carboxy-terminal portion of the protein. These domains are affected in all known naturally occurring mutant RB proteins, suggesting an important role for this region of the protein in tumor suppression.

Thus far, much effort in RB research has been focused on both the viral oncoprotein-binding domains and the C-terminal domain, whereas little is known about the N-terminal moiety of the protein. Recent reports demonstrate that an N-terminal truncated RB protein of 94 kDa (pRB^{94}) exerts more potent cell growth suppression than the full-length pRB protein in a diversity of tumor cell lines, including those having a normal endogenous *RB* gene. Tumor cells transfected with the pRB^{94} expressing plasmid displayed multiple morphological changes frequently associated with cellular sequences and/or apoptosis. They failed to enter S phase and rapidly died. The pRB^{94} expressed in recipient tumor cells had a longer half-life than the full-length pRB protein and tended to remain in an active un- or hypophosphorylated form. Since it has also been found that N-terminal truncated RB proteins often accumulate in growth-arrested and/or differentiated tumor cells, it is suggested that N-terminal truncation of pRB may be one of the cellular mechanisms modulating the RB protein function in cell cycle control.

RB mRNA is expressed in various tissues during mouse embryonic development, as early as day 10 of gestation. A high level of *RB* expression is detected in megakaryocytes, blood-forming cells of fetal liver, osteoblasts, skeletal muscle, and neurons and glial cells in the central nervous system. Thus, *RB* is expressed by cells

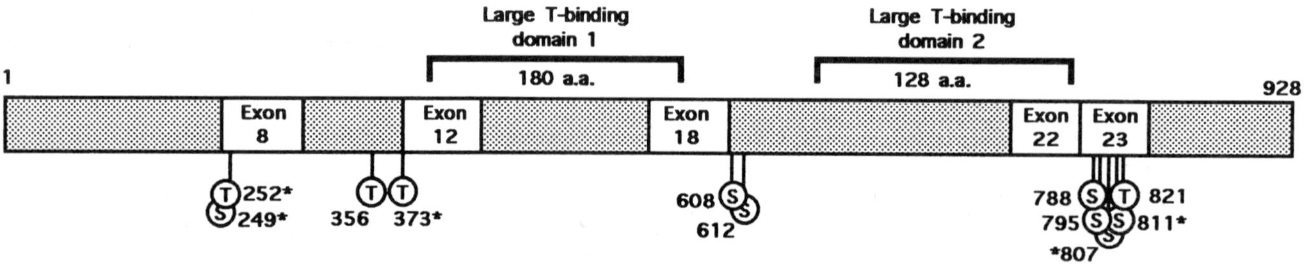

Figure 1. The human retinoblastoma protein. The 4.7 kb transcript of *RB* encodes a protein of 928 amino acids; specific exons of *RB* are indicated for reference. Binding of $p110^{RB}$ by viral transforming proteins T_{ag} or E1A requires two domains (indicated by brackets), one of 180 (large T-binding domain 1; amino acids 393–572) and another of 128 amino acids (large T-binding domain 2; amino acids 646–773). Naturally occurring mutations in *RB* disrupt one of these two domains. Considering conserved sequences between the human and murine proteins, the human protein has 10 Thr or Ser residues: seven are preceded by basic or polar residues (Thr 252, Thr 356, Thr 373, Ser 612, Ser 788, Ser 795, Ser 811) and three by nonpolar residues (Ser 608, Ser 807, Thr 821). Five phosphorylation sites have been identified and are indicated by asterisks. It is interesting that all these sites are outside the T_{ag}/E1A-binding domains.

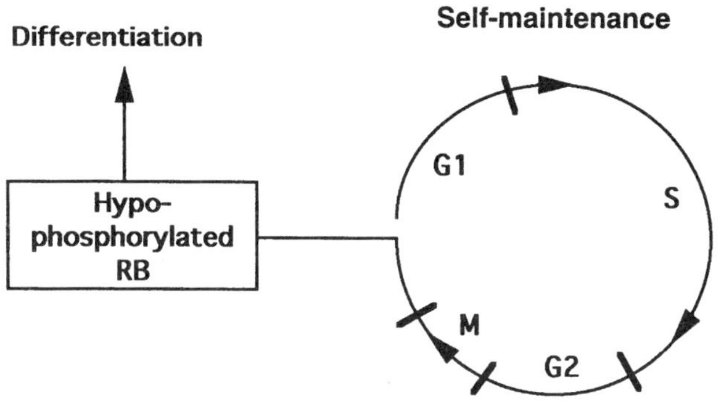

Figure 2. Model for the role of *RB* in regulation of cell differentiation. Phosphorylation of *RB* allows progression through G_1 and continuation of a self-maintenance pathway. Unphosphorylated or hypophosphorylated *RB* blocks cell cycle progression in the G_1 phase.

undergoing differentiation to produce nondividing cells or giving rise to daughter cells different from the parental cell. *RB* gene knockout mice die in utero around day 14 of gestation. The embryos show disturbed neural and hematopoietic differentiation, indicating that *RB* is vitally important for these processes. This notion is further supported by studies demonstrating that *RB* expression in mouse embryo tissues is highest in cells undergoing differentiation and that *RB* is required for MyoD-induced muscle differentiation.

In summary, the RB protein acts as a cell cycle checkpoint at the G_1 phase and plays a vital role in the differentiation process, at least in certain tissues. According to the model depicted in Figure 2, the hypophosphorylated form of *RB* is needed for permanent withdrawal from the cell cycle and initiation of a differentiation pathway.

4 THE *P53* GENE

The human *p53* gene spans about 20 kb of genomic DNA and contains 11 exons. The gene for *p53* is localized on the short arm of human chromosome 17p13 and encodes a 393 amino acid (50–55 kDa) protein (Figure 3). The *p53* gene is a tumor suppressor gene, negatively regulating the cell cycle by blocking the transit of cells through G_1, or from G_1 into S. Mutations in the *p53* gene are the most common gene mutations identified in carcinomas to date. Human families with Li–Fraumeni syndrome, an inherited predisposition to cancer, have a mutated germ line *p53* gene, a rare disorder in which affected patients are at high risk for several different types of malignant lesion at early age. In breast cancer, point mutations and/or small deletions in the *p53* gene are found clustered in four "hot spots" located in exons 5, 7, and 8 (codons 130–290/domains II–V: Figure 3), which coincide with the four most highly conserved regions of the gene. Three codons, 175, 248, and 273, stand out as extreme hot spots in colon and breast tumors. The wild-type (wt) *p53* can efficiently inhibit the in vitro transformation of primary rodent cells (rat embryo fibroblasts, REF) induced by a variety of oncogenes. In contrast, tumor-derived *p53* mutants exhibit a complete loss of this inhibitory capacity. Mutant p53 proteins differ from each other and from the wild-type p53 in conformation, localization, and transforming potential. Mutations in codons 143, 175, and 275 confer strong transforming potential. A serine 135 *p53* mutant has an intermediate transforming potential, while the histidine codon 273 allele transforms weakly, if at all. Many mutant p53 proteins are more stable than the normal or "wild-type" p53 protein. Because of their higher levels, mutant p53 proteins can be detected more easily than the wild type.

The mRNA transcribed from the *p53* gene varies between 1.8 and 3.0 kb, depending on the species. The 3′-untranslated region is long and varies in size from 800 nucleotides for mice to 1800 bp for *Xenopus levis*. The level of *p53* mRNA in normal cells is highest in undifferentiated cells, spleen cells, cells undergoing rapid embryonic development (particularly during midgestation in the mouse), and other rapidly proliferating cell types. *In vitro,* the level of *p53* mRNA appears to be down-regulated in tumor cell lines induced to differentiate and up-regulated in quiescent fibroblasts induced to divide by serum stimulation, suggesting that proliferative state of the cell is correlated with *p53* mRNA levels. Overexpression of wild-type *p53* in tumor cells can have one of three results: suppression of cell proliferation, induction of apoptosis (programmed cell death), or induction of differentiation. Which pathway the tumor cell enters is probably dependent on the tumor cell type and levels of wild-type *p53* expressed. Recently, the role for the *p53* tumor suppressor gene in the execution of apoptosis has been demonstrated in cells produced from animals with homozygous inactivation of the *p53* gene.

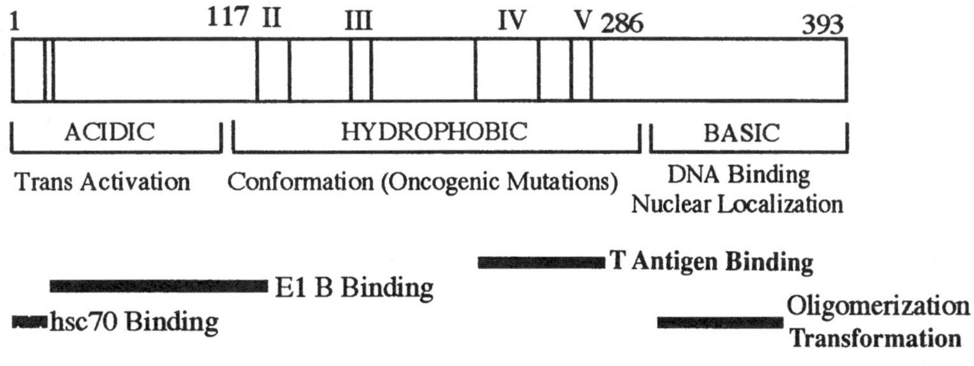

Figure 3. Structural features of the murine p53 protein. The 387 amino acid murine p53 protein contains five highly conserved domains labeled I–V. The protein is also divided into three major regions based on charge. The domains of the protein involved in oligomerization and binding to other viral and cellular proteins are also indicated.

The mechanism underlying *p53* growth suppression is still undefined. Recent efforts to further discern the function of *p53* have centered on the underlying molecular basis for this tumor suppressor. In particular, research has focused on the identification of cellular molecules (specifically DNA and proteins) with which p53 protein associates. Several biochemical features of p53 have been elucidated, and at least two of these are currently of much interest. First, p53 has been shown to repress transcription of *RB*, c-*fos*, β-actin, *p53*, *hsc70*, c-*jun*, *MDR1*, and *PCNA* from viral and cellular promoters. This suppression is apparently sequence independent and may involve p53 binding to the TATA-binding protein or to other transcription factors. Second, p53 can bind to DNA in a sequence-specific manner. Kern et al. in 1991 demonstrated that p53 binds specifically to DNA fragments containing two or three repeats of TGCCT sequences, and the G residue of a 5 bp repeat is critical for this interaction. Proteins encoded by *p53* mutants (e.g., $273Arg^{his}$, $175Arg^{his}$ lose this binding ability. More recently, a 20 bp consensus-binding site, consisting of two copies of the 10 bp sequence 5′-PuPuPuC(A/T)(T/A)GPyPyPy-3′, separated by up to 13 bp, was identified. Both copies of the 10 bp sequence are required for efficient binding by p53. In addition, p53 contains a strong transcriptional activation sequences near its amino terminus and can stimulate the expression of genes downstream of its binding site. Such stimulation has been demonstrated in both mammalian and yeast cells, as well as in in vitro systems.

The ability of p53 to activate transcription from specific sequences suggests that genes induced by p53 may mediate its biological role as a tumor suppressor. To date, several genes containing p53-binding sites have been identified. These include muscle creatine kinase *(MCK)*, *GADD45*, *MDM2*, *WAF1* and *CIP1*. The relation of any of these genes to suppression of cell growth by p53 remains unclear. It has been suggested that *MDM2* may be a feedback regulator of p53 action by being transcriptionally induced and then inhibiting p53 function. In this regard, *MDM2* functions as an oncogene rather than as a tumor suppressor gene. Another gene, the wild-type p53-activated fragment 1 gene, named *WAF1*, is highly induced by p53. Introduction of *WAF1* cDNA suppressed the growth of human brain, lung, and colon tumor cells in culture. A p53-binding site was identified 2.4 kb upstream of *WAF1* coding sequences. *WAF1* is directly regulated by p53 and can itself suppress tumor cell growth in culture. *WAF1* may thus be an important component of the p53 growth suppression pathway.

Thus, p53 emerges as a DNA-binding protein with an important but undefined role in cell cycle regulation, functioning in the G_1–S time frame. It seems likely that new insights into its activities will be published in the near future. Additional target genes that p53 binds and activates or represses transcription will likely develop. Likewise, it is highly likely that either by biochemical means or by means of yeast screens, new proteins that bind to p53 will be identified. The challenge will then be to determine which of these interactions are physiologically relevant. Some of the more intriguing questions to be illuminated include what controls the levels of p53 in normal cells and why mutant p53 proteins frequently are much more stable in cells than is wild-type form.

5 *BRCA1* AND *BRCA2* GENES AND BREAST CANCER

The first association of a tumor suppressor gene with inherited breast cancer was demonstrated for *p53* on chromosome 17p13. Mutations in this locus are inherited in families with the rare breast cancer and sarcoma syndrome termed Li–Fraumini. A large group of patients with an inherited pattern of breast and ovarian cancer has also been studied. The location of specific gene(s) responsible was initially established on chromosome 17q21 and was termed *BRCA1*.

BRCA1 has now been cloned and sequenced; it contains more than 1800 amino acids and encodes a zinc finger protein of unknown function. Of great interest was the observation that the *BRCA1* gene is not commonly mutated in sporadic breast tumors. Work from various laboratories demonstrate that either the BRCA1 protein is often not made in breast tumors or, if the protein is produced, it may end up in the wrong place in the cell, or the BRCA1 protein is located in the nucleus of normal breast epithelial cells, but in the cytoplasm of cells derived from tumors. Exactly what this set of possibilities means is not yet known, but it suggests that something has gone wrong in BRCA1 function in the tumor cells.

A second locus, termed *BRCA2*, at 13q 12-113, is associated with a smaller proportion of familial breast cancers, in particular those arising at a younger age. Northern analysis has demonstrated that *BRCA2* is encoded by a transcript of 10–12 kb that is present in normal breast epithelial cells, placenta, and the breast cancer cell line MCF7. Very recently, a 7.3 kb long, partial cDNA (including 300 bp of 3′-untranslated sequence) of *BRCA2* gene, encoding 2329 amino acids, was isolated. The BRCA2 protein shows a weak similarity to BRCA1 protein over a restricted region (amino acid 1394–1474 in BRCA1, and 1783–1863 in the portion of BRCA2). The significance of this is not clear.

Loss of heterozygosity on chromosome 13q has been observed in sporadic breast and other cancers, suggesting that there is a somatically mutated tumor suppressor gene in this region. *BRCA2* is a strong candidate for this gene, and analysis of a large series of cancers is ongoing to investigate whether *BRCA2* is somatically mutated during oncogenesis. The identification of *BRCA2* should allow more comprehensive evaluation of families at high risk of developing breast cancer.

6 WILMS' TUMOR GENE, *WT1*

Wilms' tumor or nephroblastoma is the most common solid pediatric malignancy, affecting 1 in 10,000 young children. This malignancy is an example of development gone awry; tumors arise from mesenchymal stem cells that should normally differentiate into the epithelial components of the nephron. Morphologically, Wilms' tumors are characterized by the disorganized presence of most of the components of normal embryonic kidney (i.e., stroma, undifferentiated blastema, dysplastic tubules, and pseudoglomeruloid structures). Additional non-kidney-derived, mesenchymal elements such as muscle or bone are also frequently involved. The genetics and histopathology of this cancer indicate a complicated and heterogeneous etiology involving genes mapping to at least three different locations (11p13, 11p15, and an unmapped locus) in the human genome. A fourth location (16q) contains a gene probably involved in tumor progression, rather than initiation.

One of these genes, the Wilms' tumor (*WT1*) gene, was cloned from chromosomal region 11p13, from the smallest region of overlap in 1990. It was shown to be specifically expressed in the developing kidney. Mapping and isolation of this gene were achieved because deletions of 11p13 were identified in children with the WAGR syndrome, which is characterized by Wilms' tumor, aniridia, genitourinary abnormalities (in at least 50% of boys with the syndrome), and mental retardation. Although tumors with gross deletions in

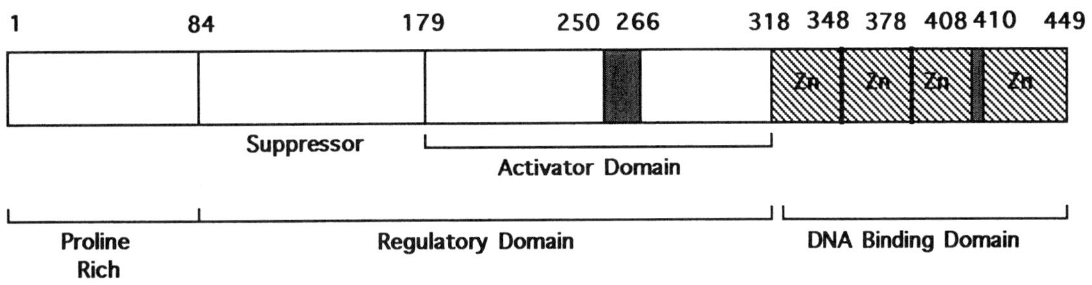

Figure 4. Structure of *WT1* gene showing various functional domains. Four forms of the WT1 protein are produced by alternative splicing. The different forms comprise 429–449 amino acids and range in size from 49 to 54 kDa. Two alternative splice are indicated by shaded rectangles. One leads to a 17 amino acid insertion (250–266) upstream of the four zinc fingers (Zn). The other splice inserts three amino acids between the third and fourth zinc fingers. The most common form includes both inserts. The proline/glutamine-rich region, the transcription regulatory domain, and the DNA-binding domain of *WT1* also are indicated.

chromosome 11p13 were instrumental in localizing and molecular cloning of the *WT1* gene, subsequent work has revealed that the more frequently observed alterations in this locus are missense point mutations and both small and large deletions. Like the *p53* tumor suppressor gene, many tumors retain the wild-type allele.

Spanning 50 kb of genomic sequence, the *WT1* gene is organized into 10 exons. The predicted WT1 polypeptide consists of 449 amino acids and shows several features characteristic of transcription factors, such as nuclear localization, four contiguous Cys^2/His^2 zinc finger domains in the carboxy-terminal DNA-binding domain, and an amino acid terminus rich in proline and glutamine residues (Figure 4). Alternative splicing at two sites, one of which inserts three amino acids (± KTS) between the third and fourth zinc fingers, and the second inserts 17 amino acids of exon 5 (± 17 amino acids), produces four discrete transcripts. Cell culture studies indicate that alternative splicing within the *WT1* zinc finger region determines whether the protein localizes mainly with splicing factors or with DNA in transcription factor domains in the nucleus. It has been proposed that *WT1* plays roles in transcriptional processing of RNA as well as in transcription. The proximal promoter region of *WT1* is GC rich and contains CCAAT, SP1, NF1, and WT1/EGR1 sites. The mRNA is about 3.5 kb; in some tissues, however, significantly different sized transcripts have been observed, indicating the presence of *WT1*-related genes.

The expression pattern of *WT1* observed in a number of species both support its role as a tumor suppressor gene in the kidney and extends its possible functions to differentiation events in many other organs. During embryonic development, *WT1* transcripts are found in the kidney, where the condensing metanephric mesenchyme and primitive renal vesicles are formed, called the gonadal ridge mesothelia, and the mesothelial lining of the celomic cavity and the organs it contains. *WT1* mRNA is also expressed in the spleen and in hematopoietic cell lines. Surprisingly, *WT1* expression is also observed in the spinal cord ventral horn motor neurons and in the area postrema of the rat brain. The latter pattern of expression in the central nervous system persists in the adult animal. The area postrema lacks a blood–brain barrier and may be a chemoreceptor trigger zone for circulation molecules such as angiotensin II. This raises the intriguing possibility that *WT1* participates in the renin–angiotensin cascade by integrating peripheral and central vascular control signals in the area postrema.

The WT1 transcript undergoes RNA editing in rat in which U^{839} is converted to C, resulting in the replacement of leucine 280 in WT1 by proline. RNA editing at the same nucleotide is also observed in human testis. In in vitro assay, the WT1-leucine^{280} polypeptide represses the EGR1 promoter 30% more efficiently than the WT1-proline^{280}. Edited WT1-C^{839} mRNA is barely detectable in neonatal kidney, whereas adult rat kidney contain both U^{839} and C^{839}-WT1 mRNA, suggesting a role for the two protein isoforms in growth and differentiation.

The zinc finger domains of WT1 have sequence similarity with those of the early growth response (EGR) family of transcription factors (e.g., EGR1 and EGR2, and others such as Sp1, krox-2, GLI, and Kruppel). A number of studies have shown that the WT1 isoforms lacking the KTS splice site can bind in vitro to the same target 9 bp sequence 5′-GCGGGGGCG-3′ as the EGR1 protein, whereas the KTS isoforms do not. In another study, a 12 bp sequence GCG-TGG-GCG-(T/G) (G/A/T) (T/G) has been found to bind to WT1(-KTS) with a fourfold higher affinity than the 9 bp EGR1 consensus sequences. David Housman's group has identified a 10 bp motif 5′-GCGTGGGAGT-3′, termed WTE, which functions as a binding site for WT1 with 20–30-fold higher affinity than the EGR1-binding motif. This binding results in transcriptional repression of *EGR1* gene and may repress the transcription of genes activated by EGR1. A second functional WT1-binding sequence (5′-TCCTCCTCCTCCTCTCC-3′), 3′ to the transcription initiation site of the platelet-derived growth factor A chain *(PDGF-A)* gene, has been recently identified by DNase I footprinting and gel mobility shift assays. Similar sequences with some divergence have also been identified within the promoters of other growth-related genes, including Ki-*ras,* epithelial growth factor receptor, insulin receptor, c-*myc,* and tumor growth *b3.*

Given the evidence of mRNA overexpression of *IGF2, PAX2, MK,* the PDGF, N-*myc,* and the insulin-like growth factor receptor 1 gene *(IGF1-R),* the notion of WT1 as a transcriptional regulator of growth-promoting genes in the kidney has recently been examined by several groups. WT1 protein binds multiple sitets in the promoter sequences of insulin-like growth factor 2 (IGF2), platelet-derived growth factor A (PDGF-A), insulin-like growth factor 1 receptor (IGF1-R), colony-stimulating factor 1 (CSF1), insulin-like growth factor binding protein 2 (IGFBP-2), and WT1 itself. Reporter constructs harboring minimal promoters from these genes have shown that they can be negatively or positively regulated by WT1 in transient transfection systems to function as potent repressors and/or activators of transcription in vivo.

WT1 may function in one of two pathways. The first consists of promoting proliferation leading to growth of the kidney substructural components during development. Second, when growth is completed, it may function to limit growth factor production and to enforce a program of differentiation on the proliferating kidney cells. Supporting this line of reasoning is the recent report that maturation of the kidney as well as the urogenital system in recombinant mouse fetuses lacking functional WT1 is seriously defective. On the other hand, functional loss of WT1 transcriptional repressor activity, or a gain of trans-activation function by mutation, may result in untimely synthesis of the growth factors in kidney blastemal cells. Unrestrained autocrine growth stimulation could lead to the genesis of Wilms' tumors. Among other functions of the WT1 protein are inhibition of *ras*-mediated transformation. WT1 protein can also stabilize p53, modulate its trans-activational properties, and inhibit its ability to induce apoptosis.

Because of the rarity of Wilms' tumors and their early onset, it has been difficult to decipher the genetic events operative during the initiation and progression of the disease. We have established an animal model system that fulfills the criteria of both the phenotype and genotype of embryonal kidney tumors and would be an valuable tool for furthering the understanding of this type of cancer. Newborn rats administered the direct-acting chemical carcinogen nitrosomethyl urea (NMU) frequently develop kidney tumors with a latency of 4–8 months. Point mutations were found in the *WT1* gene in 7 of 18 tumors. The *WT1, IGF2, Pax2,* and *MK* mRNAs, which are all overexpressed in Wilms' tumors, were expressed at levels two- to sevenfold higher in 60% of the tumors compared to newborn kidney. The histopathology of the rat kidney tumors and the presence of genetic alterations in multiple loci are reminiscent of those in Wilms' tumors. These similarities establish this rat model as a relevant system for studying the human disease. Since the process of tumor formation in these animals mimics tumorigenesis in humans, these animals provide valuable resources for dissecting the multiple steps in carcinogenesis. They also serve as tumor models for testing new treatment strategies such as gene therapy.

In summary, WT1 serves as a paradigm for how alterations of transcription factor function can lead to oncogenesis. The picture that is emerging suggests that unlike *p53* and *RB, WT1* is a highly tissue-specific and developmentally regulated transcription factor that functions as a tumor suppressor gene. *WT1* probably plays a primary role in initiating or maintaining a mesenchymal–epithelial differentiation program in kidney. Loss of *WT1* function leads to lack of or aberrant differentiation, leading to unrestrained growth of metanephric blastemal cells in the embryonic kidney, leading to Wilms' tumor. Further work on *WT1* and Wilms' tumor must address the following questions:

- Is WT1 involved in metabolic pathway that contains other Wilms' tumor susceptibility genes?
- What are the natural downstream target genes of the WT1 protein?
- What are the critical protein–protein interactions in the cell that dictate whether WT1 will activate or repress transcription?

7 THE NEUROFIBROMATOSIS TYPE 1 AND 2 (NF1 AND NF2) GENES

Neurofibromatosis is a term used to describe two major human genetic disorders, types 1 and 2 (NF1 and NF2), both of which display autosomal dominant inheritance and involve tumors of the nervous system, but are distinct clinical entities. One of the most common dominantly inherited human diseases, NF1, or von Recklinghausen NF, affects 1 in 4000 children, half of whom are new germ line mutants. Neurofibromatosis is characterized by the highly variable expression of an array of features that include neurofibromas, cafe-au-lait macules, Lisch nodules of the iris, and a predisposition to certain malignant tumors such as schwannoma, glioma, pheochromocytoma, and, at low frequency, leukemia, rhabdomyosarcoma, and Wilms' tumor. It is a hereditary disease probably caused by the germ line mutations in one allele of the *NF1* gene. The *NF1* gene mapped to chromosome 17q11 has been cloned and sequenced. The gene is large, encoding a 13 kb transcript and a ubiquitously expressed protein, neurofibromin, of at least 2485 amino acids, with a region of considerable homology to the catalytic domain of $p120^{GAP}$. The gene's function is similar to that of the GAP (GTPase-activating protein) gene product. Loss of activity results in failure of hydrolysis of GTP to GDP by the RAS protein. This loss of neurofibromin function is thought to result in elevated levels of the GTP-bound RAS protein that transduces signals for cell division whose disruption can lead to tumor formation.

The high germinal mutation rate of *NF1* may be attributed to its heavy methylation. The mutations observed in the gene frequently produce truncated, functionless proteins. In a significant fraction of the characteristic malignant tumors of NF1, the second (normal) copy of the gene appears to be mutated or absent. Mutations of the *p53* are also observed in some cases of neurofibrosarcomas, as often noted in other sarcomas. Wild-type *NF1* again exemplifies the inhibition of an oncogene by an antioncogene. The tissue-specific existence of both GAP and NF1 suggests redundancy in the regulation of RAS protein. It is possible that the tissues that are susceptible to tumorigenesis have little GAP function, leaving NF1 to be sole regulator.

Neurofibromatosis 2 (NF2), by contrast, is a much less common disorder than NF1; it occurs in about 1 of 40,000 live births and shows a high penetrance. This dominantly inherited disorder is characterized by the occurrence of bilateral vestibular schwannomas and other central nervous system tumors including multiple meningiomas. Juvenile cataracts are also frequently seen in NF2. Genetic linkage studies and investigations of both sporadic and familial tumors suggest that NF2 is caused by inactivation of a tumor suppressor gene in chromosome 22q12. Sporadic, nonhereditary forms of these tumors are often monosomic for chromosome 22. A candidate NF2 gene has recently been cloned that encodes a protein consisting of 587 amino acids. Its protein product is unusual for tumor suppressor genes in that it is homologous with proteins found at the interface between the plasma membrane and the cytoskeleton interface. Mutations found in the tumor typically causes truncation of the protein product; *NF2* clearly qualifies as an antioncogene, although its precise mechanism of action is not yet known.

8 THE VON HIPPEL–LINDAU (VHL) SYNDROME AND RENAL CELL CARCINOMA

VHL disease is a familial cancer syndrome that is dominantly inherited and predisposes affected individuals to a variety of tumors. The most frequent tumors are hemangioblastoma of the central nervous system and retina; less common are the bilateral pheochromocytoma, or renal cell carcinoma (RCC). The minimum birth incidence of the disease is 1 in 36,000, penetrance is almost complete by 65 years of age, and median actuarial life expectancy is reduced to 49 years, with RCC being the most common cause of death. The

VHL gene, localized to chromosome 3p25 by linkage analysis, was cloned recently. The *VHL* gene is evolutionarily conserved and encodes two widely expressed transcripts of approximately 6 and 6.5 kb. The protein product of the *VHL* gene is an approximately 30 kDa cytoplasmic protein. The partial sequence of the inferred gene product contains an acidic repeat domain found in the procyclic surface membrane glycoprotein of *Trypanosoma brucei*. Its protein product appears to be a cell surface molecule involved in cell adhesion and signal transduction. Abnormalities of the *VHL* gene are observed in a high percentage of nonhereditary renal cell carcinomas, making it the principal RCC gene. Both these nonhereditary tumors and the tumors found in the syndrome typically reveal mutation or loss of the second copy of the gene, as expected for an antioncogene. The renal cell carcinoma cell line 786-o is known to harbor a *VHL* mutation, and it fails to produce a wild-type VHL protein. Reintroduction of wild-type, but not mutant, *VHL* into these cells had no demonstrable effect on their growth in vitro but inhibited their ability to form tumors in nude mouse.

9 FAMILIAL ADENOMATOUS POLYPOSIS AND COLON CANCER

The classic syndrome of familial colorectal cancer is that associated with dominant, autosomally inherited premalignant syndrome, adenomatous polyposis (FAP), caused by germ line mutation in the adenomatous polyposis coli (APC) gene. FAP has a prevalence of around 1 in 10,000, whereas various lines of evidence suggest that other major genes may be responsible for a significant proportion of colorectal cancers, and familial studies show a threefold increased risk in first-degree relatives of colorectal probands.

FAP, which is characterized by the development of hundreds, or carpets of thousands, of colorectal adenomas, some of which develop into carcinomas, is transmitted by the *APC* gene. Mapping of the *APC* gene to 5q21-22 by linkage studies in families allowed its identification and the characterization of mutations in affected individuals. In virtually all cases, without prophylactic colonectomy, progression to colorectal cancer occurs and death ensues at an average age of 42 years. In addition to almost universal upper gastrontestinal neoplasia, extracolonic manifestations such as osteomas, epidermoid cycts, desmoid tumors, and congenital hypertrophy of the retinal pigment epithelium (CHRPE) are common. Ocular examination revealed that patients expressing CHRPE tend to cluster in specific families. The extent of CHRPE is dependent on the position of the mutation along the coding sequence. CHRPE lesions are almost always absent if the mutation occurs before exon 9, but are systematically present if it occurs after this exon. Thus, the range of phenotypic expression observed among affected patients may result in part from different allelic manifestations of *APC* mutations.

The *APC* mRNA is expressed in normal human and rodent colorectal mucosa and a variety of other tissues. The *APC* gene contains an 8538 bp open reading frame and is predicted to encode a 2843 amino acid polypeptide with a few homologies to other proteins. Its protein product ($\approx$300 kDa) is located in the cytoplasm. The *APC* germ line mutations described to date are notable in that they are almost exclusively either single base pair changes leading to termination codons or small deletions, insertions, or splicing mutations that cause translational frameshifts and subsequent downstream stops.

A less severe form of familial polyposis, attenuated adenomatous polyposis coli (AAPC), is characterized by a low number of adenomatous polyps (usually <100), yet patients sustain an elevated risk of colon cancer. The mutant alleles responsible for this attenuated phenotype have been mapped in several families to the adenomatous polyposis coli (*APC*) locus on human chromosome 5q. Four distinct mutations in the *APC* gene have now been identified in seven AAPC families. These mutations, which predict truncation products, either by single base pair changes or by frameshifts, are similar to mutations identified in families with classical APC. However, they differ in that the four mutated sites are located very close together and nearer the 5′ end of the *APC* gene that any base substitution or small deletions yet discovered in patients with classical APC.

The *MCC* (mutated in colon cancer) gene, a candidate gene located at or near the *APC* locus and found to be mutant in some colon carcinomas, has been shown to map outside the segments deleted in some APC patients. Three candidate genes (*DP1, DP2, DP3*) and the *MCC* genes are located within the deleted region. *MCC* encodes an 829 amino acid protein with a short region bearing similarity to a G protein (cf. *NF1*). In the absence of conformation of a constitutional mutation in the *MCC* gene in FAP patients, however, the question of whether the *MCC* gene is indeed the *APC* gene is still open.

Mutation at the *APC* locus is a necessary though not sufficient condition for carcinoma of the colon. Other genetic events are the rule for this tumor, the targets being the **K**-*ras* oncogene, the *p53* antioncogene, and the *DCC* (deleted in colon cancer) antioncogene. Mutations in **K**-*ras* are featured in larger polyps, where their incidence is 40%. They are rare in very small polyps, and their incidence in carcinomas is about the same as in large polyps. On the other hand, *p53* and *DCC* mutations occur uncommonly in polyps but in the majority of carcinomas. It has not been established whether there is a necessary sequence for mutation in these two genes.

The *DCC* gene on chromosome 18q, a region that shows heterozygous loss in more than 80% of colorectal carcinomas, is expressed in most normal tissues but with a greatly reduced or absent expression in colorectal carcinomas. The *DCC* gene encodes a putative translation start site, a signal peptide, and a hydrophobic transmembrane region dividing the predicted protein into extracellular (1100 amino acid) and intracellular (324 amino acid) domains. Several tissue-specific *DCC* transcripts may be generated through alternative splicing of the primary transcript. The coding region is composed of at least 28 exons and is flanked by lengthy 5′- and 3′-untranslated regions, accounting for a significant portion of the 10–12 kb *DCC*-encoded message. The deduced amino acid sequence of *DCC* shows that it encodes a 190 kDa transmembrane phosphoprotein having attributes of a cell surface receptor.

Mutations are detected in the *DCC* gene, and the predicted amino acid is closely similar to neural cell adhesion molecules (N-CAMs) and other cell surface glycoproteins. The characterized mutations usually produce termination codons that in turn lead to truncated, presumably nonfunctional, proteins. No normal copies of *DCC* remain in the tumor cells, as expected for an antioncogene. This is the only antioncogene that has been cloned without the availability of constitutional mutations. So far there is no hereditary condition in which *DCC* is mutant. *DCC*, for example, is expressed in neural tissues as well as in the gastrointestinal tract. One patient with a constitutional deletion of 18q, the arm that contains the *DCC* gene, developed a brain tumor. Could it be that germ line mutations of *DCC* are associated with brain tumors rather than with colon cancer?

Several lines of evidence suggest that cancers require multiple oncogene mutations to produce a metastatic tumor. For example, six independent mutations have been documented for the development of colon tumors. This number of mutations appear to be excessively high to be accounted for by the normal frequency of spontaneous mutations in cells. Recently, a human homologue of the bacterial MutS and *Saccharomyces cerevisiae* MSH proteins, hMSH2, was identified. hMSH2 maps to human chromosome 2p22-21 near a locus implicated in hereditary nonpolyposis colon cancer (HNPCC). A T-to-C transition mutation has been detected in the −6 position of a splice acceptor site in sporadic colon tumors and in affected individuals of two small HNPCC kindreds. The similarity between *S. cerevisiae msh2* mutants and HNPCC patients with regard to instability of dinucleotide repeat sequences, the correspondance of map locations of HNPCC and the hMSH2 gene, and the detection of a specific mutation in small HNPCC kindreds, as well as sporadic colorectal tumors, suggest that mutations in the hMSH2 gene are responsible for HNPCC, perhaps as a result of a defect in the mismatch repair system, which leads to an increased rate of accumulation of spontaneous mutations.

10 *KAI1* GENE AND PROSTATE CANCER

KAI1, a gene from human chromosome 11p11.2, has been isolated and is shown to suppress metastasis when introduced into rat AT6.1 prostate cancer cells. Expression of this gene is reduced in human cell lines derived from metastatic prostate tumors. *KAI1* cDNA has a single open reading frame (nucleotide positions 166–966) that predicts a protein of 267 amino acids with a molecular mass of 29,610 Da. KAI1 protein contains four hydrophobic and presumably transmembrane domains with three potential N-glycosylation sites. *KAI1* is evolutionarily conserved, is expressed in many human tissues, and encodes a member of a structurally distinct family of leukocyte surface glycoproteins. Decreased expression of this gene may be involved in the malignant progression of prostate and other cancers.

11 SEARCH FOR NEW ANTIONCOGENES

So far 11 antioncogenes (*BRCA1, BRCA2, RB1, WT1, VHL, APC, DCC, KAI1, TP53, NF1,* and *NF2*) have been cloned. All except *DCC* have been found to be mutated in the germ lines of persons predisposed to one or more tumor types. Another six putative antioncogenes (*RCC, NB1, MLM, MEN1, BCNS,* and *LC1*) have been mapped but not cloned (Table 2). Germinal mutations in all except *LC1* are known to predispose those who carry them to tumors. Obviously the analysis of hereditary cancer provides excellent entry to the world of antioncogenes.

Table 2 Uncloned Human Suppressor Genes: Chromosomal Location, Protein Products, and Characteristic Neoplasm

Gene	Chromosomal Location	Neoplasm
NB1	1p36	Neuroblastoma
MLM	9p21	Melanoma
MEN1	11q13	Pituitary adenoma
BCNS	9q31	Medulloblastoma, skin carcinoma
RCC	3p14	Kidney carcinoma

12 TRANSGENIC ANIMAL MODELS

The recent development of animal models to knock out tumor suppressor genes further substantiated the idea of cancer suppression. Either *p53* or *RB* gene disruption results in mice that are prone to cancer formation. Targeted inactivation of the *WT1* locus in mice leads to death in utero of homozygous mutant *WT1* embryos, with a failure to develop kidneys and gonads. However, mice heterozygous for the mutant *WT1* allele undergo normal development and do not develop any malignancies. Obviously, the loss of one allele of *WT1* in the mouse is insufficient for kidney tumor initiation or the urogenital anomalies typical of WAGR patients.

13 GENE THERAPY

Somatic gene therapy is defined as the insertion of an exogenous normal gene into somatic cells to correct an abnormality or deficiency of a specific protein. This can be carried out by transfecting the new gene in the presence of an abnormal gene (additional therapy) or by attempting to replace a defective gene by inserting a new one at the same site, using homologous recombination. The genetic information will not be passed on to future generations. Formal selection criteria for gene therapy have been established, include the following: (1) the condition selected is a life-threatening disease; (2) the gene responsible is available in cloned form; (3) precise regulation of the gene is not required; and (4) a suitable delivery system is available. Considerable effort is going into somatic gene therapy for single-gene disorders such as adenosine deaminase immunodeficiency and thalassemia. These diseases might seem to be more amenable to this approach than a complex polygenic disorder such as cancer. However, several systems have been explored in malignancy to enhance the selective destruction of tumor versus normal cells.

The other type of gene therapy, called germ cell transfection or germ cell therapy, poses considerable ethical dilemmas and probably has a lesser part to play in cancer therapy. However, at some point in the future it might be used to prevent cancers in individuals carrying defective tumor suppressor genes (e.g., the *RB* gene in retinoblastoma, *p53* in Li–Fraumeni syndrome). In addition, gene transfer techniques can be applied to target prodrug activation specifically to tumor cells and also to protect normal tissues against toxic chemotherapy.

14 CONCLUDING COMMENTS

Finally, it is evident that the label "tumor suppressor" is applicable to a wide variety of genes involved in normal cellular functioning. The unifying characteristic appears to be the necessity of having one functioning wild-type allele, to prevent abnormal proliferation or differentiation at particular stages of cell growth. It is this feature that permits normal activity into a hemizygous state and underlies the association of some of these genes with certain inherited cancer predispositions. To date, the genes identified in this category are of diverse function. Some may act within a narrow range of cell types and consequently may be involved in the etiology of relatively few tumor types, as with *WT1*. Others, of fundamental importance to the growth and maintenance of a broad range of cells, will be found to be implicated in many tumors, as with *p53* and *RB*. Since many of these genes appear to act as negative regulators of cellular proliferation, and their presence in a single copy is sufficient for the normal control of proliferation, they may well offer an important approach for the future therapeutic control of abnormal growth.

The continuous efforts in cloning and characterizing new tumor suppressor genes should yield exciting and worthwhile results. In combination with investigation of dominant-acting oncogenes, a much clearer picture of the mechanisms involved in positive and negative signal interactions should emerge. Understanding these mechanisms will provide the basis for development of novel methods of clinical intervention. The hope, of course, is that the use of tumor suppressor genes and/or their products will emerge as an effective clinical strategy in the effort to cure and prevent many forms of human cancer.

See also CANCER; CANCER CHEMOTHERAPY, THEORETICAL FOUNDATIONS OF; DNA IN NEOPLASTIC DISEASE DIAGNOSIS; HUMAN GENETIC PREDISPOSITION TO DISEASE; ONCOGENES; RETINOBLASTOMA, MOLECULAR GENETICS OF; STEROIDS, ANTITUMOR.

Bibliography

Campbell, C. E., Huang, A., Gurney, A. L., Kessler, P. M., Hewitt, J. A., and Williams, B. R. G. (1994) Antisense transcripts and protein binding motifs within the Wilms' tumor *(WT1)* locus. *Oncogene,* 9:583–595.

Cawthon, R. M., Weiss, R., Xu, G., Viskochil, D., Culver, M., Stevens, J., Robertson, M., Dunn, D., Gesteland, R., O'Connell, P., and White, R. (1990) A major segment of the neurofibromatosis type I gene: cDNA sequence, genomic structure, and point mutations. *Cell,* 62:193–201.

Cooper, G. M. (1992) *Elements of Human Cancer.* Jones and Barntlett Publishers, Boston.

Dong, J.-T., Lamb, P. W., Rinker-Schaeffer, C. W., Vukanovic, J., Ichikawa, T., Isaacs, J. T., and Barrett, J. C. (1995) *KAI1,* a metastasis suppressor gene for prostate cancer on human chromosome 11p11.2. *Science,* 268:884–886.

El-Deiry, W. S., Tokino, T., Velculescu, V. E., Levy, D. B., Parsons, R., Trent, J. M., Lin, D., Mercer, E., Kinzler, K. W., and Vogelstein, B. (1993) WAF1, a potential mediator of p53 tumor suppressor. *Cell,* 75:817–825.

Futreal, P. A., Liu, Q., Shattuck-Edens, D., Cochran, C., Harshman, K., Tavtigian, S., Bennett, L. M., Haugen-Strano, A., Swensen, J., Miki, Y., et al. (1994) *BRCA-1* mutations in primary breast and ovarian carcinomas. *Science,* 266:120–122.

Kreidberg, J. A., Sariola, H., Loring, J. M., Madea, M., Pelletier, J., Housman, D., and Jaenisch, R. (1993) WT-1 is required for early kidney development. *Cell,* 74:679.

Latif, F., Troy, K., Gnarra, J., Yao, M., Duh, F.-M., Orcutt, M. L., Stackhouse, T., Kuzamin, I., Modi, W., Geil, L., Schmidt, L., Zhou, F., Li, H., Wei, M. H., Chen, F., Glenn, G., Choyke, P., Walther, M. M., Weng, Y., Duan, D.-S. R., Dean, M., Glavac, D., Richards, F. M., Crossey, P. A., Ferguson-Smith, M. A., Paslier, D. L., Chumakov, I., Cohen, D., Chinault, C., Maher, E. R., Linehan, M., Zbar, B., and Lerman, M. I. (1993) Identification of the von Hippel–Lindau disease tumor suppressor gene. *Science,* 260:1317–1320.

Malkin, D., Li, E. P. Strong, L. C., Fraumeni, J. F., Jr., Nelson, C. E., Kim, D. H., Kassel, J., Gryka, M. A., Bischoff, F. Z., Tainsky, M. A., and Friend, S. H. (1990) Germ line p53 mutations in a familial syndrome of breast cancer, sarcomas, and other neoplasms. *Science,* 250:1233–1238.

Sharma, P. M., Yang, X., Bowman, M., Roberts, V., and Sukumar, S. (1992) Molecular cloning of rat Wilms' tumor complementary DNA and a study of messenger RNA expression in the urogenital system and the brain. *Cancer Res.* 52:6407–6412.

Spirio, L., Olschwang, S., Gorden, J., Robertson, M., Samowitz, W., Joslyn, G., Gelbert, L., Thliveris, A., Carlson, M., Otterud, B., Lynch, H., Watson, P., Lynch, P., Laurent-Puig, P., Burt, R., Hughes, J. P., Thomas, G., Leppert, M., and White, R. (1993) Alleles of the APC gene: An attenuated form of familial polyposis. *Cell,* 75:951–957.

Toguchida, J., McGee, T. L., Paterson, J. C., Eagle, J. R., Tucker, S., Yandell, D. W., and Dryja, T. P. (1993) Complete genomic sequence of the human retinoblastoma susceptibility gene. *Genomics,* 17:535–543.

Wallace, M. R., Marchuk, D. A., Andersen, L. B., Letcher, R., Odeh, H. M., Saulino, A. M., Fountain, J. W., Brereton, A., Nicholoson, J., Mitchell, A. L., Brownstein, B. H., and Collins, F. S. (1990) Type 1 neurofibromatosis gene: Identification of a large transcript disrupted in three NF1 patients. *Science,* 249:181–186.

Weinberg, R. A. (1991) Tumor suppressor genes. *Science,* 254:1138–1145.

Wooster, R., Bignell, G., Lancaster, J., et al. (1996) Identification of the breast cancer susceptibility gene *BRCA2. Nature,* 378:789–792.

Tyrosine and Tryptophan Hydroxylase: *see* Brain Aromatic Amino Acid Hydroxylases.

U

Ultratin Gels, High Speed DNA Sequencing in: *see* DNA Sequencing in Ultrathin Gels, High Speed

Ultratin Gels, High Speed DNA Sequencing in: *see* DNA Sequencing in Ultrathin Gels, High Speed

ULTRAVIOLET RADIATION DAMAGE TO DNA

David L. Mitchell

1 **Effects of UVR on Living Systems**

2 **The Electromagnetic Spectrum**

3 **DNA Photoproducts**
- 3.1 Pathways for Dissipation of Absorbed Energy
- 3.2 Genomic Distribution of DNA Damage

4 **Analysis of DNA Photodamage and Repair**
- 4.1 DNA Damage
- 4.2 DNA Repair

5 **DNA Damage Tolerance Mechanisms**
- 5.1 Photoprotection
- 5.2 Photodependent Repair
- 5.3 Excision Repair

6 **Spatial and Temporal Distribution of DNA Repair**

7 **Developmental Regulation of DNA Repair**

8 **Genetic Regulation of DNA Repair**

9 **Biological Consequences of UVR**
- 9.1 UVR Carcinogenesis
- 9.2 Environmental Effects of Stratospheric Deozonation

Key Words

Dimerization Covalent bonding of adjacent purine or pyrimidine bases resulting from the resolution of the unstable electronic configuration created by absorption of a photon within the UVR range of light.

Endonuclease An enzyme, often associated with the excision repair process, that cleaves a phosphodiester bond to either the 3′ or 5′ side of a base or abasic site in DNA.

Excision Repair Replacement of damaged DNA with nascent, corrected DNA involving recognition, unwinding, incision, excision, resynthesis, and ligation.

Photoenzymatic Repair (PER) A light-dependent process involving the enzyme-catalyzed reversal of cyclobutane dimers.

Photohydrate A type of monobasic photoproduct to which a hydroxyl (—OH) group has been added to the carbon in the 5- or 6-position of a pyrimidine base.

Photoproduct A stable change in DNA structure involving a single base or adjacent bases and resulting from the dissipation of absorbed UVR.

Ultraviolet Radiation (UVR) Wavelengths of the electromagnetic spectrum ranging from 190 nm to 400 nm; UVR is subdivided into vacuum UV (190–240 nm), UVC (240–280 nm), UVB (280–320 nm), and UVA (320–400 nm).

Xeroderma Pigmentosum A rare autosomal recessive disease characterized by sun sensitivity resulting in progressive degenerative changes in sun-exposed portions of the skin and eyes, often leading to neoplasia.

It has been known for more than a hundred years that ultraviolet radiation (UVR) can have an adverse effect on the growth and development of living organisms. Degradation of stratospheric ozone and the resultant increase in UVB radiation at the earth's surface have focused our attention on the public health and environmental effects of solar DNA damage. With the advent and advancement of molecular biology over the past two decades, our understanding of the biochemical foundation of the cytotoxic, mutagenic, and carcinogenic effects of UVR has greatly increased.

Armed with an arsenal of powerful techniques, today's researchers are able to probe the intricate structure of UVR damage induction and repair in DNA. Indeed, recent cloning of DNA repair genes from bacteria, yeast, and humans has set the stage for resolving the molecular mechanisms underlying this complex and essential process. In concert with public awareness and political action, the adverse effects of global stratospheric deozonation can be mitigated by the development and application of basic photobiological research.

1 EFFECTS OF UVR ON LIVING SYSTEMS

UVR damage in cellular DNA can be resolved in several ways, depending on the type of lesion produced, the genomic location of the damage, the type of cell affected, and developmental state of the cell (Figure 1). It is plausible that some lesions are not perceived as different from normal DNA by the cell. It is more probable, however, that unless the damage is repaired, it will disrupt the normal operation of vital cellular processes such as DNA replication or transcription. In this event, cell proliferation will cease or, if the lesion is situated in a gene required for an essential metabolic function, the cell will die.

Some types of photodamage are more effective than other types at blocking the progression of DNA or RNA polymerase. Photoproducts that allow lesion bypass by the DNA replication machinery may produce misincorporation of an incorrect complementary base, that is, a mutation. A mutation may have several outcomes.

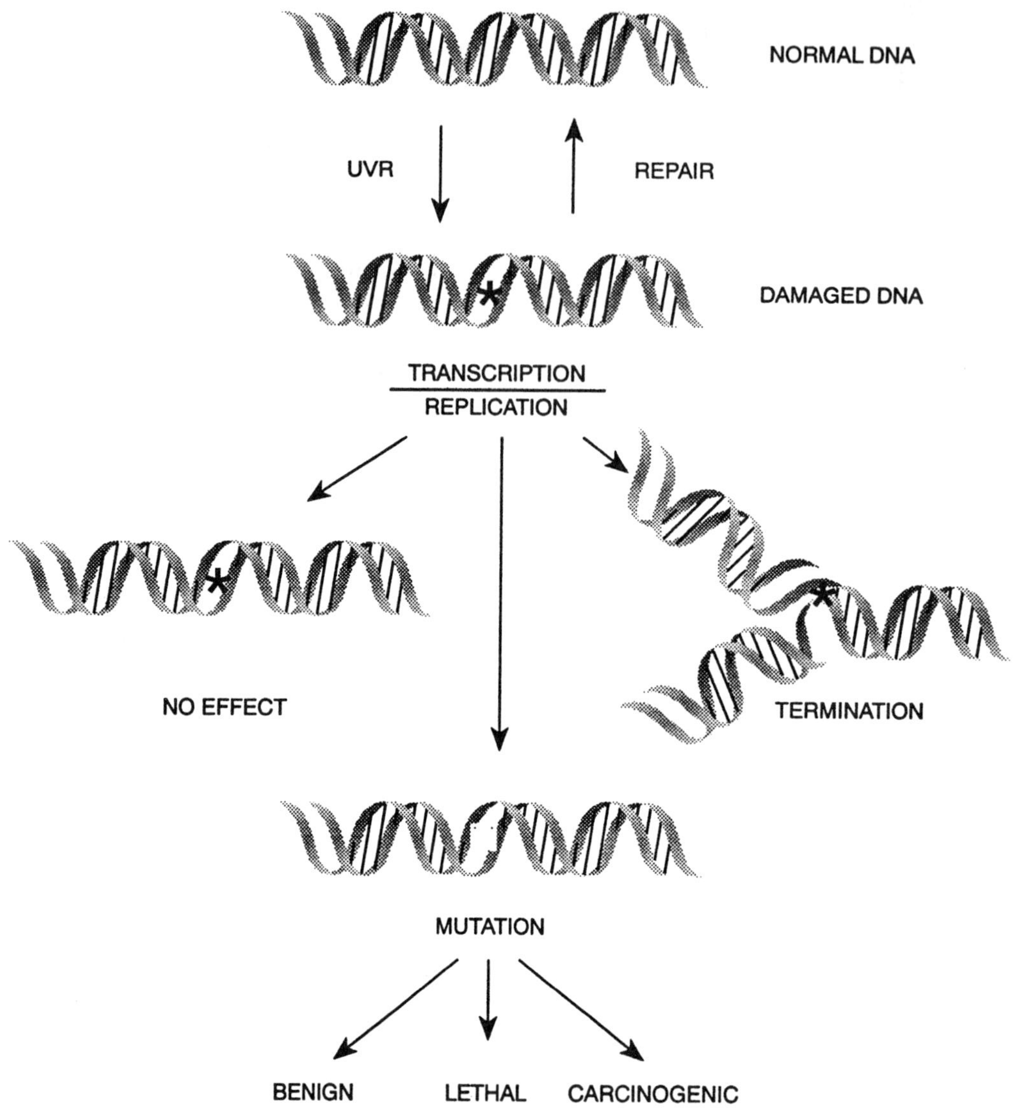

Figure 1. Biological consequences of UVR: molecular events leading to cell killing, mutation induction, and carcinogenesis.

First, it may be benign, neither altering the genetic code nor affecting normal metabolism. Second, the mutation may create a truncated or partial RNA transcript and a dysfunctional protein. If the protein is essential, then the mutation is lethal. Finally, it may result in activation of an oncogene or inactivation of a tumor suppressor gene, leading to the initiation of the carcinogenic process.

2 THE ELECTROMAGNETIC SPECTRUM

The sun emits a heterogeneous array of energies at wavelengths ranging through 11 orders of magnitude from 1 nm to 100 m (Figure 2). Much of this energy is biologically irrelevant; long wavelength, low energy emissions such as far-infrared light and microwaves are highly unlikely to impart sufficient energy to influence biochemical reactions. Likewise, visible and near-infrared light, although photobiologically very important, especially in plants, are not considered to be hazardous to living organisms. Finally, ionizing radiations such as high energy particles, X-rays and γ-rays are expended by atomic collisions in the upper atmosphere. Hence, UVR emerges as the most biologically relevant component of the solar spectrum.

The UV spectrum is divided into four regions; vacuum UV (190-240 nm), UVC (240–290 nm), UVB (290–320 nm), and UVA (320–400 nm). Absorption of UVC irradiation by stratospheric ozone greatly attenuates these wavelengths, and as a result, very little light shorter than 300 nm reaches the earth's surface. Although the UVA and UVB constitute a negligible portion of the sun's energy, they are primarily responsible for the pathological effects of solar radiation. Because of their high energies and availability, germicidal (UVC) lamps have historically been used for photochemical and photobiological studies. However, since the earth's atmosphere absorbs UV wavelengths below 300 nm, other components of the solar spectrum, such as UVB and UVA, are of more concern. For this reason, recent research has been directed more toward characterizing DNA damages produced by these biologically relevant solar wavelengths.

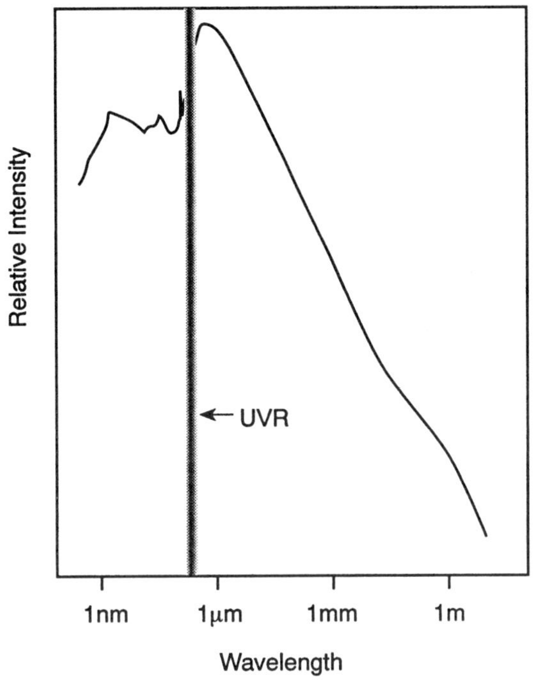

Figure 2. The electromagnetic spectrum: energy emitted by the sun for wavelengths between 1 Å and 100 m. The ultraviolet spectrum (190–400 nm) is indicated by the heavy vertical line.

3 DNA PHOTOPRODUCTS

3.1 Pathways for Dissipation of Absorbed Energy

Photon absorption rapidly converts a pyrimidine base to an excited state (Figure 3). This event promotes an electron in a filled bonding π orbital in the singlet ground state into a higher energy, empty π^*-antibonding orbital, thus initiating photoproduct formation. Formation of an excited base occurs within 10^{-12} s after photon absorption. Various pathways are then available for resolution of this unstable electronic configuration. The major pathway involves rapid dissipation of the energy of the excited singlet base to the ground state (10^{-9} s) by nonradiative transition or by fluorescence, yielding heat or light in the process. Second, the excited base can react with other molecules to form unstable intermediates (i.e., free radicals) or stable photoproducts. Finally, there is a low probability that intersystem crossing, a nonradiative pathway, can transfer a base from the excited singlet state to the excited triplet state. The lifetime of the triplet state is several orders of magnitude longer than the excited singlet state (10^{-3} s), increasing the chance of photoproduct formation. Formation of the triplet state can be greatly enhanced in the presence of sensitizers such as acetone or acetophenone. The cyclobutane pyrimidine dimer (I in Figure 4) forms through the excited triplet state, whereas other photoproducts form by other mechanisms.

The predominant photoproduct induced by UVR is the cyclobutane pyrimidine dimer (I) linked by a cyclobutyl ring between the carbons in the 5- and 6-positions of adjacent thymine (T) or cy-

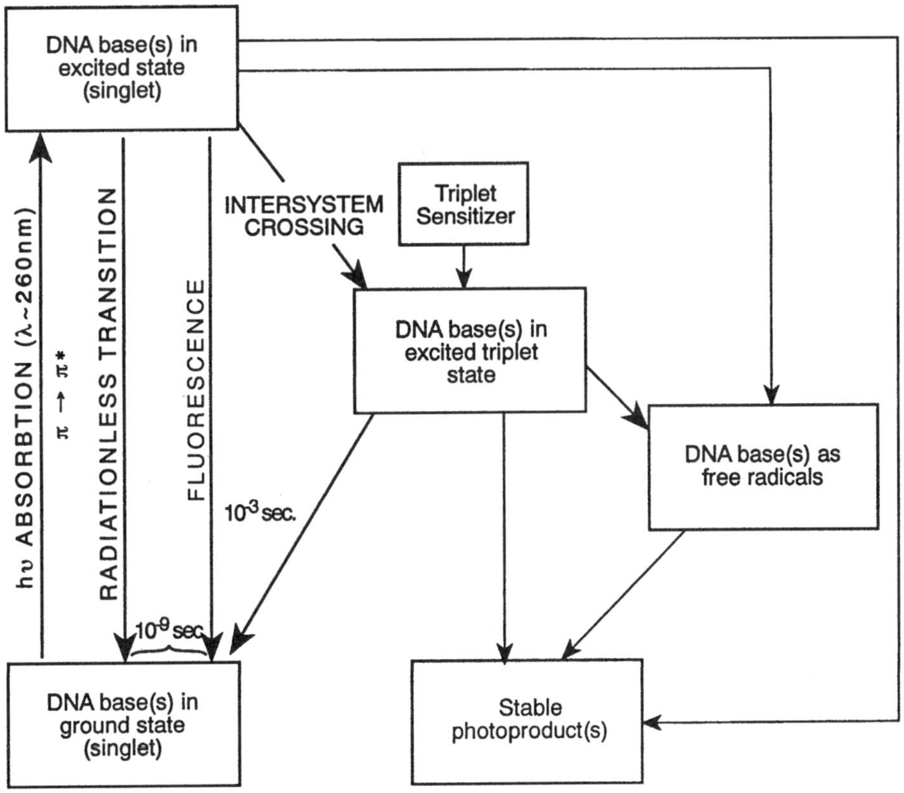

Figure 3. Pathways of dissipation of energy absorbed by UVR: schematic of the routes of UVR absorption and resolution leading to photoproduct formation.

tosine (C) bases (Figure 5). Of the four possible cyclobutane dimer combinations (T<>T, C<>T, T<>C, and C<>C) the thymine–thymine dimer occurs with greatest frequency at 40–60% of the sum of cytosine-containing dimers. At wavelengths above 300 nm, however, photochemical alterations of cytosine, rather than thymine, may account for most biological effects. The primary reason for this is that at wavelengths below 300 nm, saturation of cyclobutane dimers containing cytosine (C-Pyr) occur at a lower fluence and at a lower level than of thymine homodimers (T<>T). Hence, whereas T<>T homodimers form at 1.5-fold *greater* frequency than cytosine-containing dimers after UVC irradiation, they occur at one-third *less* the frequency of cytosine-containing dimers after UVB irradiation.

The (6-4) photoproduct (II in Figure 4) is the second most prevalent type of UV damage, occurring at 10–30% of the frequency of the cyclobutane dimer. This dimer links C-6 of a 3′-pyrimidine (either cytosine or thymine) with C-4 of a 5′-pyrimidine (primarily cytosine) by way of an unstable oxetane or azetidine intermediate. Un-

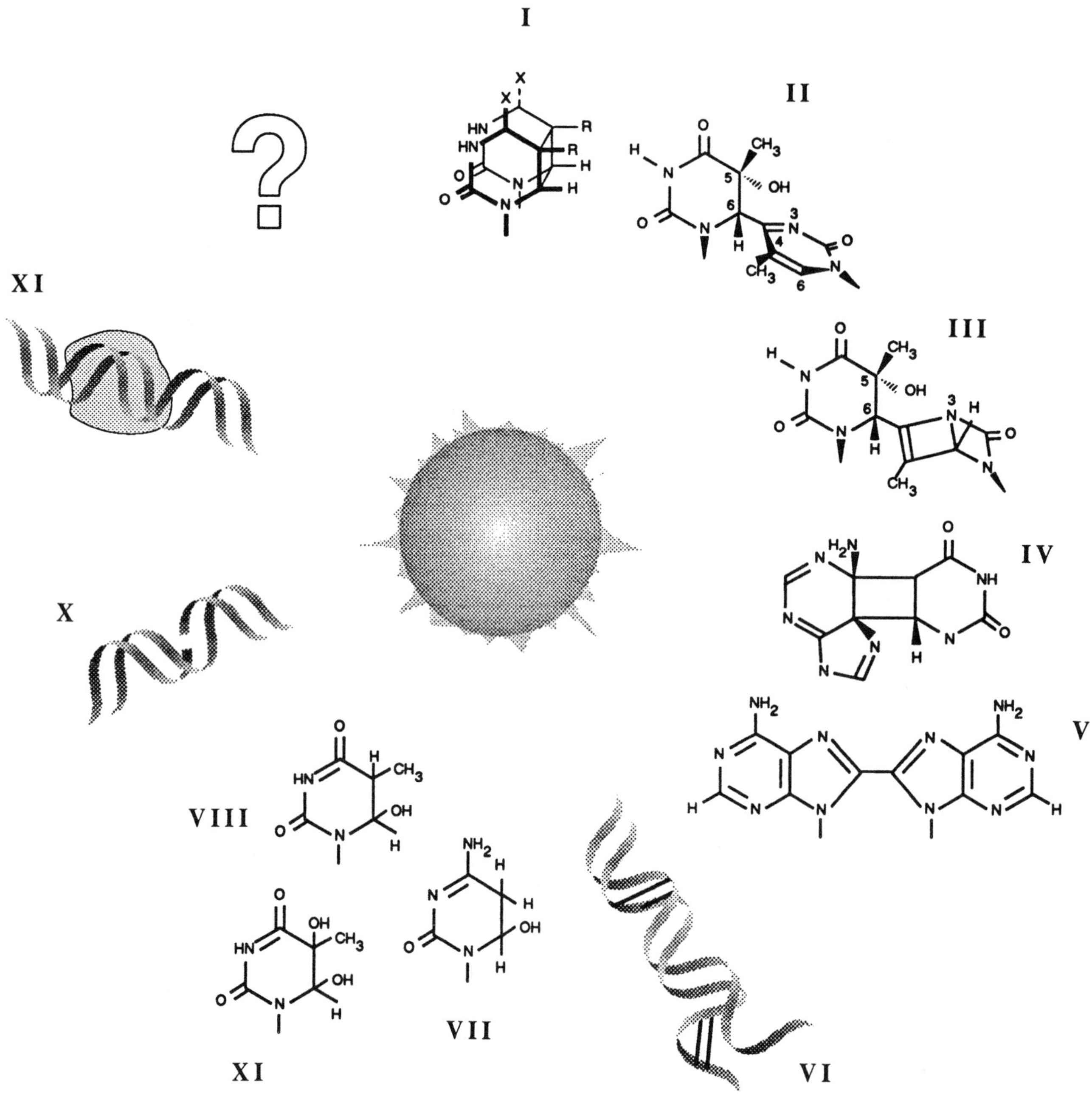

Figure 4. DNA photoproducts: types and structures of known photoproducts induced in DNA by solar UV radiation. I, cyclobutane dimer; II, (6-4) photoproduct; III, Dewar pyrimidinone; IV, thymine–adenine dimer; V, 8,8-adenine dehydrodimer; VI, inter- and intrastrand DNA cross-links; VII, 6-hydroxy-5,6-dihydrocytosine; VIII, 6-hydroxy-5,6-dihydrothymine; IX, 5,6-dihydroxy-5,6-dihydrothymine (thymine glycol); X, single-strand break; XI, DNA—protein cross-link.

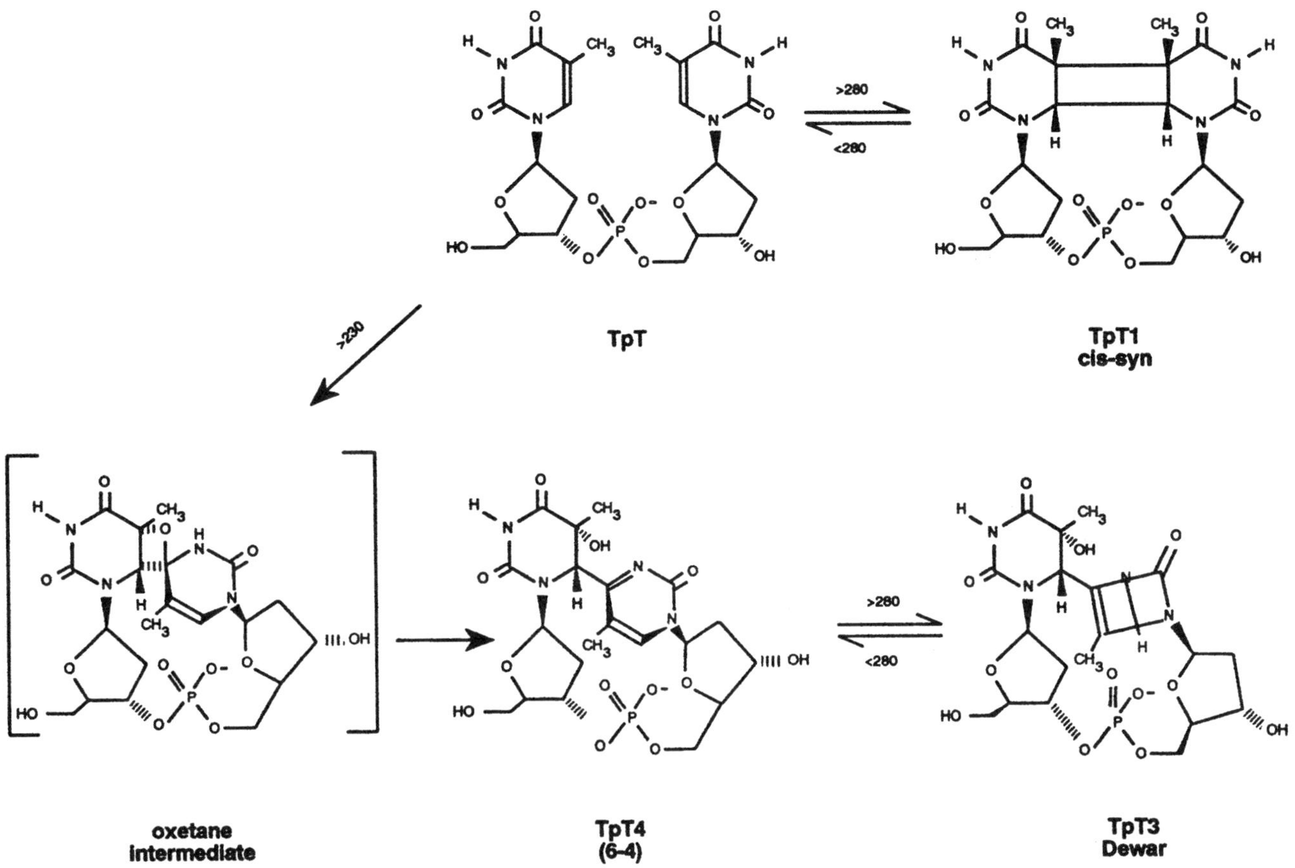

Figure 5. Reaction scheme for the formation of the major UV photoproducts: photochemical reactions in a dipyrimidine DNA sequence leading to the formation of the cyclobutane dimer (TpT1), (6-4) photoproduct (TpT4), and Dewar pyrimidinone (TpT3). [Redrawn from Taylor and Cohrs, *J. Am. Chem. Soc.* 109:2834–2835 (1987), with permission.]

like cyclobutane dimers, the formation of (6-4) photoproducts does not proceed via an excited triplet state (Figure 3), hence is less dependent on the photochemical behavior of thymine bases in DNA. Infrared spectroscopic and nuclear magnetic resonance analyses of the (6-4) photoproduct indicate that the plane of the 3′-cytosine base is shifted 90° relative to the 5′-thymine.

As shown in the reaction scheme for dimer formation (Figure 5), absorption of UVB light by the (6-4) photoproduct results in formation of its photoisomer, the Dewar pyrimidinone (III in Figure 4). Rare dimers containing purine, such as thymine–adenine dimer (IV) and the 8,8-adenine dehydrodimer (V) have also been isolated from heavily irradiated DNA. Dimerizations between nonadjacent pyrimidines either on the same strand or on opposite strands can produce intra- or interstrand cross-links (VI). It should be kept in mind that the relationship between photoproduct frequency and its biological effect depends on the intrinsic capability of that lesion to kill cells or induce mutations. Hence, even though a photoproduct may be very rare, its structure and location may elicit a potent biological effect.

Compared to photoproducts involving two pyrimidines, monobasic damage occurs in UV-irradiated DNA at very low frequencies (≈1–2% of dimers). Because of its instability and lability to the analytical procedures used to quantify DNA damage, the photobiology of cytosine photohydrates (VII) has been difficult to study. It should be kept in mind, however, that the half-life of the cytosine photohydrate is not trivial (i.e., 24 h at room temperature). Hence, this lesion may indeed pose a viable biological threat to a proliferating cell system. Recent data using sequencing analysis indicate that saturated thymine photoproducts (VIII and IX), such as the thymine glycol, do not occur in UVC- or UVB-irradiated DNA at a significant rate, hence do not appear to contribute significantly to the biological effects of UVR.

In contrast to direct induction of DNA damage by UVC light, UVA and, to a lesser extent, UVB radiation produce damage indirectly through highly reactive chemical intermediates. Similar to ionizing radiation, UVA generates oxygen and hydroxyl radicals, which in turn react with DNA to form monomeric photoproducts such as pyrimidine photohydrates and photo-oxidation products, as well as strand breaks and DNA–protein cross-links (X and XI in Figure 4). Although the number of cytosine photohydrates compared to cyclobutane dimers and (6-4) photoproducts remains constant throughout the UVR, the relative proportion of single-strand breaks and protein–DNA cross-links increases significantly in DNA irradiated with UVB and, especially, UVA light.

3.2 Genomic Distribution of DNA Damage

Cell killing by UV light is thought to result from DNA damage that inhibits DNA synthesis or transcription of genes essential to cell growth and metabolism. Recent advances in biotechnology have greatly increased our understanding of the diverse nature of DNA damage in mammalian systems, particularly regarding the spatial and temporal distribution of UV photoproducts in mammalian chromatin. Photochemical and biochemical data indicate that the distribution of UVR damage is nonrandom, being determined by DNA sequence and DNA–protein interactions.

Using biochemical and immunological analyses, (6-4) photoproducts have been quantified at 23% the level of cyclobutane dimers in purified DNA irradiated in vitro. However, the relative occurrence of (6-4) photoproducts and cyclobutane dimers varies widely, depending on base content of DNA and chromatin structure. Photoproduct yield increases with thymine content (i.e., increased AT:GC ratio); for instance, the induction of (6-4) photoproducts in *Clostridium perfringens* DNA (70% AT:GC) > *Escherichia coli* DNA (50% AT:GC) > *Micrococcus luteus* DNA (30% AT:GC) (Figure 6). Using DNA sequencing techniques, the incidence of (6-4) photoproducts and cyclobutane dimers in fragments of the *lacI* and *supF* genes was shown to be highly variable. Particularly, although thymine–thymine cyclobutane dimers were the major lesions detected, (6-4) photoproducts occurred with 100-fold greater frequency than dimers at certain dipyrimidine sites. In sequenced DNA fragments, (6-4) photoproducts were seen to form preferentially at thymine–cytosine dipyrimidines, followed by cytosine–cytosine and thymine–thymine sites.

The distribution of UV photoproducts is influenced by secondary structure mediated through DNA–protein interactions; chromatin structure modulates sites of cyclobutane dimer induction in core nucleosome DNA. Using the $3' \rightarrow 5'$ exonuclease activity of T4 DNA polymerase to map UV photoproducts, it was shown that these products do not occur randomly along the DNA helix, but are situated with a periodicity of 10.3 bases. Sites of maximum photoproduct induction in core DNA correspond to sites where the phosphate backbone is farthest from the core histone surface. The limited flexibility of the DNA helix resulting from histone–DNA interactions is thought to inhibit photoproduct formation.

The relative induction of (6-4) photoproducts and cyclobutane dimers has been measured in nucleosomal (core) and internucleosomal (linker) DNA in UV-irradiated human chromatin. Cyclobutane dimers form in equal amounts per nucleotide in core and linker DNA, whereas (6-4) photoproducts occur with six-fold greater frequency per nucleotide in linker DNA. These data indicate that (6-4) photoproduct formation is strongly influenced by DNA–protein interactions, occurring more readily in DNA that is less tightly bound to nucleosomal proteins.

Distribution of (6-4) photoproducts in chromatin may also be de-

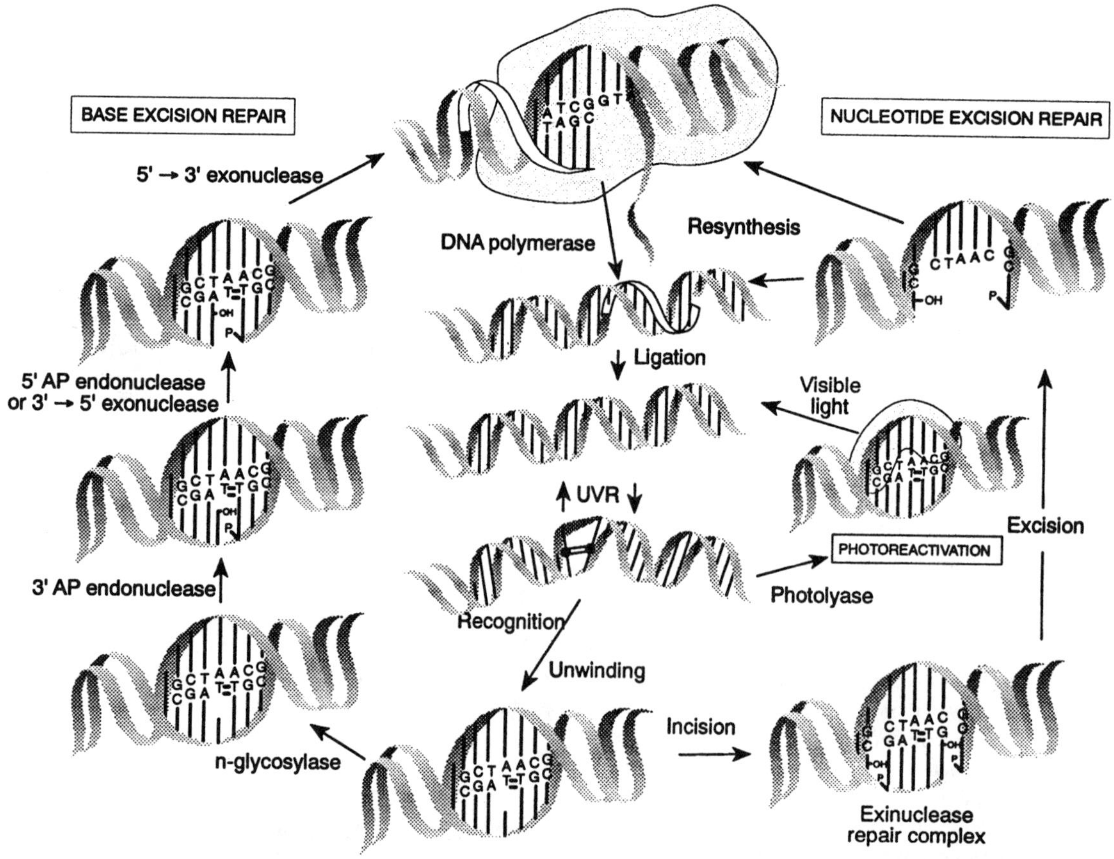

Figure 6. DNA repair pathways: enzymatic pathways for the repair of UVR damage in DNA. Shown are enzymatic photoreactivation, base excision repair, and nucleotide excision repair: T=T, cyclobutane dimer; G, guanine; A, adenine; C, cytosine; T, thymine.

termined by base modifications associated with metabolic activity. Sequencing analysis of DNA irradiated in vitro has shown that methylation of the 3′-cytosine of a (6-4) photoproduct inhibits its formation in isolated DNA. These data have recently been supported by studies of genomic DNA irradiated in vivo using the ligation-mediated polymerase chain reaction (LMPCR) to map (6-4) photoproducts at single-nucleotide resolution. Analysis of the human phosphoglycerate kinase gene promoter region showed that the frequency of (6-4) photoproducts differs in DNA derived from active and inactive X chromosomes. It was suggested that inactivation of the X chromosome by cytosine methylation results in inhibition of (6-4) photoproduct formation.

Recent studies indicate that (6-4) photoproducts may be more readily induced in actively transcribing genes compared to nonexpressed genes or the overall genome in human fibroblasts. Preferential induction of (6-4) photoproducts in functionally active regions of chromatin would serve to increase the lethal potential of this type of photodamage. Indeed, as discussed later, repair of these lesions may be facilitated by their predominance in nonnucleosomal (metabolically active) regions of the genome; DNA damage in more open chromatin conformations may be more accessible to excision repair enzymes than damage induced in more compact regions.

4 ANALYSIS OF DNA PHOTODAMAGE AND REPAIR

4.1 DNA Damage

Methods for quantifying UV damage in DNA rely on (1) separation of modified bases by chromatography, enzymatic or biochemical incision of DNA at sites of photoproducts, or antibody binding to structural damage in DNA. Damage caused by UVR was first detected in DNA nearly 30 years ago by means of paper chromatography. Chromatographic measurements require that DNA photoproducts be labeled during cell proliferation by incorporation of a radioactive pyrimidine precursor, such as [^{3}H]thymidine. After UV irradiation, the DNA is extracted, purified, and reduced to individual bases by acid or enzymatic hydrolysis. Photoproducts are then separated from unmodified bases by two-dimensional paper, thin-layer, or high performance liquid chromatography.

Photodamage can also be measured by incising DNA at sites of damage and quantifying the resultant single-strand breaks by alkaline gradient centrifugation or gel electrophoresis. Sites of cyclobutane dimers can be cleaved with UV endonuclease V from T4 phage, an enzyme that specifically recognizes this type of damage. UvrABC exinuclease, a partial excision repair complex purified from *E. coli* (Figure 7), cleaves DNA a few bases to either side of

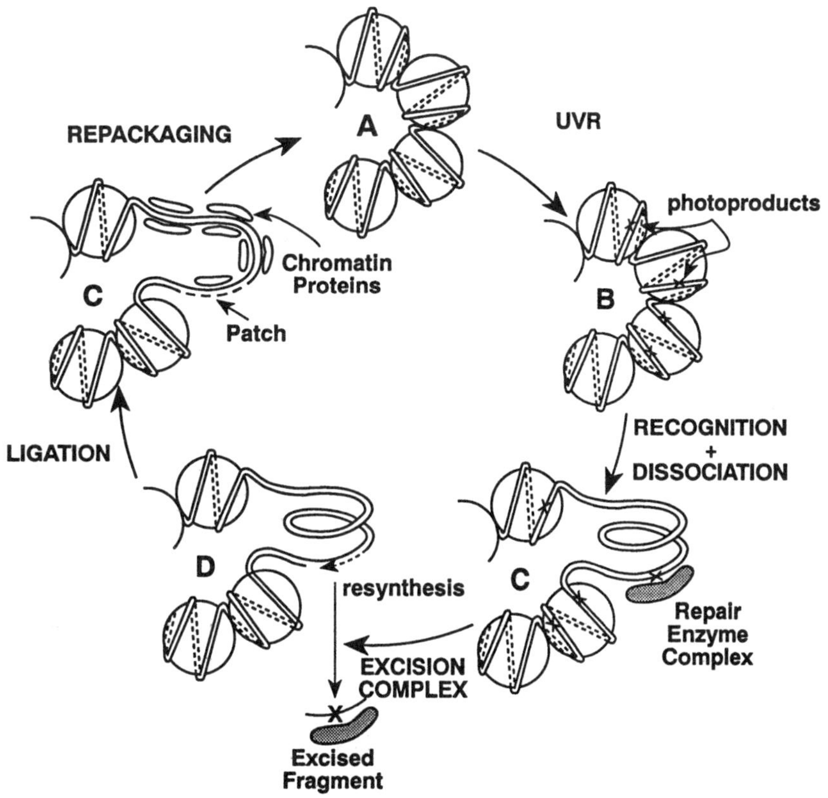

Figure 7. Excision repair in mammalian chromatin: repair is initiated by photoproduct recognition and dissociation of damaged DNA from nucleoproteins to make the DNA more accessible to repair enzymes. This is followed by the sequential incision by an endonuclease, excision and resynthesis by a DNA polymerase, and sealing of the patch by a polynucleotide ligase. The repaired segment is finally reassembled and repackaged in the original nucleosomal structure. [Redrawn from Cleaver and Mitchell (1993).]

the photoproduct. This broad-spectrum enzyme has been used to measure overall levels of DNA damage in UV-irradiated DNA and noncyclobutane dimer damage remaining after removal of cyclobutane dimers by enzymatic photoreactivation. Endonuclease III purified from *E. coli* has been used to cleave DNA at cytosine and thymine photohydrates as well as unidentified photoproducts occurring at sites of (6-4) photoproducts and modified purines.

Nonenzymatic procedures have also been used to incise DNA at sites of photodamage. The (6-4) photoproduct can be converted to the alkali-labile Dewar pyrimidinone by extended secondary irradiation with UVA light (>320 nm) (Figure 5). Prolonged incubation of the Dewar photoisomer in mild alkali results in a strand break at the site of the original damage. Photoinduced alkali-labile sites (PALS) can then be quantified by the same techniques used to measure sites of enzyme digestion. As with the chromatographic techniques, analyses of photodamage as strand breaks requires radioactive labeling to visualize and quantify DNA on alkaline gradients or agarose gels. Although these techniques are more sensitive than chromatography, their sensitivity depends on the maintenance of high molecular weight DNA throughout the purification and analysis.

Recently, the following techniques have been developed to explore the fine structure of UV damage induction and repair using strand scission.

1. Sequencing gels reveal sites of photoproducts manifested as strand breaks after enzymatic or biochemical cleavage. Such approaches have permitted the mapping of the sequence specificities of cyclobutane dimers, (6-4) photoproducts, and cytosine photohydrates induced by UVC and UVB light.
2. Southern blot hybridization of DNA restriction fragments separated by agarose gel electrophoresis can be used to measure photodamage in specific sequences, such as genes. Incision at sites of photodamage results in reduced intensity of gel bands hybridized to specific radioactive probes. This technique has revealed important information regarding the distribution of DNA repair in active and inactive regions of the mammalian genome.
3. Ligation-mediated polymerase chain reaction (LMPCR) has been used to analyze the distribution of photoproduct formation and repair at the sequence level in a gene promoter irradiated in vivo with low UV fluences. This technique has revealed important information regarding the link between DNA photodamage and the induction of skin cancer. Analysis of DNA repair rates at specific sites in the *p53* tumor suppressor gene has shown that seven of eight hot spots for mutation coincide with "slow spots" for photoproduct excision.

Over the past 20 years, antisera raised against UV-irradiated DNA have proven to be powerful and versatile reagents for the study of various photochemical and photobiological phenomena. Polyclonal and monoclonal antibodies have been produced which bind a variety of forms of photodamage, including the cyclobutane dimer, (6-4) photoproduct, Dewar pyrimidinone, and thymine glycol. A variety of immunological assays have been developed and have provided unique insights into photochemical and photobiological problems. Prominent within the extensive arsenal of immunological approaches adapted to the analysis of DNA damage and repair are quantitative immunofluorescence, immunoprecipitation, enzyme-linked immuoassays, radioimmunoassays, and mobility shift immunoassays. Each technique has its own unique attributes and applications. Unlike chromatographic techniques, immunological assays do not require chemical or enzymatic degradation of DNA prior to analysis; unlike endonucleolytic assays, their sensitivities do not depend on the molecular weight (or purity) of sample DNA.

4.2 DNA Repair

Removal of individual photoproducts can be assayed directly using the techniques described earlier. Techniques have also been developed to monitor various stages of the excision repair process. Transient incisions associated with the initial step of excision repair (Figure 6) can be measured by the technique of alkaline elution. Alkaline lysis of cells on filters results in denaturation of the DNA, which then elutes at a rate dependent on its molecular weight. Smaller DNA fragments containing more incisions elute faster than larger fragments; by measuring the molecular weight of the eluting fragments, the rate of repair incision can be estimated. Incorporation of base analogues that result in premature termination of repair synthesis (e.g., 1-β-D-arabinofuranosylcytosine) causes strand break accumulation. To discriminate between strand breaks associated with DNA repair and with DNA replication, semiconservative DNA synthesis can be inhibited by specific inhibitors (e.g., hydroxyurea), or nonreplicating senescent (G_0) or terminally differentiated nonproliferating cell populations can be used. Studies with normally UV-resistant mammalian cells show a rapid accumulation of strand breaks within the first few (0–15) minutes after UV irradiation followed by gradual restoration to the original molecular weight within an hour.

As with incision assays, protocols for measuring the resynthesis stage of excision repair (Figure 6) utilize cells in which normal DNA replication has been inhibited. Unscheduled DNA synthesis (UDS) provided the first direct evidence for the DNA excision repair process in mammalian cells. Cells were grown for a short period (pulsed) in [^{3}H]thymidine and then examined autoradiographically (as silver grains on a photographic plate). Cells undergoing DNA synthesis displayed intense labeling of their nuclei, whereas nonreplicating cells undergoing repair (unscheduled) synthesis showed much reduced labeling. Repair synthesis can also be determined using buoyant density centrifugation of DNA. In this technique the base analogue 5-bromodeoxyuridine (BUdR) is incorporated into DNA in place of thymidine. Regions containing this base have a higher density than the surrounding DNA. Hence, DNA that has replicated semiconservatively (i.e., produced one entire new strand) migrates as a hybrid density band on cesium chloride gradients. Repair synthesis occurring in very small patches (≈30 bases) migrates to a position of normal density DNA. The amount of radioactivity (^{3}H-BUdR) incorporated into unreplicated DNA is a measure of the total amount of repair synthesis.

5 DNA DAMAGE TOLERANCE MECHANISMS

The amount of photodamage present in cellular DNA at any given time depends not only on the amount of UV light absorbed by the DNA but also on the ability of the organism to repair the damage. UV damage tolerance mechanisms vary and may include combinations of photoprotection and DNA repair. The DNA repair strategy utilized depends on the type of photoproduct encountered and the organism at risk. DNA repair is composed in whole or part of at least two well-studied mechanisms: photoenzymatic repair and nucleotide or base excision repair (Figure 7).

5.1 Photoprotection

Nearly all organisms have behaviors or natural features that lower exposure of DNA to solar UVR and reduce the amount of photodamage. Human behaviors include wearing clothes, hats, sunglasses, and sunscreens, and the general habit of living and working indoors. Many plants and animals have similar strategies. Biological components such as bark, cuticles, skin, fur, feathers, scales, and shells can strongly attenuate or eliminate the absorption of UVR by internal areas of cells or organisms. Habitat selection is also a significant factor. Shielding by terrain, foliage, water, or another organism can reduce UVR exposure. Actively seeking shade to avoid direct sunlight is also a common biological response.

Morphology may also affect the amount of radiation absorbed by cellular DNA. Small cells and organisms are more susceptible to the physical damage caused by ionizing radiation. Recent studies on marine plankton indicate that UVR damage may likewise be mediated by morphological factors. The amount of DNA damage induced in Antarctic phytoplankton by UVB light correlates with the size of the individual species. In smaller cells the distance between the cell surface and nuclear DNA is less than in larger cells. The shortened light path reduces refraction and absorption by cytoplasmic components and increases the amount of UVR reaching the nucleus. Similar principles can be applied to multicellular organisms. In humans, for example, the ratio of exposed surface area to internal tissues and organs is reduced compared to smaller animals.

Several mechanisms for biochemical protection from UVR occur in nature. Pigmentation in the outer layers of plants and animals modulates the intensity and wavelength distribution of light reaching internal components. In some cases, light exposure induces the production of pigments. For example, the UVR-absorbing compounds melanin and anthocyanin are produced in human skin and plants, respectively, after exposure to solar UV. Melanin has multiple functions that can be either harmful or beneficial to the organism. On the one hand, it can be photoprotective, acting as a natural sunscreen, although not a very good one, and as a scavenger of active chemical species produced by solar UVR, although again not a very effective one. On the other hand, melanin can itself produce active radical species that can damage DNA, and it has the capacity to bind to drugs in ways that can either benefit or harm the cell. The beneficial role of melanin in humans is best illustrated by the observed protection against sunburn and reduction in skin cancer in dark-skinned peoples. In black skin, melanosomes, the cellular compartments in which melanin is sequestered, are oval, single and densely packed with melanin and clustered around the nucleus. UVR is blocked from the nucleus and free radicals are absorbed. In contrast, in white skin, melanosomes are round, aggregated, and lightly melanotic. UVR can bypass the clusters to enter the nucleus.

In addition to pigments, colorless UV-absorbing compounds have also been identified as possible UV-protective chemicals. Concentrations of flavenoids in terrestrial plants, mycosporine amino acids (MAAs) in fungi, and mycosporine-like compounds in marine organisms all correlate with UV exposure history. More than 25 different mycosporine-like compounds have been identified in marine algae and invertebrates. These molecules display different UVA and UVB absorption profiles, with several compounds present in a single species. Combinations of these "natural sunscreens" provide broadband UV coverage and optimize protection from solar UVR. Because of these attributes, the compounds resembling mycosporins are being investigated for human applications.

5.2 Photodependent Repair

Photoenzymatic repair (PER) is a light-dependent process involving the enzyme-catalyzed reversal of cyclobutane dimers. The first DNA repair mechanism recognized, PER was discovered simultaneously by Albert Kelner and Renato Dulbecco in 1949. PER involves binding of a low molecular weight protein to a cyclobutane dimer followed by absorption of photons within the UVA/visible range of light, reversal of the damage, and, finally, release of the enzyme. PER appears to be a ubiquitous repair mechanism, widespread throughout the plant and animal kingdoms (Table 1).

Although it is a relatively simple enzymatic process, PER displays much diversity in various biological systems. There are numerous organisms in which it has not been found, including several bacteria, such as *Haemophilus influenzae, Diplococcus pneumoniae, Bacillus subtilis,* and *Micrococcus radiodurans,* some algae

Table 1 Plants and Animals Capable of Removihng Cyclobutane dimers by Enzymatic Photoreactivation (PER)

Group	Species	Common Name
Cyanobacteria	*Plectonema boreanum*	Blue-green alga
	Anacystis nidulans	Blue-green alga
Schizomycophyta	*Escherichia coli*	Colon bacteria
	Streptomyces griseus	Soil actinomycete
Eumycophyta	*Saccharomyces cerevisiae*	Baker's yeast
	Neurospora crassa	Bread mold
Euglenophyta	*Euglena gracilis*	[a]
Angiospermae	*Phaseolus vulgaris*	Pinto bean
	Phaseolus lunatus	Lima bean
	Arabidopsis thaliana	[a]
	Zea mays	Maize
	Nicotiana tabacum	Tobacco
Gymnospermae	*Gingko biloba*	Gingko tree
Protozoa	*Paramecium aurelia*	Paramecium
	Tetrahymena pyriformis	[a]
Mollusca	*Physa* sp.	Pond snail
Echinodermata	*Arbacia punctulata*	Sea urchin
	Echinarachnius parma	Sand dollar
Arthropoda	*Angasta kiihniella*	Flower moth
	Gecarcinus lateralis	Land crab
	Artemia salina	Brine shrimp
	Homarus americanus	Lobster
Insecta	*Trichoplusia ni*	Cabbage looper
	Drosophila melanogaster	Fruit fly
Teleosts	*Haemulon sciuris*	Blue-striped grunt
	Pimephales promelas	Fathead minnow
	Anoptichthys jordani	Blind cave fish
	Tautoga onitis	Tautog
	Tautogolabrus adsperus	Cunner
	Xiphopherus variatus	Variable platyfish
	Carassius auratus	Goldfish
Amphibia	*Bufo marinus*	Cowflop toad
	Xenopus laevis	African clawed toad
	Rana pipiens	Leopard frog
Reptilia	*Terrapene carolina*	Box turtle
	Iguana iguana	Iguana lizard
	Gekko gekko	Gecko lizard
Aves	*Gallus gallus*	Domestic chicken
Marsupialia	*Didelphis marsupialis*	American opossum
	Caluromys derbianus	Woolly opossum
	Potorous tridactylis	Rat kangaroo
Mammalia	(*Homo sapiens ?*)	Human

[a]No common name

and higher plants, such as *Chaetoceros convolutus, Thalassiosira australis, Phaseolus aureus,* and *Haplopappus gracilis,* and the nematode *Caenorhabditis elegans.* The existence of PER in placental mammals, particularly in human skin, has been a matter of some debate for the past two decades. For most researchers, the demonstration of PER in humans has been elusive. What evidence exists appears to rely on unique media requirements in tissue culture experiments or on specific irradiation conditions in human skin. Some data suggest that the efficiency of PER in humans may vary widely between individuals and may be dependent on the presence and activity of specific cofactors required for either DNA binding or damage reversal.

Differences in action spectra, constitutive levels of photolyase, and the presence of required cofactors may account for the diverse PER efficiencies among organisms. The wavelength dependence for maximum PER efficiency is about 380 nm in *E. coli* and *Euglena,* about 400 nm in *Neurospora,* and about 440 nm in *Streptomycea griseus.* The lack of PER in many organisms and its evolutionary conservation in systems never exposed to sunlight (e.g., soil or enteric bacteria, blind cave fish) suggest that photolyase as well as PER may function in other molecular processes, such as excision repair. In support of this idea, it is been shown that *E. coli* photolyase stimulates excision repair in vitro, and a DNA-binding protein associated with excision repair in human cells has been identified with properties similar to a yeast photolyase. It has been shown that PER is inducible in fish cells both in culture and in vivo (in caudal fins). Preillumination with visible light for 8 hours prior to damage induction significantly predisposes cells to cyclobutane dimer removal by PER. These studies provided the first clear demonstration of an inducible repair system in eukaryotes. It has been suggested that the mechanism of enhanced photorepair may involve conversion of photolyase from an inactive to an active form.

With the development and application of immunological methods for detecting (6-4) photoproducts in DNA, questions have been raised about the specificity of PER for the cyclobutane dimer. It is evident that removal of cyclobutane dimers by PER can increase excision of noncyclobutane dimer damage. This phenomenon, termed *photoenhanced excision repair,* was first observed in frog cells and has since been shown to occur in diatoms, in fish, and in marsupial epidermis. The manner in which PER influences (6-4) photoproduct excision in these systems is unknown. A putative pathway involves removal of a competing substrate (i.e., cyclobutane dimers) followed by commitment of a common repair system to the task of (6-4) photoproduct removal. Another possibility, which has been recently demonstrated in insects, is that PER systems exist for the repair of damage other than the cyclobutane dimer, such as the (6-4) photoproduct.

5.3 Excision Repair

DNA excision repair is a genetically complex and phenotypically diverse system; its competency is dependent on the species and developmental state of the individual, as well as the type of photoproduct encountered and its genomic distribution. The excision repair pathway is thought to proceed by the following steps (Figures 6 and 7):

- recognition of the lesion as a structural distortion in the DNA helix
- unwinding or other activity to disassociate the lesion from chromatin proteins and provide accessibility to repair enzymes
- incision of the DNA backbone at or near the site of the lesion
- excision and resynthesis of the DNA around the damaged site
- ligation of the single-strand nick remaining after disengagement of the DNA polymerase complex

This process is very versatile, correcting various classes of chemical damage in addition to those induced by UV radiation.

Excision repair may proceed by one of two routes (Figure 6). *Base excision repair* is initiated by enzymatic recognition of the lesion and scission of the bond between the damaged base and its associated deoxyribose sugar, a process called aglycosylation. Examples of *n*-glycosylases include endonuclease III from *E. coli,* which repairs photohydrates, and endonuclease V from T4 phage, which has the unique attribute of recognizing the 5′-pyrimidine of a cyclobutane dimer as a modified base. In the latter case, cleavage of the *n*-glycosyl bond leaves a "dangling" dimer and an abasic site. After removal of the base damage, a 3′-AP endonuclease cleaves the phosphodiester bond 3′ to the abasic site. To prepare the strand for resynthesis, a 5′-AP endonuclease or 3′ → 5′ exonuclease associated with DNA polymerase digests the remaining abasic site. The damaged strand is removed and resynthesized predominantly by DNA polymerase α or δ, and the remaining nick is ligated to return the DNA duplex to its original state. The resultant repair "patch" of newly incorporated DNA is about 30 bases long in mammalian cells.

The *nucleotide excision repair* pathway recognizes a broad spectrum of UVR-induced photoproducts as well as other "bulky adducts" that may be induced by genotoxic agents. This process has been well studied in *E. coli* and is thought to be the primary excision repair pathway in eukaryotes as well. As with base excision repair, the nucleotide excision repair process is initiated by enzymes that recognize and bind the helical distortion created at the damaged site. An exinuclease complex is assembled around the lesion and cleaves the DNA on both sides of the lesion, leaving a gap. This gap is filled by a DNA polymerase, and the strand is ligated to restore the DNA duplex to its original integrity.

Figure 7 shows excision repair of damaged sites in mammalian chromatin. The first step involves mechanisms that recognize damage and dissociate nucleoproteins to make the DNA accessible to repair enzymes. After recognition and dissociation, the excision repair pathway proceeds as in Figure 6, followed by reassembly and repackaging of the chromatin complex. The actual number of damaged sites involved in excision repair at any one time is very small, comprising only about 1% of the lesions produced in DNA irradiated with a low UV fluence. Excision is, therefore, rate-limited by the enzymes involved in the early stages of DNA repair, which dissociate the DNA from the nucleosome scaffolding and provide accessibility to excision enzymes.

6 SPATIAL AND TEMPORAL DISTRIBUTION OF DNA REPAIR

The efficiency of the excision repair process is dependent on the organism, the type of damage encountered, and its location in the genome. Figures 8 and 9 present repair curves for cyclobutane dimer and (6-4) photoproduct excision for several organisms representing the plant and animal kingdoms. The initial rates and extent of cyclobutane dimer repair (Figure 8) show great diversity, ranging from very rapid removal in primitive eukaryotes, such as bread mold and yeast, to much slower repair in vertebrates such as fish,

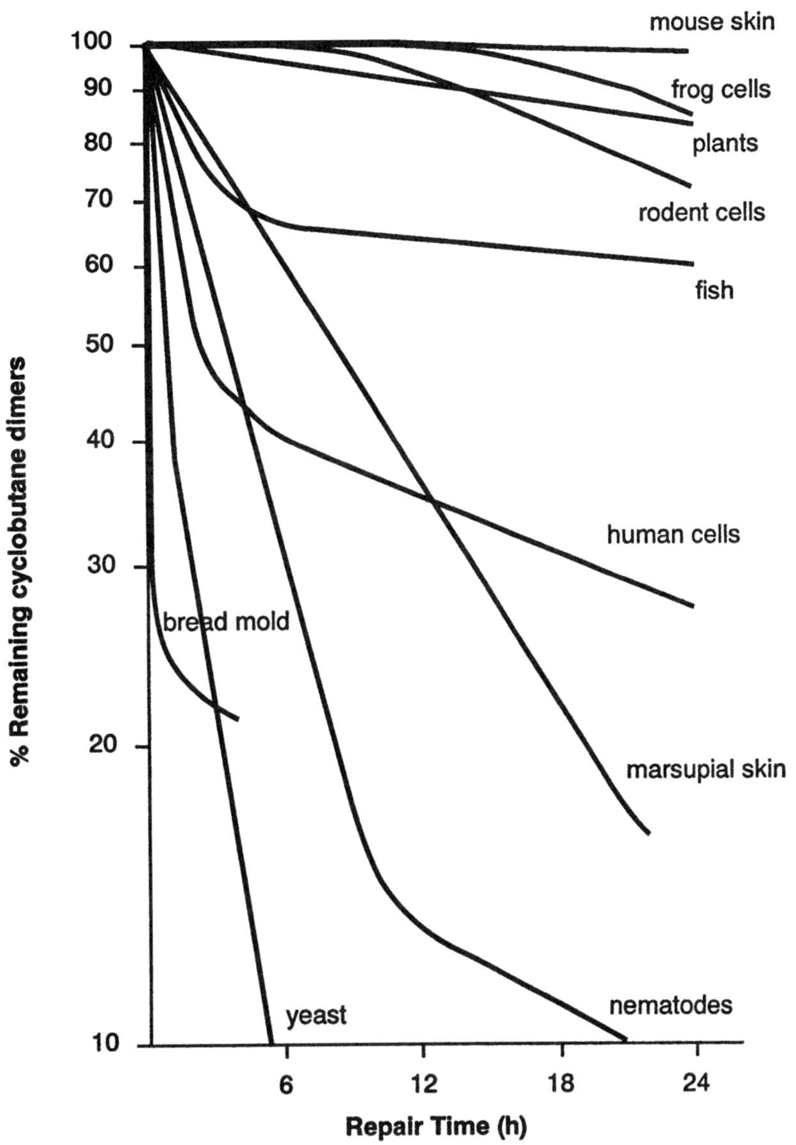

Figure 8. Kinetics of cyclobutane dimer repair: radioimmunoassay of DNA repair of cyclobutane dimers in various organisms, including *Neurospora crassa* (bread mold), *Saccharomyces cerevisiae* (yeast), *Caenorhabditis elegans* (nematode), *Monodelphis domestica* (marsupial), *Xiphophorus variatus* (fish), *Arabidopsis thaliana* (plant), and *Xenopus laevis* (frog).

frogs, and mammals. With few exceptions, the rate of photoproduct excision is much faster for the (6-4) photoproduct than for the cyclobutane dimer (Figure 9). Notable exceptions include the common bread mold (*Neurospora crassa*), the nematode *Caenorhabditis elegans,* and the epidermis of the marsupial *Monodelphis domestica.* Rapid repair of the (6-4) photoproduct may be a consequence of its structure and its nonrandom distribution in chromatin. The 90° orientation of the 3′ base of the (6-4) photoproduct (II in Figure 4) produces a segment of DNA with considerable distortion in the phosphodiester backbone, local denaturation, and disruption of normal base pairing. This structure is bound with 10-fold greater affinity by the uvrABC exinuclease of *E. coli* and blocks the progression of DNA polymerase about 20-fold more frequently than the cyclobutane dimer.

Rapid repair of the (6-4) photoproduct may also belie the structural and functional role of chromatin in DNA repair (Figure 9). Insight into the relationship between DNA repair and chromatin structure has come from studies using a yeast minichromosome as a model system. Regions of this chromosome containing unstable nucleosomes are more readily repaired than regions in which the nucleoprotein structure is more constant. These data imply that "open" chromatin configurations are more accessible and more repairable than compact, heterochromatic regions of the genome. As mentioned, (6-4) photoproducts are induced with sixfold greater frequency in linker DNA versus nucleosomal core DNA. The nonrandom induction of the (6-4) photoproducts in more accessible regions of chromatin may, therefore, facilitate its rapid removal by the excision repair complex.

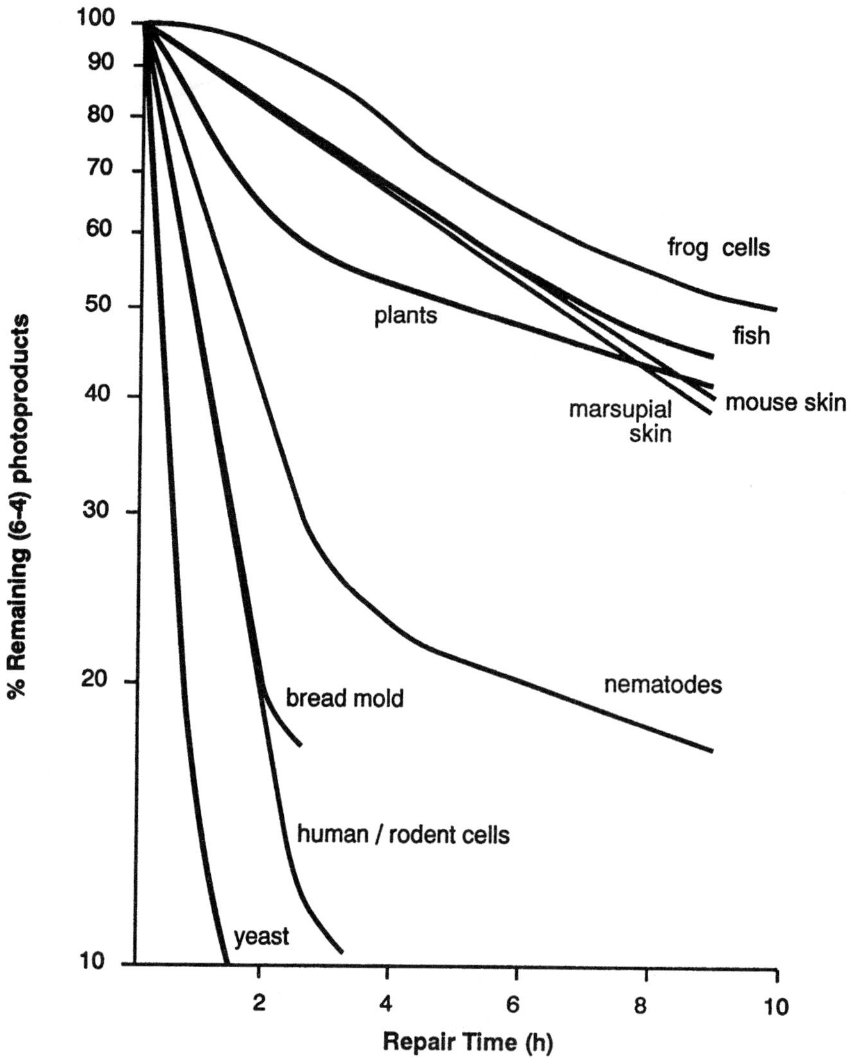

Figure 9. Kinetics of (6-4) photoproduct repair: radioimmunoassay of DNA repair of (6-4) photoproducts in various organisms (as in Figure 8).

In addition to its effect on photoproduct distribution, chromatin determines the efficiency of repair. The distribution of cyclobutane dimer excision in and around the dihydrofolate reductase (DHFR) gene is shown in Figure 10. The percentage repair at 8 hours is greatest at the 5′ end of the gene and decreases upstream and downstream from the transcribed region. The general conclusions are that actively transcribing genes are repaired more rapidly than nontranscribed genes and that the transcribed strand is repaired more rapidly than the nontranscribed strand. Several mechanisms have been proposed to account for these observations, the most recent of which involves displacement of a stalled RNA polymerase from the transcribed gene by transcription-coupled, helicase-like proteins (TFIIH, or transcription factor H for polymerase II, and Cockayne syndrome gene products for eukaryotes, TRCF for *E. coli*). The removal of the transcription complex by this factor thus renders the site of damage on the transcribing strand more accessible to the excision repair complex. In addition to transcription-coupled repair, changes in local chromatin structure that coincide with transcription may facilitate the repair process.

Preferential repair of cyclobutane dimers has been invoked to explain an outstanding enigma in the DNA repair field. Termed the "rodent paradox" (or "repairadox"), this mystery involves the ability of rodent cells to survive the lethal effects of UVR as well as human cells, even though few cyclobutane dimers are excised from total cellular DNA; DNA repair in rodent cells appears to be targeted largely to actively transcribing genes. This hypothesis presupposes that photodamage in essential genes is lethal and that damage in nontranscribing regions (i.e., the bulk of the genome) is benign. Indeed, the DNA replication machinery in rodent cells readily bypasses cyclobutane dimers. The distance between UV-induced termination sites in rodent cells correlates with the induction of (6-4) photoproducts, lesions that have been shown to be considerably more effective blocks to DNA polymerase in vitro. The observation that rodent cells repair the (6-4) photoproduct as efficiently as human cells may also explain the "rodent paradox," assuming that this lesion is the primary lethal lesion after UVR. It is probable that both preferential repair of essential genes and removal of blocks to DNA replication are responsible for cell survival.

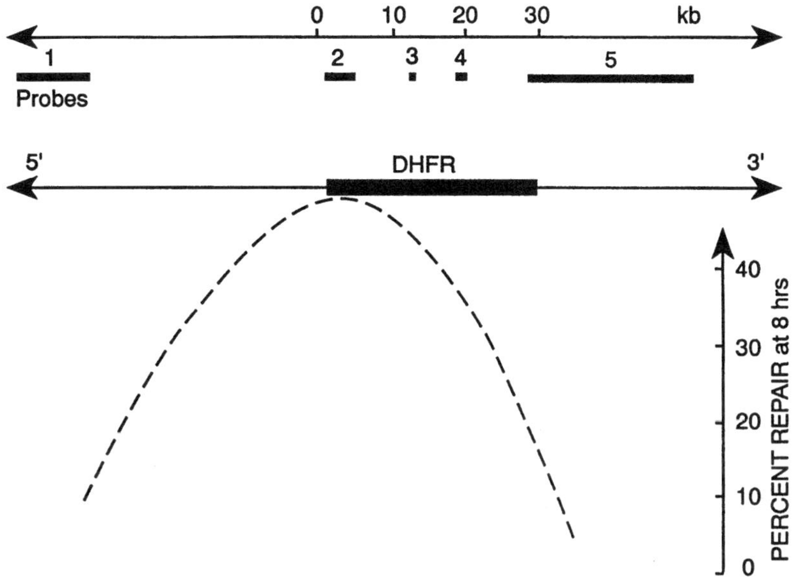

Figure 10. Preferential repair of cyclobutane dimers: schematic of the efficiency of DNA repair in relation to the dihydrofolate reductase (DHFR) gene in rodent cells. Various probes used to assay repair by Southern blot analysis are shown with respect to their location in and around the DHFR gene. [Redrawn from V. A. Bohr, *Dan. Med. Bull.* 34:309 (1989), with permission.]

7 DEVELOPMENTAL REGULATION OF DNA REPAIR

The efficiency of DNA repair depends on the developmental state of a cell, tissue, or organism. Both developmental and genetic attributes have made the nematode *C. elegans* a popular organism for probing many biological problems, including DNA repair. A striking phenotype of certain *C. elegans* mutants is the manifestation of stage-specific variation in UV hypersensitivity. For example, *rad-3* mutant nematodes are moderately hypersensitive when irradiated as embryos, but larvae are highly hypersensitive. Unlike wild-type nematodes, whose repair capacity remains relatively constant, the excision repair capacity of *rad-3* nematodes varies substantially throughout development. Although *rad-3* embryos display a reduced capacity for excision repair, larvae assayed 24 hours later are almost totally deficient in excision repair. The correlation between UV hypersensitivity and reduced excision repair capacity suggests that DNA repair in the nematode is developmentally regulated.

Evidence also suggests that DNA repair is developmentally regulated in higher eukaryotes. Excision repair of UV damage in cultured fish cells is very low, although significant excision repair is observed in primary embryonic cells. Similar regulation is seen in mice: exposure to photoreactivating light has no effect on cyclobutane dimer removal in adult mouse epidermal DNA or mouse fibroblasts in culture; yet such exposure significantly reduces the number of these lesions in the epidermis of neonatal mice, and excision (dark) repair of cyclobutane dimers is much more efficient in early embryos than in adult mice.

DNA damage resulting from UVR is more efficiently repaired in proliferating basal cells in rodent and human epidermis than in terminally differentiated keratinocytes. A compelling generality one can derive from the developmental pattern of DNA repair in organisms and tissues is that more rapidly proliferating systems (i.e., stem cells, embryos, fetuses, larvae) have increased capacities for self-correction. Hence, organisms most vulnerable to the lethal and mutagenic effects of genetic change are best able to tolerate it. DNA repair exhibits different developmental profiles in different organisms; thus, both ontogeny and phylogeny of DNA repair can vary substantially.

8 GENETIC REGULATION OF DNA REPAIR

Selection of UV-sensitive rodent cells and the study of human disorders symptomatized by sensitivity to sunlight have identified a large array of genetic loci that control the response of mammalian skin to UVR damage (Table 2). These loci are all characterized by significant increases in sensitivity to UVC or UVB radiation and include the following groups: excision repair cross-complementing (ERCC) series in rodent cells (currently 10 complementation groups), xeroderma pigmentosum (XP) (8 complementation groups), Cockayne syndrome (CS) (3 complementation groups), trichothiodystrophy (TTD) (3 distinct phenotypes), and basal cell nevus syndrome (BCNS).

Except for BCNS, all these disorders represent increased sensitivity to UVB and UVC wavelengths due to recessive mutations. A characteristic of BCNS, which is a dominant disorder, is a unique sensitivity to UVB radiation. The recessive disorders can be considered to be subsets within a large family of genes that regulate human cell DNA repair. These subsets are not mutually exclusive: CS complementation group C is identical to XP complementation group B, and TTD overlaps with group D. The XP group B corresponds to ERCC3 as well as to CS group B; XP group D corresponds to ERCC2.

The ERCC genes were identified by selection of UV-sensitive rodent cells. *ERCC1, ERCC2, ERCC3, ERCC5,* and *ERCC6* have been cloned (Table 2). *ERCC1* was the first human repair gene cloned by complementation of the UV-hypersensitive phenotype in a rodent mutant cell line. The *ERCC1* gene spans 15 kb of DNA, contains 10 exons, and encodes a protein of 297 amino acids. The protein contains a putative helix–turn–helix DNA binding motif and shares extensive sequence homology with the *RAD10* gene of the yeast *Saccharomyces cerevisiae* and with the *E. coli* UvrA and C proteins.

The *ERCC2* gene is 20 kb in length, encodes a protein of 760 amino acids, and corrects the excision repair defect in XP group D. It is the human homologue of the *S. cerevisiae RAD3* gene, which codes for an ATP-dependent helicase.

The *ERCC3* gene is approximately 45 kb in length, contains at least 14 exons, and encodes a protein of 782 amino acids. It complements the repair defect in XP group B and Cockayne syndrome and has been shown to be identical to the yeast gene homologue, *RAD25.* Like *ERCC2, ERCC3* contains sequence motifs suggesting that it functions as a helicase. All the *ERCC* genes appear to regulate the excision of DNA photoproducts, and *ERCC1* and *ERCC2* appear to correspond to important enzymatic functions of photoproduct recognition and removal.

The *ERCC5* gene is approximately 32 kb in length and encodes a 4.6 kb mRNA. An incomplete 450 amino acid ORF (open reading frame) is highly acidic, and sequence comparison has disclosed significant similarity with the yeast *RAD2 NER* gene. Microinjection of the *ERCC5* cDNA into fibroblasts of various XP complementation groups led to complete recovery of UV resistance in XP group G cells.

The *ERCC6* gene is approximately 85 kb in length and encodes two mRNAs of 5 and 7 kb. It encodes a protein of 1493 amino acids with putative chromatin- and nucleotide-binding domains and seven helicase "signatures." Like *ERCC2* and *ERCC3, ERCC6* contains helicase regions, making it the third putative helicase involved in mammalian excision repair. Transfection of *ERCC6* cDNA into Cockayne syndrome group B cells showed recovery of the UV resistance phenotype. CS group B cells lose the ability to repair DNA damage in transcribing strands, even though the ability to repair UV photodamage in the remainder of the DNA is unaffected.

Xeroderma pigmentosum is a rare autosomal recessive disease that occurs at a frequency of 1:250,000 in the United States. Affected individuals have sun sensitivity resulting in progressive degenerative changes in sun-exposed portions of the skin and eyes, often leading to neoplasia. Some XP patients have, in addition, progressive neurological degeneration. Cells from patients with XP excise DNA photoproducts at reduced rates of 0–90% of normal, except for the variant group, which has near-normal rates (Table 2). The existence of eight complementation groups in XP probably indicates that the initial step of excision repair involves cooperative action between multiple proteins that locate and bind DNA damage, unwind and modify the site before endonucleolytic cleavage, and are then displaced by helicase and polymerase.

Compared with normal cells, XP groups A and D are very sensitive to the lethal effects of UV light and are unable to excise the two major photoproducts in DNA, the cyclobutane dimer and (6-4) photoproduct. The XP group A protein has been partially purified and has an apparent molecular weight of 40–45 kDa. The XPAC (XP group A complementing) gene has been cloned and encodes a protein of 273 amino acids with a very acidic region, a zinc finger motif, and a putative nuclear location signal. Group C is the largest group and is often referred to as the common or classic form of the disease. Cells have low levels of repair (10–20% normal) and are less sensitive to killing by UVR than cells in groups A and D. As mentioned, XP group C shows normal preferential repair. The XP group C gene has been cloned *(XPCC)* and encodes a 3.8 kb mRNA and a highly hydrophilic protein of 823 amino acids. The amino acid sequence of the *XPCC* cDNA shows significant homology to the yeast *RAD4* gene product. XP group E cells display an intermediate phenotype, both in their UV sensitivity and their ability to excise UV damage.

Table 2 Complementation Groups in UV-Sensitive Human and Rodent Cell Lines

Origin	Cell Line	Group	Relative Repair	Cloned Gene[a]	Human Chromosome Location[a]	Size of Gene (kb)[a]	Activity of Gene Product
Human	Xeroderma pigmentosum						
		A	<5	*XPAC*	9q34	*	DNA-binding protein?
		B	<5	*ERCC 3*	2q21	*	DNA helicase
		C	5–20		*	*	DNA helicase
		D	25–50	*ERCC 2*	19q13	*	DNA helicase
		E	50	*	*	*	DNA-binding protein
		F	20	*	15	*	Excision nuclease
		G	<5	*ERCC 5*	*	*	Single-strand nuclease
		Variant	100	*	*	*	DNA replication protein?
Rodent	Chinese hamster ovary						
		1	<5	*ERCC 1*	19q13	15–17	Excision nuclease?
		2	<5	*ERCC 2*	19q13	20	DNA helicase
		3	<5	*ERCC 3*	2q21	45	DNA helicase
		4	<5	*ERCC 4*	16	*	Excision nuclease?
		5	<5	*ERCC 5*	13q	32	Single-strand nuclease
		6	*	*ERCC 6*	10q11	100	DNA helicase

[a]Asterisk indicates data not available.

The cloning of DNA repair genes from yeast, rodent, and human cells has greatly increased our understanding of the functional relationships between the various components of the nucleotide excision repair process in eukaryotes. The initial recognition of the DNA damage involves at least two DNA-binding proteins, XPA and XPE. Structural alterations in chromatin required before excision can proceed are performed by DNA helicases. Which gene products are involved in this step depends on whether the damage is located in a transcribing or nontranscribing region of the genome. Transcription-dependent repair requires CSA, CSB, and a transcription factor complex containing several repair genes, namely, TFIIH. The protein XPA is involved solely in excision repair in nontranscribing regions, since these cells repair damage in transcribing genes as well as normal cells. The damaged region is excised in a 30-nucleotide patch by the action of two nucleases: one is a complex of *ERCC1, ERCC4, ERCC11,* and *XPF* that act on both single-stranded and duplex, supercoiled DNA; the other is the *XPG* gene product, which can digest only single-stranded DNA. After the damaged patch has been excised, the vulnerable single-stranded region is protected by single-strand binding protein RPA and a complementary strand is then synthesized by proliferating cell nuclear antigen (PCNA), DNA polymerase, and ligase.

9 BIOLOGICAL CONSEQUENCES OF UVR

Air pollution has resulted in global decreases in stratospheric ozone concentrations and an increase in the amount of harmful solar radiation reaching the earth's surface. The effects of increased UVB radiation on the human population are complex: the obvious and direct consequences include increased skin cancer and accelerated aging; less obvious and more indirect effects include deterioration of world food crops and natural ecosystems.

9.1 UVR Carcinogenesis

Sunlight is a ubiquitous carcinogen, responsible for much of the skin cancer in the human population today. The evidence for this is compelling: (1) basal and squamous cell carcinomas appear on parts of the body exposed to the sun (e.g., the face and trunk in men, face and legs in women); (2) tumor incidence and mortality increase with decreasing latitude, corresponding to exposure; (3) skin cancers are lower in dark-skinned populations as opposed to lighter skinned peoples; and (4) tumor incidence increases with occupational exposure, such as that experienced by ranchers and fishermen.

Carcinogenesis often appears to proceed by a multistep process, the first being an initiation event, followed by promotional events that can often occur much later (10–20 years). One view of carcinogenesis correlates initiation with the induction of somatic mutations, and promotion with alteration in the expression of these mutations. Three molecular mechanisms are currently considered to be important in the initiation of some cancers: activation of proto-oncogenes, inactivation of tumor suppressor genes, and disruption of "programmed cell death," or apoptosis. Alterations in DNA sequence, such as point mutations or deletions, can cause any of these events.

Recently, mutation analyses have been performed on a series of patients with invasive squamous cell carcinomas of the skin. More than half the tumors analyzed contained mutations in the *p53* tumor suppressor gene. The role of UVR in these mutations is indicated by the occurrence of a CC $\rightarrow$ TT double-base change in the DNA sequence, a change that is produced only by UV light. In addition, a high frequency of T $\rightarrow$ C transitions was seen, a common mutation after UVR.

The lethality of a particular type of DNA damage (photoproduct) may ultimately influence its mutagenicity and tumorigenicity. This relationship has been considered in terms of a "pass/fail" rule: that is, that lethal lesions that block DNA polymerization are nonmutagenic. In other words, the more cytotoxic a lesion, the less likely that it will allow translesion synthesis and mutation induction at that site. In terms of this model, the mutagenicities of the major photoproducts in mammalian cells are determined by their cytotoxicities; their cytotoxicities are modulated by repair.

9.2 Environmental Effects of Stratospheric Deozonation

Global decreases in stratospheric ozone have been observed in recent years as a consequence of air pollution from natural and anthropogenic sources, such as volcanic eruptions and hydrocarbon (fuel) combustion. There is widespread concern about the effects of such atmospheric change on the biosphere. Reduced ozone concentrations affect the intensity and spectral distribution of UVR at the earth's surface and to at least 20 m deep in the oceans by increasing the amounts of biologically harmful wavelengths of UVB.

An increase in solar UV, as postulated from models of stratospheric ozone depletion, could significantly alter populations of marine and terrestrial plant and animal communities. Selection by UVR would result in reductions of UV-sensitive species and expansion of more UV-resistant species or clones. Trends of this sort would inevitably produce a more UV-resistant community; however, the consequence of changes in taxonomic structure on other parameters relevant to the ecosystem, such as primary production and trophic dynamics, are problematic. Decreases in total primary production have been documented during Antarctic ozone depletion and are expected at other latitudes.

The annual occurrence of springtime ozone depletion over the Antarctic provides a unique setting for monitoring the effects of increased UVB radiation on natural populations. Studies have been conducted on the induction and repair of UVB damage in various Antarctic phytoplankton species. The results indicate that there is great diversity in the photobiology of individual phytoplankton species; induction of photodamage in individual species varies nearly 100-fold. Comparable differences in the ability of individual species to repair UV damage is also observed; the ability of individual species to excise DNA photodamage ranges from negligible to near complete removal within 6 hours. Parallel to this diverse molecular response, the ability of phytoplankton to survive the effects of UVB light is also extremely variable, correlating with DNA repair capacity. It is apparent that, within a cross section of the Antarctic marine community, relatively sensitive and resistant phytoplankton species coexist.

Marine phytoplankton are an important component of the oceanic food chain and provide much of the world's oxygen. Knowledge of the molecular and biological responses to UVB light at the species-specific level will enable us to predict changes in the biomass and species composition of marine ecosystems. Using phytoplankton as a model, insight into the ultimate effects of stratospheric deozonation on other marine and terrestrial ecosystems can be gained. Knowledge of the molecular biology of DNA damage and repair in selected plants and animals at different stages of their development will allow us to identify species at risk and to predict the consequences of deozonation on man and the earth's environment.

See also DNA DAMAGE AND REPAIR; ENVIRONMENTAL STRESS, GENOMIC RESPONSES TO; IONIZING RADIATION DAMAGE TO DNA; MUTAGENESIS, MOLECULAR BASIS OF; YEAST, DNA REPAIR IN.

Bibliography

Björn, L. O., Moan, J., and Young, A. R., Eds. (1993) *Environmental UV Photobiology.* Plenum Press, London.

Cleaver, J. E., and Mitchell, D. L. (1993) Ultraviolet radiation carcinogenesis. In *Cancer Medicine,* 3rd ed., J. F. Holland, E. Frei III, R. C. Bast, D. W. Kufe, D. L. Morton, and R. R. Weichselbaum, Eds., pp. 245–254. Lea & Febiger, Philadelphia.

Friedberg, E. C. (1985) *DNA Repair.* Freeman, New York.

———, and Hanawalt, P. C., Eds. 1981 (Vol. I), 1983 (Vol. II), 1988 (Vol. III) *DNA Repair: A Laboratory Manual of Research Procedures,* Vols. 1–3. Dekker, New York.

Hanawalt, P. C. (1989) Preferential repair of damage in actively transcribed DNA sequences in vivo. *Genome,* **31:**605–611.

Hill, H. Z. (1992) The function of melanin, or Six blind people examine an elephant, *BioEssays,* 14:49–56.

Mitchell, D. L., and Nairn, R. S. (1989). The biology of the (6-4) photoproduct. Annual review. *Photochem. Photobiol.* 49:805–819.

Smith, K. C. (1977) *The Science of Photobiology.* Plenum Press, New York.

Wang, S. Y., Ed. (1976) *Photochemistry and Photobiology of Nucleic Acids,* Vols. I and II. Academic Press, New York.

V

Vaccination, Genetic Approaches to: *see* Genetic Vaccination, Gene Transfer Techniques for Use in.

Vaccination, Genetic Approaches to: *see* Genetic Vaccination, Gene Transfer Techniques for Use in.

VACCINE BIOTECHNOLOGY

Tilahun D. Yilma

1 Introduction

2 Recombinant Vaccines
- 2.1 Vectors
- 2.2 Vaccinia Virus
- 2.3 Genetic Engineering

3 Vaccinia Virus Recombinant Vaccines
- 3.1 A Vaccinia Virus Recombinant Vaccine for Rinderpest
- 3.2 Vaccines for Peste des Petits Ruminants
- 3.3 A Vaccinia Virus Recombinant Vaccine for VSV
- 3.4 Recombinant Vaccinia Virus Vaccines for SIV
- 3.5 Safety and Efficacy of Live Recombinant Vaccines
- 3.6 Potential Recombinant Vaccines for HIV

4 Baculovirus Expression Vectors and Rapid Diagnostic Kits

5 Advantages of Vaccinia Virus Recombinants

Key Words

Adjuvant Genes Genes (often lymphokine genes) that are used for the enhancement of efficacy of recombinant vaccines.

Attenuating Genes Genes reducing the virulence of agents to enhance safety for use as vaccines or vectors.

Baculovirus A DNA insect virus widely used for abundant and inexpensive generation of recombinant proteins in insect cells or larvae.

Chimeric Describing a product of fusion of two or more nucleic acids or polypeptides.

Recombinant An organism, nucleic acid, or protein generated by genetic engineering.

Vaccinia Virus An orthopox virus of uncertain origins that provides protective immunity against a number of orthopox virus infections, including smallpox.

Vector A bacterium, virus, or plasmid that is used for the introduction of DNA molecules or the generation of recombinants.

Using the techniques of molecular biology, effective modern vaccines are being developed by a variety of innovative methods that circumvent some of the problems associated with conventional vaccines. Live recombinant vaccines, which are prepared by isolating an immunogenic gene from a disease organism and inserting it into a vector, are most promising. Such vaccines can provide immunity to agents that cannot be reliably attenuated or inactivated without destroying immunogenicity.

1 INTRODUCTION

Over the past century, the gradual accumulation of knowledge concerning contagion and immunity has led to great advances in our ongoing struggle against infectious disease. In an effort to gain protection from the most serious infections, we have progressed from the earliest type of quarantine (the shunning of the obviously ill) to deliberate, though controlled and limited, exposure to disease manifestations. One of the earliest examples of such exposure is seen in the battle against the dreaded scourge, smallpox. The dangerous and sometimes deadly practice of variolation—deliberate exposure to the lesions of human cases, in hopes of contracting a mild form of the disease—was practiced sporadically for centuries before the fortuitous discovery of the protection afforded by related, nonvirulent strains of virus. By 1792, Dr. Edward Jenner was carefully introducing pox pustular material from cows into his patients' skin—a process called vaccination—to protect against smallpox. The latter half of the nineteenth century brought an explosion of discoveries in immune protection, with specific vaccines produced for a wide variety of human and animal diseases. Over time, many of the most serious of human disease threats have retreated before the advances of classical vaccine therapy.

Classical vaccines have generally been of two types: live, attenuated agents and inactivated agents. Vaccinia virus, the poxvirus strain that protects human beings from smallpox, was attenuated naturally during its adaptation to different species. From the time of Pasteur and Koch, laboratories have worked to produce other attenuated disease agents (by varying their conditions of growth and maintenance) or to find nonvirulent, protective strains in nature. To produce effective vaccines of some disease agents, it has been necessary to attempt inactivation of the agent by heat or chemicals, destroying the capacity for growth or replication and attenuating virulence while retaining the protective, immunity-inducing antigens.

Live, attenuated, and inactivated vaccines each have their advantages and disadvantages. Inactivated vaccines are usually easier to produce and are the only protection available against certain diseases. Protection against diseases caused by excreted toxins can be similarly accomplished by using heat or chemicals to convert these to toxoids. Thorough inactivation relieves safety concerns, although there have been instances of disease resulting from incomplete inactivation of an agent. Even safe, thoroughly inactivated vaccines may be less effective than live attenuated vaccines, however, because live vaccines have the capacity to reproduce or replicate in the host organism, amplifying the antigen. This amplification effect increases the stimulus to the immune system and may elicit a better immune response by providing optimal amounts of antigen. Maintaining live cultures and producing live vaccines, however, is much more demanding, requiring special laboratory and transport facilities. Many live vaccines are heat labile and must be refrigerated, further complicating their use in remote areas. Other safety concerns also arise with live vaccines because attenuated strains may revert to virulence in the vertebrate host or an arthropod vector, new strains of the agent may arise from the vaccine strain, or the vaccine may actually induce the disease in pregnant or debilitated animals.

A further, serious disadvantage of classical vaccines is that whether living or killed, they contain whole disease agents that elicit a full range of immune response. Thus, neither live attenuated nor inactivated vaccines give the capacity for serologic tests that differentiate vaccinees from recovered carriers. This is an important consideration in both epidemiologic studies and veterinary practice. It can cause special problems in animal disease diagnosis and control, and bans on importation of infected livestock and their products can result in substantial economic losses.

Advances in development processes in this century have given rise to a remarkable new generation of vaccines that includes subunit, synthetic peptide, and live recombinant vaccines. Many of these advances have their origins in techniques in molecular biology that allow the identification of an agent's specific, immunity-producing molecules, the synthesis of all or active portions of these molecules, and the incorporation of the responsible immunogenic genes into the genome of a nonvirulent carrier or vector agent. A variety of vaccines of this new generation are now being readied for use and may greatly improve our ability to protect against infectious disease. Live recombinant vaccines in particular appear to hold great promise for continuing improvements in disease control.

2 RECOMBINANT VACCINES

2.1 VECTORS

Recombinant DNA technology now makes it possible to retain most of the advantages of live vaccines, while overcoming many of their inherent safety hazards and logistical disadvantages, by inserting and expressing heterologous genes in a variety of different viruses. Thus far, simian virus 40, bovine papillomavirus, adenovirus, herpesvirus, and members of the retrovirus family have been used most extensively as expression vectors. Although the relatively small size of most of these viruses facilitates genetic engineering, this property also restricts the amount of DNA that can be inserted into the virus particle. In addition, many of these virus vectors have a limited host range that does not allow their replication in the vaccine's target species. They may be defective, requiring either a helper virus or special cell lines for their replication. Furthermore, many of these vectors are oncogenic and therefore may be unsuitable for use as vaccines in humans and domestic animals.

2.2 VACCINIA VIRUS

The disadvantages of the vector systems can be overcome by using vaccinia virus (VV) as a vector, creating a live recombinant vaccine with the immunogenic genes of disease agents incorporated into the VV genome. When the VV vector is introduced into an animal, the virus replicates in the host's cells and the foreign genes are expressed along with those of VV; thus the animal becomes immune both to VV and to the agent from which the foreign gene was taken. The history of VV in human hosts attests to its safety: during this past century, the VV vaccination procedure has protected hundreds of millions of people worldwide from smallpox, with minimal side effects and no adverse environmental impact despite global release of the VV under varying hygienic conditions. Vaccination with this virus has been effected by scarification, by subcutaneous or intraperitoneal injection, and by intranasal or even oral administration. It can be grown outside the laboratory by scarification of the abdomen of a single calf to prepare more than 200,000 vaccine doses—a great advantage for use in developing countries. The virus is heat stable in the lyophilized state, and it can be reconstituted easily. Modern vaccine technology now gives us, through genetic engineering, the capability of using VV as a live vector to protect animals and people against other, unrelated, disease agents.

A member of the poxvirus family, VV is a large, double-stranded DNA virus that packages a complete transcription system and replicates in the cytoplasm of infected cells. The virus has a wide host range that includes humans, cattle, horses, swine, sheep, goats, mice, and monkeys, making it useful for both human and a variety of animal vaccines. Its large genome can accept more than a dozen heterologous genes, opening the possibility of producing a multivalent vaccine that could protect against several diseases with one administration. The molecular biology of the virus is well documented and supports the engineering techniques necessary for producing, isolating, and identifying recombinants. The genome contains more than a dozen nonessential regions that can be used for incorporation of foreign antigen genes and reporter genes that assist in identifying the desired recombinants, in addition to the commonly used marker, the *TK* (thymidine kinase) gene. Using recombinant DNA technology, we have developed convenient marker rescue techniques and plasmid vectors containing homologous sequences to the virus DNA that allow recombination within infected cells.

2.3 GENETIC ENGINEERING

The application of recombinant DNA technology utilizes convenient marker rescue techniques and plasmid vectors containing homologous sequences to the virus DNA, allowing recombination within infected cells. The strategy for using VV as a vector involves first constructing a plasmid carrying a chimeric gene. This unit consists of a VV promoter fused to a foreign-protein-coding sequence that is flanked by DNA from a nonessential region of the VV genome. The chimeric gene is incorporated into the VV genome by homologous recombination in tissue culture cells that have been infected with wild-type VV and then transfected with the plasmid. Although any nonessential region of the VV genome can be used as the site of gene insertion, the thymidine kinase gene locus provides some advantages because recombinants are then TK^-. The TK^- phenotype serves to attenuate viral pathogenicity, and also can be easily distinguished from wild-type TK^+ virus by bromouracildeoxyriboside (BUdR) selection. Rapid screening of recombinants of interest can also be facilitated with the coexpression of reporter genes, such as the *Escherichia coli* β-galactosidase *(lacZ)* gene. Expression of the *lacZ* gene lends a blue coloration in the presence of 5-bromo-4-chloro-3-indolyl-β-D-galactoside; one is thereby able to distinguish recombinants from wild-type virus by their blue plaque phenotype.

3 VACCINIA VIRUS RECOMBINANT VACCINES

A description of three VV recombinants that we have constructed and used with varying degrees of success as vaccines will illustrate the potential of these agents: a vaccine for rinderpest that has induced 100% protection against a challenge greater than 1000 times the normally lethal dose, a vaccine for vesicular stomatitis that has been moderately successful in protecting against the disease, and a vaccine aimed at simian immunodeficiency disease that has not prevented infection but has significantly reduced the virus load from control levels. In the last case, we have not determined whether the reduction translates into prevention or delayed onset of clinical disease.

3.1 A Vaccinia Virus Recombinant Vaccine for Rinderpest

Our most successful vaccinia virus recombinant vaccine is one developed for rinderpest virus (RPV), the single most important viral disease of livestock in developing countries. An acute, febrific, highly contagious viral disease of ruminants (particularly cattle and buffalo) with a rapid course and a mortality rate exceeding 90%, rinderpest is characterized by inflammation, hemorrhaging, necrosis, and erosion of the gastrointestinal tract accompanied by bloody diarrhea, wasting, and death. This disease was controlled in Africa in the 1970s by an international effort called the Joint Project/15 (JP/15), which vaccinated 124 million cattle, but recent resurgences have necessitated a second Pan-African Rinderpest Campaign, currently in progress.

Continuing outbreaks of rinderpest have pointed up the logistical difficulties of reliance on live vaccines in the regions where this disease is prevalent, but the three vaccines currently employed—the caprinized, the lapinized, and the Plowright tissue culture vaccines for rinderpest—are all live attenuated vaccines with the limitations common to such preparations. All are extremely heat labile and require maintenance of a cold chain; in most instances, vaccines must be used within 30 minutes of reconstitution to ensure potency. The caprinized and lapinized vaccines also have varying degrees of pathogenicity for some animals. The Plowright tissue culture rinderpest vaccine (TCRV) is effective, but the requirement of skilled personnel for vaccine preparation and administration and a need for refrigeration facilities—often lacking in the hot and isolated areas where the disease is prevalent—are obstacles to its use.

A VV recombinant vaccine for rinderpest, many felt, would provide an alternative and circumvent many of the obstacles limiting the use of current preparations. The heat-stable lyophilized form of VV is easily produced, stored, and transported; and it can be administered by scarification. It is possible, as was shown in the smallpox eradication campaign, for each independent area to carry on vaccine production by collecting and emulsifying scabs from vaccinated animals for use as needed. The self-sufficiency thus afforded to local animal handlers is a rare and valuable endowment for developing countries that frees them from reliance on international assistance. For these reasons, a VV recombinant vaccine for rinderpest is a valuable addition to the arsenal of defense and raises the possibility of eradicating this disease. Our efforts to construct such a vaccine have had a highly successful conclusion: we have produced one of the two most effective recombinant vaccines to date. Our rinderpest vaccine, which was designated as a prototype for genetically engineered vaccines for export, has been approved for field trials in Kenya and Ethiopia by the U.S. Department of Agriculture/Animal and Plant Health Inspection Service and the U.S. Agency for International Development. The other successful agent is a recombinant vaccine for rabies currently in field trials.

The causative virus of rinderpest is enveloped and has a single-stranded RNA genome with a minus polarity. It is in the family *Paramyxoviridae* and is a member of the morbillivirus group along with measles virus of human beings, distemper virus of dogs, and peste des petits ruminants virus (PPRV) of goats and sheep. We propagated the highly virulent Kebete "O" strain of RPV in primary bovine kidney cells and characterized eight viral proteins. We made cDNA copies of the phosphoprotein (P), hemagglutinin (H), fusion (F), nucleoprotein (N), and matrix genes (M), and determined the complete nucleotide sequence of these genes. In all paramyxoviruses, the H and F surface proteins elicited protective immunity. Standard procedures were used to construct VV recombinants expressing the *H* gene (vRVH) and the *F* gene (vRVF) of RPV.

Using these VV recombinants (vRVH and vRVF), protective immune response studies in cattle were conducted in the high containment facility at the Plum Island Animal Disease Center. Each animal was shown to be seronegative to RPV prior to vaccination, and 10^8 plaque-forming units (pfu) of the VV recombinant vaccine was administered in a single vaccination by intradermal inoculation and scarification. Groups of animals (five each) received either one recombinant (vRVH or vRVF) or a cocktail of both recombinants (vRVH + vRVF). Two animals were vaccinated with TCRV for a positive control, and two were left unvaccinated for a negative control. For a VV control, two animals were vaccinated with v50, a VV recombinant expressing the glycoprotein gene of the vesicular stomatitis virus (VSV).

Pock lesions developed as early as 4 days in all animals vaccinated with the recombinants, but were limited to the site of inoculation and healed completely by 2 weeks. Humoral responses to the vaccines were assessed by serum neutralization (SN) assay. All animals vaccinated with the recombinants or TCRV produced SN antibodies to RPV. As expected, control animals had no detectable antibody titer.

To evaluate protection, all animals were challenged subcutaneously with a heavy dose (10^3 $TCID_{50}$—tissue culture infectious growth for 50% of subjects) of RPV in the prescapular lymph node region on day 35 following primary immunization. As low as one $TCID_{50}$ induced clinical rinderpest with 100% mortality. Cattle vaccinated with the recombinants or TCRV were completely protected when challenged, even though the inoculation was greater than 10^3 times a normally lethal dose of RPV. Some of the cattle, most of these vaccinated with the *F* recombinant (vRVF), had strong anamnestic responses to RPV after challenge inoculation; this indicates replication of the challenge virus. No anamnestic response at all could be demonstrated in the groups vaccinated with a cocktail of both recombinants or TCRV.

Viral infection and spread are mediated by receptor binding (to H) and membrane fusion (by F), so it is reasonable to expect a vaccine preparation containing both antigens to be superior to one containing the H or the F glycoprotein alone. Vaccinating with the vRVF + vRVH mixture in the field, however, is cumbersome and expensive, and particularly burdensome for developing countries. For this reason as well as to obtain greater standardization and more reliable results, we decided to attempt the inclusion of both genes in a single recombinant. Using techniques similar to those described earlier, we developed a VV (Wyeth strain) double recombinant that includes both the *F* and the *H* genes (vRVFH) of RPV; this recombinant expresses the authentic RPV F and H proteins. The *F* and *H* genes were inserted into the VV *TK* and *H* gene regions, respectively, thus inactivating these VV genes and giving an even more highly attenuated VV recombinant vaccine. Cattle vaccinated with this double recombinant (vRVFH) showed no pock lesions at the site of inoculation, indicating a high degree of attenuation.

Cattle were vaccinated with 10^8 pfu of vRVFH by intradermal inoculation in the neck region. Two groups of cattle were vaccinated with VV recombinants and kept in separate isolation rooms. In the first group, five animals were vaccinated with vRVFH. In the second group, four animals were vaccinated with a mixture of vRVF and vRVH. In addition, two unvaccinated animals were housed with each group, to assess the transmissibility of VV recombinants from vaccinated to contact animals. Unlike the cocktail vaccinees, animals vaccinated with vRVFH developed no detectable pock lesions; a thorough examination failed to demonstrate pock lesions in the contact animals in both groups. Further, serum samples taken from contact animals on the days of vaccination (day 0) and challenge (day 28) were negative to VV by SN and plaque reduction assays (data not shown). All cattle vaccinated with the recombinants produced SN antibody to RPV as early as 8 days after vaccination. However, all contact and control animals lacked detectable SN antibody to RPV during the course of the experiment.

One month postvaccination, all animals in both groups were challenge-inoculated with 10^3 $TCID_{50}$ of the pathogenic Kebete "O" strain of RPV. Cattle vaccinated with VV recombinants (both groups) were completely protected from rinderpest, exhibiting no detectable illness and a normal temeprature of 38°C. The four unvaccinated contacts developed high fever (42°C) by day 2 and died by day 6 after challenge, showing lesions typical of severe rinderpest. Daily monitoring for 2 weeks showed no detectable clinical disease in vaccinated animals, and the experiment was terminated.

We have also tested vRVFH and the parental Wyeth strain for pathogenicity by inoculating nude mice intraperitoneally, and no clinical disease was detected in 12 test mice. In addition, we have assessed the effects of vRVFH and the mixture of single recombinants (vRVF + vRVH) on the immune systems of rhesus macaques infected with the simian immunodeficiency virus (SIV). These monkeys suffer from an immunodeficiency disease virtually identical to the human acquired immunodeficiency syndrome (AIDS). Hence, SIV is a legitimate model for HIV. Four normal macaques and four SIV-infected macaques were vaccinated with vRVFH. Apparently, the lack of development of pock lesions in cattle vaccinated with vRVFH is host specific; both groups of monkeys developed small pock lesions at the site of inoculation, but these healed completely by 2 weeks postvaccination. No clinical disease or dissemination of VV was observed in either normal or SIV-infected macaques during a one-month observation following vaccination.

Rinderpest appears to be an excellent candidate for eradication using the VV recombinant vaccine. There is only one serotype of RPV, although there are different strains manifesting different degrees of pathogenicity in the field. A vaccine against one strain will immunize against all, including PPR of sheep and goats. Use of the rinderpest double recombinant VV (vRVFH) in areas of the world where PPRV is endemic would also aid in the control and eradication of PPR.

3.2 Vaccines for Peste des Petits Ruminants

PPR is a disease of small ruminants caused by a virus (PPRV) serologically and ultrastructurally related to RPV of cattle. The disease is characterized by necrotizing and erosive stomatitis, enteritis, and pneumonia. Morbidity and mortality are very high, and the virus is a serious health problem in small ruminants in Africa and the Middle East. Although cattle can be infected with PPRV, they do not exhibit clinical disease. Here we report that vaccination of goats with vRVFH provides complete protection against a challenge-inoculation with 10^3 $TCID_{50}$ of PPRV.

All goats used in these experiments were determined to be seronegative to PPRV and RPV at the start of the study. Eight goats were vaccinated with 10^8 pfu of vRVFH by intradermal inoculation and scarification. For controls, six goats were vaccinated with the parental VV (Wyeth strain). On day 35 postvaccination, goats were challenge-inoculated with 10^3 $TCID_{50}$ of PPRV, strain Mieliq-Sudan, by subcutaneous injection just anterior to the prescapular lymph node. Serum was collected on days 0, 35, 42, and 56, and titers to RPV and PPRV were determined by SN and by ELISA (enzyme-linked immunosorbent assay). Antibody directed to the F protein was measured by ELISA only, because the glycoprotein is not the antireceptor of the virus and does not usually induce SN antibody. Goats vaccinated with vRVFH had SN (32–512) and ELISA titers (2–16) to RPV and undetectable SN titers to PPRV by day 35 postvaccination. Anti-PPRV ELISA titers of vaccinated goats were undetectable in four goats and low (2, 2, 2, and 8) in the remaining four goats on the day of challenge.

All control animals lacked detectable SN and ELISA titers to either PPRV or RPV on the day of challenge. After challenge inoculation, however, there was a strong anamnestic response, as indicated by high SN and ELISA titers. In contrast, the postchallenge ELISA titers to PPRV and RPV of goats vaccinated with vRVFH ranged from 64 to 2048 and from 64 to 1024, respectively. Although SN titers to PPRV could not be detected in the animals vaccinated with vRVFH on the day of challenge inoculation, all vaccinees were completely protected against PPRV, and no clinical signs (fever or diarrhea, e.g.) were exhibited over the following period. In contrast, all the control animals vaccinated with the parental Wyeth strain of VV were febrile (>39.7°C) and suffered from sever PPR by 4 days postchallenge. Three animals died; the rest were euthanized.

Clearly, this vaccine—a VV double recombinant vaccine expressing the *F* and *H* genes of RPV—completely protects goats against PPR, even though no neutralizing antibody to PPRV could be detected before challenge inoculation. Although no anti-PPRV ELISA titers were found in four of the vaccinated animals, and the remaining four had very low titers, all vaccinated goats were equally protected. It has been reported that in cattle administered classic live viral vaccines to RPV, any detectable SN titer is associated with complete protection; this is, however, the first report of protection in the morbillivirus group in the complete absence of SN antibody to the homologous virus. In this regard, we have recently provided evidence that cell-mediated immunity may be the major determinant in protection against morbilliviruses. Cattle vaccinated with subunit antigens (F and H) of RPV expressed in baculovirus produced viral neutralizing antibody but were not protected.

Control of PPR is of great economic importance in countries engaged in large-scale rearing of small ruminants. Currently, there is no safe and efficacious vaccine against PPR. Until one is developed, probably by expressing the homologous *F* and *H* genes of PPRV, vRVFH may be used for the control of PPR. Replication of VV is unaffected by antibody to PPRV, and vRVFH can be used to immunize young animals that still have passive (maternal) antibody to PPRV. Also, one can employ a serological test to distinguish vRVFH-vaccinated goats from those exposed to the whole virus, either by vaccination or infection. We have developed such a rapid, inexpensive diagnostic kit based on the *N* gene of RPV (see Section 4).

3.3 A Vaccinia Virus Recombinant Vaccine for VSV

One of the earliest VV recombinants prepared and tested in our laboratory was a vaccine for vesicular stomatitis (VS), a contagious disease of horses, cattle, and pigs characterized by vesicular lesions on the tongue and other areas of the oral mucosa. Interestingly, the virus can be used to infect mice in a model system, where it causes neuropathy and death in 100% of the subjects. A serious cause of economic loss throughout North and South America, this disease is notable for the confusion of diagnosis between VS and foot-and-mouth disease, a pathological epizootic condition with a 100% morbidity rate.

Experimental attenuated and inactivated VSV vaccines have been developed, but their use for vaccination is generally prohibited by U.S. Department of Agriculture because vaccinees cannot be distinguished from naturally infected animals. It is possible to distinguish animals vaccinated with a subunit preparation by serology, however, and this is an essential consideration for epidemiological studies, disease control, and establishment of quarantine programs. Advantages of subunit and live vaccines could theoretically be combined in a VV recombinant expressing specific VSV proteins, and such recombinants were constructed by the methods described for rinderpest.

Of all the antibodies to various proteins of VSV, only those directed to G, the external glycoprotein, have been shown to be protective in mice. When DNA copies of mRNAs for the G or N proteins of VSV were linked to a VV promoter and inserted into the genome of VV, the recombinants retained infectivity and synthesized VSV polypeptides. We have demonstrated that the polypeptides expressed by these recombinants are similar or identical in size and antigenicity to the natural VSV proteins, and that the G glycoprotein is transported normally to the cell surface (perhaps important for antigenicity). After intradermal vaccination with live recombinant virus expressing the G protein (v50), mice produced VSV-neutralizing antibodies and were protected against lethal encephalitis upon intravenous challenge with VSV. VSV SN titers were detected by day 14 and increased over a 42-day period. Half the mice were given a booster vaccination on day 28, which resulted in a seven- to eightfold increase in SN titers over that obtained with a single vaccination.

Challenge was carried out by intravenous injection of VSV, which produces lethal encephalitis in mice within 6–12 days. For control purposes, a group of mice was vaccinated with vHBs4, a VV recombinant that expresses the hepatitis B virus surface antigen gene. When this group was challenged, 7 of 11 animals died of encephalitis. By contrast, only 1 of 15 mice that received a primary vaccination with v50 died, and none of 16 that received two vaccinations died.

To further evaluate the immunogenicity of the recombinant virus, cattle were vaccinated intradermally and later given intralingual injections of virulent VSV; the degree of protection was correlated with the neutralizing antibody produced following vaccination. After vaccination, typical pox lesions appeared within 4 days. SN titers to VSV were detected on day 7 and reached values of 80–480 by day 29. At that time a second vaccination was given, following which the titers increased severalfold. On day 44, cattle were challenged by intralingual injection of 10^2 and 10^3 pfu of VSV. Preliminary experiments indicated that most unvaccinated cattle challenged in this manner developed vesicular lesions within 2 days. All controls (unvaccinated cows and cows vaccinated with vHBs4) developed typical vesicular lesions at the 10^2 and 10^3 pfu VSV injection sites. In contrast, only two of six vaccinated cows developed vesicular lesions at the lower challenge dose of VSV. The remaining four cows had no lesions at the 10^2 pfu VSV injection sites, although all animals developed lesions at the 10^3 pfu sites.

Good correlation was observed between antibody titers and protection against challenge with 10^2 pfu of VSV. All animals that were protected had SN antibody titers of 1280 or greater, whereas those that were sensitive had titers of 640 or lower. The degree of protection obtained is particularly noteworthy in view of the severity of the experimental challenge, compared to that of a natural infection.

The VV recombinants protected mice against a VSV challenge that would normally kill all mice within 6 to 12 days. In cattle, two successive vaccinations with the VV recombinant vaccine for VSV provided protection for two-thirds of the animals against a challenge of 100 pfu; this protection was closely correlated with neutralizing antibody titers. The lack of complete protection for all vaccinees may indicate a need for cell-mediated immunity, which is not induced by this vaccine. It also may be possible to construct recombinants synthesizing considerably more G glycoprotein, which would provide more effective protection.

3.4 Recombinant Vaccinia Virus Vaccines for SIV

Experimental infection of rhesus macaques with simian immunodeficiency virus provides a valuable animal model for AIDS vaccine research. Several lines of inquiry recently have brought out the following information:

1. Inactivated whole-virus vaccines have provided limited protective immunity against SIV in macaques challenged with virus propagated in human but not rhesus peripheral blood mononuclear cells.
2. Recently, evidence has been presented to indicate that human cellular antigens play a role in protection against SIV in macaques immunized with whole inactivated virus propagated in human cells.
3. Complete protection from challenge with pathogenic rhesus-grown SIV was demonstrated in monkeys vaccinated with nonpathogenic (attenuated) nef-deleted SIV mutants.

However, whether a live retroviral vaccine could ever meet the safety requirements for human use is highly questionable; this virus integrates into the host genome and may induce disease at some later date.

Thus, as an alternative to inactivated and live attenuated classical vaccines, we and others have evaluated the potential of priming rhesus macaques with a live recombinant VV vaccine expressing the glycoproteins of SIV, and then boosting with the same recombinant subunit antigen expressed in mammalian cells, Chinese hamster cells (CHO), or a baculovirus expression vector (see Section 4). Protection in these animals was evaluated by intravenous challenge with pathogenic SIV isolates. Although binding antibodies were detected in all animals following immunization with these recombinants, neutralizing antibodies were undetectable one week prior to virus challenge—results that differed from those for macaques vaccinated with whole, inactivated SIV. However, all animals became infected after intravenous inoculation with 1–10 AID_{50} of challenge virus.

There was no protection afforded by any of our recombinant vaccines, even though in vitro suppression of virus was demonstrated, and cell-free infectious virus load in plasma (plasma viremia) of macaques primed and boosted with certain preparations of SIV surface glycoprotein was considerably lower than in unimmunized controls at 2 weeks postchallenge. These studies suggest that immunization with surface glycoprotein may not necessarily provide protective immunity against immunodeficiency virus infection, even though the relationship between reduction in virus load and delay in the onset of disease remains to be evaluated. One protocol did have notable success in heightening the immune response: animals primed with a VV recombinant vaccine and then boosted with a baculovirus subunit product showed greatly enhanced antibody responses compared to those boosted with the original vaccine.

3.5 Safety and Efficacy of Live Recombinant Vaccines

With the threat of untoward effects always in mind, we have investigated possibilities of further attenuation of our VV recombinant vaccines by the insertion of lymphokine genes like interferon gamma (IFN-γ) or interleukin 2 (IL-2). This gives us a new attenuating approach that should be applicable to other live recombinant vaccines. We constructed fusion proteins of IFN-γ (human and murine species) with various immunogens and demonstrated reduced pathogenicity for immunodeficient (athymic nude) mice of several VV recombinants expressing these proteins. Immunodeficient mice were able to clear the recombinant VV at doses as high as 10^8 pfu, while as few as 10^2 pfu of VV is normally 100% lethal for nude mice. Mixtures of recombinant vaccinia viruses were always lethal to the mice if one of the recombinant viruses did not express the IFN-γ gene, showing that IFN-γ-induced attenuation in VV is operational in a "cis" fashion only.

We have further provided evidence that the antiviral and attenuating activities of IFN-γ have different mechanisms of action. We demonstrated that unlike the antiviral activity, the attenuating activity of IFN-γ is not species specific. Moreover, we have shown attenuating activity of IFN-γ in the absence of any detectable antiviral activity. Although human IFN-γ did not have antiviral activity on murine cells, expression of this cytokine did increase the attenuation of recombinant vaccinia viruses for athymic mice. It is interesting to note that the anti-inflammatory genes of VV do not appear to block the attenuating activity of IFN-γ.

The development of an adjuvant system for VV recombinant vaccines may be crucial if the technology is to be considered an effective alternative to classical forms of vaccination, such as inactivated or live attenuated vaccines. We have demonstrated that IFN-γ mixed with G glycoprotein of VSV duplicates the adjuvant effect of complete Freund's adjuvant and futher increases the anamnestic response on booster administration. Others have demonstrated the adjuvant effects of interleukin 2.

The use of immunoregulatory genes like IFN-γ and IL-2 as adjuvant genes in multivalent VV recombinant vaccines must be further evaluated. If these lymphokine genes work as adjuvants to augment the immune response, the concept of vaccination will be revolutionized. Such natural products could eliminate the need for harsher adjuvants (e.g., Freund's) and thus avoid the untoward effects presently observed. The human IFN-γ gene could be incorporated into vaccinia vectors for human pathogens, and similar vectors could be constructed for other domestic species.

3.6 Potential Recombinant Vaccines for HIV

Several laboratories are investigating VV recombinants as vaccines for human immunodeficiency virus (HIV) infection. It has been shown that the HIV antigens expressed by some of these recombinants are correctly processed and transported to the appropriate cellular compartments. Moreover, experimental animals as well as human volunteers that were vaccinated were shown to develop detectable cell-mediated immunity to the HIV proteins expressed. There is considerable interest and some optimism that an effective vaccine might be prepared using VV as a vector.

Innovative solutions to the safety problems occasioned by putting such a vaccine into general use are gaining attention, especially following the report of a disseminated VV infection caused by smallpox vaccination of an army recruit with subclinical HIV infection. It has been demonstrated that vaccinia virus is attenuated by insertional inactivation of its *TK* gene. Also, the expression of IL-2 and murine IFN-γ has prevented disseminated VV infection in nude mice.

We have now reported the development of a novel approach for the attenuation of a live recombinant vaccine for HIV. We have constructed chimeric genes coding for IFN-γ and the structural proteins of HIV-1, and inserted them into VV to develop recombinants that express these genes. These fusion proteins retained the antigenic characteristics of both IFN-γ and HIV. The biologic antiviral activity of IFN-γ could not be demonstrated for all the fusion proteins, but the in vivo attenuating activity of IFN-γ for nude mice was retained at variable rates for all the recombinants.

We illustrated the validity of this concept by demonstrating that recombinants expressing IFN-γ have reduced pathogenicity for nude mice. We have provided evidence that the antiviral and attenuating activities of IFN-γ have different mechanisms of action, and that the attenuating activity, unlike the antiviral activity, is not species specific and may occur in the absence of any antiviral activity.

4 BACULOVIRUS EXPRESSION VECTORS AND RAPID DIAGNOSTIC KITS

The VV recombinant vaccines prepared for RP, PPR, VS, and SIV all lend themselves to employment of a new rapid diagnostic technique that our laboratory has prepared in kit form. The baculovirus expression vectors used in our laboratory give us the capacity to produce large quantities of different virus proteins that may be used as coating antigens for ELISAs. When an internal protein gene such as the *N* gene of RPV or VSV, or the *p27* gene of SIV, is incorporated into the genome of this insect virus strain before infection of a larval host, the virus polyhedrin promoter drives a very high level of expression of crude recombinant protein (up to 67% of the weight of the larva). The protein from one infected and homogenized larva (0.2–0.3 g) provides antigen sufficient for coating 200 standard 96-well ELISA plates, which can be used to diagnose up to 10,000 serum samples in duplicate. Purification of the N or p27 protein appears to be unnecessary; there was no interference or loss of sensitivity using the crude larval lysate for coating ELISA plates.

Used in conjunction with a recombinant vaccine incorporating a gene for a major immunogen of the disease agent, this ELISA allows distinction of vaccinated animals from those that have been naturally infected. Vaccination with the recombinant produces antibody to the incorporated gene's product only, while natural infection brings about responses to many different antigens. This test differentiates animals showing responses to a variety of proteins (naturally infected by the disease agent) from those that respond only to the vaccine's foreign gene product (vaccinees).

We have also introduced a novel approach for the production of monoclonal antibodies to one specific protein of a virus or other agent consisting of several proteins, without the use of purified antigen in either the immunization or screening phase of the procedure. This system has been effective in producing monoclonal antibodies to rare antigens like Alzheimer's protein, or to those—the VSV nucleocapsin (N) protein, for example—that cannot be easily screened for biological activities such as hemagglutination or neutralization.

5 ADVANTAGES OF VACCINIA VIRUS RECOMBINANTS

Experience with mass production, storage, and administration of smallpox vaccine suggests that VV or related poxvirus recombinants could be the bases for veterinary and human vaccines offering advantages over current preparations. Lyophilized VV is heat stable and can be effectively administered by a variety of methods. During the global eradication of smallpox, scarification or jet gun administration was used to vaccinate millions of people. We have vaccinated animals by intradermal inoculation and scarification, and we plan to investigate the effectiveness of oral administration for wildlife and cattle.

It is possible to make a polyvalent vaccine using VV as the vector, by expressing several genes that represent various serotypes of the same agent or even a number of unrelated agents. While the manipulations for including several genes are increasingly difficult with each addition, new methods are being developed rapidly. The possibility of developing a polyvalent vaccine, with a single VV recombinant expressing a number of heterologous genes, is especially important for nomadic regions of Africa and Asia. Theoretically, one stock of VV could carry genes from most of the important livestock diseases in a region, and one inoculation could vaccinate an animal against half-a-dozen or more diseases.

The VV recombinant vaccine contains very few antigens in common with disease agents, and so we have found it possible to develop diagnostic kits for rinderpest, vesicular stomatitis, and simian AIDS that can distinguish vaccinated animals from those exposed to the whole agent either by infection or vaccination. Currently, we are expanding this technology to provide rapid diagnosis for a variety of livestock diseases.

We recognize the potential safety problems with widespread use of VV as a vector for different vaccines. However, the strain used for vaccine construction is the Wyeth strain of vaccinia virus, which was used worldwide without serious problems in the successful eradication of smallpox. Further, it has been demonstrated that insertional inactivation of the *TK* gene attenuates the virus. In constructing a rinderpest double recombinant, the TK region was inactivated by insertion of the RPV *H* gene; insertional inactivation of the VV HA region by incorporating the *F* gene of RPV has led to further attenuation of the recombinant virus. The absence of detectable pock lesions at the site of vaccination by the double recombinant is an indication of its greater attenuation, even though it is still completely protective. Moreover, the double recombinant VV vaccine we developed for rinderpest is highly attenuated: it is apathogenic for both athymic nude mice and SIV-infected rhesus macaques.

It is important that safety considerations be thoroughly addressed before live recombinant viruses are introduced into the environment. The rapidly expanding epidemic of HIV, particularly in Africa, demands that live vaccines not pose a threat to indigenous populations. The safety of VV recombinant vaccines has been confirmed by the absence of transmission of VV from vaccinated to contact animals. No pock lesions or antibody to vaccinia virus, as measured by SN and plaque reduction assays, could be demonstrated in control animals housed in contact with cattle vaccinated with our VV double recombinant vaccine for rinderpest. We believe that these outstanding safety features of the double recombinant make it suitable for field use. We believe also that this vaccine is only the first to bring the benefits of modern molecular biology and biotechnology to disease control. As knowledge and expertise in these fields expand, recombinant vaccines will be constructed for many of the more stubborn afflictions of human beings and animals, providing creative solutions to disease problems old and new.

See also Electroporation and Electrofusion; Expression Systems for DNA Processes; Genetic Immunization; Molecular Genetic Medicine; Plasmids; Transgenic Livestock as Bioreactors.

Bibliography

Brown, F., Ed. (1994) *Recombinant Vectors in Vaccine Development: Developments in Biological Standardization.* Karger, New York.

Chanock, R. M., Giensburg, H., Lerner, R., and Brown, F., Eds. (1987) *Vaccines 87. Modern Approaches to New Vaccines Including Prevention of AIDS.* Cold Spring Harbor Laboratory, Cold Spring Harbor, NY.

National Academy of Sciences & National Research Council (1988) *Science & Technology for Development: Prospects Entering the 21st Century.* NAS/NRC, Washington, DC.

Quinnan, G. V., Ed. (1985) *Vaccinia Viruses as Vectors for Vaccine Antigens.* Elsevier-North Holland, New York.

Talwar, G., Rao, K., and Chauhan, V., Eds. (1994) *Recombinant and Synthetic Vaccines.* Narosa Publishing House, New Delhi.

Yilma, T. (1987) Adjuvant genes for infectious virus recombinant vaccines: Use of lymphokines as adjuvants. *Biotechnol. U.S.A.:* 238–246.

———. (1989) Prospects for the total eradication of rinderpest. *Vaccine,* 7:484–485.

———. (1990) A modern vaccine for an ancient plague: Rinderpest. *BioTechnology,* 8:1007–1009.

———. (1991) The role of biotechnology in tropical diseases. In *Tropical Veterinary Medicine: Current Issues and Perspectives,* J. Williams, K. Kocan, and E. P. Gibbs, Eds., *Ann. N.Y. Acad. Sci.,* 653:1–6.

———. (1991) A vaccinia virus recombinant vaccine for rinderpest. *World Animal Review* (a Quarterly Journal of the FAO/UN on Animal Health, Production, and Products) 69:2–6.

Varicella-Zoster Virus and Cytomegalovirus Vaccines

Lawrence A. Hunt

1 Introduction to Varicella-Zoster Virus and Human Cytomegalovirus
- 1.1 Human Diseases and Protective Immunity
- 1.2 Efficacy and Safety of Live Attenuated Vaccines
- 1.3 Alternate Strategies for Vaccine Development

2 VZV and CMV Structure and Composition
- 2.1 Immunogenic Proteins
- 2.2 Viral Gene Sequences

3 Development of Protein Subunit Vaccines
- 3.1 Purified Protein Versus Inactivated Virus
- 3.2 Expression Vectors for Recombinant *E. coli* Bacteria
- 3.3 Recombinant Baculoviruses and Insect Cells
- 3.4 Mammalian Cell Lines Expressing Cloned Genes
- 3.5 Synthetic Peptides and Carrier Proteins
- 3.6 Anti-Idiotype Vaccines

4 Development of Live Recombinant Vaccines
- 4.1 Recombinant Vaccinia Viruses
- 4.2 Recombinant Adenoviruses

5 Plasmid DNA Vaccines for VZV and CMV

6 Animal Systems for Testing Vaccine Safety and Efficacy

7 Quantitation of Immune Responses to VZV and CMV Vaccines

Key Words

Adjuvant A natural or chemical substance that enhances antigen-stimulated antibody production by B lymphocytes.

Antigen A foreign substance that can stimulate a specific immune response, usually a specific antibody; specific determinants on the antigen which are bound by antibody are known as epitopes.

Live Vaccine A replication-competent viral or bacterial vaccine; either an attenuated form of natural infectious agent or a recombinant agent expressing a foreign gene.

Monoclonal Antibody A molecularly pure population of antibody produced by a single clone of B cells and with single specificity of antigen binding.

Neutralizing Antibody An antibody to surface protein of virus particles that renders the virus noninfectious.

Subunit Vaccine A purified component of an infectious agent that is incapable of replication in the host but can stimulate a protective immune response.

Classical live attenuated vaccines were developed several decades ago for two important members of the herpesvirus family, varicella-zoster virus and human cytomegalovirus. The varicella vaccine is now licensed for normal children in Japan, Korea, and the United States, and the cytomegalovirus vaccine has been tested in a limited number of kidney transplant recipients and healthy adult volunteers. A number of new cytomegalovirus vaccines are being developed using molecular biology and recombinant DNA technology, because of theoretical concerns about the long-term safety and effectiveness of the live cytomegalovirus vaccines. Some of the same techniques are also being applied to varicella-zoster virus to develop improved vaccines which could supplement or replace the current varicella vaccine. Most of the high technology vaccine development is based on the products of the viral glycoprotein genes and includes both recombinant protein subunit vaccines and live recombinant virus vaccines.

1 INTRODUCTION TO VARICELLA-ZOSTER VIRUS AND HUMAN CYTOMEGALOVIRUS

1.1 Human Diseases and Protective Immunity

Varicella-zoster virus (VZV) and human cytomegalovirus (CMV) are both members of the herpesvirus family. Along with other human members, such as herpes simplex virus types 1 and 2 and Epstein–Barr virus, these viruses have a common clinical feature of viral latency and reactivation following primary virus infection. Medical complications of reactivation are most severe in immunodeficient or immunosuppressed individuals.

VZV is the cause of the common childhood illness known as chickenpox or varicella, and this disease is characterized by a distinct vesicular rash on the trunk, head, and limbs. Complications of varicella, which include pneumonia and encephalitis, may be life threatening in immunocompromised persons. Passive immunization with high-titer human antibody against VZV, known as varicella-zoster immune globulin (VZIG), can prevent or significantly reduce disease if administered within a few days after exposure to VZV. Immunocompetent individuals acquire life-long immunity to varicella/chickenpox following this initial infection, even though the viral genetic information persists in a latent form in nerve tissue known as sensory ganglia.

The same virus also causes the disease known as shingles or herpes zoster, which is most common in the elderly. This disease results from reactivation of virus that has been latent in sensory ganglia for decades, with the skin lesions and accompanying pain usually localized to an area of skin innervated by the particular sensory ganglion. The major complication of herpes zoster is the "postherpetic neuralgia," which may persist for months in elderly patients. Whereas specific antiviral chemotherapy can reduce the severity of primary infections of at-risk children and adults, the same chemotherapy is relatively ineffective in treating reactivated VZV.

In contrast to chickenpox/varicella, most primary CMV infections of children are mild or inapparent, with no clinical symptoms. The site of viral latency following primary CMV infection is not as well characterized as for VZV, but presumably involves white blood cells known as monocytes and polymorphonuclear leukocytes. Serious infections are limited to two major groups: infants infected before birth when virus from women infected during pregnancy crosses the placental barrier, and immunocompromised persons such as immunosuppressed organ transplant recipients or persons with AIDS. Interuterine infection of the fetus during the primary infection of the mother carries the highest risk of serious disease and involves multiple organ systems. Antibodies and cell-mediated immunity induced by prior CMV infection of the mother appear insufficient to prevent virus reactivation during pregnancy, but the resulting congenital infections are almost always asymptomatic. Transplacental antibody can also protect newborns from infection after birth until this maternal antibody disappears from circulation.

Similarly, the risk of serious and life-threatening CMV disease in immunosuppressed organ transplant recipients is lower for persons undergoing CMV reactivation and reinfection than for persons with primary CMV infection following transplantation. Uninfected, CMV-seronegative transplant recipients can be passively immunized with high-titer human antibody against CMV to reduce the severity of CMV infection transmitted from the donor tissue. CMV infection is also common among adults infected with human immunodeficiency virus, and in this case even the secondary, reactivated infection can be serious. Specific antiviral chemotherapy can help reduce the severity of CMV infections in immunocompromised patients, but these drugs often display significant toxicity. In summary, acquired antibody and cell-mediated immune responses to natural CMV infection are insufficient to prevent viral reactivation or reinfection, but usually they can reduce the severity of the infection and prevent serious disease.

1.2 Efficacy and Safety of Live Attenuated Vaccines

The live attenuated viral vaccines for VZV and CMV were developed in the 1970s from human clinical isolates of virus by the classical technique of long-term propagation with multiple passages in human diploid fibroblast cell cultures. By unknown mechanisms, this passaging in cell culture led to VZV and CMV that were attenuated for replication in humans, and thereby caused infection without disease in most healthy persons when administered by intramuscular or subcutaneous infection. The licensed OKA vaccine strain of VZV and the candidate AD-169 and Towne vaccine strains of CMV can be distinguished from corresponding clinical isolates of VZV or CMV based on restriction endonuclease analysis of the viral DNA genome. However, clinical isolates of CMV usually also differ from each other. The significance of the difference between some freshly isolated strains and the candidate vaccines strains of CMV, AD-169, and Towne is not clear. Further, AD-169 and Towne strains are different from each other. Since the technology for restriction endonuclease analysis was not available at the time of the isolation of AD-169 or Towne strains, and very early passages of these virus strains are not available for analysis today, the genetic alterations responsible for the attenuation have not been identified. Thus, there is a theoretical risk of reversion to virulence during replication in humans.

Extensive studies of the varicella vaccine in large numbers of healthy children have demonstrated a high degree of safety and efficacy in inducing neutralizing antibody to VZV. In immunocompromised children, who are most at risk for dangerous disease and are therefore a primary target for vaccination, the live attenuated varicella vaccine is somewhat less effective and may actually cause a mild form of varicella. The varicella vaccine virus has the capacity to establish latency, and cases of zoster due to vaccine virus have been documented.

The candidate live attenuated CMV vaccines have been tested in smaller numbers of adults, including both CMV-seronegative transplant recipients and healthy individuals, with vaccination resulting in CMV-neutralizing antibody in the majority of individuals and reduced disease severity in transplant patients receiving organs from CMV-seropositive donors. These CMV vaccines appear to be safe, and latent infection and reactivation have not been demonstrated with these attenuated CMV strains. The primary targets for a safe and effective CMV vaccine would be CMV-seronegative females of childbearing age, to prevent symptomatic congenital infection, and immunocompromised children or adults who are CMV-seronegative.

The major concerns about live attenuated vaccines for VZV and CMV include the following: (1) the potential for latent infection and subsequent reactivation and disease, (2) limited duration of immunity compared to infection with "wild-type" virus, especially in immunocompromised individuals, (3) reversion to virulence during replication in humans, and (4) potential oncogenicity in humans. There is no direct evidence for a role of VZV or CMV in human cancers, although another human herpesvirus, Epstein–Barr virus, is important in the development of some B-cell malignancies and nasopharyngeal carcinoma. These major concerns also apply to the development of live attenuated vaccines for other human herpesviruses, but are not significant with respect to commonly used live attenuated vaccines for measles, mumps, and rubella viruses.

1.3 Alternate Strategies for Vaccine Development

Because of the theoretical concerns about live attenuated herpesvirus vaccines, there is great interest in new approaches to developing human VZV and CMV vaccines with higher levels of safety and effectiveness in preventing infection or disease. These approaches are based on biochemical understanding of the important protein components of these viruses and molecular understanding of the viral genes encoding these potential vaccine components. One major and current approach for VZV and CMV vaccines is the production of recombinant protein subunit vaccines in recombinant bacteria, insect cells, or mammalian cell culture. A second major approach is the production of recombinant human viruses that express VZV or CMV genes and can be used as live recombinant vaccines. Other modern technologies, such as anti-idiotype vaccines and plasmid DNA vaccines, can also be applied to the development of safe and effective VZV and CMV vaccines.

2 VZV AND CMV STRUCTURE AND COMPOSITION

All the members of the human herpesvirus family have a common structure for the virus particle, or virion, as shown in Figure 1. From the inside to the outside, there are four structures: an inner core containing the viral genetic information, or DNA genome, associated with viral protein; a protein shell with icosahedral symmetry known as an icosahedral capsid; an amorphous layer containing viral proteins known as the tegument or matrix; and a surface envelope composed of a cellular lipid bilayer and multiple species of viral glycoprotein. The envelope is the most fragile component of the virion, and removal or destruction of the envelope with detergents or organic solvents such as ether or chloroform renders the virion noninfectious. The corresponding capsid, tegument/matrix, and envelope glycoproteins of VZV and CMV serve similar functions in virus attachment to cells, intracellular replication, and virus assembly and egress. However, the corresponding VZV and CMV proteins have distinct amino acid sequences and can be distinguished by reactivity with VZV-specific versus CMV-specific antibody.

The viral genome, which encodes all these virion-associated proteins as well as additional intracellular enzymes and regulatory proteins, is a linear molecule of double-stranded DNA. As shown in Figure 2, herpesviral genomes have the unusual characteristic of re-

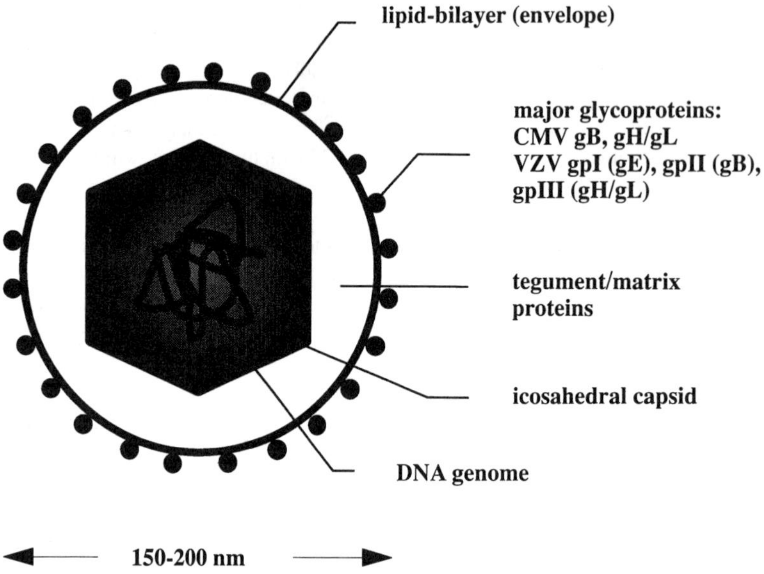

Figure 1. Structure and composition of VZV and CMV: a cross-section of the spherical virion for VZV and CMV. Major VZV glycoproteins in the envelope are identified by both old (gpI–gpIII) and new (gE, gB, gH/gL) nomenclatures.

peated DNA sequences at one or both ends and at an internal site (TR and IR in Figure 2), which results in recombinational inversion of unique long (U_L) and/or unique short (U_S) sequences during intracellular DNA replication and equal proportions of four different DNA isomers for CMV and two different isomers for VZV. Each of the isomeric forms of the linear DNA genome can be distributed equally into new virions, and all will exhibit similar infectivity. In addition, the purified DNA genome is infectious when artificially introduced into human cells in culture, resulting in the production of new infectious virions.

The DNA genome of CMV, at 229,000 base pairs (229 kb) is among the largest of all human viruses. The complete nucleotide sequence has been determined for both the AD-169 and Towne strains, and this sequence contains approximately 200 distinct open reading

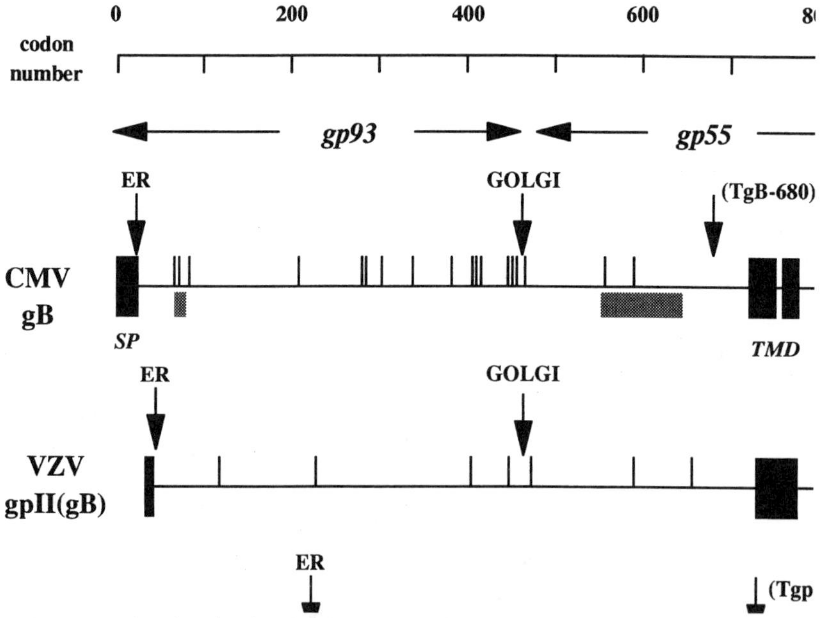

Figure 2. Structure of the DNA genome and location of major glycoprotein genes for CMV and VZV: one of four possible isomers for CMV and two possible isomers for VZV. The top scale of this genome diagram corresponds to genetic map units (0.0–1.0). Open boxes represent inverted repeat sequences at the ends of the unique long (U_L) and short (U_S) genome segments; arrows indicate the orientation of coding sequences for glycoprotein genes, which are identified by gene number and gene product.

frames (ORFs) capable of coding for approximately 200 distinct polypeptides. In contrast, the VZV DNA genome is 125 kb in length, and the complete nucleotide sequence of a laboratory strain other than the OKA vaccine strain predicts 71 ORFs for potential viral polypeptides.

2.1 Immunogenic Proteins

A general principle of viral immunology is that the surface proteins of the complete, infectious virion are the target antigens for protective antibody responses. In the case of lipid-enveloped viruses such as VZV and CMV, the neutralizing antibodies are elicited and bound by the envelope glycoproteins, whereas antibody produced to internal virion proteins such as tegument and capsid proteins is not protective and is useful only for confirming VZV or CMV infection. Some of these internal proteins, however, are important targets for specific helper T lymphocyte (T_H) and cytotoxic T lymphocyte (CTL) responses in natural CMV and VZV infection, including one of the major tegument/matrix proteins. The T_H cells assist B cells in the production of specific antibody, and the CTLs are capable of recognizing and destroying virus-infected host cells.

The major glycoprotein species for CMV are gB and the gH/gL heterodimer, and the major species for VZV are gE, gB, and the gH/gL heterodimer (Figure 1). These individual glycoproteins may exist in oligomeric forms in their native conformation on the surface of the virion, such as a dimer or tetramer for gB. The small gL glycoprotein is anchored in the envelope by its association with gH, rather than being independently anchored. The characteristics of these viral gene products are summarized in Table 1.

The original electrophoretic analyses of purified VZV and CMV had detected a large number of electrophoretically distinct envelope glycoproteins, but subsequent immunological analyses with monoclonal antibodies demonstrated that these multiple species actually represent a more limited set of gene products with heterogeneity due to oligosaccharide modifications and posttranslational proteolytic cleavage. For example, the CMV gB gene encodes a precursor glycoprotein of approximately 130–160 kDa and processed subunits of 93–116 and 55 kDa (gp93 and gp55 proteins in Figure 3). These glycoproteins contain multiple asparagine-linked oligosaccharides, which are added and processed by host cell machinery, and some of the proteins may also contain some serine- or threonine-linked oligosaccharides. Both viruses contain at least two additional minor glycoprotein species, but their contribution to the induction of neutralizing antibody in natural VZV and CMV infections is less well characterized. The OKA vaccine strain of VZV is deficient in the minor gpV(gC) protein compared to clinical isolates, but this deficiency is not thought to be responsible for the attenuation of the vaccine strain or the reduced immunogenicity of the vaccine compared to natural infection.

2.2 Viral Gene Sequences

Determination of the location and nucleotide sequences of individual glycoprotein genes in the large DNA genomes of VZV and CMV (Figure 2) permitted detailed structural analysis of the deduced amino acid sequences (Figure 3) and subsequent expression using recombinant DNA technology. In contrast to typical cellular genes for membrane glycoproteins, there are no intervening sequences (introns) within the coding sequences for the VZV and CMV glycoproteins. As a result, DNA fragments can be directly isolated by restriction endonuclease digestion from viral genome DNA, rather than having to isolate viral mRNA and then produce complementary DNA (cDNA) prior to recombinant expression.

The deduced sequences for the predominant CMV and VZV glycoproteins are indicative of type I integral membrane proteins, with an amino-terminal hydrophobic signal peptide (SP) necessary for transport of the protein into the lumen of the endoplasmic reticulum (ER) and another hydrophobic, transmembrane domain (TMD) closer to the carboxy-terminus, which functions as a membrane anchor (Figure 3). The removal of the SP is usually a cotranslational event, whereas the cleavage of CMV and VZV gB into mature subunits occurs late in processing, after transport from the ER to Golgi membranes. The positions of potential sites for cotranslational addition of asparagine-linked oligosaccharides in the ectodomain between the SP and TMD correspond to Asn-X-Ser/Thr, where X is any amino acid except proline. CMV gB, especially the gp93 subunit, is among the most highly glycosylated of all viral glycoproteins, and these oligosaccharides may help the virus evade host immune responses by preventing access of neutralizing antibodies to antigenic determinants on the viral polypeptide. Multiple cysteine residues in the ectodomain of these major VZV and CMV glycoproteins form intramolecular disulfide bonds, which are presumed to be very important in the overall structure and expression of antigenic determinants in the native molecules.

The important "linear" neutralizing epitopes of the CMV gB protein have been specifically mapped using synthetic or recombinant peptides based on the deduced amino acid sequence: the immunodominant domain is within the gp55 subunit, and an additional epitope is located near the amino-terminus of the gp93 subunit (Figure 3). "Conformational" or "discontinuous" epitopes, unlike "linear"

Table 1 Characteristics of Major Envelope Glycoproteins of CMV and VZV

Protein	Size(s)[a]	Viral Function	Immunogenicity
CMV gB	130–160 kDa precursor; 94–116 and 55 kDa mature subunits	Adsorption, penetration; major protein in viral envelope	Principal target of neutralizing antibody; target for T_H and CTL
VZV gB	125–140 kDa precursor; 62–65 and 57 kDa mature subunits	Adsorption, penetration; second most abundant protein in viral envelope	Important target of neutralizing antibody; target for T_H and CTL
VZV gE	81 kDa precursor; 98 kDa mature	Fc receptor; major protein in viral envelope	Major target of neutralizing antibody; target for T_H and CTL
CMV gH/gL	86/32 kDa	Cell fusion, viral maturation and egress	Target of neutralizing antibody, T_H and CTL
VZV gH/gL	118/22 kDa	Cell fusion, viral maturation and egress	Target of neutralizing antibody, T_H and CTL

[a] Size polymorphism is due to oligosaccharide heterogeneity and posttranslational proteolytic cleavage into subunits.

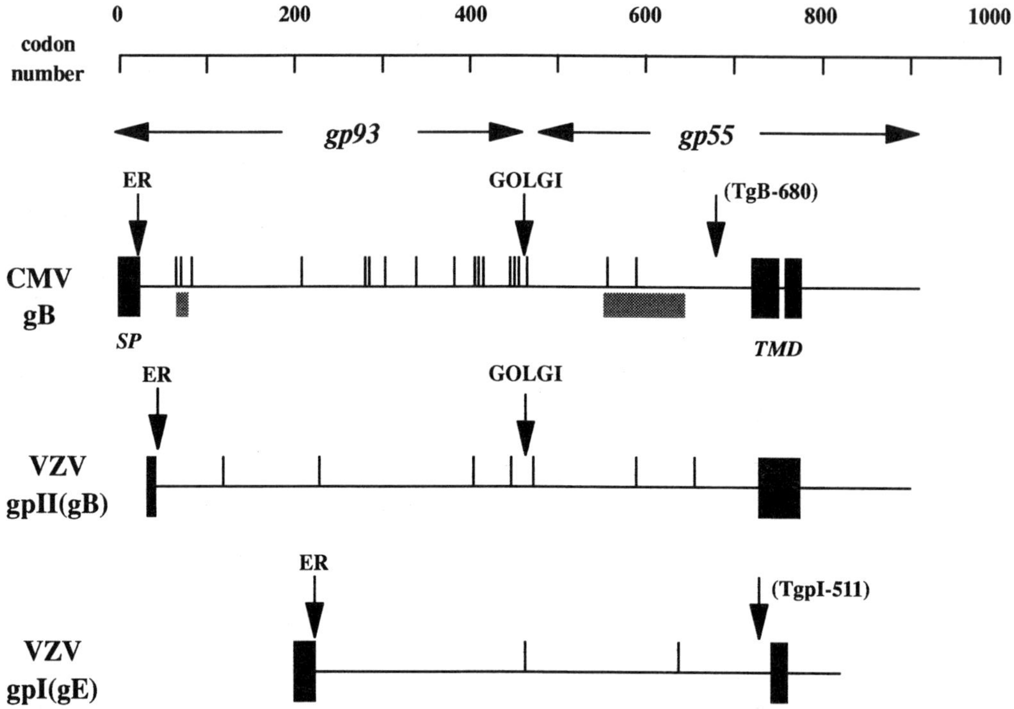

Figure 3. Linear structure of major CMV and VZV envelope glycoproteins: glycoprotein diagrams based on the derived amino acid sequences from the nucleic acid sequences of the corresponding genes. The solid black boxes represent hydrophobic domains: amino-terminal signal peptide (SP) and membrane anchor/transmembrane domain (TMD). Short vertical lines indicate the positions of potential asparaginyl-oligosaccharides. Vertical arrows indicate the positions of cotranslational cleavage by the cellular signal peptidase (ER), postranslational cleavage of gB into mature gp93 and gp55 subunits by a cellular endoprotease (Golgi), and removal of the membrane anchor/TMD by gene truncation after codon 680 of CMV gB (TgB-680) or codon 511 of VZV gpE (TgpI-511). Cross-hatched rectangles below the CMV gB polypeptide indicate the positions of an amino-terminal linear neutralization epitope in the gp93 subunit and an immunodominant linear neutralization domain in the gp55 subunit.

epitopes, cannot be duplicated by synthetic or recombinant peptides and usually are lost after protein denaturation or protein misfolding. Neutralization epitopes for the other VZV and CMV envelope glycoproteins are less well characterized.

3 DEVELOPMENT OF PROTEIN SUBUNIT VACCINES

Because of potential concerns with the long-term safety and effectiveness of live attenuated herpesvirus vaccines, there are major efforts under way in developing and testing protein subunit vaccines for CMV, and more limited studies are under way for VZV. Nonreplicating and inactivated protein vaccines have the major advantage of safety compared to live attenuated viral vaccines, but the disadvantage of requiring multiple vaccine doses plus a chemical or biological adjuvant to obtain helper T-cell responses and levels of antibody production by B cells comparable to those obtained with a single inoculation of a live attenuated vaccine. Also, inactivated protein vaccines do not usually stimulate specific CTL responses.

3.1 Purified Protein Versus Inactivated Virus

Chemical inactivation of whole virus particles has been used in the production of a number of licensed human vaccines, including vaccines for rabies, influenza, polio (Salk vaccine), and hepatitis A; whole-inactivated VZV and CMV have also been tested as vaccines in limited studies with animals and humans. Purified protein antigens are more expensive to produce, but they have less risk of incomplete inactivation of virion infectivity or transmission of infectious DNA genomes.

One technique for purification of soluble protein from clarified extracellular media or detergent lysates of cells is immunoaffinity chromatography, where specific antigen is bound to columns containing monoclonal antibody covalently linked to beads, and subsequently eluted. Another technique for viral envelope antigens is lentil lectin-affinity chromatography, which enriches for proteins with asparagine-linked oligosaccharides. Further purification based on the size or ionic charge of the antigen can be accomplished with gel filtration or ion exchange chromatography.

A number of recombinant techniques are being used to produce the envelope glycoproteins of CMV or VZV because of the difficulties in producing large quantities of the native proteins from virions or virus-infected cells. These also include production of the live recombinant human viruses discussed in Section 4, in the cases where the recombinant protein is purified from cells or culture media for use as a nonreplicating subunit vaccine. The characteristics of natural and recombinant protein subunit vaccines for CMV and VZV are summarized in Table 2.

3.2 Expression Vectors for Recombinant *E. coli* Bacteria

A number of plasmid vectors are available for high level, inducible expression of eukaryotic gene sequences in recombinant *E. coli* as

Table 2 Characteristics of Protein Subunit Vaccines for CMV and VZV

Gene Product	Source of Protein	Immune Responses
CMV gB	Immunoaffinity- or lectin-affinity-purified from CMV-infected human cells	Neutralizing antibody and T_H response in guinea pigs and humans
CMV gH	Immunoaffinity-purified from CMV-infected human cells	Neutralizing antibody in guinea pigs
CMV gB (nonglycosylated)	Purified from recombinant *E. coli* bacteria	Neutralizing antibody in mice
CMV gB (nonglycosylated)	Purified from insoluble fraction of recombinant yeast	Nonneutralizing antibody to denatured gB, but no antibody to native gB
CMV gB	Purified from insoluble fraction of recombinant baculovirus-infected insect cells	Neutralizing antibody in mice
CMV gB (soluble, truncated)[a]	Purified from extracellular medium of recombinant CHO mammalian cell line	Neutralizing antibody in guinea pigs and humans
CMV gB (soluble truncated/deleted)[b]	Purified from extracellular medium of recombinant baculovirus-infected insect cells	[Studies in progress]
VZV gE (soluble, truncated)[c]	Purified from extracellular medium of recombinant vaccinia virus-infected mammalian cells	Neutralizing antibody in rabbits

[a]Truncated TgB-680 protein in Figure 3.
[b]Truncated/deleted proteins in Figure 5.
[c]Truncated TgpI-511 protein in Figure 3.

fusion proteins, with many of these vectors using the *lac* operator/repressor system and induction by isopropyl-β-D-thiogalactopyranoside (IPTG). There are, however, several potential problems in expressing viral glycoprotein genes into immunogenic protein with the same neutralizing epitopes as the natural protein: (1) glycosylation at the multiple asparagine sites does not occur in bacteria, (2) the amino-terminal signal peptide (SP) may be toxic to the bacteria, and (3) the intramolecular and intermolecular disulfide bonds of the native protein do not form within the cytoplasm of the bacteria. A lysate of recombinant *E. coli* containing a nonglycosylated, SP-deleted form of CMV gB was able to elicit low levels of neutralizing antibody in mice. However, immunological analysis with monoclonal antibodies to native gB indicated that the *E. coli*-produced protein lacked some of the important antigenic determinants of the native CMV gB and might not be the most appropriate vaccine against CMV infection and disease in humans. Similar immunization results were obtained when CMV gB was expressed as a nonglycosylated and insoluble protein in recombinant yeast.

Recombinant fusion proteins containing CMV gB-specific peptide sequences at their carboxy-terminus have also been expressed in *E. coli,* to map linear neutralizing epitopes recognized by human antibody in natural CMV infection, but these fusion proteins have not been tested in vaccination studies in animals. One specific example is an IPTG-inducible gene encoding the maltose-binding protein linked to the amino-terminal neutralizing epitope of the gp93 subunit (Figure 3). Other bacterial plasmid expression systems are available which allow easy purification of the fusion protein based on the amino-terminal peptide or polypeptide, such as a histidine hexamer that can be absorbed to nickel or a glutathione *S*-transferase polypeptide, which can be bound to glutathione-Sepharose 4B beads. These expression plasmids also encode proteolytic cleavage sites downstream of the amino-terminal "tags," permitting the isolation of the desired viral peptide or polypeptide, free of the amino-terminal sequences prior to immunization studies.

3.3 Recombinant Baculoviruses and Insect Cells

The baculovirus and insect cell system is a popular eukaryotic expression system for production of viral glycoproteins because large quantities of glycosylated protein result when the viral coding sequences replace the coding sequences for the polyhedrin protein. The polyhedrin gene is expressed in huge quantities late in viral infection, but the gene product is unnecessary for virus replication in cell culture. The initial step in the construction of recombinant baculoviruses is to clone the viral coding sequences into an *E. coli* bacterial plasmid that contains a multiple restriction site for cloning downstream of the polyhedrin promoter and is flanked by baculovirus sequences. As shown in Figure 4, the recombinant baculoviruses are produced by cotransfection of full-length viral genome DNA and the adaptor plasmid containing the herpesviral gene sequences, followed by selection of rare recombinant viruses that result from intracellular reciprocal recombination. This process is simplified by the use of commercially available adaptor plasmids that also contain the *E. coli* β-galactosidase gene *(lacZ)* under control of another baculovirus promoter, for selection of blue viral plaques in cultures of baculovirus-infected insect cells. For scale-up production of recombinant protein, recombinant virus can be propagated in insect larvae rather than in tissue culture.

The major difference in posttranslational processing in insect cells versus mammalian cells is in the final stages of asparaginyl–oligosaccharide processing in Golgi vesicles: high-mannose, precursor oligosaccharide structures ($Man_{5\text{-}9}GlcNAc_2$-Asn) can be converted into complex, acidic structures [(NeuNAc ± Gal-Glc-

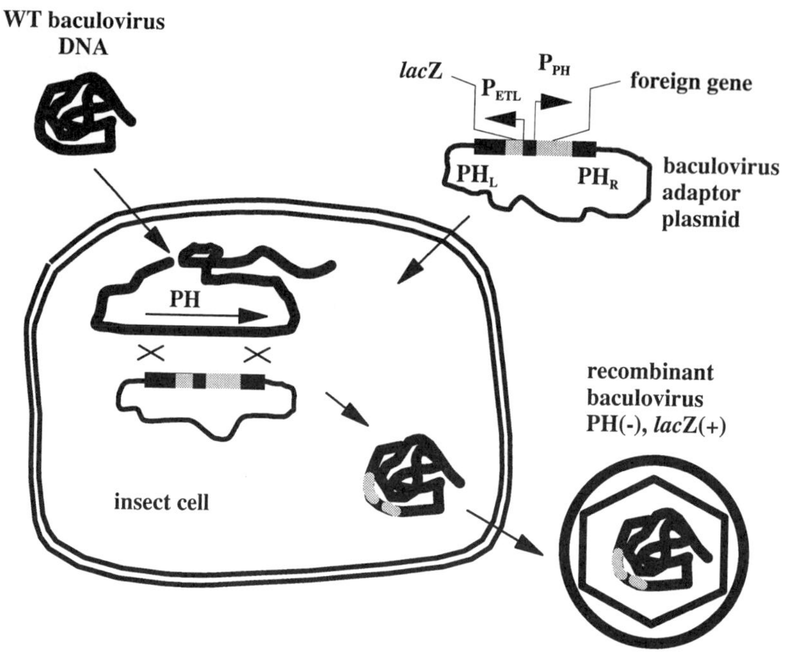

Figure 4. Production of recombinant baculoviruses to express CMV and VZV glycoprotein genes: the rare reciprocal recombination event that occurs in cells cotransfected with the full-length, wild-type (WT) DNA genome and adaptor plasmid DNA containing the foreign herpesviral gene. PH refers to the polyhedrin gene, PH_L and PH_R to left and right baculovirus recombination sequences in the plasmid, P_{ETL} and P_{PH} to specific baculovirus promoters for gene transcription; *lacZ* is the coding sequence for the bacterial β-galacosidase protein used in the blue-color selection for recombinant virus.

NAc)$_{2-4}$Man$_3$GlcNAc$_2$(±Fuc)-Asn] in mammalian cells, but are converted into more simple mature structures (Man$_3$GlcNAc$_2$-Asn) in insect cells. This oligosaccharide difference does not appear to adversely affect the structure and conformation of the polypeptide, and may actually make some of the linear or conformational epitopes even more accessible. In contrast, asparagine-linked oligosaccharides in yeast can be processed to large oligomannose structures known as mannan, and this modification can lead to insolubility and loss of antigenic determinants for recombinant glycoproteins.

The glycosylated, full-length gB protein of CMV has been purified from recombinant baculovirus-infected insect cells and demonstrated to induce neutralizing antibodies in mice. To obtain soluble and secreted gB protein, which can be more easily purified for testing as a subunit vaccine, we expressed truncated versions of gB lacking the membrane anchor (Figure 5); this was done, as well, in the laboratory of R. R. Spaete at Chiron Corporation (TgB-680 in Figure 3). Cleavage into gp93 and truncated gp55 subunits is inefficient in the baculovirus-infected insect cells, even when modified

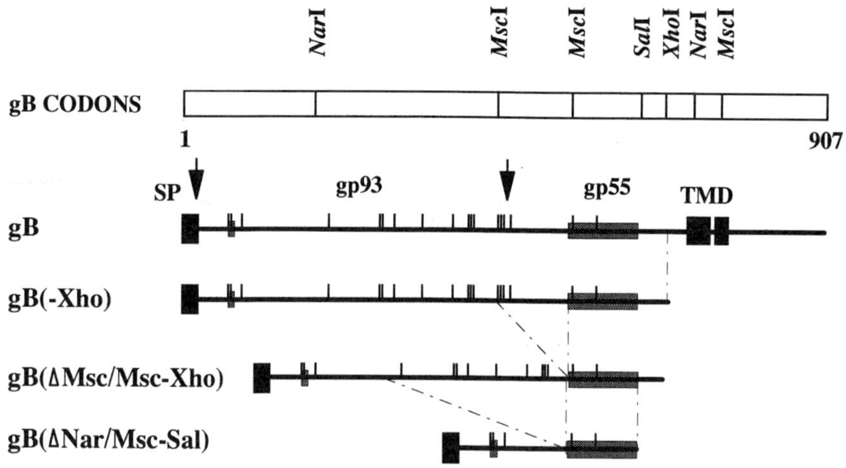

Figure 5. Structure of truncated and deleted CMV gB gene products expressed by recombinant baculoviruses; symbols as described in the legend to Figure 3. Restriction endonuclease sites within DNA sequences which were used for terminal truncation or internal deletion of gB coding sequences are displayed at the top, and the names of the three modified gene products also indicate the restriction endonuclease cleavages used for deletion or truncation.

gB proteins are efficiently transported through Golgi membranes. We have also demonstrated that internally deleted versions of gB (Figure 5) are efficiently secreted into the extracellular media and still maintain the antigenic determinants of both the immunodominant gp55 domain and the amino-terminal gp93 linear neutralization epitope. The smallest of these secreted products has less than one-third of the polypeptide and glycosylation of the full-length gB glycoprotein. These deleted/truncated gB proteins will be compared with full-length gB in animal immunization studies to determine whether the immunogenicity can be improved by the elimination of heavily glycosylated polypeptide domains.

Truncated versions of the VZV gB and gE glycoproteins can also be produced as soluble antigens in the baculovirus–insect cell system for testing as subunit vaccines. A number of different envelope glycoproteins of human herpes simplex virus (HSV) have been produced by recombinant baculovirus technology, and vaccination of mice induced neutralizing antibodies and protection against lethal challenge with HSV.

3.4 Mammalian Cell Lines Expressing Cloned Genes

Recombinant viral glycoproteins expressed in mammalian cells are usually similar, if not identical, to the corresponding natural protein with respect to co- and posttranslational modifications and antigenic structure. This advantage in production of glycoprotein subunit vaccines is balanced by the higher cost of mammalian cell culture compared to the baculovirus–insect cell system. Chinese hamster ovary (CHO) cell lines are commonly used in genetic studies because of the ease of manipulation and growth properties in selective media. CHO cell lines expressing the desired viral gene are obtained using methotrexate drug selection following cotransfection of a eukaryotic expression plasmid containing the viral gene with heterologous transcription signals and another plasmid encoding dihydrofolate reductase (DHFR) into a DHFR-negative CHO cell line. The drug selection leads to cell lines with high copy numbers of the plasmid DNA permanently incorporated into chromosomal DNA, and individual cell clones can be assayed for their level of viral gene expression.

CHO cell lines have been utilized for the production of secreted forms of human immunodeficiency virus gp120 and human CMV gB (TgB-680 in Figure 3) for testing as subunit vaccines. The membrane-anchor-deleted CMV gB is secreted as a mixture of cleaved subunits and uncleaved glycoprotein, and this secretion still occurs when the gp93/gp55 endoproteolytic cleavage site is eliminated by site-specific mutagenesis. This secreted gB has induced neutralizing antibody in animal studies and in limited human clinical trials. Analogous truncated forms of the VZV gB and gE genes might also be expressed as secreted and immunogenic proteins in CHO cell lines, but for intracellular transport and secretion of the gH/gL heterodimer to occur, truncated forms of the gH gene of VZV or CMV would need to be coexpressed with the corresponding gL gene in the same CHO cell.

3.5 Synthetic Peptides and Carrier Proteins

Peptides that mimic linear epitopes of native proteins, such as peptides corresponding to the immunodominant domain and amino-terminal neutralization epitopes of CMV gB (top of Figures 3 and 5), are not immunogenic and capable of stimulating significant antibody levels unless they are coupled to a carrier protein. This carrier protein functions as a target for specific T_H cells after proteolytic processing and presentation by major histocompatibility complex (MHC) molecules on the surface of "antigen-presenting cells." Such peptides could be covalently coupled to a protein such as the diphtheria toxoid currently used as a human bacterial vaccine.

This vaccine approach is limited to well-defined linear neutralization epitopes versus discontinuous conformational epitopes, and therefore further characterization and mapping of such epitopes for VZV gE and gB is necessary before development of peptide-based vaccines for glycoproteins other than CMV gB. Peptide-based vaccines might be useful in booster inoculations following primary immunization with protein subunit or live attenuated vaccines, to enhance the antibody response to specific antigenic determinants.

3.6 Anti-Idiotype Vaccines

The unique amino acid sequence and structure of the antigen-binding site of antibody molecules from a B-cell clone is known as the idiotype. Since secondary antibody elicited by this idiotype determinant on the primary antibody can sometimes mimic the original antigenic determinant, anti-idiotype antibodies can be used as surrogate subunit vaccines without the necessity of propagating infectious agents. Monoclonal antibody that neutralizes VZV or CMV, such as the human C23 monoclonal antibody to the amino-terminal linear epitope of CMV gB (Figures 3 and 5), is the ideal reagent for producing anti-idiotype antibodies in humans. A disadvantage to this approach is that these anti-idiotype antibodies are most easily obtained in mice using hybridoma technology, but would be inappropriate for human use unless the constant region domains were replaced by human immunoglobulin constant regions. A large number of murine monoclonal antibodies with specificity for CMV and VZV glycoproteins and virion neutralization activity have been produced by standard hybridoma technology. Recombinant anti-idiotypic antibody for use as a human subunit vaccine can be produced by ligating DNA sequences encoding the variable region of the immunoglobulin heavy and light chains from these murine anti-idiotypic antibodies to constant region sequences from human heavy and light chain immunoglobulin genes.

4 DEVELOPMENT OF LIVE RECOMBINANT VACCINES

Live recombinant vaccines have the advantage of inducing neutralizing antibody plus cytotoxic T lymphocyte responses, and single inoculations are usually as effective as multiple doses of a nonreplicating subunit vaccine. Vaccinia virus, a member of the poxvirus family, has been most extensively developed as a live recombinant vector because this naturally attenuated virus has been used as a smallpox vaccine in humans for two centuries. Vaccinia virus does cause serious complications in humans in rare instances, especially in immunocompromised persons. Avian poxviruses are also being engineered to express human viral proteins, because the incomplete replication of these viruses in human cells makes them safer than vaccinia virus.

Enteric-coated human adenoviruses, used extensively by the military since 1969 as vaccines to prevent acute respiratory disease, have more recently been adapted as live recombinant vectors. The characteristics of live recombinant vaccinia virus and adenovirus vaccines for CMV and VZV are summarized in Table 3. Attenuated bacterial strains that replicate in the human intestinal tract, such as

Table 3 Characteristics of Live Recombinant Virus Vaccines for CMV and VZV

Gene Product	Recombinant Vector	Immune Responses
CMV gB	Vaccinia virus	Neutralizing antibody in mice
CMV gB	Adenovirus	Neutralizing antibody in guinea pigs, CTL response in mice
VZV gE	Vaccinia virus	Neutralizing antibody in rabbits
VZV gE (soluble, truncated)[a]	Vaccinia virus	Neutralizing antibody in rabbits
VZV tegument protein (ORF62)	Vaccinia virus	Protection against VZV challenge in guinea pigs (CTLs and/or T_H cells ?)

[a]Truncated TgpI-511 protein in Figure 3.

Salmonella typhi, have been genetically engineered to express foreign genes, but these live recombinant bacterial vaccines would be most appropriate as vectors for nonglycosylated proteins of enteric human viruses.

4.1 Recombinant Vaccinia Viruses

The construction and isolation of recombinant vaccinia viruses for expression of human viral genes is similar in some respects to the process for recombinant baculoviruses: the foreign gene sequences are cloned downstream of a vaccinia promoter in an adaptor plasmid, this plasmid DNA is transfected into mammalian cells in culture, which have been infected several hours before with wild-type vaccinia virus, and rare recombinant virus is produced by reciprocal recombination and replacement of the vaccinia thymidine kinase (TK) gene with the foreign gene plus the *lac*Z indicator gene (Figure 6). TK^-*lac*Z^+ recombinant virus can be negatively selected by using a TK^- human cell line and 5-bromodeoxyuridine (Brdu), and positively selected by the appearance of blue plaques in the presence of 5-bromo-4-chloro-3-indolyl-β-D-galactoside (X-gal) or halogenated indolyl-β-D-galactoside (bluo-gal). Elimination of the vaccinia TK gene activity, which is unnecessary for virus replication in cell culture, further attenuates the vaccinia virus; thus complications of vaccinia in humans may be lessened.

The large vaccinia DNA genome (185 kb) can accommodate at least 25 kb of additional foreign DNA sequences without significant deleterious effects on virus replication, with the result that multiple foreign virus sequences can potentially be expressed in the same recombinant vaccinia virus. For example, one virus could be designed to express the VZV gE and gB genes, whose products are most important for induction of neutralizing antibodies, and also the VZV ORF62 tegument protein gene, to induce high levels of VZV-specific CTLs.

Since membrane-anchored, cell surface antigens are usually more immunogenic than the corresponding secreted or intracellular forms, full-length rather than truncated glycoprotein genes may be

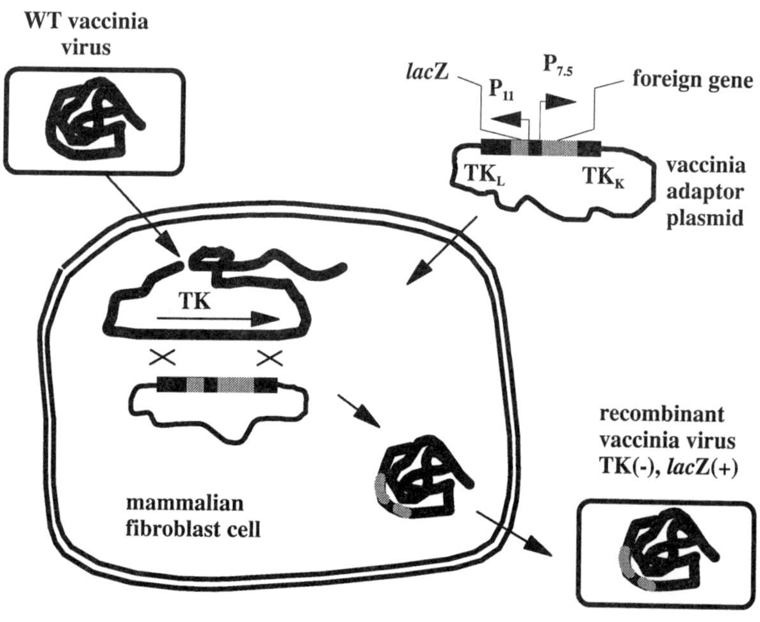

Figure 6. Production of recombinant vaccinia viruses to express CMV and VZV genes: the rare reciprocal recombination event that occurs in cells infected with wild-type vaccinia virus and cotransfected with adaptor plasmid DNA containing the foreign herpesviral gene; abbreviations and symbols as in the legend to Figure 4, except that TK refers to the vaccinia thymidine kinase DNA sequences and enzymatic activity, and P_{11} and $P_{7.5}$ refer to transcriptional promoters that are specific to vaccinia virus.

most appropriate when recombinant vaccinia or adenoviruses are to be used as live recombinant vaccines. The gB genes of both CMV and VZV have been cloned into vaccinia virus, although the short signal peptide coding sequence for the VZV gB (middle of Figure 3) had to be replaced with a longer sequence to obtain expression of the recombinant protein. Immunization of mice or rabbits with these live vaccines resulted in neutralizing antibodies and/or CTL responses. Recombinant vaccinia viruses expressing either the full-length or truncated VZV gE genes elicited neutralizing antibody in rabbits, but the relative levels of immunity induced by the two recombinant viruses were not compared in detail.

4.2 Recombinant Adenoviruses

In contrast to the large capacity of recombinant vaccinia virus for additional foreign DNA, the 36 kb DNA genome of human adenovirus can accommodate only approximately 1.8 kb additional DNA without genetic instability and inefficient packaging of the genome into the adenoviral icosahedral capsid. Because 1.9 or 2.7 kb deletions can be made in the viral E3 genes, which are nonessential for replication, foreign gene inserts of up to 3.7 or 4.5 kb can be accommodated during the construction of replication-competent, recombinant adenoviruses. Construction of these recombinant adenoviruses utilizes an adaptor plasmid containing the final 40% of the genome sequences for human adenovirus type 5, with the foreign coding sequences inserted downstream of the viral E3 promoter (Figure 7). These recombinant viral sequences are cotransfected or electroporated into human epithelial cells in culture along with an adenoviral DNA fragment containing the first 76% of the genome, and infectious adenovirus is isolated after several days.

A replication-competent adenovirus that expressed the full-length CMV gB gene was capable of eliciting both neutralizing antibody and CTL responses following intraperitoneal inoculation into mice or intranasal inoculation into hamsters. These immune responses occurred even though adenovirus infection is incomplete in these nonhuman hosts, presumably because the adenoviral E3 promoter is expressed at an early time after infection even when intracellular replication is deficient. Deletion of the viral E3 gene sequences may actually enhance CTL responses because one of the E3 gene products of the wild-type adenovirus down-regulates the processing and presentation of viral antigens by host cells. An adenovirus expressing the gB gene of human herpes simplex virus (HSV) was able to protect mice against a lethal challenge with HSV. The construction and testing of recombinant adenoviruses expressing other CMV or VZV glycoproteins has not yet been reported.

Most of these recombinant adenoviruses are derived from human adenovirus type 5, but analogous systems are being developed with the type 4 and 7 viruses, which are currently in use in an enteric-coated form for military recruits. The enteric-coated adenoviruses have the advantage of oral administration in humans without the threat of respiratory disease. In addition, these vaccines have the potential to induce mucosal IgA antibody, which can block virus soon after entrance into the body, rather than just the conventional IgG antibody in serum, which blocks infection at a later, generalized stage.

5 PLASMID DNA VACCINES FOR VZV AND CMV

The most revolutionary development in vaccines against human virus infection and disease utilizes direct inoculation with plasmid DNA containing foreign gene sequences under the control of eu-

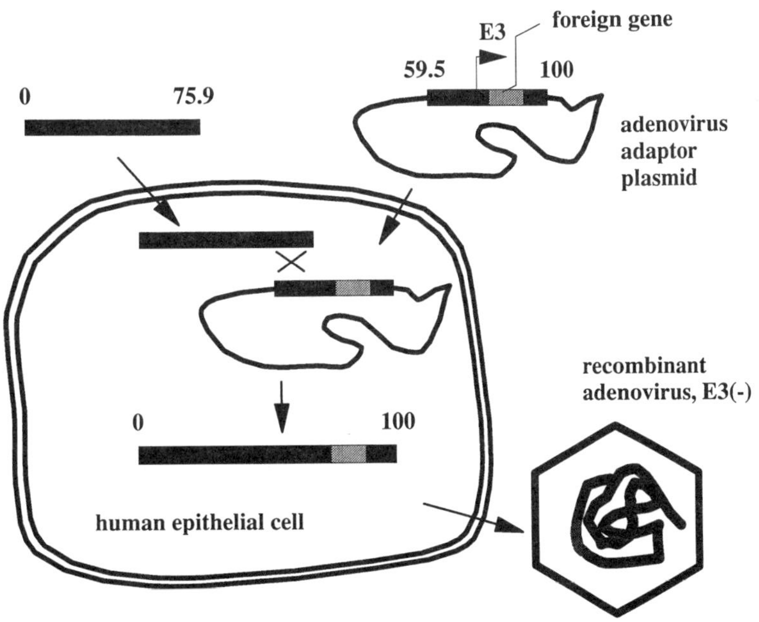

Figure 7. Production of recombinant adenoviruses to express CMV and VZV genes: the rare reciprocal recombination event that occurs in cells cotransfected with overlapping fragments of the complete adenovirus DNA genome (0–100 map units), with one plasmid-associated fragment (59.6–100 map units) containing herpesviral gene sequences replacing adenoviral E3 coding sequences downstream of the E3 promoter. The infectious icosahedral virion (nonenveloped) of the recombinant adenovirus is displayed at the bottom right.

karyotic transcriptional signals. The seminal discovery leading to these new nucleic acid vaccines was the observation in 1990 that naked plasmid DNA injected into the muscle of a mouse could be expressed into proteins with the correct antigenic or enzymatic structure. Prior to that report, the transfection of naked plasmid DNA into animal cells was believed to be a very inefficient process unless conducted in cell culture under very special laboratory conditions. The initial reports of successful plasmid DNA vaccination against virus infection appeared in 1993 and utilized cloned influenza A virus genes. This technology has since been applied to a large number of other viral genes and would be most applicable to viral genes that induce protective immune responses when expressed by live recombinant vaccinia and adenovirus vaccines: the gB gene of human CMV and the gE and ORF62 tegument protein genes of VZV.

The basic requirements for the plasmid DNA vaccines are the same for plasmids used for transient gene expression in transfected mammalian cells in culture: upstream promoter and enhancer sequences recognized by host cell transcriptional machinery, coding sequences with in-frame ribosomal initiation and termination codons, and downstream polyadenylation and RNA termination sequences. An intron sequence either upstream or downstream of the insertion site of the foreign coding sequences can enhance the processing and transport of the expressed mRNA.

The amount of plasmid DNA required for effective immune responses varies with the specific gene sequences and the route of inoculation (from <1 μg to >100 μg). Use of a "gene gun" for epidermal inoculation of gold beads coated with plasmid DNA has been much more effective at low doses of DNA than more conventional intramuscular or intravenous inoculation routes in mice and ferrets for protection against challenge with influenza virus.

These DNA vaccines are safer than live recombinant vaccines because they do not insert into chromosomal DNA and do not replicate once they have entered the cell nucleus, but retain the potential of eliciting CTL responses. In addition, the same plasmid DNA vector can be used multiple times with either the same or different foreign gene sequences, without the immunized host developing immunity to the vector. In contrast, reimmunization with recombinant adenovirus or vaccinia viruses containing a new foreign gene is less effective in stimulating antibody and cellular immune responses to the new gene product because of acquired immunity to the virus vector. The production and purification of bacterial plasmid DNA is also much less expensive than the purification of protein for subunit vaccines or the production and purification of live recombinant virus vaccines. This new plasmid DNA technology is also applicable to human gene therapy, such as replacement of the defective cystic fibrosis transmembrane conductance regulator gene in the epithelial tissue of affected humans.

6 ANIMAL SYSTEMS FOR TESTING VACCINE SAFETY AND EFFICACY

Development of suitable animal models has been difficult for both VZV and CMV because of the limited host range of these human viruses. A model system with newborn guinea pigs that was developed with a guinea pig adapted strain of VZV can be used for VZV challenge following immunization with protein subunit, live recombinant, or plasmid DNA vaccines. Most of the immunological studies with the VZV–guinea pig model have involved immune responses to natural VZV infection rather than responses to VZV vaccines. A live recombinant vaccinia vaccine expressing the ORF62 tegument protein of VZV was able to protect guinea pigs against a challenge with the guinea pig adapted strain of VZV, but the published studies with recombinant VZV gE vaccines have utilized rabbits and neutralizing antibody assays (summarized in Tables 2 and 3).

Since there is no animal host for human CMV replication and disease, immunized animals cannot be directly challenged with human CMV. There are mouse and guinea pig cytomegaloviruses (mCMV and gpCMV), and infections of the corresponding hosts with these two animal herpesviruses results in diseases with some similarity to human CMV disease. Transplacental transmission and congenital disease have been demonstrated for gpCMV. The gB gene product encoded by gpCMV is heavily glycosylated and has been recently shown to have some sequence similarity with human CMV gB, but there is no antibody cross-reactivity between the two proteins. The gpCMV–guinea pig system is probably a better biological and clinical model system than the mCMV–mouse system, but it cannot replace direct testing of human CMV-derived vaccines.

Immunization studies with protein subunit or live attenuated vaccines for human CMV have utilized guinea pigs or mice for animal studies and humans in limited trials of subunit vaccines. Whereas the live recombinant vaccinia or adenovirus vaccines require only a single dose for neutralizing antibody or CTL responses in most vaccine trials, the nonreplicating subunit vaccines generally require the addition of an adjuvant to the viral protein plus booster immunizations to induce significant titers of antibody in the majority of immunized animals or humans. The complete or incomplete Freund's adjuvant, derived from inactivated bacteria, can be utilized in animal studies with protein vaccines, but the local inflammation at the site of immunization is not acceptable for human vaccines. Alum (aluminum hydroxide) can be used as an adjuvant in human vaccines, but it is less effective than Freund's adjuvant in animal studies. Recent studies with a baculovirus-produced CMV gB subunit vaccine in mice demonstrated that QS-21 is an effective adjuvant and helps stimulate high titers of neutralizing antibody to human CMV. QS-21 is a natural product from the bark of the *Quillaja saponaria* tree.

The animal studies with live recombinant vaccinia viruses can be conducted in a wide range of animal hosts because of the broad host range for vaccinia virus, but the only nonhuman hosts that allow complete replication of human adenoviruses are chimpanzees and cotton rats. Other rodents have been used in animal studies of the live recombinant adenovirus expressing the CMV gB gene, but replication is incomplete in guinea pig and mouse tissues and does not mimic the complete adenovirus replication that would occur in human epithelial tissues. Animal studies with live recombinant adenoviruses have utilized intranasal or intramuscular inoculation, but to mimic the inoculation route used in human vaccination with adenovirus types 4 and 7, future studies should involve oral administration with enteric-coated adenovirus.

Replication-defective adenoviruses have been used in human gene therapy trials, such as adenoviruses encoding the cystic fibrosis transmembrane conductance regulatory protein. These defective viruses have deletions in several viral regulatory protein genes, and viral particles must be produced in a human cell line that constitutively synthesizes these adenovirus proteins. Because these defective particles do not replicate their genomes and reproduce infectious particles once they have entered human epithelial tissue cells in the human respiratory or gastrointestinal tract, their recombinant gene expression would be lower than with replication-competent adenoviruses. However, these defective adenovirus particles might induce immune responses similar to single doses of a plasmid DNA vaccine.

7 QUANTITATION OF IMMUNE RESPONSES TO VZV AND CMV VACCINES

The ideal assay for quantitation of protective immune responses following immunization with human VZV or CMV vaccines is the challenge of immunized animals or humans with a defined amount of infectious natural virus, with determination of whether infection and disease is reduced in vaccinated individuals. This challenge is possible only in the guinea pig animal system with VZV, and experimental challenges of humans with VZV or CMV are usually not allowed for ethical reasons. Thus, most of the quantitation of immune responses to protein subunit, live recombinant, and DNA vaccines for CMV and VZV requires indirect assays for protective immune responses: quantitation of neutralizing antibody and quantitation of helper/cytotoxic T-cell responses.

Neutralizing antibody titers are more reliable than titers of antigen binding from simple enzyme-linked immunosorbant assays (ELISAs). Neutralizing antibody assays are performed by mixing serial dilutions of serum from immunized animals and control, unimmunized animals with a defined amount of infectious CMV or VZV, and then inoculating human fibroblast cells in culture and quantitating a reduction in viral plaques, viral antigen production, or viral cytopathic effects. These herpesviral neutralization assays are usually performed in the presence and absence of complement, because many antibodies to CMV and VZV glycoproteins require complement for neutralization activity. Neutralizing antibody from human clinical trials with CMV or VZV vaccines can be quantitated in the same manner.

Specific helper T cell (T_H) responses are usually determined by removing peripheral blood mononuclear cells (PBMCs) from immunized humans or spleen cells from immunized animals; then, following incubation in the presence or absence of specific viral antigen, cell proliferation is quantitated by radioactive thymidine incorporation into chromosomal DNA. Quantitation of cytotoxic T cell responses to processed VZV or CMV antigens in the immunized animals or humans requires antigen-presenting cells from the same host or genetically identical host which are synthesizing VZV or CMV proteins. For example, recombinant vaccinia virus infected host cells expressing CMV gB could be utilized in cell culture as target cells for analyzing CTL responses in animals receiving gB protein subunit vaccines or DNA vaccines encoding CMV gB. Cytotoxicity is usually monitored by release of ^{51}Cr from target cells in the presence of spleen or PBMCs from the immunized animals or humans.

Protective immune responses in humans following immunization with new vaccines for CMV or VZV can also be quantitated in long-term studies by following the natural incidence of virus infection and disease in the immunized population versus a corresponding nonimmunized population. Conclusive evidence of long-term immunological protection and safety is necessary before a human CMV vaccine can be licensed for general use, or a new human VZV vaccine is licensed to replace or supplement the current live attenuated varicella vaccine.

See also VIROLOGY, MOLECULAR; VIRUS–RECEPTOR INTERACTIONS.

Bibliography

Chee, M., Bankier, A., Beck, S., Bohni, R., Brown, C., Cerny, R., Horsnell, T., Hutchison III, C., Kouzarides, T., Martignetti, E., Preddie, E., Satchwell, S., Tomlinson, P., Weston, K., and Barrell, B. (1990) Analysis of the protein coding content of the sequence of human cytomegalovirus AD169. *Curr. Top. Microbiol. Immunol.* 154:125–169.

Committee on Infectious Diseases. (1995) Recommendations for the use of live attenuated varicella vaccine. *Pediatrics,* 95:791–796.

Davison, A. J., and Scott, J. E. (1986) The complete DNA sequence of varicella-zoster virus. *J. Gen. Virol.* 67:1759–1816.

O'Reilly, D. R., Miller, L. K., and Luckow, V. A. (1992) *Baculovirus Expression Vectors: A Laboratory Manual.* Freeman, New York.

Pardoll, D. M., and Beckerleg, A. M. (1995) Exposing the immunology of naked DNA vaccines. *Immunity,* 3:165–169.

Plotkin, S. A. (1994) Cytomegalovirus vaccines. In *Vaccines,* 2nd ed., S. A. Plotkin and E. A. Mortimer, Eds., pp. 803–807. Saunders, Philadelphia.

Roizman, B., Whiteley, R. J., and Lopez, C., Eds. (1993) *The Human Herpesviruses.* Raven Press, New York.

Takahashi, M., and Gershon, A. A. (1994) Varicella vaccine. In *Vaccines,* 2nd ed., S. A. Plotkin and E. A. Mortimer, Eds., pp. 387–417. Saunders, Philadelphia.

White, D. O., and Fenner, F. J. (1994) *Medical Virology,* 4th ed. Academic Press, San Diego, CA.

VIRAL ENVELOPE ASSEMBLY AND BUDDING

Milton J. Schlesinger

1 The General Problem of Virus Release from Cells; Models of Assembly

2 Components of Enveloped Viruses Critical to Assembly and Budding

3 Experimental Methods for Studying Assembly and Budding

4 Specific Virus Families and Assembly Processes
- 4.1 Alphaviruses
- 4.2 Flaviviruses
- 4.3 Rhabdoviruses
- 4.4 Rotaviruses
- 4.5 Retroviruses
- 4.6 Herpesviruses

Key Words

Lipid Bilayer The chemical structure that forms cellular membranes and is composed of phospholipids, sphingolipids, and cholesterol organized in two apposing monolayers.

Matrix Protein The viral-encoded polypeptide that links the nucleoprotein with the membrane and binds viral-encoded, membrane-associated glycoproteins.

Membrane Curvature The bending of the lipid bilayer that accompanies virus budding and results in an asymmetry of lipid bilayer composition.

Nucleoproteins The viral genetic material (i.e., RNA or DNA) and viral-encoded proteins organized in either stable icosahedral or helical structures.

Transmembranal Glycoproteins The viral-encoded polypeptides that form spikes on the outer surfaces of enveloped viruses but are anchored to the membrane by hydrophobic sequences that span the lipid bilayer and contain domains that are topologically oriented to the interior of the virus particle or to the cytoplasmic face of the membrane of the infected host cell.

The assembly and budding of enveloped viruses comprise the interactions that occur between viral nucleoprotein complexes and membrane-embedded, viral-specific glycoproteins at infected cell membranes and lead to the envelopment of the nucleoproteins followed, in most cases, by release of newly replicated, infectious viruses from host cells.

1 THE GENERAL PROBLEM OF VIRUS RELEASE FROM CELLS; MODELS OF ASSEMBLY

To continue their cycle of infection, reproduction, and transmission, newly assembled viruses must be able to spread from the cell, which serves as the host for virus replication. For most "naked" viruses (i.e., those containing only a genome packaged in a protein shell and growing in a eukaryotic cell), newly assembled viruses form paracrystalline arrays in the cellular cytoplasm or nucleus, and release generally requires death and disintegration of the host cell. In contrast, enveloped viruses containing a lipid bilayer as part of their structure utilize a host cell membrane and a set of biochemical activities that lead to a "budding" or membrane extrusion, which often occurs in the absence of host cell disintegration. The task for the virus in this latter pathway is to bring together at the proper time and place the components essential for creating a newly replicated and infectious particle and to initiate a set of biochemical events that will ensure envelopment of the lipid bilayer around the nucleoprotein core of the virus. Under the same conditions, there is generally an exclusion of nonviral proteins. For some viruses this exclusion is virtually complete and only viral products are in the secreted particles; for many enveloped viruses, however, cellular proteins, which are present on the membranes at sites of budding, can account for as much as 5% of the total protein in the particle.

Three separate events can be depicted in this assembly and budding process (Figure 1): a nucleation, involving contacts between the nucleoprotein cores and transmembranal virus-specific polypeptides; a wrapping of the membrane around the virus core, driven by additional protein–protein interactions; and a fusion of the virus envelope, which frees the virus from the cell membrane. The curvature of the bilayer around the core may lead to as much as a 25% differential in volume between inner and outer bilayers, creating a significant asymmetry in bilayer composition. Some lipids, such as cholesterol and sphingomyelin, are probably involved in these changes. A biophysical analysis of virus budding, based on membrane thermal fluctuations, hydrodynamic interactions, and glycoprotein diffusion, concludes that the membrane deformation accompanying budding requires a lowering of interfacial tension of the outer leaflet of the bilayer, which could be provided by association of the glycoproteins or other virus proteins with specific lipids. In some cases, viruses might overcome some of the problems of membrane deformation by selecting for their sites of budding cellular membranes having high curvature (i.e., intracellular vesicles instead of the cell surface).

Direct visualization of the assembly and budding process has been obtained for a variety of enveloped viruses utilizing high resolution electron microscopy. Thin sections of fixed and embedded infected cells and surface replicas of deep-frozen, freeze-fractured preparations clearly illustrate the process of virus budding out of cells (Figure 2).

2 COMPONENTS OF ENVELOPED VIRUSES CRITICAL TO ASSEMBLY AND BUDDING

For assembly of most enveloped viruses, three distinct components are needed: (1) the nucleoprotein core, consisting of the viral genome in the form of either RNA or DNA and its protein-protective coat, (2) the viral-encoded, membrane-anchored proteins (often referred to as spikes), which are localized to the surface of the virus envelope, and (3) the lipid bilayer.

The nucleoprotein cores contain the viral genome wrapped in virus-encoded proteins, which form structures that are either helical or icosahedral. The molecular events operating at this stage of virus assembly involve a coating of the DNA or RNA with multiple copies of polypeptides or a packaging of the nucleic acid concomitant with assembly of a protein shell. Specific sequences in the nucleic acids and specific domains in the protein function as sites for binding nucleic acid and protein and for initiation of this part of the virus assembly. Nucleoprotein core formation tends to be localized in the infected cell to areas in which replication of the nucleic acid occurs—namely, the cell's nucleus for DNA and the cell's cytoplasm for RNA, although there are exceptions for specific viruses, as noted later. The packaged genome then participates in the next stage of assembly, which involves a cellular membrane.

Three families of enveloped viruses do not conform to the process just described. One is the *Retroviridae,* and most members of this group—including HIV, the agent responsible for AIDS—assemble by a series of events that *combine* the packaging of nucleic acid with membrane budding. The second is the *Hepadnaviridae*—including the hepatitis B virus—which package an RNA molecule into a partially formed protein core and then transcribe the RNA to DNA prior to final assembly and release from the infected cell. The third family is the *Orthomyxoviridae*—including influenza viruses—which contain within the virus particle eight separate helical RNA–protein complexes. These genomic RNAs are formed into their helical protein cores in the infected cell's nucleus and are then transported to the plasma membrane, where budding occurs.

Different families of enveloped viruses use different host cell membranes for their sites of assembly and budding (Figure 3). The specific site is determined by the cellular membrane to which the virus-encoded glycoproteins are directed. For example, some coronaviruses assemble and bud from the Golgi vesicles in which their glycoproteins accumulate; in contrast, the influenza viruses bud from the cell's plasma membrane, where their glycoproteins accumulate. In addition, cells of certain kinds—when organized into their normal physiological state—contain gap junctions that lead to a separation of their plasma membranes into apical and basal surfaces. The site of virus budding in these cells is determined also by the site to which the virus glycoprotein migrates. Thus the influenza virus glycoprotein is found at the apical surface of such cells and virus buds from this surface, whereas the vesicular stomatitis virus

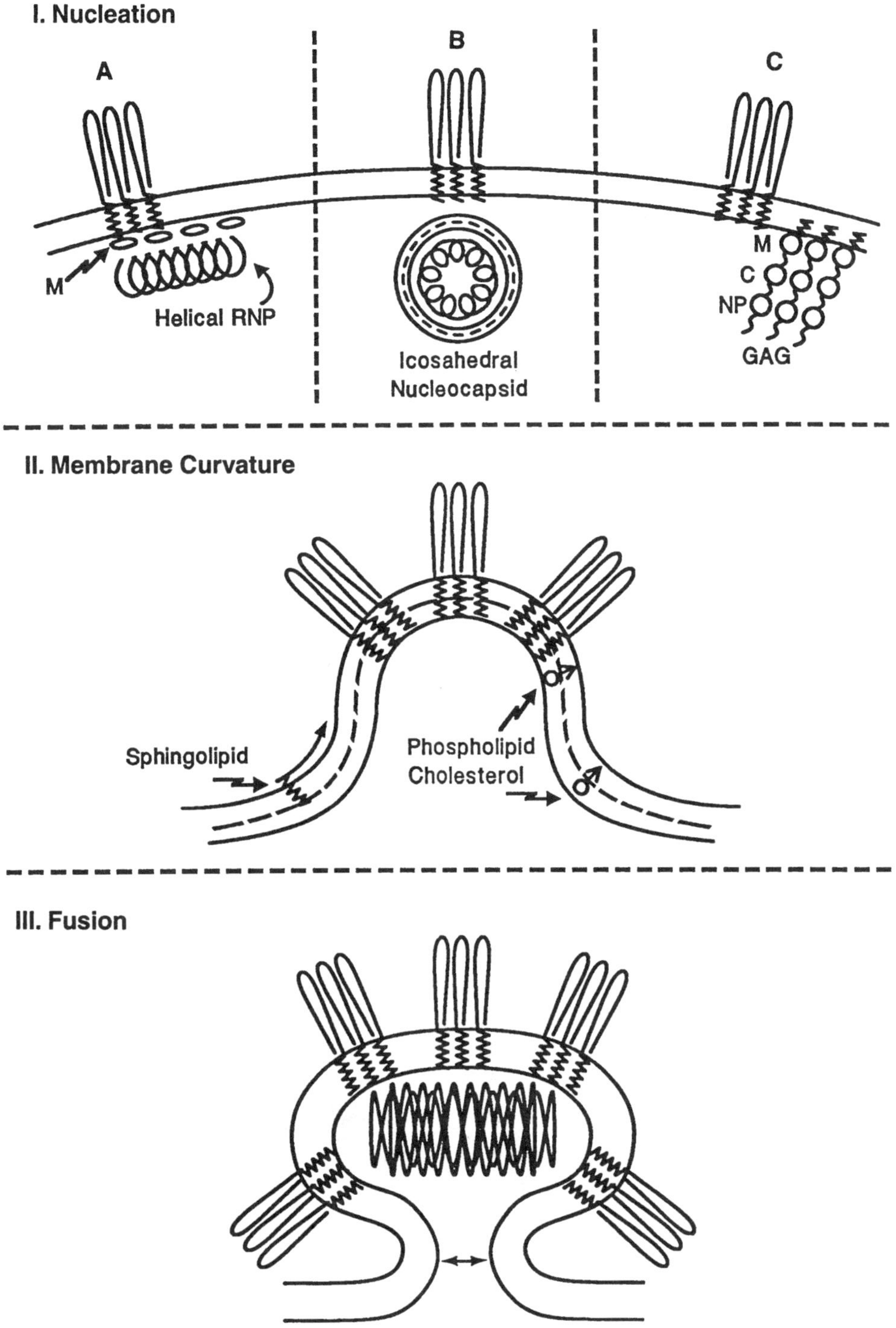

Figure 1. Events in the assembly and budding of enveloped viruses. (IA) A helical ribonucleoprotein complex (RNP) interacts with a matrix protein (M), binding to the membrane and also binding a transmembrane glycoprotein. (IB) A preformed icosahedral nucleoprotein core binds to transmembrane glycoproteins. (IC) A polyprotein (GAG) containing domains that include a matrix protein (M), a core protein (C), and a nucleic acid binding protein (NP) binds to a membrane, and transmembrane glycoproteins migrate into these assembly sites. (II) The deformation of the membrane as a result of virus budding. For small viruses, the volume difference between inner and outer leaflets of the bilayer can be greater than 25%, which requires changes and results in asymmetry in lipid composition of the bilayer. (III) Membrane fusion, which may be driven by the highly deformed membrane at this site.

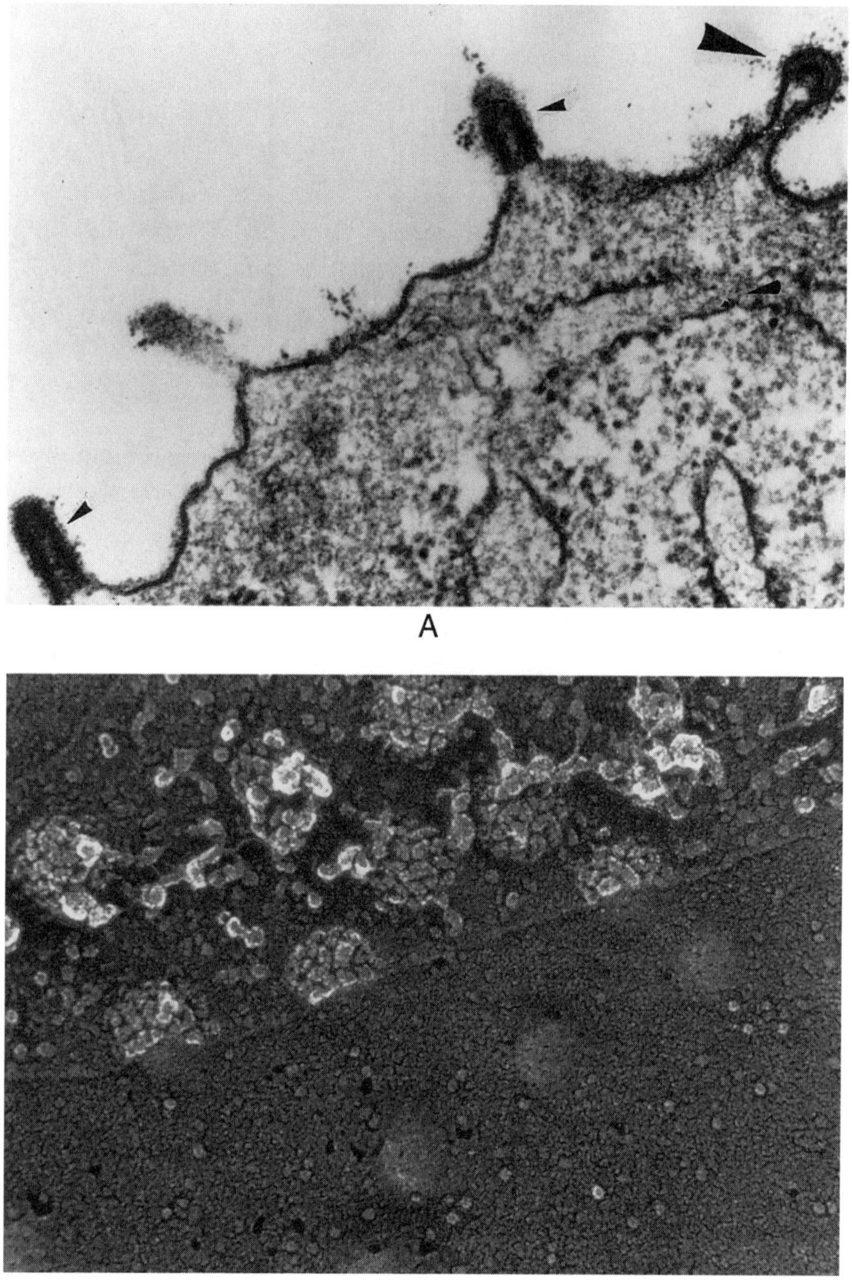

A

B

Figure 2. Electron micrographs of budding viruses. (A) A thin section of a chicken embryo fibroblast cell with vesicular stomatitis virus (the bullet-shaped particle: small arrowhead) and Rous sarcoma virus (large arrowhead) budding from the surface membrane. (Courtesy of Susan Pegrum, Imperial Cancer Research Fund, London.) (B) A replica of a freeze-fracture preparation of the surface of a chicken embryo fibroblast cell infected with Sindbis virus. (Courtesy of J. Heuser, Department of Cell Biology and Physiology, Washington University Medical School.) The fracture plane cuts the membrane between the inner and outer leaflets of the lipid bilayer and shows the outer surface of the cell on the top with circular arrays of the virus spikes at budding sites. On the bottom are imprints of the virus cores pushing up into the membrane from the cell cytoplasm.

(VSV) glycoprotein migrates to the basal lateral surface and the VSV buds from this side of the cell. When modifications are made to the glycoprotein that change its location in these polarized cells, the site of virus budding also changes.

The major factor that directs a virus-encoded glycoprotein to a particular cellular membrane is, in general, the sequences of amino acids that constitute the transmembrane and cytoplasmic domains of the protein. Virtually all enveloped virus glycoproteins are so structured that most of their protein mass is on the surface of the virus; however, these proteins contain additional sequences that form helices embedded in the lipid bilayer (the transmembrane domain) and extend beyond the membrane to form structures lying inside the virus or in the cell's cytoplasm (Figure 4). These proteins are found in three types. Type 1 has a single transmembrane domain

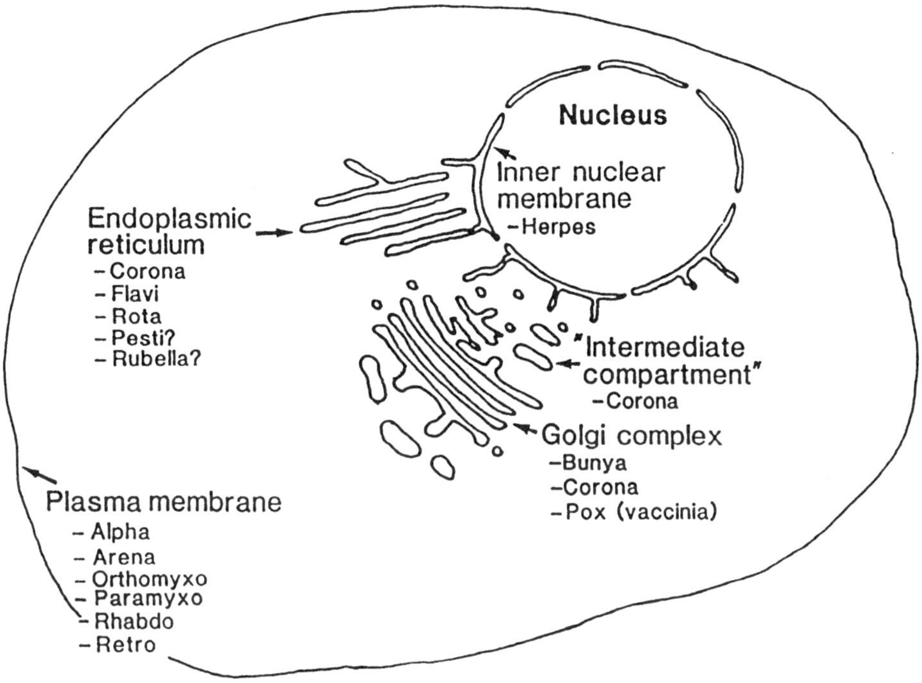

Figure 3. Cellular compartments used for virus assembly and budding. Names refer to various virus families. (Courtesy of R. Pettersson, Karolinska Institute.)

and the carboxyl terminus of the protein in the cytoplasm; type 2 also has a single transmembrane domain but has its amino terminus as the cytoplasmic domain; type 3 has multiple transmembrane domains. The protein is anchored to the membrane by the cytoplasmic domain together with the transmembrane portion. These regions of the polypeptide also provide structures and arrays of amino acids that bind cellular proteins, whose function is to retain the membrane-embedded protein to a particular vesicle of the cell. In the absence of these sequences, the glycoproteins flow to the plasma membrane. Importantly, regions in the cytoplasmic domain of the virus glycoprotein bind the virus's nucleoprotein core and can initiate budding. They also regulate the recycling of the protein from the surface back into cellular endosomes.

With regard to their synthesis, posttranslational modifications, and transport through the cell's secretory vesicles, virus glycoproteins resemble the glycoproteins secreted by the host cells. These proteins are formed from messenger RNAs bound to the cellular endoplasmic reticulum (ER). A portion of the polypeptide sequence

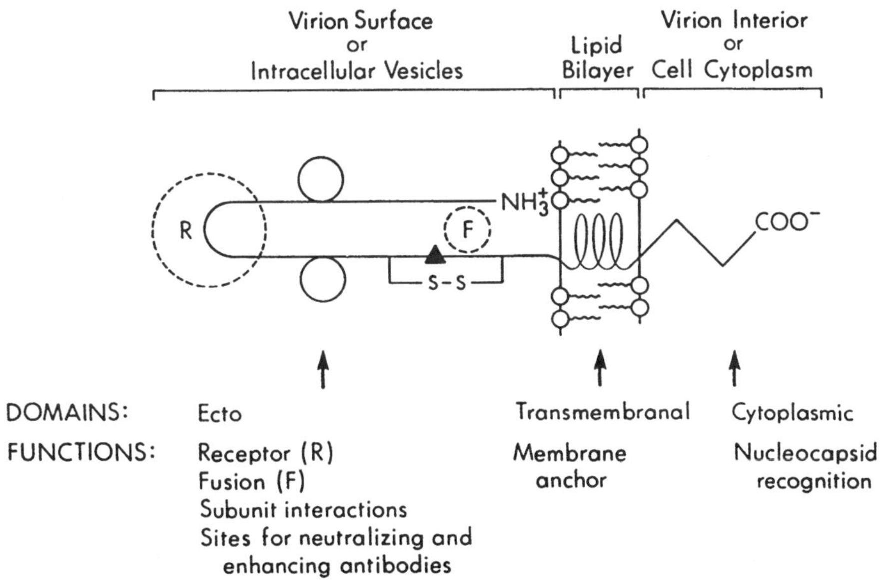

Figure 4. Domains of a typical type 1 transmembrane virus glycoprotein. [Reproduced from Schlesinger and Schlesinger (1987).]

encodes a signal sequence that translocates products translated subsequently across the membrane into the ER lumen. For type 1 glycoproteins the signal sequence is the amino-terminal residues. Near the carboxyl terminus is a hydrophobic stretch of about 25 amino acids that functions as a stop-transfer signal, allowing this portion of the protein to remain in the membrane. For type 2 glycoproteins, the signal sequence is close to but not at the amino terminus, and there is no stop-transfer signal. Type 3 glycoproteins will have several signal and stop-transfer sequences. Shortly after translocation to the ER, the glycoproteins are modified by large preformed oligosaccharide groups that become covalently bound to asparagines. There may be more than 20 of these groups attached to a virus glycoprotein, or none, depending on the protein and on the host cell in which the virus replicates.

In the ER, many virus glycoproteins form trimers and some interact also with cellular proteins that function as "chaperones." Subsequent passage into the Golgi vesicles leads to loss of mannose from the oligosaccharides and addition of other sugars. For some glycoproteins, carbohydrate is linked via the oxygen atom of serines and theonines of the protein. In an intermediary compartment between the ER and Golgi, the C-16 fatty acid palmitate is covalently attached to cysteine groups located in the cytoplasmic domain of several virus glycoproteins. In a trans-Golgi compartment, a cellular protease clips certain virus glycoproteins at sites containing several lysines or arginines. This latter event primes the virus protein for subsequent activities essential for infectivity.

In addition to the virus glycoprotein and nucleoprotein, many enveloped viruses utilize in their assembly and budding an intermediate membrane-associated polypeptide, referred to as a matrix protein. Matrix proteins are not generally glycosylated, nor do they penetrate deeply into the membrane. In some cases matrix protein provides a binding site for the transmembranal glycoproteins, and they also can bind nucleoprotein cores. For retroviruses like the HIV, matrix protein alone can form a protein shell, and for the Mason–Pfizer retrovirus, a single amino acid change in the matrix protein sequence changes the normal site of protein shell assembly in the cytoplasm to one occurring at the plasma membrane.

The matrix protein of vesicular stomatitis virus has been studied in great detail and not only participates in the assembly and budding of virus but is a critical component in regulating virus replication. This matrix protein is the major structural component of infectious virus and may function as a scaffold for the helical nucleoprotein core in addition to its role in assembly and budding at the plasma membrane. For the paramyxoviruses such as the measles and respiratory syncyctial viruses, association of the matrix protein to the membrane requires the presence of the glycoprotein, and matrix accompanies the latter during intracellular transport through the vesicles of the secretory pathway.

In influenza type C viruses, the genetic trait responsible for viruses of filamentous shape is attributable to the gene-encoding matrix protein, and the protein also binds to two other influenza virus proteins that are part of the virus's ribonucleoprotein complex.

Virus-encoded proteins other than the glycoproteins and matrix protein are important in enveloped virus assembly and budding. Many of these are embedded in the membrane and function as enhancers or catalysts in the process. Table 1 lists a few of these.

The lipid composition of enveloped viruses is similar to the cellular membrane used for budding but differs in the precise composition of this membrane, indicating that viral proteins select certain lipids in the process of assembly and budding. For example, the membranes of vesicular stomatitis virus and herpes simplex virus have higher phosphatidylserine and sphingolipid and lower phosphatidylcholine composition than do the plasma and nuclear membranes, respectively, from which these viruses bud. The glycoprotein and matrix protein of the former virus have affinity for negatively charged lipids and together create a lipid domain highly enriched in phosphatidic acid. Cholesterol is also enriched in the membrane of this virus, in contrast to the membrane of the infected cell and together with aminophospholipid is responsible for the higher viscosity of the viral envelope compared to that of the plasma membrane of the cell. The membrane viscosity of alphaviruses is close to that of the host cell plasma membrane and is much lower for virus replicating in insect cells, which also have low viscosities, than for virus formed in vertebrate cells, which are higher in viscosity. The human immunodeficiency viruses (HIV-1 and HIV-2) have significantly higher ratios of cholesterol to phospholipid than do the host cell plasma membranes, and the matrix protein of this virus has properties that lead to a selection of lipid that results in domains enriched for specific phospholipids.

Table 1 Accessory Virus Proteins Involved in Assembly and Building

Virus	Protein	Function
Alphaviruses	6K	Affects membrane curvature; may alter interactions between some glycoproteins and lipid
Retroviruses	P6 of HIV	Regulates and increases efficiency of budding
Ortho- and paramyxoviruses	M2	Forms ion channels
Lentiviruses	VPU	Processing of the envelope protein

3 EXPERIMENTAL METHODS FOR STUDYING ASSEMBLY AND BUDDING

Morphological information obtained from high resolution electron microscopy has provided the most thorough picture of virus assembly and budding. Immunofluorescence tagging of viral antigens with detection by optical microscopy and immunoelectron microscopy is used to identify the cellular membranes in which budding occurs and to detect other cellular or virus components at sites of assembly and budding. Higher resolution of virus structures is obtained by using cryoelectron microscopy, which allows viruses to be examined in vitrified ice rather than as dehydrated samples. But these experimental procedures are limited in defining events at the molecular level.

An important biochemical approach to the study of virus assembly and budding employs inhibitors that can selectively block virus replication at the assembly stage. For example, two calcium channel blocking drugs—chlorpromazine and verapamil—interfere with both influenza virus and Sindbis virus particle formation at doses that do not affect intracellular production of the viral proteins and replication of the RNA genome. These results indicate that calcium levels and cytoskeletal structures play important roles in the budding process. In fact, calcium alone can induce reversible shape changes in cell membranes leading to a crinkling (echinocytosis) of human erythocytes. Inhibitors such as the ionophore, monensin, and the glycosylation blocker tunicamycin affect movement of cellular proteins through the secretory vesicles, perturbing normal virus assembly and creating in some cases new sites of budding as virus glycoproteins accumulate in different compartments of the cell.

Genetic techniques are used widely to study virus assembly and currently are the method of choice for studying molecular events in virus budding. Initially, for viruses of several kinds, mutants that were temperature sensitive for replication were isolated and their phenotypes exhaustively analyzed to identify specific virus genes required for assembly and budding. The most prevalent genetic techniques now utilize recombinant DNA technology and reverse genetics in which site-specific mutations are produced in genes encoding proteins that function in virus assembly and budding. The process of cloning of cDNAs encoding various virus proteins and inserting them into expression vectors offers a means for investigating how a specific virus polypeptide or a set of these can function in a cell in the absence of total virus replication. These techniques enable the investigator to delete, modify, or add sequences to a particular polypeptide and in this way determine domains important to the protein's function in the assembly process. Studies with closely related virus strains have employed the construction of chimeric proteins in which the cDNA contains portions of sequences from the different strains assorted in various combination. From the latter, it has been possible to detect various kinds of complementarity required for protein–protein interaction during virus assembly.

Monoclonal antibodies directed against selective epitopes of the protein have also proved highly effective in probing the molecular biology of virus assembly. These reagents are sensitive detectors of changes in protein conformation, and results of their application have shown that proteins in most viruses undergo structural changes during the budding process. These kinds of antibody have also been used to prepare anti-idiotypic antibodies, and the latter may in some cases help to probe regions of complementarity during protein–protein binding.

Table 2 is a partial listing of results from recombinant DNA experiments carried out for several retroviruses and one alphavirus, the Semliki Forest virus. In contrast to the latter, where no budding occurs unless *both* glycoproteins and nucleoprotein cores are present, retrovirus particles are formed and secreted from infected cells in the absence of the cognate glycoprotein. In fact, for the retroviruses, only the matrix-type protein has been reported to be necessary to form a cell-free particle, suggesting that self-association of this protein is sufficient to form an icosahedral shell. Particles formed with only the virus subcomponents are, of course, not infectious.

These various reconstructed systems clearly indicate the minimal components needed for membrane budding; however, they fail to address the problem of "efficiency" and kinetics of virus release. For example, in the cases of the Semliki Forest and Sindbis alphaviruses, mutants lacking or highly defective in a gene encoding a small hydrophobic, membrane-associated protein form infectious viruses, which are indistinguishable morphologically from the normal; however, they are produced at only 1–5% of the rate that is measured for the wild-type virus. For this latter example, the virus appears to have evolved a gene that selectively enhances the efficiency of the assembly and budding process. There are other examples, including the selective truncation of a part of the transmembranal envelope gene of the SIV_{mac} retrovirus, which shows that the mutations have different effects in different hosts, with some cells selectively assembling viruses with shorter transmembranal polypeptides. The latter results offer strong evidence that host cell, membrane-associated polypeptides are also important components in the assembly and budding process. Their identification and function will be essential in any thorough analysis of the molecular basis for enveloped virus assembly.

4 SPECIFIC VIRUS FAMILIES AND ASSEMBLY PROCESSES

Individual virus families assemble and bud in a characteristic manner with features somewhat unique to that virus. Sections 4.1 through 4.6 offer examples of these events for a few specific viruses.

4.1 ALPHAVIRUSES

The alphaviruses, which comprise the major genera of the *Togaviridae,* are probably the simplest of all enveloped animal viruses, both in structure and in assembly. An alphavirus has a single strand of (+) RNA as its genome, which is coiled within a protein shell

Table 2 Viral Proteins Identified as Components in Assembly and Budding, Based on Genetic Manipulations

Virus	Components	Type of Data
HIV-1	Cytoplasmic domain of gp41	Mutation/truncation of gene leads to release of noninfectious particle lacking glycoprotein
	Glycoprotein	Location of budding in polarized cells changes to basal surface
	Matrix protein p19	Small deletions allow for release of particles without glycoprotein.
Rous sarcoma virus	GAG Envelope	Particles detected when GAG is expressed alone; no particles when envelope gene is expressed alone; coexpression forms particles with glycoproteins
Semliki Forest virus	Nucleocapsid envelope proteins	Expression of viral structural components with deletions in either the nucleocapsid or enveloped genes shows no budding unless both are expressed in the same cell

made up of 240 copies of a single polypeptide chain of 260–270 amino acids. Tissue culture cells of many kinds are susceptible to infection by these viruses, where they grow rapidly to release as many as 10,000 new infectious virus particles within 8–10 hours. In the process, normal host cell macromolecular synthesis stops as the virus coopts for its own use all the cells' synthetic machinery—including RNA, protein, and lipid biosynthesis.

Much information has accumulated about the assembly and budding of these viruses from studies with two members of the family called Sindbis and Semliki Forest as prototypic viruses. Prior to development of recombinant DNA technology, electron microscopy and isolation of temperature-sensitive mutants provided data that led to a model for virus assembly and budding. In this model a preformed nucleocapsid interacts with virus-encoded glycoproteins that have migrated through the vesicles of the cellular secretory pathway to the plasma membrane of the infected cell. The initial nucleation event is followed by further nucleoprotein–glycoprotein contacts, and these protein–protein interactions bend the lipid bilayer around the icosahedral nucleocapsids (see Figure 1, IB). The final release of virus follows from fusion of the lipid bilayer, which has become highly deformed in curvature. The molecular interactions are extremely specific, and only viral glycoproteins are found in the highly symmetrical virus structure, which has recently been analyzed to a resolution of 20 Å, based on images from cryoelectron microscopy. The surface glycoproteins, denoted as E1 and E2, are arrayed as trimeric spikes of heterodimer polypeptides. Near the carboxy terminus both have helical hydrophobic domains, which anchor the proteins in the membrane bilayer. In addition, the E2 has an extended "tail" of 33 amino acids that is topologically oriented inside the bilayer and appears to contact the surface of the nucleocapsid shell.

Biochemical analysis of several temperature-sensitive mutants showed that assembly of viruses of these kinds depended on an appropriate glycoprotein array and structure. Several mutations in the ectodomain of the Sindbis virus E2 glycoprotein perturb the assembly process: one allows for nucleocapsids to bind to membrane regions containing the glycoproteins, but no budding occurs; another interferes with the normal bending of the bilayer around the cores, producing particles with highly aberrant structures that contain several nucleocapsids. A temperature-sensitive mutant with two amino acid substitutions in the Sindbis virus E1 protein alters its structure and blocks its transport to the plasma membrane. Cells infected with this mutant and grown at the nonpermissive temperature show no nucleocapsids binding to the membrane. A temperature-sensitive mutation in the Semliki Forest virus E1 protein also blocks its transport out of the endoplasmic reticulum at the nonpermissive temperature. Shifting cells infected with this mutant to permissive temperature leads to budding of virus particles into the ER lumen, a result consistent with other data, noted earlier, indicating that the localization of the glycoprotein determines the site of virus budding.

Results based on site-directed mutations, obtained by manipulating cDNA clones constructed from the virus RNA genome, showed that a small hydrophobic protein of 6 kDa encoded by a gene located between the two virus glycoproteins can affect the morphology of the budded virus. Mutations in this gene lead to larger particles, with many cores within a single envelope bilayer. Suppressors of the 6 kDa mutant phenotype contain mutations in the E2 or E1 glycoproteins, indicating that virus budding includes interactions between the latter and the 6 kDa protein. Mutations in the 33 amino acid sequence composing the E2 cytoplasmic domain can also affect virus assembly. A role for the latter in the budding of these viruses was most clearly demonstrated from results with a chimeric strain that had almost its entire genome derived from the Ross River strain, although its capsid gene was from the Sindbis virus strain. This chimera was nonviable; however, replacement of several Sindbis virus amino acids for those of the Ross River virus in the E2 cytoplasmic domain restored infectious particle production. This result clearly shows that complementarity between capsid protein structure and E2 cytoplasmic domain sequences is critical for virus assembly and formation.

4.2 Flaviviruses

The yellow fever virus is the prototypic virus of the flaviviruses, an enveloped RNA-virus family, and molecular biological studies identify a set of proteolytic processing events crucial to assembly and budding of this virus. A single protein, C, forms the shell that packages the RNA into a nucleoprotein core, but the form of C synthesized initially contains sequences that anchor the protein to the ER membrane. This ensures localization of budding to this compartment and also leads to assembly of the core at the membrane. No cores are found free of membranes in infected cells. Processing of the anchored C is catalyzed by a protease encoded in viral genes. Two transmembrane proteins are associated with infectious virus: one is termed M (membrane protein) and initially appears in the infected cell as a larger glycosylated form, but this portion of the polypeptide is proteolytically released by host cell proteases and does not appear in infectious virus; the second is a larger glycoprotein, E, that is localized to the surface of the virus. There is speculation that the glycosylated form of M binds to E and prevents the latter from oligomerizing into a fusogenic state activatable during the initial stages of replication of cell free virus. The heterodimers of glycosylated M and E are the forms involved in the assembly process. After assembly, the E protein forms dimers that are oriented head to tail and lie like logs along the virus surface instead of in spikelike projections.

4.3 Rhabdoviruses

Rabies virus is the important member of the rhabdovirus family, but information about the assembly and budding of these viruses comes from studies with the vesicular stomatitis virus, VSV. The core of this virus is a helical ribonucleoprotein that appears to contain many molecules of the matrix protein possibly functioning as a scaffold. Some of these proteins also bind to cell membranes and interact with the single glycoprotein, G, encoded by the RNA genome of the virus. Assembly occurs at the cell plasma membrane and requires, at the least, the transmembrane and cytoplasmic domain of G. Small amounts of cellular proteins normally found in the membrane are detected in the released virus particle.

Data employing cells transfected with cDNAs coexpressing the VSV M and G proteins as well as from in vitro studies that involve viral subcomponents and nucleoproteins with membranes loaded with different VSV polypeptides support a model in which positively charged domains of membrane-bound M protein bind to negatively charged lipids in the membrane and act as the nucleation site for assembly of nucleoprotein cores—perhaps via M–M protein oligomerization. G glycoproteins that have migrated to the plasma membrane move laterally along this membrane to patches of bound nucleoproteins where they bind, through their cytoplasmic sequences, primarily to the N proteins that form the surface of the helical nucleoprotein complexes. In addition, data obtained with protein cross-linking reagents show that G binds to N and that equal numbers of copies of N and G but not M and G are found in particles, thus supporting a notion that it is the N protein rather than the M protein that interacts with G during budding. In genetically reconstructed systems with chimeric proteins, the cytoplasmic domain of G was found to be essential for the incorporation of significant amounts of a foreign protein into the VSV particle.

Association of the matrix protein with the membrane of the intact virion has been demonstrated by its labeling with various hydrophobic probes introduced into the membrane and by studies showing membrane perturbations upon interaction with matrix.

4.4 Rotaviruses

An unusual assembly process occurs with the rotaviruses, which are nonenveloped in their extracellular form yet contain a membrane structure during maturation in the virus-infected cell. During the initial stages of assembly, the major viral structural proteins collect into amorphous organelles in the infected cell cytoplasm and single-shelled particles bud into the endoplasmic reticulum through an interaction between one of the core proteins and a virus-encoded protein inserted into the ER membrane. As a result, the virus is transiently enveloped by a segment of the ER membrane; upon transit through the ER and vesicles of the secretory pathway, however, this envelope is shed, and rearrangements occur in the virus structure. The virus then takes on "surface" proteins, which form an outer capsid of the virion. Significant changes occur in topology of some viral proteins during this budding and shedding.

4.5 Retroviruses

For most retroviruses, the normal assembly process begins with alignment of the large GAG along the cell surface membrane and is then followed by condensation of the polypeptides, which in turn leads to the proteolytic processing that forms the individual structures (Figure 5). All this occurs concomitant with lipid envelop-

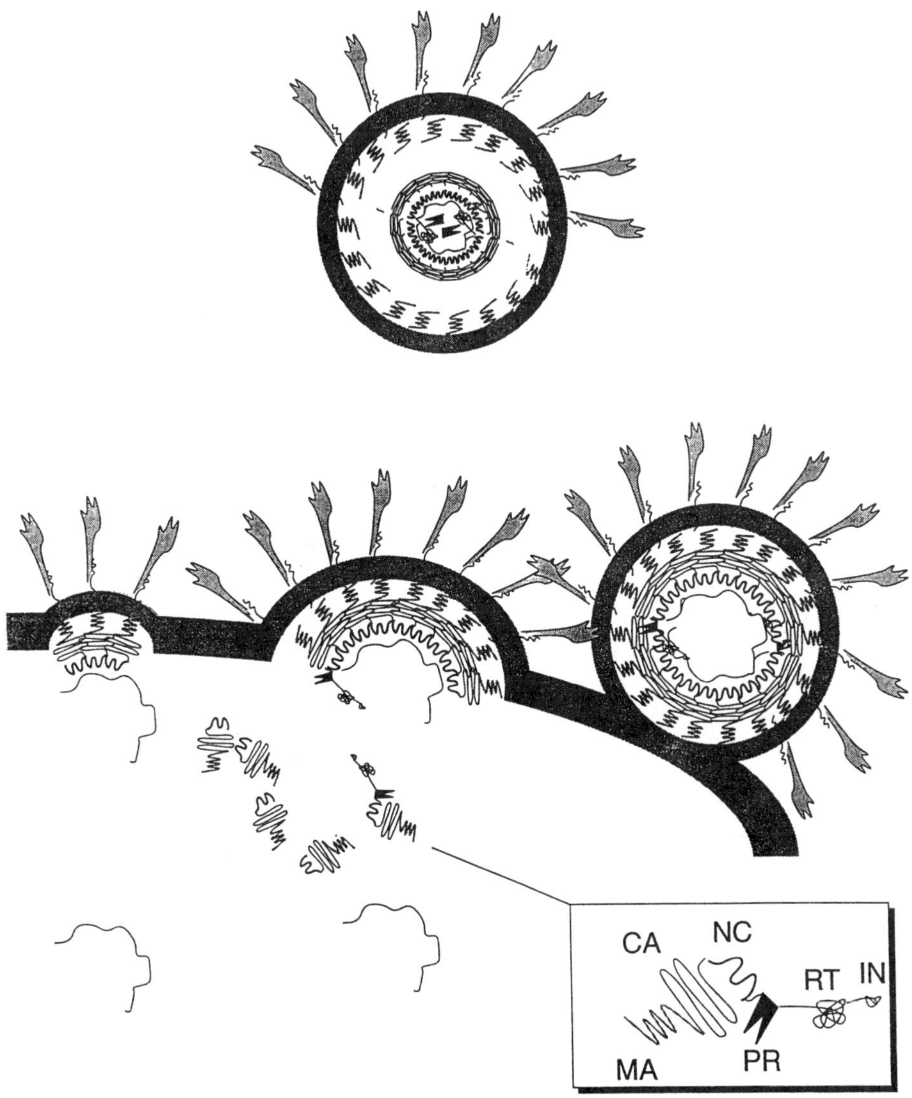

Figure 5. Assembly and budding of a retrovirus: MA, matrix protein; CA, capsid protein, NC, nucleoprotein; PR, protease; RT, reverse transcriptase, IN, intergrase. The genomic RNA is shown as a light wiggly line. Lightly shaded spikes are the virus glycoproteins. (Reproduced from J. Coffin, The *Retroviridae:* Their viruses and replication. *Fundamental Virology,* 3rd ed., Fields et al., Eds. Lippincott-Raven Press, New York, 1996.)

ment. Bound to the amino-terminal portion of the intact GAG is the 14-carbon myristyl fatty acid, which interacts with the membrane. Virus-encoded glycoproteins move into this portion of the membrane—most likely interacting in some way with the membrane-associated matrix protein to become part of the particles; however, the latter are not required for effective budding, and the GAG polyprotein alone is capable of forming viruslike particles of 100–120 nm diameter inside a cell expressing only this protein. It is estimated that about 2000 copies of GAG are in the virus particle. Intact but noninfectious particles can be formed and released using only the 180 amino acids of the matrix sequences and a small part of the capsid region in place of the normal 800 amino acids in the intact GAG; thus, the matrix structure most likely contains motifs involved in protein self-assembly.

It is the virus-specific protease that processes the polyprotein to the individual subunits that form a membrane-bound structure (matrix), the core shell (capsid); the nucleoprotein binds the RNA genome via sequences of cysteine-histidine that resemble "zinc finger" domains for DNA-binding proteins. Some groups of retroviruses (viz., the mouse mammary tumor virus and the Mason–Pfizer monkey virus) form nucleoprotein shells within the infected cell cytoplasm, and these move to the surface membrane for the final process of budding. For these two types of retrovirus, noted subgroups B and D, respectively, expression of their *GAG* gene alone—in a bacterial cell transfected with the appropriate cDNA clones—leads to protein shells whose structures closely resemble those found in cells infected with the intact virus. The protease remains inactive until the assembly process, and activation probably requires both the increased concentration resulting from the assembly itself as well as ionic conditions associated with budding from the membrane. Activation of only a few protease molecules would lead to an exponential increase in activity, since more protease molecules would be freed from the polyprotein in an active state.

Smaller polypeptides freed during the processing of the GAG polyprotein play roles in the budding of these viruses, but the mechanisms remain obscure. The HIV P6, for example, is needed to completely free virions from the cell.

4.6 Herpesviruses

Members of the *Herpesviridae* family contain a double-stranded DNA composed of more than 200 genes within a complex protein shell wrapped by an envelope containing 10 to 11 different glycoproteins. Core formation occurs in the infected cell nucleus, where the DNA is replicated and packaged into preformed protein shells. Budding occurs primarily at the inner membrane of the nuclear envelope at patches of viral membrane proteins, detectable in the outer nuclear membrane. Particles are transported through the secretory vesicles to emerge from the cell. There may be additional events in the cytoplasm of the infected cell that involve an uncoating and recoating of some forms of these viruses in particular cells. Many studies have been carried out with viruses containing mutations in many of the genes encoding glycoprotein which perturb the normal assembly and budding. The complexity of these viruses, however, has thus far precluded the development of a clear picture of events in assembly and budding.

See also Membrane Fusion, Molecular Mechanism of; Virology, Molecular; Viruses, DNA Packaging of; Viruses, RNA Packaging of.

Bibliography

Caplan, M., and Matlin, K. S. (1989) The sorting of membrane and secretory proteins in polarized epithelial cells. In *Functional Epithelial Cells in Culture,* K. S. Matlin and J. D. Valentich, Eds., pp. 71–127. Liss, New York.

Doms, R. W., Lamb, R. A., Rose, J. K., and Helenius, A. (1993) Folding and assembly of viral membrane proteins. *Virology,* 193:545.

Dubois-Dalcq, M., Holmes, K. V., and Rentier, B. (1984) *Assembly of Enveloped RNA Viruses.* Springer-Verlag, New York.

Fields, B. N., et al., eds. (1996), *Fundamental Virology,* 3rd ed. Lippincott-Raven Press, New York.

Harrison, S. C. (1990) Principles of virus structure. In *Fundamental Virology,* 2nd ed., B. N. Fields, D. M. Knipe, R. M. Chanock, M. S. Hirsch, J. L. Melnick, T. P. Monath, and B. Roizman, Eds., p. 37. Raven Press, New York.

Hunter, E. (1994) Macromolecular interactions in the assembly of HIV and other retroviruses. *Semin. Virol.* 5:71.

Lerner, D. M., Deutsch, J. M., and Oster, G. F. (1993) How does a virus bud? *Biophys. J.* 65:73.

Pettersson, R. F. (1991) Protein localization and virus assembly at intracellular membranes. *Curr. Top. Microbiol. Immunol.* 170:67–106.

Schlesinger, M. J., and Schlesinger, S. (1987) Domains of virus glycoproteins. *Adv. Virus Res.* 33:1.

Virology, Molecular

Andrew P. Rice

1 **Introduction**
2 **Types of Animal Virus**
3 **Viral Infectious Cycle**
  3.1 Entry of Virus Via Cellular Receptor
  3.2 Internalization and Uncoating of Virus
  3.3 Transcription of Viral RNA
  3.4 Translation of Viral mRNA
  3.5 Replication of Viral Genome
  3.6 Assembly and Release of Virion
4 **Virus Host–Cell Interactions**
  4.1 Infections in Individual Organisms
  4.2 Infections in Individual Cells
5 **Use of Viruses as Biological Tools**
6 **Perspectives**

Key Words

Cellular Receptor An attachment site for a virus particle, located on the surface of a cell.

Enveloped Virus A virus particle containing a lipid bilayer plus associated protein(s).

Fusion and Endocytosis Pathways of viral entry into cell: fusion occurs when the lipid envelope of a virus fuses with cell membrane; in endocytosis, a virus bound to a cellular receptor enters the cell through a normal pathway for uptake of materials bound to cell surface receptors.

Host Range Species that are permissive for replication of a specific virus.

Infection In acute infection, the infecting virus undergoes active viral replication; in the chronic form, it replicates persistently at a relatively low level. In latent infection, the infecting virus is present but does not undergo active replication.

LTR, Long Terminal Repeats (LTRs) Sequences duplicated at the 5′ and 3′ ends of all retrovirus genomes.

Nonenveloped Virus A virus particle containing only protein coat.

Nucleocapsid In a virus particle, a shell composed of viral DNA or RNA complexed with viral proteins.

Provirus For retroviruses, the viral genome inserted as DNA into host cell chromosome.

Tissue Tropism Designates the existence of specific tissues in infected host where viral replication occurs.

Viral Attachment Protein A protein on the surface of a viral particle that binds to the cellular receptor.

Virion An infectious particle that contains the viral genetic material.

A virus is a subcellular entity composed of genetic material, either RNA or DNA, that is packaged in a particle called a virion. Viruses require a host cell for their replication. Molecular virology can be thought of as using the increasingly powerful techniques of molecular biology to study and manipulate viruses. Molecular virologists are presently examining the roles of viruses as the causative agents of diseases. Additionally, molecular virologists are manipulating viruses to develop biological tools for beneficial purposes such as vectors for novel vaccines and for human gene therapy.

1 INTRODUCTION

The impact of viruses on mankind is enormous: plant viruses can ruin crops; animal viruses can destroy livestock; variola virus, the cause of smallpox, has devastated human societies; human immunodeficiency viruses, the cause of AIDS, are likely to result in the death of at least tens of millions of people. The number of different viruses on earth is huge: more than 1000 distinct viruses have been described, and this likely represents only a fraction of the total number. This article deals with animal viruses, especially those that are related to human diseases, which has been the most active research area in molecular virology owing to its medical importance.

We begin by comparing the different types of animal virus. Rather than attempting a comprehensive overview of the molecular virology of animal viruses, however, we discuss in detail only two viruses—vaccinia virus and human immunodeficiency virus type 1 (HIV-1). A presentation of the infectious cycles of vaccinia virus and HIV-1 is followed by a treatment of interactions between vaccinia virus and HIV-1 and the infected host cell. Finally, we turn to current efforts to use molecular virology to harness these or related viruses as biological tools. Vaccinia virus and HIV-1 are very different viruses, different in both physical makeup and biological properties. A comparison between these two virus will allow discussion of some of the major principles of molecular virology.

Vaccinia virus, which has been of paramount medical importance for more than two centuries, is related to the deadly variola virus, the cause of smallpox. Vaccinia has been used effectively since the 1790s as a vaccine for smallpox, hence the name. Because of the widespread use of this vaccine, the earth has been free from smallpox infection since 1979. The origins of present-day strains of vaccinia virus are unknown; vaccinia virus may be attenuated variola virus, attenuated cowpox virus (a close relative of variola virus), or a unique virus no longer found outside the laboratory.

HIV-1 was first identified in 1985 as the causative agent of acquired immunodeficiency syndrome, or AIDS. Presently, two different human immunodeficiency viruses have been identified, HIV-1 and HIV-2. HIV-1 infection is found throughout the world and is the cause of most of the AIDS cases reported to date. HIV-2 infection has been restricted to eastern Africa, although infection has recently been spreading in Asia. HIV-2 may be less pathogenic than HIV-1. It may be that HIV-2 does not cause AIDS in all infected individuals; or, the period of time required to cause disease in infected individuals may be considerably longer for HIV-2 than for HIV-1.

2 TYPES OF ANIMAL VIRUS

Viruses can be classified into families according to their virion structure and type of genetic material. Virions are the particles that contain the viral genetic material. Table 1 lists some virus families having members that are human pathogens. Table 1 is not a com-

Table 1 Some Types of Animal Viruses

Virus Family Name	Genetic Material	Human Virus (Example/*Disease*)	Virion	Genome Size (nucleotides)
Picornaviridae	Single-stranded RNA	Poliovirus/*Poliomyelitis (paralysis)*	Nonenveloped, icosahedral; diameter 30 nm	7,500
Togaviridae	Single-stranded RNA	Rubella virus/*Rubella*	Enveloped, icosahedral; diameter 65 nm	10,000
Reoviridae	Double-stranded RNA (segmented)	Rotavirus/*Diarrhea*	Nonenveloped, icosahedral; diameter 70 nm	15,000
Retroviridae	Single-stranded RNA (DNA step in replication)	Human immunodeficiency virus (HIV)/*AIDS*	Enveloped, spherical; diameter 100 nm	9,000
Hepadnaviridae	Single/double strands of DNA (RNA step in replication)	Hepatitis B virus/*Hepatitis*	Enveloped, spherical; diameter 45 nm	3,300
Parvoviridae	Single-stranded DNA	Parvovirus B-19/*Aplastic crisis in children with sickle-cell anemia*	Nonenveloped, icosahedral; diameter 25 nm	5,000
Adenoviridae	Double-stranded DNA	Adenovirus/*Respiratory illness*	Nonenveloped, icosahedral; diameter 70 nm	36,000
Poxviridae	Double-stranded DNA	Variola virus/*Smallpox* vaccinia virus	Enveloped, brick shaped; 200 mm × 400 nm	190,000

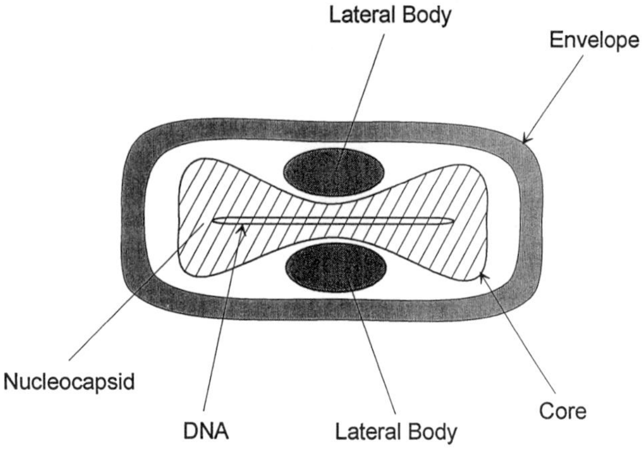

Figure 1. Morphological structure of vaccinia virus. The vaccinia virus double-stranded DNA genome is packaged in the nucleocapsid of the enveloped virion. Next to the nucleocapsid are lateral bodies of unknown composition and function. The vaccinia virus particle is approximately 200 nm × 400 nm.

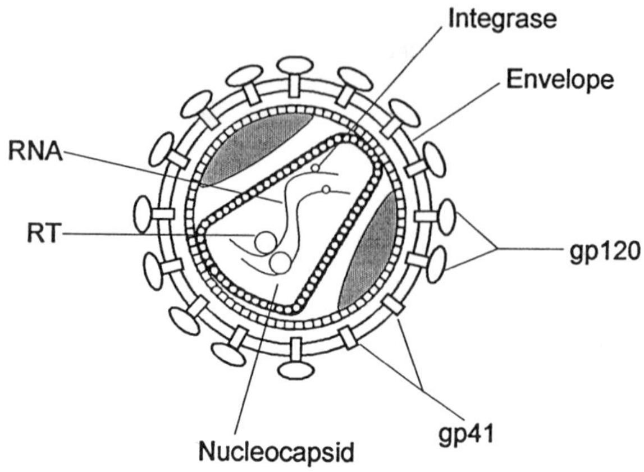

Figure 2. Morphological structure of HIV-1. Two copies of the viral RNA genome are packaged in the nucleocapsid of the enveloped virion. Also contained in the virion are the viral integrase and reverse transcriptase enzymes. The viral gp120 is the HIV-1 attachment protein.

prehensive list of virus families, but it illustrates that a wide range of viruses infect humans and can be of medical importance. As shown in Table 1 under the heading "Virion," viruses can be classified as either enveloped or nonenveloped. The enveloped viruses have virions composed of a lipid bilayer plus associated proteins that package the viral genetic material. The nonenveloped viruses have virions composed of only a protein coat that packages the viral genetic material. Vaccinia virus is an enveloped virus and is a member of the *Poxviridae* family; HIV-1 is also an enveloped virus and is a member of the *Retroviridae* family.

The virion size of different viruses varies considerably. Virions of the *Parvoviridae* family are among the smallest animal viruses; the nonenveloped virion of human parvovirus B-19 is approximately 25 nm in diameter. The *Poxviridae* family are among the largest viruses; the brick-shaped vaccinia virus virion has dimensions of 200 nm by 400 nm (Figure 1). The HIV-1 virion is roughly spherical, with a diameter of about 100 nm (Figure 2).

As shown in Table 1, viruses contain genetic material that is single-stranded or double-stranded nucleic acid, either DNA or RNA. The *Hepadnaviridae* family, whose members contain virions with both single- and double-stranded DNA, is an exception. Additional diversity is found in the RNA viruses, where the genome can be a lone single-stranded RNA molecule or, alternatively, a number of segments of double-stranded RNA. Rotaviruses, members of the *Reoviridae* family, contain 11 segments of double-stranded RNA packaged into the virion. The genome sizes of animal viruses varies over a wide range: the smallest virus in Table 1 is hepatitis B virus, a *Hepadnaviridae* family member, with a genome size of 3300 nucleotides; the largest virus in Table 1 is vaccinia virus, with a genome size of 190,000 nucleotides (Figure 3), or more than 50 times as large as hepatitis B virus. The HIV-1 genome is about 9000 nucleotides (Figure 4).

A general principle of molecular virology is that with increasing genome size, viruses gain greater independence from the host cell

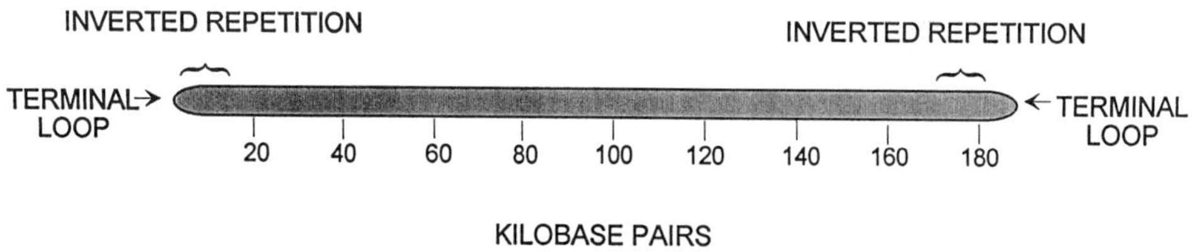

Figure 3. The vaccinia virus genome. The viral double-stranded DNA genome is 190,000 nucleotides. The ends of the genome are connected by a single-strand DNA terminal loop. The ends of the genome contain 10,000 nucleotides of inverted terminal repeat sequences.

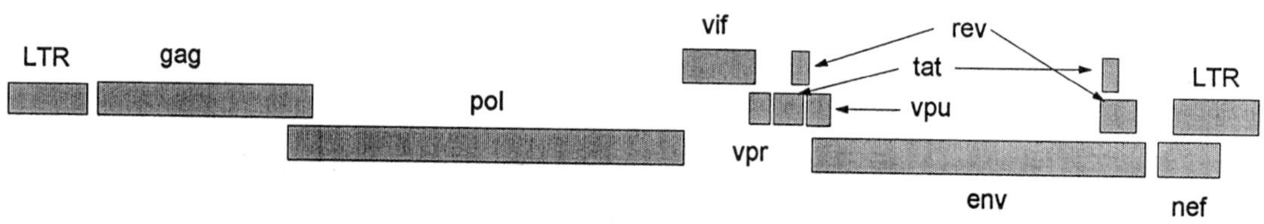

Figure 4. The HIV-1 genome. The HIV-1 genome of 9000 nucleotides encodes nine genes. The *tat* and *rev* genes are encoded in noncontiguous segments of the genome; their mRNAs are produced by splicing.

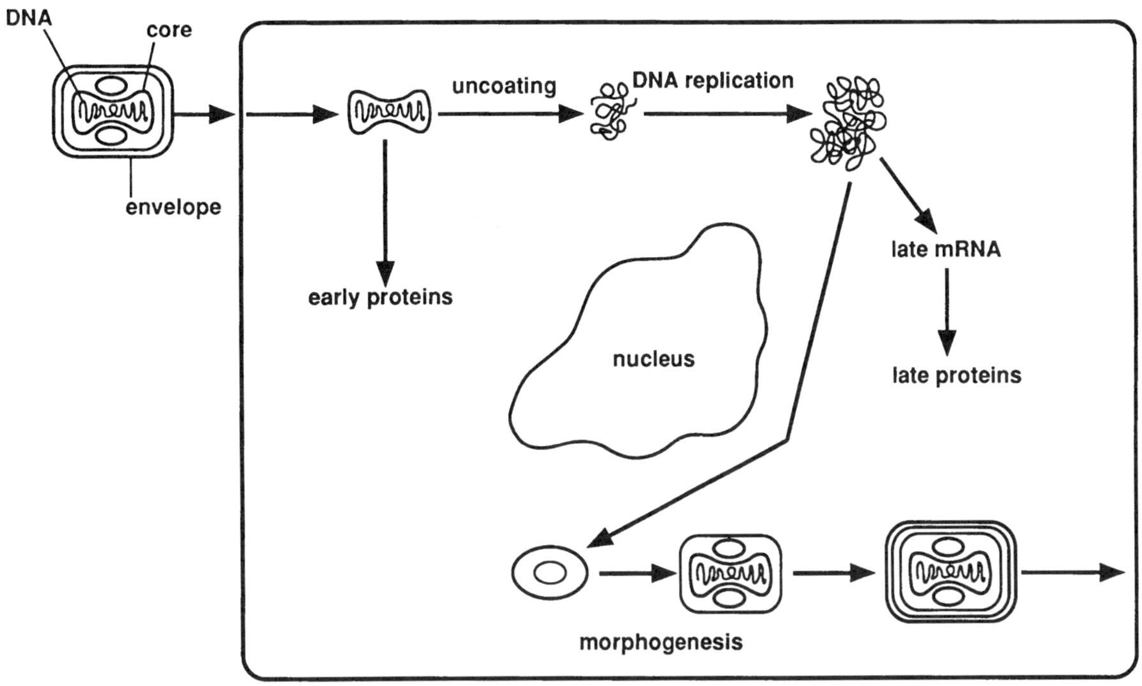

Figure 5. Replication cycle of vaccinia virus.

for functions required for viral replication. An example of this is the contrast between vaccinia virus and HIV-1. The 190,000-nucleotide vaccinia virus genome encodes about 200 genes. Vaccinia virus is therefore able to encode a large number of viral functions, including all the enzymes required for synthesizing viral RNA and replicating viral DNA. In fact, vaccinia virus replicates in the host cell cytoplasm with few requirements for any specific nuclear host cell factors. In contrast, the 9000-nucleotide HIV genome encodes only nine genes and, consequently, HIV requires a number of cellular functions for replication, including the cellular enzymes that synthesize viral RNA.

3 VIRAL INFECTIOUS CYCLE

The viral infectious cycle is the period required for a virus to enter its host cell via a cellular receptor, replicate the viral genome, and assemble infectious progeny viruses. For vaccinia virus, the infectious cycle (Figure 5) takes about 18 hours. For HIV-1, the infectious cycle (Figure 6) takes about 36 hours.

3.1 Entry of Virus Via Cellular Receptor

Each virus is able to replicate only in host cells from a limited number of species, the so-called virus *host range.* Additionally, viruses exhibit *tissue tropism*—that is, they are able to infect only certain tissues of their susceptible host. For example, HIV-1 infects and replicates predominantly in a subset of human lymphocytes, either T cells, or monocyte/macrophage cells. In contrast, vaccinia virus infects a number of different mammals and has a broad tissue tropism.

The first requirement for a virus to successfully infect a cell is the presence on the cell's surface of a cellular receptor for a component of the virion. If the cell has the appropriate receptor, the virion is able to form a stable, high affinity complex with the cell, whereupon it enters the cell and begins the infectious cycle. The protein on the surface of the virion that binds to the cellular receptor is known as the *viral attachment protein.* However, entry of a virus into a cell via the proper receptor does not guarantee viral infection, because cells may lack one or more factors required for viral replication. The techniques of molecular virology have allowed the identification and molecular cloning of cellular receptors for a number of viruses. In recent years, cellular receptors have been identified for members of several viral families, including the *Adenoviridae, Picornaviridae,* and *Retroviridae.*

Investigators have not yet identified the cellular receptor(s) for vaccinia virus. Because the large vaccinia virus virion bears a number of different viral proteins on its surface, the possibility exists that several proteins on the surface of cells will be able to act as receptors for different viral attachment proteins. The broad host range and tissue tropism of vaccinia virus suggest that may be the case.

The protein called CD4, which is found on the surface of some human cells, has been shown to be the receptor for HIV-1. This has clearly been demonstrated by an experiment in which the CD4 protein was expressed in human fibroblast cells that do not normally have the CD4 molecule on their surface. These $CD4^+$ fibroblasts created in the laboratory could then be infected with HIV-1, whereas the parental $CD4^-$ fibroblasts could not. The HIV-1 viral attachment protein on the surface of the enveloped virion is called gp120 (Figure 2). The CD4 and gp120 molecules form a high affinity interaction with a binding constant of about 10^{-9} M.

3.2 Internalization and Uncoating of Virus

After binding of the viral attachment protein to the cellular receptor, the virus may enter the cell by one of two distinct pathways: *fusion* or *endocytosis.* In the fusion pathway, the lipid envelope of the

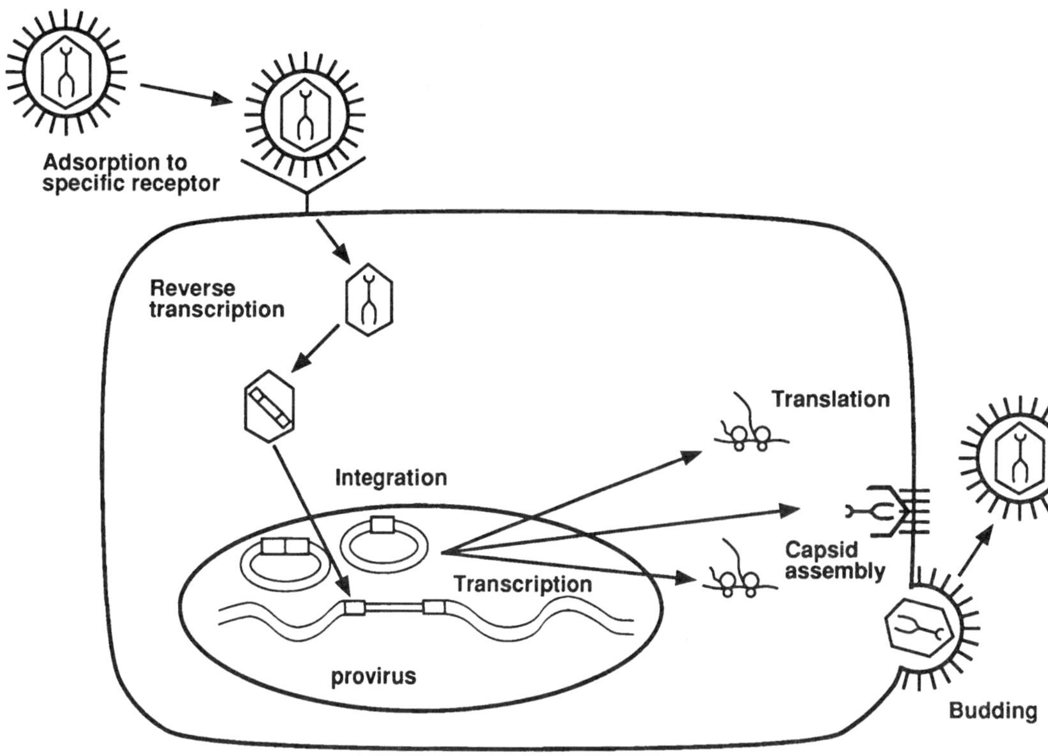

Figure 6. Replication cycle of HIV-1.

virion fuses with the cell membrane, leading to release of the viral *nucleocapsid* into the cytoplasm. The nucleocapsid is composed of the viral DNA or RNA complexed with viral proteins. In the endocytosis entry pathway, the virion bound to the receptor enters the cell through a normal cellular pathway for uptake of materials bound to cell surface receptors. During endocytosis, the cell membrane at the site of virus attachment invaginates and forms a vesicle containing the virion. This vesicle then enters the cytoplasm where the viral nucleocapsid is released.

3.2.1 Vaccinia Virus Uncoating

Because agents that inhibit endocytosis have little effect on vaccinia virus entry, fusion is believed to be the most important entry pathway for this virus. Vaccinia virus particles lose their outer lipid envelope after they enter cells (see Figure 1). The virus then undergoes a second uncoating step, in which the inner membrane is lost and the nucleocapsid containing the viral DNA with associated proteins is released. Vaccinia virus then begins to express its genetic information.

3.2.2 HIV-1 Uncoating and Reverse Transcription

Following binding of the HIV-1 gp120 attachment protein to the cellular CD4 receptor, a virion protein known as gp41 (Figure 2) mediates fusion of the viral envelope with the cell membrane. The HIV-1 nucleocapsid then enters the host cell cytoplasm, where infection can proceed. The next stage in the life cycle of HIV-1 and all members of the *Retroviridae* family is the reverse transcription of the virion RNA into double-stranded DNA. Two identical copies of viral genomic RNA are present in the HIV-1 virion. These RNA molecules are complexed with the viral enzyme known as RNA-dependent DNA polymerase or reverse transcriptase. The reverse transcriptase in the nucleocapsid that has entered the cell cytoplasm copies the two viral RNA molecules into a single double-stranded DNA molecule. This DNA copy of the HIV-1 genome is called the *provirus.* The proviral DNA remains complexed to viral proteins and is transported into the nucleus of the infected cell. One of the viral proteins, known as integrase, acts to precisely integrate a single copy of the proviral DNA into a cellular chromosome at a site chosen at random. The integrated HIV-1 provirus is now a new genetic complex irreversibly inserted into the human genome.

3.3 Transcription of Viral RNA

For both vaccinia virus and HIV-1, the next stage in the infectious cycle is the transcription of viral RNA from the viral DNA template. The molecular mechanisms of transcription for both viruses have been studied in great detail by molecular virologists. For HIV-1, some of the transcribed RNA serves as messenger RNA (mRNA) which is translated into protein, while other RNA serves as genomic RNA, which is packaged into the virion as infectious progeny viruses are assembled. For vaccinia virus, all the transcribed RNA serves as mRNA. For both viruses there is a precise temporal order in which subsets of viral mRNAs are transcribed. However, the mechanisms by which vaccinia virus and HIV-1 regulate this temporal order are profoundly different.

3.3.1 Vaccinia Virus Transcription

For vaccinia virus, two general classes of mRNA are transcribed in the host cell cytoplasm from the viral DNA template. These classes are the "early" and "late" mRNAs. As the names imply, the early mRNAs are synthesized first in infection, while the late mRNAs are not synthesized until vaccinia virus DNA replication has commenced. If DNA synthesis is inhibited during infection, late mRNAs are not synthesized. There are about 50 viral polypeptides encoded in the early mRNAs and about 150 encoded in the late mRNAs. For vaccinia virus, the enzyme responsible for synthesizing viral mRNA, the RNA polymerase, is encoded in the viral genome. The vaccinia virus RNA polymerase is a multisubunit complex, containing eight individual polypeptides that are encoded in separate genes. The vaccinia virus RNA polymerase is packaged in the virus particle and enters the infected cell in the nucleocapsid. The molecular mechanisms whereby the vaccinia virus RNA polymerase synthesizes viral mRNAs are related to those used by the cellular RNA polymerase II, since the two largest viral RNA subunits are related in sequence to the two largest subunits of human RNA polymerase II.

Specific DNA sequences in the vaccinia virus genome control the synthesis of mRNAs. Sequences called promoters determine where mRNA synthesis initiates. The vaccinia virus early mRNA promoters are very rich in A and T nucleotides, and they extend from the site of initiation to about 30 nucleotides upstream of this site. A viral early protein known as a transcription factor binds directly to these early promoter sequences. The binding of this early transcription factor then facilitates the binding of the viral RNA polymerase, after which synthesis of early mRNAs begins. Similar to cellular mRNAs, the 5′ ends of viral mRNAs (both early and late) are modified by methylation by the viral encoded capping enzyme. All vaccinia virus mRNAs are also modified at the 3′ ends by polyadenylation; the enzyme that carries out this modification, poly(A) polymerase, is also viral encoded. To control the termination of synthesis of viral mRNAs, a consensus sequence found at the 3′ ends of early genes serves as a termination signal. This sequence is recognized as RNA, rather than DNA, by a viral transcription termination factor. Interestingly, this 3′ termination factor is part of the viral capping enzyme that modifies the 5′ ends of viral transcripts.

When vaccinia virus DNA replication begins, the synthesis of early mRNAs is shut off and the synthesis of the late vaccinia virus mRNAs begins. Although this switch in viral transcription has long been known to occur, the mechanisms that regulate the process are poorly understood. The vaccinia virus DNA sequences that constitute promoters for late mRNAs are distinct from the early promoter sequences. The late promoters extend from the site of initiation of late mRNAs to about 20 nucleotides upstream of this site. Like the early promoters, the vaccinia virus late promoters are rich in A and T nucleotides. Termination of transcription of the late mRNAs at specific sites is inefficient and, consequently, late mRNAs encoding specific proteins are heterogeneous in size.

3.3.2 HIV-1 Transcription

The HIV-1 proviral DNA integrated in the cellular chromosome serves as the template for transcription of viral RNA. During the act of reverse transcription and integration into the chromosome, DNA sequences of a few hundred nucleotides are duplicated at the 5′ and 3′ ends of the viral genome. These sequences are called the *long terminal repeats (LTRs)*. The 5′ LTR contains the HIV-1 promoter, which regulates initiation of transcription. the 3′ LTR contains the sequences that are responsible for the termination and polyadenylation of viral transcripts. Like vaccinia virus, HIV-1 transcription entails both early and late classes of mRNAs. However, the mechanisms that regulate early and late mRNAs are different for the two viruses. As discussed earlier, vaccinia virus DNA replication regulates a switch in transcriptional initiation from the early to late viral mRNAs. At the end of this subsection, we discuss how an HIV-1 posttranscriptional process regulates the switch from early to late transcription.

The single HIV-1 promoter is considerably more complex than the multiple early and late promoters of vaccinia virus. Unlike the simple vaccinia virus promoters, the HIV-1 promoter contains a number of distinct cis-regulatory DNA sequences that bind cellular proteins which are transcription factors. Moreover, some of these transcription factors are multiprotein complexes. Because some of the subunits of these multiprotein complexes belong to families of related proteins, there are numerous combinations of transcription factors that can bind and regulate transcription of the HIV-1 promoter. The HIV-1 core promoter extends from the site of initiation to about 100 nucleotides upstream. Under some circumstances, LTR sequences located up to several hundred nucleotides upstream of this core promoter can also regulate HIV-1 transcription.

To initiate transcription of the HIV-1 genome, a transcription complex is assembled on the HIV-1 promoter. This complex contains cellular RNA polymerase II, which is composed of 10 subunits, plus about 20 loosely associated proteins that are known as the basal transcription factors. The cis-regulatory sequences in the HIV-1 promoter regulate the rate at which this transcription complex is assembled. At this point, basal transcription of the HIV-1 genome commences. The HIV-1 transcription complex, however, is extremely inefficient in transcribing the entire 9000 nucleotides of the HIV-1 genome. Consequently, basal transcription directed by the HIV-1 LTR terminates within a few hundred nucleotides of the site of initiation. When a transcription complex does transcribe the entire HIV-1 genome, the primary RNA transcript in the nucleus is processed by multiple RNA splicing events. These multiply spliced early viral mRNAs are then exported to the cytoplasm, where they are translated into the early viral proteins.

One of these early proteins is the viral transcriptional activator known as the Tat protein. Tat functions to increase transcription of the HIV-1 genome up to several hundredfold. The Tat protein acts by entering the nucleus and binding directly to a cis-regulatory RNA element found at the 5′ ends of nascent viral transcripts. By binding to this RNA element, called TAR RNA, the Tat protein acts to stimulate the ability of transcription complexes to efficiently transcribe the entire HIV-1 genome. Regulation of transcription through a 5′ RNA element such as TAR is an unusual mechanism of gene expression; this type of regulation has been observed only in a few retroviruses. In addition to acting at the level of transcriptional elongation, Tat may also act to increase the initiation of transcription of the HIV-1 genome.

Another of the early multiply spliced HIV-1 mRNAs encodes a regulatory protein known as Rev. The Rev protein acts posttranscriptionally to activate the switch from expression of early to late viral mRNAs. After being translated from its early mRNA, the Rev protein enters the nucleus and, like the Tat protein, functions by binding directly to a cis-regulatory RNA element. This element, called the Rev response element (RRE), is located toward the middle of the full-length HIV-1 primary transcript. Binding to the RRE prevents splicing of the primary transcript, and unspliced or singly spliced viral RNAs are exported to the cytoplasm. In general, these incompletely spliced mRNAs encode viral proteins that are assembled into the virion. Unspliced HIV-1 RNA can also serve as the genomic RNA that is packaged in the virion. The consequence, therefore, of Rev action is to activate a switch from the nuclear export of early or multiply spliced viral RNAs to the export of late or incompletely spliced viral RNAs.

3.4 Translation of Viral mRNA

Some viral infections can result in profound changes in the profile of proteins being synthesized in the infected cell. Vaccinia virus infection has such an impact on protein synthesis. In other viral infections, such as that of HIV-1, there might be only modest alterations in protein synthesis.

Vaccinia virus infection results in a phenomenon called the shutoff of host cell protein synthesis. This means that during vaccinia virus infection the synthesis of cellular mRNAs is inhibited and protein synthesis in the cell becomes exclusively viral. Thus large amounts of viral proteins are certain to accumulate in the infected cell, allowing the production of high levels of progeny virus. A major contributing factor to host shutoff during vaccinia virus infection is the apparent indiscriminate degradation of all mRNAs in the cell, both viral and cellular. Within the first few hours of vaccinia virus infection, transcription of cellular mRNAs in the nucleus is inhibited. Therefore, the degraded mRNA in vaccinia virus infected cells can be replaced only with newly synthesized viral mRNAs. As a consequence, only viral mRNAs are found in the cell late in infection. The rapid degradation of all mRNAs during infection also assures that the early viral mRNAs are replaced by the late mRNAs and, therefore, only late viral proteins are synthesized at the later stages of the infectious cycle.

HIV-1 infection, on the other hand, does not result in the shutoff of host cell protein synthesis. Rather, viral mRNAs appear to be translated into proteins by similar mechanisms as do cellular mRNAs. There exists evidence, however, that under some circumstances the viral Tat and/or Rev protein may influence viral mRNA translational efficiency or the stability of viral mRNA.

3.5 Replication of Viral Genome

An extremely unusual property of vaccinia virus and all members of the *Poxviridae* family is the presence of a cytoplasmic site for DNA replication. Discrete sites of viral DNA replication, called factory areas, have been observed in the cytoplasm by autoradiographic and microscopic techniques. Vaccinia virus DNA replication occurs during a discrete time in the infectious cycle, in general beginning at about 1.5 hours postinfection and ending at 6 hours postinfection. Replication results in the production of about 10,000 genomic copies of each viral genome that has entered the cell. Because of the large size (190,000 nucleotides) of the vaccinia virus genome, the mechanisms of viral DNA replication have been difficult to investigate and are poorly understood. There is some evidence that replication begins at each end of the double-strand DNA genome. Interestingly, the ends of each of the two strands of the vaccinia virus genome are covalently connected by a short single strand of DNA. Efforts to identify one or more vaccinia virus DNA sequences serving as origin(s) of replication have been unsuccessful. Most if not all of the enzymes required to replicate vaccinia virus DNA are viral encoded. These enzymes include DNA polymerase, thymidine kinase, ribonucleotide reductase, and perhaps DNA ligase.

In the replication of HIV-1, as already discussed, the viral RNA that enters the infected cell is converted to DNA and integrated into a cellular chromosome. The viral genome replicates by the process of transcribing the DNA proviral DNA into viral RNA. The proviral DNA is subject to normal cellular mechanisms of DNA replication, and the integrated provirus replicates only if the infected cell replicates its own DNA.

3.6 Assembly and Release of Virion

Electron microscopic analysis has shown that vaccinia virus virions are assembled in the infected host cell cytoplasm in areas apart from cellular membranes. The first morphologically distinct viral structures seen are viral-specific bilayer membranes. After this, a dense nucleocapsid mass surrounded by membranes is observed, indicating that the viral DNA and associated proteins have entered the immature viral envelope. Additional complex morphological steps occur which involve the organization of the internal and external components of the vaccinia virus virion. Proteolytic processing of several viral proteins occurs as they are assembled into the virion. Mature vaccinia virus particles move to the cell periphery, where they are enveloped by additional membranes and released from the cell. However, large amounts of vaccinia virus particles do not receive this final envelope and remain in the cell.

The HIV-1 virion is assembled on the inner surface of the infected cell plasma membrane. The viral envelope protein is synthesized as a 160 kDa precursor, which is inserted into the cell membrane and subsequently cleaved by a cellular protease into the gp120 and gp41 proteins found in the virion (see Figure 2). The viral proteins found in the viral nucleocapsid are also transported to the inner membrane surface as polyprotein precursors. One of these precursors, known as the GAg precursor, appears to bind HIV-1 genomic RNA and aid in transport of the RNA to the site of virion assembly. The nucleocapsid precursor polypeptides plus two genomic RNA molecules are assembled into a virus particle. The virion then buds out from the cell membrane, incorporating the viral gp120 and 41 polypeptides. Upon final assembly of the HIV-1 particle or during the budding process, the viral precursor polypeptides are cleaved by a viral protease into the mature viral proteins. These proteins include reverse transcriptase and integrase.

4 VIRUS HOST–CELL INTERACTIONS

4.1 Infections in Individual Organisms

Viral infections in individuals can be broadly categorized into three types: acute, chronic, and latent. In *acute infection,* the infecting virus undergoes active expression of its genes and viral replication. An example of acute infection is that of Epstein–Barr virus, a member of the *Herpesviridae* family, which actively replicates in B lymphocytes and causes infectious mononucleosis. In *chronic infection,* the infecting virus replicates persistently at a relatively low level over a prolonged length of time. An example of chronic infection is that of hepatitis B virus, a member of the *Hepadnaviridae* family, which causes liver disease or hepatitis. In *latent infection,* the infecting virus is present but does not undergo active expression of its genes and does not replicate except during intermittent episodes. The herpesviruses, members of the *Herpesviridae* family, can persist in latent form in neuronal cells without causing illness except during periodic bursts of viral gene expression and replication.

Both viral and cellular factors influence the course of infections and contribute to whether an infection will be acute, chronic, or latent. Investigations into cellular and viral mechanisms that regulate these factors is extremely difficult as a result of the complexity of infection in an individual. The infected individual has a number of different tissues, each with unique biological properties. The individual also has antiviral mechanisms, such as the immune system and the interferon system, which act to limit or stop infection. The virus itself, which during replication in the individual can generate mutations in a number of viral genes, entails further complexity. Specific mutations can influence replication in different tissues, leading to viral variants with special properties in different tissues in the individual. Research on this topic has largely been carried out in molecular virology by means of experimental approaches that utilize tissue culture cells and isolated viruses. This research has yielded information and insight into viral infections, but the lessons learned from these simple experimental systems are limited and give only a glimpse of the complexity of an authentic infection. A major challenge for molecular virology is to understand the cellular and viral factors that regulate infection in individuals.

4.1.1 Vaccinia Virus Infection in Individuals

During smallpox vaccination, individuals are deliberately infected with live vaccinia virus by puncture or scarification. The draining lymph node at the inoculation area becomes enlarged, and acute infection is almost always restricted to cells and tissues at this site. In rare circumstances, usually in individuals with immunological deficiencies, vaccinia infection spreads throughout the body and can affect tissues such as spleen, lymph nodes, and nerve cells. Normally, cell-mediated and humoral forms of immunity act to limit and clear vaccinia virus infection within about 3 weeks. A scar remains at the site of inoculation, acting as a lasting record of the smallpox vaccination. Because smallpox has been eradicated, vaccinia virus vaccinations are no longer carried out.

4.1.2 HIV-1 Infection in Individuals

HIV-1 infection in individuals almost invariably leads to the development of AIDS and ultimately death. During initial HIV-1 infection, the individual may have no symptoms, or a slight fever or general malaise over the course of a few weeks. During this time, HIV-1 is thought to replicate acutely before the infected individual's immune system can act to suppress infection. From this point, HIV-1 infection can be considered both persistent and latent.

HIV-1 appears to replicate persistently in the lymph nodes, producing a continuous level of virus that can infect CD4+ cells, especially the helper T-cell population. Because of the high mutation rate of HIV-1, viral variants are continually being generated, each with potential special tropism and ability to evade the immune system. Additionally, infected cells may contain an integrated HIV-1 provirus that does not express any viral proteins and, therefore, no viral replication occurs. If some event stimulates this latent provirus, however, the infectious cycle will resume. Eventually, HIV-1 replication acts to debilitate the infected individual's immune system, leading to illness due to opportunistic infections from various human pathogens. Additionally, the failure of an active immune system may allow the unrestricted growth of transformed or cancerous cells in an individual. Thus, many who are infected with HIV-1 develop fatal forms of cancer, often lymphomas.

4.2 Infections in Individual Cells

Viral infections in individual cells can vary tremendously in their outcomes. An example of two extremes are those of vaccinia virus and HIV-1. Vaccinia virus infection results in drastic changes in cellular metabolism, and infection usually leads to cell death. In contrast, HIV-1 infection results in introduction of the proviral DNA into the cellular DNA, which under some circumstances leads to only minor changes in cellular metabolism and has only minor effects on cellular growth.

Vaccinia virus infection leads to profound changes in cellular metabolism. Cellular transcription is inhibited early during infection and as discussed earlier, cellular protein synthesis is also inhibited. With its large genome, vaccinia virus is able to encode several functions that are devoted to perturbation of specific cellular regulatory mechanism. Many of these viral functions act to inhibit cellular antiviral mechanisms. One major antiviral system is the interferon system.

Interferons are a family of secretory proteins that bind to specific cellular receptors. The binding of interferons to their receptors activates a signal transduction pathway that leads to the induction of transcription of certain mRNAs. These interferon-inducible mRNAs encode proteins that have antiviral properties. The best characterized interferon-induced protein is a protein kinase known as PKR. PKR's kinase activity is activated by double-stranded RNA, which is often produced during viral infection. Activated PKR phosphorylates a cellular protein called eIF-2α, which is a key component of the translational machinery. Phosphorylation of eIF-2α leads to the shutoff of all protein synthesis, both cellular and viral. The role of PKR during viral infection, therefore, is that of a suicide mechanism to terminate all protein synthesis and prevent the production of viral progeny. However, vaccinia virus encodes a protein, known as K3L, that acts to inhibit the activation of PKR during infection. Thus, vaccinia virus is able to overcome this cellular antiviral mechanism. An additional interferon-induced antiviral mechanism is a ribonuclease system known as 2-5A; the 2-5A system is also activated by double-stranded RNA produced during viral infection. Vaccinia virus appears to encode a function to inhibit the activity of the 2-5A antiviral system.

Vaccinia virus also encodes genes that perturb regulation of the host immune system. Vaccinia virus encodes a protein that is homologous to a cellular protein that inhibits the complement pathway. Thus, vaccinia virus can block this immunological control mechanism. Vaccinia virus also encodes a protein that is secreted from infected cells and binds and inhibits the IL-1β cytokine; the interleukin IL-1β normally is involved in regulation of the inflammatory and immune response. Additionally, vaccinia virus encodes a growth factor, known as vaccinia growth factor (VGF), which is related to epidermal growth factor (EGF). VGF binds to the cellular EGF receptor and activates EGF-mediated pathways in the cell. The role of VGF may be to stimulate uninfected cells near infected cells, allowing increased viral replication upon the infection of these stimulated cells. There are additional gene products encoded by vaccinia virus that interact with other cellular regulatory mechanisms. It appears that the large vaccinia virus genome has enabled the virus to acquire sophisticated mechanisms for inhibiting or usurping cellular functions for its own purposes.

5 USE OF VIRUSES AS BIOLOGICAL TOOLS

Through the use of recombinant DNA technology, molecular virologists are able to manipulate viral genomes and use the engineered viral genomes as biological tools. For example, foreign genes can be inserted into the vaccinia virus genome and subsequently expressed at high levels during infections of the viral vector. Remarkably, vaccinia virus vectors can accommodate up to 25,000 base pairs of foreign DNA. If the foreign gene encodes a protein from a pathogen, the vaccinia virus vector can be used as a vaccine to elicit a strong immune response against the foreign gene. In many cases, vaccination of animals with vaccinia virus vectors protects the vaccinees from subsequent infection by the pathogen. Such vaccinia virus vectors have been used successfully to protect animals against infections by a number of viruses, including hepatitis B virus (chimpanzees), herpes simplex virus (mice), and rabies virus (mice, foxes).

For members of the *Retroviridae* family such as HIV-1, molecular virologists are able to construct molecular clones of the relatively small viral genome. The cloned genome can then be extensively manipulated; viral genes that are essential for causing disease can be removed and replaced with foreign genes. Retroviral vectors are much more limited than vaccinia virus vectors in the foreign DNA that they can accommodate, being restricted to only a few thousand base pairs. The biological properties of retroviruses make them well suited as vectors for purposes such as gene therapy. A retroviral vector can be irreversibly inserted into the host cell chromosome with no adverse effects on cellular metabolism or function. The foreign gene expressed from the retroviral vector can be a protein that corrects a metabolic defect that may arise from a genetic deficiency. A murine retroviral vector was used to introduce and express the gene for the essential enzyme adenosine deaminase (ADA) into the T lymphocytes of two children who suffer from severe combined immunodeficiency. A rare genetic disease prevents these children from expressing ADA. To date, both patients appear to have responded positively to this gene therapy. Retroviral vectors have potential for other genetic diseases such as sickle-cell anemia and thalassemia. Retroviral vectors also have potential for treatment of pathological nongenetic conditions, such as cancer or infectious diseases.

6 PERSPECTIVES

The recent pandemic of human immunodeficiency viruses has dramatically reaffirmed that viral pathogens will remain a major human health concern. It seems certain that new viruses will continue to appear, causing much human suffering. The lack of success in developing an effective vaccine against human immunodeficiency viruses also emphasizes the formidable challenges facing molecular virologists in understanding mechanisms involved in the development of viral diseases. The major goal facing molecular virologists is an understanding of both viral and cellular factors that control viral disease in individuals. With increased knowledge about these factors, effective means to combat viral diseases can be developed.

See also AIDS, Inhibitor Complexes of HIV-1 Protease in; AIDS, Transcriptional Regulation of HIV in; AIDS HIV Enzymes, Three-Dimensional Structures of; Poliovirus, Molecular Biology of; Varicella-Zoster Virus and Cytomegalovirus Vaccines; Viral Envelope Assembly and Budding; Viruses, DNA Packaging of.

Bibliography

Cullen, B. R. (1992) Mechanism of action of regulatory proteins encoded by complex retroviruses. *Microbiol. Rev.* 56:375–394.

Fields, B. N., Knipe, D. M., Chanock, R. M., Hirsch, M. S., Melnick, J. L., Monath, T. P., and Roizman, B., Eds. (1990) *Fundamentals of Virology,* 2nd ed. Raven Press, New York.

Morgan, R. A., and Anderson, W. F. (1993) Human gene therapy. *Annu. Rev. Biochem.* 62:191–217.

Moss, B., and Flexner, C. (1987) Vaccinia virus expression vectors. *Annu. Rev. Immunol.* 5:305–324.

Vaishnav, N., and Wong-Staal, F. (1991) The biochemisry of AIDS. *Annu. Rev. Biochem.* 60:577–630.

Viruses, DNA Packaging of

Philip Serwer

1 **Basic Phenomenology and Rationale**
 1.1 The Use of Genetics to Analyze Pathways
 1.2 The Use of Gene Cloning to Purify Bacteriophage Gene Products
2 **The Life Cycle of Double-Stranded DNA Bacteriophages**
 2.1 Structure of the Bacteriophages
 2.2 Adsorption to a Host Cell; Injection of DNA
 2.3 Early Metabolic Events
 2.4 Assembly of the Bacteriophage
3 **The DNA Packaging Pathway**
 3.1 Data for the Analysis of DNA Packaging Pathway
 3.2 Transduction of Energy During DNA Packaging
4 **DNA Packaging by Animal Viruses**
5 **Contribution to the Intellectual Mainstream of Molecular Biology and Biotechnology**

Key Words

Bacteriophage A virus that infects bacteria.

Capsid, Viral The protein component of a virus, within which the viral nucleic acid is packaged.

Concatemer, DNA An end-to-end aggregate or polymer of DNA

Energy, Transduction of The conversion of energy from one form to another.

Mutant, Conditional-Lethal An organism genetically altered so that it is nonviable in some, but not all, circumstances in which the nonmutant organism is viable.

Procapsid, Viral The pre-DNA packaging precursor of the mature capsid of a virus.

Protein, Scaffolding A protein that is both part of and necessary for assembly of a procapsid, but is lost during conversion of the procapsid to mature capsid.

The packaging of DNA in preformed viral procapsids is an event of DNA metabolism that is studied to develop understanding of biological specificity, energy transduction, and structural dynamics. DNA packaging is also used for the purposes of biotechnology. To assist in achieving these objectives, integrated multidisciplinary analysis is performed. The disciplines are genetics, physical chemistry, and biochemistry. Genetic alteration of DNA packaging is used to determine functions of proteins, to determine specificity of DNA–protein binding, and to assist in the isolation of particles in the DNA packaging pathway. Physical chemistry is used to understand both the structural transitions and the accompanying mechanisms of energy transduction during DNA packaging. Biochemistry is used to develop in vitro systems for performing DNA packaging; these systems provide improved control over conditions of experimentation.

1 BASIC PHENOMENOLOGY AND RATIONALE

Of the characteristics that distinguish a living organism from nonliving matter, purposeful behavior is the most obvious. To determine how purposeful behavior has evolved from interactions of macromolecules that are inanimate when isolated from each other, systems are needed that both mimic purposeful behavior and have sufficient simplicity for experimental analysis. To provide such systems, researchers have analyzed the processes by which a bacterial virus (also called a bacteriophage, or phage) multiplies after it has infected a host cell. Among reasons for choosing bacteriophage systems was the possession by bacteriophages of the following properties that assist analysis by the techniques of genetics, physical chemistry, and biochemistry:

1. The bacteriophages have a comparatively short life cycle: 13 minutes at 37°C in the case of the related bacteriophages T3 and T7.
2. Because of the properties of their prokaryotic hosts, both complete and partially assembled bacteriophages can be isolated in milligram quantities at comparatively low cost.
3. Most of the events of the bacteriophage life cycle can be made to occur in cell-free, but physiological, mixtures of the components of bacteriophage-infected cells (in vitro).

1.1 The Use of Genetics to Analyze Pathways

In the bacteriophage life cycle, each of the events described in Section 2 can be prevented by making nonfunctional the protein product of a bacteriophage gene. For the analysis of the bacteriophage life cycle, genetic alteration (mutation) of bacteriophage-encoded proteins has been used to block specific events, while often minimally changing the remaining events. Investigators use this type of mutation to determine both the pathways of metabolism in bacteriophage-infected cells and the role of bacteriophage gene products in each of the various metabolic events.

Mutations that cause premature termination of protein synthesis (polypeptide chain termination mutations) have been the most useful for preventing all processes that require a bacteriophage gene product. To analyze the effects of such a mutation, initially, growth of mutant bacteriophage must be performed. If, as is usually the case, the mutated gene is essential for growth of the bacteriophage, a condition must be found in which the effects of the mutation are suppressed. If such a condition exists, the mutation is called conditional-lethal. That is, the mutant grows in one (permissive) condition, but not in another (nonpermissive condition). In the case of polypeptide chain termination mutants, the permissive condition is a permissive host that reads a chain termination signal as an amino acid. The genetic code has three polypeptide chain termination signals, named (unscientifically) amber, ochre, and opal. The mutants are named by the type of polypeptide chain termination signal (amber mutant, etc.).

Temperature-sensitive mutants (nonpermissive conditions, high temperature; permissive conditions, low temperature) are a second type of conditional-lethal mutant. Isolation of temperature-sensitive mutants can be performed by growing bacteriophages, first at 30°C and then at 42°C, in host cells that are spread in a gelled support made of agar. Temperature-sensitive mutants (arrows in Figure 1) form a circular cleared zone of killed bacterial cells (plaque) that is smaller than the plaque formed by nonmutant T7.

After production of chain termination mutants at random on a bacteriophage genome, the essential genes of double-stranded DNA bacteriophages can be identified by mapping the mutants and determining which protein is not synthesized when a chain termination mutant infects a nonpermissive host. Although these mutants have been obtained by mutating the entire bacteriophage genome, more recent studies use either site-specific or region-specific mutagenesis to selectively, without chain termination, alter a protein segment required for only one of several activities of the protein.

To determine both the pathways of bacteriophage-infected cell metabolism and the activities of bacteriophage gene products in these pathways, a nonpermissive host is initially infected with a bacteriophage chain termination mutant. After the infected cells have burst, the effect of the mutation on metabolism is determined by analysis of the contents of the burst cells. For example, typically, an amber mutation in any one of several genes will prevent synthesis of DNA. Analysis of mutant proteins reveals, as expected, that one of these genes encodes a DNA-synthesizing enzyme, DNA polymerase. However, other such genes typically encode not a DNA polymerase but, instead, accessory proteins that assist the DNA polymerase. This genetic analysis identified a multiprotein DNA replication assembly that was much more complex than that expected before the genetic analysis. Like DNA replication, bacteriophage DNA packaging requires genetically identified accessory proteins that act while part of a multiprotein assembly.

1.2 The Use of Gene Cloning to Purify Bacteriophage Gene Products

Researchers isolate bacteriophage-encoded gene products in large (milligram) quantities during studies of DNA packaging for at least two purposes: for the physical characterization of components that participate in DNA packaging, and for the development of in vitro systems that mimic in vivo DNA packaging, by use of purified components only. Researchers will eventually use these in vitro systems to physically (e.g., by means of spectroscopy) monitor the events that occur during DNA packaging.

To purify the proteins of in vitro systems that are normally present in small amounts in vivo, the following procedure is used to obtain cells that make these proteins in greater amount. The gene that encodes a protein is, first, removed from its normal position in a bacteriophage genome (Figure 2a). Then the gene is placed next to a piece of controlling DNA (indicated by ϕ in Figure 2) that will cause

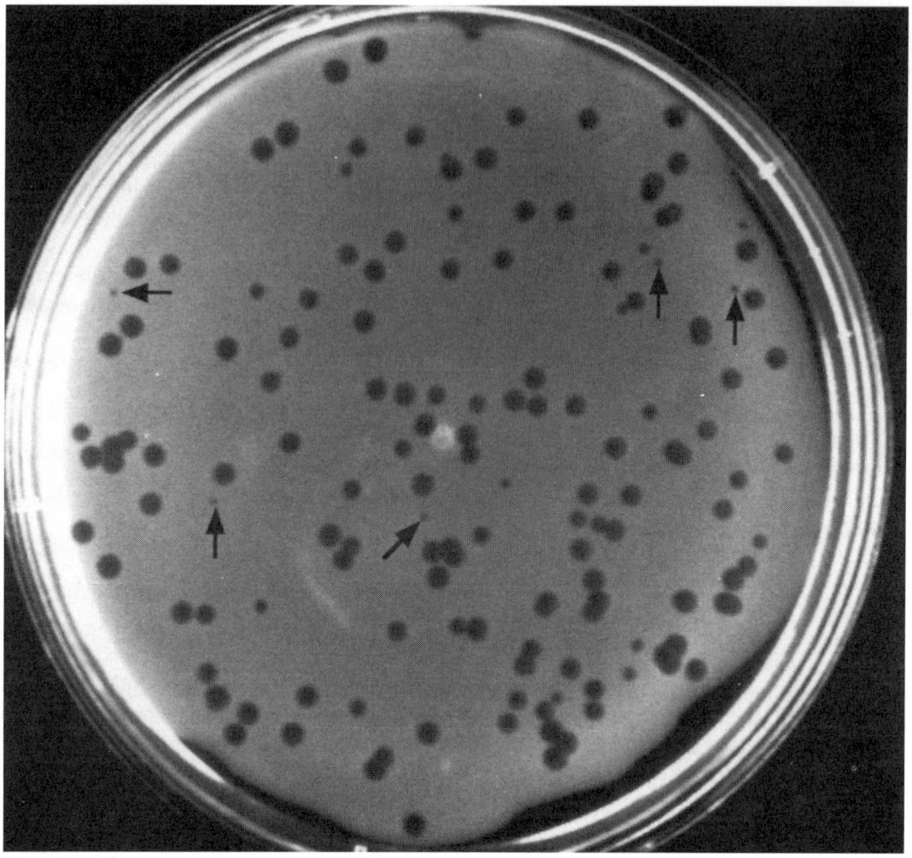

Figure 1. Isolation of temperature-sensitive bacteriophage T7 mutants by growth in a host spread in gelled growth medium.

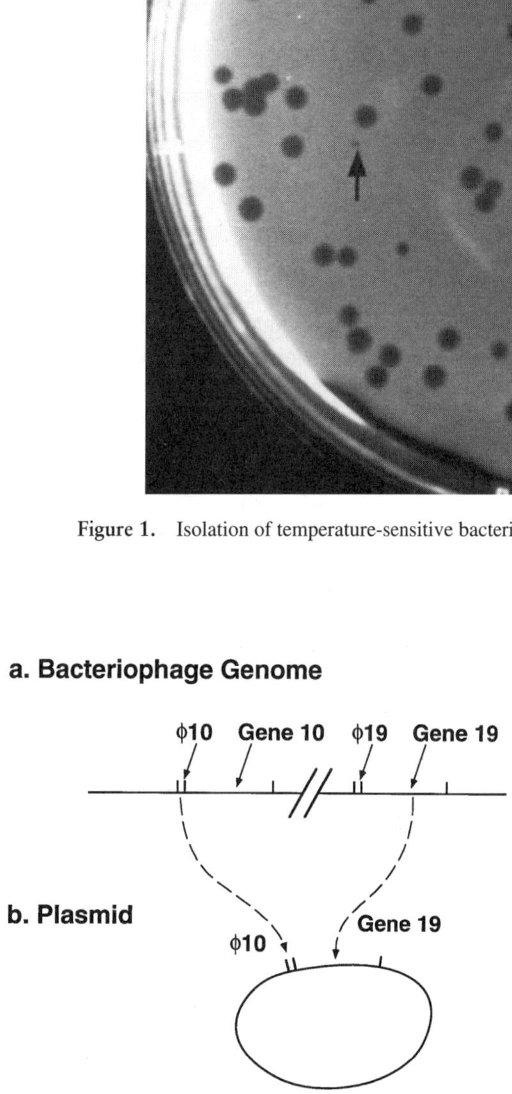

Figure 2. High level expression of a bacteriophage gene. (a) For gene 19, the controlling element (ϕ19) causes comparatively low levels of RNA synthesis during a bacteriophage infection of its host. In contrast, ϕ10 causes high levels of RNA synthesis for gene 10. (b) To obtain high levels of RNA synthesis from gene 19, this gene is moved to a plasmid, together with the controlling element, ϕ10. To obtain large amounts of the protein encoded by gene 19, the plasmid-containing host is initially grown with the ϕ10 controlling element in a nonexpressing condition. When the concentration of bacterial cells is optimal, the researcher adds a compound that activates the ϕ10 to an expressing condition, thereby causing synthesis of both the messenger RNA and the protein encoded by gene 19.

the gene to be expressed much more actively than normally would be the case. Expression consists of directing the synthesis of messenger RNA that, in turn, directs the synthesis of protein.

Typically, the gene, together with its new controlling element, is inserted in a circular piece of extrachromosomal DNA, called a plasmid (Figure 2b); plasmids are often found in bacteria. After growing plasmid-containing bacteria to high density, the researcher adds a compound that activates the controlling DNA, thereby causing production of large amounts of the protein. This process usually also kills the host cell. After induction of protein synthesis, the cells are lysed (burst), and they release the induced protein in large amount. Finally, the protein is purified, usually by standard procedures of protein purification.

2 THE LIFE CYCLE OF DOUBLE-STRANDED DNA BACTERIOPHAGES

2.1 Structure of the Bacteriophages

Structurally, the double-stranded DNA bacteriophages have two features in common: double-strandedness, and genome packaging in an outer shell of protein.

2.1.1 The Genome Is a Double-Stranded DNA

The genome ends can be either blunt (i.e., the two polynucleotide chains of the DNA double helix have the same length) or cohesive-

Mature Genome	Concatemer
ABCDEF...XYZABC	ABCDEF...XYZABCDEF...XYZABCDEF...

Figure 3. Comparison of a mature bacteriophage genome (nonpermuted) and a concatemer formed by end-to-end joining of these genomes. Strings of letters indicate blocks of nucleotides.

ended (i.e., complementary single-stranded regions protrude at both ends of the double helix). The blunt-ended genomes can have ends that are either in a unique position in the nucleotide sequence (nonpermuted) or in more than one position in the nucleotide sequence (permuted), as though cut from a precursor that was either circular or formed of linear genomes joined end-to-end (concatemeric; Figure 3). The nonpermuted genomes sometimes have a protein covalently joined to the 5′ end of both DNA strands (terminal protein). When blunt-ended genomes do not have a terminal protein, the nucleotide sequence at one end is repeated at the other (terminally repetitious). The genomes of some of the more frequently studied bacteriophages are illustrated in the second column of Figure 4.

2.1.2 The Genome Is Packaged Within an Outer Shell of Protein

The outer protein shell is the most massive component of the protein of the bacteriophage. In addition to the outer shell, the bacteriophage has protein in an external projection (tail) that varies in length among the different bacteriophages; the tail usually has fibers attached (third column in Figure 4). Greater detail is shown for the bacteriophage T7 capsid in Figure 5a. Internal proteins are also present; the organization of internal proteins varies among the different bacteriophages. For all double-stranded DNA bacteriophages, the outer shell, tail, and internal proteins are joined at a 12- 13-mem-

Name of Bacteriophage	Type of Genome	Type of Capsid
φ29	5' Pro 5'●———3' 3'———●5'	 38nm
λ	NP-Coh 5'———3' 3'———5'	 64nm
P1	P-B 5'———3' 3'———5'	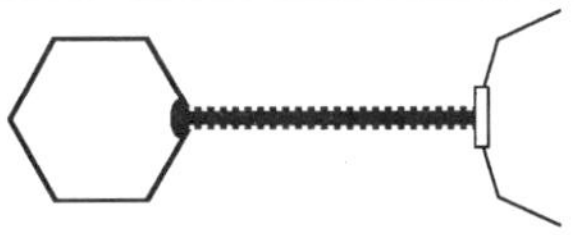 85nm
P2	NP-Coh 5'———3' 3'———5'	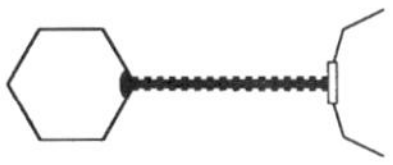 58nm
P22	P-B 5'———3' 3'———5'	 61nm
T2 (T4)	P-B 5'———3' 3'———5'	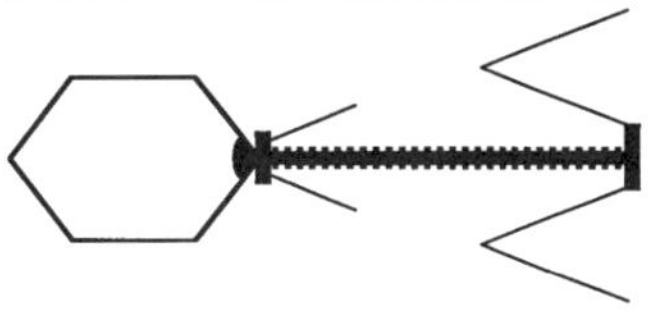 85nm
T3 (T7)	NP-B 5'———3' 3'———5'	 60nm
T5	NP-B 5'———3' 3'———5'	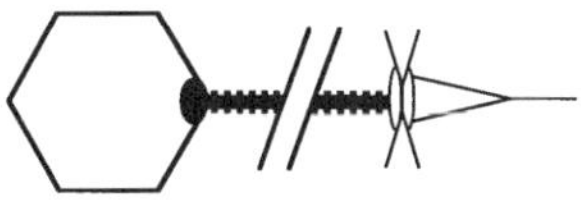 90nm

Figure 4. Structure of bacteriophage genomes and capsids. In the third column, measurements (nm) indicate width of outer shell. Other conventions: 5′ Pro, protein covalently joined to 5′ ends of DNA strands; NP, nonpermuted; Coh, cohesive-ended (i.e., single-stranded ends that are complementary, hence adhere to each other); P, permuted; B, blunt-ended.

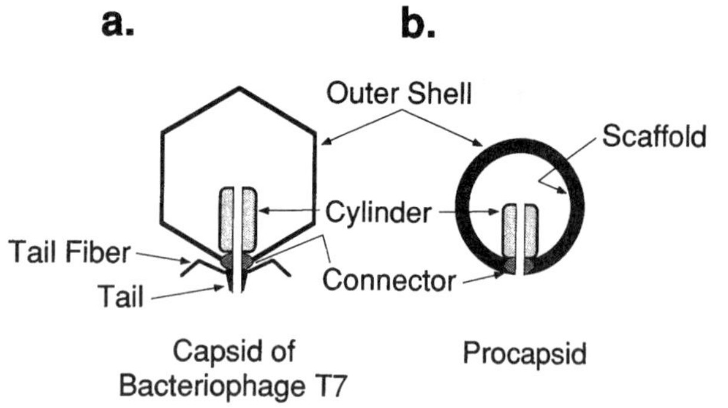

Figure 5. The bacteriophage T7 capsid and its procapsid: (a) the capsid of mature bacteriophage T7 and (b) the T7 procapsid.

bered ring of subunits, called the connector (Figure 4; see also Figure 5a). Together, these protein components are called a capsid. When sufficiently studied, the outer shell has been found to be either an icosahedrally symmetric array of subunits (e.g., λ, P1, P2, P22, T3, and T5 in Figure 4) or two hemispheres of icosahedrally arranged subunits, separated by a cylinder of the same subunits (ϕ29 and T4 in Figure 4).

2.2 Adsorption to a Host Cell; Injection of DNA

Initiation of infection by a double-stranded DNA bacteriophage occurs by adsorption to a host cell. Most, if not all, of these bacteriophages make the first contact via the fibers attached to the tail. Subsequently, the genome is injected through the tail into the host cell; injection is illustrated in Figure 6 for bacteriophage T7. In the case of bacteriophages with linear, nonpermuted genomes, the first end injected is specified by events that occurred during an earlier round of DNA packaging. In the case of T7, the left end (L in Figure 6) is injected first. This polar injection is part of the mechanism for controlling the expression of bacteriophage genes. That is, the first genes injected are the first expressed. In the case of permuted genomes, obviously, this mechanism will not work; other mechanisms are used for controlling the order of gene expression.

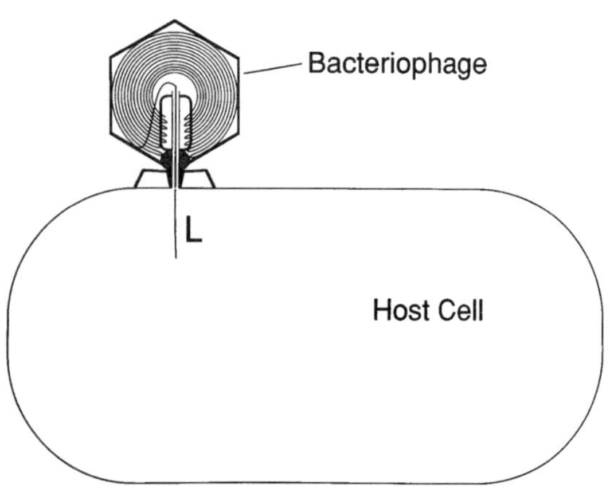

Figure 6. Injection of bacteriophage DNA into a host cell. The left-hand DNA end (L) is injected before the right-hand end of bacteriophage T7 DNA.

2.3 Early Metabolic Events

If a bacteriophage is to reproduce, logically, the first tasks after injection are to inactivate host defense mechanisms and convert host metabolism to bacteriophage metabolism. After a bout of molecular "star wars," either the bacteriophage genome is destroyed by a DNA-digesting enzyme of the host or the bacteriophage enters one of the following pathways:

1. The first (lysogenic) pathway leads to a state (called lysogeny) in which the bacteriophage DNA is in peaceful coexistence within the host bacterium. A lysogenic DNA can either become part of the host chromosomal DNA (example: bacteriophage λ) or it can exist separately from the host chromosomal DNA, like a plasmid (example: bacteriophage P1).
2. The second (lytic) pathway leads to both production of progeny bacteriophages and destruction of the host cell.

When conditions are unfavorable for a lysogenized bacterium, often the bacteriophage genome will abandon lysogeny and enter the lytic pathway. For some bacteriophages, such as T2, T3, T4, T5, and T7, the lysogenic pathway does not exist; these bacteriophages are, therefore, called lytic bacteriophages.

After entering the lytic pathway, the bacteriophage genome will express proteins in the order that is required to produce progeny bacteriophages. After the inactivation of host defense systems and the substitution of bacteriophage RNA synthesis for host RNA synthesis, this order is synthesis of DNA, followed by synthesis of protein components of the bacteriophage. For some bacteriophages (e.g., T3 and T7), substitution of bacteriophage RNA synthesis for host RNA synthesis is accomplished by expression of a bacteriophage-encoded RNA polymerase that recognizes signals for initiating synthesis of messenger RNA that come from the bacteriophage, but not the host. The bacteriophage RNA polymerase is expressed via RNA synthesis by the host RNA polymerase, during the earliest phase of infection. For other bacteriophages (e.g., T4), this task is accomplished by replacing one of several subunits of the host RNA polymerase with a bacteriophage-encoded protein.

2.4 Assembly of the Bacteriophage

In the final stage of a bacteriophage infection, progeny bacteriophages are assembled from components. In the case of all double-stranded DNA bacteriophages that have been studied, this occurs via the following pathway: a DNA-free capsid, called a procapsid, is assembled, whereupon the procapsid first binds newly synthesized DNA and then packages it. For some bacteriophages, the newly synthesized DNA is a concatemer of the mature bacteriophage genome (Figure 3); during packaging, this concatemer is cut to genomes of mature length. The procapsid has invariably been found to differ from the mature bacteriophage capsid in the following respects. First, the procapsid is less angular; for icosahedral mature capsids, the procapsid appears in the electron microscope to be a sphere. The procapsid is also smaller, less stable, more sensitive to proteases, and more electrically charged at its surface than the mature bacteriophage capsid. The procapsid of bacteriophage T7 is sketched in Figure 5b.

Although bacteriophage metabolism has several interesting aspects, all of which can be studied by the procedures of genetics, biophysics, and biochemistry, Section 3 describes only work done on

DNA packaging. Parallel studies on the packaging of DNA by double-stranded DNA animal viruses are discussed in Section 4.

3 THE DNA PACKAGING PATHWAY

The DNA packaging pathway starts with both the bacteriophage procapsid and its DNA substrate. Initially, the procapsid binds to its DNA substrate. When the substrate is concatemeric, the first end to be packaged is then produced by cleavage of the concatemer. In addition to both procapsid and DNA, two bacteriophage-encoded accessory proteins are needed to bind DNA. For those sufficiently studied bacteriophages (ϕ29, λ, and T3), initiation of packaging occurs in a complex that has the following sequence of spatial connections: procapsid→larger accessory protein→smaller accessory protein→DNA.

Although not proven, the packaging of concatemerized genomes appears to occur progressively on the concatemer, from one end to another. In the case of bacteriophage T7, a model proposed for initiation of packaging is shown in Figure 7. In this model, the procapsid binds the concatemer at two identical nucleotide sequences, one on each of two adjacent genomes within a concatemer (left-hand side of Figure 7). The binding occurs via a projection from the procapsid at a DNA site removed from the site at which the concatemer will be cleaved; the projection is made of the two accessory proteins for DNA packaging. After procapsid–concatemer binding, cleavage of the concatemer forms the mature right DNA end (Figure 7, middle) that subsequently initiates entry into the capsid, through an axial hole in the connector. The right-hand side of Figure 7 represents two possible DNA conformations during entry of DNA into the capsid. A conversion from procapsid to bacteriophage-like capsid occurs soon after entry has begun. Note that the pathway in Figure 7 does not show either the terminal phase of DNA packaging or the terminal cleavage required to separate a genome from a concatemer. The reason is that the data are not yet sufficient data to develop a reasonable model. For packaging/cleavage of nonpermuted, linear genomes, the following puzzle has yet to be solved: to produce a terminally repetitious genome, either two copies of the terminal repeat must be present at the joint between genomes in a concatemer, or the terminal repeat must be selectively replicated during packaging, or one genome must be destroyed for every genome packaged. The first possibility is known to be incorrect, and the other two are difficult to rationalize.

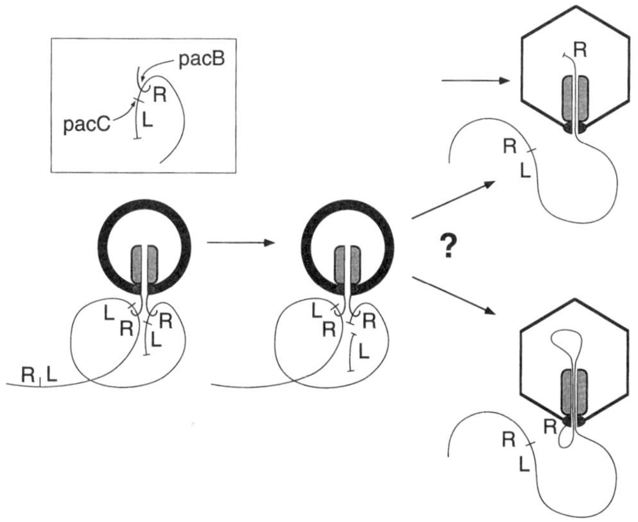

Figure 7. A model pathway for the packaging of concatemeric DNA, indicating proposed pathways for bacteriophage T7: L, left-hand end; R, right-hand end of a T7 genome within a concatemer. Inset: pacB and pacC sites.

3.1 Data for the Analysis of DNA Packaging Pathway

We turn now to the procedures that were used for developing the model in Figure 7. Both this model and models of other processes are derived from the types of data described in Sections 3.1.1 through 3.1.4.

3.1.1 Nucleotide Sequences Are Either Altered in or Moved from Regions of the Genome That Are Required for Specificity of Packaging

For example, to help identify the nucleotide sequences (called pac sequences) that determine the specificity of the initiation of non-permuted DNA packaging, regions of the bacteriophage genome are transferred to plasmids. If a region contains a pac sequence, its presence is revealed by packageability of the plasmid during infection by the bacteriophage. By progressively reducing the region transferred, this type of study has shown that, in general, two nucleotide sequences are necessary for initiation of the packaging/cleavage of concatemeric bacteriophage DNA to form a nonpermuted genome: a sequence at which the procapsid binds (pacB) and a second sequence at which an accessory protein cleaves (pacC) (see inset, left, Figure 7). In the future, more detailed information will be obtained by mutating nucleotides within either pacB or pacC.

3.1.2 The DNA Packaging Pathway Is Characterized After Bacteriophage Mutants Have Been Used to Infect in Nonpermissive Conditions

For example, in the case of most studied double-stranded DNA bacteriophages, two genes have been found to yield the following result when an amber mutant infects a nonpermissive host: both normal procapsids and a normal DNA substrate for packaging are made, but DNA packaging does not occur. This experiment was the first to identify the accessory proteins needed for the initiation of bacteriophage DNA packaging. In the future, further mutagenesis of accessory proteins will be used in attempts to arrest DNA packaging complexes at the various stages of DNA packaging.

3.1.3 The DNA Packaging Pathway Is Analyzed by Both Isolating and Characterizing Particles in the Pathway

The roles of both procapsids and DNA concatemers were initially determined by means of isolating and characterizing particles in the pathway. Ideally, one could establish the entire DNA packaging pathway in this manner. Although some success has been achieved in isolating DNA–capsid complexes that are in DNA packaging pathways, many of these particles are either too short-lived (in the pathway) or too unstable (during isolation) to characterize.

3.1.4 The Effects of Modifying Components Present Are Determined During In Vitro DNA Packaging

The roles proposed earlier for both procapsids and concatemers were determined during in vitro packaging. This approach also demonstrated that a blunt end is the substrate for DNA packaging for at least some bacteriophages, including T7 (see Figure 7). In vitro DNA packaging was first achieved by bursting bacteriophage-infected cells and incubating the lysate (i.e., the material released from burst cells) with DNA. To avoid in vivo packaging during preparation of a DNA packaging lysate, the infected cells used were nonpermissive cells infected with an amber mutant. For example, one type of lysate is actually a mixture of two lysates of infected, nonpermissive cells: one prepared by use of an amber mutant that does not form procapsids and a second prepared by use of an amber mutant that does not synthesize bacteriophage DNA. Mixed, these two lysates have all the components necessary for DNA packaging. When prepared carefully, this mixture of lysates does package DNA. However, neither of the component lysates packages DNA in the absence of the other.

The components needed for activity of in vitro DNA packaging lysates have been purified in the case of ϕ29, λ, and T3. Removing nonessential components of the lysates produces the following advantages:

1. When characterizing the lysates by means of optical, spectroscopic procedures, background from unrelated events is reduced.
2. The minimal components necessary for DNA packaging are determined during purification.
3. Uniformity of the DNA substrate for packaging is achieved (lysates have several forms of bacteriophage DNA).
4. Synchronization of packaging among the different capsids can be improved.

The major disadvantage is that the process of DNA packaging may be distorted. For example, P22, T3, and T7 in vitro DNA packaging does not occur efficiently (i.e., the pathway may be distorted) in the absence of high concentrations of a neutral compound (sugar, polyol, polysaccharide, other polymer). Presumably, this neutral compound mimics some aspect of the crowded condition of the inside of a cell.

3.2 Transduction of Energy During DNA Packaging

During in vitro packaging in a system of purified components, energy from the cleavage of ATP is required for DNA packaging. For ϕ29 and T3, quantification reveals that one ATP molecule is cleaved for every two nucleotide pairs packaged. The large accessory protein, an ATPase, probably is responsible for ATP cleavage during DNA packaging. In the case of T3 and T7, the rate of packaging at 30°C in vitro is roughly 400 base pairs per second. The free energy of DNA condensation measured for the T3-related T7 is 0.47 ± 0.05 kcal per mole of nucleotide pair. The mechanism of energy transduction is, thus far, an unsolved puzzle, as it is for most other energy-transducing, physiological events (e.g., muscle contraction).

Because of the scale of biological motors, such as the DNA packaging apparatus of double-stranded DNA bacteriophages, random thermal motion (Brownian motion) must be an integral part of any mechanism of energy transduction. Brownian motion might be the major source of motion for entry of DNA into a bacteriophage capsid. That is, by one hypothesis, DNA diffuses into the capsid by Brownian motion; an energy-requiring gate prevents the DNA from diffusing out of the capsid. In this case, the scale of Brownian fluctuations in the absence of packaging should also be the scale of fluctuations in the rate of DNA entry.

Alternative hypotheses propose that protein motors either push or pull the DNA into the capsid. These hypotheses include worm gear–like pushing derived from ATP-driven rotation of the connector, and reach-and-pulling motions by accessory proteins bound to the connector. In the latter case, Brownian fluctuations of DNA entry will be on a shorter time scale. Perhaps measurement of fluctuations in the rate of DNA entry will help to constrain theories that describe the mechanism of energy transduction during the entry of DNA into a bacteriophage capsid.

4 DNA PACKAGING BY ANIMAL VIRUSES

Of the double-stranded DNA animal viruses, some package DNA that is bound to the histone proteins that also bind chromosomal DNA in eukaryotic cells. In analogy with double-stranded DNA bacteriophages, other animal viruses package DNA that has not bound either histones or detectable amounts of other host proteins. The current evidence suggests that the histone-containing double-stranded DNA animal viruses (including polyoma and SV-40 viruses) do not package DNA in a preformed procapsid.

In contrast, however, at least two histone-free DNA animal viruses do form a procapsid in vivo; the procapsid subsequently packages DNA. The first is adenovirus, a virus that, during DNA replication, produces a linear DNA with a 5′ covalently bound protein. This linear product of replication undergoes polar entry of DNA, in analogy with both the terminal protein-containing bacteriophage ϕ29 genome and the terminally repetitious nonpermuted genomes of bacteriophages λ, T3, and T7. The adenovirus DNA also has a pacB-like DNA sequence near the end that first enters a capsid during packaging.

Herpes simplex virus is the second DNA animal virus that has histone-free packaged DNA. When observed by electron microscopy, packaged herpesvirus DNA resembles packaged bacteriophage T7 DNA, observed by the same procedure, in both density and texture. Like double-stranded DNA bacteriophages and adenovirus, herpes simplex virus also assembles a procapsid that subsequently packages DNA in vivo. Like bacteriophages λ, P22, T3, T4, and T7, herpes simplex virus packages a concatemeric product of DNA replication.

Although the DNA packaging of both adenoviruses and herpesviruses has several features in common with the packaging of the double-stranded DNA bacteriophages, a connector has not yet been found in either of these animal viruses. The omnipresence of connectors in bacteriophage capsids suggests that they are essential for some event of either DNA packaging or DNA injection that does not occur for the animal viruses.

5 CONTRIBUTION TO THE INTELLECTUAL MAINSTREAM OF MOLECULAR BIOLOGY AND BIOTECHNOLOGY

For studies of basic molecular biology, bacteriophage systems have often been favored for the reasons discussed in Section 1. Historically, bacteriophage systems have been used as exploratory experimental material for understanding both the fundamentals of biological information transfer and the enzymology of metabolic events (e.g., DNA and RNA synthesis). They have also been used as tools in accomplishing the goals of biotechnology. These goals include cloning of genes, high efficiency expression of genes, sequencing of DNA, site-specific mutagenesis, specimen preparation for electron microscopy, and fractionation of multimolecular particles. For example, DNA packaging can be used in the transfer of plasmid-cloned genes to a host cell. In this case, pac signals are incorporated in the plasmid; the plasmid is packaged in vitro in a bacteriophage capsid. To introduce the packaged plasmid into a host cell, the cell is infected with the bacteriophage that has packaged the plasmid. By use of either bacteriophage λ or bacteriophage P1, this procedure is in common use for the cloning of eukaryotic genes.

Although the future of science is always hard to predict, the author risks the prediction that a productive new direction in the analysis of DNA metabolism will be observation of DNA metabolism (including DNA packaging) at the level of a single DNA molecule. This would be achieved by use of light microscopy. For example, single DNA molecules have been observed by use of fluorescence light microscopy of DNA stained with fluorescent dyes. Therefore, might it not be possible to use fluorescence light microscopy to observe a single DNA molecule undergoing DNA packaging?

The information obtainable by light microscopy includes spatial coordinates at a resolution of 0.5 μm, effective radius of a particle (based on the magnitude of its Brownian motion), and any spectroscopic information that can be obtained from either the molecules involved or from probes attached to the molecules. Thus, for molecular biology in general and DNA packaging in particular, this field of endeavor depends on the quality of both the probes developed and the techniques for performing spectroscopy through a light microscope. Bacteriophage systems seem to be as appropriate for the further development of these aspects of light microscopy as they have been for developing other areas of molecular biology.

See also Bacterial Growth and Division; Cellular Energetics, Biochemical Basis of; Viral Envelope Assembly and Budding; Virology, Molecular.

Bibliography

Black, L. W. (1989) DNA packaging in dsDNA bacteriophages. *Annu. Rev. Microbiol.* 43:267–292.

Casjens, S. (1985) Nucleic acid packaging by viruses. In *Virus Structure and Assembly,* S. Casjens, Ed., pp. 75–147. Jones and Bartlett, Boston.

Earnshaw, W., and Casjens, S. (1980) DNA packaging by the double-stranded DNA bacteriophages. *Cell,* 21:319–331.

Fujisawa, H., and Hearing, P. (1994) Structure, function, packaging signals and specificity of the DNA in double-stranded DNA viruses. *Semin. Virol.* 5:5–13.

Serwer, P. (1989) Double-stranded DNA packaged in bacteriophages: Conformation, energetics and packaging pathway. In *Chromosomes: Eukaryotic, Prokaryotic and Viral,* K. W. Adolph, Ed., pp. 203–223. CRC Press, Boca Raton, FL.

Viruses, Photodynamic Inactivation of

James L. Matthews and Millard M. Judy

1 **Mechanism of Action of Photosensitizers**
2 **Effects of Irradiation on Nucleic Acid Pyrimidine Bases with and Without Photosensitizers**
3 **Protection Against Photochemical Damage**
4 **Virus Form and Susceptibility to Photokilling**
5 **Properties and Sources of Light Used for Photochemistry**
6 **Photosensitizers Used for Virus Eradication**
 6.1 Porphyrin Derivatives
 6.2 Phthalocyanines and Naphthalocyanines
 6.3 Merocyanines
 6.4 Naphthalimides
 6.5 Psoralen Photosensitizers
7 **Potential Uses and Prospects for Photosensitizers**

Key Words

Envelope A lipid-containing membrane that surrounds some viral particles. It is acquired during viral mutation by a budding process through a cellular membrane. Virus-encoded glycoproteins are exposed on the surface of the envelope.

Photodynamic Describing light-mediated damage to biologic systems usually involving the following sequence: light absorption by photosensitizers, energy transfer to oxygen, attack of singlet oxygen on biologic structure.

Photosensitizer A dye that reacts to light to form excited states.

Singlet Oxygen Oxygen with paired electronic configuration; may be produced by reaction of photosensitizer triplet with ground state oxygen.

Virion The complete viral particle, which may or may not have an envelope, and serves to transfer the viral nucleic acid from one cell to another.

Attempts to develop chemotherapeutic methods to treat viral diseases or to eradicate free virus or virus-infected cells from biological fluids have been disappointing. In spite of the remarkable successes of antibacterial, antifungal chemicals and antibiotics, antiviral compounds remain to be developed, for reasons that include the diversity of viral types, mutations to new forms, and the integral relationship that exists between the virus and host cell—that is, the dependence of the virus on host cells for metabolism and replication. Chemotherapeutic drugs were characterized as possessing a high affinity between drug and target receptor and a toxic potential capable of destroying the target once drug–receptor binding had occurred. It is presently recognized that several different direct or indirect mechanisms of target killing may be involved, but the premise of selective binding or uptake and ultimate target toxicity pertain. Antiviral chemicals are often too toxic to general host cells to be used as chemotherapeutic agents. For example, actinomycin D and puromycin are too toxic for in vivo use. Nevertheless, some drugs or biological compounds (halogenated pyrimidines, adenine arabinoside, interferons, etc.) are available for a limited number of viruses.

In recent years, an increasing number of investigations into the potential use of photochemicals as antiviral agents have been initiated. It is generally agreed that study of photobiology (the role of endogenous and exogenous photochemicals in biologic systems) started with the research performed in 1900 by Oscar Raab. He found that low concentrations of certain colored compounds such as acridine and eosin, which had no effect in the dark, led to the rapid killing of protozoans and paramecia upon illumination; only the wavelengths of light absorbed by the photosensitizer were effective. This phenomenon was called "photodynamic action." Subsequently, researchers reported that enzymes could be inactivated and that certain types of bacteria, animal and plant cells, and viruses could be killed by photodynamic action, raising the possibility that these compounds might offer a new class of agents for therapeutic use.

1 MECHANISM OF ACTION OF PHOTOSENSITIZERS

Molecules of many kinds can act as photosensitizers, including naturally occurring materials (e.g., chlorophyll, iron-free porphyrins, flavins, hypericin and other polycyclic hydrocarbons, a number of psoralens) and synthetic organic compounds (e.g., acridine, anthraquinone, thiazine and xanthene dyes), and some inorganic substances. In general, following absorption of light, these sensitizers (D) are excited to a short-lived singlet excited state (^1D) by the absorption of a photon ($h\nu$), and then they cross over into a long-lived triplet excited state (^3D), which mediates the photobiological damage.

Photosensitized reactions can be categorized into three types of reaction mechanism as summarized by Spikes. Type I and type II reactions both require the presence of molecular oxygen, and type III reactions are anaerobic.

Type I reactions proceed by radical mechanisms (reactions 1–7); the triplet sensitizer ^3D extracts an electron (or hydrogen atom) from the biological substrate S to give free radicals according to:

$$^3D + S \rightarrow D^{\cdot -} + S^{\cdot +} \qquad (1a)$$

or

$$^3D + S \rightarrow DH^{\cdot} + S^{\cdot} \qquad (1b)$$

The free radicals produced are very reactive because they have an odd number of electrons. Substrate free radicals can react with molecular oxygen to produce peroxides, which react further in antioxidation processes to give oxidized substrate products. Additionally, sensitizer free radicals may react with ground state molecular oxygen ($^3O_2$) according to:

$$D^{\cdot -} + {}^3O_2 \rightarrow D + O_2^{\cdot -} \qquad (2a)$$

in which the general state sensitizer D is regenerated and the chemically reactive superoxide radical anion are produced, or

$$DH^{\cdot} + {}^3O_2 \rightarrow D + HO_2^{\cdot} \qquad (2b)$$

in which $HO_2^{\cdot -}$ is the chemically reactive conjugate acid of $O_2^{\cdot -}$. Some semireduced sensitizer radicals $DH^{\cdot}$ can undergo dismutation according to:

$$DH^{\cdot} + DH^{\cdot} \rightarrow + DH_2 \qquad (3a)$$

Further, the fully reduced DH_2 may react with $^3O_2$ to give ground state sensitizer and hydrogen peroxide according to:

$$DH_2 + {}^3O_2 \rightarrow D + H_2O_2 \qquad (3b)$$

Additionally, superoxide radical anion can be produced in low yields according to:

$$^3D + {}^3O_2 \rightarrow D^{\cdot +} + O_2^{\cdot -} \qquad (4)$$

The semioxidized sensitizer $D^{\cdot +}$ further can react with the substrate according to:

$$D^{\cdot +} + S \rightarrow D + S^{\cdot +} \qquad (5)$$

from which the semioxidized substrate may further react with oxygen to give oxidized products or may react with other available free radicals to give covalent bonding of sensitizer to substrate according to:

$$D^{\cdot -} + S^{\cdot +} \rightarrow \dot{D}\text{—}S \qquad (6)$$

Additionally, the free radical reaction

$$S^{\cdot} + S^{\cdot} \rightarrow S\text{—}S \qquad (7)$$

gives covalent bonding of substrate by molecular type I reactions, which are favored by high substrate and low oxygen concentrations, since oxygen competes with substrate for interaction with ^3D.

Type II photosensitizer reactions are initiated by direct energy transfer from excited triplet state sensitizer to triplet ground state oxygen to give an electronically excited oxygen singlet state $^1O_2$ according to:

$$^3D + {}^3O_2 \rightarrow D + {}^1O_2 \qquad (8)$$

The excited singlet oxygen species $^1O_2$ is much more reactive than ground state $^3O_2$ because most organic substrates are also in a singlet ground state and reaction with the triplet state is spin-forbidden.

Type III anaerobic photosensitized reactions proceed in the absence of oxygen. In some cases the reaction is even inhibited by oxygen. The excited photosensitizer is converted into a toxic photoproduct. For example, psoralens form photoadducts of cross-linked DNA with high efficiency in the absence of oxygen. The photosensitized tranquilizer chlorpromazine, under anaerobic conditions, gives a stable, detergent-like product that is toxic to red blood cells and macrophages.

To elucidate mechanisms of photodynamic action in biological systems, a large amount of work has been carried out with cells and viruses, as well as their component biomolecules such as lipids, purines, pyrimidines, nucleic acids, amino acids, and proteins. Photodynamic degradation of these biomolecules results in damage to living systems.

Lipids, including unsaturated fatty acids, unsaturated neutral fats, unsaturated phospholipids, and steroids, are susceptible to photosensitized oxidation. Unsaturated fatty acids and their esters can be photooxidized with a variety of sensitizers.

With very few exceptions, proteins are photooxidized rapidly upon illumination in the presence of oxygen and photodynamic sensitizers. Five amino acids, all of which have electron-rich side chains, can be rapidly photooxidized in sensitized reactions: cysteine, histidine, methionine, tryptophan, and tyrosine. Therefore, the mechanism of the photooxidation depends on the amino acid, the sensitizer, the solvent, the reactant concentrations, and, in particular, the pH.

Many types of alteration occur in proteins as a result of photodynamic treatment, including physicochemical changes in charge, conformation, heat sensitivity, electrophoretic mobility, optical rotation, solubility, viscosity, and other properties. Changes in the biologic activity include loss of antibody binding, loss of activity of enzymes and protein (e.g., loss of toxicity of bacterial toxins and snake venoms), and loss of activity of insulin to bind cofactors. Some of these changes result from the photodynamic cross-linking of proteins to nucleic acids, and various other molecules can also occur.

Studies of the sensitized photooxidation of nucleic acids reveal that guanine and its derivatives are photooxidized most rapidly. Thymine and its derivatives are oxidized more slowly, while the other bases and their derivatives are oxidized very slowly with most sensitizers. The rates of oxidation of the nucleic acid bases increase sharply with increasing pH, suggesting that the anions are much more sensitive than the neutral bases. The sensitized photooxidation of purines and pyrimidines typically leads to a complex mixture of reaction products. The array of guanosine reaction products sensitized by methylene blue is different from the products resulting from oxidation with singlet oxygen produced by a radio frequency discharge. Thus it may be that both type I and type II processes are involved in the photodynamic reaction.

During photooxidation of nucleic acids in solution, guanine residues in DNA are preferentially destroyed with most sensitizers. In many cases, binding of the sensitizers to nucleic acids increases the efficiency of photodynamic damage. Free radicals are produced in the DNA moiety of DNA–sensitizer complexes upon illumination. The physicochemical changes include spectral shifts, chain cleavage, production of alkali-labile bonds, and changes in conformation, flexibility, sedimentation coefficients, and so on. For example, photodynamic treatment of DNA with methylene blue followed by piperidine specifically cleaves the chain at each guanine residue. Photodynamic treatment also alters the biologic properties of nucleic acids, such as destruction of bacterial DNA-transforming principle and loss of template activity of both prokaryotic and eukaryotic DNA.

2 EFFECTS OF IRRADIATION ON NUCLEIC ACID PYRIMIDINE BASES WITH AND WITHOUT PHOTOSENSITIZERS

Irradiation with ultraviolet light (220–400 nm) causes the formation of cyclobutane pyrimidine dimers in cellular DNA. These reactions can be lethal, mutagenic, or carcinogenic in various cells and organisms. DNA damage by UV is believed to be one possible initial event in the induction of cancer. Generally, UV primarily induces pyrimidine lesions while ionizing radiation causes purine damage.

DNA photolyase (PRE) directly reverses the pyrimidine dimers into pyrimidine nucleotides. This repair process is called photoreactivation, a different mechanism from excision repair, which involves removing the modified base as a base or as a nucleotide, usually an oligonucleotide.

Irradiation of DNA and RNA in the presence of photosensitizers may amplify the damage due to irradiation alone. Psoralens are classic enhancers of DNA lesion production. Psoralens and other furocoumarins have been isolated from numerous plants and have biologic activity when activated with UVA 320–400 nm. Additional synthetic derivatives have been made and evaluated. Psoralens are able to intercalate between base pairs of double-stranded nucleic acids. The prime mechanism of photodamage is adduct formation with double-stranded DNA, although other mechanisms are established. For example, DNA cross-linking in viral inactivation is reported, as are cross-links between DNA and protein and protein–protein linkages.

The initial reaction is a [2 + 2] cycloaddition between a 5:6 double bond of a pyrimidine base and a 3:4 or 9′, 5′-ethylenic double bond of the intercalated psoralen producing a monoadduct. Following monoadduct formation on the furan side, further irradiation (absorption of a second photon) produces a cycloadduct on the pyrone side, generating an interstrand cross-linked DNA. Reactions between psoralen and free thymidine yield numerous psoralen base adducts.

In addition to photoadduct formation, psoralen also generates reactive oxygen species, especially singlet oxygen species, that are capable of producing lesions that may result in DNA cleavage as well as lipid oxidation, leading to cell membrane degradation. Many other photosensitizers have been shown to produce similar oxidative damage of DNA. For example, Kagan et al. and Gasparro have reviewed psoralen–DNA–light interactions. Still other mechanisms are involved as several different psoralens cause DNA breaks even in the absence of oxygen. This is especially evident in supercoiled DNA, where breaks are more readily shown.

DNA damage is also reported as a result of some nonpsoralen photosensitizers, including those whose action is believed to be primarily effective at membranes. It is reported that DNA breaks occur with porphyrins and phthalocyanines. The resultant number of breaks is comparable to what is observed following X-radiation, but the breaks are restricted in location. This limited DNA damage is attributed to the duration of the singlet oxygen species generated at light excitation and to the short diffusion path of singlet oxygen. Proliferating cell nuclear antigen (PCNA) has been used to mark DNA damage sites, since the PCNA protein participates in excision repair of the damaged DNA. Even some porphyrins may intercalate into the DNA if their membrane binding is reduced as a result of their lower hydrophobicity. *meso*-Tetramethylpyridylporphine (TMPP) is an example of a porphyrin that binds with high avidity to DNA.

3 PROTECTION AGAINST PHOTOCHEMICAL DAMAGE

It is interesting to note that some photochemicals that sensitize the formation of singlet oxygen (e.g., bacteriochlorophylls, porphyrins, phthalocyanines, naphthalocyanines) are active quenchers of singlet oxygen in the ground state and thus may be used as protective compounds. The potential for a combination of photosensitizers and protective agents (quenchers) is recognized because of the need for selective damage, with minimal cytotoxic effects on the surrounding host tissue, on a preferred target such as tumor cells or infectious organisms. It is possible that sensitizer and quencher can influence both viricidal action and specificity of the reaction, particularly since viral inactivation may be achieved by different mechanisms. It is likely that addition of compounds to reaction media may either interfere with the viricidal actions or increase the protection of normal cells without compromising viricidal action. Addition of albumin to herpes simplex virus suspensions, for example, reduces collateral damage to blood platelets, but at the expense of loss of viricidal action. In contrast, adding mannitol, glycerol, reduced glutathione (GSH), or superoxide dismutase (SOD), known quenching agents of type I mechanisms, has no effect on the inactivation rate of vesicular stomatitis virus (VSV) in the presence of phthalocyanine sensitizers. Inhibition of VSV kill was shown with type II quenchers, such as trytophan or sodium azide, suggesting that if properly selected, sensitizers and quenchers may be useful adjuncts to viral kill systems. If one accepts that singlet oxygen is an important radical involved in virus inactivation, and that free radicals such as hydroxyl radicals are primarily responsible for cell damage, it makes sense to employ agents such as mannitol and thiols as anti-hydroxyl radicals. Superoxide dismutase (a scavenger of superoxide anions) or ascorbates or quercetin may be less useful because they scavenge both radicals and singlet oxygen. Selenium, a component of selenoperoxidase, may also have selective protective value for collateral normal cells, as may fluoride ions when used with some photosensitizers.

4 VIRUS FORM AND SUSCEPTIBILITY TO PHOTOKILLING

In general, viruses are very small particles ranging in size from 18 nm for parvovirus to 250 nm for poxvirus. All viruses contain a central core of nucleic acid and a protein coat or capsid that is specific for each virus, and there may be lipid and protein envelope surrounding the capsid and nucleic acid core. Most viruses also contain enzymes that are related to replication of their nucleic acid components, including reverse transcriptase activity for RNA viruses. Either DNA or RNA makes up the genetic material. Most clinically important DNA viruses have double-stranded DNA, although single-stranded circular or linear strands exist. RNA viruses exist primarily in single-stranded forms, although double-strand forms of RNA exist also. The related capsid proteins may be closely apposed to the helical structure, but are usually seen as a shell around the nucleic acid in an icosohedral form. There are many stages involved in virus–cell interactions, starting with viral attachment to target cell surface and proceeding to penetration of cell, uncoating (loss of protein), replication, assembly, and release. In the case of some bacterial viruses (bacteriophage), only the phage DNA but not the protein enters the cell, to initiate replication events. The viral replication events occur independently of the host cell replication, enabling multiple replications of the viral genome.

It should be noted that not all viruses that enter a cell immediately enter replication, maturation, and release. Some viruses exhibit a latent form in which the viral genomes are not replicated initially, but, rather, integrate into the host DNA chromosome to become a provirus. These conditions impose several problems for the person attempting to eradicate virus from biologic solutions, such as blood, for the virus may exist in three forms: free in the extracellular milleau in suspension, present within an infected cell, and in a provirus form within a cell.

Bacteriophages that integrate their DNA into bacterial chromosomes are termed lysogenic bacteriophages. Provirus forms may later leave the host chromosome to complete several replication cycles prior to assembly, maturation, and release from the infected cell. The ideal photochemical would be capable of eradicating all these forms; that is, the photosensitizer would be taken up by infected cells with replicating virus, provirus cells, and cell-free virus. Some likely candidate photosensitizers are reported to have an affinity for the viral membrane.

Since the membrane of enveloped viruses is derived from the membrane of the infected cell, a strong probability exists that the lipid/protein composition of the viral envelope is very similar to that of the infected cell. Since the provirus-containing cell was initially infected by virus, and thus has the membrane composition and protein features essential for viral infection, it is possible, but not necessarily probable, that the provirus cell also has some features common to the viral envelope. If such were the case, hydrophobic photosensitizers with membrane avidity might be common for all three forms. In the case of singlet oxygen generating sensitizers, lipid oxygenation could be expected to eliminate both cells and enveloped free virus, since the latter requires the envelope components for docking on a specific cell surface prior to penetration and infection.

Once sensitizers have been taken up by the infected cell membrane or virion envelope (or capsid in virions without envelopes), further diffusion into the interior can occur. Some cell organelles may exhibit preferential uptake varying as a consequence of the sensitizer's anionic or cationic charge, its hydrophobicity, and the amino acid composition of target proteins. Sensitizers that can cross the nuclear membrane or are not bound to capsid protein can interact with the nucleic acid components. Thus photosensitizers have several potential targets for their cytocidal or viricidal activity. Fortunately, many of the human viral pathogens are enveloped, so cell kill and viral kill are possible with the same sensitizer.

One use for this photochemical technology is the eradication of virus from blood to achieve a safer blood supply (even though donated blood is tested for a variety of pathogens, the tests presently used do not exclude 100% of infected blood). The two important formed elements of greatest clinical value are erythrocytes (RBC) and platelets, the latter serving to block vascular leaks and to participate in coagulation. Platelets are special cytoplasmic fragments from marrow-bound megakaryocytes and have no nuclear material. Since mature human erythrocytes, as well, have no nucleus, it is possible to treat blood with photosensitizers that damage virus nucleic acids without concern for nuclear damage to the important cells. The blood leukocytes (WBC) are not important to ordinary blood product use and can be removed by filtration or may also be killed by photochemical attack. Photochemical elimination is important because lymphocytes, for example, may be infected by retroviruses such as human immunodeficiency virus (HIV). Optimally, 100% viral inactivation can be achieved with little or no RBC and platelet damage. Again, augmentation with antioxidants, or temperature control, or reaction conditions in reduced oxygen environments may help achieve this goal.

Other biologic fluids and tissues also may be treated for viral inactivation. Fluids and tissues that are used, such as semen, cornea, and skin, also introduce the risk of transfer of infection from donor to recipient. One major limitation of the application of photochemicals to viral inactivation rests with the need for delivery of activating light to all parts of the medium or tissue being treated to achieve viral inactivation. Another obstacle is the potential for selective absorption of the activating light by some component of the normal target tissue, potentially resulting in thermal damage. Blood, for example, with its RBCs containing the chromophobe hemoglobin, absorb light over a range of 400–650 nm, which could produce thermal damage. Additionally, light scattering coupled with absorption imposes limits on the depth of penetration of the exciting light through the tissue; for whole blood, this depth is calculated to be less than 2 mm. In general, wavelengths in red range (630–690 nm) are able to be transmitted significantly deeper than others, and as a result, many photosensitizers with absorption excitation maxima in the red range have been synthesized.

5 PROPERTIES AND SOURCES OF LIGHT USED FOR PHOTOCHEMISTRY

Each photochemical sensitizer is maximally responsive to one or more wavelengths that constitute a physical characteristic absorp-

Table 1 Commonly Used Light Source for Photosensitization

Pathogen	Photosensitizer/ Dye Concentration (μM)/ Peak Absorbtion Wavelength	Light Source	Incident Power Density* (w/cm²)	Energy Density** (Joules/cm²)/ Surviving Fraction	Ref
HSV-1	MC540/ 26.3/520–570	Incandescent Lamp	3×10^{-3}	3.0×10^{-2}/ 3×10^{-3}	Lytle et al. 1991
HSV-1	Photofrin II/ 2.0/630	Fluorescent Lamp	9×10^{-4}	1×10^{0}	Lytle et al. 1989
HSV-1	Photofrin II/ 10/630	Filtered Xenon Arc Lamp	1.2×10^{-2}	1.0×10^{1}/ 4×10^{-4}	Judy et al. 1991
HSV-1	Sapphyrin 10/690	Dye Laser (Kiton Red Dye)	4.77×10^{-2}	1×10^{1}/ 4×10^{-4}	Judy et al. 1991

*POWER DENSITY = Power (watts) divided by area (cm²) over which irradiating beam is incident
**ENERGY DENSITY = Power density (w/cm²) multiplied by the exposure time (sec); One joule = one(watt)(sec)

tion spectrum unique to that compound. The wavelengths presently found in biologically effective sensitizers range from 350 nm for the psoralens to 950 nm for some extended ring porphyrins. Typically a variety of light sources is required to cover this wide range at the power densities needed to achieve complete reactivity. Table 1 summarizes light sources of different types that have been used for sensitized inactivation of viruses and other pathogens. Useful sources have included incandescent, fluorescent, and high pressure vapor arc lamps, dye lasers, and light-emitting photodiode arrays.

Incandescent and fluorescent lamps, individually or in arrays, offer the advantages of low cost and a broad spectral range of output optical radiation, which includes the wavelength range of 400–840 nm encompassed by often-used photosensitizers. Disadvantages include relatively low output power, usually necessitating longer treatment times, and the need, as a rule, to filter out potentially damaging ultraviolet and infrared wavelengths. Additionally, their highly diverging light patterns typically limit these lamps to irradiation of in vitro targets or cutaneous in vivo targets, as opposed to targets of limited extent that can be reacted endoscopically.

High pressure vapor discharge lamps such as the xenon arc lamp offer much higher total output power levels than incandescent or fluorescent lamps. Additionally, the visible spectrum emitted by the high power vapor discharge lamp tends to be richer, relatively, in the blue–green wavelength range than is the incandescent lamp, thus offering a flatter spectral power output. Typically, high pressure vapor discharge lamps incorporate collimating reflectors and lenses, ensuring that the emitted beam is slowly diverging, and relatively higher illumination intensities (ten to hundreds of milliwatts per square centimeter) are available over areas a few centimeters in extent. A disadvantage of the high pressure vapor discharge lamp is an unusual requirement for separating out the visable, near-infrared spectrum used in photosensitization from the high levels of UV and IR wavelengths typical of these lamps. Typically, dichroic mirrors (which pass only visible radiation), often in conjunction with a water or glass filter for IR filtration, are used. Additionally, since the high pressure vapor discharge lamp requires a specialized electrical power supply a lamp system costs much more than the setup for incandescent or fluorescent lamps.

Dye lasers, typically pumped by use of an argon ion laser, have also been used in studies of photosensitized inactivation of pathogens. These lasers can be tuned to emit radiation in the wavelength range of 500–900 nm. Choice of the identity of the light-emitting dye molecules fixes the lasing wavelength within the dye laser range. Dyes are readily available commercially. Output power of nonchromatic radiation up to 2–3 W are readily attained. Power density values ranging from tens of milliwatts to watts per square centimeter are readily attained over regions with dimensions ranging from centimeters to tens of centimeters. Dye laser systems are very costly, yet they offer very high output power. Furthermore, the output beam is very small in diameter ($\leq$ 1 mm diameter) and can be focused easily into a flexible optical fiber for endoscopic delivery for internal photosensitization therapy.

Semiconductor diodes, when conducting electrical current, emit light in the wavelength range of roughly 410–690 nm according to their chemical composition. These diodes are compounds of the chemical group III elements gallium and aluminum and one or more of the group IV elements, nitrogen, phosphorus, arsenic, and antimony. By adjustment of the relative ratios of the constituent elements, the emission wavelength can be moved over the emission range. Output optical powers of individual diodes currently available range between milliwatts and hundreds of milliwatts, with diodes emitting in the red range (630–650 nm) typically offering higher output power levels. Notably, the large number of porphyrin photosensitizers absorb light in this wavelength range. Planar arrays of semiconductor diodes offer the ability to illuminate targets of large area for photosensitization.

6 PHOTOSENSITIZERS USED FOR VIRUS ERADICATION

A variety of photochemicals or photosensitizers have been developed to address variation in virus chemistry, structure, and replication events, and to spare the normal cells in their environs. Only a few of the several hundred available photosensitizers are detailed here, as representatives of different types.

6.1 Porphyrin Derivatives

One of the first porphyrin photochemicals to be tested for antiviral action was Photofrin II, a complex mixture of porphyrins in oligomeric and monomeric form at different degrees of aggregation. In plasma, the majority of this mixture is associated with high den-

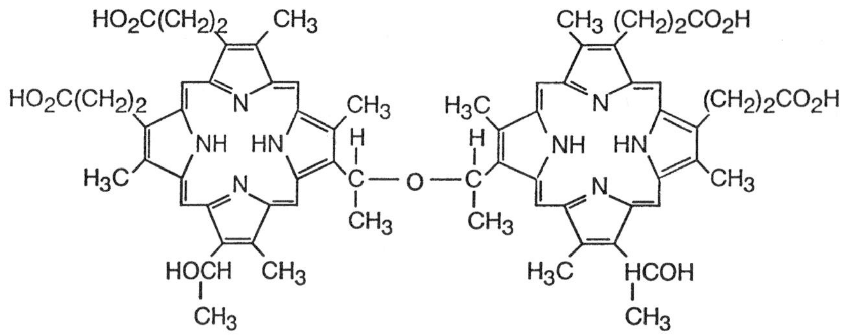

Figure 1. Hematoporphyrin derivative: molecular structure.

sity lipoproteins (HDL), with low density lipoproteins (LDL) and albumin accounting for the rest.

That porphyrins given to tumor-bearing animals accumulate in the tumors has been known for at least 50 years. Their typical red fluorescence, used for tumor localization as well as for therapeutic purposes, was reported first by Auler and Banzer in 1943. Present-day tumor therapy exploits this property widely. It is interesting to note that many of the photosensitizers found to be efficacious for tumor therapy are also effective in virus kill. Interestingly, many photosensitizers are nontoxic in the dark, at relatively high concentrations, whereas light-irradiated local sensitizer concentrations are lethal to the fluorescing cells, supporting their potential as photochemotherapeutic agents.

Over time, several isolated porphyrins have been tested and several pure porphyrins have been synthesized. The hematoporphyrin derivative (HpD), whose structure is shown in Figure 1, exhibits an action peak around 630 nm in the red spectrum. Upon irradiation, molecular singlet oxygen is generated (type II reaction) with a relatively high quantum yield of singlet oxygen production or moles of O_2 produced per mole of absorbed photon $\phi_\Delta = 0.275$. Some skin sensitivity to light is demonstrated by animals that received intravascular sensitizer followed by exposure of skin to light, since high levels of light and dye concentration must be used to compensate for the lack of light penetration and absorbance at lower wavelengths.

Viral inactivation with these compounds and their derivatives is impressive. Table 2 shows data obtained from tests in HSV-1 (a herpesvirus) using dihematoporphyrin ether (DHE) as the photosensitizer. The viral suspensions were held static in optically transparent containers or were passed by means of a transparent tubular flow system through an illuminated region. The virus was suspended in aqueous buffer or blood and either irradiated (with or without dye) or nonirradiated (with or without dye). A xenon light source was equipped with a dichroic heat mirror and a 630 ± 5 nm band-pass filter (Figure 2). Typical intensities of the light source were 1.04×10^{-2} W/cm^2. HSV-1 was propagated in vero cells to obtain stock concentrations of 10^6–10^7 (pfu/mL). Viral inactivity was assessed by determining the number of plaque-forming units (pfu) per milliliter. DHE was used at concentrations of 2.5 mg/mL.

Table 2 Photoinactivation of Herpes Simplex Virus Type 1 (HSV-1) in a Flow Cell System

Concentration of DHE μg/mL	PFU/mL in Dark	PFU/mL in Light	Log_{10} reduction in PFU-Light to Dark
0	3.8×10^6	2.7×10^6	0.2
2.5	3.1×10^6	$<1.0 \times 10^1$	>5.5

HSV-1 photoirradiated in the absence of dye showed a reduction in infectivity of less than 1 log_{10} PFU/mL. (From Newman JT, Matthews JL, Sogandares-Bernal F, et al: Photodynamic Inactivation of Viruses and Its Application for Blood Banking, *Baylor University Medical Center Proceedings* 1(2):8, 1988. Permission granted by *Proceedings.*)

Photodynamic inactivation of human immunodeficiency virus (HIV-1) in cell culture medium was also assessed in the flow cell system. An initial concentration of 4×10^4 infectious units (iu) was diluted, incubated with 2.5, 10, and 20 mg/mL of DHE in the dark for approximately 30 minutes, and irradiated in the system at 5 J/cm^2. Dark controls were also run in the system. After photoirradiation, tenfold dilutions of each sample were prepared (near to 10^{-4}), and 0.1 mL of each dilution was inoculated in to the human CEM T-cell line A301. On days 7, 10, 14, and 18, supernatants from the infected cultures were assessed for reverse transcriptase (RT) activity. Figure 3 shows a typical result of this experiment.

In a similar manner, HIV-infected H9 cells at a concentration of 2×10^6 cells/mL, were photoirradiated through the flow cell at a light fluence of 5 J/cm^2 with DHE concentrations of 2.5 or 10 mg/mL. All aliquots of infected cells were counted after treatment and resuspended at a density of 10^3, 10^4, or 10^5 cells/mL. The cultures were maintained for 21 days and supernatants harvested at 3–21 days for RT activity as described by Chanh. Both in cell-free HIV virus suspensions and in HIV-infected cells, viral inactivation was achieved in samples treated with dye and light, but not in those treated with dye alone or light alone. Several other porphyrins excited at higher wavelengths, tested subsequently, were found to be equally or more effective. One example of these longer wavelength porphyrin antivirals is benzoporphyrin derivative (BPD) whose structural formula is shown in Figure 4. Several analogues have been made of this basic structure to obtain improved antiviral performance. One analogue monoacid ring A (BPD-MA) is currently in phase II clinical trials for the treatment of cutaneous, psoriatic lesions. It is also useful in leukemia and in viral sterilization of blood. This second-generation photosensitizer is more advantageous than HPD and DHE because of its optimal absorbance at 690 nm, which permits readier tissue penetration and its efficient generation of singlet oxygen with a quantum yield of $\phi_\Delta = 0.76$. Interestingly, BPD has been conjugated with antibodies using a modified polyvinyl alcohol that couples several molecules of photosensitizer, offering the possibility of a more targeted delivery.

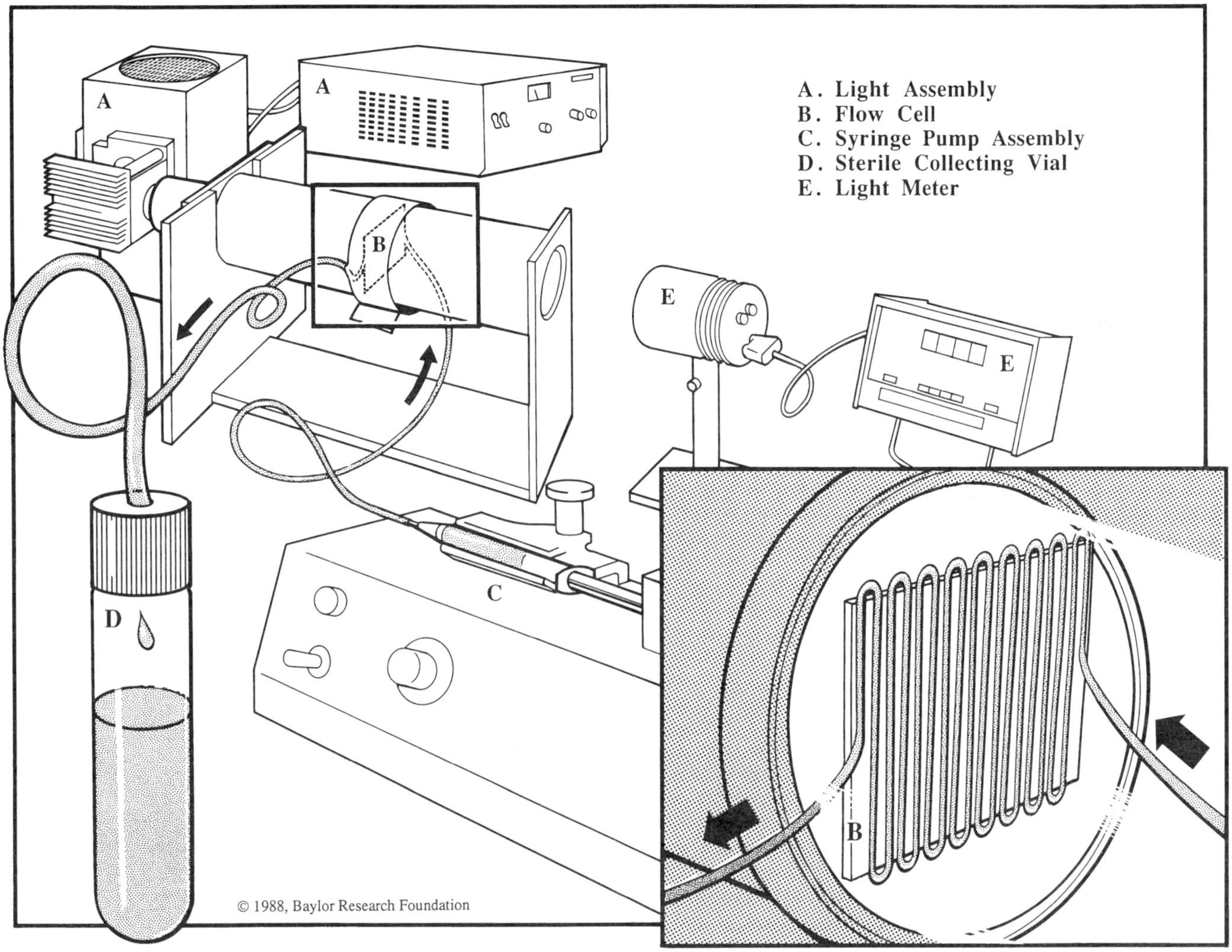

Figure 2. Diagram of flow cell apparatus for irradiation of virus plus photosensitizer suspensions. [From J. T. Newman, F. Sogandares-Bernal, et al., Photodynamic inactivation of viruses and its application for blood banking. *B.U.M.C. Proc.* *1*(2):3 (1988). Permission granted by *B.U.M.C. Proceedings.*]

Other porphyrins with potential antiviral use include cationic porphyrins and extended ring porphyrins.

6.2 Phthalocyanines and Naphthalocyanines

Another group of second-generation sensitizers are the phthalocyanines and naphthalocyanines. The latter offer strong red adsorption maxima in the range of 755–780 nm. The phthalocyanines not only show great potential for cancer therapy but also are effective against enveloped virus. Use of the photosensitizer aluminum phthalocyanine tetrasulfonate ($AlPcS_4$) in whole blood spiked with vesicular stomatitis virus (VSV) and irradiated with 25 mW/cm^2 light resulted in the rapid and complete inactivation of the virus in the presence of known quenchers of type I reactions, but was less effective in the presence of sodium azide, a quencher of type II reactions. This observation confirmed the singlet oxygen action of the dye and suggested the potential use of type I quenchers to protect the blood elements. A typical structural morphology of phthalocyanines is shown in Figure 5. Several analogues of these phthalocyanines and naphthalocyanines are available, and the properties of several of these were reviewed by Zuk et al.

6.3 Merocyanines

Merocyanines also offer good antiviral potential, although their excitation maxima are in the range of 520–570 nm and they are not appropriate for sterilization of RBC. Numerous analogues of merocyanines with higher excitation maxima are being investigated. The basic molecular structure of the parent MC540 is shown in Figure 6.

An approach for using merocyanine 540 as an antiviral photosensitizer developed by Gulliya et al. uses the same photosensitizers as in the standard dye–light–substrate system, but the temporal sequence of irradiation is modified: here, the photosensitizer is irradiated prior to exposure to the virus suspension or infected cells. In this case, singlet oxygen is generated but decays long before ex-

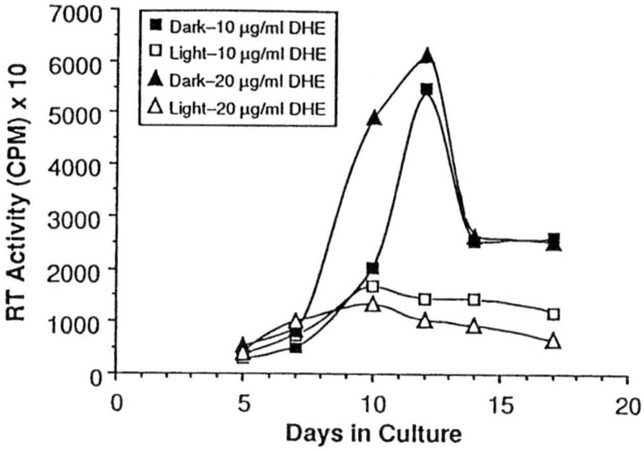

Figure 3. Reverse transcriptase activity in counts per minute (cpm) resulting from the inoculation of A3.01 cells with undiluted HIV samples. Samples of HIV at a concentration of 2×10^4 iu/mL (untreated) were passed through the flow cell in the dark (flow control) or photoirradiated with a light energy density of 5 J/cm^2. HIV samples were also photoirradiated at 5 J/cm^2 in the presence of either 2.5, 10, or 20 μg/mL of dihematoporphyrin ether (DHE). Tenfold dilutions of each sample were prepared (neat to 10^{-4}), and 0.1 mL of each dilution was inoculated into the human CEM T-cell line A3.01. Supernatants were obtained from the infected cultures at days 7, 10, 14, and 18 and assessed for reverse transcriptase (RT) activity as described by Chanh et al. RT activity observed with uninfected cell cultures with our system usually ranges from 1000 to 2500 cpm. Each point in the graph is the mean of triplicate assays. [From J. L. Matthews, J. T. Newman, F. Sogandares-Bernal, et al. Photodynamic therapy of viral contaminants with potential for blood banking applications. *Transfusion,* 28(1):82 (1988). Permission granted by *Transfusion.*]

posure to the virus. Thus, any virus toxicity or cytocidal activity must be attributed to some other mechanism. Analysis of the post-irradiated photosensitizer disclosed a family of new chemical species, which arose from photochemical reactions involving the photosensitizer itself. With some photosensitizers, such as merocyanine 540, the irradiation of photosensitizer alone yields a mixture of products that inactivate virus and viral-infected cells. These products maintain viricidal properties for at least 30 days, if refrigerated. Three of these irradiation products have now been isolated (Figure 7). A typical experiment using "preirradiated" merocyanine 540 (PMC540) is shown in Table 3.

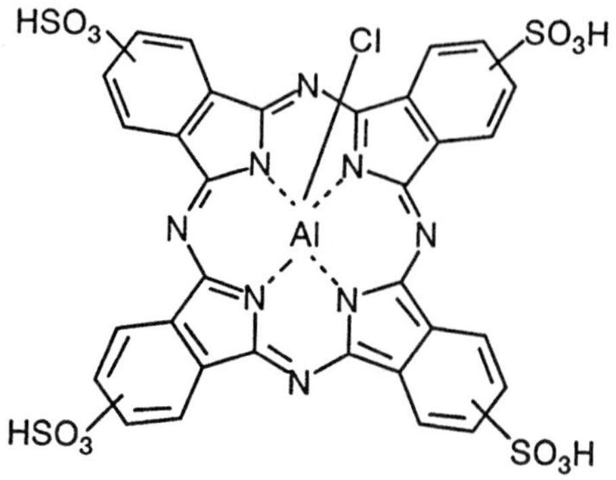

Figure 5. Phthalocyanine: molecular structure.

The analysis of mechanisms of viral kill using these products is incomplete, but membrane (envelope) damage has been evoked, and apoptosis of cells treated in this manner has been noted. The addition of radical quenchers such as glutathione reduces the efficacy of the system.

6.4 Naphthalimides

A fourth class of photosensitizers has recently been developed. These agents act independently of oxygen and are based on the 3-bromo-4-alkylamino-*N*-alkyl-1,8-naphthalimide skeleton shown in Figure 8. In fact, some of these new photosensitizers proved successful in neutralizing HIV-1 and inhibiting HIV-1-induced syncytium formation in vitro following light irradiation, while retaining relatively low toxicity toward cells and extremely low toxicities in the absence of light.

The naphthalimide photochemical effect results in formation of covalent bonds with amino acids (e.g., methionine, cysteine, tryptophan, tyrosine) in and between intramembranous proteins accessed by the lipophilic photosensitizer. This protein-binding mechanism is supported by bleaching experiments with the amino acids in appropriate solvents and in the loss of tryptophan in analysates of gramicidin A, a polypeptide known to form helical channels in membranes following irradiation of liposomes containing the polypeptide. The order of reactivity was found to correlate with the ease with which each reactive compound can act as a single electron donor, suggesting that a single electron transfer (s.e.t.) step is

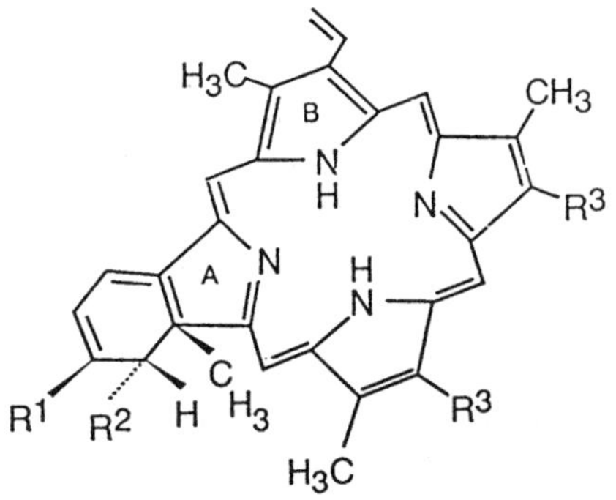

Figure 4. Benzoporphyrin derivative: molecular structure.

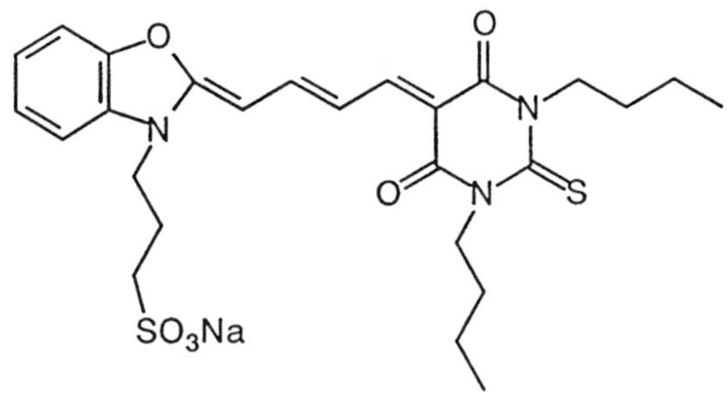

Figure 6. Merocyanine 540: molecular structure.

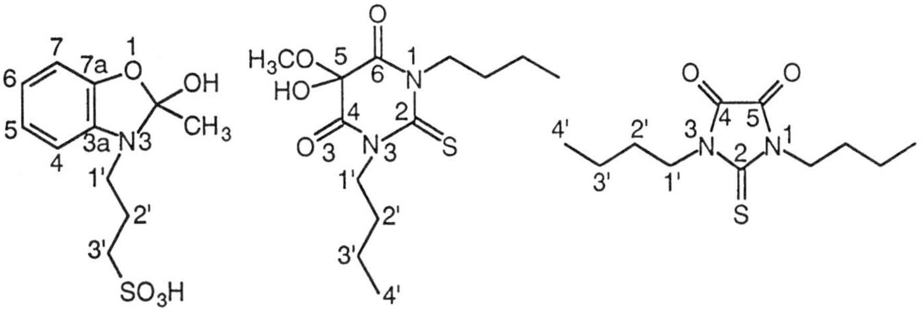

Figure 7. Three derivatives of light-activated merocyanine 540: molecular structures.

Table 3 Preactivated MC540 inactivates cell-free HIV-1 and SIV

MT-4 cell infection		P-MC540 Concentration	Viral antigens, (pg/ml)[a]			
			HIV-1		SIV	
HIV-1	SIV	(μg/ml)	Day 4	Day 7	Day 5	Day 10
+	−	—	735.3	956.7	—	—
+	−	80.0	9.5	12.4	—	—
+	−	160.0	7.6	6.8	—	—
+	−	200.0	8.1	8.0	—	—
−	+	—	—	—	73.0	65.9
−	+	80.0	—	—	10.4	12.9
−	+	160.0	—	—	4.2	5.0
−	+	200.0	—	—	3.5	2.8

HIV, human immunodeficiency virus; SIV, simian immunodeficiency virus; P-MC540, preactivated merocyanine 540.

[a]Mean of duplicate determinations using the antigen-capture enzyme-linked immunosorbent assay kits for p24 antigen HIV-1 and core antigen SIV.

(From Chanh TC, Allan JS, Pervaiz S, Matthews JL, et al: Preactivated merocyanine 540 inactivates HIV-1 and SIV: Potential therapeutic and blood banking applications, *J. AIDS* 5:191, 1992. Permission granted by Raven Press Publishers.)

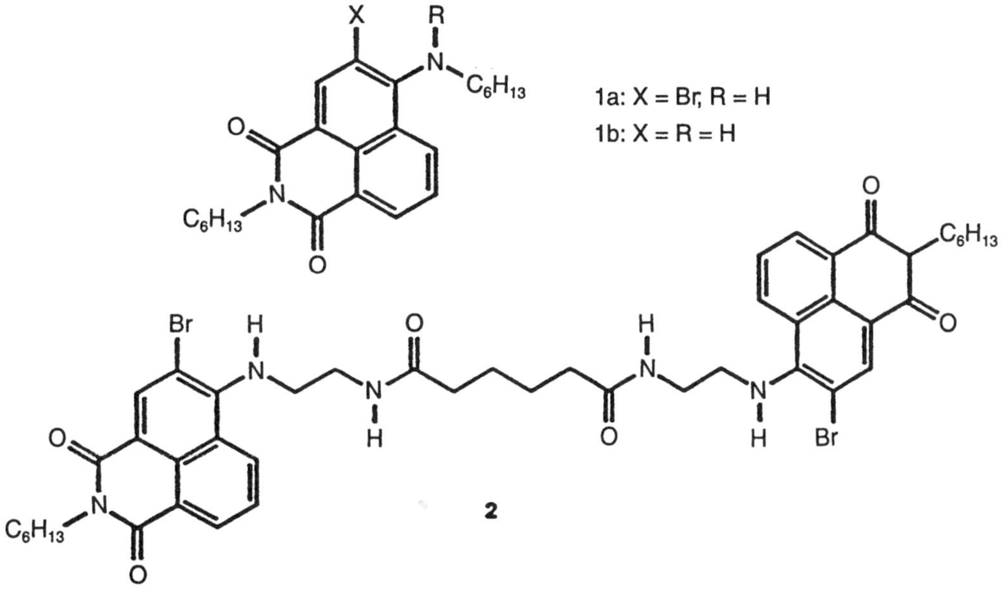

Figure 8. Naphthalimide: molecular structure.

involved in the mechanism. Covalent linking of the photosensitizer to transmembrane protein would prevent the needed conformational change (as occurs during viral invasion of cells in receptor sites), thus precluding delivery of the nucleocapsid or contents to the cell. Gel electrophoresis of photosensitizer-irradiated membrane proteins of HIV indeed show binding of these proteins. Of the compounds tested to date, the monomeric naphthalimides represented by 3-bromo-4-(hexylamino)-*N*-hexyl-1,8-naphthalimide are effective against HSV-1 at concentrations around 250 nM, and the dimeric naphthalimides represented by the bisamide are effective against the same pathogen at concentrations below 100 nM. Both classes of dye are also effective in neutralizing vesicular stomatitis virus, bovine herpes virus, type 1 (BHV-1), and HIV-I.

6.5 Psoralen Photosensitizers

With the increased interest in the potential for employing photodynamic methods for viral eradication, additional photosensitizers that bind to DNA and RNA are being evaluated. In this context, a fifth group of photosensitizers, which can act with or without oxygen, is being examined. These are the psoralen dyes (Figure 9), substances that have been used extensively for decades in the clinical care of psoriasis and for other medical applications. In the presence of oxygen, these substances are photodynamically active and singlet oxygen is generated. However, these substances, excitable in the UVA spectrum (320–400 nm), can also act in the absence of oxygen as a consequence of their capability to form photoadducts with DNA and RNA. Presumably this chemistry reflects the nature of psoralens—that is, planar tricyclic furocoumarins, which, in turn, can intercalate between stacked base pairs as reviewed in Section 2 on photosensitizer mechanisms.

The foregoing observations suggest that psoralen photosensitizers, when irradiated with UV, could inactivate several types of virus with either RNA and DNA constructs. Not surprisingly, therefore, numerous analogues of psoralen have been synthesized. Many of these have shown efficacy in viral kill, with their antiviral effects depending on their affinity for single- or double-stranded RNA. Cross-linking of viral nucleic acids with UVA excitation of psoralen photosensitizers is significantly higher than that typically required in proteins for virus inactivation. This "overkill" precludes the opportunity for survival or repair of the affected virus. Thus, psoralens have been studied for their potential use in eradicating virus from blood. As with the other sensitizers already described, much of the motive here is to reduce the possibility of transmission of disease via transfusion. This procedure inactivates HIV, CMV, and other viral pathogens, with only minor effects reported on the red blood cells and platelets. The major concern for utilizing this system is the mutagenic potential for photosensitizer/UVA-treated virus. However, since both the RBCs and the platelets are anuclear, this mutagenic potential is not as high as it would be in other applications.

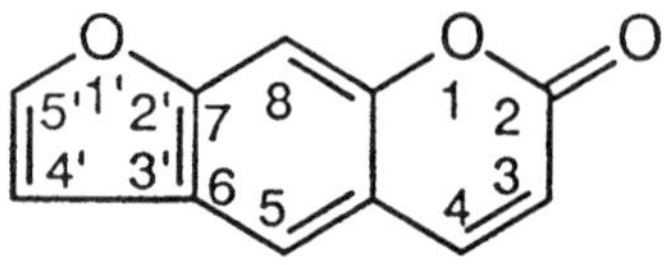

Figure 9. Psoralen: molecular structure.

7 POTENTIAL USES AND PROSPECTS FOR PHOTOSENSITIZERS

Although photochemical approaches have not yet been employed in the blood bank, or for clinical use for viral infections, there is increasing use of photodynamic therapy (PDT) for treating various tumors, with Photofrin (HPD) being the first to be used in clinical trials and to receive regulatory approval. With the toxicity clearance for humans and the demonstrated efficacy for tumor kill and the excellent in vitro data supporting the efficacy of viral inactivation of free virus and infected cells, and, with the increasing number of photosensitizers to choose from, direct PDT and "preactivation of sensitizers" appear to be poised for future exploitation. Photosensitizers may indeed become a useful new tool against viral infection.

See also Ultraviolet Radiation Damage to DNA; Virology, Molecular; Virus–Receptor Interactions.

Bibliography

Auler, H., and Banzer, G. (1943) Untersuchungen über die Rolle der Porphyrine be qeschwulstranken Menschen and Tieren. *Z. Krebsforsch.* 53:65–68.

Chanh, T. C., Allan, J. S., Matthews, J. L., Sogandares-Bernal, F., and Newman, J. T. (1989) Photodynamic inactivation of simian immunodeficiency virus. *J. Viral. Methods,* 26:125–132.

Davis, N., Liu, D., Ashok, K. J., Jiang, S. Y., Richter, A., and Levy, J. G. (1993) Modified polyvinyl alcohol–benzporphyrin derivative conjugates as phototoxic agents. *Photochem. Photobiol.* 57:641–647.

Gasparro, F. P. (1988) Psoralen DNA interactions: Thermodynamics and photochemistry. In *Psoralen DNA Photobiology,* Vol. 1, F. P. Gasparro, Ed., pp. 5–36. CRC Press, Boca Raton, FL.

Gulliya, K. S., Pervaiz, S., Dowben, R. M., and Matthews, J. L. (1990) Tumor cell specific dark cytotoxicity of light-exposed merocyanine 540: Implications for system therapy without light. *Photochem. Photobiol.* 52:831–838.

Kagan, J., Chen, X., Wang, T. P., and Forlot, P. (1992) Psoralens cleave pBB322 DNA under ultraviolet radiation. *Photochem. Photobiol.* 56:185–193.

Krasnovsky, A. A. Jr., Rogers, M. A. J., Galpern, M. E., Rihter, B., Kenney, M. E., and Lukjanetz, E. A. (1992) Quenching of singlet molecular oxygen by phthalocyanines and naphthalocyanines. *Photochem. Photobiol.* 55:691–696.

Kuratomi, K., and Kobayashi, Y. (1976) Photodynamic action of lumiflavin on the template DNA of RNA polymerase. *FEBS Lett.* 72:295–298.

Lewis, D. E., Utecht, R. E., Judy, M. M., and Matthews, J. L. (1993) Photochemical neutralization of HIV-1 and inhibition of HIV-1-induced syncytium formation by halogenated 1,8-naphthalimides. *The Spectrum,* pp. 8–14. Center for Photochemical Sciences, Bowling Green, OH.

Raab, O. (1900) Über die Wirkung fluorescierender Stoffe and Infusorien. *Z. Biol.* 39:524.

Rywkin, S., Lenny, L., Goldstein, J., Geacintou, N. E., Margolis-Nunno, H., and Horowitz, B. (1992) Importance of type I and type II mechanisms in the photodynamic inactivation of viruses in blood with aluminum phthalocyanine derivatives. *Photochem. Photobiol.* 56:463–469.

Spikes, J. D. (1989) In *The Science of Photobiology,* K. C. Smith, Ed. Plenum Press, New York, London.

Zuk, M. M., Rihter, B. D., Kenney, M. E., Rogers, M. A. J., and Kreimer-Birnbaum, M. (1994) Pharmacokinetic and tissue distribution studies of the photosensitizer *bis* (di-isobutyl octadecylsiloxy) silicon 2,3-naphthalocyanine (iso-BOSINC) in normal and tumor bearing rats. *Photochem. Photobiol.* 59:66–72.

VIRUSES, RNA PACKAGING OF

Dennis H. Bamford

1 **Background**
2 **Tobacco Mosaic Virus (TMV), the Classic Case**
3 **Small, Icosahedral, Positive-Strand RNA Viruses**
4 **Double-Stranded RNA Viruses**
4.1 Bacteriophage φ6
4.2 Yeast Virus L-A
5 **RNA Packaging Signaling in Animal Viruses**
5.1 Sindbis Virus
5.2 Mouse Hepatitis Virus
5.3 Influenza Virus
5.4 Hepadnaviruses and Retroviruses
6 **Perspectives**

Key Words

Packaging, Encapsidation The process by which the virus genome is translocated and condensed into the virus capsid.

Virus An intracellular parasite that utilizes the cellular machinery to synthesize its components, which assemble to a particle containing the genome and protected by protein or protein and lipid layers.

Virus Capsid Protective symmetrical protein shell formed by a large number of protein subunits.

Virus Genome A nucleic acid molecule. Chemically the genome can be either ribonucleic or deoxyribonucleic acid (RNA or DNA). These molecules can exist in single- or double-stranded form. Any of these four possible molecules can serve as a virus genome.

A key event in the virus assembly is the specific recognition of the genome with the capsid precursor. Particularly with RNA viruses, this interaction must be very selective, to be able to recognize the viral genome among the number of nucleic acid molecules in the cell. The understanding of this event might give an opportunity to combat viral diseases. Since, however, RNA and protein interactions are one of the most important macromolecular recognition events in biology, RNA virus packaging offers a good model system to study this largely unknown phenomenon.

1 BACKGROUND

Viruses are intracellular parasites that are fully dependent on the metabolic apparatus of the cell. The viral genome has the information for its multiplication in the cell as well as for directing the synthesis of structural components of the virion. The virus genome can be either DNA or RNA, double-stranded (ds) or single-stranded (ss). The nucleic acid interacts with the viral structural proteins to form a nucleic acid–protein complex. This event is referred to as nucleic acid packaging or encapsidation, and the formed protein coat protecting the nucleic acid is the virus capsid. The capsid can be organized either helically or icosahedrally. In the helical nucleocapsid, an elongated rod-shaped structure, the coat proteins form a helix around the genome. The icosahedral capsid is a spherical shell-like structure enclosing the condensed viral genome (Figure 1). In some viruses these structures are additionally surrounded by a membrane.

Double-stranded nucleic acids are antiparallel molecules. In the case of RNA the strand that can serve as the messenger in protein synthesis is known as the (+)-strand and the complementary strand as the (−)-strand. There are RNA viruses that have either (+), (−), or dsRNA as their genome. The genome can be a single RNA molecule or segmented. The genome segments can reside in a single capsid or in different capsids (multipartite). The (+)-strand RNA genomes, upon introduction to the cell, can serve directly as messengers, whereas the (−)-strand and dsRNA genomes must be transcribed to (+)-strands in the cell. This (+)-strand synthesis is carried out by virion-associated polymerases either directly or via a DNA intermediate (retroviruses).

One of the key events during virion assembly is the encapsidation (packaging) of the viral genome into a protein capsid. Based on the great variety of genome strategies in RNA viruses, there evidently must also exist variation in the mechanisms used for RNA packaging. It is obvious that only very specific interactions between the RNA and the capsid protein(s) can produce the distinction of the viral RNA molecules in the cell and their subsequent packaging. The formation of a functional metastable virion structure which, upon response to stimuli during the virus entry, results in virion disassembly and genome activation. Our understanding of the RNA packaging events relies mostly on a few model systems, but information from a great variety of viruses is rapidly accumulating. This article describes various examples of RNA virus systems about which packaging information is available.

2 TOBACCO MOSAIC VIRUS (TMV), THE CLASSIC CASE

TMV is a helical virus composed for a (+)-sense, single-stranded RNA molecule 6395 nucleotides long, associated with some 2140 identical capsid protein molecules (Figure 1A). As long ago as the early 1950s, a reconstitution of infectious virus particles from puri-

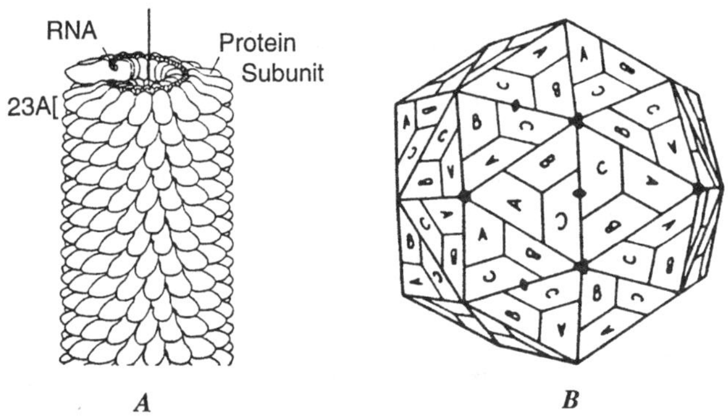

Figure 1. Principal structural arrangements of virus capsids. (A) A helical capsid, with the coat proteins organized helically around the viral RNA (e.g., tobacco mosaic virus). (B) An icosahedral capsid in which the coat proteins (A, B, C) form a spherical structure enclosing a condensed viral RNA molecule [e.g., a plant (+)-strand RNA virus].

fied RNA and dissociated coat proteins was achieved. This self-assembly reaction of purified components allowed a detailed study of the packaging reaction.

The protein structure initiating virus assembly is most probably a short two- to three-turn helix of the coat protein containing 39–49 subunits (the so-called 20S particle). A disordered loop in the protein prevents further elongation of this initiation complex. An internal region of the genomic RNA called the "origin of assembly" (OAS), located in the region between nucleotides 5443 and 5518, interacts with the nucleating aggregate. The binding of the RNA bases to the protein causes the folding of the disordered loop in the coat protein, leading to the helix elongation to both directions of the RNA molecule. The current packaging (assembly) model postulates that the 5′ direction encapsidation is rapid whereas the 3′ direction encapsidation is slow. Furthermore the 5′ direction RNA passes through the lumen of the growing rod as it is incorporated into the virion. It is not clear how large building blocks are added to the growing rod. It might be that the 5′ direction assembly uses short helices, whereas protein subunits or small aggregates assemble to 3′ direction. The assembly process is greatly dependent on the pH and ionic conditions, as is the disassembly upon infection. The genome length determines the length of the virion.

3 SMALL, ICOSAHEDRAL, POSITIVE-STRAND RNA VIRUSES

Small (+)-strand RNA viruses have been found infecting bacteria, plants, and animals. These viruses are among the simplest replication systems. High resolution structural data, which are available for a number of such viruses, have greatly advanced our understanding of the assembly of these small viruses. The overall structural principles are surprisingly similar in ssRNA viruses infecting plants and animals (Figure 1B), whereas a different arrangement is found in the corresponding bacterial viruses. In spite of the structural similarities of the capsids, the condensed RNA molecule inside the capsid can be organized differently. It can be either randomly organized (so that it does not contribute significantly to the diffraction pattern obtained from virus crystals) or parts of the RNA can be associated with the capsid proteins in a symmetric fashion (making the nucleic acid visible in the electron density map).

Bacteriophage assembly initiation of ssRNA is probably the best understood among small, icosahedral (+)-strand viruses. The viruses (MS2, R17, etc.) have one coat protein that exists as a dimer. Initially coat proteins bind specifically to an asymmetric RNA hairpin loop, 21 nucleotides long (Figure 2), which is located in the translation initiation region of the replicase gene. All unpaired nucleotides are necessary. There are two A residues in the loop which are base specific. There is also an unpaired A in the stem which cannot be changed without the loss of the activity. The paired stem nucleotides, however, do not show any base specificity (compare also to yeast virus L-A packaging signal in Section 4.2: Figure 4). Later in the infection, when the coat protein concentration in the cell increases, this binding leads to translational repression of the replicase synthesis and is most probably also responsible for the initiation of the genome packaging. The initial binding stimulates the binding of additional coat protein dimers to nonspecific sequences, resulting in a highly cooperative assembly of coat proteins, which form a shell that encloses simultaneously the viral RNA.

An intact ssRNA bacteriophage capsid also contains a single copy of a maturase protein, which is necessary for infection. It has been

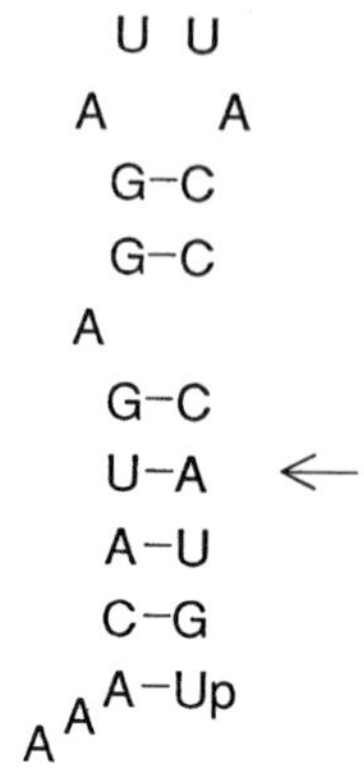

Figure 2. The bacteriophage R17 coat protein binding site just upstream of the replicase gene in the R17 genome; arrow indicates start of the replicase gene.

shown to bind RNA, but capsid proteins can specifically package virus specific RNA without this protein. In the model represented by ssRNA bacteriophages, the RNA genome is condensed to the capsid simultaneously with the protein shell assembly. Both picornaviruses and certain plant viruses [e.g., southern bean mosaic virus, tomato bushy stunt virus, turnip crinkle virus (TCV)] assemble via discrete intermediate units.

In the TCV assembly model, three coat protein dimers bind to a specific viral RNA sequence. The N-termini of these proteins locate themselves at the threefold symmetry axis. Additional dimers are then added to this structure. A preformed capsid packaging is not yet, however, completely ruled out. Anyway, the packaging is very specific for the (+)-sense genomic RNA. The picornavirus genomes contain also a 5′-terminally covalently linked protein, but it is not the specificity factor for packaging.

Ordered ssRNA has been observed in the crystal structure of bean pod mottle virus, a comovirus. Some 20% of the RNA is icosahedrally oriented with no sequence specificity. Whether this RNA–capsid protein interaction has implications for RNA packaging remains to be seen. These viruses, however, can form stable empty capsids, indicating that correct capsid protein–protein interactions take place without the genomic RNA. On the other hand, a nodavirus (flock house virus) capsid has been shown to bind portions of the genomic ssRNA in a dsRNA form. About 10 bases in the dsRNA form bind to the coat proteins at the icosahedral twofold axis. The dsRNA is in the A form and is formed from the secondary structures of nonunique sequences. This RNA is an integral part of the capsid quaternary structure, suggesting a capsid condensation around the RNA. This model is also supported by the absence of reports of empty particles in nodavirus-infected cells.

4 DOUBLE-STRANDED RNA VIRUSES

Since cellular polymerases are not capable of operating on dsRNA templates, all dsRNA viruses have a virion-associated, RNA-dependent RNA polymerase. Although in most cases the dsRNA viruses have segmented genomes, many fungal viruses have only a single genome segment. The viruses with segmented genomes, in addition to having the specificity for selecting viral RNA for packaging, must have a mechanism ensuring that one copy of each genome segment is included in the virion. Depending on the virus, the number of segments can vary from one to more than 10. Here I concentrate on two dsRNA virus systems (φ6 and L-A), where in vitro assembly experiments have shed light on the RNA encapsidation reaction.

4.1 Bacteriophage φ6

Pseudomonas syringae virus φ6 has three dsRNA genome segments (L, 6374 bp; M, 4057 bp; S, 2948 bp). These are packaged into a dodecahedral polymerase complex particle made of four viral proteins (P1, P2, P4, P7), which are the products of the L genomic segment. This structure is surrounded by an additional protein layer composed of protein P8. The structure containing protein P8 is called the nucleocapsid (NC). The nucleocapsid resides in a virus-specific membrane envelope.

The NC enters the cell via membrane fusion and endocytosis-type events. In the cytoplasm, NC is activated by removal of the protein P8 layer. This activation leads to the production of (+)-strand transcripts from all genomic segments. The L-segment-encoded polymerase complex proteins are synthesized early in infection and self-assemble to empty dodecahedral particles. The synthesized (+)-strands serve also as genomic strands, which are packaged to these polymerase complex (procapsid) particles. The packaging signal extends 200–300 nucleotides from the 5′ end of each segment. These signals are in the noncoding region of the segments, extending to a few tens of nucleotides from the first ribosome-binding site. The 18 first 5′ nucleotides are conserved in all segments, and there is also a 12-nucleotide homologous sequence some 70 nucleotides from the terminus. However, the location of this nucleotide region is slightly different in the three segments. The RNA-folding algorithm shows extensive secondary structure within the packaging regions, but the prediction is different for each of the three segments.

The polymerase complex particle recognizes the viral (+)-strand packaging signal and translocates and condenses the RNA molecules into the particle interior in an energy-requiring process. The energy is provided by NTP hydrolysis carried out by the procapsid. The genome segments act separately and independently. The next step in the virus life cycle is the synthesis of the (−)-strand, which occurs inside the polymerase complex particle using the (+)-strands as the templates. The (−)-strand synthesis signal (four hairpin loops) resides at the very 3′ end of the (+) strands. The (−)-strand synthesis is coupled to the packaging in such a way that the initiation does not occur before all three segments have been packaged. The particles with synthesized dsRNA genomes move to (+)-strand synthesis mode, which takes place at low Ca^{2+} concentration. At high Ca^{2+} concentration (≈ 0.5 mM), (+)-strand synthesis ceases and protein P8 is assembled on the particle. Also the P8 assembly is a calcium-driven reaction. The envelopment of the NC leads to particle maturation inside the host cell.

Figure 3 depicts the NC activation, procapsid packaging, and NC maturation. It remains to be seen whether members of the reovirus family follow this general scheme for their RNA packaging. It is intriguing to address the specific RNA–protein interactions leading to the highly selective packaging of all the viral genomic RNA segments.

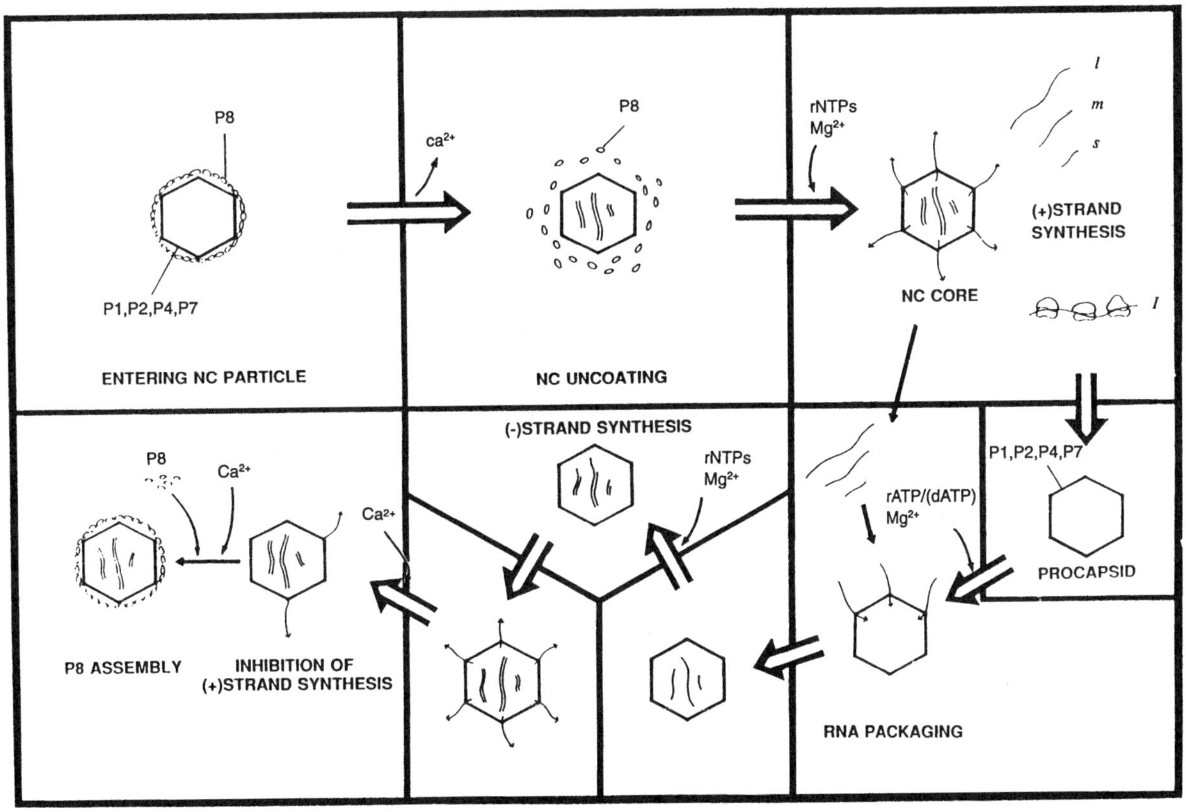

Figure 3. Bacteriophage φ6 nucleocapsid activation, synthesis of the procapsid (RNA polymerase complex), procapsid packaging, and maturation of the nucleocapsid. (From the thesis of Päivi M. Ojala, Department of Genetics, University of Helsinki, with permission.)

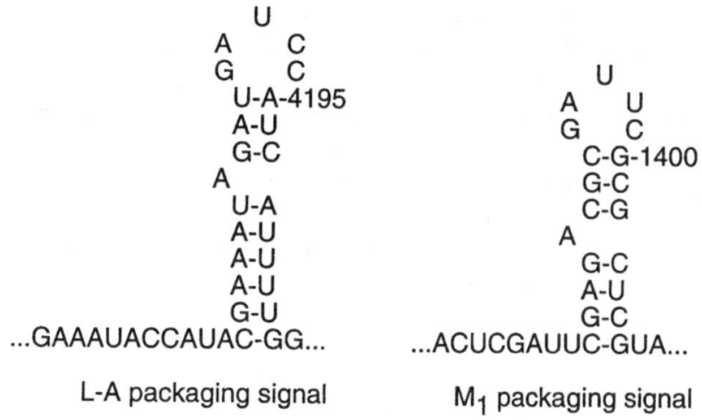

Figure 4. Packaging signal of the L-A virus and that of its satellite RNA M_1. The structures are very similar and the loop sequences nearly identical. (Figure kindly provided by Dr. Reed B. Wickner, U.S. National Institutes of Health.)

4.2 Yeast Virus L-A

Yeast dsRNA virus L-A is a degenerate virus, as are all viruses found in fungi. They exist intracellularly only, no external virus particles having been observed. L-A is found in *Saccharomyces cerevisiae.* The L-A system also supports the replication of several satellite dsRNAs. The L-A genome is a 4.6 kb long, single dsRNA molecule, whereas all the satellite dsRNAs are smaller than the L-A genome. The satellites code for a secreted toxic protein and immunity to that toxin.

The L-A genome codes for the major capsid protein (gag) and a fusion protein that has a gag as its N-terminal part and an RNA polymerase domain as its C-terminal portion (gag-pol). The gag-pol fusion protein is formed by a −1 ribosomal frameshift event. The frameshifting efficiency controls the formation of the gag-pol protein. In vitro, 1.8% of the gag transcripts are read to gag-pol fusion protein transcripts. The ratio for gag and gag-pol is very close to that estimated from the isolated virus particles; 120 gag proteins, 2 gag-pol proteins (and one RNA molecule).

About 400 nucleotides from the 3′ end of the (+)-strand resides a 24-base stem–loop structure. This structure and ten 5′ adjacent bases form the RNA packaging site (Figure 4). The RNA packaging site binds to the pol domain of the gag-pol fusion protein. The stem structure, but not the sequence of the stem, is necessary for the binding. An A residue protruding from the 5′ side of the stem, as well as the loop sequence, however, are critical.

The packaging model involves the formation of an initiation complex where two gag-pol proteins dimerize and bind one (+)-strand RNA molecule. The gag domain of the fusion protein primes the assembly of additional gag proteins to form an icosahedral shell condensing the RNA inside (Figure 5). The possibility of packaging of a preformed capsid is, however, not completely ruled out.

Following packaging, the (−)-strand synthesis takes place inside the particle. The particle size is designed to give room to a single L-A dsRNA genome. If, however, the particle packages a short satellite RNA using a mechanism and a mode of signaling similar to those of the L-A genome, the RNA synthesis produces progeny RNA molecules that stay inside the particle as long as the RNA mass is equal to that of the L-A genome. This system results in particles with multiple dsRNA segments, the number depending on the size of the packaged molecule.

5 RNA Packaging Signaling in Animal Viruses

5.1 Sindbis Virus

Sindbis virus, a togavirus, has icosahedral symmetry. The assembly of the virion requires two steps. First, the genomic (+)-strand ssRNA interacts with the capsid protein, leading to the formation of the nucleocapsid. Second, the nucleocapsid specifically interacts with the viral membrane proteins, which are translocated to the cellular membranes. This event results in the release (budding) of viruses from the plasma membrane. The genomic RNA is used to translate the nonstructural proteins from the 5′ portion of the molecule. Structural proteins are translated from a subgenomic RNA (identical to one-third of the 3′ end of the genomic RNA). Only the genomic RNA, not the subgenomic fragment, is packaged. The packaging signal is located to a 132-nucleotide region (between nucleotides 945 and 1076 in the Sindbis virus genome).

The binding site lies within the coding region of a nonstructural protein nsP1. Strong secondary structure is predicted for the RNA in the binding region. The binding of this region to the capsid protein is supposed to trigger the protein–protein interactions between the capsid proteins leading to the nucleocapsid formation. The crystal structure of the Sindbis virus capsid protein reveals a dimer. However, monomeric capsid protein binds RNA. The RNA binding site is a region some 30 amino acids long around amino acid 100 of the protein which is about 246 amino acids long. The binding site is located between two domains of this protein. The N-terminal domain is largely unstructured (flexible) and basic, a good candidate

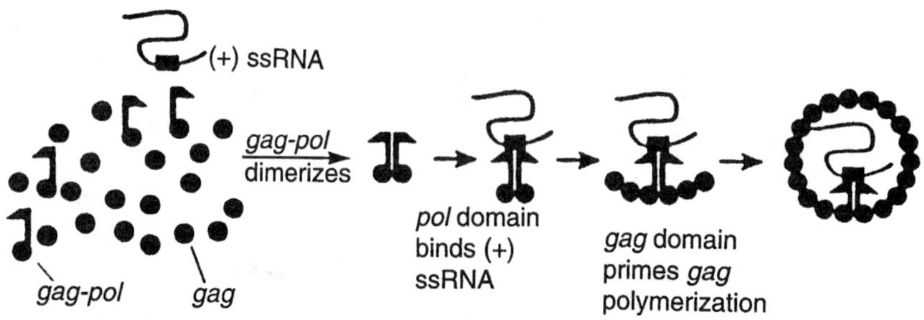

Figure 5. RNA packaging model of the L-A family of dsRNA viruses. (Figure kindly provided by Dr. Reed B. Wickner, U.S. National Institutes of Health.)

for an RNA interacting protein region in the viral capsid. The other domain forms the capsid structure.

5.2 Mouse Hepatitis Virus

Mouse hepatitis virus, a coronavirus, is an enveloped virus with a helical capsid. The genome is a positive-sense ssRNA molecule. Also in this case there are subgenomic RNA species in addition to the genome length one. Only the genome length RNA is packaged. Deletion mapping has located the packaging signal to a 61-nucleotide internal sequence close to the 3′ end of gene 1. There are implications of a stem–loop RNA structure in this region. It is assumed that the nucleocapsid protein binds to this packaging site. Such an arrangement would regulate the expression of the viral RNA polymerase (gene 1), as is the case in ssRNA bacteriophages.

5.3 Influenza Virus

The influenza virus, an orthomyxovirus, has eight different ssRNA genome segments of negative polarity which are associated with four polypeptides forming a ribonucleoprotein core. This core exhibits RNA-dependent RNA polymerase activity and is enclosed into the viral envelope. About 15 of the terminal untranslated 3′ and 5′ nucleotides of the genomic RNAs are involved in forming circles via a panhandle structure. Some 20 nucleotides at the termini of the segments are sufficient signals for transcription, replication, and packaging of the RNAs into the virus particles. It is assumed that the packaging of the genomic segments is not accurate, but more than eight segments are packaged. This random packaging leads to different assortments of genomic segments within the viral particles. Thus, with only a few particles having complete sets of genomic segments, specific infectivity should be low, as in fact has been observed. However, a more specific packaging system has not been ruled out. Figure 6 depicts the genome regions of the Sindbis, mouse hepatitis, and influenza viruses that have been implicated in the genome packaging.

5.4 Hepadnaviruses and Retroviruses

In both the hepadna- and retroviruses, the packaged nucleic acid is positive-sense RNA, but the virus particle contains an RNA-dependent DNA polymerase (reverse transcriptase) which converts the packaged RNA to DNA form.

Hepadnaviruses are small enveloped icosahedral viruses. Upon infection, the nucleocapsid is transported to host cell nucleus. The 3.2 kb genome is converted to a covalently closed circular molecule, which is transcribed to a genome-length RNA molecule, as well as to two subgenomic RNAs. Only the genome-length (+)-strand RNA is encapsidated to the nucleocapsid along with the reverse transcriptase. At this stage, the packaged RNA molecule acts as a template for the synthesis of the complementary (−)-DNA strand, which is further synthesized to a partially duplex DNA molecule. The nucleocapsid acquires the virus-specific envelope from the endoplasmic reticulum, after which the viral particle is removed from the cell by the vesicular transport pathway.

The hepadnavirus genome-length RNA contains a region desig-

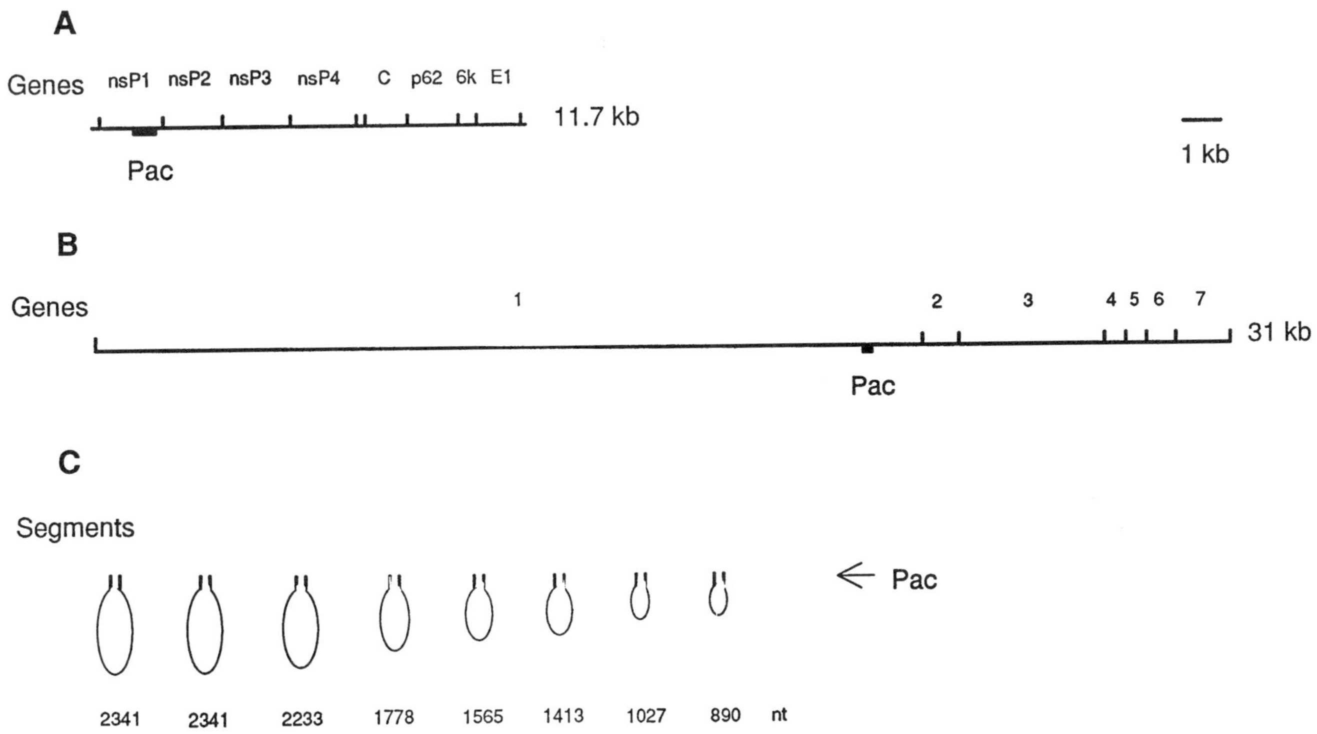

Figure 6. Location of the packaging sites (pac) found in the genomes of Sindbis virus (A), mouse hepatitis virus (B), and influenza virus (C). The scale for the influenza virus genome segments is about twice to that of the other two; actual size of each genome segment is indicated below the respective segment. The paired nucleotides in the panhandle structures at the segment termini are indicated with thicker line. (Figure kindly provided by A. Marika Grahn, Department of Genetics, University of Helsinki.)

nated as ϵ; it is a sequence about 90 nucleotides long, located near the 5′ end of the molecule, and it is necessary and sufficient for genome packaging. The same signal is also found at the 3′ end of the genome but does not have packaging activity at this position. The ϵ sequence forms a hairpin loop, which also binds the reverse transcriptase and is the initiation site for the reverse transcription. Even though the packaging is coupled to the reverse transcription, these two events are independent. Nonfunctional reverse transcriptase does not prevent the packaging. However, if there are no reverse transcriptase proteins available, only empty capsids are formed. The capsid (core) protein has a basic C-terminus that might have nonspecific nucleic acid binding functions. It is not, however, clear how ϵ, reverse transcriptase, and the capsid proteins interact to initiate the encapsidation reaction. An interesting feature in hepadnavirus packaging is a cis preference. When wild-type genomes are used to complement reverse transcriptase mutants, the progeny genomes packaged are preferentially of wild-type origin.

Retroviruses are also enveloped polyhedral viruses with (+)-strand 8–11 kb RNA genomes. The retroviruses contain two copies of genomic RNA molecules. Upon entry to the host cell, the retroviral RNA genome is converted to a DNA molecule by the virion-associated reverse transcriptase. After the virus particle has entered the nucleus, the viral genome integrates to the host genome. The integrated provirus acts as a template for transcriptase, and genome length and subgenomic RNAs are produced.

Again only full genome length RNA molecules are packaged together with RNA-associated reverse transcriptase. The virus capsid obtains the envelope by budding through the plasma membrane. Retroviral genomes code for polyprotein precursors (gag, pol, env). The reverse transcriptase (pol) is synthesized as a gag-pol fusion protein by a ribosomal frame shifting (see also Section 4.2 on yeast virus L-A). The gag region produces the virion capsid proteins. After virus maturation, the reverse transcriptase is activated by a proteolytic cleavage, which leads to the binding of the reverse transcriptase to the RNA. It is assumed that the packaging of the reverse transcriptase is based on the gag (capsid) domain of the fusion protein. As in hepadnaviruses, the retrovirus RNA contains a packaging signal (ψ) near the 5′ end of the genomic RNA. This sequence is about 200 nucleotides long and is sufficient to direct the packaging of heterologous RNA molecules as well. Sequences near the 3′ end of the genomic RNA have also been implicated in genome encapsidation.

Even though the reverse transcriptase has a role in the virus life cycle, it is not needed for packaging, since the gag proteins alone with the genomic RNA do form RNA-containing particles. One of the gag-encoded proteins (NC) binds RNA but with no sequence specificity. Also, the proteolytic cleavage of the gag is not necessary for packaging, since the gag proteins are in certain conditions capable of empty particle formation. It is not yet clear what triggers the specific encapsidation reaction. The hepadna- and retrovirus genomes, with the RNA packaging sites, indicated are depicted in Figure 7.

6 PERSPECTIVES

Virus RNA packaging is a highly selective event in which where almost exclusively, viral RNA is encapsidated. The selectivity is obtained by means of specific RNA–capsid protein interactions. These interactions may also have translational regulatory effects. There is a vast amount of information about DNA–protein interactions, but our understanding of RNA–protein interactions is still relatively limited. Surely RNA viruses will be in the forefront in helping us to elucidate the interactions between the two most active cellular components, RNA and protein.

The development of recombinant DNA technologies (including reverse transcription) and the understanding of the packaging signaling in many virus systems have opened the opportunity to package heterologous RNA into viruses. These advances are of crucial importance in helping us to understand the packaging mechanism.

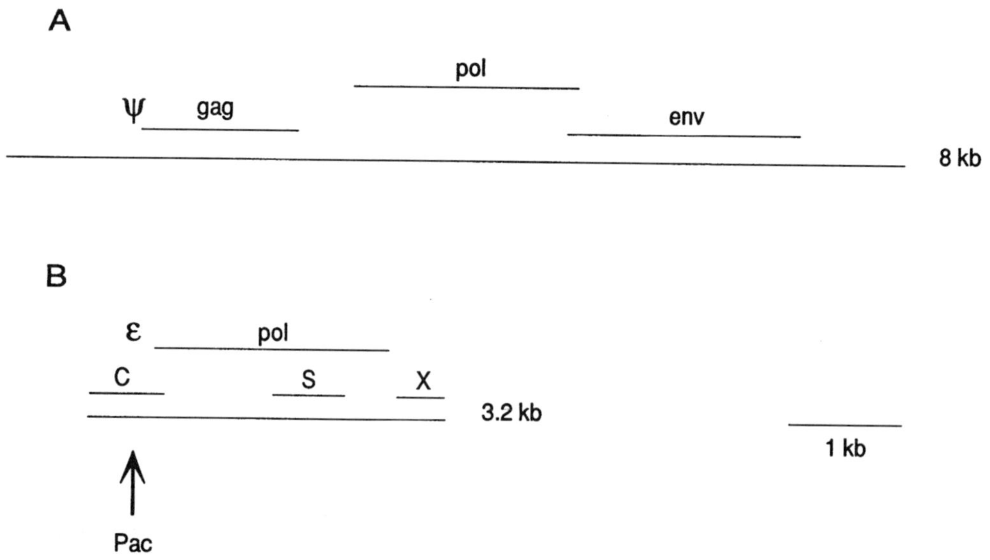

Figure 7. Encapsidation signals indicated by the "pac arrow" in hepadna- and retroviruses. (A) Murine leukemia virus (packaging site designated by ψ) and (B) human hepatitis virus B (signal is designated by ϵ). (Figure kindly provided by A. Marika Grahn, Department of Genetics, University of Helsinki.)

Also they have allowed the construction of a number of vector systems that are widely used in biological and medical investigations.

See also RNA Three-Dimensional Structures, Computer Modeling of; Viral Envelope Assembly and Budding; Virology, Molecular; Viruses, DNA Packaging of.

Bibliography

Bamford, D. H., and Wickner, R. B. (1994) *Semin. Virol.* 5:61–69.

Caspar, D. L. D., and Namba, K. (1990) *Adv. Biophys.* 26:157–185.

Chen, Z., Stauffacher, C., Schmidt, T., Fisher, A., and Johnson, J. E. (1990) In *New Aspects of RNA Positive-Strand Viruses,* M. A. Brinton and F. X. Heinz, Eds., pp. 218–226. American Society for Microbiology, Washington DC.

Fox, J. M., Johnson, J. E., and Young, J. Y. (1994) *Semin. Virol.* 5:51–60.

Ganem, D. (1991) *Curr. Top. Microbiol. Immunol.* 168:61–63.

Linial, M. L., and Miller, A. D. (1990). *Curr. Top. Microbiol. Immunol.* 157:125–152.

Rossmann, M. G., and Johnson, J. E. (1989) *Annu. Rev. Biochem.* 58:533–573.

Schlesinger, S., Makino, S., and Linial, M. L. (1994) *Semin. Virol.* 5:39–50.

Virus–Receptor Interactions

Richard L. Crowell and Ruliang Xu

1 **Principles**

2 **Cellular Receptor Specificity**

3 **Purification, Characterization, and Cloning of Receptor Genes**

4 **Receptors as Determinants of Virus Tropism in Pathogenesis**

5 **Virus–Receptor Interactions as Targets for Chemotherapy**

6 **Perspectives**

Key Words

Attachment or Binding The first step in the life cycle of a virus is attachment to host cells by the interaction of a virion attachment protein (VAP) or a virion attachment site (VAS) with a specific cellular receptor. This reaction is considered to be temperature independent and reversible (by low pH or by high salt concentrations).

Cellular Receptor Specificity Specific cellular receptors are identified by demonstrating that they can be saturated with the ligand (unlabeled virus), and that the same virus strain or a virus sharing the same receptor specificity, when labeled with radioactivity to permit measurement of binding, cannot bind to the saturated cells. Nonspecific binding is not saturable and must be subtracted from the amount of bound virus to determine the amount of virus that is bound specifically.

Eclipse Virus eclipse is the third step in infection (after virus attachment and penetration) and is irreversible. It is accomplished by the virus–receptor interaction in the case of picornaviruses. The virus undergoes a major conformational change and cannot be recovered as an infectious virus particle.

Tropism Viruses have a predilection for certain cell types during infection of humans or animals. This homing of viruses to these cells is called tropism and is considered to be controlled by the presence of specific receptors for a given virus. A neurotropic virus has a tropism for neurons in the pathogenesis of infection.

Uncoating Uncoating of the virus genome is the final step in the early events of the virus life cycle. The capsid has been disassembled and the viral nucleic acid becomes susceptible to applied nucleases when the cell is artificially disrupted.

Virion Attachment Protein (VAP) or Virion Attachment Site (VAS) A virion surface protein (capsid or envelope protein) that serves to attach the virion to a specific cellular receptor. Each virion possesses multiple copies of the VAP to permit multivalent binding to multiple copies of cellular receptors.

The initiation of virus infection begins with attachment of virions to specific receptors on the surface of host cells. Viruses of plants, bacteria, insects, and animals, including humans, come in many different kinds, which are classified according to taxonomic principles. This review focuses on the different receptors used by the different families of viruses that infect humans and animals. Specific receptors for the different viruses serve multiple functions for the cell, and progress on the identification of these cell surface molecules has increased dramatically over the past few years. Viruses have evolved to take advantage of these cellular molecules for attachment and entry into cells, and X-ray crystallography has contributed to our understanding of the structural basis for the binding of receptor molecules to the pocket or "canyon" located on virions or on virion attachment proteins.

The genes for a number of receptor molecules have been cloned and expressed in receptor minus cells to provide evidence of their role for virus entry. Transgenic mice (engineered by molecular biological techniques to carry a gene or genes from another species) have been developed, which express the human receptor for polioviruses, and upon infection with wild-type polioviruses, these animals develop paralytic disease mimicking that produced in humans. Thus, there is strong evidence that specific receptors are major determinants of virus tropism in the pathogenesis of infection. The development of antiviral reagents that will block virus replication during the early events in the virus life cycle continues to be an important strategy for control of virus infections.

1 PRINCIPLES

A very large number of viruses of plants and animals, including insects, have been classified into groups (families) based on structure, nucleic acid type, sites of replication in the cell, presence or absence of an envelope, and other characteristics that provide the basis for the taxonomy of viruses. All viruses, which are obligate intracellular parasites, have evolved to take advantage of cellular molecules for their cycle of replication. Viruses have found that specific molecules on the cell surface, which serve assorted cellular functions, also serve as specific receptors for virus attachment. Plant viruses, however, appear to be uniquely transferred to host cells by the bite of insects or by some other form of cell injury. This article focuses on virus–receptor interactions of human and animal viruses, about which a number of reviews and books have been written over the past decade.

The first step in the life cycle of a virus is attachment to host cells by the interaction of a virion attachment protein (VAP) or, when more than one virion protein serves to bind with a specific cellular receptor, a virion attachment site (VAS). This reaction is considered to be temperature independent (although reaction rates are generally reduced at 0°C, reflecting the lower mobility of receptor molecules in the lipid-containing cell membrane, where multivalent binding is accomplished) and reversible [by low pH or by high salt concentrations, as obtainable by treatment with sodium dodecyl sulfate (SDS)]. The virus–receptor binding is mediated by the long-range electrostatic (ionic) forces and subsequently by hydrophobic bonds and, once the virus-binding pocket has received the receptor molecule, by other shorter range bonds (hydrogen bonds, etc.).

Following virus attachment to the cell surface, the virus particle (virion) may be disassembled at the cell surface to permit the viral nucleic acid to enter the cell (possibly for some picornaviruses), or the virion may be engulfed (receptor-mediated endocytosis: picornaviruses, togaviruses, myxoviruses, adenoviruses, etc.) or, in the case of some enveloped viruses (paramyxoviruses, poxviruses, herpesviruses, lentiviruses, etc.), the envelope of the virion may fuse with the plasma membrane to enter the cell. This second step in virus infection is temperature dependent (i.e., it will not occur at 0°C), and the pathway followed is dependent on the type of virus.

Upon entry of the virion into the cell, the virion must be disassembled (the virus is eclipsed and the genome is uncoated) to release the viral genome for transcription or, in the case of positive-strand RNA viruses, for translation. Virus eclipse is the third and irreversible step in infection, which is accomplished by the virus–receptor interaction in the case of picornaviruses. Evidence for the role of receptors in virus eclipse is derived from the finding that solubilized receptor preparations will begin the disassembly of the virion in cell-free suspension at 37°C at neutral pH. Some viruses (influenza, rhinoviruses, foot-and-mouth disease viruses, etc.) require the lower pH of the endosome for conformational changes in the virion proteins to facilitate the uncoating of the genome. Other viruses, like enterovirusees (polio, coxsackie, echo) are stable at low pH (i.e., 3.0), and are not affected by the endosomal pH of 5. These viruses must rely mainly on the action of receptors for uncoating. Thus, the early events in the life cycle of a virus are centered around the activities of specific receptors that both bind and eclipse virus infectivity (infectious virus cannot be recovered from the preparation by artificially disrupting the infected

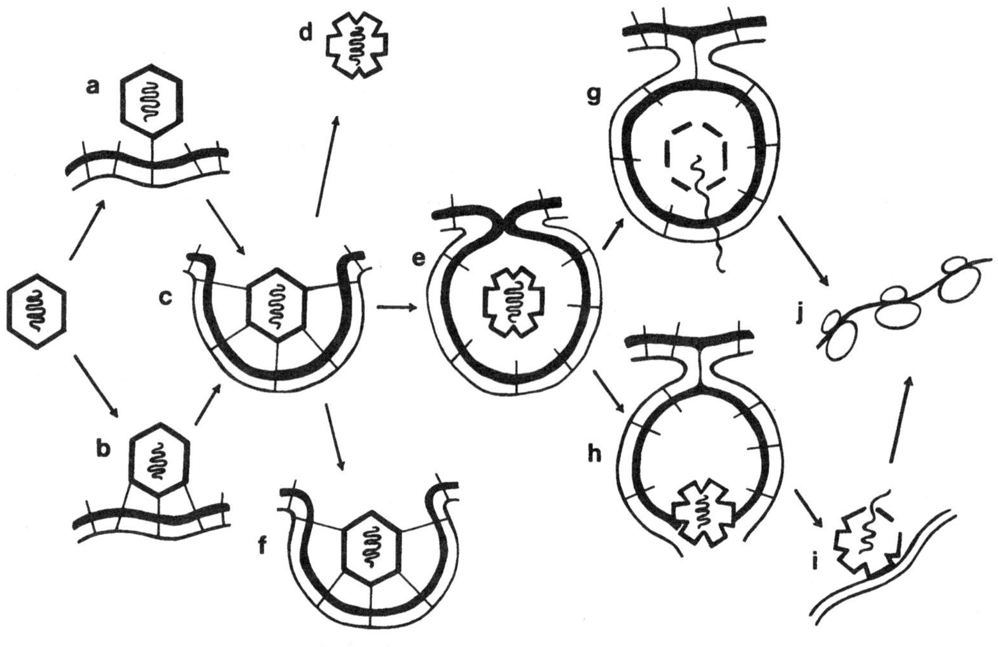

Figure 1. Scheme for early events in picornavirus infection at 37°C. (a) Virion attached by a virion attachment protein (VAP) to a cellular receptor unit (CRU) (loose binding, reversible with EDTA). (b) Virion attached by multiple VAPs to CRUs (tight binding, reversible at pH 2). (c) Virion penetration (virion inaccessible to neutralizing antibodies, but recoverable by pH 2 or SDS). (d) Eluted virus = "A particle" (the lowest molecular weight capsid protein, VP4, has become dissociated from the virion; the viral genome remains inside the altered particle, however). Virus is now eclipsed and not recoverable by pH 2 or by SDS. Most of the eluted virus is dissociated from the cells (≈ 50–80%), contributing to the low ratio of virus particle to infectivity found for picornaviruses. (e) Engulfed "A particle" (virus is eclipsed and not recoverable by pH 2 or by SDS). The "A particle" no longer will bind to receptors and is shown floating free in the vesicle or endosome. (f) Virion that will remain static at the cell surface (recoverable by pH 2 or SDS). Approximately 20% of attached virus will remain in this state without being eclipsed or uncoated. (g) Disassembly of "A particle" to release the RNA genome (uncoating). The RNA can pass across the membrane of the vesicle (RNA is susceptible to hydrolysis by applied RNase following disruption of cells by sonication or by SDS) and attach to ribosomes. (h) "A particle" is transported across membrane of vesicle, since it is more lipophilic than virions. (i) "A particle" may attach to endoplasmic reticulum prior to disassembly and release of RNA (uncoating). No specific receptors for "A particles" have been demonstrated on intracellular membranes. (j) RNA combines with ribosomes for translation. (Reprinted from R. L. Crowell, et al., *Methods in Enzymology,* Vol. 96, pp. 443–452, 1983, by permission from Academic Press, New York.)

cell); these activities, in turn, are designed to help liberate the genome for replication (Figure 1).

Further evidence for the role of receptors in uncoating of the viral genome was provided by the results of experiments in which the delivery into cells of poliovirions encapsulated inside liposomes yielded only minimal virus replication, since the receptors were bypassed. In a control experiment, liposome delivery of infectious viral RNA into the cells was highly efficient for virus replication. In the first case the interaction of receptors with virions was prevented by liposome encapsulation. In the second case, receptors were not needed for processing of naked RNA to become infectious, since naked RNA (or even viral cDNA) can enter receptor negative cells and initiate virus replication. In separate experiments, a "surrogate" receptor for poliovirus (PV) and foot-and-mouth disease virus (FMDV) was provided by attaching antiviral antibodies to the respective viruses. The viruses were allowed to infect Chinese hamster ovary cells, which had been engineered to express the Fc receptor for binding immunoglobulins. Both viruses entered the cells, but only FMDV was capable of replication. It is likely that FMDV was uncoated by the acid pH of the endosomes, whereas PV, in the absence of specific virus receptors, was not uncoated because it was stable to the pH 5 environment of the endosomes. Thus, the specificity of cellular receptors for binding and uncoating certain viruses, in large measure, determines the host range of the virus and at the cellular level is a major determinant of virus tropism (which cell types will become infected) in the pathogenesis of infection.

In contrast to enteroviruses of the *Picornaviridae* family, the alpha viruses of the *Togaviridae* family are uncoated in the cytoplasm of the cell by the action of ribosomes, following receptor-mediated endocytosis of the virion (there appear to be more than one receptor specificity for these viruses too, see Table 1). The low pH of the endosome provides encouragement for the virion spike proteins to fuse with the endosomal membrane, releasing the nucleocapsids into the cytoplasm, where they are rapidly disassembled on ribosomes. Thus, there appear to be multiple ways for viral genomes to be uncoated. Furthermore, papovaviruses, adenoviruses, and herpesviruses usually have their capsids altered in some way prior to entry of the viral genome into the nucleus, where transcription of the viral genes begins the replicative events leading to new virus synthesis.

A number of specific receptors for viruses have been identified (Tables 1 and 2) and some of their genes cloned within the past few years. The cloning of genes for receptors also has allowed for the development of transgenic mice. For example, the human poliovirus receptor gene *(PVR)* has been introduced into the germ line of mice, which then produces the specific human protein, PVR, and renders the mice susceptible to infection by wild-type polioviruses. The PV-infected transgenic mice develop a classical disease typical of human poliomyelitis, and furthermore, PV infects only selected cell types within the brain and spinal cord, leading to paralysis and death of the animal. This example of the role of specific receptors in controlling the species specificity for infection by polioviruses (usually limited to humans and primates) also demonstrates the selected neurotropism of these viruses. That is, the introduction of a single receptor gene converts a resistant animal into a susceptible one and, in addition, provides the histotropism associated with paralytic poliovirus infection of humans.

A number of structural studies of purified virions and viral proteins examined by X-ray crystallography have identified the virion attachment site (VAS), which binds to the receptor, at the atomic structure level. These studies have shown that the VAS is sequestered from the activity of antibody molecules; that is, the antibody "footprint" spans the receptor binding pocket (Figure 2a). Certainly, evolution has endowed viruses with a mechanism to evade preexisting antibodies and to gain a foothold in the host to initiate virus replication. The elimination of viruses from the host is dependent on neutralizing antibodies and cell-mediated immune responses, which react with other sites on the virus surface or with viral proteins expressed on the surface of infected cells. In the case of picornaviruses, a number of them (rhinoviruses, polioviruses, and coxsackieviruses of group B) have "canyons" encircling the fivefold axis of symmetry (vertices of the icosahedron that are devoid of envelopes) into which bind the specific receptor molecules (Figure 2c). Enhanced electron micrographs have revealed the binding of a soluble form (domains 1 and 2; D1D2) of intercellular adhesion molecule 1 (ICAM-1), which is the receptor for the major group of human rhinoviruses (HRV), to the "canyons" of the virions. It has been shown that 60 receptor molecules can bind to each virion (five at each pentameric vertex) (Figure 2b). Thus through ultrastructural studies, both virions and, in some cases, their specific receptors, have been delineated at the atomic level.

2 CELLULAR RECEPTOR SPECIFICITY

Specific cellular receptors are identified by demonstrating that they can be saturated with the ligand (unlabeled virus) and that the same virus strain, when labeled with radioactivity to permit measurement of binding, cannot bind to the saturated cells. Nonspecific binding is not saturable and must be subtracted from the amount of bound virus to determine the amount of virus that is bound specifically. When possible, receptor negative cells are used to help determine the amounts of nonspecific binding of virus. The number of receptors on cells varies from about 1×10^4 to 1×10^6 per cell, depending on the virus–cell system. These values are approximated by analysis of receptor saturation curves (Figure 3) in which viruses are presumed to be multivalent, whereas monoclonal antibodies are divalent. Thus, 3–10 times more receptors per cell will be found when the receptor saturation studies are performed with monoclonal antibodies than with virions. Analysis of the number of receptors on cells by Scatchard plots is not recommended, since that method assumes that ligand binding is monovalent, which clearly is not the case for virions or bivalent antibodies. Figure 3C reveals that there are approximately 5×10^4 receptors per cell for a divalent monoclonal antibody with specificity for coxsackie virus B3 (CVB3) receptors, as determined with radiolabeled antibody. This is about three times the number estimated for labeled virus and by assays of plaque-forming units (pfu) for virus. For the pfu assay, one can assume that approximately 100 virions are required to score one pfu, in this system. Thus, there is good agreement among the different methods used to determine the number of specific receptors per cell.

The results of virus competition for cellular receptors have sorted the picornavirus species into the taxonomic groupings in which they were placed originally, by disease production and histopathological responses (Table 3). Thus, the three serotypes of polioviruses compete for the same receptor on HeLa cells (human carcinoma cells), which is different from the receptor used by the group B coxsackieviruses, which also is different from that used by echoviruses, and so on. The grouping of viruses by receptor specificity (receptor families) suggests that receptors are important determinants of virus tropism in the pathogenesis of disease.

Table 1 Receptors Identified for Enveloped Human and Animal Viruses Listed According to Virus Classification

Virus	Receptor	
	Function[a]	Size of Glycoprotein, gp (kDa)
Arenaviridae		
Arenavirus: Lymphocytic-choriomeningitis virus	(?)	gp120–140
Bunyaviridae		
Bunyavirus		
Bunyamwera virus	(?)	
LaCrosse virus	(?) G_1-binding glycoprotein	
Phelebovirus: Rift Valley fever virus	(?)	
Hantavirus: Hantaan virus	(?)	
Coronaviridae		
Coronavirus		
Human coronavirus 229E	Aminopeptidase N (metalloprotein)	
OC43	9-*O*-Acetylated sialic acid	
Animal coronavirus		
Murine hepatitis viruses	Human CEA-related; biliary gp	gp110
Bovine corona virus	9-*O*-Acetylated sialic acid	
Transmissible gastroenteritis virus (TGEV)	Aminopeptidase N	gp150
Flaviviridae		
Flavivirus		
Group B arboviruses		
Yellow fever virus	Fc receptor	
Japanese encephalitis virus	(?)	
Dengue viruses	Fc receptor for neutral antibody	
Tick-borne encephalitis virus	(?)	gp110
Hepadnaviruses		
Hepatitis B virus (human)	Endonexin II; lipid-binding receptor; asialoglycoprotein	
Hepatitis B-like (animal)	(?)	gp180
Herpesviridae		
Simplexvirus: Herpes simplex virus	Heparan sulfate; other (?)	
Varicellavirus		
Varicella-zoster virus	Mannose 6-phosphate receptor; FcR	
Pseudorabies virus	Heparan sulfate	
Cytomegalovirus: Cytomegalovirus	CD13; HLA-B27; phospholipid	
Lymphocryptovirus: Epstein–Barr virus	Complement receptor 2 (CD21)	
Bovine herpesvirus 1	(?)	gp60
Orthomyxoviridae		
Influenzavirus		
Influenza A	Sialyloligosaccharides (*N*-acetylneuraminic acid); glycophorin A (erythrocyte MN antigen); Fc receptor	
Influenza B	*N*-Acetylneuraminic acid	
Influenza C	9-*O*-Acetylated sialic acid	

(continued)

Table 1 *(Continued)*

Virus	Receptor: Function[a]	Receptor: Size of Glycoprotein, gp (kDa)
Paramyxoviridae		
Paramyxovirus		
Parainfluenzaviruses	Sialic acid (?)	
Newcastle disease virus	Sialic acid	
Sendai virus	Gangliosides	
Morbillivirus		
Measles virus	CD46 (membrane cofactor protein)	gp57–67
	Moesin (membrane-organizing extension spike protein)	gp75
Canine distemper virus	(?)	
Pneumovirus		
Respiratory syncytial virus	(Resembles receptor for Sendai virus)	
Pneumonia virus of mice	(?)	
Poxviridae		
Orthopoxvirus: Vaccinia virus	Epidermal growth factor (?)	gp170
Hepadnaviridae		
Retroviridae		
Oncovirus type C		
Murine leukemia virus A	Phosphate transporter protein	
Murine leukemia virus E	Sodium-dependent basic amino acid transporter (permease)	
Gibbon ape leukemia virus	Permease, or GLVR-1	
Avian Rous sarcoma virus	Homologue of low density lipoprotein receptor (LDLR)	
Oncovirus type B: Murine mammary tumor virus	(?)	
Lentivirus		
Human immunodeficiency virus 1	CD4 (T4); galactosyl ceramide; FcR	
HIV-2	(?)	gp260
Human T-cell leukemia virus 1	Interleukin 2 receptor B p75;	gp31, 45, 55
Simian immunodeficiency virus	(?)	
Feline immunodeficiency virus	Feline homologue of CD9	gp70
Visna virus of sheep	MHC Class II antigens and (?)	gp30–33; gp50
Rhabdoviridae		
Vesiculovirus: Vesicular stomatitis virus	Phospho- or glycolipid carbohydrate moiety	
Lyssavirus: Rabies virus	Acetylcholine receptor, fibronectin, other (?)	
Togaviridae		
Alphavirus		
Group A arboviruses		
Semliki Forest virus	Class I MHC antigens	gp44
Sindbis virus	Catecholinergic receptor	gp90
	Laminin receptor	gp67, 74
Lactate dehydrogenase virus	Ia antigen (?); Fc receptor	gp30
Rubivirus: Rubella virus	Membrane lipid moiety	

[a]MHC, major histocompatibility complex; CEA, carcinoembryonic antigen.

Table 2 Receptors Identified for Nonenveloped Human and Animal Viruses Listed According to Virus Classification

Virus	Receptor: Function	Receptor: Size of Glycoprotein, gp (kDa)
Adenoviridae		
Adenovirus (human)		
Hemagglutination group A: Ad12	Integrins $\alpha\nu\beta3$ and $\alpha\nu\beta5$	
Hemagglutination group B: Ad3 and 7	(?)	
Hemagglutination group C		
Ad2 and 5 fiber	(?)	gp34, 42, 49, 60, 78, 88, 115, etc.
(Penton base)	Integrins $\alpha\nu\beta3$ and $\alpha\nu\beta5$	
Papovaviridae		
Papillomaviruses (human)	(?)	
Polyomavirus		
Polyomavirus (murine)	Sialyloligosaccharides	gp25, 30, 50, 95 (?)
Simian virus 40	MHC class I (complex) (?) Oligosaccharides	gp30, 54, 58, 90 (?)
JC (human)	(?)	
Parvoviridae		
Parvovirus B19 (human)	Erythrocyte P antigen (globoside)	
Parvovirus canine (animal)	Sialic acid ($\alpha_{2,3}$-linked)	gp40–42
Picornaviridae		
Enterovirus (human)		
Polioviruses T1–3	NCAM (?)	gp67 or 80
Coxsackievirus A9	Integrin ($\alpha\nu\beta3$)	gp100
Coxsackieviruses A13, 15, 18, 21	ICAM-1	gp90
Coxsackieviruses B1–6	(?)	gp46–49
Coxsackievirus B3-RD	CD55 (DAF)	gp60–70
Echoviruses 1, 8	Integrin (VLA-2)	gp145
Echoviruses 3, 6, 7, 11, 12, 20, 21	CD55 (DAF)	gp70
Hepatovirus: Hepatitis A virus	α_2-macroglobulin ("surrogate")	
Enterovirus (animal)		
Bovine enterovirus	(?)	gp180
Theiler's virus	(?)	gp34
Swine vesicular disease	(?) (virus related to CBV5)	
Cardiovirus (animal): Encephalomyocarditis virus	VCAM-1 sialoglycoprotein 70 glycophorin A (on erythrocytes)	
Rhinovirus (human)		
Major receptor group	ICAM-1	gp90
Minor receptor group	LDLR	gp120
Aphthovirus (animal): Foot-and-mouth disease	Integrin ($\alpha\nu\beta3$)	gp100
Reoviridae		
Reovirus T3	β-Adrenergic-like	gp67
	Epidermal growth factor	gp150–170
	Multiple sialic acid proteins	gp26–200
Rotaviruses (human and simian)	Sialic acid	
Rotaviruses (bovine)	Monosialoganglioside	

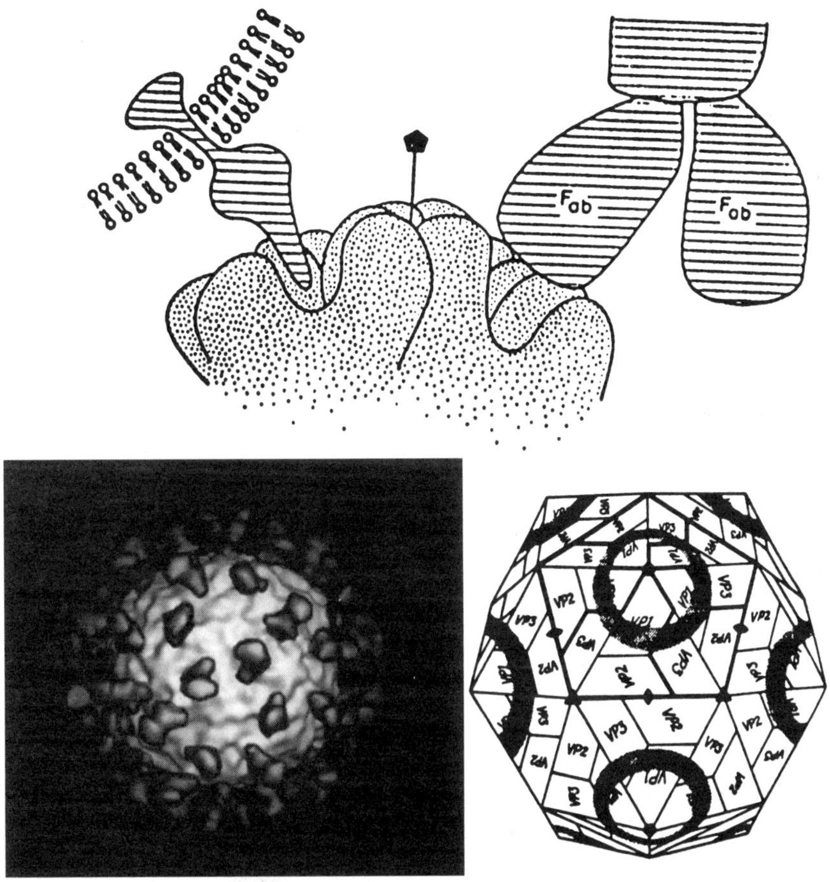

Figure 2. (a) The presence of depressions on the picornavirus surface suggests a strategy for the evasion of immune surveillance. The dimensions of the receptor binding sites (the canyon in HRV14, the pit in Mengovirus) sterically hinder an antibody's (top right) recognition of residues at the base of the site, while still allowing recognition and binding by a smaller cellular receptor (top left). This would allow for receptor specificity while at the same time permitting evolution of new serotypes by mutating residues about the rim of the canyon or pit. [Reprinted with permission from M. G. Rossman and A. C. Palmenberg, *Virology,* 164:373–382 (1988). Academic Press, Inc., New York.]
(b) Cryoelectron microscopy of human rhinovirus (HRV): 16 particles and their complex with the soluble, truncated receptor fragment, domain 1 and domain 2 (D1D2) of ICAM-1. Reconstruction of the HRV-16–D1D2 complex, viewed along an icosahedral two-fold axis in approximately the same orientation shown in (c) for HRV. The D1D2 molecules saturated the 60 available sites on the virion, with each D1D2 molecule having an approximate shape of a dumbbell, consistent with the presence of a two-domain structure. This micrograph and other evidence confirmed the canyon hypothesis.
(c) The icosahedral symmetry, subunit organization, and canyon (shaded) for HRV. Heavy lines encircle five protomers of VP1–VP3. The fourth viral protein, VP4, is inside the capsid. [(b) and (c) reprinted from N. H. Olson et al., *Proc. Natl. Acad. Sci. U.S.A.* 90:507–511 (1993), with permission from the National Academy of Sciences.]

The finding that monoclonal antibodies can be developed against specific cellular receptors has confirmed and extended the grouping of viruses into receptor families. Cells that possess specific viral receptors can be saturated with monoclonal antibodies that block the subsequent binding of given related viruses that share the same receptor specificity. In many cases, the monoclonal antibodies will also protect the cells from infection by viruses that would have used the blocked receptors. This latter assay of monoclonal antibody is a major challenge for the antibody, since if only one receptor is not blocked by the antibody in the cell preparation being challenged with virus, the virus may enter a cell and initiate replication. To reveal the location of the receptors on the cells, the monoclonal antibodies can be labeled with fluorescent dyes and examined under ultraviolet light or, after binding to cells, labeled with colloidal gold and examined with the electron microscope. The use of this type of reagent or the use of labeled viruses, or their VAP, has revealed that receptors are located on the cell surface on microvilli or on the smooth surface of susceptible cells. Receptors also may exist on the apical or basolateral surfaces of polarized cells (SV-40, VSV, TGEV, etc.); under certain circumstances, when cross-linked at 37°C, the receptors can migrate to the poles of the cells (receptor capping). These reagents also have permitted identification, by means of in situ methods, of receptor location in tissues.

It is axiomatic that all cells that are susceptible to a given virus possess specific receptors for that virus, however, all cells expressing these receptors are not necessarily susceptible to infection, since other factors may influence whether the cell will be permissive to virus replication. Thus, for example, it has been found that all cell types in the PVR transgenic mice express messenger RNA for the PVR, and yet only selected cell types (neurons, etc.) serve to replicate virus, which leads to disease. The reason for this lack of correlation between receptor distribution and cell susceptibility remains

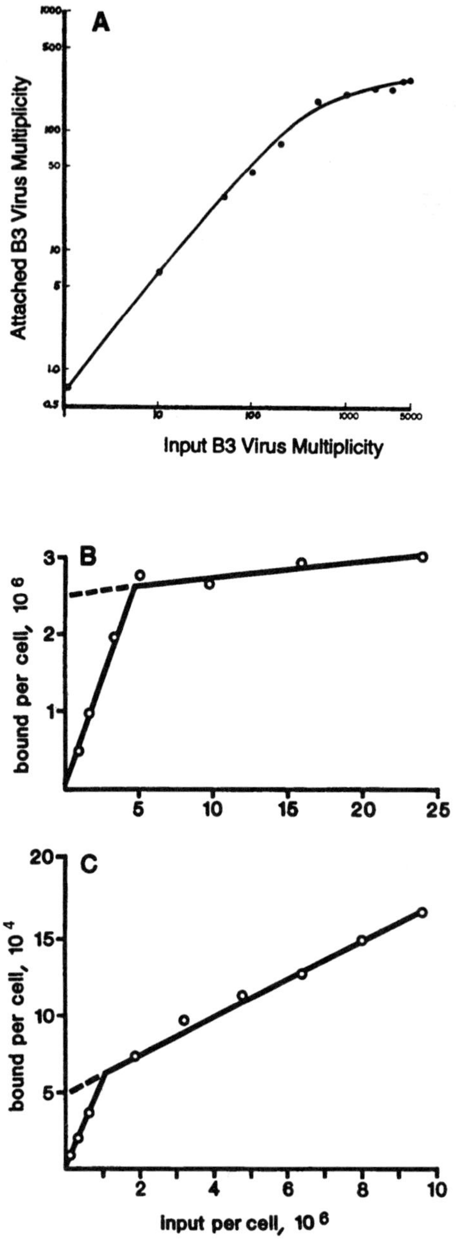

Figure 3. (A) Influence of input virus multiplicity (pfu) on the amount of coxsackievirus B3 that will attach to HeLa cells within 1 hour at 37°C. This result is the first demonstration of saturation of host cell receptors by a human virus. [Reprinted from R. L. Crowell, *J. Bacteriol.* 91:198–204 (1966), with permission from the American Society for Microbiology, Washington, DC.]
(B) and (C) Titration of binding sites for monoclonal antibodies on cells. Dilutions of ^{125}I-labeled monoclonal antibodies were incubated with cell suspensions for 60 minutes at room temperature. The cells were washed and counted for radioactivity. The number of molecules bound per cell was determined and plotted versus input multiplicity of antibody molecules. Plots show the binding of RmcA (receptor monoclonal anti-DAF, CVB-RD receptor) to HeLa cells (B) and (C) the binding of RmcB (anti-49 kDa receptor molecule for CVB1-6) to HeLa cells. The binding of CVB3 to HeLa cells has a very low reversibility; thus it was assumed that the adsorption isotherms for antibodies or coxsackieviruses on cells are composed of two components. The first is a specific saturable attachment; superimposed on this is a nonspecific attachment, which appears not to be saturable, but proportional to input multiplicity. The intercept of the slope of the nonspecific binding at zero input was considered to be the approximate number of specific sites per cell. [Reprinted from K.-H. Ksu, et al., *J. Virol.* 62:1647–1654 (1988), with permission, from the American Society for Microbiology, Washington, DC.]

to be determined, but it probably will be found to lie in the capacity of receptors to eclipse and uncoat the viral genome. In some cases, a receptor specific cofactor is required for initiating infection. For example, the cofactor fusin and beta chemokine receptor serve as cofactors for T-lymphotropic and macrophage tropic HIV virus strains, respectively.

A number of investigators have attempted to use monoclonal antibodies as immunogens, to develop anti-idiotypic (anti-ID) antibodies to search for receptors in cell extracts. It was reasoned that the anti-ID antibodies would have the mirror image of the reactive antibody site and therefore would be able to detect the specific virus receptor. Despite the occasional exception (reoviruses), however, these studies have not been very successful, since the virus site (VAS or VAP) that combines with the receptor is secluded from the antibody "footprint" in a receptor-binding pocket, canyon, or depression in the virus particle; coxsackieviruses of group B, Sindbis virus, and rubella virus have failed to develop anti-ID antibodies under experimental manipulation. In these latter viruses, consequently, there is no mirror image available for recognition of the cellular receptor.

Several studies have now demonstrated that a given virus may utilize more than one receptor specificity to initiate infection of different cell types. For example, variants of the group B coxsackieviruses, which were selected for growth in human rhabdomyosarcoma cells (RD cells) from prototype strains of virus, were found to have acquired the capacity to agglutinate erythrocytes (hemagglutination). The prototype viruses from which the variants were selected were hemagglutination negative. The receptor on erythrocytes for binding these viruses has recently been shown to be the decay-accelerating factor (DAF), a complement regulatory protein,

Table 3 Relationship Between the Receptors for Picornavirus Species and Virus Attachment Interference (Receptor Competition)[a]

Competing Virus	Interference with Attachment of Virus Immunotypes[b] Polio 1–3	Cox B 1–6	Cox A 13, 21	HRV 14	HRV 2	Echo 6	FMDV SAT	EMC
Polio 1–3	+	−	−	−	−	−	−	−
Cox B 1–6[c]	−	+	−	−	−	−	−	−
Cox A 13, 21[d]	−	−	+	+	−	−	−	−
HRV 14[e]	−	−	+	+	−	−	−	−
HRV 2[f]	−	−	−	−	+	−	−	−
ECHO 6[g]	−	−	−	−	−	+	−	−
FMDV[h]	−	−	−	−	−	−	+	−
EMC[i]	−	−	−	−	−	−	−	+

[a] Cells that have been saturated with competing (interfering) virus are washed and tested for capacity to attach a second type of virus. These experiments usually are carried out at room temperature or colder, and the second virus is usually radiolabeled for ease of assay. Every serotype may not have been tested in both directions.

[b] +, interference; −, lack of virus attachment competition; Polio, poliovirus types 1–3; Cox B, coxsackievirus group B types 1–6; Cox A, coxsackievirus group A types 13 and 21; HRV, human rhinovirus type 14 or 2; ECHO, echovirus type 6; FMDV, foot-and-mouth disease virus South African Territories types; EMC, encephalomyocarditis virus of mice.

[c] Coxsackievirus B3-RD variant, which is hemagglutination positive, shares one of two receptor specificities (CD55) with echoviruses 3, 6, 7, 11, 12, 20, 21, which also are hemagglutination positive. The other receptor specificity for CVB3-RD, a molecule of 49 kDa, is yet to be defined. CVB3-RD will block binding of CVB3 to HeLa cells, but CVB3 will not block binding of CVB3-RD, because the latter virus can utilize the second receptor (CD55) and CVB3 cannot.

[d] Coxsackieviruses A13, 15, 18, and 21 share a receptor specificity and coxsackieviruses A2 and A5 share a receptor of different specificity from the others. CVA9 uses an integrin, vitronectin, for its receptor, although following proteolysis, CVA9 uses another yet undefined receptor. Receptors for other CVAs are unknown.

[e] Human rhinovirus type 14 shares a receptor, ICAM-1, with about 90 other serotypes of HRV and is in the "major" receptor grouping. Interestingly, this receptor also is used by CVA 13, 15, 18, and 21.

[f] Human rhinovirus type 2 shares a receptor, LDLR, with about 10 other serotypes of HRV and is in the "minor" receptor grouping.

[g] Echoviruses 3, 6, 7, 11, 12, 20, and 21 share the receptor CD55 (DAF), whereas echoviruses 1 and 8 use the integrin VLA-2 for their receptor. The receptors for other echovirus serotypes remain to be determined.

[h] FMDV serotypes have the amino acid sequence RGD (arginine, glycine, and aspartic acid) in their virions and use the vitronectin integrin as their receptor.

[i] EMC virus strains of mice do not use the same receptor, although they have only one serotype. They are hemagglutination positive and use glycophorin A on erythrocytes for their receptor. A different receptor, VCAM-1 or another molecule, serves to bind virus to host cells.

and this DAF molecule also serves to bind virus to initiate infection of RD cells. Thus, these coxsackieviruses acquired a VAS, located on the surface of the virion, to bind to a second receptor with specificity that differed from the primary receptor that occurs on monkey kidney and on HeLa cells. The primary receptor is considered to bind to the VAS located in the canyon of the CVB virion. It is of interest that the RD cells are devoid of the primary CVB receptor molecule (a 50 kDa protein), whereas monkey kidney cells possess the 50 kDa receptor and lack the DAF molecule. HeLa cells, however, possess both receptor molecules on their surface.

Prior to the recognition that viruses may have more than one receptor specificity, investigators would debate the authenticity of receptor identification for a given virus. Thus, for example, the human immunodeficiency virus (HIV) may infect cells via its usual receptor on human T lymphocytes, the CD4 molecule or, in the absence of CD4, on certain cell types (viz., neurons and macrophages) the route to infection may entail alternative receptor molecules, such as gangliosides (neutral lipids). In addition, immunoglobulins that bind to virus particles may also bind to the Fc receptor on macrophages or other cell types, and thereby facilitate infection of those cells (sometimes referred to as "surrogate" receptors) by this virus-linked molecule.

Dengue viruses appear to be enhanced in their infection of host cells by the presence of virus-neutralizing antibodies. This would appear to be anomalous behavior of viruses, since neutralizing antibodies are generally considered to protect cells against infection. Furthermore, some viruses that have a very wide host range (i.e., infect almost all cell types), like herpes simplex, have been found to use as a specific receptor heparan sulfate, a molecule very widely distributed in nature. Rabies virus, likewise, has a wide host range, infecting all warm-blooded animals.

Viruses with a wide host range have presented problems to investigators, who require cells that are insusceptible to infection for use as control cells when they are looking for a specific receptor. Rabies virus is a classical case of a virus for which more than one type of molecule serves as the virus receptor. For rabies, these molecules include fibronectin and the acetylcholine receptor; hence it has been very difficult to prove receptor specificity for this virus.

3 PURIFICATION, CHARACTERIZATION, AND CLONING OF RECEPTOR GENES

The recognition in the 1940s that influenza virus causes erythrocytes to aggregate (the hemagglutination phenomenon) placed this virus–cell system in the forefront for detailed studies of virus–receptor interaction. It has been shown that the VAP is the hemagglutinin glycoprotein, which occurs in multiple copies as homotrimers embedded in the virus envelope and extending from the surface of the virion. The hemagglutinin was purified, and the amino

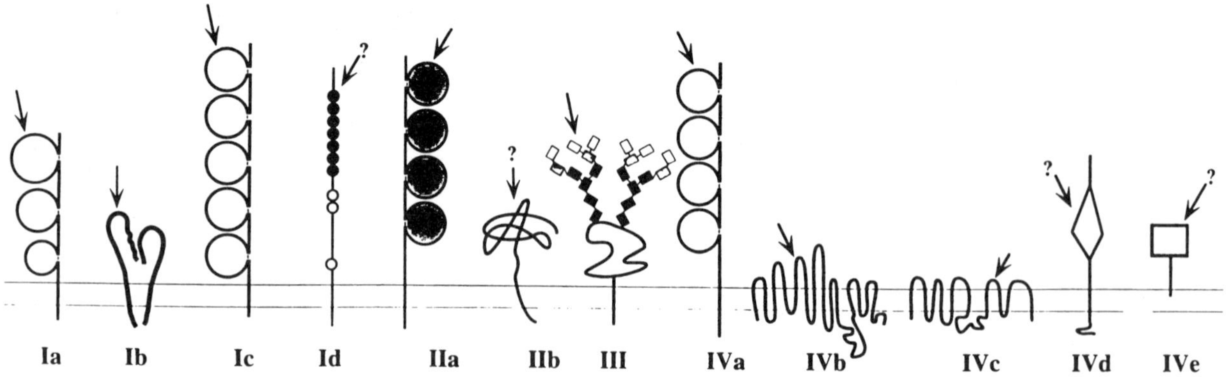

Figure 4. Schematic representation of some plasma membrane molecules that serve as receptors for viruses. Roman numerals indicate that the receptors are used by some members of the following virus families: I, *Picornaviridae,* II, *Coronaviridae,* III, *Orthomyxoviridae* and *Paramyxoviridae;* IV, *Retroviridae;* Ia, poliovirus receptor (PVR; an Ig-like protein of unknown function; polioviruses); Ib, integrin VLA-2 (echoviruses 1 and 8); Ic, ICAM-1 (major group of human rhinoviruses); Id, LDL receptor (minor group of human rhinoviruses); IIa, murine biliary glycoprotein (related to the human carcinoembryonic antigen family; murine hepatitis virus); IIb, aminopeptidase N (a metalloprotease; porcine and human coronaviruses); III, sialic acid residues, glycoproteins of which interact with the virion (orthomyxoviruses and paramyxoviruses); IVa, CD4 (T-cell surface glycoprotein; HIV-1); IVb, mouse cationic amino acid transporter (MCAT-1; ecotropic murine leukemia virus); IVc, sodium-dependent phosphate transporter (gibbon ape leukemia virus); IVd, protein of unknown function (bovine leukemia virus); IVe, an LDL-related protein with an extracellular domain of only 83 amino acids (subgroup A avian leukosis and sarcoma viruses). Arrows mark proposed virus-binding sites. [Reprinted from E. Wimmer, Cellular receptors for animal viruses, in Wimmer, Ed. (1994), pp. 1–14, with permission from Cold Spring Harbor Laboratory Press, Cold Spring Harbor, NY.]

acids sequenced and subjected to X-ray crystallographic studies for determination of atomic structure. It was found that the distal end of the stalklike molecule, which terminates in a globular form, contains the receptor-binding pocket; sialic acid from the receptor glycoprotein on the erythrocyte was shown to fit snugly into this pocket, accounting for virus attachment to the cell. A neighboring viral protein, neuraminidase, was found to dissociate (elute) the virion from the erythrocyte and in the process destroy the cellular receptor (glycophorin) by release of sialic acid. Consequently, the neuraminidase-treated receptor no longer can bind virus. It is remarkable that more than 50 years of analysis of this virus–receptor system went into the resolution of the structure–function relationships at the atomic structure level. The difficulty of the task reflects the high level of complexity of the system and the need for development of newer technologies to decipher this virus–receptor interaction.

Other receptor systems have been even more elusive, since the plasma membranes of somatic cells are more complex than those of red blood cells. Nevertheless, considerable progress for the identification of other virus receptors was facilitated by the use of specific antibodies to identify a number of antigenic sites on lymphocytes and other cell types as virus receptors (see Tables 1 and 2). The specific cellular receptors for picornaviruses have been purified, only with great difficulty, in amounts sufficient for identification but precluding any significant biochemical characterization.

Use of receptor-specific monoclonal antibodies in affinity chromatography, however, has permitted the purification of amounts of the receptor for the major group of human rhinoviruses sufficient for partial amino acid sequencing. This accomplishment led to the cloning of the receptor gene and its identification as ICAM-1. This molecule belongs to the immunoglobulin superfamily of genes, which includes the gene for the CD4 molecule, which serves as the receptor for HIV-1, and the receptor genes for the polioviruses and the murine hepatitis virus (MHV), respectively (Figure 4). Studies have revealed that the functional sites of the receptors for binding to their specific viruses are located in the N-terminal disulfide-bonded loop of each molecule. Use of recombinant DNA technology has also shown that chimeric molecules produced between two different receptors reflect the virus specificity that belongs to the virus binding to the N-terminal loop, and that the length of the molecule is of little importance in determining whether it can serve as a functional molecule to initiate infection. Indeed, there seemed to be no specific requirement for the cytoplasmic tail of the receptor for full functional activity. In general, the carbohydrate moieties of the receptor play little role in virus binding; however, they enhance the magnitude of their binding activity when present, perhaps by providing a unique conformation to the molecule. Thus, the central theme in numerous studies designed to demonstrate the specificity of viral receptors is to transfect receptor negative cells, or prepare transgenic animals, with the receptor gene and with different constructs of the gene to determine what elements of the receptor molecule are essential to initiate infection by the virus in question.

4 RECEPTORS AS DETERMINANTS OF VIRUS TROPISM IN PATHOGENESIS

The early studies of Holland, McLaren, and Syverton in 1959 showed that the restriction of poliovirus replication in nonprimates was due to the lack of a specific cellular receptor for poliovirus. The nonprimate cells, nevertheless, could be infected with naked poliovirus RNA to demonstrate that the resistant cells were capable of poliovirus replication. These classical studies paved the way to the concept that specific cellular receptors for poliovirus were the major determinants of virus tropism in the pathogenesis of poliovirus infection. Although polioviruses are considered to be neurotropic viruses because they produce human paralysis with destruction of motor neurons in the brain and spinal cord, they also are capable of replication in a variety of other cell types including lymphocytes, skeletal and heart muscle cells, and endothelial and epithelial cells. Thus, as mentioned earlier, transgenic mice producing the human poliovirus receptor (PVR) were found to produce mRNA for the PVR in all cell types studied, and yet the animals developed a classical paralytic disease following inoculation of wild-type polioviruses.

Researchers wondered whether receptors could really be the determinants of virus tropism in the poliovirus system, since a restricted number of motor neurons were the major cell types destroyed in this infection, instead of a wide range of cells putatively expressing specific PVRs. Further studies will be required to resolve this apparent discrepancy in the role of receptors in determining virus tropism. It is very likely, however, that the number of receptors per cell, or their ability to cluster together in sufficient number to eclipse virus and initiate the viral RNA uncoating, will be found to be directly related to the distribution of virus infection in this model system. Perhaps the terminal stage of differentiation of motor neurons provides these cells with a unique receptor configuration that distinguishes them from other cell types that express poliovirus receptors but do not participate in the pathogenesis of the infection.

The Epstein–Barr virus (EBV), one of the human herpesviruses, causes infectious mononucleosis and is associated with nasopharyngeal carcinoma. This virus also has been linked to the etiology of a number of human lymphoproliferative diseases including Burkitt's lymphoma. The receptor for EBV has been identified as the molecule that binds the third component of complement, C3dg, and is a single 145 kDa membrane glycoprotein known as CD21 (complement receptor type 2). This receptor occurs mainly on B lymphocytes and on some T cells. The receptor also must occur on epithelial cells, since a number of epithelial cells in culture serve as host cells for EBV, although the epithelial cell receptor may not be identical to that on lymphocytes. Unlike herpes simplex virus, which uses for its receptor a widely distributed molecule, heparan sulfate, EBV appears to have a much more limited host cell range. Consequently, EBV presents a more practical problem for potential strategies for antiviral therapies at the receptor level.

5 VIRUS–RECEPTOR INTERACTIONS AS TARGETS FOR CHEMOTHERAPY

The inhibition of virus replication by antiviral chemotherapeutic or chemoprophylactic reagents is a continuing challenge to pharmaceutical companies worldwide. At least three major problem areas must be addressed before an effective reagent can be produced for the benefit of humans and animals. The reagent must be:

1. Rapidly acting, with appropriate pharmacodynamic properties, and able to block virus replication in its tracks (since a very large amount of replication has occurred prior to the development of significant clinical signs).
2. Not too toxic to host cells, tissues, or organs, since viral replication utilizes many host cell components.
3. Able to discourage the virus from rapidly developing drug-resistant mutations.

Relatively few antiviral reagents have met all these requirements, although there are available a few somewhat effective antivirals that have reduced morbidity and mortality due to virus infections. Nevertheless, the early events in virus replication have not yet provided adequate targets for effective antiviral reagents. Inasmuch as neutralizing antibodies are naturally occurring antiviral reagents that combine with the surface of virions and with viral proteins that are produced on the surface of virus-infected cells, neutralizing antibodies should be considered to be reagents that interfere with virus replication by blocking an early event in the virus life cycle. It is likely that in addition to blocking virus attachment (which may occur under rare conditions of high antibody concentrations), the neutralizing antibodies interfere with the productive uncoating of virions by combining with viral surface epitopes. These antibodies would preclude a close fit of the multivalent binding of receptors, considered to be necessary for successful uncoating of the viral genome. Hopefully this mechanism of virus neutralization will be given further consideration as another target for development of antivirals.

Some major advances in the design of drugs that inhibit early events in virus replication have been made. The most notable is the licensed drug amantadine, which blocks the uncoating of the influenza A strains, by combining with a region on the hemagglutinin molecule proximal to the site near the virion envelope. More recent data suggest, however, that the site of action of amantadine is to block a pore in the M2 protein of the virion to inhibit acidification of the interior of the virion and thereby block virus–membrane fusion. Amantadine has been an important reagent in the prevention of influenza A infections of the elderly and in immunologically incapacitated persons. Unfortunately, strains of virus emerge that become resistant to amantadine upon prolonged usage.

The use of soluble CD4 for the prevention of HIV-1 infection had at one time appeared to be quite promising. The soluble receptor protein was engineered to prolong its circulation in the body by coupling it to an immunoglobulin chain. As in many attempts at chemotherapy, however, the capacity to provide nearly 100% efficacy by blocking attachment of all HIV-1 virions could not be met. In addition, HIV-1 may use alternative receptors for infection. Even if there were no alternative receptors, this strategy would likely fail because such a receptor decoy approach would not be able to block 100% of invading virions. In this infection, however, all virions must be blocked, since successful infection by only a few virions would lead to viral genome integration in host cells and the capacity to initiate disease. Nevertheless, in acute diseases such as those due to rhinovirus infections, administration of the soluble form of ICAM-1 might delay the rapid replication of virus. This would permit the immune responses to develop, and disease manifestations would be lessened.

There have been numerous attempts to develop a series of chemicals that would occupy an internal pocket lying just below the receptor-binding pocket of enteroviruses and rhinoviruses. These reagents have the property of stabilizing the virions against receptor-mediated uncoating. The studies required to develop these reagents were elegant in design and required knowledge of the atomic structure of the respective viral protein coats. In clinical trials, however, these compounds were not as effective as in tissue culture, and drug-resistant mutants rapidly emerged. Further studies of the mutants revealed that only one amino acid change in the region of the drug-binding pocket could inhibit the activity of the drug, which was designed to block either virus binding to receptors or uncoating of the viral genome.

Without question, the development of virus mutations that provide drug resistance is the major threat to successful viral chemotherapy or prophylaxis. The future trend in viral chemotherapy is toward combinations of drugs that have different modes of action, so that the emergence of virus resistance will be reduced significantly. Thus, the development of molecular constructs of soluble receptors to serve as decoys for viruses, and the development of virus-binding drugs to help stabilize them against the uncoating action of receptors, in combination with drugs that block viral nucleic acid replication, together hold promise for future attempts to control viral diseases.

6 PERSPECTIVES

Studies of cellular receptors for viruses have progressed rapidly over the past 10 years. The primary impetus for attention to this area of research stems from the importance of receptors as determinants of virus tropism in the pathogenesis of viral infections. These studies have been advanced by the emergence of new technologies and their application to the elucidation of the details of virus–receptor interactions, such as X-ray crystallographic analysis of viral binding pockets and the fit of soluble receptor domains into these pockets. These studies have provided a basis for computer modeling of potentially useful drugs that will serve to distort the virus–receptor bonding and either block the interaction or stabilize the capsid structure against receptor-mediated uncoating of the viral genome.

The development of transgenic mice, which can express human receptor proteins, has expanded the capability of evaluating the role of receptors in viral pathogenesis. In the case of poliovirus infection of transgenic mice, it has been found that receptors are expressed on most cell types, and yet poliovirus replication is restricted to limited cell types in causing paralytic disease. The next level of study of this interesting animal model of a human disease should reveal the contribution made by receptor configuration and density in the plasma membrane of the cell, together with possible cofactors, in virus eclipse and uncoating (the second function of virus receptors). Further transgenic murine models expressing other human receptor glycoproteins (e.g., the CD46 molecule used by measles virus) should be equally exciting. Transgenic animals provide potentially useful models for the study of virus pathogenesis for viruses (such as polioviruses and measles virus) which have a host range restricted to humans or to subhuman primates.

It appears that all receptors (specific or "surrogate") accomplish binding of their respective viruses to the cell surface. In the next few years, the exact mechanisms involved in the presentation of the viral capsid for fusion with cell membranes or for fusion of conformationally altered viral particles for the different types of viruses should be better understood. For some viruses these alterations may occur at the cell surface and for others within endosomes or phagocytic vesicles. In addition, some types of infecting viral particle must travel toward the nuclear membrane and find ways to insert their genomes into cell nuclei for transcription.

The emerging field of gene therapy has utilized viruses of a number of different types (adenoviruses, retroviruses, poxviruses, herpesviruses, etc.) as vectors for the delivery of selected genes to specified target cells. The cell targets for receipt of genes may be better defined and achieved when the distribution and activities of specific receptors become known for each virus vector delivery system. Moreover, since cell receptors can be altered by addition of gangliosides or carbohydrate molecules, the cells can be made to serve as targets for specific virus variants. Future developments in gene delivery systems promise remarkable achievements in control of human diseases, including cancer therapies.

The alternative pathways by which different viruses enter cells must be elucidated for the rational development of reagents that may prove effective in blocking the entry of viruses and the uncoating of their genomes. It would appear that unique therapies or prophylactic reagents will need to be developed for each of the separate groups of viruses, which utilize as receptors diverse cell surface glycoproteins, gangliosides, or carbohydrates. At present we do not understand the reason for receptor diversity among viruses, especially among closely related viruses. Presumably viruses have undergone continuous evolution to adapt to the structures that are presented to them at the cell surface.

Study of viral inhibitors has revealed all too predictably that virus variants will emerge that are resistant to these inhibitors. Furthermore, some virus variants can utilize more than one receptor specificity for initiating infection. This latter capability has made it more difficult to block virus infection at the different stages of virus entry into cells.

The specificity of virus–receptor interactions has proven to be an important revelation of nature. This specificity coincides with many other specific interactions of biological significance, such as enzyme–substrate, antigen–antibody, and hormone–receptor interactions. Progress will continue to be made in our understanding of the early events in virus infection as we apply future technologies that are developed in related fields. The early events of virus–cell interactions remain as potential targets for intervention in virus infections.

See also VIROLOGY, MOLECULAR.

Bibliography

Bentz, J., Ed. (1993) *Viral Fusion Mechanisms.* CRC Press, Boca Raton, FL.

Crowell, R. L., and Lonberg-Holm, K., Eds. (1986) *Virus Attachment and Entry into Cells.* American Society for Microbiology Press, Washington, DC.

Fields, B. N., Knipe, D. M., Chanock, R. M., Hirsch, M. S., Melnick, J. L., Monath, T. P., and Roizman, B., Eds. (1990) *Virology,* 2nd ed. Raven Press, New York.

Haywood, A. M. (1994) Virus receptors: Binding, adhesion strengthening, and changes in viral structure. *J. Virol.* 68:1–5.

Marsh, M., and Helenius, A. (1989) Virus entry into animal cells. *Adv. Virus Res.* 36:107–151.

Nomoto, A. (1992) Cellular Receptors for Virus Infections. In *Seminars in Virology,* Saunders Scientific Publications/Academic Press, London.

Norkin, L. C. (1995) Virus receptors: Implications for pathogenesis and the design of antiviral agents. Clin. Microbiol. Rev. 40: 8(2):293-315.

Sauve, G. J., Saragovi, H. U., and Greene, M. I. (1993) Reovirus receptors. *Adv. Virus Res.,* 42:325–341.

Tucker, S. P., and Compans, R. W. (1993) Virus infection of polarized cells. *Adv. Virus Res.* 42:187–247.

Wiley, D. C., and Skehel, J. J. (1987) The structure and function of the hemagglultinin membrane glycoprotein of influenza virus. *Annu. Rev. Biochem.* 56:365–394.

Wilson, T. M. A. (1992) *Early Events in RNA Virus Infection.* In *Seminars in Virology.* Saunders Scientific Publications/Academic Press, London.

Wimmer, E., Ed. (1994) *Cellular Receptors for Animal Viruses.* Cold Spring Harbor Laboratory Press, Cold Spring Harbor, NY.

VITAMIN RECEPTORS

Krishnamurti Dakshinamurti

1 Introduction

2 Vitamin A
- 2.1 Absorption and Transport
- 2.2 Retinoid Receptors: Cellular Binding Proteins
- 2.3 Nuclear Retinoic Acid Receptors

3 Vitamin D
- 3.1 General
- 3.2 Vitamin D Receptors
- 3.3 Vitamin D Responsive Genes

4 Vitamin E
- 4.1 General
- 4.2 Transport of α-Tocopherol
- 4.3 Binding of α-Tocopherol to Erythrocyte Membranes
- 4.4 Liver Cytosolic α-Tocopherol-Binding Proteins
- 4.5 Nonantioxidant Effects of Vitamin E

5 Ascorbic Acid
- 5.1 General
- 5.2 Ascorbic Acid Transport and Transporters
- 5.3 Ascorbic Acid and Collagen Production

6 Biotin
- 6.1 Noncovalent Biotin-Binding Proteins
- 6.2 Role of Biotinidase as a Biotin Carrier Protein
- 6.3 Nonprosthetic Group Functions of Biotin
- 6.4 Effect of Biotin on the Regulation of the Synthesis of Specific Proteins

7 Vitamin B_6
- 7.1 General
- 7.2 Pyridoxal Phosphate Binding Proteins
- 7.3 Nuclear Pyridoxal Phosphate Binding
- 7.4 Function of Pyridoxal Phosphate in the Nucleus

Key Words

Receptor Proteins with a recognition function for a vitamin or its derivative, that is, proteins that bind to the vitamin ligand.

Vitamer Derivatives of a vitamin that function as a vitamin.

Vitamins Accessory factors of diverse chemical structure required for the maintenance and growth of protozoans and metazoans.

Transcription Synthesis of mRNA or mRNA precursor.

Transcription Factor Direct-acting factors that control transcription by interacting with DNA sites near the target gene.

Response Element The characteristic DNA-binding sites for hormones, vitamins, or their active forms.

Interlocking ligand–receptor systems constitute an intra- and intercellular communication system in which exogenous or endogenous molecules such as vitamins, hormones, cytokines, and foreign macromolecules convey signals through association with membrane structures (i.e., receptors). Sequential interactions between ligands and receptors on membranes of cells or organelles coordinate metabolism within the same and between different cells and organs. Cell communication is regulated by activation events and by processes that limit such activation.

This article presents current information on the ligand (vitamin)–receptor (binding protein) interaction as it applies to vitamins, from the site of enteric absorption to the site of vitamin function. For the absorption and transport of most vitamins, specific proteins, which have a recognition factor incorporated into them, are involved. At the plasma membrane of the cell, a second recognition factor has to function before the vitamin is internalized. Intracellular vitamin-binding proteins and their interaction with cellular systems resulting in the biological action of the vitamins provides the last step in the successive stages of communication. Thus, we have the concept of a ligand (vitamin), an extraorganismic molecule, reacting with a specific protein entity, the receptor, facilitating absorption and transport of the ligand to its site of action with the resultant sequence of physiological events to which we refer as "vitamin function." The term "receptor," in the context of vitamins, refers to proteins with a recognition function, not to receptors in the traditional pharmacological sense. It includes proteins that bind to the vitamin ligand and function as extracellular or intracellular transporters as well as the smaller number of vitamin-binding nuclear proteins. In the latter case, vitamins might act as small molecular modulators of protein–DNA interaction, a situation analogous to that of some hormones. The designation "receptors" for these categories of vitamin-binding proteins appears to be appropriate inasmuch as it connotes a function more than ligand binding.

The transcriptional patterns of selected genes are often altered by external signals that the cell receives. Thus, changes in nutritional supplies can either induce or repress the transcription of specific genes or gene sets. In many cases the regulation of transcription (up or down) is mediated by alterations in the availability of essential transcription factors or by change in the factors' ability to bind to specific regulatory sequence motifs.

1 INTRODUCTION

Vitamins are required by all protozoans and metazoans that have been studied. They do not conform to any single chemical classification. Some vitamins, such as retinol (vitamin A), cholecalciferol (vitamin D), α-tocopherol (vitamin E) and phylloquinone (vitamin K), are lipid soluble: others, such as thiamine (vitamin B_1), riboflavin (vitamin B_2), niacin, pyridoxine (vitamin B_6), panthothenic acid, lipoic acid, biotin, folic acid, cyanocobalamin (vitamin B_{12}), and ascorbic acid (vitamin C), are water soluble. While the unlimited availability of vitamins may have permitted the evolutionary loss of the ability of animals to synthesize vitamins, the ability to absorb vitamins from the environment or from digested food is essential. The modes of uptake and intra- and intercellular transport of these two broad categories of vitamins are different, based on their solubility characteristics. Lipoprotein binders are implicated in the transport of the lipid-soluble vitamins. Investigators recognized early, among the water-soluble vitamins, the bindings of cyanocobalamin and folate, respectively, to specific proteins. Increasing evidence has accumulated to indicate that the absorption of many water-soluble vitamins in the gastrointestinal tract is through saturable mechanisms. In addition, the concentrations of many water-soluble vitamins are much higher in the brain than in plasma. This is true of placental transport as well, implicating specific transporters for individual vitamins across the blood–brain and placental barriers, respectively.

Absorption of vitamins by most cells occurs through membrane-associated transport systems, which are present in low concentrations. These receptor-like proteins may consist of an extracellular vitamin-binding domain linked to a transmembrane anchor. The fertilized egg and embryo obtain vitamins from the external environment using cell surface vitamin receptor/transporter proteins. The vitamins are used directly without storage. In contrast, the eggs of birds and reptiles contain both the vitamins and their respective vitamin-binding proteins deposited in the yolk for transport to the oocyte.

The physiological effects of vitamins follow their entry into the cell, and subsequent metabolic transformation and association with specific apoenzymes, resulting in accentuation of a metabolic pathway. In this respect vitamins are metabolic initiators. Where the vitamin function is other than as a coenzyme or prosthetic group of an enzyme, its effect is through regulation of cellular protein synthesis. The mechanisms of action of lipid-soluble vitamins are similar to those of the steroid hormones. Water-soluble vitamins such as biotin, ascorbic acid, and vitamin B_6 also seem to have a modulatory influence on cellular protein synthesis acting through similar mechanisms.

2 VITAMIN A

2.1 Absorption and Transport

Vitamin A, an essential nutrient in the diet, plays a significant role in diverse physiological processes such as development, differentiation, and vision. Its role in embryogenesis is evident from certain known effects of a deficiency or excess of vitamin A, namely, malformation or teratogenicity. Vitamin A deficiency has profound effects on the immune system, and retinoids seem to affect differentiation, the stimulation of phagocytosis, and the modulation of cytokines and eicosanoids. Several clinical studies have demonstrated beneficial effects of retinoids on skin diseases such as acne, psoriasis, keratodermas, and skin cancers. Retinoids play a role in the development of the central body axis and limbs of vertebrates. High levels of endogenous retinoids are found in proximity to the developing axis in a variety of species. They exert their wide spectrum of biological functions through their ability to regulate gene expression at specific targets. The term "retinoid" includes the natural forms of vitamin A as well as synthetic analogues, which may or may not have the biological activity of vitamin A. The natural forms of vitamin A include retinol, retinal, and retinoic acid.

Ingested retinyl esters are hydrolyzed to retinol by the lipases in the intestinal lumen. Retinoid production from carotenoids occurs by oxidative or eccentric cleavage of the central double bond of carotenoids. The transport of retinol across the brush border of enterocytes is the first step in the absorption of vitamin A into the organism. A carrier-mediated mechanism specific for all-*trans*- and 3-dehydroretinol with passive diffusion of the 13-*cis* compound and retinal has been suggested.

In the intestinal mucosal cell, free retinol is esterified and incorporated into chylomicrons for transport. Chylomicron remnants are primarily cleared by liver, although extrahepatic uptake occurs in other tissues such as bone marrow and spleen. In the liver, retinyl esters from chylomicron remnants are taken up by the parenchymal cells, hydrolyzed to retinol, and bound to retinol-binding protein (RBP). This series of events is followed by paracrine transfer of retinol from hepatocytes to the perisinusoidal stellate cells for storage. Retinol is mobilized from the stellate cells bound to RBP and released into circulation. Retinol–RBP complex binds to transthyretin in the plasma compartment and thus, thyroid hormone and RBP–retinol complex are transported together. Retinol reaches target cells bound to the RBP transport protein. Transport of retinol across cell membrane follows receptor-mediated endocytosis, internalization, degradation of RBP, and liberation of retinol within the cells.

2.2 Retinoid Receptors: Cellular Binding Proteins

Retinol is oxidized to retinoic acid (RA) and other compounds such as 9-*cis* RA. Within the cell, retinol and RA, respectively, bind to ligand-specific cellular binding proteins: cellular retinol binding protein (CRBP) and cellular retinoic acid binding protein (CRABP). These proteins belong to the superfamily of fatty acid binding proteins. CRABP-I, CRABP-II, and CRBP have been cloned from human and rat sources. CRABP-I and CRABP-II differ in their tissue distribution and affinity for RA. CRABP-II, the major form expressed in human epidermis, is up-regulated by RA during differentiation. The promoter of the CRABP-II gene has been shown to have an RA response element. Retinoid translocation into target cells is regulated by binding to specific high affinity cellular receptors.

2.3 Nuclear Retinoic Acid Receptors

Almost all the biological activity of vitamin A is due to the ability of its metabolite RA (all-*trans* RA, or atRA) to regulate gene expression, a receptor-mediated event. By analogy with lipid-soluble hormones, the RA receptor is a nuclear-acting transcription factor. These receptors belong to a highly conserved superfamily of proteins with similar structural and functional domains.

Following the isolation of the human RA receptor (hRARα), it was shown that its DNA-binding and ligand-binding domain was highly conserved with other members of the nuclear receptor family. There are three subtypes of RAR (RAR_α, RAP_β, RAR_γ) in various species, and their ligand-binding domains are highly conserved. A second class of retinoid-responsive transcription factor—the human retinoid X receptor ($hRXR_\alpha$), isolated later—had a sequence dissimilar to that of RAR, although both had the ability to respond to RA. RAR and RXR significantly differ in terms of their binding to tritiated all-*trans* RA, leading to the suggestion that a metabolite of RA might be the ligand for RXR. An isomer of RA, 9-*cis*-retinoic acid (9cRA) was identified as this RA metabolite, which was 40 times more potent than RA in transacting RXR. It is also active with RAR. RXRs bind and respond to 9cRA, and they are able to activate target organs in the absence of RAR gene products.

The presence of two receptor systems suggests that target genes responsive to each receptor system are distinct. RA seems to modulate positively the expressions of CRBP and CRABP. CRBP binds retinol with high affinity and is involved in its esterification to retinyl esters. Similarly, CRABP binds to RA and seems to be involved in the oxidative inactivation of RA. An additional regulatory step is the isomerization of RA to 9cRA.

Gene expression is modulated by members of the nuclear receptor superfamily by binding as homo- or heterodimers to specific DNA sequences, the hormone response elements (HREs). Sequence analysis of the response elements activated by RARs and RXRs indicates that similar to the HREs for other nonsteroid members of the nuclear receptor family, the retinoid response elements consist of a minimal half-site consensus sequence (AGGT-CA) configured into a variety of structured motifs, including direct repeats and palindromes. Unlike the heterogeneity of RAREs, RXREs are exclusively of the DR-1 type (a direct repeat separated by one nucleotide). Most of the RXREs discovered seem to occur in promoters of genes involved in vitamin A and lipid metabolism, suggesting a role for RXRs in several distinct but coregulated metabolic pathways. RXR has a role both as a ligand-dependent transcription factor (as homodimer) and as a partner (as heterodimer) to several nuclear receptors. It is interesting that 9cRA, which promotes RXR homodimer formation, is derived from RA, which activates only the RAR heterodimer. RAR-RXR heterodimer can repress RXR homodimer function. Thus under limiting concentration of RXR, the target genes for RAR will be preferentially activated. When the concentration of 9cRA is high, the RXR homodimer is stabilized. Also, the RAR-RXR heterodimer has a higher affinity for DNA than the RXR homodimer and might compete more favorably for binding to RXREs. A combinatorial regulation of receptor expression and ligand availability seems to increase the diversity of the signaling pathways. Experiments using knockout mice and cultured cells indicate that each subtype of RAR and RXR has a distinct effect on target gene expression. The cellular retinoid-binding proteins also play a significant role in the retinoid signaling pathway by regulating the intracellular transport and metabolism of retinoids.

3 VITAMIN D

3.1 General

Vitamin D is grouped along with vitamins A, E, and K as one of the fat-soluble vitamins; it is formed in the skin of mammals by ultraviolet irradiation, upon conversion of the precursor 7-dehydrocholesterol to cholecalciferol (vitamin D_3). Its physiological function is related to whole-body calcium homeostasis, inasmuch as a deficiency of this vitamin leads to rickets in children and osteomalacia in adults. Ultraviolet irradiation prevents or heals bone disorders induced by a deficiency.

In the vitamin-deficient state the absorptive capacity of small intestine for calcium is decreased. One of the physiological functions of vitamin D is the maintenance of a constant extracellular concentration of calcium acting in concert with parathyroid hormone and possibly, calcitonin, thus preventing hypo- or hypercalcemia. Cholecalciferol is converted to 1,25-dihydroxycholecalciferol, the biologically active form of the vitamin. This is a two-step process. The product of the initial hydroxylation, which takes place in the liver, is 25-hydroxycholecalciferol (25[OH]D_3). This is secreted into blood, where it circulates bound to a specific vitamin D binding globulin. The final step takes place in the kidney through a 1-hydroxylation reaction, and the product 1,25-dihydroxycholecalciferol (1,25[OH]$_2$$D_3$) is the active form of vitamin D. In addition to the role in calcium homeostasis, vitamin D is now recognized to be involved in regulating growth and differentiation of a variety of cell types.

3.2 Vitamin D Receptors

Most of the biological actions of vitamin D occur through a 1,25(OH)$_2$$D_3$-mediated regulation of the transcription of specific genes in target tissues. This active form of vitamin D binds to a specific nuclear receptor (VDR), which belongs to the superfamily of steroid, thyroid hormone (TR), and vitamin A (RAR) receptors. Ligand binding is necessary for the high affinity binding of VDR to specific vitamin D DNA response elements (VDREs) located in the promoter region of the target genes. This is similar to the binding of TR and RAR to their cognate receptors, since TR, RAR, and VDR belong to the same subfamily of receptors. Interaction between the ligand–receptor complex with VDREs modulates the transcription of genes specific for the biological action of vitamin D.

Osteoporosis is an age-related loss of bone mineral. Bone mineral mass is affected by both genetic and nongenetic factors. A correlation between alleles of VDR and bone density has been identified. In bone, 1,25(OH)$_2$$D_3$ recruits bone-resorbing osteoclasts from hematopoietic precursor cells. It also modulates, through VDR, the bone-forming osteoblasts. Restriction fragment length polymorphism analysis indicates the presence of two alleles of VDR. In a study of 250 Caucasian twins and a second study of 311 unrelated healthy women, it was shown that common allelic variants of the VDR gene can be used to predict differences in bone mineral density, hence susceptibility to osteoporosis. The magnitude of the VDR allele difference between the two genotypes could mean a 10-year difference for postmenopausal women in the onset of vertebrate fractures.

3.3 Vitamin D Responsive Genes

Vitamin D acting through 1,25(OH)$_2$$D_3$ enhances intestinal calcium absorption by stimulating transcellular calcium transport. The stimulation is achieved by both nongenomic and genomic actions of vitamin D. One such genomic action is the induction of the calcium-binding protein calbindin-D, which could facilitate the diffusion of calcium from the brush border to the basolateral membrane of the enterocyte. Other vitamin D responsive calcium-binding proteins such as Ca^{2+}-ATPase and the integral membrane calcium-binding protein have been characterized. A list of 51 gene products that are reported to be 1,2(OH)D_3 sensitive in terms of altered mRNA levels has been assembled. This list includes genes associated in the mineral homeostasis, vitamin D metabolism, cell differentiation and proliferation, bone matrix protein, extracellular matrix proteins, oncogenes, chromosomal proteins, growth factors, signal transduction proteins, peptide hormones, and energy metabolism. In clonal rat pituitary tumor (GH) cells in culture, 1,25(OH)$_2$$D_3$ selectively increased prolactin production, prolactin mRNA, and agonist-induced prolactin release. In primary cultures of rat anterior pituitary cells, 1,25(OH)$_2$$D_3$ selectively enhanced TRH-induced thyrotropin (TSH) release. In studies using primary dispersed and clonal pituitary cells, 1,25(OH)$_2$$D_3$ has been shown to stimulate pituitary hormone production and agonist-induced hormone release. There was an upregulation of thyrotropin-releasing hormone (TRH)-binding sites as shown by an increase in TRH receptor specific mRNA. Incubation of pituitary GH4C1 cells with vitamin D markedly reduces the response of rat growth hormone (GH) mRNA to thyroid hormone (T3) and retinoic acid. Rat GH gene expression is stimulated by T3. The 5′-flanking region of rat GH gene contains a TRE that mediates T3 stimulation. RA also increases the transcription of thc GH gene. The RA response was mapped to the same sequence that mediated regulation by T3, indicating that TRE also functioned as RARE. Vitamin D inhibited the T3 and RA transactivation of the rat GH gene by interference with the common response element present in the 5′-flanking region of the gene. Vitamin D also inhibited activation of other T3- and RA-responsive promoters. The formation of TR-VDR and RAR-VDR heterodimers, which have low affinities for these response elements, can inhibit binding of more active heterodimers with RXR. These results highlight the repressive role of VDR and indicate that the biological response to T3 and RA could be selectively inhibited in cells that contain receptors for vitamin D.

4 VITAMIN E

4.1 General

"Vitamin E" is the generic term for two groups of naturally occurring fat-soluble compounds, tocopherols and tocotrienols. The members of each group are designated α, β, γ, or δ, depending on the numbers and position of the methyl groups attached to the chroman nucleus. The most biologically active of these compounds is α-tocopherol, which functions as an antioxidant in membranes to prevent lipid peroxidation. The intake of γ-tocopherol is two to four times that of α-tocopherol because of its prevalence in many diets. Both α- and γ-tocopherol are absorbed equally by the intestine and transferred to the bloodstream by chylomicrons, which are taken up by the liver. However, α-tocopherol is present in human plasma at a concentration many times higher than that of γ-tocopherol, suggesting a mechanism of discrimination in the liver. This accounts for the generally recognized lower biopotency of γ-tocopherol.

4.2 Transport of α-Tocopherol

There is a similarity between the uptakes of tocopherol and cholesterol. Tocopherol exchanges between different lipoproteins and between lipoproteins and erythrocytes. Intravenous administration of chylomicrons or very low density lipoprotein (VLDL) containing radioactive tocopherol results in its rapid distribution among all the plasma lipoproteins, red blood cells, and tissues. In normal human subjects, plasma low density lipoprotein (LDL) fraction, which contains the largest amount of total lipid, also contains the largest amount of tocopherol. Following oral administration of radioactive α-tocopherol, the absorbed α-tocopherol was associated with the lipoprotein fraction of the intestinal mucosa cytosol. This lipoprotein is similar to the VLDL of serum.

4.3 Binding of α-Tocopherol to Erythrocyte Membranes

α-Tocopherol has been shown to decrease the hemolytic damage to erythrocytes by oxidative stress. Since the disruption of erythrocyte membrane is the likely locus of oxidative stress, and since α-tocopherol has been proposed to have a role in membrane structure and function, it is possible that α-tocopherol binding proteins or receptors are present on these membranes. Red blood cell membranes have specific binding sites for α-tocopherol. These binding sites have the properties expected of membrane receptors (i.e., specificity; saturability; moderate affinity; time-, protein-, and temperature-dependence, reversibility). The binding of α-tocopherol to the membrane is highly specific, inasmuch as only γ-tocopherol competes with α-tocopherol for binding to the membrane.

4.4 Liver Cytosolic α-Tocopherol-Binding Proteins

A rat liver cytosolic protein that specifically binds to α-tocopherol was indicated to function in the intracellular transport of α-tocopherol between membranes. This α-tocopherol transfer protein [α-TTP] has been purified and characterized and shown to consist of two closely similar isoforms of molecular mass 30,500 kDa, with high specificity for binding to the α isomer of tocopherol. This protein is present in rat liver and absent in other rat tissues such as heart, spleen, testes, and lung. The α-TTP gene was sequenced and shown to have homology with bovine and human 11-*cis*-retinaldehyde-binding protein of the retina. These proteins do not belong to the family of cellular retinoid-binding proteins (CRBP, CRABP) or the fatty acid binding proteins. It has been suggested that the liver α-TTP, in view of its specificity for the α isomer, discriminates against the γ-tocopherol that is excreted into the bile. The α-tocopherol is transferred to VLDL. A familial vitamin E deficiency, which results in spinocerebellar lesions with progressive development of ataxia and areflexia, has been described. Vitamin E is almost absent in the plasma of these patients, although lipid absorption, pancreatic, and intestinal functions, as well as lipoprotein transport, were normal. The only abnormality was in the transfer of α-tocopherol into VLDL, with the suggestion of an abnormality in the α-TTP of these patients.

4.5 Nonantioxidant Effects of Vitamin E

Among the factors that retard the development of arteriosclerosis is the ingestion of the antioxidant vitamin α-tocopherol, which scavenges peroxide radicals in the hydrophobic phase of cellular membranes. Various observations have indicated additional mechanisms of action. The plasma concentration of α-tocopherol is regulated by the liver cytosolic α-tocopherol transfer protein, indicating that α-tocopherol is more important than other antioxidants for specific cellular functions. The highly tissue-specific pathology associated with vitamin E deficiency would suggest a tissue target of action rather than a general antioxidant function. Recent observations indicate that α-tocopherol acts as a negative regulator of smooth muscle proliferation. The mechanism of action was studied in smooth muscle cells of rat (line A7r5) and human aorta. α-Tocopherol and not β-tocopherol inhibited smooth muscle cell proliferation in a dose-dependent manner. The results indicate the possible existence of a tocopherol-binding protein in smooth muscle cells. This protein would bind both α- and β-tocopherol at the same site.

α-Tocopherol but not the β isomer is able to induce a modification of the binding protein at the beginning of the cascade of cellular events such as modulation of the transcription factor AP-1, protein kinase C inhibition, and growth inhibition. α-Tocopherol may induce or activate a protein phosphatase that is capable of dephosphorylating and inhibiting the action of protein kinase C. The following sequence of events has been suggested in the inhibition of smooth muscle cell proliferation: binding of α-tocopherol to a receptor protein, activation of AP-1, activation (or expression) of a protein phosphatase, dephosphorylation, and inhibition of protein kinase C, resulting in the inhibition of proliferation. The functional significance of the effect of vitamin E on smooth muscle cell proliferation is highlighted by epidemiological studies indicating that a 25% decrease of plasma α-tocopherol is associated with a fourfold increase in the frequency of ischemic heart diseases.

5 ASCORBIC ACID

5.1 General

The antioxidant activity of ascorbic acid in all living organisms is the best-recognized function of this vitamin. Vitamin C and vitamin E eliminate cytotoxic free radicals by redox cycling. Ascorbate and vitamin E function at physiological pH as donors of single hydrogen atoms, enabling these vitamins to react efficiently with free radicals and not with molecular oxygen. Ascorbic acid is a cofactor in the synthesis of collagen and proteoglycans, the two major constituents of the extracellular matrix. The synthesis of collagen by fibroblasts involves enzymes that hydroxylate lysine and proline in the procollagen molecule. Ascorbate has also been shown to be a cofactor of the enzyme peptidylglycine α-aminating monooxygenase, which amidates neuroendocrine peptides. The concentration of ascorbate in brain and endocrine glands is high. In addition, the presence of a complex system to regulate its intracellular and intravesicular availability indicates a critical role for ascorbate in these structures. Ascorbate is released from brain and endocrine tissues and can affect cellular functions by acting on plasma membrane receptors.

5.2 Ascorbic Acid Transport and Transporters

Most animals, exception primates and guinea pig, synthesize ascorbate from glucose in the liver. Hence, intestinal transport of ascorbate has special significance in primates and the guinea pig as the source of ascorbic acid. Vitamin C (L-dihydroascorbic acid) is a stereospecific compound. The L stereoisomer is the biologically active compound: the D form has only one-twentieth its biological activity. Once inside the cell, the D epimer (D-isoascorbic acid) is as efficient as the L form, at least in one aspect of its biological action, that of collagen synthesis. Thus, the cell membrane is the primary site of stereoselectivity. The carrier proteins in the cell membrane are stereospecific in terms of binding and transporting the vitamin across the cell membrane to its intracellular site of action.

Dietary vitamin C is a mixture of ascorbate and dehydroascorbic acid. Both compounds are antiscorbutic. Dehydroascorbic acid, the oxidized form, is not an acid, does not ionize, and is more hydrophobic than ascorbic acid. A net transepithelial absorption and transport of this form was shown in everted and noneverted loops of guinea pig intestine. The vitamer reaching the serosal side is in the reduced state. Such simultaneous absorption and reduction is characteristic of species that do not synthesize ascorbic acid. On the contrary, there is no net absorption of dehydroascorbic acid in the intestine of the rat, a species capable of synthesizing ascorbic acid in the liver.

In contrast to the intestinal transport of ascorbate by various mammalian species, renal handling of this substance seems to be very similar. Ascorbate is filtered in the glomerulus and is reabsorbed in the proximal convoluted tubule. The uptake was saturable, inhibited by D-iso-ascorbate and dehydroascorbic acid, temperature dependent, and susceptible to metabolic inhibitors. Ascorbate is reabsorbed by the kidney in a sodium-dependent active process not shared by other acid anions.

When incubated in a medium containing physiological amounts of ascorbic acid, human neutrophils accumulate millimolar amounts using two transporters. The high affinity transporter with an apparent Michaelis constant K_m of 2–5 μM is saturated at normal extracellular ascorbic acid concentrations. The low affinity transporter has an apparent K_m of 6–7 mM and is in the linear phase of uptake at normal extracellular concentrations of ascorbic acid. The intracellular concentration of ascorbic acid in neutrophils, which is dependent on the extracellular concentration, functions to preserve neutrophil integrity. It also protects host tissue by inactivating free radicals and oxidants produced by the phagolysosomes.

Ascorbic acid is essential for the formation of bone by osteoblasts. In vivo studies indicate that administered ^{14}C-labeled ascorbic acid accumulates preferentially at sites of active osteogenesis. In vitro studies have shown that ascorbic acid is essential for the formation of mineralized extracellular matrix by rabbit long bone cells and rat calvaria cells. Two cell systems, a clonal osteoblast-like cell line ROS 17/2.8 and a primary culture of rat calvaria cells, have been used as in vitro osteoblast models. Both cell cultures accumulate ascorbate. Consistent with a carrier-mediated mechanism, transport was saturable, temperature dependent, stereoselective, and dependent on external Na^+ concentration. Ascorbate taken up is rapidly compartmentalized in the lumen of the rough endoplasmic reticulum and is bound or metabolized to a nonpermeant species, in keeping with its putative role as a cofactor in the posttranslational modification of procollagen. Using the clonal osteosarcoma cell line UMR-106, it was shown that transforming growth factor β (TGFB), an important modulator of cell proliferation and differentiation, stimulates carrier-mediated ascorbate transport. Stimulation of ascorbate uptake was seen at TGFB concentrations similar to that required for stimulating the production of collagenous extracellular matrix. The enhancement by TGFB of ascorbate uptake was associated with an increase in the apparent V_{max} of the rate of transport without any change in the affinity of the transporter for ascorbic acid, indicating that either the number of functional transporters or the rate at which individual transporters translocate ascorbate is increased.

The effect of TGFB is completely blocked by cycloheximide. This would suggest that TGFB increases the rate of synthesis of the Na^+ ascorbate cotransporter or a regulatory protein that interacts with the existing transporters to increase their turnover number.

Ascorbic acid is transported into 3T6 fibroblasts, which synthesize collagen, by a carrier-mediated saturable active Na^+ cotransporter process that appears to be the mechanism of transport in most cell types. It was observed that serum from various animals contained a heat-labile factor that interacted with bacterial endotoxin and inhibited ascorbate transport. The inhibitor may have been generated during endotoxin-induced activation of complement in unheated serum. The inhibitor has been purified from insulin-activated human serum. This heat-stable serum protein factor, which is produced when serum complement is activated, has a molecular mass of 9000 D. The reactivity of the purified inhibitor with antiserum to C3a suggests that the inhibitor is C3a or a very similar molecule. The inhibitor of ascorbate transport increased the apparent K_m for ascorbate without affecting the V_{max}. In view of the similarity of ascorbate transport in many cell types, it is possible that this protein can inhibit ascorbate transport in other cell types as well, resulting in depletion of tissue ascorbate during infection or autoimmune process when serum complement is activated.

Following intracardiac injection of [^{14}C]ascorbic acid into guinea pigs, the neurohypophysis takes up the label to a high concentration (exceeded only by the adrenals). The uptake of [^{14}C]ascorbic acid by isolated nerve endings from ox neurohypophysis was demonstrated to be a saturable process, suggesting the presence of both glucose-dependent and sodium-dependent uptakes. Cortisol and triiodothyronine inhibited the uptake. Posttranslational processing of vasopressin and oxytocin precursors involves a trypsinlike enzyme, a carboxypeptidase, and an amidating enzyme [peptidylglycyl-α-amino monooxygenase (PAM)]. PAM, which requires copper, molecular oxygen, and ascorbic acid, may be limiting in the processing of amidated peptides. In a study of pancreatic islet cells, in which several peptides (e.g., pancreastatin, thyrotropin-releasing hormone, pancreatic polypeptide, amylin) are

C-terminally amidated, the presence of PAM activity has been reported. Research has demonstrated the presence of PAM, a high content of ascorbic acid, as well as a carrier-mediated process for its uptake in the islet cells. As in neurohypophyseal nerve terminals, ascorbate uptake was inhibited by triiodothyronine and glucocorticoids.

5.3 Ascorbic Acid and Collagen Production

Ascorbate modulates collagen production through its effect on prolyl hydroxylation. It has been reported that ascorbate may have an additional role. After prolonged exposure to ascorbate, collagen synthesis in cultured human fibroblasts increased eightfold with no significant change in the synthesis of noncollagen protein. This effect of ascorbate was unrelated to its cofactor function in collagen hydroxylations. Thus perhaps there is a shift in the distribution of collagen-synthesizing ribosomes toward polysome in the presence of ascorbate. In further work it was shown that ascorbic acid can stimulate collagen production with no apparent change in the intracellular degradation of newly synthesized procollagen in cultured skin fibroblasts. It was found also that the steady state levels of type I (which accounts for $\approx$85% of the collagenous protein produced by these cells) and type III procollagen mRNAs are increased in cells treated with ascorbic acid. A three- to fourfold increase in collagen synthesis was associated with a two- to threefold increase in the levels of mRNAs for both type I and type III procollagens. A translational control linked either to procollagen gene transcription or mRNA degradation has been suggested.

Human foreskin fibroblasts have been cultured for up to 6 weeks in a medium supplemented with ascorbic acid (50 μg/mL). The cells produced a new connective tissue matrix with properties, as seen in vivo, which included complete processing of procollagen to collagen α-chains, covalent cross-linking of collagen, and assembly of collagen fibrils into bundles. The high degree of molecular and topographical differentiation, a unique feature of long-term cultures, was not seen in short-term cultures or in long-term cultures in the absence of ascorbate.

An ascorbate-binding protein has not yet been isolated, although all evidence indicates its presence. The roles of ascorbic acid in prolyl hydroxylation in collagen synthesis, in dopamine hydroxylation, in norepinephrine synthesis, and in amidation of peptides are well understood. However, there are indications that ascorbate may be involved in the synthesis of the procollagens acting possibly at the translation step as well as in the macromolecular assembly—for example, in the fabrication of connective tissue matrix. Cellular ascorbate has to interact with organelles involved in such assembly and the mechanism of action of ascorbate will be an area of future study.

6 BIOTIN

6.1 Noncovalent Biotin-Binding Proteins

Biotin in most biological materials is covalently bound to a lysine residue of proteins (carboxylases). The noncovalent binding of biotin to various proteins seems to be crucial for the uptake, transport, and intracellular functions other than the prosthetic group functions of the vitamin.

A group of proteins bind to biotin noncovalently. Both avidin (the biotin-binding protein of rat egg white) and strepavidin (the extracellular biotin-binding protein of *Streptomyces avidinii*) have exceedingly high affinities for biotin, with a dissociation constant K_d of 10^{-15} M. These are the strongest noncovalent bindings known between a protein and a small molecular weight ligand. Other biotin-binding proteins (e.g., the egg yolk biotin-binding proteins, biotinidase, biotin-holocarboxylase synthetase) have progressively lower affinities for biotin. Biotin-binding proteins vary considerably along the evolutionary scale. In *E. coli,* the biotin-binding protein biotin-holocarboxylase synthetase has been shown to regulate the synthesis of the enzymes of biotin biosynthesis pathway.

In the laying hen there are three biotin-binding proteins. In addition to avidin, which is present in albumen of egg, there are two biotin-binding proteins in egg yolk. The presence of distinct biotin-binding proteins in the yolk suggests a specialized role for biotin in the regulation of the development of the embryo. In mammals only two proteins have so far been shown to bind noncovalently to biotin.

6.1.1 Avidins

Avidin has been isolated from hen egg white and the oviduct of the avian species. It is a basic glycoprotein with a molecular mass of 67 kDa and an isoelectric pH of 10.5. It is a tetramer of four identical subunits, each with 128 amino acid residues and one biotin-binding site. The composition and structural arrangement of the carbohydrate moiety of avidin has been shown to be heterogeneous. Strepavidin is a nonglycosylated neutral protein secreted by *Streptomyces avidinii.* Its binding to biotin is remarkably similar to that of avidin, with a dissociation constant of 4×10^{-14} M. Both avidin and strepavidin are tetrameric, with each subunit containing one biotin-binding site. However, they lack immunological cross-reactivity and evolutionary relatedness. Avidin is a secretory protein produced by the oviduct and concentrated in the egg white. It is induced by progesterone. In addition to being an oviductal secretory protein, avidin is synthesized in areas of tissue damage and inflammation caused by diverse agents such as toxic doses of actinomycin D, septic infection, or thermal injury in chicken. This progesterone-independent synthesis of avidin is inhibited by anti-inflammatory steroids.

The avidin–biotin system has been used in a wide range of biological studies including isolation, localization, cytochemistry, immunochemistry, and diagnostics, as well as for probing genes. Since there is some nonspecific adsorption of avidin, strepavidin generally is used instead. However, strepavidin has been shown to bind with high affinity to cell surfaces. Strepavidin itself can displace fibronectin from cell surfaces. The RYD (Arg-Tyr-Asp) sequence of strepavidin mimics the RGD (Arg-Gly-Asp) sequence, which is a universal recognition domain for adhesion receptors, as well as for the receptors for fibronectin, fibrinogen, and other adhesive proteins.

6.1.2 Nonavidin Biotin-Binding Proteins

In addition to avidin and strepavidin, the two examples of proteins with exceptionally strong affinity for biotin, other biotin-binding proteins have lower affinities for biotin. These include the egg yolk biotin-binding proteins: biotin-holocarboxylase synthetase and biotinidase.

Biotin-Holocarboxylase Synthetase Biotin carboxylases are synthesized in the form of the apoproteins that undergo posttranslational covalent modification by the addition of biotin, the prosthetic group, to the ε-amino group of lysine of the apoprotein. This covalent attachment of biotin to a specific lysine in the apoenzyme is catalyzed by biotin-holocarboxylase synthetase in a two-step reaction. The synthetase catalyzes the biotination of apocarboxylase from heterologous sources.

The biotin locus in *E. coli* was shown to be located in the same region as *gal* and *att,* and the order was determined to be *gal, att, bio.* The biotin locus is composed of five consecutive genes in the order *A B F C D* (representing the genes coding for various enzymes of biotin biosynthesis) and three unlinked genes *bioH, bioR* (a regulatory gene coding for the biotin repressor protein), and *bioP* (a permeability gene). Transcription of this cluster of genes is from two "face-to-face," partially overlapping promoters. One transcript includes the *BioB F C D* genes, where as the second transcript includes *bioA* gene, and an open reading frame of unknown function. Transcription from both promoters was blocked in vivo upon addition of high levels of biotin to the medium of *E. coli* cultures. Starvation of *E. coli* biotin auxotrophs resulted in increased transcription. In the biotin operon the repressor is the biotin–protein ligase and the corepressor is biotinyl AMP. Maximal rates of biotin operon transcription are associated with conditions of biotin limitation.

Biotinidase Acid hydrolysis of biotin proteins releases free biotin, whereas proteolytic hydrolysis yields biotin peptides. The smallest among these is biocytin, ε-*N*-biotinyl-L-lysine. The name biotinidase (biotin-amide amidohydrolase, EC 3.5.1.12) was proposed for this enzyme, which also hydrolyzes biocytin. The rat liver enzyme is associated with membranes, although not an integral part of the membrane structure. The enzymes extracted from diverse sources such as the rat liver and human serum are similar in many of their properties (e.g., optimal pH and K_m for substrate, acidic nature, inhibition by chaotropic agents and sulfhydryl agents). The major difference between these two proteins is in the heat stability of the rat liver enzyme. It appears that the liver is the site of synthesis of the serum enzyme. Human serum biotinidase has been purified to homogeneity. The enzyme is a glycoprotein monomer with a molecular mass of 68 kDa. It contains sialic acid residues. However, these are not required for enzyme activity. Although the predicted amino acid sequence from the isolated biotinidase clone did not share extensive homology with avidin and strepavidin, there is conservation of sequence around tryptophan residues that has been shown to be critical for biotin binding. It is possible that the hydrophobic binding of biotin to biotinidase and egg yolk biotin-binding proteins requires tryptophan residues in the active center.

Egg Yolk Biotin-Binding Proteins The proteins of egg yolk differ from those of egg white in regard to time and place of synthesis. The entire egg white is secreted by the magnum of the oviduct within 4–6 hours of ovulation. The yolk proteins, in contrast, are serum proteins synthesized by the liver and contain high affinity binding proteins for riboflavin, thiamine, cholecalciferol, retinol, vitamin B_{12}, and iron. Vitamins and minerals are transported protein bound to the oocyte. The transport proteins such as the riboflavin and thiamine-binding proteins and transferrin present in both egg yolk and egg white are products of the respective single genes.

The exception seems to be the biotin-binding proteins. A biotin-binding protein (BBP-I) isolated from hen egg yolk was shown to be distinct from avidin. Avidin binds to biotin very tightly and does not exchange, whereas the yolk biotin-binding protein exchanges biotin at the body temperature of the bird. Thus, biotin could be made available to the developing embryo from its reserve in the yolk. Only about 15% of the biotin of the whole egg is associated with avidin, under normal conditions of biotin nutrition of the bird. The rest is associated with the yolk BBP. The plasma of laying hens and egg yolk contains another biotin-binding protein (BBP-II), which is stable to 45°C, whereas BBP-I was stable to 65°C. Both are tetrameric isoproteins, normally saturated with biotin, and together they account for most of the biotin of hen plasma and yolk. Yolk BBPs have a nutritional role, that of transporting biotin to the developing embryo. BBP-II is preferentially deposited in the yolk and is specifically involved in the transport of biotin to the oocyte. Both BBP-I and BBP-II cross-react with anti-BBP-II serum. The maximal production of BBP-I is attained at a lower level of dietary biotin (≈50 μg/kg) than for BBP-II (≈250 μg/kg). At this higher level of dietary biotin, the production of BBP-II is severalfold higher than that of BBP-I. The observation that the amount of these proteins is limited by dietary biotin within the normal dietary range would suggest that biotin is required for their synthesis, stability, or secretion. The synthesis of both BBP-I and BBP-II is stimulated during egg laying, indicating that the gene(s) for these two proteins are induced by estrogen. The difference in their response to dietary biotin content as well as in their distribution between hen and chick plasma would suggest that they are either products of different genes or differentially modified products of the same gene. Both estrogen and the specific ligand for BBP-I and BBP-II are involved in the regulation of their synthesis, although the mechanisms of action of biotin in this sequence is not known yet.

A biotin-binding protein analogous to BBP-II has been characterized in the serum of the pregnant or estrogenized female rat. This protein is not present in the sera of normal male rats. The molecular mass of this protein is 66 kDa and the isoelectric pH is 4.1. It exhibits immunological cross-reactivity with the purified egg yolk biotin-binding protein, BBP-II. The biological role of this protein is not known. The physiological relevance of the biotin-binding protein during rodent embryonic development was demonstrated by selective immunoneutralization of the maternal vitamin carrier, an intervention that led to a drastic decrease in the transport of biotin to the embryo, which in turn resulted in early embryonic mortality—without affecting maternal biotin nutrition.

6.2 Role of Biotinidase as a Biotin Carrier Protein

Based on investigations with HeLa cells and human fibroblasts, we suggested that biotin transport in mammalian cells may involve pinocytosis through the functioning of specific circulating proteins. Both biotinidase and biotin-binding activities coincided during serum fractionation, suggesting that biotinidase is the only protein that binds biotin in human serum and might be the biotin carrier protein in human serum. We determined the binding of biotin to pure human serum biotinidase. The Scatchard plot as well as displacement curve analysis showed two linear parts, indicating that two classes of sites with high and low affinities for biotin (0.5 and 50 nM, respectively), were present. The high capacity, low affinity binding corresponds to that of other vitamin-binding proteins.

The decreased level of plasma biotin in biotinidase-deficient patients was tentatively attributed to decreased levels of biotinidase or a specific biotin-binding protein in plasma. Other indirect evidence for the existence of a biotin-binding protein comes from the study of epileptic patients on long-term therapy with anticonvulsants. Significantly low levels of serum biotin were observed in these patients, suggesting that these anticonvulsants probably compete with biotin for binding to a circulatory protein. These anticonvulsants share with biotin the common structural feature of a cyclic carbamide group. Protein binding involves this group of biotin and the anticonvulsants. We have shown that all these anticonvulsants compete with biotin for the binding to biotinidase, indicating that biotin in human serum is essentially noncovalently bound to protein(s) and that biotinidase is the only protein that exchanges with $[^3H]$*d*-biotin.

The uptake of biotin by various human cell lines was reported by us to be saturable. Based on the binding of avidin and avidin–biotin complex to rat liver plasma membrane, we suggested that avidin in these systems was mimicking a natural biotin-binding carrier involved in biotin transport. At concentrations of biotin less than 5 μM, absorption of biotin by rat jejunal segments proceeded largely by a saturable process. The biphasic transport of biotin and biocytin in the rat small intestine suggests that at biotin concentrations in the gut below 50 nM, the saturable uptake mechanism would operate. Such a system would have a tremendous advantage in the context of fluctuating amounts of biotin ingested in the diet. This is born out by the rarity of primary biotin deficiency in humans. The late onset type of multiple carboxylase deficiency was shown to be due to the deficiency of biotinidase, the patients lacking the system for absorbing biotin in the nanomolar range. Biotinidase is the only protein in brush border membrane of rat intestine that binds to biotin, indicating that biotinidase in vivo may have a role in the transport of biotin.

6.3 Nonprosthetic Group Functions of Biotin

Biotin has been implicated in areas of metabolism where its role is not explained on the basis of its function as the prosthetic group of known biotin-containing enzymes. Biotin has been shown to have an important role in the regulation of the biotin operon in *E. coli.* Biotinyl AMP–holocarboxylase synthetase functions as the repressor for the biotin operon. In the avian system, biotin has been shown to regulate the synthesis of both the egg yolk BBP-I and BBP-II by transcriptional control that overrides hormonal control. In mammals the concentration of various proteins and enzymes are regulated by the biotin status of the animal. It is likely that transcriptional control, as for some hormones, is involved in these.

Biotin is essential for the expression of the asialoglycoprotein receptor in HepG2 cells. We have shown that testicular function in biotin-deficient rats responds to luteinizing hormone or biotin administration: specifically, there is an increase in testosterone production. The morphological abnormalities of the biotin-deficient rat testes were reversed by biotin treatment alone, whereas continuous testosterone supply through implantation throughout the biotin-deficient period was without any effect. The stimulation of testicular protein synthesis by testosterone required normal biotin status.

Congenital malformations have been reported in chick embryos of domestic fowl maintained on a biotin-deficient diet. Maternal biotin deficiency was strongly teratogenic in mice, even when dams did not exhibit any typical clinical signs of biotin deficiency. In a recent study of the development of palatal process in culture, we found that after 72 hours of organ culture (from 12.5 days to 15.5 days of embryo age) more than 90% of the explants from normal mouse embryos (biotin sufficient) were at stage 6 of development. The corresponding figure for explants from biotin-deficient embryos cultured in a biotin-deficient medium was 6.5%. If the deficient explants were cultured in a medium containing biotin (10^{-8} M), the proportion at stage 6 of development rose above 30%. Administration of biotin (20 μg) to biotin-deficient dams 24 hours prior to removal of the embryos resulted in 33% of the explants at stage 6 of development when cultured in a biotin-deficient medium and more than 50% at stage 6 if cultured in medium containing biotin (10^{-7} M). These results highlight the continuous requirement of biotin during the proliferation of the mesenchyme. Embryological malformations could not be ascribed to a deficient activity of the biotin carboxylases, and these results emphasize the role of biotin in the synthesis of proteins, including growth factors, during organogenesis.

6.4 Effect of Biotin on the Regulation of the Synthesis of Specific Proteins

The asialoglycoprotein receptor is characteristic of fully differentiated hepatocytes. The human hepatoblastoma line HepG2 expresses maximum receptor activity only in confluent cultures. HepG2 cells grown to confluency in minimal essential medium made 10% with respect to fetal bovine serum (FBS) demonstrate an asialoglycoprotein receptor with ligand-binding characteristics and molecular mass comparable to those of the receptor purified from human liver. However, dialysis or ultrafiltration, with removal of the low molecular weight fraction of FBS, dramatically reduced expression of asialoglycoprotein receptor. The addition of biotin or biocytin to medium supplemented with dialyzed FBS during logarithmic growth supported normal expression of asialoglycoprotein receptor by confluent HepG2 cells. Biotin was most effective at a concentration of 10^{-8} M. Isolation of mRNA from HepG2 revealed no difference in asialoglycoprotein receptor transcripts when cells were grown in a dialyzed or undialyzed medium. Thus, a biotin-dependent posttranscriptional event seems to permit the ultimate expression of asialoglycoprotein receptor by HepG2 cells.

Glucokinase is specifically induced by biotin. Biotin is not a part of this protein. Earlier work from this laboratory demonstrated that liver glucokinase activity was altered in response to the biotin status of rats. Biotin also plays a role in the precocious development of glucokinase in young rats. In all these experiments increase in enzyme activity was associated with protein synthesis. The synthesis of glucokinase is under developmental, nutritional, and hormonal control. The in vivo regulation of glucokinase mRNA by biotin was studied by us in starved rats. An hour after the injection of biotin, the concentration of glucokinase mRNA of biotin-injected starved rats had increased 3.9-fold in comparison to normal-fed rats. Following this increase there was rapid decay of glucokinase mRNA, which was hardly detectable 4 hours after biotin administration. The induction of glucokinase mRNA at 1 hour following biotin administration to starved rats was 19-fold in comparison to starved rats not receiving the biotin injection. The induction of glucokinase mRNA by biotin is quite marked, relatively rapid, and correlated with the increase in glucokinase activity. In "run-on" transcription experiments with isolated liver nuclei, biotin administration increased glucokinase gene transcription by about 6.7-fold. The increased transcription of glucokinase gene was not simply due to an increase in overall transcriptional efficiency, inasmuch as the transcription of the β-actin gene was unaffected by biotin. The difference between induction of mRNA and rate of transcription could also be due to the stabilization of glucokinase mRNA by biotin.

In both fasted and diabetic rats, hepatic phosphoenol pyruvate carboxykinase (PEPCK) activities are markedly increased. Refeeding a high carbohydrate diet to fasted rats decreased PEPCK activity markedly. The decrease in PEPCK synthesis in animals fed a high carbohydrate diet is directly due to a decrease in PEPCK mRNA which, in turn, is due to the repression of transcription of PEPCK gene by insulin. Since we had shown that the administration of biotin causes a rapid and marked induction of glucokinase mRNA in starved rats, we investigated the effect of biotin on the regulation of PEPCK in the liver and kidney of diabetic rats. After administration of biotin, hepatic PEPCK mRNA decreased to 15% of the non-biotin-injected control level at 3 hours after injection. The level of hepatic mRNA had recovered significantly by 5 hours after biotin injection. In parallel studies on insulin administration to diabetic rats, the inhibition was nearly fully reversed by 7 hours after insulin injection. The relative rates of liver PEPCK gene transcription at various time intervals following biotin administration to diabetic rats was estimated in "run-on" transcription experiments. The transcription of the actin gene, included as internal control, was not influenced by biotin administration. Biotin suppressed the transcription rates of hepatic PEPCK gene by 55% at 30 minutes, and gradually, the rate of transcription increased back to the original level.

It is to be noted that the inhibition of transcription by biotin is dominant over other stimulatory effects. The elevated plasma glucagon characteristic of the fasting or diabetic condition induces PEPCK synthesis through enhanced transcription of the PEPCK gene. There are many similarities between biotin and insulin in their action on the enzymes of glucose metabolism. Both induce the mRNA that encodes glucokinase, a key glycolytic enzyme and repress the mRNA that encodes PEPCK, a key gluconeogenic enzyme. This was the first demonstration that a water-soluble vitamin, biotin, exerted both positive and negative transcriptional control over key enzymes of glucose metabolism.

7 VITAMIN B_6

7.1 General

The term "vitamin B_6" refers to a group of naturally occurring pyridine derivatives that include pyridoxine, pyridoxal, pyridoxamine, and their phosphorylated derivatives. Vitamin B_6 is essential for normal growth, development, and biological function. Pyridoxal phosphate, generally, is the active form of this vitamin and is a cofactor for many enzymes involved with the metabolism of amino acids and lipids. In addition to a role in general metabolism, its functions in the nervous and immune systems are well documented.

7.2 Pyridoxal Phosphate Binding Proteins

Pyridoxal phosphate in plasma is bound to albumin with high affinity and exists as a Schiff base. Virtually none of the pyridoxal phosphate in plasma is dialyzable. This binding protects against the renal glomerular filtration of pyridoxal phosphate. Pyridoxal is less tightly bound to albumin than pyridoxal phosphate, hence is susceptible to glomerular filtration. However, pyridoxal appearing in the glomerular filtrate is reabsorbed. The binding of pyridoxal phosphate to albumin also protects it against hydrolysis by alkaline phosphatase. Liver is the primary source of pyridoxal phosphate in plasma and possesses a unique transport mechanism that possibly releases pyridoxal phosphate as an albumin complex. The complex contains pyridoxal phosphate and albumin in a molar ratio of 2:1. The binding of pyridoxal to hemoglobin results in the accumulation of the vitamer in the erythrocyte. This makes it possible for pyridoxal as well to be a transport form of vitamin B_6. The role of protein binding in the regulation of the liver cellular pyridoxal phosphate has also been indicated. When hepatic cytosol is dialyzed exhaustively, about 50% of the pyridoxal phosphate is still bound to proteins. This binding protects pyridoxal phosphate from hydrolysis by the liver plasma membrane associated phosphatase activity. Thus, protein binding within cells as well as in plasma is a significant modifier of the hydrolysis of pyridoxal phosphate by phosphatase. It has been proposed that the distribution of pyridoxal phosphate between protein-bound and free forms within the cell, together with the enzymatic hydrolysis of pyridoxal phosphate, maintains the concentration of this vitamer form. In liver, cytosol pyridoxal phosphate is found as five major protein-bound forms.

Skeletal muscle is the major repository of vitamin B_6 in the body. About 60–90% of the total vitamin B_6 present in muscle is associated with glycogen phosphorylase, an enzyme that comprises about 5% of the total soluble protein of muscle. Pyridoxal phosphate is bound principally to glycogen phosphorylase. Phosphorylase requires pyridoxal phosphate for catalytic activity. Pyridoxal phosphate binds to Lys-679 in rabbit muscle glycogen phosphorylase with the 5′ phosphate adjacent to the substrate-binding site. The function of muscle phosphorylase as a storage depot for pyridoxal phosphate has been demonstrated by the effect of vitamin B_6 deficiency on muscle phosphorylase activity.

7.3 Nuclear Pyridoxal Phosphate Binding

In rats fed a diet adequate in vitamin B_6 the fraction of total pyridoxal phosphate found in the nuclei of liver cells was 21%, and this increased to 39% in rats fed a vitamin B_6 deficient diet, indicating a conservation of the vitamin in the nuclear compartment. Pyridoxal phosphate is the only significant form of vitamin B_6 found in the nucleus, and there is one particularly prominent pyridoxal phosphate binding protein in the nucleoplasmic fraction of nuclear preparations obtained from HTC cells. Results of analysis by sodium dodecyl sulfate/polyacrylamide gel electrophoresis (SDS-PAGE) indicate that this protein has an apparent molecular mass of 50–55 kDa. The presence of pyridoxal phosphate in the nucleus could be of physiological significance. It is not known how the nucleus acquires pyridoxal phosphate or whether the phosphorylation of pyridoxal takes place in the nucleus, given the general assumption that phosphorylated vitamin B_6 vitamers do not cross membranes of mammalian cells.

7.4 Function of Pyridoxal Phosphate in the Nucleus

The function of the protein-bound pyridoxal phosphate in cell nucleus is of much interest. Various reports indicate that pyridoxal phosphate affects steroid hormone activity by altering the interaction of steroid receptor complexes with DNA, chromatin, and nuclei. Rat hepatoma cells when grown in presence of 5 mM pyridoxine have a significantly reduced glucocorticoid-dependent induction of tyrosine aminotransferase, and the enzyme activity is increased when the cells are grown in a medium deficient in pyridoxine. The addition of pyridoxal phosphate to the medium containing mouse mammary gland explants, however, results in a significant decrease of both dexamethasone binding to nuclear steroid receptor and dexamethasone-stimulated casein mRNA synthesis. The possibility that pyridoxal phosphate might act as a modulator of protein–DNA interaction was suggested.

The possibility that vitamin B_6 might regulate transcriptional activation by glucocorticoid receptors as well as by other members of the steroid hormone receptor family has been investigated. Elevated vitamin B_6 concentrations suppressed transcription mediated through glucocorticoid receptor. Conversely, such transcription was enhanced in mild vitamin B_6 deficiency. Under these conditions the levels of hormone-independent constitutive gene expression was not affected. It was also shown that transcriptional activities mediated by androgen, progesterone, and estrogen were reduced by elevated vitamin B_6 levels and enhanced under conditions of vitamin deficiency. Similar results were obtained using different cell lines, thus highlighting the modulation of multiple steroid hormone mediated transcriptional activation by vitamin B_6. The level of steroid-induced gene expression from simple promoters containing only hormone response elements and a TATA sequence was not affected by changes in the vitamin B_6 status. However, the modulatory influence of vitamin B_6 status was restored when a binding site for a transcription factor, nuclear factor (NF1), was included within the hormone responsive promoter. These results suggest that vitamin B_6 modulates steroid hormone mediated gene expression through its influence on a functional interaction between the steroid hormone receptors and the transcription factor NF1. Thus, the vitamin B_6 status of the cells introduces an additional level of cell-specific control over the regulation of gene expression by steroid hormones.

Bibliography

Chauhan, J., and Dakshinamurti, K. (1991) Transcriptional regulation of the glucokinase gene by biotin in starved rats. *J. Biol. Chem.* 266:10035–10038.

Dakshinamurti, K., Ed. (1994) *Vitamin Receptors.* Cambridge University Press, Cambridge.

———. (1997) Neurobiology of vitamin B_6. *Prog. Neurobiol.*

———, and Triggs-Raine, B. (1997) Biotin and multiple carboxylase deficiency. In *Clinical Studies in Medical Biochemistry,* R. H. Glow, Ed. Oxford University Press, Oxford.

Diliberto, E. J., Daniels, A. J., and Viveros, O. H. (1991) Multicompartmental secretion of ascorbate and its dual role in dopamine β-hydroxylation. *Am. J. Clin. Nutr.* 54:1163S–1172S.

Hannah, S. S., and Norman, A. W. (1994) 1α,25$(OH)_2$ Vitamin D_3-regulated expression of the eukaryotic genome. *Nutri. Rev.* 52:376–382.

Mangelsdorf, D. J. (1994) Vitamer A receptors. *Nutri. Rev.* 52:S32–S44.

Sporn, M. B., Roberts, A. B., and Goodman, D. S., Eds. (1994) *The Retinoids: Biology, Chemistry and Medicine,* 2nd ed. Raven Press, New York.

Tully, D. B., Allgood, V. A., and Cidlowski, J. A. (1994) Modulation of steroid receptor-mediated gene expression by vitamin B_6. *FASEB J.* 8:343–349.

Walters, M. R. (1992) Newly identified actions of vitamin D endocrine system. *Endocr. Rev.* 13:719–764.

Vitamins, Structure and Function of

Donald B. McCormick

1 **Classification**
 1.1 Discovery
 1.2 Nomenclature

2 **Individual Vitamins**
 2.1 Fat-Soluble Cases
 2.2 Water-Soluble Cases

3 **Perspectives**
 3.1 Quasi-Vitamins
 3.2 Pharmacologic Uses

Key Words

Apoenzyme A protein moiety of an enzyme that requires a coenzyme.

Avitaminosis A disease condition, described as a deficiency syndrome, resulting from lack of a vitamin.

Coenzyme An organic molecule, generally derived from a vitamin, that functions catalytically in an enzyme system.

Cofactor A natural reactant, usually either a metal ion or coenzyme, required in an enzyme-catalyzed reaction.

Holoenzyme A catalytically active enzyme constituted by coenzyme bound to apoenzyme.

Hypervitaminosis An unhealthy condition resulting from excess of a vitamin.

Hypovitaminosis An unhealthy condition resulting from too little of a vitamin; interchangeable with avitaminosis.

A vitamin is an organic compound that occurs as a natural component of foods and must be supplied exogenously in small amounts to maintain growth, health, and reproduction of an organism. Hence, a vitamin is essential because it cannot be made in the organism that requires it; yet it functions, usually after metabolic alteration, in indispensable ways. Since only small quantities, usually microgram to milligram amounts per human adult per day, are required to avoid a deficiency disease, vitamins are classified as micronutrients along with the trace elements.

1 CLASSIFICATION

1.1 Discovery

Symptoms of diseases that were likely the result of insufficient dietary intake of one or more vitamins are described in early writings. For example, signs of beri-beri (lack of thiamine) are described in the ancient Chinese herbals (ca. 2600 B.C.), and the Eber papyrus (ca. 1150 B.C.) gives the signs of scurvy (lack of vitamin C). That diet could influence such diseases was also suspected in early writings from ancient Greek, Roman, and Arab physicians, who recognized that ingestion of liver would both prevent and cure night blindness, which today we know to be an early symptom of hypovitaminosis A. Such empirical relationships set the stage for the experimental phase of vitamin discovery that became scientifically based around the beginning of this century.

1.2 Nomenclature

The historical manner in which vitamins have been named is first based on capital Arabic letters (A, B, C, etc.). When it was realized that some vitamins (e.g., B) were mixtures more complex than initially perceived, subscript numbers were added to the individually isolated compounds (e.g., vitamins B_1, B_2). In some instances, subscripts were used as it became evident that members of a group are functionally similar but differ in one or more particulars of structure [e.g., vitamin A, which is required for vision, exists as A_1 (retinol) and A_2 (3-dehydroretinol)]. Common chemical names, which are receiving greater usage, give a better indication of the types of compound involved. In some cases, these have replaced earlier alphanumeric designations. Thus vitamin H is now called biotin, and vitamin B_3 is known as niacin. The common names often reflect the presence of a specific atom (e.g., *thia*mine denotes sulfur). Chemically functional groups are also indicated (e.g., pyridox*amine* with the amine function). Even larger portions of the molecular structure can be reflected (e.g., *ribo*flavin with a ribityl chain). Parts of some names reflect a biofunctional property as with chole*calciferol,* which is an alcohol that "carries" calcium, or reflect source/location, as in *phyllo*quinone from plants.

2 INDIVIDUAL VITAMINS

2.1 Fat-Soluble Cases

The relative solubilities of vitamins continue to provide a generally useful division. Members of the "fat-soluble" group (i.e., A, D, E, K) are more soluble in organic solvents, which also relates generally to their fatlike absorption, transport, and storage. Hence, they are rather well retained in the body and in some cases become toxic at high dosage. Table 1 lists the fat-soluble vitamins required by humans and many animals. Aspects of chemistry, metabolism, and function of individual groups are considered in Sections 2.1.1 through 2.1.4.

2.1.1 Vitamin A

The structure of retinol (A_1) as one form of vitamin A is shown in Figure 1. 3-Dehydroretinol (A_2) has an additional double bond between carbons 3 and 4. These alcohol compounds are yellowish oils or low-melting solids that are practically insoluble in water but soluble in organic solvents and mineral oil. Vitamin A is sensitive to oxygen and ultraviolet light. Vitamin A_1 predominates, especially as long chain fatty acid esters, in the liver of mammals and saltwater fish (e.g., cod), whereas the biologically less active vitamin A_2 is found in freshwater fish oils. Higher animals are unable to synthesize the β-ionone-type ring structure but can derive the aldehyde retinal from provitamin carotenes and cryptoxanthin of plants. Retinal is then reversibly reduced to retinol. Retinoic acid is the irreversible oxidation product from retinal.

The structure for the most common and effective provitamin A, β-carotene, is also given in Figure 1. The other carotenes yield less vitamin A_1 activity. Such carotenoid compounds, which constitute the yellow-to-orange pigments of most vegetables and fruits, vary considerably in their availability to the human; they often constitute the main dietary source of what ultimately becomes vitamin A.

The emulsification to micelles of forms of vitamin A and provitamin A by bile salts enhances their uptake by mucosal cells of the small intestine. Absorbed free retinol is reesterified with long chain fatty acids within the cell. Retinyl esters in association with chylomicrons then pass via the lymphatic system to the liver and are

Table 1 Fat-Soluble Vitamins

Group/Vitamers	Adult RDA[a]	Dietary Sources	Physiologic Roles	Deficiency	Excess
A Retinol Retinal	800–1000 μg as retinol equivalents	Provitamins are carotenoids with a β-ionone ring in green and yellow/orange vegetables; retinol in liver, whole milk products	Visual pigments, cellular differentiation and proliferation	Night blindness, xerophthalmia, keratomalacia	Headache, vomiting, diplopia, alopecia, desquamation, bone abnormalities
D Cholecalciferol Ergocalciferol	5–10 μg as cholecalciferol (= 400 IU of D)	Provitamins are 7-dehydrocholesterol in skin and ergosterol from ergot; fortified milk and dairy products, eggs, margarine	Adsorption of calcium and mineralization of bones and teeth	Rickets (children) and osteomalacia (adults)	Nausea and vomiting, anorexia, diarrhea and loss of weight; calcification of soft tissue, especially kidney and heart
E Tocopherols Tocotrienols	8–10 mg as α-tocopherol equivalents	Seeds/vegetable oils, margarine and shortening; green leafy vegetables	Antioxidant for cell membrane unsaturated lipids and low density lipoprotein	Anemia (premature infants), nerve and muscle degeneration	Relatively nontoxic, but may antagonize vitamin K in those on anticoagulant therapy
K Phylloquinones Menaquinones	55–80 μg	Green leafy vegetables; lesser amounts in milk and dairy products, meats, eggs, cereals, fruits	Formation of blood clotting proteins that bind calcium ions	Hemorrhagic disease	Relatively nontoxic, but synthetic menadione can lead to jaundice

[a]The lower range is for women, but needs are modestly higher in pregnancy or lactation.

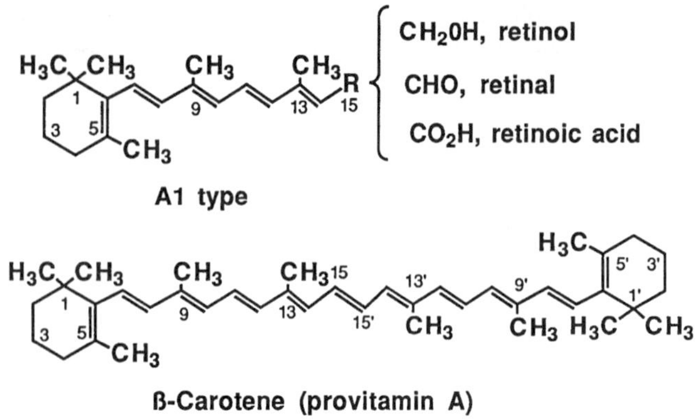

Figure 1. Vitaminic forms of A_1, A_2, and β-carotene.

stored in a lipoglycoprotein complex. Retinol associated with plasma retinol-building protein is released from hepatocytes to form, with circulating prealbumin, a molecular aggregate of sufficient size to avoid loss through glomerular filtration.

The best-understood physiological function of vitamin A is its participation as retinal in vision. Retinol and retinal are involved in a reversible cycle with opsin and rhodopsin, resulting in optic nerve transmission. Other broader functions of vitamin A occur in reproduction and growth. Systemic effects of an optimal level of vitamin A are the stabilization of cellular and intracellular membranes, maintenance of the integrity of epithelial tissue, and synthesis of glycoproteins. Retinoic acid also functions in cell development and differentiation.

2.1.2 Vitamin D

The two forms of vitamin D, ergocalciferol (D_2) and cholecalciferol (D_3), are steroid-derived compounds. D_2 does not occur naturally but is produced by ultraviolet irradiation of ergosterol, which occurs in molds, yeast, and higher plants. Vitamin D_3 in our bodies derives from irradiation of provitamin 7-dehydrocholesterol in skin. As shown by structures in Figure 2, the only chemical difference between the two vitamin and provitamin forms is in the side chain. There are at least 10 compounds known to yield vitamin D–active compounds upon irradiation. Vitamin D forms white, odorless crystals that are soluble in fats and organic solvents. It is generally stable to heat, alkalies, acids, and oxidation.

The extent to which ultraviolet irradiation of 7-dehydrocholesterol in skin provides vitamin D_3 varies with seasonal changes in sunlight, amount of clothes worn, degree of skin pigmentation, and other factors. The availability and consumption of such vitamin D_3–containing foods as fatty fish, eggs, liver, and butter also varies. Hence, widespread fortification of food has been adopted to ensure an adequate supply. Because dairy foods in the United States supply about 75% of dietary calcium in a calcium/phosphorus ratio near optimal for bone growth, milk and infant formulas are desirable vehicles for vitamin D fortification. Though a poor source of vitamin D itself, milk now marketed contains 10 μg (400 IU) of added vitamin D per quart.

Vitamin D is efficiently absorbed from the gastrointestinal tract unless there is fat malabsorption due to lack of bile salts, pancreatic insufficiency, or a defect of intestinal mucosal function. When absorbed, the vitamin is bound directly to chylomicrons. Much of the vitamin initially supplied to the liver is hydroxylated at the terminal side chain position to yield 25-hydroxy-D. This 25-hydroxycholecalciferol, which represents the major metabolite of D_3 in plasma, circulates bound to a vitamin D–binding α-globulin having a molecular weight (MW) of approximately 52,000. In kidneys, a second hydroxylation occurs at the 1α-position. This change is regulated by vitamin D status and by calcium/phosphate levels, which are controlled by parathormone (PTH) and calcitonin.

The 1α,25-dihydroxycholecalciferol formed in the kidney complexes with vitamin D–binding protein for transfer via blood to tar-

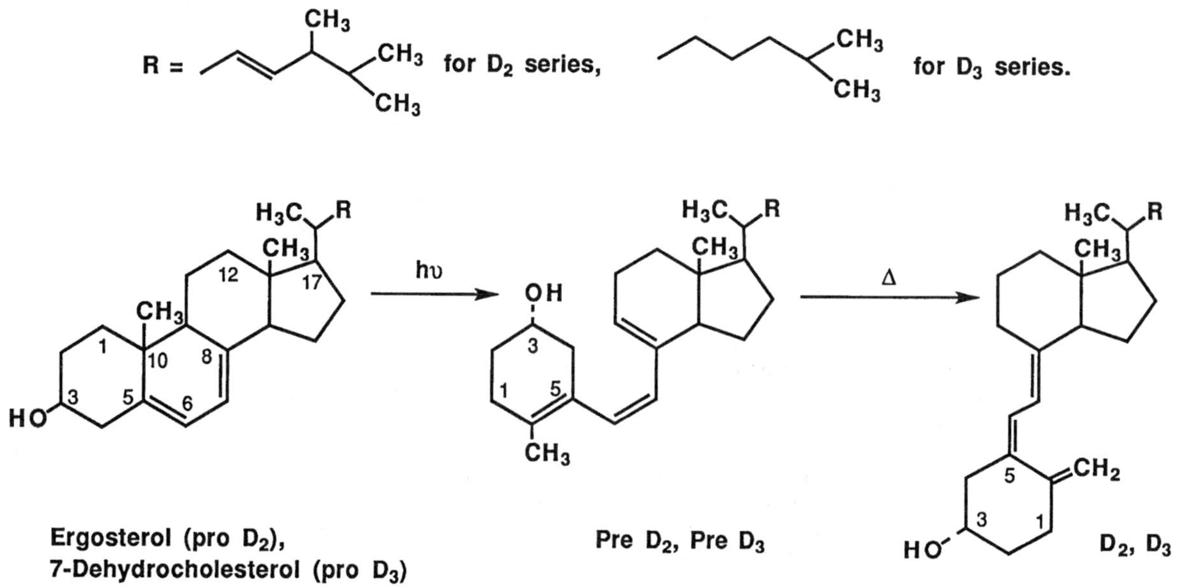

Figure 2. Provitamins, previtamins, and vitamins D_2 and D_3.

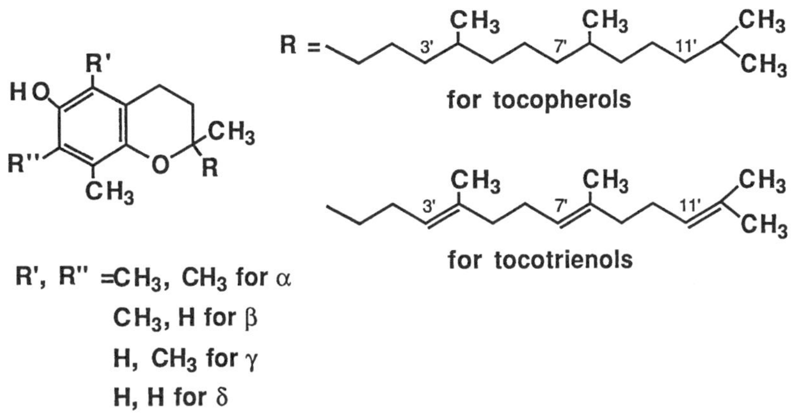

Figure 3. Vitaminic forms of E.

get tissues (e.g., small intestine, bone). In the intestinal mucosal cell, the hormonally active dihydroxy-D mediates synthesis of a specific calcium-binding protein.

2.1.3 Vitamin E

The form of vitamin E that is biologically most active, and on which units are based, is *d*-α-tocopherol. There are eight related natural compounds in this vitaminic group, which is biosynthesized in plants and is especially abundant in vegetable oils. Structures are shown in Figure 3. These compounds are soluble in fat solvents and insoluble in water, stable in acid and heat in the absence of oxygen, but labile to oxygen in alkaline solutions and to ultraviolet light.

Vitamin E is absorbed from the small intestine in the presence of bile. Most tocopherol enters the bloodstream via lymph, where the tocopherol is associated with chylomicrons and very low density lipoproteins. The vitamin is stored in most tissues, with the largest amount stored in adipose tissue. Rapid exchange of tocopherol occurs between the erythrocyte membranes and plasma lipoproteins. When physiological amounts are administered, only a small fraction of the dose appears in urine.

The best-defined role for vitamin E is as an antioxidant for unsaturated fatty acyl moieties of lipids within membranes. Oxidative damage to polyunsaturated fatty acyl parts of membrane phospholipids can occur as a result of hydrogen peroxide production by flavoprotein oxidases. Concurrent free radical damage can ensue.

2.1.4 Vitamin K

Compounds in the vitamin K series, shown in Figure 4, are the phylloquinones (K_1 type) synthesized in plants and the menaquinones (K_2 type) of bacterial origin. Several synthetic analogues and derivatives have been used; most relate to or derive from menadione (K_3), which lacks a side chain substituent at position 3. The K vitamins are insoluble in water but dissolve in organic fat solvents. They are destroyed by alkaline solutions and reducing agents and are also sensitive to ultraviolet light.

The absorption of natural vitamin K from the small intestine is facilitated by bile. Vitamins K_1 and K_2 are bound to chylomicrons for transport from mucosal cells to the liver. Menadione (K_3) is more rapidly and completely absorbed. Vitamin K in the bloodstream is associated with circulating β-lipoproteins for transport. Whereas only traces of urinary metabolites of vitamins K_1 and K_2 appear in urine, a considerable portion of vitamin K_3 (menadione hydroquinone) is conjugated to β-glucuronide and sulfate esters, which are excreted.

Initially, vitamin K was recognized as a dietary antihemorrhagic factor. It has been found to be necessary for liver syntheses of plasma clotting factors II (prothrombin), VII (proconvertin), IX (plasma thromboplastin component), and X (Stuart factor). It is now known that reduced K (hydroquinone) participates in the oxygen-dependent incorporation of CO_2 into specific L-glutamyl residues of prothrombin and other plasma proteins. The formation of bone Gla

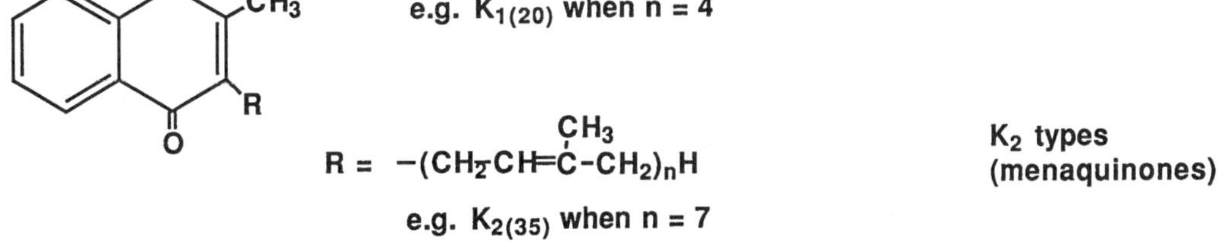

Figure 4. Vitaminic forms of K.

Table 2 Water-Soluble Vitamins

Group/Vitamers	Adult RDA[a]	Dietary Sources	Physiologic Roles	Deficiency
B_1: Thiamine	1.0–1.5 mg	Pork, organ meats, whole grains, legumes	Coenzyme functions in metabolism of carbohydrates and branched-chain amino acids	Beri-beri, polyneuritis, Wernicke-Korsakoff syndrome
B_2: Riboflavin	1.2–1.8 mg	Milk and dairy products, meats, green vegetables	Coenzyme functions in numerous redox reactions	Cheilosis, angular stomatitis, dermatitis
Niacin Nicotinic acid Nicotinamide	13–20 mg as niacin equivalents	Liver, lean meats, grains, legumes; can be formed from tryptophan	Cosubstrates/coenzymes for H-transfer with numerous dehydrogenases	Pellagra with diarrhea, dermatitis, and dementia
B_6 Pyridoxine Pyridoxamine Pyridoxal	1.6–2.0 mg	Meats, vegetables, whole-grain cereals	Coenzyme functions in metabolism of amino acids, glycogen, and sphingoid bases	Nasolateral seborrhea, glossitis, peripheral neuropathy (epileptiform convulsions in infants)
Folacin Folic acid Folyl polyglutamates	180–200 μg	Liver, yeast, leafy vegetables, legumes, whole-wheat products	Coenzyme functions in single-carbon transfers in metabolism of nucleic and amino acids	Megaloblastic anemia, mental disorder, gastrointestinal disturbances, glossitis, pallor
B_{12}: Cyanocobalamin	2 μg	Meats, eggs, dairy products (not in plant foods)	Coenzyme functions in metabolism of odd-number fatty acids and methyl transfers	Pernicious anemia, neurological disorders, pallor
Biotin	30–100 μg (ranges recommended)	Liver, yeast, egg, yolk, soy flour, cereals	Coenzyme functions in bicarbonate-dependent carboxylations	Fatigue, depression, nausea, dermatitis, muscular pains
Pantothenic acid	4–7 mg (ranges recommended)	Animal tissues, whole-grain cereals, legumes; widely distributed	Constituent of CoA and phosphopantetheine involved in fatty acid metabolism	Fatigue, sleep disturbances, impaired coordination, nausea
C *l*-Ascorbic acid Dehydroascorbic acid	60 mg	Vegetables and fruits, especially broccoli and citrus	Formation of collagen and carnitine; important as antioxidant	Scurvy with petechiae, ecchymoses, perifollicular hemorrhage, spongy and bleeding gums

[a]The lower range is for women, but needs are modestly higher in pregnancy or lactation.

protein (γ-carboxylglutamyl proteins) or osteocalcin, possible regulators of mineralization, also depends on vitamin K.

Bis-4-hydroxycoumarin (dicumarol), the anticlotting compound from spoiled sweet clover, and synthetic 4-hydroxycoumarins such as warfarin, act as anticoagulants by interfering with the conversion of provitamin K to the functional hydroquinone. Hence, 4-hydroxycoumarins are able to suppress the formation of prothrombin and other vitamin K–dependent Ca^{2+}-binding proteins.

2.2 Water-Soluble Cases

Vitamin C and the vitamins in the B-complex group are "water soluble" and as such share the fate of other solutes more compatible with an aqueous, physiological medium. Characteristics include a lesser tendency to be retained for long times in the body and a greater loss by way of urinary excretion. Table 2 lists the water-soluble vitamins required by the human and many plants and animals.

The eight vitamin groups that comprise the B complex function after metabolic conversions to coenzymes, which are involved in holoenzymic systems responsible for catalyzing numerous reactions. In the case of vitamin C, a cosubstrate-like function is seen with its oxidation coupled to certain hydroxylations. Aspects of chemistry, metabolism, and function of individual groups are considered in Sections 2.2.1 through 2.2.9.

2.2.1 Thiamine

Thiamine (vitamin B_1) is a pyrimidyl-substituted thiazole (Figure 5). The principal if not sole coenzyme form is the pyrophosphate ester (TPP). Monophosphate and triphosphate esters occur naturally. Small amounts of thiamine and its phosphates are present in most plant and animal tissues, but more abundant sources are unrefined cereal grains, liver, heart, kidney, and lean cuts of pork. Thiamine is somewhat heat labile, particularly in alkaline solutions.

Thiamine is readily absorbed in the small intestine by an active transport process when intake is less than 5 mg/day; at higher levels of intake, passive diffusion increasingly contributes to absorption. Phosphorylation takes place in the jejunal mucosa to yield thiamine pyrophosphate. Thiamine is carried by the portal blood to the liver. The free vitamin occurs in the plasma, but the coenzyme, TPP, predominates in the cellular components. About half the body stores are found in skeletal muscles, with much of the remainder in heart,

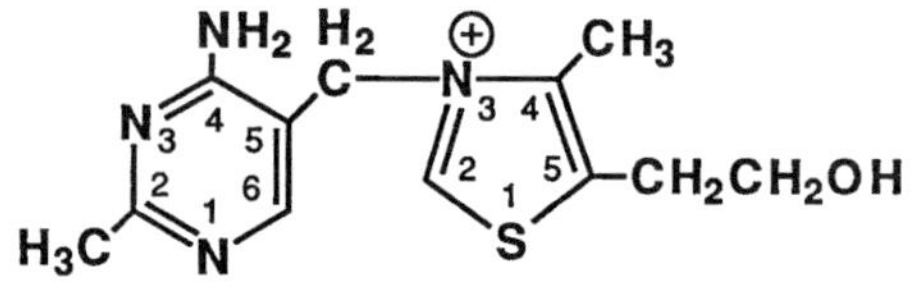

Figure 5. Structure and ring numbering for thiamine.

liver, kidneys, and nervous tissue, including brain. Thiamine, as well as several of its catabolites, is excreted into the urine by the renal tubules.

As a thiamine deficiency develops, there is a rather rapid loss of the vitamin from all tissues except the brain. The decrease of TPP in the erythrocytes roughly parallels the decrease of this coenzyme in other tissues. During this time, the thiamine in urine falls to near zero; the urinary metabolites remain high for some time before decreasing.

TPP functions as the Mg(II)-coordinated coenzyme for so-called active aldehyde transfers in the oxidative decarboxylation of α-keto acids and the formation of α-ketols (ketoses) catalyzed by transketolase.

2.2.2 Riboflavin

Riboflavin (vitamin B_2) is a ribityl-substituted isoalloxazine (Figure 6). It and its natural derivatives are yellow fluorescent compounds that are widely distributed throughout the plant and animal kingdoms. Principal among these flavins are the coenzymes flavin mononucleotide (FMN) and flavin adenine dinucleotide (FAD). Such coenzyme forms are rich sources of the vitamin in liver, kidney, and heart. Many vegetables are also good sources, but cereals are rather low in flavin content. Raw milk is a good source of the vitamin, but considerable loss can occur from exposure to light during pasteurization and bottling or as a result of irradiation to increase the vitamin D content. Flavins are stable to heat but are decomposed by light.

The process by which riboflavin and lesser amounts of natural derivatives are released by digestion of complexes with food proteins involves hydrolysis of coenzyme forms of the vitamin (mainly FAD; less FMN) as a consequence of gastric acidification. The vitamin is primarily absorbed in the proximal small intestine by a saturable transport system that is rapid and proportional to intake before leveling off at higher doses. Bile salts appear to facilitate the uptake. Much of the riboflavin circulating in blood plasma is loosely associated with albumin; lesser amounts complex with other proteins. One subfraction of immunoglobulin G (IgG) avidly binds a small portion of the total flavin in blood. Conversion of riboflavin to the coenzymes FMN and FAD occurs within the cell cytoplasm of most tissues, but particularly in the small intestine, liver, heart, and kidneys. Thyroxine and triiodothyronine stimulate FMN and FAD synthesis in mammalian systems. FAD is the predominant flavocoenzyme present in tissues, where it is mainly complexed with numerous flavoprotein dehydrogenases and oxidases. Since there is little storage of riboflavin as such, the urinary excretion reflects dietary intake.

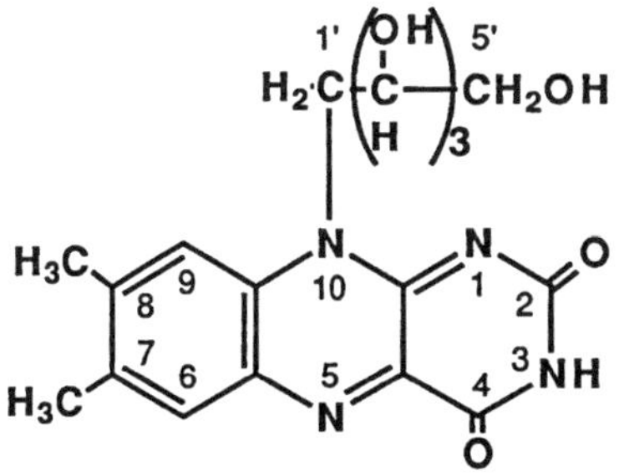

Figure 6. Structure and numbering for riboflavin.

In bound coenzymatic form, riboflavin participates in oxidation–reduction reactions in numerous metabolic pathways and in energy production via the respiratory chain. Flavins serve as redox carriers upon differential binding to proteins, participating, as well, in both one- and two-electron transfers. In reduced (1,5-dihydro) form, flavins react rapidly with oxygen.

2.2.3 Vitamin B_6

The three natural members of the vitamin B_6 group—pyridoxine (pyridoxol), pyridoxamine, and pyridoxal—are all 2-methyl-3-hydroxy-5-hydroxymethyl pyridines, as shown in Figure 7. Both pyridoxamine-5′-phosphate and pyridoxal-5′-phosphate (PLP) interconvert as coenzyme forms during reactions catalyzed by aminotransferases (transaminases). Vitamin B_6 is widely distributed in animal and plant tissues, where the phosphorylated forms, and particularly PLP, predominate. Meats, poultry, and fish are good sources, as are yeast, certain seeds, and bran; somewhat more limited sources are milk, eggs, and green leafy vegetables. The common commercial form of the vitamin is pyridoxine hydrochloride, which is a water-soluble, white, crystalline solid. Solutions of the B_6 vitamins are decomposed by light, especially in the UV region, at neutral to alkaline pH. There is significant loss of PLP during thermal processing of foods.

The three B_6 vitamers are readily absorbed by the mucosal cells. PLP binds for catalytic function with numerous specific apoenzymes throughout the cell. The erythrocytes, in addition, trap PLP as a conjugate Schiff base with hemoglobin. Glycogen phosphorylase contains most of the PLP in skeletal muscle.

Free vitamin is released when the phosphates are hydrolyzed by nonspecific alkaline phosphatase located on the plasma membrane of cells or when pyridoxine-5′-β-D-glucoside is hydrolyzed by nonspecific glucosidase within cells. The vitamin released from animal derived foods is mainly pyridoxal; from plants, pyridoxine is preponderant. Some PLP is also released into the circulation by the liver. Although PLP is the principal tissue form of vitamin B_6 and pyridoxal constitutes much of the circulating vitamin, the main catabolite excreted in urine is 4-pyridoxic acid.

As coenzyme PLP, vitamin B_6 functions in the metabolism of proteins, carbohydrates, lipids, and amino acids. PLP coenzyme-dependent reactions include formation of amines such as epinephrine and norepinephrine, formation of aminolevulinate in heme synthesis, and phosphorolysis of the α-1,4-linkages of glycogen.

2.2.4 Niacin and Niacinamide

Though the term "niacin" is chemically synonymous with nicotinic (pyridine-3-carboxylic) acid, it is now used as the generic name for the specific compound as well as for derivatives exhibiting qualitatively the biological activity of niacinamide (nicotinamide, nico-

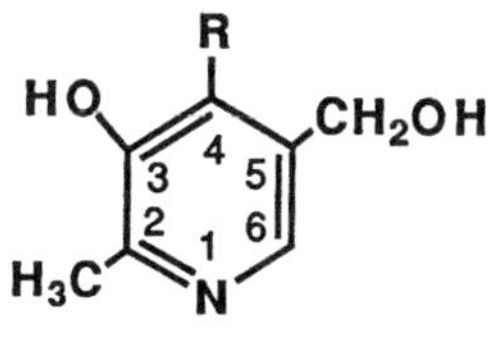

R = CH_2OH for pyridoxine
CH_2NH_2 for pyridoxamine
CHO for pyridoxal

Figure 7. Natural forms of vitamin B_6.

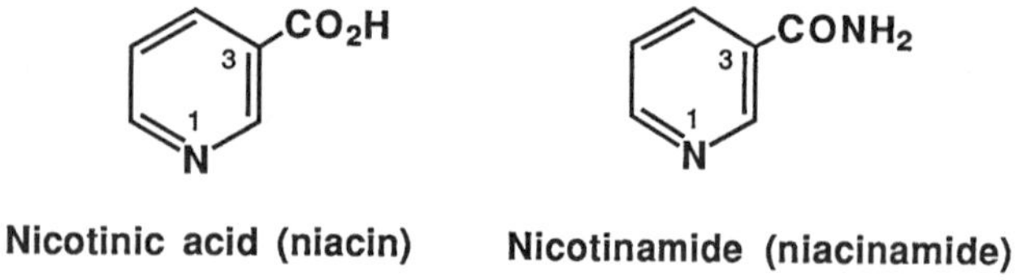

Figure 8. Structures for niacin and niacinamide.

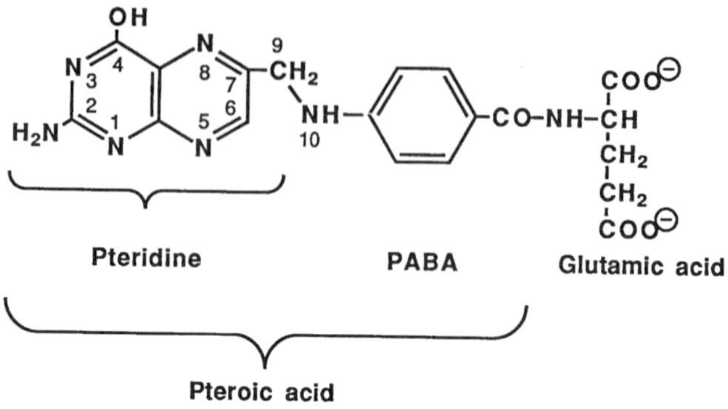

Figure 9. Structure with numbering for folic acid.

tinic acid amide). Thus "niacin activity" and "niacin deficiency" carry this broader meaning in nutritional literature. Figure 8 gives the structures of both vitamers.

The coenzymes nicotinamide adenine dinucleotide (NAD; diphosphopyridine nucleotide, DPN) and nicotinamide adenine dinucleotide phosphate (NADP; triphosphopyridine nucleotide, TPN) represent most of the niacin activity found in food sources, which include yeast, lean meats, liver, and poultry. Milk, canned salmon, and several leafy green vegetables have lesser amounts. Protein provides a considerable portion of niacin equivalent because of the tryptophan content. As much as two-thirds of niacin required by adults can be derived from tryptophan metabolism via nicotinic acid ribonucleotide to NAD and NADP. Free forms of the vitamin are white, stable solids that are quite soluble in water. The oxidized coenzymes are labile to alkali, whereas the reduced (dihydro) coenzymes are labile to acid. Reduction of the oxidized coenzymes commonly occurs by addition of a hydride ion to the para (4) position of the nicotinamide ring, with simultaneous formation of a solvated proton. NADH and NADPH (but not NAD and NADP) absorb light in the near-ultraviolet region (339 nm).

The coenzymes are hydrolyzed in the intestinal tract, and both the acid and amide forms of the vitamin are readily absorbed. Nicotinic acid and nicotinamide are both present in the blood and move between blood and cerebrospinal fluid. Both compounds are converted to the coenzyme forms in blood cells, kidneys, brain, and liver. In the tissues most of the vitamin is present as nicotinamide in NAD and NADP, although liver may contain a significant fraction of the free vitamin. There is little storage of niacin as such. Although nicotinamide can be converted to nicotinic acid by a rather widespread microsomal deamidase, there is no direct reamidation of nicotinic acid. Humans excrete 1-methylnicotinamide and 1-methyl-3-carboxamido-6-pyridone, which are primary urinary metabolites.

There are hundreds of enzymes that require the nicotinamide moiety within either NAD or NADP. Most of these oxidoreductases function as dehydrogenases and catalyze such diverse reactions as the conversion of alcohols (often sugars and polyols) to aldehydes or ketones, hemiacetals to lactones, aldehydes to acids, and certain amino acids to keto acids. Generally, most NAD-dependent enzymes are involved in catabolic reactions, whereas NADP systems are more common to biosynthetic reactions.

2.2.5 Folacin

Folacin is the term covering both folic acid (pteroyl monoglutamate) and its natural polyglutamyl forms. The structure of the unconjugated acid, given in Figure 9, comprises a pterin linked through a *p*-aminobenzoyl residue to the glutamate terminus. Reduction proceeds through the 7,8-dihydro to the 5,6,7,8-tetrahydro level before coenzymic activity is generated. Green leafy vegetables (e.g., spinach) are excellent sources of the vitamin.

Folacin is digested in the upper small intestine by cleavage of excess glutamyl residues by "conjugases": substances that hydrolyze the predominant coenzymic forms present in foods to liberate the free folate, which is then absorbed. Folate-binding proteins occur in plasma. In tissues, folate is reduced by the NADPH-dependent dihydrofolate reductase to tetrahydrofolate, which is then polyglutamylated by an ATP-dependent synthetase.

Coenzymic cycling is generally at the tetrahydro level for one-carbon transfer reactions, though an oxidation–reduction shift from the tetrahydro to dihydro occurs with thymidylate synthetase. In this case the transfer of a methylene function from 5,10-methylene tetrahydrofolyl coenzyme to deoxyuridylate involves its reduction to a methyl on the thymidylate product concomitant to oxidation of the pterin portion to generate the dihydrofolyl coenzyme that must be recycled. Other important roles of tetrahydrofolyl (THF) coenzymes relating to biosynthesis of nucleic acid precursors include the transformylase-catalyzed conversions of glycinamide ribonucleotide to the *N*-formyl compound with participation of 5,10-methenyl-THF, and 5-aminoimidazole-4-carboxamide ribonucleotide to the 5-formamido compound with participation of 10-formyl-THF.

With respect to amino acid metabolism, enzyme systems include the hydroxymethyltransferase-catalyzed conversion of serine with THF to glycine and 5,10-methylene-THF, the formiminotransferase reaction with formiminoglutamate and THF to yield glutamate plus 5-formimino-THF, and the biosynthesis of methionine from homocysteine plus 5-methyl-THF to regenerate THF.

2.2.6 Vitamin B_{12}

The commercial form of vitamin B_{12} is cyanocobalamin, shown in Figure 10, but the tissue-predominate B_{12} has water or the hydroxyl ion in place of the coordinate cyanide. Primary structural features include a central cobinamide with a corrin ring system and an affixed dimethylbenzimidazole nucleotide with coordination of this nitrogen base to the corrin cobalt. The cobalamins are photolabile and also unstable at extremes of pH. Cyanocobalamin exhibits a distinctive absorption spectrum with a long wavelength maximum at 550 nm for the reddish aqueous solutions. Best food sources are meats and dairy products; higher plants, which cannot synthesize this vitamin, provide little or none. Hence, true vegetarians are at risk for deficiency.

Upon ingestion vitamin B_{12} (extrinsic factor) forms a complex with a glycoprotein called intrinsic factor, which is absorbed in the distal ileum. Additional B_{12}-binding proteins (transcobalamins, R

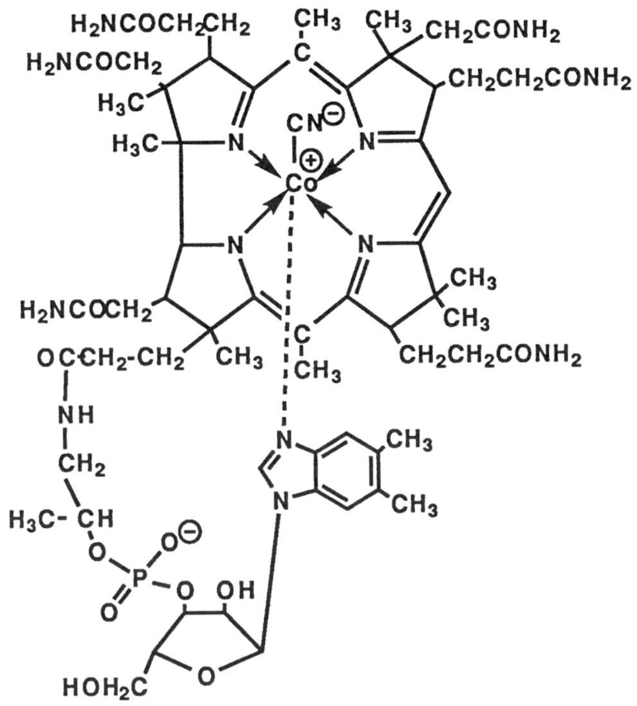

Figure 10. Structure of cyanocobalamin (B_{12}).

proteins) help vector the vitamin through plasma and into cells. For conversion to functional coenzymes, the aquocobalamin (B_{12a}) is reduced through the radical (B_{12r}) to the Co(I) reactive B_{12s} that can react with ATP to add the 5′-deoxyadenosyl moiety of coenzyme B_{12} in a transferase-catalyzed step or with methyl-THF to form methyl B_{12}, which participates in methionine biosynthesis.

Besides the essential function of B_{12} in formation of methionine in mammals including the human, the coenzyme form is required for the methylmalonyl–coenzyme A mutase that routes L-methylmalonyl–CoA to succinyl–CoA and the Krebs cycle. More broadly in other organisms, B_{12} coenzymes function in ribonucleotide reductase, diol dehydratase, ethanolamine ammonia lyase, L-β-lysine mutase, glutamate mutase, and so on. In some cases (e.g., the adenylcobamide coenzyme used in a bacterial glutamate mutase), the N base that is a dimethyl-benzimidazole in the Co-B_{12} in our bodies is replaced by a different N base.

2.2.7 Biotin

Biotin (vitamin H) is a fused biocyclic system. In most organisms this vitamin, shown in Figure 11, occurs mainly bound to protein. Good sources of biotin include liver, kidney, pancreas, eggs, yeast, and milk.

Digestion of dietary proteins containing bound biotin yields considerable biocytin (ε-N-biotinyl lysine). Biocytin and biotin are readily absorbed. Biotinidase releases biotin from biocytin. Biotin is cleared from the circulating blood and is taken up by such tissues as liver, muscle, and kidney. Careful balance studies show that in humans, the urinary excretion of biotin often exceeds dietary intake and fecal excretion is always more than dietary intake because of microfloral biosynthesis. For the average adult, total body content of biotin is perhaps only 1 mg.

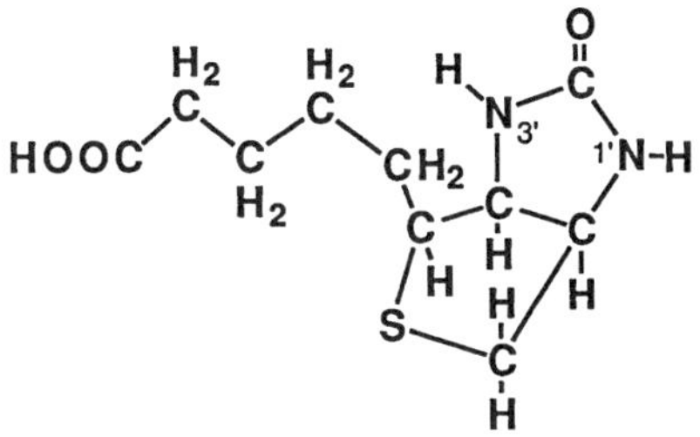

Figure 11. Structure of biotin.

Four biotin-dependent enzymes have been found in human tissues. They are carboxylases for acetyl–CoA, propionyl–CoA, β-methylcrotonyl–CoA, and pyruvate.

2.2.8 Pantothenic Acid

Pantothenic acid is the amide formed from pantoic acid and β-alanine, as shown in Figure 12. This vitamin is ubiquitous in nature and is synthesized by most microorganisms and plants. The vitamin is an integral part of a covalently attached prosthetic group of acyl carrier proteins and is within the structure of coenzyme A. The vitamin is widely distributed in foods and is particularly abundant in animal sources, legumes, and whole-grain cereals. Excellent food sources include egg yolk, kidney, liver, and yeast. Fair sources include broccoli, lean beef, skimmed milk, sweet potatoes, and molasses. More than half the pantothenate in wheat may be lost during the manufacture of flour, and up to a third is lost during the cooking of meat. The most common commercial synthetic form is the calcium salt.

Coenzyme A, the form in which much of the pantothenic acid is ingested, is hydrolyzed by intestinal pyrophosphatase and phosphatase to pantetheine, which together with pantothenate is absorbed into the portal circulation. Within cells, resynthesis of coenzyme compounds occurs. About 80% of the vitamin in animal tissues is in CoA form. Cleavage enzymes operate during turnover and release of the vitamin, which is excreted in the urine. Only a small fraction of pantothenate is secreted in milk, and even less into colostrum.

The myriad acyl thiol esters of CoA, of which pantothenic acid is a constituent, are central to the metabolism of numerous compounds (especially lipids) and the ultimate catabolic disposition of carbohydrates and ketogenic amino acids. For example, acetyl–CoA, which derives from the metabolism of carbohydrates, fats, and amino acids, can acetylate compounds such as choline and hexosamines to produce essential biochemicals; it can also condense with other metabolites such as oxaloacetate or with itself to supply citrate and cholesterol, respectively. Another essential role of pantothenic acid is its participation in the 4′-phosphopantetheine moiety of acyl carrier protein (ACP), where it facilitates chain elongation during fatty acid biosynthesis. Although the reactive thiol

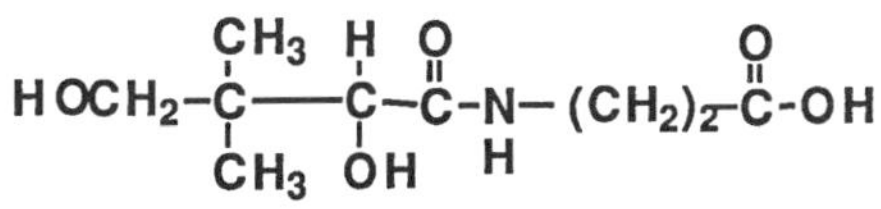

Figure 12. Structure of pantothenic acid.

function of CoA and ACP is not an integral part of pantothenate, the steric and chemical properties conferred by the vitamin structure are important for enzymic recognition.

2.2.9 Vitamin C

Vitamin C (L-ascorbic acid) is a white, crystalline solid that is readily soluble in water. Ascorbic acid is a relatively strong reductant with an E_0' (pH 7) of +0.58 V. Ascorbic acid is reversibly oxidized to dehydroascorbic acid (ascorbone); both compounds are shown in Figure 13. The dehydro form is more labile than the reduced form. Some mammals, including the human, lack the enzyme L-gulonolactonase that catalyzes the formation of 2-keto-L-gulonolactone, which spontaneously tautomerizes to L-ascorbic acid. The best sources of the vitamin are citrus fruits, berries, melons, tomatoes, green peppers, raw cabbage, and leafy green vegetables. Losses during processing, especially with heat and aerobic conditions, can be considerable.

Absorption of vitamin C occurs readily, mostly from the stomach, where some of the ascorbic acid is converted to the dehydro form. At physiological pH, the uncharged dehydroascorbic acid passes across cell membranes faster than the anionic L-ascorbate. Passive diffusion of vitamin C accounts significantly for entry into some cells, such as leukocytes and erythrocytes, but an active transport mechanism also operates, especially for platelets, liver, adrenals, and retina. Vitamin C is found in most tissues, but glandular tissues such as the pituitary, adrenal cortex, corpus luteum, and thymus have the highest amounts, and the retina has 20–30 times the plasma concentration. The half-life for vitamin C in humans is only about 16 days. Ascorbate, dehydroascorbic acid, and lesser amounts of a number of catabolites are present in urine.

The most clearly established and critical functional role for ascorbic acid is a cofactor for protocollagen hydroxylase, the enzyme responsible for hydroxylation of prolyl and lysyl residues within nascent peptides in connective tissue proteins. Among these are collagen and related proteins that comprise the intercellular material of cartilage, dentin, and bone. Vitamin C may also be involved in carnitine biosynthesis, tyrosine metabolism, microsomal drug metabolism, synthesis of epinephrine and anti-inflammatory steroids by the adrenals, folic acid metabolism, and leukocyte functions. Absorption of Fe(II) is enhanced by simultaneous ingestion of the vitamin.

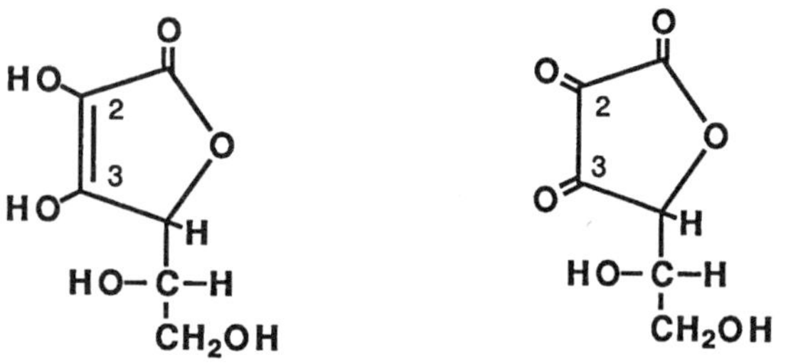

L-Ascorbic acid   Dehydroascorbic acid

Figure 13. Structure for L-ascorbic and dehydroascorbic acids.

3 PERSPECTIVES

3.1 Quasi-Vitamins

Some compounds not listed in Tables 1 and 2 are legitimate vitamins for certain species, though not the human. For examples, lipoic and *p*-aminobenzoic acids for certain microorganisms or carnitine for the common mealworm represent essential organic micronutrients that cannot be synthesized adequately by the organisms that depend on dietary supply of these compounds. More than micro amounts of choline and myoinositol are required at early, growth stages for rodents. There are, furthermore, several compounds that at one time or another have been falsely claimed to be vitamins with reputed benefit to humans. These include such ineffective factors as B_{15} or pangamic acid and B_{17} or laetrile. Only the 13 vitaminic groups listed in Tables 1 and 2 presently account for human needs.

3.2 Pharmacologic Uses

As mentioned in connection with fat-soluble vitamins, particularly A and D, toxicity can result from the ingestion of too much, especially chronically. Excessive amounts of at least a couple of water-soluble vitamins can also pose a health hazard. Hepatoxicity can result from high amounts of nicotinic acid, and neurologic dysfunction has been associated with chronic excess of pyridoxine.

There is growing certainty, however, that amounts of L-ascorbate (C), and α-tocopherol (E) that are modestly above the present recommended daily allowance (RDA) levels, have beneficial effects as natural antioxidants. The decrease in some specific cancers has been associated epidemiologically with higher intakes of foods rich in C, E, and carotenoids, including the provitamin A β-carotene. Extra folic acid also seems to protect the fetuses in a subset of women who have a history of neural tube defects in their newborns. More specific and direct experimentation will be needed to clarify such pharmacologic roles of vitamins.

See also Coenzymes, Biochemistry of; Nutrition; Retinoids; Trace Element Micronutrients.

Bibliography

Brown, M. L., Ed. (1990) *Present Knowledge in Nutrition,* 6th ed. International Life Science Institute–Nutrition Foundation, Washington, DC.

Combs, G. E., Jr. (1992) *The Vitamins, Fundamental Aspects in Nutrition and Health.* Academic Press, San Diego, CA.

Machlin, L. J., Ed. (1991) *Handbook of Vitamins,* 2nd ed. Dekker, New York.

McCormick, D. B., and Greene, H. L. (1994) Vitamins, In *Tietz Textbook of Clinical Chemistry,* 2nd ed. C. Burtis and E. Ashwood, Eds. Saunders, Philadelphia.

National Research Council (1989) *Recommended Dietary Allowances,* 10th ed. National Academy Press, Washington, DC.

Sauberlich, H. E., and Machlin, L. J. (1992) Beyond deficiency. New views on the function and health effects of vitamins. *Ann. N.Y. Acad. Sci.* 669.

Shils, M. E., Shike, M., and Olson, J. A., Eds. (1993) *Modern Nutrition in Health and Disease,* 8th ed. Lea & Febiger, Philadelphia.

W

Whole Chromosome Complementary Probe Fluorescence Staining

Heinz-Ulrich G. Weier, Daniel Pinkel, and Joe W. Gray

1 **Probes**
 1.1 Species-Specific Hybridizations
 1.2 Chromosome-Specific Probes

2 **Applications**
 2.1 Species-Specific Staining of Human DNA in Human × Hamster Hybrid Cells
 2.2 Translocation Detection Using FISH with WCPs
 2.3 Reversed Painting
 2.4 Demonstration of Syntenic Regions in Distantly Related Species
 2.5 Chromosomal Organization of Cell Interphase Nuclei
 2.6 Detection of Aneusomy in Metaphase and Interphase Cells
 2.7 Gene Mapping

3 **Perspectives**

Key Words

In Situ Hybridization The binding of nucleic acid probes to a target nucleic acid sequence inside a mostly intact cell or chromosome.

Label A reporter molecule attached to a hybridization probe that is normally absent in the hybridization target and can be detected by its fluorescent or immunogenic properties.

Locus-Specific Probe (LSP) A single probe molecule or collection of labeled nucleic acid fragments that bind to a limited region of the genome typically containing a gene, a translocation breakpoint, or other relevant DNA sequence.

Probe A collection of one or many nucleic acid molecules, each a few hundred base pairs long, carrying covalently bound labels.

Probe Complexity A measure of the number of different nucleic acid sequences in a probe.

Whole-Chromosome Probe (WCP) A collection of labeled nucleic acid fragments that have sequence homology with regions of the genome distributed over an entire chromosome.

Fluorescence in situ hybridization (FISH) with nucleic acid probes that are complementary to specific chromosomes or parts thereof delineates and fluorescently "stains" the region targeted by the probes in the interphase and metaphase cells. Probes for FISH typically are labeled enzymatically by nick translation or random primer extension in the presence of deoxynucleoside triphosphates attached to reporter molecules such as biotin, digoxigenin, or fluorescent molecules (fluorescein, rhodamine, Spectrum Orange, Spectrum Green, coumarine derivatives, CY3, etc.). FISH with these probes is accomplished by denaturing the probe and target nucleic acid sequences so that substantial fractions of the probe and target molecules become single-stranded. The probe and target are then mixed under conditions ensuring that the probe molecules hybridize to complementary sequences in the target cell.

Probe molecules that carry nonfluorescent molecules are detected after hybridization by incubation with fluorochrome-conjugated avidin or antibodies that have high affinity for the reporter molecules. Targets as small as 1 kb or as large as whole chromosomes can be stained using FISH. The staining of whole chromosomes or substantial portions thereof has proved useful for detection and characterization of mammalian or plant chromosomes in interspecies hybrids, for visualization of chromosomal rearrangements or for chromosome enumeration in tumor cells or in cells exposed to clastogenic agents, for investigation of the chromosomal organization of interphase nuclei, and for demonstration of syntenic regions in distantly related mammals.

1 PROBES

1.1 Species-Specific Hybridizations

The first demonstration of whole-chromosome staining was accomplished using genomic DNA from one species to stain chromosomes of that species in mixed species hybrid cells (e.g., human chromosomes in human × hamster hybrid cells). Species-specific hybridization is accomplished using whole genomic DNA from the target species. Hybridization occurs primarily to interspersed repeated sequences in the chromosomes of the target species, with the result that all DNA from the target species is stained more or less uniformly. The divergence between nucleic acid sequences in humans and rodents is high enough to permit species-specific staining to be easily achieved: that is, interspersed repeats in human genetic material will not bind to interspersed repeats in nucleic acid sequences from rodents. However, the use of genomic DNA as a probe for species-specific staining works less well when the DNA sequence homology is high between the two species (e.g., between human and chimpanzee), and it does not enable staining of specific chromosome types.

The same approach has been used to differentially stain parental genomes in plant hybrids. Biotinylated probes are typically detected by conjugation with fluorescent avidin for fluorescence microscopy. Avidin coupled to horseradish peroxidase or alkaline phosphatase, on the other hand, is used to generate insoluble precipitate at the location of bound probe and allows high resolution electron microscopic analysis.

1.2 Chromosome-Specific Probes

Chromosome-specific staining requires whole-chromosome probes (WCPs), that are composed of a large number of different, labeled DNA sequences that have homology throughout the target region. In general, increasing the probe complexity (i.e., the fraction of the target DNA sequence that has homology to sequences in the probe) will result in more intense and homogeneous staining. Generation of WCPs begins with enrichment of DNA fragments from the target chromosome, as described in the subsections that follow. The enriched DNA fragments are then amplified for use in in situ hybridization. The amplified DNA, however, contains repetitive sequences that are distributed throughout the genome. These sequences will hybridize to all chromosomes. To achieve chromosome-specific hybridization, these repeats must be removed prior to hybridization (quite a labor-intensive process) or inhibited from binding by including unlabeled repetitive DNA in excess during hybridization. Enrichment of chromosome-specific DNA has been accomplished using several different techniques, three of which are described in Sections 1.2.1 through 1.2.3.

1.2.1 Chromosome Sorting

Chromosome sorting was the first technique used for enrichment of chromosome-specific DNA for production of WCPs. In this approach, aqueous suspensions of chromosomes are isolated from metaphase cells and stained with one or more DNA-specific dyes. Chromosomes of one type are enriched using fluorescence-activated sorting. Sorter-enriched DNA can be immortalized and amplified by cloning restriction enzyme fragments into plasmid, cosmid, or phage vectors, followed by propagation in a bacterial host, or by in vitro DNA amplification using the polymerase chain reaction (PCR). The number of chromosomes required for the establishment of high complexity probes depends on the technique used for immortalization (see Table 1). In general, cloning requires up to 10^5–10^7 flow-sorted chromosomes, depending on the efficiency of the cloning vector. Far fewer chromosomes are needed if PCR is used to amplify the specific DNA fragments. In fact, high complexity probes have been prepared by PCR amplification of hundreds or thousands of flow-sorted chromosomes.

Two approaches to PCR amplification have been used. In one, chromosomal DNA is digested using a frequently cutting endonuclease, and adapter oligonucleotides are ligated to both ends of the restriction fragments. The resulting fragments are amplified using PCR with primers to the adapter oligonucleotides (linker–adapter PCR). In another approach, chromosomal DNA is PCR-amplified using highly degenerate, mixed-base primers (DOP-PCR).

DOP-PCR is particularly simple: it requires almost no manipulation of the chromosomal DNA. Common protocols include initial incubation in the presence of proteinase K or topoisomerase I to facilitate access of oligonucleotide primers to chromosomal DNA, followed by a brief heat inactivation of the enzyme. The PCR reaction is then performed for four to eight cycles at low annealing temperature to allow the 3′ ends of the mixed-base primers to anneal to the template DNA. The annealing temperature is then raised and another 25–40 PCR cycles are performed for efficient amplification of the primer extension products generated during the low annealing temperature cycles.

1.2.2 Interspersed Repeat Sequence (IRS) DNA Amplification

An alternative technique for generation of WCPs takes advantage of the presence of repeated sequences mentioned earlier, which are interspersed with single-copy DNA in intervals of 10 to 100 kilobase pairs throughout primate and rodent genomes. The short and long interspersed repeat DNA sequences (SINEs and LINEs, respectively) provide species-specific primer annealing sites to accommodate the amplification of the intervening single-copy DNA sequences by PCR (IRS-PCR). IRS-PCR amplification of human DNA from human–rodent cells carrying single human chromosomes (or portions thereof) has been used to produce high quality WCPs.

1.2.3 Microdissection

Microdissection allows collection of DNA fragments from whole chromosomes or from chromosomal subregions. DOP-PCR amplification of the microdissected DNA from 10–20 chromosomes generates WCPs that can be used to stain specific chromosomes or chromosome subregions, if only that portion of a chromosome is dissected. These probes are useful for detection and mapping of deletions, gene amplifications, and other structural rearrangements.

1.2.4 Availability

WCPs have now been produced for each normal human chromosome, for many chromosomal subregions, and for several rodent chromosomes. WCPs for all human chromosomes are available

Table 1 Generation of Whole-Chromosome Probes

Enrichment Procedure	DNA Immortalization Procedure	Required Number of Chromosomes	Library Amplification	Ref.
Flow sorting	Endonuclease digestion and molecular cloning	10^5–10^6	Bacterial host	Gray et al. (1987); Collins et al. (1991)
Flow sorting	Linker-adaptor PCR	10^4–10^5	PCR	Vooijs et al. (1993)
Flow sorting	DOP-PCR	10^2–10^3	PCR	Telenius et al. (1992); Weier et al. (1994)
Cell hybrids	IRS-PCR	10^2–10^3	PCR	Liu et al. (1993)
Microdissection	DOP-PCR	10–30	PCR	Meltzer et al. (1992)

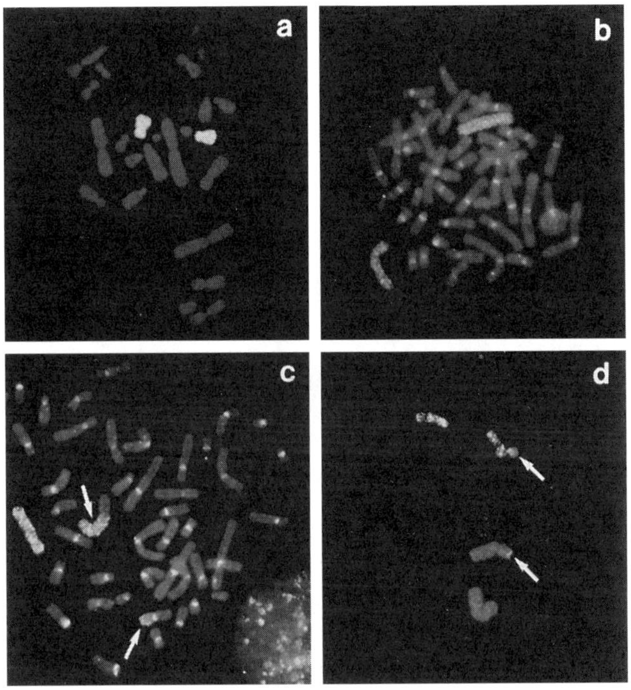

Figure 1. Chromosome-specific staining using FISH. (a) Human chromosomes specifically stained in metaphase spreads from human × hamster hybrid cells by hybridization with human genomic DNA. Hybridization signals appear yellow. Chromosomes to which the probe did not hybridize appear red. (b) Both copies of human chromosome 4 in a normal human metaphase spread stained by hybridizing a WCP for chromosome 4; staining the same as for (a). Centromeric regions were simultaneously stained with a probe that hybridizes to repeated DNA in the centromeric region of all chromosomes. (c) FISH to a metaphase spread from an irradiated cell population using a WCP to chromosome 4. A radiation-induced exchange between one copy of chromosome 4 and another chromosome is apparent. The derivative chromosomes appear red and yellow (arrows). The centromeres were stained as described in (b) to allow discrimination between translocations (one centromere) and dicentrics (two centromeres) (d) FISH with WCPs to chromosomes 7 (red) and 12 (green) to a metaphase spread prepared from human tumor. Derivative chromosomes resulting from a t(7;12) translocation are visible (arrows) as red and green chromosomes (Photomicrograph d courtesy of M. Vooijs.) (See color plate 3.)

commercially. Table 1 summarizes approaches to WCP production that have been reported and provides references to each.

2 APPLICATIONS

2.1 Species-Specific Staining of Human DNA in Human × Hamster Hybrid Cells

Figure 1a (see color plate 3) shows the results of fluorescence in situ hybridization (FISH) with human genomic DNA labeled with biotin and hybridized to a metaphase spread prepared from a human × hamster hybrid cell line. After hybridization, the slides were incubated with fluorescein-labeled avidin and counterstained with propidium iodide. The human chromosomes appear yellow because of the superposition of green fluorescence from the hybridized probe (detected with fluorescein) and red fluorescence (propidium iodide), while the hamster chromosomes appear red (propidium iodide only). The technique allows study of the chromosomal composition of human × hamster cell lines, the organization of the human chromosomes in the hybrid nuclei, and sensitive detection of small rearrangements between human and hamster chromosomes. It provides no information about the identity of the involved chromosomes.

2.2 Translocation Detection Using FISH with WCPs

2.2.1 Biological Dosimetry

FISH with WCPs allows analysis of the structural integrity of the target chromosomes in metaphase spreads. Figure 1b, for example, shows FISH to a normal human metaphase spread using a WCP to chromosome 4. The two intact number 4 chromosomes are stained yellow by the hybridization process, while the other chromosomes are stained red by the counterstain propidium iodide. Figure 1c shows FISH with a WCP for chromosome 4 to a metaphase spread prepared from a cell carrying a translocation involving one copy of chromosome 4. The derivative chromosomes resulting from the translocation are visible as chromosomes that have juxtaposed red and yellow segments. Translocations and other structural aberrations involving exchanges between nonhomologous chromosomes can be scored rapidly using FISH. As a result, this technique has been useful for assessment of structural aberration frequencies as a measure of the extent of radiation-induced genetic damage.

2.2.2 Analysis of Subtle or Complex Rearrangements

FISH with WCPs has proved to be a useful adjunct to conventional banding analysis for classification of complex chromosome rearrangements and for detection of subtle chromosomal exchanges. In one approach, FISH with WCPs is used to verify chromosome aberrations suggested by banding analysis. Figure 1d, for example, shows dual-color FISH with WCPs for chromosomes 7 and 12 to a metaphase spread prepared from cells reported to carry a t(7:12) translocation. The dual-color chromosomes resulting from the translocation are clearly visible.

Marker chromosomes in cytological preparations from tumors are often composed of fragments that are derived from two or more nonhomologous chromosomes. WCPs for chromosomes that are suspected to be involved in marker chromosome generation can be hybridized to tumor cell metaphase spreads. Figure 2a (see color plate 4), for example, shows the painting of a marker chromosome with a Spectrum Orange (red) labeled WCP specific for chromosome 3, which is frequently found rearranged in certain types of human solid tumor. Red signals that are specific for chromosome 3 material were found on both arms of the marker chromosome. The preparation was counterstained with the DNA-specific dye DAPI (blue). The banding pattern of DAPI DNA counterstain (blue) and red hybridization signals suggested that the chromosome 3 material on both arms of the marker chromosome was derived from the long arm of human chromosome 3. The hybridization mixture also contained a chromosome 17 specific, digoxigenin-labeled WCP, which was visualized with antibodies labeled with fluorescein isothiocyanate (FITC) against digoxigenin (green). This probe revealed aberrations of chromosome 17 in this cell line and demonstrated that the centromeric region of the marker chromosome (green) originated from chromosome 17. This approach requires either some indication of which probes to use or the performance of a large series

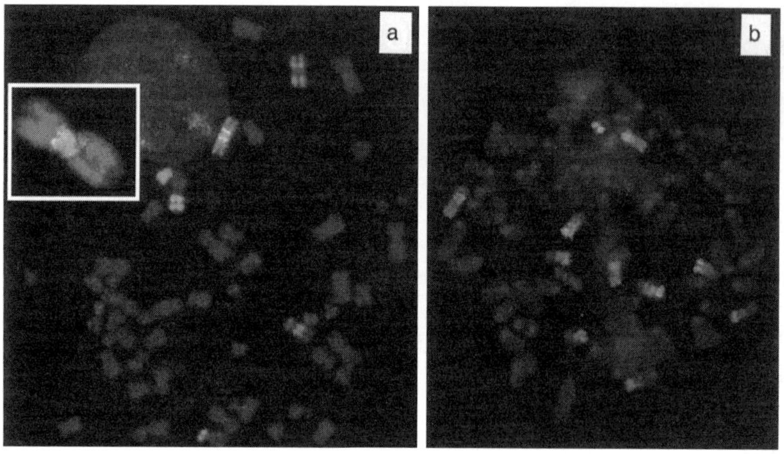

Figure 2. Application of chromosome painting for identification of marker chromosomes and probe mapping. (a) Combined hybridization of two or more chromosome painting probes allows definition of marker chromosome composition: a WCP for human chromosome 3 (Spectrum Orange, red) was combined with a WCP for chromosome 17 (FITC, green) and hybridized to metaphase spreads from the thyroid cancer cell line FTC236. The marker chromosome (insert) contains substantial parts derived from chromosome 3q and chromosome 17. Both painting probes were generated by DOP-PCR. Chromosomal DNA was counterstained with DAPI (blue). (Photomicrograph a courtesy of B. O'Brien.) (b) Assignment of a locus-specific probe (LSP) to mouse chromosome 11 by two-color FISH. Three chromosome-specific WCPs were hybridized simultaneously to expedite the mapping process. The WCPs specific for mouse chromosomes 9, 11, and 15, respectively, were generated by DOP-PCR from flow-sorted chromosomes and labeled by enzymatic incorporation of biotin during the last PCR cycles. Bound probes are visualized with avidin-FITC (green) following hybridization, and chromosomes are identified by their size as well as the pattern of chromosome painting. Metaphase spreads from a mouse 3T3 cell line showed three normal and one rearranged copy of chromosome 11. The LSP to be mapped was digoxigenin-labeled by random priming and detected with rhodamine-conjugated sheep-anti-digoxigenin antibodies (red). All four chromosomes containing the distal part of mouse chromosome 11 also showed the LSP hybridization signal. Chromosomal DNA was counterstained with DAPI (blue). (See color plate 4.)

of hybridizations. These obstacles can be overcome by the "reversed painting" approach described in Section 2.3.

2.3 Reversed Painting

The difficulty in discovering the chromosomal components of marker chromosomes can be overcome by preparing WCPs from the aberrant chromosomes (enriched either by chromosome sorting or by microdissection). FISH with these WCPs to normal metaphase spreads, termed *reversed painting,* immediately shows which chromosomes are involved. FISH with WCPs to the involved chromosomes can then be applied to metaphase spreads from the cells carrying the aberration to determine its structural organization. The major limitation to this analytic procedure is the need to prepare WCPs for each derivative chromosome. However, the development of DOP-PCR as described has greatly simplified this process, for the probe can be prepared by means of microdissecting or flow sorting only a few copies of the marker chromosome.

2.4 Demonstration of Syntenic Regions in Distantly Related Species

Comparative cytogenetics has provided a powerful tool to establish interspecies chromosome homologies and to test models for karyotype evolution. Recent improvements in FISH protocols have enabled a hybridization-based comparative chromosome painting capable of detecting homologous segments in species representing different mammalian orders (primate, rodents, even-toed ungulates, and whales). Comparative mapping is presently restricted by the limited number of chromosome-specific painting probes that are available for species other than mouse and human and by the relatively low DNA sequence homology between highly divergent species.

2.5 Chromosomal Organization of Cell Interphase Nuclei

FISH with WCPs to interphase nuclei coupled with optical sectioning microscopy allows investigation of the chromosomal organization of interphase nuclei. Such studies in somatic cells show clearly that chromosomes are organized in discrete, reasonably compact domains in interphase and that these domains remain well defined during progression through the cell cycle. Investigation of the extent to which the organization is the same in cells of the same type is now being pursued in several laboratories.

The high degree of chromatin condensation in particular cell types such as spermatocytes has hampered the investigation of chromosome organization in these cells in the past because accessibility for probes and bulky detection reagents was limited. The application of fluorochrome-labeled WCPs circumventing the immunocytochemical detection steps has shown promising first results, and the organization of chromosomes in germ cells is being studied intensively.

2.6 Detection of Aneusomy in Metaphase and Interphase Cells

Chromosomal territories marked by chromosomal painting can be scored to detect gain or loss of chromosomes in somatic cells in interphase. The number of a particular chromosome type can be determined easily in metaphase spreads by FISH, and thus large numbers of cells can be scored, facilitating detection of minimum residual disease. FISH with WCPs to interphase nuclei produces hybridization signal domains that often overlap, complicating accurate scoring. However, the mean number of chromosomal domains in a population of cells with an abnormal number of copies of a chromosome may still differ from values obtained for diploid controls, thus allowing detection of the abnormality if it is present in a high frequency of cells in the population under study.

2.7 Gene Mapping

The cytogenetic mapping of gene- or locus-specific probes (LSPs) is facilitated by the use of WCPs to counterstain chromosomes of a particular type. Chromosome identification using WCPs circumvents the need for chromosome banding and high quality metaphase spreads. Different WCPs can be combined to label several chromosomes simultaneously. The example shown in the Figure 2b demonstrates the use of three WCPs to specifically paint mouse chromosomes 9, 11, and 15, respectively. These mouse chromosomes differ slightly in size and are distinctly painted by the WCPs, permitting LSP probes to be mapped unambiguously. Figure 2 shows LSP-specific hybridization signals (red) on the distal long arm of chromosome 11 stained (green) by FISH with a WCP for chromosome 11. FISH with WCPs also facilitates detailed analysis of rearranged chromosomes or chromosomal fragments using LSPs by distinctly staining the derivative or marker chromosomes.

3 PERSPECTIVES

Directly labeled whole-chromosome probes are now available for almost all human chromosomes and some rodent chromosomes. Furthermore, WCPs labeled directly with fluorochromes greatly simplify application of whole-chromosome staining by eliminating time-consuming immunocytochemical detection steps. Multiple probes labeled with different fluorochromes can be analyzed simultaneously. WCPs for most of the human chromosomes are now commercially available, and the availability of probes for chromosome subregions is increasing rapidly. Thus, this powerful technique is rapidly approaching the point of being a candidate for routine use in the clinical laboratory.

See also Fluorescence Spectroscopy of Biomolecules; Gene Mapping by Fluorescence In Situ Hybridization; Gene Order by FISH and FACS; Partial Denaturization Mapping.

Bibliography

Carter, N. P. (1994) Cytogenetic analysis by chromosome painting. *Cytometry,* 18:2–10.

Collins, C., Kuo, W. L., Segraves, R., Fuscoe, J., Pinkel, D., and Gray, J. W. (1991) Construction and characterization of plasmid libraries enriched in sequences from single human chromosomes. *Genomics,* 11:997–1006.

Dauwerse, J. G., Wiegant, J., Raap, A. K., Breuning, M. H., and van Ommen, G. J. B. (1992) Multiple colors by fluorescence in situ hybridization using ratio-labeled DNA probes create a molecular karyotype. *Hum. Mol. Genet.* 1:593–598.

Gray, J. W., Dean, P. N., Fuscoe, J. C., Peters, D. C., Trask, B., van den Engh, G., and Van Dilla, M. A. (1987) High-speed chromosome sorting. *Science,* 238:323–329.

Leitch, A. R., Mosgoller, W., Schwarzacher, T., Bennett, M. D., and Heslop-Harrison, J. S. (1990) Genomic in situ hybridization to sectioned nuclei shows chromosome domains in grass hybrids. *J. Cell Sci.* 95:335–341.

Liu, P., Siciliano, J., Seong, D., Craig, J., Zhao, Y., de Jong, P. J., and Siciliano, M. J. (1993) Dual *Alu* polymerase chain reaction primers and conditions for isolation of human chromosome painting probes from hybrid cells. *Cancer Genet. Cytogenet.* 65:93–99.

Manuelidis, L. (1990) A view of interphase chromosomes. *Science,* 250:1533–1540.

Meltzer, P. S., Guan, X.-Y., Burgess, A., and Trent, J. M. (1992) Rapid generation of region specific probes by chromosome microdissection and their application. *Nature Genet.* 1:24–28.

Pinkel, D., Straume, T., and Gray, J. W. (1986) Cytogenetic analysis using quantitative, high-sensitivity, fluorescence hybridization. *Proc. Natl. Acad. Sci. U.S.A.* 83:2934–2938.

———, Landegent, J., Collins, C., Fuscoe, J., Segraves, R., Lucas, J., and Gray, J. W. (1988) Fluorescence in situ hybridization with human chromosome–specific libraries: Detection of trisomy 21 and translocations of chromosome 4. *Proc. Natl. Acad. Sci. U.S.A.* 85:9138–9142.

Scherthan, H., Cremer, T., Arnason, U., Weier, H.-U., Lima-de-Faria, A., and Froenicke, L. (1994) Comparative chromosome painting for demonstration of syntenic groups in evolutionary distant mammalian species. *Nature Genet.* 6:342–347.

Telenius, H., Pelmear, A. H., Tunnacliffe, A., Carter, N. P., Behmel, A., Ferguson-Smith, M. A., Nordenskjold, M., Pfragner, R., and Ponder, B. A. (1992) Cytogenetic analysis by chromosome painting using DOP-PCR amplified flow-sorted chromosomes. *Genes, Chromosomes Cancer* 4:257–263.

Trask, B., and Pinkel D. (1990) Fluorescence in situ hybridization with DNA probes. *Methods Cell Biol.* 33:383–400.

Vooijs, M., Yu, L. C., Tchachuk, D., Pinkel, D., Johnson, D., and Gray, J. W. (1993) Libraries for each human chromosome constructed from sorter-enriched chromosomes using linker-adapter PCR. *Am. J. Hum. Genet.* 52:586–597.

Wang, M. L., Leitch, A. R., Schwarzacher, T., Heslop-Harrison, J. S., and Moore, G. (1992) Construction of a chromosome-enriched *Hpa* II library from flow-sorted wheat chromosomes. *Nucleic Acids Res.* 20:1897–1901.

Weier, H.-U., Polikoff, D., Fawcett, J., Lee, K.-H., Cram, S., Chapman, V., and Gray, J. W. (1994) Generation of five high complexity painting probe libraries from flow sorted mouse chromosomes. *Genomics,* 24:641–644.

X

Xenobiotic Metabolism

Bernard Testa

1 Introduction
1.1 Definitions and Concepts
1.2 Factors Affecting Xenobiotic Metabolism

2 Monooxygenases and Their Reactions
2.1 Monooxygenases
2.2 Reactions of Carbon Oxidation
2.3 Reactions of Nitrogen Oxygenation and Oxidation
2.4 Reactions of Oxidation of Sulfur, Oxygen, and Other Atoms

3 Other Oxidoreductases and Their Reactions
3.1 Oxidoreductases
3.2 Reactions of Oxidation

4 Peroxidases and Their Reactions
4.1 Peroxidases
4.2 Reactions of Oxidation and Oxygenation

5 Reductions Catalyzed by Cytochrome P450 and Other Reductases
5.1 Reductions at Carbon Atoms
5.2 Reductions at Other Atoms

6 Hydrolases and Their Reactions
6.1 Hydrolases
6.2 Reactions of Hydration and Hydrolysis

7 Transferases and Their Reactions
7.1 Methylation
7.2 Sulfoconjugation
7.3 Glucuronidation
7.4 Acetylation and Other Reactions of Acylation
7.5 Acyl–Coenzyme A Thioesters as Metabolic Intermediates
7.6 Conjugation with Glutathione
7.7 Other Reactions of Conjugation

8 Concluding Remarks

Key Words

Chemospecificity The specificity of an enzyme in terms of the type(s) of reaction it catalyzes.

Conjugation, Reactions of Metabolic reactions by which a xenobiotic or a metabolite thereof is coupled with a usually polar endogenous molecule (an "endocon"). Also less adequately called phase II reactions.

Detoxication The result of metabolism by which a toxic chemical (be it a xenobiotic or metabolite thereof) is rendered nontoxic or nonreactive. In the first half of the twentieth century, all metabolic reactions were considered as a rule to result in detoxication.

Functionalization Reactions Metabolic reactions that introduce or unmask a functional group in the substrate. Also called phase I reactions.

Genetic Polymorphism (in Xenobiotic Metabolism) Genetically determined differences in the activity of a given xenobiotic-metabolizing enzyme, leading to the classification of individuals into extensive ("normal") and slow ("defective") metabolizers of the enzyme's substrates.

Interindividual Factors Factors affecting xenobiotic metabolism that by definition remain constant during the life span of the organism or individual.

Intraindividual Factors Factors affecting xenobiotic metabolism that can change rapidly or slowly in any given individual. These are related to pathological states, physiological changes, or external agents affecting, for example, the hormonal balance, immunological reactions, or enzymic activity of individuals.

Macromolecular Conjugates A special type of metabolite resulting from the covalent binding of reactive metabolites (or, on occasion, unchanged, chemically reactive xenobiotics) to biomacromolecules such as proteins.

Metabolism In the broad sense, metabolism is synonymous with biodisposition: that is, the sum of the processes (absorption, distribution, storage, excretion) affecting the fate of a chemical substance in the body. In the narrow sense, it means biotransformation.

Product Specificity (and Product Selectivity) The property of an enzyme (and the characteristic of a metabolic reaction) to transform a substrate into one product faster than into another. For example, the product enantioselectivity of a metabolic reaction is observed when a prochiral substrate is metabolized at different rates to two enantiomeric products.

Selectivity The detectable expression of the specificity of an enzyme: that is, the property of a metabolic reaction to select some substrate(s) or produce some metabolite(s).

Specificity A property or set of properties specifying the catalytic behavior of an enzyme (e.g., its substrate specificity and product specificity).

Substrate Specificity (and Substrate Selectivity) The property of an enzyme (and the characteristic of a metabolic reaction) to transform some substrate(s) faster than other(s). For example, the substrate enantioselectivity of a metabolic reaction is observed when one enantiomer is metabolized faster than the other.

Toxication The result of metabolism by which a xenobiotic is activated to a toxic metabolite, or to a reactive intermediate able to leave the catalytic site and elicit chemical damage.

Xenobiotics Compounds that are foreign to the body (i.e., have no nutritional or other physiological role to fulfill). Examples include drugs, food additives, agrochemicals, industrial chemicals, and pollutants.

Xenobiotic metabolism is an old science, born in 1841 with the isolation from horse urine of hippuric acid as a metabolite of benzoic acid. The major metabolic pathways were discovered in the decades that followed. By the first half of the twentieth century, the role of enzymes in the biotransformation of xenobiotics was recognized, and the belief was firmly held that metabolism could only mean detoxication.

The second half of the twentieth century introduced many concepts and generalizations, and it was recognized that a large number of substrates undergo biotransformation reactions that generate toxic intermediates or metabolites. Thus xenobiotic metabolism may also result in toxication, a major reason for the great growth of this science in the last decades. In recent years, the contributions of molecular biology and computational chemistry have been particularly impressive.

1 INTRODUCTION

1.1 Definitions and Concepts

Xenobiotics (i.e., "foreign compounds") are substances that enter the body but have no nutritional or physiological value. They include drugs, food constituents devoid of physiological roles, food additives, chemicals of "pleasure" or abuse (e.g. ethanol, caffeine, nicotine, hallucinogens), agrochemicals, industrial and technical chemicals, and pollutants of natural, microbiological, or synthetic origin.

The study of the disposition—or fate—of xenobiotics in living systems includes the consideration of their absorption into the organism, how and where they are distributed and stored, the chemical and biochemical transformations they may undergo, and how and by what route(s) they are finally excreted and returned to the environment. The word "metabolism" has acquired two meanings, one is synonymous with disposition (i.e., the sum of the processes affecting the fate of a chemical substance in the body), and the second with biotransformation, as understood in this article.

The metabolism of drugs and other xenobiotics is typically a biphasic process in which the compound first undergoes a reaction of functionalization (also called phase I reaction) involving oxidation, reduction, or hydrolysis. This introduces or unmasks a functional group, typically a hydroxyl or amino group, suitable for linkage with an endogenous molecule or moiety in a second metabolic step known as a reaction of conjugation (also called phase II reaction). In a number of cases, phase I metabolites may be excreted prior to conjugation, while many xenobiotics can be directly conjugated. Alternatively, reactions of functionalization may follow some reactions of conjugation.

Xenobiotic biotransformation thus produces two types of metabolites, namely, functionalized products and conjugates. In addition, biochemists and pharmacologists have progressively come to recognize the existence of a third class of metabolites, the macromolecular conjugates (also called xenobiotic–macromolecule adducts), which are formed when a xenobiotic binds covalently to a biological macromolecule, usually following metabolic activation. Such adducts may have particular toxicological relevance.

Reactions of xenobiotic metabolism, like all other enzymatic reactions, are qualified as selective or specific. In biochemistry, the specificity of an enzyme often covers an ensemble of properties, the description of which makes it possible to list the components of the enzyme's behavior. In contrast, the term "selectivity" is applied to metabolic processes, indicating that a given metabolic reaction or pathway is able to select some substrates or generate some products from a larger set. In other words, the selectivity of a metabolic reaction is the detectable expression of the specificity of an enzyme.

Selectivities (or specificities) of different types are encountered in biochemistry generally and in xenobiotic metabolism in particular. What characterizes an enzyme from a catalytic viewpoint is first its chemospecificity, that is, its specificity in terms of the type(s) of reaction it catalyzes. When two or more substrates are metabolized at different rates by a single enzyme under identical conditions, substrate selectivity is observed. Such a definition does not consider the nature of the product(s) and their isomeric relationship. Substrate selectivity is distinct from product selectivity, which is observed when two or more metabolites are formed at different rates by a single enzyme from a single substrate. Thus, substrate-selective reactions discriminate between different compounds, while product-selective reactions discriminate between different groups or positions in a given compound.

The substrates being metabolized at different rates may share various types of structural relationship. They may be analogues, in which case the term of "substrate selectivity" is used in a narrow sense. Alternatively, the substrates may be isomers of various kinds, resulting in substrate regioselectivity or substrate stereoselectivity. Products formed at different rates in product-selective reactions may also share various types of relationship. Thus, they may be analogues, regioisomers, or stereoisomers, resulting in product selectivity (narrow sense), product regioselectivity, or product stereoselectivity. Note that the product selectivity displayed by two distinct substrates in a given metabolic reaction may be different; in other words, the product selectivity may be substrate selective. The term "substrate–product selectivity" can be used to describe such complex cases.

1.2 Factors Affecting Xenobiotic Metabolism

A variety of physiological and pathological factors influence xenobiotic metabolism and may thus affect the wanted and unwanted activities of drugs. Such factors are usually subclassified as interindividual and intraindividual, depending on whether they vary between or within given organisms or individuals, respectively (Table 1).

Table 1 Factors Affecting Xenobiotic Metabolism

Interindividual Factors	Intraindividual Factors
Animal species	Age
Genetic factors	Biological rhythms
Sex	Pregnancy
	Stress
	Diseases
	Nutritional factors
	Enzyme induction
	Enzyme inhibition

The interindividual factors by definition remain constant during the life span of an organism or individual. Species differences in xenobiotic metabolism are an important issue pertaining to the extrapolation of animal data to humans. Genetic differences result from the deficiency of some xenobiotic-metabolizing enzymes among a number of individuals. The consequences of this genetic polymorphism are an impaired metabolism of the drugs that are substrates of such enzymes, with a marked risk of overdose or therapeutic failure in affected individuals. Sex may also be a factor of significance in a number of cases, which may reflect hormonal influences on enzyme activities.

Intraindividual factors are related to pathological states, physiological changes, or external agents affecting, for example, the hormonal balance, immunological reactions, or enzymic activity of individuals. The age of a person may markedly influence his or her ability to metabolize drugs, especially at the extremes of life. The study of biological rhythms is the realm of chronopharmacology. Pregnancy affects drug metabolism as a result of the profound hormonal and physiological changes in the mother, while the intrinsic activity of the placenta also comes into play. Much information is available on the influence of disease, but relatively few rationalizations and explanatory mechanisms have been proposed. Nutritional factors such as diet, nutrients, and starvation influence xenobiotic metabolism. As far as drug therapy and toxicology are concerned, factors of even greater significance are enzyme induction and enzyme inhibition. Enzyme inducers act by increasing the concentration, hence the activity, of some enzymes, while inhibitors decrease the activity of some enzymes by reversible inhibition or irreversible inactivation.

2 MONOOXYGENASES AND THEIR REACTIONS

2.1 Monooxygenases

Monooxygenation reactions involve the splitting of molecular oxygen into two atoms, one of which forms H_2O while the other oxygenates the substrate [XH in equation (1)]:

$$XH + O_2 + 2e^- + 2H^+ \rightarrow XOH + H_2O \qquad (1)$$

All monooxygenases must bind the substrate and O_2 and need two electrons for O_2 reduction. A variety of different enzymes have evolved which activate oxygen according to equation (1) and transfer it to a substrate. Despite this superficial mechanistic uniformity, the various types of monooxygenases differ markedly in their structure and properties. As far as xenobiotic metabolism is concerned, the major monooxygenases are the cytochromes P450 (EC 1.14.14.1, also EC 1.14.15.1, 1.14.15.3, 1.14.15.4–1.14.15.6), the flavin-containing monooxygenases [dimethylaniline monooxygenase (N-oxide forming), EC 1.14.13.8], and dopamine β-hydroxylase (dopamine β-monooxygenase, EC 1.14.17.1).

Cytochromes P450 (CYP) are by far the most important xenobiotic-and endobiotic-metabolizing monooxygenases. They form a very large group of hemoproteins encoded by the *CYP* gene superfamily and are classified in families and subfamilies as summarized in Table 2. A cytochrome P450 enzyme consists of a protein and a heme moiety, iron-protoporphyrin IX, which is common not only to all cytochrome P450 enzymes but also to other hemoproteins and enzymes such as hemoglobin, myoglobin, catalase, and most peroxidases. In contrast to the constant porphyrin, the protein part of

Table 2 The Humsan P450 Gene Superfamily: A Table of Gene Products

Families	Subfamilies	Gene Products
P450 1 (mammalian aryl hydrocarbon hydroxylases; xenobiotic metabolism inducible by polycyclic aromatic hydrocarbons)	P450 1A	CYP1A1 CYP1A2
P450 2 (mammalian; xenobiotic and steroid metabolism; constitutive and xenobiotic-inducible)	P450 2A	CYP2A6 CYP2A7
	P450 2B (includes phenobarbital-inducible forms)	CYP2B6
	P450 2C (constitutive forms; includes sex-specific forms)	CYP2C8 CYP2C9 CYP2C10 CYP2C18 CYP2C19
	P450 2D	CYP2D6
	P450 2E (ethanol-inducible)	CYP2E1
	P450 2F	CYP2F1
P450 3 (mammalian; xenobiotic and steroid metabolism; steroid-inducible)	P450 3A	CYP3A3 CYP3A4 CYP3A5
P450 4 (mammalian fatty acid ω- and (ω-1)-hydroxylases; peroxisome proliferator-inducible)	P450 4A	CYP4A9 CYP4A11
	P450 4B	CYP4B1
	P450 4F	CYP4F2
P450 5 (TXA synthetase)		CYP5
P450 7 (mammalian cholesterol 7α-hydroxylase)		CYP7
P450 11 (mammalian mitochondrial steroid hydroxylases)	P450 11A (cholesterol side-chain cleavage)	CYP11A1
	P450 11B (steroid 11β-hydroxylases)	CYP11B1 CYP11B2
P450 17 (mammalian steroid 17α-hydroxylase)		CYP17
P450 19 (mammalian steroid aromatase)		CYP19
P450 21 (mammalian steroid 21-hydroxylases)	P450 21A	CYP21A2
P450 27 (mammalian steroid hydroxylase; mitochondrial)		CYP27

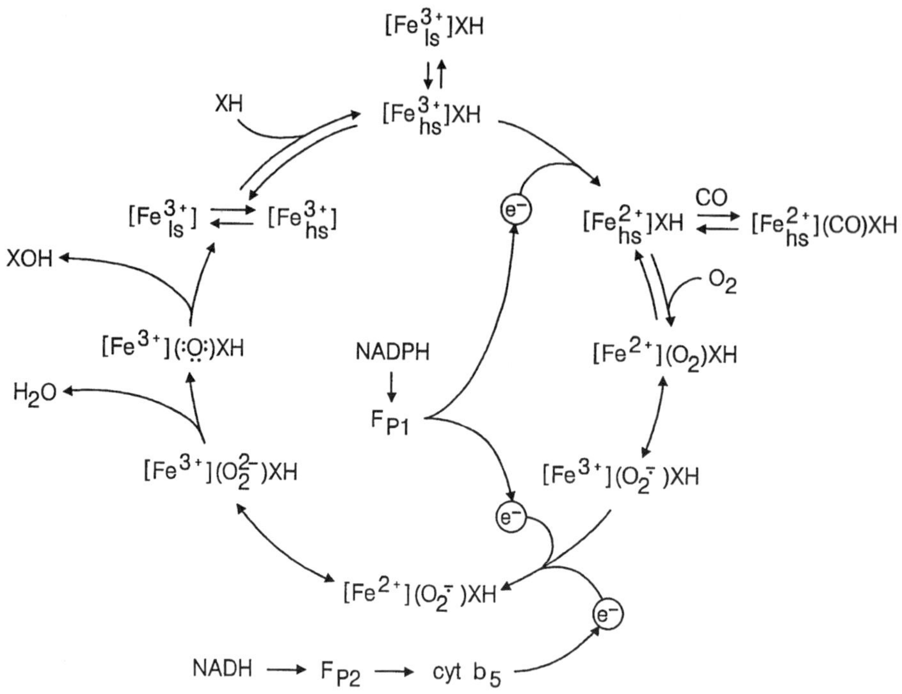

Figure 1. The catalytic cycle of cytochrome P450, starting with the binding of the substrate (XH) and ending with its oxygenation.

these enzymes varies markedly from one enzyme/isozyme to the other and accounts for the differences in their properties, including molecular weight (approximate range 45–60 kDa), substrate and product specificities, and sensitivity to inhibitors. The catalytic cycle of cytochrome P450, shown in Figure 1, reveals:

- how the substrate (XH) binds to the high spin form of the enzyme
- how the enzyme–substrate complex undergoes a single-electron reduction
- how molecular oxygen binds to form a ternary complex
- how this ternary complex undergoes the second single-electron reduction step
- how molecular oxygen, which now exists as O_2^{2-}, can undergo cleavage to O^{2-} (which protonates to H_2O) and an electrophilic oxene atom
- how oxene will attack the substrate, either oxygenating or oxidizing it.

In the endoplasmic reticulum (microsomes), NADPH–cytochrome P450 reductase (NADPH:ferrihemoprotein oxidoreductase, EC 1.6.2.4) is the major oxidoreductase which catalyzes the transfer of the first and second electron from NADPH to cytochrome P450. A second electron transport system exists, namely, NADH–cytochrome b_5 reductase (NADH:ferricytochrome b_5 oxidoreductase, EC 1.6.2.2); this is a flavoprotein that takes its reduction equivalents from NADH and reduces cytochrome b_5, which in turn can supply the second electron to cytochrome P450.

Cytochrome P450 is not only the major drug-metabolizing enzyme system, using most drugs as substrates and playing a key role in detoxication and toxication—it is also of significance in medicinal chemistry because several CYP enzymes are drug targets [e.g., thromboxane synthase (CYP5) and steroid aromatase (CYP19)].

2.2 Reactions of Carbon Oxidation

The most common reactions of oxidation involving carbon atoms in substrate molecules are summarized in Figure 2. The reactions include attacks at sp^3-carbons (pathways 1–6), sp^2-carbons (pathways 7 and 8), and sp-carbons (pathway 9).

In the simplest cases, nonactivated sp^3-carbons in alkyl groups undergo cytochrome P450 catalyzed hydroxylation, the terminal and penultimate positions being the preferred but not exclusive sites of attack (pathways 1 and 2, respectively). Such hydroxylated metabolites, but also xenobiotic alcohols, can then be further oxidized by dehydrogenases as discussed in Section 3.2. In addition, recent evidence indicates that for a number of substrates, the oxidation of primary and secondary alcohols and of aldehydes can also be catalyzed by cytochrome P450. A typical example is the C(10)-demethylation of androgens and analogues catalyzed by aromatase (CYP19).

Desaturation is a special case of carbon oxidation that produces an olefinic group (pathway 3). An interesting example is provided by testosterone, which, among many cytochrome P450 catalyzed reactions, undergoes allylic hydroxylation of 6β-hydroxytestosterone and desaturation to 6,7-dehydrotestosterone.

There is a known regioselectivity in cytochrome P450 catalyzed hydroxylations of carbon atoms adjacent (alpha) to an unsaturated system (pathway 4) or a heteroatom such as N, O, or S (pathway 5). In the former cases, hydroxylation can easily be followed by dehydrogenation (not shown). In the latter cases, however, the hydroxylated metabolite is usually unstable and undergoes a rapid, postenzymatic hydrolytic cleavage. Depending on the substrate, this pathway produces a secondary or primary amine, an alcohol or phenol, or a thiol, while the alkyl group is cleaved as an aldehyde or a ketone. Pathway 5 is a most frequent one as far as drug metabolism

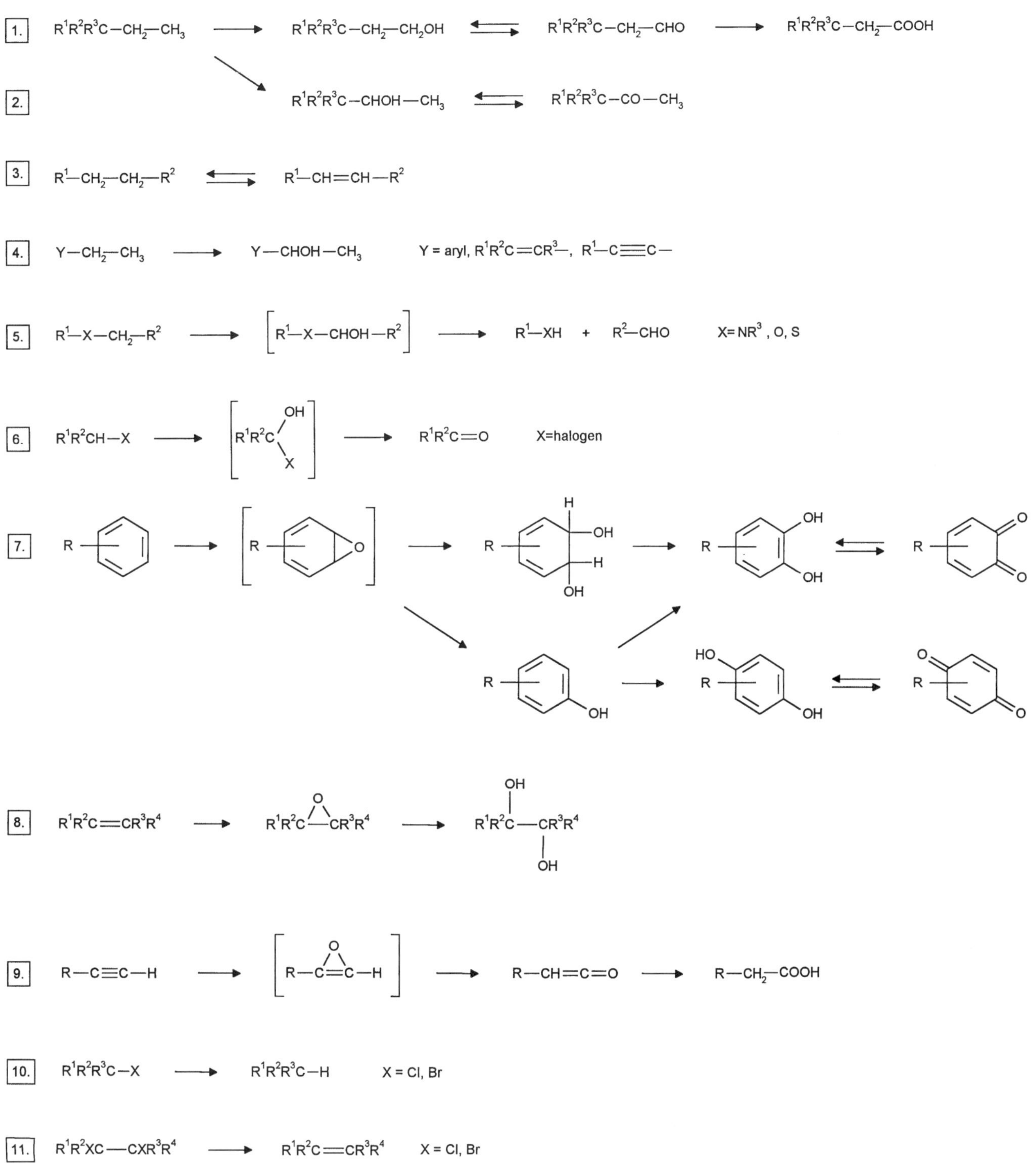

Figure 2. Major reactions of functionalization involving sp^3-, sp^2-, or sp-carbons in substrate molecules: mainly oxidations (oxygenations and dehydrogenations) and reductions (hydrogenations), plus some reactions of hydrolysis and postenzymatic rearrangement.

is concerned, since it accounts for many well-known metabolic reactions of N—C cleavage such as N-dealkylations and deaminations, O—C cleavage (e.g., O-dealkylations, O-demethylations), and S—C cleavage (S-dealkylations). Note that the actual mechanism of such reactions is usually more complex than that shown here and may involve intermediate oxidation of the heteroatom.

Aliphatic carbon atoms bearing one or more halogen atoms (mainly chlorine or bromine) can be similarly metabolized. Thus, oxidative dehalogenation as catalyzed by cytochrome P450 proceeds by hydroxylation followed by loss of HX (pathway 6).

Reactions at sp^2-carbons are characterized by their own pathways, catalytic mechanisms, and products (pathways 7 and 8). Thus, the oxidation of aromatic rings generates a variety of (usually stable) metabolites. Their common precursor is often a reactive epoxide (pathway 7), which may be hydrolyzed by epoxide hydrolase to a dihydrodiol (Section 6.2); alternatively, it may be rearranged under proton-catalysis to a phenol. Dehydrogenation of dihydrodiols by dihydrodiol dehydrogenase restores aromaticity with production of a catechol metabolite. The further oxidation of phenols and phenolic metabolites is also possible, the rate of reaction and the nature of products depending on the ring and on the nature and position of the substituents. Although catechols are commonly formed, hydroquinones are sometimes also produced.

Olefinic bonds in xenobiotic molecules can also be targets of cytochrome P450 catalyzed epoxidation (pathway 8). In contrast to arene oxides, the resulting epoxides are fairly stable and have often been isolated and characterized. Acetylenic bonds also are targets for cytochrome P450 catalyzed oxidation. Oxygenation of the triple bond (pathway 9) yields an intermediate that, depending on the substrate, can react in a number of ways, forming, for example, a highly reactive ketene whose hydration produces a substituted acetic acid.

2.3 Reactions of Nitrogen Oxygenation and Oxidation

The main metabolic reactions of oxygenation and oxidation nitrogen atoms in organic molecules are summarized in Figure 3. These reactions are catalyzed by cytochrome P450 and/or flavin-containing monooxygenases. The principal functional groups involved are amines and amides and their oxygenated metabolites. A few other functional groups are considered, as well.

Nitrogen oxygenation is an (apparently) straightforward metabolic reaction of tertiary amines (pathway 1), be they aliphatic or

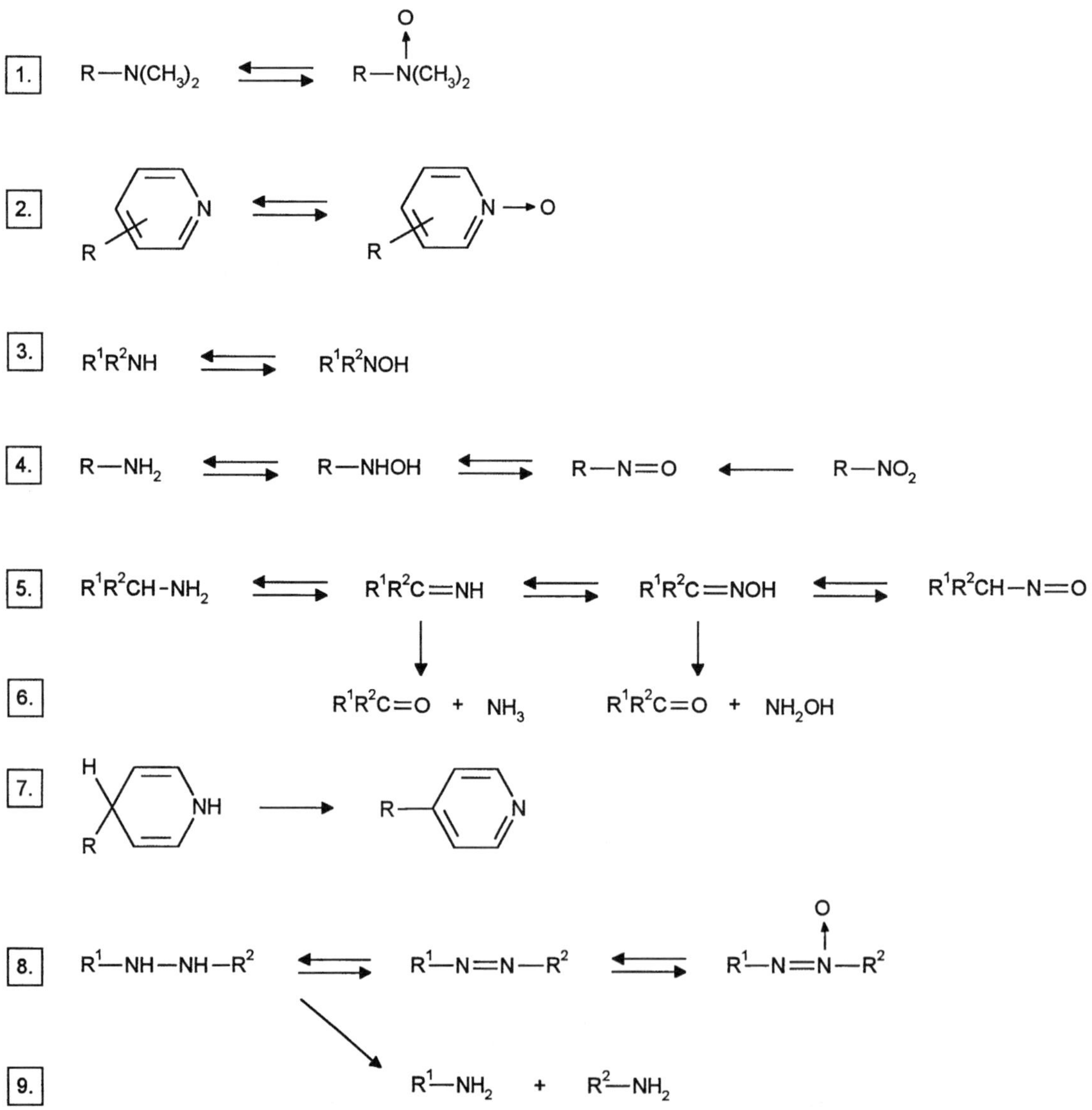

Figure 3. Major reactions of functionalization involving nitrogen atoms in substrate molecules: mainly oxidations (oxygenations and dehydrogenations) and reductions (deoxygenations and hydrogenations).

aromatic. Identical considerations apply to pyridines and analogous aromatic azaheterocycles (pathway 2).

Secondary and primary amines also undergo N-oxygenation, the first isolable metabolites being hydroxylamines (pathways 3 and 4). These compounds can be aliphatic or aromatic amines, but they can also be secondary or primary amides (i.e., R = acyl). As expected, tertiary amides are resistant to N-oxygenation. The oxidation of secondary amines and amides usually stops at the hydroxylamine/hydroxylamide level, but formation of short-lived nitroxides (not shown) has been reported.

As opposed to secondary amines and amides, their primary analogues can be oxidized to nitroso metabolites (pathway 4). Further oxidation of the latter compounds to nitro compounds does not seem to occur in vivo. Note that primary aliphatic amines having a hydrogen on the alpha carbon can display additional metabolic reactions, as shown in pathways 5 and 6. Indeed, N-oxidation also yields imines whose further oxidation produces oximes which are in equilibrium with their nitroso tautomer. Hydrolytic cleavage of imines and oximes are pathways of deamination.

1,4-Dihydropyridines, and particularly the calcium channel blockers, are effectively oxidized by cytochrome P450. The reaction is one of aromatization (pathway 7) to yield the corresponding pyridine. Dinitrogen moieties are also targets of oxidoreductases. Depending on their substituents, hydrazines are oxidized to azo compounds (pathway 8), some of which can be oxygenated to azoxy compounds.

2.4 Reactions of Oxidation of Sulfur, Oxygen, and Other Atoms

The major redox reactions occurring at sulfur atoms in organic compounds are summarized in Figure 4.

Thiol compounds can be oxidized to sulfenic acids, then to sulfinic acids, and finally to sulfonic acids (pathway 1). Depending on the substrate, the pathway is mediated by cytochrome P450 and/or flavin-containing monooxygenases. Another route of oxidation of thiols leads to disulfides either directly (pathway 2, via thiyl radicals), or by dehydration between a thiol and a sulfenic acid. However, our understanding of sulfur biochemistry is largely incomplete, and much remains to be understood.

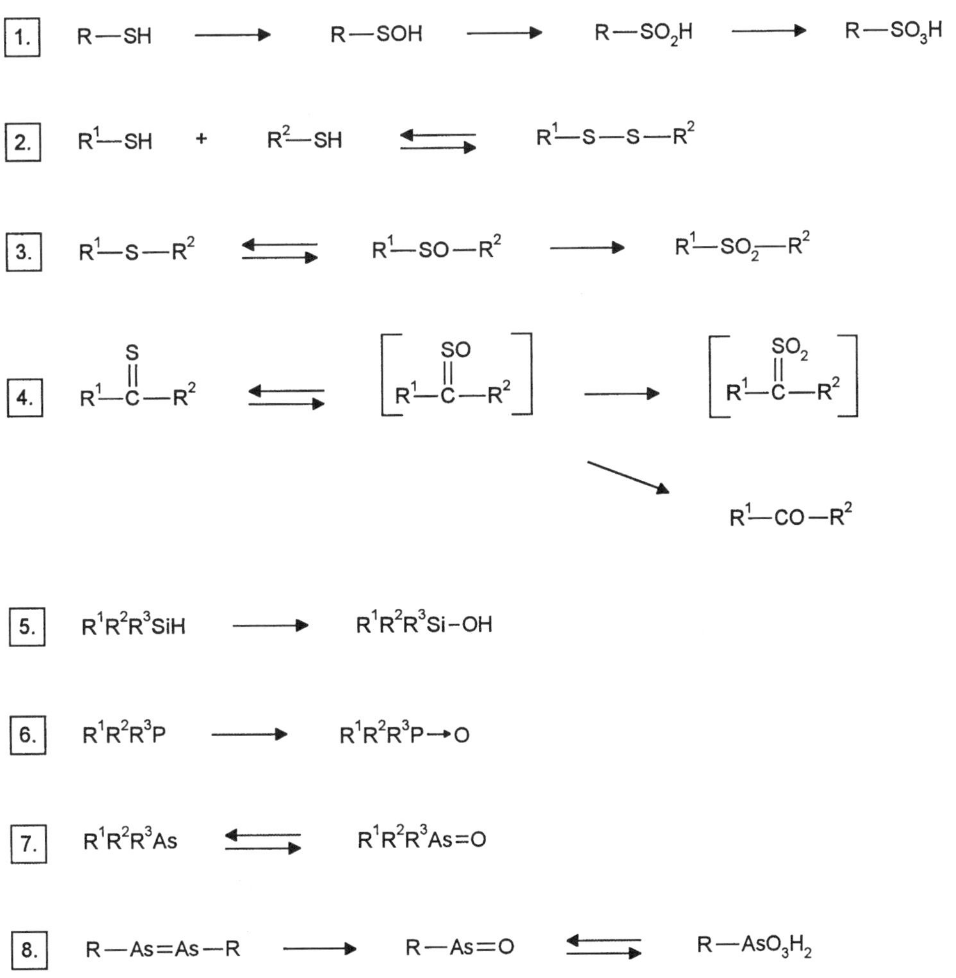

Figure 4. Major reactions of functionalization involving sulfur and a few other atoms in xenobiotics: oxidations (oxygenations and dehydrogenations) and reductions (deoxygenations and hydrogenations).

The metabolism of sulfides (thioethers) is rather straightforward. Besides reactions of S-dealkylation mentioned in Section 2.2, these compounds can also be oxygenated by monooxygenases to sulfoxides and then to sulfones (pathway 3). Thiocarbonyl compounds are also substrates of monooxygenases, forming S-monoxides (sulfines) and then S-dioxides (sulfenes) (pathway 4). As a rule, the reactivity of these metabolites is so high that they cannot be identified as such. Thus, S-monoxides rearrange to the corresponding carbonyl by expelling a sulfur atom, a reaction known as oxidative desulfuration, which is observed with thioamides and thioureas (e.g., thiopental). As for the S-dioxides, they react very rapidly with nucleophilic sites in biological macromolecules. This covalent binding results in the formation of adducts, a mechanism believed to account for the carcinogenicity of a number of thioamides.

Other elements of lesser quantitative importance in medicinal chemistry that can enter redox reactions are compounds of silicon, phosphorus, arsenic, and selenium, among others. Note, however, that the enzymology and mechanisms of these reactions are insufficiently understood. For example, a few silanes have been shown to yield silanols in vivo (pathway 5). The same applies to some phosphines, which can be oxygenated to phosphine oxides by monooxygenases (pathway 6).

Arsenicals have received more attention as a result of their therapeutic significance. Both inorganic and organic arsenic compounds display an As(III)–As(V) redox equilibrium in the body. This is illustrated with the arsine–arsine oxide and arsenoxide–arsonic acid equilibria shown in pathways 7 and 8, respectively. Another reaction of interest is the oxidation of arseno compounds to arsenoxides (pathway 8), a reaction of importance in the bioactivation of a number of chemotherapeutic arsenicals.

3 OTHER OXIDOREDUCTASES AND THEIR REACTIONS

3.1 Oxidoreductases

Alcohol dehydrogenases (ADHs, alcohol:NAD$^+$ oxidoreductase, EC 1.1.1.1) are zinc-containing enzymes found in the cytosol of the mammalian liver and in various extrahepatic tissues. Mammalian liver alcohol dehydrogenases (LADHs) are dimeric enzymes. The human enzymes belong to three different classes: class I, comprising the various isozymes that are homodimers or heterodimers of the α, β, and γ subunits (e.g., the $\alpha\alpha$, $\beta_1\beta_1$, $\alpha\beta_2$, and $\beta_1\gamma$ isozymes); class II, comprising the $\pi\pi$ enzyme; and class III, comprising the $\chi\chi$ enzyme.

Enzymes categorized as aldehyde reductases include alcohol dehydrogenase (NADP$^+$) [alcohol:NADP$^+$ oxidoreductase; aldehyde reductase (NADPH), EC 1.1.1.2], aldehyde reductase [alditol: NAD(P$^+$) 1-oxidoreductase; aldose reductase, EC 1.1.1.21], and many others. Aldehyde reductases occur in a considerable number of mammalian tissues. Their subcellular location is cytosolic and sometimes also mitochondrial. The so-called ketone reductases include α- and β-hydroxysteroid dehydrogenases (e.g., EC 1.1.1.50 and EC 1.1.1.51), various prostaglandin ketoreductases (e.g. prostaglandin-F synthetase, EC 1.1.1.188; prostaglandin-E_2 9-reductase, EC 1.1.1.189), and many others that are comparable to aldehyde reductases. One group of particular importance consists of the carbonyl reductases (NADPH, EC 1.1.1.184). Furthermore, the many similarities between monomeric, NADPH-dependent aldehyde reductase (AKRI), aldose reductase (AKR2), and carbonyl reductase (AKR3) have led to their designation as aldo-keto reductases. Other reductases that have a role to play in xenobiotic metabolism include glutathione reductase (NADPH:oxidized-glutathione oxidoreductase, EC 1.6.4.2) and quinone reductase [NAD(P)H:(quinone acceptor) oxidoreductase; DT-diaphorase, EC 1.6.99.2].

Aldehyde dehydrogenases [aldehyde:NAD(P)$^+$ oxidoreductases, EC 1.2.1.3 and EC 1.2.1.5] exist in multiple forms in the cytosol, mitochondria, and microsomes of various mammalian tissues. Aldehyde dehydrogenases form a superfamily of related enzymes consisting of class 1 enzymes (cytosolic), class 2 enzymes (mitochondrial), and class 3 enzymes (tumor-associated, and other isozymes). In all three major classes, constitutive and inducible isozymes exist. Dihydrodiol dehydrogenases (*trans*-1,2-dihydrobenzene-1,2-diol:NADP$^+$ oxidoreductases, EC 1.3.1.20) are cytosolic enzymes whose preferred cofactor is NADP$^+$.

Other oxidoreductases that can play a role in xenobiotic metabolism are hemoglobin, monoamine oxidases (MAO-A and MAO-B, EC 1.4.3.4), which are essentially mitochondrial enzymes, the cytosolic molybdenum hydroxylases (xanthine oxidase, EC 1.1.3.22; xanthine dehydrogenase, EC 1.1.1.204; aldehyde oxidase, EC 1.2.3.1), and the broad group of copper-containing amine oxidases (EC 1.4.3.6).

3.2 Reactions of Oxidation

A number of reactions catalyzed by the enzymes named in Section 3.1 can be found in Figures 2 and 3. Thus, the dehydrogenation of primary or secondary alcohols by dehydrogenases yields aldehydes or ketones, respectively (see pathways 1 and 2 in Figure 2). These dehydrogenations are reversible, meaning that the enzymes catalyze the reactions in both directions. And while a ketone can hardly undergo further enzymatic oxidation, aldehydes are good substrates for aldehyde dehydrogenases or other enzymes and lead irreversibly to carboxylic acid metabolites. A classical example is that of ethanol, which in the body exists in redox equilibrium with acetaldehyde, a metabolite that is rapidly and irreversibly oxidized to acetic acid.

The dehydrogenation of dihydrodiols to catechols (pathway 7, Figure 2) is typically mediated by dihydrodiol dehydrogenase. Monoamine oxidase catalyzes the oxidative deamination of primary amines, mainly endogenous but also exogenous, in a reaction involving imine formation and hydrolysis (pathways 5 and 6, Figure 3). The role of these oxidoreductases in mediating other reactions of oxidation in Figures 3 and 4 remains to be investigated.

4 PEROXIDASES AND THEIR REACTIONS

4.1 Peroxidases

Various peroxidases are progressively being recognized as important enzymes in drug metabolism. Thus, protaglandin–endoperoxide synthetase (EC 1.14.99.1) is able to use a number of xenobiotics as cofactors in a reaction of co-oxidation. A variety of other peroxidases may oxidize drugs (e.g., catalase, EC 1.11.1.6; myeloperoxidase, donor:hydrogen-peroxide oxidoreductase, EC 1.11.1.7). Interestingly, some cytochrome P450 enzymes have been shown to have peroxidase activity.

4.2 Reactions of Oxidation and Oxygenation

Catechols and hydroquinones can undergo further oxidation to quinones (pathway 7, Figure 2). Such reactions occur by two single-electron steps and can be either enzymatic or nonenzymatic (i.e., resulting from autoxidation and yielding as by-product the superoxide anion radical $O_2^{\cdot -}$). The intermediate in this reaction is a semiquinone. Both quinones and semiquinones are reactive, in particular toward biomolecules, and have been implicated in many reactions of toxication. For example, the high toxicity of benzene for bone marrow is believed to be explainable by the myeloperoxidase-catalyzed oxidation of catechol and hydroquinone.

Besides phenols and alcohols, other xenobiotics or metabolites can be substrates of peroxidase activities. Thus, aromatic amines can undergo nitrogen oxygenation to yield nitro compounds (pathway 4, Figure 3). Aromatic amines can also be N-oxidized by removal of one electron to yield an aminium radical cation that reacts in a number of ways, leading, for example, to dimers and adducts (not shown). Similarly, thiols are oxidized to thiyl radicals and oxygenated metabolites.

5 REDUCTIONS CATALYZED BY CYTOCHROME P450 AND OTHER REDUCTASES

A number of reduction reactions have been demonstrated in the metabolism of xenobiotics. From a quantitative viewpoint, they are less important than oxidations, since our *milieu intérieur* is mostly an aerobic one. From a qualitative viewpoint, however, reactions of reduction may be of great pharmacological or toxicological significance when they generate active metabolites or toxic metabolic intermediates.

5.1 Reductions at Carbon Atoms

A number of reactions of reduction at carbon atoms can be found in Figure 2. Thus, aldehydes and ketones are readily reduced to primary and secondary alcohols, respectively (pathways 1 and 2, Figure 2). The oxidation of diphenols to quinones is also reversible (pathway 7, Figure 2), a variety of cellular reductants being able to mediate the reduction of quinones either by a two-electron mechanism or by two single-electron steps. The two-electron reduction can be catalyzed by carbonyl reductase and quinone reductase, while cytochrome P450 and some flavoproteins act by single-electron transfers. The nonenzymatic reduction of quinones can occur, for example, in the presence of superoxide or some thiols such as glutathione.

Reduction of olefinic groups—that is, the reverse of desaturation—is documented for a few drugs bearing an α,β-ketoalkene function (pathway 3, Figure 2). The reaction is thought to be catalyzed by various NAD(P)H oxidoreductases. Dehalogenation reactions can also proceed reductively. Reductive dehalogenations involve replacement of a halogen by a hydrogen (pathway 10, Figure 2), or vicinal bisdehalogenation (pathway 11). Some radical species formed as intermediates may have toxicological significance.

5.2 Reductions at Other Atoms

Various reactions of N-oxidation are reversible, cytochrome P450 and other reductases being able to deoxygenate N-oxides back to the amine (pathways 1 and 2, Figure 3). The same is true for aromatic nitro compounds, aromatic nitroso compounds and hydroxylamines, and imines and oximes, which can ultimately be reduced to primary amines (pathways 3–5, Figure 3). Azo and azoxy compounds can be reduced to hydrazines (pathway 8, Figure 3). An important pathway of hydrazines is their reductive cleavage to primary amines (pathway 9, Figure 3). A toxicologically significant pathway thus exists for the reduction of some aromatic azo compounds to potentially toxic primary aromatic amines.

Other reductions can be found in Figure 4 and involve sulfur and a few other atoms. Thus, disulfides are reduced to thiols (pathway 2). There are numerous examples of the reduction of sulfoxides (pathways 3 and 4). In contrast, the reduction of sulfones has never been found to occur. The case of pathway 1 is unclear, the reduction of sulfenic, sulfinic, and sulfonic acids appearing unlikely. Note also some possible reductions of arsenic-containing compounds (pathways 7 and 8).

6 HYDROLASES AND THEIR REACTIONS

6.1 Hydrolases

Hydrolases constitute a very complex ensemble of enzymes, many of which are known or suspected to be involved in xenobiotic metabolism. Relevant enzymes among the serine hydrolases include carboxylesterases (carboxylic-ester hydrolase, EC 3.1.1.1), arylesterases (aryl-ester hydrolase, EC 3.1.1.2), cholinesterase (acylcholine acylhydrolase, EC 3.1.1.8), and a number of serine endopeptidases (EC 3.4.21). The role of arylsulfatases (EC 3.1.6.1), aryldialkylphosphatases (EC 3.1.8.1), β-glucuronidases (EC 3.2.1.31), and epoxide hydrolases (EC 3.3.2.3) is also worth noting. Some cysteine endopeptidases (EC 3.4.22), aspartic endopeptidases (EC 3.4.23), and metalloendopeptidases (EC 3.4.24) are of potential interest, as well.

6.2 Reactions of Hydration and Hydrolysis

Hydrolases catalyze the addition of a molecule of water to a variety of functional groups. Thus, epoxide hydrolase hydrates epoxides to yield *trans*-dihydrodiols (pathway 7, Figure 2). This reaction is documented for many arene oxides and olefinic epoxides. Here, a molecule of water has been added to the substrate without loss of a molecular fragment; hence the use of the term "hydration" sometimes found in the literature.

The main reactions of hydrolytic cleavage (hydrolysis) are shown in Figure 5. They are frequent for organic esters (pathway 1), inorganic esters such as nitrates (pathway 2), and amides (pathway 3). These reactions are catalyzed by esterases, peptidases, or other enzymes. Nonenzymatic hydrolysis is also known to occur for sufficiently labile compounds under biological conditions of pH and temperature. Such reactions are of particular significance in the activation of ester prodrugs.

7 TRANSFERASES AND THEIR REACTIONS

In reactions of conjugation, an endogenous molecule or moiety (sometimes called an "endocon") is bound covalently to the substrate, which may be either the unchanged xenobiotic or a metabolite. Reactions of conjugation have several defining characteristics: (1) they are catalyzed by enzymes known as transferases; (2) involve a cofactor that binds to the enzyme in the close proximity of the substrate and carries the endocon to be transferred; and (3) this endocon is highly polar (hydrophilic), and usually of a molecular size comparable to that of the substrate.

The foregoing defining characteristics, or criteria, are not unique to conjugation reactions, since in reactions of hydrogenation (i.e., typical reactions of functionalization) the hydride is also transferred from a cofactor (NADPH or NADH). In addition, the criteria suffer

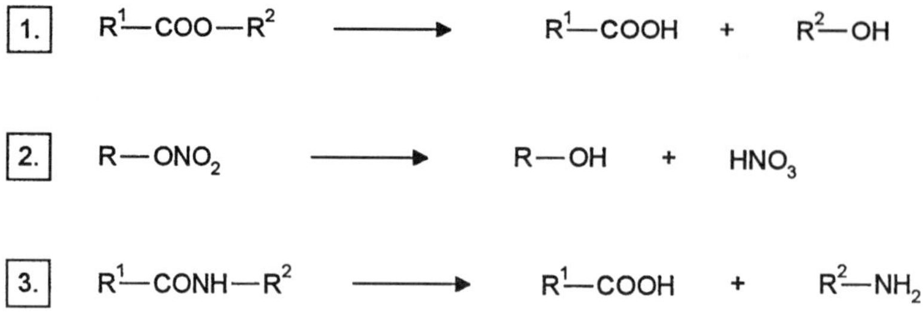

Figure 5. Major reactions of hydrolysis of xenobiotics involving esters and amides.

from important exceptions, as exemplified in the sections that follow. As a result of this fuzziness, reactions of hydration are sometimes classified as conjugations.

7.1 Methylation

Reactions of methylation imply the transfer of a methyl group from the cofactor *S*-adenosyl-L-methionine (SAM) to the substrate, a reaction catalyzed by methyltransferases. Reactions of methylation satisfy only two of the three criteria just defined, since the methyl group is hydrophobic and comparatively small. The main enzyme responsible for methylations of oxygen atoms is catechol *O*-methyltransferase [*S*-adenosyl-L-methionine:catechol *O*-methyltransferase (COMT), EC 2.1.1.6], which is mainly cytosolic but also exists in membrane-bound form. Reactions of xenobiotic N-methylation are catalyzed by several enzymes having different substrate specificities. Examples include nicotinamide *N*-methyltransferase (*S*-adenosyl-L-methionine:nicotinamide *N*-methyltransferase, EC 2.1.1.1), histamine methyltransferase (*S*-adenosyl-L-methionine: histamine *N*-methyltransferase, EC 2.1.1.8), phenylethanolamine *N*-methyltransferase (*S*-adenosyl-L-methionine:phenylethanolamine *N*-methyltransferase; noradrenaline *N*-methyltransferase, EC 2.1.1.28), and nonspecific amine *N*-methyltransferase-(*S*-adenosyl-L-methionine:amine *N*-methyltransferase; arylamine *N*-methyltransferase; tryptamine *N*-methyltransferase, EC 2.1.1.49). Reactions of xenobiotic S-methylation are mediated by the membrane-bound thiol methyltransferase (*S*-adenosyl-L-methionine:thiol *S*-methyltransferase, EC 2.1.1.9) and the cytosolic thiopurine methyltransferase (*S*-adenosyl-L-methionine:thiopurine methyltransferase, EC 2.1.1.67).

The methyl group in *S*-adenosylmethionine is bound to a sulfonium center, giving it a marked electrophilic character and explaining its reactivity toward the three types of functionality undergoing biomethylation, namely, hydroxyl (phenolic), amino, and thiol groups. Figure 6 summarizes the main reactions of methylation seen in xenobiotic metabolism. O-Methylation is common for catechol-containing compounds and shows a usual regioselectivity for the meta position (pathway 1). Frequently, O-methylation is seen as a late event in the metabolism of aryl groups, after they have been oxidized to catechols, as shown in Figure 2, pathway 7. This sequence was seen, for example, in the metabolism of many drugs containing a phenyl ring.

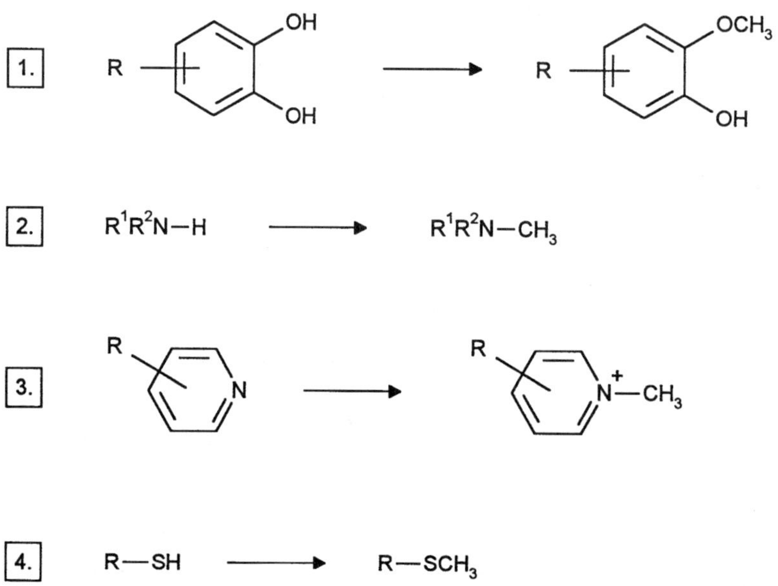

Figure 6. Major reactions of methylation of xenobiotics involving catechol, amino, and thiol groups.

Reactions of N-methylation involve a variety of functional groups (pathways 2 and 3, Figure 6). Some xenobiotic primary and secondary amines have been shown to be in vitro substrates of amine *N*-methyltransferase, while some phenylethanolamines and analogues are methylated by phenylethanolamine *N*-methyltransferase (pathway 2). However, such reactions are seldom of significance in vivo, presumably as a result of effective oxidative N-demethylation. N-Methylation of pyridine-type nitrogen atoms (pathway 3) appears to be of greater in vivo pharmacological significance. First, the resulting metabolites, being quaternary amines, are more stable than tertiary or secondary amines toward N-demethylation. And second, these metabolites are also more polar than the parent compounds. S-Methylation of thiol groups (pathway 4) is documented for a few drugs. Other substrates are metabolites (mainly thiophenols) resulting from the S—C cleavage of (aromatic) glutathione and cysteine conjugates. Once formed, such methylthio metabolites can be further processed before excretion as sulfoxides and sulfones.

7.2 Sulfoconjugation

In reactions of sulfoconjugation, a sulfate moiety is transferred from the cofactor (3′-phosphoadenylylsulfate, PAPS) to the substrate by cytosolic enzymes known as sulfotransferases. These include aryl sulfotransferase (3′-phosphoadenylylsulfate:phenol sulfotransferase; phenol sulfotransferase, EC 2.8.2.1), alcohol sulfotransferase (3′-phosphoadenylylsulfate:alcohol sulfotransferase; hydroxysteroid sulfotransferase, EC 2.8.2.2), amine sulfotransferase (3′-phosphoadenylylsulfate:amine *N*-sulfotransferase; arylamine sulfotransferase, EC 2.8.2.3), estrone sulfotransferase (3′-phosphoadenylylsulfate:estrone 3-sulfotransferase, EC 2.8.2.4), tyrosine–ester sulfotransferase (3′-phosphoadenylylsulfate:L-tyrosine-methylester sulfotransferase, EC 2.8.2.9), steroid sulfotransferase (3′-phosphoadenylylsulfate:phenolic steroid-sulfotransferase, EC 2.8.2.15), and cortisol sulfotransferase (3′-phosphoadenylylsulfate:cortisol 21-sulfotransferase; glucocorticosteroid sulfotransferase, EC 2.8.2.18). The first three enzymes are of particular significance in the metabolism of xenobiotics.

The sulfate moiety in PAPS will bind to a nucleophilic —OH or —NH— site in the substrate, forming a sulfate of sulfamate conjugate (Figure 7). Sulfoconjugation of alcohols (pathway 1) leads to metabolites of different stabilities. Endogenous hydroxysteroids (i.e., cyclic secondary alcohols) form relatively stable sulfates, while sulfate conjugates of some secondary alcohol metabolites of xenobiotics (e.g., safrole and estragole) form genotoxic carbocations. Primary alcohols (e.g., methanol and ethanol) can also form sulfates with well-documented alkylating capacity. Particular mention must be made of polycyclic hydroxymethylarenes, which yield reactive sulfates believed to account for the carcinogenicity of these compounds. In contrast to some alcohols, phenols form stable sulfate esters. The reaction is usually of high affinity, but the limited availability of PAPS restricts the amounts of conjugate being produced.

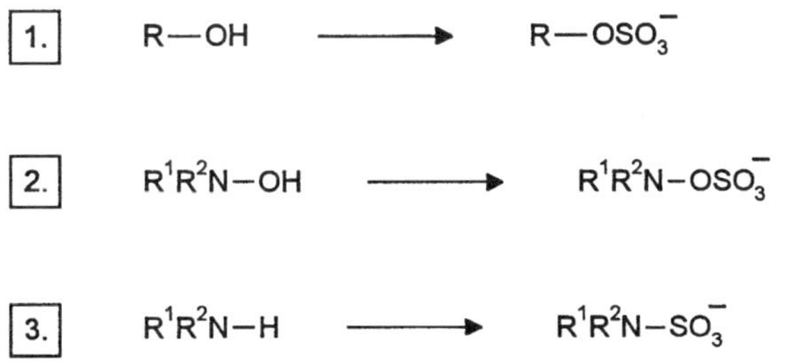

Figure 7. Major reactions of sulfation of xenobiotics involving alcohol, phenol, hydroxylamino, hydroxylamido, and amino groups.

Aromatic hydroxylamines and hydroxylamides are good substrates for some sulfotransferases and yield unstable sulfate esters (pathway 2). Indeed, heterolytic N—O cleavage produces highly electrophilic nitrenium ions believed to account for part or all of the cytotoxicity of arylamines and arylamides. In contrast, significantly more stable products are obtained upon formation of sulfamates from amines such as primary and secondary alkyl- and arylamines (pathway 3).

7.3 Glucuronidation

In glucuronidation, one molecule of glucuronic acid is transferred to the substrate from the cofactor uridine-5′-diphospho-α-D-glucuronic acid (UDPGA). The reaction is catalyzed by glucuronyltransferase [UDP-glucuronate β-D-glucuronosyltransferase (acceptor-unspecific); UDPGA-glucuronosyltransferase, EC 2.4.1.17; UDPGT]. This enzyme consists in a number of products of the *UGT* gene superfamily. According to recent evidence, the human UDPGT isozymes are the products of two gene families, *UGT1* and *UGT2*, whose products are the phenol/bilirubin UGTs and steroid UGTs, respectively. Individual enzymes include UGT1*1 and UGT1*4, UGT1*6, UGT2B4, UGT2B7, UGT2B8, UGT2B9, and UGT2B10.

The mechanism of glucuronidation involves a nucleophilic substitution with inversion of configuration of α-D-glucuronic acid in UDPGA to form β-D-glucuronides. As shown in Figure 8, the functional groups able to undergo glucuronidation are nucleophiles, a common characteristic they share despite their great chemical variety. As a consequence of this diversity, the products of glucuronidation are classified as O-, N-, S-, and C-glucuronides. The O-glucuronidation of phenolic xenobiotics or metabolites (pathway 1) is often in competition with O-sulfation, with the latter reaction predominating at low concentrations and the former at high concentrations. Alcohols form another major group of substrates, be they primary, secondary, or tertiary (pathway 2). An interesting example is that of morphine, which is conjugated on its phenolic and secondary alcohol groups to form the 3-*O*-glucuronide (a weak opiate antagonist) and the 6-*O*-glucuronide (a strong opiate agonist), respectively.

Another important pathway of O-glucuronidation is the formation of acyl glucuronides (pathway 3). Good substrates are arylacetic acids and aliphatic acids. These metabolites are relatively reactive, rearranging to positional isomers and binding covalently to plasma and seemingly also tissue proteins. Thus, acyl glucuronide formation cannot be viewed solely as a reaction of inactivation and detoxication. A special class of acyl glucuronides is that formed by carbamic acids (pathway 4), which themselves are not stable enough to be characterized. An increasing number of primary and secondary amines are found to yield this type of glucuronide, whose chemical and biochemical reactivities remain to be fully understood. Hydroxylamines and hydroxylamides may also form O-glucuronides (pathway 5). Thus, a few drugs and a number of aromatic amines are known to be N-hydroxylated and then O-glu-

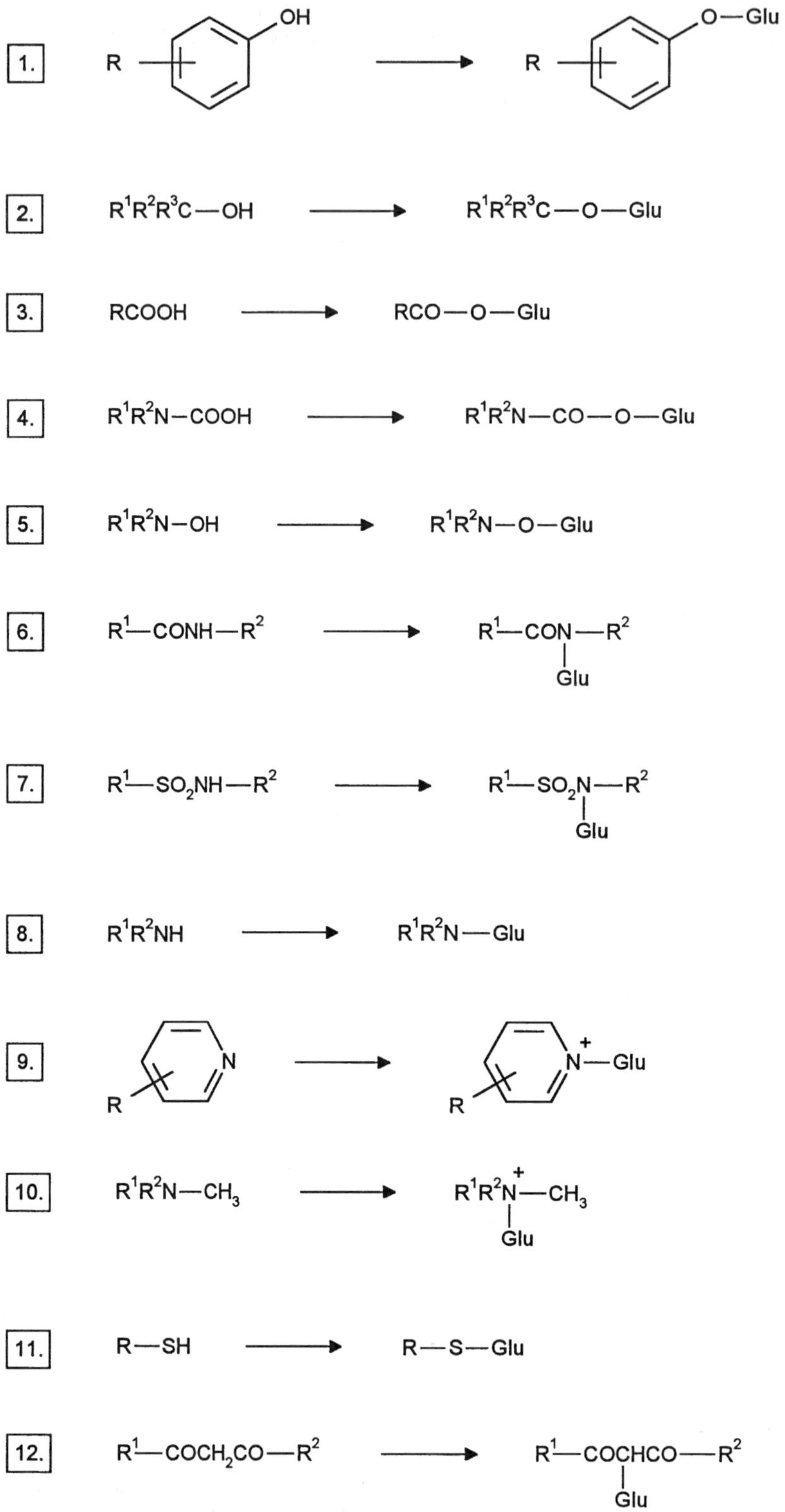

Figure 8. Major reactions of glucuronidation of xenobiotics involving phenol, alcohol, carboxyl, carbamoyl, hydroxylamino, hydroxylamido, carboxamido, sulfonamido, amino, thiol, dithiocarboxyl, and 1,3-dicarbonyl groups: Glu, glucuronosyl moiety.

curonidated. The susceptibility of N-O-glucuronides to undergo heterolytic cleavage and form nitrenium ions is well characterized.

Second in importance to O-glucuronides are the N-glucuronides formed from carboxamides (pathway 6), sulfonamides (pathway 7), and various amines (pathways 8–10). The reaction has special significance for antibacterial sulfanilamides, producing highly water-soluble metabolites that do not crystallize in the kidneys. N-Glucuronidation of aromatic and aliphatic amines (pathway 8) has been observed in a few cases only. Similarly, there are a few observations that pyridine-type nitrogens can be N-glucuronidated (pathway 9). As far as human drug metabolism is concerned, a reaction of much greater significance is the N-glucuronidation of lipophilic, basic tertiary amines containing one or two methyl groups (pathway 10).

Third in importance are the S-glucuronides formed from aliphatic thiols, aromatic thiols, and dithiocarboxylic acids (pathway 11). As for C-glucuronidation (pathway 12), this reaction has been seen in humans for 1,3-dicarbonyl drugs such as sulfinpyrazone.

7.4 Acetylation and Other Reactions of Acylation

Reactions of acetylation and more generally acylation involve the transfer of an acetyl or acyl moiety to an acceptor group, which may be an amino or hydroxyl function. The major enzyme system catalyzing reactions of acetylation is arylamine *N*-acetyltransferase [acetyl–CoA:arylamine *N*-acetyltransferase (NAT); arylamine acetylase, EC 2.3.1.5]. Two enzymes have been characterized, NAT1 and NAT2, the latter having two closely related isoforms NAT2A and NAT2B. The cofactor of *N*-acetyltransferase is acetyl–coenzyme A (CoA-S-Ac), where the acetyl moiety is bound by a thioester linkage.

Two other activities are also involved in the acetylation of aromatic amines and hydroxylamines, namely, aromatic hydroxylamine *O*-acetyltransferase (*N*-hydroxy-4-acetylaminobiphenyl:*N*-hydroxy-4-aminobiphenyl *O*-acetyltransferase, EC 2.3.1.56) and *N*-hydroxyarylamine *O*-acetyltransferase (acetyl–CoA:*N*-hydroxyarylamine *O*-acetyltransferase; arylhydroxamate *N,O*-acetyltransferase, EC 2.3.1.118). Other acetyltransferases exist: for example, diamine *N*-acetyltransferase (acetyl–CoA:alkane-α,ω-diamine *N*-acetyltransferase; putrescine acetyltransferase, EC 2.3.1.57) and aralkylamine *N*-acetyltransferase (acetyl–CoA:aralkylamine *N*-acetyltransferase; serotonin acetyltransferase, EC 2.3.1.87), but their involvement in xenobiotic metabolism does not appear to be documented.

The substrates of acetylation shown in Figure 9 are mainly amines of medium basicity. Very few basic amines (primary or secondary) form N-acetylated metabolites (pathway 1), and then in low yields. In contrast, a large variety of primary aromatic amines are N-acetylated (pathway 2). Examples are sulfonamide drugs and various carcinogenic amines such as benzidine. Besides amines, other nitrogen-containing functionalities undergo N-acetylation, hydrazines and hydrazides being particularly good substrates (pathway 3).

Arylhydroxylamines (e.g., the N-hydroxy metabolites of a number of arylamines) can also be acetylated, but the reaction is one of O-acetylation (pathway 4). This is the reaction formally catalyzed by EC 2.3.1.118 (with acetyl–CoA acting as the acetyl donor), but also by EC 2.3.1.5 and EC 2.3.1.56. The same conjugates can be

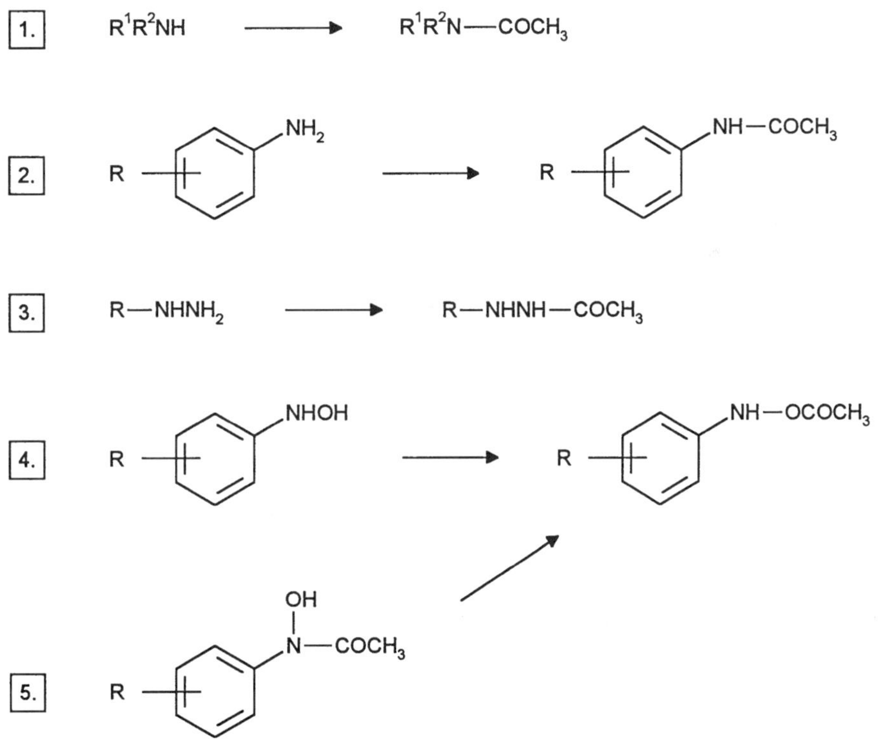

Figure 9. Major reactions of acetylation involving aliphatic amines, aromatic amines, hydrazines and hydrazides, and arylhydroxylamines.

formed by intramolecular *N,O*-acetyl transfer, when an *N*-aryl-*N*-hydroxyacetamide is substrate of EC 2.3.1.56 (pathway 5). In addition, such an arylhydroxamic acid can transfer its acetyl moiety to an acetyltransferase, which can then acetylate an arylamine or an arylhydroxylamine (intermolecular *N,N*- or *N,O*-acetyl transfer).

N-Formylation is also a genuine route of conjugation for some arylalkylamines and arylamines, and particularly polycyclic aromatic amines. The reaction is postulated to be catalyzed by arylformamidase (aryl-formylamine amidohydrolase; kynurenine formamidase, EC 3.5.1.9) in the presence of *N*-formyl-L-kynurenine.

A very different type of reaction, represented by the conjugation of xenobiotic alcohols with fatty acids, yields highly lipophilic metabolites that may accumulate in tissues. Thus, ethanol forms esters with, for example, palmitic acid, oleic acid, linoleic acid, and linolenic acid. These reactions are catalyzed by sterol esterase (steryl-ester acylhydrolase; cholesteryl ester synthetase, EC 3.1.1.13) and fatty acyl-ethyl-ester synthetase (long chain fatty acyl-ethyl-ester acylhydrolase, EC 3.1.1.67). Larger xenobiotics such as tetrahydrocannabinols and codeine are also acylated with fatty acids, possibly by sterol *O*-acyltransferase (acyl–CoA:cholesterol *O*-acyltransferase; sterol-ester synthetase, EC 2.3.1.26).

7.5 Acyl–Coenzyme A Thioesters as Metabolic Intermediates

All the reactions described in this section have in common the participation of xenobiotic carboxylic acids (R—COOH) coupling with coenzyme A (CoA—SH) to form an acyl–CoA metabolic intermediate:

$$\text{R—COOH} \rightarrow \text{R—CO—S—CoA}$$

The reaction requires ATP and is catalyzed by various acyl—CoA ligases of overlapping substrate specificity, including acetate—CoA ligase [acetate:CoA ligase (AMP-forming); acetyl—CoA synthetase, EC 6.2.1.1], butyrate—CoA ligase [butanoate:CoA ligase (AMP-forming); fatty acid thiokinase (medium chain), EC 6.2.1.2], long chain fatty acid–CoA ligase [acid:CoA ligase (AMP-forming); fatty acid thiokinase (long chain); acyl–CoA synthetase, EC 6.2.1.3], benzoate–CoA ligase [benzoate:CoA ligase (AMP-forming); EC 6.2.1.25], and phenylacetate–CoA ligase [phenylacetate:CoA ligase (AMP-forming), EC 6.2.1.30].

The acyl–CoA conjugates thus formed are seldom excreted but undergo further transformation by one of a variety of pathways:

- hydrolysis (by thiolester hydrolases, EC 3.1.2)
- formation of amino acid conjugates
- formation of hybrid triglycerides and phospholipids
- formation of cholesteryl esters and bile acid esters
- formation of acyl–carnitines
- protein acylation
- unidirectional chiral inversion of arylpropionic acids (possibly involving methylmalonyl–CoA epimerase, EC 5.1.99.1)
- dehydrogenation and β-oxidation
- two-carbon chain elongation

Amino acid conjugation, a major route for many xenobiotic acids, involves the formation of an amide bond between the xenobiotic acyl–CoA and the amino acid. Glycine is the amino acid most frequently used for conjugation (pathway 1 in Figure 10). A few glutamine conjugates have been characterized in humans. The enzymes catalyzing these transfer reactions are various *N*-acyltransferases—for example, glycine *N*-acyltransferase (acyl—CoA:glycine *N*-acyltransferase, EC 2.3.1.13), glutamine *N*-phenylacetyltransferase (phenylacetyl–CoA:L-glutamine α-*N*-phenylacetyltransferase, EC 2.3.1.14), glutamine *N*-acyltransferase (acyl–CoA:L-glutamine *N*-acyltransferase, EC 2.3.1.68), and glycine *N*-benzoyltransferase (benzoyl–CoA:glycine *N*-benzoyltransferase, EC 2.3.1.71). The

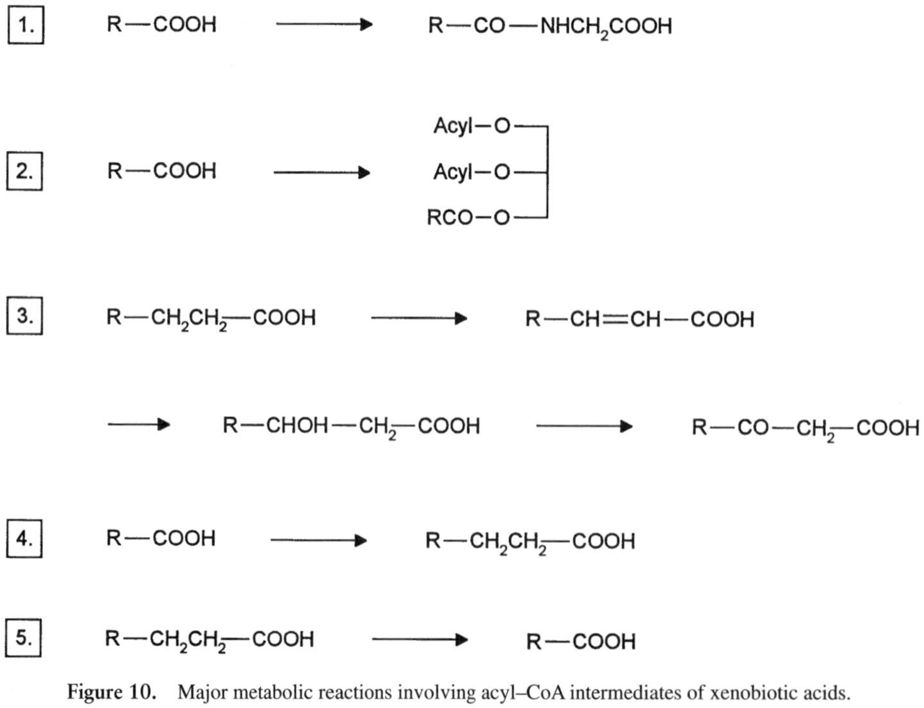

Figure 10. Major metabolic reactions involving acyl–CoA intermediates of xenobiotic acids.

xenobiotic acids undergoing amino acid conjugation are mainly benzoic acids such as benzoic acid itself and salicylic acid, which form hippuric acid and salicyluric acid, respectively. Phenylacetic acid derivatives can yield glycine and glutamine conjugates. In addition, other amino acids can be used for conjugation in various animal species (e.g., alanine and taurine), as well as some dipeptides.

Incorporation of xenobiotic acids into lipids has been characterized rather recently. The resulting highly lipophilic metabolites may burden the body as long-retained residues. In the majority of cases, triacylglycerol analogues (pathway 2, Figure 10) or cholesterol esters are formed, but some phospholipid analogues have also been characterized. The enzymes catalyzing such reactions include diacylglycerol *O*-acyltransferase (acyl–CoA:1,2-diacylglycerol *O*-acyltransferase, EC 2.3.1.20), 2-acylglycerol *O*-acyltransferase (acyl–CoA:2-acylglycerol *O*-acyltransferase, EC 2.3.1.22), and sterol *O*-acyltransferase [acyl–CoA:cholesterol *O*-acyltransferase (ACAT); cholesterol acyltransferase; sterol-ester synthetase, EC 2.3.1.26].

In some cases, acyl–CoA conjugates formed from xenobiotic acids can also enter the physiological pathways of fatty acid catabolism or anabolism. Thus, intermediate metabolites of β-oxidation may be observed (pathway 3, Figure 10), as exemplified by valproic acid. The enzymes catalyzing this pathway are clearly those involved in the metabolism of fatty acids, such as acyl–CoA ligases (see beginning of this section), acyl–CoA:carnitine *O*-acyltransferases (e.g., EC 2.3.1.7, 2.3.1.21, 2.3.1.137), acyl–CoA dehydrogenases (e.g., EC 1.3.99.2, 1.3.99.3), enoyl–CoA hydratase (EC 4.1.2.17), and 3-hydroxyacyl–CoA dehydrogenase (EC 1.1.1.35). In addition, a few examples are known of xenobiotic alkanoic and arylalkanoic acids undergoing two-carbon chain elongation (pathway 4), or chain shortening by two carbons (pathway 5), or even four or six carbons, as catalyzed by 3-ketoacyl–CoA thiolases (e.g., EC 2.3.1.9, 2.3.1.16).

7.6 Conjugation with Glutathione

Glutathione (GSH, γ-glutamyl-cysteinyl-glycine) is a thiol-containing tripeptide of capital significance in the detoxication and toxication of drugs and other xenobiotics. Glutathione reacts with endogenous and exogenous compounds in a variety of manners. First, the nucleophilic properties of the thiol group make it an effective conjugating agent. And second, glutathione can act as a reducing or oxidizing agent depending on its redox state (i.e., GSH or GSSG). Furthermore, the reactions of glutathione can be enzymatic (e.g., conjugations catalyzed by glutathione *S*-transferases, peroxide reductions catalyzed by glutathione peroxidase) or nonenzymatic (e.g., some conjugations and various redox reactions).

The glutathione transferase [RX:glutathione *R*-transferase; glutathione *S*-transferase (GST); EC 2.5.1.18] comprises multifunctional proteins coded by a multigene family. These proteins act both as enzymes and as binding proteins. These enzymes are mainly localized in the cytosol as homodimers and heterodimers and exist as four classes in mammals. The human enzymes comprise the following dimers: A1-1, A1-2, A2-2, A3-3 (alpha class); M1a-1a, M1a-1b, M1b-1b, M1a-2, M2-2, M3-3 (mu class); P1-1 (pi class); T1-1 (theta class); and a microsomal enzyme (MIC). The GST A1-2, A2-2, and P1-1 display selenium-independent glutathione peroxidase activity, a property also characterizing the selenium-containing enzyme glutathione peroxidase (glutathione:hydrogen peroxide oxidoreductase, EC 1.11.1.9). The GST A1-1 and A1-2 are also known as ligandin when they act as binding or carrier proteins, a property also displayed by M1a-1a and M1b-1b. In the latter function, these enzymes bind and transport a number of active endogenous compounds (e.g., bilirubin, cholic acid, steroid and thyroid hormones, hematin), as well as some exogenous dyes and carcinogens.

The nucleophilic character of glutathione is due to its thiol or rather thiolate group. Indeed, an essential component of the catalytic mechanism of glutathione transferase is the marked increase in acidity experienced by the thiol group upon binding of glutathione to the active site of the enzyme. As a result, GSTs transfer glutathione to a very large variety of electrophilic groups; depending on the nature of the substrate, these nucleophilic reactions can be categorized as substitutions or additions. And with compounds of sufficient reactivity, these reactions can also occur nonenzymatically.

Once formed, glutathione conjugates (GS-R) are seldom excreted as such but usually undergo further biotransformation. Cleavage of the glutamyl moiety by glutamyl transpeptidase (EC 2.3.2.2) and of the cysteinyl moiety by cysteinylglycine dipeptidase (EC 3.4.13.6) or aminopeptidase M (EC 3.4.11.2) leaves a cysteine conjugate (Cys-*S*-R), which is further N-acetylated by cysteine-*S*-conjugate *N*-acetyltransferase (acetyl–CoA:S-substituted L-cysteine *N*-acetyltransferase, EC 2.3.1.80) to yield an *N*-acetylcysteine conjugate (CysAc-*S*-R). Conjugates of the latter type are known as mercapturic acids, a name clearly indicating that they were first characterized in urine. This however does not imply that the degradation of unexcreted glutathione conjugates must stop at this stage, since cysteine conjugates can be substrates of cysteine-*S*-conjugate β-lyase [L-cysteine-*S*-conjugate thiol-lyase (deaminating), EC 4.4.1.13] to yield thiols (R-SH). These in turn can rearrange as discussed next or be S-methylated and then S-oxygenated to yield thiomethyl conjugates (R-S-Me), sulfoxides (R-SO-Me) and sulfones (R-SO_2-Me).

The major reactions of glutathione are shown in Figure 11: the first reaction is nucleophilic addition to epoxides to yield a nonaromatic conjugate (pathway 1), which may undergo further transformation as just described. This reaction is well documented for the arene oxide metabolites of numerous drugs and xenobiotics containing an aromatic moiety. The same reaction can also occur readily for epoxides of olefins. An important pathway of substitution at sp^3-carbons is shown in pathway 2. Various electron-withdrawing leaving groups X may be involved—for example, the chlorine atom at the NCH_2CH_2Cl group of anticancer alkylating agents.

Addition at activated olefinic groups (e.g., β-γ-unsaturated carbonyls) are quite varied (pathway 3). A typical substrate is acrolein ($CH_2{=}CH{-}CHO$). Quinones (ortho- and para-) and quinone imines (not shown) react with glutathione by two distinct and competitive routes, namely, nucleophilic addition to form a conjugate (pathway 4), and reduction to the hydroquinone (pathway 5) or the aminophenol. The conjugates produced by pathway 4 may undergo reoxidation to *S*-glutathionylquinones or *S*-glutathionylquinone imines of considerable reactivity.

Haloalkenes may react with GSH either by substitution (pathway 6) or by addition (pathway 7). Formation of mercapturic acids occurs as for other glutathione conjugates, but in this case S—C cleavage of the *S*-cysteinyl conjugates by the renal β-lyase yields thiols of significant toxicity, since they rearrange to form highly reactive thioketenes (XRC=C=S) and/or thioacyl halides. With a good leaving group and adequate substituents, nucleophilic aromatic substitution reactions also occur at aromatic rings (pathway 8). A good example of the detoxication of acyl halides with glutathione (pathway 9) is provided by phosgene ($O{=}CCl_2$), an extremely toxic

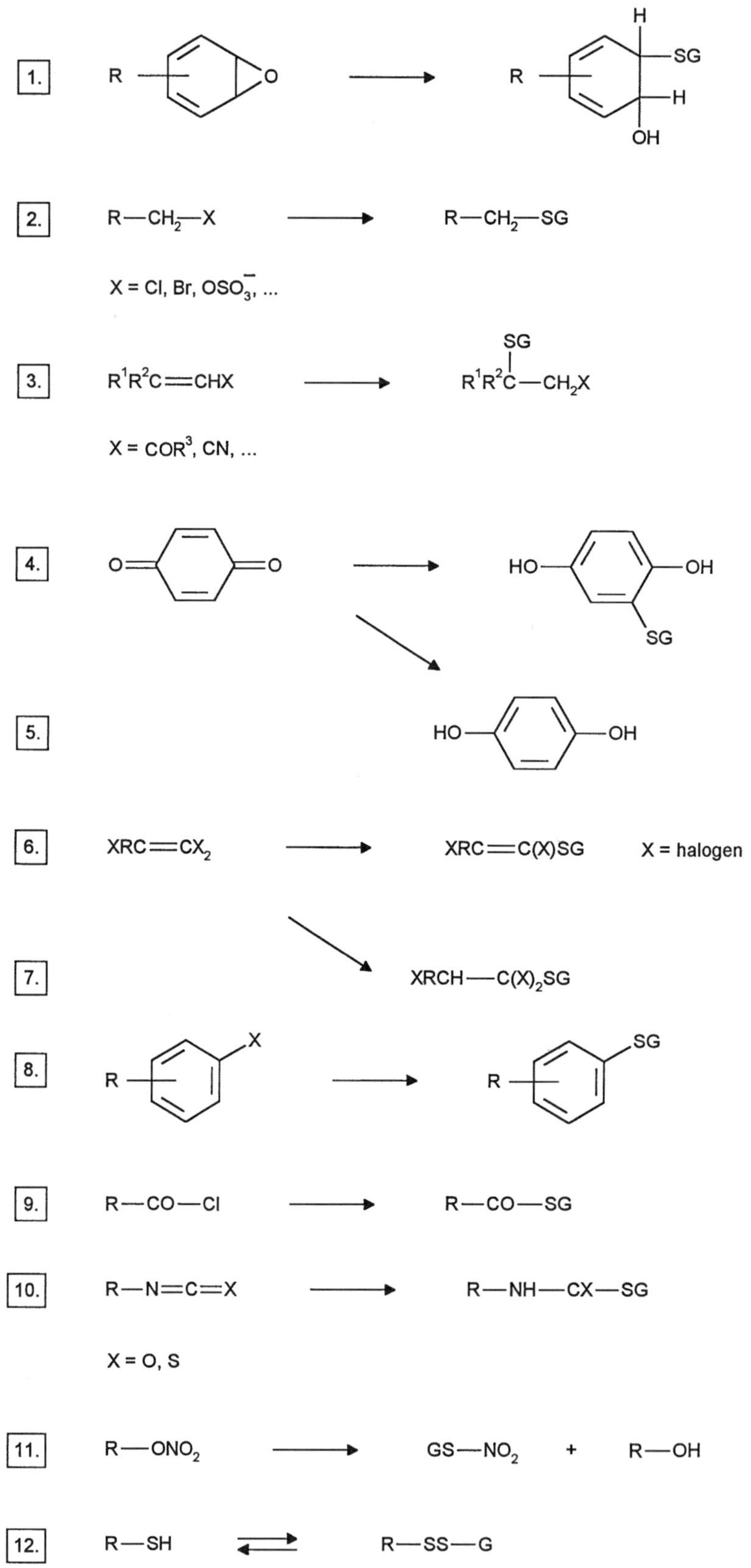

Figure 11. Major metabolic reactions involving glutathione (GSH) as a nucleophilic or reducing agent.

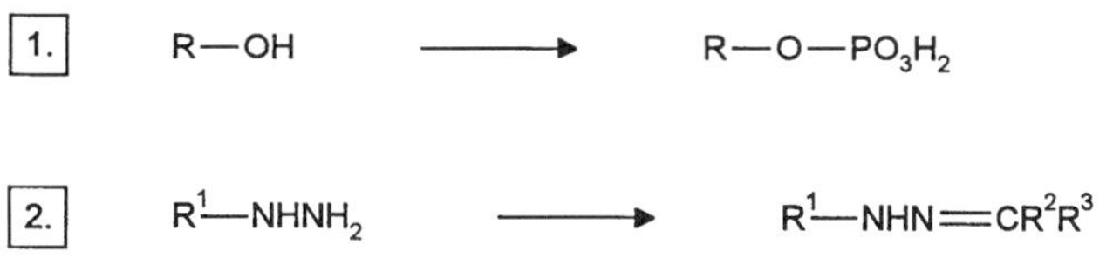

Figure 12. Reactions of phosphorylation and hydrazone formation.

metabolite of chloroform, which is inactivated to the diglutathionyl conjugate O=(SG)$_2$. The addition of glutathione to isocyanates and isothiocyanates is significant because of its reversible character (pathway 10). Substrates of the reaction are xenobiotics such as the well-known toxin methyl isocyanate, whose glutathione conjugate behaves as a transport form able to carbamoylate various macromolecules, enzymes, and membranes structures.

Organic nitrate esters such as nitroglycerine are vasodilators whose action results from their reduction to nitric oxide (NO). Glutathione and other thiols play an important role in this activation, a thionitrate being formed in the first step (pathway 11). Nitrogen reduction may then proceed by various routes having nitrate (NO_2^-) or *S*-nitrosoglutathione (GS-NO) as an intermediate. The formation of mixed disulfides between GSH and a xenobiotic thiol has been observed in a few cases (pathway 12).

7.7 Other Reactions of Conjugation

Several other routes of xenobiotic conjugation have been reported, but their importance is restricted to a few exogenous substrates. Here, we consider briefly phosphorylation and carbonyl conjugation (Figure 12).

While phosphorylation is of great significance in the processing of endogenous compounds and macromolecules, relatively few xenobiotics form phosphate esters (pathway 1, Figure 12). The enzymes involved are various phosphotransferases with a hydroxyl group as acceptor (EC 2.7.1), phosphotransferases with a phosphate group as acceptor (EC 2.7.4), diphosphotransferases (EC 2.7.6), and possibly also nucleotidyltransferases (EC 2.7.7). Thus, a number of antiviral nucleoside analogues yield the mono-, di-, and triphosphates in vitro and in vivo [e.g., zidovudine (AZT)].

The reaction of hydrazines with endogenous carbonyl-containing compounds (pathway 2) occurs nonenzymatically and involves a variety of aldehydes (mainly acetaldehyde) and ketones (e.g., acetone, pyruvic acid, α-ketoglutaric acid). The products thus formed are hydrazones, which either are excreted as such or undergo further transformation.

8 CONCLUDING REMARKS

This chapter focuses on two major aspects of xenobiotic metabolism, namely, a systematic consideration of biochemical reactions and a presentation of the enzymes involved. The reader cannot fail to marvel at the impressive variety of substrates, reactions, and enzymes whose study constitutes the science of xenobiotic metabolism. The unveiling of such a broad picture, a slow and demanding endeavor, owes much to the extraordinary development of analytical techniques and molecular biology in the last decades. Much is now known about the structure and function of enzymes, the exquisite stereoelectronic details of their catalytic mechanisms, the complexity of their regulation and expression, and their molecular evolution.

But beyond these facts and concepts, the void of our ignorance is infinite. This is what gives sense and meaning to scientific research and, as we progress in our understanding of life, the new questions we formulate are more important than the facts we discover and the concepts we create. These questions are the lights that guide us in our study of ourselves.

See also Biochemical Genetics, Human; Biotransformation of Drugs and Chemicals; Cytochrome P450; Enzymes, Energetics and Regulation of Biological Catalysis by; Pharmacogenetics.

Bibliography

Burchell, B., Nebert, D. W., Nelson, D. R., Bock, K. W., Iyanagi, T., Jansen, P. L. M., Lancet, D., Mulder, G. J., Chowdhury, J. R., Siest, G., Tephly, T. R., and Mackensie, P. I. (1991) *DNA Cell Biol.* 10:487–494.

International Union of Biochemistry and Molecular Biology (1992) *Enzyme Nomenclature 1992.* Academic Press, San Diego, CA.

Jenner, P., and Testa, B., Eds. (1980 and 1981) *Concepts in Drug Metabolism,* Parts A and B. Dekker, New York.

Mulder, G. J., Ed. (1990) *Conjugation Reactions in Drug Metabolism.* Taylor & Francis, London.

Nelson, D. R., Kamataki, T., Waxman, D. J., Guengerich, F. P., Estabrook, R. W., Feyereisen, R., Gonzalez, F. J., Coon, M. J., Gunsalus, I. C., Gotoh, O., Okuda, K., and Nebert, D. W. (1993) *DNA Cell Biol.* 12:1–51.

Testa, B. (1995) *The Metabolism of Drugs and Other Xenobiotics—Biochemistry of Redox Reactions.* Academic Press, London.

———, and Jenner, P. (1976) *Drug Metabolism. Chemical and Biochemical Aspects.* Dekker, New York.

———, and Mayer J. M. (1996) *The Metabolism of Drugs and Other Xenobiotics—Biochemistry of Hydration and Conjugation Reactions.* Academic Press, London, in preparation.

X-Ray Diffraction of Biomolecules

K. Ravi Acharya and Anthony R. Rees

1 Principles

2 Stages of Structure Determination

2.1 Crystallization

2.2 Characterization and Data Collection

2.3 Phase Determination

3 Model Building and Refinement

4 Assessing the Structure

5 Perspectives

Key Words

Crystallography The science of crystal structures and the most widely used method to obtain accurate three-dimensional information on biomolecules.

Crystals Regular three-dimensional arrays of atoms or molecules.

Diffraction Specific patterns generated by the scattering of X-rays from atoms or molecules organized in a crystalline array.

Chemical reactions involve interactions between atoms or molecules, which are three-dimensional objects. To understand fully the process of interaction, it is necessary to know the structures of the atoms or molecules involved. This is particularly true with respect to large biological macromolecules such as proteins and nucleic acids, the major constituents of living tissue. From such structural knowledge two important molecular requirements for interaction can be seen: complementarity of shape and charge. Satisfying these requirements is essential to the effective design of, for example, inhibitor or activator molecules for an enzyme reaction, or to the design and construction of new proteins using protein engineering techniques. Thus, the development of "rules" for molecular recognition and the concomitant progress in rational drug design rely more and more on three-dimensional structural information. In recent years considerable effort has been put into devising more efficient methods of obtaining such information, of which X-ray diffraction is the most commonly used.

X-ray diffraction of a molecule that forms a single crystal can give its three-dimensional architecture at or near atomic resolution. The resolution of a study critically affects how much we can learn about the structure of the biomolecule. For example, a protein structure at 2 Å resolution will provide detailed information about atomic positions, specific atomic interactions (intra- and intermolecular hydrogen bonds, ionic interactions, etc.), and solvent accessibility. Some indication of flexibility or mobility within the molecule can also be obtained, although changes in the conformations of flexible regions can also be induced by packing forces within the crystal. Comparative structural studies in the crystal (using X-ray crystallography) and in solution (using nuclear magnetic resonance techniques) for the same macromolecule, as well as studies of the same molecule in different crystallization conditions, demonstrate that the intrinsic dynamic behavior of macromolecules is not obscured in the solid state. This is most convincingly illustrated by experiments describing enzymes that, after crystallization, are seen to exhibit the identical activity measured in solution. An exciting consequence is that experimental crystallographer can carry out time-resolved structural studies, thus providing structural information at various points along the enzymatic reaction pathway.

Over the last 10 years there has been a considerable increase in the rate of determination of three-dimensional structures of macromolecules by X-ray crystallography, due in large part to developments in diffraction data collection and computer technology. The greater availability of purified macromolecules, through advances in protein chemistry and molecular biology, has also contributed and has made possible structural studies of complex biological assemblies. Thus, macromolecular crystallography can now be applied successfully to small peptides, proteins, multiprotein complexes, and even viruses.

1 PRINCIPLES

Crystals are three-dimensional arrays of molecules in a regular, repeating arrangement. The minimal repeating unit within the crystal is known as the "unit cell" and is the basic building block from which the whole volume of the crystal is constructed. The unit cell may contain multiple copies of the same molecule whose positions within the cell are governed by symmetry rules. This allows compact packing of the molecules within the crystal.

In X-ray crystallography, a crystal is bombarded with electromagnetic radiation of a wavelength comparable to the desired resolution. For example, X-rays must be used if a resolution of 1–2 Å is required, since X-radiation has wavelengths between 1 and 1.5 Å. The diffraction pattern that is obtained, which arises mainly from the scattering of X-rays by the electrons of atoms in the crystal lattice, was first explained by Lawrence Bragg in 1913. Using Bragg's law, it is possible to translate the information present in the diffraction patterns of molecules into atomic structures. The process is akin to the formation of an image in an optical microscope, where light rays are scattered by an object and then recombined by means of a system of lenses to form an image.

In X-ray crystallography, although there is no lens capable of refocusing the X-rays, molecular image can be produced by computational means. Since the diffracting power of a single biomolecule is rather weak, three-dimensional crystals that possess billions of molecules are used. As a result of this averaging, diffraction intensities from protein crystals can be measured very precisely. The resultant scattered radiation is then recombined to form the image of the object, a task that calls for knowledge of the "amplitudes" of the scattered waves with respect to the others and of the "phases" of the scattered radiation. The amplitudes can be experimentally measured but the phases cannot, a phenomenon that is often referred to in macromolecular crystallography as the "phase problem." With reasonable estimates for the phases of all the scattered X-ray beams, obtained by any of several methods described in the sections that follow, it is possible to calculate the electron density at any point within the crystal using a mathematical technique involving the "Fourier transform."

When estimates for the phases of all the scattered X-ray beams have been obtained, it is straightforward to calculate the electron density ρ at any position x,y,z within the crystal by the relationship:

$$\rho(x, y, z) = \frac{1}{V} \sum_h \sum_k \sum_l |F_{(hkl)}| \exp\left[-2\pi i\left(\frac{hx}{a} + \frac{ky}{b} + \frac{lz}{c}\right) - \alpha(hkl)\right]$$

where hkl identifies the index of the scattered beam, $|F_{(hkl)}|$ is the amplitude, $\alpha(hkl)$ is the phase of the diffracted beam of index hkl (the structure factor), $a,b,$ and c describe the dimensions of the unit cell of the crystal, and V is the volume of the unit cell.

Note that the electron density at any point x,y,z in the crystal is determined by all the observations, which makes crystallography an essentially definitive science. Since the scattering occurs simultaneously from all molecules within the crystal lattice, we see an average picture of the entire volume of the lattice that was illuminated by the X-rays. Crystallography also provides a large number of frozen snapshots of the unit cell, since the frequency of X-rays is far higher than the fastest chemical vibrations within a biomolecule. In general there can be many intensity observations in comparison to the number of atomic parameters. This "overdetermination" is somewhat diminished in the case of macromolecules because of the exponential fall off with scattering angle that depends on atomic mobility, which is often quite large in macromolecular crystals.

The information at a particular point within the biomolecule is "synthesized" from a complete set of diffraction measurements that obeys Bragg's law of X-ray diffraction,

$$2d \sin\theta = \lambda$$

where λ is the wavelength of the radiation, d is the spacing within the object being examined (commonly called *resolution* in defining the quality of macromolecule structures), and θ is the half-angle of scattering. The beams that are deflected by only a small angle (low resolution data, quite often due to the scattering contribution of water molecules surrounding the protein molecules within the crystal lattice) contribute to the broad features of the molecule under study, while the quality and the extent of the fine detail present in the electron density map depend directly on the higher angle diffraction data (high resolution data). The resolution is an important term in defining the quality of the structure of a macromolecule. It should be noted, however, that the linear inverse relationship implies that to improve the resolution by a factor of 2, eight times as much data must be measured.

The main features of a macromolecule that can be observed at different resolutions are:

6 Å resolution: outline of the molecule, features such as helices can be identified.
3 Å resolution: course of the polypeptide chain can be traced and topology of folding can be established. With the aid of the amino acid sequence it is possible to place the side chains within the electron density map.
2 Å resolution: the main chain conformation can be established with high accuracy. Details of side chain conformations can be interpreted easily without amino acid sequence data; bound water molecules can be identified.
1.5 Å resolution: individual atoms are almost resolved; detailed description of water structure.
1.2 Å resolution: hydrogen atoms may become visible.

In practice, only a few macromolecules diffract to the very highest resolution. This is due to "smearing" of the structure, obscuring the fine details and weakening the high resolution data beyond the possibility of measurement. The smearing may arise from packing defects (spatial) or from thermal motion of atoms (temporal). Using X-ray diffraction, direct information about the spatial dynamics can also be obtained, in contrast to a technique, that probes only the temporal dynamics of a biomolecule (e.g., NMR spectroscopy).

2 STAGES OF STRUCTURE DETERMINATION

A typical crystallographic analysis of a macromolecular structure normally proceeds through a sequence of experimental or analytical operation that generate objects (Figure 1). Crystals are grown, diffraction data are measured, phases for the structure factors are evaluated, an electron density map is calculated, an atomic model is fitted into this electron density map, and finally this model is refined. This flow of procedures often includes cycles that repeat some of the operations, such as in refinement, or set up shunts as alternative attempts to find a solution are tried. Thus the process can sometimes be tedious, although using standard protocols, some of which are described here, the repetitive elements of the procedure can be automated. For the reader who requires more detail on the various stages of structure determination using X-ray crystallography, several excellent reviews are available in the literature. In this article we shall highlight only the practical aspects of the method, as used in the determination of protein structures.

The process of crystallization begins with a requirement for several milligrams of pure (> 95%), homogeneous protein. Where necessary, methods such as site-directed mutagenesis can be used to engineer changes that improve crystallization or provide specific sites for the binding of heavy atoms.

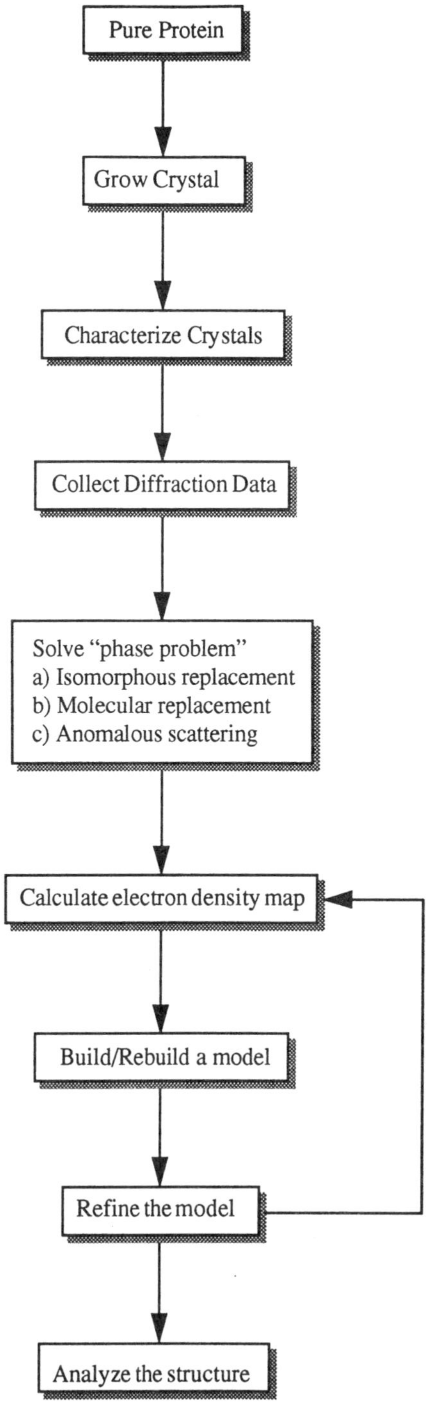

Figure 1. Flow diagram showing the basic stages in the determination of a protein structure.

2.1 CRYSTALLIZATION

One of the most difficult steps in X-ray crystallography, crystallization requires the experimentalist to establish precise conditions under which weak intermolecular forces between the protein molecules will be directed to produce highly ordered crystal packing rather than random aggregation as a precipitate. This procedure depends on achieving:

supersaturation of the protein is solution
nucleation at a few sites for crystal growth
sustained growth of single crystals

Despite significant improvements in the general methodology of crystallization, most notably by the introduction of robotic methods, the crystallization of a protein is still largely a black art. In any event, the single most important requirement is that the purity and homogeneity of the protein be as high as possible.

Standard protocols such as the hanging drop, sitting drop, dialysis, micro- and macroseeding, batch, free capillary, and temperature gradient methods are commonly used, depending on the amount of material available. Quite often it is necessary to explore a wide range of conditions, such as sample concentration, pH, concentration of the precipitant (e.g., ammonium sulfate or other salt, polyethylene glycol or other alcohol), added ions or effectors and their concentrations, and temperature. More recently the use of nonionic detergents has widened the scope of the crystallographic method to include membrane proteins. Since several crystals may often be required to obtain a full set of structural data, many crystallization trials may be needed to produce enough crystals of high quality. Thus, to stand a reasonable chance of success, an experimenter should have available at least 10 mg of pure protein.

2.2 Characterization and Data Collection

Once crystals have been obtained, the next step is to characterize their diffraction behavior. If the crystals diffract well, it is necessary to determine the crystallographic and noncrystallographic symmetry they exhibit. Crystallographic symmetry is usually detected uniquely from the diffraction pattern. However, the presence of noncrystallographic symmetry sometimes may be hard to detect and, where present, can make structure determination quite difficult. On the positive side, multiple copies of molecules (e.g., "promoter" structures of a virus capsid) governed by noncrystallographic symmetry can provide a large amount of phasing information, resulting, perhaps, in an electron density map that is superior in quality to that obtained from a simple packing involving only crystallographic symmetry operations.

In most experimental assemblies the intensities of the diffracted X-rays are recorded by rotating the crystal within the X-ray beam. These intensities provide the amplitudes of the diffracted beam, or the structure factors, directly. Often, crystallographers face two difficulties:

1. The crystals are usually sensitive to irradiation, with the result that the diffraction pattern fades as the crystal is exposed to the X-ray beam.
2. A weak diffraction pattern is seen because the diffracted energy is a very small proportion of the total energy in the incident beam and is spread over a large number of reflections simulated simultaneously.

X-rays for protein crystallography experiments are produced by one of two methods. In laboratory sources electrons are accelerated at high voltage against a metal target. Crystals with small unit cells ($<$ 50 Å in each dimension) are often suited for a diffractometer that records one diffraction spot at a time. Crystals with unit cells of intermediate dimensions (50–150 Å) are often suited for routine data collection on an electronic area detector that can record several diffraction spots at once. For crystals with large unit cells (such as are seen with viruses) or well-ordered crystals with dimensions $0.1 \times 0.1 \times 0.1$ mm^3 "synchrotron radiation" is often used. Synchrotron radiation, which extends from X-rays to the wavelengths of visible light, is produced at high flux in a special facility that accelerates charged particles at velocities close to the speed of light, whereupon a magnetic field forces them to move along a curved path. Where X-rays are used, the diffraction spots may be recorded either on a photographic film or on an electronic area detector. Since the synchrotron produces an extremely intense and mainly parallel beam of X-rays, when used in combination with electronic area detectors it is possible to increase the data collection rate considerably, and so reduce the radiation damage during data collection.

More recently, image plates have become an often preferred state-of-the-art alternative to area detector systems for data collection. They have a high spatial resolution (100–150 μm) and subtend a large angle, permitting the collection of more data in one image. During exposure of the image plate to X-rays, the X-ray photon causes a chemical change in the plate coating, resulting in fluorescence that is detected by a photomultiplier when scanned with light of the proper wavelength. This process of "reading" the image plate is carried out by a laser beam on a scanner.

The protein crystal is usually mounted in a glass or quartz capillary tube of appropriate diameter. Fragile platelike crystals may be mounted in specially flattened capillary tubes. More recently a method has been suggested whereby crystals are mounted suspended in a thin film confined within a thin-wire loop by surface tension, which has been used routinely in flash-freezing data collection experiments. Flash-freezing of crystals to liquid nitrogen temperatures is currently recommended as a means of prolonging crystal lifetime.

For time-resolved structural studies, such as are required to probe enzyme reactions, data collection must be fast in comparison to the speed of the reaction. For some systems, standard modes of data collection may be sufficiently rapid, possibly aided by lowering the working temperature using cooling techniques. Alternatively, very fast data collection using the Laue technique may be employed; here the crystal usually remains stationary but data are collected using a continuous spectrum of X-ray wavelengths. Depending on the crystal system, and assuming favorable conditions, it is possible to collect about 70% of the data on one photograph. The potential increase in the rate of data collection with this method is enormous. Most of the recently reported time-resolved experiments using this technique involve data collection on the millisecond timescale.

2.3 Phase Determination

To determine the structure of a molecule after three-dimensional X-ray intensity data have been collected, it is necessary to address the phase problem. There are various methods to evaluate, or approximate, the phases required to reconstruct a three-dimensional structural image from experimental diffraction data. All of them are somewhat indirect, since there is no straightforward method of obtaining phases from diffraction measurements on a single crystalline species. Quite often it is possible to derive valuable information on the molecular packing from calculations based on native amplitudes. Determination of phases is a critical step in macromolecular crystallography, and it is important to evaluate them as accurately as possible, since errors in phases usually have a much greater impact on the accuracy of the electron density map than do errors in the diffraction measurements themselves. The three major methods for obtaining phase information—namely, isomorphous replacement, molecular replacement, and anomalous scattering—are described in Sections 2.3.1 through 2.3.3.

2.3.1 Isomorphous Replacement

Isomorphous replacement has played a central role in macromolecular crystallography since its introduction during the X-ray structure determination of hemoglobin by Green, Ingram, and Perutz in 1954, in which mercury atoms were attached to the crystals. This method, now in general use, provides indirect experimental estimates of the protein phase angles by observing the interference effects on the intensities of the scattered beams when heavy atoms at very low concentrations (typically in the range of 1 to 10 mM) are added to the protein. Normally, heavy atom derivatives are prepared by soaking crystals in heavy atom compounds, which diffuse into the crystals and bind to specific site(s) in the protein. The binding can be either covalent or ionic. If the heavy atom substitution does not affect the surrounding protein structure—that is, if the derivative is isomorphous with the native structure—and if more than one pattern of heavy atom binding can be produced, it is possible to obtain estimates of the protein phase angles.

Certain chemical groups within the protein are good candidates for combining with heavy atoms; these include free sulfydryls and single or exposed disulfide bridges, which can incorporate a mercury atom. It is also possible to mutate proteins to provide favorable site(s) for heavy atom binding—for example, by the introduction of free cysteines via site-directed mutagenesis of, say, glutamine residues. Usually it is necessary to screen a wide range of putative heavy atom derivatives at relatively low (3.5 Å) resolution. A good single isomorphous derivative can be sufficient for a structure determination, but quite often it is necessary to obtain multiple derivatives to obtain phase information. It is possible to use anomalous scattering signals, although for resolving phase ambiguities such signals are often small (see Section 2.3.3). The information, however, allows the experimenter to make the correct choice of "handedness" (enantiomer).

2.3.2 Molecular Replacement

Frequently in macromolecular crystallography it is desirable to determine the crystal structure of a molecule for which the structure of a similar or homologous molecule is already available. For example, the structures of several inhibitors or substrates complexed with the same enzyme, or the structures of different forms of a protein mutated by protein engineering experiments, may be required. In most cases, the crystal of experimental interest is isomorphous with the native structure, and the analysis can proceed directly using the difference Fourier methods.

Where isomorphism does not pertain, however, the new structure must be determined de novo, or other ways of obtaining phase information must be sought which allow a reasonable estimation of starting phases for the new structure. The only successful technique for this purpose, originally described by Rossmann, is called the "molecular replacement" method. It makes use of a known three-dimensional structure as an approximate starting model to provide phase angles for the observed structure factor amplitudes from the unknown structure. In summary, it involves generating a starting model of the target crystal structure by orienting and positioning the search molecule within the unit cell of the target crystal so as best to account for the observed diffraction pattern. From this model it is possible to estimate the phase angles, which can be combined with the observed amplitudes to give an approximate Fourier synthesis of the unknown structure. This resultant model can then be subjected to any of the available refinement procedures to provide a final, accurate structure of the target crystal. It is applicable only if there is significant amino acid sequence identity (a minimum of 30% but preferably 50% or higher) to a molecule of known three-dimensional structure. Where a family of homologous structures is available, it is preferable to calculate an "average structure" for use as a starting model. Quite often the presence of more than one copy of the unknown structure in the asymmetric unit complicates the search unless the relative positions of the copies are known. In such cases, it is necessary to use an oligomeric model as the search model to obtain phase information. Many structural studies now use this technique, either alone or in combination with the method of isomorphous replacement.

2.3.3 Anomalous Scattering

When the protein is rather small, it is sometimes difficult to prepare isomorphous heavy atom derivatives. In such situations, the scattering properties of atoms change when the incident radiation is of an energy close to an absorption edge of the atom, and it is possible to determine the phase angles from the native data. For the usual wavelengths used, this means that quite significant "anomalous scattering" effects occur for some atoms (e.g., Fe, Cd, Se), while the effect is much smaller for sulfur and virtually nonexistent for carbon, nitrogen, oxygen, and hydrogen. There are two variations in using this method. Hendrickson's method of resolved anomalous scattering requires very accurate intensity measurements at one wavelength, while the multiple wavelength anomalous dispersion method (mainly used for larger molecules) requires intensity measurements at several wavelengths. A tunable X-ray source such as synchrotron radiation is ideal for performing such experiments. This method is of obvious use for metalloproteins and would be the method of choice for proteins when mutant proteins containing metal atoms can be generated.

It is possible to improve the accuracy of the initial set of phases when there is redundancy of information in the asymmetric unit, either because there are several copies of the molecule, giving rise to noncrystallographic symmetry, or more generally due to the solvent content of the crystal, which varies between 30 and 90%. "Solvent flattening" is a standard technique for crystals with an average solvent content (≥50%). When the level density in the solvent region is observed, this method can be sufficient to resolve the phase ambiguity for a single isomorphous derivative, or in general to improve the quality of the multiple isomorphous replacement phases to yield an interpretable electron density map. If there are multiple copies of the molecule in the asymmetric unit, or more than one type of crystal, molecular averaging may also be used to improve the quality of the electron density map. Although computationally demanding, this method has proved to be useful in reducing the phase errors due to averaging procedure and has been used routinely in multimeric proteins and in large complex structures such as viruses.

3 MODEL BUILDING AND REFINEMENT

Based on a good set of phases, it is possible to calculate an electron density map at 3 Å resolution. This map can be used to build an atomic model with the aid of the known amino acid sequence, the most exciting part of macromolecular structure determination. The quality of the electron density map will depend on the phase errors. Thus, it is important to start with a medium resolution electron density map. A computer graphics workstation is now a necessity for model-building work. At this stage, depending on the resolution and quality of the electron density map, it is possible to construct a three-dimensional structure of the protein. From this structure it is possible to obtain a starting set of atomic coordinates. However these coordinates may not be accurate, in part because of the phase errors and the resolution. Even for models derived from the best electron density maps, the accuracy of the atomic coordinates is likely to be no better than 0.5 Å, and localized regions of maps derived from average quality maps may suffer more serious errors. Such models are accurate enough to show the overall feature of the molecule and the tertiary fold, but not accurate enough to describe the inter- and intramolecular interactions that are important to understand the biological function of the molecule.

Thus, it is necessary at this stage to refine the initial model against the X-ray observations. "Refinement" refers to the process of adjusting the parameters of the models, mainly the positions (x,y,z) and the thermal parameters (B factor) of atoms within the unit cell of the structure, to improve the agreement between the structure factor amplitudes of the observed reflections and the values calculated from the model parameters. A commonly used measure of this match is the R value,

$$R = \frac{\Sigma \, ||F_{obs}| - |F_{calc}||}{\Sigma \, |F_{obs}|}$$

where $|F_{obs}|$ and $|F_{calc}|$ are observed and calculated structure factor amplitudes, respectively. One current standard refinement protocol uses a "restrained molecular dynamics" procedure in which the starting structure is initially energy minimized, to reduce the steric clashes, then heated to 2000–3000 K for a restrained dynamics run, cooled slowly [at, say, intervals of 1 picosecond ($= 10^{-12}$ s)] over a fixed period, and the final structure energy minimized to a global minimum. Finally, the restrained B factors are refined, a step that leads to greatly improved phase information. Usually, the quality of the new, refined map enables the resolution of any remaining ambiguities in the structure.

Thus, it can be seen that these late processes require the alternate application of model building and refinement until all parts of the molecule are well explained in terms of their electron densities and stereochemistry. It is important to note that a full refinement of individual (x,y,z) coordinates and the isotropic B factors for each of the nonhydrogen atoms of a protein is not meaningful at less than about 3 Å resolution because the observation-to-parameter ratio is usually too low.

4 ASSESSING THE STRUCTURE

Structural models based on X-ray crystallographic data produce a wealth of information. Before making detailed analysis of the structure, however, it is important to assess its accuracy. It is easy to convince a noncrystallographer of the beauty of a new structure, even where gross errors are present within it. Depending on the resolution, the quality of the final structure can be assessed from two factors:

1. The reliability index, or R factor, which should have values between 0.15 and 0.20 for a well-refined structure.
2. Root mean square deviation (RMSD) from covalent bond lengths, a measure of the accuracy of the stereochemistry; typically each one should be in the range of 0.01–0.02 Å.

When the experimental model is correct, both the R value and the stereochemical identity will be well within these values. Phi-psi (also known as Ramachandran) plots are good indicators of the accuracy of the protein model. The values of the main chain backbone dihedral angles in a protein can have only a limited range of values (except glycine, which does not have side chains). It is important that all amino acid residues within the protein structure be within the allowed regions. If there are significant numbers of residues lying outside the allowed ranges, the structure will require further improvement.

Atomic mobility and structural disorder are quite difficult to model in crystal structures. The dynamic character of proteins is such that atoms possess large thermal fluctuations in the crystal. These movements can be highly anisotropic, requiring the experimentalist to derive multiparameter definitions for B factors to model the reality. Newer sophisticated methods for treatment of both atomic mobility and B factors are under development. However, even isotropic B factors have been used to derive functional information.

The solvent structure is also an important component in refined macromolecular structures. Protein crystals usually contain about 50% solvent of crystallization. It is because of this high solvent content that molecules in crystals retain the identical activities they exhibit in solution. Quite often it is possible to visualize water-mediated interactions at the active site of, say, an enzyme, or within the crystal lattice. For this reason it is important to pay special attention to accurately designating water molecules during the final stages of structure refinement.

As an example of how the methods of macromolecular crystallography operate in practice, the different stages involved in the structure determination of the protein toxic shock syndrome toxin 1 (TSST-1), a microbial superantigen, are summarized in Figure 2 (see color plate 5). The majority of protein structures published in the literature are held by the Protein Data Bank (Chemistry Department, Brookhaven National Laboratory, Upton, New York 11973, U.S.A.).

5 PERSPECTIVES

X-ray crystallography is a highly specialized field, and we have attempted to give only a bird's-eye view of the technique, which, when used effectively, is an extremely powerful tool for the visualization of biomolecular interactions so critical to many areas of modern biomedical research. To be able to use X-ray crystallography to reveal clues to function, it is essential to understand molecular structure. For example, if a certain protein is known to play a key role in a particular disease state, knowledge of its three-dimensional structure might enable the "rational design" of therapeutic drugs that act to modify the protein function. The combination of an increased understanding of the rules governing protein structure and the power of molecular genetics will soon make the production of "tailor-made" molecules routine. At the level of structural techniques, the recent advances in fast data collection (in the range of milliseconds and submilliseconds using the Laue method), it is possible to dissect in detail the molecular events associated with catalysis, ligand binding, and structural transitions in crystals. These time-resolved X-ray diffraction experiments have brought the crys-

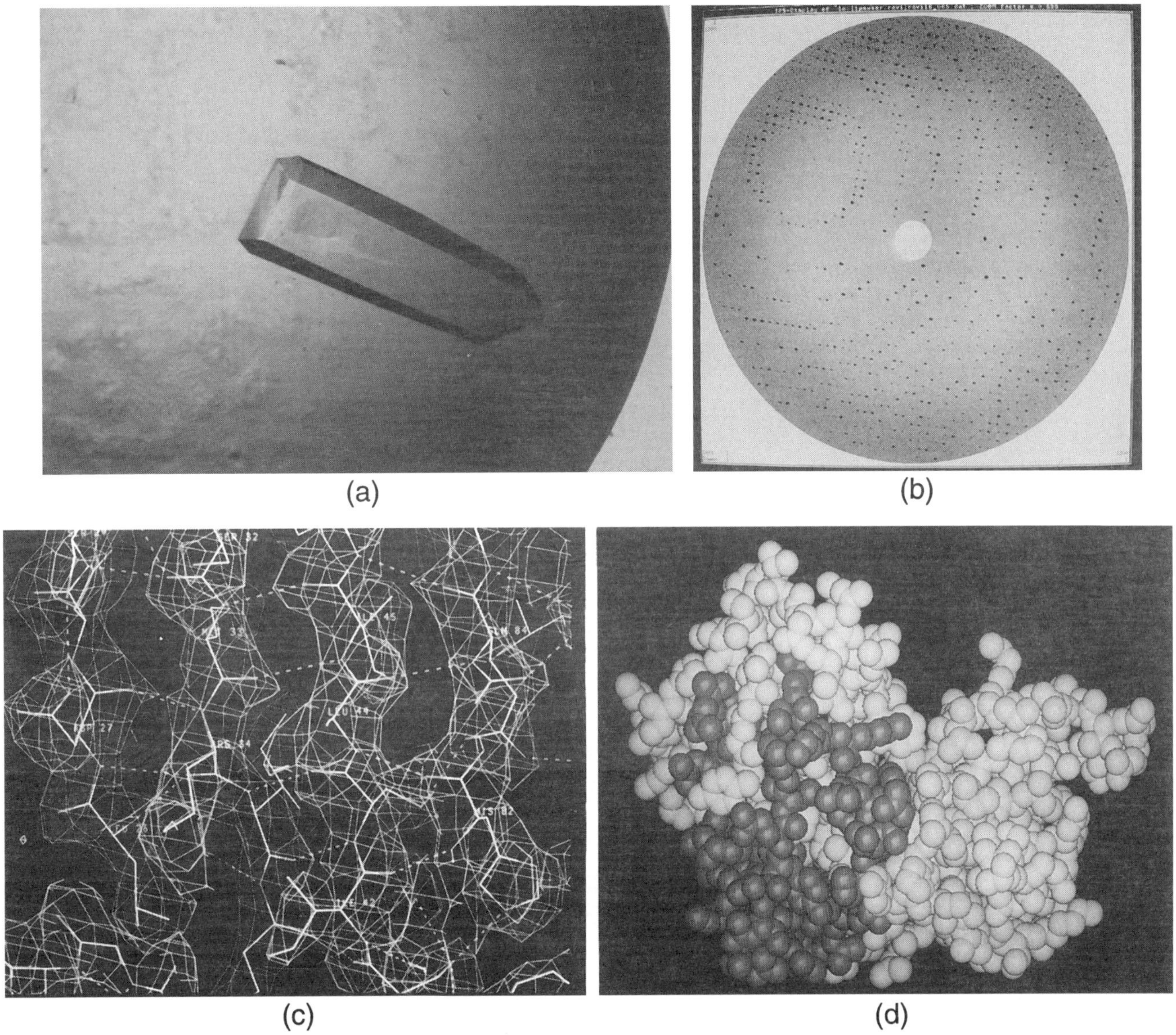

Figure 2. Stages involved in the structure determination of a protein molecule, toxic shock syndrome toxin 1 (TSST-1), a microbial superantigen. (a) A single crystal of TSST-1. (b) Diffraction pattern of TSST-1 crystal recorded using MAR Research Image Plate system (wavelength, 1.5418 Å). The data extend to 2.5 Å resolution. (c) Portion of the 2.5 Å resolution electron density map of TSST-1 with atomic positions superimposed. The region displayed shows four strands of a β-sheet. (d) Space-filling diagram of TSST-1 (drawn using the 2.5 Å crystal structure). (See color plate 5.)

tallographer a step closer to looking at biological events in real time, until recently an unfulfilled dream. For these and other reasons, X-ray crystallography will continue to play a central role in modern biology.

See also PROTEIN ANALYSIS BY X-RAY CRYSTALLOGRAPHY; PROTEIN MODELING; SUPERANTIGENS.

Bibliography

Acharya, K. R., et al., (1994) *Nature,* 367:94–97.
Bernstein, F. C., et al., (1977) *J. Mol. Biol.* 112:535–542.
Blundell, T. L., and Johnson, L. N. (1976) *Protein Crystallography.* Academic Press, London.
Bricogne, G. (1976) *Acta Crystallogr.* A32:832–847.
Brünger, A. T., Kuriyan, J., and Karplus, M. (1987) *Science,* 235:458–460.
Ducruis, A., and Giege, R., Eds. (1992) *Crystallization of Nucleic Acids and Proteins—A Practical Approach.* IRL Press, Oxford.
Gilliland, G. L., and Davies, D. R. (1984) *In Methods in Enzymology,* Vol. 104, W. B. Jacoby, Ed., pp. 370–391. Academic Press, San Diego, CA.
Green, D. W., Ingram, V. M., and Perutz, M. F. (1954) *Proc. R. Soc. London,* A225:287–307.
Hendrickson, W. A. (1985) In *Methods in Enzymology,* Vol. 115, H. W. Wyckoff, C. H. W. Hirs, and S. N. Timasheff, Eds., pp. 252–270. Academic Press, San Diego, CA.
———. (1991) *Science,* 254:51–59.
———, Smith, J. L., and Sheriff, S. (1985) In *Methods in Enzymology,* Vol. 115, H. W. Wyckoff, C. H. W. Hirs, and S. N. Timascheff, Eds., pp. 41–55. Academic Press, San Diego, CA.
Hope, H. (1989) *Acta Crystallogr.* B45:190–199.
Johnson, L. N. (1992) *Protein Sci.* 1:1237–1242.

Jones, E. Y., and Stuart, D. I. (1992) In *Protein Engineering—A Practical Approach,* A. R. Rees, M. J. Sternberg, and R. Wetzel, Eds., pp. 3–32. IRL Press, Oxford.

Jones, T. A. (1985) In *Methods in Enzymology,* Vol 115, H. W. Wyckoff, C. H. W. Hirs, and S. N. Timascheff, Eds., pp. 157–171. Academic Press, San Diego, CA.

McPherson, A. (1982) *Preparation and Analysis of Protein Crystals.* Wiley, New York.

McRee, D. E. (1993) *Practical Protein Crystallography.* Academic Press, San Diego, CA.

Michel, H., Ed. (1991) *Crystallization of Membrane Proteins.* CRC Press, Boca Raton, FL.

Miyahara, J., et al. (1986) *Nuclear Instrum. Methods,* A245:572–578.

Pai, E. F. (1992) *Curr. Opinion Struct. Biol.* 2:821–827.

Petsko, G. A. (1985) In *Methods in Enzymology,* Vol. 114, H. W. Wyckoff, C. H. S. Hirs, and S. N. Timasheff, Eds., pp. 147–156. Academic Press. San Diego, CA.

Ramachandran, G. N., and Sasisekharan, V. (1968) *Adv. Protein Chem.* 23:283–437.

Rossmann, M. G., Ed. (1972) *The Molecular Replacement Method,* International Science Review, No. 13. Gordon & Breach, New York.

Schlichting, I., et al. (1990) *Nature,* 345:309–315.

Wang, B. C. (1985) In *Methods in Enzymology,* Vol. 115, H. W. Wyckoff, C. H. W. Hirs, and S. N. Timascheff, Eds., pp. 90–111. Academic Press, San Diego, CA.

Wyckoff, H. W., Hirs, C. H. W., and Timascheff, S. N., Eds. (1985) *Methods in Enzymology,* Vols. 114 and 115. Academic Press, San Diego, CA.

X-Ray Crystallography in Protein Analysis: *see* Protein Analysis by X-Ray Crystallography.

Y

Yeast, DNA Repair in

Errol C. Friedberg

1 Yeast as a Eukaryotic Model

2 DNA Repair by Reversal of Base Damage

2.1 Enzymatic Photoreactivation (EPR)

2.2 Repair of Alkylation Damage by O^6-Methylguanine–DNA Methyltransferase

3 DNA Repair by Excision of Base Damage

3.1 Base Excision Repair

3.2 Nucleotide Excision Repair

3.3 Mismatch Repair

Key Words

Base Excision Repair A DNA excision repair mode whereby damaged bases are excised as the free base.

Cyclobutane Pyrimidine Dimers Covalently joined adjacent pyrimidines in DNA produced by exposure of DNA to ultraviolet radiation.

DNA Photolyase The enzyme that catalyzes enzymatic photoreactivation of pyrimidine dimers in DNA.

Enzymatic Photoreactivation A process in which pyrimidine dimers produced by ultraviolet irradiation of DNA are repaired by their enzyme-catalyzed monomerization.

Mismatch Repair A DNA repair mode whereby misrepaired bases are excised from DNA.

O^6-Methylguanine–DNA Methyltransferase An enzyme that specifically transfers methyl groups from the O^6 position of guanine in DNA to itself.

Nucleotide Excision Repair A DNA excision repair mode whereby damaged bases are excised as nucleotide structures.

DNA repair comprises a series of biochemical processes whereby alterations of the chemical structure of DNA are eliminated and the DNA is restored to its native state. By strict definition, the term does not include cellular responses by which DNA damage is functionally tolerated but is not removed from the genome. DNA repair occurs by a variety of distinct biochemical pathways, which vary with the particular type of damage under consideration. Such damage can affect either the nitrogenous bases or the sugar–phosphate backbone of DNA, or both. This discussion is confined to a consideration of the repair of damage to the nitrogenous bases, the informational components of genetic material. The detailed mechanisms of the known multiple DNA repair pathways vary in different organisms, usually as a function of their level of biological organization. This article considers DNA repair pathways in eukaryotic cells, using the yeast *Saccharomyces cerevisiae* as a model organism.

1 YEAST AS A EUKARYOTIC MODEL

Yeasts are eukaryotes because within each cell the genetic material is contained in a discrete cellular organelle, the nucleus. Since, however, yeast cells are not organized into tissues and organs; yeasts are considered to be lower eukaryotes. Nonetheless, the essential parameters of nucleic acid metabolism in yeast cells are very similar to those in the cells of higher eukaryotes, so yeasts are excellent models for exploring the molecular biology of eukaryotic DNA transactions. Interest in yeast as an experimental organism began early in this century, when it was discovered that alcoholic fermentation could be carried out with yeast cell-free extracts. Since then yeast has been shown to be a highly accessible organism for genetic studies and more recently for recombinant DNA technology. Yeast cells are also very easy and inexpensive to grow in the laboratory and have a short life cycle, which provides the further advantage of very rapid multiplication relative to mammalian cells. For all these reasons yeast has been extensively studied with respect to multiple and diverse aspects of cellular and molecular biology. At present the two yeasts that are most widely used in research laboratories are the budding yeast *Saccharomyces cerevisiae* and the fission yeast *Schizosaccharomyces pombe.* These two eukaryotes, although both are yeasts, are evolutionarily as distant from one another as each is from *Homo sapiens.* This review is largely confined to a consideration of DNA repair in the yeast *Saccharomyces cerevisiae.* Thus, the terms "yeast" and "*Saccharomyces cerevisiae*" are often used interchangeably.

The term "DNA repair" has come to have a rather broad general use in the literature and frequently encompasses cellular events in response to genomic insults that mitigate against phenotypic end points such as lethality, without physically correcting damage in DNA. For the purposes of this discussion, the term "DNA repair" is applied only to the consideration of several distinct biochemical pathways by which chemical or sequence alterations of the bases in DNA are physically removed and the genetic information is restored to its proper native state.

2 DNA REPAIR BY REVERSAL OF BASE DAMAGE

Several types of base damage can be repaired by single-step biochemical reactions during which particular chemical modifications are directly reversed without other perturbations of the chemical structure of the DNA molecule. The two best-studied examples of this mode of DNA repair in yeast are the enzymatic photoreactivation of cyclobutane pyrimidine dimers and the repair of methylation of the sixth and fourth oxygen atoms of guanine and thymine, respectively, in DNA.

2.1 Enzymatic Photoreactivation (EPR)

When living cells are exposed to ultraviolet (UV) light at wavelengths of approximately 260 nm, which is the absorption maximum for DNA, the absorbed radiation can produce several chemical alterations of the bases, resulting in the formation of distinct photoproducts. The quantitatively major photoproduct in DNA consists of the covalent joining of adjacent pyrimidines through saturation of their 5,6 double bonds. These altered bases are called cyclobutane pyrimidine dimers (CPDs), because the covalent linkage between them comprises a cyclobutane ring structure. Two distinct repair enzymes are known that specifically and uniquely recognize CPD in DNA. One of these is called *photoreactivating enzyme* or *DNA photolyase*. It is widely distributed in nature, but a recognizable functional homologue seems to have been lost during the evolution of some placental mammals.

DNA photolyases catalyze the reversal of CPD in DNA to yield native monomeric pyrimidines. This enzymatic reaction is a photochemical one that requires the presence of light at wavelengths ranging from approximately 350 to 450 nm, depending on the particular enzyme. The absorption of light at photoreactivating wavelengths and the photochemistry required to monomerize CPD are effected by two chemical chromophores that are noncovalently bound components of all known DNA photolyases. One of the chromophores is flavin adenine dinucleotide. In some DNA photolyases, including that in yeast, the second chromophore belongs to the folate family of compounds, while in others the second chromophore belongs to the deazaflavin family of compounds. The folate moiety of yeast DNA photolyase is thought to function as the primary light-harvesting entity in the enzyme, while the flavin adenine dinucleotide moiety is believed to be the photocatalyst that utilizes the energy of the absorbed light to effect the reactions resulting in the monomerization of CPD. The specific binding of DNA photolyase to CPD sites in DNA does not require the presence of chromophores or photoreactivating light. Hence, the complete EPR reaction can be written as follows:

$$\mathrm{E + DNA\,[Py\text{-}Py]} \underset{}{\overset{\text{Dark reaction}}{\rightleftharpoons}} \mathrm{E\text{-}DNA\,[Py\text{-}Py]} \xrightarrow{\text{Light reaction}} \mathrm{E + DNA\,[PyPy]}$$

The enzyme from *S. cerevisiae* is encoded by the *PHR1* gene, which maps to the right arm of chromosome XV. It encodes a polypeptide having a calculated molecular weight of 66.2 kDa, consistent with the measured molecular weight ($\approx$ 60 kDa) determined for the purified enzyme. The *PHR1* gene is regulated in living yeast cells and increased steady state levels of *PHR1* mRNA are present in cells exposed to UV radiation at about 254 nm. Curiously, gene expression is also up-regulated when yeast cells are exposed to forms of DNA damage that are not repaired by EPR. Mutants that are defective in the *PHR1* gene are unable to carry out EPR, hence they are abnormally sensitive to the lethal and mutagenic effects of UV radiation under photoreactivating conditions. EPR thus constitutes a fundamental DNA repair mode for UV radiation damage to DNA in prokaryotes, lower eukaryotes, and many, but not all, higher eukaryotes.

2.1.1 The Utility of DNA Photolyases

The absolute substrate specificity of DNA photolyases for CPD has facilitated the use of this class of DNA repair enzymes as diagnostic probes for the presence of these photoproducts in DNA. Since DNA exposed to UV radiation contains multiple types of photoproducts, it is possible to correlate a particular biological end point with the specific presence of CPD by showing that the end point is mitigated by the removal by EPR of these photoproducts. This correlation is particularly useful for studies in living cells. The removal of CPD from UV-irradiated DNA by incubating the DNA in the presence of DNA photolyase in vitro enriches for other types of photoproducts that may be the primary focus in DNA repair studies. Finally, when DNA photolyase is used under appropriate conditions, the tight binding affinity of the enzyme for CPD in DNA allows for the detection and quantitation of these lesions.

2.2 Repair of Alkylation Damage by O^6-Methylguanine–DNA Methyltransferase

A second ubiquitous example of the repair of specific types of base damage in DNA is that effected by the enzyme O^6-methylguanine–DNA methyltransferase (MGMT). This novel enzyme removes methyl (and, much less efficiently, ethyl or butyl) groups specifically from the O-6 position of guanine and the O-4 position of thymine in DNA. It is an unusual enzyme in that it acts stoichiometrically rather than catalytically and is consumed in the reaction in which it participates. The methylation of oxygen and nitrogen atoms in the nitrogenous bases in DNA is a frequent consequence of the exposure of DNA to reactive alkylating agents. Many such sites in each of the four bases are sensitive to the covalent addition of small alkyl groups. However, such alkylations at the O-6 position of guanine and O-4 of thymine are uniquely available for repair by MGMT. During the course of this repair reaction, the offending methyl groups (which have mutagenic potential because they can alter the normal base-pairing properties of DNA during replication) are transferred and covalently attached to a specific cysteine in the protein, resulting in the formation of an *S*-methylcysteine residue in the polypeptide. This moiety apparently inactivates the enzyme. Hence, each molecule of MGMT can repair only a single alkylated base and all known MGMTs are "suicide enzymes."

MGMT was originally discovered in *Escherichia coli* during the study of a phenomenon called "the adaptive response to alkylation damage." When *E. coli* and various other prokaryotes are exposed to very low levels of alkylating agents, they become less sensitive (adapt) to the potentially lethal and mutagenic effects of subsequent exposure to higher doses. The adaptation to lethality results from upregulation of the gene that encodes MGMT, providing markedly increased levels of the protein. Interestingly, the transcriptional activation of the gene that encodes MGMT is effected by the methylated protein itself, once it has been specifically modified by the removal of methyl groups from DNA methylated at a site other than guanine or thymine. The details of the adaptive response need not concern as here, since the phenomenon is restricted to prokaryotes. Though organisms such as yeast are endowed with this enzyme, it is not regulated in response to exposure to alkylating agents.

The yeast enzyme is encoded by a gene called *MGT1*. Not unexpectedly, *mgt1* mutants are hypersensitive to the lethal and mutagenic effects of alkylating agents. Interestingly, such mutants also manifest an increased rate of spontaneous mutations, suggesting that alkylation of guanine and thymine at O-6 and O-4, respectively, occurs naturally in yeast. Some natural sources of such spontaneous alkylation damage to DNA have in fact been identified. However, since most known sources of such damage are not natural to the environment, there has been considerable conjecture as to why living cells should have evolved specific mechanisms for the repair of alkylation damage to DNA in the first place.

Extracts of yeast contain very low levels of MGMT activity. Consistent with the absence of an adaptive response to alkylation damage, the yeast protein accepts methyl groups only from the O-6 position of guanine and to a lesser extent the O-4 position of thymine. Hence, it is not modified to a form that can function as a transcriptional activator, as is the case in many prokaryotes. Nonetheless, the amino acid sequence of the yeast enzyme is highly conserved. In all MGMT proteins studied, including that from yeast, the same cysteine residue is used as the methyl acceptor species. This cysteine residue is present in the highly conserved penta-amino acid sequence Pro-Cys-His-Arg-Val/Ile. It is not obvious what structural features of alkylated guanine and thymine provide the exquisite substrate specificity of MGMT, although it has been noted that both the O-6 position of guanine and the O-4 position of thymine occupy the major groove of double-stranded DNA. However, MGMT does not catalyze the removal of alkyl groups in other positions on bases that also project into the major groove. Thus, enzyme specificity is apparently also conferred by oxygen atoms of the nitrogenous bases carrying bound alkyl groups.

3 DNA REPAIR BY EXCISION OF BASE DAMAGE

The fundamental paradigm that operates in all excision repair modes is the physical removal of damaged, incorrect (with respect to base pairing) or inappropriate (with respect to normal DNA function) bases and the restoration of normal sequence information by using the opposite intact DNA strand as a template for DNA synthesis. Hence, all excision repair modes require template-directed repair DNA synthesis (i.e., double-stranded DNA). Considering that free-living forms probably have had to contend with base damage from UV radiation since the beginning of evolution, it may not be unreasonable to consider the teleological argument that DNA evolved as a double-stranded molecule to allow excision repair, and that coincidentally this structure provides an obvious mechanism for its replication.

All excision repair modes share in common the requirement for the enzyme-catalyzed removal of damaged bases. In some cases the damaged (or inappropriate) bases are excised as free bases. This mode of repair is called *base excision repair* (BER). In other cases the damaged bases are removed as part of a larger entity, typically an oligonucleotide fragment, which in yeast is believed to be about 30 nucleotides long. This mode of repair is called *nucleotide excision repair* (NER). There are indications from studies on the yeast *S. pombe* that NER can occur by several biochemically distinct modes. A form of NER that operates specifically on mispaired bases and is specifically directed to the strand containing the incorrect base is called *mismatch repair* or *mismatch correction.*

3.1 Base Excision Repair

The substrates amenable to base excision repair are determined by the total range of specificity exhibited by the DNA glycosylases present in any given cell type. In general, BER operates on bases that are chemically altered in ways that are believed to result in minimal helix distortion, if any. Many such forms of base damage result from the presence of chemically reactive oxygen species in cells. Hence, oxidative base damage is heavily dependent on BER for its repair. Such damage is believed to be an important form of spontaneous base damage in mammalian cells and an important contributor to the pathogenesis of spontaneous cancer.

3.1.1 DNA Glycosylases

During BER, the excision of damaged bases is achieved by the enzyme-catalyzed hydrolysis of the *N*-glycosyl bond that joins the base to the rest of the DNA molecule (Figure 1). Enzymes endowed with this specific catalytic function are called *DNA glycosylases.* Yeast and most other living organisms have multiple DNA glycosylases, each of which is more or less specific for different types of base damage. DNA glycosylases are monomeric proteins with no requirement for divalent cations or other cofactors. The first DNA glycosylase discovered is the most specific. It is called uracil–DNA glycosylase because it exclusively recognizes uracil residues in DNA, but not in RNA. Thus, DNA glycosylases are endowed with the ability not only to recognize particular bases but also to recognize these only in the context of deoxyribose linkages. The structural basis for this substrate specificity remains to be determined. It is equally mysterious how some DNA glycosylases have achieved a relaxed substrate specificity. For example, a DNA glycosylase in *E. coli* called 3-methyladenine–DNA glycosylase II recognizes not only 3-methyladenine residues in DNA but also 7-methylguanine and 3-methylguanine, and a similar enzyme in mammalian cells additionally recognizes 1-N^6-ethenoadenine in DNA. A 3-methyladenine–DNA glycosylase activity has been identified in yeast cells. The protein is encoded by a gene called *MAG,* which is up-regulated when yeast cells are exposed to alkylating agents. Curiously, as is the case with the yeast *PHR1* gene, induction of *MAG* expression also occurs following DNA damage by agents such as UV radiation that clearly does not result in the formation of alkyl groups in DNA.

3.1.2 Pyrimidine Dimer–DNA Glycosylase

E. coli cells infected with bacteriophage T4 and the bacterium *Micrococcus luteus* have evolved a novel mechanism for the repair of pyrimidine dimers in DNA. These cells exploit a DNA glycosylase that specifically hydrolyzes the 5′-glycosyl bond in the dimerized pyrimidine pair. The apyrimidinic site generated is then processed further, as discussed later, eventually leading to excision of the dimer from the DNA. This enzyme merits attention here for two reasons. First, this particular enzyme apparently occupies a very limited evolutionary niche. Aside from the two biological sources just mentioned, pyrimidine dimer–DNA glycosylase has not been identified in any other organism. Second, like DNA photolyase, the exquisite specificity of the pyrimidine dimer–DNA glycosylase for CPD in DNA has made the phage T4 and *M. luteus* enzymes extremely useful experimental tools for the detection and quantitation of such lesions in DNA. In vitro, the pyrimidine dimer–DNA glycosylase catalyzes hydrolysis of a phosphodiester bond 3′ to the apyrimidinic site generated by the glycosylase, considered shortly. While the biological relevance of this activity is controversial, the fact is that incubation of pyrimidine dimer–DNA glycosylase with DNA containing CPD results in nicking of the DNA. Nicks in DNA (hence CPD) can be detected with exquisite sensitivity by a variety of techniques.

3.1.3 AP Endonucleases

The release of free bases catalyzed by DNA glycosylases results in sites of base loss [apurinic or apyrimidinic (AP) sites] in the genome

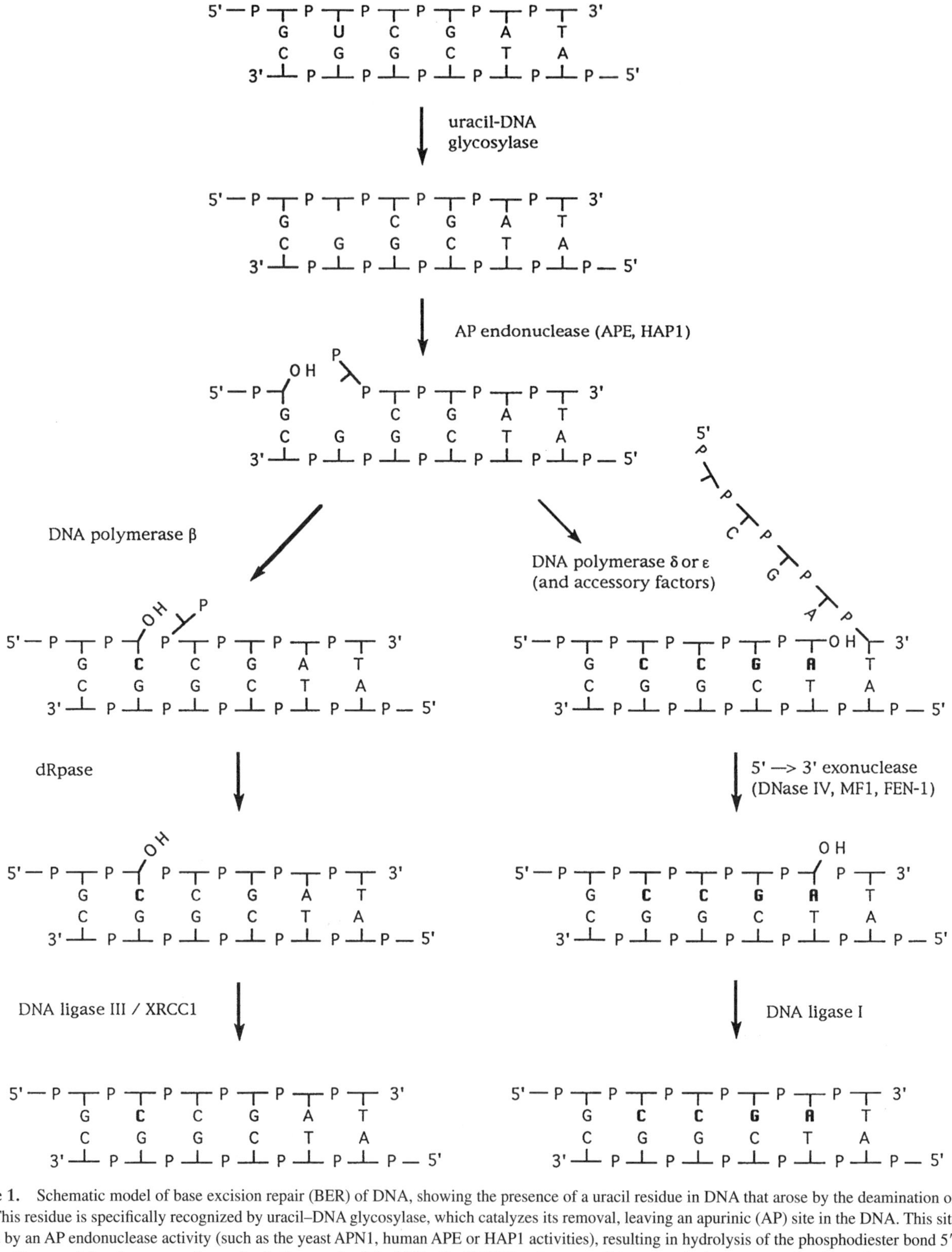

Figure 1. Schematic model of base excision repair (BER) of DNA, showing the presence of a uracil residue in DNA that arose by the deamination of cytosine. This residue is specifically recognized by uracil–DNA glycosylase, which catalyzes its removal, leaving an apurinic (AP) site in the DNA. This site is attacked by an AP endonuclease activity (such as the yeast APN1, human APE or HAP1 activities), resulting in hydrolysis of the phosphodiester bond 5′ to the AP site. Two possible schemes are shown for further repair of the DNA. (Left) DNA polymerase β inserts a single correct nucleotide and at the same time a dRpase activity excises the 3′-deoxyribose–phosphate moiety generated by the AP endonuclease. DNA ligase (which in human cells is believed to act in concert with XRCC1 protein) ligates the DNA, thereby completing DNA repair. (Right) A scheme suggesting that in some cases a different DNA polymerase may begin synthesis at the site of the nick created by the AP endonuclease, peeling back a 5′ single-stranded region and eventually generating a flap structure. This flap is specifically recognized by an enzyme called Rad27 protein in yeast and DNase IV, MF1, or FEN-1 in mammalian cells. DNA ligation then completes repair as before. [Reproduced from Dianov and Lindahl (1994), with permission from T. Lindahl, © Copyright Civorent Biology Ltd.]

(Figure 1). In theory the simplest mode of addressing these would be the reverse of the DNA glycosylase reaction, that is, the direct insertion of the correct base into the DNA. Several laboratories have searched for DNA insertases, but there is no evidence for the existence of such an enzyme activity in nature. The repair of AP sites, hence the completion of the process of BER, requires the participation of several other enzymes. Sites of base loss generated by the action of DNA glycosylases or by spontaneous hydrolysis of *N*-glycosyl bonds are attacked by a specific class of endonucleases called apurinic/apyrimidinic (AP) endonucleases. All known AP endonucleases catalyze the hydrolysis of the phosphodiester bond 5′ to the site of base loss, generating nicks with 3′-OH and 5′-P termini (Figure 1). In vitro some DNA glycosylases (such as the pyrimidine dimer–DNA glycosylase discussed earlier) are able to effect cleavage of the 3′-phosphodiester bond at AP sites through a so-called AP lyase activity. However, it is uncertain whether this activity operates in living cells.

The AP endonuclease that has been most extensively characterized and appears to be the quantitatively major activity in yeast has been variously designated as yeast AP endonuclease I, yeast DNA diesterase, yeast 3′-phosphoglycoaldehyde (3′-PGA) diesterase, or yeast apurinic endonuclease/3′-diesterase. This enzyme, which has a molecular weight of about 40.5 kDa, is a metalloenzyme. Atomic absorption spectroscopy has revealed the presence of 3.3 atoms of Zn^{2+}/molecule of protein. Nuclease activity is also stimulated by cobalt ions, though the enzyme does not have an absolute requirement for this metal cofactor. In addition to its AP endonuclease activity, yeast AP endonuclease can excise a variety of 3′-esters in DNA, including 3′-phosphoglycoaldehyde, 3′-phosphoryl groups, and 3′-α,β-unsaturated aldehydes. Like all AP endonucleases, the enzyme hydrolyzes DNA 5′ to AP sites, generating 3′-OH and 5′-deoxyribose-phosphate termini.

The structural gene that encodes this AP endonuclease is designated *APN1*. Strains carrying disruptions or deletions of the *APN1* gene are hypersensitive to killing by a variety of oxidative and alkylating agents that damage DNA. Additionally, such mutants have a spontaneous mutation rate about an order of magnitude greater than that observed in wild-type yeast cells. The predicted amino acid sequence of the *APN1* gene is homologous with that of an *E. coli* gene that encodes an AP endonuclease. In fact, the yeast *APN1* gene can complement sensitivity to killing by oxidants and alkylating agents in *E. coli* mutants that are defective in AP endonuclease activity.

3.1.4 Repair Synthesis, Excision, and DNA Ligation

The incision of DNA 5′ to sites of base loss provides a normal 3′-OH nucleotide terminus for the initiation of repair synthesis of DNA (Figure 1). There are indications that in yeast, repair synthesis during BER is effected by an enzyme called DNA polymerase ϵ, although there are suggestions that another related enzyme, DNA polymerase δ, may also carry out this function. It is not known what specific conditions dictate the preferential utilization of a particular DNA polymerase. The repair synthesis reaction is limited. Only about two to four nucleotides are inserted. To achieve this, the DNA strand with the 5′-P terminal AP site must be displaced by limited nick translation. As shown in Figure 1, displacement by repair synthesis of a single strand with a 5′ terminus would generate a three-stranded flap structure. A yeast enzyme has been identified which specifically recognizes flap structures with 5′ single-strand tails and removes the single-stranded flap. This protein is encoded by a gene called *RAD27*. The substrate specificity of the Rad27 nuclease has in fact driven the notion that repair synthesis precedes excision of the AP site during BER as shown in Figure 1. However, an equally plausible model is that the 5′-terminal AP site is excised first and repair synthesis then follows this event. While the properties of the Rad27 nuclease are consistent with a role in BER, there are indications that this protein is primarily involved in DNA replication.

Figure 1 shows an alternative repair synthesis/excision reaction for which there is substantial evidence in mammalian cells. In this reaction a single nucleotide is believed to be inserted by DNA polymerase β. The 5′-terminal AP site is then excised by an enzyme called DNA deoxyribophosphodiesterase (dRpase). Hence, the repair synthesis patch during this putative mode of BER is one nucleotide long. A dRpase activity has not been identified in yeast. A gene that encodes a polypeptide with the properties of mammalian DNA polymerase β has been cloned from yeast. However, deletion of this gene produces no recognizable phenotype. This does not necessarily mean that the polymerase has no role in DNA repair, since in its absence the alternative longer patch repair synthesis mode may operate exclusively.

3.1.5 DNA Ligation

The final event in all excision repair reactions is the joining of newly inserted nucleotides with 3′-OH termini to the 5′-P terminus of the extant DNA. This joining reaction is catalyzed by an enzyme called *DNA ligase*. Mammalian cells contain multiple DNA ligases, most (but possibly not all) of which are encoded by distinct genes. As is the case with the multiple DNA polymerases present in eukaryotic cells, it remains to be determined which DNA ligase is used for which specific function. At present a single DNA ligase encoded by the yeast *CDC9* gene has been purified and characterized. However, there are indications that yeast cells contain at least one other DNA ligase activity.

3.2 Nucleotide Excision Repair

Like BER, nucleotide excision repair is a ubiquitous repair process. Unlike BER in which different repair enzymes (DNA glycosylases) recognize different forms of base damage, NER is in general characterized by the recognition of multiple diverse types of base damage by a single repair apparatus. A fundamental issue yet to be resolved is what determinant(s) set up the broad substrate specificity of the NER apparatus. Yeast has been a particularly informative model for deciphering the many genetic and biochemical complexities of NER in eukaryotes. Parallel studies in yeast and mammalian cells have shown that the genes and proteins involved in NER are highly conserved, and that the detailed biochemistry of this process in these organisms is quite different from that in prokaryotes. This structural and functional conservation has allowed for reliable extrapolations from yeast to man and vice versa, and at the risk of some inaccuracy I will continue to make such extrapolations in the ensuing discussions. Very recent studies in the yeast *S. pombe* have also provided evidence that NER can occur by more than one biochemical pathway in eukaryotic cells (see later discussion).

3.2.1 NER Mediated by Dual Damage-Specific Endonucleases

The nucleotide excision repair mode is characterized by enzyme-catalyzed incisions in the affected DNA strand both 5′ and 3′ to sites

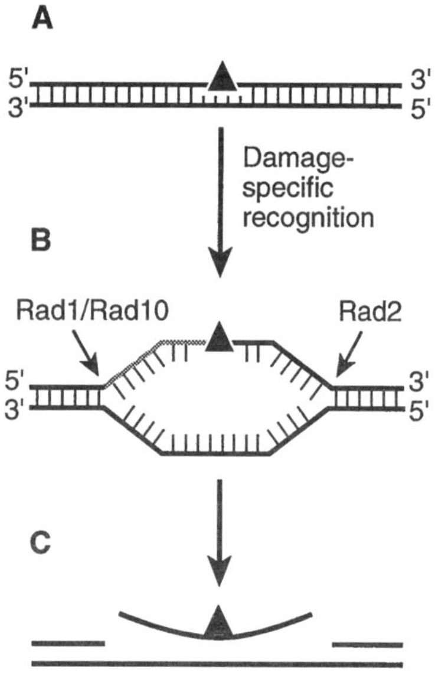

Figure 2. Cartoon showing the postulated mechanism of bimodal incision of DNA during GNER in yeast. Following the specific recognition of base damage by the GNER machinery (A) a "bubble" structure (B) is somehow generated. The Rad1/Rad10 enzyme recognizes the junction between double-stranded and single-stranded DNA, and cleaves the 3′ single-stranded DNA. The Rad2 endonuclease similarly recognizes the junction between double-stranded and single-stranded DNA, and cleaves the 5′ single-stranded DNA. The opposite (undamaged) strand (C) must somehow be protected, since *in vivo* incisions are only made on the damaged strand. [Reproduced with permission from E. C. Friedberg, G. C. Walker, and W. Siede (1995).]

of base damage at relatively precise distances from such sites, generating oligonucleotide fragments roughly 30 nucleotides long which are subsequently excised (Figure 2). I refer to this NER mode as general NER (GNER) to distinguish it from a second mode of NER, which is characterized by a *single* damage-specific incision event 5′ to sites of base damage in UV-irradiated DNA. Evidence from studies on mammalian cells indicates that during GNER the 5′ incisions are located about 22 nucleotides from the sites of damage, while the 3′ incisions are located just five nucleotides from these sites (Figure 2). Presumably the same holds true in yeast and other eukaryotes. At present we have no understanding of what determines this precise spacing of nicks. In one working model, however, events intrinsic to the recognition of base damage by the GNER machinery process the DNA locally, such that a discrete "repair bubble" of approximately 30 nucleotides is identified for bimodal incision.

This model derives from the discovery in yeast and human cells of proteins known to be required for GNER, which specifically recognize and cleave double-strand/single-strand junctions in DNA with polarities that are specified by the directionality of the single-stranded DNA. Hence, the putative 5′-endonuclease for GNER in yeast, consisting of a heterodimer of the products of the *RAD1* and *RAD10* genes, specifically cleaves the 3′ single-stranded DNA at double-strand/single-strand junctions in vitro (Figure 2). On the other hand, the putative 3′-endonuclease for GNER, consisting of the monomeric product of the human *XPG* gene (homologous to the yeast *RAD2* gene), specifically cleaves the 5′ single-stranded DNA at double-strand/single-strand junctions in vitro (Figure 2). So it would appear that the substrate specificity for incision of damaged DNA during GNER is determined by the presence of duplex/single-strand DNA junctions rather than by the damaged bases themselves.

In yeast one of the many genes that is indispensable for GNER, called *RAD3*, encodes a 5′ → 3′ DNA helicase. Rad3 protein is an integral component of a large nucleotide excision repairosome comprising at least 10 other proteins (including the Rad1/Rad10 and Rad2 nucleases just discussed), which has been identified in extracts of yeast cells not exposed to DNA damage. In vitro, the translocation of purified Rad3 helicase on duplex DNA is arrested at sites of base damage located on the strand on which it translocates, but not by damage on the opposite strand. Collectively these facts have led to the suggestion that the preassembled nucleotide excision repairosome loads onto DNA and processively patrols the genome (for unspecified distances) in search of base damage, by the action of the Rad3 helicase (Figure 2). The yeast repairosome includes a second DNA helicase encoded by the *SSL2* (*RAD25*) gene, with a directionality opposite to that of the Rad3 protein. Conceivably this helicase participates in scanning the other DNA strand for base damage. Proteins encoded by the *RAD7, RAD16,* and *RAD23* genes may be required for the processing of chromatin during this patrolling function, a notion that is consistent with the known requirement for these genes for GNER in transcriptionally silent DNA. When a site of base damage is encountered, the arrested translocation of the Rad3 helicase may position the repairosome in the immediate proximity of the damage (Figure 3; see color plate 6). Further biochemical events requiring other subunits of the repairosome may stabilize this binding, resulting in the formation of a tight repairosome–DNA complex, the precise topology of which places the catalytic domain of the Rad1/Rad10 endonuclease in the immediate vicinity of a double stranded/3′ single-strand junction, and the catalytic domain of the Rad2 endonuclease in the immediate vicinity of a double-stranded/5′ single-strand junction (Figure 2).

Besides the Rad1, Rad10, Rad2, Rad3, and Ssl2 proteins already alluded to, the nucleotide excision repairosome is known to include at least six other proteins designated Rad4, Rad14, Ssl1, Tfb1, Tfb2, and Tfb3. Rad14 and its homologue in human cells, XPA protein, bind preferentially to damaged DNA in vitro and are possibly involved in the postrecognition protein–DNA stabilization events postulated earlier. Conceivably the Rad4, Rad7, Rad16, and Rad23 proteins are involved in these events as well. Based on studies with human cells, the yeast nucleotide excision repairosome may contain other polypeptides, including single-stranded binding protein, replication factor A, and proliferating cell nuclear antigen.

3.2.2 Proteins for GNER Are Also Required for Transcription by RNA Polymerase II

The Rad3, Ssl2, Ssl1, Tfb1, Tfb2, and Tfb3 proteins are components of a protein complex called TFIIH, which in concert with several other transcription factors is required for the initiation of basal transcription by RNA polymerase II. It has been directly demonstrated that Rad3, Ssl2, Ssl1, and Tfb1 are indispensable for GNER in vitro, and it is likely that the same is true of the Tfb2 and Tfb3 proteins. Hence, all six of these proteins are probably bifunctional, with obligatory roles in GNER and in RNA polymerase II transcription.

The discovery that some GNER proteins are also involved in transcription provides a compelling biochemical correlate of the observation that in vivo this DNA repair mode operates more rapidly on the transcribed (template) strand of transcriptionally active genes

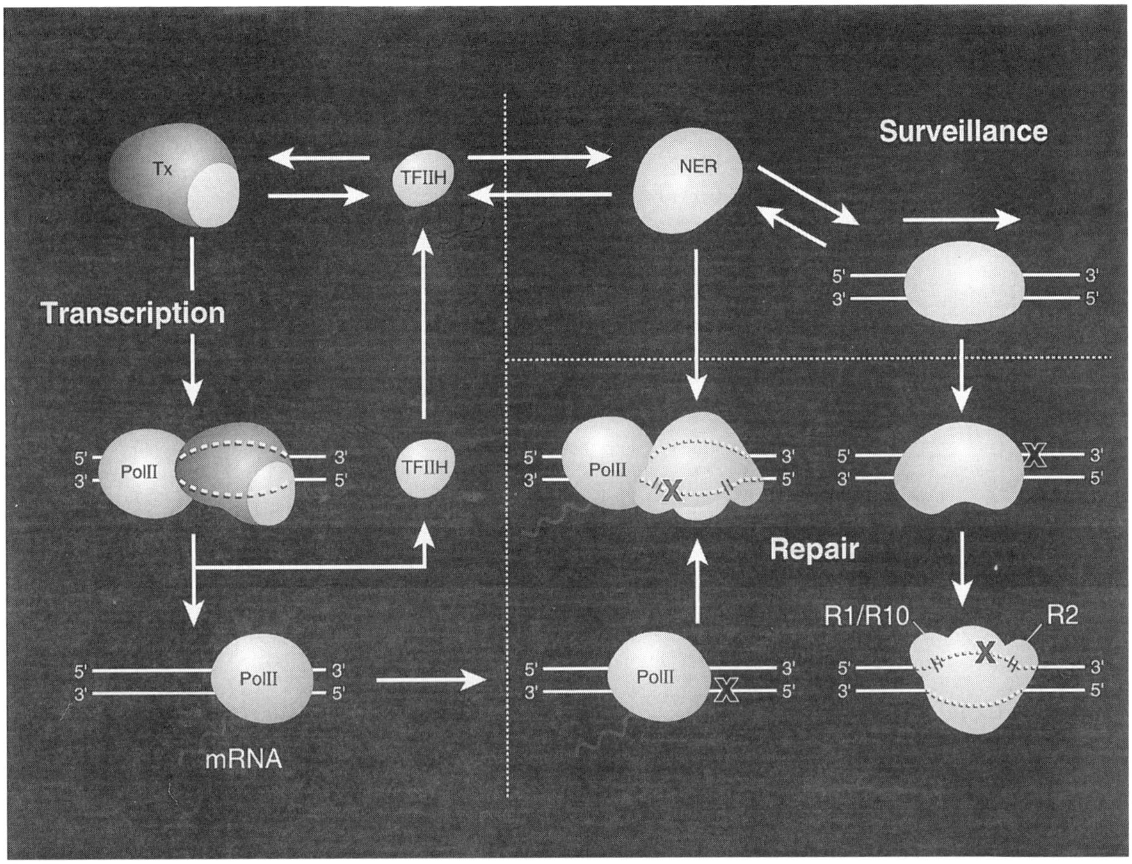

Figure 3. Cartoon summarizing the relationship between GNER and RNA polymerase II transcription in yeast. Transcription factor IIH (TFIIH) is common to both the holoTFIIH transcription (Tx) initiation complex and the repairosome for GNER (NER) and may shuffle between these two supercomplexes with varying equilibrium states. The preassembled repairosome may constantly survey the genome for base damage in transcriptionally silent DNA (surveillance), and when such damage (X) is encountered, repair it (bottom right). HoloTFIIH is believed to be loaded onto promoter sites during RNA polymerase II transcription initiation, where it associates with the polymerase and with other transcription factors. When transcription has been successfully initiated TFIIH is released. When the transcription elongation complex encounters sites of base damage (X) in the template strand, transcription stalls. The repairosome is recruited to sites of stalled transcription through interactions that specifically require the TFIIH subunits. The recruited repairosome effects repair of the damage, as in transcriptionally silent DNA. (See color plate 6.)

than it does on the nontranscribed (coding) strand. A logical means of reconciling these two sets of experimental findings is to postulate that TFIIH is obligatorily loaded onto the template DNA strand during transcription initiation by RNA polymerase II and remains associated with the transcription machinery during mRNA synthesis. When the transcription machinery encounters a site of base damage, transcription is arrested and the TFIIH complex provides a nucleation site for the assembly of a complete functional nucleotide excision repairosome, thus accounting for the rapid kinetics of repair of the template strand.

However, if a nucleotide excision repairosome is indeed assembled on the core TFIIH complex following arrested transcription, what is the function of the preassembled repairosome observed in extracts of yeast cells in the absence of exposure to DNA damage, that includes all the TFIIH components? One possibility is that the preassembled repairosome is uniquely required for GNER in transcriptionally silent regions of the genome. A more likely explanation is that TFIIH operates during GNER at sites of arrested transcription as a component of a preassembled repairosome, which specifically associates with base damage at sites of arrested transcription. This logic suggests that the TFIIH complex that is loaded onto promoter DNA during transcription initiation dissociates from the transcription machinery during the elongation phase. In fact, experiments with mammalian cells suggest that TFIIH does *not* participate in the process of transcript elongation. Hence, conceptually the requirement for TFIIH both for transcription and for GNER may operate through its selective affinity for a transcription complex at transcription initiation (promoter) sites in the former case, and for sites of arrested transcription by base damage in the latter case. There is experimental evidence in both yeast and mammalian cells that TFIIH does indeed associate with two physically and functionally distinct multiprotein complexes.

The model just described implies that in the absence of DNA damage, core TFIIH exists in a steady state equilibrium between a transcription initiation complex (called TFIIH holocomplex in yeast) and a repairosome. When cells are exposed to DNA damage, the equilibrium state shifts to repairosome assembly. This shift may be triggered by the increased synthesis of repair proteins, which are limiting in the absence of such damage. It is known, for example, that when yeast cells are exposed to a variety of types of DNA damage, expression of both the *RAD2* and *RAD7* gene is induced. Hence, the steady state levels of Rad2 and Rad7 proteins may limit repairosome formation in the undamaged state. The increased repairosome levels could facilitate efficient association of these complexes with sites of base damage generated in regions of active transcription. The preferential assembly of TFIIH into repairosomes would be expected to limit transcription initiation, thereby reducing gene expression in the presence of DNA damage, a teleologically expected and experimentally observed response to genomic insult.

3.2.3 NER Mediated by a Single Damage-Specific Endonuclease Activity

Recent studies in the yeast *S. pombe* and in *Neurospora crassa* have identified a novel endonuclease that apparently specifically recognizes CPD and another bulky photoproduct (the so-called 6-4 photoproduct) in DNA. Since both these lesions result from exposure of cells to UV radiation, this endonuclease represents yet another cellular defense that has evolved in some organisms against this potent naturally occurring genotoxin. The enzyme nicks DNA uniquely 5′ immediately adjacent to sites of damage. The events that follow the 5′ incision event are not yet known. It is likely that the physical excision of the base damage (presumably as part of an oligonucleotide fragment) is effected by events associated with repair synthesis and nick translation, without a requirement for a second endonuclease.

3.3 Mismatch Repair

Mismatch repair (MMR) refers to biochemical events that result in the excision of mismatched bases from DNA. As is the case with general excision repair, the repair of mismatches can occur in either a BER mode or an NER mode; that is, the mismatched bases can be excised as free bases or as oligonucleotide fragments. The BER mode of MMR is limited to certain mismatches and has not been identified in yeast. In contrast, the NER mode of MMR is ubiquitous and appears to be mechanistically conserved from prokaryotes to *H. sapiens.* During semiconservative DNA synthesis in *E. coli,* newly replicated DNA strands are methylated at GATC tetranucleotide sequences (Figure 4). The kinetics of this process results in a period of transient undermethylation of these strands. The MMR system exploits this chemical difference to provide strand specificity for the excision of mispaired bases.

In *E. coli* the presence of both mismatched bases and undermethylated (at GATC sequences) DNA results in the binding of proteins encoded by the *mutS* and *mutL* genes. The formation of a bound MutS/MutL complex activates a latent endonuclease in a protein encoded by the *mutH* gene, resulting in incision of the newly replicated DNA strand at an unmethylated GATC sequence (Figure 4). This nick can occur either 5′ or 3′ to a mismatch and initiates degradation of the DNA between the nick and the mismatch, resulting in excision of the latter. The excision reaction, which is dependent on the MutS and MutL proteins, a DNA helicase, and an exonuclease, results in the excision of a tract of DNA that extends from the site of the mismatch to the site of incision at the nonmethylated GATC sequence (Figure 4).

Homologous systems for MMR have been identified in yeast and

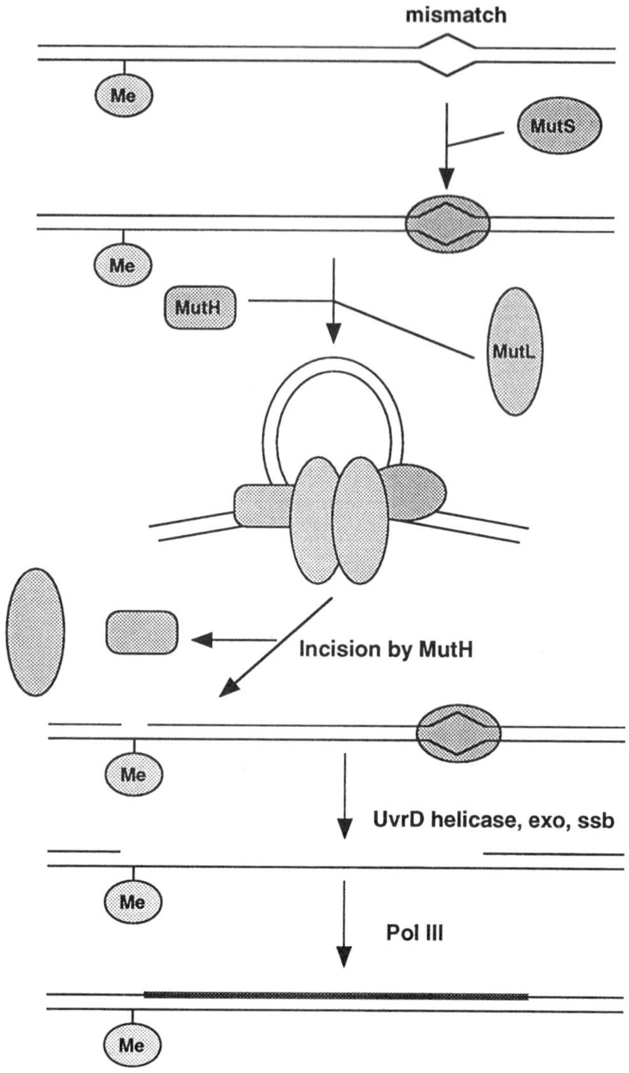

Figure 4. General (long patch) mismatch repair of DNA in *E. coli.* The eukaryotic process is believed to share many of the features shown here. See text for details [Reproduced with permission from E. C. Friedberg and R. D. Wood, (1996).]

in higher eukaryotes. Extracts of human cells support an excision reaction that is remarkably similar to that described in *E. coli.* Yeast DNA is not methylated; hence the molecular basis for the strand specificity during MMR in this organism is unknown, and a cell-free system that supports strand-specific mismatch correction in yeast has not yet been achieved. A yeast gene called *PMS1* has been shown to be homologous to the *E. coli mutL* gene. A second *mutL* homologue called *MLH1* (for MutL homologue) has also been cloned, and four homologues of the *E. coli mutS* gene have been isolated, designated *MSH1, MSH2, MSH3,* and *MSH4.*

See also DNA Damage and Repair; DNA Repair in Aging and Sex; Free Radicals in Biochemistry and Medicine; Ionizing Radiation Damage to DNA; Mutagenesis, Molecular Basis of; Ultraviolet Radiation Damage to DNA.

Bibliography

Aboussekhra, A., and Wood, R. D. (1994) Repair of UV-damaged DNA by mammalian cells and *Saccharomyces cerevisiae*. *Curr. Biol.* 4:212–220.

Dianov, G., and Lindahl, T. (1994) Reconstitution of the base excision-repair pathway. *Curr. Biol.* 4:1069–1076.

Friedberg, E. C., Siede, W., and Cooper, A. J. (1991) Cellular responses to DNA damage in yeast. In *The Molecular and Cellular Biology of the Yeast Saccharomyces. Genome Dynamics, Protein Synthesis, and Energetics*, J. R. Broach, J. R. Pringle, and E. W. Jones, Eds., pp. 147–192. Cold Spring Harbor Press, Plainview, NY.

———, Walker, G. C., and Siede, W. (1995) *DNA Repair and Mutagenesis*. ASM Publications, Washington DC.

———, and Wood, R. D. (1996). In *DNA Replication in Eukaryotic Cells*, M. L. DePamphilis, Ed., pp. 249–269. Cold Spring Harbor Laboratory Press, Plainview, NY.

Modrich, P. (1991) Mechanisms and biological effects of mismatch repair. *Annu. Rev. Genet.* 25:229–254.

Prakash, S., Sung, P., and Prakash, L. (1993) DNA repair genes and proteins of *Saccharomyces cerevisiae*. *Annu. Rev. Genet.* 27:33–70.

Yeast Artificial Chromosome (YAC) Construction and Applications

Christine Loehrlein and Ronald W. Davis

1 **Principles of YAC Construction**
  1.1 YAC Vectors: Description and Logic of Design
  1.2 Current Technical Limitations
2 **Techniques for Constructing Yeast Artificial Chromosomes**
  2.1 Outline of the Procedure
  2.2 In Vitro Construction of YACs
  2.3 Transforming YACs into Yeast
  2.4 Working with YACs in Yeast
3 **Using YACs as Tools and Devices**
  3.1 Genomic Mapping
  3.2 Finding and Building Genes
  3.3 Constructing Transgenic Cell Lines and Organisms
4 **Concluding Remarks**

Key Words

CHEF Gel *C*ontour-clamped *h*omogeneous *e*lectric *f*ield gel—a derivative of an alternating electric field gel used for separation of large DNA molecules (up to 2 megabase pairs).

Chimera A clone containing two or more pieces of DNA that are not adjacent to each other in their normal genetic locations.

Physical Map of a Genome The information describing the order of sequence information in the genome, established using structural data such as the order of sites of a restriction enzyme or regions of known sequence called sequence-tagged sites (STSs).

Saccharomyces cerevisiae **(yeast)** A fast-growing, single-celled eukaryotic microorganism commonly used in biological studies.

Yeast Artificial Chromosome (YAC) A large foreign fragment of DNA propagated as a linear molecule in yeast, containing all the essential elements required for maintenance of natural yeast chromosomes.

Yeast artificial chromosomes (YACs) are fragments of DNA constructed in vitro that contain all the elements required for their stable propagation as an independent chromosome in yeast. Fragments of DNA sometimes exceeding a megabase in length have been cloned by means of YACs. Because YAC clones are extremely large, they are an important resource for generating maps of large genomes, for providing long-range continuous coverage of a complex genome contained within contiguously cloned DNA, and for subcloning and maintaining very large eukaryotic genes. This technology has been critical in bridging fine-scale and large-scale mapping data to produce an integrated map of the entire human genome. YACs have allowed cloning of several disease genes, including the *BCL2* proto-oncogene, the gene responsible for cystic fibrosis, and the breast cancer susceptibility loci *BRCA1* and *BRCA2*. YACs have also proven useful in generating stable transgenic mammalian cells, promising a valuable application in gene therapy using very large fragments of cloned DNA.

1 PRINCIPLES OF YAC CONSTRUCTION

1.1 YAC Vectors: Description and Logic of Design

For a linear chromosome to be stably maintained in yeast, it must contain a centromere for proper segregation in mitosis, telomere sequences at each end to protect the ends from degradation, at least one origin of replication, and selectable genes. Vectors designed for building yeast artificial chromosomes contain all these essential elements, as well as additional components that facilitate use of a YAC clone once it has been obtained. Some of the vectors that have been constructed for making YACs are shown in Figure 1A. The structure of a linear YAC made with the vectors pRML1 and pRML2 is illustrated in Figure 1B.

In studies of the replicative stability of YACs, we have found that a single ARS (*a*utonomously *r*eplicating *s*equence—a yeast origin of replication) provides for stable replication of only about 100 kb of DNA in each direction. To increase the stability of clones generated from regions of DNA that may have insufficient origin sequences, vectors should be designed to provide an ARS on both ends of the linear YAC.

YACs generated with pCL99 and pCL100 contain the *ADE2* gene on the acentric arm, providing a visual screen for monitoring stability and rearrangement of individual YAC clones. As shown in Figure 2 (see color plate 7), yeast strains having a mutation in the *ADE2* gene form red colonies on solid media containing limiting adenine (Figure 2A), while wild-type strains form white colonies (Figure 2B). If a yeast strain with a mutation in the *ADE2* gene is transformed with a YAC containing the wild-type gene, it will form white colonies only as long as it maintains the YAC. The tendency of a transformant to lose a particular YAC clone can therefore be assessed visually by the frequency of red sectors arising within a white colony. The more red the colony, the more unstable the YAC clone (Figures 2C and 2D). Unstable clones are prone to rearrangements that remove DNA causing the instability. Such events can be seen as sectors arising within a colony that have a rate of sectoring dif-

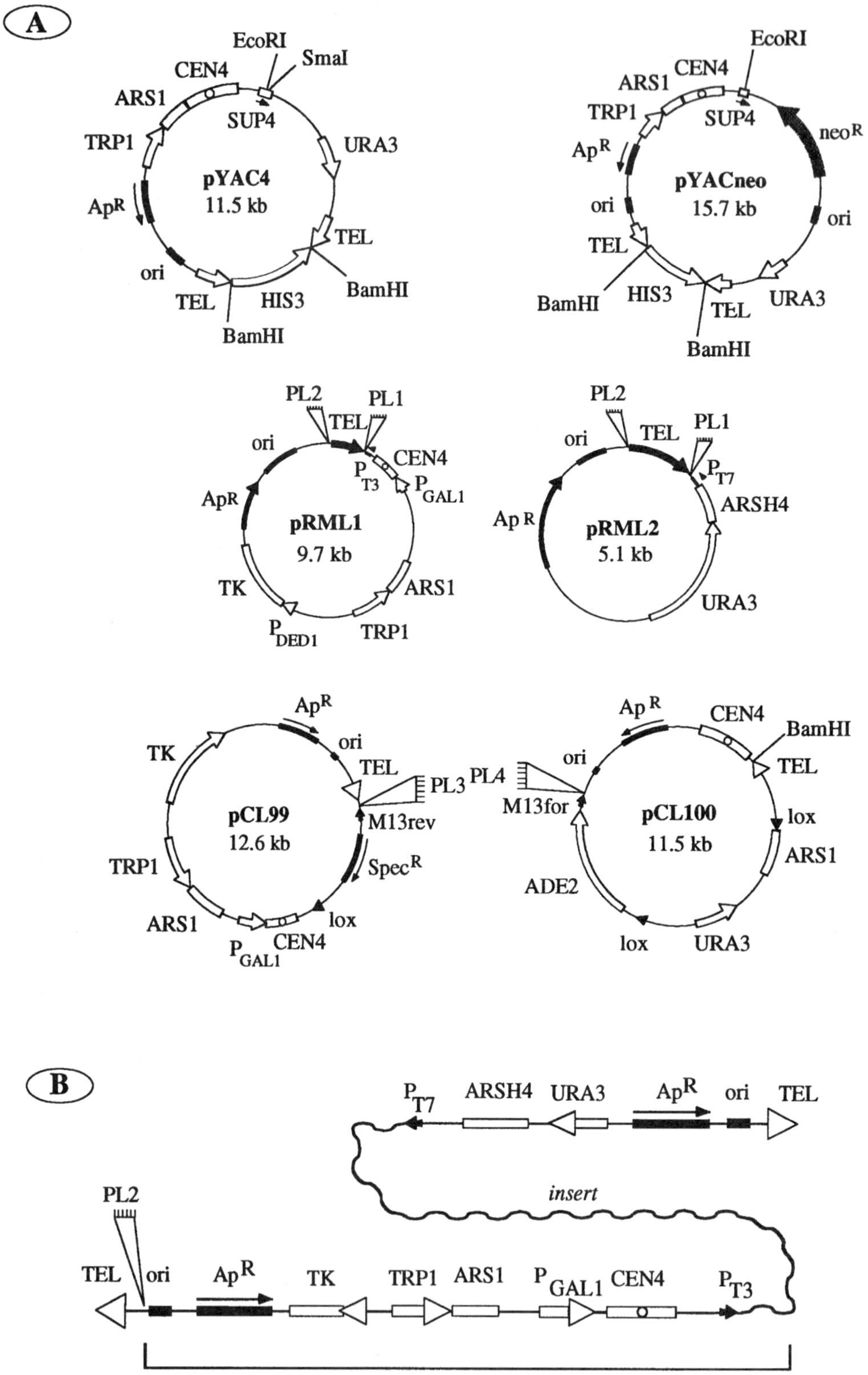

Figure 1. YAC vectors and structure. (A) Maps of plasmid vectors. pYAC4 and pYACneo each provide both vector arms of a YAC clone. With the pRML and pCL vectors, each arm is constructed on a separate plasmid. *PL,* polylinker; P_{T3}, P_{T7}, *M13for,* and *M13rev,* sequences that anneal to primers for sequencing ends of the inserted fragment; *Spec*^{R}, *Ap*^{R}, antibiotic resistance genes; neo^R, G418 drug resistance gene; *CEN4,* centromere; *P-GAL1, P-DED1,* transcriptional promoters; *ARS1, ARSH1,* yeast origins of replication; *TRP1, ADE2, HIS3, SUP4, URA3,* yeast selectable metabolic markers; *TK,* chromosome copy number amplification marker; *ori,* bacterial origin of replication; *TEL,* telomere; *lox,* recognition sequence for the site-specific recombinase Cre. (B) Structure of a YAC made with the pRML vectors; brackets indicate region illustrated in Figure 4.

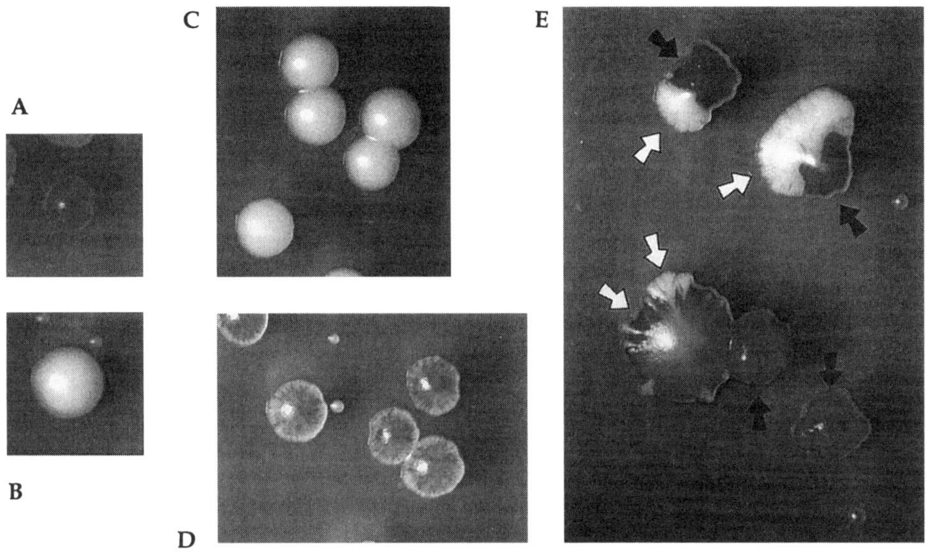

Figure 2. Use of the *ADE2* gene to assess YAC stability. (A) Yeast colony formed by an *ade2* mutant strain. (B) Colony formed by a wild-type strain. (C) A very stable YAC clone. (D) A YAC clone showing intermediate instability. (E) Very unstable clone, prone to rearrangement at a high frequency: closed arrows, phenotype of the unrearranged clone; open arrows, rearranged derivatives of varying lengths and stabilities. (See color plate 7.)

ferent from the original phenotype (open arrows, Figure 2E). (Changes are generally seen as a reduction in sectoring; but if the rearrangement removes part of the *ADE2* gene, a solid red sector will result.) If colonies of such a mixed phenotype are streaked out to single cells on solid medium and allowed to re-form colonies, a mixture of colony phenotypes is observed: those showing the original sectoring phenotype (with sectors from rearrangement events arising within the colony once again) and more stable subclones (forming colonies with a lower rate of sectoring and showing a lower frequency of sectors resulting from rearrangement events). Because an unstable clone can be conveniently separated from its rearranged products, the sectoring phenotype provides for maintenance of unstable regions of DNA that are prone to rearrangement, regions of DNA that would be lost in libraries that do not provide such a stability screen (Section 2.4.1).

The pRML and pCL vectors shown in Figure 1 also contain elements that allow amplification of the copy number of a YAC in yeast from a single copy typical of centromere-containing chromosomes to copies as high as 20 per cell. Increasing copy number increases the amount of cloned DNA that can be obtained from a yeast cell, greatly facilitating characterization and use of the YAC clones. The vector elements required for amplification are the thymidine kinase cell as a selectable marker and a strong regulatable transcriptional promoter placed adjacent to the centromere. Accurate segregation of replicated chromosomes from each other relies on functionality of the centromere, which has been shown to be repressed by transcription. When transcription from the promoter adjacent to the centromere is induced, cells can missegregate the YAC and subsequently grow more vigorously, as extra copies of the *TK* gene provide for gradually improved growth (Section 2.4.4).

Obtaining the sequences of DNA at the ends of the insert in a YAC clone greatly facilitates construction of a physical map of a genome by providing a probe to search for other YAC clones having overlapping sequences. The pRML and pCL vectors contain multiple cloning sites that allow isolation of the ends of the inserted DNA (Section 2.4.5). The vectors also have sequences adjacent to the insert DNA that hybridize to sequencing primers (P_{T3} and P_{T7}, Figure 1B), allowing determination of the sequences at the ends of the insert (Section 3.1).

The potential uses of YAC libraries extend well beyond studies of physical maps and sequence information of the genomes of complex organisms. If vectors used to construct a YAC library are appropriately designed, the resulting clones can be used for generating transgenic cell lines or organisms. The pYACneo and pCL vectors illustrate this application. pYACneo contains the neomycin phosphoribosyltransferase gene (*neo*), a selectable marker conferring resistance to the drug G418 in higher eukaryotic cells. YACs made from the pCL vectors contain *spec,* a gene encoding resistance to the antibiotic spectinomycin in bacteria and plant cells. Generation of transgenic cells using YACs is discussed further in Section 3.3.

1.2 Current Technical Limitations

1.2.1 Chimera Formation

Perhaps the most problematic artifact to arise in the development of artificial chromosome technology is the presence of chimeric clones. Such clones contain two or more fragments of DNA that are not adjacent to each other at their normal genetic loci. When such clones are used for mapping or reassembling a complete gene, they result in a jump in the physical map from one area of the genome to another. The difficulty is that it is impossible to detect such an event without an independent source of information about the organization of a genome. Methods for identifying chimeric clones before they are used are generally very laborious. Some of these methods are discussed in Section 2.4.3.

Rates of chimerism, which vary enormously among various libraries, have ranged from a few percent to greater than 60%. The rates of chimera formation are generally proportional to the average insert size of a library. Chimeric clones can arise in a variety of ways, both in construction of YAC clones and after introduction into yeast.

When YACs are constructed, if the insert is not thoroughly mixed with vector (which is present in excess), two separate insert fragments can ligate to each other. Incomplete mixing can also cause the vector to be ligated to insert on only one end; shearing of insert DNA may create ends incompatible with the vector. Such fragments of DNA, when transformed into yeast, will be forced to recombine with a second YAC to ensure that the required metabolic markers are present on each end. This recombinational event, which would give rise to a chimeric product, is facilitated by the tendency of the transformation process to generate cells capable of taking up more than one fragment of DNA.

Chimeras can also arise from correctly assembled clones once transformed into yeast. If the insert DNA is physically damaged, it may induce in yeast DNA repair enzymes that remove the damage through recombinational pathways. If a specific segment of DNA is unstable in yeast for any reason, rearrangements may occur that remove the cause of instability.

There are no protocols for artificial chromosome construction that specifically and consistently eliminate formation of chimeras, but there are several factors that can reduce their frequency. When libraries are made from human/rodent hybrid cell lines, the vast majority of insert DNA is from the rodent source, so human–human chimeras are unlikely. YAC libraries can be established in various strains deficient in homologous recombination, but such mutants are also defective in carrying out many possible manipulations of a YAC's structure once it has been established. Transformants can be monitored for instability using the *ADE2* gene, as described in Section 2.4.1. This approach, which allows characterization of the instability of a clone, provides a means to maintain an unstable clone in an unrearranged configuration. Finally, because the rate of chimera formation has been shown to be proportional to the average insert size of a library, libraries can intentionally be made with smaller insert sizes. The extra labor involved in using smaller pieces of DNA to assemble a map can be offset by the circumvention of mapping errors that result from chimeras.

1.2.2 Isolation of Large Quantities of Cloned DNA

Once a YAC library has been established, it is essential in some applications to obtain large quantities of the insert DNA in a highly concentrated form, purified away from yeast DNA. This is difficult to obtain from yeast; methods described in Sections 2.4.4 and 2.4.5, however, provide enough DNA for most purposes.

2 TECHNIQUES FOR CONSTRUCTING YEAST ARTIFICIAL CHROMOSOMES

2.1 Outline of the Procedure

The in vitro construction of YACs requires a series of manipulations of extremely long fragments of DNA. The most difficult aspect of artificial chromosome construction is the minimization of random shearing of the genomic DNA used for the insert. Not only does shearing reduce the average length of the clones, but it may also promote rearrangements when the clones are in yeast. Long fragments of DNA become more prone to shearing as their concentration (and consequently viscosity) is reduced. The protocol described here carries out all DNA manipulations essentially in a solid phase, since the genomic DNA has been prepared in agarose. This allows construction of artificial chromosomes with a very small quantity of insert DNA and a minimum of shearing.

The manipulations most prone to causing shearing of genomic DNA are pipetting and mixing with reagents. Shearing due to pipetting can be minimized by cutting the ends off disposable plastic pipette tips with a sharp razor blade to leave a larger opening. Shearing that results from mixing reagents into the DNA can be minimized by bearing in mind that DNA is a long, linear polymer. Reagents are added by holding the tube containing the DNA at an angle, and adding reagents by slowly expelling them from the tip of the pipette. To mix, the tip is slowly drawn through the DNA in parallel lines, taking care to not draw solution up into the tip.

Electrophoretic analysis of very large fragments of DNA has been made possible by the development of pulsed field gel systems in which two separate electric fields are generated across the gel at different angles, alternating with each other for defined periods of time. It is thought that the continual changing of the direction of the electric field enables very large molecules of DNA to move through the gel without becoming entangled, thus allowing their resolution. The times chosen for switching between the two electric fields can vary from a few seconds up to half an hour, with longer switch times resolving fragments of greater length. An innovation of the pulsed field gel developed in this laboratory is called the *c*ontour-clamped *h*omogeneous *e*lectric *f*ield (CHEF) gel. This design provides a much more uniform electric field, thereby reducing distortions in DNA migration and improving resolution.

The basic steps for construction of YACs are described in subsequent numbered sections as indicated below. DNAs at various stages of YAC construction are illustrated in Figure 3.

2.2.1	Prepare genomic DNA in low melting point agarose
2.2.2	Digest with restriction enzyme
2.2.2	Size-select DNA fragments on CHEF gel
2.2.3	Ligate to vector arms in low melting point agarose
2.2.3	Size-select a second time on CHEF gel
2.2.3	Convert DNA sample to liquid phase with agarase
2.3.2	Transform into yeast spheroplasts
2.4.3	Characterize and use the library

2.2 In Vitro Construction of YACs

2.2.1 Preparation in Agarose of High Molecular Weight Genomic DNA from Tissue Culture Cells

The following protocol has been developed for cultured mammalian cells. It can be adapted for any cell type by modifying steps 2 and 5 as appropriate.

Protocol

1. Harvest about 5×10^7 cells. Pellet at 3000 rpm for 10 minutes in a clinical benchtop centrifuge; decant supernatant.
2. Wash cells with serum-free medium; pellet; decant.
3. Warm cells to 40°C. Add an equal volume of 1.5% low melting point agarose containing 125 mM EDTA/NaOH, pH 8, which has been melted and cooled to 50°C. Mix quickly by pipetting, being careful to avoid making bubbles.
4. Aliquot mixture into small-volume (50–100 μL) Plexiglas or plastic molds (see Comments). Allow to solidify for 5 minutes at 4°C.

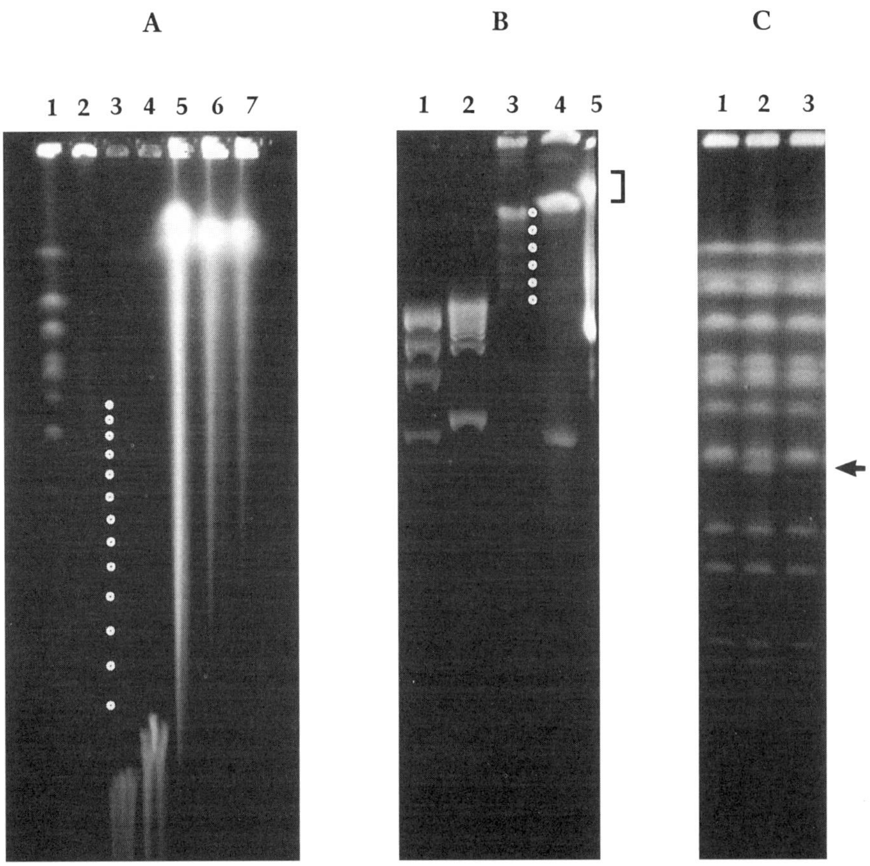

Figure 3. DNAs at different stages of YAC construction. (A) Partial *Xba* I digestion of human DNA: lane 1, an untransformed yeast strain; lane 2, ladder of 48.5 kb λ concatemers (dots indicate positions of the multimers); lanes 3–7, partial digestions of human genomic DNA incubated 10 minutes with *Xba* I at 50, 25, 2.5, 0.25, and 0.025 units/μg DNA. (B) Size selection of ligation products: lane 1, λ *Hin*dIII digest; lane 2, 5 kb ladder; lane 3, 48.5 kb lambda ladder; lane 4, yeast chromosomes; lane 5, YAC ligation products. Bracket indicates region of the gel used for transforming yeast. DNA molecules of 300 kb and greater remain unresolved under these conditions. Excess ligated and unligated vector fragments can be seen as the faster migrating species. (C) A human YAC and the natural chromosomes of yeast; arrow indicates the position of a 555 kb human YAC in lane 2. Untransformed strains are in lanes 1 and 3.

5. Immerse solidified samples in CDP solution (about 1 mL/block). Incubate overnight at 50°C.
6. Decant CDP and dialyze against THE + 0.1 mM phenylmethylsulfonyl fluoride (PMSF) (1 mL/sample) twice for 1 hour at 50°C, followed by dialysis against THE three times for 30 minutes.
7. Store samples in THE at 4°C.

Media and Reagents

Agarose (for 5 mL):
 0.075 g low melting point agarose
 3.75 mL H_2O
 Tare, melt in microwave oven, bring to original volume with sterile H_2O, then add:
 1.25 mL 0.5 M EDTA/NaOH, pH 8

CDP: 0.5 M EDTA/NaOH, pH 8
 1% N-lauroylsarcosine
 2 mg/mL proteinase K

THE: 10 mM Tris/HCl, pH 8
 50 mM EDTA/NaOH, pH 8

Comments

1. Molds are made of plastic or Plexiglas and can be of any dimension, as long as the volume is small enough to avoid aliquoting the sample after preparation. Molds are often custom-made; one commercial supplier is Bio-Rad (e.g., model 170-3713). Samples can also be made by simply pipetting aliquots of the cell/agarose mixture onto a clean, flat plastic surface.
2. Wear gloves during all manipulations. Samples that have been prepared carefully are stable indefinitely in storage.
3. The final concentration of DNA in the blocks is about 0.3 μg/μL.

2.2.2 Digestion of Genomic DNA in Agarose Blocks; Size Fractionation of Products by CHEF Gel Electrophoresis; Preparation of Vector Fragments

When genomic DNA is cut to completion by restriction enzymes with rare recognition sequences, a good representation of the genome can be obtained with a relatively small number of clones.

In addition, much less DNA is required for preparation of the insert DNA. However, assembly of a contiguous physical map will require an independent source of overlapping clones, such as YACs, made with partially digested DNA.

Protocol for Partial Digestion of Insert DNA

1. Dialyze agarose blocks in 8 × volume of the appropriate restriction enzyme buffer containing at least 50 mM NaCl and without $MgCl_2$, twice for 15 minutes at 37°C. Remove buffer; add 2 × volume of the same. Cool on ice.
2. Add restriction enzyme varying from 0.025 to 50 units per microgram of DNA. Incubate on ice for 4 hours to diffuse enzyme into the agarose.
3. Add $MgCl_2$ to 10 mM final concentration. Digest for 10 minutes at 37°C.
4. Stop reaction by removing all the buffer, adding 0.1 × volume of 0.5 M EDTA/NaOH, pH 8, then incubating for 15 minutes at 70°C.
5. To characterize the extent of digestion, load an aliquot of the liquid sample directly into a dry well of a 1% agarose gel, using a pipette tip with the end cut off. Electrophorese on a CHEF gel apparatus for 20 hours at 6 V/cm, 14°C, with switch times ramping linearly from 0.5 to 94 seconds. Such conditions will resolve fragments up to 2 megabases in length.
6. To remove small molecular weight products, load as just described in step 5 the remainder of the sample into a dry well of a 1% low melting point agarose gel. Electrophorese on a CHEF gel apparatus for 5–10 hours at 6 V/cm, 14°C, with a switch time of 15 seconds. Such conditions will focus DNA fragments over 300 kb into a single band.
7. Cut marker lanes into the gel and remove a very thin edge of the sample lane. Stain with ethidium bromide and visualize by UV. Cut notches in the gel flanking the focused region of YAC DNA. Reconstruct the gel (without UV light), and cut the area containing the unresolved YAC DNA from the gel in the smallest volume possible. Stain the remaining parts of the gel and visualize with UV to ensure that the desired ligation products have been removed.
8. Store sample at 4°C in 10 × volume of THE.

Protocol for Complete Digestion of Insert DNA

1. Dialyze agarose blocks in 8 × volume of the appropriate restriction enzyme buffer with $MgCl_2$, twice for 15 minutes at 37°C. Remove buffer; add 2 × volume of the same.
2. Add 10 units of restriction enzyme per microgram of DNA. Digest overnight at 37°C.
3. Add NaCl to 50 mM if not included in step 1. Allow to diffuse into agarose for 15 minutes at 37°C, remove all the buffer, then incubate for 15 minutes at 70°C to melt the sample.
4. Proceed with steps 6–8 as described.

Protocol for Preparing Vector Fragments A large quantity of each vector arm is needed as the fragments are added in equal weight to insert DNA for the ligation. When preparing vector fragments, digest with the restriction enzyme that cuts adjacent to the telomere, treat with phosphatase, then digest with the enzyme that cuts at the site to be ligated to the insert. The vector fragment is purified from contaminating fragments and undigested plasmid by standard electrophoresis of the digestion products on a low melting point agarose gel, followed by gel purification.

Media and Reagents Use THE, as described in Section 2.2.1.

Comments

1. To prevent denaturation of the DNA when heated to 70°C, the digest buffer must contain at least 50 mM NaCl.
2. In the partial digestion protocol, no digestion of DNA occurs until step 3. This allows restriction enzyme to be evenly mixed with DNA without stirring, minimizing shearing of the DNA. Mg^{2+} dialyzes into solid agarose very rapidly and does not limit the reaction.
3. CHEF gels used for analysis are made with 1% agarose containing 0.5× TBE and no ethidium bromide. They are typically run at 6 V/cm, at 14°C, and stained with ethidium bromide after electrophoresis. Preparatory CHEF gels are made with 1% low melting point agarose.
4. An optimal partial digest will show digestion products ranging from 100 to 2000 kb, with the amount of DNA seen by ethidium bromide staining increasing with length, and 0.3–0.5 of the DNA appearing undigested. The results of a typical partial digestion of genomic DNA are shown in Figure 3A. An example of size selection of ligation products by CHEF gel electrophoresis is shown in Figure 3B.
5. Vector fragments are often treated at both ends with phosphatase to reduce vector-to-vector ligation. We have found, however, that treatment with phosphatase of the end of the vector to be ligated to insert yields DNA that does not ligate well.

2.2.3 Ligation of Insert to YAC Vectors; Size Selection of Ligation Products; Agarase Treatment to Liquefy Sample

Protocol

1. Dialyze the size-selected insert DNA against 10 volumes of TS for 15 minutes at 37°C; remove buffer and melt at 65°C.
2. Add each vector arm in a weight equal to insert DNA. Cool to 50°C, then add 0.1× sample volume of 10× ligation buffer and 1.5 Weiss units T4 DNA ligase per microgram of insert DNA. Stir together carefully but as thoroughly as possible, drawing the pipette tip through the mixture in parallel lines. Ligate overnight at 16°C.
3. Melt sample at 65°C and load directly into a dry well of a 1% low melting point agarose gel, using a pipette tip with the end cut off. Load appropriate markers in adjacent wells.
4. Run the gel first by conventional electrophoresis at 25 V for 8–15 hours. Transfer to a CHEF gel apparatus and electrophorese as described in Section 2.2.2, step 6.
5. Remove the region of the gel containing the unresolved region of large molecular weight YAC DNA, as described in Section 2.2.2, step 7.
6. Dialyze against 2 volumes TS for 15 minutes at 37°C, followed by 2 volumes TSP, twice for 15 minutes, at 37°C.
7. To liquefy the sample, remove buffer, melt at 65°C, cool to 40°C, then add 1 unit agarase/100 μL sample, stirring gently in parallel lines. Incubate for 1 hour at 40°C. Store at 4°C.

Media and Reagents

TS:	10 mM Tris/HCl, pH8
	50 mM NaCl
TSP:	10 mM Tris/HCl, pH 8
	50 mM NaCl
	1 mM EDTA/NaOH, pH 8
	0.75 mM spermidine
	0.3 mM spermine
Agarase:	β-Agarase I from New England BioLab, no. 392

Comments

1. A typical ligation reaction contains 7–12 μg of insert DNA and an equivalent amount of each vector in a total volume of 100 μL.
2. In the size selection step, the sample is run first by conventional electrophoresis to allow circular ligation products and some of the vast molar excess of vector fragments to migrate beyond the area of the gel in which the desired YAC ligation products will be after CHEF gel electrophoresis.

2.3 Transforming YACs into Yeast

2.3.1 Considerations in Choosing a Yeast Strain as Host

An enormous amount of work goes into making and characterizing a YAC library. The effort justifies a careful consideration of the yeast strain to be used, since it cannot easily be changed after the library has been established. Some important general features of a strain are as follows: it does not tend to clump together; it grows well on media of all types, sporulates well, and transforms well; it has several useful genetic metabolic markers (mutations in genes that allow selection for addition of the wild-type gene); and it is from a well-characterized genetic background. Some of the most useful strains include YPH274 (*MAT***a**/α *trp1-Δ1 ura3-52 ade2-101 lys2-801 his3-Δ200 leu2-Δ1* ψ^-) and its haploid derivatives, and CGY2516 (*MAT***a** *ura3-52 trp1-Δ63 lys2-Δ202 his3-Δ200 leu2-Δ1* ψ^-). A more detailed discussion of yeast strains and their characteristics can be found in the volume edited by Nelson and Brownstein, cited in the bibliography.

Establishment of a YAC library requires a number of genetic markers. YACs made from the vectors illustrated in Figure 1 all require their host to have mutations in the *TRP1* and *URA3* genes. The pCL vectors also require a mutation in the *ADE2* gene to select for and maintain the correct YAC structure and to allow visual screening of YAC stability. It is useful to have additional genetic markers to facilitate subsequent manipulation of the YAC or the strain containing it (see, e.g., Section 3.3). Other commonly used metabolic marker genes include *LEU2, HIS3, LYS2,* and *CAN1.*

Radiation-sensitive mutants have been used as hosts for YAC libraries. Such strains have been shown to reduce the number of chimeric clones in a library, which is thought to be due to their defects in homologous recombination and repair of damaged DNA. However, virtually all manipulations of a YAC that can be done after it has been established in yeast are not possible in these mutant strains.

A final consideration is the mating type of the host. Yeast grows stably in both a haploid and a diploid state. Haploid yeast exists as one of two mating types, called **a** and α, which are capable of mating to form an **a**/α diploid cell. Many manipulations of a YAC can be carried out by introduction of a new mutation to the host or introduction of a second YAC. Such experiments are easily carried out by mating, but this calls for the establishment of YAC libraries in a haploid strain. Finally, when making the initial library, YACs should be transformed into isogenic haploid strains of both mating types. As described in Section 3.2.2, extremely long eukaryotic genes can be completely reconstructed by patching together parts of the gene by homologous recombination between two individual YAC clones having overlapping sequence. If a YAC library is established in both mating types, YAC substrates for these recombinational steps will already be in hand.

2.3.2 Yeast Spheroplast Transformation; Plating Cells with Low Melting Point Agarose

Although there are several methods for transforming DNA into yeast cells, using yeast spheroplasts is the only method proven reproducibly to provide for transformation by high molecular weight DNA. Because the protocol is a standard methodology of yeast molecular biology, it is not presented in detail here. Rather, the modifications that are particular to transformation with large DNA molecules are described, as well as a method for plating the transformed cells that allows the initial transformant colony to grow on the surface of the solid media. This in turn makes possible the transfer of colonies to hybridization membranes directly from the transformation plate. Details of the standard spheroplast protocol are given by Nelson and Brownstein.

Transforming with Large Molecular Weight DNA To introduce the YAC DNA to spheroplasted cells, 5 μg of carrier DNA is mixed with 100 μL of cells, followed by addition of 10 μL of agarose-digested, gel-purified YAC DNA in liquid phase. The DNA must be added very carefully by means of a pipette tip with the end cut off, and stirred with the cells as carefully yet thoroughly as possible by drawing the pipette tip through the mixture in parallel lines.

Plating Transformants Pipette cells and 700 μL of plating agarose onto selective SORB plates prewarmed to 37°C. Spread as a thin film lightly over the surface of the plate. Transformants start to appear after 2–3 days but continue to arise for 5–7 days. Because the overlay of low melting point agarose is thin, colonies will break through to the surface with very little growth.

Reagents

Plating agarose:	2% low melting point agarose
	1 M sorbitol

Tare solution; melt in a microwave oven; bring back to volume with sterile water. Cool to 48°C.

Comments

1. We obtain zymolyase-20T from ICN ImmunoBiologicals (32-092-1). A freshly made stock should always be retitrated for digestion of yeast. Aliquots frozen at −20°C generally will give reproducible kinetics when used with the same yeast strain. Use a fresh aliquot for each transformation.

2. The rate of spheroplasting may vary among different strains.
3. The manufacturers, lots, and molecular weights of polyethylene glycol (PEG) can have a strong impact on transformation efficiency. We have obtained best results with PEG 10,000 from Sigma (P-6667).
4. When optimized, spheroplast transformation will typically yield 10^5–10^6 transformants per microgram of plasmid DNA, and 10^2–10^3 transformants per microgram of YAC DNA.

2.4 Working with YACs in Yeast

2.4.1 Screening for Unrearranged Clones

As described in Section 1.1 and illustrated in Figure 2, the presence of the *ADE2* gene on the acentric arm of the YAC provides a visual screen to assess instability and rearrangement of the YAC based on red sectoring within a white colony. Medium used to visualize this sectoring phenotype must have adenine present, but in limiting quantities, to allow red sectors to grow yet still develop the red pigment. The untransformed strain should be grown with a nonlimiting concentration of adenine to prevent selection for revertants. Generally, adenine in limiting concentration is 5 μg/mL, while complete medium contains 20 μg/mL. Some variability in optimal adenine concentrations may be observed between different strains. When YACs are transformed into yeast, they can be grown on limiting medium from the very start, allowing identification of unstable clones before any rearrangement occurs.

2.4.2 Preparation in Agarose of Yeast Genomic DNA and YACs

Protocol

1. Inoculate 1.5 mL of YPAD or 6 mL of YMglu (selective medium) with a yeast colony freshly grown on solid selective medium.
2. Grow with aeration at 30°C until saturation.
3. Pellet cells by centrifugation.
4. Wash with 1 mL of TE9.5. Repellet and remove supernatant completely. Pellet will have a volume of approximately 25–35 μL.
5. Add 1 μL of zymolyase to each sample.
6. Add 45 μL of 1% low melting point agarose containing 125 mM EDTA, pH 8, which has been melted and cooled to 50°C. Mix quickly by pipetting, being careful to avoid making bubbles.
7. Aliquot mixture into small-volume (50–100 μL) Plexiglas or plastic molds (see Comments). Solidify for 5 minutes at 4°C.
8. Immerse samples in TEβz (1 mL/block). Incubate for 5 hours at 37°C.
9. Decant buffer; incubate in BCDP (1 mL/block) for 12 hours at 50°C.
10. Dialyze against THE (1 mL/block), 8 times, for 15 minutes at 50°C.
11. Store samples in THE at 4°C.
12. Electrophorese samples on a CHEF gel to visualize chromosomes. (The gel in Figure 3C was run for 24 hours at 6 V/cm, 14°C, with switch times ramping linearly from 7 seconds to 84 seconds).

Media and Reagents

YPAD:
1% Bacto-yeast extract
2% Bacto-peptone
2% dextrose
20 μg/mL adenine

YMglu:
0.67% yeast nitrogen base without amino acids
2% glucose
metabolic supplements as appropriate for YAC selection

TE9.5:
10 mM Tris/HCl, pH 9.5
0.5 M EDTA/NaOH, pH 8

Zymolyase-20T (ICN ImmunoBiologicals, 32-092-1):
10 mg/mL stock in 10 mM $NaPO_4$ buffer, pH 7.5.

Store aliquots at −20°C. Use a fresh aliquot for each preparation.

TEβz:
Add to TE9.5:
5 μL β-mercaptoethanol/mL
150 μg zymolyase 20T/mL

Make fresh before use.

BCDP:
10 mM Tris/HCl, pH 9.5
0.5 M EDTA/NaOH, pH 8
1% N-lauroylsarcosine
0.5 mg/mL proteinase K

THE: see Section 2.2.1

Comments

1. Figure 3C shows the profile of the 16 natural yeast chromosomes and a transformant containing a human YAC.
2. To maximize the amount of YAC DNA in a culture, transformants should be grown in selective medium.
3. Yeast cells become more difficult to digest with zymolyase as the time the cultures have been in stationary phase increases. Cultures should not remain stationary for more than about 6 hours before proceeding with the preparation.
4. See Comments 1 and 2 in Section 2.2.1.

2.4.3 Double Transformants and Chimeras

Transformation using spheroplasts tends to generate cells containing more than one molecule of transforming DNA. As much as 10–20% of YAC transformants may have obtained more than one artificial chromosome. But because they have the same selectable markers, mitotic missegregation will eventually cause them to arise individually in subclones of the original transformant. To segregate the YACs, streak the strain out to single cells on appropriate solid selective medium. Allow colonies to form, and examine the chromosomal profile of a few subclones by CHEF gel to identify those containing a single YAC.

Chimeric YACs, clones containing more than one insert fragment of DNA, are much more problematic because they can cause time-consuming errors when used for mapping or cloning genes. One of the ways in which chimeras are thought to arise is by recombination between insert DNAs once transformed into yeast. The tendency of spheroplasted cells to take up multiple fragments of DNA would contribute to this problem. One of the more common ways of determining whether a YAC is chimeric is to use it as a probe in an in situ hybridization experiment. Most chimeras will give more than one hybridization signal, each arising from sequences from different parts of the genome contained within the single clone.

Another method makes use of human–rodent somatic cell hybrids that contain individual human chromosomes or fragments of chromosomes. The insert DNA in a YAC is amplified with the PCR, using *Alu* sequences as primers. The resulting DNA is then hybridized to a panel of "dot-blot" DNAs representing a collection of the somatic cell hybrids. A nonchimeric clone would be expected to hybridize to a single spot. Although this method has lower resolution, it is a more accessible technology, and enables testing a much greater number of clones.

2.4.4 Amplification of YAC Copy Number

Protocol

1. Inoculate 5 mL of YMglu medium with a YAC transformant colony freshly grown on solid selective medium.
2. Grow approximately 8 hours with aeration at 30°C.
3. Dilute cells into 6 mL of MSTgal medium to a density of 10^5 cells/mL. Grow to saturation with aeration at 30°C.
4. Recover cells and process to make purified DNA, or inoculate about 200 μL of culture into YMglu medium and grow with aeration to saturation overnight at 30°C (about five cell doublings).
5. Recover cells and process to make purified DNA in an agarose block as described in Section 2.4.2.

Media and Reagents

YMglu:	see Section 2.4.2
MSTgal:	0.67% yeast nitrogen base without amino acids
	2% galactose
	metabolic supplements as appropriate for YAC selection
	50 μg/mL methotrexate
	2 mg/mL sulfanilamide
	0.8 mg/mL thymidine

Comments

1. Cells grow very slowly in MSTgal (≈ 6–7 h/doubling).
2. The amplified copy number of a YAC relative to natural yeast chromosomes can be as high as 20. However, cells in the amplification medium can be fairly resistant to digestion, resulting in reduced yields of DNA released from agarose blocks. When cells are transferred to YMglu medium after amplification, the copy number of the YAC drops only about 50% after 10 cell doublings, and cells are easily digested. Consequently, the yield of YAC DNA made from samples of cells regrown in YMglu medium may be improved even though the copy number of the YAC relative to natural chromosomes is reduced.
3. If a YAC comigrates with a natural chromosome on a CHEF gel, it should be isolated from samples taken directly from amplification medium. This is because the copy number of the YAC relative to the contaminating chromosome will be higher and therefore will yield a more pure preparation.

2.4.5 Isolation of Insert DNA for Subcloning and Sequencing

Use of the following method requires that a recognition sequence for a rare-cutting restriction enzyme be in the vector adjacent to or at the site of insertion. The pRML vectors shown in Figure 1 contain *Not* I in the polylinkers, and the pCL vectors have both *Pme* I and *Not* I in their polylinkers.

Protocol

1. Load agarose blocks containing yeast chromosomes and YAC DNA on a 1% regular agarose CHEF gel. Electrophorese as described in Section 2.4.2, step 12.
2. Cut a very thin edge of the sample lane off the gel and process as in the first protocol in Section 2.2.2, step 7.
3. Dialyze twice against 2 volumes of appropriate restriction enzyme buffer for 15 minutes at 37°C.
4. Remove buffer; add 2 volumes fresh buffer plus restriction enzyme (see Comments) at 10 units per microgram of DNA. Incubate for 2 hours at 37°C.
5. Load sample on a second CHEF gel cast with 1% low melting point agarose. Electrophorese with conditions identical to those used in step 1.
6. Recover sample as in step 2.
7. Sample may be liquefied as described in Section 2.2.3, steps 6 and 7, dialyzing only against TS.

Comments

1. If the YAC is from a library in which the insert DNA was digested to completion, use the same restriction enzyme used to generate the insert fragments. If the YAC is from a library in which the insert was generated by partial digestion, cut with either *Not* I or *Pme* I, as appropriate for the vectors used. These rare-cutting enzymes are unlikely to have a recognition sequence in the insert of most clones, and extremely unlikely to have two. Thus, even if the insert is cut internally, it usually will generate products much longer than the vector fragments.
2. This protocol provides for purification of insert DNA in three ways. First, DNA at a given position is contaminated by species with faster mobilities (since individual molecules often are trapped in the gel); running a second CHEF gel eliminates most contaminants of this type. Second, any chromosome with the same mobility as the YAC will be separated in the second gel because the mobility of the YAC species will be increased as a result of the removal of 20 kb of vector sequence. Finally, insert DNA is purified away from vector.

3 USING YACs AS TOOLS AND DEVICES

YAC libraries have been a valuable resource in the establishment of integrated, detailed maps of the human genome. Because of their extremely large clone sizes (as large as 1 Mb of insert DNA), they provide a critical bridge between the fine-structure maps supplied by restriction digests of smaller fragments of cloned DNA such as that made in cosmids and the large structure whole-chromosome maps acquired from genetic mapping and cytology. Once a physical map that contains an ordered set of overlapping clones has been established, YACs offer the best tools for use of this information.

Because their insert sizes can be so large, they provide the best means for obtaining large genes within a single clone, and subsequently generating transgenic cell lines by transformation with the clone. Although only a brief overview of these applications is possible here, more detailed descriptions of the many uses of YAC libraries can be found in the literature, cited at the end of this article.

3.1 Genomic Mapping

3.1.1 Mapping by In Situ Hybridization

In situ chromosomal hybridization has been an indispensable tool for construction of physical maps of the human genome by cytological observation. The current limit of resolution is a few megabases for metaphase chromosomes. This high degree of resolution is achieved in part by development of fluorescent detection systems that have replaced radioactively labeled probes, and in part from the strength of the fluorescent signal observed with a very long probe.

Because of their length, YACs provide ideal probes for *f*luorescent *i*n *s*itu *h*ybridization (FISH) studies. In addition, the same piece of DNA used as the probe for hybridization mapping also provides a subclone of the region. A valuable technical application that is made possible by the use of very large clones is the study of chromosomal aberrations. Because the inserts in YACs are so large, they provide ideal probes for identification of structural changes such as insertions and deletions, inversions, translocations, and amplifications, as well as a means for obtaining the DNA spanning the regions of these rearrangements.

As described in Section 3.2.1, YACs can be used for building sets of overlapping clones called contigs that provide contiguous coverage of a region of a genome. Mapping by FISH using the YACs from the ends of a contig as the probes provides a very facile method for linking the contigs together. To use FISH for routine linkage, however, the contigs must have an average separation no greater than the limit of resolution of FISH, currently about 5 Mb.

3.1.2 Generating End Probes and Using Them to Build YAC Contigs

A library can be used to generate a physical map by searching for overlapping sequence information between individual clones. In this way, a map for a specific region of a genome can be defined as the order of overlapping clones in a library. The most useful probes for searching for sequence overlap come from the ends of the insert in a clone because this will identify clones with an end point that is the greatest possible distance from the end of the original clone.

Figure 4 illustrates a method for determining the sequence at the ends of inserted DNA using YACs constructed with the pRML vectors shown in Figure 1A; the centric arm and its junction with insert [shown in brackets in Figure 1B] is at the top. The only *Bam*HI site in the vector lies between the telomere and a bacterial origin of replication. Because restriction sites for this enzyme are fairly frequent, it can be expected to cut within the insert fairly close to the vec-

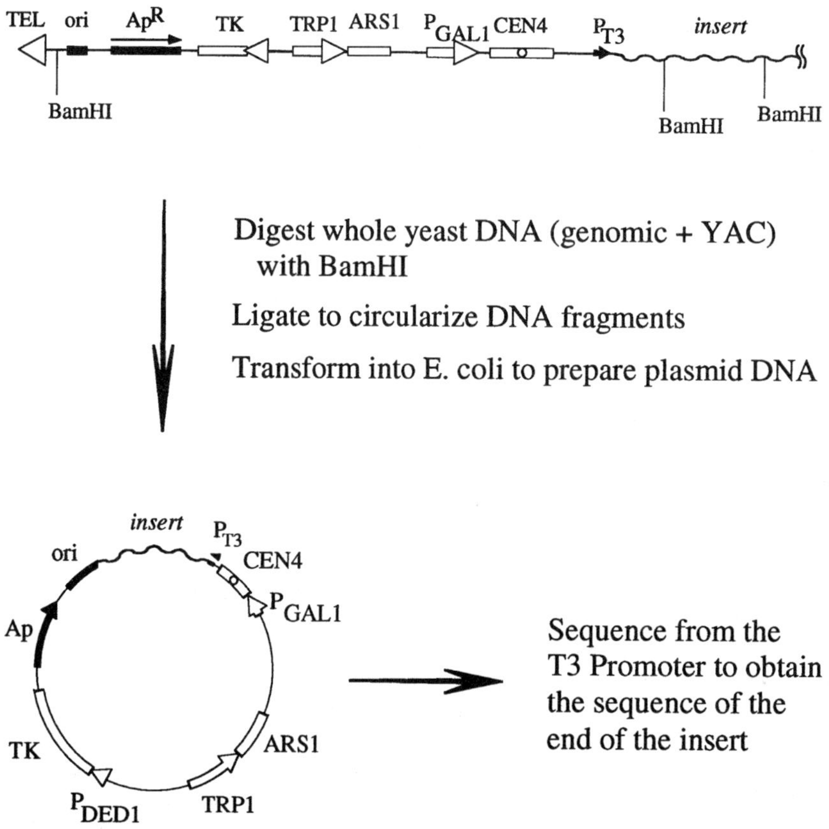

Figure 4. Recovering sequences of insert ends. The region of the YAC shown is from the bracketed region of Figure 1B.

tor–insert junction. Genomic DNA of a YAC transformant is prepared in agarose, then digested with *Bam*HI. The resulting fragments are isolated, ligated to form closed circles, and transformed into *E. coli* to obtain a circularized plasmid derivative of the YAC containing the vector–insert junction. If this plasmid is digested with *Bam*HI and the restriction enzyme used to make the library, the insert end fragment will be released with no attached vector sequence and can be used directly as a probe to search a library by Southern hybridization for other clones containing this sequence. Numerous rounds of library screening, end fragment isolation, and rescreening allow one to "walk" along a chromosome and collect clones, called contigs, that span a region without interruption.

The vector–insert junction can also be sequenced using the T3 promoter as a priming site on the plasmid template. When the sequence is known, it can function as a genetic marker (called a sequence-tagged site, or STS) for this region of the genome. The presence of an STS marker is scored by designing primers for the polymerase chain reaction (PCR) from the sequence and using them to screen for the presence of the site by successful PCR amplification. Because PCR is amenable to automation and requires very little template to screen for a sequence, STS markers have become central in the assembly of physical maps of complex genomes. (See Section 3.1.3.)

3.1.3 Integration of the Physical Map of a Complex Genome

To sequence an entire genome, a physical map is needed to provide ready access to specific chromosomal regions. Physical maps with a high density of markers have become available which are defined as collections of sequence-tagged sites. This type of physical map has been made possible by the development of the PCR because an STS is defined by PCR primers, and screened for by PCR amplification. STS-based maps have the added advantage of being made up of a collection of sequences rather than clones; therefore, unlike biological reagents, they are not subject to degradation and can be used to screen any library. The required density of markers on a physical map is determined by the average length of clones used to assemble a contig with those markers. For this reason, YAC libraries are extremely useful, for they have by far the largest average insert sizes of any cloning vehicle.

Methods that allow rapid screening to identify regions of overlapping sequence among thousands of clones are essential in efforts to obtain a clone collection that will provide complete coverage of a genome. Such a method is provided by STS screening, which defines clone overlap on the basis of the STS content of individual YACs. YAC-containing yeast strains are typically screened for STSs in pools, since the frequency of occurrence of any given STS in a YAC is very low. YACs sharing a given STS can be compared for their overall STS content. Definition of shared STSs, which define overlapping sequence, allows ordered assembly of YACs into a contig that faithfully represents genomic DNA. This approach has been successfully used to generate a map of the entire human genome.

3.2 Finding and Building Genes

3.2.1 Identifying Expressed Sequences Contained in YACs

The gene content of a YAC can be defined by directly screening a cDNA library with the entire YAC. To do so, the YAC is purified from yeast chromosomes on a CHEF gel, radiolabeled, and prehybridized with human DNA to block repetitive sequences. The probe is then used for hybridization with a cDNA library represented as plaques on filter disks.

Genes can also be identified in a YAC based on STS content. Databases containing sequences from the end of expressed genes (called *e*xpressed-*s*equence *t*ags, or ESTs) have been used to design PCR primer pairs that allow amplification of these regions as STSs. This subset of STSs serves the dual purposes of defining a genetic locus and serving as a marker for a coding sequence.

3.2.2 Modifying Cloned DNA by Homologous Recombination

Homologous recombination can be used to manipulate YAC DNA in vivo in a variety of ways. For example, if a gene is known to be contained within a clone, the gene can be located and its end points roughly defined by homologous recombination between repetitive *Alu* sequences within the clone and a linear recombination vector having an *Alu* sequence at one end and a telomere at the other. When the structure or sequence of a gene is known, defined mutations can be created, including point mutations and null mutations. The vector itself can be modified by introduction of new selectable markers or reporter genes to test the cloned DNA in other organisms.

Perhaps one of the most important applications of recombinational modification of YACs is in reconstruction of human genes for which YAC clones containing the entire gene have not been found, or genes that are too long to be isolated in a single YAC. Once two YACs containing overlapping sequence information have been identified by mapping, the two strains can be mated to form a diploid. A recombinational event within the region of homology will produce a hybrid clone containing all the sequences of each individual clone.

3.3 Constructing Transgenic Cell Lines and Organisms

Once established in yeast, YAC clones can be transferred to other organisms by a variety of methods, allowing functional tests of the cloned sequences. To date, experiments have been limited to transferring individual clones of well-characterized structure. The efficiency required for extending these studies to allow screening of a complex YAC library has yet to be developed.

Ideally, to permit the identification of transgenic cells that have taken up DNA, the YAC should contain a selectable marker for the cell line being transformed. Some of the most widely used mammalian selectable markers include neomycin phosphoribosyltransferase (*neo*), thymidine kinase, and hygromycin B phosphotransferase (*hyg*). If the YAC vectors used to construct a library do not contain one of these markers, a researcher may select from a number of plasmids that facilitate the addition of the markers to the YAC by homologous recombination in yeast.

Several different methods for introduction of YAC DNA into mammalian cells have been developed. Mammalian cells can be fused with yeast spheroplasts using polyethylene glycol. In this case the entire yeast genome is introduced, and a variable amount of natural yeast DNA can be found cointegrated with the YAC. Remarkably, the yeast DNA does not seem to have any observable effect on the transformed cell lines.

YACs have been introduced into cells by microinjection after they have been purified from contaminating natural yeast chromosomes by CHEF gel electrophoresis. This method has allowed for transformation of YACs greater than 0.5 Mb and bypasses any unknown complications that may arise from the presence of yeast DNA in the transgenic cell.

Purified total yeast DNA has also been transformed into cells by lipofection, calcium phosphate coprecipitation, or electroporation. The size of YACs successfully introduced in these ways appears to be rather small relative to other methods, perhaps because of the shearing of large molecular weight DNA during isolation from yeast and transformation.

4 CONCLUDING REMARKS

Technology development for the study of complex genomes is evolving rapidly. Since the development of YACs and of general methods for handling large pieces of DNA, we have seen the advent of artificial chromosome cloning systems in bacteria (BACs), assembly of a physical map of the human genome with a high density of markers (which will allow rapid mapping of YAC clones), and successful efforts in transforming YACs into mammalian genomes in which they are stably maintained. The potential applications of YAC clones in the establishment of transgenic organisms have only begun to be explored. Further dramatic improvements in artificial chromosome technology are to be expected. But as is generally true for complex problems, it is unlikely that any single approach will address every technical difficulty or be useful for every application. Yeast artificial chromosomes will always be an indispensable tool in the understanding and description of complex genomes.

See also EXPRESSION SYSTEMS FOR DNA PROCESSES; RECOMBINATION, MOLECULAR BIOLOGY OF; YEAST, DNA REPAIR IN; YEAST GENETICS.

Bibliography

Burke, D. T., and Olson, M. V. (1991) Preparation of clone libraries in yeast artificial chromosome vectors. *Methods Enzymol.* 194:251–270.

Guthrie, C., and Fink, G. R., Eds. (1991) *Guide to Yeast Genetics and Molecular Biology.* Academic Press, San Diego, CA.

Huxley, C., and Gnirke, A. (1991) Transfer of yeast artificial chromosomes from yeast to mammalian cells. *BioEssays,* 13:545–550.

Jakobobits, A. (1994) Humanizing the mouse genome. *Curr. Biol.* 4:761–763.

Nelson, D. L., and Brownstein, B. H., Eds. (1994) *YAC Libraries: A User's Guide.* Freeman, New York.

Schlessinger, D., et al. (1991) Yeast artificial chromosome-based genome mapping: Some lessons from Xq24-q28. *Genomics,* 11:783–793.

Spencer, F., Ketner, G., Connelly, C., and Hieter, P. (1993) Targeted recombination based cloning and manipulation of large DNA segments in yeast. *Methods* (companion to *Methods Enzymol.*) 5:161–175.

YEAST GENETICS

Fred Sherman

1 Yeast as a Model Eukaryote

2 Information on Yeast

3 Yeast Strains

4 Growth and Life Cycles

5 The Yeast Genome

6 Genetic Nomenclature
- 6.1 Chromosomal Genes
- 6.2 Mitochondrial Genes
- 6.3 Non-Mendelian Determinants

7 Genetic Analyses
- 7.1 Overviews with Examples
- 7.2 Tetrad Analysis
- 7.3 Non-Mendelian Inheritance

8 Transformation
- 8.1 Yeast Vector and DNA Fragments
- 8.2 Synthetic Oligonucleotides
- 8.3 Mitochondrial Transformation

9 Yeast Vectors
- 9.1 YIp Vectors
- 9.2 YEp Vectors
- 9.3 YCp Vectors

10 Genes Important for Genetic Studies
- 10.1 *URA3* and *LYS2*
- 10.2 *ADE1* and *ADE2*
- 10.3 *GAL1* Promoter
- 10.4 *lacZ* and Other Reporters

11 Manipulating the Genome In Vitro with Plasmids
- 11.1 Cloning by Complementation
- 11.2 Mutagenesis In Vitro
- 11.3 Two-Step Gene Replacement
- 11.4 Gene Disruption and One-Step Gene Replacement
- 11.5 Plasmid Shuffle
- 11.6 Recovering Mutant Alleles

12 Interactions of Genes
- 12.1 Heterozygosity and Dominant-Negative Mutations
- 12.2 Intragenic Complementation
- 12.3 Nonallelic Noncomplementation
- 12.4 Suppressors
- 12.5 Synthetic Enhancement and Epistatic Relationships

13 Analyses with Yeast Systems
- 13.1 Two-Hybrid Systems
- 13.2 Yeast Artificial Chromosomes (YACs)
- 13.3 Expression of Heterologous Proteins in Yeast

Key Words

Ascus (plural, asci) A saclike structure containing a tetrad of four spores (or ascospores).

Heterothallic Describing strains of yeast that have cross-compatible mating types and are stable both as haploids and diploids.

Homothallic Describing strains of yeast that give rise to tetrads containing four potentially self-fertile members, because the transient haploid cells switch their mating types and thus have only a stable diplophase.

Plasmid Shuffle A procedure for screening of mutations, derived from a mutagenized plasmid, requiring the loss of a second plasmid to assay for the recessive mutations.

Shuttle Vectors Vectors that can be propagated in both yeast and *E. coli.*

Tetrad The four products of meiosis.

Two-Hybrid System A genetic assay used in yeast for detection of protein–protein interactions.

The yeast *Saccharomyces cerevisiae* is clearly the most ideal eukaryotic microorganism for biological studies. The "awesome power of yeast genetics" has become legendary and is the envy of those who work with higher eukaryotes. The complete sequence of its genome, which recently became available, will be extremely useful as a reference toward the sequences of human and other higher eukaryotic genes. Furthermore, the ease of genetic manipulation of yeast allows its use for conveniently analyzing and functionally dissecting gene products from other eukaryotes.

1 YEAST AS A MODEL EUKARYOTE

This article deals only with the yeast *Saccharomyces cerevisiae,* and related interbreeding species. The fission yeast *Schizosaccharomyces pombe,* which is only distantly related to *S. cerevisiae,* has equally important features but is not as well characterized. The general principles of the numerous classical and modern approaches for investigating *S. cerevisiae* are described, with emphasis on the explanation of terms and nomenclature used in current yeast studies. The article should be particularly useful to those who are exposed for the first time to experimental studies of yeast. Detailed protocols are described in the primary literature and in a number of reviews in the books listed in the bibliography. The original citations for the material covered here also can be found in these comprehensive reviews.

Although yeasts have greater genetic complexity than bacteria, containing 3.5 times more DNA than *Escherichia coli* cells, they share many of the technical advantages that permitted rapid progress in the molecular genetics of prokaryotes and their viruses. The properties that make yeast particularly suitable for biological studies include rapid growth, dispersed cells, the ease of replica plating and mutant isolation, a well-defined genetic system, and most important, a highly versatile DNA transformation system. Unlike many other microorganisms, *S. cerevisiae* is viable with numerous markers. Being nonpathogenic, yeast can be handled with few precautions. Large quantities of normal baker's yeast are commercially available and can provide a cheap source for biochemical studies.

Unlike most other microorganisms, strains of *S. cerevisiae* have both stable haploid and diploid states. Thus, recessive mutations can be conveniently isolated and manifested in haploid strains, and complementation tests can be carried out in diploid strains. The development of DNA transformation has made yeast particularly accessible to gene cloning and genetic engineering techniques. Structural genes corresponding to virtually any genetic trait can be identified by complementation from plasmid libraries. Plasmids can be introduced into yeast cells either as replicating molecules or by integration into the genome. In contrast to most other organisms, integrative recombination of transforming DNA in yeast proceeds exclusively via homologous recombination. Exogenous DNA with at least partial homologous segments can therefore be directed at will to specific locations in the genome. Also, homologous recombination, coupled with yeast's high levels of gene conversion, has led to the development of techniques for the direct replacement of genetically engineered DNA sequences into their normal chromosome locations. Thus, normal wild-type genes, even those having no previously known mutations, can be conveniently replaced with altered and disrupted alleles. The phenotypes arising after disruption of yeast genes have contributed significantly toward understanding of the function of certain proteins in vivo. Many investigators who disrupted genes previously assumed to be essential were shocked to find viable mutants with few or no detrimental phenotypes. Also unique to yeast, transformation can be carried out directly with synthetic oligonucleotides, permitting the convenient production of numerous altered forms of proteins. These techniques have been extensively exploited in the analysis of gene regulation, structure–function relationships of proteins, chromosome structure, and other general questions in cell biology. The overriding virtues of yeast are illustrated by the introduction of mammalian genes into yeast for systematic analyses of the functions of the corresponding gene products.

In addition, yeast has proved to be valuable for studies of other organisms, including the use of the two-hybrid screening system for the general detection of protein–protein interactions, the use of YACs for cloning large fragments of DNA, and expression systems for the laboratory and commercial preparation of heterologous proteins. Many of these techniques are described herein.

During the last two decades, an ever-increasing number of molecular biologists have taken up yeast as their primary research system, resulting in a virtually autocatalytic stimulus for continuing investigations of all aspects of molecular and cell biology. Most significantly, members of an international consortium of laboratories have collaborated on the DNA sequencing of the entire yeast genome, which was completed in 1996. A knowledge of the DNA sequence of the genome has altered the way molecular and cell biologists approach and carry out their studies. In addition, plans are under way to systematically investigate the possible functions of all yeast genes by examining the phenotypes of strains having disrupted genes.

2 INFORMATION ON YEAST

Chapters in *Molecular Biology of the Gene,* by Watson et al. offer a general introduction to a few selected topics on yeast. Comprehensive and excellent reviews of the genetics and molecular biology of *S. cerevisiae* are contained in two volumes edited by Broach, Jones, and Pringle. An important source for methods used in genetics and molecular biology of yeast is contained in a book edited by Guthrie and Fink. Overviews of numerous subjects are also covered in other sources, and a more comprehensive listing of earlier reviews can be found in Sherman's work. Hall and Linder present interesting and amusing accounts of developments in the field in *The Early Days of Yeast Genetics.* The journal *Yeast* publishes original research articles, reviews, short communications, sequencing reports, and selective lists of current articles on all aspects of *Saccharomyces* and other yeast genera.

Current and frequently updated information and databases on yeast can be conveniently retrieved on the Internet through the World Wide Web, including the "Saccharomyces Genome Database" (http://genome-www.stanford.edu/saccharomyces) and linked files containing DNA sequences, lists of genes, news groups, bulletin boards, and other useful information concerning yeast.

3 YEAST STRAINS

Although genetic analyses and transformation can be performed with a number of taxonomically distinct varieties of yeast, extensive studies have been limited primarily to the many freely interbreeding species of the budding yeast *Saccharomyces* and to the fission yeast *Schizosaccharomyces pombe.* Although "*Saccharomyces cerevisiae*" is commonly used to designate many of the laboratory stocks of *Saccharomyces* used throughout the world, it should be pointed out that most of these strains originated from the interbred stocks of Winge, Lindegren, and others who employed fermentation markers not only from *S. cerevisiae* but also from *S. bayanus, S. carlsbergensis, S. chevalieri, S. chodati, S. diastaticus,* and so on. Nevertheless, it is still recommended that the interbreeding laboratory stocks of *Saccharomyces* be denoted as *S. cerevisiae,* to conveniently distinguish them from the more distantly related species of *Saccharomyces.*

Care should be taken in choosing strains for genetic and biochemical studies. Unfortunately there are no truly wild-type *Saccharomyces* strains that are commonly employed in genetic studies. Also, most domesticated strains of brewer's yeast and probably many strains of baker's yeast and true wild-type strains of *S. cerevisiae* are not genetically compatible with laboratory stocks. It is usually not appreciated that many "normal" laboratory strains, which were in fact derived from pedigrees involving mutagenized strains, contain mutant characters. The haploid strain S288C is often used as a normal standard because it gives rise to well-dispersed cells and is widely used, and because many isogenic mutant derivatives are available. However, S288C contains mutations making it undesirable to use in mitochondrial studies. An other strain, D273-10B, has been extensively used as a typical normal yeast, especially for mitochondrial studies. (These laboratory strains should be denoted as "normal" or "standard," but not "wild type.") One should examine the specific characters of interest before initiating a study with any strain.

Many strains containing characterized auxotrophic, temperature-sensitive, and other markers can be obtained from the Yeast Genetics Stock Culture Center [Department of Molecular and Cell Biology, 229 Stanley Hall, University of California, Berkeley, CA 94720-3206; (510) 642-0815; fax (510) 642-8589; e-mail ygsc@violet.berkeley.edu]. Other sources of yeast strains include American Type Culture Collection [12301 Parklawn Drive, Rockville, MD 20852; (301) 881-2600; (800) 638-6597; e-mail request@atcc.org; http://www.atcc.org]; National Collection of Yeast Cultures (Food Research Institute, Colney Lane, Norwich NR4 7UA, U.K.); Centraalbureau voor Schimmelcultures (Yeast Division, Julianalaan 67a, 2628 BC Delft, Netherlands); Slovak Collection of Yeasts (Institute of Chemistry, Slovak Academy of Sciences, Dubravaska cesta, 809 33 Bratislava, Slovak Republic).

4 GROWTH AND LIFE CYCLES

Vegetative cell division of yeast characteristically occurs by budding, in which a daughter is initiated as an outgrowth from the mother cell, followed by nuclear division, cell wall formation, and finally cell separation. The sizes of haploid and diploid cells vary with the phase of growth and from strain to strain. Typically, diploid cells are 5 μm × 6 μm ellipsoids and haploid cells are 4 μm diameter spheroids. The volumes and gross composition of yeast cells are listed in Table 1. During exponential growth, haploid cultures tend to have higher numbers of cells per cluster compared to diploid cultures. Also haploid cells have buds that appear adjacent to the previous one, whereas diploid cells have buds that appear at the opposite pole. Each mother cell usually forms no more than 20–30 buds, and its age can be determined by the number of bud scars left on the cell wall.

Table 1 Size and Composition of Yeast Cells

Characteristic	Haploid Cell	Diploid Cell
Volume (μm^3)	70	120
Composition (10^{-12} g)		
Wet weight	60	80
Dry weight	15	20
DNA	0.017	0.034
RNA	1.2	1.9
Protein	6	8

In addition, certain diploid strains of *S. cerevisiae* can assume a markedly different cell and colony morphology, denoted pseudohyphae, in agar medium that is limiting for nitrogen sources. These pseudohyphal cells are significantly elongated, and mother–daughter pairs remain attached to each other. This characteristic pseudohyphal growth causes extended growth of branched chains outward from the center of the colony, and invasive growth under the surface of agar medium.

"Normal" laboratory haploid strains have a doubling time of approximately 90 minutes in complete YPD (1% yeast extract, 2% peptone, 2% glucose) medium and approximately 140 minutes in synthetic media during the exponential phase of growth at the optimum temperature of 30°C. However, strains with greatly reduced growth rates in synthetic media are often encountered. Usually strains reach a maximum density of 2×10^8 cells/mL in YPD medium. Titers 10 times this value can be achieved with special conditions that can be delivered in fermentors, such as pH control, continuous additions of balanced nutrients, filtered–sterilized media, and extreme aeration.

S. cerevisiae can be stably maintained as either heterothallic or homothallic strains, as illustrated in Figure 1. Both heterothallic and homothallic diploid strains sporulate under conditions of nutrient deficiency, and especially in special media, such as potassium acetate medium. During sporulation, the diploid cell undergoes meiosis yielding four progeny haploid cells, which become encapsulated as spores (or ascospores) within a saclike structure called an ascus. The percentage of sporulation varies with the particular strain, ranging from no or little sporulation to nearly 100%. Many laboratory strains sporulate to more than 50%. The majority of asci contains four haploid ascospores, although varying proportions asci with three or fewer spores are also observed.

Because the **a** and α mating types are under control of a pair of *MAT***a**/*MAT*α heterozygous alleles, each ascus contains two *MAT***a** and two *MAT*α haploid cells. Upon exposure to nutrient condition, the spores germinate, vegetative growth commences, and mating of the *MAT***a** and *MAT*α can occur. However, if the haploid spores are mechanically separated by micromanipulation, the haplophase of heterothallic strains can be stably maintained, thus allowing the preparation of haploid strains. In contrast, the presence of the *HO* allele in homothallic strains causes switching of the mating type in growing haploid cells, such that *MAT***a** cells produce *MAT*α buds and *MAT*α cells produce *MAT***a** buds. As a consequence, mating oc-

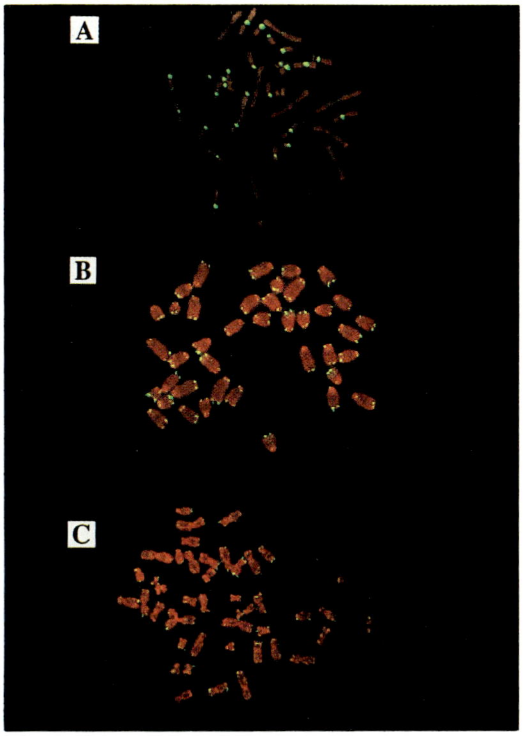

Plate 1. Identifying $(TTAGGG)_n$ DNA sequences on chromosomes by fluorescence in situ hybridization: (A) a Chinese hamster fibroblast cell line, (B) a mouse embryonic stem cell line, and (C) human peripheral blood lymphocytes. The yellow fluorescence identifies $(TTAGGG)_n$ arrays. The chromosomes are counterstained with propidium iodide and appear red. (Courtesy of R. Critcher, Department of Genetics, Cambridge University.) (See Telomeres, Mammalian, Figure 3.)

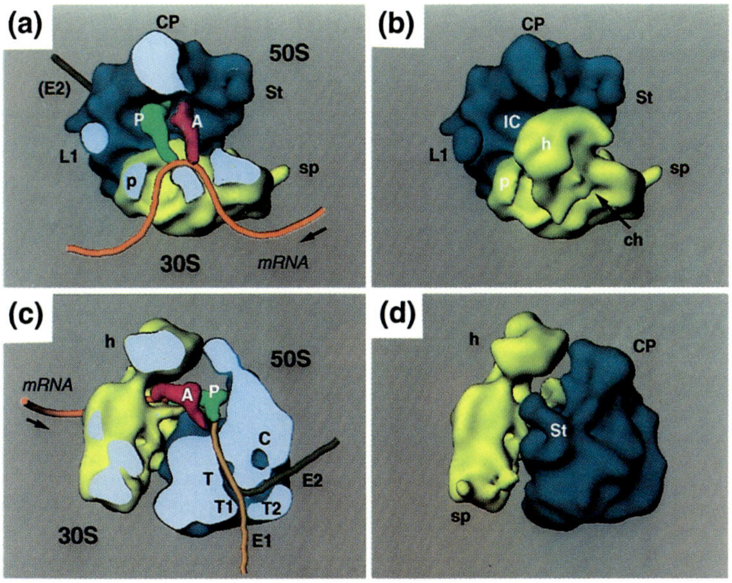

Plate 2. A model of elements involved in translating mRNA to protein based on 2.5 nm reconstructions of images of the *E. coli* ribosome obtained by cryoelectron microscopy, according to Frank et al. in Matheson et al. (1996) and in *Nature* [376, 441–444 (1995)]. The small subunit (yellow) has prominent features such as the head (h), platform (p), a "spur" (sp) at the base, and a channel (ch) that penetrates the "neck." The main features of the large subunit (blue) include the L1 arm (L1), the central protuberance (CP), the stalk (St), and the interface canyon (IC). The subunits are joined by five bridges across the subunit–subunit interface and enclose a space that accommodates mRNA, aminoacyl-tRNA (A) and peptidyl-tRNA (P). Sections through two views of the ribosome are shown in (a) and (c) to reveal the proposed locations of mRNA and tRNAs. The corresponding views of the whole ribosome are shown in (b) and (d). (a) A section through the top view of the ribosome (b) revealing the channel and the cleft into which it leads. The mRNA (orange) is shown after it has entered from the cytosol through the channel, encircling the back portion of the neck. The arrow points away from the 3' end of the mRNA and indicates the direction of movement relative to the ribosome. The complex of A (pink) and P (green) is placed with the anticodons facing the cleft region immediately next to the channel. (c) A section through a side view of the ribosome (d) revealing the bifurcating tunnel (T), its first branch (T1) and second branch (T2), and the compartment (C) contiguous with T2. E1 and E2 are alternative exit sides for the nascent polypeptide chain shown in yellow (E1) and yellow then olive green (E2). The tRNA complex is placed with the aminoacyl (or peptidyl) ends at the entry to the tunnel, indicating the location of the peptidyl transferase center. The location of this center is not known precisely but ranges from near the aminoacyl ends of the AP complex, as shown to the limit where the aminoacyl ends of the AP complex are placed close to the roof formed by the head of the small subunit. (Reproduced by permission from an original photograph by Dr. J. Frank, Wadsworth Center for Laboratories and Research, New York State Health Department, Albany, NY 12201, U.S.A.) (See Translation of RNA to Protein, Figure 8.)

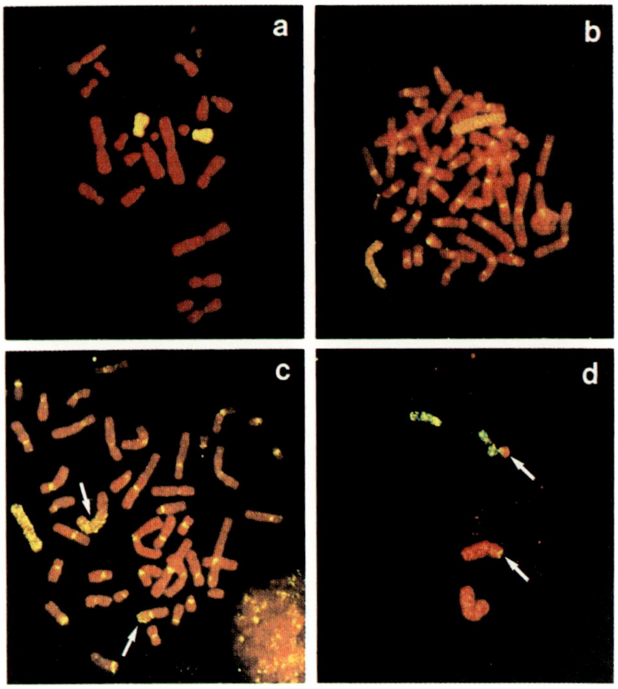

Plate 3. Chromosome-specific staining using FISH. (a) Human chromosomes specifically stained in metaphase spreads from human × hamster hybrid cells by hybridization with human genomic DNA. Hybridization signals appear yellow. Chromosomes to which the probe did not hybridize appear red. (b) Both copies of human chromosome 4 in a normal human metaphase spread stained by hybridizing a WCP for chromosome 4; staining the same as for (a). Centromeric regions were simultaneously stained with a probe that hybridizes to repeated DNA in the centromeric region of all chromosomes. (c) FISH to a metaphase spread from an irradiated cell population using a WCP to chromosome 4. A radiation-induced exchange between one copy of chromosome 4 and another chromosome is apparent. The derivative chromosomes appear red and yellow (arrows). The centromeres were stained as described in (b) to allow discrimination between translocations (one centromere) and dicentrics (two centromeres) (d) FISH with WCPs to chromosomes 7 (red) and 12 (green) to a metaphase spread prepared from human tumor. Derivative chromosomes resulting from a t(7;12) translocation are visible (arrows) as red and green chromosomes (Photomicrograph d courtesy of M. Vooijs.) (See Whole Chromosome Complementary Probe Fluorescence Staining, Figure 1.)

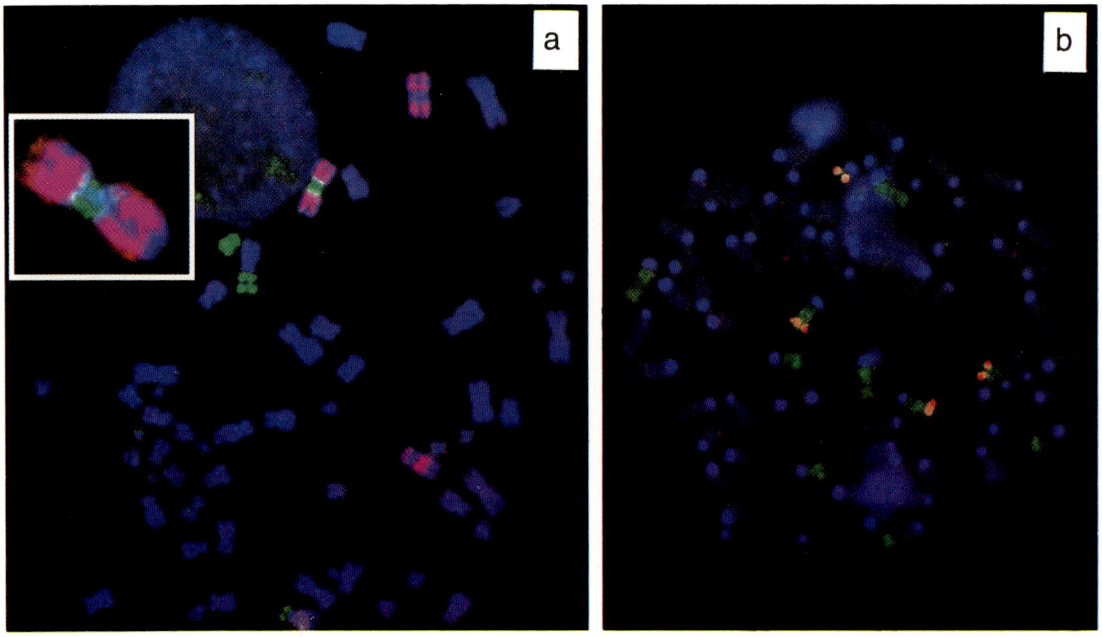

Plate 4. Application of chromosome painting for identification of marker chromosomes and probe mapping. (a) Combined hybridization of two or more chromosome painting probes allows definition of marker chromosome composition: a WCP for human chromosome 3 (Spectrum Orange, red) was combined with a WCP for chromosome 17 (FITC, green) and hybridized to metaphase spreads from the thyroid cancer cell line FTC236. The marker chromosome (insert) contains substantial parts derived from chromosome 3q and chromosome 17. Both painting probes were generated by DOP-PCR. Chromosomal DNA was counterstained with DAPI (blue). (Photomicrograph a courtesy of B. O'Brien.) (b) Assignment of a locus-specific probe (LSP) to mouse chromosome 11 by two-color FISH. Three chromosome-specific WCPs were hybridized simultaneously to expedite the mapping process. The WCPs specific for mouse chromosomes 9, 11, and 15, respectively, were generated by DOP-PCR from flow-sorted chromosomes and labeled by enzymatic incorporation of biotin during the last PCR cycles. Bound probes are visualized with avidin-FITC (green) following hybridization, and chromosomes are identified by their size as well as the pattern of chromosome painting. Metaphase spreads from a mouse 3T3 cell line showed three normal and one rearranged copy of chromosome 11. The LSP to be mapped was digoxigenin-labeled by random priming and detected with rhodamine-conjugated sheep-anti-digoxigenin antibodies (red). All four chromosomes containing the distal part of mouse chromosome 11 also showed the LSP hybridization signal. Chromosomal DNA was counterstained with DAPI (blue). (See Whole Chromosome Complementary Probe Fluorescence Staining, Figure 2.)

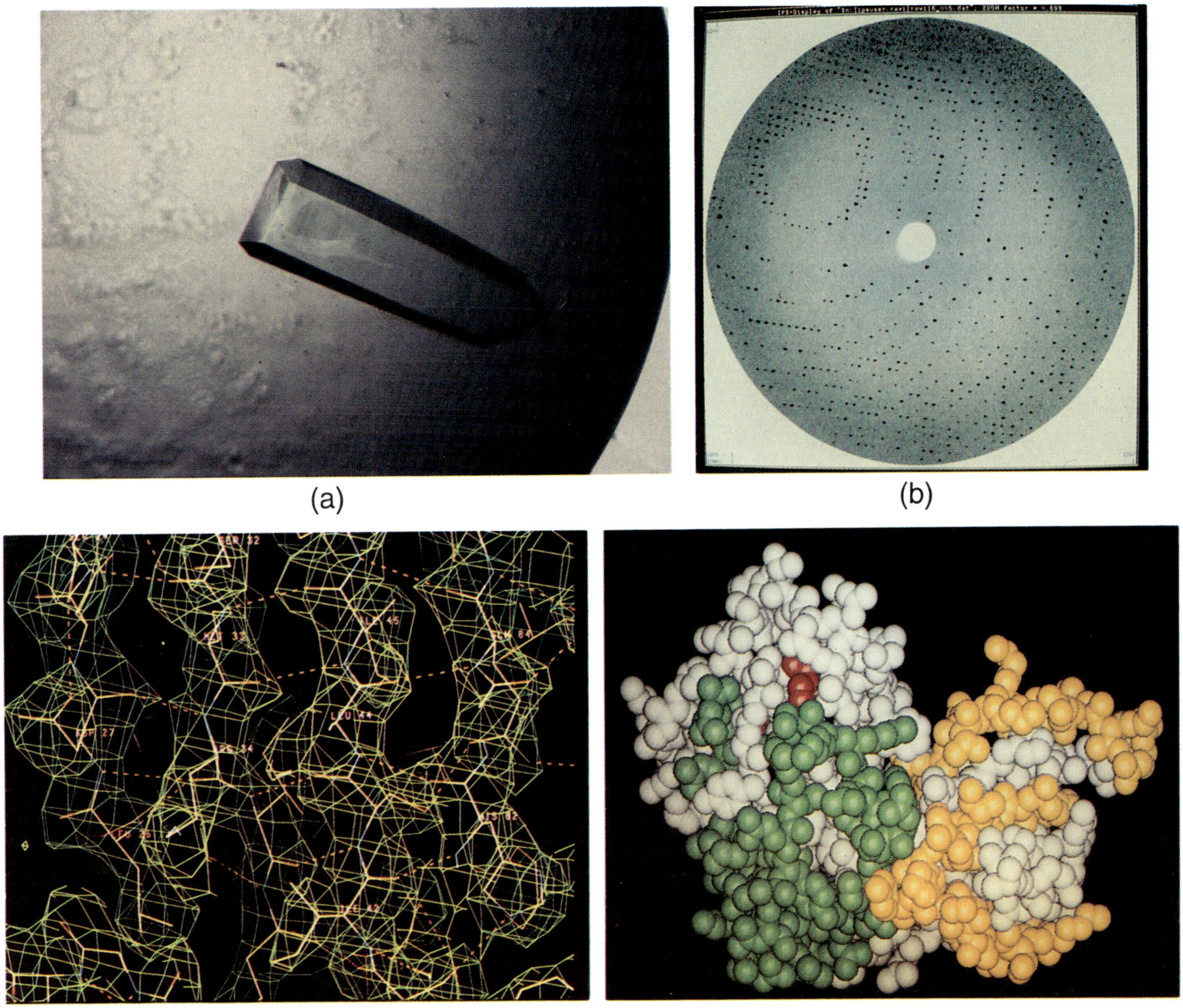

(a) (b) (c) (d)

Plate 5. Stages involved in the structure determination of a protein molecule, toxic shock syndrome toxin 1 (TSST-1), a microbial superantigen. (a) A single crystal of TSST-1. (b) Diffraction pattern of TSST-1 crystal recorded using MAR Research Image Plate system (wavelength, 1.5418 Å). The data extend to 2.5 Å resolution. (c) Portion of the 2.5 Å resolution electron density map of TSST-1 with atomic positions superimposed. The region displayed shows four strands of a β-sheet. (d) Space-filling diagram of TSST-1 (drawn using the 2.5 Å crystal structure). (See X-Ray Diffraction of Biomolecules, Figure 2.)

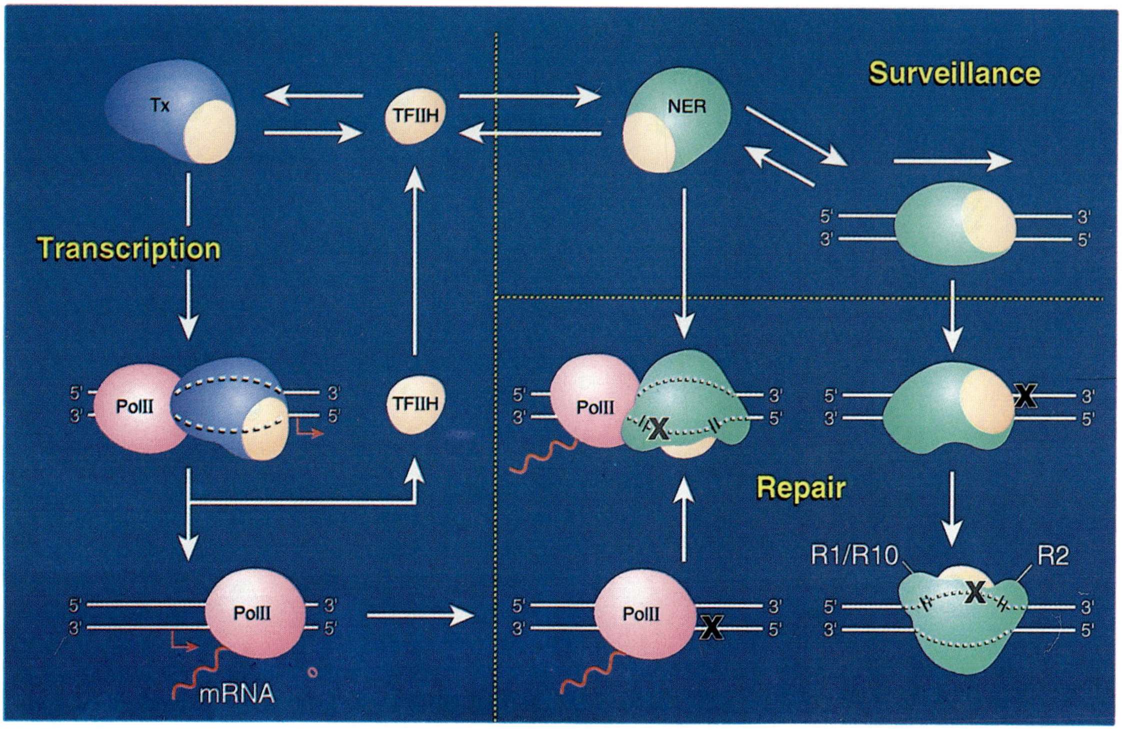

Plate 6. Cartoon summarizing the relationship between GNER and RNA polymerse II transcription in yeast. Transcription factor IIH (TFIIH) is common to both the holoTFIIH transcription (Tx) initiation complex and the repairosome for GNER (NER) and may shuffle between these two supercomplexes with varying equilibrium states. The preassembled repairosome may constantly survey the genome for base damage in transcriptionally silent DNA (surveillance), and when such damage (X) is encountered, repair it (bottom right). HoloTFIIH is believed to be loaded onto promoter sites during RNA polymerase II transcription initiation, where it associates with the polymerase and with other transcription factors. When transcription has been successfully initiated TFIIH is released. When the transcription elongation complex encounters sites of base damage (X) in the template strand, transcription stalls. The repairosome is recruited to sites of stalled transcription through interactions that specifically require the TFIIH subunits. The recruited repairosome effects repair of the damage, as in transcriptionally silent DNA. (See Yeast, DNA Repair in, Figure 3.)

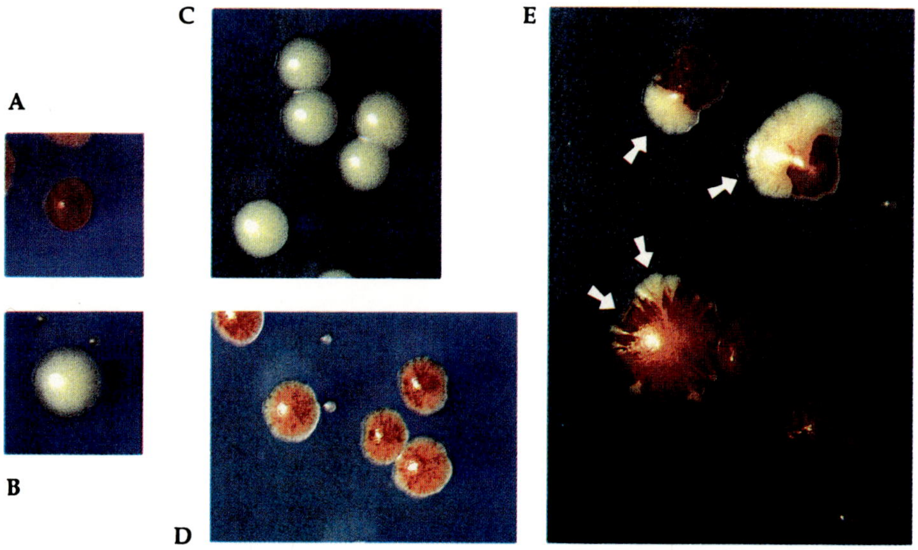

Plate 7. Use of the *ADE2* gene to assess YAC stability. (A) Yeast colony formed by an *ade2* mutant strain. (B) Colony formed by a wild-type stain. (C) A very stable YAC clone. (D) A YAC clone showing intermediate instability. (E) Very unstable clone, prone to rearrangement at a high frequency: closed arrows, phenotype of the unrearranged clone; open arrows, rearranged derivatives of varying lengths and stabilities. (See Yeast Artificial Chromosome (YAC) Construction and Applications, Figure 2.)

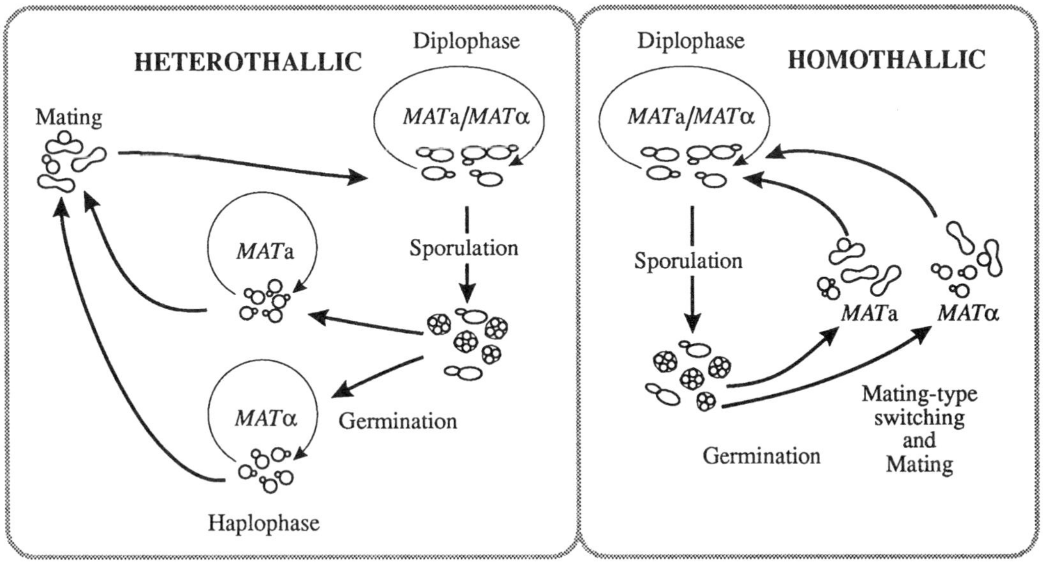

Figure 1. Life cycles of heterothallic and homothallic strains of *S. cerevisiae*. Heterothallic strains can be stably maintained as diploids and haploids, whereas homothallic strains are stable only as diploids, because the transient haploid cells switch their mating type, and mate.

curs and there is only a transient haplophase in homothallic strains (Figure 1).

Controlled crosses of *MAT***a** and *MAT*α haploid strains are simply carried out by mixing approximately equal amounts of each strain on a complete medium and incubating the mixture at 30°C for at least 6 hours. Prototrophic diploid colonies can then be selected on appropriate synthetic media if the haploid strains contain complementing auxotrophic markers. If the diploid strain cannot be selected, zygotes can be separated from the mating mixture with a micromanipulator. Zygotes are identified by a characteristic thick zygotic neck, and are best isolated 4–6 hours after incubating the mixture when the mating process has just been completed.

5 THE YEAST GENOME

S. cerevisiae contains a haploid set of 16 well-characterized chromosomes, ranging in size from 200 to 2,200 kb. The total sequence of chromosomal DNA, constituting 12,052 kb, was released in April, 1996. A total of 6,183 open-reading-frames (ORF) were reported, and approximately 5,800 of them were predicted to correspond to actual protein-coding genes. In contrast to the genomes of multicellular organisms, the yeast genome is highly compact, with genes representing 72% of the total sequence. The average size of yeast genes is 1.45 kb, or 483 codons, with a range from 40 to 4,910 codons. A total of 3.8% of the ORF contains introns. Approximately 30% of the genes already have been characterized experimentally. Of the remaining 70% with unknown function, approximately one half either contain a motif of a characterized class of proteins or correspond to genes encoding proteins that are structurally related to functionally characterized gene products from yeast or from other organisms.

Ribosomal RNA is coded by approximately 120 copies of a single tandem array on chromosome XII. The DNA sequence revealed that yeast contains 262 tRNA genes, of which 80 have introns. In addition, chromosomes contain movable DNA elements, called retrotransposons, which vary in number and position in different strains of *S. cerevisiae,* although most laboratory strains have approximately 30.

Other nucleic acid entities, presented in Figure 2, also can be considered to be part of the yeast genome. Mitochondrial DNA encodes components of the mitochondrial translational machinery and approximately 15% of the mitochondrial proteins. Completely lacking in mitochondrial DNA, ρ° mutants are deficient in the respiratory polypeptides synthesized on mitochondrial ribosomes (i.e., cytochrome *b* and subunits of cytochrome oxidase and ATPase complexes). Even though ρ° mutants are respiratory deficient, they are viable and still retain mitochondria, which, however, are morphologically abnormal.

The 2 μm circle plasmids, present in most strains of *S. cerevisiae,* apparently function solely for their own replication. Generally cir° strains, which lack 2 μm DNA, have no observable phenotype. However, a certain chromosomal mutation, *nib1,* causes a reduction in growth of cir^+ strains, as a result of the presence of 2 μm DNA that is abnormally high in copy number.

Similarly, almost all *S. cerevisiae* strains contain double-stranded (ds) RNA viruses, which constitute approximately 0.1% of total nucleic acid. RNA viruses include three families with dsRNA genomes, L-A, L-BC, and M. Two other families of dsRNA, T and W, replicate in yeast but so far have not been shown to be viral. M dsRNA encodes a toxin, and L-A encodes the major coat protein and components required for the viral replication and maintenance of M. The two dsRNAs, M and L-A, are packaged separately with the common capsid protein encoded by L-A, resulting in viruslike particles that are transmitted cytoplasmically during vegetative growth and conjugation. L-B and L-C (collectively denoted L-BC), similar to L-A, have an RNA-dependent RNA polymerase and are present in intracellular particles. *KIL*-**o** mutants, lacking M dsRNA and consequently the killer toxin, are readily induced by growth at elevated temperatures, and by chemical and physical agents.

Yeast also contains a 20S circular single-stranded RNA (not shown in Figure 2) that appears to encode an RNA-dependent RNA

Inheritance	Mendelian	Non-Mendelian		
Nucleic acid	Double-stranded DNA			Double stranded RNA
Location	Nucleus		Cytoplasm	
Genetic determinant	Chromosomes	2-μm plasmid	Mitochondrial DNA	RNA Viruses: L-A, M, L-BC, T, W
Relative amount	85%	5%	10%	80%, 10%, 9%, 0.5%, 0.5%
Number of copies	2 sets of 16	60-100	~50 (8-130)	103, 170, 150, 10, 10
Size (kb)	13,500 (200-2,200)	6.318	70-76	4.576, 1.8, 4.6, 2.7, 2.25
Deficiencies in mutants	All kinds	None	Cytochromes $a{\cdot}a_3$ & b	Killer toxin (L-A, M); None (L-BC, T, W)
Wild-type	*YFG1*$^+$	cir$^+$	ρ^+	*KIL*-k$_1$
Mutant or variant	*yfg1-1*	cir^o	ρ^-	*KIL*-o

Figure 2. The genome of a diploid cell of *S. cerevisiae* (see text). A wild-type chromosomal gene is depicted as *YFG1*$^+$ (your favorite gene) and the mutation as *yfg1-1*.

polymerase, which acts as an independent replicon and is inherited as a non-Mendelian genetic element.

Only mutations of chromosomal genes exhibit Mendelian 2:2 segregation in tetrads after sporulation of heterozygous diploids; this property is dependent on the disjunction of chromosomal centromeres. In contrast, non-Mendelian inheritance is observed for the phenotypes associated with the absence or alteration of other nucleic acids described in Figure 1.

6 GENETIC NOMENCLATURE

6.1 Chromosomal Genes

The genetic nomenclature for chromosomal genes of the yeast *S. cerevisiae* is now more or less universally accepted, as illustrated in Table 2, using *ARG2* as an example. Whenever possible, each gene, allele, or locus is designated by three italicized letters (e.g., *ARG*, which is usually a describer, followed by a number, e.g., *ARG2*). Unlike most other systems of genetic nomenclature, dominant alleles are denoted by using uppercase italics for all letters of the gene symbol (e.g., *ARG2*), whereas lowercase letters denote the recessive allele (e.g., the auxotrophic marker *arg2*). Wild-type genes are designated with a superscript "plus" (*sup6*$^+$ or *ARG2*$^+$). Alleles are designated by a number separated from the locus number by a hyphen (e.g., *arg2-9*). The symbol Δ can denote complete or partial deletions (e.g., *arg2-Δ1*). Insertion of genes follows the bacterial nomenclature by using the symbol ::. For example, *arg2::LEU2* denotes the insertion of the *LEU2* gene at the *ARG2* locus, in which *LEU2* is dominant (and functional) and *arg2* is recessive (and defective).

Phenotypes are sometimes denoted by cognate symbols in roman type and by the superscripts + and −. For example, the independence and requirement for arginine can be denoted by Arg$^+$ and Arg$^-$, respectively. Proteins encoded by *ARG2*, for example, can be denoted Arg2p, or simply Arg2 protein. However, gene symbols are generally used as adjectives for other nouns (*ARG2* mRNA, *ARG2* strains, etc.).

Although most alleles can be unambiguously assigned as dominant or recessive by examining the phenotype of the heterozygous diploid crosses, dominant and recessive traits are defined only with pairs, and a single allele can be both dominant and recessive. For example, because the alleles *CYC1*$^+$, *cyc1-717*, and *cyc1-Δ1* produce, respectively, 100, 5, and 0% of the gene product, the *cyc1-717* allele can be considered to be recessive in the *cyc1-717/CYC1*$^+$ cross and dominant in the *CYC1-717/cyc1-Δ1* cross. Thus, sometimes it is less confusing to denote all mutant alleles in lowercase letters, especially when considering a series of mutations having a range of activities.

Although superscript letters should be avoided, it is sometimes expedient to distinguish genes conferring resistance and sensitivity by superscript R and S, respectively. For example, the genes controlling resistance to canavanine sulfate (*can1*) and copper sulfate (*CUP1*) and their sensitive alleles could be denoted, respectively, as *can*^{R}*1*, *CUP*^{R}*1*, *CAN*^{S}*1*, and *cup*^{S}*1*.

Table 2 Genetic Nomenclature, Using *ARG2* as an Example

Symbol	Definition
ARG$^+$	All wild-type alleles controlling arginine requirement
ARG2	A locus or dominant allele
arg2	A locus or recessive allele confering an arginine requirement
arg2$^-$	Any *arg2* allele confering an arginine requirement
ARG2$^+$	The wild-type allele
arg2–9	A specific allele or mutation
Arg$^+$	A strain not requiring arginine
Arg$^-$	A strain requiring arginine
Arg2p	The protein encoded by *ARG2*
Arg2 protein	The protein encoded by *ARG2*
ARG2 mRNA	The mRNA transcribed from *ARG2*
arg2-Δ1	A specific complete or partial deletion of *ARG2*
ARG2::LEU2	Insertion of the function *LEU2* gene at the *ARG2* locus; *ARG2* remains functional and dominant
arg2::LEU2	Insertion of the functional *LEU2* gene at the *ARG2* locus; *arg2* is or became nonfunctional
art2–10::LEU2	Insertion of the functional *LEU2* gene at the *ARG2* locus; the specified *arg2–10* allele, is nonfunctional
cyc1-arg2	A fusion between the *CYC1* and *ARG2* genes; both are nonfunctional
P_{CYC1}-*ARG2*	A fusion between the *CYC1* promoter and *ARG2*; *ARG2* is functional

Wild-type and mutant alleles of the mating-type locus and related loci do not follow the standard rules. The two wild-type alleles of the mating-type locus are designated *MAT***a** and *MAT*α. The wild-type homothallic alleles at the *HMR* and *HML* loci are denoted *HMR***a**, *HMR*α, *HML***a**, and *HML*α. The mating phenotypes of *MAT***a** and *MAT*α cells are denoted simply **a** and α, respectively. The two-letter combination *HO* denotes the gene encoding the endonuclease required for homothallic switching.

Dominant and recessive suppressors should be denoted, respectively, by three uppercase or three lowercase letters, followed by a locus designation (e.g., *SUP4, SUF1, sup35, suf11,* etc.). In some instances UAA ochre suppressors and UAG amber suppressors are further designated, respectively, **o** and **a** following the locus. For example, *SUP4-***o** refers to suppressors of the *SUP4* locus that insert tyrosine residues at UAA sites; *SUP4-***a** refers to suppressors of the same *SUP4* locus that insert tyrosine residues at UAG sites. The corresponding wild-type locus that encodes the normal tyrosine tRNA and lacks suppressor activity can be referred to as $sup4^+$. Intragenic mutations that inactivate suppressors can be denoted, for example, $sup4^-$ or *sup4-***o***-1*. Frameshift suppressors are denoted as *suf* (or *SUF*), whereas metabolic suppressors are denoted with a variety of specialized symbols, such as *ssn* (suppressor of *snf1*), *srn* (suppressor of *rna1-1*), and *suh* (suppressor of *his2-1*).

Capital letters are also used to designate certain DNA segments whose locations have been determined by a combination of recombinant DNA techniques and classical mapping procedures (e.g., *RDN1,* the segment encoding ribosomal RNA).

The general form YCRXXw is now used to designate genes uncovered by systematically sequencing the yeast genome, where Y designates yeast; C (or A, B, etc.) designates the chromosome III (or I, II, etc.); R (or L) designates the right (or left) arm of the chromosome; XX designates the relative position of the start of the open-reading frame from the centromere; and w (or c) designates the Watson (or Crick) strand. For example, YCR5c denotes *CIT2,* a previously known but unmapped gene situated on the right arm of chromosome III, fifth ORF from the centromere on the Crick strand.

E. coli genes inserted into yeast are usually denoted by the prokaryotic nomenclature (e.g., *lacZ*).

A list of gene symbols is tabulated in the book edited by Wheals *et al.,* whereas a current list can be found in the Internet file "Saccharomyces Genome Database."

6.2 Mitochondrial Genes

Special consideration should be made of the nomenclature describing mutations of mitochondrial components and function that are determined by both nuclear and mitochondrial DNA genes. The growth on media containing *non*fermentable *s*ubstrates (Nfs) as the sole energy and carbon source (such as glycerol or ethanol) is the most convenient operational procedure for testing mitochondrial function. Lack of growth on nonfermentable media (Nfs^- mutants), as well as other mitochondrial alterations, can be due to either nuclear or mitochondrial mutations as outlined in Table 3. Nfs^- nuclear mutations are generally denoted by the symbol *pet;* however,

Table 3 Mitochondrial Genes and Mutations with Examples

Wild Type	Mutation (with examples)	Mutant Phenotype or Gene Product[a]
Nuclear Genes		
PET^+	pet^-	Nfs^-
	pet1	Unknown function
	cox4	Cytochrome *c* oxidase subunit IV
	hem1	δ-Aminolevulinate synthetase
	cyc3	Cytochrome *c* heme lyase
Mitochondrial DNA		
	Gross aberrations	
ρ^+	ρ^-	Nfs^-
	ρ°	ρ^- mutants lacking mitochondrial DNA
	Single-site mutations	
ρ^+	mit^-	Nfs^-, but capable of mitochondrial translation
[*COX1*]	[*cox1*]	Cytochrome *c* oxidase subunit I
[*COX2*]	[*cox2*]	Cytochrome *c* oxidase subunit II
[*COX3*]	[*cox3*]	Cytochrome *c* oxidase subunit III
[*COB1*]	[*cob1*] or [*box*]	Cytochrome *b*
[*ATP6*]	[*atp6*]	ATPase subunit 6
[*ATP8*]	[*atp8*]	ATPase subunit 8
[*ATP9*]	[*atp9*] or [*pho2*]	ATPase subunit 9
[*VAR1*]		Mitochondrial ribosomal subunit
ρ^+	syn^-	Nfs^-, deficient in mitochondrial translation
	$tRNA^{Asp}$ or M7-37	Mitochondrial $tRNA^{Asp}$ (CUG)
	antR	Resistant to inhibitors
[ery^S]	ery^R or [*rib1*]	Resistant to erythromycin, 21S rRNA
[cap^S]	cap^R or [*rib3*]	Resistant to chloramphenicol, 21S rRNA
[par^S]	par^R or [*par1*]	Resistant to paromomycin, 16S rRNA
[oli^S]	oli^R or [*oli1*]	Resistant to oligomycin, ATPase subunit 9

[a] Nfs^- denotes lack of growth on nonfermentable substrates.

Table 4 Nomenclature of Presumptive Prions Exhibiting non-Mendelian Inheritance

Prion State: Positive	Prion State: Negative	Putative Gene Product	Phenotype of Negative State
ψ^+	ψ^-	Sup35p	Decreased efficiency of certain suppression
ξ^+	ξ^-	Sup35p	Decreased efficiency of certain suppression
[*URE3*]	[*ure3*$^-$]	Ure2p	Deficiency in ureidosuccinate utilization

more specific designations (e.g., *cox4, hem1*) have been used instead of *pet* when the gene products were known.

The complexity of nomenclatures for mitochondrial DNA genes, outlined in Table 3, is due in part to complexity of the system, polymorphic differences of mitochondrial DNA, complementation between exon and intron mutations, the presence of intron-encoded maturases, diversed phenotypes of mutations within the same gene, and the lack of agreement among various workers. Unfortunately, the nomenclatures for most mitochondrial mutations do not follow the rules outlined for nuclear mutations. Furthermore, confusion can occur between phenotypic designations, mutant isolation number, allelic designations, loci, and cistrons (complementation groups).

6.3 Non-Mendelian Determinants

In addition to the non-Mendelian determinants described in Figure 2 (2 μm plasmid, mitochondrial genes, and RNA viruses) and discussed in Section 5 ("The Yeast Genome"), yeast contains elements that have been proposed to be prions (i.e., infectious proteins) on the bases of their genetic properties. Table 4 presents the nomenclature of these putative prions, representing alternative protein states.

7 GENETIC ANALYSES

7.1 Overviews with Examples

There are numerous approaches for the isolation and characterization of mutations in yeast. Generally, a haploid strain is treated with a mutagen, such as ethylmethanesulfonate, and the desired mutants are detected by any one of a number of procedures. For example, if Yfg$^-$ (your favorite gene) represents an auxotrophic requirement, such as arginine, or temperature-sensitive mutants unable to grow at 37°C, the mutants could be scored by replica plating. Once identified, the Yfg$^-$ mutants could be analyzed by a variety of genetic and molecular methods. Figure 3 illustrates three major methods: complementation, meiotic analysis, and molecular cloning.

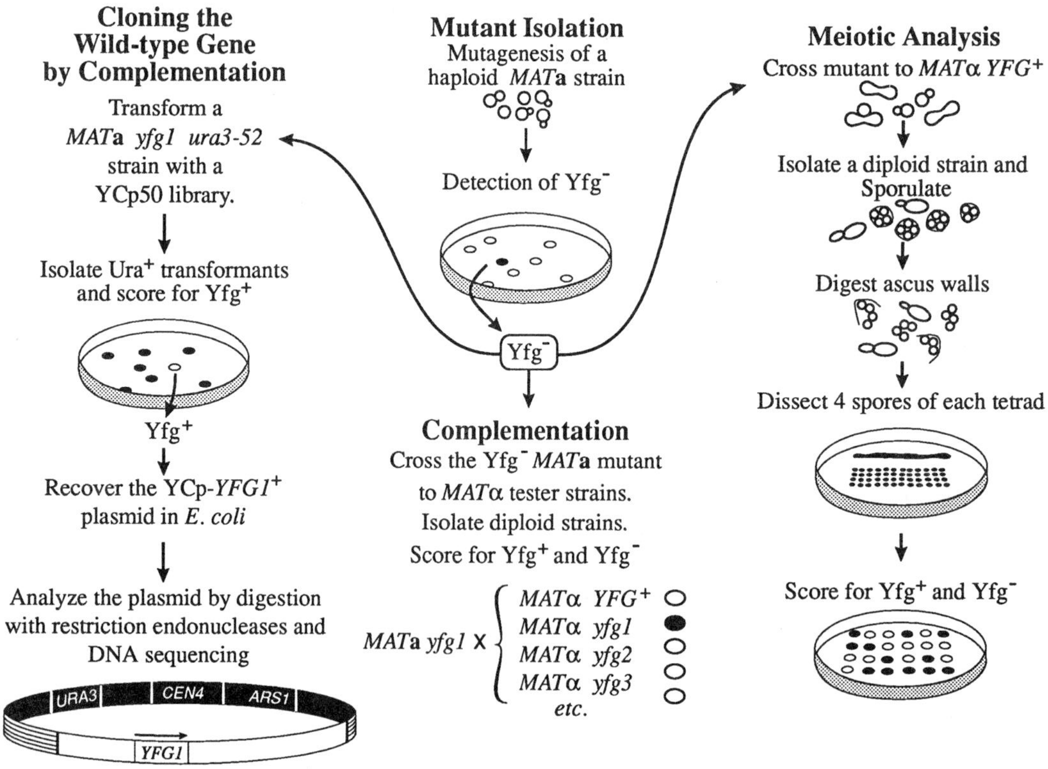

Figure 3. General approaches for genetic analysis. As an example, a *MAT***a** strain is mutagenized and a hypothetical trait Yfg$^-$ is detected. The Yfg$^-$ mutant is analyzed by three methods: complementation, meiotic analysis, and molecular cloning (see text).

Genetic complementation is carried out by crossing the Yfg$^-$ *MAT***a** mutant to each of the tester strains (*MAT*α *yfg1*, *MAT*α *yfg2*, etc.) as well as the normal control strain *MAT*α. These strains (*yfg1*, *yfg2*, etc.) contain previously defined mutations causing the same phenotype. The diploid crosses are isolated and the Yfg trait is scored. The Yfg$^+$ phenotype in the heterozygous control cross establishes that the Yfg$^-$ mutation is recessive. The Yfg$^-$ phenotype in *MAT*α *yfg1* cross, and the Yfg$^+$ phenotype in the *MAT*α *yfg2*, *MAT*α *yfg3*, etc., crosses reveals that the original Yfg$^-$ mutant contains a *yfg1* mutation.

Meiotic analysis can be used to determine whether a mutation is an alteration at a single genetic locus and to determine genetic linkage of the mutation both to its centromere and to other markers in the cross. As illustrated in Figure 3, the *MAT***a** *yfg1* mutant is crossed to a normal *MAT*α strain. The diploid is isolated and sporulated. Typically, sporulated cultures contain the desired asci with four spores, as well as unsporulated diploid cells and rare asci with fewer than four spores. The sporulated culture is treated with snail extract, which contains an enzyme that dissolves the ascus sac but leaves the four spores of each tetrad adhering to each other. A portion of the treated sporulated culture is gently transferred to the surface of a petri plate or an agar slab. The four spores of each cluster are separated with a microneedle controlled by a micromanipulator. After separation of the desired number of tetrads, the ascospores are allowed to germinate and form colonies on complete medium. The haploid segregants can then be scored for the Yfg$^+$ and Yfg$^-$ phenotypes. Because the four spores from each tetrad are the product of a single meiotic event, a 2:2 segregation of the Yfg$^+$:Yfg$^-$ phenotypes is indicative of a single gene. If other markers are present in the cross, genetic linkage of the *yfg1* mutation to the other markers or to the centromere of its chromosome could be revealed from the segregation patterns.

The molecular characterization of the *yfg1* mutation can be carried out by cloning the wild-type *YFG1*$^+$ gene by complementation, as illustrated in Figure 3 and described later, in Section 11.1.

7.2 Tetrad Analysis

Meiotic analysis is the traditional method for genetically determining the order and distances between genes of organisms having well-defined genetics systems. Yeast is especially suited for meiotic mapping because the four spores in an ascus are the products of a single meiotic event, and the genetic analysis of these tetrads provides a sensitive means for determining linkage relationships of genes present in the heterozygous condition. It is also possible to map a gene relative to its centromere if known centromere-linked genes are present in the cross. Although the isolation of the four spores from an ascus is one of the more difficult techniques in yeast genetics, requiring a micromanipulator and practice, tetrad analysis is routinely carried out in most laboratories working primarily with yeast. Even though linkage relationships are no longer required for most studies, tetrad analysis is necessary for determining whether a mutation corresponds to an alteration at a single locus, for constructing strains with new arrays of markers, and for investigating the interaction of genes.

There are three classes of tetrads from a hybrid that is heterozygous for two markers, *AB* × *ab*: PD (parental ditype), NPD (nonparental ditype), and T (tetratype), as shown in Figure 4. The following ratios of these tetrads can be used to deduce gene and centromere linkage:

	PD	NPD	T
	AB	*aB*	*AB*
	AB	*aB*	*Ab*
	ab	*Ab*	*ab*
	ab	*Ab*	*aB*
Random assortment	1 :	1 :	4
Linkage	>1 :	<1	
Centromere linkage	1 :	1 :	<4

There is an excess of PD to NPD asci if two genes are linked. If two genes are on different chromosomes and are linked to their respective centromeres, there is a reduction of the proportion of T asci. If two genes are on different chromosomes and at least one

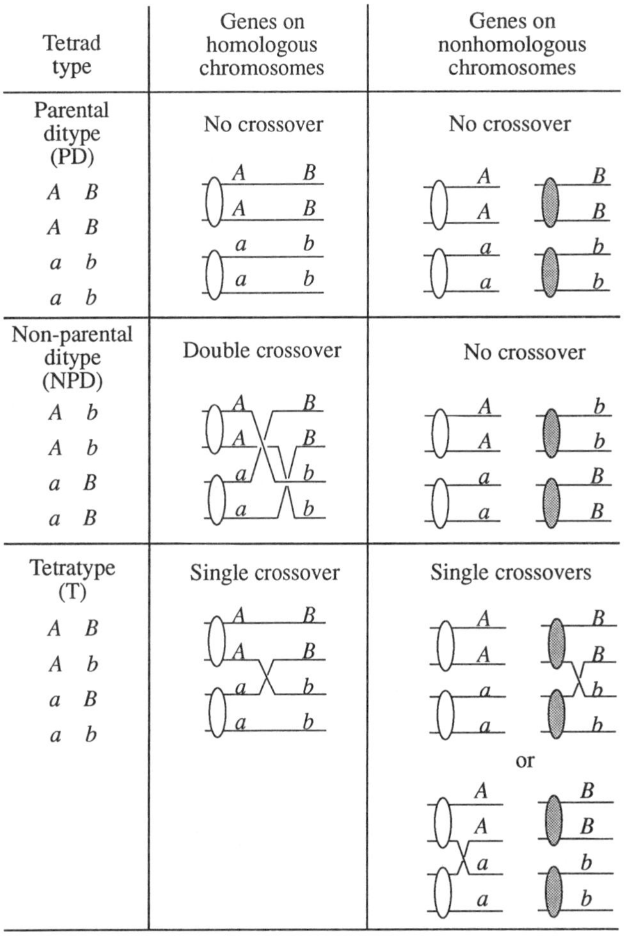

Figure 4. Origin of different tetrad types. Different tetrad types (left) are produced with genes on homologous (center) or nonhomologous (right) chromosomes from the cross *AB* × *ab*. When PD > NPD, the genes are on homologous chromosomes, because of the rarity of NPDs, which arise from four-strand double crossovers. The tetratype (T) tetrads arise from single crossovers. See text for the method of converting the percentages of T and NPD tetrads to map distances when genes are on homologous chromosomes. If gene are on nonhomologous chromosomes, or if they greatly separated on the same chromosome, PD = NPD, because of independent assortment, or multiple crossovers. Tetratype tetrads of genes on nonhomologous chromosomes arise by crossovers between either of the genes and their centromere (lower right). The percentage of T can be used to determine centromere distances if it is known for one of the genes (see text).

gene is not centromere-linked, or if two genes are widely separated on the same chromosome, there is independent assortment and the PD:NPD:T ratio is 1:1:4. Figure 4 illustrates the origin of different tetrad types.

The frequencies of PD, NPD, and T tetrads can be used to determine the map distance in centimorgans (cM) between two genes if there are two or lesser exchanges within the interval:

$$\mathrm{cM} = \frac{100}{2}\left[\frac{\mathrm{T} + 6\mathrm{NPD}}{\mathrm{PD} + \mathrm{NPD} + \mathrm{T}}\right]$$

The equation for deducing map distances is accurate for distances up to approximately 35 cM. For larger distances, up to approximately 75 cM, the value can be corrected by the following empirically derived equation:

$$\mathrm{cM\ (corrected)} = \frac{(80.7)(\mathrm{cM}) - (0.883)(\mathrm{cM})^2}{83.3 - \mathrm{cM}}$$

Similarly, the distance between a marker and its centromere cM′, can be approximated from the percentage of T tetrads with a tightly linked centromere marker, such as *trp1:*

$$\mathrm{cM}' = \frac{100}{2}\left[\frac{\mathrm{T}}{\mathrm{PD} + \mathrm{NPD} + \mathrm{T}}\right]$$

7.3 Non-Mendelian Inheritance

The inheritance of non-Mendelian elements can be revealed by tetrad analysis. For example, a cross of ρ^+ *MAT***a** and ρ^- *MAT*α haploid strains would result in ρ^+ *MAT***a**/*MAT*α and ρ^- *MAT***a**/*MAT*α diploid strains, the proportion of which would depend on the particular ρ^- strain. Each ascus from a ρ^+ diploid strain contains four ρ^+ segregants or a ratio of 4:0 for ρ^+:ρ^-. In contrast, a cross of *pet1 MAT***a** and *PET1*$^+$ *MAT*α strains would result in a *PET1*$^+$/*pet1 MAT*α/*MAT***a** diploid, which would yield a 2:2 segregation of *PET1*$^+$/*pet1*. Similarly, the other non-Mendelian determinants also produce primarily 4:0 or 0:4 segregations after meiosis.

Another means for analyzing non-Mendelian elements is cytoduction, which is based on the segregation of haploid cells, either *MAT***a** or *MAT*α, from zygotes. Haploid cells arise from zygotes at frequencies of approximately 10^{-3} with normal strains, and nearly 80% with *kar1* crosses (e.g., *kar1 MAT***a** × *KAR1*$^+$ *MAT*α). While the haploid segregants from a *kar1* cross generally retains all the chromosomal markers from either the *MAT***a** or *MAT*α parental strain, the non-Mendelian elements can be reassorted. For example, a *MAT***a** *can*^{R}*1 kar1* [ρ^- ψ^- *kil*-**o**] × *MAT*α *CAN*^{S}*1* [ρ^+ ψ^+ *kil*-k] cross can yield *MAT***a** *can*^{R}*1 kar1* haploid segregants that are [ρ^+ ψ^+ *kil*-k], [ρ^- ψ^+ *kil*-k], and so on. In addition, high frequencies of 2 μm plasmids and low frequencies of chromosomes can leak from one nucleus to another.

Also, the mating of two cells with different mitochondrial DNAs results in a heteroplasmic zygote containing both mitochondrial genomes. Mitotic growth of the zygote usually is accompanied by rapid segregation of homoplasmic cells containing either one of the parental mitochondrial DNAs or a recombinant product. The frequent recombination and rapid mitotic segregation of mitochondrial DNAs can be seen, for example, by mating two different *mit*$^-$ strains, and observing both Nfs$^-$ parental types as well as the Nfs$^+$ recombinant (see Table 3).

8 TRANSFORMATION

8.1 Yeast Vector and DNA Fragments

In general, transformation is the introduction into cells of exogenously added DNA and the subsequent inheritance and expression of that DNA. The most important advances in the molecular characterization and controlled modification of yeast genes have relied on the use of shuttle vectors, which can be used to transform both yeast and *E. coli.*

The following three main methods are currently used to transform yeast: those using spheroplasts, those using cells treated with lithium salts, and those using electroporation.

Spheroplasts for transformations are prepared by the action of hydrolytic enzymes to remove portions of the cell wall in the presence of osmotic stabilizers, typically 1 M sorbitol. Cell wall digestion is carried out either with a snail gut extract, usually denoted glusulase, or with zymolyase, an enzyme from *Arthrobacter luteus.* DNA is added to the spheroplasts, and the mixture is coprecipitated with a solution of polyethylene glycol (PEG) and Ca^{2+}. Subsequently, the cells are resuspended in a solution of sorbitol, mixed with molten agar, and then layered on the surface of a selective plate containing sorbitol. Although this protocol is particularly tedious, and efficiency of transformation can vary by more than four orders of magnitude with different strains, very high frequencies of transformation ($> 10^4$ transformants/μg DNA) can be obtained with certain strains.

Most investigators use cells treated with lithium salts for transformation. After treating the cells with lithium acetate, which apparently permeabilizes the cell wall, DNA is added and the cells are coprecipitated with PEG. The cells are exposed to a brief heat shock, washed free of PEG and lithium acetate, and spread on plates containing ordinary selective medium. Increased frequencies of transformation are obtained by using specially prepared single-stranded carrier DNA and certain organic solvents.

A commonly used method for transforming a wide range of different species of cells is based on the induced permeability to DNA by exposure to electrical fields. The interaction of an external electric field with the lipid dipoles of a pore configuration is believed to induce and stabilize the permeation sites, resulting in cross-membrane transport. Freshly grown yeast cultures are washed and suspended in an osmotic protectant, such as sorbitol; DNA is added, and the cell suspension is pulsed in an electroporation device. Subsequently, the cells are spread on the surface of plates containing selective media. The efficiency of transformation by electroporation can be increased over 100-fold by using PEG, single-stranded carrier DNA, and cells that are in late log phase of growth. Although electroporation procedures are simple, the specialized equipment and the required cuvettes are costly.

8.2 Synthetic Oligonucleotides

A convenient procedure has been described for producing specific alterations of chromosomal genes by transforming yeast directly with synthetic oligonucleotides. This procedure is easily carried out by transforming a defective mutant and selecting for at least partially functional revertants. Transformation of yeast directly with synthetic oligonucleotides is thus ideally suited for producing a large number of specific alterations that change a completely nonfunctional allele to at least a partially functional form. The oligonucleotide should contain a sequence that would correct the defect and produce the desired additional alterations at nearby sites. The method is apparently applicable to all mutant alleles whose functional forms can be selected. Although it is a general procedure, so far it has been extensively used only with mutations of *CYC1*, which encodes iso-1-cytochrome *c*, and *CYT1*, which encodes cytochrome c_1. The transformation is carried out by the usual lithium acetate procedure, using approximately 50 μg of oligonucleotides that are approximately 40 nucleotides long.

8.3 Mitochondrial Transformation

Standard methods for transformation of nuclear genes are ineffective for mitochondrial DNA genes. However, DNA can be delivered to the mitochondrial matrix by high velocity bombardment of yeast cells with tungsten microprojectiles carrying mitochondrial DNA. Several high velocity microprojectile bombardment devices are commercially available, and these are powered by gunpowder charge or compressed gas.

This method was used to demonstrate that ρ° strains can be converted to stable "synthetic ρ^{-}" strains by transformation with bacterial plasmids carrying mitochondrial genes (see Table 3). Similar to natural ρ^{-} mitochondrial DNA, the synthetic ρ^{-} mitochondrial DNA can recombine with ρ^{+} mitochondrial DNA, thus providing means to replace ρ^{+} wild-type genes with mutations generated in vitro.

Synthetic ρ^{-} strains are isolated by bombarding a lawn of ρ° cells on the surface of a petri plate with YEp or YCp plasmids carrying both a selectable marker, such as *URA3*, and the mitochondrial gene of interest. The nuclear and mitochondrial genes may be either on separate plasmids or on the same plasmid. Ura^{+} colonies, for example, are then screened for the presence of the mitochondrial gene by crossing the colonies to an appropriate *mit*$^{-}$ tester strain and scoring the diploids for Nfs^{+} (see Table 3). The efficiency of mitochondrial transformation, which varies from experiment to experiment, can be from 2×10^{-3} to less than 10^{-4} mitochondrial transformant per nuclear transformant.

9 YEAST VECTORS

A wide range of vectors are available to meet various requirements for insertion, deletion, alteration, and expression of genes in yeast. Most plasmids used for yeast studies are shuttle vectors, which contain sequences permitting them to be selected and propagated in *E. coli*, thus allowing for convenient amplification and subsequent alteration in vitro. The most common yeast vectors originated from pBR322 and contain an origin of replication (*ori*), promoting high copy number maintenance in *E. coli*, and the selectable antibiotic markers—the β-lactamase gene *bla* (or *Amp*^{R}), and sometimes the tetracycline resistance gene *tet* or (*Tet*^{R})—conferring resistance to, respectively, ampicillin and tetracycline.

In addition, all yeast vectors contain markers that allow selection of transformants containing the desired plasmid. The most commonly used yeast markers include *URA3, HIS3, LEU2, TRP1,* and *LYS2,* which complement specific auxotrophic mutations in yeast, such as *ura3-52, his3-Δ1, leu2-Δ1, trp1-Δ1,* and *lys2-201.* These complementable yeast mutations have been chosen because of their low reversion rate. Also, the *URA3, HIS3, LEU2,* and *TRP1* yeast markers can complement specific *E. coli* auxotrophic mutations.

Table 5 Components of Common Yeast Plasmid Vectors: YIp, YEp, YRp, YCp

Component	YIp	YEp	YRp	YCp
Plasmid				
E. coli genes or segments: *ori, bla; tet*	+	+	+	+
Yeast genes or segments				
URA3; HIS3; LEU2; TRP1; LYS2; etc.	+	+	+	+
leu2-d	0	+	+	0
2 μm; 2 μm-*ori REP3;*	0	+	0	0
ARS1; ARS2; ARS3; etc.	0	0	+	+
CEN3; CEN4; CEN11; etc.	0	0	0	+
Host (yeast) markers				
ura3-52; his3-Δ1; leu2-Δ1; trpl-Δ1; lys2-201; etc.	+	+	+	+
Stability	++	+	±	+

The *URA3* and *LYS2* yeast genes have an additional advantage because both positive and negative selections are possible, as discussed in Section 10.1 for *URA3* and *LYS2*.

Although there are numerous kinds of yeast shuttle vector, those used currently can be broadly classified in either of following three types as summarized in Table 5: integrative vectors, YIp; autonomously replicating high copy number vectors, YEp; or autonomously replicating low copy number vectors, YCp. Vectors of another type, yeast artificial chromosomes (YACs), for cloning large fragments, are discussed in Section 13.2.

9.1 YIp Vectors

The YIp integrative vectors do not replicate autonomously; rather, they integrate into the genome at low frequencies by homologous recombination. Integration of circular plasmid DNA by homologous recombination leads to a copy of the vector sequence flanked by two direct copies of the yeast sequence, as illustrated in Figure 5 (top). The site of integration can be targeted by cutting the yeast segment in the YIp plasmid with a restriction endonuclease and transforming the yeast strain with the linearized plasmid. The linear ends are recombinogenic, and they direct integration to the site in the genome that is homologous to these ends. In addition, linearization increases the efficiency of integrative transformation from 10- to 50-fold.

The YIp vectors typically integrate as a single copy. However multiple integrations do occur at low frequencies, a property that can be used to construct stable strains overexpressing specific genes. YIp plasmids with two yeast segments, such as *YFG1* and *URA3* marker, have the potential to integrate at either of the genomic loci, whereas vectors containing repetitive DNA sequences, such as Ty elements or rDNA, can integrate at any of the multiple sites within genome. Strains constructed with YIp plasmids should be examined by polymerase chain reaction (PCR) analysis, or other methods, to confirm the site of integration.

Strains transformed with YIp plasmids are extremely stable, even in the absence of selective pressure. However, plasmid loss can occur at approximately 10^{-3}–10^{-4} frequency by homologous recombination between tandemly repeated DNA, leading to looping out

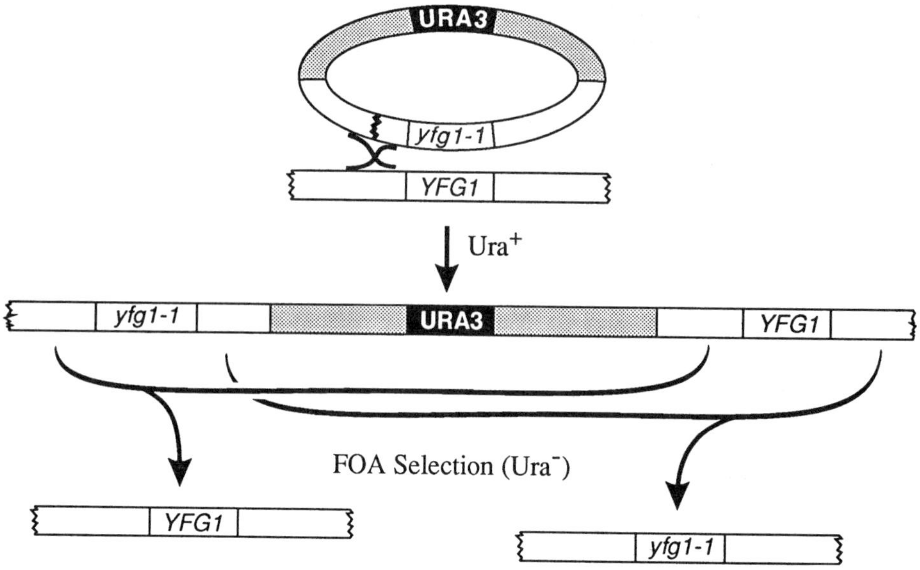

Figure 5. Two-step gene replacement. The wild-type chromosomal *YFG1*$^+$ allele can be replaced by the mutant *yfg1-1* allele from a YIp integrating plasmid. The plasmid is first integrated in the chromosome corresponding to the site on the plasmid that was cleaved by a restriction endonuclease. Strains that have excised the *URA3* marker in vivo by homologous recombination are selected on FOA medium. Either the original *YFG1*$^+$ allele, or the *yfg1-1* allele remains in the chromosome, depending on the site of the crossover.

of the vector sequence and one copy of the duplicated sequence, as illustrated in Figure 5 and discussed in Section 11.3.

9.2 YEp Vectors

The YEp yeast episomal plasmid vectors replicate autonomously because of the presence of a segment of the yeast 2 μm plasmid that serves as an origin of replication (2 μm *ori*). The 2 μm *ori* is responsible for the high copy number and high frequency of transformation of YEp vectors.

YEp vectors contain either a full copy of the 2 μm plasmid, or, as with most vectors of these kinds, a region that encompasses the *ori* and the *REP3* gene. The *REP3* gene is required in cis to the *ori* for mediating the action of the trans-acting *REP1* and *REP2* genes, which encode products that promote partitioning of the plasmid between cells at division. Therefore, the YEp plasmids containing the region encompassing only *ori* and *REP3* must be propagated in cir$^+$ hosts containing the native 2 μm plasmid (Figure 2).

Most YEp plasmids are relatively unstable, being lost in approximately 10^{-2} or more cell after each generation. Even under conditions of selective growth, only 60–95% of the cells retain the YEp plasmid.

The copy number of most YEp plasmids ranges from 10 to 40 per cell of cir$^+$ hosts. However, the plasmids are not equally distributed among the cells, and there is a high variance in the copy number per cell in populations.

Several systems have been developed for producing very high copy numbers of YEp plasmids per cell, including the use of the partially defective mutation *leu2-d,* whose expression is several orders of magnitude less than the wild-type *LEU2*$^+$ allele. The copy number per cell of such YEp *leu2-d* vectors ranges from 200 to 300, and the high copy number persists for many generations after growth in leucine-containing media without selective pressure. The YEp *leu2-d* vectors are useful in large-scale cultures with complete media, where plasmid selection is not possible. The most common use for YEp plasmid vectors is to overproduce gene products in yeast.

9.3 YCp Vectors

The YCp yeast centromere plasmid vectors are autonomously replicating vectors containing centromere sequences, CEN, and autonomously replicating sequences, ARS. The YCp vectors are typically present at very low copy numbers (1–3 per cell, and possibly more) and are lost in approximately 10^{-2} cell per generation without selective pressure. In many instances, the YCp vectors segregate to two of the four ascospores from an ascus, indicating that they mimic the behavior of chromosomes during meiosis, as well as during mitosis. The *ARS* sequences are believed to correspond to the natural replication origins of yeast chromosomes, and all of them contain a specific consensus sequence. The *CEN* function is dependent on three conserved domains, designated I, II, and III; all three of these elements are required for mitotic stabilization of YCp vectors. YRp vectors, containing *ARS* but lacking functional *CEN* elements, transform yeast at high frequencies but are lost at too high a frequency (> 10% per generation), making them undesirable for general vectors.

The stability and low copy number of YCp vectors make them the ideal choice for cloning vectors, for construction of yeast genomic DNA libraries, and for investigating the function of genes altered in vivo. *ARS1,* which is in close proximity to *TRP1,* is the most commonly used *ARS* element for YCp vectors, although others have been used. *CEN3, CEN4,* and *CEN11* are commonly used centromeres that can be conveniently manipulated. For example, the vector YCp50 contains *CEN4* and *ARS1.*

10 GENES IMPORTANT FOR GENETIC STUDIES

Several genes and promoters are commonly employed for genetic manipulations and studies with yeast. Some of these genes have special properties, whereas others were originally chosen arbitrarily and are conveniently available in strains and plasmids. Several of the genes most commonly used for a variety of purposes are described in Sections 10.1 through 10.4.

10.1 *URA3* AND *LYS2*

The *URA3* and *LYS2* yeast genes have a marked advantage because both positive and negative selections are possible. Positive selection is carried out by auxotrophic complementation of the *ura3* and *lys2* mutations, whereas negative selection is based on specific inhibitors, 5-fluoro-orotic acid (FOA) and α-aminoadipic acid (αAA), respectively, which prevent growth of the prototrophic strains but allow growth of the *ura3* and *lys2* mutants, respectively.

URA3 encodes orotidine-5′-phosphate decarboxylase, an enzyme that is required for the biosynthesis of uracil. *Ura3*$^-$ (or *ura5*$^-$) cells can be selected on media containing FOA. The *URA3*$^+$ cells are killed because FOA appears to be converted to the toxic compound 5-fluorouracil by the action of decarboxylase, whereas *ura3*$^-$ cells are resistant. The negative selection on FOA media is highly discriminating, and usually less than 10^{-2} FOA-resistant colonies are Ura$^+$. The FOA selection procedure can be used to produce *ura3* markers in haploid strains by mutation, and, more importantly, for expelling *URA3*-containing plasmids, including YCp- and YEp-replicating plasmids, and integrated YIp plasmids, as discussed in Section 11 for a number of genetic strategies. Because of this negative selection and its small size, *URA3* is the most widely used yeast marker in yeast vectors. The specific allele, *ura3-52,* which is the most commonly used host marker, contains a Ty1 insertion, is not revertible, and prohibits integration of YIp-*URA3* plasmids at the *URA3* chromosomal locus in most, but not all strains.

LYS2 encodes α-aminoadipate reductase, an enzyme that is required for the biosynthesis of lysine. The mutants *lys2*$^-$ and *lys5*$^-$, but not normal strains, grow on a medium lacking the normal nitrogen source but containing lysine and αAA. Apparently, *lys2* and *lys5* mutations cause the accumulation of a toxic intermediate of lysine biosynthesis that is formed by high levels of αAA, but these mutants still can use αAA as a nitrogen source. Numerous *lys2* mutants and low frequencies of *lys5* mutants can be conveniently obtained by simply plating high densities of normal cells on αAA medium. Similar to the FOA selection procedure, *LYS2*-containing plasmids can be conveniently expelled from *lys2* hosts. Because of the large size of the *LYS2* gene and the presence of numerous restriction sites, the FOA selection procedure with *URA3* plasmids is more commonly used.

10.2 *ADE1* AND *ADE2*

The *ADE1* and *ADE2* yeast genes encode phosphoribosylaminoimidazole-succinocarbozamide synthetase and phosphoribosylaminoimidazole-carboxylase, respectively, two enzymes in the biosynthetic pathway of adenine. *Ade1* and *ade2* mutants, but no other *ade*$^-$ mutants, produce a red pigment that is apparently derived from the polymerization of the intermediate phosphoribosylaminoimidazole (denoted AIR). Furthermore, the formation of AIR is prevented by blocks preceding the *ADE1* and *ADE2* steps. For example, *ade2* strains are red, whereas both *ade3* and the double mutant *ade2 ade3* are white, similar to the color of normal strains. Red colonies and red–white sectored colonies are easily detected by visual inspection.

Two characteristics—the *ade1* and *ade2* red pigmentation, and the prevention of the coloration by *ade3* or other *ade*$^-$ mutation—have been incorporated as a detection scheme in a number of diversed genetic screens. Also, the combination of the *ade2-1* UAA mutation and the suppression of formation of the red pigment by the *SUP4-***o** suppressor has been used in a variety of genetic screens. Most of the screens are based on the preferential loss, or the required retention, of a plasmid containing a component involved in the formation of the red pigment.

Examples of *ade*$^-$ red genetic screens include the detection of conditional mutations (Section 11.5), the isolation of synthetic lethal mutations (Section 12.5), the detection of YAC transformants (Section 13.2), and the isolation of mutations that effect plasmid stability.

10.3 *GAL1* PROMOTER

Cloned genes can be expressed with constitutive or regulatable promoters. The most commonly used regulated promoter for yeast studies is P_{GAL1}.

There are two regulatory proteins, Gal4p and Gal80p, which effect the transcription of the following structural genes: *GAL1,* a kinase; *GAL2,* a permease; *GAL7,* a transferase; *GAL10,* an epimerase; and *MEL1,* a galactosidase. Gal3p appears to be required for the production of the intracellular inducer from galactose. In presence of the inducer, Gal4p binds to sites in the upstream activation sequence (UAS) and activates transcription. In the absence of the inducer (e.g., when cells are grown in media containing nonfermentable carbon sources), Gal80p binds to the carboxyl-terminal region of Gal4p, masking the activation domain. Expression is repressed in cells exposed to glucose-containing media for several reasons in addition to the absence of the inducer; these include the action of repressors at sites between the UAS and the TATA box and the inhibition of galactose uptake. Therefore, the addition of glucose to cells growing in galactose medium causes an immediate repression of transcription. The UAS of galactose structural genes contains one or more 17 base pair palindromic sequences to which Gal4p binds, with the different levels of transcription determined by the number and combinations of the palindromes.

The UAS of the divergently transcribed *GAL1* and *GAL10* is contained within a 365 bp fragment, denoted P_{GAL1}, that is sufficient for maximal galactose induction and thorough glucose repression. P_{GAL1} can rapidly induce the expression of downstream-fused genes more than 1000-fold after the addition of galactose to cells growing in media with a nonfermentable carbon source. Furthermore, P_{GAL1} can be turned off by the addition of glucose to the galactose-containing medium.

P_{GAL1} has been used in a wide range of studies with yeast, including the overproduction of yeast proteins as well as heterologous proteins (Section 13.3). Most importantly, the strong glucose repression of P_{GAL1} has been used to investigate the terminal phenotype of essential genes, in much the same way that temperature shifts are used to control the activity of temperature-sensitive mutations (see Section 11.2). Also, the P_{GAL1} system has been used to investigate suppression (Section 12.4) and growth inhibition by overexpressed normal or mutant genes (dominant-negative mutations, Section 12.1). P_{GAL1} is also an important component of one of the two-hybrid systems (Section 13.1).

10.4 *lacZ* and Other Reporters

Activities of promoters, and protein–protein and protein–DNA interactions involving promoter regions, can be readily converted into selectable and quantifiable traits by fusing the promoter regions to reporter genes. Reporter genes can be used to determine the levels of transcription, or the levels of translation of the transcript, under various physiological conditions. The most common use of reporter genes has been to identify elements required for transcription by systematically examining series of mutations in promoter regions. Similarly, reporter genes have been used to identify trans-acting factors that modulate expression by transcription or translation.

The *E. coli lacZ* gene, which encodes β-galactosidase, is the most commonly used reporter with yeast and other systems, because its activity can be assayed semiquantitatively on plates and fully quantitatively by enzyme assay of liquid cultures. Rare events can be detected by the differential staining of colonies using X-gal (5-bromo-4-chloro-3-indolyl-β-D-galactoside).

For positive selection, the reporter gene could include, for example, the translated region of the *HIS3* gene, lacking a UAS. His^+ colonies arise when active promoters are formed, such as in the cloning of heterologous components required for the activation of a defined DNA segment. Combining the *HIS3* selection with a *lacZ* screen is a commonly used strategy; this approach of using two different reporters in parallel with the same promoter region is an efficient means for identifying trans-acting factors.

11 MANIPULATING THE GENOME IN VITRO WITH PLASMIDS

The greatest virtues of using yeast have been the ease with which genes can be retrieved, deleted, inserted, and modified in a controlled manner. These methods rely on the combined use of recombinant DNA techniques, transformation, and classical yeast genetics procedures. Overviews of some of these major approaches are described in Sections 11.1 through 11.6.

11.1 Cloning by Complementation

Molecular cloning and DNA analysis is the most definitive way of characterizing a gene that corresponds to a mutation. Cloning by complementation is usually carried out with a library of a YCp vector containing inserts of a more or less random series of genomic fragments, as illustrated in Figure 3 with the hypothetical *yfg1* mutation.

The *yfg1* strain is transformed with the YCp library, and the transformants are examined for the Yfg^+ trait. YCp vectors are generally used because each transformant contains a single plasmid per cell, or only a few. The method of screening transformants for complementation varies according to the specific phenotype that is to be scored. Direct selection can be used in some instances. However, if the original mutation reverts, as is often the case, a high frequency of false positives occurs among the transformants. Thus, an alternative method of indirect selection by replica plating is preferred. Thus, by this method, the transformant containing the desired YCp-*YFG1*$^+$ plasmids appear as homogeneous Yfg^+ colonies, whereas the colonies containing *yfg1* revertants appear as heterogeneous Yfg^+ and Yfg^- mixtures after replica plating. Most importantly, the true transformants will be dependent on the YCp-*YFG1*$^+$ plasmid for the Yfg^+ phenotype. In almost all studies, plasmid dependency is conveniently determined with the *ura3* system and usually with the nonreverting allele *ura3-52.* Because *ura3* mutants can be selected on FOA medium, plasmid-free strains can be recovered and subsequently tested for the loss of complementation. For example, the *yfg1 ura3* YCp-*YFG1*$^+$ strain would be Yfg^+Ura^+, while the *yfg1 ura3* strain, lacking the plasmid, would be Yfg^-Ura^-. Furthermore, the authenticity of the plasmid can be confirmed by first recovering the plasmid in *E. coli,* then retransforming the *yfg1* strain.

It is also desirable to confirm that the cloned segment truly encompasses the *YFG1*$^+$ gene. Even though the transformants may contain only a single copy of the putative gene, there are cases of two wild-type copies of a gene, one on the chromosome and the other on the plasmid, sometimes suppressing a mutation situated at a different locus. An independent test, based on homologous recombination, relies on YIp vector containing the insert. If the insert contains a unique restriction site, cleavage at this site will enhance integration of the plasmid at the homologous chromosomal site. Without cleavage, the plasmid could integrate at the site of other yeast markers on the plasmid, as well as at the *YFG1*$^+$ locus. After the integrant has been obtained, the site of integration can be investigated by meiotic analysis. For example, integration of the p[*YFG1*$^+$ *URA3*$^+$] plasmid at the site of *YFG1*$^+$ locus would result in a *YFG1*$^+$::[*YFG1*$^+$ *URA3*$^+$] *ura3* strain. After crossing to a *yfg1 ura3* strain and carrying out a meiotic analysis, the segregants should show a 2:2 segregation for both Yfg^+/Yfg^- and Ura^+/Ura^- and both markers would segregate as parental ditypes. On the other hand, if the plasmid integrated at a site other than the *YFG1* locus, an excessive number of Yfg^+ segregants would be recovered, indicating that the normal chromosomal *YFG1*$^+$ allele and the integrated plasmid *YFG1*$^+$ allele were not linked, or were at least not in close proximity.

If the sequence of the *YFG1* gene and flanking regions are known, the site of integration could be confirmed solely by PCR analysis.

After the desired plasmid has been demonstrated to encompass the *YFG1* gene, restriction fragments can be analyzed to narrow down the region of the locus, which can be subsequently sequenced and studied by a variety of other methods.

11.2 Mutagenesis In Vitro

Two common experimental goals are to produce either specific or "random" mutations within a gene. DNA alterations are required for investigating, for example, structure–function relationships and essential regions of proteins, and for producing conditional mutations, such as temperature-sensitive mutation. Specific alterations are carried out by the general procedure of oligonucleotide-directed mutagenesis, which is applicable to any cloned DNA segment, including those used for yeast studies. Oligonucleotide-directed mutagenesis has been used to systematically replace amino acids within proteins, especially in the substitution of alanine residues for charged amino acids. Such alanine replacements have resulted in a multitude of effects, including proteins that were unaffected, inactive, and temperature sensitive.

Also, numerous general procedures for producing "random" point mutations are available, including treating plasmid DNA with hydroxylamine and misincorporation by PCR mutagenesis. Most importantly, a simple procedure has been developed for the localized mutagenesis of yeast genes, as illustrated in Figure 6B. The region to be altered is first amplified under mutagenic PCR conditions, resulting in the generation of fragments containing "random" *yfg1-x* mutations. A *yfg1*-Δ strain is then cotransformed with these PCR products and with a gapped YCp plasmid containing homology to both ends of the PCR products. Repair of the gap with the PCR products (see Section 11.6) results in a series of strains with YCp plasmids containing the altered *yfg1-x* alleles. The yeast strains containing the *yfg1*-Δ chromosomal deletion can then be individually scored for the phenotype of each of the *yfg1-x* mutants. This procedure is particularly effective for targeting "random" mutations in specific regions and does not require subcloning steps in *E. coli.*

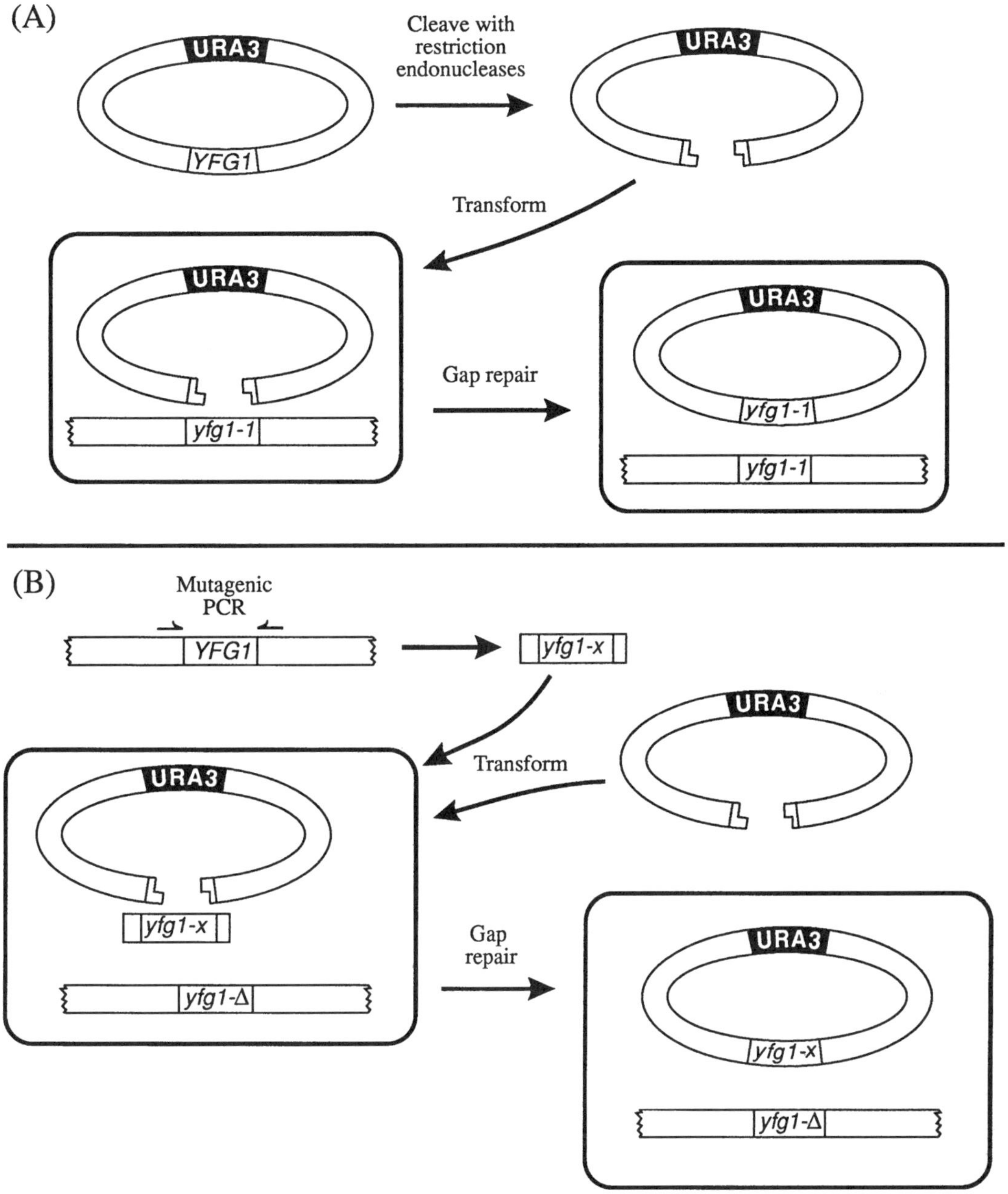

Figure 6. (A) Retrieval of a chromosomal *yfg1-1* mutation by transforming a mutant with a gapped plasmid. Only the repaired circular plasmid containing the mutation is stably maintained in yeast. (B) Generation of a series of *yfg1-x* mutations by PCR mutagenesis and gap repair. The *yfg1*-Δ mutant is cotransformed with the mutagenized PCR fragments and the gapped plasmid. The phenotypes of the *yfg1-x* mutants can be directly assessed in the strain containing the *yfg1*-Δ deletion.

11.3 Two-Step Gene Replacement

After a gene has been cloned, the most efficient means for obtaining mutations in the gene is by mutagenesis in vitro of the cloned DNA segment as just described. The effects of the mutations can then be examined in vivo by introducing the altered gene in yeast by transformation. A simple procedure, and the most common one, is to transform a yeast strain that lacks a functional copy of the chromosomal gene with a YCp plasmid that contains the altered gene. This can be accomplished directly in a single step if the gene in question is not essential. However, the best procedure, eliminating the problems of copy number and vector sequences, is to replace the chromosomal copy of the gene with the altered plasmid copy. This can be accomplished by the two-step gene replacement procedure illustrated in Figure 5. A YIp plasmid, containing the altered *yfg1-1* gene, is integrated in the chromosome in the region containing the *YFG1*$^+$ normal gene. Homologous recombination results in two copies of the gene, *yfg1-1* and *YFG1*$^+$, separated by the plasmid sequences. The second step involves homologous crossing over in the repeated DNA segment to loop out the plasmid, along with the *URA3* gene. Such desired Ura$^-$ strains can be selected on FOA medium. The resulting plasmid is lost during growth of the cells because the plasmid lacks an origin of replication.

The second crossover can occur in either of two regions as depicted in Figure 6: the region to the left or to the right of the *yfg1-1* alteration. The crossover at the left results in the regeneration of normal *YFG1* allele, whereas a crossover at the right results in the introduction of only the desired *yfg1-1* mutation.

The position of the crossover in the second step is approximately random, resulting in recovery of both YFG1 and *yfg1-1* strains. However, the relative frequencies of crossovers in the two regions are probably related to their lengths. To recover the altered *yfg1-1* allele, the second crossover must occur at the opposite side of the site of integration. Therefore, it is desirable to force the initial integration at the smaller region by cutting the plasmid in this region with a restriction endonuclease.

In addition to the *URA3* marker, the *LYS2* can be also used for both positive and negative selection (see Section 10.1). However, if neither *URA3* or *LYS2* can be used, loop-out recombination is often sufficiently high (10^{-3}–10^{-4}), making it possible to detect the loss of the marker by replica plating. If enough altered replacements are contemplated, an additional marker could be introduced into the *YFG1*$^+$ locus, allowing for the convenient scoring of the desired loop-out.

11.4 Gene Disruption and One-Step Gene Replacement

One of the most important and widely used methods of characterizing yeast genes is gene disruption. The complete disruption of a gene unambiguously reveals its function and can be helpful for generating additional mutations. Several methods can be used to produce deletions and null mutations, including the two-step gene replacement described in the preceding section.

The one-step gene disruption procedure, usually preferred because of its simplicity, is based on the use of a linear fragment of DNA containing a selectable marker flanked by 5′ and 3′ homologous regions (Figure 7A). The free ends of the fragment, prepared by digestion with restriction endonucleases, are recombinogenic, resulting in the integration of *URA3* marker and the loss of wild-type *YFG1* allele.

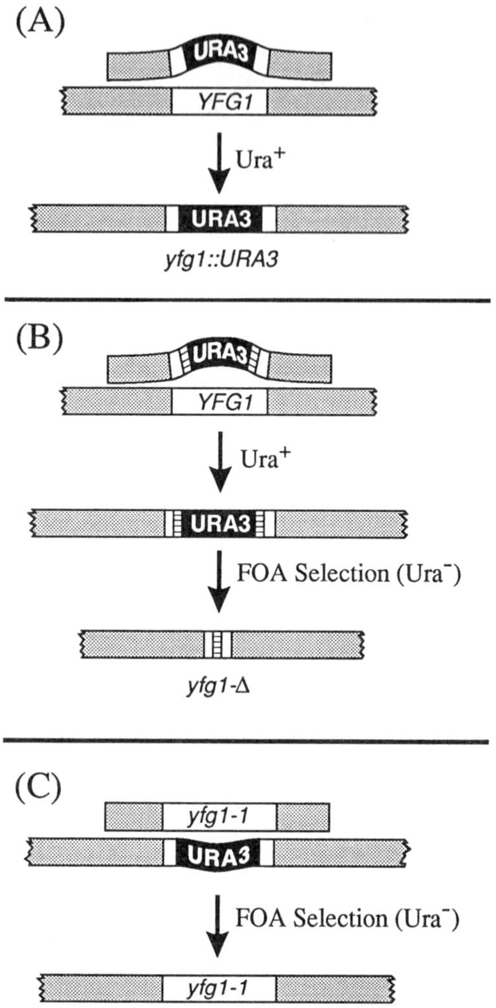

Figure 7. Gene disruption and single-step gene replacement. (A) The *YFG1*$^+$ gene is disrupted by transforming the strain with a linear fragment containing a *URA3* selectable marker flanked by homologous sequences. The chromosomal segment is replaced by this *URA3*-containing fragment after integration by homologous recombination at the two ends. (B) The *URA3* marker introduced in the *YFG1* locus can be excised if *URA3* is also flanked by direct repeats of DNA, preferably not originating from yeast. Homologous recombinants, selected on FOA medium, lack the *URA3* marker and retain a single copy of the repeated DNA. (C) Single-step gene replacement of mutant alleles, such as *yfg1-1*, can be carried out by first replacing the *YFG1* gene by *URA3*, transforming the strain with linear fragment encompassing the *yfg1-1* mutation, and selecting transformants on FOA medium, in which *URA3* is replaced by *yfg1-1*.

It should be noted that if the gene encodes an essential function, the transformation must be carried out in a diploid strain. Also, the disruption of the desired genes should be verified by PCR or Southern blot analysis. The fragment required for single-step disruptions can be also conveniently generated by PCR, eliminating the need to clone the *YFG1* gene.

Because the one-step gene disruption procedure results in a *URA3*$^+$ strain, the method has been modified, as illustrated in Figure 7B. In this method, *URA3* is flanked by identical copies of the bacterial *hisG* gene (or any nonyeast DNA segment). The *hisG-URA3-hisG* is first used to produce the gene disruption; subsequently, recombination between the direct repeats and selection on FOA produces a single copy of *hisG* at the site of the disruption. Thus, multiple rounds of gene disruption can be carried out in the same strain.

A similar procedure has been developed for conveniently replacing mutant alleles in a single step, as illustrated in Figure 7C. The *YFG1* gene is first disrupted with the *URA3* gene as described. Replacements of the disrupted *YFG1* by altered alleles can be selected on FOA medium among transformants or after minimal growth of transformants on complete medium. This and related procedures are particularly useful when many replacements are required.

Another method for producing gene disruptions, as well as simultaneously testing for the promoter activity, is based on a dominant resistant module consisting almost entirely of heterologous DNA. Transformants resistant to geneticin (G418) are selected and examined for *lacZ* activity. To allow for repeated use of the G418 selection, the module is flanked by short direct repeats, promoting excision in vivo.

11.5 Plasmid Shuffle

As mentioned in Section 11.2, the most common procedure for isolating and characterizing a series of altered alleles, *yfg1-x*, is simply to transform a strain lacking the gene, *yfg1*-Δ, with YCp plasmids containing the altered forms and to examine the phenotype of the transformants, *yfg1*-Δ p[*yfg1-x*]. However, the characterization of mutations of an essential gene poses an additional technical difficulty because of the inviability of strains containing the *yfg1*-Δ null mutation, as well as of those containing nonfunctional *yfg1-x* alleles.

Although the two-step gene replacement procedure (Section 11.3) could be used to generate condition *yfg1-x* mutations of an essential gene, a method that more clearly reveals the nature of the *ygf1-x* mutation has been developed, the "plasmid shuffle" procedure, illustrated in Figure 8.

A haploid strain is first prepared, which contains a YCp plasmid with the $URA3^+$ and $YFG1^+$ genes and in which the chromosomal copy of the $YFG1^+$ gene has been deleted or disrupted (*yfg1*-Δ). Such a strain could be prepared by transforming a diploid strain that is hemizygous ($YFG1^+$/yfg1-Δ) and choosing the appropriate meiotic segregant with the plasmid.

The $YFG1^+$ gene on YCp-*LEU2* plasmid, for example, is mutagenized, and the resulting p[*yfg1-x LEU2*] plasmids containing the *yfg1-x* alterations are introduced into the *yfg1*-Δ p[*YFG1 URA3*] strain by transformation. Tests on FOA medium then can be used to determine the nature of the *yfg1-x* mutation. If the *yfg1-x* allele was not altered or was completely functional, the p(*YFG1 URA3*) plasmid may be lost without preventing growth of the strain, which would appear to be FOA resistant (see Section 10.1). On the other hand, if the *yfg1-x* allele was completely nonfunctional, the strain will not grow on FOA medium. Furthermore, if the *yfg1-x* allele was conditional, the growth on FOA medium would correspond to the condition. For example, a temperature-sensitive mutation would be revealed by growth of the replica-plated colony on FOA medium at 22°C but not at 37°C, whereas the strain would grow at both temperatures on complete medium lacking FOA.

The major disadvantage of the plasmid shuffle procedure, and other procedures using YCp plasmid, is that the copy number varies from one to three or possibly more copies per cell. Thus, different phenotypes may arise because of the different levels of expression of the altered gene product. Overexpression of some altered alleles can produce a nearly wild-type phenotype, although a single copy produces a mutant phenotype. A more exact evaluation of a mutant allele may require the integration of a single copy with a YIp plasmid.

Several variations of the plasmid shuffle procedure have been developed, and these rely on the production of a red pigment by certain adenine mutations. Strains having mutations in the *ADE2* gene (e.g., *ade2-1*) accumulate a red pigment and form red colonies. However, if the strain also contains the *ade3*-Δ deletion, the adenine biosynthetic pathway is blocked before the step encoded by *ADE2* and the strain is white (see Section 10.2). If *ADE3* is carried by a $YFG1^+$ plasmid, then the *ade2-1 ade3*-Δ strains are red, but they produce white sectors when the *ADE3* plasmid is lost. The procedure does not require replica plating and is useful for detecting rare events. However, there are numerous other mutations that also produce white colonies, including ρ^- mutations, resulting in relatively high numbers of false positives.

11.6 Recovering Mutant Alleles

A convenient method for recovering chromosomal mutations involves transformation with gapped YCp plasmids, as illustrated in Figure 6A. A double-stranded gap is produced by cleavage at two restriction sites within the cloned segment. The gapped plasmid is then used to transform a strain containing the desired mutation that is encompassed in the chromosomal region corresponding to the gap. The gapped plasmid is repaired with the homologous chromosomal region, resulting in the capture of the *yfg1-1* mutation by the plasmid. The gapped plasmid is preferentially maintained, because only the circular form is replicated. The plasmid with the *yfg1-1* mutation can be subsequently recovered by transforming *E. coli* with DNA from the yeast strain. As little as 100 base pairs of homology on either side of the gap is sufficient to allow gap repair, although larger regions increase the efficiency of the process.

12 INTERACTION OF GENES

Yeast genetics has been particularly amenable to identifying and characterizing gene products that directly or indirectly interact with each other, especially when two mutations alleviate or enhance each other's defects. Information on a protein sometimes can be inferred simply by examining the phenotypes of haploid and diploid strains containing two or more mutations. In addition, these genetic properties can be used for isolating novel genes whose products interact. The genetic terms used to denote the interaction of genes are summarized in Table 6, using $YFG1^+$, and so on as hypothetical examples.

12.1 Heterozygosity and Dominant-Negative Mutations

When two recessive mutants are crossed in a standard complementation test, the phenotype of the resulting diploid strain usually reveals whether the two mutations are allelic and encode the same gene product. For example, if the *yfg1-1* and *yfg1-2* mutations produce inactive Yfg1 proteins, the diploid cross will be Yfg^-. On the other hand, if the two recessive mutations, *yfg1-1* and *yfg2-1,* are in

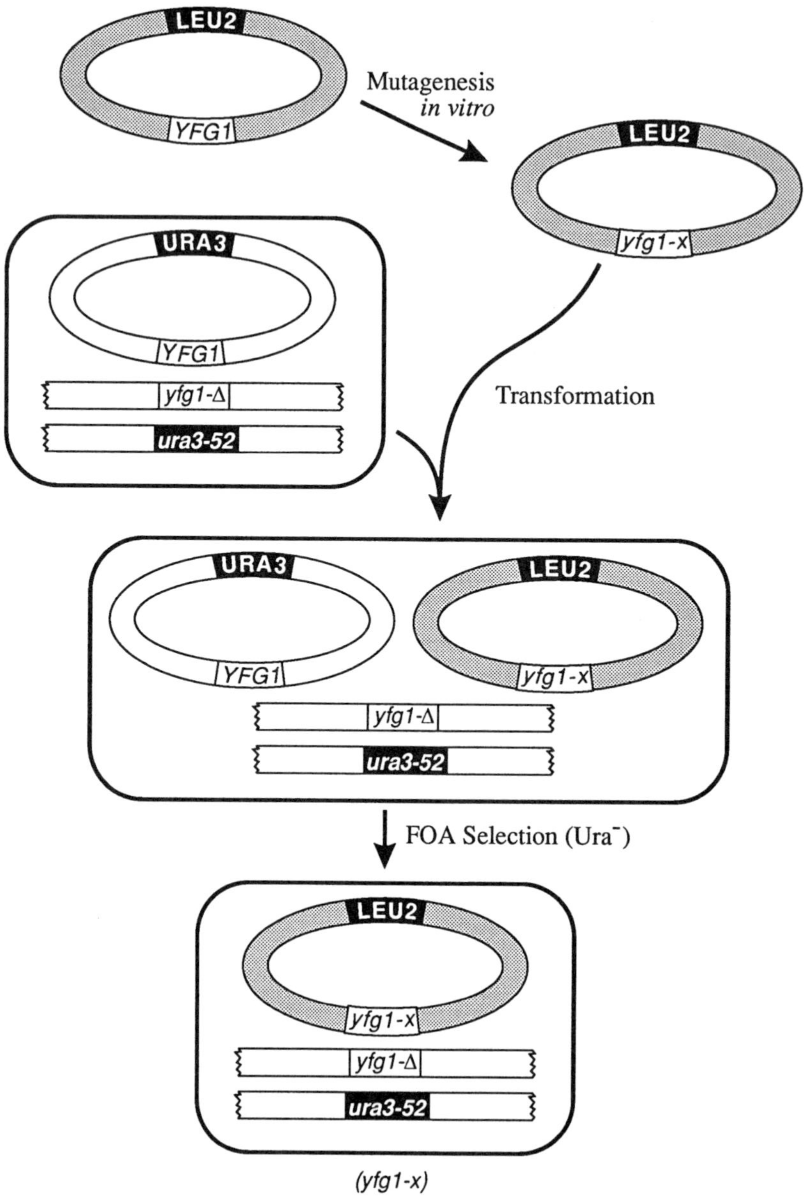

Figure 8. Plasmid shuffle. The chromosomal copy of *YFG1* is replaced by the *yfg1*-Δ deletion, but the Yfg$^+$ phenotype is maintained by the YCp plasmid containing *YFG1* and *URA3*. The strain is transformed with a mutagenized *LEU2* plasmids having the *YFG1* gene. Recessive *yfg1-x* mutations are manifested by selecting for strains on FOA medium. The strain will not grow on FOA medium if *YFG1* is an essential gene and if the *yfg1-x* mutation is not functional.

two different genes, encoding two different polypeptide chains, then the diploid cross, *yfg1-1 YFG2*$^+$ × *YFG1*$^+$ *yfg2-1* would be Yfg$^+$, because both Yfg1p and Yfg2p are produced by the wild-type alleles in the doubly heterozygous diploid strain.

As expected, mutations that inactivate a function are usually recessive. However, rare nonfunctional mutations can be dominant. Such dominant-negative mutations are particularly important because they can be used to identify nonfunctional forms of the protein that retain their proper structure and associate with other cellular components.

As illustrated in Table 6, dominant-negative mutations can be revealed either by overexpressing the mutation in a haploid (or diploid) strain, such as *YFG1*$^+$ p[*yfg1-4*]$_N$, or by a single copy in heterozygous strains, such as *YFG1*$^+$/yfg1-4. Most studies use multicopy YEp plasmids for overexpressing mutations to uncover dominant-negative mutations. Similarly, the P_{GAL1} promoter fused to mutant alleles, P_{GAL1}-*yfg1-4*, could be used for the controlled overexpression in tests for dominant-negative mutations (see Section 10.3).

Dominant-negative mutations act by a variety of mechanisms. For example, a mutationally altered transcriptional activator that retains DNA-binding activity, but lacks the ability to transactivate, could complex with the DNA-binding sites and displace the wild-type protein. While most recessive missense mutations produce an overall misfolding of proteins, dominant-negative mutations retain at least portions of the structure, thus revealing specific critical regions.

Table 6 Interactions of *YFG* (Your Favorite Gene) Genes

Genotype	Ploidy	Phenotype	Description
$YFG1^{+}$	1n	Yfg^{+}	Wild-type dominant allele
yfg1-1	1n	Yfg^{-}	Nonfunctional, recessive mutation
$YFG1^{+}$ / *yfg1-1*	2n	Yfg^{+}	Heterozygous diploid
yfg1-1 / *yfg1-2*	2n	Yfg^{-}	Heteroallelic diploid
yfg1-1 / *yfg1-Δ1*	2n	Yfg^{-}	Hemizygous diploid
$YFG1^{+}$ / *yfg1-4*	2n	$Yfg^{\pm}$	Dominant-negative *yfg1-4* mutation
$YFG1^{+}$ p[*yfg1-4*]$_N$	1n	Yfg^{-}	Dominant-negative overexpressed *yfg1-4* mutation
yfg1-1 / *yfg1-3*	2n	$Yfg^{\pm}$	Intragenic complementation
$YFG1^{+}$ *yfg2-1* / *yfg1-1* $YFG2^{+}$	2n	Yfg^{+}	Double heterozygous diploid
$YFG1^{+}$ *yfg3-1* / *yfg1-1* $YFG3^{+}$	2n	$Yfg^{\pm}$	Noncomplementation of a double heterozygous diploid
suy1-1	1n	Yfg^{+}	Suppressor of *yfg1-1*
yfg1-1 suy1-1	1n	Yfg^{+}	Suppression of *yfg1-1* by *suy1-1*
yfg1-1 p[$YFG2^{+}$]$_N$	1n	Yfg^{+}	Suppression of *yfg1-1* by overexpression of $YFG2^{+}$
yfg1-1 P_{GAL1}-$YFG2^{+}$	1n	Yfg^{+}	Suppression of *yfg1-1* by overexpression of $YFG2^{+}$
yfg1-4	1n	$Yfg^{\pm}$	Partially functional mutation of $YFG1^{+}$
yfg2-2	1n	$Yfg^{\pm}$	Partially functional mutation of $YFG2^{+}$
yfg1-4 yfg2-2	1n	Yfg^{-}	Synthetic enhancement

Dominant-negative mutations can also act in heterozygous diploid strains with one copy of each allele. Such mutant proteins generally have a higher than normal affinity for a cellular component and displace the wild-type protein. For example, numerous nonfunctional *CYC7* mutations were at least partially dominant because the altered forms of cytochrome *c* were arrested at one of the steps in mitochondrial import or heme attachment, preventing entry of the normal form.

12.2 Intragenic Complementation

One common exception in which heteroallelic diploids have a wild-type or near-wild-type phenotype is intragenic complementation (also denoted intraallelic, allelic, or intracistronic complementation) (Table 6). If large numbers of pairwise crosses of independent mutations of a gene are analyzed, complex complementation patterns are often encountered, with some alleles showing complementation while others do not. For example, a *yfg1-Δ* deletion would not show intragenic complementation with other *yfg1-1* and *yfg1-3,* although the *yfg1-1/yfg1-3* cross could. Also, intragenic complementation does not always restore the activity to the normal level, and heteroallelic diploid strains even can have conditional phenotypes.

There are at least two mechanisms for intragenic complementation, one involving proteins with two or more functional domains, and the other involving proteins composed of two or more identical polypeptide chains.

If a protein has two or more functional domains that act independently, then a missense mutation (an amino acid replacement) could inactivate one domain without greatly affecting the others. Thus, partial or complete restoration could be observed in crosses of two such mutations, when each affected a different functional domain. As expected, intragenic complementation of this type is often observed with missense mutations, not with deletions, and only with special subsets of nonsense mutations. Numerous examples of intragenic complementation of this type occur with genes encoding amino acid biosynthetic enzymes, including *HIS4,* which has three functional domains.

Intragenic complementation is also observed with genes encoding proteins that are composed of two or more identical polypeptide chains. A haploid mutant could produce an abnormal polypeptide that assembles into an inactive homomultimeric protein. On the other hand, if two different abnormal polypeptide chains are produced in a heteroallelic diploid strain, the abnormal peptides could assemble in certain combination to produce a catalytically active multimeric protein. In such cases, each abnormal polypeptide chain in some way mutually compensates for the other's defect. Intragenic complementation of genes encoding multimeric proteins is surprisingly frequent among missense mutations of certain genes. For example, 5 out of 10 genes controlling histidine biosynthesis show extensive complementation, and the nature of the genetic complementation maps suggests that multimeric proteins may be involved.

12.3 Nonallelic Noncomplementation

There are rare exceptions to the complementation of nonallelic genes, and these exceptions are denoted "nonallelic noncomplementation" or "unlinked noncomplementation." As illustrated in Table 6, certain recessive mutations of two different genes, *yfg1-1* and *yfg3-1,* fail to complement in the doubly heterozygous diploid $YFG1^{+}$/*yfg1-1 yfg3-1*/$YFG3^{+}$, even though they are clearly functional in each of the singly heterozygous diploid strains *YFG1*$^+$/*yfg1-1* and *YFG3*$^+$/*yfg3-1*. Noncomplementation of recessive nonallelic mutations is only rarely encountered and may be the property of only certain proteins.

Several explanations can account for nonallelic noncomplementation. One can consider two proteins Yfg1p and Yfg3p, which carry out related functions and are near their limiting concentration in the cell. In normal strains, both Yfg1p and Yfg2p are, by definition, at the normal 100% level, producing 100 units of each of the hypothetical proteins, with a total of 200 units; in singly heterozygous strains, the total would be 150 units, whereas in doubly heterozygous strains the total would be 100 units. If the total level of 100 units for both proteins is below a critical threshold, a mutant phenotype will be manifested in the doubly heterozygous strains. Noncomplementation of nonallelic genes has been observed for mutant genes encoding cytoskeletal proteins whose cellular concentrations are critical for normal growth.

The lack of complementation of nonallelic genes has also been explained by the formation of inactive heteroligomers or protein complexes, in which, for example, both altered Yfg1p and Yfg3p independently assemble and inactivate the same protein complex, reducing its level below a critical concentration. Also, although recessive by themselves, these mutations may act like dominant-negative mutation when in the doubly heterozygous condition, encoding abnormal proteins that actively compete or replace wild-type subunits.

12.4 Suppressors

A suppressor is generally defined as a mutation that completely or partially restores the mutant phenotype of another mutation. In one of the last examples in Table 6, the Yfg$^-$ phenotype of *yfg1-1* mutation is restored by the suppressor *suy1-1*. Suppressors can either have or not have a phenotype by themselves.

Suppressors can be broadly assigned to two major groups: informational and metabolic. Informational suppressors encode either altered tRNAs or other components of the translational machinery, and they act by misreading mRNAs. For example, the nonsense suppressor *SUP4*-**o** encodes an altered tyrosine tRNA that inefficiently inserts tyrosine residues at UAA chain-terminating codons because of an altered anticodon. Another class of informational suppressors consists of the so-called omnipotent suppressors, which cause ribosomal misreading because of alterations in any one of a number of proteins of the translational apparatus, including ribosomal proteins, elongation factors, and release factors. Informational suppressors characteristically act on certain mutations of most, if not all genes; that is, they are allele specific but not gene specific.

On the other hand, metabolic suppressors usually act on genes common to the same pathway or to a single metabolic function. While there are many possible mechanisms for metabolic suppression, these suppressors are now routinely isolated and investigated as a means for identifying novel gene products of a pathway, as well as proteins that directly interact with each other.

The suppressors that act by direct physical interactions between two mutant proteins are expected to be both gene and allele specific; that is, the suppressors should act on only one or a few genes, and on only a restricted subclass of mutants. For example, a temperature-sensitive mutation of actin gene was used to identify a suppressor that was subsequently shown to encode an actin-binding protein.

In contrast, the suppressors that act indirectly on the same pathway are expected to be gene specific, but allele nonspecific; that is, the suppressor could act on any allele, including null mutations, of a specific group of genes. Some of these are denoted as "bypass" suppressors if they replace the function of the initial mutation. For example, *cyc1-Δ* mutations, which cause the deficiency of the major isocytochrome *c*, can be replaced by *CYC7-H* mutations, which act as suppressors by overproducing the minor isocytochrome *c*, and allow growth on lactate medium. Another mechanism for pathway suppression involves the loss of regulation of successive steps, such as is encountered, for example, in amino acid biosynthesis or the signal transduction pathways. If an upstream gene that normally causes the activation of a downstream gene is destroyed by mutation, mutant forms of the downstream gene can act as suppressors by rendering it independent of signaling from the upstream gene.

Another often-used means for uncovering interacting components of common functions is suppression by overexpression of wild-type alleles. As in the case for suppression due to mutation, suppression due to overexpression can occur by many different mechanisms. Overexpression can be brought about with YEp multicopy plasmids, denoted as *yfg1-1* p[*YFG2*$^+$]$_N$ in Table 6; or by fusion to a strong inducible promoter such as P_{GAL1}, denoted as *yfg1-1* P_{GAL1}-*YFG2*$^+$ in Table 6. Furthermore, P_{GAL1} has the advantage that the activity can be conveniently turned off by the addition of glucose to the galactose-containing medium, resulting in the loss of suppression (see Section 10.3).

12.5 Synthetic Enhancement and Epistatic Relationships

In some instances, the combination of two different mutant genes in a haploid strain can enhance the severity of the phenotype more than can be achieved by either of the mutant genes by itself. This exacerbation by the combination of two genes, illustrated at the end of Table 6 with the *yfg1-4* and *yfg2-2* alleles, was denoted by a variety of terms in the early literature of genetics; currently, this phenomenon is denoted "synthetic lethality" when cell growth is involved, or more generally, "synthetic enhancement" or "synthetic phenotypes." For example, two different mutant genes individually could cause temperature-sensitive growth, but the double mutant would be inviable at any temperature.

Synthetic enhancement is conceptually identical to epistatic relationships. If having two mutant genes in a haploid condition confers a phenotype that is quantitatively identical to that conferred by each of the single mutant genes alone, the two genes are said to be epistatic with respect to each other. If, however, the doubly mutant strain has an enhanced phenotype, the two genes are said to be in two separate epistasis groups. For example, detailed study of epistatic relationships of more than 30 mutants sensitive to UV or ionizing radiation has revealed three nonoverlapping epistasis groups.

As with suppression, synthetic enhancement can be caused by a number of mechanisms, but often the genes are associated with parallel or related pathways controlling the same function, and some encode proteins that physically interact with each other.

There are numerous examples of synthetic lethality arising when two proteins are functionally redundant, such that neither nonfunctional form causes inviability, but both together are inviable. Synthetic enhancement of redundant genes is illustrated with strains having deletions of either of the *CYC1* or *CYC7* genes encoding, respectively, the two isocytochromes *c;* single deletion strains are respiratory competent, while strains carrying both deletions are respiratory deficient. Synthetic enhancement can be indirect, as exemplified by mutations of *ARG4,* encoding an arginine biosynthetic enzyme, and *CAN1,* encoding the arginine permease; the *arg4 can1* double mutant does not grow on synthetic medium containing arginine, while each of the *arg4* and *can1* single mutants does grow.

As with suppressors, genes causing synthetic enhancement are commonly isolated as means to identify genes in the same or related pathways and genes encoding interacting proteins. Several methods have been devised for this aim, and one is illustrated in Figure 9. A *yfg1-1 ura3-52* strain, carrying a YCp plasmid, p[*YFG1*$^+$ *URA3*$^+$], is mutagenized and tested on FOA medium. A mutation, *yfg2-1,* causing synthetic enhancement, is manifested in the Ura$^-$ strain because of the loss of the *YFG1*$^+$ allele; in contrast, the other Ura$^-$ strains would show only the *yfg1-1* phenotype. If the *yfg2-1* mutation causes synthetic lethality and the *yfg1-1 yfg2-1* double mutant is inviable, the colony would not grow on FOA medium (see Section 10.1).

Other methods have been developed for isolating mutations that confer synthetic lethality; these are similar to the detection schemes used with the plasmid shuffle procedures (Section 11.5). One of these other methods relies on the *ade2 ade3* host mutations and a *YFG1 ADE3* plasmid; when this combination is lost, red sectoring results (see Section 10.2). Another method is based on fusion of the *GAL1* promoter, and screening P_{GAL1}-*YFG1* colonies for those that grow on galactose but not on glucose medium, presumably because of the presence of the desired *yfg2-1* mutations (see Section 10.3).

13 ANALYSES WITH YEAST SYSTEMS

The accessibility of the yeast genome for genetic manipulation and the availability of techniques to introduce exogenous DNA into yeast cells have led to the development of methods for analyzing and preparing DNA and proteins not only from yeast itself, but also from other organisms. For example, many mammalian homologues of yeast genes have been cloned by using heterologous cDNA expression libraries in yeast expression vectors. Also, yeast is being used to investigate the detailed functions of heterologous proteins, such as mammalian transcription factors and nuclear hormone receptor. In fact, like *E. coli,* yeast has become a standard microorganism for carrying out special tasks, some of which are described in Sections 13.1 through 13.3.

13.1 Two-Hybrid Systems

Powerful new methods, denoted two-hybrid systems, have been designed for screening and investigating interacting proteins. Because of the ease of the assay, exploratory two-hybrid screens are usually the first method of choice when information of interacting proteins are desired.

Some two-hybrid systems are based on the properties of certain eukaryotic transcription factors, usually Gal4p, which have two separate domains: one for DNA binding and the other for transcriptional activation. While the two domains are normally on the same polypeptide chain, the transcription factor also functions if these two domains are brought together by noncovalent protein–protein interactions. In practice, gene fusions are constructed so that the DNA-binding domain is linked to one protein, Yfg1p, and the activation domain is linked to another protein, Yfg2p, as illustrated in Figure 10. Interactions of Yfg1p and Yfg2p brings the DNA-binding and activation domains close together, leading to the expression of a reporter gene that is regulated by the transcription factor. Another two-hybrid system is based on the use of the lexA repressor protein and the *lexA* operator sequences from *E. coli.* These assays are almost always carried out in yeast, although mammalian cells have been used.

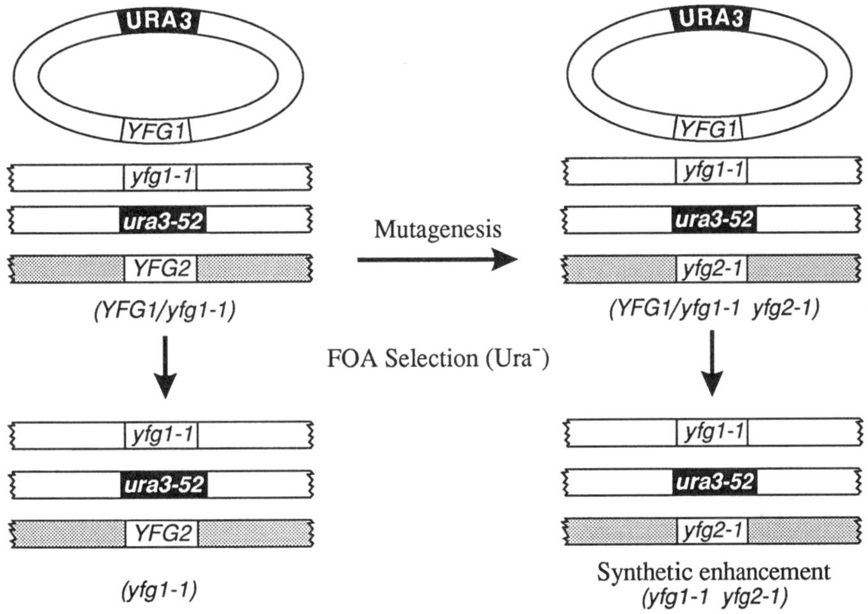

Figure 9. Synthetic enhancement and synthetic lethality. Novel chromosomal genes mutations (*yfg2-1,* etc.) that enhance the phenotype of a *yfg1-1* mutations can be uncovered by mutagenizing a *yfg1-1* strain containing a YCp plasmid p[*YFG1 URA3*], and subsequently selecting for the loss of the plasmid. If *yfg1-1 yfg2-1* strain are inviable, they will not grow on FOA medium.

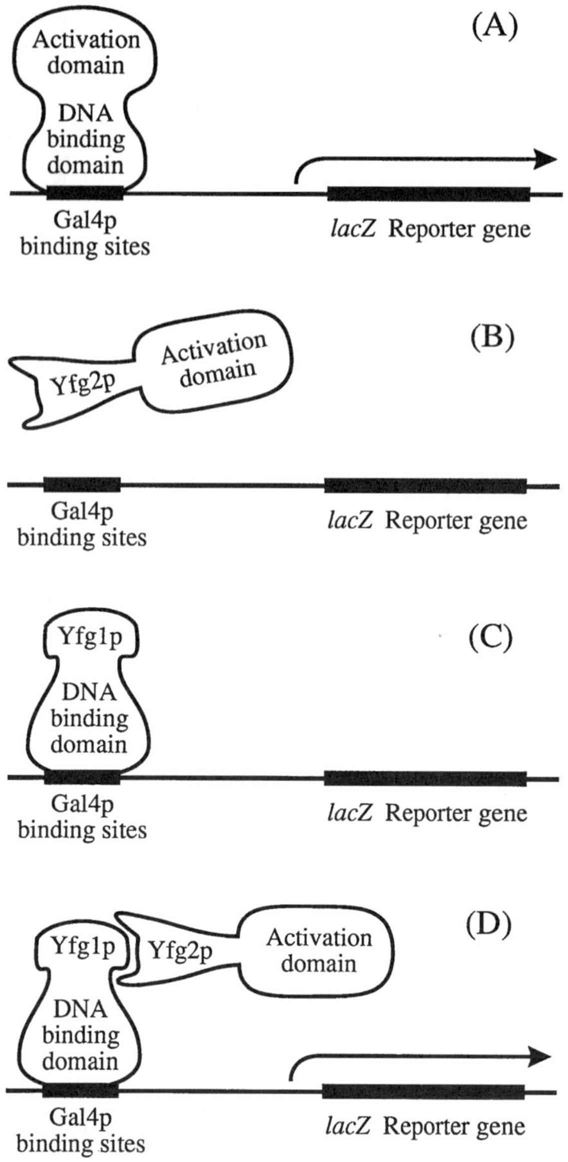

Figure 10. The two-hybrid system. (A) Normally, the Gal4 transcription activator binds to DNA at the Gal4p-binding sites and activates transcription of the *lacZ* reporter gene. (B) A hybrid of the Gal4 activation domain with the Yfg2 protein does not activate transcription because it does not localize at the Gal4-binding site. (C) A hybrid of the Gal4 DNA-binding site domain with the Yfg1 protein does not activate transcription of the reporter gene because a transcriptional activation domain is lacking. (D) Protein–protein interaction between Yfg1p and Yfg2p reconstitutes Gal4p function and activates transcription of the reporter gene.

Yeast plasmid vectors are available in which the *GAL4* DNA-binding domain and the *GAL4* activation domain are on separate plasmids, with convenient restrictions sites and with selectable yeast markers. These plasmids are used in conjunction with reporter yeast strains, in which upstream activation sequences from the *GAL1–GAL10* region are used to promote transcription of the *E. coli lacZ* gene (the P_{GAL1}-*lacZ* reporter gene) (see Sections 10.3 and 10.4). Complete or partial genes are fused in frame with the *GAL4* DNA-binding domain and the *GAL4* transcription activation domains. If these two hybrid proteins interact, the *lacZ* reporter gene is transcribed, leading to the blue color of the strain on medium containing the chromogenic substrate X-gal. In addition, yeast strains having not only the P_{GAL1}-*lacZ* but also the P_{GAL1}-*HIS3* reporter genes are available. It is advantageous to select directly for expression of the P_{GAL1}-*HIS3* reported gene, followed by a screen for P_{GAL1}-*lacZ* expression.

Another version of the two-hybrid system uses the *lexA* operator sequence and the DNA-binding domain from the *E. coli* lexA repressor protein. In this system, the activator domain is a segment of *E. coli* DNA that expresses an acidic peptide, which acts as a transcriptional activator in yeast when fused to a DNA-binding domain. As with the *GAL4* system, *lexA* transcriptional activator also contains a nuclear localization signal that directs the protein into the nucleus. Yeast strains having *lexA* operators upstream of both the *E. coli lacZ* and yeast *LEU2* gene have served as reporter genes.

In addition, epitope tags have been built into the constructs of both the *GAL4* and *lexA* systems, allowing for the detection of expressed hybrid proteins. Although false positive and false negative results can be obtained, a substantial number of protein combinations have been successfully uncovered with the two-hybrid system, and its use has become widely accepted. Because of its sensitivity, moreover, relatively low affinity interactions can be detected. Also, the cloned genes encoding proteins that interact with the target protein become immediately available when the two-hybrid system is used in a screen with libraries of fused genes.

The two-hybrid system has been mainly used for three applications: testing proteins that are believed to interact on the basis of other criteria, defining domains or amino acids critical for interactions of proteins that are already known to interact, and screening libraries for proteins that interact with a specific protein. The two-hybrid system has been successfully used to identify diverse sets of interacting proteins in yeast and mammalian cells, and it has been particularly successful in studies of oncogenes, tumor suppressors, protein kinases, and cell cycle control. Some examples of interacting proteins uncovered with the two-hybrid system in mammalian cells include Jun and Fos; Ras and the protein kinase Raf; the retinoblastoma protein and the SV-40 large T antigen; p53 and p53; and other oncoproteins.

13.2 Yeast Artificial Chromosomes (YACs)

The initial step in the molecular characterization of eukaryotic genomes generally requires cloning of large chromosomal fragments, which is usually carried out by digestion with restriction endonucleases and ligation to specially developed cloning vectors. Usually 200–800 kb fragments are cloned as yeast artificial chromosomes (YACs), and 100–200 kb fragments are cloned as bacterial artificial chromosomes (BACs or PACs). The importance of YAC technology has been heightened by the recently developed methods for transferring YACs to cultured cells and to the germ line of experimental animals.

YAC cloning systems are based on yeast linear plasmids, denoted YLp, containing homologous or heterologous DNA sequences that function as telomeres *(TEL)* in vivo, as well as containing yeast *ARS* (origin of replication) and *CEN* (centromere) segments. Manipulating YLp linear plasmids in vitro is complicated by the inability of these DNA segments to be propagated in *E. coli.* However, specially developed circular YAC vectors have been developed for amplification in *E. coli.* For example, a circular YCp vector, containing a head-to-head dimer of *Tetrahymena* or yeast telomeres, is resolved in vivo after yeast transformation into linear molecules with the free ends terminated by functional telomeres. One common type of YAC vector that can be propagated in *E. coli* contains in inverted orientation telomeric sequences that flank a DNA cassette containing the HIS3 gene (Figure 11). After amplification in *E. coli* and before the transformation of the yeast, the plasmid is digested with a restriction endonuclease, usually *Bam*HI, which excises the *HIS3* cassette and generates a linear form in vitro. Yeast are transformed by this linear structure at high frequencies, although the transformants are unstable. Despite the presence of a *CEN* sequence, the YLp is present at high copy numbers and is lost at high frequency because of its small size. Increasing the size of the YLp by homologous integration in vivo or by ligation in vitro increases the stability of the plasmid and reduces the copy number to approximately one per cell.

The developed highly efficient YAC cloning vectors also contain *TRP1* and *URA3* markers and a *SUP4*-**o** gene flanked by the *Not* I and *Sfi* I rare restriction sites, as shown in Figure 11. The *SUP4*-**o** suppressor also harbors a naturally occurring *Sma* I site. The host strain contains the *ade2-1* UAA mutation, causing the formation of a red pigment, unless the mutation is suppressed by *SUP4*-**o** (see Section 10.2). The YAC vector is cleaved with *Bam*HI and *Sma* I and treated with alkaline phosphatase, whereupon the two arms are ligated to the exogenous DNA fragments desired to be cloned. An *ade2-1 ura3-52 trp1*-Δ host strain is transformed with the ligated mixture. Both arms are anticipated to be present in Ura$^+$ and Trp$^+$ transformants, and inserts should be present in the Ura$^+$ Trp$^+$ Ade$^-$ (red) transformants. YACs present in these transformants are then subjected to pulsed field electrophoresis to estimate the size of the inserts.

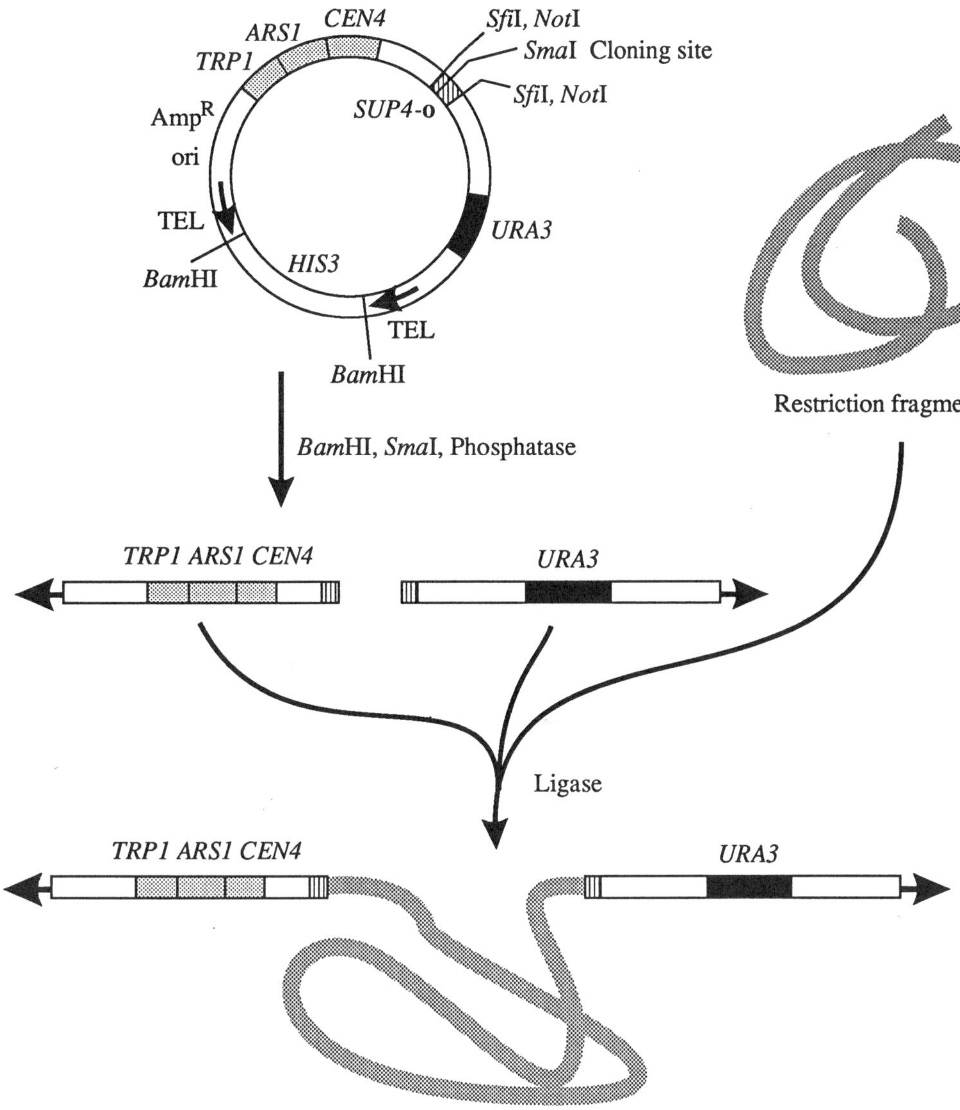

Figure 11. A yeast artificial chromosome (YAC) cloning system. The YAC vector contains telomeric ends (black arrowheads). The vector also contains a unique *Sma* I cloning site flanked by *Sfi* I and *Not* I restriction sites having 8 base pairs. The vector can be used to clone 50–500 kb restriction fragments (see text).

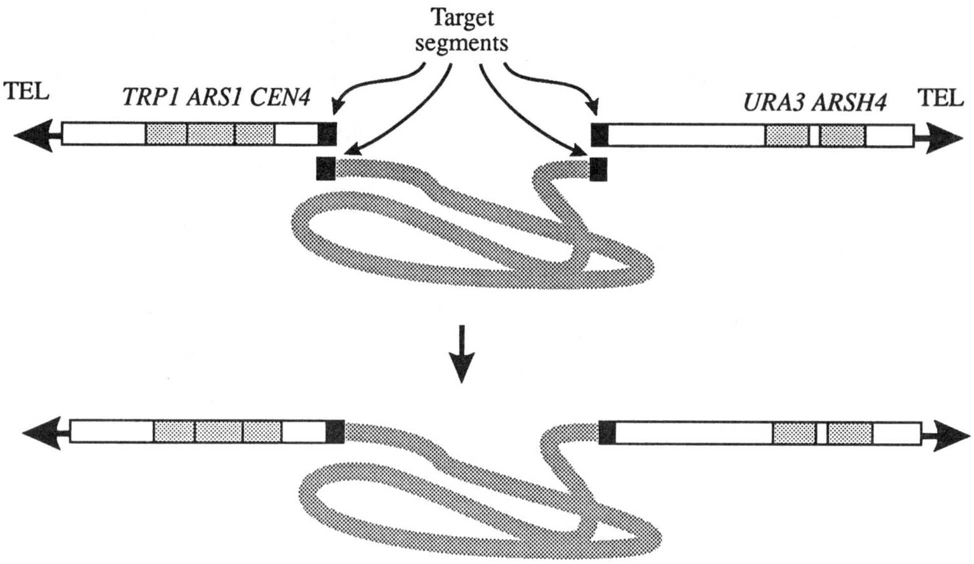

Figure 12. Recombinational targeted cloning with YAC vectors. A yeast strain is transformed with a mixture of the two YAC vector arms and large fragments of DNA. Recombination in vivo results in the formation of a specific YAC clone. The two YAC vector arms are derived from linearized plasmids that contain targeting segments homologous to the termini of the DNA segment to be cloned.

The potential value of YAC cloning technology has been enhanced by the ability to use homologous recombination for manipulating exogenous DNA in the yeast host. In recombinationally targeted YAC cloning, YACs are assembled in vivo, by recombination, and not by ligation in vitro. Recombination takes place between a target segment of the exogenous DNA, and the YAC vector that contain sequences homologous to these targets as illustrated in Figure 12. Transformation of the two YAC vector arms and the exogenous segment, flanked by the target segments, followed by recombination, results in the formation of the desired stable YACs. The specific target DNA segments for the YAC vector can be obtained from the exogenous DNA as restriction fragments or PCR products.

Also YACs can be modified after cloning by "retrofitting," using homologous recombination with yeast plasmids having targeting sequences. For example, a neomycin-resistant gene has to be incorporated into a YAC that will be transferred to mammalian cells using selection, as was done for transferring the entire human β-globin gene in embryonic stem cells. Also, overlapping YAC clones can be recombined, resulting in larger clones encompassing more extensive regions. Furthermore, special YAC vectors have been designed for generating terminal and internal deletions of cloned YAC inserts.

YACs have been useful for cloning not only genes but also mammalian telomeric and centromeric regions, and chromosomal origins of replication.

13.3 Expression of Heterologous Proteins in Yeast

Although *E. coli* is still the first choice for the producer of heterologous proteins, the yeast *S. cerevisiae* has some attractive features. Proteins produced in yeast, unlike those produced in *E. coli,* lack endotoxins. In certain special cases, such as hepatitis B core antigen, the products produced in yeast have higher activity than those produced in *E. coli.* In contrast with using *E. coli,* several posttranslational processing mechanisms available in yeast have allowed the expression of several human or human pathogen associated proteins with appropriate authentic modifications. Such posttranslational modifications include particle assembly, amino-terminal acetylation, myristylation, and proteolytic processing. In addition, heterologous proteins secreted from specially engineering strains are correctly cleaved and folded and are easily harvested from yeast culture media. The use of either homologous or heterologous signal peptides has allowed authentic maturation of secreted products by the endogenous yeast apparatus.

To underscore the importance of yeast for the production of protein products by recombinant DNA methods, one can point out that the first approved human vaccine, hepatitis B core antigen, and the first food product, rennin, were produced in yeast.

The cloning of specific cDNAs from other organisms and the study of their function using yeast as a surrogate does not necessarily require high level expression of the foreign protein. In these instances, the aim is just to produce physiological quantities of the protein in a form that is correctly modified and localized in the cell such that the activity accurately reflects the activity in the original organism. However, commercial and laboratory preparations of proteins generally require high expression vectors.

There are numerous varieties of expression vectors currently available for producing heterologous proteins in yeast, and these are derivatives of the YIp, YEp, and YCp plasmids described in Section 9. The cDNA, synthetic DNA, or genomic DNA lacking introns is inserted in the vector. Promoters used in expression vectors include a transcription initiation site and variable amounts of DNA encoding the 5′-untranslated region. Because most of the yeast expression vectors do not contain an ATG in the transcribed region of the promoter, the heterologous gene must provide an ATG that establishes the correct reading frame corresponding to the amino terminus of the protein. It is essential that this ATG correspond to the first AUG of the mRNA, because translation almost always initiates at the first AUG on mRNAs from yeast as well as from other eukaryotes. The 5′-untranslated region of the vector also should be similar to the naturally occurring leader region of abundant mRNAs by lacking secondary structures and being A rich, and G deficient, and by having an A at position −3 relative to the ATG translational initiator codon.

Many of the expression vectors include a known signal for 3′ end formation of yeast mRNA, although vectors lacking such defined signals synthesize transcripts until encountering a 3′ end forming signal from another gene or a fortuitous signal on the plasmid.

Numerous normal and altered yeast promoters have been used, and these are chosen because of their high activity and sometimes because of their regulatory properties. Some of the promoters have been derived from genes encoding alcohol dehydrogenase I, enolase, glyceraldehyde 3-phosphate dehydrogenase, phosphoglycerate kinase, triose phosphate isomerase, galacokinase (P_{GAL1}, see Section 10.3), repressible acid phosphatase, and α-mating factor. These promoters almost always produce high levels of transcription of heterologous gene, but there is wide variation in the amount of the corresponding proteins finally produced in the yeast strain, depending on the specific heterologous gene.

See also Expression Systems for DNA Processes; Mitochondrial DNA; Plasmids; Yeast, DNA Repair in; Yeast Artificial Chromosome (YAC) Construction and Applications.

Bibliography

Broach, J. R., Jones, E. W., and Pringle, J. R., Eds. (1991) *The Molecular and Cellular Biology of the Yeast Saccharomyces,* Vol. 1, *Genome Dynamics, Protein Synthesis, and Energetics.* Cold Spring Harbor Laboratory Press, Cold Spring Harbor, NY.

Fields, S., and Johnson, M., Eds. (1993) Yeast methods for analyzing proteins from other organisms. *Methods,* 5:77–187.

Guthrie, C., and Fink, G. R., Eds. (1991) *Methods in Enzymology,* Vol. 194, *Guide to Yeast Genetics and Molecular Biology.* Academic Press, New York.

Hall, M. N., and Linder, P., Eds. (1992) *The Early Days of Yeast Genetics.* Cold Spring Harbor Laboratory Press, Cold Spring Harbor, NY.

Heslot, H., and Gaillardin, C. (1992) *Molecular Biology and Genetic Engineering of Yeast.* CRC Press, Boca Raton, FL.

Johnston, J. R., Ed. (1994) *Molecular Genetics of Yeast, A Practical Approach.* IRL Press, Oxford.

Jones, E. W., Pringle, J. R., and Broach, J. R., Eds. (1992) *The Molecular and Cellular Biology of the Yeast Saccharomyces,* Vol. 2, *Gene Expression.* Cold Spring Harbor Laboratory Press, Cold Spring Harbor, NY.

Sherman, F. (1991) Getting started with yeast. In *Methods in Enzymology,* Vol. 194, *Guide to Yeast Genetics and Molecular Biology,* C. Guthrie and G. R. Fink, eds., pp. 3–21. Academic Press, New York.

Watson, J. D., Hopkins, N. H., Roberts, J. W., Steitz, J. A., and Weiner, A. M. (1987) *Molecular Biology of the Gene,* Chapters 18 and 19. Benjamin/Cummings, Menlo Park, CA.

Wheals, A. E., Rose, A. H., and Harrison, J. S., Eds. (1995) *The Yeast,* Vol. 6, *Yeast Genetics,* 2nd ed. Academic Press, London.

Z

Zinc Finger DNA-Binding Motifs

Louise Fairall and John W. R. Schwabe

1 **Introduction**
2 **Classical Zinc Fingers (TFIIIA Type)**
3 **Nuclear Hormone Receptors and Related Motifs**
4 **Gal4 and Related Proteins**
5 **Nucleocapsid Cysteine-Rich Domain**
6 **Other Zinc-Binding Motifs**
6.1 TFIIS
6.2 LIM Domain
6.3 RING Finger
6.4 Cysteine-Rich Domain of Protein Kinase C (PKC CRD)
6.5 Methionyl–tRNA Synthetase Zinc-Binding Domain (MetRS)
7 **Perspectives**

Key Words

α-Helix and β-Sheet Regions of regular structure in proteins where the peptide backbone traces, respectively, a right-handed helical path or a straight extended path with one or more adjacent strands.

Extended X-Ray Absorption Fine Structure (EXAFS) Analysis A technique that explores the environment of a metal ion by comparing the X-ray absorption profile of a sample with that of a model compound.

Major and Minor Grooves of DNA Spiral "grooves" in the DNA that arise through the helical nature of the molecule. At any one position, the two grooves are on opposite faces of the helix. Each base pair is accessible from both grooves. The minor groove is generally the narrower.

Mutagenesis The use of genetic engineering to produce proteins whose amino acid sequence differs from that found in vivo.

Proteolysis The use of enzymes to digest the exposed flexible regions of a protein, leaving a structured core.

Transcription Factor (TF) A protein involved in the regulation of the transcription of DNA into messenger RNA (often involves binding to DNA).

Xenopus laevis The African clawed toad—an amphibian species used widely both as a tool and as a model system in molecular biology research.

The term "zinc finger" was first used to describe a 30-residue, repeated-sequence motif that has been found to be extraordinarily widespread in eukaryotic DNA-binding proteins. This is now known as the classical zinc finger motif. Consistent with early predictions, it has emerged that these motifs form independent minidomains folded around a central zinc ion. Adjacent zinc fingers are strung together to make up modular DNA-binding domains.

In the same way that the classical zinc finger motif was identified as a repeated pattern of cysteine and histidine residues, similar patterns of conserved potential metal ligands were found in a number of other families of proteins, some of which are also DNA-binding proteins. It was suggested that these might be variants or different classes of the zinc finger domain. We now know the structure of the classical zinc fingers along with the structures of many of these other zinc-binding domains. Furthermore, for those that bind to DNA, the details of their interaction with DNA have been established. From these structures it is clear that whereas some of the domains are structurally related, others are distinct from each other.

1 INTRODUCTION

For many years biologists have sought to understand the mechanisms by which genes are specifically switched on and off during development and in response to external stimuli. It is now established that the specificity of this regulation is largely brought about by transcription factors binding, highly selectively, to specific DNA-binding sites associated with their target genes. The selective recognition of these short DNA sequences is often achieved via discrete DNA-binding domains within these transcription factors. It has emerged that nature has employed comparatively few different designs of DNA-binding domains. Rather, it has tinkered with these few designs to produce families of related proteins, different members of which bind to distinct sequences. We focus on a number of DNA-binding motifs that have in common the use of zinc to stabilize their folded structure.

Interest in zinc as an integral part of small protein domains began in 1985 with the discovery of the classical zinc finger motif (see Section 2). Following this, other miniprotein domains were identified that apparently also required zinc as a structural accessory. These discoveries spurred structural biologists to apply to these systems the biophysical tools of NMR spectroscopy and X-ray crystallography. NMR spectroscopists have rapidly exploited the small size of these domains to facilitate the structure determination of many of these proteins in solution. Crystallographers have focused on the complexes of several of the proteins with their DNA targets. More recently, ambitious NMR studies have contributed in the latter area with structures of two additional complexes.

2 CLASSICAL ZINC FINGERS (TFIIIA TYPE)

The classical (Cys_2His_2) zinc finger motif was first observed in 1985 as a repeated 30 amino acid sequence motif in the protein tran-

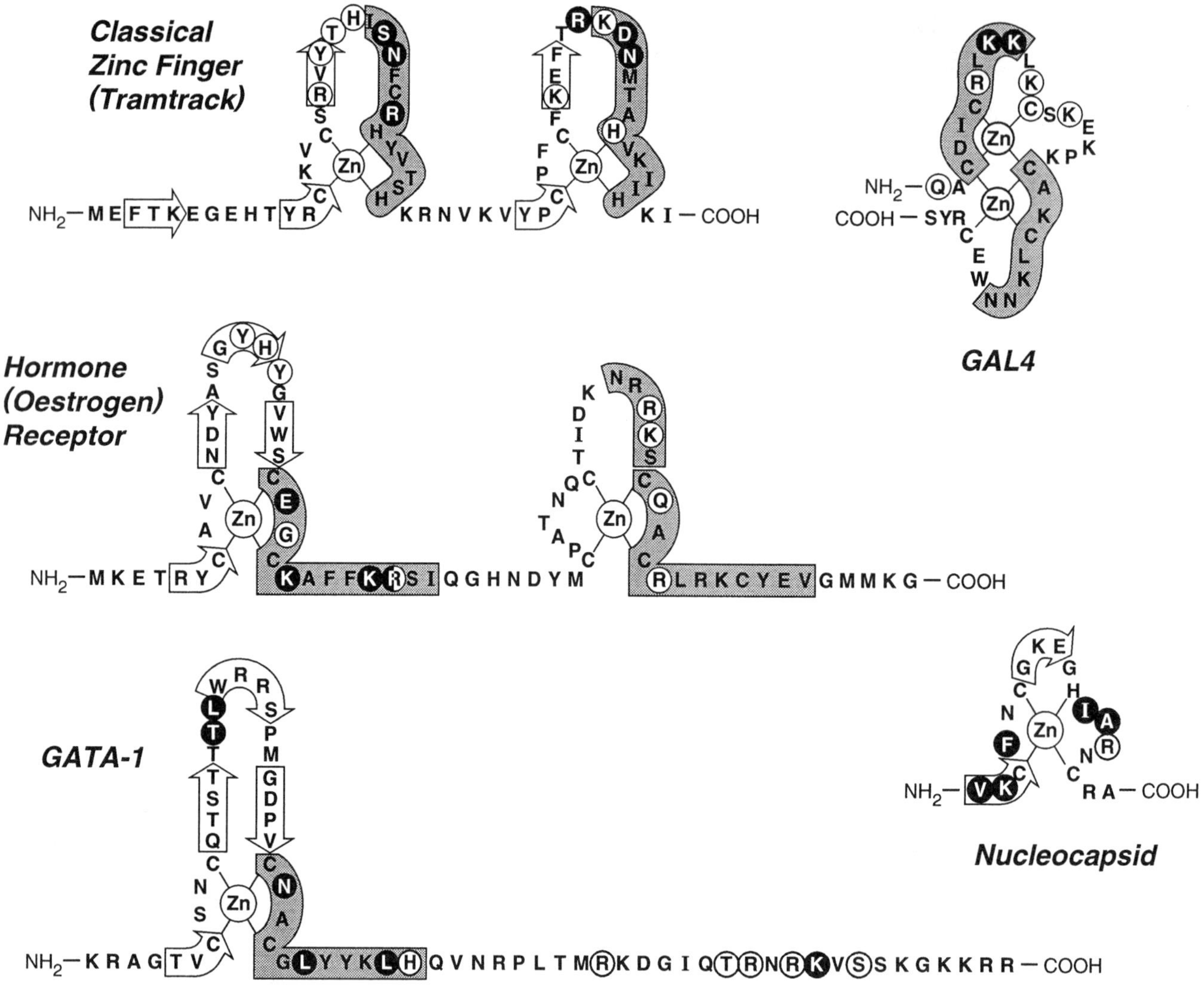

Figure 1. Classes of zinc-containing DNA-binding motif, showing the pattern of zinc coordination. The protein sequence is indicated by the one-letter code. Shaded regions fold to form α-helices. Arrows indicate β-strands. Residues that contact DNA are circled (phosphate contacts in white; base contacts in black) Structures shown in the following figures are generated as cartoons in the program Molscript [P. J. Kraulis, *J. Appl. Crystallogr.* 24:946–950 (1991).]

scription factor TFIIIA from the toad *Xenopus laevis.* This zinc finger motif is defined by two cysteine and two histidine residues that ligate a zinc ion (arranged: $CysX_{(2-5)}CysX_{12}HisX_{(2-5)}His$, where X represents any amino acid) and three conserved hydrophobic residues (Figure 1). The observation of this motif, combined with proteolysis studies and analysis of metal content, led to the proposal that this sequence would form an independent minidomain folded around a central zinc ion. This proposal was quickly substantiated by extended X-ray absorption fine structure (EXAFS) analysis of the environment of the zinc ion, which showed that the ion was tetrahedrally ligated by the two cysteine and two histidine residues of the motif.

The repeated nature of this motif is a feature common to all proteins containing classical zinc fingers, but the number of repeats varies widely among the proteins ranging from 2 (in, e.g., the *Drosophila* protein Tramtrack) to 37 (in the *Xenopus* protein Xfin). Since these motifs have been observed in a very large number of eukaryotic protein sequences, the zinc finger appears to be a widespread structural motif for nucleic acid recognition. To date, more than 200 different cDNA sequences have been found to encode zinc finger motifs, amounting to more than 1200 individual zinc fingers. Indeed it has been estimated that between 300 and 700 human genes (nearly 1% of the human genome) encode zinc finger containing proteins.

Prior to the completion of structural studies, model building was used to predict (with hindsight, successfully) the structure of the zinc finger domain based on the structures of other zinc-binding proteins. However, it was not until the structure of a single zinc finger from the *Xenopus* protein Xfin was solved, using NMR spectroscopic techniques, that direct evidence revealed that each 30-residue motif folds to form an independent domain. The folded structure of this domain consists of an irregular antiparallel, two-stranded β-sheet packed against a short α-helix held together by a single zinc ion and the three conserved hydrophobic residues (Figure 2a).

More recently, NMR studies of constructs with two adjacent zinc

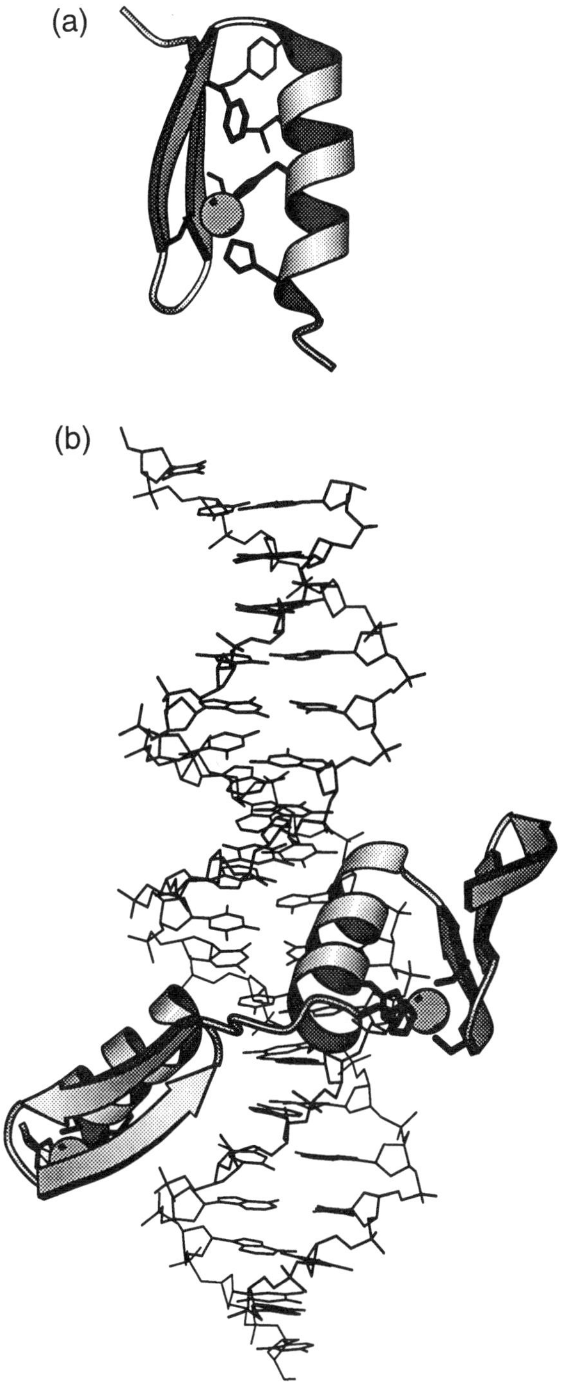

Figure 2. (a) The solution structure, determined by NMR spectroscopy, of the second of three classical zinc finger motifs in the yeast transcription factor SWI5. The zinc ion and ligands are shown. [D. Neuhaus, Y. Nakaseko, J. W. R. Schwabe, and A. Klug, *J. Mol. Biol.* 228:637–651 (1992).] (b) The crystal structure of the two classical zinc finger motifs of the fruit fly transcription factor Tramtrack in complex with DNA. [L. Fairall, J. W. R. Schwabe, L. Chapman, J. T. Finch, and D. Rhodes, *Nature,* 366:483–487 (1993).]

finger domains have demonstrated that there is little or no interaction between adjacent domains in solution and that the linker between domains is flexible. Nevertheless, there is evidence that the sequence of the linker is important for DNA binding by TFIIIA. Furthermore, there is increasing evidence that the ββα structure in certain zinc fingers (especially the first finger in a run) may require additional structural elements for stability. For example, in the yeast protein SWI5 and in the *Drosophila* protein Tramtrack, an additional N-terminal β-strand is observed in the N-terminal zinc finger of the run. Thus the ββα structure may represent just the minimal core zinc finger domain.

Three crystal structures of the zinc finger DNA-binding domains in complex with their target sites have now been solved (the three zinc fingers of Zif268 from mice, the five zinc fingers of GLI from humans, and the two zinc fingers of Tramtrack from fruit flies: Figure 2b). These structures illustrate a modular mode of binding, with each zinc-finger domain binding with the N-terminus of its helix oriented directly into the major groove of the DNA. Amino acids on the surface of this helix make specific contacts to the base pairs, but primarily to only one of the two strands of the DNA.

The first of these structures (Zif268) suggested that there might be a recognition code that could be generally applicable to all classical zinc finger proteins. In this structure, one amino acid immediately preceding the α-helix and two amino acids on adjacent turns of the α-helix have the potential to contact the bases at three adjacent positions on one strand of the binding site (Figure 3a positions labeled −1, 3, and 6). The importance of these three positions was supported by a statistical analysis of many zinc finger sequences. The interactions seen in the structure of the Tramtrack zinc finger–DNA complex are very similar to those seen in the Zif268–DNA complex. One difference is that a conserved phosphate contact from the first histidine–zinc ligand is replaced by a phosphate contact from a tyrosine. The structure also demonstrates that given small distortions of the DNA structure, amino acids at other positions could contact the bases of the binding site (Figure 3a). The GLI–DNA structure shows considerable deviations from the interactions seen in the Zif268–DNA complex. GLI has five zinc fingers of which the N-terminal zinc finger makes no contact with the DNA. The second finger makes only one base contact and a few phosphate contacts, while the third makes only phosphate contacts. The fourth and the fifth fingers make extensive base contacts, using between them most of the positions in the α-helix. However the contacts made by finger 4 are primarily to the opposite strand of the DNA compared with those made by Zif268. Hence the recognition of DNA by zinc fingers is more complex than was perceived from the first crystal structure. However, the consensus of opinion seems to be that the GLI structure is likely to be the exception rather than the rule. The three structures do have two features in common: (1) in contrast to other DNA-binding domains, water-mediated intermolecular contacts appear to play little or no role in the sequence specificity, and (2) there are rather few contacts made to the sugar–phosphate backbone of the DNA.

Although zinc finger DNA recognition is more complicated than was first perceived, the simplicity of the first structure Zif268 suggested that the zinc finger might represent a candidate template for designing proteins to recognize DNA. Residues at the three key positions (−1, 3, and 6 as just described) have been the targets for mutagenesis experiments aimed at deriving a recognition code. More recently a catalog of recognition patterns for zinc fingers has been established using phage display techniques. This involves incorpo-

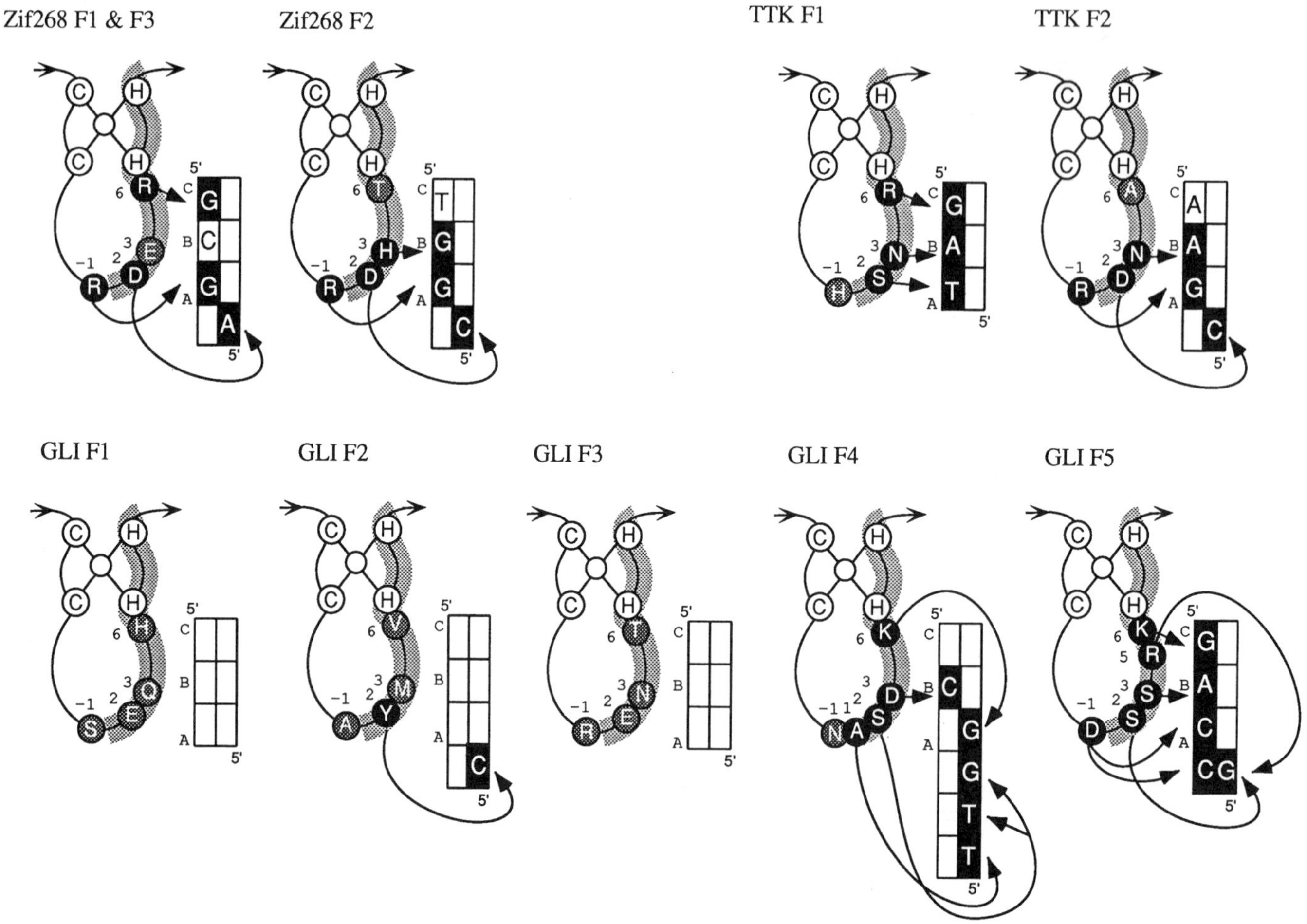

Position

Base	C	B	A
G	Arg(6) Ser(6)/*Asp(2*)* Thr(6)/*Asp(2*)*	His(3)	Arg(-1)/Asp(2)
A		Asn(3)	Gln(-1)/Ala(2)
T	Ser(6)/*Asp(2*)* Thr(6)/*Asp(2*)*	Ala(3) Ser(3) Val(3)	Asn(-1) Gln(-1)/Ser(2)
C		Asp(3) Leu(3) Thr(3) Val(3)	Asp(-1)

Figure 3. (a) The amino acid–base contacts seen in the crystal structures of Zif268 [N. P. Pavletich and C. O. Pabo, *Science* 252: 809–817 (1991)], GLI [N. P. Pavletich and C. O. Pabo, *Science* 261: 1701–1707 (1993)], and TTK. Each finger is illustrated separately (F1, F2, etc.) except for fingers 1 and 3 of Zif268, which are essentially identical. Amino acid positions relative to the start of the helical region (shaded) are numbered. The bases at the three positions thought to be most important for zinc finger recognition are labeled A, B, and C. Note that the first three fingers of GLI make virtually no base contacts. (b) Consensus zinc finger recognition code derived from screening a zinc finger library. The triplet position A, B, or C ($3' \rightarrow 5'$) is given at the top of the table and the base, at that position, down the side. Amino acids that can recognize a particular base at a particular position are given in the body of the table, with the helical position of that residue in brackets. An asterisk indicates a residue in the adjacent finger. [Y. Choo and A. Klug, *Proc. Natl. Acad. Sci. U.S.A.* 91:11168–11172 (1994).]

rating zinc finger sequences with randomized amino acids into the DNA recognition helix as fusions with the minor coat protein of bacteriophage fd. In this way, classical zinc finger proteins that recognize specific target sites of a particular sequence can be selected from a large library. Thus a catalog of zinc finger sequences can be generated to recognize specific DNA targets (Figure 3b). A potential use of the phage display technique has been demonstrated by the ability to select zinc fingers that recognize and block transcription of an oncogenic gene.

A number of important issues regarding zinc fingers remain unresolved. It is clear that the wrap-around mode of DNA binding would pose a topological problem for proteins containing long runs of zinc fingers. Since three zinc fingers wrap round a complete turn of the DNA double helix, it seems unlikely that proteins with more than five zinc fingers that are embedded in a larger protein could achieve such a mode of binding without crossing the minor groove. The biochemical data for TFIIIA suggest that the nine fingers might not bind equivalently, that groups of fingers may bind partially independently, and that the protein does cross the minor groove. The finding that the first finger of GLI does not bind to DNA is consistent with the idea that some zinc fingers in a protein might act as spacers, allowing the protein to span the minor groove.

A final question is that of how zinc fingers bind to RNA. TFIIIA binds to both RNA and DNA, and it has been suggested that proteins such as the *Xenopus* protein Xfin (with 37 zinc fingers) might be used to store RNA (perhaps maternal mRNA transcripts) and not to bind DNA. It is possible that the mode of binding to double-stranded RNA is not so different from that of binding to double-stranded DNA, since the DNA in all three crystal structures has some characteristics of A-DNA, and double-stranded RNA has a structure that is similar to A-DNA. But RNA usually exists folded into more complicated tertiary structures, and it has been established that the mode of TFIIIA binding to 5S RNA is different from binding to DNA. It appears that the protein has multifunctional fingers, with fingers 1–3 defining the sequence-specific binding to DNA and fingers 4–6 the specificity for 5S RNA.

3 NUCLEAR HORMONE RECEPTORS AND RELATED MOTIFS

The nuclear hormone receptors form a superfamily of ligand-activated transcription factors. This family includes receptors for the steroid hormones (e.g., estrogens and glucocorticoids), thyroid hormones, retinoids, and many others. Members of this family range from just over 400 to 1000 amino acids, but they all share a highly conserved DNA-binding domain (70–80 residues) that is characterized by eight conserved cysteine residues. The spacing of these residues is similar to that of the cysteine and histidine metal ligands in the classical zinc fingers (Figure 1). Consequently, it was suggested that the DNA-binding domain of the nuclear hormone receptors might contain two variant zinc finger domains.

Like the classical zinc fingers, it was found that the glucocorticoid receptor DNA-binding domain requires two zinc ions for DNA-binding activity. Each of these is bound by a tetrahedral arrangement of cysteine ligands. However, NMR spectroscopic analyses of the DNA-binding domains from the estrogen and glucocorticoid receptors in solution revealed that the two "zinc finger"–like motifs fold to form a single structural domain, hence are quite distinct from the classical modular zinc fingers. The structure of the estrogen receptor DNA-binding domain in solution is illus-

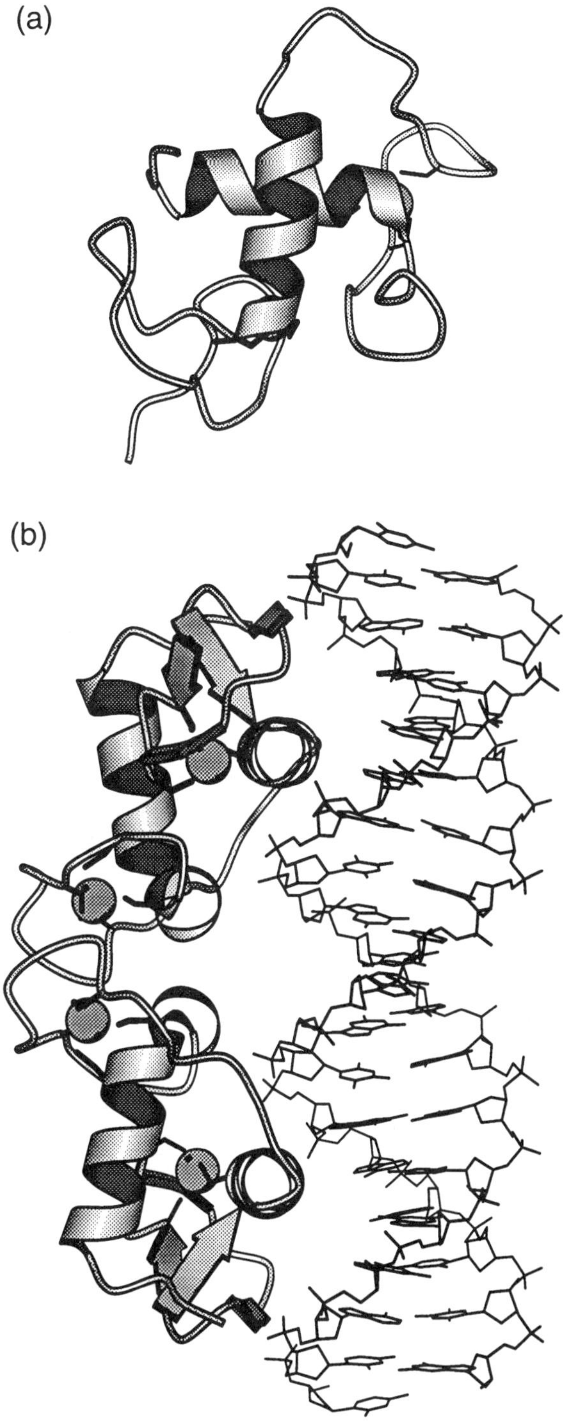

Figure 4. (a) The solution structure, determined by NMR spectroscopy, of the DNA-binding domain from the human estrogen receptor containing two class II zinc-binding motifs that are folded to form a single structural unit. The zinc ions and ligands are shown. [J. W. R. Schwabe, D. Neuhaus, and D. Rhodes, *Nature,* 348:458–461 (1990).] (b) The crystal structure of a dimer of the DNA-binding domain from the human estrogen receptor in complex with DNA. [J. W. R. Schwabe, L. Chapman, J. T. Finch, and D. Rhodes, *Cell,* 75:567–578 (1993).]

trated in Figure 4a. The dominant feature of this structure is an arrangement of two helices oriented perpendicularly to each other and crossing at their midpoints. At the N-terminus of each of these helices a zinc-binding pocket is formed from four conserved cysteines, two of which lie in the helix. One region of the protein is comparatively poorly ordered in solution. This region encompasses residues from the first ligands of the second zinc-binding site to the beginning of the second helix.

The two zinc-binding motifs have some similarities, but are clearly distinct. This seems to suggest that the two motifs have evolved as a duplication, but they appear to require each other for structural stability, which is different from the modular nature of the classical zinc finger. However the core structure of GATA-1, described shortly, is similar to the N-terminal zinc-binding motif of the steroid hormone receptors and yet is a stable entity, implying that the two zinc-binding motifs of the steroid hormone receptors probably arose as a duplication of a single domain resembling GATA-1, followed by diversification to perform different functions.

The DNA-binding domains of both the estrogen and glucocorticoid receptors are monomeric in solution yet bind highly cooperatively to DNA, forming symmetrical homodimers. Their DNA target sites consist of two conserved six base pair half-sites arranged as a palindrome with three intervening base pairs. The consensus half-sites recognized by the two receptors differ by only two out of the six base pairs (e.g., estrogen half-site, AG**GT**CA; glucocorticoid half-site, AG**AA**CA). The details of the protein–DNA interaction have been revealed in the crystal structures of the glucocorticoid and estrogen receptor DNA-binding domains in complex with DNA.

Figure 4b illustrates the structure of the estrogen receptor complex. Two molecules of the protein bind to adjacent major grooves, from one side of the DNA double helix. In common with many other DNA-binding domains, but in contrast to the classical zinc fingers, each of the proteins in the dimer straddles the major groove of the DNA double helix. The protein makes extensive contacts to the phosphate backbone on either side, orienting the DNA-binding domain such that the recognition helix ("reading head") enters the major groove, allowing surface side chains to make sequence-specific contacts to the base pairs. Comparison of the complexes of the estrogen and glucocorticoid receptor DNA-binding domains with their target sites reveals differences in the protein–DNA interface that are responsible for the discrimination of the two target sites. These differences are largely consistent with mutagenesis experiments demonstrating that residues at just three positions can switch the specificity of the estrogen receptor to that of the glucocorticoid receptor [glutamate, glycine, and alanine (E, G, and A) at positions 2, 3, and 6 from the N-terminus of the first helix (Figure 1) changed to glycine, serine, and valine]. Interestingly, the region of the protein that was disordered in solution forms the interface between the two proteins of the dimer when bound to DNA, as well as making phosphate contacts. Thus, with these isolated DNA-binding domains, a structured dimer interface is formed only upon binding to DNA.

The DNA targets for members of the nuclear receptor family are not all like those of the estrogen and glucocorticoid receptors. Some receptors bind as monomers to a single half-site, others bind as heterodimers. It appears that those that bind as heterodimers are able to distinguish the spacing between the two half-sites. It is not yet known how this is achieved, but the structure of the retinoid X receptor, obtained recently by means of NMR techniques, reveals an additional helix C-terminal to the structured domain in Figure 4a. It has been suggested that this may be part of a head-to-tail dimer interface.

GATA-1 is a member of a family of proteins characterized by the recognition of the DNA sequence GATA and by the presence of two copies of the sequence Cys-X_2-Cys-X_{17}-Cys-X_2-Cys separated by 29 residues. GATA-1 is an erythroid-specific transcription factor responsible for the regulation of transcription of erythroid-specific genes. Its target site is found in the regulatory elements of all globin genes and other erythroid-specific genes that have been analyzed.

Mutagenesis of GATA-1 has demonstrated that only one of the motifs described earlier is required for sequence-specific binding (Figure 1). This motif and 28 amino acids C-terminal to the last cysteine residue have been solved in complex with its DNA target site using multidimensional NMR techniques. The cysteine ligands are in the core of the domain, which is composed of two small, irregular, two-stranded, antiparallel β-sheets and an α-helix followed by a long loop that leads into the carboxyl-terminal tail (Figure 5a). The helix and the loop that joins the two β-sheets interact with the major groove of the DNA. The carboxyl-terminal tail, which contributes to the sequence-specific binding of the domain, wraps around and interacts in the minor groove (Figure 5b). The base and phosphate contacts for this domain are shown in Figure 1. Unlike other zinc-binding motifs, the DNA contacts are predominantly

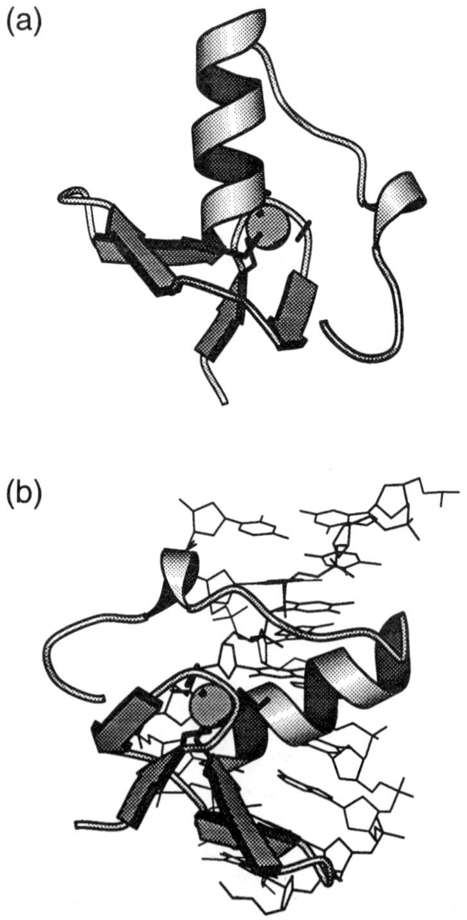

Figure 5. (a) The structure of the DNA-binding domain of GATA-1 taken from (b) the NMR structure of the domain complexed with DNA. [J. G. Omichinski, G. M. Clore, O. Schaad, G. Felsenfeld, C. Trainor, E. Appella, S. J. Stahl, and A. M. Gronenborn, *Science,* 261:438–446 (1993).]

from hydrophobic residues; in addition, there are extensive sugar contacts. The core structure of GATA-1 is similar to the N-terminal zinc-binding motif of the steroid hormone receptors described earlier in this section, a circumstance that has evolutionary implications.

4 GAL4 AND RELATED PROTEINS

GAL4 is the best-studied member of a group of more than 10 fungal transcription factors with conserved DNA-binding domains containing a 28-residue, cysteine-rich motif with the sequence Cys-X_2-Cys-X_6-Cys-X_6Cys-X_2-Cys-X_6-Cys (Figure 1). This motif ligates two zinc ions that are required for the structural integrity of the protein, hence for DNA-binding activity. However, an additional 26 amino acids C-terminal to the cysteine-rich motif are also required for DNA-binding activity. Although this motif has not been observed outside fungi, GAL4 has been found to be a potent activator of transcription in many heterologous systems.

The structure of the cysteine-rich domain from Gal4 was solved by NMR spectroscopy. As had been predicted, the conserved six cysteines of the motif ligate two zinc ions in a single cluster (Figure 1), with the first and fourth cysteines shared between the two zinc ions. The structure consists of two short α-helices each capped at their N-terminus by a pair of cysteine ligands: the first and second cysteines in the first helix and the fourth and fifth in the second helix (Figure 6a). These helices are held in a rigid conformation with respect to each other by the shared zinc ions. Each helix is followed by an extended peptide strand forming two helix–strand motifs with approximate twofold structural symmetry, although with rather low sequence homology.

X-ray structures of Gal4 and PPR1 have shown that the cysteine-rich domain acts as a DNA-binding module. Each module binds in the major groove of the DNA with the C-terminus of the first α-helix oriented toward the bases of its binding site. This is in contrast to the classical zinc fingers and the steroid hormone receptors, where the N-terminus of the helix is oriented toward the bases. The base-specific contacts are made by two adjacent lysines at the C-terminus of the first helix. Phosphate contacts are made by residues in the first helix–strand motif.

In solution, the DNA-binding domain of GAL4 is monomeric and the residues C-terminal to the cysteine-rich motif are flexible and disordered. In the crystal structures, however, the protein is bound to DNA as a dimer. Each DNA-binding module interacts in the major groove with a CGG base triplet. The spacing between the CGG triplets in the Gal4- and PPR1-binding sites is different, 11 and 6 bp, respectively. In Gal4 each DNA-binding module binds 1.5 turns of the double helix apart from its partner (Figure 6b) and is joined by an extended linker (which makes a number of phosphate contacts to an amphipathic α-helical dimerization domain) that forms a parallel coiled-coil structure with its partner in the dimer. The Gal4 coiled-coil structure is oriented perpendicular to the DNA, with the N-termini of the two helices positioned over the minor groove making additional contacts to the phosphate backbone. In PPR1 a similar coiled-coil dimer interface is formed, but its disposition relative to the DNA is significantly different as a consequence of the different spacing between the CGG base triplets. Comparison of the Gal4 and PPR1 structures shows that the different specificity of spacing is determined by the folds of the linkers and by the residues at the amino terminus of the coiled coils. In PPR1 the coiled coils do not contact the phosphate backbone. Hence, like the DNA-binding domain of the steroid hormone receptors, this DNA-binding motif is doubled up as a dimer when bound to DNA, such that the protein interacts with a greater number of base pairs as well as measuring the spacing between the two halves of the binding site.

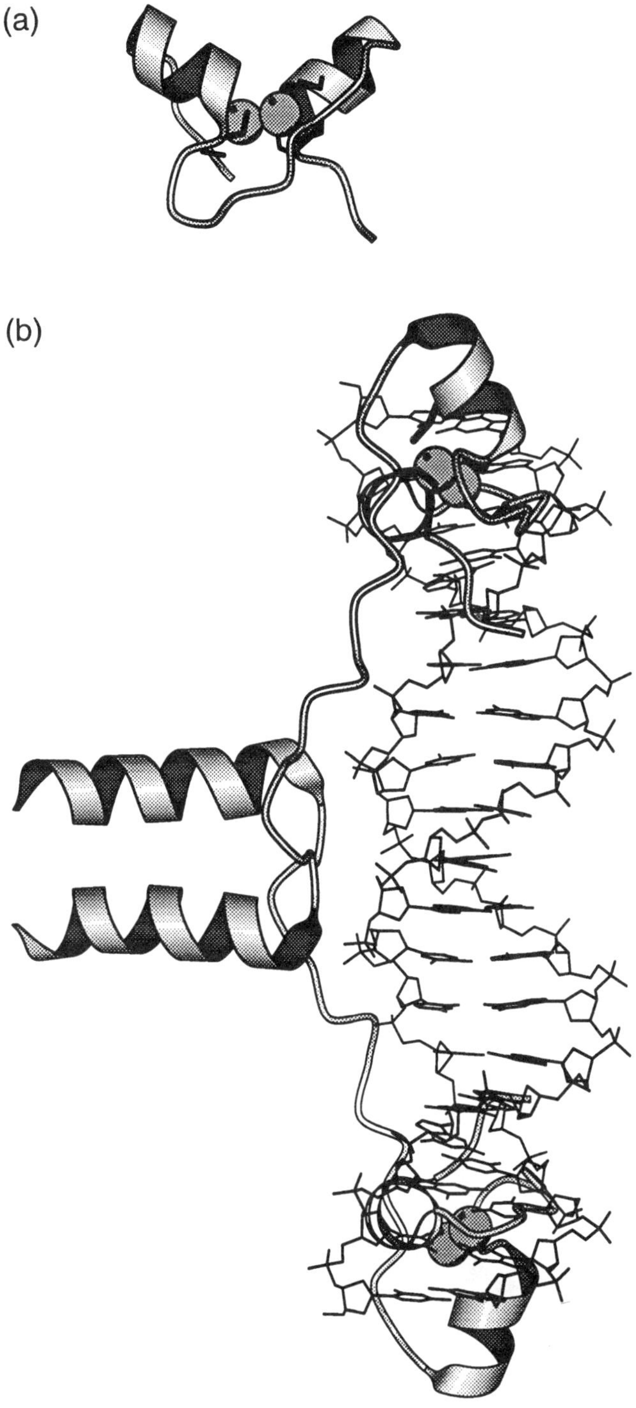

Figure 6. Crystal structures of (a) the cysteine-rich DNA-binding module of the yeast transcription factor GAL4, showing the zinc ions and ligands, and (b) a dimer of the complete DNA-binding domain from the yeast transcription factor GAL4 in complex with DNA. [R. Marmorstein, M. Carey, M. Ptashne, and S. C. Harrison, *Nature,* 356:408–414 (1992).]

5 NUCLEOCAPSID CYSTEINE-RICH DOMAIN

One or two copies of the zinc-binding motif Cys-X_2-Cys-X_4-His-X_4-Cys are found in all retroviral nucleocapsid proteins. Nucleocapsid proteins not only are involved in sequence-specific packaging of two copies of the single-stranded retroviral RNA but apparently have nonspecific RNA-binding activity to different targets during the course of the retroviral life cycle. Proper RNA packaging requires a signal in the viral genome of about 44 nucleotides known as the psi packaging signal. This site may serve as a primary signal for sequence-specific nucleocapsid RNA binding. The nucleocapsid protein also appears to be involved in viral RNA dimerization, tRNA primer hybridization, and nucleic acid unwinding during reverse transcription.

The structure of the nucleocapsid protein has been solved on its own and in complex with DNA using NMR techniques. In both structures the protein has an identical simple looped structure (Figure 7a). Interestingly this simple fold is structurally very similar to a small zinc-binding domain within the Met tRNA synthetase. The structure of the complex has been solved with a single-stranded piece of DNA (dACGCC) corresponding to part of the psi packaging signal of the RNA genome. In this structure the DNA has been modeled as A-DNA with fixed idealized geometry; hence the structure does not determine the details of the nucleic acid component of the complex. The structure was also modeled with a B-DNA conformation, but the A-DNA provided a better model in the calculations. Like the binding of GATA-1, the binding of the nucleocapsid protein involves many hydrophobic interactions as well as hydrogen bonds from some backbone atoms of the protein (Figure 1). Part of the nucleic acid recognition seems to be shape recognition involving a hydrophobic cleft (Figure 7b).

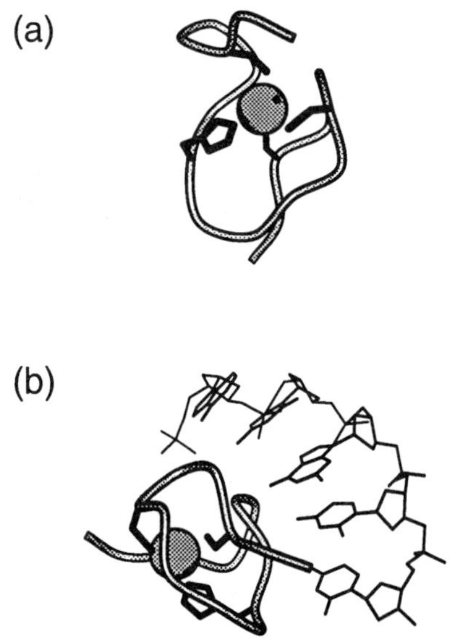

Figure 7. (a) The structure of the HIV nucleocapsid protein RNA-binding domain taken from (b) the NMR structure of the domain complexed with DNA. [T. L. South and M. F. Summers, *Protein Sci.* 2:3–19 (1993).]

6 OTHER ZINC-BINDING MOTIFS

The discovery of the classical zinc finger domain stimulated great interest in the role of zinc ions in stabilizing small structural domains, leading in turn to a recent explosion of new zinc-binding motifs, not all of which are involved in nucleic acid binding and some of which have unknown functions (Figure 8). The domains for which some structural information is available are described in Sections 6.1 through 6.5.

6.1 TFIIS

Transcription factor IIS (TFIIS) is an elongation factor required for polymerase II transcription. It contains a zinc-binding motif with four cysteines in the form Cys-X_2-Cys-X_{24}-Cys-X_2-Cys. The NMR structure of this domain shows that the two N-terminal cysteine residues are located in a loop, which is packed against a three-stranded, antiparallel β-sheet. The other two ligands are in another loop between the second and third β-strands (Figure 1). While intact TFIIS has not been shown to bind to nucleic acid, the C-terminus, which contains the zinc-binding motif described earlier, does bind to ssDNA, dsDNA, ssRNA, and RNA/DNA hybrids. It is proposed that the native protein folds in a way that serves to block nucleic acid binding by the zinc-binding motif, but in the elongation complex, the protein undergoes a conformational change to enable the zinc-binding motif to recognize nucleic acid.

6.2 LIM Domain

The LIM motif is a cysteine-rich domain that was discovered by comparison of the sequences of Lin-11, Isl-1, and Mec-3, hence the name. It has now been found in more than 20 proteins and tends to occur as two tandem copies in the N-terminal region of these proteins. While many of these proteins are DNA-binding transcription factors (containing the homeo-domain), this motif has not been shown to bind to nucleic acid. Biochemical studies suggest that it is involved in protein–protein interactions.

The LIM domain consists of seven conserved cysteine residues and a histidine arranged as follows:

$$\text{Cys-X}_2\text{-Cys-X}_{16-23}\text{-His-X}_2\text{-Cys-X}_2\text{-}$$
$$\text{Cys-X}_2\text{-Cys-X}_{16-21}\text{-Cys-X}_{2-3}\text{-Cys}$$

The structure of the domain from the avian cysteine-rich protein (CRP) has been solved by NMR spectroscopy. The domain folds to form a single structural unit containing two independent zinc-binding sites. Each domain contains two two-stranded β-sheets packed perpendicular to each other. The C-terminal zinc-binding motif contains an α-helix at its C-terminus (Figure 8). The global fold of this C-terminal zinc-binding site is very similar to that of GATA-1 and the N-terminal zinc-binding site of the steroid hormone receptors. The two zinc-binding domains of the LIM domain are very similar to each other, leading researchers to speculate that the two zinc-binding sites may have arisen as an early duplication of a single motif resembling the GATA-1 domain.

6.3 RING Finger

A number of eukaryotic and viral proteins have been found to contain a conserved cysteine-rich domain that is known as the RING

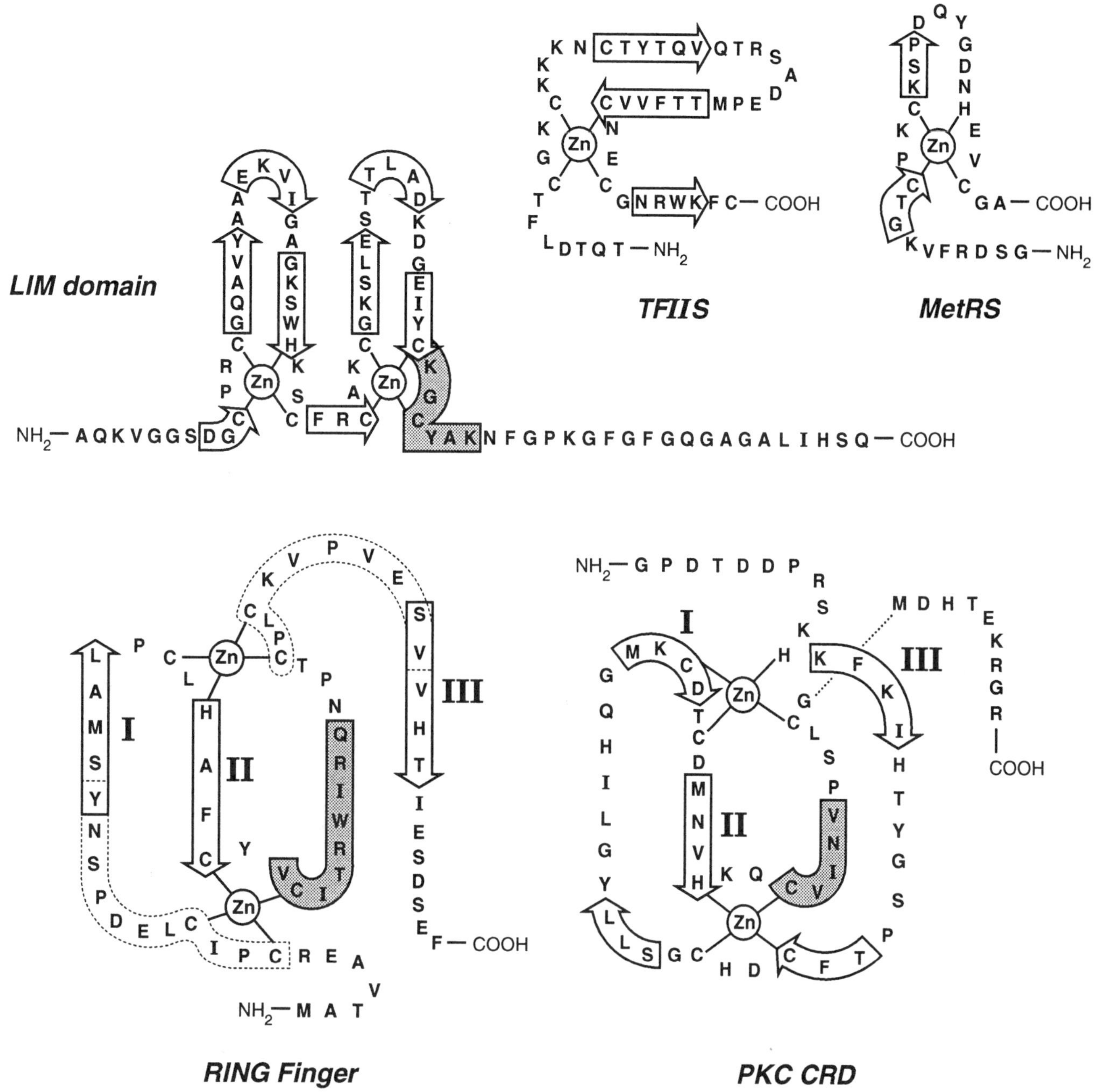

Figure 8. More classes of zinc-binding motif, showing the pattern of zinc coordination. The protein sequence is indicated by the one-letter code. Shaded regions fold to form α-helices. Arrows indicate β-strands.

(Really Interesting New Gene) finger. The spacing of the cysteines in this domain is as follows:

$$\text{Cys-X}_2\text{-Cys-X}_{11-27}\text{-Cys-X-His-X}_2\text{-}$$
$$\text{Cys-X}_2\text{-Cys-X}_{6-17}\text{-Cys-X}_2\text{-Cys}$$

Accordingly, this domain is also known as the C_3HC_4 zinc finger.

The structure of the RING finger domain from the equine herpes virus type I immediate–early gene product, EHV-63, which has been solved by NMR analysis, consists of an amphipathic α-helix that lies along one surface of a triple-stranded β-sheet. Unlike all the domains described hitherto, the zinc ligands are interleaved: that is, the first zinc is bound to the first and third pair of ligands and the second zinc is bound to the second and fourth pair of ligands (Figure 8). The RING finger domain does not appear to bind specifically to DNA or RNA. It has been demonstrated that the polar side chains of the α-helix are required for trans-activation of gene expression by an intact herpes virus regulatory protein.

6.4 Cysteine-Rich Domain of Protein Kinase C (PKC CRD)

Protein kinase C plays a major role in the transfer of extracellular signals into intracellular signals by phosphorylation. It contains a cysteine-rich, zinc-binding domain (CRD) that is homologous to the proteins of the *raf-mos* family of oncogenes, which are the receptors for tumor-promoting phorbolesters. The NMR structure of this domain from the rat protein kinase C shows that it has some similarities to the RING finger motif in that both motifs consist of an α-helix packed against a three-stranded β-sheet. However the positions of the entry and exit points of the N- and C-termini are significantly different (Figure 8).

6.5 Methionyl–tRNA Synthetase Zinc-Binding Domain (MetRS)

Methionyl–tRNA synthetase (MetRS) from *E. coli* contains a zinc-binding domain. The 28 amino acids that constitute the domain were expressed in *E. coli* and the structure solved by NMR spectroscopy. This zinc-binding domain has four cysteine zinc ligands with a simple structure consisting of four tight turns (Figure 8) and similar to that structure of the HIV nucleocapsid zinc-binding domain (Figure 7). However, it is not known whether the Met tRNA synthetase zinc-binding domain has any nucleic acid binding activity.

7 PERSPECTIVES

The role of zinc as an electrophilic catalyst to activate water molecules has been studied extensively in the zinc-containing enzymes. In addition to its catalytic role, zinc has also been observed to act to stabilize parts of the enzyme structure (e.g., in alcohol dehydrogenase). Similarly, in some DNA-binding proteins with largish domains, zinc is incorporated as part of the structural framework of the protein. This is seen in the crystal structures of the tumor suppressor protein p53 and adenovirus single-stranded DNA-binding protein. The special feature of zinc fingers, however, is that the zinc stabilizes a small modular functional domain that would otherwise be unstable. This is perhaps the best definition of a zinc finger motif.

At one time or another, all the motifs in this article have been described as zinc finger motifs; yet in fact, only the classical zinc finger protein TFIIIA has been shown to ligate zinc in vivo. Both GAL4 and the hormone receptors are found to contain zinc ions when overexpressed in bacteria, but it is known that other metals ions (e.g., Cd, Co) can substitute for the zinc. Nevertheless, it is a fairly safe assumption that zinc is used in vivo, since it is the most abundant metal in cells that is suitable for such a structural role.

Nature has presumably employed the services of zinc coordination as a structural buttress in intracellular proteins, since the highly reducing environment inside the cell makes disulfide bonds (common in extracellular proteins) unsuitable for this task. This is particularly advantageous in small protein domains, which are inherently unstable and cannot sequester disulfides in the interior of a large protein.

Remarkably, despite the general diversity of the zinc finger motifs, the local stereochemistry around the zinc-binding sites is very similar in a number of the proteins that utilize a pair of cysteine ligands arranged as follows: CXXCXX. In this motif there is a type I turn between the two cysteine ligands. Hydrogen bonds orient the cysteine side chains through fixing the sulfur atoms in relation to backbone amides. Particularly favorable are hydrogen bonds between Cys-SG(i) and X-NH($i + 2$). The motif has been termed the rubredoxin knuckle from the iron-binding protein where it was first observed. The rubredoxin "knuckle" requires that the backbone be extended before and after the zinc-binding site. This type of structure appears to occur in every one of the zinc-binding folds with the exception of Gal4.

The zinc-binding motifs described in this article represent some quite distinct structural solutions to the same biological task of directing specific DNA recognition. Most of the zinc-binding, DNA-binding domains make use of an α-helical element as a "reading head" in the major groove of the double helix. The α-helix is a rigid structure, on which to build the various amino acid side chains that confer the specificity of binding; hence there is little or no entropic cost in DNA binding by an α-helix. The orientation of this reading head is markedly different in the classes of zinc-binding motif. In the classical zinc finger, the N-terminus points down into the major groove. In GAL4 the helix has an almost opposite orientation, with the C-terminus of the helix pointing into the major groove. The recognition helix in the hormone receptors shows an orientation between these two extremes. The variation in recognition helix orientation is a consequence of the different background structures in these protein families. Since this in turn means that the surface amino acids are presented to DNA in different ways in these classes of protein, it is difficult to envisage any recognition code that could be generally applicable across these different classes of domain.

See also Chromatin Formation and Structure; Gene Expression, Regulation of; Protein Designs for the Specific Recognition of DNA.

Bibliography

Berg, J. M., and Shi, Y. G. (1996) The galvanization of biology—a growing appreciation for the roles of zinc. *Science* 271: 1081-1085.

Coleman, J. E. (1992) Zinc proteins: Enzymes, storage proteins, transcription factors, and replication proteins. *Annu. Rev. Biochem.* 61:897–946.

Klug, A., and Rhodes, D. (1987) "Zinc fingers": A novel protein motif for nucleic acid recognition. *Trends Biochem. Sci.* 12:464–469.

———, and Schwabe, J. W. R. (1995) Protein motifs. 5. Zinc fingers. *FASEB J.* 9: 597–604.

Pabo, C., and Sauer, R. (1992) Transcription factors: Structural families and principles of DNA recognition. *Annu. Rev. Biochem.* 61:1053–1095.

Rhodes, D., and Klug, A. (1993) Zinc fingers. *Sci. Am.* 268:56–65.

Glossary of Basic Terms

The most basic terms in molecular biology are defined below. These, in combination with the key words listed at the head of each article, provide definitions of all essential terms in this desk reference.

Alleles Alternative forms of a given gene, inherited separately from each parent, differing in nucleotide base sequence and located in a specific position on each homologous chromosome, affecting the functioning of a single product (RNA and/or protein).

Amplification (gene) The process of replication of specific DNA sequences in disproportionately greater amounts than are present in the parent genetic material, e.g., PCR is an in vitro amplification technique.

cDNA (complementary DNA) A DNA copy of an RNA molecule synthesized from an mRNA template in vitro using an enzyme called reverse transcriptase; often used as a probe.

Chromatin The complex of nucleic acids (DNA and RNA) and proteins (histones) comprising eukaryotic chromosomes.

Chromosome In prokaryotes, the usually circular duplex DNA molecule constituting the genome; in eukaryotes, a threadlike structure consisting of chromatin and carrying genomic information on a DNA double helix molecule. A viral chromosome may be composed of DNA or RNA.

Cloning Asexual reproduction of cells, organisms, genes or segments of DNA identical to the original.

Cloning vector See: *vector*.

Complementary base pairing Nucleic acid sequences on paired polymers with opposing hydrogen bonded bases adenine (designated A) bonded to thymine (T), guanine (G) to cytosine (C) in DNA and adenine to uracil (U) replacing adenine to thymine in RNA.

DNA (deoxyribonucleic acid) The molecular basis of the genetic code consisting of a poly-sugar phosphate backbone from which project thymine, adenine, guanine and cytosine bases. Usually found as two complementary chains (duplex) forming a double helix associated by hydrogen bonds between complementary bases.

DNA polymerase Enzymes that catalyze the replication of DNA from the deoxyribonucleotide triphosphates using single- or double-stranded DNA as a template.

E. coli (Escherichia coli) A colon bacillus which is the most studied of all forms of life and whose genome is presently the best sequenced and mapped.

Eukaryote An organism (or cell) whose cells contain a true nucleus; all living matter except viruses, bacteria and blue-green algae.

Expression The process of making the product of a gene, which is either a specific protein giving rise to a specific trait or RNA forms not translated into proteins (e.g., transfer ribosomal RNAs).

Gene A DNA sequence, located in a particular position on a particular chromosome, which encodes a specific protein or RNA molecule.

Heterozygous Having two different alleles for a given trait in the homologous chromosomes.

Homologies Similarities in DNA or protein sequences between individuals of the same species or among different species.

Homologous chromosomes Chromosome pairs, each derived from one parent, containing the same linear sequence of genes, and as a consequence, each gene is present in duplicate (e.g., humans have 23 homologous chromosome pairs but the toad has 11 pairs, the mosquito has three pairs, and so on).

Homozygous Having two identical alleles for a given trait in the homologous chromosomes.

Hybridization The formation of a double-stranded polynucleotide molecule when two complementary strands are brought together at moderate temperature. The strands can be DNA or RNA or one of each; a technique for assessing the extent of sequence homology between single strands of nucleic acids.

Ligation The formation of a phosphodiester bond to join adjacent terminal nucleotides (nicks) to form a longer nucleic acid chain (DNA of RNA); catalyzed by ligase.

Marker A gene or a restriction enzyme cutting site with a known location on a chromosome and a clear-cut phenotype (expression), or pattern of inheritance, used as a point of reference when mapping a new mutant.

mRNA (messenger RNA) RNA used to translate information from DNA to ribosome where the information is used to make one or several proteins.

Nucleotide The monomer which, when polymerized, forms DNA or RNA. It is composed of a nitrogenous base bonded to a sugar (ribose or deoxyribose) bonded to a phosphate.

Oligonucleotide A polynucleotide 2 to 20 nucleotide units in length.

Operon A series of prokaryote genes encoding enzymes of a specific biosynthesis pathway and transcribed into a single RNA molecule.

Plasmid An extrachromosomal circular DNA molecule found in a variety of bacteria encoding "dispensable functions," such as a resistance to antibiotics. Often found in multiple copies per cell and reproduces every time the bacterial cell reproduces. May be used as a cloning vector.

Polymorphism Difference in DNA sequence among individuals expressed as different forms of a protein in individuals of the same interbreeding population.

Polynucleotide The polymer formed by condensation of nucleotides.

Probe A radioactively, fluorescent or immunologically labeled oligonucleotide (RNA or DNA) used to detect complementary sequences in a hybridization experiment (e.g., identify bacterial colonies that contain cloned genes or detect specific nucleic acids following separation by gel electrophoresis.

Prokaryote An organism that lacks a true nucleus, a bacterium, virus or blue-green algae.

Replication The copying of a DNA molecule duplex yielding two new DNA duplex molecules, each with one strand from the original DNA duplex. Single-stranded DNA replication results in a single-stranded DNA molecule.

Repressor A protein that binds to a specific location (operator) on DNA and prevents RNA transcription from a specific gene or operon.

Restriction mapping Uses restriction endonuclease enzymes to produce specific cuts (cleavage) in DNA, allowing preparation of a genome map describing the order and distance between cleavage sites.

Reverse transcription The synthesis of cDNA from an RNA template as catalyzed by reverse transcriptase.

RFLP (restriction fragment length polymorphism) DNA fragment cut by enzymes specific to a base sequence (restriction endonuclease) generating a DNA fragment whose size varies from one individual to another. Used as markers on genome maps and for screening for mutations and genetic diseases.

Ribosomes Small cellular components composed of proteins plus ribosomal RNA that translate the genetic code into synthesis of specific proteins.

RNA (ribonucleic acid) A single-stranded polynucleotide with a phosphate oxyribose backbone and four bases that are identical to those in DNA, with the exception that the base uracil is substituted for thymine.

RNA polymerase The enzyme (peptide) that binds at specific nucleotide sequences, called promoters, in front of genes in DNA, catalyzing transcription of DNA to RNA.

Transcription Synthesis of an RNA molecule from a DNA template (gene) catalyzed by RNA polymerase.

tRNA (transfer RNA) RNA molecules that transport specific amino acids to ribosomes into position in the correct order during protein synthesis.

Vector DNA molecule originating from a virus, a plasmid or cell of a higher organism into which another DNA fragment can be integrated without loss of the vector's capacity for self-replication. Vectors introduce foreign DNA into host cells where it can be reproduced in large quantities.

Cumulative Contents

Volume 1

Preface vii

Acknowledgments ix

Contributors xi

Achilles' Cleavage 1
Michael D. Koob

Aging, Genetic Mutations in 5
James E. Fleming

AIDS, Inhibitor Complexes of HIV-1 Protease in 14
Alexander Wlodawer

AIDS, Transcriptional Regulation of HIV in 19
Melanie Ruth Adams and Boris Matija Peterlin

AIDS HIV Enzymes, Three-Dimensional Structures of 27
Zuzana Hostomska and Zdenek Hostomsky

AIDS Therapeutics, Biochemistry of 34
Christine Debouck and Martin Rosenberg

Alzheimer's Disease 42
Frank Ashall and Alison M. Goate

Amino Acid Synthesis 52
Robert M. Williams

Annexins 59
Carl E. Creutz

Antibody Molecules, Genetic Engineering of 62
Sherie L. Morrison

Antisense Oligonucleotides, Structure and Function of 69
Eugen Uhlmann and Anusch Peyman

Arabidopsis Genome 76
Vivian F. Irish

Autoantibodies and Autoimmunity 84
K. Michael Pollard, Edward K. L. Chan, Robert L. Rubin, and Eng M. Tan

Bacterial Growth and Division 95
Stephen Cooper

Bacterial Pathogenesis 103
Luciano Passador and Barbara H. Iglewski

Bacteriorhodopsin 108
Janos K. Lanyi

Biochemical Genetics, Human 113
Frits A. Hommes

Biodegradation of Organic Wastes 117
Duane Graves

Bioinorganic Chemistry 125
R. Bruce Martin

Biomimetic Materials 134
Robert J. Campbell

Biomolecular Electronics and Applications 148
Elias Greenbaum

Bioorganic Chemistry 159
Simon H. Friedman and George L. Kenyon

Bioprocess Engineering 172
Kimberly L. Ogden

Biosensors 177
Anthony E. G. Cass

Biotransformations of Drugs and Chemicals 192
Paul R. Ortiz de Montellano

Body Expression Map of the Human Genome 203
Kousaku Okubo and Kenichi Matsubara

Brain Aromatic Amino Acid Hydroxylases 207
Seymour Kaufman and Paula Ribeiro

Breast Cancer, Genetic Analysis of 221
Patricia Margaritte-Jeannin

Caenorhabditis elegans, Molecular Biology of 227
Robert J. Barstead

Calcium Biochemistry 237
Joachim Krebs

Cancer 250
Thomas G. Pretlow and Theresa P. Pretlow

Cancer Chemotherapy, Theoretical Foundations of 252
Robert C. Jackson

Carbohydrate Analysis 266
Martin F. Chaplin

Carbohydrate Antigens 278
David R. Bundle

Cardiovascular Diseases 283
Todd Leff and Peter J. Gruber

Cell-Cell Interactions 292
Geoffrey M. W. Cook

Cell Death and Aging, Molecular Mechanisms of 299
Michael Hengartner

Cellular Energetics, Biochemical Basis of 303
Stephen E. Specter and Barbara A. Horwitz

Chaperones, Molecular 315
Elizabeth A. Craig

Chemiluminescence and Bioluminescence, Analysis by 322
Larry J. Kricka

Chirality in Biology 332
Ronald Bentley

Chlamydomonas 347
Jean-David Rochaix

Chromatin Formation and Structure 360
Timothy J. Richmond and Fritz Thoma

Chromosome Microdissection and Microcloning 370
Nabil G. Hagag and Michael V. Viola

Circular Dichroism in Protein Analysis 384
Nicholas C. Price

Coenzymes, Biochemistry of 396
Donald B. McCormick

Colon Cancer 407
Joanna Groden

Combinatorial Phage Antibody Libraries 415
R. Anthony Williamson and Dennis R. Burton

Cystic Fibrosis 425
Mitchell L. Drumm

Cytochrome P450 435
Michael R. Waterman

Cytokines 441
Alan G. Morris

Cytoskeleton-Plasma Membrane Interactions 451
Frederick M. Pavalko

Glossary of Basic Terms 461

VOLUME 2

Preface vii

Acknowledgments ix

Contributors xi

Denaturation of DNA 1
Richard D. Blake

Diabetes Insipidus 19
Walter Rosenthal, Anita Seibold, Daniel G. Bichet, and Mariel Birnbaumer

Diabetes Mellitus 28
Catherine H. Mijovic, David Jenkins, and Anthony H. Barnett

DNA Damage and Repair 40
George J. Kantor

DNA Fingerprint Analysis 48
Lorne T. Kirby, Ron M. Fourney, and Bruce Budowle

DNA in Neoplastic Disease Diagnosis 62
David Sidransky

DNA Markers, Cloned 66
Eugene R. Zabarovsky

DNA Repair in Aging and Sex 78
Carol Bernstein and Harris Bernstein

DNA Replication and Transcription 86
Hisaji Maki, Yusaku Nakabeppu, and Mutsuo Sekiguchi

DNA Sequencing in Ultrathin Gels, High Speed 96
Robert L. Brumley, Jr. and Lloyd M. Smith

DNA Structure 102
C. R. Calladine and Horace R. Drew

Down Syndrome, Molecular Genetics of 118
Charles J. Epstein

Drosophila Genome 121
Robert D. C. Saunders

Drosophila Mating, Genetic Basis of 130
Jeffrey C. Hall

Drug Addiction and Alcoholism, Genetic Basis of 146
Judith E. Grisel and John C. Crabbe

Drug Synthesis 155
Daniel Lednicer

Drug Targeting and Delivery, Molecular Principles of 165
Alfred Stracher

E. coli Genome 169
Hirotada Mori and Takashi Yura

Electric and Magnetic Field Reception 172
Eberhard Neumann

Electron Microscopy of Biomolecules 181
Claus-Thomas Bock, Hanswalter Zentgraf, and John Sommerville

Electron Transfer, Biological 193
Christopher C. Moser, Ramy S. Farid, and P. Leslie Dutton

Electroporation and Electrofusion 198
Donald C. Chang

Endocrinology, Molecular 206
Franklyn F. Bolander, Jr.

Environmental Stress, Genomic Responses to 216
John G. Scandalios

Enzymes, Energetics and Regulation of Biological Catalysis by 222
Stephen W. Raso, Jon A. Christopher, Miriam M. Ziegler, and Thomas O. Baldwin

Enzymes, High Temperature 240
Michael W. W. Adams

Enzymology, Nonaqueous 249
Sudipta Chatterjee, Shabbir Bambot, Richard Bormett, Sanford Asher, and Alan J. Russell

Epidemiology, Molecular 258
Paul A. Schulte, Frederica P. Perera, and Nathaniel Rothman

Epidermal Keratinocytes, Molecular Biology of 262
Miroslav Blumenberg

Expression Systems for DNA Processes 271
Ka-Yiu San and George N. Bennett

Extracellular Matrix 278
Linda J. Sandell

Female Reproductive System, Molecular Biology of 285
J. K. Findlay and J. E. Mercer

Fluorescence Spectroscopy of Biomolecules 294
Joseph R. Lakowicz

Food Proteins, Interactions of 306
Nicholas Parris and Charles Onwulata

Fragile Sites 313
Grant R. Sutherland, Elizabeth Baker, and Robert I. Richards

Fragile X Linked Mental Retardation 319
A. J. M. H. Verkerk and B. A. Oostra

Free Radicals in Biochemistry and Medicine 330
Barry Halliwell

Fungal Biotechnology 337
Brian McNeil

Gaucher Disease 347
Ernest Beutler

Gel Electrophoresis of Proteins, Two-Dimensional Polyacrylamide 352
Peter J. Wirth

Gene Distribution in the Human Genome 360
Giorgio Bernardi

Gene Expression, Regulation of 364
Göran Akusjärvi

Gene Mapping by Fluorescence In Situ Hybridization 377
Amanda C. Heppell-Parton

Gene Order by FISH and FACS 386
Malcolm A. Ferguson-Smith

Genetic Analysis of Populations 391
A. Rus Hoelzel

Genetic Diversity in Microorganisms 402
Werner Arber

Genetic Intelligence, Evolution of 407
David S. Thaler and Bradley T. Messmer

Genetics 415
D. Peter Snustad

Genetic Testing 419
Frank K. Fujimura

Genetic Vaccination, Gene Transfer Techniques for Use in 430
Stephen G. Kayes and Jon A. Wolff

Genomic Imprinting 439
Wolf Reik, Robert Feil, and Nicholas D. Allen

Glycobiology 446
Akira Kobata

Glycoproteins, Secretory 458
Alistair G. C. Renwick

Growth Factors 463
Antony W. Burgess

Glossary of Basic Terms 473

VOLUME 3

Preface vii

Acknowledgments ix

Contributors xi

Heart Failure, Genetic Basis of 1
Frank Kee and Alun E. Evans

Hemoglobin 2
Gino Amiconi and Maurizio Brunori

Hemoglobin, Genetic Engineering of 20
Jeremy Tame

Hemophilia 28
Francesco Giannelli

Heterotrimeric G Proteins, Signal Transduction by 41
Anya Harry and Ravi Iyengar

Histones 48
Gary S. Stein, Janet L. Stein, and André J. van Wijnen

Homeodomain Proteins 55
Thomas R. Bürglin

HPLC of Biological Macromolecules 76
Karen M. Gooding

Human Chromosome 1, Genetic and Physical Map Status of 84
Ben Carritt

Human Chromosome 9, Genetic and Physical Map Status of 97
David J. Kwiatkowski and Sue Povey

Human Chromosome 10, Genetic and Physical Map Status of 112
Nancy L. Carson

Human Chromosome 11, Genetic and Physical Map Status of 125
Marcel Mannens

Human Chromosome 14, Genetic and Physical Map Status of 142
Sunil D. Pandit and Helen Donis-Keller

Human Chromosome 15, Genetic and Physical Map Status of 160
Mansoor Sarfarazi and Michael W. Kilpatrick

Human Chromosome 16, Genetic and Physical Map Status of 174
Helen Kozman and John Mulley

Human Chromosome 18, Genetic and Physical Map Status of 190
Gary A. Silverman and Joan Overhauser

Human Chromosomes, Physical Maps of 215
Nikolay Bukanov, Denan Wang, and Cassandra L. Smith

Human Disease Gene Mapping 221
Scott R. Diehl

Human Genetic Predisposition to Disease 227
Belina J. F. Rossiter and C. Thomas Caskey

Human Linkage Maps 233
Jurg Ott

Human Repetitive Elements 240
Jerzy Jurka a5nd Mark A. Batzer

Huntington's Disease 246
James F. Gusella and Marcy E. MacDonald

Hybridization for Sequencing of DNA 252
William Bains

Hydrogen Bonding in Biological Structures 261
Thomas Steiner

Immunology 275
Jiri Novotny and Antony Nicholls

Immuno-PCR 288
Takeshi Sano, Cassandra L. Smith, and Charles R. Cantor

Immunotoxicological Mechanisms 295
Rodney R. Dietert and Karen A. Golemboski

Infectious Disease Testing by Ligase Chain Reaction 302
Helen H. Lee, Gregor W. Leckie, and John J. Carrino

Inorganic Solids, Biomolecules in the Synthesis of 309
Trevor Douglas and Stephen Mann

Interferons 317
Anthony Meager

Interleukins 331
Anthony Meager

Intracellular Protein Degradation, Mechanisms of Regulated 346
Stuart M. Arfin and Ralph A. Bradshaw

Ionizing Radiation Damage to DNA 354
Clemens von Sonntag and Heinz-Peter Schuchmann

Isoenzymes 366
Adrian O. Vladutiu and Georgirene D. Vladutiu

Kallikrein-Kininogen-Kinin System 377
Michael E. Rusiniak and Nathan Back

Labeling, Biophysical 385
Gertz Likhtenshtein

Ligase-Mediated Gene Detection 391
Ulf Landegren, Martina Samiotaki, Marek Kwiatkowski, and Mats Nilsson

Ligation Assays 400
Claire Delahunty, Pui-Yan Kwok, and Deborah Nickerson

Lipid and Lipoprotein Metabolism 403
Clive R. Pullinger and John P. Kane

Lipid-Based Gene Vectors 415
Christopher J. Kirby

Lipids, Microbial 426
Colin Ratledge

Lipids, Structure and Biochemistry of 442
Donald M. Small and R. A. Zoeller

Lipopolysaccharides 462
Ilkka M. Helander, P. Helena Mäkelä, Otto Westphal, and Ernst T. Rietschel

Lipoprotein Analysis 471
E. Roy Skinner

Liver Cancer, Molecular Biology of 480
Mehmet Ozturk

Lung Cancer, Molecular Biology of 491
Jack A. Roth

Major Histocompatibility Complex 499
James Driscoll

Male Reproductive System, Molecular Biology of 500
K. A. L. Loveland and D. M. de Kretser

Mammalian Genome 510
John Schimenti and Elizabeth M. Simpson

Glossary of Basic Terms 521

VOLUME 4

Preface vii

Acknowledgments ix

Contributors xi

Mass Spectrometry High Speed DNA Fragment Sizing 1
C. H. Chen

Mass Spectrometry of Biomolecules 12
Raymond E. Kaiser, Jr.

Medicinal Chemistry 28
David J. Triggle

Membrane Fusion, Molecular Mechanism of 42
Koert N. J. Burger

Memory and Learning, Molecular Basis of 53
Timothy E. Kennedy and Eric R. Kandel

Metalloenzymes 58
Walther R. Ellis, Jr. and Gregory M. Raner

Methanogens and Archaea, Molecular Biology of 66
Madeline E. Rasche and James G. Ferry

Mitochondrial DNA 79
Ulf Gyllensten

Mitochondrial DNA, Evolution of Human 84
Linda Vigilant

Molecular Diversity, Theoretical Aspects of 90
Jack D. Keene

Molecular Genetic Medicine 100
Bernice E. Morrow and Raju Kucherlapati

Motor Neuron Disease 105
Michael Sendtner

Motor Proteins 112
Ravindhra G. Elluru, Janet L. Cyr, and Scott T. Brady

Mouse Genome 120
Stephen D. M. Brown

Muscle, Molecular Genetics of Human 129
Prem Mohini Sharma

Mutagenesis 148
Raymond Devoret

Mycobacteria 156
Johnjoe McFadden and Neil G. Stoker

Nematodes, Neurobiology and Development of 167
Darryl MacGregor, Ian A. Hope, and R. Elwyn Isaac

Neurofibromatosis 1 172
David H. Viskochil

Neurofibromatosis 2 180
Jaime O. Claudio, Michael Belliveau, and Guy A. Rouleau

Neuropeptides, Insect 188
Edward P. Masler

Neuropsychiatric Diseases 198
F. Owen

Nitric Oxide in Biochemistry and Drug Design 200
Larry K. Keefer, Carl F. Nathan, and William J. Payne

Nonribosomal Peptides 207
Juan F. Martin

Nuclear Magnetic Resonance of Biomolecules in Solution 218
Brendan C. Maguire and Betty J. Gaffney

Nucleic Acid Enzymes, Evolutionary Engineering of 226
Michael Famulok

Nucleic Acid Hybrids, Formation and Structure of 235
James G. Wetmur

Nucleic Acid-Protein Complexes, Laboratory Techniques in 243
Arnold Revzin

Nucleic Acid Sequencing Techniques 254
Christopher J. Howe

Nutrition 257
C. J. K. Henry

Oligonucleotide Analogues, Phosphate-Free 269
Rajender S. Varma

Oligonucleotides 281
Fritz Eckstein, Olaf Heidenreich, Mabel Ng, and Tom Tuschl

Oncogenes 285
Anthony Byrne and Desmond N. Carney

Origins of Life, Molecular Biology of 288
James P. Ferris

Paleontology, Molecular 299
Rob DeSalle

Parasite Genome Projects 306
Steven A. Williams and David J. Kemp

Partial Denaturation Mapping 312
Ross B. Inman

Pathology, Molecular 323
Daniel H. Farkas

PCR Technology 333
Henry A. Erlich

Peptide Hormones, Biosynthesis, Processing, and Secretion of 340
Iris Lindberg

Peptides 347
Tomi K. Sawyer, Wayne L. Cody, Daniele M. Leonard, and Mac E. Hadley

Peptides, Synthetic 361
Gregory A. Grant

Peptides and Mimics, Design of Conformationally Constrained 371
Victor J. Hruby and Lakmal W. Boteju

Peptide Synthesis, Solid Phase 383
R. C. Sheppard

Pesticidal Compounds from Prokaryotes 393
Michael David, John T. Turner, A. Lawrence Christy, Jeffrey L. Kelly, and Peter S. Carlson

Pharmacogenetics 399
Ann K. Daly and Jeffrey R. Idle

Phenylketonuria, Molecular Genetics of 405
Randy C. Eisensmith and Savio L. C. Woo

Phospholipases 415
Mary F. Roberts and Chun Zhou

Phospholipids 432
Dennis E. Vance

Photosynthetic Energy Transduction 440
Neil R. Baker

Plant Cells, Genetic Manipulation of 450
Roland Bilang and Martin Schrott

Plant Cell Transformation, Physical Methods for 458
John J. Finer, Kim R. Finer, and Eliane R. Santarem

Plant Gene Expression Regulation 465
Eric Lam

Plant Pathology, Molecular 475
Sarah Jane Gurr and Sally A. Francis

Plants, Molecular Biology of 485
Donald E. Fosket

Plasma Lipoproteins 495
Thomas L. Innerarity

Glossary of Basic Terms *507*

VOLUME 5

Preface *vii*

Acknowledgments *ix*

Contributors *xi*

Plasmids 1
Kimber Hardy

Poliovirus, Molecular Biology of 13
Stephen Todd and Bert L. Semler

Polymers, Genetic Engineering of 26
Maurille J. Fournier, Thomas L. Mason, David A. Tirrell, and Jane G. Tirrell

Polymers for Biological Systems 31
Robert C. Thomson, Susan L. Ishaug, Antonios G. Mikos, and Robert Langer

Population-Specific Genetic Markers and Disease 44
L. B. Jorde

Post- and Co-translational Modifications of Proteins, Enzyme Catalyzed 49
Bruce L. Martin

Prions 57
Stanley B. Prusiner

Protein Aggregation 73
Jeannine M. Yon

Protein Analysis by Raman Spectroscopy 93
George J. Thomas, Jr.

Protein Analysis by X-Ray Crystallography 110
Gordon V. Louie

Protein Architecture and Analysis 115
Robert A. Copeland

Protein Designs for the Specific Recognition of DNA 127
Aaron Klug

Protein Expression by Expansion of the Genetic Code 135
Lance E. Steward and A. Richard Chamberlin

Protein Folding 147
Thomas E. Creighton

Protein Modeling 158
Thomas R. Defay and Fred E. Cohen

Protein Phosphorylation 169
Clay W Scott and Theodora W. Salcedo

Proteins and Peptides, Isolation for Sequence Analysis of 173
Richard J. Simpson, Gavin E. Reid, Robert L. Moritz, and Larry D. Ward

Protein Sequencing Techniques 185
Jeffrey N. Keen and John B. C. Findlay

Protein Targeting 194
Brian Austen and David Stephens

Radioisotopes in Molecular Biology 209
Robert James Slater

RecA Protein, Structure and Function of 220
Michael M. Cox

Receptor Biochemistry 229
Tatsuya Haga

Recognition and Immobilization, Molecular 243
Irwin M. Chaiken, Jun Li, and Robert Miceli

Recombination, Molecular Biology of 251
Hannah L. Klein

Renal System 264
Samir S. El-Dahr, Lisa M. Harrison-Bernard, R. Ariel Gomez, and L. Gabriel Navar

Repressor–Operator Recognition 274
Peter G. Stockley and Simon E. V. Phillips

Respiratory Syncytial Virus Vaccine, Approaches to 285
Edward E. Walsh and Caroline B. Hall

Restriction Endonucleases and DNA Modification Methyltransferases for the Manipulation of DNA 296
Eric W. Fisher and Richard I. Gumport

Restriction Landmark Genomic Scanning Method 304
Yoshihide Hayashizaki, Shinji Hirotsune, Yasushi Okazaki, Masami Muramatsu, and Jun-ichi Asakawa

Retinoblastoma, Molecular Genetics of 319
Brenda L. Gallie

Retinoids 322
Robert R. Rando

Ribosome Preparation and Protein Synthesis Techniques 327
Gary Spedding

Ribozymes 338
Joyce E. Heckman and John M. Burke

RNA Methodologies 342
Robert E. Farrell, Jr.

RNA Replication 352
Claude A. Villee

RNA Secondary Structures 360
John A. Jaeger

RNA Structure, Nonhelical 368
Brian Wimberly

RNA Three-Dimensional Structures, Computer Modeling of 371
François Major and Daniel Gautheret

Robotics and Automation in Genome Research 388
Edward Theil

Scanning Probe Microscopy in Sequencing of DNA 401
T. L. Ferrell, D. P. Allison, T. Thundat, and R. J. Warmack

Scleroderma Diagnosis with Recombinant Protein 410
Kenji Sugimoto, Yoshinao Muro, Masaru Ohashi, and Michio Himeno

Sequence Alignment of Proteins and Nucleic Acids 419
William R. Taylor

Sequence Analysis 428
Martin Bishop

Sex Determination 434
Stephen S. Wachtel and Lee P. Shulman

Sexual PCR and Assembly PCR 447
Willem P. C. Stemmer

Single-Stranded DNA Binding Proteins (SSBs) 457
Michael Mitas

Steroid Analysis 466
H. L. J. Makin and D. B. Gower

Steroid Hormones and Receptors 478
Robin Leake

Steroids, Antitumor 485
Robert T. Blickenstaff

Streptomycetes, Molecular Biology of 501
Dimitrina Lyutzkanova and Iain Hunter

Superantigens 508
Monique Lafon

Synapses 512
Hui-Quan Han and Robert A. Nichols

Synthetic Peptide and Nonpeptide Combinatorial Libraries 516
Kit S. Lam

Glossary of Basic Terms 525

VOLUME 6

Preface vii

Acknowledgments ix

Contributors xi

Tandemly Repeated Noncoding DNA Sequences 1
Wolfgang Stephan

Taxol® and Taxane Production by Cell Culture 10
Arthur G. Fett-Neto and Frank DiCosmo

Telomerase 17
Silvia Bacchetti and Calvin B. Harley

Telomeres, Mammalian 27
Andrew H. Palin and Christine J. Farr

Theoretical Molecular Biology 37
Andrzej K. Konopka

Tomatoes, Gene Alterations of 53
Belinda Martineau and William R. Hiatt

Trace Element Micronutrients 58
Robert J. Cousins

Transgenic Animal Modeling 63
Carl A. Pinkert, Michael H. Irwin, and R. Jeffrey Moffatt

Transgenic Animal Patents 74
W. Lesser

Transgenic Fish 84
Thomas T. Chen, Jenn-Kan Lu, Mike J. Shamblott, Clara M. Cheng, Chun-Mean Lin, Jane C. Burns, Renate Reimschuessel, N. Chatakondi, and Rex A. Dunham

Transgenic Livestock as Bioreactors 99
Michael A. Dalrymple and Ian Garner

Translation of RNA to Protein 108
R. A. Cox and H. R. V. Arnstein

Transport Proteins 125
Milton H. Saier, Jr.

Transposons in the Human Genome 133
Abram Gabriel

Tumor Suppressor Genes 138
Prem Mohini Sharma and S. R. Dev Sharma

Ultraviolet Radiation Damage to DNA 149
David L. Mitchell

Vaccine Biotechnology 165
Tilahun D. Yilma

Varicella-Zoster Virus and Cytomegalovirus Vaccines 172
Lawrence A. Hunt

Viral Envelope Assembly and Budding 183
Milton J. Schlesinger

Virology, Molecular 192
Andrew P. Rice

Viruses, DNA Packaging of 200
Philip Serwer

Viruses, Photodynamic Inactivation of 207
James L. Matthews and Millard M. Judy

Viruses, RNA Packaging of 217
Dennis H. Bamford

Virus-Receptor Interactions 223
Richard L. Crowell and Ruliang Xu

Vitamin Receptors 235
Krishnamurti Dakshinamurti

Vitamins, Structure and Function of 244
Donald B. McCormick

Whole Chromosome Complementary Probe Fluorescence Staining 253
Heinz-Ulrich G. Weier, Daniel Pinkel, and Joe W. Gray

Xenobiotic Metabolism 259
Bernard Testa

X-Ray Diffraction of Biomolecules 275
K. Ravi Acharya and Anthony R. Rees

Yeast, DNA Repair in 283
Errol C. Friedberg

Yeast Artificial Chromosome (YAC) Construction and Applications 291
Christine Loehrlein and Ronald W. Davis

Yeast Genetics 302
Fred Sherman

Zinc Finger DNA-Binding Motifs 327
Louise Fairall and John W. R. Schwabe

Glossary of Basic Terms 337

Cumulative Contents 339

Cumulative Contributors 347

Index 369

Cumulative Contributors

K. Ravi Acharya
School of Biology and Biochemistry
University of Bath
Bath, BA2 7AY, United Kingdom
X-Ray Diffraction of Biomolecules

Melanie Ruth Adams
Scientific Services
Irwin Memorial Blood Centers
San Francisco, CA 94118, USA
AIDS, Transcriptional Regulation of HIV in

Michael W. W. Adams
Department of Biochemistry
University of Georgia
Athens, GA 30602, USA
Enzymes, High Temperature

Göran Akusjärvi
Department of Medical Immunology and Microbiology
Uppsala University
S-751 23 Uppsala, Sweden
Gene Expression, Regulation of

Nicholas D. Allen
Laboratory of Developmental Genetics and Imprinting
The Babraham Institute
Cambridge CB2 4AT
United Kingdom
Genomic Imprinting

D. P. Allison
Oak Ridge National Laboratory
Oak Ridge, TN 37381, USA
Scanning Probe Microscopy in Sequencing of DNA

Gino Amiconi
Department of Biochemical Sciences
University of Rome La Sapienza
00185 Rome, Italy
Hemoglobin

Werner Arber
Department of Microbiology
Biozentrum
University of Basel
CH-4056 Basel, Switzerland
Genetic Diversity in Microorganisms

Stuart M. Arfin
Department of Biological Chemistry
College of Medicine
University of California
Irvine, CA 92717, USA
Intracellular Protein Degradation, Mechanisms of Regulated

H. R. V. Arnstein
King's College, London
London WC2R 2LS
United Kingdom
Translation of RNA to Protein

Jun-ichi Asakawa
Tsukuba Life Science Center
Institute of Physical and Chemical Research (RIKEN)
Tsukuba, Ibaraki 305, Japan
Restriction Landmark Genomic Scanning Method

Frank Ashall
St. Louis University School of Medicine
St. Louis, MO 63102, USA
Alzheimer's Disease

Sanford Asher
Department of Chemistry
University of Pittsburgh
Pittsburgh, PA 15261, USA
Enzymology, Nonaqueous

Brian Austen
Department of Surgery
St. George's Hospital Medical School
London SW17 0RE, United Kingdom
Protein Targeting

Silvia Bacchetti
Cancer Research Group
and
Institute for Molecular Biology and Biotechnology
McMaster University
Hamilton, Ontario L8N 3Z5, Canada
Telomerase

Nathan Back
Department of Biochemical Pharmacology
School of Pharmacy
State University of New York
Buffalo, NY 14260, USA
Kallikrein-Kininogen-Kinin System

William Bains
PA Consulting Group
Melbourn, Royston
Hertfordshire SG8 6DP
United Kingdom
Hybridization for Sequencing of DNA

Elizabeth Baker
Department of Cytogenetics and Molecular Genetics
Centre for Medical Genetics
Women's and Children's Hospital
Adelaide 5006, Australia
Fragile Sites

Neil R. Baker
Department of Biological and Chemical Sciences
University of Essex
Colchester CO4 3SQ, United Kingdom
Photosynthetic Energy Transduction

Thomas O. Baldwin
Department of Biochemistry and Biophysics
Texas A&M University
College Station, TX 77843, USA
Enzymes, Energetics and Regulation of Biological Catalysis by

Shabbir Bambot
Department of Chemical Engineering
Center for Biotechnology and Bioengineering
University of Pittsburgh
Pittsburgh, PA 15261, USA
Enzymology, Nonaqueous

Dennis H. Bamford
Institute of Biotechnology
and
Department of Biosciences Division Genetics
University of Helsinki
FIN-00014 Helsinki, Finland
Viruses, RNA Packaging of

Anthony H. Barnett
Department of Medicine
University of Birmingham and Birmingham Heartlands Hospital
Birmingham B9 5SS
United Kingdom
Diabetes Mellitus

Robert J. Barstead
Program in Molecular and Cellular Biology
Oklahoma Medical Research Foundation
Oklahoma City, OK 73104, USA
Caenorhabditis elegans, *Molecular Biology of*

Mark A. Batzer
Department of Pathology
Stanley S. Scott Cancer Center
Louisiana State University Medical Center
New Orleans, LA 70112, USA
Human Repetitive Elements

Michael Belliveau
Center for Research in Neuroscience
McGill University and Montreal General Hospital
Montreal, Quebec H3G 1A4, Canada
Neurofibromatosis 2

George N. Bennett
Department of Biochemistry and Cell Biology
Rice University
Houston, TX 77251, USA
Expression Systems for DNA Processes

Ronald Bentley
Department of Biological Sciences
University of Pittsburg
Pittsburgh, PA 15260, USA
Chirality in Biology

Giorgio Bernardi
Laboratoire de Génétique Moléculaire
Institut Jacques Monod
75005 Paris, France
Gene Distribution in the Human Genome

Carol Bernstein
Department of Microbiology and Immunology
University of Arizona College of Medicine
Tucson, AZ 85724, USA
DNA Repair in Aging and Sex

Harris Bernstein
Department of Microbiology and Immunology
University of Arizona College of Medicine
Tucson, AZ 85724, USA
DNA Repair in Aging and Sex

Ernest Beutler
Department of Molecular and Experimental Medicine
The Scripps Research Institute
La Jolla, CA 92037, USA
Gaucher Disease

Daniel G. Bichet
Department of Medicine
Centre de Recherche
Hôpital du Sacré-Coeur de Montréal
Montreal, Quebec H4J 1C5
Canada
Diabetes Insipidus

Roland Bilang
Institute of Plant Sciences
Swiss Federal Institute of Technology
ETH-Zentrum
CH-8092 Zurich
Switzerland
Plant Cells, Genetic Manipulation of

Mariel Birnbaumer
Department of Anesthesiology
University of California School of Medicine
Los Angeles, CA 90095, USA
Diabetes Insipidus

Martin Bishop
UK Human Genome Mapping Project Resource Centre
Hinxton
Cambridge CB10 7KQ
United Kingdom
Sequence Analysis

Richard D. Blake
Department of Biochemistry, Microbiology, and Molecular Biology
University of Maine
Orono, ME 04469, USA
Denaturation of DNA

Robert T. Blickenstaff
Department of Veterans Affairs
Richard L. Roudebush VA Medical Center
and
Department of Biochemistry and Molecular Biology
Indiana University School of Medicine
Indianapolis, IN 46202, USA
Steroids, Antitumor

Miroslav Blumenberg
Ronald O. Perelman Department of Dermatology and
Department of Biochemistry
New York University Medical Center
New York, NY 10016, USA
Epidermal Keratinocytes, Molecular Biology of

Claus-Thomas Bock
Abteilung Gastroenterologie und Hepatologie
Medizinische Hochschule Hannover
D-30625 Hannover, Germany
Electron Microscopy of Biomolecules

Franklyn F. Bolander, Jr.
Department of Biological Sciences
University of South Carolina
Columbia, SC 29208, USA
Endocrinology, Molecular

Richard Bormett
Department of Chemistry
University of Pittsburgh
Pittsburgh, PA 15261, USA
Enzymology, Nonaqueous

Lakmal W. Boteju
Department of Chemistry
University of Arizona
Tucson, AZ 85721, USA
Peptides and Mimics, Design of Conformationally Constrained

Ralph A. Bradshaw
Department of Biological Chemistry
College of Medicine
University of California
Irvine, CA 92717, USA
Intracellular Protein Degradation, Mechanisms of Regulated

Scott T. Brady
Department of Cell Biology
and Neuroscience
University of Texas
Southwestern Medical Center
Dallas, TX 75235, USA
Motor Proteins

Stephen D. M. Brown
Medical Research Council
Mouse Genome Centre
Harwell
Didcot
Oxfordshire OX11 ORD
United Kingdom
Mouse Genome

Robert L. Brumley, Jr.
Department of Chemistry
University of Wisconsin
Madison, WI 53706, USA
DNA Sequencing in Ultrathin Gels, High Speed

Maurizio Brunori
Department of Biochemical
Sciences
University of Rome
La Sapienza
00185 Rome, Italy
Hemoglobin

Bruce Budowle
Laboratory Division
Forensic Science Research
and Training Center
Federal Bureau of Investigation
Academy
Quantico, VA 22135, USA
DNA Fingerprint Analysis

Nikolay Bukanov
Center for Advanced Biotechnology and
Departments of Biomedical
Engineering, Biology, and
Pharmacology
Boston University
Boston, MA 02215, USA
Human Chromosomes, Physical Maps of

David R. Bundle
Department of Chemistry
University of Alberta
Edmonton, Alberta T6G 2G2
Canada
Carbohydrate Antigens

Koert N. J. Burger
Department of Cell Biology
University of Utrecht
Medical School AZ
3584 CX Utrecht
The Netherlands
Membrane Fusion, Molecular Mechanism of

Antony W. Burgess
Ludwig Institute for Cancer Research
Melbourne
Victoria 3050, Australia
Growth Factors

Thomas R. Bürglin
Department of Cell Biology
Biozentrum
University of Basel
CH-4056 Basel, Switzerland
Homeodomain Proteins

John M. Burke
Department of Microbiology
and Molecular Genetics
Markey Center for Molecular
Genetics
University of Vermont
Burlington, VT 05405, USA
Ribozymes

Jane C. Burns
Department of Pediatrics
School of Medicine
University of California
San Diego, CA 92093, USA
Transgenic Fish

Dennis R. Burton
Departments of Immunology and
Molecular Biology
The Scripps Research Institute
La Jolla, CA 92037, USA
Combinatorial Phage Antibody Libraries

Anthony Byrne
Department of Clinical Pharmacology
Royal College of Surgeons in Ireland
Dublin 2, Ireland
Oncogenes

C. R. Calladine
Department of Engineering
University of Cambridge
Cambridge CB2 1PZ
United Kingdom
DNA Structure

Robert J. Campbell
Chemistry and Biological Sciences
Division
U.S. Army Research Office
Research Triangle Park, NC 27709
USA
Biomimetic Materials

Charles R. Cantor
Center for Advanced Biotechnology and
Departments of Biomedical Engineering
and Pharmacology
Boston University
Boston, MA 02215, USA
Immuno-PCR

Peter S. Carlson
Oceanix Biosciences Corp.
Hanover, MD 21076, USA
Pesticidal Compounds from Prokaryotes

Desmond N. Carney
Department of Medical Oncology
Mater Misericordiae Hospital
Dublin 7, Ireland
Oncogenes

John J. Carrino
Probe Diagnostics Business Unit
Abbott Laboratories
Abbott Park, IL 60064, USA
Infectious Disease Testing by Ligase Chain Reaction

Ben Carritt
Medical Research Council
Human Biochemical Genetics Unit
The Galton Laboratory
University College London
London NW1 2HE, United Kingdom
Human Chromosome 1, Genetic and Physical Map Status of

Nancy L. Carson
Department of Clinical Chemistry
Health Sciences Centre
Winnipeg, Manitoba R3A 1R9, Canada
Human Chromosome 10, Genetic and Physical Map Status of

C. Thomas Caskey
Department of Molecular and Human
Genetics
Baylor College of Medicine
and Merck Research Laboratories
West Point, PA 19486, USA
Human Genetic Predisposition to Disease

Anthony E. G. Cass
Department of Biochemistry
Imperial College of Science, Technology and Medicine
South Kensington
London SW7 2AZ, United Kingdom
Biosensors

Irwin M. Chaiken
Rheumatology Division
University of Pennsylvania School of Medicine
Philadelphia, PA 19104, USA
Recognition and Immobilization, Molecular

A. Richard Chamberlin
Department of Chemistry
University of California
Irvine, CA 92717, USA
Protein Expression by Expansion of the Genetic Code

Edward K. L. Chan
Department of Molecular and Experimental Medicine
W. M. Keck Autoimmune Disease Center
The Scripps Research Institute
La Jolla, CA 92037, USA
Autoantibodies and Autoimmunity

Donald C. Chang
Department of Biology
Hong Kong University of Science and Technology
Clear Water Bay, Kowloon, Hong Kong
Electroporation and Electrofusion

Martin F. Chaplin
School of Applied Science
South Bank University
London SE1 0AA, United Kingdom
Carbohydrate Analysis

N. Chatakondi
Department of Fisheries and Allied Aquacultures
Auburn University
Auburn, AL 36849, USA
Transgenic Fish

Sudipta Chatterjee
Department of Chemical Engineering
Center for Biotechnology and Bioengineering
University of Pittsburgh
Pittsburgh, PA 15261, USA
Enzymology, Nonaqueous

C. H. Chen
Health and Safety Research Division
Oak Ridge National Laboratory
Oak Ridge, TN 37831, USA
Mass Spectrometry High Speed DNA Fragment Sizing

Thomas T. Chen
Biotechnology Center
University of Connecticut
Storrs, CT 06269, USA
Transgenic Fish

Clara M. Cheng
Center for Marine Biotechnology
Maryland Biotechnology Institute
Baltimore, MD 21202, USA
Transgenic Fish

Jon A. Christopher
Department of Chemistry
Texas A&M University
College Station, TX 77843, USA
Enzymes, Energetics and Regulation of Biological Catalysis by

A. Lawrence Christy
AgriTech, Inc.
Crofton, MD 21114, USA
Pesticidal Compounds from Prokaryotes

Jaime O. Claudio
Center for Research in Neuroscience
McGill University and Montreal General Hospital
Montreal, Quebec H3G 1A4
Canada
Neurofibromatosis 2

Wayne L. Cody
Department of Chemistry
Parke-Davis Pharmaceutical Research
Ann Arbor, MI 48206, USA
Peptides

Fred E. Cohen
Department of Cellular and Molecular Pharmacology
University of California
San Francisco, CA 94143, USA
Protein Modeling

Geoffrey M. W. Cook
Department of Anatomy
Cambridge University
Cambridge CB2 3DY, United Kingdom
Cell-Cell Interactions

Stephen Cooper
Department of Microbiology and Immunology
University of Michigan Medical School
Ann Arbor, MI 48109, USA
Bacterial Growth and Division

Robert A. Copeland
The DuPont Merck Research Laboratories
Wilmington, DE 19880, USA
Protein Architecture and Analysis

Robert J. Cousins
Food Science and Human Nutrition Department
Center for Nutritional Sciences
University of Florida
Gainesville, FL 32611, USA
Trace Element Micronutrients

Michael M. Cox
Department of Biochemistry
University of Wisconsin
Madison, WI 53706, USA
RecA Protein, Structure and Function of

R. A. Cox
National Institute for Medical Research
London NW7 1AA, United Kingdom
Translation of RNA to Protein

John C. Crabbe
Veterans Affairs Medical Center and Department of Medical Psychology
Oregon Health Sciences University
Portland, OR 97201, USA
Drug Addiction and Alcoholism, Genetic Basis of

Elizabeth A. Craig
Department of Biomolecular Chemistry
University of Wisconsin
Madison, WI 53706, USA
Chaperones, Molecular

Thomas E. Creighton
European Molecular Biology Laboratory
D-69012 Heidelberg, Germany
Protein Folding

Carl E. Creutz
Department of Pharmacology
Health Sciences Center
University of Virginia
Charlottesville, VA 22908, USA
Annexins

Richard L. Crowell
Department of Microbiology
and Immunology
Allegheny University of the
Health Sciences
Philadelphia, PA 19129, USA
Virus-Receptor Interactions

Janet L. Cyr
Laboratory of Sensory Neuroscience
Rockefeller University
New York, NY 10021, USA
Motor Proteins

Krishnamurti Dakshinamurti
Department of Biochemistry and
Molecular Biology
Faculty of Medicine
University of Manitoba
Winnipeg, Manitoba R3E OW3
Canada
Vitamin Receptors

Michael A. Dalrymple
PPL Therapeutics Ltd
Roslin, Edinburgh
Scotland EH25 9PP
United Kingdom
Transgenic Livestock as Bioreactors

Ann K. Daly
Department of Pharmacological
Sciences
University of Newcastle upon Tyne
The Medical School
Pharmacogenetics Research Unit
Newcastle upon Tyne NE2 4HH
United Kingdom
Pharmacogenetics

Michael David
Intellectual Property
Foley & Lardner
Attorneys
Washington, DC 20007, USA
Pesticidal Compounds from Prokaryotes

Ronald W. Davis
Department of Biochemistry
Beckman Center
Stanford University
School of Medicine
Stanford, CA 94305, USA
Yeast Artificial Chromosome (YAC) Construction and Applications

Christine Debouck
Biopharmaceutical Research and
Development
SmithKline Beecham Pharmaceuticals
King of Prussia, PA 19406, USA
AIDS Therapeutics, Biochemistry of

Thomas R. Defay
Graduate Group in Biophysics
University of California
San Francisco, CA 94143, USA
Protein Modeling

D. M. de Kretser
Institute of Reproduction and
Development
Monash University
Clayton, Victoria 3168
Australia
Male Reproductive System, Molecular Biology of

Claire Delahunty
Department of Molecular
Biotechnology
University of Washington
Seattle, WA 98195, USA
Ligation Assays

Rob DeSalle
Department of Entomology
American Museum of Natural History
New York, NY 10024, USA
Paleontology, Molecular

Raymond Devoret
Institut Curie
Centre Universitaire
F-91405 Orsay, France
Mutagenesis

Frank DiCosmo
Department of Botany
Institute of Biomedical Engineering
and
Department of Surgery
University of Toronto
Toronto, Ontario M5S 3B2
Canada
Taxol® and Taxane Production by Cell Culture

Scott R. Diehl
Molecular Epidemiology and Disease
Indicators Branch
National Institute of Dental Research
National Institutes of Health
Bethesda, MD 20892, USA
Human Disease Gene Mapping

Rodney R. Dietert
Institute for Comparative and
Environmental Toxicology and
Department of Microbiology
and Immunology
College of Veterinary Medicine
Cornell University
Ithaca, NY 14853, USA
Immunotoxicological Mechanisms

Helen Donis-Keller
Departments of Surgery, Genetics,
and Psychiatry
Washington University
School of Medicine
St. Louis, MO 63110, USA
Human Chromosome 14, Genetic and Physical Map Status of

Trevor Douglas
Department of Biological and
Physical Sciences
Montana State University
Billings, MT 59101, USA
Inorganic Solids, Biomolecules in the Synthesis of

Horace R. Drew
Laboratory of Molecular Biology
CSIRO Division of Biomolecular
Engineering
North Ryde, New South Wales
2113 Australia
DNA Structure

James Driscoll
Department of Molecular Biology
Harvard Medical School
Boston, MA 02115, USA
Major Histocompatibility Complex

Mitchell L. Drumm
Departments of Pediatrics and Genetics
Case Western Reserve University
Cleveland, OH 44106, USA
Cystic Fibrosis

Rex A. Dunham
Department of Fisheries and
Allied Aquacultures
Auburn University
Auburn, AL 36849, USA
Transgenic Fish

P. Leslie Dutton
Department of Biochemistry
and Biophysics
Johnson Research Foundation
University of Pennsylvania
Philadelphia, PA 19104, USA
Electron Transfer, Biological

Fritz Eckstein
Max-Planck-Institut für
Experimentelle Medizin
D-37075 Göttingen, Germany
Oligonucleotides

Randy C. Eisensmith
Department of Cell Biology
Baylor College of Medicine
Houston, TX 77030, USA
Phenylketonuria, Molecular Genetics of

Samir S. El-Dahr
Department of Pediatrics
Tulane University School of Medicine
New Orleans, LA 70112, USA
Renal System

Walther R. Ellis, Jr.
National Center for the Design
of Molecular Function
Utah State University
Logan, UT 84322, USA
Metalloenzymes

Ravindhra G. Elluru
Department of Cell Biology
and Neuroscience
University of Texas Southwestern
Medical Center
Dallas, TX 75235, USA
Motor Proteins

Charles J. Epstein
Department of Pediatrics
University of California
San Francisco, CA 94143, USA
Down Syndrome, Molecular Genetics of

Henry A. Erlich
Human Genetics Department
Roche Molecular Systems, Inc.
Alameda, CA 94501, USA
PCR Technology

Alun E. Evans
Department of Epidemiology and
Public Health
The Queens University of Belfast
Belfast VT12 6BJ, Northern Ireland
Heart Failure, Genetic Basis of

Louise Fairall
Medical Research Council
Laboratory of Molecular Biology
Cambridge CB2 2QH, United Kingdom
Zinc Finger DNA-Binding Motifs

Michael Famulok
Institut für Biochemie der
Universität München
Genzentrum
D-81375 München
Germany
Nucleic Acid Enzymes, Evolutionary Engineering of

Ramy S. Farid
Department of Chemistry
Rutgers University
Newark, NJ 07102, USA
Electron Transfer, Biological

Daniel H. Farkas
Department of Clinical Pathology
William Beaumont Hospital
Royal Oak, MI 48073, USA
Pathology, Molecular

Christine J. Farr
Department of Genetics
University of Cambridge
Cambridge CB2 3EH
United Kingdom
Telomeres, Mammalian

Robert E. Farrell, Jr.
Exon-Intron, Inc.
Columbia, MD 21045, USA
RNA Methodologies

Robert Feil
Laboratory of Developmental Genetics
and Imprinting
The Babraham Institute
Cambridge CB2 4AT
United Kingdom
Genomic Imprinting

Malcolm A. Ferguson-Smith
Department of Pathology
Cambridge University
Cambridge CB2 1QP, United Kingdom
Gene Order by FISH and FACS

T. L. Ferrell
Oak Ridge National Laboratory
Oak Ridge, TN 37381, USA
Scanning Probe Microscopy in Sequencing of DNA

James P. Ferris
Department of Chemistry
Rensselaer Polytechnic Institute
Troy, NY 12180, USA
Origins of Life, Molecular Biology of

James G. Ferry
Department of Biochemistry
and Molecular Biology
Eberly College of Science
Pennsylvania State University
Paul M. Althouse Laboratory
University Park, PA 16802, USA
Methanogens and Archaea, Molecular Biology of

Arthur G. Fett-Neto
Department of Botany
University of Toronto
Toronto, Ontario M5S 3B2, Canada
Taxol® and Taxane Production by Cell Culture

J. K. Findlay
Prince Henry's Institute of Medical
Research
Clayton, Victoria 3168, Australia
Female Reproductive System, Molecular Biology of

John B. C. Findlay
Department of Biochemistry
and Molecular Biology
University of Leeds
Leeds LS2 9JT, United Kingdom
Protein Sequencing Techniques

John J. Finer
Department of Horticulture
and Crop Science
Plant Molecular Biology
and Biotechnology Program
Ohio Agricultural Research
and Development Center
The Ohio State University
Wooster, OH 44691, USA
Plant Cell Transformation, Physical Methods for

Kim R. Finer
Department of Biological Sciences
Kent State University, Stark Campus
Canton, OH 44720, USA
Plant Cell Transformation, Physical Methods for

Eric W. Fisher
Department of Biochemistry
University of Illinois at
Urbana-Champaign
Urbana, IL 61801, USA
Restriction Endonucleases and DNA Modification Methyltransferases for the Manipulation of DNA

James E. Fleming
Institute of Molecular Medical Sciences
Palo Alto, CA 94306, USA
Aging, Genetic Mutations in

Donald E. Fosket
Department of Developmental
and Cell Biology
University of California
Irvine, CA 92717, USA
Plants, Molecular Biology of

Ron M. Fourney
Biology Research
and Development Section
Central Forensic Laboratory
Royal Canadian Mounted Police
Ottawa, Ontario K1G 2M3, Canada
DNA Fingerprint Analysis

Maurille J. Fournier
Department of Biochemistry
and Molecular Biology
University of Massachusetts
Amherst, MA 01003, USA
Polymers, Genetic Engineering of

Sally A. Francis
Department of Plant Sciences
Oxford University
Oxford OX1 3RB, United Kingdom
Plant Pathology, Molecular

Errol C. Friedberg
Laboratory of Molecular Biology
Department of Pathology
University of Texas Southwestern
Medical Center
Dallas, TX 75235, USA
Yeast, DNA Repair in

Simon H. Friedman
Department of Pharmaceutical
Chemistry
School of Pharmacy
University of California
San Francisco, CA 94143, USA
Bioorganic Chemistry

Frank K. Fujimura
Myriad Genetic Laboratories, Inc.
Salt Lake City, UT 84108, USA
Genetic Testing

Abram Gabriel
Department of Molecular Biology
and Biochemistry
Rutgers University
Piscataway, NJ 08854, USA
*Transposons in the
Human Genome*

Betty J. Gaffney
Department of Chemistry
Johns Hopkins University
Baltimore, MD 21218, USA
*Nuclear Magnetic Resonance of
Biomolecules in Solution*

Brenda L. Gallie
Department of Ophthalmology
and Division of Immunology
and Cancer Research
The Hospital for Sick Children
and
Department of Medical Genetics
and Ophthalmology
University of Toronto
Toronto, Ontario M5G 1X8
Canada
*Retinoblastoma, Molecular
Genetics of*

Ian Garner
PPL Therapeutics Ltd.
Edinburgh EH25 9PP
United Kingdom
*Transgenic Livestock as
Bioreactors*

Daniel Gautheret
Structural and Genetic Information,
CNRS-EP91
13402 Marseille, France
*RNA Three-Dimensional
Structures, Computer
Modeling of*

Francesco Giannelli
Division of Medical and Molecular
Genetics
United Medical and Dental Schools
Guy's and St. Thomas Hospitals
London SE1 9RT, United Kingdom
Hemophilia

Alison M. Goate
Department of Psychiatry
Washington University School
of Medicine
St. Louis, MO 63110, USA
Alzheimer's Disease

Karen A. Golemboski
Department of Microbiology and
Immunology
College of Veterinary Medicine
Cornell University
Ithaca, NY 14853, USA
Immunotoxicological Mechanisms

R. Ariel Gomez
Department of Pediatrics
University of Virginia Health
Sciences Center
Charlottesville, VA 22908, USA
Renal System

Karen M. Gooding
Micra Scientific, Inc.
Lafayette, IN 47903, USA
*HPLC of Biological
Macromolecules*

D. B. Gower
St. Bartholomew's and The Royal
School of Medicine and Dentistry
London E1 2AD
United Kingdom
Steroid Analysis

Gregory A. Grant
Department of Molecular Biology
and Pharmacology
Washington University
School of Medicine
St. Louis, MO 63110, USA
Peptides, Synthetic

Duane Graves
Biotechnology Applications Center
IT Corporation
Knoxville, TN 37923, USA
*Biodegradation of
Organic Wastes*

Joe W. Gray
Cancer Genetics Program
Cancer Center
University of California
San Francisco, CA 94143, USA
*Whole Chromosome
Complementary Probe
Fluorescence Staining*

Elias Greenbaum
Chemical Technology Division
Oak Ridge National Laboratory
Oak Ridge, TN 37831, USA
*Biomolecular Electronics and
Applications*

Judith E. Grisel
Veterans Affairs Medical Center and
Department of Medical Psychology
Oregon Health Sciences University
Portland, OR 97201, USA
*Drug Addiction and Alcoholism,
Genetic Basis of*

Joanna Groden
Department of Molecular Genetics, Biochemistry, and Microbiology
University of Cincinnati
College of Medicine
Cincinnati, OH 45267, USA
Colon Cancer

Peter J. Gruber
Department of Surgery
Johns Hopkins Hospital
Baltimore, MD 21287, USA
Cardiovascular Diseases

Richard I. Gumport
Department of Biochemistry and College of Medicine
University of Illinois at Urbana-Champaign
Urbana, IL 61801, USA
Restriction Endonucleases and DNA Modification Methyltransferases for the Manipulation of DNA

Sarah Jane Gurr
Department of Plant Sciences
Oxford University
Oxford OX1 3RB, United Kingdom
Plant Pathology, Molecular

James F. Gusella
Harvard Medical School and Molecular Neurogenetics Unit
Massachusetts General Hospital
Charlestown, MA 02129, USA
Huntington's Disease

Ulf Gyllensten
Department of Medical Genetics
University of Uppsala Biomedical Center
S-751-23 Uppsala, Sweden
Mitochondrial DNA

Mac E. Hadley
Department of Cell Biology and Anatomy
University of Arizona
Tucson, AZ 85724, USA
Peptides

Tatsuya Haga
Institute for Brain Research
University of Tokyo
Tokyo 113, Japan
Receptor Biochemistry

Nabil G. Hagag
University Medical Center
State University of New York
Stony Brook, NY 11794, USA
Chromosome Microdissection and Microcloning

Caroline B. Hall
Department of Pediatrics and Medicine
School of Medicine and Dentistry
University of Rochester
Rochester, NY 14621, USA
Respiratory Syncytial Virus Vaccine, Approaches to

Jeffrey C. Hall
Department of Biology
Brandeis University
Waltham, MA 02254, USA
Drosophila Mating, Genetic Basis of

Barry Halliwell
Neurodegenerative Disease Research Centre
Pharmacology Group
King's College
University of London
London SW3 6LX, United Kingdom
Free Radicals in Biochemistry and Medicine

Hui-Quan Han
Laboratory of Molecular and Cellular Neuroscience
The Rockefeller University
New York, NY 10021, USA
Synapses

Kimber Hardy
Geneva Biomedical Research Institute
Glaxo Wellcome Research and Development S.A.
Case Postale 674
Geneva, Switzerland
Plasmids

Calvin B. Harley
Geron Corporation
Menlo Park, CA 94025, USA
Telomerase

Lisa M. Harrison-Bernard
Department of Physiology
Tulane University School of Medicine
New Orleans, LA 70112, USA
Renal System

Anya Harry
Department of Pharmacology
Mount Sinai School of Medicine
City University of New York
New York, NY 10029, USA
Heterotrimeric G Proteins, Signal Transduction by

Yoshihide Hayashizaki
Tsukuba Life Science Center
Institute of Physical and Chemical Research (RIKEN)
Tsukuba, Ibaraki 305, Japan
Restriction Landmark Genomic Scanning Method

Joyce E. Heckman
Department of Microbiology and Molecular Genetics
Markey Center for Molecular Genetics
University of Vermont
Burlington, VT 05405, USA
Ribozymes

Olaf Heidenreich
Max-Planck-Institut für Experimentelle Medizin
D-37075 Göttingen, Germany
Oligonucleotides

Ilkka M. Helander
Laboratory of Vaccine Development
Department of Vaccines
KTL, National Public Health Institute
FIN-00300 Helsinki, Finland
Lipopolysaccharides

Michael Hengartner
Cold Spring Harbor Laboratory
Cold Spring Harbor, NY 11724, USA
Cell Death and Aging, Molecular Mechanisms of

C. J. K. Henry
School of Biological and Molecular Science
Oxford Brookes University
Gipsy Lane Campus, Headington
Oxford OX3 0BP, United Kingdom
Nutrition

Amanda C. Heppell-Parton
Clinical Oncology and Radiotherapeutics Unit
Medical Research Council Centre
Cambridge CB2 2QH, United Kingdom
Gene Mapping by Fluorescence In Situ Hybridization

William R. Hiatt
Calgene, Inc.
Davis, CA 95616, USA
Tomatoes, Gene Alterations of

Michio Himeno
Department of Applied Biological Chemistry
College of Agriculture
University of Osaka
Sakai, Osaka 593, Japan
Scleroderma Diagnosis with Recombinant Protein

Shinji Hirotsune
Tsukuba Life Science Center
Institute of Physical and Chemical Research (RIKEN)
Tsukuba, Ibaraki 305, Japan
Restriction Landmark Genomic Scanning Method

A. Rus Hoelzel
Laboratory of Viral Carcinogenesis
National Cancer Institute
Frederick, MD 21702, USA
Genetic Analysis of Populations

Frits A. Hommes
Human Genetics Program
Department of Pediatrics
New York University Medical Center
New York, NY 10016, USA
Biochemical Genetics, Human

Ian A. Hope
Department of Biology
University of Leeds
Leeds LS2 9JT, United Kingdom
Nematodes, Neurobiology and Development of

Barbara A. Horwitz
Division of Biological Sciences
Section of Neurobiology, Physiology, and Behavior
University of California
Davis, CA 95616, USA
Cellular Energetics, Biochemical Basis of

Zuzana Hostomska
Agouron Pharmaceuticals, Inc.
San Diego, CA 92121, USA
AIDS HIV Enzymes, Three-Dimensional Structures of

Zdenek Hostomsky
Agouron Pharmaceuticals, Inc.
San Diego, CA 92121, USA
AIDS HIV Enzymes, Three-Dimensional Structures of

Christopher J. Howe
Department of Biochemistry
Cambridge University
Cambridge CB2 1QW, United Kingdom
Nucleic Acid Sequencing Techniques

Victor J. Hruby
Department of Chemistry
University of Arizona
Tucson, AZ 85721, USA
Peptides and Mimics, Design of Conformationally Constrained

Lawrence A. Hunt
Department of Microbiology and Immunology
University of Louisville School of Medicine
Louisville, KY 40292, USA
Varicella-Zoster Virus and Cytomegalovirus Vaccines

Iain Hunter
Department of Bioscience and Biotechnology
University of Strathclyde
Glasgow G1 1QW, United Kingdom
Streptomycetes, Molecular Biology of

Jeffrey R. Idle
Institute for Cancer Research and Molecular Biology
Medisinsk Teknisk Senter
Norwegian University of Science and Technology
7005 Trondheim, Norway
Pharmacogenetics

Barbara H. Iglewski
Department of Microbiology and Immunology
University of Rochester School of Medicine
Rochester, NY 14642, USA
Bacterial Pathogenesis

Ross B. Inman
Institute for Molecular Virology and
Department of Biochemistry
University of Wisconsin
Madison, WI 53706, USA
Partial Denaturation Mapping

Thomas L. Innerarity
Gladstone Institute of Cardiovascular Disease
Cardiovascular Research Institute
Department of Pathology
University of California
San Francisco, CA 94141, USA
Plasma Lipoproteins

Vivian F. Irish
Department of Biology
Osborn Memorial Laboratory
Yale University
New Haven, CT 06520, USA
Arabidopsis Genome

Michael H. Irwin
Department of Comparative Medicine
University of Alabama at Birmingham
Birmingham, AL 35294, USA
Transgenic Animal Modeling

R. Elwyn Isaac
Department of Biology
University of Leeds
Leeds LS2 9JT, United Kingdom
Nematodes, Neurobiology and Development of

Susan L. Ishaug
Institute of Biosciences and Bioengineering
Rice University
Houston, TX 77251, USA
Polymers for Biological Systems

Ravi Iyengar
Department of Pharmacology
Mount Sinai School of Medicine
City University of New York
New York, NY 10029, USA
Heterotrimeric G Proteins, Signal Transduction by

Robert C. Jackson
Agouron Pharmaceuticals, Inc.
San Diego, CA 92121, USA
Cancer Chemotherapy, Theoretical Foundations of

John A. Jaeger
Genta, Inc.
San Diego, CA 92122, USA
RNA Secondary Structures

David Jenkins
The Diabetic Centre
Worcester Royal Infirmary
Worcester WR5 1HN, United Kingdom
Diabetes Mellitus

L. B. Jorde
Department of Human Genetics
University of Utah Health
Sciences Center
Salt Lake City, UT 84112, USA
Population-Specific Genetic Markers and Disease

Millard M. Judy
Baylor Research Institute
Baylor University Medical Center
Dallas, TX 75226, USA
Viruses, Photodynamic Inactivation of

Jerzy Jurka
Genetic Information Research Institute
Los Altos, CA 94024, USA
Human Repetitive Elements

Raymond E. Kaiser, Jr.
Lilly Research Laboratories
Eli Lilly & Company
Indianapolis, IN 46285, USA
Mass Spectrometry of Biomolecules

Eric R. Kandel
Center for Neurobiology and Behavior
Howard Hughes Medical Institute
College of Physicians and Surgeons
of Columbia University
New York, NY 10032, USA
Memory and Learning, Molecular Basis of

John P. Kane
Cardiovascular Research Institute
University of California
San Francisco, CA 94143, USA
Lipid and Lipoprotein Metabolism

George J. Kantor
Department of Biological Sciences
Wright State University
Dayton, OH 45435, USA
DNA Damage and Repair

Seymour Kaufman
Laboratory of Neurochemistry
National Institute of Mental Health
Bethesda, MD 20892, USA
Brain Aromatic Amino Acid Hydroxylases

Stephen G. Kayes
Department of Structural and Cellular
Biology
University of South Alabama
Mobile, AL 36688, USA
Genetic Vaccination, Gene Transfer Techniques for Use in

Frank Kee
Department of Epidemiology and
Public Health
The Queens University of Belfast
Belfast VT12 6BJ
Northern Ireland
Heart Failure, Genetic Basis of

Larry K. Keefer
National Cancer Institute
Frederick Cancer Research
and Development Center
Frederick, MD 21702, USA
Nitric Oxide in Biochemistry and Drug Design

Jeffrey N. Keen
Department of Biochemistry
and Molecular Biology
University of Leeds
Leeds LS2 9JT
United Kingdom
Protein Sequencing Techniques

Jack D. Keene
Department of Microbiology
Duke University Medical Center
Durham, NC 27710, USA
Molecular Diversity, Theoretical Aspects of

Jeffrey L. Kelly
Biosys, Inc.
Columbia, MD 21046, USA
Pesticidal Compounds from Prokaryotes

David J. Kemp
Menzies School of Health Research
Darwin, Northern Territory 0811
Australia
Parasite Genome Projects

Timothy E. Kennedy
Department of Anatomy
School of Medicine
University of California
San Francisco, CA 94143, USA
Memory and Learning, Molecular Basis of

George L. Kenyon
Department of Pharmaceutical
Chemistry
School of Pharmacy
University of California
San Francisco, CA 94143, USA
Bioorganic Chemistry

Michael W. Kilpatrick
Department of Clinical Genetics
University of Birmingham
Birmingham BI5 2TG
United Kingdom
Human Chromosome 15, Genetic and Physical Map Status of

Christopher J. Kirby
Cortecs Research Laboratory
University of London
London School of Pharmacy
London WC1N 1AX
United Kingdom
Lipid-Based Gene Vectors

Lorne T. Kirby (retired)
Department of Pathology
University of British Columbia
Vancouver, British Columbia V6H 3V4
Canada
DNA Fingerprint Analysis

Hannah L. Klein
Department of Biochemistry
New York University Medical Center
New York, NY 10016, USA
Recombination, Molecular Biology of

Aaron Klug
Laboratory of Molecular Biology
Medical Research Council
Cambridge CB2 2QH
United Kingdom
Protein Designs for the Specific Recognition of DNA

Akira Kobata
Tokyo Metropolitan Institute
of Gerontology
35–2 Sakaecho, Itabashi-Ku
Tokyo 173, Japan
Glycobiology

Andrzej K. Konopka
BioLingua Research, Inc.
Gaithersburg, MD 20879, USA
Theoretical Molecular Biology

Michael D. Koob
Department of Neurology
University of Minnesota
Medical School
Minneapolis, MN 55455, USA
Achilles' Cleavage

Helen Kozman
Department of Cytogenetics and Molecular Genetics
Centre for Medical Genetics
Women's and Children's Hospital
North Adelaide, South Australia 5006
Australia
Human Chromosome 16, Genetic and Physical Map Status of

Joachim Krebs
Laboratory of Biochemistry
Swiss Federal Institute of Technology (ETH)
CH-8092 Zurich, Switzerland
Calcium Biochemistry

Larry J. Kricka
Department of Pathology and Laboratory Medicine
University of Pennsylvania
Philadelphia, PA 19104, USA
Chemiluminescence and Bioluminescence, Analysis by

Raju Kucherlapati
Department of Molecular Genetics
Albert Einstein College of Medicine
Bronx, NY 10461, USA
Molecular Genetic Medicine

David J. Kwiatkowski
Division of Experimental Medicine and Hematology-Oncology
Brigham and Women's Hospital
Harvard Medical School
Boston, MA 02115, USA
Human Chromosome 9, Genetic and Physical Map Status of

Marek Kwiatkowski
Department of Medical Genetics
Uppsala Biomedical Center
University of Uppsala
S-75123 Uppsala, Sweden
Ligase-Mediated Gene Detection

Pui-Yan Kwok
Division of Dermatology
Washington University School of Medicine
St. Louis, MO 63110, USA
Ligation Assays

Monique Lafon
Département de Virologie
Institut Pasteur
75724 Paris, France
Superantigens

Joseph R. Lakowicz
Department of Biological Chemistry
University of Maryland School of Medicine
Baltimore, MD 21201, USA
Fluorescence Spectroscopy of Biomolecules

Eric Lam
Department of Plant Sciences
Ag Biotech Center
Rutgers University
New Brunswick, NJ 08903, USA
Plant Gene Expression Regulation

Kit S. Lam
Arizona Cancer Center
Department of Medicine, Microbiology, and Immunology
College of Medicine
University of Arizona
Tucson, AZ 85724, USA
Synthetic Peptide and Nonpeptide Combinatorial Libraries

Ulf Landegren
Department of Medical Genetics
Uppsala Biomedical Center
University of Uppsala
S-75123 Uppsala, Sweden
Ligase-Mediated Gene Detection

Robert Langer
Department of Chemical Engineering
Massachusetts Institute of Technology
Cambridge, MA 02139, USA
Polymers for Biological Systems

Janos K. Lanyi
Department of Physiology and Biophysics
University of California
Irvine, CA 92717, USA
Bacteriorhodopsin

Robin Leake
Department of Biochemistry
University of Glasgow
Glasgow G12 8QQ
United Kingdom
Steroid Hormones and Receptors

Gregor W. Leckie
Probe Diagnostics Business Unit
Abbott Laboratories
Abbott Park, IL 60064, USA
Infectious Disease Testing by Ligase Chain Reaction

Daniel Lednicer (Retired)
826 Bowie Road
Rockville, MD 20852, USA
Drug Synthesis

Helen H. Lee
Medical Research Council
Department of Hematology
University of Cambridge
Cambridge, CB2 2QH, United Kingdom
Infectious Disease Testing by Ligase Chain Reaction

Todd Leff
Department of Signal Transduction
Parke-Davis
Ann Arbor, MI 48105, USA
Cardiovascular Diseases

Daniele M. Leonard
Department of Chemistry
Parke-Davis Pharmaceutical Research
Ann Arbor, MI 48206, USA
Peptides

W. Lesser
Department of Agricultural Economics
New York State College of Agriculture and Life Science
Ithaca, NY 14853, USA
Transgenic Animal Patents

Jun Li
Department of Molecular Immunology
SmithKline Beecham Pharmaceuticals
King of Prussia, PA 19406, USA
Recognition and Immobilization, Molecular

Gertz Likhtenshtein
Department of Chemistry
Ben-Gurion University of the Negev
Beer-Sheva 84105, Israel
Labeling, Biophysical

Chun-Mean Lin
Biotechnology Center
University of Connecticut
Storrs, CT 06269, USA
Transgenic Fish

Iris Lindberg
Department of Biochemistry and Molecular Biology
Louisiana State University Medical Center
New Orleans, LA 70112, USA
Peptide Hormones, Biosynthesis, Processing, and Secretion of

Christine Loehrlein
Department of Biochemistry
Beckman Center
Stanford School of Medicine
Stanford, CA 94305, USA
Yeast Artificial Chromosome (YAC) Construction and Applications

Gordon V. Louie
Department of Crystallography
University of London
Birbeck College
London WC1E 7HX
United Kingdom
Protein Analysis by X-Ray Crystallography

K. A. L. Loveland
Institute of Reproduction and Development
Monash University
Clayton, Victoria 3168, Australia
Male Reproductive System, Molecular Biology of

Jenn-Kan Lu
Biotechnology Center
University of Connecticut
Storrs, CT 06269, USA
Transgenic Fish

Dimitrina Lyutzkanova
Department of Bioscience and Biotechnology
University of Strathclyde
Glasgow G1 1QW, United Kingdom
Streptomycetes, Molecular Biology of

Marcy E. MacDonald
Harvard Medical School and Molecular Neurogenetics Unit
Massachusetts General Hospital
Charlestown, MA 02129, USA
Huntington's Disease

Darryl MacGregor
Department of Biology
University of Leeds
Leeds LS2 9JT, United Kingdom
Nematodes, Neurobiology and Development of

Brendan C. Maguire
Department of Chemistry
Johns Hopkins University
Baltimore, MD 21218, USA
Nuclear Magnetic Resonance of Biomolecules in Solution

François Major
Départment d'Informatique et de Recherce Opérationnelle
Université de Montréal
Montreal, Quebec H36 3J7, Canada
RNA Three-Dimensional Structures, Computer Modeling of

P. Helena Mäkelä
Department of Vaccines
KTL, National Public Health Institute
FIN-00300 Helsinki, Finland
Lipopolysaccharides

Hisaji Maki
Department of Prokaryotic Molecular Genetics
Graduate School of Biosciences
Nara Institute of Science and Technology
Ikoma, Nara 630–01, Japan
DNA Replication and Transcription

H. L. J. Makin
St. Bartholomew's and The Royal School of Medicine and Dentistry
London E1 2AD, United Kingdom
Steroid Analysis

Stephen Mann
School of Chemistry
University of Bath
Bath BA2 7AY, United Kingdom
Inorganic Solids, Biomolecules in the Synthesis of

Marcel Mannens
Institute of Human Genetics
Academic Medical Center
University of Amsterdam
1105 AZ Amsterdam (ZO)
The Netherlands
Human Chromosome 11, Genetic and Physical Map Status of

Patricia Margaritte-Jeannin
INSERM U155
Unité de Recherches d'Epidémiologie Génétique
Château de Longchamp
75016 Paris, France
Breast Cancer, Genetic Analysis of

Bruce L. Martin
Department of Biochemistry
University of Tennessee
Memphis, TN 38163, USA
Post- and Co-translational Modifications of Proteins, Enzyme Catalyzed

Juan F. Martin
Area of Microbiology
Faculty of Biology
University of León
24071 León, Spain
Nonribosomal Peptides

R. Bruce Martin
Department of Chemistry
University of Virginia
Charlottesville, VA 22901, USA
Bioinorganic Chemistry

Belinda Martineau
Calgene, Inc.
Davis, CA 95616, USA
Tomatoes, Gene Alterations of

Edward P. Masler
U.S. Department of Agriculture
Agricultural Research Service
Plant Sciences Institute
Insect Neurobiology and Hormone Laboratory
Beltsville, MD 20705, USA
Neuropeptides, Insect

Thomas L. Mason
Department of Biochemistry and Molecular Biology
University of Massachusetts
Amherst, MA 01003, USA
Polymers, Genetic Engineering of

Kenichi Matsubara
Institute for Molecular and Cellular Biology
Osaka University
Osaka 565, Japan
Body Expression Map of the Human Genome

James L. Matthews
Baylor Research Institute
Baylor University Medical Center
Dallas, TX 75226, USA
Viruses, Photodynamic Inactivation of

Donald B. McCormick
Department of Biochemistry
Emory University School of Medicine
Atlanta, GA 30322, USA
Coenzymes, Biochemistry of; Vitamins, Structure and Function of

Johnjoe McFadden
School of Biological Sciences
University of Surrey
Guildford, Surrey GU2 5XH, United Kingdom
Mycobacteria

Brian McNeil
Department of Bioscience
and Biotechnology
University of Strathclyde
Glasgow G1 1XW, United Kingdom
Fungal Biotechnology

Anthony Meager
Division of Immunobiology
National Institute for Biological
Standards and Control
Herts, England EN6 3QG, United
Kingdom
Interferons; Interleukins

J. E. Mercer
Department of Obstetrics
and Gynaecology
Monash Medical Centre
Monash University
Clayton, Victoria 3168, Australia
Female Reproductive System, Molecular Biology of

Bradley T. Messmer
Laboratory of Molecular Genetics
and Informatics
Rockefeller University
New York, NY 10021, USA
Genetic Intelligence, Evolution of

Robert Miceli
Research and Development
Gull Laboratories
Salt Lake City, UT 84117, USA
Recognition and Immobilization, Molecular

Catherine H. Mijovic
Department of Medicine
University of Birmingham
Clinical Research Block
Queen Elizabeth Hospital
Birmingham B15 2TH, United
Kingdom
Diabetes Mellitus

Antonios G. Mikos
Institute of Biosciences and
Bioengineering
Rice University
Houston, TX 77251, USA
Polymers for Biological Systems

Michael Mitas
Department of Biochemistry
and Molecular Biology
Oklahoma State University
Stillwater, OK 74078, USA
Single-Stranded DNA Binding Proteins (SSBs)

David L. Mitchell
M.D. Anderson Cancer Center
Science Park, Research Division
University of Texas
Smithville, TX 78957, USA
Ultraviolet Radiation Damage to DNA

R. Jeffrey Moffatt
Department of Comparative Medicine
and the Transgenic Animal/ES Cell
Resource
University of Alabama at Birmingham
Birmingham, AL 35294, USA
Transgenic Animal Modeling

Hirotada Mori
Nara Institute of Science and
Technology
Ikoma, Nara 630–01, Japan
E. coli *Genome*

Robert L. Moritz
Joint Protein Structure Laboratory
Ludwig Institute for Cancer Research
(Melbourne Branch)
and
The Walter and Eliza Hall Institute
of Medical Research
Parkville, Victoria 3050, Australia
Proteins and Peptides, Isolation for Sequence Analysis of

Alan G. Morris
Department of Biological Sciences
University of Warwick
Coventry CV4 7AL, United Kingdom
Cytokines

Sherie L. Morrison
Department of Microbiology and
Molecular Genetics
The Molecular Biology Institute
University of California
Los Angeles, CA 90095, USA
Antibody Molecules, Genetic Engineering of

Bernice E. Morrow
Department of Molecular Genetics
Albert Einstein College of Medicine
Bronx, NY 10461, USA
Molecular Genetic Medicine

Christopher C. Moser
Department of Biochemistry
and Biophysics
Johnson Research Foundation
University of Pennsylvania
Philadelphia, PA 19104, USA
Electron Transfer, Biological

John Mulley
Department of Cytogenetics and
Molecular Genetics
Centre for Medical Genetics
Women's and Children's Hospital
North Adelaide, South Australia 5006
Australia
Human Chromosome 16, Genetic and Physical Map Status of

Masami Muramatsu
Tsukuba Life Science Center
Institute of Physical and Chemical
Research (RIKEN)
Tsukuba, Ibaraki 305, Japan
Restriction Landmark Genomic Scanning Method

Yoshinao Muro
Department of Dermatology
Nagoya University School of Medicine
Nagoya 466, Japan
Scleroderma Diagnosis with Recombinant Protein

Yusaku Nakabeppu
Department of Biochemistry
Medical Institute of Bioregulation
Kyushu University
Fukuoka 812, Japan
DNA Replication and Transcription

Carl F. Nathan
Department of Medicine
Division of Hematology-Oncology
Cornell University Medical College
New York, NY 10021, USA
Nitric Oxide in Biochemistry and Drug Design

L. Gabriel Navar
Department of Physiology
Tulane University School of Medicine
New Orleans, LA 70112, USA
Renal System

Eberhard Neumann
Faculty of Chemistry
University of Bielefeld
D-33501 Bielefeld, Germany
Electric and Magnetic Field Reception

Mabel Ng
Max-Planck-Institut für
Experimentelle Medizin
D-37075 Göttingen
Germany
Oligonucleotides

Antony Nicholls
Department of Biochemistry and Molecular Biophysics
Columbia University
New York, NY 10032, USA
Immunology

Robert A. Nichols
Departments of Pharmacology and of Neurobiology and Anatomy
MCP-Hahnemann School of Medicine
Allegheny University of the Health Sciences
Philadelphia, PA 19129, USA
Synapses

Deborah Nickerson
Department of Molecular Biotechnology
University of Washington
Seattle, WA 98195, USA
Ligation Assays

Mats Nilsson
Department of Medical Genetics
Uppsala Biomedical Center
University of Uppsala
S-75123 Uppsala, Sweden
Ligase-Mediated Gene Detection

Jiri Novotny
Department of Macromolecular Structure
Bristol-Myers Squibb Research Institute
Princeton, NJ 08543, USA
Immunology

Kimberly L. Ogden
Department of Chemical Engineering
University of Arizona
Tucson, AZ 85721, USA
Bioprocess Engineering

Masaru Ohashi
Department of Dermatology
Nagoya University School of Medicine
Nagoya 466, Japan
Scleroderma Diagnosis with Recombinant Protein

Yasushi Okazaki
Tsukuba Life Science Center
Institute of Physical and Chemical Research (RIKEN)
Tsukuba, Ibaraki 305, Japan
Restriction Landmark Genomic Scanning Method

Kousaku Okubo
Institute for Molecular and Cellular Biology
Osaka University
Osaka 565, Japan
Body Expression Map of the Human Genome

Charles Onwulata
U.S. Department of Agriculture
Agricultural Research Service
Eastern Regional Research Center
Philadelphia, PA 19118, USA
Food Proteins, Interactions of

B. A. Oostra
Department of Clinical Genetics
Erasmus University
3000 DR Rotterdam, The Netherlands
Fragile X Linked Mental Retardation

Paul R. Ortiz de Montellano
Department of Pharmaceutical Chemistry
School of Pharmacy
University of California
San Francisco, CA 94143, USA
Biotransformations of Drugs and Chemicals

Jurg Ott
Laboratory of Statistical Genetics
Rockefeller University
New York, NY 10021, USA
Human Linkage Maps

Joan Overhauser
Department of Biochemistry and Molecular Biology
Thomas Jefferson University
Philadelphia, PA 19107, USA
Human Chromosome 18, Genetic and Physical Map Status of

F. Owen
Division of Neuroscience
School of Biological Sciences
University of Manchester
Manchester M13 9PT, United Kingdom
Neuropsychiatric Diseases

Mehmet Ozturk
Laboratoire d'Oncologie Moléculaire
INSERM CJF 9302
Centre Leon Berard
69373 Lyon, France
Liver Cancer, Molecular Biology of

Andrew H. Palin
Department of Genetics
University of Cambridge
Cambridge CB2 3EH
United Kingdom
Telomeres, Mammalian

Sunil D. Pandit
Division of Human Molecular Genetics
Department of Surgery
Washington University School of Medicine
St. Louis, MO 63110, USA
Human Chromosome 14, Genetic and Physical Map Status of

Nicholas Parris
U.S. Department of Agriculture
Agricultural Research Service
Eastern Regional Research Center
Philadelphia, PA 19118, USA
Food Proteins, Interactions of

Luciano Passador
Department of Microbiology and Immunology
University of Rochester School of Medicine
Rochester, NY 14642, USA
Bacterial Pathogenesis

Fredrick M. Pavalko
Department of Physiology and Biophysics
Indiana University School of Medicine
Indianapolis, IN 46202, USA
Cytoskeleton-Plasma Membrane Interactions

William J. Payne
Department of Microbiology
University of Georgia
Athens, GA 30602, USA
Nitric Oxide in Biochemistry and Drug Design

Frederica P. Perera
Division of Environmental Health Sciences
School of Public Health
Columbia University
New York, NY 10032, USA
Epidemiology, Molecular

Boris Matija Peterlin
Departments of Medicine,
Microbiology, and Immunology
Howard Hughes Medical Institute
University of California
San Francisco, CA 94143, USA
AIDS, Transcriptional Regulation of HIV in

Anusch Peyman
Hoechst Aktiengesellschaft
General Pharma Research G838
D-65926 Frankfurt am Main, Germany
Antisense Oligonucleotides, Structure and Function of

Simon E. V. Phillips
Department of Biochemistry
and Molecular Biology
University of Leeds
Leeds LS2 9JT, United Kingdom
Repressor-Operator Recognition

Daniel Pinkel
Department of Laboratory Medicine
Cancer Genetics Program
Cancer Center
University of California
San Francisco, CA 94143, USA
Whole Chromosome Complementary Probe Fluorescence Staining

Carl A. Pinkert
Department of Comparative Medicine
and the Transgenic Animal/ES Cell
Resource
University of Alabama at Birmingham
Birmingham, AL 35294, USA
Transgenic Animal Modeling

K. Michael Pollard
Department of Molecular and
Experimental Medicine
W. M. Keck Autoimmune
Disease Center
The Scripps Research Institute
La Jolla, CA 92037, USA
Autoantibodies and Autoimmunity

Sue Povey
Medical Research Council
Human Biochemical Genetics Unit
The Galton Laboratory
University College London
London NW1 2HE
United Kingdom
Human Chromosome 9, Genetic and Physical Map Status of

Theresa P. Pretlow
Department of Pathology
Case Western Reserve University
School of Medicine
Cleveland, OH 44106, USA
Cancer

Thomas G. Pretlow
Department of Pathology
Case Western Reserve University
School of Medicine
Cleveland, OH 44106, USA
Cancer

Nicholas C. Price
Department of Biological and
Molecular Sciences
University of Stirling
Stirling FK9 4LA
United Kingdom
Circular Dichroism in Protein Analysis

Stanley B. Prusiner
Departments of Neurology
and of Biochemistry
and Biophysics
University of California
San Francisco, CA 94143, USA
Prions

Clive R. Pullinger
Cardiovascular Research Institute
University of California
San Francisco, CA 94143, USA
Lipid and Lipoprotein Metabolism

Robert R. Rando
Department of Biological Chemistry
and Molecular Pharmacology
Harvard Medical School
Boston, MA 02115, USA
Retinoids

Gregory M. Raner
Department of Biological Chemistry
University of Michigan
Medical School
Ann Arbor, MI 48109, USA
Metalloenzymes

Madeline E. Rasche
Department of Chemistry and
Biochemistry
Utah State University
Logan, UT 84322, USA
Methanogens and Archaea, Molecular Biology of

Stephen W. Raso
Department of Chemistry
Texas A&M University
College Station, TX 77843, USA
Enzymes, Energetics and Regulation of Biological Catalysis by

Colin Ratledge
Department of Biological Sciences
University of Hull
Hull HU6 7RX, United Kingdom
Lipids, Microbial

Anthony R. Rees
School of Biology and Biochemistry
University of Bath
Bath, BA2 7AY, United Kingdom
X-Ray Diffraction of Biomolecules

Gavin E. Reid
Joint Protein Structure Laboratory
Ludwig Institute for Cancer Research
(Melbourne Branch)
and
The Walter and Eliza Hall Institute of
Medical Research
Parkville, Victoria 3050, Australia
Proteins and Peptides, Isolation for Sequence Analysis of

Wolf Reik
Laboratory of Developmental Genetics
and Imprinting
The Babraham Institute
Cambridge CB2 4AT, United Kingdom
Genomic Imprinting

Renate Reimschuessel
Department of Pathology
Aquatic Pathobiology Center
University of Maryland at Baltimore
Baltimore, MD 21201, USA
Transgenic Fish

Alistair G. C. Renwick
Institute of Health
and Community Medicine
University of Malaysia Sarawak
94300 Kota Samarahan
Sarawak, Malaysia
Glycoproteins, Secretory

Arnold Revzin
Department of Biochemistry
Michigan State University
East Lansing, MI 48824, USA
Nucleic Acid-Protein Complexes, Laboratory Techniques in

Paul Ribeiro
Institute of Parasitology
McGill University
Ste. Anne-de-Bellevue
Montreal, Quebec H9X 3V9
Canada
Brain Aromatic Amino Acid Hydroxylases

Andrew P. Rice
Division of Molecular Virology
Baylor College of Medicine
Houston, TX 77030, USA
Virology, Molecular

Robert I. Richards
Department of Cytogenetics
and Molecular Genetics
Centre for Medical Genetics
Women's and Children's Hospital
Adelaide 5006, Australia
Fragile Sites

Timothy J. Richmond
Institute for Molecular Biology and Biophysics
Swiss Federal Institute of Technology (ETH)
CH-8093 Zurich, Switzerland
Chromatin Formation and Structure

Ernst T. Rietschel
Research Center Borstel
Center for Medicine and Biosciences
D-23845 Borstel, Germany
Lipopolysaccharides

Mary F. Roberts
Merkert Chemistry Center
Boston College
Chestnut Hill, MA 02167, USA
Phospholipases

Jean-David Rochaix
Departments of Molecular Biology and Plant Biology
University of Geneva
CH-1221 Geneva 4, Switzerland
Chlamydomonas

Martin Rosenberg
Biopharmaceutical Research and Development
SmithKline Beecham Pharmaceuticals
King of Prussia, PA 19406, USA
AIDS Therapeutics, Biochemistry of

Walter Rosenthal
Rudolf-Buchheim-Instut für
Pharmakologie der Justus-Liebig-
Universität Gießen
D-35392 Gießen, Germany
Diabetes Insipidus

Belina J. F. Rossiter
Human Genome Center and
Department of Molecular and
Human Genetics
Baylor College of Medicine and
4421 Darsey Street
Bellaire, TX 77401, USA
Human Genetic Predisposition to Disease

Jack A. Roth
M. D. Anderson Cancer Center
Department of Thoracic and
Cardiovascular Surgery
University of Texas
Houston, TX 77030, USA
Lung Cancer, Molecular Biology of

Nathaniel Rothman
School of Hygiene and Public Health
Johns Hopkins University
Baltimore, MD 21202, USA
Epidemiology, Molecular

Guy A. Rouleau
Center for Research in Neuroscience
McGill University and Montreal
General Hospital
Montreal, Quebec H3G 1A4, Canada
Neurofibromatosis 2

Robert L. Rubin
Department of Molecular and
Experimental Medicine
W. M. Keck Autoimmune
Disease Center
The Scripps Research Institute
La Jolla, CA 92037, USA
Autoantibodies and Autoimmunity

Michael E. Rusiniak
Departments of Molecular and Cellular Biology
Roswell Park Cancer Institute
Buffalo, NY 14263, USA
Kallikrein-Kininogen-Kinin System

Alan J. Russell
Department of Chemical Engineering
Center for Biotechnology
and Bioengineering
University of Pittsburgh
Pittsburgh, PA 15261, USA
Enzymology, Nonaqueous

Milton H. Saier, Jr.
Department of Biology
University of California at San Diego
La Jolla, CA 92093, USA
Transport Proteins

Theodora W. Salcedo
RIN Department
Zeneca Pharmaceuticals
Wilmington, DE 19850-5437, USA
Protein Phosphorylation

Martina Samiotaki
Department of Medical Genetics
Uppsala Biomedical Center
University of Uppsala
S-75123 Uppsala, Sweden
Ligase-Mediated Gene Detection

Ka-Yiu San
Department of Chemical Engineering
Rice University
Houston, TX 77251, USA
Expression Systems for DNA Processes

Linda J. Sandell
Veterans Affairs Medical Center
and Departments of Orthopaedics
and Biochemistry
University of Washington
Seattle, WA 98108, USA
Extracellular Matrix

Takeshi Sano
Center for Advanced Biotechnology and
Department of Biomedical Engineering
Boston University
Boston, MA 02215, USA
Immuno-PCR

Eliane R. Santarem
Department of Horticulture
and Crop Science
Plant Molecular Biology
and Biotechnology Program
Ohio Agricultural Research
and Development Center
The Ohio State University
Wooster, OH 44691, USA
Plant Cell Transformation, Physical Methods for

Mansoor Sarfarazi
Department of Surgery
Surgical Research Center
University of Connecticut Health Center
Farmington, CT 06030, USA
Human Chromosome 15, Genetic and Physical Map Status of

Robert D. C. Saunders
Department of Anatomy and Physiology
University of Dundee
Dundee DD1 4HN, United Kingdom
Drosophila *Genome*

Tomi K. Sawyer
Department of Chemistry
Parke-Davis Pharmaceutical Research
Ann Arbor, MI 48206, USA
Peptides

John G. Scandalios
Department of Genetics, Box 7614
North Carolina State University
Raleigh, NC 27695, USA
Environmental Stress, Genomic Responses to

John Schimenti
Jackson Laboratory
Bar Harbor, ME 04609, USA
Mammalian Genome

Milton J. Schlesinger
Department of Molecular Microbiology
Washington University School of Medicine
St. Louis, MO 63110, USA
Viral Envelope Assembly and Budding

Martin Schrott
Institute of Plant Sciences
Swiss Federal Institute of Technology
ETH-Zentrum
CH-8092 Zurich, Switzerland
Plant Cells, Genetic Manipulation of

Heinz-Peter Schuchmann
Max-Planck-Institut für Strahlenchemie
D-45470 Mülheim an der Ruhr, Germany
Ionizing Radiation Damage to DNA

Paul A. Schulte
National Institute for Occupational Safety and Health
Centers for Disease Control and Prevention
Cincinnati, OH 45226, USA
Epidemiology, Molecular

John W. R. Schwabe
Medical Research Council
Laboratory of Molecular Biology
Cambridge CB2 2QH, United Kingdom
Zinc Finger DNA-Binding Motifs

Clay W Scott
RIN Department
Zeneca Pharmaceuticals
Wilmington, DE 19850-5437, USA
Protein Phosphorylation

Anita Seibold
Graduate School of Biomedical Sciences
University of Texas Health Science Center
Houston, TX 77030, USA
Diabetes Insipidus

Mutsuo Sekiguchi
Department of Biochemistry
Medical Institute of Bioregulation
Kyushu University
Fukuoka 812, Japan
DNA Replication and Transcription

Bert L. Semler
Department of Microbiology and Molecular Genetics
College of Medicine
University of California
Irvine, CA 92717, USA
Poliovirus, Molecular Biology of

Michael Sendtner
Department of Neurology
University of Würzburg
D-97080 Würzburg, Germany
Motor Neuron Disease

Philip Serwer
Department of Biochemistry
University of Texas Health Science Center
San Antonio, TX 78284, USA
Viruses, DNA Packaging of

Mike J. Shamblott
The Johns Hopkins University School of Medicine
Baltimore, MD 21287, USA
Transgenic Fish

Prem Mohini Sharma
Department of Medicine
University of California at San Diego
La Jolla, CA 92093, USA
Muscle, Molecular Genetics of Human; Tumor Suppressor Genes

S. R. Dev Sharma
PrimeX General Imaging
Carlsbad, CA 92008, USA
Tumor Suppressor Genes

R. C. Sheppard
Medical Research Council
Molecular Biology Laboratory
Cambridge CB2 2QH, United Kingdom
Present Address:
15 Kinnaird Way
Cambridge CB1 4SN
United Kingdom
Peptide Synthesis, Solid Phase

Fred Sherman
Department of Biochemistry and Biophysics
University of Rochester Medical Center
Rochester, NY 14642, USA
Yeast Genetics

Lee P. Shulman
Department of Obstetrics and Gynecology
Division of Reproductive Genetics
University of Tennessee
Memphis, TN 38103, USA
Sex Determination

David Sidransky
Department of Otolaryngology—Head and Neck Surgery
Head and Neck Cancer Research Division
Johns Hopkins University School of Medicine
Baltimore, MD 21205, USA
DNA in Neoplastic Disease Diagnosis

Gary A. Silverman
Department of Pediatrics
Harvard Medical School
Boston, MA 02115, USA
Human Chromosome 18, Genetic and Physical Map Status of

Elizabeth M. Simpson
Jackson Laboratory
Bar Harbor, ME 04609, USA
Mammalian Genome

Richard J. Simpson
Joint Protein Structure Laboratory
Ludwig Institute for Cancer Research (Melbourne Branch)
and
The Walter and Eliza Hall Institute of Medical Research
Parkville, Victoria 3050, Australia
Proteins and Peptides, Isolation for Sequence Analysis of

E. Roy Skinner
Department of Molecular and Cell Biology
University of Aberdeen
Marischal College
Aberdeen AB9 1AS
United Kingdom
Lipoprotein Analysis

Robert James Slater
Division of Biosciences
University of Hertfordshire
Hatfield, Hertfordshire AL10 9AB
United Kingdom
Radioisotopes in Molecular Biology

Donald M. Small
Department of Biophysics
Boston University Medical Center
Boston, MA 02118, USA
Lipids, Structure and Biochemistry of

Cassandra L. Smith
Center for Advanced Biotechnology and Departments of Biomedical Engineering, Biology, and Pharmacology
Boston University
Boston, MA 02215, USA
Human Chromosomes, Physical Maps of; Immuno-PCR

Lloyd M. Smith
Department of Chemistry
University of Wisconsin
Madison, WI 53706, USA
DNA Sequencing in Ultrathin Gels, High Speed

D. Peter Snustad
Department of Genetics and Cell Biology
University of Minnesota
St. Paul, MN 55108, USA
Genetics

John Sommerville
School of Biological and Medical Sciences
University of St. Andrews
St. Andrews, Fife KY16 9TS
United Kingdom
Electron Microscopy of Biomolecules

Clemens von Sonntag
Max-Planck-Institut für Strahlenchemie
D-45470 Mülheim an der Ruhr
Germany
Ionizing Radiation Damage to DNA

Stephen E. Specter
Department of Nutrition
University of California
Davis, CA 95616, USA
Cellular Energetics, Biochemical Basis of

Gary Spedding
Department of Chemistry
Butler University
Indianapolis, IN 46208, USA
Ribosome Preparation and Protein Synthesis Techniques

Gary S. Stein
Department of Cell Biology
University of Massachusetts Medical Center
Worcester, MA 01655, USA
Histones

Janet L. Stein
Department of Cell Biology
University of Massachusetts Medical Center
Worcester, MA 01655, USA
Histones

Thomas Steiner
Institut für Kristallographie
Freie Universität Berlin
D-14195 Berlin, Germany
Hydrogen Bonding in Biological Structures

Willem P. C. Stemmer
Maxygen, Inc.
Palo Alto, CA 94304, USA
Sexual PCR and Assembly PCR

Wolfgang Stephan
Department of Zoology
University of Maryland
College Park, MD 20742, USA
Tandemly Repeated Noncoding DNA Sequences

David Stephens
Department of Surgery
St. George's Hospital Medical School
London SW17 0RE, United Kingdom
Protein Targeting

Lance E. Steward
Department of Chemistry
University of California
Irvine, CA 92717, USA
Protein Expression by Expansion of the Genetic Code

Peter G. Stockley
Department of Genetics
University of Leeds
Leeds LS2 9JT, United Kingdom
Repressor-Operator Recognition

Neil G. Stoker
Department of Clinical Sciences
London School of Hygiene & Tropical Medicine
London WC1E 7HT, United Kingdom
Mycobacteria

Alfred Stracher
Department of Biochemistry
Health Science Center at Brooklyn
State University of New York
Brooklyn, NY 11203, USA
Drug Targeting and Delivery, Molecular Principles of

Kenji Sugimoto
Laboratory of Applied Molecular Biology
Department of Applied Biological Chemistry
College of Agriculture
University of Osaka
Sakai, Osaka 593, Japan
Scleroderma Diagnosis with Recombinant Protein

Grant R. Sutherland
Department of Cytogenetics and Molecular Genetics
Centre for Medical Genetics
Women's and Children's Hospital
Adelaide 5006, Australia
Fragile Sites

Jeremy Tame
Department of Chemistry
University of York
York YO1 5DD, United Kingdom
Hemoglobin, Genetic Engineering of

Eng M. Tan
Department of Molecular and Experimental Medicine
W. M. Keck Autoimmune Disease Center
The Scripps Research Institute
La Jolla, CA 92037, USA
Autoantibodies and Autoimmunity

William R. Taylor
Division of Mathematical Biology
National Institute of Medical Research
London NW7 1AA, United Kingdom
Sequence Alignment of Proteins and Nucleic Acids

Bernard Testa
Institut de Chimie Thérapeutique
Section de Pharmacie
Université de Lausanne
CH-1015 Lausanne, Switzerland
Xenobiotic Metabolism

David S. Thaler
Laboratory of Molecular Genetics and Informatics
Rockefeller University
New York, NY 10021, USA
Genetic Intelligence, Evolution of

Edward Theil
Center for Human Genome Studies
University of California
Lawrence Berkeley Laboratory
Berkeley, CA 94720, USA
Robotics and Automation in Genome Research

Fritz Thoma
Institute for Cell Biology
Swiss Federal Institute of Technology (ETH)
CH-8093 Zurich, Switzerland
Chromatin Formation and Structure

George J. Thomas, Jr.
Division of Cell Biology and Biophysics
School of Biological Sciences
University of Missouri
Kansas City, MO 64110-2499, USA
Protein Analysis by Raman Spectroscopy

Robert C. Thomson
Institute of Biosciences and Bioengineering
Rice University
Houston, TX 77251, USA
Polymers for Biological Systems

T. Thundat
Oak Ridge National Laboratory
Oak Ridge, TN 37831, USA
Scanning Probe Microscopy in Sequencing of DNA

David A. Tirrell
Department of Polymer Science and Engineering
University of Massachusetts
Amherst, MA 01003, USA
Polymers, Genetic Engineering of

Jane G. Tirrell
Biotechnology Program
University of Massachusetts
Amherst, MA 01003, USA
Polymers, Genetic Engineering of

Stephen Todd
Department of Microbiology and Molecular Genetics
College of Medicine
University of California
Irvine, CA 92717, USA
Poliovirus, Molecular Biology of

David J. Triggle
School of Pharmacy
State University of New York
Buffalo, NY 14260, USA
Medicinal Chemistry

John T. Turner
United States Department of Agriculture
Foreign Agricultural Service
Washington, DC 20250, USA
Pesticidal Compounds from Prokaryotes

Tom Tuschl
Max-Planck-Institut für Experimentelle Medizin
D-37075 Göttingen, Germany
Oligonucleotides

Eugen Uhlmann
Hoechst Aktiengesellschaft
General Pharma Research G838
D-65926 Frankfurt am Main
Germany
Antisense Oligonucleotides, Structure and Function of

Dennis E. Vance
Department of Biochemistry and Lipid and Lipoprotein Research Group
University of Alberta
Edmonton, Alberta T6G 2S2, Canada
Phospholipids

Rajender S. Varma
Department of Chemistry
Sam Houston State University
Huntsville, TX 77341, USA
and
Office of Research
Houston Advanced Research Center
The Woodlands, TX 77381, USA
Oligonucleotide Analogues, Phosphate-Free

A. J. M. H. Verkerk
Department of Clinical Genetics
Erasmus University
3000 DR Rotterdam
The Netherlands
Fragile X Linked Mental Retardation

Linda Vigilant
Department of Anthropology
Pennsylvania State University
University Park, PA 16802, USA
Mitochondrial DNA, Evolution of Human

Claude A. Villee
Department of Biological Chemistry and Molecular Pharmacology
Harvard Medical School
Boston, MA 02115, USA
RNA Replication

Michael V. Viola
University Medical Center
State University of New York
Stony Brook, NY 11794, USA
Chromosome Microdissection and Microcloning

David H. Viskochil
Department of Pediatrics
Division of Medical Genetics
University of Utah School of Medicine
Salt Lake City, UT 84112, USA
Neurofibromatosis 1

Adrian O. Vladutiu
Departments of Pathology, Microbiology, and Medicine
State University of New York
School of Medicine and Biomedical Sciences
Buffalo, NY 14214, USA
Isoenzymes

Georgirene D. Vladutiu
Departments of Pediatrics and Neurology
State University of New York
School of Medicine and Biomedical Sciences
Buffalo, NY 14214, USA
Isoenzymes

Stephen S. Wachtel
Department of Obstetrics and Gynecology
Division of Reproductive Genetics
University of Tennessee
Memphis, TN 38103, USA
Sex Determination

Edward E. Walsh
University of Rochester
School of Medicine and Dentistry
Infectious Disease Unit
Rochester, New York 14621, USA
Respiratory Syncytial Virus Vaccine, Approaches to

Denan Wang
Center for Advanced Biotechnology and Departments of Biomedical Engineering, Biology, and Pharmacology
Boston University
Boston, MA 02215, USA
Human Chromosomes, Physical Maps of

Larry D. Ward
Joint Protein Structure Laboratory
Ludwig Institute for Cancer Research (Melbourne Branch)
and
The Walter and Eliza Hall Institute of Medical Research
Parkville, Victoria 3050, Australia
Proteins and Peptides, Isolation for Sequence Analysis of

R. J. Warmack
Oak Ridge National Laboratory
Oak Ridge, TN 37831, USA
Scanning Probe Microscopy in Sequencing of DNA

Michael R. Waterman
Department of Biochemistry
Vanderbilt University School of Medicine
Nashville, TN 37232, USA
Cytochrome P450

Heinz-Ulrich G. Weier
Division of Life Sciences
Lawrence Berkeley National Laboratory
Berkeley, CA 94720, USA
Whole Chromosome Complementary Probe Fluorescence Staining

Otto Westphal
Chemin de Ballalaz 18
CH-1820 Montreux, Switzerland
Lipopolysaccharides

James G. Wetmur
Department of Microbiology
Mount Sinai School of Medicine
New York, NY 10029, USA
Nucleic Acid Hybrids, Formation and Structure of

André J. van Wijnen
Department of Cell Biology
University of Massachusetts Medical Center
Worcester, MA 01655, USA
Histones

Robert M. Williams
Department of Chemistry
Colorado State University
Fort Collins, CO 80523, USA
Amino Acid Synthesis

Steven A. Williams
Department of Biological Sciences
Clark Science Center
Smith College
Northampton, MA 01063, USA
Parasite Genome Projects

R. Anthony Williamson
Departments of Immunology and Molecular Biology
The Scripps Research Institute
La Jolla, CA 92037, USA
Combinatorial Phage Antibody Libraries

Brian Wimberly
Department of Molecular Biology
Scripps Research Institute
La Jolla, CA 92037, USA
RNA Structure, Nonhelical

Peter J. Wirth
Laboratory of Experimental Carcinogenesis
National Cancer Institute
National Institutes of Health
Bethesda, MD 20892, USA
Gel Eletrophoresis of Proteins, Two-Dimensional Polyacrylamide

Alexander Wlodawer
Macromolecular Structure Laboratory
National Cancer Institute
Frederick Cancer Research and Development Center
Frederick, MD 21702, USA
AIDS, Inhibitor Complexes of HIV-1 Protease in

Jon A. Wolff
Department of Pediatrics
University of Wisconsin
Madison, WI 53705, USA
Genetic Vaccination, Gene Transfer Techniques for Use in

Savio L. C. Woo
Department of Cell Biology and Molecular Genetics
Baylor College of Medicine
Houston, TX 77030, USA
Phenylketonuria, Molecular Genetics of

Ruliang Xu
Department of Microbiology
School of Dental Medicine
University of Pennsylvania
Philadelphia, PA 19104, USA
Virus-Receptor Interactions

Tilahun D. Yilma
Department of Veterinary Pathology, Microbiology, and Immunology
International Laboratory of Molecular Biology for Tropical Disease Agents
University of California
Davis, CA 95616, USA
Vaccine Biotechnology

Jeannine M. Yon
Laboratoire d'Enzymologie, Physiochimique et Moléculaire
Unité associé du Centre National de la Recherche Scientifique
Université de Paris-Sud
91405 Orsay, France
Protein Aggregation

Takashi Yura
HSP Research Institute
Kyoto Research Park
Kyoto 600, Japan
E. coli *Genome*

Eugene R. Zabarovsky
Microbiology and Tumor Biology Center
Karolinska Institute
Stockholm S-171 77, Sweden
DNA Markers, Cloned

Hanswalter Zentgraf
Applied Tumor Virology
German Cancer Research Center
D-69120 Heidelberg, Germany
Electron Microscopy of Biomolecules

Chun Zhou
Merkert Chemistry Center
Boston College
Chestnut Hill, MA 02167, USA
Phospholipases

Miriam M. Ziegler
Department of Biochemistry and Biophysics
Texas A&M University
College Station, TX 77843, USA
Enzymes, Energetics and Regulation of Biological Catalysis by

R. A. Zoeller
Department of Biophysics
Boston University Medical Center
Boston, MA 02118, USA
Lipids, Structure and Biochemistry of

Index

Main entries are printed in CAPITAL LETTERS with the page ranges given in **bold face.** The letter f following a page number indicates a figure; the letter t following a page number indicates a table. Volume numbers are indicated by **bold face** numbered followed by a colon.

A band
  C protein of, **4:**142
  myosin fibers, **4:**132
ABC excinuclease, **2:**45
ABC proteins (ATP-binding-cassette proteins), **1:**429
  superfamily of, **6:**130
Abdominal-b (abd-b) class of genes, HOX cluster, **3:**67
Abetalipoproteinemia, **3:**404, 412
Abscisic acid, **4:**492
Absorption, of vitamin A, **6:**236
Absorption spectroscopy
  biomolecular spectra, **2:**295f
  for following DNA denaturation, **2:**2, **2:**6t
Abundance, of essential and toxic elements, **1:**127
Ac (acetyl group), defined, **5:**485
Acceptor molecule, defined, **5:**516
Acceptors, in hydrogen bond interactions, **3:**262, 262f
  directionality of, **3:**269, 269f
Accessory pigments, for energy transfer, **4:**442
Accuracy, of biological marker identification, **2:**261
ACEDB (A *Caenorhabditis elegans* database), defined, **4:**167
Acetamidate linkages, in modified oligonucleotides, **4:**272, 272f
Acetaminophen
  conductimetric enzyme electrode measurement of, **1:**187, 187f
  toxicity of, **1:**200, 202
Acetate, methane production from, **4:**72, 73f
Acetate activation, genes for, in archaea, **4:**77
Acetic anhydride, reaction with nucleophiles, **1:**160
Acetoacetyl-CoA reductase, **3:**434f
N-α-Acetylated proteins, degradation of, ubiquitin-dependent, **3:**352
Acetylation
  of peptide hormones, **4:**345–46
  and posttranslational modification, **5:**52
  in xenobiotic metabolism, **6:**271–72, 271f
Acetylcholine
  activation of tyrosine hydroxylase by, **1:**215
  genes encoding synthesis and breakdown, in *Drosophila*, **2:**135–36
  of motor neurons, **4:**106
  of nematodes, **4:**169
Acetylcholinesterase
  inhibition of, by diisopropylfluorophosphate, **2:**236
  mutation of gene controlling, in *Drosophila*, **2:**136, 136f
N-Acetylglucosaminephosphotransferase, defect in, causing I-cell disease, **5:**203
N-Acetylneuraminic acid, defined, **1:**278
N-Acetyltransferase (NAT), **4:**404; **6:**271
ACHILLES' CLEAVAGE, **1:1–5**; **3:**219
Achiral compounds
  defined, **4:**269
  with dephospho internucleoside linkages, **4:**270
  glycine, **1:**336
  nuclease-resistant sulfamate linkage, **4:**277
  polyamide linkage in, polyamide nucleic acid, **4:**273
  sulfones, **4:**277
Acholinesterasemia, *Alu* insertion in cholinesterase gene, **6:**136
Acid-base catalysis, **2:**232
Acidemias, detection of, **1:**114
Acid-fastness, defined, **4:**156
Acid phosphatase, isoenzymes of, to diagnose damage to bone, prostate, or erythrocytes, **3:**375
Acids, hard and soft, **1:**129–30
Aconitase, **1:**344, 345f
  iron in, **4:**61–62
Acoustic wave biosensors, **1:**191–92, 191f
Acquired immunity, **3:**296, 297f
Acridine, photodynamic action of, **6:**207
Acridinium esters, as chemiluminescent labels, **1:**326
Acrocentric chromosome
  defined, **2:**118; **3:**160
  example, chromosome 21, **2:**119
Actin, **4:**137
  α-cardiac, marker of muscle differentiation, **4:**146
  cardiac muscle, gene localization to chromosome 15, **3:**167
  condensation of, response to *Listeria monocytogenes* infection, **1:**106
  and endocytosis, **4:**51
  F, crosslinker for, **1:**454
  filaments of, **4:**138f
    cytoskeleton, **1:**452
  pseudogene, mapping to chromosome 18, **3:**201
  in thin myofilaments, **4:**115, 130
α-Actin
  upregulation in heart failure, **1:**290
β-Actin
  cytoskeletal, mapping to chromosome 7, **3:**199–200
Actin-binding protein (ABP), **1:**453
α-Actinin, **1:**453–54, 454f
α-Actinin
  defined, **1:**451
  homodimer actin-binding domain, **4:**142f
  structure of, **4:**142
Actinomycin D, effect of, on cell death, **1:**299
Actinophage, ϕC31, **5:**502
Actinorhodin biosynthesis, **5:**507
  pathway for, **5:**503
Action potential, **5:**513–14
  defined, **5:**512
Activated keratinocytes
  defined, **2:**263
  markers of, **2:**266

Activating transcription factor (ATF), **3**:54
Activation
of enzymes, **2**:237
of oncogenes, **4**:286
Activation domain
factor IX, **3**:37
Tat transactivation, HIV-1, **1**:21–23
Activation energy
defined, **2**:223
of nucleation, **3**:310
and rate of reaction, **2**:227–28
Activator/activator protein
for CaM PK II, **1**:215
in enzyme-directed posttranslational modification, **5**:51
Active site
defined, **2**:223
of HIV-1 protease, **1**:27
of HIV integrase, **1**:33
Active transport
ATP-dependent, **6**:130
primary, **6**:131–32
secondary, **6**:126, 127
Activin, **2**:468; **5**:242
follicle-stimulating hormone regulation by, **2**:288
Activity, physical, and energy requirements, **4**:260t
Activity coefficient, and rate constant in an electric field, **2**:178
Acute lymphoid leukemia (ALL), translocation in, **2**:64, 64t
Acute myeloblastic leukemia, monosomy 18 associated with, **3**:209
Acute promyelocytic leukemia (APL), translocation in, **3**:160, 164
Acyl-CoA/cholesterol acyltransferase (ACAT), role in lipoprotein metabolism, **3**:410
Acyl compounds
acylamino acids, as coupling reagents in solid phase peptide synthesis, **4**:390
acylglycerols, structure and biochemistry of, **3**:456
acylglycerophospholipids, **3**:435
derivatives of carboxylic acids, **1**:160–61, 161f
N-Acyltransferases, **1**:198–99
Adaptins, **5**:204
Adaptive mutation, **2**:409
defined, **2**:407
mechanisms for, proposed, **2**:410
Adaptive systems, **4**:93–95
Adaptor proteins, **5**:241
AP1, role in budding from the trans-Golgi network, **4**:51
AP2, **4**:51
in endocytosis, **5**:204
defined, **2**:207
Grb2, in Ras-GDP activated complex, **2**:209–10
Addison's disease, familial form of, **3**:214
Addresses
electronic newsgroup, *Arabidopsis thaliana*, **1**:84
internet
Centraalbureau voor Schimmelcultures of, **6**:304
for chromosome 14 information sources, **3**:159
of the Yeast Genetic Stock Culture Center, **6**:304
Adenine (A), **2**:415
solubility of, and denaturation of DNA by solvents, **2**:10
Adenine nucleotides, concentration of, and control of respiration, **1**:312
Adenine nucleotide translocator protein (ANT), **1**:394
of mitochondria, **1**:309
Adenine radical ion, chemistry of, **3**:359f
Adeno-associated virus (AAV), gene transfer mediated by, **2**:436–37
Adenocarcinoma
ErbB2 amplification in, **4**:286
in IGF I transgenic medaka, **6**:96, 98f
lung, in smokers and nonsmokers, **3**:493
Adenomas
colon, **1**:408
defined, **1**:407
Adenomatous polyposis, familial (FAP). *See* Familial adenomatous polyposis
Adenomatous polyposis coli (APC), **6**:145
and colon cancer, **1**:408–9
Adenosine 5′-triphosphate, phosphorylation by, **1**:160
Adenosine 5′-(γ-thio)triphosphate (ATPγS), Achilles' cleavage protocol using, **1**:5
Adenosine deaminase (ADA) deficiency, gene therapy for, experimental, **4**:40–41, 101
Adenosine deaminase negative severe combined immunodeficiency syndrome, **3**:244
Adenosine dinucleotide phosphate-ribosyl transferase, **1**:401
Adenosine triphosphate (ATP), **1**:305–6
synthesis of, controls on, **1**:311–13
Adenosine triphosphate synthetase, **1**:308
S-Adenosylmethionine (SAM)
activity of, in Achilles' cleavage, **1**:4
binding to MetJ, **5**:277
damage to DNA caused by, **4**:302–3
methyl donor for posttranslational modification, **5**:54
Adenoviruses
defenses against antiviral reactions, **3**:325
DNA and RNA of, **5**:357, 357t
DNA packaging by, **6**:206
enteric-coated, for vaccination against respiratory disease, **6**:179–180
for genetic immunization, **2**:436
recombinant, **6**:181, 181f
in phenylketonuria therapy, **4**:413–14, 414f, 415f
replication-defective, for human gene therapy, **6**:182
transcription factor E1A, **5**:463
Adenylate charge, regulation of adenosine triphosphate by, **1**:311–12
Adenylate cyclase
activation by G-protein-linked receptors, **2**:21; **4**:355; **5**:235
interaction with G proteins, **5**:237
Adenylate cyclase activating polypeptide (*PACAP*) gene, pituitary, mapping to chromosome 18, **3**:208
Adenylate-forming enzyme family, **4**:213f
defined, **4**:207
Adenylate-forming region, of peptide synthetases, **4**:215
Adenylyl cyclase, **3**:46, 47f
regulation of, by G proteins, **3**:45
Adenylylimidodiphosphate (AMP-PNP)
effect on motor proteins, variation among families, **4**:118
effect on organelle motility in microtubles, **4**:117–18
Adherence, of pathogens to host cells, **1**:104
Adherens-type junctions, **1**:452, 457–58
Adhesins
of bacteria, **1**:104
defined, **1**:103
Adhesion
of cells, **1**:175, 293–97
calcium ion mediation of, **1**:239
carbohydrate mediation of, **2**:292
leukocyte adhesion deficiency, **1**:459
and membrane fusion, **4**:45–47, 47f
Adhesion molecules
in atherosclerotic lesions, **1**:288
CD4 and CD8, **2**:431
neural cell, **4**:56f
primary, **1**:292, 295
substrate, **1**:293
in T-cell interaction with antigen-presenting cells, **1**:446
Adhesion protein
paxillin, **1**:452, 454
phosphorylation of, **1**:458
phosphorylation of, in embryonic development, **1**:459
Adhesives, bone cement, from poly(methyl methacrylate), **5**:41
Ad hoc frequency counts, **6**:41

Adipokinetic hormones, **4:**195
I, **4:**197
cyclic AMP enhancement by, **4:**196
II, **4:**197
Adipose tissue, brown, thermogenesis by, **1:**313
Adjuvant
defined, **4:**156; **6:**172
for subunit vaccines, safety of, **6:**182
for vaccinia virus recombinant vaccines, **6:**170
Adjuvant genes, defined, **6:**165
ADP-ribosylation, **5:**52–53
ADP ribosylation factor (ARF), transport vesicle, **5:**201
ADP-ribosylation factor (ARF) proteins, as enzyme-directed activators, **5:**51
Adrenal androgens, synthesis of, **5:**480
Adrenal hyperplasia, congenital (CAH), **1:**437; **2:**214, 424; **5:**468
Adrenaline (epinephrine), lead for drug synthesis, **2:**157–58
β-Adrenergic, defined, **2:**155
Adrenergic receptors
types of, **1:**289–90
β-Adrenergic receptors, **5:**237
Adrenocortical carcinoma, mapping of, to chromosome 11, **3:**127
Adrenocorticotropic hormone (ACTH)
isolated congenital deficiency of, **2:**214
and P450, **1:**439
Adrenoleukodystrophy, **5:**206
Adriamycin, as an anticancer drug, **1:**254
Adsorption, of double-stranded DNA bacteriophage to a host cell, **6:**204
Adsorptive sequencing, **5:**187, 187t
Adult polycystic kidney disease, **2:**426
Aequorin, as a bioluminescence label, **1:**326
Aerobic organisms, bacteria, **1:**119–21
Afferent conduction, defined, **4:**53
Affine model, to adjust sequence alignment for gap size, **5:**422–23
Affinity chromatography
analytical, **4:**251; **5:**246f
for high density lipoprotein subfraction measurement, **3:**478
for molecular recognition, **5:**245
defined, **5:**243
screening with, **5:**249
Affinity maturation, **4:**95
Affinity selection, with synthetic peptide libraries, **5:**519
Affinity sensors, **1:**178–79, 179t
Aflatoxins
defined, **3:**480
and liver cancer, **3:**485
oxidation by cytochromeP450, to a carcinogen, **1:**202
and *p53* mutations, **3:**488–89
Agammaglobulinemia, mutation in, **4:**126–27, 126f
Agaricus bisporus, **2:**338
Agarose gel techniques
in DNA fingerprint analysis, **2:**55–56
with polymerase chain reaction, **4:**326
for preparing genomic DNA, **1:**4
in restriction landmark genomic scanning, **5:**307
Aggrecan (AGC1)
in cartilage, **2:**457
gene for, localization on chromosome 15, **3:**165
Aggregates
defined, **5:**73
inclusion body
β-lactamase, **5:**84
characteristics of, **5:**83
lipid, for assays of phospholipases, **4:**419f
protein, detection and characterization of, **5:**80–81
Aggregation theory, of cell growth, **1:**96
AGING
and Alzheimer's disease, **1:**42
incidence of, **1:**42; **4:**199
and capacity for DNA repair, **2:**46
cell
during the division cycle, **1:**96
and racemization of amino acids, **1:**336
and cell death
molecular mechanisms, **1:**299–303
role of telomerase in, **6:**26, 26f
of cell preparations for chromosome microdissection, **1:**372
defined, **1:**5; **2:**78
DNA damage theory of, **2:**79–81
GENETIC MUTATIONS IN, **4:**151, **1:5–14**z
and loss of synapses, **5:**516
maternal, and incidence of Down's syndrome, **2:**118
and mitochondrial mutations, **4:**84, 90
and onset and severity of Gaucher disease, **2:**349
organismal, **1:**302
and oxygen reactions, **1:**133
and posttranslational changes in enzymes, **3:**376
and protein synthesis, **5:**336
and telomerase, **6:**34–35
telomere hypothesis of, **6:**34
and telomere length, **3:**242
Aglycan, defined, **2:**446
Agonists, **3:**43
β-adrenergic, **2:**157–58
defined, **2:**155; **3:**42; **5:**229, 478
Agretope, defined, **3:**275
Agricultural applications
aquaculture using transgenic fish, **6:**92
of electrofusion, **2:**205
of molds and mold products, **2:**343–44
of patented transgenic animals, **6:**77
Agrobacterium rhizogenes, vector for gene transfer, **4:**451
Agrobacterium tumefaciens
defined, **4:**485; **6:**53
modification of tomato genes using, **6:**55, 55f
Ti plasmid of, **5:**9
tumors produced by, **5:**11–12
vector for gene transfer, **2:**277; **4:**451, 469
in dicotyledonous plants, **4:**459
in tomatoes, **6:**54
vector for plant genomes, **1:**80
AIDS
cytokine role in pathogenesis of, **1:**449
defined, **1:**14, 19
HIV ENZYMES, THREE-DIMENSIONAL STRUCTURE OF, **1:27–34**
INHIBITOR COMPLEXES OF HIV-1 PROTEASE IN, **1:14–19**
and mycobacteria, **4:**157
reduction in CD4 and CD8 T cells in, **3:**297
THERAPEUTICS, BIOCHEMISTRY OF, **1:34–42**
TRANSCRIPTIONAL REGULATION OF HIV IN, **1:19–26**
Ajmalicine, production of, from cell cultures, **6:**17
Alamethicin, nonribosomal peptide, **4:**207
Alanine, derivatization of the β-carbon, **1:**54
Alanine-glyoxylate transaminase (AGT), deficiency of, genetic disorder associated with, **5:**206
Alanine racemase, **1:**343
Albinism, oculocutaneous, mapping to chromosome 15, **3:**163
Albright hereditary osteodystrophy 2, deletion in chromosome 15, **3:**164
Albumin
bovine, matrix-assisted laser desorption ionization spectrum, **4:**4f
steroid hormones bound to, **5:**471, 481
Albuterol, **2:**158
structure of, **2:**159f
Alcohol
and liver cancer, **3:**485
pharmacogenetics of metabolism of, **4:**403–4
withdrawal from, mice, genetic studies, **2:**148–49
Alcohol dehydrogenases, **1:**195–96, 196f
in an organic solvent, EPR study, **2:**252
oxidation of allyl alcohol by, **1:**202

Alcohol dehydrogenases (*cont.*)
- secondary structure, from circular dichroism measurements, **1**:390
- from *Thermococcus litoralis*, **2**:245

Aldehyde dehydrogenases, **1**:195–96
- polymorphism in, and alcohol metabolism, **4**:403, 403t

Aldehyde oxidoreductase (AOR), of *Pyrococcus furiosus*, **2**:245, 248

Aldo/ketoreductase superfamily, 3α-HSD of, **5**:470

Aldolase
- hydroxyoxoglutarate, reaction at chiral center of the substrate, **1**:342
- staphylococcal, reversible thermodenaturation of, **5**:74

Aldosterone, control of sodium and potassium levels by, **5**:482

Alfalfa mosaic virus (AlMV), **6**:57

Algae, unicellular, lipids of, **3**:440–42

Alglucerase, for treating Gaucher disease, **2**:351

Algorithm
- backtracking, search tree generated by, **5**:380
- defined, **5**:420
- Needleman-Wunsch-Sellers, **5**:432–33

Alignment
- defined, **5**:420, 428
- domain for binding of DNA to telomerase, **6**:19
- significance of, **5**:423–24

Alkaline phosphatase
- analysis by anion exchange chromatography, **3**:80f
- chemiluminescence assay for, **1**:327, 329f
- isoenzyme separation by heat stability difference, **3**:374–75

Alkaline Southern transfer, of DNA, **2**:56

Alkaloids
- enantiomers of, **1**:338
- vinca, as anticancer drugs, **1**:254–55

Alkanes, bacterial metabolism of, **1**:120

Alkaptonuria, **1**:113

N-Alkylation, to modify backbone amides, **4**:372

Alkylation damage to DNA, **2**:41
- repair by O^6-methylguanine-DNA methyltransferase, **6**:284–85

Alkylsilyl linkages, in phosphate-free oligonucleotide analogues, **4**:277–78, 278f

Allatostatins, juvenile hormone regulation by, **4**:198

Alleles
- complementation test for, **2**:415–16
- defined, **3**:246
- in fragile X linked mental retardation, **2**:323
- for hemophilia A and hemophilia B, **3**:29
- heterogeneity of, **3**:223
- modifying oncoprotein function, **4**:286–87
- mutant
  - classes of, **2**:125
  - recovery of, **6**:317
  - single-step replacement of, yeast studies, **6**:317
- null
  - on chromosome 16, **3**:176
  - from gene targeting, **4**:103
- replacement of, **4**:163

Allele-specific ligation, **2**:423

Allele-specific oligonucleotides (ASOs), **2**:421–22
- for sequence polymorphism typing, **2**:57

Allelic loss, defined, **2**:62

Allelotyping of colonic tumor genes
- to follow tumor development, **1**:410–11
- for mapping, **1**:411

Allelozymes, **3**:366

Allenes, optical activity of, **1**:335

Allergic disease, kinins associated with, **3**:379

Allogeneic extrapolation, of immunotoxicity evaluations, **3**:300

Allostatins, **4**:194
- juvenile hormone production control by, **4**:190

Allosteric protein, defined, **3**:3

Allostery
- and cooperative binding of oxygen by hemoglobin, **3**:9–12, 9f
- defined, **2**:223
- in enzyme regulation, **2**:237
- and hormone activity, **2**:212
- study with spin labels, **3**:389

Allotropins, juvenile hormone production control by, **4**:190, 190f

Allotype, defined, **3**:275

Allyl alcohol, oxidation of, to toxic acrylaldehyde, **1**:202

Alphabet, genetic, **2**:415

Alpha-fetoprotein, **1**:261

Alpha-helix, **5**:119f
- amide bands corresponding to, **5**:97
- assignments for histone proteins, **1**:362f
- defined, **5**:128; **6**:327
- as a domain subunit, **5**:150
- in proteins, **2**:307
- reverse transcriptase, **1**:31
- Rnase H domain of HIV-1 reverse transcriptase, **1**:31

Alphaviruses, assembly and budding of, **6**:189–90

Alport syndrome, **2**:426

Alternating current/direct current effects, **2**:180–81

Alternative splicing
- *APC* gene, in colon cancer, **1**:412
- in *Drosophila melanogaster*, sex determination pathway, **2**:374f
- *FMR-1* genes, **2**:323
- of genes for fast-twitch and slow-twitch muscle forms, **1**:247
- HDL receptor gene, **3**:408
- isoforms of the plasma membrane calcium ion pump due to, **1**:246f
- and protein diversity, **2**:372–73
- RNA, defined, **1**:407

Aluminum, and incidence of Alzheimer's disease, **1**:42

Alu-PCR, for cloning DNA markers, **2**:76

Alu repeats, **3**:217–18, 243–44, 243f, 512; **5**:430
- defined, **3**:215
- LDL receptor gene, **3**:408
- polymorphic markers associated with, **2**:68

Alu sequences, **1**:371, 380; **3**:242; **6**:136
- defined, **6**:133

Alu source gene, **3**:244, 244f

ALZHEIMER'S DISEASE, **1:42–52**
- age of onset, **3**:223
- and apoE, **5**:46
- and apoE allele ϵ_4, **3**:414, 479–80
- contigs for, chromosome 14, **3**:151
- defined, **4**:199
- in Down syndrome, **2**:119, 333
- mapping to chromosome 14, **3**:155, 157
- mitochondrial mutations associated with, **4**:83
- synaptic connections in, **5**:516

Amadori rearrangement, **2**:310, 310f

Amantadine, inhibition of influenza A virus replication by, **6**:233

Amberization, **4**:300–1
- defined, **4**:299

Amber termination codon, **5**:145–46, 145f; **6**:201

Ambiguity in hybridization, **3**:253, 253f
- defined, **3**:252
- reconstruction of sequences of different lengths, **3**:254–55, 254f

American Type Culture Collection, addresses of, **6**:304

Amerindians, mtDNA markers, **4**:86–87

Amidases, hydrolysis by, **1**:196

Amidation
- of peptide hormones, **4**:345, 345f
- of peptides, **5**:53

Amide band
- defined, **5**:93
- I, **5**:96
  - Raman profiles of polypeptide chain conformations, **5**:100f
- I and III, in Raman analysis of polypeptides and proteins, **5**:99t

Amide bond torsion angle (ω), **4**:352, 363, 363f
Amide chain, hydrogen bond motifs decomposed into, **3**:269
Amides, **1**:160
  hydrolysis of, **1**:196
Amiloride, treatment for nephrogenic diabetes insipidus induced by Li^+, **2**:24
Amination, nucleophilic, of α-substituted acids, **1**:56
Amines, biogenic, of nematodes, **4**:170
Amino acids, **4**:263t; **5**:116–17, 117t
  β-alkyl, in peptide design, **4**:379
  analysis of
    for protein identification, **5**:180
    by HPLC, **5**:188t
  artificial, proteins containing, **5**:30–31
  chirality of, **1**:336–38
  defined, **4**:347; **5**:516
  dietary patterns of consumption, **4**:264t
  essential, **2**:307, 309
  metal binding by, **1**:132
  for polypeptide synthesis, **6**:124
  relatedness of, **5**:421–22
  sequence for PrP, **5**:58
  stabilizing and destabilizing, in ubiquitin- dependent protein degradation, **3**:350–51, 350t
  study of, in molecular paleontology, **4**:301, 301f, 303
  symbols for, **4**:364t
  α,β-unsaturated, in peptide design, **4**:379
α-Amino acids
  chiral, **1**:52–58
  α-hydrazido-, **1**:56
Amino acid sequences
  conserved, in P450s, **1**:439
  homology among integrin β subunits, **1**:459
  of human integrin β_1 and β_2 cytoplasmic domains, **1**:455f
D-Amino acid substitutions, in peptides, **4**:374
AMINO ACID SYNTHESIS, **1:52–58**
Aminoacylase 1, **3**:201
Aminoacyl-tRNA synthetases, defined, **5**:135, 327
α-Aminoadipyl-cysteinyl-valine (ACV)
  and the β-lactam biosynthetic pathway, **4**:211–12, 212f
  biosynthesis of, model, **4**:214–15f
α-Aminoadipyl-cysteinyl-valine synthetase, **4**:211–12
γ-Aminobutyric acid (GABA), **5**:230–33
  of nematodes, **4**:169–70
  potentiation by avermectins, **4**:396
  receptors for, **5**:233
    localization to chromosome 15, **3**:168
3-Aminocarboxydiaza-cyclodecadione, **4**:377
1-Aminocyclopropane-1-carboxylate synthetase (ACC synthetase), ethylene biosynthesis catalyzed by, **4**:474
1-Aminocyclopropane-1-carboxylic acid (ACC), ethylene biosynthesis from, in tomatoes, **6**:58
Aminoglycosides, and antibiotic resistance conferred by R plasmids, **5**:10
Amino groups, peptide, protecting during synthesis, **4**:383–84, 384f
Aminomethyl/coumarin (AMCA), label for FISH, **2**:384, 384t
Aminomethyl naphthoic acid, in peptides, **4**:377
3-Aminopiperidonecarboxylic acid, **4**:378f
Aminotransferase, function of pyridoxal 5′-phosphate, **1**:401–2, 402f
Amperometric biosensors, **1**:182–86
  defined, **1**:177
  electrodes, **1**:179
AMP-FLP (amplified fragment length polymorphism), **2**:51
  characteristics of loci, **2**:59t
  defined, **2**:48
Amphetamine, effects on rodents, genetic studies, **2**:150
Amphiphile, defined, **3**:442
Amphoteric compound, defined, **2**:352
Amplicon
  carryover contamination by, **4**:328
  defined, **3**:302; **4**:323
  detection of, LCR, **3**:306–7, 307f
  unit length, in polymerase chain reaction, **4**:326
Amplified polymorphic PCR (AP-PCR), defined, **2**:48
Amplification
  of c-*myc* in small-cell lung cancer, **3**:492–93
  detecting defect in hemophilia A, **3**:35f
  for detecting hemophilia mutations, **3**:34f
  in FISH, **2**:384
  of genes
    in cancers, **4**:286
    in hepatocellular carcinomas, **3**:487–88
  of a genomic library, **2**:69
  LCR, **3**:396, 396f
  linear and exponential techniques, **3**:400–1
  mechanisms of, in electric and magnetic fields, **2**:178–81
  in molecular diversity experiments, **4**:96
  nucleic acid, in vitro technologies, **4**:327t
  PCR, **4**:335f
  specific, **4**:339
  of tandem repeats, **6**:6
Amylase
  defined, **2**:240
  gene cluster, **6**:137
  isoenzyme separation to identify pancreatitis, **3**:375
  of *Pyrococcus woesei*, **2**:244
Amyloidosis, Dutch-type, **1**:43
Amyloid peptide
  circular dichroism measurements, **1**:389
  gene for, **3**:223
β-Amyloid peptide, **1**:42, **1**:43
  in Alzheimer's disease, **4**:199
  biochemistry of formation, **1**:46–47
  soluble, **1**:49
Amyloid plaques
  defined, **4**:198
  in prion disease, **5**:69–70
  PrP 27–30, **5**:61, 61f
Amyloid precursor protein (APP), **1**:42, 46–47
  amino acid substitutions in, **1**:45f
  functional domains of, **1**:47f
β-Amyloid protein, toxicity of, to neuronal cells, **4**:109
Amylopectin, **5**:33–34, 33f
Amyloplasts, **4**:488
Amylopullulanase, **2**:244
Amyotrophic lateral sclerosis (ALS), **2**:425
  defined, **4**:105
  pathology of, **4**:107
  sporadic, **4**:105
Anaerobic metabolism, bacterial, **1**:121
Anaerobic reactions, photosensitized, **6**:208
Analgesic, μ opioid, **4**:380–381
Analysis of variation, in population genetic studies, **2**:401
Analyte, defined, **1**:177
Analyte matrix, **1**:177
Analytical methods
  carbohydrate assays, **1**:267t
  for characterizing peptides, after solid state synthesis, **4**:388
  for cloned DNA evaluation, **1**:381–82
  for detecting UV damage to DNA, **6**:155–56
  for determining metal ions in enzymes, **4**:60
  for DNA denaturation assessment, **2**:2–5
  DNA fingerprint analysis, specimen processing and analysis, **2**:50–51
  electron microscopy, **2**:182–90
  ELISA, for ACA detection, **5**:418–19
  for evaluating telomerase, **6**:21–23, 23t
  for evaluating transgene integration and expression, **6**:71–72
  for foreign DNA determination, **4**:455–56

Analytical methods (*cont.*)
for foreign gene product determination, **4:**454–55
gel electrophoresis, **2:**353–60
double- and triple-label, **2:**356, 357f
for genetic analysis in populations, **2:**394–97
for genetic analysis of mtDNA, **4:**88
for genetic testing, **2:**420–21, 421t
for homologous recombination event detection, **4:**103
for immuno-PCR product identification, **3:**290–95
for LCR amplicon detection, **3:**307–08
for molecular paleontology, **4:**300, 305
for molecular recognition, with immobilized ligands, **5:**245–49
for mtDNA assessment, **4:**85
for neuropeptide identification, **4:**193–95
for nitric oxide identification, **4:**201
in PCR, **4:**337–38
for peptide synthesis control, **4:**391
phospholipase assay, **4:**418–19
for protein quantity determination, **5:**121–23, 121t
for protein sequence determination, **5:**186–88
for quantitation of VZV and CMV vaccine immune responses, **6:**183
reagent requirements in protein synthesis assays, **5:**331t
for RNA three-dimensional structure determination, **5:**378–79
scanning-tunneling microscopy, **5:**402–3
for synthetic peptide evaluation, **4:**368t
for following taxol^REG and taxane production, **6:**13–15
in transgene integration and expression, **6:**88–89
in vivo assays of promoter functions, **4:**470
for yeast genetic studies, **6:**308–10, 308f, 321–25
Anandamide, **4:**29
structure of, **4:**30f
Ancestors
common mtDNA, **4:**87–89
recency of, **4:**88–89
Ancient sources
quality of material from, **4:**301–2, 301t
study of, in molecular paleontology, **4:**305–6
Ancylostoma duodenale, **4:**168f
Androgenetic embryos
defined, **2:**439
development of, **2:**440, 441f
β-Androgenic blockers, propranolol, **4:**31
Androgen-insensitivity syndromes, **5:**471
Androgen receptor gene, in breast cancer in males, **3:**223–24
Androgens
adrenal, synthesis of, **5:**480
role of, **5:**482
steroid subgroup, **5:**479
testicular, synthesis of, **5:**480
Androstenedione, **5:**480
Anesthetics, fluorocarbon, antagonizing vasopressin, **2:**24
Aneuploidy
chromosome 16, lethal mutant, **3:**186
defined, **3:**191
syndromes involving, **3:**211–13
Aneusomy, detection of, in metaphase and interphase cells, **6:**256–57
Angelman syndrome (AS)
deletion in, **3:**161–62
genomic imprinting in, **2:**424, 445; **3:**160, 517
physical map of the chromosome 15 region involved in, **3:**161f
Angiogenesis, **1:**261, 450
cytokine role in, **1:**448
defined, **5:**485
fibroblast growth factors as agents of, **2:**469
Angiotensin, in essential hypertension, **2:**214
Angiotensin-converting enzyme (ACE), **1:**167–68, 168f; **5:**266
deletion in the gene for, **3:**2
inhibitor for, development as a drug, **4:**31
Angiotensin II receptor antagonist, losartan, **4:**31
Angiotensinogen (Ao), **5:**265, 266
Angiotensinogen gene (ATG), mapping of, to chromosome 1, **3:**92
Anilinothiazolinone (ATZ), **5:**174
Animal bioreactors, transgenic livestock as. *See* Transgenic livestock as bioreactors
Animal models
for antibody production, **1:**91t
for autoimmune disorders, **1:**85f, 90–92
genetic, for study of drugs of abuse, **2:**147–50
for heart failure, **1:**289
for human disease, **4:**103–4
cystic fibrosis, **1:**434
insulin-dependent diabetes mellitus, **2:**36–37
for immunotoxic evaluations, **3:**299, 300–1
for Krabbe disease, **3:**158
for motor neuron disease, **4:**109
and neurofibromatosis 1, **4:**179–80
for noninsulin-dependent diabetes mellitus, **2:**38
transgenic, **6:**66–69
Animal rights
defined, **6:**74
ethical issues in transgenic research, **6:**82–83
Animal welfare
defined, **6:**74
ethical issues transgenic research, **6:**82
Animal Welfare Act of 1985, **6:**82
Anions
nucleobase radical, **3:**359, 359f
sensitivity of DNA to denaturation in the presence of, **2:**10, 12
Aniridia gene, *PAX6*, mapping to chromosome 11, **3:**127
Anisotropy
defined, **2:**294
fluorescence, **2:**299
defined, **2:**294
Ankyrin, **1:**452
Annealing, defined, **3:**302; **4:**281, 333
ANNEXINS, **1:59–62**, 242
gene localization to chromosome 15, **3:**169
membrane fusion roles of, **4:**50
Annuloaortic ectasia, gene for, mapping to chromosome 15, **3:**163
Anode, defined, **2:**96
Anomalous scattering, for phase determination, X-ray crystallography, **6:**279
Anomeric structure, defined, **1:**266
Anomers, defined, **2:**446
Antagonistic pleiotropy, **1:**10
defined, **1:**5
Antagonists
adrenergic, **2:**157–58
bradykinin receptor, **3:**384
defined, **2:**155; **3:**42; **5:**229, 478
G protein receptor, **3:**43
peptide, therapeutic use of, **4:**359–60
Antennae
defined, **4:**440
pigments of, **4:**441–42
pigments of, modification in response to light, **4:**449–50
Antennapedia class of genes, **3:**66–67, 66f
Antennary structure, defined, **1:**266
Anterior pituitary peptides, **4:**341t
Anthocyanin, protection against UV damage by, **6:**157
Antiangiogenesis, by antitumor agents, **5:**498–99, 499f
Antibiotics
amino acids of, **1:**337
antagonism to antidiuretic actions of vasopressin, **2:**24
defined, **2:**155; **4:**393
hybrid, development using streptomycetes, **5:**506–7

ion-translocating, **6:**128
β-lactam, manufacture of, **2:**163–64, 164f
modification of, by plasmids, **5:**9
peptide, **4:**207, 354
resistance to
bacterial, **5:**9
cloning genes for, **5:**503
defined, **2:**271
plasmids encoding, **5:**9
sensitivity to, of archaea, **4:**76
Antibodies
biotinylated, in immuno-PCR, **3:**289
catalytic, **1:**170; **2:**239–40; **3:**288
chimeric, **1:**65–66
defined, **1:**84, 148; **5:**243
diversity of, dependence on somatic recombination, **3:**513
genes coding for, **3:**281–82, 282f
against HIV-1 envelope, **1:**36
against HIV-1 virus, **1:**420–21
identification of ligands for, **5:**523
isolation of, from peripheral blood lymphocytes, **1:**64, 416
maternally derived, against respiratory syncytial virus, **5:**291
monoclonal
electroporation for introducing into cells, **2:**205
production of, using electrofusion, **2:**205
neutralizing, **6:**233
analysis using, **6:**183
defined, **6:**172
from nonimmune libraries, **1:**422–24
sensors using, **1:**142
specificity of, **3:**283–85, 285f
and clinic utility, **1:**415
structure of, **1:**62–63, 63f; **3:**276–79, 276f
therapeutic applications of engineered molecules, **3:**286
against therapeutic factor VIII, in hemophilia A, **3:**35
in vitro evolution of, with sexual PCR, **5:**455, 457
Antibody-antigen assay, **1:**190
Antibody binding site
engineering of, **3:**286, 286f
structure of, **3:**284f
Antibody-dependent cell-mediated cytotoxicity (ADCC), **1:**65; **3:**327
Antibody-forming cell (AFC) assay, **3:**301
Antibody libraries
combinatorial phage, **1:**415–24
semisynthetic, **1:**424
ANTIBODY MOLECULES
in electron microscopy, **2:**190
GENETIC ENGINEERING OF, **1:62–69**
Antibody repertoire, defined, **1:**415
Anticentromere antibodies (ACAs), **5:**411
defined, **5:**410
in patients with limited scleroderma, **5:**417
Anticentromere autoantibodies, **5:**414–15, 414f
Antichaotropic salts, for hydrophobic interaction chromatography, **3:**81
α_1-Antichymotrypsin, gene for, mapping to chromosome 14, **3:**155
Anticipation, in an inherited disorder, **2:**328, 427
defined, **2:**419
Anticodons
defined, **5:**352; **6:**109
flexible loop, **5:**371
hairpin loop, tRNA, **5:**376
of tRNA-cognate synthetases, **5:**370, 370f
Antidiabetic, defined, **2:**155
Antidiuretic hormone (ADH). *See* Vasopressins
Antidromic hydrogen bonding, **3:**266
Antiestrogens
antitumor activity of, **5:**496
for cancer therapy, **5:**486
Antifreeze protein (AFP), gene for, introduction into fish, **6:**96
Antigen-antibody reactions, association rate constant, **1:**180
Antigen-binding fragments (Fab), of immunoglobulins, **3:**277
Antigen capture, **1:**421, 421f
Antigenicity
and cell-type tropism, **5:**16
cellular, **3:**276, 285
defined, **3:**275, 366
humoral, **3:**276
versus immunogenicity, **3:**285
Antigen-presenting cells (APCs), **1:**446; **2:**31; **5:**508
in autoantibody response, **1:**86
interaction with T and B lymphocytes, **3:**296
Antigens
bacterial, carbohydrate, **1:**278–79
CENP-B, in scleroderma, **5:**412
colonization factor, **5:**11
and cytokine production, **1:**444–45
defined, **1:**84, 148; **3:**499; **6:**172
diversity of, among respiratory syncytial virus isolates, **5:**288, 288f
epitope on the functional structure of, **5:**416–17
human blood group ABO(H), **1:**281
La, interaction with picornaviruses, **5:**19
MIC12, gene localization to chromosome 15, **3:**167
MIC12, mapping the gene for, to chromosome 15, **3:**167
molecular properties of, **3:**285
mycobacterial, **4:**159–61
oncofetal, **1:**261
O-polysaccharide, **1:**279; **3:**467–68
PM-Scl, **5:**414–15
presentation of, genes involved in, **2:**31
processing, genes involved with, **2:**31
purified protein, for vaccines, **6:**176
synthetic peptides corresponding to, **4:**368–71, 371f
transition state analogues, **1:**170, 170f
Antigen-specific antibodies, in response to MHC class II, **2:**431
Antihormones
as anticancer drugs, **1:**256
steroidal, **5:**484, 484f
Anti-idiotype vaccines, **6:**179
Antimetabolites
as anti-cancer drugs, **1:**254
defined, **1:**252
Antimicrobial activity, of interferons, **3:**326
Antimüllerian hormone (AMH), **5:**435
Antinuclear antibodies, in scleroderma, **5:**411
Antioncogenes, **6:**146
NF2, **6:**144
See also Tumor suppressor genes
Antioxidant response element (ARE), **2:**220
Antioxidants
defense mechanisms using, **2:**219–220, 333
defined, **2:**330
as enzyme defense mechanisms, **2:**220
vitamin C, **6:**238–40
vitamin E, **6:**238
Antiparallel base pairing
defined, **6:**109
mRNA and tRNA, **6:**112
Antiport, **6:**127
defined, **6:**125
example of, maintenance of the Na^+, K^+ gradient, **6:**131–32
Antiproliferative activity, defense against viruses, interferon involvement in, **3:**325–26
Antirepressor, Tat role as, **1:**25
Antisense blocker, enzymatic, **1:**173
Antisense genes, expression of, tomato plants, **6:**55
Antisense message, of the *lin-4* gene, in *Caenorhabditis elegans*, **1:**235
Antisense methodology, **4:**282
for studying lung cancer, **3:**493–94
ANTISENSE OLIGONUCLEOTIDES, **4:**269, 270

ANTISENSE OLIGONUCLEOTIDES (*cont.*)
for alcohol and drug sensitivity studies, **2:**152
analogue of, defined, **1:**69
cationic liposomes for delivery of, **3:**421
defined, **1:**69; **4:**281
DNA-DNA, with formacetal linkages, **4:**279
phosphorothioates, **4:**283
STRUCTURE AND FUNCTION OF, **1:69–76**
for therapy in neurological disorders, **5:**512
therapy using, **3:**418
with pH-sensitive liposomes, **3:**419
for treating neointimal hyperplasia, **3:**424
Antisense peptides, synthesis of, **4:**370
Antisense polynucleotides, effect on translation of RNA to protein, **6:**124
Antisense RNA
defined, **4:**465; **6:**53
effect on translation of RNA to protein, **6:**124
Antisense strands, **5:**355–56
defined, **5:**352
Antisense therapy, **1:**165; **5:**336
for glioma, **1:**252
Antisocial personality, tendency toward, in alcoholics and drug abusers, **2:**147
Antitemplate strategies, for developing antiviral drugs, **1:**40–41
α_1-Antitrypsin (AAT)
deficiency of, **2:**424
expression of the human gene in mice, **6:**105
gene for, mapping to chromosome 14, **3:**155
production of, in the milk of transgenic animals, **6:**106–7, 106t
Antiviral activity
association with interferons, **3:**325
mechanisms of, **3:**325t
Antiviral cytokines, list, **1:**444t
Aortic aneurysms, gene for, mapping to chromosome 15, **3:**160
Aplysia, sensitization in, **4:**54
Apocarbonic anhydrase, zinc ion binding to, **4:**60
Apocytochrome *c*, **5:**206
Apoenzyme, defined, **1:**397; **6:**244
Apolipoprotein A-I, **3:**410–11
Apolipoprotein A-II, **3:**411
Apolipoprotein A-IV, **3:**411
Apolipoprotein B, **1:**286; **3:**411–12
degradation of, **3:**353–54
Apolipoprotein C-I, **3:**412
Apolipoprotein C-II, **3:**412–13
Apolipoprotein D, **3:**413
Apolipoprotein E, **1:**286; **3:**413–14
in Alzheimer's disease, **1:**46
expression of, **4:**505
ϵ4 allele, **1:**46
gene for, in Alzheimer's disease, **3:**223
Apolipoprotein E4, in Alzheimer's disease, **1:**27, 50
Apolipoprotein J, **3:**414
Apolipoprotein regulatory protein (ARP1)
gene for
localization to chromosome 15, **3:**170
mapping to chromosome 15, **3:**165, 167
Apolipoproteins, **1:**284, 285t; **3:**410–15
apo(a), **1:**286; **3:**414–15, 472
characterization of, **4:**496, 498
concentration of, measuring, **3:**475–76
defined, **3:**404, 471; **4:**495
expression in transgenic mice, **4:**503t
properties and functions of, **3:**473t
regulatory protein, **3:**165
transport by, **4:**496
Aponitrous oxide reductase, copper incorporation, **4:**60
Apoptosis, **1:**257, 300–1f, 448–49
association with overexpression of *p53* tumor cells, **6:**141
Bcl-2 gene involved in, **2:**65
defined, **1:**253, 299; **2:**463
and G_1 cell cycle block, **1:**263
transforming growth factor β involvement in, **3:**482
Apparatus design, horizontal ultrathin gel electrophoresis, **2:**96–98, 97f
Approximation, catalysis by, **2:**230
AP sites (apurinic or apyrimidinic sites), **2:**43f
result of DNA damage, **2:**41
Aptamers
defined, **4:**226
DNA, selection of, **4:**227
nucleic acid, examples of, **4:**227–30
RNA
modeling of, **5:**385–86
selection of, **4:**227
Aquaporin, **5:**273
gene for, in atypical congenital nephrogenic diabetes insipidus, **2:**27
Arabidopsis Biological Resource Center (ABRC), **1:**84
ARABIDOPSIS GENOME, **1:76–84**
mapping of, **4:**491
retrotransposons of, **4:**488
Arabidopsis thaliana
defined, **1:**76; **4:**485
and plant molecular biology, **4:**490
Arachidonic acid
defined, **4:**432
eicosatetraenoic acid, **3:**453f
metabolism of
eicosanoids formed in, **4:**438
pathways involved in endotoxin reactions, **3:**471
metabolites of, release at inflammation sites, **1:**447
synthesis of
by algae, **3:**441
commercial, from molds, **3:**439
ARCHAEA
defined, **2:**240
domain of, defined, **4:**66
hyperthermophilic, **2:**241
lipids of, **3:**428–29
MOLECULAR BIOLOGY OF, **4:66–78**
Archaebacteria, **1:**109
Archaeology
human and plant studies, **4:**305
molecular, **4:**299
Architecture, three-dimensional, from X-ray diffraction, **6:**276
Argentine hemorrhagic fever, treatment with pooled human sera, **1:**420
Arginine, binding of, effect on HIV TAR, **1:**21
Arginine-rich motif (ARM), **5:**385
Arginine vasopressin, **2:**20–21, 20f
Aromatase
deficiency of, **3:**169
inhibitors of, **5:**491–92, 491f
testololactone, **5:**490
in treatment for breast cancer, **1:**256
P450, **3:**502
Aromatic compounds
bacterial metabolism of, **1:**120f, 121
hydrocarbons, induction of P450 CYP1A1 by, **4:**402–3
Aromatic side chains, azo-linked, in peptides, **4:**377
Arrestin
interaction with phosphorylated rhodopsin, **5:**234
interaction with rhodopsin, **5:**324
Arrhenius plot, for protein folding, **5:**155
Arrhythmias, genetic susceptibility to, **3:**2
Arsenical resistance pumps, **6:**132–33
Arteriosclerosis, effect of vitamin E on development of, **6:**238
Arthritis, role of nitric oxide in, **4:**205
Arthrogyroposis type I (AMCD1), mapping to chromosome 9, **3:**104
Articular cartilage replacement
with poly(2-hydroxyethyl methacrylate), **5:**42
with poly(vinyl alcohol), **5:**43
Artificial heart, polyurethane in, **5:**38
Ascaris suum, locomotory muscle of, **4:**171
Ascorbic acid, **1:**338
receptors for, **6:**238–40
structure and function of, **6:**252

Ascus
 defined, **6**:302
 of spores, **6**:304
As low as reasonably achievable (ALARA) dose of radiation, **5**:211
Asparagine amidohydrolase, **3**:351
Asparagine turn, substrate-directed posttranslational modification of, **5**:52
N-Asparaginyl-tRNA synthetase (*Nars*), localization to chromosome 18, **3**:201
Aspartame, amperometric electrodes for determination of, **1**:185
Aspartic proteases, **1**:14
Aspartyl hydroxylation, posttranslational, **5**:54
Aspartyl phosphate, site of, catalytic domain of the calcium ion pump, **1**:244
Aspartyl proteases
 HIV, **1**:27
 HIV-1, **1**:38
Aspartyl residues
 bacteriorhodopsin, **1**:111
 HIV protease, structure of the binding region, **1**:28
Aspergillus niger, **1**:184; **2**:338
 for citric acid production, **2**:338–40
Aspirin
 antiinflammatory effect, **1**:168
 production of, **2**:161, 161f
Assembly PCR, **5**:448
 applications, **5**:451–52
 defined, **5**:447
Assessment, of X-ray diffraction models, **6**:280
Association
 allelic, defined, **3**:222
 studies of, in gene mapping, **3**:225
Associative learning, defined, **4**:53–54
Ast, mitochondrial, isoenzyme pattern in heart or liver damage, **3**:375
Astaxanthin, synthesis by *Phaffia rhodozyma*, **3**:437
Asymmetric gap ligase chain reaction (AG-LCR), **3**:302
 mechanism of, **3**:305, 305f
Asymmetric rolling circle replication, **5**:4, 4f
Ataxia
 genes involved in, mapping to chromosome 11, **3**:127
 Machado-Joseph disease, mapping to chromosome 14, **3**:158
 spinocerebellar type 3, mapping to chromosome 14, **3**:158
Ataxia telangiectasia (AT), **1**:8
 chromosome breakage in, **1**:413
 excision repair defects in, **2**:46
 premature aging in, **2**:80
Atenolol, **2**:158
Atherogenesis
 and ACAT activity, **3**:410
 and lipoprotein(a) expression, **4**:504
Atheroma, **1**:284, 288
Atherosclerosis, **1**:283–88
 cytokine role in pathogenesis of, **1**:450
 defined, **1**:283; **3**:472
 in knockout mice, deficient in apoE, **4**:505
 and lipoproteins, **3**:414; **4**:495–96
  high density, **4**:502
  levels of, **3**:479
 premature, and apoE deficiency, **3**:414
 in transgenic mice, **4**:504, 504f
  relationship with apoA-II expression, **4**:503
Atherosclerotic lesions, **1**:287–88
Atmosphere, of the early earth, **4**:291
Atomic force microscopy (AFM), **1**:158, 158f; **4**:251; **5**:404–6
 for transistor fabrication, **1**:158–59, 159f
Atomic model, defined, **5**:110
ATP (adenosine triphosphate)
 for activation of the cystic fibrosis transmembrane conductance regulator, **1**:430–32
 cofactor for ligation of nucleic acids, **3**:392, 392f
 defined, **4**:129
 fluorescent analogue of, **2**:295
 hydrolysis of
  and muscle contraction, **4**:143–45
  RecA filament, **5**:223–24, 226
 role in membrane fusion, **4**:45
 synthesis of, and bacteriorhodopsin, **1**:109
ATPases
 F_1/F_0, **6**:132
 ion-translocating, **6**:131–32
 motor proteins as, **4**:118
 myosin as, **4**:115, 134
 Na^+-K^+, **5**:273
 oxyanion-specific (A-type), **6**:132
 traffic, **1**:429
 vacuolar (V-type), **6**:132
ATP synthetase
 defined, **4**:440
 of methanogens, genes encoding, **4**:78
 in photophosphorylation, **4**:448–49, 448f
Atrial natriuretic factor (ANF), and heart failure, **1**:290
Atropine, **1**:338
Attached cell antigen, gene for, mapping to chromosome 15, **3**:167
Attachment, by viruses, defined, **6**:223
Attachment protein (G), respiratory syncytial virus, **5**:288–89
Attenuated vaccines, **6**:165–66
 interferon-γ activity and, **6**:170
Attenuating genes, defined, **6**:165
Attenuator region, tryptophan transcription, **2**:371, 372f
Auger electrons, **5**:210
AUTOANTIBODIES
 AND AUTOIMMUNITY, **1**:**84–93**
 versus monoclonal antibodies (Mabs), **5**:417
 in scleroderma, **5**:410–11
 See also Autoimmune disease
Autoantigens, nuclear, functions of, **1**:89–90, 90t
Autocatalysis, self-replicating systems, **4**:92
Autocrine, **2**:463
Autographa californica, **2**:277
Autoimmune defects, induced by deficiency in TGF-β1 gene, **2**:468
Autoimmune disease, **1**:84–93
 endogenous retroviruses associated with, **6**:137
 insulin-dependent diabetes, **2**:28
 interferons associated with, **3**:328
 motor neuron disease as, **4**:108
 and programmed cell death, **1**:302
 protein kinase C associated with, **5**:172
 role of cytokines in, **1**:441, 449
 superantigen in, **5**:511
Autoinducer synthase, LasI gene for, **1**:108
Autoinduction, **1**:107
Autoinhibition
 defined, **5**:49
 protein kinase domain, **5**:51
Automated sequencer, defined, **5**:388
Automation
 of amplicon detection, LCR, **3**:307
 defined, **5**:389
 of DNA fingerprinting, **2**:58, 60, 61
 of Edman degradation reactions, **5**:185
 in genome research, **5**:388–99
 of ligation assays, **3**:403
 of nucleic acid sequencing, **4**:255–56, 256f
 of oligodeoxyribonucleotide synthesis, **4**:282
 of oligonucleotide ligation amplification, **3**:401
 in PCR technology, **4**:337; **5**:305
 of peptide synthesis, **4**:390
  solid phase, **1**:171
 of protein sequencing, **5**:174
 and recognition biosensors, **5**:245
 of secondary structure prediction, RNA, **5**:363
 of sequencing, **5**:187, 187t
 of solid state synthesis of peptides, **4**:364–65

Automation (*cont.*)
  use of HUGE, **2**:96
  of DNA sequencing, **2**:99–102, 99f
Automodification
  defined, **5**:49
  in enzyme-directed posttranslational modification, **5**:51
Autonomous elements, transposon, **4**:487
Autonomously replicating sequence (ARS), yeast, **6**:291
Autoradiograph, horizontal ultrathin gel electrophoresis, **2**:98f
Autoradiography
  defined, **5**:115, 209
  in HUGE, **2**:96
  in restriction landmark genomic scanning, **5**:307
  use and sensitivity of, **5**:219–20, 220t
Autosomal dominant disorders, **2**:425–26
  adenomatous polyposis coli, **1**:409
  Alzheimer's disease, early onset, **1**:43
  central diabetes insipidus, **2**:23–24
  Huntington's disease, **4**:199
  retinoblastoma, **2**:27; **5**:319–22
Autosomal dominant polycystic kidney disease (ADPKD), localization to chromosome 16, **3**:188–90
Autosomal dominant trait, defined, **1**:42; **4**:173
Autosomal recessive disorders
  colon cancer, **1**:413
  nephrogenic diabetes insipidus, **2**:27
  phenylketonuria, **4**:406
Autosomal recessive trait, **2**:424–25
  lethal, **2**:124, 125, 125f
Autosomes
  defined, **2**:118, 313, 419; **3**:160, 227; **4**:198
  sex-determining genes of, **5**:443
Autosynthesis, of layered structures, **1**:153–54
Autotrophic complementation, *URA3* and *LYS2* genes, yeast, **6**:313
Autotrophs, utilization of light energy by, **1**:306
Auxins, **4**:492
  phyto-oncogene, expression of, **4**:490
Auxotrophs
  for cloning experiments, **5**:502
  defined, **5**:135
  novel protein synthesis by, **5**:137
  for in vivo novel protein synthesis, **5**:30
Available Chemical Directory, **4**:99
Avermectins, **4**:396, 397f
  of nematodes, **4**:170
Avidin
  biotin affinity for, **6**:240
  fluorescent, for species-specific hybrid identification, **6**:253–54
Avirulence genes, **4**:483–84
Avitaminosis, defined, **6**:244
Axonemal dyneins, **4**:116
Axons
  length of, **4**:106
  structure and movement of organelles in, **4**:117
Azetidone, **2**:157
Azidometmyoglobin, FTIR spectra of, in water and in organic solvents, **2**:253–54, 255f
Aziridines
  antitumor activity of, **5**:497
  for synthesis of β-substituted amino acids, **1**:54
Azospermia factor (AZF), **3**:503
AZT (3′-azido-2′,3′-dideoxythymidine), **1**:14, 37
  binding of, to HIV reverse transcriptase, **2**:236
  mutation after exposure to, in reverse transcriptase, **1**:31
Aztreonam, structure of, **2**:158f
Azurin, reorganization energy, electron transfer, **2**:195

Bacillus brevis, D-specific hydantoinase, **1**:58
Bacillus Calmette-Guerin (BCG), vaccination with, **4**:157, 163
Bacillus cereus, structure of PI-PLC from, **4**:424, 426, 426f, 429f
Bacillus subtilis
  host, **2**:275–76
  sporulation in, **2**:369, 369f
Bacillus thuringiensis (Bt)
  δ-endotoxins of, **4**:393
  PI-PLC enzymes from, **4**:424, 424f
Backbones
  DNA, exogenously induced damage to, **2**:41
  oligonucleotide, phosphate-free, **4**:270–80, 280f
  peptide, modifying amides of, **4**:372–74, 374f
  phosphate-free, defined, **4**:269
  phosphodiester, **1**:70–71; **2**:1
  polyamide, in DNA analogues, **1**:166
  polypeptide, **4**:361; **5**:149f
Backcross
  defined, **4**:121
  for mapping the mouse genome, **4**:123f
Backtracking algorithm, search tree generated by, **5**:380
Bacteria
  anaerobic, photosynthetic, **1**:310
  directed evolution of, **1**:125
  domain of, defined, **4**:67
  energy-transducing membrane systems of, **1**:307f
  genetic immunization against, **2**:437
  genetic investigations with, **2**:403
  gram-negative, cell envelope, **3**:463f
  lipids of, **3**:429–34
  nitric oxide formation in, **4**:200
  pathogenic, lipopolysaccharide architecture of, **3**:468f
  protein targeting in, **5**:206–7, 207f
  ribosomes of, characterization of, **5**:328
  S layers of, **1**:141
Bacterial cells
  extracts from as protein-synthesizing systems, **5**:333–34
  preparing ribosomes from, **5**:329
BACTERIAL GROWTH AND DIVISION, **1:95–102**
Bacterial infections, cytokine role in immune response to, **1**:447
BACTERIAL PATHOGENESIS, **1:103–8**
  role of cytokines in, **1**:449
Bacteriochlorophyll proteins, absorption maxima of, **4**:441
Bacteriophage
  assembly of, **6**:204–5
  defined, **1**:62; **2**:271; **4**:254; **6**:200
  fd, DNA of, **2**:185f, 191f
  filamentous, **1**:64
  immunoglobulin gene expression in, **1**:64
  life cycle, double-stranded DNA, **6**:202–5
  186, partial denaturation mapping, **4**:318
  P1, for inserting apo B genes in mice, **4**:504
  P2, alignment of physical and genetic maps of, **4**:314
  polarity of DNA injection by, **4**:321–22
  receptors for, O-specific chains as, **3**:467
  replication of, **5**:359
  T4
    intron homing in, **1**:351
    single-stranded DNA-binding protein of, **5**:460–61
    source of DNA ligase, **3**:392
  T5, electron microscope image, **2**:191f
  T7, **6**:202f
    capsid of, **6**:203–4, 204f
    promoter for genetically engineered silk formation, **5**:27
  translational control of synthesis, **6**:123, 123f
Bacteriophage κ, **1**:1–2
  A+T regions of, **4**:313, 313f
  DNA replication, role of hsp70s, **1**:318–19, 319f
  electron microscope image, **2**:191f
  N protein, **1**:21
  N protein antiterminator, **1**:25
  partial denaturation mapping, **4**:312, 313, 313f, 319–21f

for producing catalytic antibodies, **1:**68–69
Bacteriophage ϕ
6, **6:**219, 219f
X 174, **1:**9
BACTERIORHODOPSIN, **1:108–13**, 137–39, 149–50
defined, **1:**108
films of, for photochromic bistable optical devices, **1:**150–51
as a light-driven ion pump, **6:**131
Baculoviruses, **1:**174–75
defined, **6:**165
for genetic control of insects, **4:**198
as hosts for glycosylated protein production, **2:**275; **6:**177–78, 178f
for producing antibody molecules, **1:**63–64
Baculovirus expression vectors, **2:**277; **6:**170–71
Baker's muscular dystrophy, **4:**147
Balanced growth, in cell replication, **1:**96
Balanced lethal stocks, defined, **2:**121
Balance method, for determining protein requirements, **4:**265
Balancer chromosomes, following genetic crosses with, **2:**125
Ballistic methods, for transformation, **4:**453–54, 454f
Band 4.1 superfamily
defined, **4:**180
place of schwannomin in, **4:**187, 187f
Band assignments, Raman spectroscopy, **5:**96
Banding patterns, FISH gene mapping, **2:**379–80
Barbiturates, withdrawal from, genetic studies, **2:**148–49
Barnase, anther-specific expression of, **4:**472–73
Barr body, **3:**511, 517
Basal cell carcinoma, nevoid, mapping to chromosome 9, **3:**104
Basal cell nevus syndrome (BCNS), excision repair disorder in, **6:**161
Basal-cell-specific markers, keratin, **2:**266
Basal layer, keratinocytes in, **2:**263–65
Basal metabolic rate (BMR)
defined, **4:**257
factors affecting, **4:**260–61, 260t
Base excision repair (BER), **2:**44; **6:**155f, 156, 158
defined, **6:**283, 285
substrates amenable to, **6:**285–7, 287f
Baseline genetic linkage map, chromosome 14, **3:**143, 143f
Basement membrane
proteoglycans in, **2:**283
skin, attachment of, **2:**264–65
Base pairing, defined, **5:**372
Base pairs
adenine/thymine, **1:**3
cytosine/guanosine and isocytosine/isoguanosine, **5:**146f
formation between a codon and an anticodon, **6:**114f
Hoogsteen and Watson-Crick, **1:**70
substitutions, **2:**418
substitutions in hemophilia B, **3:**37
view along the DNA helix axis, **5:**280f
Watson-Crick
in the B form of DNA, **2:**105–7, 105f
and second-level structure of DNA, **2:**103
Bases
damage to
reversal by DNA excision repair, **6:**285–90
reversal by DNA repair, **6:**283–85
hard and soft, **1:**129–30
modifying
in oligodeoxyribonucleotides, **4:**283
in oligoribonucleotides, **4:**284
replacement by a base analogue, **4:**152
Base-specific cleavage, for nucleotide sequencing, **4:**256
Base stacking, in RNA, **5:**360, 371
Base step geometry, of DNA, **2:**107–11
Base substitutions, synonymous, from point mutations, **4:**149
Base triples, in tRNA, **5:**377
Basonuclin, transcription factor for epidermal keratinocytes, **2:**266
Batch culture, **2:**340–41
defined, **2:**337
Batimastat, inhibition of matrix metalloproteases by, **1:**260
Batten disease, localization to chromosome 16, **3:**187–88
Battleboro rat, model for autosomal dominant central diabetes insipidus, **2:**27
Bayes's theorem, application in breast cancer analysis, **1:**224
B-cell cancers, **2:**65
B-cell chronic lymphocytic leukemia, *bcl-2* gene rearrangement in, **4:**329
B-cell lymphoma, translocation between chromosome 18 and chromosome 14 associated with, **3:**209–10
B-cell lymphoma 2 (*BCL2*) proto-oncogene, on chromosome 18, **3:**201
B cells
activation of, role of cytokines, **1:**446–47
differentiation of, **1:**447
BCR/*abl* gene rearrangement, for identifying chronic myelogenous leukemia, **4:**328–29
Beals' syndrome, **3:**163
Beauveria bassiana, source of mycoinsecticides, **2:**343
Becker muscular dystrophy (BMD), **1:**289; **2:**426
Beckwith-Wiedemann syndrome (BWS), **2:**429, 445
cosmid contig identifying, **3:**140f
gene for, mapping to chromosome 11, **3:**127
imprinting in, **3:**127
Becquerel (Bq) (unit), defined, **5:**211
Behavior
effects of neuropeptides on, **4:**195
impairments in phenylketonuria, **4:**407
Behavioral mutants, of *Caenorhabditis elegans*, **4:**171
Bending, of DNA, **2:**111–14
Benzoate, defined, **5:**485
Benzodiazepines
chlordiazepoxide, **2:**160
derivatives reacting with receptors, **4:**35
inhibition of reverse transcriptase, HIV-1, **1:**38
sensitivity to, point mutation in GABA receptor complex, **2:**149
structure, **1:**39f
synthesis of, **2:**162f
withdrawal from, genetic studies, **2:**148–49
Benzoporphyrin, derivatives of, as photosensitizers, **6:**212
Benzothiazepines, binding to L-type calcium channels, **1:**243
Benzo[*a*]pyrene, association with carcinogenesis, **1:**197, 197f
Beri-beri, thiamine deficiency, **6:**244
Bernoulli text, **6:**39–40, 41
Best disease, mapping of, to chromosome 11, **3:**127
Best location alignment, **5:**433
Bestrabucil, for cancer therapy, **5:**486
Best region(s) of local similarity, **5:**433
Beta decay, **5:**209–10
Beta-pleated sheet, **2:**307
amide bands corresponding to, **5:**97
amphipathic, in apoB, **3:**411–12
channels formed by, **1:**140
defined, **1:**27; **5:**128, 159; **6:**327
design of polymers based on, **5:**29–30
as a domain subunit, **5:**150
immunoglobulin structure involving, **3:**278–79
model of, **5:**168f
reverse transcriptase, **1:**30
ribonuclease H, HIV-1, **1:**30
structure of, **5:**118–19, 119f
Beta-turn
mimics, dipeptides of modified amino acids as, **4:**377

Beta-turn (*cont.*)
modeling, **5**:164
secondary structure component, **5**:119
substrate-directed posttranslational modification of, **5**:52
BH_4 (tetrahydrobiopterin)
cofactor for tryptophan hydroxylase, **1**:218
defined, **1**:207
specific requirement for, tyrosine and tryptophan hydroxylases, **1**:208
Bialaphos
bioherbicides, **4**:395
structure of, **4**:396f
Bicistronic, defined, **5**:13
Bicistronic mRNAs, for studying cap-independent translation, in picornaviruses, **5**:17–19
Bidirectional replication, defined, **1**:95
Bifunctional antibodies (BFAs), **1**:68
Bifurcation loop, defined, **5**:360
Bilayer vesicles, kinetic analyses of phospholipases using, **4**:418
Bile acids
biosynthesis of, **1**:439
synthesis of, **3**:461
in liver, **3**:409–10
Bilharzia, **4**:307
Bilirubin, glucuronidation of, **1**:198
Binding
actin filaments, by talin, **1**:453
antigen, antibody sites for, **1**:62
catecholamines, by tyrosine hydroxylase, **1**:213
characteristics of T4gp@ interaction with ssDNA, **5**:460
conformation changes induced by, **2**:175
curves for specific DNA-protein complexes, **4**:246
DNA
by RecA protein, **5**:224–25
sites for Achilles' cleavage, **1**:2
by small molecules, **1**:166, 167f
of drugs, to HIV-1 reverse transcriptase, **1**:30–31
homophilic, in cell adhesion, **1**:295
of inhibitor with enzyme, **1**:167–68
metal, RNase H, **1**:30, 32
oxygen, structural changes accompanying, **3**:10–11
peptide nucleic acid, to RNA and DNA, **1**:73
promoter/enhancer module, **1**:19
of interleukin-1 genes, **3**:334
protein, **1**:162, 162f
and RNA secondary structure, **5**:363
Tat domain, **1**:21
by viruses, defined, **6**:223
See also Antibody binding site
Binding affinity
defined, **1**:69; **5**:478
of oligonucleotide analogues, **1**:72–73
Binding constant, apparent, effect of cooperativity on, **5**:458
Binding domain
animo acid variations, **1**:145
TAR RNA binding, HIV-1, **1**:21
Binding elements, determination of, in protein-binding nucleic acids, **4**:228–29
Binding energy, enzyme-substrate, **2**:229
Binding proteins (BiP)
heavy chain, **5**:199
insulin-like growth factor modulation by, **2**:466
Binding site size
defined, **5**:458
of *Eco*DSSB, **5**:461–62
and *Eco*SSB function, **5**:462
Bioassay, defined, **4**:188
Bioavailability
defined, **2**:165
of metal ions, **4**:59–60
Biobreeding rat, diabetes mellitus model, **2**:29, 37
Biochemical approach, for studying virus assembly and budding, **6**:188
BIOCHEMICAL GENETICS, HUMAN, **1:113–17**
bases of phenylketonuria, **4**:406–7
Gaucher disease, **2**:349
of replication protein A, **5**:463–64
Biochemical modulators, in combination chemotherapy, **1**:266
Biochemical pathways, and selectivity of drugs, **1**:253
Biochemistry, of DNA replication, **2**:87–90
Biocomputing, **1**:157–58, 157f
Biocontrol, using molds, **2**:343
Biodegradable polymers, **5**:32–36
defined, **5**:31
BIODEGRADATION OF ORGANIC WASTES, **1:117–25**
Bioenergetics
cellular, **1**:304
defined, **1**:303
of methanogenesis, **4**:72–73
Biofields, internal, **2**:174
Biofungicides, blasticidin S and kasugamycin, **4**:396
Biogenesis, of telomerase, **6**:26–27
Biogeochemical nitrogen cycle, nitric oxide in, **4**:204
Bioherbicides, bialaphos, **4**:395
BIOINORGANIC CHEMISTRY, **1:125–34**
Bioinsecticides, avermectins, **4**:396
Biolistic method
defined, **4**:465
for DNA delivery, **4**:466
helium-driven device, **4**:461
for promoter analyses, **4**:470
transformation using, **4**:459–61, 461f
in chloroplasts, **4**:489
defined, **4**:485
in plants, **4**:490
Biological clock
globin evolution, **3**:18f
norpA gene of *Drosophila*, mutant, **2**:133
period gene of *Drosophila*, **2**:139f
mutant, **2**:137, 138
Biological clock hypothesis, for investigating evolution, **3**:17
Biological distribution, of isoenzymes, **3**:373
Biological evolution, defined, **2**:402
Biological markers
defined, **2**:258
defined as molecular fossils, **4**:299
of effect, **2**:259
defined, **2**:258
of exposure, **2**:259
defined, **2**:258
of susceptibility, **2**:259
defined, **2**:258
Biological methods, for plant cell transformation, **4**:459
Biological properties, of steroid hormones, **5**:481–82, 481f
Biological response modifiers (BRMs)
for cancer treatment, **1**:256–57
production of, fungal biotechnology, **2**:340
Biological rhythms, defined, **2**:130
Biological systems
activities of interleukins, **3**:340–44, 344t
responses to ultraviolet radiation, **6**:149–50, 150f
Bioluminescence, **1**:323
defined, **1**:322
genes for, **1**:107–8
tracer for biodegradation reactions, **1**:122–23
Biomembrane fusion
models of, **4**:48–53
morphology of, **4**:43–44, 44f, 44f
topology of, **4**:42, 43f
See also MEMBRANE FUSION
BIOMIMETIC MATERIALS, **1**:134–48, 148t
Biomineralization, **1**:147; **3**:309, 316
model for, crystallization in phospholipid vesicles, **3**:314
Biomolecular arrays, **1**:139–41
BIOMOLECULAR ELECTRONICS AND APPLICATIONS, **1:148–59**
Biomolecules, in solution, nuclear magnetic resonance measurements, **4**:218–26

BIOORGANIC CHEMISTRY, **1:159–71**
Biophysical properties, of the msSSB gene, **5:**465
Biopolymers
  colloid aggregation prevented by, **3:**316
  prebiotic formation of, **4:**294–96, 294f
BIOPROCESS ENGINEERING, **1:172–77**
Bioprocessing, defined, **1:**172
Bioreactors
  transgenic animals as, **6:**69, 70–71
  transgenic livestock as, **6:**102
  whole-cell, **1:**174–75
Bioremediation, **4:**68
  defined, **1:**118
Bioresorbable polymers, **5:**35–36
  defined, **5:**31
BIOSENSORS, **1:177–92**; **5:**247f
  bioluminescent, **1:**123
  defined, **1:**118; **5:**243
  enzymatic, **1:**173
  enzymes as, **1:**272
  poly(2-hydroxyethyl methacrylate) used in, **5:**42
  recognition, **5:**245
  surface plasmon resonance, **5:**245
Bioseparations, **1:**175–76
Biospecificity, defined, **5:**243
Biosurfactant lipids, **3:**434
Biosynthesis
  of glycoproteins, **2:**459–60
    control of, **2:**461
  of nitric oxide, **4:**202–3, 202f
    in plants and microbes, **4:**204, 204f
  of peptides, **4:**353–54
  of phospholipids, **4:**433–38, 438f
  of proteins, folding during, **5:**158
  of steroid hormones, **5:**466, 468, 468f
  of steroids, **5:**479–80, 479f
Biotin (vitamin H)
  coenzymic form of, **1:**405
  defined, **3:**288
  hapten for FISH, **2:**382
  receptors for, **6:**240–43
  structure and function of, **6:**251, 251f
Biotin-binding protein 1 (BBP-1), from egg yolk, **6:**241
Biotinidase, **6:**241
Biotin-streptavidin-biotin bridge, **1:**154
Biotinylation
  defined, **3:**288
  and phage binding to antigen, **1:**418
BIOTRANSFORMATIONS
  fungal, **2:**344–45
  OF DRUG AND CHEMICALS, **1:192–202**
Biotrophs
  defined, **4:**476
  plant defense against, **4:**478
Biphasic column technology, **5:**187
Biphenyls, polyhalogenated, **1:**193
Bisheteroarylpiperazines, inhibition of reverse transcriptase, HIV-1, **1:**38
Bislactim ethers, **1:**53–54, 53f
Bisphosphoglycerate (BPG), **3:**8
  in blood substitutes, **3:**20
  defined, **3:**20
Bistability, defined, **1:**148
Biuret assay, **5:**121–22
Bladder cancer, **2:**65
Blasticidin S, **4:**396–97, 398f
Blastomere, analysis of transgenic, **6:**69
BLAST program, **5:**434
  for sequence alignment, **5:**423, 425, 433
Bleomycin, **1:**166
  as an anticancer drug, **1:**257
Blocking
  in immuno-PCR, **3:**295
  in RLGS, **5:**306, 307
    defined, **5:**304
    techniques for, **5:**308–10
Block model, for determining DNA structure, **2:**106–11, 106f
Blocks, database of, **5:**434
Blood
  coagulation system, **3:**377
  extraction of DNA from, **2:**52, 53
  local flow of, regulation by the kallikrein-kininogen-kinin system, **3:**378
Blood group antigens, **2:**458
  ABO(H), **1:**281–82, 282f
Blood pressure
  drug for controlling, **1:**167–68
  regulation by glucocorticoids, **5:**482
Blood substitutes, hemoglobin-based, **3:**19–20, 26–27
Bloom's syndrome
  chromosome breakage in, **1:**413
  excision repair defects in, **2:**46
  gene for, mapping to chromosome 15, **3:**160
Blot, defined, **4:**235
Blue light responses, in plant development, **4:**493
Blue-white selection, **2:**70
  defined, **2:**66
B lymphocytes
  defined, **3:**295
  immune responses of, **3:**296
  interactions with interferons, **3:**327
  interleukin-4 as a competence factor for, **3:**341–42
  interleukin-6 role in differentiation of, **3:**342
Boc-benzyl-polystyrene technique, for peptide synthesis, **4:**386–88
Boc group, **4:**384
  for protecting Nα-amino groups, **4:**387
Boc method for peptide synthesis, **4:**366, 366f, 384
BODY EXPRESSION MAP OF THE HUMAN GENOME, **1:203–6**
Body fluid stains, extracting DNA from, **2:**52
Body mapping
  defined, **1:**203
  human genes, adult lung, **1:**206, 206t
Bohr effect, **3:**12
  defined, **3:**3, 8, 8f, 20
  in fish, **3:**27
  in hemoglobin regulation, **2:**237
Bombesin, and lung cancers, **3:**494
Bombyx mori, silk fibroins of, **5:**27
Bond angles, hydrogen bonds, **3:**269, 269f
Bond cooperativity, in hydrogen bonding, **3:265–66,** 270, 270f
Bonded phase, defined, **3:**76
Bond energies, hydrogen bonds, moderate and weak, **3:**264, 264t
Bond strength, hydrogen bonds, defined as moderate, weak and strong, **3:**261
Bone
  extraction of DNA from, **2:**53
  formation of, role of ascorbic acid, **6:**239
  regulation of mineralization, by vitamin K, **6:**248
  role of calcium in, **1:**248
Bone cement, poly(methyl methacrylate), **5:**41
Bone marrow
  for antibody library construction, **1:**420
  interleukin-07 production in, **3:**342
  interleukin-11 production in, **3:**343
  transplantation of, granulocyte/macrophage-colony stimulating factor used with, **2:**469
Bone morphogenic proteins (BMP), **2:**468
Bordetella pertussis, transmission of, **1:**103
Bosswell and McLachlan's method, for local alignment of sequences, **5:**424
Bottom-up design
  defined, **5:**388
  evolutionary, **5:**390–91
Botulism, transmission of, **1:**103
Boundaries, and nucleosome positioning, **1:**368
Bovine papilloma virus (BPV), electron microscope image, **2:**192f
Bovine serum albumin (BSA), antibodies to, in diabetics, **2:**36
Bradford assay, **5:**122
Bradykinin, **3:**378, 378f
  activation of PKC and MAP kinases by, **1:**216
  generation and destruction of, **3:**379f

Bradykinin (*cont.*)
liberation from high molecular weight kininogen, **3**:380f
receptors for, **5**:268
antagonists to, **3**:384
structure of, **3**:379f
Bragg's law, **6**:276–77
Brain
accessibility of drugs to, **1**:260
active histone gene transcription in, **3**:52
artificial dura from poly(tetrafluoroethylene), **5**:39
concentration of water-soluble vitamins in, **6**:235–36
glioma of, **1**:252
mammalian, DNA damage to, **2**:79
neurotransmitter receptor, effect of antisense oligonucleotides, **1**:75
neurotransmitters in, **5**:238
P-type calcium channels in, **1**:243
targeting for drug therapy, **1**:67–68
tryptophan hydroxylase in, **1**:218
tumors of, associated with hereditary colon cancer, **1**:410
BRAIN AROMATIC AMINO ACID HYDROXYLASES, **1:207–20**
BRCA1 mutation, associated with breast cancer, **1**:223
Breakage-fusion-bridge cycle, **6**:35
Breakpoint maps, chromosome 11, **3**:137, 138f
BREAST CANCER
antihormone treatment for, **1**:256
BRCA1 and *BRCA2* genes of, **6**:142
chromosome 18 genes involved in, **3**:210
FGF-4 gene amplification in, **2**:469
GENETIC ANALYSIS OF, **1:221–25**; **2**:428; **3**:223–24; **4**:286
HER/2/*neu* oncogene, prognostic indicator in, **4**:329
p53 gene, **6**:141
response to tamoxifen, **5**:484
Breast Cancer Linkage Consortium, **1**:223
Breast implants
poly(2-hydroxyethyl methacrylate) hydrogel, limitations of, **5**:42
silicones in, **5**:37
Breeding, selective, for sensitivity to drugs, **2**:147–50
o-Bromobenzyl derivatives, for protecting side chains, during peptide synthesis, **4**:387
Bromodeoxyuridine (BrdU), for synchronizing cell growth, **2**:379
Bromouracildeoxyriboside (BUdR), for screening recombinant viruses, **6**:166
Bromoxynil, herbicide, **6**:57, 57f
Brown fat, **1**:313
Brownian motion, and DNA packaging by bacteriophages, **6**:206
Brugia malayi, genome construction, **4**:309–10
B/T-cell gene rearrangement, test for lymphoproliferative disorder diagnosis, **4**:328
Budding, of enveloped viruses, **6**:183–92, 192f, 222
Building blocks, for combinatorial syntheses, **4**:34f
Bulge loop, **5**:361
defined, **5**:360
and quaternary interactions of RNA with proteins, **5**:362
TAR, predicting the structure of, **5**:383–85, 383f
tRNA structure, **5**:371, 376–77, 376f
Bungarotoxin, effects on nicotinic receptors, **5**:232
Burkitt's lymphoma
association with Epstein-Barr virus, **6**:233
translocation in, **2**:65; **3**:159; **4**:286
Butoxycarbonyl (Boc) group, **4**:384
t-Butoxycarbonyl (Boc) group
for protecting Nα-amino groups, **4**:387
N-Butyldeoxynojirimycin, in vitro inhibition of HIV-1 binding, **1**:36

CAAT box, genes for skeletal muscle actin, **4**:138
CAAT consensus sequence, defined, **4**:129
Cadherins, **1**:295
cytoskeletal interactions with, **1**:452–53, 457–58
defined, **1**:292
desmosomal, mapping to chromosome 18, **3**:201
E-, **1**:452, 458
gene for, **3**:487
N-, **1**:294f
role in cytoskeleton-plasma membrane interactions, **1**:451
Cadmium
response of genes to, in vivo, **6**:60–61
stability of metal-ligand complexes, **1**:131, 131t
CAENORHABDITIS ELEGANS, **4**:167–69
cell death studies, **1**:299, 301
delayed senescence in, **1**:10–11, 10f
HOX clusters of, **3**:64, 64f
MOLECULAR BIOLOGY OF, **1:227–37**
mtDNA of, **4**:79
Cahn-Ingold-Prelog system, for naming prochiral compounds, **1**:344
Cairnsian mutation, defined, **2**:407
Calbindin 9K, **1**:242
Calcineurin
role in immune response, **3**:279
target of immunosuppressive agents, **1**:242
Calcisequestrin, **1**:292
Calcite, crystal morphology, **3**:312–13f
Calcitonin (CT), **1**:238
regulation of vitamin D status by, **6**:246
Calcium
binding by annexins, **1**:60f
binding by selectins, **1**:296
binding by troponin-C, **4**:140
Ca^{2+}-dependent serine-threonine protein kinase C, **1**:459
comparison with magnesium ion, **1**:127–28
cytoplasmic, role in muscle contraction, **4**:130
defined, **1**:237
dependence of stimulated peptide hormone secretion on, **4**:346
effect on cell membranes, **6**:188
fluorescence detection of, **2**:295–96
handling of, in heart failure, **1**:290, 292; **3**:2
homeostasis of, role of vitamin D in, **6**:237–38
increase intracellular, as a result of oxidative stress, **2**:334
mitochondrial, and respiratory control, **1**:312
regulatory role in oxidative phosphorylation, **1**:312
requirement for, in peptide hormone reactions, **4**:354
role in cell adhesion, **1**:295
sarcoplasmic reticulum, role in muscle contraction, **4**:144
second messenger, **2**:210–11
in tryptophan hydroxylase activation, **1**:220
and tyrosine hydroxylase regulation, **1**:217
Calcium-binding motifs, of α-actinin, **4**:142
Calcium-binding proteins (ICaBP), intestinal, cell cycle control by, **1**:242
CALCIUM BIOCHEMISTRY, **1:237–50**
Calcium channels, **5**:512
annexin V as, **1**:242
dihydropyridine-sensitive, effect of ethanol on, in mice, **2**:149
and exocytosis, **5**:514
ligand-gated, **2**:210–11
membrane, in eye-pigment mutant *Drosophila*, **2**:134
types of, **1**:292
Calcium phosphate precipitation, for gene insertion into mammalian cells, **2**:432

Calcyclin
  association of annexin XI with, **1**:61
  cell cycle control by, **1**:242
Caldarchaeol, **3**:429
Calelectrin (annexin VI), structure, **1**:60f
Calibration curve, protein separation, size exclusion chromatography, **3**:78–79, 78f
Callus, defined, **6**:10
Calmodulin, **2**:211; **3**:46
  calcium-binding sites of, **1**:240
    relationship with learning and memory, **2**:137
  calcium ion binding sites, **1**:241
  and calcium pump of plasma membranes, **1**:243–45
  cell cycle control by, **1**:240
  defined, **1**:237
  movement of, between cytosol and nucleus, **1**:248
Calmodulinlike domain protein kinase (CDPK), **1**:249
Calorimetry
  for measuring energy content, of foods, **4**:259
  for protein measurements, thermodynamic stability, **5**:127
Calpactin (annexin II)
  structure, **1**:60f
  tetramer of, in chromaffin granule aggregation, **1**:61
Calpains, **1**:249; **3**:353
Calpectin (annexin II), role in membrane fusion, **4**:50
Calreticulin, and long-term sensitization, **4**:56
Calusterone, for cancer therapy, **5**:486
Cameras, CCD, for signal visualization, FISH, **2**:385
cAMP (cyclic adenosine monophosphate)
  and calcium ion functions in cells, **1**:238–39
  defined, **1**:207
  as a second messenger, **3**:44
  and tyrosine hydroxylase regulation, **1**:217
cAMP-dependent chloride permeability, **1**:426
cAMP-dependent phosphorylation, and regulation of L-type calcium channels, **1**:243
Camphor, regulation of P450 enzymes by, **1**:440
CaM PK (Ca^{2+}/calmodulin-dependent protein kinase), **3**:46
  II, **1**:210
    defined, **1**:207
    and regulation of tyrosine hydroxylase, **1**:215
Campomelic dysplasia, XY sex reversal associated with, **5**:443
cAMP-response element binding protein (CREB), **2**:291
  phosphorylation of, and gene transcription, **5**:233
Camptothecin, as an anticancer drug, **1**:254
CaMs
  binding domain, myosin light chain kinase, **1**:241f
  hormonal control of, in the nucleus, **1**:248
CAMs (cell adhesion molecules), **1**:293–95
  defined, **1**:292
  L-, **1**:295
  N-, neural tissue, **1**:294, 294t
  P-, **1**:295
CANCER, **1:250–52**
  B-cell, **2**:65
  bladder, **2**:65, 65
  defined, **1**:250; **2**:463
  growth factors in development of, **1**:450
  growth factors in therapy for, **2**:463
  hormone-sensitive, steroids for treating, **5**:483
  immunological control of, **2**:438–439
  imprinting syndromes, **2**:439
  isoenzyme expression in, **3**:375
  susceptibility to
    heritable, **2**:428
    multiple primary cancers of the aerodigestive tract, **3**:496
    in slow acetylators, **4**:404
  taxol for treatment of, **6**:11
  telomere length in tumor cells, **6**:34
  two-hit model of gene mutation, **3**:232f
CANCER CHEMOTHERAPY, THEORETICAL FOUNDATIONS OF, **1:252–66**
Cancer family syndrome, **1**:410
Cancer predisposition syndromes, **2**:47, 62–63, 63t
Candida curvata, mutation by deletion in, **3**:434
Candidate genes, **3**:226
  defined, **3**:222
5′-Cap, defined, **4**:129
Capacitor discharge (CD) pulse, for transfection with reporter genes, **2**:202
Capacity, for DNA repair in human cells, **2**:46
Capillary electrophoresis, **5**:189
  defined, **1**:266
Capillary liquid chromatography, **5**:184f
Capillary zone electrophoresis mass spectrometer (CZE-MS)
  defined, **4**:1
  for DNA sizing, **4**:11, 11f
Cap-independent translation, defined, **5**:13
Capped peptides, **4**:367
Capsid
  bacteriophage T7, **6**:203–4, 204f
  picornavirus, **5**:14–17
  viral
    defined, **6**:200, 217
    structure of, **6**:217, 217f
Capsid antigens, poliovirus, **5**:16
Cap site, MLC1f gene, **4**:135
Capsular antigens, **1**:278–79
Capsule, defined, **1**:103
Captopril
  development of, **4**:31
  structure of, **4**:32f
Carbamate linkages, in modified oligonucleotides, **4**:272f, 272–73, 273f
Carbamates, antitumor activity of, **5**:495
Carbocyclic compounds, drugs based on, **2**:161
CARBOHYDRATE ANALYSIS, **1:266–77**
CARBOHYDRATE ANTIGENS, **1:278–83**
Carbohydrates
  chirality of, **1**:338
  complexes with protein, **2**:309–10
  crystal structures, hydrogen bonds in, **3**:271, 271f
  surface, and cell-cell interactions, **1**:296–97
  use by hyperthermophiles, **2**:242
Carbon
  inorganic, **1**:126
  metabolic requirement of microbes, **1**:119
Carbonate linkages, in modified oligonucleotides, **4**:270
Carbon dioxide (CO_2), methanogenesis using, **4**:69–71, 71f
Carbonic anhydrase
  of methanogens, genes encoding, **4**:77
  neoprontosil binding to, **1**:394
  refolding of, **5**:77, 85
  specificity constant of, **5**:301
  tertiary structure, **2**:308f
  zinc in, **4**:64, 64f, 65f
α-Carbon modification, peptides, **4**:374
Carbon monoxide (CO)
  binding to hemoglobin, **3**:8–9
  as a second messenger, **1**:239
Carbonmonoxyhemoglobin, resonance Raman spectrum, **3**:25
Carbon tetrachloride, free radical formation during metabolism of, **2**:334–35
Carbon wires, in nanometer-sized channels, **1**:155–56, 156f
Carboplatin, as an anti-cancer drug, **1**:254
γ-Carboxyglutamic acid, **1**:239

Carboxy groups, peptide, protecting during synthesis, **4**:383–84, 384f
Carboxylases, **1**:403–4
phosphoribosylaminoimidazole, in yeast, **6**:313
Carboxylation, posttranslation modification by, **5**:53
Carboxylesterases, **1**:196
liver, acyl CoA-cholesterol acyltransferase activity of, **3**:410
Carboxyl terminal domain (CTD), essential role in transcription, **2**:92
Carboxymethyl linkages, in modified oligonucleotides, **4**:270–72, 272f
Carboxypeptidase
dipeptidyl, ACE as, **5**:266
E, **4**:345
Carcinoembryonic antigen (CEA), **1**:261
level after cell infection with Harvey murine sarcoma, **3**:493
vaccination against, in mice, **2**:439
Carcinogenesis, and ultraviolet radiation, **6**:163
Carcinogens
mutagens as, **4**:155
point mutations induced by, **4**:152–53
ras in tumors induced by, **3**:493
Carcinoid tumor syndrome (CTS), **3**:379
interferon-α therapy for, **3**:330
Carcinomas
adrenocortical, mapping to chromosome 11, **3**:127
basal cell, **3**:104
colon, loss of heterozygosity in chromosome 5q, **1**:411
colorectal
DCC gene loss in, chromosome 18, **3**:201
deletions from chromosome 18 associated with, **3**:210
defined, **1**:407
See also Adenocarcinoma; Renal cell carcinoma
Cardiac hypertrophy, defined, **1**:283
Cardiac muscles, **4**:130
in familial hypertrophic cardiomyopathy, gene mapping to chromosomes, 1, 11, 14, and 15, **3**:157
isoforms of myosin light chains in, **4**:135–37
Cardiolipin, **4**:433
circular dichroism measurements, **1**:390
Cardiolipin synthetase, **4**:435
Cardiomyopathies, genetic diseases causing, **1**:289
CARDIOVASCULAR DISEASES, **1:283–92**
Cardiovascular system, regulation by kinins, **3**:378
Carmustine, as an anti-cancer drug, **1**:254
Carotenes
provitamin A, **6**:245–46, 246f
β-Carotenes
structure of, comparison with squalene, **3**:451f
synthesis by *Blakeslea trispora*, **3**:440
synthesis by *Dunaliella salina*, **3**:441
Carotenoids, **4**:441–42
lipid, synthesis by yeasts, **3**:437
retinoid production from, **6**:236
synthesis by algae, **3**:441
synthesis by fungi, **3**:440
Carp
F_2 transgenic, growth and weight changes from growth hormone insertion, **6**:94–95, 94f
transgenic, F_1 progeny of, **6**:93, 93t
Carrier, defined, **2**:419
Carrier ampholytes, defined, **2**:353
Carrier protein, defined, **6**:125
Cartilage
histogenesis of, **2**:457
replacement of, with silicone rubber, **5**:37
Cartilage-hair hypoplasia (CHH), gene for, mapping to chromosome 9, **3**:103
Cartilage oligomeric protein (COMP), **2**:284
Cascade
chemical amplification, **2**:178–81
defined, **5**:49
Case-case studies, in molecular epidemiology, **2**:262
Case-control studies, in molecular epidemiology, **2**:261–62
Casein, **6**:103–5
Casein kinase II, activation by polyamines, **2**:211
Cassette mutagenesis, **5**:453
defined, **5**:447
Cassette permutation, in assembly PCR, **5**:451
Catabolism
chylomicron, **3**:404–6
of low density lipoprotein, **3**:408
of steroid hormones, **5**:470
of very low density lipoprotein, **3**:406–7
Catabolite activator protein (CAP)
footprint of, **4**:248f
homology with histone H1, **1**:361
nonspecific DNA binding of, **4**:245
X-ray crystallography for studying, **4**:252
Catalase (CAT), **4**:62
enzymatic antioxidant defense mechanism, **2**:220
gene for, circadian regulation of, **2**:222f
iron of, **2**:232
mutants for, effects on life span, **1**:12
and mycobacterial virulence, **4**:160
in plant response to infection, **4**:480–1
removal of peroxides by, **2**:333
Catalysis
bioengineered, **1**:143
bioorganic chemistry of, **1**:161–62
by engineered antibodies, **1**:68–69
by enzymes, **2**:223
by metalloenzymes, **6**:59
by RNAs, **4**:284
scooting mode and hopping mode, phospholipases, **4**:417–18
Catalytic antibodies, **3**:288
biomimetic, **1**:144–45
defined, **1**:159
Catalytic domain
calcium ion pump, **1**:244
factor IX, substitutions in, **3**:38
Catalytic sensors, **1**:178–79, 179t
Catalytic triad, of serine proteases, **2**:233
Catecholamines
biosynthesis of, **1**:207, 208f
defined, **1**:207
interaction with β-adrenoceptor, **4**:38f
levels of, in heart failure, **1**:290
Cathepsins
activity of
in lysosomes, **3**:347–48
around tumors, **1**:260
H, gene for on chromosome 15, **3**:165
mapping to chromosome 14, **3**:155
Catheters
from polyurethanes, **5**:38–39
from silicone rubber, **5**:37, 38
Cathode, defined, **2**:96
Cationic liposome-based complexes, **3**:419–21, 421f
Cationic nonliposomal complexes, as gene vectors, **3**:421–23
Cation-independent mannose-6-phosphate receptor, in enzyme transport to lysosomes, **5**:202–3
Cations
nucleobase radical, **3**:358–59
SSB-induced release from DNA, **5**:460
Caudal (Cad) class, of homeobox genes, Hexapeptide encoding, **3**:67
Caudina arenicola, hemoglobin of, **3**:12
C bands, chromosomal, **1**:371
Cd15
cell adhesion and, **2**:455
defined, **2**:446
cDNA
and Alzheimer's disease, **1**:51
amplification of, **5**:341–42
clone
N-CAM, **1**:294
as a sequence-tagged site, **3**:520
3-′-directed library, **1**:203
encoding β-amyloid, mutation of, **1**:49

library, **4:**127
screening of, **1:**383
randomly primed library, **1:**203
strands of, **5:**355
synthesis of, **5:**350–51
transgenic, poor expression of, **6:**102
tyrosine hydroxylase, **1:**209
CELL-CELL INTERACTIONS, **1:292–99**
adhesion requiring calcium, **1:**452
repulsion, **1:**293, 297
Cell cycle
checkpoints in, and drug selectivity, **1:**263
control of growth, by cytokines, **1:**448–49
defined, **3:**48–49
effects of drugs on, **1:**257–58
regulation of, **3:**507
Cell death
developmental, **4:**172
neuronal, implications for motor neuron disease, **4:**106
programmed, **1:**299–302
versus pathological, **1:**299
CELL DEATH AND AGING, MOLECULAR MECHANISMS OF, **1:299–303**
Cell division, and cyclin-dependent kinases, **4:**287, 287f
Cell division cycle (cdc), **1:**100–1
G_0,G_1 and G_2 stages of, defined, **6:**138
rate of division, and survival of plasmid DNA, **3:**424
role of *p53* gene in, **1:**412
roles of calcium in, **1:**240
S phase of, **3:**51
Cell-free extracts, translational systems using, **5:**137
Cell-free translation (protein-synthesizing) system, defined, **5:**327–28
Cell fusion
induced by an electric field, **2:**200–2
study of, with electroporation, **2:**205
See also MEMBRANE FUSION
Cell loss factor, **1:**257
Cell membrane
defined, **1:**95
phospholipid heterogeneity in, **3:**460
Cellobiohydrolase, **2:**244
Cell production, defined, **2:**463
Cells
as bioenergetic units, **1:**314
ciliated, movement involving dynein, **4:**116
counting, with luciferase assay, **1:**330
cultures for producing taxol and taxane, **6:**12–13, 13f
distribution of micronutrients in, **6:**59–60
growth factor action sites, **2:**464f
hypertrophy of, in heart failure, **3:**1–2
labeling for study of, **3:**389–90
methods for disrupting, **5:**344–45
sources of interleukins, **3:**337t
structures recognized by autoantibodies, **1:**88t
total RNA of, **5:**343
transplantation of, collagen-glycosaminoglycan substrates for, **5:**33
virus infections in, **6:**199
Cell surface
antigen of, detecting with PCR, **3:**293
Fas/APO-1 antigen of, **1:**302
proteins of, regulation of expression by interferons, **3:**327
receptors on, **3:**43
c-met proto-oncogene, **3:**482
interferon-specific, **3:**322
for interleukins, **3:**332
segregation of wall during division, **1:**100, 100f
synthesis during the division cycle, **1:**98–99, 98f
Cell suspension, defined, **6:**10
Cell-type tropism, and antigenicity, **5:**16
Cellular aging, **1:**303; **6:**23
Cellular antigenicity, determinants of, **3:**285
Cellular binding proteins, retinoid, **6:**236
CELLULAR ENERGETICS, BIOCHEMICAL BASIS OF, **1:303–15**
Cellular immunity
defined, **3:**499
to respiratory syncytial virus, **5:**292–93
role of the major histocompatibility complex in, **3:**500
Cellular oncogene, defined, **6:**138
Cellular pharmacokinetics, **1:**260
Cellular receptor, defined, **6:**192
Cellular receptor specificity, defined, **6:**223
Cellular receptor unit (CRU), picornavirus, **6:**224f
Cellular retinoic acid binding protein (CRBP), localization to chromosome 15, **3:**169
Cellular uptake
defined, **1:**69
of oligonucleotides, **1:**72
Cellulase
for ethanol production, **1:**173
role in plant pathology, **4:**482
Cellulose, **1:**146
as a biodegradable polymer, **5:**34
Cell wall
advent of, and speciation, **4:**78
plant, **4:**466, 480
altering the hyrolases of, **6:**57–58
Centimorgan (cM), **3:**113, 217
defined, **1:**79; **3:**142
Centraalbureau voor Schimmelcultures, addresses of, **6:**304
Central diabetes insipidus (CDI), **2:**23–24
defined, **2:**19
genetic, **2:**20
Central nervous system
role of kinins in, **3:**378
transfection of plasmids into brain, **3:**424
tyrosine hydroxylase in, **1:**209
Centre d'Étude du Polymorphisme Humain (CEPH)
database of, **3:**170, 236–37
mapping of chromosome 16, **3:**176
reference pedigrees from, **3:**143
Centrifugation
for peptide purification, **5:**189
for studying nucleic acid-protein complexes, **4:**244–45
Centrin, amino acid change in, in mutant of *Chlamydomonas reinhardtii*, **1:**359
Centromeres
Arabidopsis genome, **1:**77–78
defined, **1:**371; **2:**118
DNA-binding protein of, CENP-B, **5:**416
Centromeric sequences, pronuclei used for mapping, **2:**379
Cephaloridine, structure of, **2:**158f
Cephalosporin
C, **2:**156–57
manufacture of, **2:**164, 164f
Cephalothin, **2:**157
structure of, **2:**158f
Ceramics
biocomposite, **3:**316
biological, **1:**147
defined, **1:**134
Ceramide, **3:**457
defined, **3:**426
lipid, **3:**437
role in glycosphingolipid synthesis, **3:**458
as a second messenger, **4:**439
Cerebellum, IP_3 in the Purkinje cells of, **1:**246
Cerebral hemorrhage, in Dutch type amyloidosis, **1:**43
Cerebrosides
line model of, **3:**458f
solid to liquid transition of, **3:**460
Cerenkov counting, **5:**219
Ceroid lipofuscinosis, infantile neuronal, gene mapped to chromosome 1, **3:**92–93
C gene, **3:**281
cGMP (guanosine monophosphate)
and nitric oxide's biological actions, **4:**205

cGMP (guanosine monophosphate) (*cont.*)
as a second messenger, **4**:195–96
visual transduction pathway role of, **3**:46
Chagas disease, **4**:307
Chain extension, in peptide synthesis, **4**:387
Chain of custody (continuity), **2**:48
defined, **2**:48
Chain promiscuity, **1**:416, 420
Chain reaction, free radical, **2**:332
Chain termination
by *APC* mutations, **1**:411–12
in genetic analysis of populations, **2**:395–96, 397f
in viral inhibition, **1**:37
Chalcone synthetase (CHS)
role in plant defense against pathogens, **4**:479, 479f
suppression of, to produce commercially desirable plant variants, **4**:474
Channel-forming integral membrane protein (CHIP), **5**:273
Channel protein, defined, **6**:125
Channels, **4**:54
calcium, **1**:243
on the endoplasmic reticulum, **2**:211
voltage-dependent, **1**:247
calcium ion release, **1**:246–47
chloride, **1**:430
defined, **1**:237
in DNA structure, **2**:110
membrane, mimicking, **1**:140
water, **5**:273–74
Chaos, and complex systems, **4**:94
Chaotropic, defined, **5**:342
Chaotropic buffers, for cell lysis, **5**:345
Chaperones
intramolecular, **1**:322; **5**:92
SecB, of *E. coli*, **1**:322
CHAPERONES, MOLECULAR, **1:315–22**
defined, **5**:73
in the endoplasmic reticulum, **4**:342
hsp60, **5**:86–90
hsp70, **5**:90, 90f
role in mitochondrial import of proteins, **5**:205, 205f
targeting of proteins to the endoplasmic reticulum membrane, **5**:198
hsp90, **5**:91, 482–83
protein folding catalyzed by, **2**:460, 461; **3**:353; **5**:86–91, 158
sc60, in mitochondria, **5**:205
virus interactions with, **6**:188
Chaperonins, **1**:315, 319–20; **5**:160
and the ATPase cycle, **5**:89f
and cell division, **1**:366
chaperonin-10, **1**:319
chaperonin-60, **1**:319–20
GroEL, structure of, **5**:88
hsp70, hsp60, and hsp90, **5**:86–91
Charcot-Marie-Tooth disease, **2**:425–26
mapping of, to chromosome 1, **3**:91
Chargaff rules, **5**:353–54
Charge
density of, phosphodiester linkages, **4**:280
mobile, **2**:174
neutral and negative, of liposomes, **3**:417–18
positive, of histones, **3**:49
Charge-coupled device
image for DNA analysis, **1**:331f
for light emission measurement, **1**:326
Charge relay mechanism, of serine proteases, **2**:233
Charge separation, in photosynthesis, **1**:139, 149
Charge state, of ions in mass spectrometry, **4**:14
CHEF gel, **6**:294
defined, **6**:291
Chelate, defined, **1**:126
Chelated hydrogen bond, **3**:265
Chemical electric and magnetic field effects, defined, **2**:172
Chemical fossils, **4**:299
Chemical library, generating, for drug development, **4**:40
Chemical methods
for gene transfer, **4**:452–53
for probing RNA structure, **5**:365, 365t
for protein-ligand interaction studies, **4**:251
Chemical probes, for RNA structure determination, **5**:378
Chemical process, for cross-linking, **4**:251
Chemicals
immunomodulating, **3**:298
for preparing genomic DNA, in RLGS, **5**:307
for preparing RNA, **5**:345–46
Chemical sequencing, C-terminal, of proteins, **5**:190–92
Chemical shift
defined, **4**:218
one-dimensional NMR, **4**:219
Chemical synapses, **5**:512, 514
Chemiluminescence, **1**:323
defined, **1**:322
for DNA fingerprint analysis, **2**:57
reactions involved in, **1**:324–25f
CHEMILUMINESCENCE AND BIOLUMINESCENCE, ANALYSIS BY, **1:322–32**
ChemInform Reaction Library, **4**:99
Chemiosmotic coupling, of ATP synthesis and electron transport, **1**:309
Chemiosmotic hypothesis, **4**:448
of ATP synthesis, **1**:307–8
Chemokine, defined, **2**:463
Chemospecificity, defined, **6**:259
Chemotaxis, **1**:175
and cytokine response in inflammation, **1**:447
Chemotherapy
combination, principles of, **1**:264–66
compounds for, **1**:166
virus-receptor interactions as targets for, **6**:233–34
Chiasmata
defined, **3**:174; **6**:27
formation during meiosis, **5**:253, 253f
and genetic distance, **3**:234
of male chromosomes, **6**:31
Chicken ovalbumin upstream promoter (COUP), **1**:20f
Chickenpox (varicella), **6**:172
Chimeras
as artifacts, in yeast artificial chromosome preparation, **6**:293–94, 298–99
defined, **6**:291
for destroying anther cells, **4**:472–73
for gene regulation with trace elements, in plant, **6**:63
knockout mice, **6**:64
polypeptide or nucleic acid, defined, **6**:165
protein, **3**:286, 286f
of related genes from sexual PCR, libraries, **5**:456f
streptavidin-protein A, **3**:289
for studying lipoprotein genes, **1**:287
Chimeric compounds, antibodies, **1**:65–66, 69
Chimeric gene
antisense, in tomato plants, **6**:55, 55f
for attenuating live recombinant HIV vaccines, **6**:170
construction of, tomato plant, **6**:54–55, 54f
defined, **6**:53
Chimeric rescue, of parthenogenetic or androgenetic embryos, **2**:443–44
Chinese hamster ovary (CHO) cell lines, **6**:179
CHIP28 water channel, **5**:273–74
Chips, light-directed peptide synthesis on, **5**:519
CHIRALITY
of amino acids, and of gypsum crystals grown in their presence, **3**:314
center and axis of, **1**:335
defined, **1**:52, 332, 384; **2**:155
IN BIOLOGY, **1:332–47**
in living systems, **4**:294

of modified oligonucleotides, **1**:73
of pharmaceuticals, and manufacturing, **2**:162–63
and pharmacological activity, **4**:39–40, 39f
reactions at the center of, with chirality lost, **1**:342
Chiral methyl, defined, **1**:332
Chiral molecules, defined, **1**:148
Chiral phosphorus atom, defined, **1**:332
Chironomus tentans, silk of, **5**:27
Chi sequence, of DNA, interaction with RecBCD enzyme, **5**:228–29
Chitin, **1**:147
Chitinase
accumulation in response to plant pathogens, **4**:480
bean, overproduction of, for resistance to fungal pathogens, **4**:472
Chlamydia trachomatis
commercial detection kit, based on PCR, **4**:326
detection of
ligase chain reaction, **3**:306, 307f, 308t
PCR-based kit, **4**:330
test for
sensitivity and specificity, **3**:308t
using ligase chain reaction, **3**:306
CHLAMYDOMONAS, **1:347–60**
Chlorambucil, defined, **5**:485
Chloramphenicol acetyltransferase (CAT) gene, as a reporter gene, **2**:151
in analysis of plant promoters, **4**:490
for apolipoprotein promoter mapping, **4**:500–501
for chemiluminescence assays, **1**:329
Chloride, permeability to, in cystic fibrosis, **1**:425
Chloride channel, cystic fibrosis transmembrane conductance regulator, **1**:430
Chloride peroxidase, reactions forming a new chiral center, **1**:343
Chlorophyll, **1**:310; **4**:441
absorption maxima of protein, **4**:441
electron density of, **2**:194
Chloroplast genome
Arabidopsis, **1**:83
heteroplasmicity of, in *Chlamydomonas*, **1**:358
Chloroplasts
of *Chlamydomonas*, **1**:347
DNA of, **1**:349–53
uniparental transmission of DNA, **1**:348
energy-transducing membrane system of, **1**:307f
genetic analysis of populations using DNA, **2**:400
genetic exchanges with mitochondria, **4**:84
genetic exchange with the nucleus, **4**:489
genome of, **4**:488–89
defined, **4**:485
mitochondrial DNA of, **4**:79
photosynthesis in, **1**:310
ribosomes of, **5**:328
transformation in *Chlamydomonas reinhardtii*, **1**:354
Chloroquine
accumulation of lipofuscin associated with, **1**:49
transfection efficiency in the presence of, **3**:423
Chlorpromazine, effect on viral budding, **6**:188
Cholecalciferol
hydroxylation of, **6**:237
from irradiation of 7-dehydrocholesterol in skin, **6**:246
Cholecystokinin (CCK)
peptide hormone, and synthetic mimetics, **4**:381–82, 382f
sulfation, and bioactivity of, **4**:343
Cholera toxin, regulation of, **1**:107
Cholesterol
biosynthesis of steroid hormones from, **5**:468, 468f
cellular metabolism, enzymes involved in, **1**:288t; **3**:473
functions of, **4**:495
measurement of, **3**:475
metabolism of, **3**:409–10
as the parent molecule for steroid hormones, **5**:478
reverse transport of, **3**:408–9, 473–74, 473f
structure of, line model, **3**:461f
synthesis of, P450-dependent step, **1**:439
transport of, **1**:284
Cholesterol 7α-hydroxylase, **3**:410
Cholesterol ester transfer protein (CETP), **1**:287, **3**:409, 474
in transgenic mice, **4**:505
Cholesteryl hemisuccinate, for cancer therapy, **5**:486
Choline acetyltransferase (Chat), in Alzheimer's disease, **1**:50
Choline/ethanolamine kinase (CK/EK), **4**:434
Cholinephosphotransferase (CPT), **4**:434
Chondrocytes, site of collagen mRNA synthesis, **2**:280
Chondrodysplasia, McKusick-type metaphyseal, gene for, mapping to chromosome 9, **3**:103
Chorionic gonadotropin, **2**:454
Choroidermia, X-linked, haplotype analysis of, **5**:47
Christmas disease. *See* Hemophilia, B
Chromaffin granules
role of annexin II in aggregation of, **1**:61
role of annexin VII in binding to membranes, **1**:61
Chromatids, sister, unequal exchange between, and tandem arrays, **6**:8–9
CHROMATIN
assembly, **1**:366
constitutively condensed, in centromeres, **1**:371
Drosophila, **2**:123
electron micrography of, **2**:185, 187
excision repair of, **6**:155f, 159–60, 160f
FORMATION AND STRUCTURE, **1:360–70**
histones in, **3**:49–50, 50f
mouse P815 cells, high resolution autoradiography, **2**:189f
nucleolar, from newt oocytes, electron microscope image, **2**:188f
photoproducts from ultraviolet radiation exposure, **6**:154–55
structure of, at mammalian telomeres, **6**:29–30
structure of the apo B gene, **4**:500–501
Chromatin fiber, **1**:364–66, 366f, 370; **2**:390f
defined, **1**:360
order of cosmids along, **2**:389
Chromatographic recognition analysis, **5**:245
Chromatography
affinity
for bioseparation, **1**:176
for bioseparation of anabolic steroid, **5**:474
for isolating phage, **1**:64
for bioseparations, **1**:175–76
for carbohydrate analysis, **1**:271
gel filtration, for isolating apolipoproteins, **3**:476
perfusion, **1**:176
for protein separation, prior to analysis, **5**:176
for steroid purification, **5**:474
for studying nucleic-acid-protein complexes, **4**:245
Chromatosome, defined, **1**:360
Chromophore
circular dichroism of, **1**:385
for nucleic acid analysis, **3**:390
purple, bacteriorhodopsin, **1**:109, 109f
Chromophore membranes, **4**:441
Chromosomal band, defined, **1**:370
Chromosomal deletion, defined, **3**:160

Chromosomal positioning, and telomere repeats, **3:**243
Chromosomal translocation, defined, **3:**160
Chromosome 1
 apoAII gene, **3:**411
 gene for ACAT, **3:**410
 gene for RPA32, **5:**463
 glucocerebrosidase gene, **2:**347, 424
 human, histone gene clusters, **3:**53
 myelin protein zero gene, **2:**426
 peptidase C gene of, **3:**370
 plasma membrane calcium pump gene, **1:**245, 245t
Chromosome 1p, mapping of *MOM1* to, **1:**413
Chromosome 2
 acid phosphatase gene of, **3:**370
 apoB gene, **3:**411
 in colon cancer, **2:**428
 hereditary nonpolyposis colon cancer, **6:**146
 IL-1β gene of, **3:**334
 isocitrate dehydrogenase gene of, **3:**370
 limb girdle muscular dystrophy type 2B map to, **3:**163–64
 malate dehydrogenase gene of, **3:**370
 motor neuron disease related to genes of, **4:**106
 mouse, *Pmv-7* gene of, **2:**153
 PAX3 gene of, **3:**223
 in Waardenburg syndrome type 1, **3:**226
Chromosome 3
 apoD gene, **3:**413
 colon cancer susceptibility linked to, **2:**428
 deletion on, in lung cancer, **3:**495
 hereditary nonpolyposis colon cancer, **1:**410
 kininogen gene, **3:**381
 plasma membrane calcium pump gene, **1:**245, 245t
 renal cell carcinoma, **6:**145
 of tumors, FISH identification of, **6:**255–56, 256f
 in von Hippel-Lindau syndrome, **2:**429
 in Waardenburg syndrome, **3:**223
 YAC contig library for, **3:**149
Chromosome 4
 FISH assessment of genetic damage using WCP, **6:**255, 255f
 gene for Huntington's disease, **3:**246–47; **4:**199
 microsomal triglyceride transfer protein gene, **3:**404
 mouse, gene for interferon-α, **3:**319
Chromosome 5
 APC gene, **2:**63, 428
 APC gene in familial adenomatous polyposis, **6:**145
 cytokine genes on, **3:**334
 hexosaminidase gene, **3:**370
 HMG-CoA synthetase and reductase genes, **3:**409
 mapping to, of congenital contractural arachnodactyly, **3:**162
 MCC gene in colon cancer, **3:**224
 MCC genes in colon cancer, and *fap* genes, **1:**251
 motor neuron degenerative diseases mapped to, **4:**106–7
Chromosome 5q
 in APC and Gardner syndrome, **1:**410
 loss of heterozygosity at, **1:**411
 MCC gene in colon cancer, **1:**413
Chromosome 6
 21-hydroxylase gene, **2:**424
 apo(a) gene, **3:**415
 data model example for, **5:**397–98
 in diabetes mellitus, **2:**29
 gene for interferon type II binding protein on, **3:**322
 human
  α-apoproteins encoded in, **2:**287
  histone gene clusters, **3:**53
Chromosome 7
 CHIP locus on, **5:**273
 cytoskeletal actin mapping to, **3:**199–200
 gene for mtSSB on, **5:**465
 gene for RPA14, **5:**463
 glucokinase gene at, **2:**38–39
 human
  defect in osteogenesis imperfecta, **2:**426
  rearrangement and translocation analysis using FISH, **6:**255, 255f
 malate dehydrogenase gene of, **3:**370
 markers for cystic fibrosis on, **1:**426
 mouse *Igf-2*, *H-19*, and *Snrpn*, **2:**442–43
 mouse, imprinted genes of, **2:**442
 NOS3 gene of, **4:**202
Chromosome 8
 gene for cholesterol 7α-hydroxylase, **3:**410
 gene for c-*myc* proto-oncogene on, **3:**487–8
 gene for Langer-Giedion syndrome, **1:**379
 gene for lipoprotein lipase, **3:**404
 motor neuron disease related to genes of, **4:**106
 translocation in Burkitts's lymphoma, **4:**286
 loss of heterozygosity at, **1:**411
Chromosome 9
 deletions linked to cancers, **4:**287
 familial melanoma susceptibility gene on, **3:**93
 genes for interferons on, **3:**319
 localization of the gene for tuberous sclerosis (TSC), **3:**188
 mouse
  drug-related phenotypes correlated with locus Drd_2, **2:**153t
  QTL affecting drug responses, **2:**154f
 translocation in the Philadelphia chromosome, **4:**286
 VLDL receptor protein gene, **3:**407
Chromosome 10
 genetic mapping of, **3:**115–18, 118f
 genetic maps, male and female, **3:**116f
 mouse, gene for interferon-γ on, **3:**319
 physical mapping of, **3:**122–23
 RET gene of, **2:**429
Chromosome 11
 acid phosphatase gene, **3:**370
 *apo*AI, *apo*CIII, *apo*AIV genes, **3:**404, 410
 apolipoprotein genes, **4:**501, 502f
 β-globin gene, **4:**332
 breakpoints present in translocations associated with leukemia, **1:**382
 cyclin D1 oncogene on, **4:**287
 follicle-stimulating hormone β gene on, **2:**287
 gene for Usher syndrome, **3:**158
 genes for β-type hemoglobin subunits, **3:**13
 insulin gene, **2:**37
 in Jacobsen syndrome, **2:**318
 lactate dehydrogenase gene of, **3:**370
 mouse
  cytokine genes on, **3:**334
  for MLCla and MLCemb, **4:**136
  imprinted genes of, **2:**441
 in multiple endocrine neoplasia, **2:**429
 prostate cancer, **6:**146
 in Usher syndrome, **4:**126
 in Wilms' tumors, **2:**429, 445; **6:**142
Chromosome 12
 α_2-macroglobulin receptor gene, **3:**405
 Cdk4 gene amplification, **4:**287
 gene for aquaporin-2, **2:**27
 gene for basic-type keratin, **2:**265–66
 gene for interferon-γ on, **3:**319
 human, rearrangement and translocation analysis using FISH, **6:**255, 255f
 KRAS gene, **3:**224
 lactate dehydrogenase gene, **3:**370
 NOS1 gene of, **4:**202
 PAH gene, **4:**407
 paralogy with chromosome 1 in, **3:**84
 peptidase B gene of, **3:**370
 plasma membrane calcium pump gene, **1:**245, 245t
 TAC contigs for, **3:**149
Chromosome 13
 breast cancer gene on, **6:**142
 gene for breast cancer on, **3:**223

RB gene, **6:**139
RBI gene, **2:**429; **4:**329; **5:**320
Chromosome 14
in α1-antitrypsin deficiency, **2:**424
in Alzheimer's disease, **3:**223
early onset, **4:**199
bradykinin receptor assigned to, **5:**268
gene for transglutaminase K, **2:**268
involvement in early onset hereditary Alzheimer's disease, **1:**42
markers associated with Alzheimer's disease, **1:**46
polymorphic markers of Alzheimer,s disease, **1:**44f
sex-average unified map, **3:**150f
translocation in Burkitts's lymphoma, **4:**286
Chromosome 15
deletion in Albright hereditary osteodystrophy 2, **3:**164
fibrillin 1 gene on, **2:**426
gene for congenital contractural arachnodactyly, **3:**163
gene for hepatic triglyceride lipase, **3:**407
gene for Tay-Sachs disease, **2:**425
hexosaminidase gene, **3:**370
isocitrate dehydrogenase gene, **3:**370
in Prader-Willi syndrome and in Angelman syndrome, **2:**424
Chromosome 16
α-globin gene, **4:**332
in α-thalassemia, **6:**35
deletion in hepatocellular carcinoma, **3:**487
gene for α-thalassemia, **2:**425
gene for α-type hemoglobin subunits, **3:**13
gene for CETP, **3:**409
lecithin-cholesterol acyltransferase gene, **3:**409
linkage to adult polycystic kidney disease, **2:**426
physical map, **3:**149
Chromosome 17
campomelic dysplasia and sex reversal genes on, **5:**443
in colon cancer, **1:**411
defect in osteogenesis imperfecta, **2:**426
gene for acid-type keratin, **2:**265–66
gene for breast cancer on, **1:**223; **2:**428; **3:**223; **6:**142
gene for neurofibromatosis 1 on, **3:**226; **4:**175; **6:**144
gene for RPA70, **5:**463
in hepatocellular carcinoma, **3:**487
mouse, *Tas* gene associated with sex reversal, **5:**446
in neurofibromatosis 1, **2:**429
NOS2 gene of, **4:**202
p53 gene on, **1:**412; **6:**141
mutation in colon cancer, **1:**251
peripheral myelin protein 22 gene, **2:**426
reciprocal translocation of, in acute promyelocytic leukemia, **3:**160
role in cancer, **2:**63
TP53 gene in colon cancer, **3:**224
of tumors, FISH identification of, **6:**255
Chromosome 18
DCC gene mutation in colon cancer, **1:**251, 412; **3:**224; **6:**145
peptidase A gene of, **3:**370
18q, in colon cancer, **1:**411; **4:**329
Chromosome 19
adeno-associated virus targeted insertion to, **2:**437
in Alzheimer's disease, **3:**223
*apo*E, *apo*CI, *apo*CII genes, **3:**404, 410
apolipoprotein E gene on, **3:**414; **4:**199
apolipoprotein gene cluster, **4:**502f
apolipoprotein genes, **4:**501–2
in early onset Alzheimer's disease, **1:**46
gene for LDL receptor, **2:**426; **3:**408
kallikrein gene, **3:**384
luteinizing hormone β gene on, **2:**287
in myotonic dystrophy, **2:**427–28
peptidase D gene of, **3:**370
polymorphic markers of Alzheimer's disease, **1:**44f
prostate-specific antigen gene on, **3:**384
Chromosome20
defects in G-protein gene, **3:**164
prion protein on, **4:**199
prion protein gene on, **5:**59
Chromosome 21
age-dependent abnormalities, **1:**42; **2:**118
in Alzheimer's disease, **3:**223, 226
APP gene on, **1:**43
in early onset Alzheimer's disease, **1:**46
genes for interferon type I binding proteins, **3:**322
polymorphic markers of Alzheimer's disease, **1:**44f
structure of, **2:**119
superoxide dismutase 1 gene, **2:**425
superoxide dismutase gene on, mutant in familial ALS, **4:**108
YAC mapping of, **3:**149
Chromosome 22
gene for CYP2D6, **4:**400
L1H element on, **6:**135
LIH insertions in, **6:**135
in meningiomas, **4:**185
in neurofibromatosis 2, **2:**429; **4:**181, 182, 182f; **6:**144
and the Philadelphia chromosome, **3:**98
physical map, **3:**149
translocation in the Philadelphia chromosome, **4:**286
Chromosome bands, **1:**371
chromosome 18, **3:**191
defined, **2:**360, 377
Chromosome breakage, syndromes associated with, **1:**413
Chromosome fragment, defined, **1:**370f
Chromosome jumping, **3:**515
for identifying the cystic fibrosis gene, **1:**426
Chromosome maps, **3:**95
CHROMOSOME MICRODISSECTION AND MICROCLONING, **1:370–84**
Chromosomes
aberrant, *Drosophila*, **2:**127–28, 127f
abnormalities in, **3:**225–26
abnormalities of, in cancer cells, **1:**383
age-dependent abnormalities, **1:**8–9
artificial, generating, **6:**37
bacterial, recombination of plasmids with, **5:**6–7
in colon cancer, loss of heterozygosity, **1:**411
defined, **1:**370
Drosophila, **2:**122–23, 122f
elongation and translocation, **6:**19
Escherichia coli, **2:**169, 170f; **5:**7
and expression of cystic fibrosis, **1:**117
fragmentation of, experimentally induced, **6:**36–37
healing by telomerase, **6:**35–36
hepatitis B virus integrations on, **3:**484
human, stained with Hoechst 33258, **2:**380f
hybridization analysis of, **2:**379
interleukin genes on, assignment and exon/intron organization, **3:**336t
isoenzyme gene locations, **3:**371–72t
linear, of streptomycetes, **5:**507
locations of growth factors and receptors, **2:**465t
locations of identified tumor suppressor genes, **6:**139t
mammalian, **3:**511
map, constructing with linking and jumping clones, **2:**78
packaging of DNA into, **2:**117–18
packing of, and DNA structure, **2:**113
pairing of, *LI* element facilitation of, **3:**245
partitioning of, in *recA* mutant cells, **5:**222
Philadelphia, **1:**250; **2:**64, 64t; **4:**286
physical maps, constructing with linking and jumping clones, **2:**76
Plasmodium falciparum, **4:**308
polytene, **2:**121
drosophila, **2:**126–28f

Chromosomes (*cont.*)
prokaryote and eukaryote, replication of, **2**:86–87, 87f
rearrangement of, **2**:418
recombinant, **5**:252f
repetitive sequences of telomere, **6**:29
restriction maps, of *Plasmodium falciparum*, **4**:308
Robertsonian fusion, **2**:119
satellites in, sex chromosomes, **5**:436, 436f
sex, **3**:511
sorting, **6**:254
structure of, **1**:370–71
telomeres of, and function, **6**:18
translocation t(14;18), in follicular lymphomas, **1**:302; **2**:65
translocation t(15;17), **2**:65
whole chromosome complementary probe fluorescence staining, **6**:253–57
Xenopus, electron microscope image, **2**:192f
yeast, **6**:305–6
yeast artificial, **1**:83
Chromosome-sorted libraries, **2**:72
Chromosome-specific painting, **2**:388
defined, **2**:386
reverse, **6**:256, 256f
Chromosome translocation, defined, **6**:138
Chromosome walking, **3**:122, 220, 515; **6**:301
with *Arabidopsis*, **4**:491
with *Drosophila*, **2**:124, 128–29
for genetic mapping, **1**:231
for identifying the cystic fibrosis gene, **1**:426
Chromosome walking techniques, *Arabidopsis* gene mapping, **1**:83
Chromosome X
and breast cancer in males, **3**:223–24
and fragile X linked mental retardation, **2**:320, 320f
gene for factor IX, **3**:41
gene for factor VIII, **3**:29
inactivation of, during meiosis, **3**:503–4
Kallman's syndrome gene, **2**:292
mouse and human compared, **4**:126, 126f
mutant gene in agammaglobulinemia, **4**:127
and nephrogenic diabetes insipidus, **2**:24–25, 25–26, 26t
Chromosome Y, **5**:435
YAC library, **3**:149
Chronic external ophthalmoplegia (CEOP), **4**:82
mtDNA deletion in, **4**:90
Chronic granulomatous disease, **2**:420
Chronic myelogenous leukemia (CML)
interferon-α therapy for, **3**:330
test for translocation indicating, **4**:328–29
translocation in, **2**:64, 64t
involvement of chromosome 9, **3**:98
Chylomicronemia, in pregnancy, **3**:405
Chylomicrons, **1**:284
assembly of, **3**:404
catabolism of, **3**:404–6
defined, **3**:404
lipid transport by, **3**:472; **4**:495, 496
pathway for synthesis and catabolism of, **4**:499f
Chymosin, culture of, in filamentous fungi, **2**:342
Chymotrypsin
acid-base catalysis by, **2**:232
model, solvent effects on biocatalyst kinetic properties, **2**:256
reaction coordinate diagram for, **2**:228, **2**:229f
reaction mechanism, **2**:228
Cigarette smoke
and cytochrome P450A1 gene expression, **3**:492
free radical formation in, **2**:335, 335t
Ciliary beating, **4**:113
dynein roles in, **4**:116, 120
Ciliary neurotrophic factor (CNTF), **4**:111; **5**:241
Cinnamate-4-hydroxylase (C4H), **4**:479f
Circadian clock responsive elements (CCREs), **2**:222
Circadian clocks, **2**:221
Circadian rhythms, **2**:221–22
defined, **2**:216
Circular assembly, PCR, **5**:450, 451
Circular chromosome, **2**:86
CIRCULAR DICHROISM
defined, **1**:384
for following DNA denaturation, **2**:6t
for following DNA-protein interaction, **4**:244, 244f
IN PROTEIN ANALYSIS, **1**:**384–96**; **5**:126, 126f
Circular satellites, extrachromosomal, **6**:5
Cis-acting elements
in apolipoprotein expression, **4**:500–501, 501f
defined, **4**:465
DNA, **4**:466
picornavirus, **5**:23
in plant genes, **4**:492
Cis factors, and regulation of gene expression, **1**:234–35
Cisplatin
as an anti-cancer drug, **1**:254
structure of, **1**:255f
Citrate synthetase, **1**:344, 345f
Citric acid, from *Aspergillus niger* fermentation, **2**:338–40
Citric acid cycle, **2**:233
Clams, transgenic, electroporation for transfer to, **6**:88
Clarkia fossil beds, **4**:300
Classical conditioning vs. operant conditioning, **4**:53–54
Classification
of fragile sites, **2**:313, 313t
of lipids, biologically active, **3**:450t
of oncogenes, **4**:286
of organisms, by nutrient source, **1**:306
of steroids, by target tissue, **5**:478
See also Nomenclature
Classification codes, and functional codes, **6**:39
Class switching
defined, **3**:275
in immune recognition, **4**:95
Clathrin, cytosolic, role in endocytosis, **4**:51
Clathrin cage, disruption of, after hsp70 binding, **1**:318
Clathrin-coated pits, mediation of endocytosis by, **5**:204, 204f
Cleavage
activation sites for, hemophilia A, **3**:31
by phospholipases A_1, A_2, C, and D, **4**:417f
of side chain protecting groups and resin attachment, peptide synthesis, **4**:387–88
Clinical features
of fragile X syndrome, **2**:319
of human prion diseases, **5**:69–72
of Huntington's disease, **3**:247
of neurofibromatosis 1, **4**:173–74
See also Phenotype
Clinical trials
of HIV protease inhibitors, **1**:18
of interleukins, **3**:345–46
CLN3 (juvenile onset neuronal ceroid lipofuscinosis)
defined, **3**:174
genetic and cytogenetic maps, **3**:189f
localization to chromosome 16, **3**:187–88
Clock boxes, **2**:222
Clock genes, **2**:221
Clonality, **1**:108
defined, **1**:103
Clone libraries, positional cloning using, **4**:127
Cloning
Arabidopsis genes, **1**:80–83
of autoantigens associated with scleroderma, **5**:412, 412t
of avirulence genes, **4**:484
cleavage sites for, **2**:274–75
direct, of a gene cluster synthesizing an antibiotic, **5**:504

of DNA markers, **2**:67
for a chromosome, **1**:377–82; **2**:72–73
of DNA repair genes, **6**:163
in *Drosophila*, **2**:124
of the Duchenne muscular dystrophy gene, **5**:46
of the human phenylalanine hydroxylase cDNA, **4**:407
immune repertoire, **3**:287
of mycobacteria genes, **4**:161–63
of myosin genes, **4**:134
of plant regulatory genes, **4**:471–72
positional, **2**:127–28, 387, 420; **3**:216, 222, 515–17
of mutant genes of *Caenorhabditis elegans*, **1**:231–32
to purify bacteriophage gene products, **6**:201–2
restriction enzymes used in, **5**:302
strategies for, in streptomycetes, **5**:503–4
of streptomycetes, vectors for, **5**:502
of telomeres, **6**:28
Cloning vector, pYAC4, for constructing a *Plasmodium falciparum* library, **4**:309
Clonotype, defined, **3**:275
Clotting, role of vitamin K in, **6**:247
Clusterin, **3**:414
Clustering, sequence, defined, **5**:428
Cnidarians, HOX-cluster-related genes of, **3**:67
Coagulation, blood
factors VIII and IX in, **3**:29f
kininogen in, **3**:380
Coagulation cascade, sulfation in, **5**:56
Coagulation factor VIII
in hemophilia, **3**:29, 29f
release by desmopressin, **2**:22
Coagulation factor XIa, inhibition by amyloid precursor protein, **1**:46
Coated pits
clathrin, mediation of endocytosis by, **5**:204, 204f
defined, **3**:415
in exocytosis, **4**:51
Coatomers, **4**:51
Golgi complex, **5**:201
Coat proteins, in endocytosis, **4**:50–51
Cobalt, in enzymes, **2**:233; **4**:65
Cocaine
effects on rodents, genetic studies, **2**:150
structure of, **4**:29f
Cockayne's syndrome, **1**:8
excision repair defects in, **2**:46; **6**:161
nature of mutation in, **2**:47
premature aging in, **2**:80
Cocliths, morphology of, **3**:310f
Cocoa butter fatty acids, **3**:436t
Coconversion of alleles, **5**:254–55, 257
defined, **5**:252
in a repeated array, **5**:262, 262f
Codes
defined, **6**:38
functional and classification, **6**:39
Coding region, defined, **2**:271; **5**:128
Codominant amplified polymorphic sequence (CAPS) markers, for mapping the *Arabidopsis* gene, **1**:80
Codon-anticodon pairs, in engineered biosynthesis, **1**:137
Codon bias, nuclear genome of *Chlamydomonas reinhardtii*, **1**:348
Codons
amyloid precursor protein, encoding valine, **1**:44
in the *Arabidopsis* genome, **1**:78
of archaea, **4**:75
consistency with statistical structure of sequences, **6**:46
defined, **1**:76, 134; **2**:271, 415; **3**:227; **5**:352
initiation, **6**:116
mutant, in Alzheimer's disease, **1**:43–44
nonsense, suppression of, **5**:143f
stop, recognition of, and suppression of recognition of, **5**:143–47
Codon usage
defined, **5**:26, 428
in sequence analysis, **5**:429
and tRNA, effect on rate of elongation in protein synthesis, **6**:124
COENZYMES
A, **6**:251–252
condensation with butyric acid, **1**:160–61, 161f
medium chain acyl CoA dehydrogenase deficiency, **2**:425
B_{12}, **1**:402–3
BIOCHEMISTRY OF, **1:396–406**
defined, **2**:223; **6**:244
M and B, for methanogenesis, **4**:69
in methanogenesis, **4**:69, 70f
of redox catalysts, **2**:232
tetrahydrobiopterin, **1**:207; **4**:69
vitamin-derived, **1**:397–405
Cofactor
defined, **1**:397; **2**:223; **6**:244
enzyme, **2**:232–33
vitamin C, for protocollagen hydroxylase, **6**:252
Cognitive impairment, in phenylketonuria, **4**:407
Coiled body, reaction with human autoimmune sera, **1**:87–88
Coiled coil
basic leucine zipper, **5**:130
polymers based on proteins with, **5**:29
Coincidence indices, **6**:39
sequence analysis using, **6**:44–47
Colcemid, for accumulating cells at mitosis, for FISH, **2**:379
Colcinogenic plasmids, **5**:12
Cold acclimation, **2**:218–19
Cold-regulated genes, *cor*, **2**:219
Cold spots, for recombination, **3**:513
Colicin plasmids, **5**:2–3
Col E1, **5**:298–99
Colicins
defined, **5**:1
effects of, on bacteria, **5**:12
Colinearity principle, in peptide synthesis, **4**:213
Collagen, **1**:146
fibril of, electron microscope image, **2**:191
genes coding for, **2**:279–82
medical application of, **5**:32–33
production of, modulation by ascorbate, **6**:240
synthetic, **5**:27–28
Collagenases, secretion by bacterial pathogens, **1**:106
Collapsin, **1**:298–99
Colleotrichum gleosporoides, culture for herbicidal properties, **2**:344
Collision-induced dissociation (CID), **4**:17
Colon, anatomy of, **1**:408
COLON CANCER, **1**:251, 407-15; **6**:145–46
genetic bases of, **3**:224
loss of heterozygosity in, **4**:329
Colony picking, automatic, **5**:391–92
Colony-stimulating factors (CSFs), **1**:442; **2**:466, 469
defined, **2**:463
list, **1**:443t
Color blindness, association with hemophilia, **3**:29
Colorectal cancer
DCC gene loss in, chromosome 18, **3**:197
deletions in, chromosome 18, **3**:198f
genetic predisposition to, **3**:232, 232f
ras mutation in, **2**:65
Colorectal carcinoma
DCC gene loss in, chromosome 18, **3**:201
deletions from chromosome 18 associated with, **3**:210
Colorimetric assays, for proteins, **5**:121
Color ratio labeling, **2**:384–85
Columns, chromatography, **3**:77, 77f
selection criteria, **3**:83t
Combinatorial chemistry, **4**:32–33, 32t
Combinatorial library
addressable, **4**:97–98

Combinatorial library (*cont.*)
antibody
applications, **1**:420–24
defined, **1**:415
for bacteriophage κ, **1**:68
of biomimetic molecules, **1**:143
defined, **4**:90
for filamentous phage, **1**:68–69
and ligand selection, **4**:96–100
nonpeptide, defined, **5**:516
peptide, **4**:91, 353; **5**:520, 520f
defined, **5**:516
peptoid, **4**:97
Combinatorial mechanisms, in antibody diversity, **3**:282–83
Combinatorial multiple cassette mutagenesis, **5**:452
COMBINATORIAL PHAGE ANTIBODY LIBRARIES, **1:415–24**
Combinatorial probe labeling, **2**:384–5
Combinatorial small molecule library, defined, **5**:516
Cometabolism
defined, **1**:118
of trichloroethylene, **1**:121
Comets, organic compounds in, **4**:291–92
Commitment complex, in assembly of a spliceosome, **2**:371
Communication
among cells, **5**:230
Hartley-Shannon theory of, **6**:51–52
Comparative genomic hybridization (CGH)
defined, **3**:215
positional information from, **3**:220
Comparative map
chromosome 15, with mouse chromosomes 2, 7, and 9, **3**:170, 173t
chromosome 18, **3**:192t
mouse, chromosome 18, **3**:198–99
Compartmentalization, of pathways, due to isoenzymes, **3**:374
Compatible interaction, host-pathogen, **4**:476
Compensatory mutations, in RNA, **6**:21
Competitive binding, by nonpeptides to peptide receptors, **4**:347
Competitive elution, in affinity chromatography, **5**:246f
Competitive inhibition of enzymes, **2**:235, 235f
Competitive in situ suppression (CISS), **2**:381–82
Complement activation, identifying structures responsible for effector functions, **1**:65–66
Complementarity-determining regions (CDRs), **1**:66, 424
loops of, **3**:277
multiple ligands for, **4**:98–99
Complementation
by a cloned gene, **5**:502
cloning by, **6**:314–17
defined, **2**:78; **5**:501
groups in UV-sensitive human and rodent cell lines, **6**:162, 162t
intragenic, **6**:319
of mutants deficient in antibiotic biosynthesis, **5**:503–4
Complementation groups, genetic, of *Caenorhabditis elegans*, **1**:229
Complementation test, **1**:114–15; **2**:416, 416f
defined, **2**:415
Complement system, **3**:500
C3dg component, receptor for Epstein-Barr virus, **6**:233
Complex disease, defined, **3**:222
Complex expression, in pattern matching, **5**:425
Complexity class, defined, **6**:48
Complex-mediated hypersensitivity, **1**:105
Complex-type glycans, **2**:458
Complex-type sugar chains, **2**:452f, 459f
defined, **2**:451–53
Composites, hierarchically ordered, **1**:145–47
Compound chromosomes, **2**:128
Compound libraries, **4**:32–33, 33f
Comprehensive Medicinal Chemistry (CMC) database, **4**:99
Computation, with DNA, **1**:157–58, 158f
Computer analysis, in protein sequencing, **5**:193
Computer modeling
in drug design, **2**:160–61
genetic algorithms, **2**:412–14
of RNA three-dimensional structures, **5**:372–73, 372f
interactive, **5**:379
Concatemer, DNA, **6**:205, 205f
defined, **6**:200
Conceptualization, defined, **5**:372
Concerted evolution, **6**:4f
in kallikrein genes, **3**:383–84
phenomenon of, **6**:134
Concerted model, of allostery, **3**:10
Conductance, defined, **1**:425
Conductimetry, biosensors based on, **1**:187
Conductivity, sample, effect on STM, **5**:402–3
Cone-rod dystrophy, gene for, mapping to chromosome 18, **3**:213
Configuration
at a chiral center, **1**:335
covalent, of RNA, **5**:369f
defined, **1**:332
Confocal microscopy, for FISH signal visualization, **2**:385, 385f
Conformation
changes in, regulation by phosphoprotein phosphatase, **5**:170
defined, **1**:335; **5**:147, 368
DNA, **2**:1
labeling methods for detecting changes in, **3**:389
main chain, Raman spectroscopic determination, **5**:104
oligonucleotide, defined, **4**:281
peptide
constraints on, **4**:371–82
defined, **4**:348
phytochrome, change in response to light, **4**:493–94
protein
methods for determining, **5**:126–27, 126f
stable forms, **5**:148–50, 150f
PrP^{Sc}, in diverse strains, **5**:67
RNA, physical forces determining, **5**:369–70
state transitions, and dipole moment, **2**:176
Conformational changes
during ATP formation, **1**:307
and calcium ion binding, **1**:241
ligand binding, HIV protease, **1**:28
Conformational freedom, of oligonucleotide analogues, **4**:280
Conformational space, two-dimensional twist *T* and slide *S* for DNA, **2**:111f
Conformations, peptide, defined, **4**:371
Conformers, enzyme, **3**:367
Congenic strains
defined, **2**:146
for drug and alcohol studies, **2**:151–52
Congenital adrenal hyperplasia (CAH), **1**:437; **2**:214, 424; **5**:468
caused by $P450_{c21}$ deficiency, **5**:469
Congenital contractural arachnodactyly, **3**:162
gene for, mapping to chromosome 15, **3**:163
Congenital disorders, in infants of mothers with phenylketonuria, **4**:407
Congenital dyserythropoietic anemia type III, mapping to chromosome 15, **3**:165
Congenital hypertrophy of the retinal pigment epithelium (CHRPE), **1**:409
Coniferyl alcohol dehydrogenase (CAD), **4**:480
Conjugation
bacterial, **2**:405
defined, **5**:1
with glutathione, in xenobiotic metabolism, **6**:273–75, 275f
of marker DNA to molecular linkers, for immuno-PCR, **3**:290

metabolic, **6:**260, 267–68
defined, **6:**259
plasmid, **5:**5–6
of steroid hormones, **5:**470–71
Conjugative enzymes, phase II enzymes as, **1:**193
Connectin, **4:**143
Connective tissue
diseases involving, genetic, **2:**426
involvement in Marfan syndrome, **3:**162–63
Connectivity diagram, NOESY-COSY, apolipoprotein C-II fragment, **4:**222f
Connexins, in gap junctions, **5:**512–13
ω-Conotoxin, N-type calcium channel blockage by, **1:**243
Consensus binding sequence
for p53 binding, **6:**142
Tat, **1:**21
Consensus map
for chromosome 1, **3:**94f
for chromosome 18, **3:**195f
Consensus sequences, **2:**389
Asn-Pro-Thr-Tyr, of amyloid precursor protein, **1:**49
calcium-binding, fibrillin gene, **3:**162
defined, **5:**420, 428
for folding to a predictable protein structure, **5:**166
of genes relating to interleukin-6, **3:**339
G-protein-linked receptor, **5:**233
homeodomain, **3:**56–57, 57f
L1H, **6:**134
in multiple sequence alignment, **5:**425
palindromic regulatory, heat shock regulatory element, **2:**218
PKA and PKC, in the regulatory domain of the calcium ion pump, **1:**244, 244f
promoter
Box A and box B, **4:**74–75
protein phosphorylation, **5:**56t
recognition by protein kinases, **5:**170
search for signal sequences using, **5:**431
in secreted glycoproteins, **2:**459
TATA and CCAAT of MLC1f promoter, **4:**135
T-cell receptor, **3:**282
transcriptional, **2:**94–95
translational, **2:**274
tyrosine hydroxylase, phenylalanine hydroxylase, tryptophan hydroxylase, **1:**210
Conservation
of *Caenorhabditis elegans* genes in humans, **1:**230
of telomeric DNA sequences, **6:**18–19
Conserved sequences
Alu repeat, species specific, **2:**73
homeobox genes, **3:**60, 60f
identifying, **5:**425
Constitutive promoter, **2:**273
Constitutive proteins, cpn60 and cpn10, **5:**86
Constitutive secretion, **4:**45
Constitutive secretory pathway, defined, **4:**340
Construct, defined, for DNA fusion in vitro, **6:**99
Construction, in vitro, of yeast artificial chromosomes, **6:**294–99, 299f
Contact lenses
from poly(2-hydroxyethyl methacrylate), **5:**41
from poly(methyl methacrylate), **5:**40
Contamination
in immuno-PCR, **3:**295
in LCR, **3:**307–8
in PCR, **4:**336, 337
Contigs
construction of, **2:**74
defined, **2:**386; **3:**125, 191; **4:**127, 156, 306; **5:**388
Genome Project use of, **3:**140, 183, 520
mapping with, **1:**382–83; **5:**404–5
for physical mapping, **3:**515
in positional cloning, **2:**387
to identify the *NF2* gene, **4:**182f, 183
from the retinoblastoma gene, **6:**140
for sequencing the *Caenorhabditis elegans* genome, **1:**230–31
YAC, **4:**125f, 127
building, **6:**300–1
for chromosome mapping, **3:**122
Contiguous disease syndromes, **3:**186
Contiguous pattern, factor VIII gene, and hemophilia A, **3:**31, 32
Continuous culture, Chemostat, **2:**341
defined, **2:**337
Contour-clamped homogeneous electric field (CHEF) gel, **6:**294
Contractile system
molecular organization and genetics of, **4:**133–43
regulation and energetics of, **4:**143–45
Contracts versus patents, **6:**81
Controlled-delivery systems, **2:**166
Control region, defined, **5:**128
Conversion, gene, **5:**254–55, 258f
and repeats, **5:**261–62, 262f
Converting enzyme, defined, **5:**49
Coomassie Brilliant Blue, stain for 2D-PAGE, **2:**355–56, 355f
Cooperative Human Linkage Center (CHLC)
map for chromosome 14, **3:**144
map for chromosome 18, **3:**195
Cooperativity
in binding by hemoglobin, **3:**8, 9–12, 9f
σ-bond and π-bond, **3:**265–66
π-bond, and hydrogen bonding, **3:**269
defined, **2:**172; **5:**147
for single-stranded binding, **5:**458
of DNA-binding, zinc fingers, **5:**133
in estrogen and glucocorticoid receptor binding to DNA, **6:**332
in excision repair of DNA damage, **6:**162
of hsp60 and hsp70 in protein folding, **5:**90–91, 91f
hydrogen bond, **3:**261, 265–66
in ligand binding, **1:**131
model for protein binding, **5:**461, 461f
in oxygen binding by hemoglobin, **1:**131; **3:**6
of protein folding, **5:**88, 151–53, 153f, 157
state transition, **2:**180
and transcription control, protein-DNA interactions, **3:**53
T state, hemoglobin in animals, **3:**27
Coordination, of enzyme regulation, **2:**237
Coordination number, of calcium, **1:**239
Copolymers, properties of, and sequence of monomers, **5:**26
Copper
in enzymes, **4:**62–63, 63t
metabolism of, **6:**60
and photosynthetic gene expression, **1:**357–58
in superoxide dismutase, **2:**334
Copper-zinc superoxide dismutase (CuZn-SOD), effects of mutants on life span, **1:**12–13
Coprotease, RecA protein as, **5:**222, 226–27
Copy number, **2:**272
amplification of, in yeast artificial chromosomes, **6:**293, 299
determining with RLGS, **5:**317, 317f, 318
of DNA molecules, **5:**305
dynamics of change in, in tandem arrays, **6:**6–7
genetic mechanisms for change in, **6:**3–5
plasmid, regulation of, **5:**5
reflected by spot intensity, of restriction landmarks, RLGS, **5:**316
in transgenic animals, **6:**72
variable, in yeast episomal plasmid vectors, **6:**312
YCp plasmid, interference in plasmid shuffle utilization, **6:**317
Copy number polymorphism, defined, **6:**1
Core, human genome, H3 isochore family, **2:**363

Core histones, **1:**361
Corepressor
 binding of, **5:**279
 defined, **5:**274
Coronary arteries, defined, **1:**283
Coronary heart disease
 defined, **3:**472
 lipoprotein metabolism disorders and, **3:**472
 and plasma lipoproteins, **3:**478–79
Coronavirus
 assembly and budding in, **6:**184
 mouse hepatitis virus, **6:**221
Corpus cardiacum (CC), **4:**190, 190f, 193f
 insect, ecdysiotropic neuropeptide of, **4:**192
Correlated spectroscopy (COSY), defined, **4:**218
Corrinoid-containing enzymes, **4:**69
Corticoskeleton, actin, **4:**119f
Corticosteroid-binding globulin (CGB), gene for, mapping to chromosome 14, **3:**155
Corticosteroids, **2:**158
 preparation from stigmasterol, **2:**165, 165f
Corticotropin, enzyme defect associated with production of, **2:**214
Cortisol-binding globulin (CBG), **5:**481
Cosegregation, **3:**224
Cosin, photodynamic action of, **6:**207
Cosmid contig map, for chromosome 16, **3:**183
Cosmid library
 for chromosome 11, **3:**140
 for chromosome 9, **3:**101
 for *Mycobacterium leprae*, **4:**157
Cosmids, **2:**68
 defined, **1:**347; **2:**386; **4:**306; **5:**501
 genomic libraries constructed in, **2:**71–72, 72f
 linked, for genome sequencing, **5:**507
 for physical mapping of *Drosophila*, **2:**128–29
 probe made from, **2:**389
Cosmid subclones, FISH ordering of, **2:**389; **3:**172–73
Cost, of producing one transgenic pig, **6:**71
Cosuppression
 defined, **4:**465
 of genes, **4:**474
COSY (correlated spectroscopy), **4:**220, 220f
 spectrum of a synthetic tridecapeptide, **4:**222f
Cotranslational modification, defined, **5:**49
Cotranslational processing
 defined, **5:**194
 N-terminal acetylation, **5:**52
Cotransporters, Na^+-Cl^- and Na^+-K^+-Cl^-, **5:**273
4-Coumarate:CoA ligase (4CL), **4:**479
Coumarin, metabolism by CYP2A6, **4:**402
Counterion distribution monitoring, in peptide solid phase synthesis, **4:**391
Counterions, concentration of
 and DNA denaturation, **2:**6–7
 DNA duplex helix, **2:**1
Counterstaining, in FISH, **2:**379
Countertransport, **6:**129
Coupling, of enzyme reactions, in a metabolic pathway, **2:**228–29
Coupling agent, defined, **4:**383
Coupling factor, and rate of ATP synthesis, **4:**449–50
Coupling reaction, peptide synthesis, **4:**384–85, 384f
Coupling reagents
 peptide synthesis, **4:**390
  dicyclohexylcarbodiimide, **4:**386
Courtship
 behaviors of, relationship with other behaviors, **2:**132
 in *Drosophila* mutants, relationship with neural membrane functions, **2:**133–35
Covalent attachment, to form constrained peptides, **4:**376–77, 376f
Covalent binding, in biosensors, **1:**181
Covalent bonds
 polynucleotide, **5:**374, 374f
 protein, **2:**309, 309f
Covalent catalysis, enzymatic, **2:**231–32
Covalent interactions, bioorganic chemistry, **1:**160–61
Cowden syndrome, **1:**410
Coxsackievirus B3, attachment to host cells, assay of, **6:**230f
C period, bacterial growth, defined, **1:**95
CpG islands (cytosine guanosine pair islands), **5:**430
 defined, **2:**319; **3:**28; **4:**173; **5:**428
 in fragile X linked mental retardation, **2:**321
 for genome mapping, **2:**74
 in hemophilia A, **3:**32
 identification of, in positional cloning, **4:**128
 markers containing, **2:**68
 markers for mapping, chromosome 16, **3:**183
 in promoter regions, **2:**321–23
 screening of, restriction landmark genome scanning, **5:**316–17
 and transcription, **3:**226
CpG methylation, **2:**444, 444f
C protein, of a band, **4:**142
Crambin, engineering of, for organic biocatalysis, **2:**251
Craniofacial-deafness-hand syndrome, *Pax-3* defect in, **3:**68
Creatine kinase
 isoenzyme distribution, as a measure of myocardial damage, **3:**374, 374f
 mitochondrial, gene localization to chromosome 15, **3:**168
CRE-binding protein (CREB), testicular, **3:**506
Crem, role in spermatogenesis, **3:**506
CREST syndrome
 defined, **5:**410
 Raynaud's phenomenon with, change in antibody species, **5:**419
Creutzfeldt-Jakob disease (CJD), **4:**199
 familial, **5:**58
 iatrogenic transmission of, **5:**70, 70t
Cri-du-chat syndrome, unbalanced translocation of chromosome ends in, **6:**35
CRI-MAP program, **3:**175
Crisis point, in telomere shortening, **6:**34
Critical micelle concentration (cmc), defined, **4:**416
Critical period
 dependence of larval development on brain secretions, **4:**189
 in development, defined, **4:**188
Crocodile, hemoglobin of, **3:**27
Crops, improvement with genetic engineering, **4:**456
Cross-bridges, in the walk-along theory of muscle contraction, **4:**144–45, 144f
Cross-hybridization, **5:**504
Crossing-over
 on homologous chromosomes, **2:**419f; **3:**142; **5:**254, 254f
 between repeated sequences, **5:**262
 unequal, **6:**4, 4f
Cross-linked polymers
 defined, **5:**31
 hydrogels, **5:**41
Cross-linking
 to identify nucleic acid-protein structures, **4:**250–51
 for RNA secondary structure studies, **5:**365
 for RNA structure modeling data, **5:**378
Cross links
 defined, **3:**354; **4:**312
 DNA-DNA and DNA-protein, from ionizing radiation damage, **3:**365
 myosin bridges, **4:**134
Crossovers, **3:**225, 513; **5:**253, 253f
 defined, **3:**377; **5:**252
 between functional gene and pseudogene, in Gaucher disease, **2:**349

identifying, with CRI-MAP analysis, **3:**238
in meiosis, **3:**233–234, 234f
secondary site, in DNA inversion, **2:**404
Cross-reactivity
as neutral sequence space, **4:**98–99
species, of autoantigens, **1:**89–90
Cross talk, among interleukins, **1:**446
Crown gall disease, **5:**11–12
Cryoelectron microscopy, defined, **2:**181
Cryptanalysis tests, **6:**45
Crystal lattice, defined, **3:**309
α-Crystallins, relationship to small heat shock proteins, **1:**322
Crystallization, **3:**309–11
of HIV-1 integrase, **1:**33
for X-ray crystallography, **6:**277–78
Crystallography
defined, **6:**275
enzyme-inhibitor interactions from, **1:**14
X-ray, defined, **1:**27
Crystals
defined, **5:**110; **6:**275
protein, **5:**111
characterization of, **5:**112
growth of, **5:**112
Crystal structures
G protein, **5:**234
zinc finger binding domains, **6:**330f
C-terminal sequence analysis, **5:**175, 190–92, 192f
defined, **5:**173, 185
Culture, of yew callus, **6:**11, 12
Curcurbitacins, antitumor activity of, **5:**497–98, 498f
Curie (Ci) unit, defined, **5:**211
Curvature
determining the fourth-level structure of DNA, **2:**104
in DNA structure, **2:**111–14, 115
defined, **2:**103
Curve-fitting procedures, for Raman data, **5:**102
Cutaneous malignant melanoma (CMM1), gene for, mapping to chromosome 1, **3:**93
Cut class, homeobox genes, **3:**70, 71f
Cuticle
composite structure of, **1:**147
leaf, composite structure of, **4:**477f
plant, penetration by pathogens, **4:**481
Cutinase, in plant infections, **4:**481
Cutters, 6 bp and 8 bp, **5:**306
C-value paradox, **6:**1–2
Cyanobacteria, **1:**310
lipids of, **3:**440–41
phycobilisomes of, **4:**442
Cyanocobalamin, structure and function of, **6:**250–51, 251f
N-Cyanoguanidine linkage, **4:**279–80, 280f
α-Cyanohydroxycinnamic acid, matrix for MALDI, **4:**16
Cycle sequencing, **4:**255
defined, **5:**388
Cyclic adenosine monophosphate. *See* cAMP
Cyclic guanosine monophosphate. *See* cGMP
Cyclic nucleotides
as second messengers, **2:**209–10
second messengers regulating, **2:**209
Cyclic peptides
constrained, **5:**250
self-assembly of, into nanotubules, **1:**153
Cyclic structures, constraints imposed by, peptides, **4:**374–77
Cyclin-dependent kinases
and cell division, **4:**287, 287f
in spermatogenesis, **3:**507
Cyclins
A, integration of a viral sequence in a gene for, **3:**489
and cell division, **4:**287, 287f
degradation of, **3:**352
spermatogenesis control involving, **3:**507
Cyclization, peptide
N- and C- terminal, **4:**374–76
N-to-N type, **4:**376f, 377
side chain-to-backbone, **4:**374–76
synthetic, **5:**517
Cyclobutane dimers
enzymatic photoactivation for removal of, **6:**157t
formation by exposure to ultraviolet radiation, **6:**154–55, 154f
preferential repair of, **6:**161f
repair, kinetics of, **6:**158–59, 159f
Cyclobutane pyrimidine dimers (CPDs)
defined, **6:**283
formed by ultraviolet damage to DNA, **6:**284
from ultraviolet damage to DNA, **2:**41
ultraviolet radiation exposure and formation of, **6:**209
β-Cyclodextrin, absorption enhancement by, **2:**167
Cycloheximide, effect of, on cell death, **1:**299
Cyclopentanoperhydrophenanthrene ring, of steroids, **5:**466
Cyclophilin (CyP), **5:**92
phosphatase inhibition by a complex of, **5:**173
structure of, from NMR, **4:**223
Cyclophosphamide, B-lymphocyte susceptibility to, **3:**298
Cyclopropyl amino acid substitution, in peptides, **4:**374, 374f
Cyclosporins, **4:**210–11, 211f
A
effect on dephosphorylation, **5:**173
interaction with cyclophilin, **4:**224
from fungi, **2:**340, 344–45
T-lymphocyte susceptibility to, **3:**298
Cyclosporin synthetase, **4:**210–11, 213
Cyproterone acetate
antisteroid, **5:**484
for cancer therapy, **5:**486
Cysteine
conformation-sensitive Raman bands, **5:**100t
poliovirus proteinase 3C active site, **5:**19–20
Raman analysis, **5:**98–100
Tat, **1:**21
Cysteine β-lyase pathway, **1:**201f
Cysteine proteinase
location on chromosome 15, **3:**165
release of interleukin-1β protein by, **3:**334
Cysteine proteinase inhibitor, **3:**380f
kininogens as, **3:**380f
Cysteine sulfhydryl, accumulation in tissues, **1:**200
Cystic fibrosis, **1:**114
animal model for, **4:**103–4
mutation in, **1:**117
CYSTIC FIBROSIS (CF), **1:425–35**; **2:**420; **3:**229; **4:**101
among African Americans, independent origins of, **5:**48–49
association analysis in identifying, **3:**226
gene therapy for, specific to lung tissue, **3:**425
among Hutterites, founder effect, **5:**47
mutation analysis in, **3:**394–95; **4:**331–32; **5:**48
natural selection and, **5:**48
point mutation scanning in, **2:**424
prevalence of, **5:**45
symptomless carriers of, **3:**231, 231f
transfer the CFTR gene to lung tissue with liposomes, **2:**433
Cystic fibrosis transmembrane conductance regulator gene, **4:**338; **6:**182–83
Cystic fibrosis transmembrane conductance regulator (CFTR)
expression of, in milk of transgenic mice, **6:**105
gene for, **6:**70
therapy using, **4:**41
structure and function of, **1:**429–32
Cytidine, deamination of, mRNA editing protein (REPR) involvement in, **4:**499–500

Cytidine 5-triphosphate, biosynthesis of phospholipids using, **4**:433–34
Cytidine diphosphate-choline pathway, for biosynthesis of phospholipids, **4**:433–35
Cytidylyltransferase
  phosphoethanolamine, **4**:434, 434f
  structure of, **4**:436
Cytochrome
  defined, **1**:148
  identification of, historic, **1**:306
  matrix-assisted laser desorption ionization spectrum, **4**:4f
  transfer of electrons via, **4**:444–45
  b, gene for, *C. smithii*, **1**:353, 353f
  b2, import to mitochondria, **5**:206
  B5 (*CYB5*), mapping to chromosome 18, **3**:201
  bc-type complex, electron transfer through, **4**:447, 447f
  c, **2**:184–85, 184f
    semisynthetic, **1**:157, 157f
  *c*1, controlled oxidation of, **1**:309
  *c*6
    codon usage, in *Chlamydomonas reinhardtii*, **1**:348
    transfer of electrons via, **1**:357–58
CYTOCHROME P450, **1:435–40**
  aromatase, **3**:502
  carbon oxidation reactions involving, **6**:262–64, 262f
  carcinogenic products of catalysis by, **1**:202
  catalytic cycle, **4**:63f
  CYP2C19, role in mephenytoin metabolism, **4**:402
  CYP2D6
    role in debrisoquine metabolism, **4**:400–2, 402f
    substrates for, **4**:402t
  CYP2E1, **4**:403
  defined, **4**:58, 399; **5**:466, 478
  genes and mRNAs for variants involved in steroid biosynthesis, **5**:469t
  and hormone synthesis, **2**:289–90
  incorporation of noncoded amino acids in, **5**:143
  iron of, **4**:62, 62f
  monooxygenases, **1**:193–95; **6**:261–62, 261t
  oxidation of acetaminophen by, **1**:201f
  oxidation of benzo[*a*]pyrene by, **1**:197, 197f
  polymorphisms of, **4**:400–403
  pregnenolone synthesis catalyzed by, **5**:468–70
  production of carcinogens by, **1**:202
  reductase, **1**:194
  reduction catalyzed by, **6**:267
  reduction of nitric oxide by, **4**:204
  specificity of chemical reactions catalyzed by, **2**:224
  subfamily of genes localized to chromosome 15, **3**:169
Cytofluorimetry, for Vβ family characterization, **5**:510
Cytogenetics
  defined, **2**:319
  of Down's syndrome, **2**:119
  of fragile sites, **2**:315–16
  of fragile X linked mental retardation, **2**:320, 329
Cytogenic map
  of chromosome 16, **3**:182
  of chromosome 18, **3**:191–92
  defined, **4**:173
  of *Drosophila*, **2**:126–28
Cytokeratins, attachment sites for, cytoskeleton, **1**:458
Cytokine network, **1**:450–51
CYTOKINES, **1:441–51**; **5**:238
  communication by, role in immune responses, **3**:297, 297f
  defined, **1**:441; **2**:463; **3**:317t, 332, 341; **5**:229
  genes for, for measuring transcriptional activity, **3**:301
  glycoprotein induction by, on endothelial cells, **2**:455
  and immunity to respiratory syncytial virus, **5**:292–93
  leukemia inhibitory factor, **2**:292
  list, **1**:443–44t
  and motor neuron functioning, **4**:111
  production by helper T lymphocytes, **3**:335
  receptor binding process, **2**:470
  receptors for
    activation of the *JAK/STAT* signaling system, **2**:465
    structure and functions of, **5**:241
  regulation of extracellular matrix molecules by, **2**:279, 279t
  role in atherosclerosis, **1**:288
  as survival factors, **1**:449
  in vitro activity of antisense or triplex-forming oligonucleotides, **1**:74t
  See also Interferons; Interleukins
Cytokinetic optimization, in combination chemotherapy, **1**:264
Cytokinetic parameters, and selectivity of drugs, **1**:253
Cytokinetics, **1**:257
  defined, **1**:253
Cytokinins, plant, **4**:492
Cytology, defined, **2**:62
Cytomegalovirus (CMV)
  DNA genome of, **6**:174–75, 174f
  gB, Chinese hamster ovary cell lines expressing, **6**:176f, 178, 178f, 179
  recombinant Fabs against, **1**:421
  vaccine for, **6**:172–83
Cytoplasm
  acetylation of histones in, **5**:52
  cell division, segregation during, **1**:99–100
  defined, **1**:95, 315
  division cycle, synthesis during, **1**:96–97, 96f
  protein targeting in, **3**:348–53; **5**:196
  total RNA of, **5**:343
Cytoplasmic domain, bacteriorhodopsin, **1**:112
Cytoplasmic dynein, **4**:116
Cytoplasmic expression of genes, **2**:437
Cytoplasmic male sterility (cms), plant, **4**:489
Cytoplasmic mannosidase-α (MANA), gene for, localization to chromosome 15, **3**:167
Cytoplasmic regulation, of gene expression, **2**:374, 376–77
Cytosine (C), **2**:415
  in an oligonucleotide for triple-helix Achilles' cleavage, **1**:4
Cytoskeleton
  defined, **1**:451; **4**:112
  interaction with motor proteins, **4**:118
  movement of elements of, as an assay for motor proteins, **4**:114
  organization and polarity in, **4**:119–20
  in vitro interactions of, with integrins, **1**:453
Cytoskeleton-membrane attachments, in vivo, **1**:454–57
CYTOSKELETON-PLASMA MEMBRANE INTERACTIONS, **1:451–60**
Cytosol
  calcium in, **1**:238
  defined, **1**:315
  enzymes for phospholipid biosynthesis in, **4**:433
  eukaryotic, Tcp 1 protein in, **1**:320
  export of ATP to, from mitochondria, **1**:309
  hsp70 in, **1**:316
  hsp70s of, **1**:317f
  α-tocopherol binding protein in, **6**:238
Cytosolic phosphatidylinositol-phospholipase C (cPI-PLC), structure and function of, **4**:427–28
Cytosolic phospholipase (cPLA$_2$), calcium-ion dependent, **4**:423
Cytosolic phospholipase cPLA$_2$, **4**:423
Cytostatic factor (CSF), **3**:507
Cytotoxic agents, for cancer therapy, **1**:253
Cytotoxic effect
  cell-mediated, of macrophages, **3**:327

of oxidized low density lipoproteins, **3**:479
Cytotoxic T cell (CTL), defined, **3**:499; **5**:286
Cytoxan, as an anti-cancer drug, **1**:254

DAF, DNA amplification fingerprinting, defined, **2**:48
Dalton (unit), defined, **4**:129
Danazol, for cancer therapy, **5**:486
Dansyl chloride (DNS-Cl), fluorescent probe, **2**:295
Darcy's law, **1**:176
Dark reactions, of photosynthesis, **1**:310
Darvon, **1**:340
Data
analysis of, for tandem arrays, **6**:8–10
control of flow, automation, **5**:390
for nucleic acid sequencing, **4**:255
from Raman spectra, **5**:95–97
for Raman spectroscopy, processing of, **5**:101–2
from X-ray diffraction, **5**:114
Databases
Arabidopsis thaliana, **1**:84
for automation, **5**:397–98
Database searching
Saccharomyces Genome Database, WWW address for, **6**:303
sequence, **5**:433
Data model, **5**:397–98
Datiscoside, antitumor activity of, **5**:498
Daunomycin, antitumor activity of, **4**:99
ddC (2′,3′-dideoxycytidine), **1**:14, 37–38
binding of, to HIV reverse transcriptase, **2**:236
structure of, **1**:39f
ddI (2′,3′-dideoxyinosine), **1**:14, 37
structure of, **1**:39f
Dead-end enzyme, tyrosine hydroxylase, **1**:208
Deamination, mutation by, **4**:151
Debrisoquine
polymorphism, **4**:400–2
and susceptibility to lung cancer, **3**:492
Decarboxylase
orotidine-5′-phosphate, gene for, **6**:313
pyruvate, in reactions forming a new chiral center, **1**:343
Decay-accelerating factor (DAF), receptor for coxsackieviruses, **6**:230–31, 230f
Deconvolution
biological testing of synthetic peptides, **5**:519–20
of data from mass spectrometry, **4**:14
Fourier, of Raman spectra, **5**:102
of pooled libraries, **4**:97
Deficiency, in chromosomes, **2**:127–28
Deficiency mapping, **2**:128, 128f
Degeneracy
codon, and secondary structure of mRNA, **6**:124
defined, **4**:90
Degenerate oligonucleotide primed PCR (DOP-PCR), for probe labeling, **2**:382
Degradation, defined, DNA, **2**:1
Degree of unusuality, **6**:39–40
Degrees of freedom, of DNA structure, **2**:107
Dehydroamino acid substitution, in peptides, **4**:374, 374f
7-Dehydrocholesterol reductase, deficiency in Smith-Lemli-Opitz syndrome, **1**:116
Dehydroepiandrosterone (DHEA), **5**:480
Dehydrogenases
alcohol
A-specific, **1**:344, 344f
circular dichroism spectra of, **1**:391f
reaction at chiral center of the substrate, **1**:342
secondary structure, **1**:390
aldehyde, **1**:195–96
in amperometric biosensors, **1**:183–84
chiral specificity of, **1**:341
defined, **2**:240
hydroxysteroid, deficiency of, **2**:214
medium chain acyl coenzyme A, deficiency of, **2**:425
NADH coenzyme, **1**:310, 400–1
NAD^+-linked, calcium ion activation of, **1**:312
nicotine, reaction not involving the chiral center of substrates, **1**:342
octopine, unfolding and refolding of, **5**:76
steroid biosynthesis mediated by, **5**:469–70
sulfide, **2**:245
Dejerine-Sottas syndrome, mutation of chromosome 1 in, **3**:91
Delayed hypersensitivity (DH)
interferon-γ role in, **3**:328
interleukin role in, **3**:335
Deletion
in α-thalassemia, **2**:425
in Angelman syndrome (AS), **3**:161–62
in angiotensin-converting enzyme gene, **3**:2
detection of disease-causing, **2**:421
cystic fibrosis, **1**:426, 432–33
in Duchenne muscular dystrophy, **2**:426
of the *FMR 1* gene, **2**:326
in hemophilia A, **3**:31, 31f
mitochondrial, **4**:82–83
mutation by, **2**:392, 418; **4**:149, 149f
chromosome 15 disorders, **3**:170
mapping for *Caenorhabditis elegans*, **1**:230
oncogene activation associated with, **4**:286
in peptide synthesis, **4**:367
in phenylketonuria, prevalence of, **4**:409
single amino acid
in factor VIII gene of hemophilia patients, **3**:33–34t
in hemophilia B, **3**:38, 38f
of tandem repeats, **6**:6
use in identifying tumor suppressor genes, **1**:411
Deletion-duplication analysis, for scoring multiple local alignments of sequences, **5**:424
Delipidization, for apolipoprotein analysis, **3**:476, 476f
Delivery systems, nucleic acids to plants, **4**:451, 451t
Dementia pugilistica, beta-amyloid plaques in, **1**:43
Dementias
loss of synapses in, **5**:516
transmissible, **4**:199
See also PRIONS
Demographics, defined, **2**:391
Demography, and interpretation of mitochondrial DNA variation, **2**:399–400
Denaturant
defined, **5**:147
urea or guanidine hydrochloride, reversible unfolding induced by, **5**:74
DENATURATION
bacteriophage 186, A+T and G+C patterns, **4**:316f
OF DNA, **2:1–19**
complete, **4**:316
defined, **3**:302; **4**:312, 333
degrees of, **4**:313–14, 314f
of proteins, defined, **5**:159
Denaturation mapping
bacteriophage lambda, A+T base content, **4**:314–15, 315f
partial, **4**:312–23
Denaturing gradient gel electrophoresis (DGGE), **2**:4–5, 4f
defined, **2**:419
for mutation screening, **2**:423
for RNA analysis, **5**:347
Dengue viruses, binding to host cell, in the presence of antibodies, **6**:231
De novo modeling, **5**:161–64
Density gradient centrifugation, for analysis of low density lipoprotein subfractions, **3**:478
Dentatorubropallidoluysian atrophy (DRPLA), **3**:242

Deoxyadenosylcobalamin (coenzyme B_{12}), **1**:402–3
structure, **1**:404f
Deoxycholate, sodium, absorption enhancement by, **2**:167
Deoxyhemoglobin, **3**:5, 10–11, 11f, 20
defined, **3**:20
optical absorption spectra, **3**:7f
structure of, **3**:21
surface area between contact interfaces in, **3**:7, 7t
Deoxy-D-*manno*-octulosonic acid (Kdo), lipopolysaccharide, **3**:466
Deoxyribonucleotides (dNTPs), adding to DNA, **6**:19
Deoxyribose, free radical induced products from, **3**:363–64, 363f
Depalmitoyl-phosphatidylcholine, in lung surfactant, **4**:433
Department of Energy, U.S., Health and Environmental Research Advisory Committee, **5**:389
Dependence, defined, **2**:146
Dephosphorylation, for constructing a genomic library, **2**:71
Depolarization, defined, **1**:207
Depolarization ratios, Raman, **5**:96–97
Depurination, of DNA, as a function of temperature, **2**:2
Dermorphin, opioid activity of, and chirality, **1**:337
Design
defined, **5**:243
of enzymes, biomimetic, **1**:143–44
molecular, cycle of in biotechnology, **5**:244f
recognition molecules, **5**:250–51
Desmin
attachment sites for, cytoskeleton, **1**:458
filament, **4**:143
Desmocollins, **1**:453, 458
Desmogleins (DSGs), **1**:452, 458
mapping to chromosome 18, **3**:201
Desmopressin
effect on blood pressure and coagulation, **2**:22–23
response to, in nephrogenic diabetes insipidus, **2**:25
for treating central diabetes insipidus, **2**:23
Desmosomes, **1**:452, 458; **2**:265
cadherins of, mapping to chromosome 18, **3**:201
defined, **2**:263
Desolvation energy, in electrostatic interactions, **1**:162–63
Destabilization, by hairpin loops, **5**:361–62
Desulfovibrio desulfuricans, **1**:175
Detection
in HPLC, **3**:77–78, 78t
in PCR, **4**:337–38
Deterministic finite automaton, defined, **5**:428
Detoxification
of aflatoxins, **3**:485
defined, **6**:259
of plant pathogen products, **4**:481–82
Devaluation, of humanness, from transgenic animal work, **6**:83
Development
in *Caenorhabditis elegans*, **4**:171–72
changes during,
of hemoglobin composition, **3**:13
of isoenzymes, **3**:372
of lactate dehydrogenase and creatine kinase, **3**:372–73
effects on
of cytomegalovirus infection, **6**:172–73
of biotin, **6**:242
of neuropeptides, **4**:195
genes controlling, mapping to chromosome 11, **3**:127
homeobox genes involved in, mapping to chromosome 14, **3**:151
of liver, **3**:481
in nematodes, **4**:172
neuronal, and adult plasticity, **4**:58
regulation during,
of PrP mRNA, **5**:59
of transcription, **2**:368–69
of telomerase, **6**:24, 24f
sequence of HOX cluster gene expression during, **3**:65–66, 65f
of sex determination in the human, **5**:435, 435f
sexual, in *Drosophila* mutants, **2**:141–42
stage of, and regulation of DNA repair, **6**:161
testicular, **3**:501–3
Dexamethasone, **5**:483
Dextran sulfate, in vitro inhibition of HIV-1, **1**:36
Diabetes, defined, **2**:28
DIABETES INSIPIDUS, **2**:**19–28**
atherosclerosis associated with, **1**:288
defined, **2**:19
DIABETES MELLITUS, **2**:**28–40**
as an endocrine deficiency, **2**:213
genomic recombination between *Alu* repeats in, **3**:244
glycated hemoglobins in, **3**:5
HLA association in, **3**:225
levels of angiotensinogen in, **5**:266
mitochondrial mutations associated with, **4**:83
role of nitric oxide in, proposed, **4**:205
streptozotocin-induced, **5**:270
susceptibility locus, *IDDM2*, **6**:5
Diacetylphloroglucinol (DAPG), as a pesticide, **4**:398, 398f
Diacylglycerol (DAG), **1**:239
defined, **4**:432
from PLC action on phospholipids, **4**:424
protein kinase C activation by, **3**:46
as a second messenger, **3**:44; **4**:439
Diacylglycerol cholinephosphotransferase, **4**:434, 434f
Diagnosis
autoantibody markers used for, **1**:86–87, 87t
of colon cancer, **1**:413
of fragile X genetic defects, **2**:326, 326f; **4**:331
with gap ligase chain reaction, **3**:304–5
of genetic disease, **3**:230–31
of hematologic malignancies, **2**:64–65
of hemophilia B, direct and indirect, **3**:40f
of hemophilias, **3**:41
of human prion diseases, **5**:69–70
of Huntington's disease, **3**:249–50
with immuno-PCR techniques, **3**:295
isoenzymes used for, **3**:374–75, 374f
molecular, of genetic conditions, **2**:422t
of neoplastic disease, **2**:63–66
of neurofibromatosis 1, **4**:173–74, 173t
of neurofibromatosis 2, criteria for, **4**:181t
of solid tumors, **2**:65–66
of tumors, from telomerase activity, **6**:26
Diagnostic, defined, **5**:243
Diamidino-2-phenylindole (DAPI)
for fluorescence in situ hybridization, **2**:379
Q-banded metaphase chromosome stained with, **2**:380f
Diaminomaleonitrile (DAMN), **4**:292
Diapause hormone, **4**:197
Diaphorases, catalysis of NADH oxidation, **1**:183
Diastase, **2**:224
Diastereoisomers, **1**:334, 340
defined, **1**:52, 332; **4**:281
Diastereoselective reaction, defined, **1**:52
Diastolic dysfunction, in heart failure, **1**:292
1,4-Diazepin-2-one derivatives, in peptides, **4**:377
Dibutyryl-cAMP, and tyrosine hydroxylase regulation, **1**:214
Dicentric chromosomes, disorders involving chromosome 15, **3**:171–72

2,6-Dichlorobenzyloxycarbonyl derivatives, for protecting side chains, during peptide synthesis, **4**:387
Dichloromethane (DCM), reaction medium, Merrifield technique, **4**:390
Dictionary, sequence, defined, **5**:428
Dictyostelium, **1**:294
  annexin from, **1**:59
Dicumarol, **6**:248
Dicyclohexylcarbodiimide (DCCI), coupling reagent, peptide synthesis, **4**:386
Dideoxycytosine (ddC). *See* ddC
Dideoxynucleotide analogues, for block nucleotides in RLGS, **5**:306, 306f
Dielectrophoresis
  cell fusion using, **2**:200–201
  defined, **2**:198
Diels-Alder reactions, biomimetic systems for catalyzing, **1**:145
Diethyl pyrocarbonate (DEPC)
  defined, **5**:342
  inhibition of ribonuclease by, **5**:346
Diethylstilbestrol (DES), **2**:158–59, 159f
Diet-induced thermogenesis (DIT), **4**:261, 261t
Difference spectra, Raman, computation of, **5**:101–2
Differential adhesion hypothesis, **1**:293
Differential RNA splicing, defined, **1**:42
Differential scanning calorimetry
  for following DNA denaturation, **2**:6t, 10
  for DPPC phase transitions, **3**:448f
Differentiation
  defined, **2**:263, 463
  early markers of, keratin, **2**:266
  keratinocyte, **2**:270, 270f
  neural and hematopoietic, role of the *RB* gene in, **6**:141, 141f
  regulation of transcription during, **2**:368–69
  signal for, epidermal keratinocytes, **2**:266–67
Diffraction, defined, **6**:275
Diffusion
  defined, **1**:177
  facilitative, **6**:129
Digestion, of proteins, for sequencing, **5**:190
Digitalis, **3**:461
Digoxigenin, hapten for FISH, **2**:382
Dihydrofolate reductase (DHFR), **5**:196
  folding of, in yeast mitochondria, **5**:90
  incorporation of alanine into, **5**:147
Dihydrofolate reductase pseudogene 1 (DHFRP1), mapping to chromosome 18, **3**:201
Dihydropteridine reductase (DHPR), **1**:207
Dihydropyridines
  binding to L-type calcium channels, **1**:243
  receptors for, **1**:247
Dihydrotestosterone propionate, cancer therapy using, **5**:490
3,4-Dihydroxyphenylalanine (dopa), tyrosine hydroxylation to, **1**:207
Diisopropylcarbodiimide (DIC), coupling agent, peptide synthesis, **4**:390
Diisopropylfluorophosphate (DIPF), as trypsin inhibitor, **2**:236
Dikaryon rescue experiment, **1**:359
Diltiazem, binding to L-type calcium channels, **1**:243
Dimerization, defined, **6**:149
Dimerization domains, leucine zippers of, **4**:492
Dimethyl-1-androstene-2,17β-diol-3-one, for cancer therapy, **5**:486
Dimethylarsenoselenocholesterol, antitumor activity of, **5**:497
Dimethylene linkages, in peptides, **4**:377
Dimethylnitrosamine, mutagenic and carcinogenic actions of, **2**:41
2,4-Dinitrophenol, as a respiratory uncoupling agent, **1**:313
Dinucleotide, pdCpA, synthesis and aminoacylation of, **5**:144
Dioctadecyl-amidoglycylspermine (DOGS), for transfecting cells, **3**:421
Dioleoyloxypropyltrimethylammonium chloride (DOTMA), in cationic liposomes, **3**:419–20
Diosgenin, synthesis of, **2**:164, 164f
Dioxin
  effect on T lymphocyte, **3**:298
  induction by, of P450, **1**:440
Dipalmitoyl phosphatidylcholine (DPPC), phase changes, example, **3**:448, 448f
Dipeptides, **4**:350, 350f
Diphasmids, **2**:68–69, 69f
Diphenyl oxazinones, **1**:54
  structure, **1**:53
Diploid organisms, alleles in, **2**:416
Dipole-dipole interaction
  between chromophores, **3**:388
  defined, **3**:385
  in electron spin resonance, **3**:386
  in hydrogen bonds, **3**:261
Dipole moment, water, **1**:163
Dipole pairing, and chemical equilibrium, **2**:176–77
Dipole reactions, **2**:174, 175
  and rate constants, **2**:177
Dipolophoresis, **2**:174
Dipteran chromosomes, preparation of, for microdissection, **1**:373
Dipyridodiazepinones
  inhibition of reverse transcriptase, HIV-1, **1**:38
  structure, **1**:39f
Directed cDNA library, **1**:205
3′-Directed cDNA library, **1**:203
Directed evolution
  defined, **4**:90
  of microorganisms with novel biodegradative capability, **1**:125
Directed Hamiltonian path problem, solution with DNA manipulation, **1**:157–58, 158f
Directed mutation, defined, **2**:407
Directionality
  acceptor, and hydrogen bonding, **3**:269
  of hydrogen bonding, **3**:264
  of motor proteins, **4**:118–19
Directional selection, limiting genetic variation through, **2**:393
Direct labeling, for FISH, **2**:384
Direct mutation detection, **2**:420–21
Direct readout, **5**:279
  defined, **5**:275
Direct repeat sequences, hepatitis B virus, **3**:483
Direct screening, for cloning plant regulatory genes, **4**:472
Direct selection, of catalytic RNAs, **4**:230–31
Direct sequence analysis, **2**:51, 57
Disaccharides, **2**:448–49, 449t
  of glycosaminoglycans, **2**:453
Disassembly reactions, protein, roles of hsp70s in, **1**:318–19
Disclosure
  defined, **6**:74
  as a requirement for patentability, **6**:76
Discrete search, for RNA three-dimensional structure modeling, **5**:380–81
Discrete uniform distribution (DUD), **6**:39–40
  measuring deviations from, **6**:41–47
Disease
  genetic basis of human, **4**:101
  interferons associated with, **3**:329t
  interleukins associated with, **3**:345, 345t
  telomerase activity related to, **6**:24–26
  validation of the relationship with biologic markers of effects, **2**:260
Disease genes
  chromosome 9, **3**:101, 103–11
  chromosome 16, **3**:186–90, 190f
  See also Genetic disorders
Disease models, animal, patents for, **6**:77
Disease vectors, *Drosophila* model for, **2**:129
Disiloxane linkages, in peptides, **4**:377

Dismutase, Cu/Zn, in Down's syndrome, **2**:80
Dismutation reaction, of superoxide radicals, **2**:331
Disomy, paternal, in Angelman syndrome, **3**:161–62
Dispersants, biopolymer, **3**:316
Dissemination, of bacterial pathogens, within the host, **1**:106
Dissociation temperature, defined, **4**:235
Distal-less (Dll) class, homeobox genes, **3**:72
Distal residues, hemoglobin, **3**:23–25
Distamycin, DNA binding to, **1**:166, 167f
Distance, genetic
 relation between map distance and recombination fraction, **3**:235
 for yeast, determination using tetrads, **6**:310
Distance charts
 defined, **6**:38
 for nucleotides, **6**:47, 49f
Distance geometry (DG)
 for computer modeling of RNA, **5**:381
 for computer modeling of RNA 16S, **5**:382–83
 for RNA three-dimensional structure modeling, **5**:379–80
Distribution coefficient K_d, chromatography, **3**:78
Disulfide bond, defined, **5**:159
Disulfide linkages
 antibody, **1**:62–63, 63f; **3**:276, 279f
 circular dichroism for determining dihedral angle, **1**:386
 in core proteins, whey acidic protein, **6**:103
 folding and pairing of bonds, **5**:199
 and inclusion body formation, **5**:84
 in osteonectin, **1**:239
 in peptides, **4**:376
 in proteins
  folding coupled to formation of, **5**:156–57, 157f
  reshuffling during refolding of, **5**:74–75
Disulfiram, inhibition of acetaldehyde dehydrogenase by, **1**:196
Diterpenes, gibberellins, **3**:440
Diuretic, defined, **2**:155
Diuretic hormones, **4**:194
 in insects, **4**:196
Diversity
 human, incidence of mutations, **1**:115
 of immune phenomena, **3**:276, 282–83
 of prions, **5**:66–68
Division cycle
 bacterial, **1**:96–102
 variation in, **1**:100–101
 See also Meiosis; Mitosis
DMBA (Dimethylbenz[*a*]anthracene), defined, **5**:485
DNA
 acquisition of, **2**:404
 agents damaging, **3**:355t
  defined, **4**:148
 ancient
  defined, **4**:299
  isolation and characterization, **4**:305–06
 B and A forms of, **2**:105f
  transition between, **2**:109, 109f
 basic unit of, as a step, **2**:107
 bending by protein, analysis by cleavage and gel electrophoresis, **5**:302
 biosynthesis of, **1**:160–61
 catalytic, selection of, **4**:233–34, 234f
 categories of, **3**:511–12
 circular, in *Plasmodium falciparum*, **4**:308
 closed circle of, and underwinding, **2**:116–18, 116f
 coding, and haploid genome size, **6**:2t
 complex with HIV-1 reverse transcriptase, **1**:29
 conserved sequences, **4**:127
 damage to, by ionizing radiation, **3**:356–65
 enzymatic sequencing of, **4**:254–56
 forms of, geometric, **2**:103–4, 104f
 fragments of, mass spectrometry for analyzing, **4**:3, 6–9
 functional, **3**:511–12
 G4, **6**:30–31
 genomic
  digestion and fractionation of, **6**:295–96
  preparation of, **1**:4
 imaging of, STM, **5**:403, 403f, 407–9, 409f
 imaging of large molecules with atomic force microscopy, **5**:405, 405f
 intergenic, **3**:512
 labeled with ruthenium(II) tris(bipyridyl), **1**:327f
 labeling with radionuclides, in vitro methods, **5**:216–17f
 linear, preparation of chromosomes for microdissections as, **1**:371
 mapping of, with atomic force microscopy, **5**:404–5
 methylation of, **1**:407
  loss in colon cancers, **1**:411
 mimicking of, by PNA, **4**:273
 mitochondrial, **2**:399–400
 modified, cleaving a phosphodiester bond of a ribonucleotide, **4**:234, 234f
 nonfunctional, **3**:512
 orders of folding, in chromosomes, **1**:370–71
 phosphorothioate-containing, **1**:159
 photoproducts of, **6**:151–55, 155f, 153f
 plant genome, of unknown function, **4**:486
 plasmid, for vaccines, **6**:181–82
 preparation of
  for cloning, **5**:504
  for restriction landmark genome scanning, **5**:307–08
 radiation-induced lesions, **3**:355
 recombinant
  defined, **1**:172
  in plants, **4**:465
  for polypeptide production, **1**:137
  for transgenic livestock production, **6**:100–01
 replication of
  as an event in the division cycle, **1**:100
  and histone protein synthesis, **3**:51
 segregation of, during cell division, **1**:100, 100f
 "selfish", **3**:512
 sequence analysis of, by chemiluminescence, **1**:331f
 single-copy, **3**:511
 sources of, for chromosome microdissection, **1**:372
 strand breakage, from ionizing radiation, **3**:356
 switch between A and B forms, **2**:107
 synthesis during the division cycle, **1**:97, 97f
 telomeric, **6**:28–29
 transposable *Tc1*, of *Caenorhabditis elegans*, **1**:230
 variations in, **3**:228–29
DNA binding, nonspecific, of catabolite activator protein (CAP), **4**:245
DNA-binding motifs
 defined, **5**:275
 GATA-1, **6**:332
 glucocorticoid receptor, zinc ions of, **6**:331–33
 homeodomain, **3**:58
 in proteins regulating transcription, **5**:277
 zinc finger, GLI, **6**:329
  Zif268, **6**:329–31, 331f
DNA-binding proteins
 characterization of, **4**:470–71
 p53, **6**:142
 unique, in spermatogenesis, **3**:506
DNA chips, **1**:157, 157f; **4**:333
DNA complexity, defined, **1**:347
DNA cross-linkers, as anti-cancer drugs, **1**:254

DNA DAMAGE AND REPAIR, **2:40–47**
considerations in aging, **1:**6–7
defined, **2:**40, 78
detecting damage, **6:**155–56
pathways for repair of UVR damage, **6:**154f
RecA protein and, **5:**223–26
DNA databank, defined, **2:**169
DNA-DNA hybridization, for infectious disease assays, **4:**329
DNA extraction, **2:**395
DNA FINGERPRINT ANALYSIS, **2:48–62**; **4:**339
DNA footprinting, genomic, **1:**367
DNA helicase, Rad3, of excision repairosomes, **6:**288
DNA IN NEOPLASTIC DISEASE DIAGNOSIS, **2:62–66**
DnaJ, stimulation of ATPase activity by, **1:**319
DnaK, stimulation of ATPase activity of, **1:**319
DNA ligase, **6:**287
DNA MARKERS, CLONED, **2:66–78**
DNA methylation, defined, **2:**439
DNA microinjection, **6:**64, 65f
by bacteriophage, polarity of, **4:**321–22
defined, **6:**64, 99
DNA minor groove, nucleosome, **1:**363
DNA modification methyltransferase, defined, **5:**296
DNA pairing, models for, **5:**225–26, 225f
DNA photolyase
defined, **6:**283
reversal of cyclobutane pyrimidine dimer formation by, **6:**284
DNA polymerase
defined, **2:**86
ε, effect on base excision repair, **6:**287
high temperature, **2:**246
Taq, for polymerase chain reactions, **4:**325–26
and telomere length, **3:**242
DNA polymorphism
defined, **1:**42
in gene mapping, **3:**225
DNA-protein complexes, as vectors for transfer of phenylketonuria genes, **4:**414–15
DNA radicals, reactions of, **3:**357
DNA rearrangement, **2:**404
defined, **2:**402
DNA repair, and replication slippage of microsatellites, **6:**4
DNA REPAIR IN AGING AND SEX, **2:78–85**
DNA repair proteins, **1:**369
DNA REPLICATION AND TRANSCRIPTION, **2:86–96**; **5:**353–54, 355
DNA Resource Center, **1:**84
DNaseI footprinting, **4:**248f, 249
DNA SEQUENCING, **2:**395–96
automated, **2:**96–102, 102f
defined, **4:**1
Escherichia coli, **2:**169–70
IN ULTRATHIN GELS, HIGH SPEED, **2:96–102**
by laser mass spectrometry, **4:**9–10, 10f
in molecular pathology, **4:**328
for mutation detection, **2:**423
DNA shuffling, **5:**448–50, 450f
DNA slippage, for analysis of mutations, **2:**392f
DNA strand breakage, **3:**363–64
defined, **3:**354
DNA strand exchange, **5:**222
models for, **5:**223–26, 226f
DNA STRUCTURE, **2:**12, 102–18; **5:**353, 353f
chromatin packaging, **1:**360–61
circular, of plasmids, **5:**1
coding strand, **6:**111
double helix, **5:**354
G+C-rich regions, **4:**316, 316f
packaging, **3:**50, 50f
sequences recognized by restriction-modification enzymes, **5:**301–2
DNA transformation, **1:**358–59
defined, **1:**227
DNA turnover, defined, **2:**391
DNA typing, PCR-based, **2:**56f
dNTP (deoxyribonucleoside triphosphate), defined, **6:**17
Docosahexaenoic acid
synthesis by algae, **3:**441
synthesis by molds, **3:**439, 439f
Dogs,
nephrogenic diabetes insipidus in, **2:**27
XX male, **5:**444
Dolichols, **3:**461
Domains
actin-binding, **4:**142
antibody recognition of, **1:**88t
bacteriorhodopsin, **1:**112, 112f
calcium-binding, **1:**240–41
catalytic
calcium ion pump, **1:**244
tyrosine hydroxylase, **1:**211, 211f
cytoplasmic cadherin, **1:**458
defined, **3:**377; **4:**66; **5:**120
functional, of the CENP-B antigen, **5:**416
immunoglobulin, **1:**293
of immunoglobulin light and heavy chains, **3:**277–78
immunoglobulinlike, **1:**294
integrin-binding, of α-actinin, **1:**455
kininogen, **3:**380–81
melting of, in DNA, **2:**4–5, 8–10
organization of humain interferons, types I and II, **3:**323f
orientation of, and phase transitions, **2:**180
in peptide synthetases, defined, **4:**207
phospholipid-binding, **1:**244
polypeptide, defined, **1:**27, 62; **5:**150
regulatory
calcium ion pump, **1:**244–45
tyrosine hydroxylase, **1:**211, 211f
transcription activation, **2:**207–8
transduction, calcium ion pump, **1:**244
transmembrane receptor, **5:**121f
variable, Vα and Vβ, on T-cell receptors, **5:**510
of virus glycoproteins, **6:**187f
zinc finger, **6:**328–29, 329f
Domestic animals, transgenic, production of, **6:**69–71
Dominance/dominant trait, defined, **1:**407; **3:**227
Dominant allele, defined, **2:**419; **3:**142
Dominant mutation
Ki-ras, in colon carcinomas and adenomas, **1:**411
of oncogenes, **6:**139
Dominant-negative mutation, **1:**411
defined, **1:**407; **4:**173
p53, **1:**412
Donors, in hydrogen bond interactions, **3:**262, 262t
water molecules as, **3:**272f
Dopa decarboxylase, reaction at chiral center of the substrate, **1:**342
Dopamine
activation by addictive drugs, **2:**149–50
in alcohol-preferring HAD and P rats, **2:**150
gene controlling synthesis of, in *Drosophila*, **2:**136–37
inhibitor, for tryptophan hydroxylase, **1:**219
receptors for, **5:**238
Doping, in site-directed randomization, **4:**95–96
Dosage analysis, for physical mapping of chromosomes, **3:**119–20
Dose limits, radiation exposure, **5:**211–12, 211t
Dosimeters, personal, thermoluminescent detectors on, **5:**220
Dot blot analysis, **2:**51
defined, **5:**342
method, for RNA analysis, **5:**348
DOTMA liposomes, complex with nucleic acids, **3:**421f
Dot plot, **5:**430, 430f, 432
defined, **5:**420, 428
sequence, **5:**427
Double dynamic programming, **5:**426
Double helix
defined, **2:**103

Double helix (*cont.*)
DNA, **5:**353–54
irregular, and third-level structure of DNA, **2:**104
left-handed Z form, **2:**104
structure of, **5:**129
Watson-Crick, **2:**103
Double-lysine motif, in the transmembrane domain, **5:**201
Double-strand break
from ionizing radiation, **3:**365
recombination repair of, **2:**81
repair model, **5:**257–58, 257f, 259–60, 260f
Double-stranded DNA bacteriophages, **6:**201
Double-stranded RNA viruses, **6:**218–20
Double-tag sandwich immunoassay, **3:**294
DOWN SYNDROME
Alzheimer's disease associated with, **1:**42, 43, 51
association of telomere length with aging in, **6:**34
MOLECULAR GENETICS OF, **2:118–21**
and premature aging, **2:**80
role of superoxide dismutase in, **2:**333–34; **4:**108–9
D period, bacterial reproduction, defined, **1:**95
DROSOPHILA GENOME, **2:121–30**
DROSOPHILA MATING, GENETIC BASIS OF, **2:130–46**
Drosophila melanogaster
age-dependent changes in proteins, **1:**9
artifacts in studies, **1:**10
alternative splicing, sex determination pathway, **2:**373
antagonistic pleiotropy in, **1:**10
homeotic selector genes of, **3:**64–65, 65f
learning in, **4:**57
longevity studies, with genetic transformation, **1:**12–13
DRUG ADDICTION AND ALCOHOLISM, GENETIC BASIS OF, **2:146–55**
Drug delivery
defined, **2:**165
gene-based, **4:**41
peptide, **4:**357
polymers for controlling, **5:**35–36
with poly(2-hydroxyethyl methacrylate) hydrogels, **5:**41–42
with poly(vinyl alcohol), **5:**43
site selection, **1:**260
using poly(ortho-esters), **5:**36
Drug design
nitric oxide and, **4:**205–6
and protein structure, **4:**372; **5:**160
small-molecule libraries for, **4:**99–100
from structures of protease-inhibitor complexes, **1:**28
See also Rational drug design
Drug metabolism
and cytochrome P450, **1:**194, 440
defined, **2:**155
therapeutic aspects of, **1:**202
Drug resistance, **2:**406
genes affecting, **4:**160
HIV, **1:**18
by tumors, **1:**262–63
Drugs
carbohydrate-based, **2:**462t
cardiovascular and psychotropic, metabolism by CYP2D6, **4:**401
combination, HIV-1 treatment with, **1:**41
enantiomers, differing effects of, **1:**340
induction of autoimmune disease by, **1:**85–86
prevention of uptake, by plasmids, **5:**9–10
psychoactive, defined, **2:**146
responses to, rodent lines selected for, **2:**148t
tiazofurin, **1:**252
DRUG SYNTHESIS, **2:155–65**
Drug targeting, defined, **2:**165
DRUG TARGETING AND DELIVERY, MOLECULAR PRINCIPLES OF, **2:165–67**
dsRNA, interaction with protein kinase R, in defense against viruses, **3:**325–26
Dual-color assay, ligation analysis, **3:**394–95, 394f
Duchenne muscular dystrophy (DMD), **1:**117, 289; **2:**420; **3:**231f; **4:**101, 146–47; **6:**137
deletion-duplication analysis for, **2:**426
diagnosis of, **3:**230–31
gene size in, **5:**46
heart failure in, **3:**1
mapping on the mouse and human genomes, **4:**126, 126f
as a single-gene disorder, **3:**230
Dunaliella salina, lipids of, **3:**441
Duplication
aberrant chromosome, **2:**128
18q syndrome, **3:**212
in hemophilia A, **3:**31
Dyad symmetries, DNA, **5:**430
Dye laser, **6:**211
Dynamic effect, of DNA length on structure, **2:**115
Dynamic light scattering, for following DNA denaturation, **2:**6t
Dynamic mutation, **2:**315
defined, **2:**313
Dynamic programming
defined, **5:**420
gap penalty adjustment, **5:**422–23
for gap placement analysis, **5:**421, 421f
for RNA structure determination, **5:**378
for structural comparison, **5:**426
Dynamic Raman spectroscopy, **5:**109
Dynamics, of hydrogen bonding, **3:**272–73
Dynamin protein, genes encoding, mutant studies in *Drosophila*, **2:**133
Dyneins, **4:**113–14, 114t
characterization of, **4:**116–17, 117f
defined, **4:**112
directionality of, **4:**120
Dysautonomia, familial, mapping to chromosome 9, **3:**107
Dysgerminoma, associated with XY gonadal dysgenesis, **5:**445
Dystonia, dopa-responsive, mapping to chromosome 14, **3:**157
Dystonia, idiopathic torsion (DYTI), mapping to chromosome 9, **3:**107
Dystrophin, **2:**426; **4:**142
in Duchenne muscular dystrophy, **3:**231f; **4:**147
gene for, X-linked, **1:**117
introns in the gene for, **2:**371

E6-associated protein, gene localization to chromosome 15, **3:**170
Ear-bone replicas
from high density poly(ethylene), **5:**40
from poly(tetrafluoroethylene)-graphite composites, **5:**39
Early response protein, Zif 268, **5:**132, 132f
EC code, for enzyme nomenclature, **2:**225
Ecdysiotropic neuropeptide, release of, in the mosquito, **4:**192
Ecdysiotropins, **4:**190, 194
bioassay based on, **4:**194
Bom-PTTH, **4:**196
ecdysteroid production stimulation by, **4:**196
Ecdysis, behavioral patterns in, **4:**195
Ecdysone, **4:**190f
structure of, **4:**189f
Ecdysteroids, **4:**189–90, 190f
Eclipse, virus
defined, **6:**223
role of receptors, **6:**224
Eclosion hormone, **4:**192
second messenger production stimulation by, **4:**195–96
Ecology, of methanogenesis, **4:**69
Economic incentives, for patents, **6:**78
Ecotype, defined, **1:**76
Ectoglycosyltransferases, **1:**296
Ectomycorrhizal fungi, use in tree culture, **2:**344

Ectopia lentis, gene for, mapping to chromosome 15, **3:**160, 163
Ectoplasmic specialization, effect of testosterone on, **3:**505
Edeine synthetase, **4:**209
Edman degradation, **5:**125, 174–75, 186f
  defined, **5:**173, 185
Edmundson helical wheel, **5:**162f
Effector, **3:**286f
  allosteric, **2:**237
  defined, **3:**42
  G protein, **3:**45–46
  G protein cycle, **3:**44, 44f
Efferent conduction, defined, **4:**53
Efficiency, of transcription, in eukaryotes, **2:**95
EF-hand binding sites, **1:**239
EF-hand family, calcium-binding proteins, **1:**59
EF-hand proteins, **1:**240–42
  calcium-binding domain, α-actin, **4:**142
  defined, **1:**237
  myosin light chains, **4:**134
  osteonectin, **1:**239
  in plants, **1:**249
EGF domain, of selectins, **1:**294f, 296
Egg proteins, **2:**308
Eicosanoids, **3:**461
  as hormones, **2:**211
Eicosapentaenoic acid
  synthesis by algae, **3:**441
  synthesis by molds, **3:**439
Eicosatetraenoic acid (Arachidonic acid), **3:**453f
18p⁻ syndrome, **3:**211
18q⁻ syndrome, **3:**211–12
Einheit der Zellbiologie, **1:**102
ELAM-1 (E-selectin), **1:**296; **2:**455
Elastases
  encoded by *lasB*, **1:**108
  pancreatic, unfolding and refolding of, **5:**76–77
  secretion by bacterial pathogens, **1:**106
Elastin, **1:**146; **2:**282
  genetic engineering of, **5:**28
  incorporating in silklike fibers, **5:**27
  silk copolymers using, **5:**27
Elastomers
  defined, **5:**31
  thermoplastic, block copolymers, **5:**26
Electrical potential, membrane, **1:**307–8
Electrical synapses, **5:**512–13, 513f, 514
ELECTRIC AND MAGNETIC FIELD RECEPTION, **2:172–81**
Electric and magnetic field strength, defined, **2:**172
Electric current, **2:**174–75
Electric discharge system, for ballistic plant cell transformation, **4:**461
Electric organs, amplification in, **2:**178–81
Electroblotting, **5:**179, 179f, 188–89, 189t
  defined, **5:**185
Electrochemical biosensors, **1:**179
Electrochemiluminescence, **1:**323
  defined, **1:**322
  of ruthenium(II) tris(bipyridyl) labeled DNA, **1:**327f
Electrodes
  defined, **1:**177
  mediator-based, **1:**183
Electrofusion, **2:**198–206
Electrolyte channel modification, potential therapy for cystic fibrosis, **1:**435
Electrolyte transport
  chloride, in the airways, pancreatic ducts and sweat glands, **1:**427f
  in cystic fibrosis, **1:**426
Electromagnetic spectrum, **6:**150
Electrometric titration curves, for DNA, **2:**7f
Electron acceptors, metabolic requirement of microbes, **1:**119, 121
Electron capture, **5:**210
Electron density map, interpretation of, **5:**114
Electronic absorption spectroscopy, for following DNA denaturation, **2:**2
Electronic image, recording, for multicolor FISH, **2:**386
Electron impact ionization spectroscopy (EIMS), **4:**1
  for carbohydrate analysis, **1:**274
Electron microscope, defined, **4:**129
ELECTRON MICROSCOPY
  of biomembrane fusion, **4:**43–44
  for detecting protein aggregation, **5:**81
  for following DNA denaturation, **2:**6t
  image of connectin, **4:**143
  OF BIOMOLECULES, **2:181–93**
  for partial denaturation mapping, **4:**312
  for RNA structure determination, **5:**378
  of smooth muscle, **4:**132
  for studying nucleic acid-protein complexes, **4:**245
  for studying virus assembly and budding, **6:**188
Electron microscopy (EM), study of talin interaction with actin, **1:**456
Electron microscopic autoradiography (EM ARG), **2:**193, 193t
  high resolution, defined, **2:**181
Electron paramagnetic resonance, for observing enzymes in anhydrous organic dispersants, **2:**252
Electron scattering labels, **3:**389
Electron spectroscopic imaging (ESI), **2:**193
Electron spin resonance, **2:**330, 335–36; **3:**385–86
  for nucleic acid conformation determination, **3:**390
ELECTRON TRANSFER
  BIOLOGICAL, **2:193–97**
  contact with biomolecules, **1:**150, 150f
  cyclic and noncyclic, **4:**444–45
  donation by reduced cofactors, **1:**308–9
  long-range photoinduced, **1:**156–57, 157f
  mimicking in proteins, **1:**139
  photosynthetic, **1:**310
  and proton translocation, **1:**309
Electron transport
  cyclic and noncyclic, **4:**446f
  noncyclic, water a source of electrons, **4:**445–47
Electron transport chain
  defined, **1:**303
  in methanogenesis, **4:**69
  photosynthetic, of *Chlamydomonas reinhardtii*, **1:**355f
Electron transport system (ETS), **1:**306–9, 436; **2:**233
  of aerobic heterotrophs, **1:**306
  photosynthetic energy transduction in, **4:**444–48
  sensors for changes in, proposed diagnostic use of, **4:**333
Electropermeabilization, defined, **2:**198
Electrophile, defined, **1:**193
Electrophilic catalysis
  defined, **4:**59
  zinc ion, **6:**336
Electrophoresis
  for bioseparation, **1:**175
    of VNTR products, **2:**58
  defined, **2:**96, 392; **3:**366; **5:**388
  high resolution, for restriction landmark genome scanning, **5:**306
  horizontal ultrathin gel electrophoresis, **2:**99
  for isoenzyme separation, **3:**368
  with polymerase chain reaction, **4:**326
  for separation of plasma lipoproteins, **3:**475
  two-dimensional, of proteins, **5:**177–78
Electrophoretic mobility shift assay (EMSA), **4:**470
  for sequence-specific DNA interactions with nuclear factors, **4:**470–71
ELECTROPORATION, **2:**432; **4:**162; **6:**88
  defined, **4:**458; **6:**84
  AND ELECTROFUSION, **2:198–206**
  for gene targeting, **4:**102
  for gene transfer, **4:**452–53, 462–63, 463f; **5:**505
    in yeast, **6:**310
  whole-tissue, for plant cell transformation, **4:**463–64

Electropore, defined, **2:**198
Electrospray ionization (ESI), **4:**13–15, 15f
  defined, **4:**12
  mass analysis after, **4:**26–27
  in mass spectrometry, **5:**193, 477
  mass spectrum
    recombinant human growth hormone, **4:**15f
    thrombin, **4:**21–22, 21f
Electrospray ionization mass spectrometry (ES-MS)
  for carbohydrate analysis, **1:**275–76
  for DNA sizing, **4:**10–11
  for protein identification, **5:**125–26
  for protein measurements, **5:**175
Electrospray mass spectrometer (ESMS), defined, **4:**1
Electrostatic interactions
  in bioorganic chemistry, **1:**162–63
  in DNA, between base pairs, **2:**110
  shielding, to orient photosynthetic reaction centers, **1:**151–52
Elicitins, from *Phytophthora*, **4:**483
Elicitor-receptor model, of plant resistance, **4:**476, 481, 483
Elicitors, defined, **4:**476
Eliminases, **2:**448, 448t
Ellipticity
  defined, for plane-polarized radiation, **1:**384
  molar, **1:**388–89
Ellis van Creveld syndrome, **5:**47
Elongation
  defined, **6:**109
  in RNA replication, **2:**365, 369; **6:**117f
  telomere, telomerase-independent, **6:**35
Elongation factor
  for archaea, **4:**75
  defined, **5:**328
  eukaryotic, **6:**121t
  human, βgene localization on chromosomes 2, 5, 15, and X, **3:**165
  prokaryotic, **6:**120t
  Tu (EF-Tu), defined, **5:**135
  and virus development, **5:**337
Elongation-translocation cycle, for peptide chain synthesis, **6:**118, 120–21
Elution, defined, **3:**76
Embryo
  globin subunits of hemoglobin in, **3:**13
  hemoglobin of, **3:**13
  monoparental diploid, **2:**440
  regulation of development of, **2:**288
Embryogenesis, translational control during, studying, **5:**335
Embryonic cells
  cytokines role in proliferation of, **1:**444t
  N-cadherin role in development of, **1:**458
Embryonic development. *See* Development
Embryonic genes, and heart failure, **1:**290
Embryonic stem (ES) cell
  defined, **6:**99
  pluripotent, **3:**518
    mouse lines, **4:**103
  transfer of
    defined, **6:**64
    technology of, **6:**68–69
    transgenic livestock developed using, **6:**100–101
    transgenic mice developed using, **6:**65f
Emission spectra, **2:**298, 298f
  biomolecules, **2:**295f
  and the Stokes shift, **2:**298
Empty-spiracles (ems) class, noncluster genes, **3:**67
Emulsan, **3:**434
Emulsification, in foods, **2:**311
Emulsion
  defined, **2:**307
  formation and breakdown of, **2:**312f
Enalapril, example of a noncovalent enzyme inhibitor, **1:**167–68
Enantiomeric excess, defined, **1:**52
Enantiomers, **1:**335, 335f; **4:**294
  amino acid, **5:**116
  defined, **1:**332; **2:**155
  physiological responses to, **1:**339–40
Enantiomorphic compound, defined, **1:**52
Enantioselectivity, of medical chemicals, **4:**38–39
Encapsidation, **6:**217
  of hepadnaviruses and retroviruses, **6:**222
Encephalitis virus infection
  tick-borne, antisense oligonucleotide effect on, **1:**75
  vesicular stomatitis virus caused, vaccination against, **6:**169
Encoding, for nonpeptide libraries, **5:**522–23, 523f
End capping, of packing for reversed phase chromatography, **3:**82
End labeling, enzymatic, **2:**382
Endo-β-galactosidase, **2:**449
Endo-β-xylosidase, **2:**449
Endocon, in transferase reactions, **6:**267–68
Endocrine, defined, **2:**463
Endocrine disorders
  deficiency, **2:**213–14
  excess hormones in, **2:**214–15
Endocrine regulation, defined, **2:**285
Endocrine system
  control of calcium homeostasis, **1:**238
  growth factors as an extension of, **2:**463
ENDOCRINOLOGY, MOLECULAR, **2:206–16**
Endocytosis, **4:**45; **5:**204, 204f
  defined, **5:**194
  for internalization of a virus by a host cell, **6:**195–96
  and intracellular membrane budding, **4:**50–52
  transfer of cholesteryl esters to liver by, **3:**409
  uptake of modified LDL by, **3:**408
  uptake of oligonucleotides by, **1:**72
Endogenous agents
  defined, **2:**40; **4:**28
  DNA damage from, **2:**41–43, 79, 85
Endogenous genes
  altered regulation of, **6:**55
  suppression of, **4:**474
Endogenous ligands, and receptor functions, **5:**231t
Endogenous retroviruses, **6:**135f
  defined, **6:**133
  element 1, localization of, to chromosome 18, **3:**203
  human, **6:**136–37
  use in genetic mapping, **3:**512
Endogenous substrates
  for cytochrome P450, **1:**435, 436
  defined, **1:**435
Endoglucosaminidases, **2:**447–48
Endoglycosidase, **2:**448
  defined, **1:**266
  structural requirements of, **2:**447t
Endohydrolases, production in response to plant pathogen attack, **4:**480
Endomembrane system, **5:**195–96
  defined, **5:**194
Endonexin (annexin IV), structure, **1:**60f
Endonexin fold, **1:**60f
Endonexin II (annexin V), structure, **1:**60f
Endonucleases
  AP, **2:**82
    in base excision repair, **6:**285–87, 287f
  Ca^{2+}-dependent, role in apoptosis, **1:**302
  contaminants in Achilles' cleavage, **1:**4
  defined, **6:**149
  degradation of end-capped oligonucleotides by, **1:**71
  nucleotide excision repair mediation by, **6:**287–88
    damage-specific activity, **6:**290
  RNase L, antiviral activity of, **3:**325
  RuvC protein, **5:**228–29
  3′-5′-, processive, RNase H as, **1:**31
Endopeptidases, **4:**354, 354t
  proteasomes as, **4:**78

Endophytic bacteria, genetically modified to express *Bt* toxin, to control corn borers, **4**:395
Endoplasmic reticulum (ER)
  adrenal cortex, cytochrome P450 in, **1**:436
  bidirectional exchange of proteins with the Golgi complex, **5**:199
  biosynthesis of glycans in, **2**:459, 460f
  bridge to mitochondria, **4**:438
  calcium ion transport system, **1**:246
    in muscle, **6**:131–32
  chaperone proteins in, **2**:461
  defined, **1**:315; **3**:347; **4**:129; **5**:194, 328
  enzymes for phospholipid biosynthesis in, **4**:433
  glycoprotein formation at, **6**:187–88
  and membrane fusion, **4**:45
  P450s of, **1**:439
  packaging of lipids into chylomicrons at, **3**:404
  protein targeting at, **3**:353–54; **5**:195, 197–99, 199f
  rough, ribosomes of, **5**:198
  in the secretory pathway for peptide hormones, **4**:342, 342f
Endoreduplication, *Arabidopsis*, **1**:77
Endorhamnosidase, tailspike, in vivo folding pathway of, **5**:82–83, 82f
β-Endorphins
  blocking opioid activity of, by acetylation, **4**:346
  response to ethanol, in rats, **2**:150
Endosomal/lysosomal proteolytic degradation, of amyloid precursor protein, **1**:49
Endosome, fusion of, effect of arachidonic acid on, **4**:48
Endosymbiosis, and the origin of mitochondria, **4**:83–84
Endothelins, **2**:291
  defined, **2**:285
  endothelin-1 (ET-1), **5**:269
Endothelium
  cells of, glycoprotein induction on, **2**:455
  schwannomin in, **4**:186
Endothelium-derived relaxing factor (EDRF)
  contribution of nitric oxide to, **4**:204–5
  defined, **4**:200
Endothelium-derived relaxing factor/nitric oxide (ERDF/NO), **5**:268–69
Endotoxin/endotoxicity
  and cytokine production, **1**:444–45
  defined, **3**:462
  of lipopolysaccharides, **3**:465, 465f, 470–71, 470f
δ-Endotoxin proteins, of *Bacillus thuringiensis*, **4**:394–95
End replication problem, in DNA synthesis, **6**:31
Energetics, chemical, for enzyme reactions, **2**:225–29
Energy
  absorbed from light exposure, dissipation of, **6**:151–53, 151f
  expenditure of
    by humans, **4**:258–59
    for muscle contraction, **4**:145
  from foods, **2**:309; **4**:145
  of ionizing radiation, **3**:355
  requirements for, **4**:259–61, 261t
    during pregnancy and lactation, **4**:261
Energy balance, **4**:259–60
Energy barrier, in making and breaking hydrogen bonds, **3**:273
Energy density, of foods, **4**:261–62, 262t
Energy function, and molecular mechanics, **5**:166
Energy minimization, for RNA structure determination, **5**:379
Energy transduction
  defined, **6**:200
  during DNA packaging, **6**:206
  in oxidative phosphorylation, **1**:306
Energy transfer
  biological, thermodynamics of, **1**:304–6
  direct, in photosensitization reactions, **6**:208
Engrailed (en) class, homeobox genes, **3**:68, 70f
Enhancer/modulatory region, long terminal repeat, HIV, **1**:19–20
Enhancer/promoter sequences, and gene transfer, in fish, **6**:96
Enhancers, **2**:366; **3**:511–12
  defined, **4**:129
  H chain, **3**:282
  intronic, **3**:512
  muscle-specific, **4**:135
  prokaryotic, function of, **2**:369f
  regulation of protein synthesis by, **2**:376–77
Enhancer trap
  Drosophila, **2**:123
  for *transformer* mutant experiments in *Drosophila*, **2**:142
Enkephalins
  conformational flexibility of, **4**:381
  met, morphine as a peptidomimetic of, **4**:34f
Enniantin, structure of, **4**:210f
Enniantin synthetase, **4**:209
  domains of, **4**:212
  gene for, **4**:213
Enolates, electrophilic amination of, **1**:54–56
Enterobacteria, lipopolysaccharide of, **3**:463
Enterobactins, **4**:212
Enterococcus hirae, plasmids of, **5**:8–9
Enterotoxins, defined, **5**:508
Enthalpy, of protein unfolding, **5**:150–51, 151f
Entrapment, electrochemically polymerized, in biosensors, **1**:180–81
Entropy
  conformational, of unfolded proteins, **5**:148
  and energy transfer, **1**:305
  k-gram, **6**:43
  maximum, and local compositional complexity, **6**:47–51
  of protein unfolding, **5**:150–51, 151f
  and redundancy of nucleotide sequences, **6**:41
  of strings of length *k*, **6**:43
Envelope, defined, **6**:207
Enveloped virus, defined, **6**:192
Environment
  cellular, and protein folding, **5**:81–82
  defined, **2**:407
  of a DNA sequence, defined, **2**:411–12
  of the earth
    and exposure to ultraviolet radiation, **6**:163
    and origins of life, **4**:289–91
  and mutation frequency, **2**:408–9
  and phenotypic expression of genetic disease, **3**:223
    Gaucher disease, **2**:350
ENVIRONMENTAL STRESS, GENOMIC RESPONSES TO, **2**:**216–22**
Env protein, HIV-1, **1**:41
Enzymatic photoreactivation (EPR), **6**:284
  defined, **6**:283
Enzyme-directed change, **5**:51
  defined, **5**:49
Enzyme electrophoresis, **2**:395
Enzyme engineering, **1**:173–74
Enzyme immunoassay
  in automated ligase chain reaction testing, **3**:306
  for detection of LCR amplicons, **3**:307
Enzyme inhibitors, **1**:166–68
  defined, **1**:159
  nucleic acids, in vitro selected, **4**:228
Enzyme kinetics, **1**:173, **2**:233–37
Enzyme-linked immunosorbent assay (ELISA)
  for analyzing response to vaccination, **6**:171
  for apolipoprotein determination, **3**:476
  for assessing orientation of reaction centers, **1**:151
Enzyme-linked receptors, **2**:208, 208f
ENZYMES
  α-amino acid synthesis using, **1**:57–58

ENZYMES (*cont.*)
  amperometric electrodes using, **1**:182–86
  binding of PNA to, **4**:274
  in bioluminescence and chemiluminescence assays, **1**:328–29
  biosensor electrodes using, **1**:182–87
  for carbohydrate analysis, **1**:272, 274t
  commercial production of, by fermentation, **2**:340
  copper, **4**:62–63, 63t
  defined, **1**:113; **2**:249
  ENERGETICS AND REGULATION OF BIOLOGICAL CATALYSIS BY, **2:222**
  enhanced production of, using streptomycetes, **5**:506
  in foods, **2**:308–9
  as genetic markers, **2**:398
  HIGH TEMPERATURE, **2:240–49**
  hyperthermostable, **2**:241
  integrity of, in organic media, **2**:257
  iron, **4**:61–62t
  kinetics of reactions, **1**:173, **2**:233–37
  lipoprotein-processing, **1**:285t
  mechanism of catalysis, and structure, **5**:159–60
  mechanochemical, defined, **4**:112
  mediation of DNA rearrangements by, **2**:404–5
  metal-dependent, **4**:61, 65t
  of microbial restriction-modification systems, **5**:297, 298f
  as molecular recognition elements, **1**:141–42
  motor proteins as, **4**:112
  myosin, **4**:115
  for nucleotide sequencing, **4**:255
  oligoribonucleotide synthesis using, **4**:284
  and pharmacogenetic defects, **4**:400t
  phase I, **1**:193–97
  phase II, **1**:197–202
  polyethylene glycol modified, **2**:251
  post- and co- translational modifications of proteins catalyzed by, **5**:49–56
  precursor peptide processing by, **4**:353
  as probes, for RNA structure determination, **5**:378
  for protein sequencing, **5**:192
  proton transfer, and hydrogen bonding in, **3**:267
  RecBCD, **5**:227–28
  for regeneration of 11-*cis*-retinal, **5**:324–25
  for replacing glucocerebrosidase in Gaucher disease, **2**:351
  restriction, *Not* I, *Sgr*AI, *FseI*, and *AscI*, **3**:217
  of S^0-reducing thermophiles, **2**:243t
  specificity of, and chirality, **1**:340–43
  spin labeling analyses, **3**:389
  stereospecific, **1**:341
  steroid-transforming, molecular biology of, **5**:468–70
  for synthesis of oligoribonucleotides, **4**:284
  in vitro evolution of, with sexual PCR, **5**:455, 455f
  See also Restriction enzymes
Enzyme substitution, by plasmids, **5**:10
ENZYMOLOGY
  history of, **2**:224
  NONAQUEOUS, **2:249–58**
Eosinophiles, activation by IL-5, **1**:447
Ependymomas, *NF2* gene in, **4**:185
Epibatidine, **4**:29, 30f
EPIDEMIOLOGY
  of coronary heart disease, **3**:478–79
  of heart failure, **3**:1
  of liver cancer, **3**:488–89
  MOLECULAR, **2:258–62**
  of respiratory syncytial virus infection, **5**:291
Epidermal growth factor (EGF), **2**:465, 466, 468
  domains coding for, factor IX gene, **3**:37
  in the kidney, **5**:271–72
  transcription of keratin genes induced by, **2**:266
Epidermal growth factor receptor (EGFR), **2**:269–70; **3**:494
  tyrosine kinases, **2**:464
EPIDERMAL KERATINOCYTES, MOLECULAR BIOLOGY OF, **2:262–71**
Epigenetic modification
  defined, **2**:364, 419, 439; **3**:366
  of enzymes, **3**:376
  in genomic imprinting, **2**:444–45
  in parental imprinting, **2**:440
Epinephrine
  atenolol derived from, **2**:158
  structure of, **2**:159f
Episomes, replication of transferred nucleic acids as, **4**:450
Epistasis, **3**:223; **6**:320–21
Epithelial cells, **6**:122f
  association of α-actinin with focal adhesion in, **1**:454
  human, transformation by papilloma virus, **6**:34
  stomach, H^+-translocating enzyme in the plasma membranes of, **6**:132
Epithelial differentiation, keratins as markers of, **2**:265–66
Epithelioma, self-healing squamous (ESS1), mapping to chromosome 9, **3**:104
Epithelium
  cells of, telomeres of, **6**:34
  defined, **1**:425
  infection barrier, **1**:103
  transcytosis in, **5**:204–5
Epitope mapping, autoantigens associated with scleroderma, **5**:414–15
Epitopes
  analysis of scleroderma antigens, **5**:411–17
  in autoantibody response, **1**:86
  defined, **1**:84, 278; **2**:165; **3**:275; **4**:156, 361
  functional, **3**:283, 285
  imitation by aptamers, **4**:230
  O antigen, **3**:467–68
  quantitation of, **3**:294
Epitope tag, in yeast analyses, two-hybrid systems, **6**:322
Epitostanol, antitumor activity of, **5**:497
Epoxide hydrolases, **1**:196–97
Epoxides
  defined, **5**:485
  formation catalyzed by cytochrome P450, **1**:194–95, 195f
Epstein-Barr virus
  BRCF-1 protein of, **3**:343
  cytokine production by, **1**:449
  defenses of, against antiviral reactions, **3**:325
  disorders associated with, and host cell receptors for, **6**:233
  human B lymphocytes immortalized by, **6**:34
  malignancy associated with, **6**:173
Equilibrium
  chemical reaction, **2**:175–76
  defined, **2**:223, 225–27
Equilibrium constants (K_{eq})
  for assembly of subunits in oxyhemoglobin, **3**:12t
  defined, **2**:223
  electric and magnetic field-induced changes, **2**:175, 180
Equine infectious anemia virus (EIAV), activation domain of Tat, **1**:21
Equivalence groups, of postembryonic development in *Caenorhabditis elegans*, **4**:172
Ergocalciferol, from irradiation of ergosterol, **6**:246
Ergodic sources, entropy of, **6**:43
Error catastrophe theory, **1**:9
  defined, **1**:5
Error detection, human linkage maps, **3**:237–38
Error-prone PCR, **5**:452, 453
  defined, **5**:447
Erythrocytes
  antigens of, **1**:281–82, 282t

binding of α-tocopherol to, **6**:238
cytoskeletal-membrane interaction model, **1**:452
hemoglobin components, **3**:5
histone transcription in, **3**:53
irradiation of, to destroy viruses, **6**:210
membranes of, **1**:244; **3**:460
schwannomin in, **4**:186
Erythrocyte surface protein 4.2, gene localization to chromosome 15, **3**:168
Erythrokeratodermia variabilis (EKV), mapping of, to chromosome 1, **3**:91
Erythromycin A, **4**:99
Erythropoietin, **1**:441; **2**:465, 469
production from recombinant organisms, **1**:174
ESCHERICHIA COLI
Achilles' cleavage of genome, **1**:2
biosynthesis of lipid A in, **3**:464
composition of, **1**:101t
enzymatic α-amino acid synthesis using, **1**:58
F^+ plasmids carried by, **3**:220
GENOME, **2:169–72**
host cell features, **2**:275
k-12, F plasmid of, **5**:5–6
replication apparatus, **2**:86
RNase H from, **1**:30
RNase H domain of, versus HIV-1 reverse transcriptase RNase H domain, **1**:32
Escherichia coli SSB, defined, **5**:458
E-selectin, receptors for, **1**:297
Essential elements, for humans, **1**:126–27, 126t
Esterases
defined, **4**:399
EST-6, and mating behavior in *Drosophila*, **2**:143–44
hydrolysis by, **1**:196
Estradiol-receptor unit, **5**:483, 483f
Estramustine
antitumor activity of, **5**:495
for cancer therapy, **5**:486, 486–87
Estrogen receptor
DNA binding domain, **5**:133
isoform analysis, with high performance hydrophobic interaction chromatography, **3**:81f
regulation of, **2**:290
zinc-binding motifs, **6**:331–32, 331f
Estrogen response element (ERE), **2**:286
Estrogens
as anticancer drugs, prostate cancer, **1**:256
nuclear receptors for, **2**:208
role of, **5**:482
in insulinlike growth factor receptor gene expression, **5**:270
steroid subgroup, **5**:479
synthesis of, **2**:158, f; **5**:480
Estrone sulfate, sodium, for cancer therapy, **5**:490
Ethanol
metabolism of, **1**:196
precipitation of DNA by, **2**:49
Ether bond, in *Archaea* lipids, **3**:428–29; **4**:67
Ether-thioether formation, in peptides, **4**:377
Ethics
and genetic testing, **2**:427, 429; **3**:222, 249–50
and germ cell transfection therapy, **6**:146
of transgenic animal patents, **6**:75–76, 81–83
of transgenic animal production, **6**:66
Ethidium bromide staining, **2**:49
Ethyl acetate, hydrolysis of, **1**:160–61, 161f
Ethylene
as a plant hormone, **4**:492
production of, role in tomato development, **6**:58
N-Ethylmaleimide-sensitive factor (NSF), fusion protein, **4**:346
Etiologic studies, in molecular epidemiology, **2**:261
Etoliation, of seedlings, by light, **4**:493
Etoposide
as an anticancer drug, **1**:254
structure of, **1**:255f
Eucarya, domain of, defined, **4**:67
Euchromatin, defined, **2**:121
Eukaryotes
algae, lipids of, **3**:441–42, 441t
cells of
energetics of, **1**:304
protein synthesis and degradation in, **3**:347
replicative capacity of, **6**:23–24
senescence in, **1**:303
cytochrome P450 of, **1**:436
diploid, **2**:416
DNA replication in, **2**:87f, 89
genes of
regulation of expression, **2**:373–74
structure, **5**:431
initiation factors (eIFs), **6**:118
mRNA of, **6**:111, 111f
nucleotide excision repair in, **2**:45
origins for replication, **2**:86
protein secretion in, **2**:460f
protein-synthesizing systems extracted from cells of, **5**:334–35
recombinational repair in, **2**:82
ribosome preparation from cells of, **5**:329–30
transcription in, **2**:94–95f, 276f, 365–69
regulation of, **2**:95
Eurogem, **3**:238
European Molecular Biology Library (EMBL), **2**:70
database of, **3**:517
Escherichia coli database, **2**:169
European Patent Convention, **6**:75
Evaluation
immunotoxic, **3**:299
of peptide syntheses, **4**:366–67
of vaccination against rinderpest, **6**:167–68
Evanescent wave sensors, **1**:188–90, 188f
Even-skipped (eve) class, HOX cluster, **3**:67
Evolution
and aging theory, **1**:10
of chromosomes, **3**:84
of complementary base pairing, **5**:359
conservation of interleukins, **3**:334
constraints for dynamic programming based on, **5**:421–23
convergent, example of, **6**:137
defined, **2**:407
directed, defined, **4**:90
of domain Archaea, **4**:73–74
genetic, defined, **2**:407
and genetic diversity, **2**:405–6
enzyme patterns, **2**:398
and genetic drift, **2**:393
of genetic intelligence, **2**:407–14
genome, **3**:512–13
data from mitochondrial DNA, **4**:84–85
and hemoglobin structure, **3**:16–17, 16f
hominoid
Africa as the source of the mtDNA ancestor, **4**:87–88, 87t
estimate from mtDNA divergence, **4**:85–86, 86f, 338
implications of enzyme potential of RNA, **2**:239
and imprinting, **2**:445
of kallikrein gene families, **3**:382–83
of the kininogen domain structure, **3**:381
of kininogen genes, **3**:377
and mitochondrial DNA, **2**:399; **4**:81–82
and multiple-loci isoenzyme occurrence, **3**:369–70
and origins of life, **4**:289
phylogenetic tree, rooting of, **4**:78
and recombination, **5**:261–62
of restriction-modification enzymes, **5**:298–99
and sequence similarities, **5**:429
and synonymous mutation, **4**:150–51
of transport mechanisms, **6**:133

Evolution (*cont.*)
 vertebrate, gene distribution in, **2:**363
 in vitro, **4:**227
Exchange interaction
 defined, **3:**385
 and electron spin resonance, **3:**386
Exchanger
 defined, **1:**237
 renal Na^+-H^+, **5:**272–73
Excision repair
 of damaged DNA, **2:**44–45, 81–82, 81f
 defined, **6:**149
 of UV-damaged DNA, **6:**158–59, 159f
Excision repair cross-complementing (ERCC) series, in rodent cells, **6:**161
Excitation energy transfer
 defined, **4:**440
 light capture and, **4:**441–42, 441f
Exclusion, defined, **2:**48
Exocytosis, **5:**514
 of chylomicrons, **3:**404
 defined, **1:**59; **4:**340; **5:**194, 512
 regulated, **4:**42, 44–45
  of peptide proteins, **4:**346–47
  in secretory protein transport, **5:**203–4
 sites for, **5:**512
Exogenous agents
 defined, **2:**40
 DNA-damaging, **2:**41
Exogenous molecules, uptake during electroporation, **2:**200
Exogenous substrates
 for cytochrome P450, **1:**435
 defined, **1:**435
 P450s, **1:**436
Exoglycosidase
 analysis of oligosaccharides using, **2:**450–51
 defined, **1:**266
Exon-intron structure, of collagen genes, **2:**280
Exons, **2:**371–73; **5:**356–57
 defined, **2:**278, 364; **3:**28, 227; **4:**129, 198; **5:**338, 352
 of eukaryotes, **2:**393, 393f
 in the factor VIII gene, **3:**36
 junction, **2:**280
 NF2 gene, **4:**185f
 phenylalanine hydroxylase gene, **4:**408f
 role in illegitimate recombination, **3:**513
 of *tat*, **1:**21
Exon skipping, in alternative splicing, **2:**283
Exon trapping, **3:**226, 516
 for positional cloning, **4:**127
Exonucleases
 digestion at a double strand break, DNA, **5:**258f
 I, **5:**228
 RecJ protein, **5:**228
 removal of DNA errors by, **2:**88
 3′-, degradation of oligonucleotides by, **1:**71
Exopeptidases, **4:**354, 354t
Explants
 defined, **6:**10
 size and preparation of, yew callus culture, **6:**11
Exposure, relationship to dose, biological markers, **2:**260
Expressed sequence tags (ESTs), **1:**203, 204; **3:**159
 Arabidopsis gene mapping, **1:**82–83
 of *Caenorhabditis elegans*, **4:**169
 of chromosome 16, **3:**186
 of chromosome 18, **3:**209
 defined, **3:**142, 174, 215; **4:**167, 306
 generating in sequencing by hybridization, **3:**255
 of *Plasmodium falciparum*, **4:**308
EXPRESSION SYSTEMS FOR DNA PROCESSES, **2:271–78**
 for hemoglobin, **3:**22–23
 T7, for producing β-globin, **3:**23
Expression vectors, **2:**272, 272f; **6:**324–25
 for hemoglobin, single operon, **3:**23
 pLcIIFx, **3:**22, 22f
 for recombinant *Escherichia coli* bacteria, **6:**176–77
Extended X-ray absorption fine structure (EXAFS)
 defined, **6:**327
 for zinc finger motif characterization, **6:**328
Extension, defined, **3:**302; **4:**333
Extracellular domain, bacteriorhodopsin, **1:**112
Extracellular fluid (ECF), calcium in, **1:**238
EXTRACELLULAR MATRIX (ECM), **1:**293; **2:278–84**
 of cartilage, **2:**457
 cell adhesion to, **1:**452
 defined, **2:**278
 degradation of, **2:**290
 transforming growth factor β control of, **2:**468
Extracellular space, calcium in, **1:**239
Extrachromosomal circular satellites, **6:**5
Extraction
 for bioseparation, **1:**176
  in yew callus cultures, **6:**13–15, 15f
 for separation of lipids from cells and tissues, **3:**449–50, 451f
Eye
 development of, and granulocyte/macrophage colony-stimulating factor, **2:**464
 poly(vinyl alcohol) for use in, **5:**43
 silicone rubbers for medical treatment of, **5:**38
Ezrin-radixin-moesin (ERM) proteins, homology with schwannomin, **4:**186

F_1 transgenic fish
 defined, **6:**84
 inheritance of transgenes, **6:**90
Fab assembly, **1:**418f
Fab fragments
 binding specificities expressed as, **1:**64–65
 cloning, **1:**415–16, 416f
 defined, **1:**62, 292
 of immunoglobulins, **3:**277
 monovalent, **1:**294
 scaffold for generating recognition molecules, **5:**250–51
FAB-MS (fast atom bombardment mass spectrometry), defined, **1:**266
Facilitated diffusion, **6:**126
 in intersegment transfer, **5:**282f
Facilitative diffusion, **6:**129
Factor-binding sites, **4:**468
Factorial method, for determining protein requirements, **4:**265
Factor IX gene, **3:**36–37, 36f
Factor recognition, roles of nucleosomes in, **1:**368f
Factory areas, host cell, for viral DNA replication, **6:**198
Familial adenomatous polyposis (FAP), **1:**409; **6:**145–46
 APC gene involved in, **2:**423, 428
 and colon cancer, **3:**224
 defined, **6:**138
 screening for colon cancer in, **2:**64
Familial disorders
 amyloidotic polyneuropathy, gene for, mapping to chromosome 18, **3:**213
 CETP deficiency, **3:**409
 chylomicronemia, apoCII deficiency associated with, **3:**413
 defective apo B100, **1:**286
 expansile osteomyelitis, mapping to chromosome 18, **3:**199
 fatal insomnia, as an inherited prion disease, **5:**72
 glucocorticoid deficiency, gene for, mapping to chromosome 18, **3:**214
 HTGL deficiency, **3:**407
 human dwarfism, **2:**214
 hypercholesterolemia (FH), **1:**286; **2:**426; **3:**2, 408
  genetic mechanism, **6:**136
  as a single-gene disorder, **3:**230
 hypertrophic cardiomyopathy, point mutation in, **3:**169

hypobetalipoproteinemia (FHB), apoB mutations associated with, **3**:412
hypocalciuric hypercalcemia, **2**:215
isolated hypoparathyroidism, mutation in, **2**:213
juvenile polyposis, **1**:410
ligand-defective apoB, **3**:411
lipoprotein lipase deficiency, **3**:405
male precocious puberty, **2**:215
mechanism of, **2**:292
Mediterranean fever (MEF), localization to chromosome 16, **3**:188
neurohypophyseal diabetes insipidus, **2**:214
relationships, DNA fingerprinting to establish, **2**:60
spastic paraplegia (FSP), gene for, mapping to chromosome 14, **3**:157
Family of repeats, defined, **3**:240
Fanconi anemia
chromosome breakage in, **1**:413
excision repair defects in, **2**:46
Faraday constant, relating Gibbs free energy change and reduction potential difference, **1**:310
Farber's lipogranulomatosis, ceramide level in, **3**:457, 460
Farm structure, impact of biotechnology on, **6**:80–81
Farnesylation, **5**:55
Farnesyl pyrophosphate synthetase, **3**:409
FASTa, for local alignment assessment, **5**:424–25
Fast atom bombardment mass spectroscopy (FAB-MS), **4**:13
for carbohydrate analysis, **1**:274–75
FAST methods, **5**:434
FASTa, for local alignment assessment, **5**:424–25
FAST program family, **5**:433
Fat-soluble vitamins, **6**:245–48, 245t
vitamin D, **6**:237–38
vitamin E, **6**:238
Fatty acid binding proteins (FABPs), **3**:455
Fatty acids
biosynthesis of, **3**:453
branched chain, synthesis by bacteria, **3**:431–32
cyclopropane, synthesis by bacteria, **3**:432
defined, **3**:426
hydroxy, synthesis by bacteria, **3**:432
metabolism of, **3**:451–54, 454f
modification pathways, **3**:455f
polysaturated, synthesis by fungi, **3**:439–40, 439f
polyunsaturated, **3**:451
synthesis by molds, **3**:437
synthesis by mycobacteria, **3**:431
profiles
commercial fungal oil products, **3**:439t
oils from algae, **3**:441t
properties of, **3**:453t
synthesis by molds, **3**:437, 438t
unsaturated
production from *Mucor* fermentation, **2**:340
synthesis by bacteria, **3**:431
yeast production of, **3**:434–35, 435t
Fatty acid synthetase, **3**:432f, 454f
Fatty alcohols, metabolism of, **3**:451–54, 454f
Fc fragment, defined, **1**:62
Fc receptor, defined, **2**:446
Fed batch culture, **2**:341
defined, **2**:337
Feedback
in estradiol production, **2**:286
in respiration, **1**:312
Feedback inhibition, by tyrosine hydroxylase, **1**:213
Feline immunodeficiency virus (FIV)
model for HIV drug resistance, **1**:18
proteases encoded by, **1**:14
FEMALE REPRODUCTIVE SYSTEM, MOLECULAR BIOLOGY OF, **2:285–93**
Females
transmission by, premutation, fragile X syndrome, **2**:328
XY, **5**:445–446
Ferguson-Smith syndrome, mapping to chromosome 9, **3**:104
Fermentation
for bioprocessing, **1**:172
recombinant, **1**:174
Fermi's Golden Rule, **2**:194
Ferredoxin
defined, **2**:240
of hyperthermophiles, **2**:244
of methanogens, genes encoding, **4**:77
in photosystem I, **4**:447–48
of *Pryococcus furiosus*, characterization of, **2**:247
Ferricenium ion, electron acceptor for amperometric sensors, **1**:184–85
Ferritin, **3**:315, 315f; **6**:60
biosynthesis of, **6**:122
iron in, **4**:61
synthesis of, regulation by a translational repressor, **2**:376–77, 377f
Ferrochelatase (FECH)
mapping to chromosome 18, **3**:203
in protoporphyria, **3**:214–15
Fertility inhibition, by R plasmids, **5**:9
Fertilization, mammalian, **2**:455–56
Fetal development, of liver, **3**:481
Fetal growth factors, defined, **2**:439
Fetal hemoglobin (HbF), **3**:5
Fetal tissues, transforming growth factor α in, **5**:272
Feulgen microspectrophotometry, **4**:486
Fiber-optic biosensors, **1**:142
Fiber-optic chemical sensors (FOCS), **1**:179
Fibers
collagen, structure of, **5**:28
DNA
geometrical parameters of double helix from, **2**:104t
X-ray diffraction patterns from, **2**:103
graphite, in orthopedic prostheses, **5**:39
Fibrillarin
autoantigen, in autoimmunity induced by heavy metals, **1**:91
component of coiled body, **1**:87–88
conservation of protein sequence, **1**:90
Fibrillin gene
mapping to chromosome 5, **3**:162, 162f
mapping to chromosome 15, **3**:160, 162
Fibrinogen
as an adherence molecule, **1**:453
expression in milk of transgenic mice, **6**:105
Fibrinolysin system, **3**:377
Fibroblast growth factors (FGFs), **1**:443t; **2**:465, 469; **5**:239–40, 240t
Fibroblasts
actin filament disassembly, in response to cAMP, **1**:456
association of α-actinin with focal adhesion in, **1**:454
choline requirement of, **4**:437
defined, **1**:113
tropomyosin isoforms in, **4**:138
Fibronectin (FN), **2**:282–83, 283f
accumulation of, in *cmd* cells, **2**:457
as an adherence molecule, **1**:453, 458
repeat in N-CAM, **1**:294f, 295
silk polymer, **5**:27
Fibronectin-binding proteins, from mycobacterial cultures, **4**:159–60
Fibronectinlike receptors, **2**:208
Fibrous proteins, **5**:27–29
Field amplification, by interfacial polarization, **2**:179f
Field effect transistor (FET), all-polymer, **1**:154
Field reception
defined, **2**:173–74
electric and magnetic, **2**:172, 173, 173f
Field strength, and efficiency of electroporation, **2**:203
Field testing, of transformed tomato plants, **6**:56
Filaggrin, **2**:268
Filamentous fungi
characteristics of, **2**:338

Filamentous fungi (*cont.*)
defined, **2:**337
metabolic activities of, **2:**338
protein production using, **2:**341–43, 341t
Filamentous hemagglutinin (Fha), **1:**104
Filamentous phage, combinatorial libraries for, **1:**68–69
Filaments, formation of, RecA protein interaction with single-stranded DNA, **5:**224–25
Filarial genome project, WHO, **4:**309–10
Filariasis, **4:**307
Filter binding
of specific nucleic acid-protein complexes, **4:**246–47, 246f
assay using, defined, **4:**243
Filtration
for assay of repressor-operator complexes, **5:**283
of blood, with polyurethane devices, **5:**38
Fimbri. *See* Pili
Fimbrin, role in endocytosis, **4:**51
Fine Chemicals Directory (FCD), **4:**99
Fingerprint assays, of oxygen-derived species, **2:**336–37
Fingerprinting
clone libraries for, **3:**255
of clones, **3:**220
of mycobacterial DNA, **4:**163, 164f
of peptides, **5:**180
First law of thermodynamics, **1:**305
First messengers
defined, **4:**347
peptide hormones as, **4:**355
Fischer projection formula, **1:**336, 336f
Fish, hemoglobin of, **3:**27
Fish-eye disease, **3:**409
Fish odor syndrome, monooxygenase deficiency in, **4:**405
Fitness, defined, **5:**45
Fixation
of chromosomes for microdissection, **1:**372
of genetic change in an individual organism, defined, **2:**407
FK506-binding protein, **5:**92
Flagellar system, of *Chlamydomonas*, **1:**358–60
Flap region, HIV protease, **1:**28
Flaps, active site, HIV-1 protease, **1:**41f
Flavenoids, protection from ultraviolet radiation, **6:**157
Flavin adenine dinucleotide (FAD), **1:**398–400
fluorescence from, **2:**294
Flavin mononucleotide (FMN), **1:**398–400
as an enzyme cofactor, **2:**232–33, 232f
Flaviviruses, assembly and budding of, **6:**190
Flavocoenzymes, **1:**398–400
Flavoprotein, in P450 metabolism, **1:**436
Flavoprotein monooxygenases, **1:**195
Flavoprotein oxidases, **1:**184
Flavoquinone coenzymes, **1:**400–401, 401f
Flexibility
of folded proteins, and free energy constraints on conformation, **5:**150
segmental, of soluble antibodies, **3:**278
Flexural plate wave device, **1:**192
Flotation, for separation of plasma lipoproteins, **3:**474
Flow cytometry
cell cycle analysis using, **1:**102
chromosome sorting with, **2:**387
lifetime measurements, fluorometry, **2:**306
Flow karyotype, defined, **2:**386
Flumezanil, **2:**160
Fluorenylmethoxycarbonyl (Fmoc) derivatives, for protecting amino groups in peptide synthesis, **4:**387, 388
Fluorescein isothiocyanate (FITC)
label for FISH, **2:**384, 384t
probe, **2:**295
Fluorescence
defined, **1:**177; **2:**294; **6:**27
for identifying isoenzymes, **3:**368
polarization, **2:**299
in polymerase chain reaction analysis, **4:**337
of pyridoxal 5′-phosphate, **1:**401–2
of quinoid forms of FMN and FAD, **1:**399
Fluorescence-activated cell sorting (FACS), **2:**72, 387, 390–91
Fluorescence binding assay, **1:**189, 189f
Fluorescence detection system, DNA sequencing, **2:**100f
Fluorescence immunoassays, **1:**142
Fluorescence in situ hybridization (FISH), **2:**320, 387, 388; **3:**119; **6:**253
for analysis of cloned DNA, **1:**382
for chromosome-specific staining, **2:**321f; **6:**255f
for cytogenetic mapping, **3:**182
defined, **3:**126, 142, 191, 215
for gene mapping, **2:**378; **6:**300
for identifying the Philadelphia chromosome, **2:**65
Fluorescence lifetime, defined, **2:**294
Fluorescence microscopy, for signal visualization, **2:**385
Fluorescence polarization, **2:**299f
defined, **2:**294
Fluorescence resonance energy transfer (FRET), **2:**299–300
defined, **2:**294
FLUORESCENCE SPECTROSCOPY OF BIOMOLECULES, **2:294–306**
of tryptophan, sensitivity to environment, **5:**126–27, 127f
Fluoride-selective electrode, **1:**186
Fluorochromes
for FISH, **2:**384, 384t
for visualizing chromosome bands, **1:**371
Fluorography
defined, **5:**209
use and sensitivity of, **5:**220t
Fluorophores, **2:**294–96
defined, **2:**294
5-Fluorouracil
inhibition of thymidylate synthase by, **1:**254
structure of, **1:**255f
Fluoxymestrone, cancer therapy with, **5:**487
Fluphenazine, modification of release by esterification, **1:**196, 197f
Flutamide, prostate cancer response to, **5:**484, 484f
Flybase (computer-based project), **2:**123, 127
Fmoc method for peptide synthesis, **4:**366, 366f, 384, 389f
Foam cells, **3:**479
in atherosclerotic lesions, **1:**288
Foaming
defined, **2:**307
in foods, **2:**311–12
Focal adhesion kinase (FAK), **1:**454
and integrins, **1:**452
pp125, **1:**459
Focal adhesions, **1:**452
defined, **1:**451
and integrins, **1:**452
Folacin, structure and function of, **6:**250, 250f
Folate-sensitive fragile site, characterization of, chromosome 16, **3:**181
Folded state, protein, stability of, **5:**150–51
Folding, protein
analysis by circular dichroism, **1:**391
control by glycosylation, **2:**461
rates of reaction and equilibria, **5:**153–54, 154f
rules of, **5:**81
in vivo, aggregation during, **5:**81–93
Folding pathway, hsp70s and hsp60 in, **1:**321
Follicle cells, **2:**85
Follicle-stimulating hormone, **2:**454
feedback loop in production of, **2:**286
follistatin production influenced by, **3:**509
interaction with CREM in spermatogenesis, **3:**506
isoforms of, and infertility, **2:**292–93
in spermatogenesis, **3:**504–5

subunits of, secretion in the milk of transgenic mice, **6**:103
Follicular lymphoma variant translocation, *FVT1* gene, chromosomes 2 and 18, **3**:203
Follistatin
follicle-stimulating hormone regulation by, **2**:288
production of, by Sertoli cells, **3**:509
Food, constituents and functions, **4**:258, 258t
Food and Drug Administration (FDA), U.S.
approval of antiviral drugs, HIV-1, **1**:37
tiered testing in immunotoxicology, **3**:299
Food industry, enzymes used in, **1**:173–74
Food protein interaction, defined, **2**:307
FOOD PROTEINS, INTERACTIONS OF, **2:306–12**
Footprint analysis, **5**:284–285, 285f
Footprinting
defined, **4**:243
DNaseI, **4**:248f, 249
for apolipoprotein study, **4**:500–501
for DNA-protein interaction analysis, **4**:471
reagents for, **5**:283, 284f
RNA-protein, with primer extension, **4**:253
for studying nucleic acid-protein interactions, **4**:248–50, 248f
Forces, noncovalent, of native duplex DNA, **2**:1
Foreign gene transfer, **2**:431–32
Forensically informative nucleotide sequencing (FINS), **2**:61
Forensic applications
of DNA typing probes, **2**:59t
of PCR technology, **4**:339
Forks, multiple, defined, **1**:95
Formacetal linkages, in deoxyoligonucleotide analogues, **4**:278–79, 279f
Formal correctness of biological models, **6**:53
Formate dehydrogenase (FDH), of methanogens, genes for, **4**:76–77
Formin, gene for, localization to chromosome 15, **3**:168
Forskolin
activation of tyrosine hydroxylase by, **1**:216
to treat diastolic failure, **1**:292
Forward genetics, for isolation of genes in *Caenorhabditis elegans* nervous system, **4**:169
Fossilization
defined, **4**:299
types of, **4**:300–303
Fostriecin, as an anticancer drug, **1**:254
Founder, defined, **6**:99
Founder animals, transgenic animal modeling, **6**:67, 107
Founder cells
Caenorhabditis elegans, **4**:171
from embryonic stem cell cultures, **6**:68
Founder effect
and distribution of cystic fibrosis mutations, **1**:432–33
in an epithelioma mapped to chromosome 9, **3**:104
and genetic drift, **5**:47–48
in phenylketonuria, **4**:411–12, 411t
and hemophilia, **3**:37
Four-center hydrogen bond, **3**:265
geometry of, **3**:268
Fourier transform infrared spectroscopy, enzymes in organic solvents, **2**:253
Fourier transform mass spectrometer, defined, **4**:12
14–3-3 family, activator protein for CaM PK II, **1**:215
F plasmid, *Hfr* strain of, defined, **5**:1
F′ plasmid, **5**:8, 8f
defined, **5**:1
Fra16a, defined, **3**:174
Fra16b, defined, **3**:174
Fractionation
by agarose gel electrophoresis, in RLGS, **5**:307
of human plasma proteins, **1**:172
FRAGILE SITES, **2:313–18**
defined, **2**:319
folate-sensitive, physical map of, **3**:182f
FRAXA, **2**:327–28, 327f
FRAGILE X LINKED MENTAL RETARDATION, **2**:318, 319–30, 420; **3**:225, 231–32, 242; **6**:5
defined, **2**:313
molecular pathology of, **4**:331
parental imprint in, **2**:445
positional cloning for identifying genes in, **2**:320–21
trinucleotide repeat expansion in, **2**:427
detection procedure, **3**:397
unstable genetic sequence in, **3**:231
Fragmentation
of labeled DNA, in RLGS, **5**:307
of proteins, chemicals and enzymes for, **5**:190t
Fragment condensation, for synthesis of peptides, **4**:384–85, 385f
Fragments, antibody, fusion proteins from, **1**:68
Frames, of data, DNA sequencing with high speed gel electrophoresis, **2**:100–101
Frameshifts, **2**:418; **4**:150, 150f
in adenomatous syndromes, **1**:410
caused by pseudoknots, **5**:362
defined, **1**:407; **2**:347; **3**:28
in factor VIII genes, **3**:32t
in familial hypobetalipoproteinemia, **3**:412
in hemophilia A, **3**:31
in products from mRNA incorporating noncoding amino acids, **5**:146–47, 146f
in recessive nephrogenic diabetes insipidus, **2**:25–26
Framework construction, in homology modeling, **5**:164
Framework maps, **3**:238
Framingham study, **3**:1
Franck-Condon factors, **2**:194–95
defined, **2**:194
Free energy, **1**:305
and cellular work, **1**:304
protein binding, **1**:162
of protein conformations, **5**:161
and relative redox potentials, **1**:309–10
Free energy of reaction
cooperative mechanism of human hemoglobin, **3**:13t
intraprotein electron transfer, **2**:195–96, 196f
of monomer association, hemoglobin of some species, **3**:13
Free induction decay, in multidimensional NMR, **4**:220
FREE RADICALS
defined, **2**:330; **3**:354
IN BIOCHEMISTRY AND MEDICINE, **2:330–37**
produced in photosensitization reactions, **6**:208
Frequency
of cystic fibrosis, by population group, **2**:424
of Gaucher disease mutation, in Jewish patients and non-Jewish patients, **2**:352
Frequency counts
nucleotide, **6**:41f
and statistical modeling, **6**:40–41
Frequency domain (FD) measurement, of fluorescence, **2**:301–2, 302f
Frequency gene, **2**:221
Freundlich isotherm, **1**:176
Friedewald formula, **3**:475
Friedreich ataxia (FRDA), mapping to chromosome 9, **3**:103–4
Frozen accident theory, of the selection of enantiomers, and the origin of life, **4**:294
Fruit quality, bioengineering to enhance, **6**:57–58
FSH isoforms, defined, **2**:285

Fukuyama-type congenital muscular dystrophy (FCMD), mapping to chromosome 9, **3:**104, 107
Fumarase
  specificity constant of, **5:**301
  stereospecificity of, **1:**344, 344f
Fumarylacetoacetase (FAH), gene for, chromosome 15, **3:**170
Förster distance, **2:**299–300, 300f
Function, of mtSSB, **5:**465
Functional analysis, mammalian genome, **3:**520
Functional codes, **6:**39
Functional domains, Tat protein, **1:**24f
Functionalization reactions, **6:**263f
  defined, **6:**259
  in metabolism of xenobiotics, **6:**260
  of nitrogen, in metabolism of xenobiotics, **6:**264–65, 264f
  of sulfur, in metabolism of xenobiotics, **6:**265, 265f
Functional roles
  of nucleosomes, **1:**368–69
  of replication protein A, **5:**464–65
FUNGAL BIOTECHNOLOGY, **2:337–46**
  products of, **2:**340t
Fungi
  biodegradation of organic wastes by, **1:**125
    white rot, **1:**122
  cultivation of, **2:**341
  filamentous, lipids of, **3:**437–40, 440t
  as host cells, **2:**275
  infection of plants by, **4:**477–78
  mtDNA of, **4:**80
  polyketide metabolites of, **1:**339, 339f
Furanose, defined, **2:**446
Furin, **4:**345
Fusarium graminearum, production of mycoprotein by, **2:**340
Fusion
  bilayer, mediation by membrane-specific SNAREs, **5:**202, 202f
  chromosome, **3:**512
  genetic, **2:**404; **5:**196, 196f
    in oncogene activation, **4:**286
  protein-antibody, **1:**67–68
  of transport vesicles, **5:**201–2
  virus envelope
    with the host cell membrane, **6:**195–96
    in virus release from cells, **6:**184
  See also MEMBRANE FUSION
Fusion and endocytosis, defined, **6:**192
Fusion gene, association with synovial sarcoma, **3:**211
Fusion peptide, **4:**52–53, 52f
Fusion protein (F), **1:**67–68, 67f; **2:**274
  cIIFX-β-globin with blood clotting factor Xa, **3:**22, 22f
  defined, **2:**271
  expressing deletion mutants, CENP-B cDNA, **5:**412
  gag-pol
    reverse transcriptase synthesis as, **6:**222
    yeast virus, **6:**220, 220f
  β-galactosidase, with ubiquitin, **3:**350, 350f
  interferon-γ and immunogens, **6:**170
  NEM-sensitive, **5:**201–2
  N-ethylmaleimide-sensitive factor (NSF), **4:**346
  liver, suppression of neuroblastoma by
  of respiratory syncytial virus, **5:**289–90, 289f
    vaccines using, **5:**294
  in translation, **2:**275
  viral, **3:**418
Fusogenic lipids, **4:**47–48
Fusogenic protein, **4:**42, 50
Fv fragments
  defined, **1:**62
  production of, **3:**286, 286f

G_0 state, defined, **6:**138
G_1 and G_2 stages of the cell cycle, defined, **6:**138
GABA receptors, **5:**233
  and sensitivity to ethanol, in SS and LS mice, **2:**148
Généthon map, **3:**238
  for chromosome 14, **3:**143–44
Gag protein, HIV-1, **1:**41
Gain of function
  mutations
    in *Caenorhabditis elegans*, **1:**235
    in colorectal carcinomas, **1:**410–11
    of the *lin-14* gene, **1:**235
  proposed mechanism
    for Huntington's disease, **3:**251
    for oncogene molecular action, **4:**286–87, 287f
  transgene, **6:**98
Galactocerebrosidase deficiency (GALC), gene for, mapping to chromosome 14, **3:**158
Galactocerebroside, **3:**458f
β-Galactosidase
  fusion protein with ubiquitin, **3:**350, 350f
  gene for
    multiple cloning site within, **4:**251
    as a reporter gene, **1:**234
  lacZ encoding of, **6:**314
  quasi-irreversible unfolding-refolding of, **5:**79
  reactions catalyzed by, **5:**276f
Galns, defined, **3:**174
Gametogenesis
  change in CAG repeat lengths during, in Huntington's disease, **3:**249
  in *Chlamydomonas reinhardtii*, **1:**347–48
Gamma decay, **5:**210
Gangliosides, **3:**457
  biosynthetic pathway, **2:**454f
  brain, binding of cholera toxin by, **2:**457
  GM_1, **1:**280–82, 282f; **2:**457
Gangliotetraosylceramide, **1:**280f
Gap, accounting for, in sequence alignment, **5:**432
Gap bias, in structure-based alignment, **5:**426
Gap junctions, **6:**128
  connexins in, **5:**512–13
  field amplification at, **2:**179
  of follicle cells, **2:**85
  viral budding in cells with, **6:**184
Gap ligase chain reaction (G-LCR), **3:**304–5, 304f, 396, 396f
  for identifying disease, **4:**327
  sensitivity of, **3:**302
Gapped plasmid, for mutagenesis, yeast, **6:**315
Gap penalty, in sequence alignment, **5:**422–23
Gap repair, during integrative transformation, **5:**258f
Gardner neurofibromatosis, **4:**181
Gardner syndrome, **1:**408–9
Gas chromatography-mass spectrometry (GC-MS)
  defined, **5:**466
  for quantitation of steroid hormones, **5:**476–77
Gas-liquid chromatography
  for analysis of lipids from cells and tissues, **3:**449–50
  for carbohydrate analysis, **1:**271, 271t
  for steroid purification, **5:**475
Gas phase spectroscopy, for hydrogen bond studies, **3:**267
Gastrin-releasing peptide, *GRP* gene for, **3:**207
  mapping to chromosome 18, **3:**203
Gastrointestinal cancers, **3:**210
Gastrointestinal tract
  involvement of, in cystic fibrosis, **1:**425–26
  upper, carcinoma of, **1:**409
Gastrointestinal tract abnormalities, associated with Down's syndrome, **2:**119
Gated calcium channels, **1:**243
Gated ion channels, mimicking, **1:**140
Gate mechanism, two-electron, in photosynthetic energy transduction, **4:**444–45, 446f
Gaucher cell, **2:**350

GAUCHER DISEASE, **2:347–52**, 424
defined, **2:**347
detection of mutation in, **3:**394f
clinical manifestations of, **2:**350
G-box-binding factor (GBF), in plants, **4:**492
Gelatin, as a drug delivery matrix, **5:**33
Gelation, **2:**309
defined, **2:**307
of proteins, **2:**311
Gel electrophoresis, **4:**3
defined, **4:**1; **5:**304
for determining DNA structure, **2:**104
for genetic analysis of populations, **2:**392, 395f
for immuno-PCR, detection of products, **3:**294
ladder, from telomerase analysis, **6:**23, 23f
retardation of specific sequence repeat DNAs in, **2:**115
for studying nucleic acid-protein complexes, **4:**247–48, 247f
two dimensional, **5:**124
defined, **5:**173
for DNA analysis with RLGS, **5:**310–16, 310f
for protein separation and analysis, **5:**183–84, 183f
GEL ELECTROPHORESIS OF PROTEINS, TWO-DIMENSIONAL POLYACRYLAMIDE, **2:352–60**
Gel filtration
for apolipoprotein purification, **3:**476
for lipoprotein analysis, **3:**474
Gel mobility shift, defined, **4:**243
Gel preparation and assembly, horizontal ultrathin gel electrophoresis, **2:**98–99
Gel retardation assay, **5:**302
for in vitro study of repressor-operator recognition, **5:**282, 282f
Gels, **3:**315–16
polystyrene bead, solid support in the Merrifield technique, **4:**387
Gel shift method, for studying nucleic acid-protein complexes, **4:**247
Geminate recombination, **3:**9
defined, **3:**3
Gene acquisition, defined, **2:**402
Genealogical analysis, **5:**47
of autosomal sex-determining genes, **5:**443
duplicate section of an X chromosome, **5:**442f
human, of X-linked sex-determining genes, **5:**441–42, 441f
of true hermaphrodite incidence, **5:**444–45, 445f
Gene amplification, defined, **3:**480; **6:**138
Gene cloning, *NF1* gene, **4:**175
Gene cluster, $C\mu$, **3:**282
Gene constructs, fish, **6:**85–88, 88f
Gene conversion, **2:**424; **5:**253f, 254–55, 255f
defined, **3:**377; **5:**252
models of, **3:**515f
GENE DISTRIBUTION, IN THE HUMAN GENOME, **2:360–64**
Gene duplication
origin of, **3:**513
by unequal homologous recombination, **3:**514f
GENE EXPRESSION
in *Caenorhabditis elegans*, **1:**233–35
and the cystic fibrosis transmembrane conductance regulator, **1:**434
and heart failure, **1:**290, 291t
inhibition of
by antisense enzymes, **1:**173
by antisense oligonucleotides, **1:**75
keratin, regulation of, **2:**266
lipoprotein system, **1:**287
in neurofibromatosis 1, **4:**178
for NOS2, **4:**202, 202t
profile of, **1:**203, 204, 205f
for regulating enzyme activity, **2:**235
regulation of
in methanogens and archaea, **4:**78
during spermatogenesis, **3:**505–7
REGULATION OF, **2:364–77**
in transgenic animals, **6:**102
Gene flow, **5:**48–49
Gene-for-gene hypothesis, **4:**476
of plant-pathogen interactions, **4:**477f, 481, 483
Gene fragments, immune system antibody, cloning, **1:**145
Gene locus, defined, **2:**386
GENE MAPPING, **3:**513–15
Arabidopsis, **1:**78–80
BY FLUORESCENCE IN SITU HYBRIDIZATION, **2:377–86**
to chromosome 1, **3:**84–91, 91t
relative locations, **3:**92f
to chromosome 9, **3:**105–07t
to chromosome 11, **3:**128–32t
to chromosome 14, **3:**151–55, 152–55t
to chromosome 18, **3:**199–209
defined, **1:**370
by fluorescence in situ hybridization, using whole chromosome probes, **6:**257
to human chromosome 14, and mouse synteny, **3:**156f
mycobacteria, **4:**157
Gene maps, defined, **3:**84
Gene mutations, DNA variation due to, **3:**228–29
Gene order, by FISH, **2:**389
GENE ORDER BY FISH AND FACS, **2:386–91**
Gene probe, defined, **2:**386
Gene products, gB cytomegalovirus, expressed by recombinant baculoviruses, **6:**178, 178f
General nucleotide excision repair (GNER), **6:**288–90, 288f
Generation of diversity, experimental examination of, **2:**407–8
Gene replacement, yeast genetic studies, **6:**316–17, 316f
Genes
actin, **4:**137
activation by steroids, pathways for, **5:**473f
addition of, in tomatoes, **6:**54
affecting *Drosophila* mating, glossary, **2:**131–32
affecting programmed cell death, conservation of, **1:**302
alkali myosin light chain, **4:**134
amyloid precursor protein, **1:**44–45
analysis of, using nuclear runoff assay, **5:**350
antennapedia class of, **4:**66–67, 66f
antibiotic resistant, **4:**466
antibody and T-cell receptor, **3:**281–82, 282f
antiethylene, to prevent ripening of fruit, **6:**58
apolipoprotein, **3:**404
of *Caenorhabditis elegans*, **1:**229
cellular, regulation of expression by interferons, **3:**322–23
chromosomal, of yeast, **6:**306–7
clock, **2:**221
cloned
mammalian cell lines expressing, **6:**179
mapping of, **1:**232
for coagulation factor VIII, **3:**29–30, 30f
for collagen, **2:**279
controlling neural membrane products, in *Drosophila*, **2:**133
conversion of, **3:**513
CpG islands associated with, **2:**74
cytokine, **1:**444–45
defined, **2:**216, 415, 419; **3:**227, 233; **4:**129, 148–55
for desmin, **4:**143
for DNA repair, human, **2:**47
as drugs, **4:**40–41, 41t
dystrophin, **1:**289
elongation factor 1α (eEF-1α), **1:**12
encoding hsp70s, **1:**317f
encoding lipoprotein metabolism enzymes, **1:**288t
for extracellular matrix proteins, **2:**279
for fibronectin, **2:**282–3, 283f

Genes (*cont.*)
fusion of, **2**:404
for producing heterologous proteins from filamentous fu, **2**:342
glucocerebrosidase, **2**:347–48, 348t
glucokinase, and diabetes, **2**:28, 39
for β-glucuronidase, activation of expression by promoter, **4**:472
hemoglobin, nucleotide sequence, **1**:114
histone, **3**:49
expression of, **3**:51–53
organization of regulation of, **3**:53–55, 53f
HIV pol, **1**:33
homeobox, **3**:64–67, 64f
human LDL receptor, transgenic experiments in, **1**:286
H-Y antigen, **5**:435
hypoxanthine-guanine phosphoribosyltransferase, **1**:6
immunoglobulin, **1**:64
for immunoglobulin domains V, L, J, and C, **3**:281
imprinted, **2**:441–43
mouse and human, **2**:443t
for insulin, **2**:37
interferon, **3**:318–19
interferon type I-inducible, regulatory nucleotide sequence of, **3**:323
interleukin, **3**:334–35
involved in lipoprotein metabolism, **1**:285t
isolation by assay of products, **5**:504
kallikrein, **3**:377, 383–84
kininogen, structure of, **3**:381, 382f
knockout technique for studying, **1**:287
β-lactamase, molecular evolution of, **5**:455, 455f
ligase-mediated detection of, **3**:391–99
longevity, **1**:10–13
major histocompatibility complex, **3**:499–00
methanogen, cloned, **4**:76–78
β-MHC, and heart failure, **1**:289
mitochondrial DNA, **4**:79–81; **6**:307–8, 307t
conservation of, **4**:80
MLC1F promoter region, **4**:135, 135f
comparisons among species, **4**:136f
multilocus, isoenzymes coded by, **3**:369–70
mycobacterial, **4**:159–61
regulation of expression, **4**:161
in myogenic differentiation, **4**:145
myosin heavy chain and myosin light chain, **4**:134
nuclear
as genetic markers, **2**:398–99
Mendelian pattern of transfer, **1**:348
overlapping, in viruses, **6**:47
overproduction of desirable products, **4**:472
peptide synthetase, **4**:213
phenotypic expression of, **2**:417–18
for phenylalanine hydroxylase (PAH), **4**:407
photosynthetic mutant, **1**:355
phytochrome-encoding, **4**:494
plant resistance, **4**:481
plasma membrane calcium ion pump, **1**:245
poliovirus receptor (*PVR*), **6**:225
prion protein, **5**:58–59
expression of, **5**:59
promoter sequences of, **5**:356
for proteinase inhibitors, **4**:483, 483f
PrP, dosage and control of scrapie incubation time, **5**:66
regulation of expression of, **2**:418
by steroid hormones, **5**:479
by trace element micronutrients, **6**:61, 61f
renin, **5**:265
reporter, for studying transcription, **1**:329–30
sarcoplasmic reticulum calcium pump, **1**:247
sex-determining region Y (*SRY*), **5**:438–41
single-copy, **3**:511
stress-responsive, **2**:217
T-cell receptor, **3**:159, 281–82
techniques for isolation, tomatoes, **6**:54–55
transcription of, **5**:233
translational control of expression of, **6**:121–24
transmembrane conductance regulator, **1**:117
tropomyosin, **4**:138–40, 140f
troponin, **4**:140
tumor suppressor, **1**:407
tyrosine hydroxylase, **1**:217, 217f
vasopressin, mutation in, **2**:23–24, 24t
vasopressin precursor, **2**:20
vasopressin receptor, mutations in nephrogenic diabetes insipidus, **2**:25–26, 26t
for vimentin, **4**:143
vitamin D responsive, **6**:237–38
for YadA adhesin, **1**:104
for yeast studies, **6**:313–14
Genes, specific
age-1, *Caenorhabditis elegans*, **1**:11
APC
in colon carcinoma and adenoma, **1**:410, 411–12
in familial adenomatous polyposis, **2**:64
MCC, and *DCC*, **6**:145
APNI, for AP endonuclease, **6**:287
*apo*AI, **1**:287
*apo*CIII, **1**:287
apo E, rat, **1**:287
avr4 and *avr9*, **4**:483–84, 483f
Arabidopsis, structure of, **1**:78
bcr, RT-PCR for detecting rearrangement of, **4**:328
Btk, **4**:127
bxn, encoding nitrilase, **6**:57
ced-3, **1**:301, 301f, 302
ced-9, **1**:301, 301f, 302
ces-1 and *ces*-2, **1**:301, 301f
c-*fos*, sensitivity to ethanol in null mutants, **2**:151
CFTR, cystic fibrosis transmembrane conductance resistance, **4**:331–32
c-*kit*, effects of mutations in, **3**:508–9
CLN3, genetic and cytogenetic maps, **3**:189f
clpP, of *Chlamydomonas reinhardtii*, **1**:353
c-*myb* and c-*src*, **4**:103
c-myc, *c-fos*, *c-myb*, **1**:26
cMYC, translocation of, **3**:159
COL2AI, **2**:280
alternative splicing in, **2**:281f
cry, coding for δ-endotoxins, **4**:394
DCC, chromosome 18, **1**:412; **6**:145
dnaJ, interaction of products with hsp70, **1**:319
Drosophila, **2**:123
E6, human papilloma virus, **1**:412
EIB, adenovirus, **1**:412
distinguishing retrotransposons from retroviruses, **4**:488
from human immunodeficiency virus, for vaccine delivery, **2**:437
erb B2, **3**:494–95
ERCC, **6**:161–62
ERCC-6, in Cockayne's syndrome, **2**:47
erd2, **5**:199
esy, for enniatin synthetase, **4**:213
fas, in an animal model for SLE, **1**:92
fem-3, in *Caenorhabditis elegans*, **1**:235
FLP, plasmid, **5**:5
FOS, mapping to chromosome 14, **3**:158–59
fragile X, *FMR 1*, **2**:321–23
GAL4-UAS, **2**:123–24
grpE, interaction of products with hsp70, **1**:319
H-19, **2**:445
HLA, **3**:499
hsp70, promoter region of, **2**:218
Igf-2 and *Igf-2r*, **2**:445
Ins-2, **2**:443
IPW, in Prader-Willi syndrome, **3**:161–62

KAI1, prostate cancer metastasis suppressor, **6:**138
katG, and isoniazid resistance, **4:**160–61
lacZ
  defined, **4:**167
  for identifying recombinant vaccinia viruses, **6:**166–67
  multiple cloning site within, **4:**251
lasI, for autoinducer synthase, **1:**108
Mash-2, **2:**443
MCC, **1:**413; **6:**145
mdm2, in human sarcomas, **1:**412
mdr, multiple drug resistance, **4:**327
mec-3, of *Caenorhabditis elegans*, **1:**233–34
MLC2, regulation of interactions between myosin head and actin, **4:**137
NAT2, for *N*-acetyltransferase 2, **4:**404, 404f
NF2, **4:**182–83
  schematic representation of mutations, **4:**185f
NFI, in colonic tumors, **1:**412–13
nod and *nif*, roles in nitrogen fixation, **5:**13
nuc-1, role in programmed cell death, **1:**301
oxyR, and isoniazid resistance, **4:**161
p16, inhibitor of CDK4 kinase, **3:**103
p53, **4:**287; **6:**141–42, 141f
  in bladder cancer, **2:**65
  in colon cancer, **1:**412
  mutations in cancer, **2:**63; **3:**495–96; **4:**155
  mutations in hepatocellular carcinoma, **3:**488–89
P, in albinism, **3:**163
Pax, **3:**68
PHRI, encoding photolyase in yeast, **6:**284
Plp, **4:**126, 126f
pol, **1:**27
ras, in lung cancers, **3:**493
Rb, in retinoblastoma and in osteosarcoma, **1:**251
RBI, in retinoblastoma, **5:**319, 320–22
recA, **5:**221, 222
recG, **5:**228
rKLK1, **5:**267
ruvA and *ruvB*, **5:**228
Snrp, mouse, **2:**443
SNRPN, imprinting of, **3:**161
SOD I, implication in familial amyotrophic lateral sclerosis, **4:**108
supF, **2:**74
  genetic selection using, **2:**66
sxl (sex-lethal), **2:**373
Tla and *Qa*, MHC, **3:**500
TCL-3, **3:**67–68
Tme (*T* maternal effect), **2:**443; **3:**517
TP53, in Li-Fraumeni syndrome, **2:**423
tufA
  Arabidopsis, **1:**83
  chloroplast genome in *Chlamydomonas reinhardtii*, **1:**349
unc-17 and *cha-1*, of *Caenorhabditis elegans*, conservation in humans, **1:**230
unc-52, alternative splicing of, **1:**235
unc-86 of *Caenorhabditis elegans*, **1:**233
WT1, **6:**142–44
XPCC, in xeroderma pigmentosum, **2:**47
YY1, mapping to chromosome 14, **3:**151
ZFY, pDP1007 probe recognizing, **5:**437
ZNF46, mapping to chromosome 14, **3:**151
Gene signature, **1:**204–5
Gene structure
  of collagens, **2:**280
  defined, **2:**278
  neurofibromatosis gene, **4:**176–77, 177f
Gene switching, and heart failure, **1:**290
Gene synthesis, with assembly PCR, **5:**451
Gene tagging
  defined, **4:**485
  T-DNA, **4:**493
Gene targeting
  animal model for, **4:**104f
  defined, **1:**407; **4:**101
  with electroporation, **2:**205–6
  by homologous recombination, **4:**101–3
  in-out method for, **4:**102–3
  in mice, **4:**103–5, 105f
    prion studies, **5:**62
Gene therapy, **3:**416; **4:**101, 332–33
  application of homologous recombination in, **5:**263–64
  for colon cancer, **1:**413–14
  electroporation used in, **2:**206
  myoblasts used in, **4:**147
  potential, for cystic fibrosis, **1:**435
  somatic, for phenylketonuria, **4:**412–15
  somatic and germ cell transfection, **6:**146
Genetic algorithms (GAs), **2:**412–14
  defined, **2:**407
Genetically directed representational difference analysis (GDRDA), for alcohol and drug sensitivity studies, **2:**151
GENETIC ANALYSIS
  of breast cancer, **1:**221–25
  of colon cancer, **1:**410–11
  of the *Drosophila melanogaster* genome, **2:**124–26
  of learning in *Drosophila melanogaster*, **4:**57
  mammalian, techniques for, **3:**513–17
  OF POPULATIONS, **2:391–401**
  in vitro protein translation systems, **5:**336
Genetic change, defined, **2:**407
Genetic code, **6:**112
  mitochondrial DNA, **4:**80
Genetic control, effects of pheromones, in *Drosophila*, **2:**144–45
Genetic counseling
  DNA profile used in, **2:**60
  in hemophilias, **3:**41
  for risk of breast cancer, **1:**223–25
  for risk of Gaucher disease, **2:**352
Genetic diagnostic, defined, **3:**252
Genetic disease, defined, **3:**84
Genetic disorders
  Alzheimer's disease, **1:**43–46
  apoE deficiency, **3:**414
  atherosclerosis associated with, **1:**286
  in cardiomyopathies, **1:**289
  central diabetes insipidus, **2:**23–24
  chromosome 15 implicated in, **3:**166–67t
  chromosome 18 regions associated with, **3:**210f
  congenital contractural arachnodactyly, **3:**162
  with defects in DNA repair, **2:**46–47
  Down syndrome, **2:**118–21
  and genomic imprinting, **2:**445
  in gonadotropin releasing hormone gene expression, **2:**292
  heart failure related to, **3:**1–2
  hemoglobin mutations, **3:**19
  hemoglobin structure related to, **3:**16–17, 17t
  human, **3:**231–32
    involving chromosome 11, **3:**133f
  hypothyroidism, **4:**356
  I-cell disease, defect in mannose-6-phosphate tag formation, **5:**203
  involving aging, **1:**302; **2:**80
  involving P450-encoding genes, **1:**437
  involving peroxisome biogenesis, **5:**206
  Li-Fraumeni cancer syndrome, **6:**138
  mapping of
    to chromosome 11, **3:**128–32t
    to chromosome 14, **3:**152–55t
    to chromosome 9, **3:**111t
  mapping loci for, **3:**239
  Marfan syndrome, **3:**162
  mitochondrial point mutations in, **4:**82–83
  molecular pathology of, **4:**331–32
  motor neuron disease, **4:**105, 106–7
  nephrogenic diabetes insipidus, **2:**20
  non-insulin-dependent diabetes mellitus, **2:**38–39
  PCR for identifying mutations in, **4:**339

Genetic disorders (*cont.*)
prevalence of, **5**:45, 45f
prion disease, **5**:71
retinoblastoma, **5**:319–22
simple and complex, **3**:222, 222t
von Hippel-Lindau syndrome, **6**:144–45
See also Congenital disorders; Familial disorders
Genetic distance, **3**:233–34, 234f
GENETIC DIVERSITY
and animal patents, **6**:81
IN MICROORGANISMS, **2:402–6**
Genetic drift, **2**:393
defined, **5**:45
and the founder effect, **5**:47–48
in small populations, **6**:7
and tandem repeats, **6**:6
See also Founder effect
Genetic engineering
of bacteriorhodopsin-based compounds, **1**:138–39
in biological waste treatment, **1**:122–23, 123f
for fungal strain improvement, **2**:345–46
for improvement of crops, **4**:456
for insect control, **4**:395
of mice, **1**:64
of polymers, **5**:26–31
of recombinant vaccines, **6**:166–67
Genetic evolution, defined, **2**:407
Genetic instability, defined, **5**:26
GENETIC INTELLIGENCE, EVOLUTION OF, **2:407–14**
Genetic interference, **3**:99, 101
Genetic linkage map
for chromosome 9, **3**:99–101
for chromosome 14, **3**:142–44
for chromosome 15, **3**:170, 171t
defined, **3**:97; **4**:173
in Huntington's disease, **4**:199
in Marfan syndrome, **3**:162–63
mouse
for chromosome 18, **3**:204f
development of, **4**:123f
in PrP mutations, **5**:71–72, 71f
Genetic manipulation
for studying viral assembly and budding, **6**:189, 189t
of yeasts, **6**:311–12
Genetic map
chromosome 1, **3**:95
chromosome 10, **3**:116f
chromosome 11, **3**:132–35
chromosome 16, **3**:175–81
chromosome 18, **3**:192, 194–96f
defined, **2**:169; **3**:112, 191; **4**:121
of DNA markers flanking site FRAXA, **2**:322f
mouse, compared with physical map, **4**:125, 125f
mouse genome, **4**:122–27
nuclear, of *Chlamydomonas reinhardtii*, **1**:350–51f
versus physical map, **3**:123–24
relationship of denaturation to, 4:314
Genetic marker
defined, **1**:221
in linkage analysis, **1**:222–23
Genetic model, for muscle assembly, *Caenorhabditis elegans*, **1**:236
Genetic polymorphism, defined, **6**:259
Genetic profiles, **2**:51
and substance abuse in humans, **2**:146–47
Genetic recombination
defined, **5**:220
participation of ligases in, **3**:392, 392f
See also Homologous recombination
Genetic risk, defined, **3**:239
GENETICS, **2:415–19**
of *Caenorhabditis elegans*, **1**:228–29
defined, **2**:407; **4**:148
of δ-endotoxins, **4**:394
of drug addiction and alcoholism, **2**:146–55
of hemophilias, **3**:29
human biochemical, **1**:113–17
of Huntington's disease, **3**:247–48
of immunotoxicity, **3**:300
and immunotoxicology testing, **3**:300, 300t
of mitochondrial DNA, **4**:81–82
molecular, of hepatocellular carcinoma, **3**:486–87
of muscle-related myopathies, **4**:146–47
of neurofibromatosis 2, **4**:182
and regulation of DNA repair, **6**:161–63
Genetic selection
defined, **2**:66
for increasing recombinant proportions in a library, **2**:70
Genetic Stock Center, *Escherichia coli*, **2**:169
Genetic systems, of *Chlamydomonas*, **1**:348–53, 353t
GENETIC TESTING, **2:419–30**
Genetic tool, the mouse as, **4**:121–22
Genetic transformation, in aging studies, **1**:11–13
Genetic typing, **3**:259
GENETIC VACCINATION
defined, **2**:430
GENE TRANSFER TECHNIQUES FOR, **2:430–39**
Genetic variability, in expression of cytochrome P450, **1**:194
Gene transfer
Agrobacterium-mediated, **4**:451–52, 452f
biological methods of, **2**:436–37
defined, **4**:450; **6**:64
direct, defined, **2**:430
with electroporation, **2**:198
by electroporation, number of pulses for, **2**:204
electroporation application in, **2**:204–5
horizontal, **2**:402
from insects to plant, **1**:78
mechanisms and effectiveness, **3**:423
overview, **6**:73–74
physicochemical methods of, **2**:432–36
in the presence of an electric field, **2**:200
for treating Gaucher disease, **2**:351
using chromosome fragments, **1**:383
Gene vectors, lipid-based, **3**:415–26
Genitourinary system, involvement of, in cystic fibrosis, **1**:426
Genogram, **1**:225f
Genome Consortium, yeast artificial chromosome source, **1**:232
Genome Database (GDB), **3**:97, 126
Genome map
Arabidopsis, **4**:491
physical
chloroplast, of *Chlamydomonas reinhardtii*, **1**:352f
defined, **4**:167
mitochondrial, of *Chlamydomonas reinhardtii* and *C. smithii*, **1**:353, 353f
Genome mapping, **3**:216–17
and chromosome fragmentation, **6**:37
long-range, shotgun sequencing for, **2**:78
Genome mutation, DNA variation from, **3**:228–29
Genome Project, **3**:520
for *Caenorhabditis elegans*, **1**:230–31
defined, **3**:510
parasite, **4**:306–12
See also Human Genome Project
Genomes
aberrations of, in cancer, **1**:251
Arabidopsis, **1**:76–84
archaea, **4**:74
bacteriophage, structures of, **6**:203–4, 203f
Caenorhabditis elegans, **4**:169
chloroplast, **4**:488–89
defined, **1**:76; **2**:169, 216, 360; **3**:112, 227, 510; **4**:121, 148; **6**:138
Drosophila, **2**:121–30
Escherichia coli, **2**:169–72
evolution of, **3**:512–13
HIV-1, **1**:35f; **6**:194f

HSV, electron microscope representation, **2:**184f
human, **3:**227–28
body expression map, **1:**203–6
compositional distribution of fragments, **2:**363
fragile sites in, **2:**314t
origin of modern, **4:**89
interstitial $(TTAGGG)^n$ sequences in, **6:**36, 36f
mitochondrial, **4:**79
mutation frequency, **1:**9–10
of plants, **4:**489
mycobacterial, **4:**157–59
nuclear, *Chlamydomonas reinhardtii*, **1:**348
picornavirus, **5:**15f
plastid, **4:**466, 488–89
poliovirus, **5:**15f
respiratory syncytial virus, **5:**287, 287f
stabilization by telomeres, **6:**34
stress responses within, **2:**220–21
distribution of damage from UV radiation, **6:**153–55
tomato, altering, **6:**54, 54f
viral
defined, **6:**217
hepatitis B virus, **3:**483, 483f
hepatitis C virus, **3:**485, 485f
herpesvirus, **6:**173–74
nucleoprotein core of, **6:**184
packaging of, **6:**217
release for transcription, **6:**224
replication of, **6:**198
size of and dependence on the host cell, **6:**194–95
vaccinia DNA, **6:**180–81
vaccinia virus, **6:**194f
yeast, **6:**305–6, 306f
Genome scanning, **5:**305–6, 305f
defined, **5:**304
Genome Sequences Databank (GENBANK), **3:**517
Genomic base distributions, denaturation studies for determining, **2:**18–19, 18f
Genomic DNA, quantification of, **2:**53–54
Genomic fluidity, **2:**220
defined, **2:**216
GENOMIC IMPRINTING, **2:439–46**
Genomic inserts, replication of transferred nucleic acids as, **4:**450
Genomic instability, in colon cancer, **1:**413
Genomic library, **2:**67
construction of, **2:**69–72, 70f
Genomic mechanisms, hormone control of, **2:**212–13
Genomic restriction maps, direct construction of, **3:**217–20
Genomic sequencing, for detecting DNA-protein interactions in vivo, **4:**471
Genomic subtraction
Arabidopsis gene cloning, **1:**83
using *Not* I linking libraries, **2:**77f
Genotype
correlation with phenotype
in neurofibromatosis, **4:**178–79
in phenylketonuria, **4:**410–11, 410t
defined, **1:**113, 227; **2:**146; **3:**142, 233; **4:**173
errors, **3:**238
for identifying poor metabolizers of debrisoquine, **4:**400
and phenotype, for cytochrome P450 alleles in debrisoquine metabolism, **4:**402f
phenylalanine hydroxylase heterogeneity, **4:**409
selective breeding for specific, **2:**148
Genotype-by-environment interactions, in laboratory studies of aging, **1:**10
Genotype/phenotype map, chromosome 16, **3:**186
Geranylgeranylation, **5:**55
Germ cell transfection, **6:**146
Germ line
for gene therapy, ethical questions about use in humans, **3:**231
immortality of, **2:**80
mutation of, **1:**5
in neurofibromatosis 2, **4:**183
transformation of, *p*-element controlled, **2:**129
Gerontogenes, **1:**11
Gerstman-Sträussler-Scheinker syndrome (GSS), **4:**199
familial, **5:**58
modeling in transgenic mice, **5:**64–65
Gibberellins, **4:**492
synthesis by molds, **3:**440
Gibbs free energy, **1:**305
of antigen-antibody reactions, **3:**283
criterion for spontaneity, **2:**225–27
defined, **2:**223
for folded and unfolded proteins, **5:**150
predicting RNA secondary structure with lowest, **5:**364
of selected methanogenesis substrates, **4:**68t
Giemsa dark bands, of chromosomes, **1:**371
Gilbert's syndrome, **4:**405
Gilles de la Tourette syndrome (GTS), gene for, mapping to chromosome 18, **3:**214
G-inhibitory protein (G_i), **1:**289
Glaucoma, juvenile onset, open-angle, **3:**91
GLC (gas-liquid chromatography), **1:**266
Glial fibrillary acidic protein (GFAP), electron microscope image, **2:**190f
Glial growth factor, **2:**466
Glioblastoma, controlled drug release system for treating, **5:**36
Glioma
NF2 gene in, **4:**185
optic nerve, in neurofibromatosis 1, **4:**175
transplantable, **1:**252
Global alignment, **5:**432–33
Global transcriptional activator, regulation of the virulence genes by, **1:**108
Globin fold, **3:**3, 12–13, 15f
β-Globin gene
perpendicular DGGE pattern, **2:**4, 4f
point mutation, perpendicular DGGE, **2:**16–17
Globins, **3:**3
assembly of monomers, **3:**12–13
β-, of myoglobin, **3:**21f
gene for, biosynthesis of, **3:**13–16
genes for, and disorders associated with mutations in, **4:**332
primary structure of subunits, **3:**4–5t
Globoid cell leukodystrophy, gene for, mapping to chromosome 14, **3:**158
Globotetraosylceramide, **1:**280f
Globular proteins, defined, **5:**115
Globulins, steroid hormones bound to, **5:**471
Glomerulogenesis, PDGF-R role in, **5:**270
Glomerulonephritis
poststreptococcal, **1:**105
up-regulation of PDGF receptor in, **5:**270
Glucanase, accumulation in response to plant pathogens, **4:**480
Glucans, **5:**33–34
Glucocerebrosidase
deficiency of, in Gaucher disease, **2:**347, 424
haplotypes, **2:**348t
Glucocorticoid receptor (GR)
complex with glucocorticoid response element (GRE), structure, **5:**133
interaction with vitamin B_6, **6:**244
Glucocorticoid-remediable aldosteronism, **2:**214
Glucocorticoids
cancer chemotherapy using, **1:**256
effects on skin, **2:**269
nuclear receptor, **2:**208
role of, **5:**481–82
synthesis of, **5:**479–80
and tyrosine hydroxylase regulation, **1:**217
Glucokinase
gene for, on chromosome 7, **2:**38–39
induction by biotin, **6:**242–43

Glucokinase (*cont.*)
mutant, effect on insulin secretion, **2**:214
in the pancreatic beta cell, **2**:39f
Glucosamine, link to phosphatidylinositol, **4**:440
Glucose
biosensor for, **1**:184
metabolism of, **1**:161
production of, from glycogen in hepatic cells, **3**:48
Glucose-6-phosphate dehydrogenase, deficiency of, **4**:405
Glucose analyzer, biosensor in, **1**:141–42
Glucose oxidase (GOx), **1**:184
chiral specificity of, **1**:341
Glucosylceramide, in Gaucher disease, **2**:347
β-Glucuronidase (GUS)
for assaying plant genes, **4**:470, 470f
reporter gene, in plant transformations, **4**:490
Glucuronidation, in xenobiotic metabolism, **6**:269–71, 271f
Glucuronides, from hydroxylated steroids, **5**:471
Glucuronyltransferases, **1**:197–98; **6**:269
Glutamate binding, and long-term potentiation of memory, **4**:57f
Glutamate dehydrogenase, **1**:400
Glutamate receptors, ionotropic and metabotropic, **5**:232–33, 233t
Glutamic acid, invariant, position 12 of EF-hand proteins, **1**:241
Glutamic acid carboxymethylation, substrate-directed posttranslational modification of sites for, **5**:52
Glutamic acid decarboxylase (GAD), potential role in insulin-dependent diabetes mellitus, **2**:36
trans-Glutaminase, role in cell death, **1**:302
Glutamine synthetase (GS)
inhibition by phosphinothricin, **4**:395
reaction not involving the chiral center of substrates, **1**:342
reversible regulation of, in nucleotidylation cascades, **5**:50
γ-Glutamylpeptidase, **1**:402
γ-Glutamyltranspeptidases, **2**:456, 456f
processing of glutathione conjugates by, **1**:200
Glutathione (GSH)
as an enzyme cofactor, **4**:354
reaction with DNA radicals, **3**:357
Glutathione peroxidase (GSHPX), selenium in, **2**:333, 333f
Glutathione reductase, **2**:333, 333f
Glutathione transferases, **1**:199–202
detoxification of epoxides by, **1**:197
metabolism of methyl bromide via, **1**:201f
polymorphism in, **4**:404–5
in xenobiotic metabolism, **6**:273–75, 275f
N-Glycanases, **2**:447
Glycan release, carbohydrate analysis using, **1**:272
Glycans, **5**:33
defined, **1**:266; **2**:446
detection of, **1**:271–72
glycoprotein, **1**:270f
N-, functions in glycoproteins, **2**:462t
O-, **2**:458
core classes of, **2**:459t
functions of, in glycoproteins, **2**:462t
Glyceraldehyde 3-phosphate dehydrogenase (GAPDH), **2**:245, 247
Glyceraldehyde 3-phosphate dehydrogenase-like 11 (*GAPDHL11*), mapping to chromosome 18, **3**:207
Glyceride, metabolism of, **3**:456
Glycerol phosphatidylinositol (GPI), linkage to proteins on cell surfaces, **4**:439–40, 439f
Glyceromannoheptose, **3**:466
Glycerophospholipids, **3**:457, 457f
Glycine
asymmetric derivatization of, **1**:53–54
in collagen, **5**:28
invariant, position 6 in EF-hand proteins, **1**:241
as a neurotransmitter, **5**:237–38
Glycine-rich proteins (GRP), **4**:480
GLYCOBIOLOGY, **2**:**446–57**
Glycocholate, sodium, absorption enhancement by, **2**:167
Glycoconjugates
defined, **1**:278; **2**:446
mammalian, **1**:280–83
structural rules in, **2**:451–53
Glycogenolysis, regulation by epinephrine, **5**:171
Glycogen phosphorylase
liver, deficiency of, **3**:157
muscle
allosteric regulation of, **2**:237, 239f
deficiency of, **4**:146
deficiency of, mapping to chromosome 11, **3**:157
Glycogen storage disease type VI, gene mapping to chromosome 14, **3**:157
Glycogen storage disease Type VII, **4**:146
Glycohormones, **2**:454
defined, **2**:446
Glycoinositol phospholipids, anchors for prion protein, **5**:68f
Glycolipids
analysis of, **1**:269, 271
defined, **2**:347; **3**:426
as surfactants, **3**:434
Glycolysis, generation of ATP during, **1**:307n
Glycolytic pathway, **2**:229
phosphofructokinase of, regulation of, **2**:235
reactions of, **2**:231t
regulation by metabolites, **5**:50
Glycophorin, membrane, **1**:452
Glycoprotein hormone family, **2**:287
GLYCOPROTEINS, **1**:282–83
adherins, **1**:239
analysis of, **1**:269–70, 270f
antibodies as, **1**:66
α_1-antichymotrypsin, mapping to chromosome 14, **3**:155
apoE, **3**:414
apoJ, **3**:414
apolipoprotein(a), **4**:504
defined, **1**:292
E-cadherin, **1**:458
ELAM-1, **2**:455
envelope, for varicella-zoster virus and cytomegalovirus, **6**:175, 175t
fibronectins, **2**:282
fusogenic, incorporation in liposome vectors, **3**:418
hemagglutin in, **6**:231–32
hydroxyproline-rich, **4**:480
interleukin-15, **3**:344
lipopolysaccharide-binding, **3**:470
lipoprotein receptor, **3**:406
MIS, **2**:288–89
mucins, **1**:282–83
Müllerian inhibiting substance, **3**:509
N-CAM, **1**:295
nicotinic acetylcholine receptors, **5**:231
N-linked, **2**:447
O-linked, **2**:447
osteopontin, **1**:239
plasma kallikrein, **3**:382
prion protein, **5**:57–58
sec66p, **5**:198
SECRETORY, **2**:**458–62**
serum, antibodies as, **3**:276
single-chain transmembrane, **3**:43
thrombospondin, **3**:168–69
virus-encoded, directing to a cellular membrane, **6**:186–88
Glycosaminoglycans, **2**:448
in proteoglycans, **2**:453
Glycosidases, **2**:461
O-Glycoside, defined, **1**:267
Glycosphingolipids, **1**:280–82, 280f; **2**:448
metabolism of, **3**:458–60
tetrasaccharide cores of, **2**:453f
Glycosylases
AP sites created by, **2**:82
DNA, specificity of, **6**:285

Glycosylation
of aromatase cytochrome P450, **1**:439
cotranslational modification by, **5**:53–54
terminal, of peptide hormones, **4**:343
in yeasts, **2**:276
Glycosyl bond, conformations of, **5**:374, 374f
Glycosylphosphatidylinositol (GPI)
attachment of N-CAM via, **1**:295
defined, **4**:416
phospholipase D specific for, **4**:431
and protein covalent modification, **3**:461–62
truncated cadherin attachment via, **1**:295
Glycosylphosphatidylinositol-phospholipase C (GPI-PLC), **4**:424
link between proteins and the plasma membrane, **4**:428
Glycosyltransferases, **2**:460–61
defined, **1**:293
GMP140 (P-selectin), **1**:296
Goat, XX male, **5**:444
Gold, crystalline, substrate for imaging DNA, **5**:407–9
Goldblatt hypertension, chronic, **5**:266
Golgi apparatus, enzymes for phospholipid biosynthesis in, **4**:433
Golgi complex
biosynthesis of glycans in, **2**:459
and the endoplasmic reticulum, bidirectional exchange of proteins, **5**:201f
protein flow to, from the endoplasmic reticulum, **5**:199
vesicle budding in, **5**:201
Gomperzian growth, **1**:258
Gonad, defined, **1**:227; **5**:434
Gonadoblastoma, in XY gonadal dysgenesis, **5**:445
Gonadotropin receptors, **2**:287
Gonadotropin-releasing hormone
defined, **2**:285
and female reproduction, **2**:286
Gonadotropin-releasing hormone receptor, pituitary, **2**:286–87
Gonadotropins
defined, **2**:285
female reproduction roles of, **2**:287
Goose, hemoglobin of, **3**:27
GO repair process, **4**:151
Gorlin-Goltz syndrome, mapping to chromosome 9, **3**:104
G protein-coupled receptor kinases (GPKs), **5**:234
G protein-coupled receptors, **2**:208–9f; **5**:236f
adenylate cyclase activation by, **4**:355
classification of, **5**:237
family of, β-adrenergic receptor in, **4**:37–38, 37f
gene for, localization to chromosome 15, **3**:169
hormone, V_2, mutation in, **2**:27
for neurotransmitters, **5**:237–38
structure, function, and regulation of, **5**:233–34, 234f
G proteins, **2**:210f; **3**:2
of *Caenorhabditis elegans*, **1**:228
defined, **2**:207, 285; **3**:1, 42
gene for, chromosome 20, **3**:164
glutamate receptors coupled to, **5**:232
heterotrimeric, signal transduction by, **3**:41–48
receptors linked to, **3**:44–45, 44f; **4**:30, 31f; **5**:233–38, 238t
binding by interleukin-8, **3**:340
receptor superfamily, interleukin-8 receptor, **3**:339
as second messengers, **2**:209–10
species of, **5**:235–37
structure of, **5**:236f
substrate in prenylation, **5**:55–56
α subunit of, N-end rule-dependent degradation of, **3**:351–52
transducin, **5**:324
trimeric, and binding of coat proteins, **5**:201
Gradient
defined, **3**:76
in high performance liquid chromatography, **3**:77
Gradient gel electrophoresis
for analyzing high density lipoprotein subfractions, **3**:477–78, 478t
for analyzing low density lipoprotein subfractions, **3**:478
for detecting single base changes in DNA, **2**:397
for following DNA denaturation, **2**:4–5, 6t
Graft rejection, **3**:499
Graft-versus-host disease (GVHD), animal model of, **1**:92
Gramicidin
A, **6**:128
induction of nonbilayer lipid structures by, **4**:48
S
structure of, **4**:210f
for transfection, **3**:422–23
Gramicidin synthetase, **4**:209
gene for, **4**:213
Gram-negative bacteria
cell envelope, **3**:463f
lipopolysaccharides of, **3**:463
defined, **2**:155; **3**:462
Gram-positive bacteria
defined, **2**:155
mycobacteria, **4**:156
Granular cells, defined, **5**:264
Granulocyte colony-stimulating factor (G-CSF), **2**:464
receptor for, structure of, **3**:337
Granulocyte/macrophage-colony stimulating factor (GM-CSF), **1**:447; **2**:469
clinical use, for neutrophil recovery after chemotherapy, **2**:469
eye development and, **2**:464
receptors for, **3**:337–38, 338f
gene encoding, on the X chromosome, **3**:503
Granulocytes, defined, **3**:295
Granulomas, interferons involved in formation of, **3**:328
Graphite fibers, in orthopedic prostheses, **5**:39
Graves' disease, **1**:86
Gravimetric sensors, **1**:191–92
Green fluorescent protein (GFP), as a reporter for promoter analysis, **1**:234
Greenhouse gas, methane as, **4**:68
Greiss reaction, nitric oxide assay, **4**:201
G-rich strand asymmetry, at telomeres, **6**:29
Grooves, major and minor, **6**:327
DNA structure, **2**:106
Gross energy (GE), of foods, **4**:258
Group intron, modeling the structure of, **5**:385, 387f
Group translocation, **6**:130
defined, **6**:125
Growth
active phase of, harvesting for electroporation, **2**:204
of crystals, **3**:310–11
in gels, **3**:315–16
of *Taxus cuspidata* cell suspension, **6**:16, 16t
Growth additives, in crystal chemistry, **3**:312–14
Growth conditions, for protein expression in vitro, **5**:137
Growth cone collapse, **1**:297, 298f
Growth factor pathway, mutation in, and oncogenes, **2**:215
GROWTH FACTORS, **1**:441, 442; **2:463–71**; **5**:238
autocrine, in lung cancers, **3**:494–95
B-cell, interleukin-14 as, **3**:343
defined, **1**:441; **2**:463; **5**:229
fetal, defined, **2**:439
in hepatocyte proliferation, **3**:481–82, 482t
nomenclature of, **1**:443t
peptide, **4**:355
receptors for, **1**:445; **5**:238–40
annexin interaction with, **1**:242
tyrosine kinase activity, **5**:238–41, 241f
regulation of ECM expression by, **2**:279

GROWTH FACTORS (*cont.*)
in the renal system, **5**:269–72
stimulation of release, by oxidized low density lipoproteins, **3**:479
and testicular function, **3**:507–10
and tyrosine hydroxylase regulation, **1**:217
Growth hormone
bovine, unfolding and refolding of, **5**:77
exogenous application of to fish, **6**:92
expression in milk of transgenic mice, **6**:105
human
MALDI spectra, **4**:17f
transmission of prions in therapy using, **5**:70
isolated growth hormone deficiency type IA, **2**:213
receptor for, structure of, **3**:337
recombinant human, ESI mass spectrum, **4**:13
in transgenic fish, **6**:91–95
in transgenic mice, **6**:70
Growth hormone releasing factor (GRF), in transgenic animals, **6**:70
Growth law, bacterial, **1**:101
Growth medium, and recombinant protein yield, **2**:278
Growth rates, bacterial, **1**:101
Growth regulators
interaction with interferons, **3**:326
in plant disease, **4**:482
Gβγ subunits, protein G, **3**:45
GTG banding, for chromosomal analysis, **1**:382
Gtp, role in exocytosis, **4**:45
GTPase-activating protein (GAP), **2**:209, 209f, 209–10, 209f
GTPases, defined, **3**:42
GTP-binding proteins (G proteins)
in endocytosis and biosynthesis, **5**:204
Golgi complex, **5**:201
regulatory, **2**:209–10; **4**:31f
defined, **5**:229
See also G proteins
Guanidine hydrochloride, denaturant, for dissolving inclusion bodies, **5**:84
Guanidine linkage, *N*-substituted, oligonucleotide analogues, **4**:279–80, 280f
Guanine (G), **2**:415
methylation of, and age, **1**:7
photooxidation of, **6**:209
Guanine nucleotide binding protein, expressed in sensory olfactory neuroepithelia, **3**:207
Guanylate cyclase-activating protein, homology with recoverin, **1**:242
Guanylyl cyclase, **4**:30
binding of NO to, **5**:268–69
soluble, **4**:200
Guthrie test, for phenylketonuria, defined, **4**:405
Gyrase, DNA
recombination mediated by, **2**:90, 404
replication mediated by, **2**:87
supercoiling induced by, **5**:1

Habitat, and genetic diversity, **2**:393
Habituation, nonassociative reflex decrement, defined, **4**:54
Habrobracon juglandis, radiation studies using, **1**:6
Haemophilus influenzae, lipopolysaccharides of, **1**:279
Hageman factor
activation of the kallikrein-kininogen-kinin systems, **3**:378f, 382
defined, **3**:377
Hair, extraction of DNA from, **2**:53
Hairpin loop
antiparallel structure, HIV-1 reverse transcriptase, **1**:30
β-, HIV protease, **1**:15
defined, **5**:360, 368
formation in DNA replication, **5**:459, 459f
RNA, **5**:361–63
size and flexibility of, **5**:371
transactivation response element, predicting the structure of, **5**:383–85, 383f
tRNA anticodon, **5**:370–71, 370f
Hairpin ribozyme, **5**:340f
Hairy cell leukemia (HCL), interferon-α therapy for, **3**:330
Haldane map function, **3**:113, 235, 235f
Half-life
defined, **5**:209
of radionuclides, **5**:211
Halobacterium halobium, bacteriorhodopsin of, **1**:109, 137–39, 149
Halogenated organic compounds
defined, **1**:118
reductive dehalogenation of hydrocarbons, **1**:122
Halomethanes, conjugation with glutathione, role of glutathione S-transferase 1, **4**:405
Halophiles, **4**:69
lipids of, **3**:429, 431f
Halopropionic acid, optically active, lipase-catalyzed production of, **2**:250
Halorhodopsin, **1**:111
Halo structures, **2**:379
Hamartomatous syndromes, **1**:410
Hammerhead ribozyme, **5**:340f
three-helix loop of, **5**:362
Hammett analysis, of transition states in organic solvents, **2**:257
Haploid organisms, **2**:402
yeast, doubling time of, **6**:304
Haplotype
for analysis of the founder effect, **5**:47–48
defined, **3**:246; **4**:405; **5**:45
phenylalanine hydroxylase gene, **4**:408–9, 408f
Haptens, defined, **2**:378; **3**:275
Hardware architecture, automated genome research, **5**:395–97, 397f
Hardy-Weinberg equilibrium, **3**:236
Hardy-Weinberg rule, **2**:401
Hartley-Shannon theory of communication, **6**:51–52
Harvey murine sarcoma virus, **3**:493
Harvey's rule, in combination chemotherapy, **1**:265
Hashing, **5**:424
defined, **5**:420
HC toxin, structure of, **4**:210f
HC toxin synthetase, **4**:209–10
Heart disease
associated with Down's syndrome, **2**:119
ischemic, **1**:292
inverse correlation with α-tocopherol, **6**:238
and low density lipoproteins, **3**:479
HEART FAILURE, **1**:288–92
defined, **1**:283
GENETIC BASIS OF, **3**:**1–2**
Heart valves, from silicone rubber, **5**:37
Heat
damage to DNA from, **2**:40
removing purines and pyrimidines, **2**:41
and energy, **1**:304–5
Heat increment of feeding, **4**:258
Heat shock, defined, **1**:315
Heat shock gene, **2**:217–18
transcription patterns, **1**:26
Heat shock proteins (hsps), **2**:217–18
dissociation of, on steroid binding to receptor, **5**:478
functions of, in nuclear receptors, **2**:207–8, 207f
genes for, mapping to chromosome 14, **3**:155
hsp10, **1**:320
hsp70, **5**:90, 90f
role in mitochondrial import of proteins, **5**:205, 205f
targeting of proteins to the endoplasmic reticulum membrane, **5**:198
hsp90, **1**:321–22
role in hormone steroid binding, **5**:482–83
immunological reactivity of, **4**:159
involvement in photosynthesis, **1**:358

role in protein folding, **5**:86
small, **1**:322
steroid receptor stabilization by, **5**:471
synthesis in response to stress, **5**:336–37
targeting to the endoplasmic reticulum by, **5**:198
Heat shock regulatory element (HSE), **2**:218
Heat stability of isoenzymes
alkaline phosphatase, **3**:374–75
separating using, **3**:368
Heavy chain class switch, **3**:282
Heavy chains, gene loci coding for, **3**:282
Heavy metals
autoimmunity induced by, **1**:91–92
induction of autoantibody response by, **1**:86
Heavy metal shadowing, defined, **2**:181
Heinz bodies, **3**:17
Helical DNA, structure of, **2**:105–7
Helical motifs, scaffold for generating recognition molecules, **5**:251
α-Helical proteins, polymers based on, **5**:29
Helical RNA, defined, **5**:368
Helicases
DNA, **2**:89, 89t; **5**:352, 354
in plasmid replication, **5**:3
RecQ, **5**:228
replication protein A interaction with, **5**:464
DnaB, **2**:88
Helix
α-, **5**:117–18
defined, **1**:27, 148; **5**:159
estimation of, using circular dichroism, **1**:389
formation from nonfolded proteins, **5**:148
in interleukin structures, **3**:334, 334f
model of, top and side views, **5**:167f
amphipathic, **4**:436
left-handed, of DNA, **2**:115
twisted, in DNA, **2**:106, 106f
Helix-destabilizing proteins, defined, **5**:352
Helix-loop-helix motif
basic, in plants, **4**:467, 467t, 468–69
Ca^{2+}-binding motif, **1**:240
CNEP-B antigen, **5**:417
in muscle proteins, **4**:145
relation to basic leucine zipper group, **5**:130, 130f
See also EF-hand proteins
Helix swap experiment, **5**:275
Helix-turn-helix motif, **5**:129–30, 275
defined, **5**:128
Helminths, infection by, genetic protection against, **2**:438
Helper T cell (T_h)
CD4 protein of, **2**:431
CD4 receptor, and viral envelope binding, **1**:35
cytokines secreted by, **3**:337t
defined, **3**:499
generation of, role of interleukin-12, **3**:343
interleukins produced by, **3**:335
responses of, quantitation for evaluating vaccines, **6**:183
Th1 and Th2, **2**:431
Hemagglutinin (HA)
and agglutination of erythrocytes caused by influenza virus, **6**:231–32
role in membrane fusion, **4**:52
Hematologic malignancies, diagnosis of, **2**:64–65
Hematopoiesis
c-*myb* gene associated with, **4**:103
defined, **3**:331
interleukin-1 associated with, **3**:340
interleukin-6 stimulation of, **3**:342
regulatory activity of interferons, **3**:327
Hematopoietic growth factors, **2**:469–71
biological actions of, **2**:470t
list, **1**:443t
Hematopoietin
defined, **3**:331
interleukins classified as, **3**:334
Hematopoietin receptor
defined, **3**:331
families, **2**:465
interleukin receptors classified as, **3**:334
members of, **3**:337
structures of, **3**:338f
Hematoporphyrin derivative (HPD), as photosensitizer, **6**:212, 212f
Heme-controlled repressor (HCR), **6**:124
Heme pocket, **3**:7f
cooperative binding and allostery of, **3**:12
hemoglobin disorders related to, **3**:19
myoglobin, **3**:21f
Hemes
biosynthesis of, **1**:401–2
coenzyme functions of, **2**:233
of cytochrome P450, **1**:435
electron density of, **2**:194
protein, **4**:62
synthesis of, **6**:62–63
Heme-thiolate protein, **1**:436
Hemicellulases, role in plant pathology, **4**:482
Hemidesmosomes, **2**:265, 265f
Hemisuccinate, defined, **5**:485
Hemochromatosis, uric acid degradation products in, **2**:336
Hemochromes, **3**:5
Hemodialysis membranes
cellulose, **5**:34
poly(vinyl alcohol), **5**:43
HEMOGLOBIN, **3:2–20**
allostery of, **2**:237
analysis by reversed phase chromatography, **3**:82f
β-, expression of the gene encoding, **2**:417f
cooperativity for oxygen binding, example, **1**:131
ferric, **3**:6f
functional properties of, **3**:6–8, 8t
GENETIC ENGINEERING OF, **3:20–28**
iron in, **4**:61
oxygen-binding curve, **2**:239f
susceptibility to damage from light, **6**:210
S variant, **5**:47
T state, cooperativity model, **3**:21, 27
Hemoglobin H disease, **2**:425
Hemoglobinopathy
autosomal recessive conditions, **2**:424
defined, **2**:419
Hemolysins
α-, **1**:140
operons containing, **1**:107
HEMOPHILIA, **3:28–41**
A, **2**:426–427
L1 insertion in, **3**:245; **6**:135
as a single-gene disorder, **3**:230
X-linked, LINEs in, **6**:135
Alu insertions in, **3**:244
B, **2**:426–27
founder effect, **5**:47–48
Hemopoiesis
control of, **1**:447–48, 448f
role of cytokines in, **1**:441
Hemopoietic growth factors, nomenclature of, **1**:443t
Hemoproteins, defined, **1**:435
Hepadnaviruses, packaging of, **6**:221–22, 221f
Heparin, **2**:456–57
binding of
to poly(2-hydroxyethyl methacrylate), for biomedical coatings, **5**:42
to poly(vinyl alcohol), for biomedical coatings, **5**:42–43
enzyme sensors for measuring, **1**:186–87
as a ribonuclease inhibitor, **5**:330
and tyrosine hydroxylase regulation, **1**:213–14
Heparin-sepharose affinity chromatography, for analysis of high density lipoprotein subfractions, **3**:478
Hepatic acute phase, of the immune response, **3**:298

Hepatic chylomicron remnant receptor, **3:**405
Hepatic lipase
expression in transgenic animals, effect on HDL, **4:**505–6
gene for, localization to chromosome 15, **3:**169
Hepatic triglyceride lipase (HTGL), **3:**407
Hepatitis B core antigen, production in yeast, **6:**324
Hepatitis B virus (HBV)
cytokine role in action of, **1:**449
cytokines used in therapy for, **1:**450
defined, **3:**480
duck, effect of phosphorothioates on infection by, **1:**75
electron microscope image, **2:**192f
encapsidation signals, **6:**222f
genes for surface antigen, transfection into liver, **3:**424
integration site 7, chromosome 18, **3:**207
interferon-α therapy for infection by, **3:**330
Hepatitis C virus (HCV)
cytokines used in therapy for, **1:**450
defined, **3:**480
detection of, asymmetric gap ligase chain reaction for, **3:**305
identifying, with RT-PCR, **4:**326
interferon-α therapy for infection by, **3:**330–31
and liver cancer, **3:**481, 482–84, 484f, 485, 485f
Hepatitis virus, mouse
packaging of, **6:**221
receptor gene for, **6:**232, 232f
Hepatobiliary system, cholesterol secretion via, **3:**408
Hepatoblastoma, gene for, mapping to chromosome 11, **3:**127
Hepatocarcinogenesis, **3:**482–84
Hepatocellular carcinoma (HCC), **3:**480
genetic basis of, **3:**490f
molecular genetics of, **3:**486–87
relationship with virus integration in chromosome 18, **3:**207
Hepatocyte growth factor (HGF), **2:**469; **5:**272
liver, **3:**482
Hepatocytes, **1:**240
adult, **3:**481
growth regulation in, **3:**481
phosphatidylethanolamine *N*-methyltransferase of, **4:**437
Hepatocyte-stimulating factor, interleukin-6 as, **3:**342
Hepatomas
cells of, **3:**484
phosphatidylethanolamine *N*-methyltransferase in, **4:**437–38
Heptad repeat
in *APC* gene product, **1:**411
defined, **1:**407
Herbicides
enantiomers of, **1:**340
engineering plants for resistance to, **6:**57, 57f
Herculin, **4:**145
Hereditary, defined, **4:**148
Hereditary cerebral hemorrhage with amyloidosis, Dutch type (HCCHWA-D), in tangles associated with Alzheimer's disease, **1:**43
Hereditary disorders
angioedema, genetic mechanism of, **6:**136
colon cancer predisposition, **1:**408–10
versus genetic disorders, **4:**149
hemorrhagic telangiectasia (HHT), mapping to chromosome 9, **3:**107
hyperekplexia, **2:**214
keratinopathies, **2:**270
multiple cutaneous leiomyomas, **3:**211
nephrogenic diabetes insipidus, **2:**214
nonpolyposis colon cancer (HNPCC) syndromes, **1:**410; **2:**428; **3:**224
persistence of fetal hemoglobin, **6:**136
Hermaphrodites
Caenorhabditis elegans, **1:**228, 228f
defined, **1:**227
XX, **5:**444–45
Hermaphrodite-specific neurons (HSN), in *Caenorhabditis elegans*, **4:**170
Herpes simplex virus (HSV)
DNA of, electron microscopy of, **2:**184f
effect of methylphosphonate oligonucleotides on, in mice, **1:**75
electron microscope image, **2:**192f
lipid composition of, **6:**188
packaged DNA of, **6:**206
photodynamic therapy for infection by, and damage to platelets, **6:**209–10
photoinactivation of, **6:**212, 212t
recombinant Fabs against, **1:**421
Herpesviruses
assembly and budding of, **6:**192
structure and composition of, **6:**173–79
varicella-zoster virus and human cytomegalovirus, **6:**172
Hers disease, gene for, mapping to chromosome 14, **3:**157
Heteroatoms
defined, **3:**354
oxidation of,
catalyzed by cytochrome P450, **1:**195, 195f
catalyzed by flavin-containing monooxygenases, **1:**195, 195f
Heterochromatin
chromosomal, CENP-B antigen of, **5:**416
constitutive, in C bands, **1:**371
defined, **2:**121
effect on gene expression, **2:**125
tandem arrays of satellites in, **6:**9–10
variation in size of, chromosome 9, **3:**98–99
Heterocyclic compounds
defined, **4:**289
for drug manufacture, **2:**161–62
Heterodimer hypotheses, *DQ* heterodimer, and susceptibility to insulin-dependent diabetes mellitus, **2:**34–35
Heterodimerization, among UAS-activating transcription factors, **2:**368, 368f
Heterodimers
asymmetric, HIV-1 reverse transcriptase, **1:**29
DQ, and risk of insulin-dependent diabetes mellitus, **2:**35
steroid receptor, **5:**472
Heterodromic hydrogen bonding, **3:**266
Heteroduplex analysis, **3:**226
Heteroduplex DNA
defined, **2:**419; **5:**221, 252
electron microscopy image, **2:**186f
Heterogeneity, genetic, defined, **2:**419
Heterogeneous nuclear RNA (hnRNA), **5:**357
Heterologous proteins, from mammary glands, application of transgenic techniques, **6:**100–101
Heterologous sequences
defined, **5:**221
expression of, in yeasts, **6:**324–25
Heteromorphism, in chromosome 9, **3:**98–99
Heteronuclear hydrogen bonds, **3:**264
Heteronuclear NMR, **4:**223
Heteroplasmy, **4:**81
defined, **4:**79
and genetic disorders, **4:**89
Heteropolymer, defined, **1:**113
Heterothallic strains
defined, **6:**302
yeast, **6:**304
HETEROTRIMERIC G PROTEINS, SIGNAL TRANSDUCTION BY, **3:41–48**
Heterotrophs, **1:**119–20
dependence on autotrophs, **1:**310
energy metabolism in, **1:**306
Heterozygosity, *DR3/4*, and susceptibility to insulin-dependent diabetes mellitus, **2:**35
Heterozygote
defined, **2:**420; **3:**246
detecting, for Gaucher disease, **2:**351

and dominant-negative mutation, **6:**317–19, 319t
Heterozygote advantage
defined, **5:**45
and phenylketonuria, **4:**412
and sickle cell disease, **5:**46–47
Heuristic reasoning, **6:**52–53
Hexapeptide superclass, homeobox genes, **3:**64–65, 66–67, 66f
Hexosaminidase A, in Tay-Sachs disease, **2:**349, **3:**375
Hierarchical structure, in biological composites, **1:**145–47
High density lipoproteins (HDL), **1:**284
analysis of, **3:**477–78
apo A-I of, expression in transgenic rodents, **4:**503–4
apo A-II of, expression in transgenic mice, **4:**504
apoC-III of, expression in transgenic mice, **4:**504–5
and atherosclerosis, **1:**284
correlation with low incidence of heart disease, **3:**479
defined, **3:**404
High density operations, automation of, **5:**394
High endothelial cells (HE cells), binding of lymphocytes to, **2:**455
High mannose type sugar chains (glycans), **2:**451–53, 459f
High mobility group (HMG) proteins, **5:**439–40
box for, gene encoding, **3:**501
HIGH PERFORMANCE LIQUID CHROMATOGRAPHY (HPLC)
amino acid composition of protein hydrolysate using, **5:**122–23
for analysis of biomolecules in solution, **4:**218
OF BIOLOGICAL MACROMOLECULES, **3:76–83**
for bioseparations, **1:**175
for carbohydrate analysis, **1:**271, 271t
defined, **1:**266; **5:**115
isocratic, **3:**77
for peptide purification, **4:**367
preparative and analytical protein separations using, **5:**124
sample purification for protein sequencing, **5:**188
for steroid purification, **5:**475–76
for taxol^{REG} analysis, **6:**15
use with electrospray ionization, **4:**13–17
Hill equation, **1:**263
Hill plot, **3:**7–8, 8f
Hinges
immunoglobulin domain, **3:**277–78
myosin, **4:**134
Hippocampus, defined, **4:**54
Hip replacement, ultra high molecular weight poly(ethylene) for, **5:**40
Hirudin, inhibition of thrombin by, **5:**56
Histidine, intragenic complementation of genes for biosynthesis, **6:**319
Histidine decarboxylase (HDC), gene for, mapping to chromosome 15, **3:**169
Histocompatibility, hybridization and polymerase chain reaction for assessing, **4:**331
Histone octamer, **1:**361
defined, **1:**360
internal structure of, **1:**363–4
Histone proteins
defined, **3:**49
modification in the nucleus, **5:**53
HISTONES, **3:48–55**
in autoantibody response, **1:**85–86
in cell division, **1:**366
in chromatin, **1:**361
in chromosomes, **2:**117–18f
HIT and RUN protocol, for gene targeting, **1:**51
HIV. *See* Human immunodeficiency virus *entries*
Hlx class, homeobox genes, **3:**72
Hodgkin's lymphoma, treatment of, **1:**253
Hoechst 33258 dye, for FISH, **2:**379
Holliday intermediate, **4:**318–21, 321f
cleavage by RuvC protein, **5:**228–29
defined, **5:**221
Holliday junctions, **5:**255–56, 256f, 261, 261f
branch migration of, effect of RuvB protein on, **5:**228
defined, **5:**252
in DNA, **2:**104
double, **5:**260
resolution of, **5:**259
Holoenzyme
defined, **1:**397; **6:**244
DNA polymerase III, **2:**88–89, 91
restriction endonucleases, type I and type III, **5:**299
RNA and protein subunits, **5:**366
Holoprosencephaly, gene for, mapping to chromosome 18, **3:**213
Holorepressors, **5:**277
Homeobox, defined, **3:**56; **5:**128
Homeobox genes
of *Caenorhabditis elegans*, homology with vertebrate HOX genes, **1:**230
HOX10 and *GSC*, mapping to chromosome 14, **3:**151
HOMEODOMAIN PROTEINS, **3:55–76**
in the adult testis, **3:**506
POU (Pit-Oct-Unc) subclass, **1:**233–34
Homeodomains, **5:**129–30
atypical, **3:**74–75, 74f
classification of, **3:**58–63, 63f
defined, **3:**56; **5:**128
plant gene, **4:**492
Homeostasis
calcium, **1:**237–38, 249
energy required to maintain, **1:**304
Homeotic genes
defined, **3:**56
selector, *Drosophila melanogaster*, **3:**64
Hominoidea, place of humans in, **4:**85–86
Homodimer
protease, **1:**28, 28f
steroid receptor, **5:**472
Homodromic hydrogen bonding, **3:**266
Homogeneous assay, defined, **1:**322
Homologous chromosomes, crossing over of genes between, **2:**419f
Homologous recombination (HR), **5:**252–54, 252f, 253f
correction of gene mutations by, **4:**105
defined, **2:**407; **4:**101; **5:**447
gene targeting by, **4:**101–3
and genetic intelligence, **2:**413
for modifying cloned DNA, **6:**301
in the plant mitochondrial genome, **4:**489
and RecA protein, **5:**221, 221f
RecA protein role in, **5:**222
in TAC cloning technology, **6:**324, 324f
for transforming yeast DNA, **6:**303
Homologous sequences, defined, **5:**221
Homology
amino acid and gene-coding sequences, myosin alkali light chains, **4:**137t
cloning by, *Arabidopsis* mapping, **1:**83
defined, **3:**317; **5:**420, 428; **6:**126
among factors VII, VIII, IX, X, and protein C, **3:**36
between factor VIII and factor V, **3:**30
genome, defined, **3:**511
of human and mouse genes including the *NF2* gene, **4:**183
human chromosome 11, with nonhuman species' genes, **3:**141, 141t
human chromosome 14, and mouse chromosome 12, **3:**142
human chromosome 1, with mouse chromosomes 1, 3, and 4, **3:**84
Homology modeling, **5:**161, 164–66, 166f
Homopolymers, polysaccharide, **1:**278
Homothallic strains
defined, **6:**302
yeast, **6:**
Homozygosity mapping, **3:**188
Homozygote, defined, **2:**420; **3:**246
Hoogsteen base pairs, **1:**70–71, 71f
reverse, A-U, **5:**377, 377f
Hormone receptor, **6:**328f
zinc finger, **5:**133

Hormone response element (HRE), **2:**290; **5:**472
 activation by retinoic acid receptors, **6:**236–37
 defined, **2:**207
 glucocorticoid, **2:**208
Hormones
 classification of, **2:**207
 cytokines as, **1:**441
 deficiency disorders, **2:**213–14
 defined, **2:**155, 207, 463; **5:**229
 endogenous, leads drug synthesis, **2:**157–60
 glycohormones, **2:**454
 and hepatocellular carcinoma, **3:**485
 and immunotoxic responses, **3:**298
 interactions of
  with ascorbic acid vitamers, **6:**239
  with pyridoxal phosphate, in the nucleus, **6:**243–44
  with vitamin D, **6:**237–38
 lipid, **3:**443
 peptide, **4:**354–55
 plant, **4:**492
 steroid, **2:**158–60
 synthetic, **4:**368
 thyroid, effect on cornification, **2:**269
Hormone therapy
 anticancer drugs, **1:**256
 for central diabetes insipidus, **2:**23
 for malignancies, **2:**215
Host
 bacterial invasion of cells, **1:**105–6
 cloning, **5:**502
 factors affecting viral RNA replication, **5:**23
 transmission of pathogens to, **1:**103; **6:**204, 204f
Host cell
 features of, **2:**275–77
 macromolecular shutoff, in picornavirus infection, **5:**25
 shutoff in virus infection, **6:**198
Host-pathogen interface, in plants, **4:**484f
Host range
 defined, **6:**192
 and identification of virus receptor sites, **6:**231
 of viruses, **6:**195
Host-vector systems, **2:**271–72
 in yeast artificial chromosome construction, **6:**297–98
Hot spots
 in colon and breast tumors, **6:**141
 for male-specific meiotic recombination, **6:**31
 for mutation, **4:**151; **5:**46
  factor IX gene, **3:**37
  modeling, **2:**413
  of the *p53* gene in human hepatocellular carcinoma, **3:**485
  in retinoblastoma, **2:**429
 in plant mtDNA, **4:**81
 of recombination, **3:**124, 513; **5:**260
  HDL receptor gene, **3:**408
Hot start procedure, in polymerase chain reaction amplification, **4:**336
HOX cluster
 defined, **3:**56
 in *Drosophila*, **3:**67
 and related homeobox genes, **3:**64–67
Hox genes, defined, **3:**56
HPAEC-PAD (high performance anion exchange chromatography)
 with pulsed amperometric detection, **1:**266
  for carbohydrate analysis, **1:**271
H-phosphonate method, **4:**283
H-phosphonates, defined, **4:**281
HPLC, *See* High performance liquid chromatography
HTF islands, defined, **3:**28. *See also* CpG islands
Human antimouse response (HAMA), of immunoglobulins, **1:**65
Human β-globin gene
 perpendicular DGGE pattern, **2:**4, 4f
 point mutation, perpendicular DGGE, **2:**16–17
HUMAN CHROMOSOMES
 GENETIC AND PHYSICAL MAP STATUS OF
  1, **3:84–97**
  9, **3:97–112**
  10, **3:112–25**
  11, **3:125–41**
  15, **3:160–74**
  14, **3:142–59**
  16, **3:174–90**
  18, **3:190–15**
 PHYSICAL MAPS OF, **3:215–21**
HUMAN DISEASE GENE MAPPING, **3:221–27**
HUMAN GENETIC PREDISPOSITION TO DISEASE, **3:227–33**. *See also* Genetic disorders
Human Genome Project (HGP), **1:**117; **2:**96, 119, 386–87; **3:**112, 226; **5:**389
 cost of, **5:**401
 probes for chromosome 15, **3:**172
Human growth hormone
 MALDI mass spectra, **4:**17f
 therapy using, and transmission of prions, **5:**70
Human immunodeficiency virus (HIV), **6:**193
 antibodies against, **1:**420–21
 defined, **1:**27
 identifying, with RT-PCR, **4:**326
 infection by, role of superantigen, **5:**511
 inhibition of the reverse transcriptase of, **4:**228, 228f
 interferon-α therapy for infection by, **3:**331
 and mycobacterial infections, **4:**157
 photodynamic inactivation of, **6:**214, 214f
 recombinant vaccines for, **6:**170
 reverse transcriptase gene mutation in, **3:**304–5
 Rev protein of, **5:**385
 transcription of, **6:**197–98
Human immunodeficiency virus 1 (HIV-1), **1:**35–36, 36f
 analysis with gap ligase chain reaction, **3:**304–5
 Chinese hamster ovary cell lines expressing gp120 of, **6:**179
 infection in individuals, **6:**199
 integration into a host cellular chromosome, **6:**198
 photodynamic inactivation of, assessment of porphyrins to enhance, **6:**212
 photoinactivation of, **6:**212
 postprimary tuberculosis associated with, **4:**156–57
 receptor decoy approach for treating, **6:**233
 receptors for, multiple, **6:**231
 response to light in the presence of merocyanine 540, **6:**215f
 TAR structure of, **4:**224
 uncoating and reverse transcription of, **6:**196
Human immunodeficiency virus protease
 defined, **1:**14
 for HIV-2, **1:**14
 peptidomimetics of, **4:**357
 synthesis of enantiomers of, **1:**341
Human leukocyte antigen (HLA)
 class II
  association with insulin-dependent diabetes mellitus, **2:**29–36
  defined, **2:**28
  function of, **2:**31
  structure of, **3:**280–81, 280f
 and tissue typing using PCR analysis, **4:**334
HUMAN LINKAGE MAPS, **3:233–40**
Human pituitary gonadotropin, therapy using, and transmission of prions, **5:**70
HUMAN REPETITIVE ELEMENTS, **3:240–46**
Humoral immunity
 defined, **3:**499

against respiratory syncytial virus, **5**:292
Huntingtin (protein), **3**:250–51
HUNTINGTON'S DISEASE, **2**:328, 420, 427; **3**:226, 242, 246–52, 513; **4**:199
genetic risk in, **3**:239
mitochondrial mutations associated with, **4**:83
parental transmission in, **2**:445
repeat sequence
detection, **3**:397
trinucleotide, **2**:427
single-gene disorder, **3**:230
unstable genetic sequences in, **3**:231
Hutchinson-Gilford syndrome, premature aging in, **6**:34
Hyaluronidase, **2**:449f
H-Y antigene (HYA) gene, **5**:435–38
Hybrid
defined, **4**:235
formation of, **4**:239–42
isoenzyme, **3**:367–68
somatic cell, **3**:118
deletion map for chromosome 18, **3**:196, 196f
Hybrid cell panels, **2**:320, 321f
Hybrid cells/cell lines
for cloning markers for a chromosome, **2**:72–73
for differential Alu-PCR approach to DNA markers, **2**:73
staining of DNA in, species specific labelling, **6**:255
Hybrid dysgenesis, *Drosophila*, **2**:123–24
Hybridization
CISS, **2**:381, 381f
colony, **1**:124
conditions for, **2**:383–84
defined, **3**:391, 391f
in situ, genomic mapping by use of, **6**:300
FOR SEQUENCING OF DNA, **3:252–60**
denaturation step in, **2**:17
species-specific, **6**:253–54
subtractive, **4**:473–74, 474f
targets for, **2**:379–80
preparing, **2**:383
Hybridization assay, **1**:326–28
Hybridization protection assay, **1**:326f
for DNA, **1**:328f
Hybridoma, **3**:276
defined, **1**:172
human, electrofusion of, **2**:204
for producing monoclonal antibodies, **1**:175
production of, using electrofusion, **2**:205
Hybrid sequences, artifacts in PCR, **4**:337
Hybrid-type sugar chains, **2**:459f
defined, **2**:451–53
Hydantoins, secondary products of exposure to ionizing radiation, **3**:361
Hydra attenuata, mtDNA of, **4**:79
Hydralazine, lupus erythematosus associated with, genetic susceptibility to, **1**:199
Hydration
of biological macromolecules, **3**:272
force due to, and cell fusion, **2**:201–2
of phosphodiester linkages, oligonucleotide analogues, **4**:280
of proteins, in foods, **2**:310
in xenobiotic metabolism, **6**:267
Hydride ion, transfer of, pyridine nucleotide coenzyme operation, **1**:401f
Hydrocarbon chain
defined, **3**:442
melting temperature, effect of substituents on, **3**:448
packing in a crystalline state, **3**:443–45, 443f, 444f
Hydrocarbons
metabolism of, **3**:450–51
substituted, metabolism of, **3**:451
Hydrocephalus, silicone shunt for treating, **5**:37–38
Hydrogels
from cross-linked polymeric networks, **5**:41, 42f
from cross-linked starch, **5**:34
defined, **5**:31
Hydrogen, and methanogenesis, **4**:69–71, 71f
Hydrogenases
of archaea, genes for, **4**:77
defined, **2**:240
of hyperthermophiles, **2**:245
nickel-iron, **4**:69–70
Hydrogenation, asymmetric, of dehydro amino acids, **1**:57
Hydrogen bond cooperativity, defined, **3**:261
Hydrogen bonding, homodromic, **3**:266
HYDROGEN BONDING IN BIOLOGICAL STRUCTURES, **3:261–73**
Hydrogen bonds
acceptors, defined, **3**:261
in bioorganic compounds, **1**:163–64
configurations, **3**:265
and stability of folded proteins, **5**:151
in cyclic peptide nanotubules, **1**:153
defined, **2**:307; **3**:261, 262; **5**:128
in DNA, **2**:106, 110–11
donors, defined, **3**:261
geometry of, **3**:268–69, 268f
homonuclear, **3**:264
in immunoglobulin domain structures, **3**:277
linear, **3**:268
β-pleated sheet, **5**:118–19
of proteins, **2**:308
in RNA, **5**:371, 374
and conformation, **5**:369–70, 370f
stabilization of α-helices by, **5**:117
strength of, **3**:263–65, 263t, 264t
Hydrogen ion, biosensor utilizing, **1**:142
Hydrogen peroxide
cellular, damage from, **2**:42
generation by oxygen free radicals, **2**:331–32
Hydrogen sulfide, from sulfate reduction by anaerobic bacteria, **1**:121
Hydrolases
cell wall, altering in tomatoes, **6**:57–58
in organic solvent systems, **2**:250
restriction endonucleases as, **5**:297
and xenobiotic metabolism, **6**:267
Hydrolysis, **1**:160–61
peptide bond, enzyme catalyzed, **2**:230f
in xenobiotic metabolism, **6**:267
Hydropathy, defined, **5**:428
Hydropathy analysis, **1**:429
Hydropathy plots, for predicting protein sequence function, **5**:432
Hydrophilic substance
defined, **2**:347; **5**:159
nonbiodegradable polymers, **5**:41–44
Hydrophobic binding energy, **3**:283
Hydrophobic cleft, **6**:334, 334f
Hydrophobic interaction chromatography (HIC), **3**:81
Hydrophobic interactions, **1**:164
defined, **2**:307
in DNA, structure resulting from, **2**:106
in proteins, **2**:309, 309f
Hydrophobic substance
defined, **2**:347; **5**:159
nondegradable polymer, **5**:36–41
Hydrothermal systems, marine, and origin of life, **4**:292
Hydrothermal vent, defined, **2**:240
Hydroxyandrostenedione, for tumor suppression, **5**:487
Hydroxyapatite, **1**:238
Hydroxybutyrate (PHB), poly-β-, **3**:433–34
Hydroxydopamine, delayed response to, by tyrosine hydroxylase, **1**:216
Hydroxylamine, mutational analysis using, **5**:366
21-Hydroxylase gene, mutation in, **2**:424
Hydroxylases
aryl hydrocarbon, product of, **3**:492
cytochrome P450, **1**:438
6β-, **5**:470

Hydroxylation
aromatic, assay for free radicals, **2:**336, 336f
by cytochrome P450, **1:**194, 194f
posttranslational modification, **5:**54
Hydroxyl radicals
chemical reactivity of, **2:**330–31; **3:**359–60, 360f
footprinting with, **4:**249
oxidative damage from, in cells, **2:**80
from radiolysis of water, **3:**355–56
Hydroxymethylglutaryl (HMG) CoA-reductase, **3:**353, 409
Hydroxymethylglutaryl (HMG) CoA-synthetase, **3:**409
Hydroxyoleic acid, **3:**437
p-Hydroxy-D-phenylglycine, optically active, preparation of, **1:**58 **1:**58Hydroxyprogesterone caproate, antitumor activity of, **5:**488
Hydroxyproline, in collagen, **5:**28
Hydroxyproline-rich glycoproteins (HRGPs), **4:**480
11β-Hydroxysteroid dehydrogenase deficiency, **2:**214
Hydroxytryptamine, of nematodes, **4:**170
Hyperaldosteronism, **2:**214
Hypercholesterolemia, from mutation in apoB, **3:**411
Hyperchromicity, in T4gp32 binding to DNA, **5:**461
Hyperglycemia, management of, with poly(2-hydroxyethyl methacrylate) hydrogels for insulin delivery, **5:**42
Hyperlipidemia, type III, role of apoE in, **3:**414
Hyperlipoproteinemia
and apo E expression, **4:**505
role of apoE in, **3:**414
Hypermutation, somatic, **3:**283
Hyperoxaluria type 1, **5:**206
Hyperparathyroidism, neonatal severe, **2:**215
Hyperphenylalaninemia, **4:**406
Hyperproinsulinemia, **2:**213
Hypersensitive reactions
in bacterial infection, **1:**105
defined, **4:**476
in plant defense against pathogens, **4:**483–84
Hypersensitive response (HR), in plant defense against pathogens, **4:**478–79
Hypertension
angiotensinogen levels in, **5:**266
essential, mapping to chromosome 1, **3:**91–92
kallikrein-kininogen-kinin system in, **3:**379
role in atherosclerosis, **1:**288
Hyperthermophile, defined, **2:**240
Hyperthermophilic genera, **2:**242t
Hypertriglyceridemia
apoCII and apoCIII mutations associated with, **3:**413
and apoCIII, **4:**504–5
Hypertrophic cardiomyopathy (CMH2), gene for, mapping to chromosome 1, **3:**92
Hypertrophy, and heart failure, **1:**289
Hypertrophy of the retinal pigment epithelium (CHRPE), congenital, **1:**409
Hypervariable regions
of apoB gene, **3:**412
defined, **2:**48
of immunoglobulin light and heavy chains, **3:**278–79, 279f
Hypervitaminosis, defined, **6:**244
Hyphages, **2:**68
Hypokalemic periodic paralysis (HOKPP), gene for, mapping to chromosome 1, **3:**92
Hypothalamic-pituitary-ovarian-uterine axis, **2:**286
Hypothalamus
oxytocin gene expression in, **2:**291
secretion of gonadotropin-releasing hormone from, **2:**286
Hypothyroidism
genetic, **4:**356
neonatal, effect on Sertoli cell proliferation, **3:**504
Hypotonic solution, for chromosome preparation for microdissection, **1:**372
Hypovitaminosis, defined, **6:**244
Hypoxanthine guanine phosphoribosyltransferase (HPRT)
defective encoding of, in Lesch-Nyhan syndrome, **2:**427
gene for, **4:**103, 105
H zone, muscle, **4:**131, 131f

Ibuprofen, manufacture of, **2:**161
Icosahedral viruses, **6:**229f
Sindbis virus, **6:**220–21
ICUMSA (International Commission for Uniform Methods of Sugar Analysis), defined, **1:**266
Identification
from DNA analysis, **2:**51
of individuals, DNA fingerprints for, **2:**60
of proteins, methods for, **5:**124–26
of transgenes by PCR, **6:**90f
Ideogram
chromosome 9, **3:**98, 98f
chromosome 10, **3:**119f
chromosome 16, **3:**188f
chromosome 18, **3:**191–92, 193f, 196f, 200f
Idiopathic dilated cardiomyopathy, **1:**290
gene expression for the calcium-sodium channel in, **1:**292
Idiopathic hypertrophic subaortic stenosis (IHSS), **1:**289
Idiotype, defined, **3:**275; **6:**179
Ifosphamide, as an anti-cancer drug, **1:**254
Ig class switching, **1:**447
Ig supergene family, T cells fused to specific members of, **1:**67
Illegitimate recombination, defined, **3:**511
Image analysis, **2:**386
defined, **2:**378
Image reconstruction, **2:**193
Imidazolidinones, **1:**54
derivatives, in peptides, **4:**377
structure of, **1:**53
Immediate early genes, and heart failure, **1:**290
Immobilization, in biosensors, chemistry of, **5:**245–46
Immobilized enzymes, in organic media, **2:**251
Immobilized ligands, for analyzing molecular recognition, **5:**245–49
Immobilized pH gradient (IPG) electrophoresis, **2:**354
Immortalization
defined, **6:**27
and telomerase, **6:**34–35
Immune cells, keratinocytes as, **2:**269–70
Immune donors, antibodies from, **1:**420–22
Immune response, **2:**430–31
and cancer treatment, **1:**261–62
cytokine role in, **1:**441
interleukin-2, **3:**343
interleukins, **3:**333
hepatic acute phase, **3:**298
host, evasion of by bacteria, **1:**104–5
and interferons, **3:**327
to *Mycobacterium tuberculosis*, **4:**156–57
to proteins and derivatives, **5:**34
quantitation of, VSV and CMV vaccines, **6:**183
Immune response genes, and insulin-dependent diabetes mellitus, **2:**38
Immune systems
networks in, **4:**95
in vivo strategy of, **3:**287f, 296–98
Immunoadhesins, **1:**67, 67f
CD4, **3:**286
Immunoaffinity chromatography
for extraction of anabolic steroid, **5:**474
for purification of protein antigens for vaccines, **6:**176
for separation of plasma lipoproteins, **3:**474–75, 478

Immunoassays, **5**:466
for antigen detection, **5**:411–17
defined, **1**:323
for quantitation of steroid hormones, **5**:476
tiers of, **3**:299
Immunoelectron microscopy, defined, **2**:182
Immunofluorescence/immunoelectron microscopy, for identifying proteins involved in focal adhesion, **1**:454
Immunogen, defined, **1**:84
Immunogenicity versus antigenicity, **3**:285
Immunogenic proteins, herpesvirus, **6**:175
Immunoglobulin class switching, **1**:447
Immunoglobulin domain, defined, **1**:293
Immunoglobulins
classes of, **1**:64f
G isotypes, defined, **5**:286
heavy chain diversity region 2, localization to chromosome 15, **3**:167
IgD, genetic engineering to obtain, **1**:66
IgE, regulation of synthesis by interferons, **3**:327
IgG
regulation of synthesis by interferons, **3**:327
structure of, **3**:276–77, 277f
sugar chains of, **2**:455, 455f
IgM, membrane-anchored, **3**:282
induction of, by interleukin-13, **3**:343
regulation of synthesis by interleukins, **3**:341–42
structure of
immunoglobulin light and heavy chain fold, **3**:277
loop, **5**:164
V_H, contig for, on chromosome 14, **3**:151
Immunoglobulin superfamily
interleukin-1 receptors, **3**:338–39
MAG of, **1**:294f
N-CAM of, **1**:294
Immunoliposomes, **3**:416
Immunological assays, for protein identification, **5**:124
IMMUNOLOGY, **3:275–88**
of respiratory syncytial virus infection, **5**:291–93
Immunomodulation
defined, **3**:295
environmentally produced, **3**:297–98
IMMUNO-PCR, **3:288–95**
quantitation of PCR product, **3**:294f
Immunosorbent surface, defined, **1**:148
Immunosuppression
chemicals and drugs producing, **3**:298t
environmentally produced, **3**:298
in interferon-α therapy, **3**:330
and susceptibility to cytomegalovirus infection, **6**:173
transforming growth factor-β induction of, **2**:468
Immunotoxic evaluations, **3**:299
IMMUNOTOXICOLOGICAL MECHANISMS, **3:295–302**
Immunotoxicology, defined, **3**:296
Immunotoxins
nature of, **3**:298
production of, **3**:286
Imprinting
cancer syndromes, defined, **2**:439
in chromosome 15, **3**:161
defined, **3**:160
DNA, defined, **3**:511
effect on genetic analysis, **2**:424
genomic, mapping to chromosome 11, **3**:127
methylation involved in, **3**:517
parental, **2**:440–41
of *SNRPN*, **3**:168
Inactivated vaccines, **6**:165
Inbred strain
defined, **2**:146
recombinant, mouse, **4**:124, 124f
Inbreeding, for drug sensitivity, **2**:150
Inbreeding depression, in laboratory studies of aging, **1**:10, 11
Incidence, of phenylketonuria in human populations, **4**:411, 411t
Inclusion, defined, **2**:48
Inclusion bodies
defined, **5**:73
detection of, **5**:80–81
formation of, **5**:82–86
Incompatibility
plant-pathogen, **4**:476
plasmid, **2**:272; **5**:5
defined, **5**:1
Incrustation, **4**:300
Incubation time, for scrapie, genetic control of, **5**:66, 67t
Indel, **5**:421
defined, **5**:420
Index map, for chromosome 14, **3**:144, 149f
Indicators, fluorescent, **2**:295–98, 298f
Indirect gene transfer, defined, **2**:430
Indirect immunofluorescence (IIF), **1**:87–89
defined, **1**:84
Indirect mutation detection, with linkage analysis, **2**:421
Indirect readout, **5**:279
defined, **5**:275
Indirect selection, of catalytic RNAs, **4**:231–32, 231f
Individuality, in xenobiotic metabolism, **1**:202
Indolepyruvate oxidoreductase (IOR), **2**:244
Induced asymmetry
from binding of ligands to proteins, **1**:394
defined, **1**:384
Inducer of short transcripts (IST), **1**:20, 20f
Inducers
ganglioside GM3, **2**:457
of interleukins, **3**:337–38t
Inducible system, **2**:371
for DNA repair, **2**:45
Induction, of fragile sites, **2**:316
Infection
agents of
detection by ligase chain reaction, **3**:305–8
molecular pathology, **4**:329–31
cytokine production in, and symptoms, **1**:449
defined, **6**:193
etiology of, human prion diseases, **5**:69–72, 69t
types of, **6**:198–99
viral, cycle of, **6**:195–98
INFECTIOUS DISEASE TESTING BY LIGASE CHAIN REACTION, **3:302–9**
Inflammatory cascade, **1**:447
Inflammatory response
expression of T-kininogens in, **3**:380
in knockout mice lacking interleukin expression, **3**:344
regulation of, **1**:447
role of interferons, **3**:328
role of selectins, **1**:296–97
Influenza virus, **5**:359
assembly and budding in, **6**:184
packaging of, **6**:221
type C, matrix protein of, **6**:188
Information
chemical flow of, transforming to electrical flow, **1**:178, 178f
genetic, **2**:415–16
RNA encoding of, **5**:338
molecular, from fluorescence, **2**:298–300
pragmatic and semantic theories of, **6**:51–52
for RNA three-dimensional modeling, **5**:378–79
sources of, for linkage analysis, **3**:239–40
transfer of
DNA to DNA, **5**:354–55
DNA to RNA, **5**:355
RNA to DNA, **5**:355, 359

Information (*cont.*)
RNA to RNA, **5:**355, 358
Information processing, with biochips, hypothetical, **1:**157–58
Information theory, **6:**41
Infrared spectroscopy
for following DNA denaturation, **2:**6t
for hydrogen bond studies, **3:**267
Infringement
defined, **6:**74
of a patent, **6:**76–77
Inheritance pattern
of fragile sites, **2:**317–18, 317f
of fragile X linked mental retardation, **2:**320
of growth hormone in transgenic carp, **6:**93, 93t
of mitochondrial DNA, **4:**81
for transgenes, **6:**90
Inhibin, **2:**468; **5:**242
family of, follicle-stimulating hormone regulation by, **2:**288
receptors for, **3:**510
secretion by Sertoli cells, **3:**509
as a tumor marker, **2:**293
Inhibition of crystal growth, **3:**313
Inhibitors, enzyme, **1:**166–68; **2:**235–37
amperometric electrodes for measuring, **1:**186
asymmetric and symmetric peptidic, to HIV protease, **1:**15–16
in enzyme-directed posttranslational modification, **5:**51
of HIV protease, structure of the binding region, **1:**28
of phospholipase A enzymes, **4:**423, 423t
of protease, mapping to chromosome 14, **3:**155
of RNase, **5:**346
for studying virus assembly and budding, **6:**188
of tryptophan hydroxylase, **1:**218–19
of tyrosine and tryptophan hydroxylases, **1:**212–13
Initiation
defined, **6:**109
of DNA replication, **1:**97; **2:**90–91
codons of metazoan mitochondria, **4:**80
mechanism, **2:**90–91
regulation of, **1:**98
of protein biosynthesis, **6:**114
of RNA replication, regulation of, **2:**365
Initiation complex, eukaryotic, 80S, **6:**118
Initiation factors (eIFs)
defined, **5:**328
eukaryotic, **6:**118, 120t
initiation of protein synthesis and, **2:**376–77f
phosphorylation of, and protein synthesis, **6:**124
Initiation signal, defined, **4:**129
Initiator (INR) elements, **2:**95
Initiator codon, **6:**116
Initiator of short transcripts (IST), mutating, **1:**25
Innate immunity, **3:**296, 297f
Inner core, lipopolysaccharide, **3:**466, 466f
INORGANIC SOLIDS, BIOMOLECULES IN THE SYNTHESIS OF, **3:309–17**
Inosine monophosphate dehydrogenase, in neoplastic cells, **1:**252
Inositol, **1:**246
Inositol triphosphate
Ca^{2+} release stimulated by, **3:**46
defined, **3:**20; **4:**432
as a second messenger, **3:**44, 279; **4:**195–96, 438–39
Inositol triphosphate-3-kinase, gene localization to chromosome 15, **3:**168
IN-OUT method, for gene targeting, **1:**51
Insect cells, posttranslational processing in, **6:**177–78
Insecticidal crystal proteins (ICPs), **4:**393, 394–95
Insects, responses of plants to attack by, **4:**492–93
Insect virus, expression systems for the production of proteins, **2:**277
Insertion, mitochondrial, **4:**82–83
Insertional inactivation, **5:**502
Insertional mutagenesis, **2:**392; **3:**518; **4:**149, 149f; **6:**67
and disease, mtDNA, **4:**89–90
in hemophilia, **3:**31
in phenylketonuria, **4:**409
for tagging *Arabidopsis* gene, **1:**82
Insertion sequences (IS), **2:**405
defined, **4:**156
in mycobacteria, **4:**159
In situ hybridization (ISH), **3:**118–19
defined, **2:**121; **6:**253
gene mapping by fluorescence, **2:**378
for mapping candidate genes, **3:**226
See also Fluorescence in situ hybridization (FISH)
In situ PCR, **4:**336–37
In situ restriction enzyme digestion, for restriction landmark genome scanning, **5:**312–16
Instrumentation
molecular and biomolecular electronics, **1:**158–59, 159f
Rama spectroscopy, **5:**101
Insulin (*INS*) gene, **2:**37, 466
defect in, in diabetes mellitus, **4:**356–57
defined, **2:**28
delivery of, with poly(2-hydroxyethyl methacrylate) hydrogel, **5:**42
effect on synthesis of apoE, **3:**414
nasal delivery of, **2:**166
production from recombinant organisms, **1:**174
receptor for, tyrosine kinase, **5:**238, 241
reduced secretion of, in maturity onset diabetes mellitus of the young, **2:**38–39
Insulin-dependent diabetes mellitus (IDDM), **2:**28
comparison among animal models, **2:**29t
cytokines' roles in pathogenesis of, **1:**449
DR3-associated susceptibility to, **2:**32
DR4-associated susceptibility to, **2:**33
DR9-associated susceptibility to, **2:**33
major histocompatibility class I and class II antigens in, **3:**328
protection against, evidence for alleles contributing to, **2:**33–34
susceptibility to, gene mapping to chromosome 15, **3:**165
Insulinlike growth factor 1 receptor (IGF1R), gene localization to chromosome 15, **3:**170
Insulinlike growth factors (IGFs), **2:**466; **5:**270–71
I, in transgenic fish, **6:**95–96, 96f
production by gliomas, **1:**252
receptors for, **5:**239
in transgenic animals, **6:**70
Insulin response element, negative, **3:**404
Integral membrane calcium-binding protein, **6:**237
Integral membrane proteins, transmembrane pores formed by, **6:**128
Integral membrane transport proteins, **5:**118; **6:**126–27
Integral proteins, of reaction centers, purple bacteria, **4:**442–43, 443f
Integrase, **1:**14
comparison with RNase H, HIV reverse transcriptase, **1:**33
drugs based on, **1:**41
HIV-1, **1:**33
Integration
of DNA, gene transfer in fish, **6:**74–75
of hepatitis B virus sequences, **3:**484
of injected genes, **6:**66, 67
analysis of, **6:**71–72
in fish, **6:**93–94, 93t
pattern of, **6:**89–90
Integrins, **1:**293, 296–97
basement membrane, **2:**264–65
cytoskeletal interactions with, **1:**453–57
defined, **1:**293, 451
role of

in bacterial invasion of cells, **1**:106
in cell adhesion, **1**:175, 452
role in cytoskeleton-plasma membrane interactions, **1**:451
subunits of, **1**:459
target of bacterial invasins, **1**:106
Inteins, defined, **4**:159
Intellectual property rights, defined, **6**:74
Intelligence, genetic, defined, **2**:407
Interaction, among cytokines, **1**:442
Interaction analysis, defined, **5**:243
Interactive modeling, of a group intron, **5**:385
Inter-*Alu* PCR, defined, **3**:215
Intercellular adhesion molecule 1 (ICAM-1), as a virus receptor, **6**:225
Interfacial activation, of lipases, measuring, **4**:418
Interfacial electric polarization, **2**:179, 179f
defined, **2**:172
Interference, in crossing over, **3**:239
Interference experiment, defined, **4**:243
Interference technique, for footprinting, **4**:249–50, 250f
Interferon-α
characterization of, **3**:319–20
genes for, **3**:318–19
introns of, **2**:371
synthesis of, **3**:321
Interferon-β
β_2, relationship with interleukin 6, **2**:465
characterization of, **3**:320, 320f
gene for, **4**:319
structure, schematic, **3**:321f
synthesis of, **3**:322
Interferon-γ, **3**:321, 321f
mediator in endotoxic reactions, **3**:471
stimulation of production of, by interleukin-12, **3**:343
synthesis of, **3**:322
in vaccinia virus vaccines, **6**:170
Interferon gene regulatory elements (IREs), **3**:321
for interferon-β, **3**:322
Interferon-ω
characterization of, **3**:320
gene for, **3**:319
synthesis of, **3**:322
Interferon-τ
characterization of, **3**:320
synthesis of, **3**:322
INTERFERONS, **1**:441; **3:317–31**
biological effects of, prevention of virus replication, **1**:447
defined, **3**:317
gene cluster on chromosome 9, **3**:101
list, **1**:444t
nomenclature of, **1**:444t
properties of, **6**:199
receptors for, defined, **3**:317
tumor necrosis factor, **1**:442
type I and type II, defined, **3**:317
Interferon-stimulated gene factor (ISGF), **3**:323
Interferon-stimulated response element, **3**:323
Intergeneric recombination, **2**:414
Interindividual factors
affecting xenobiotic metabolism, **6**:260–61, 260t
defined, **6**:259
Interleukin-1 (IL-1), biological activities of, **3**:340
Interleukin-1 (IL-1) receptor antagonist (IL-Ira), **3**:334–35
Interleukin-1β, structure of, **3**:335f
Interleukin-1β converting enzyme (ICE), homology with CED-3, **1**:302
Interleukin-2 (IL-2)
adjuvant for vaccina virus vaccines, **6**:170
biological activities of, **3**:340–41
cancer therapy using, **2**:469
receptor for, **2**:464; **3**:301
Interleukin-3 (IL-3)
biological activities of, **3**:341
as a colony-stimulating factor, **2**:466
Interleukin-4 (IL-4)
biological activities of, **3**:341–42
effect on nitric oxide synthase 2, **4**:203
structure of, **5**:162–63, 163f
Interleukin-5 (IL-5), biological activities of, **3**:342
Interleukin-6 (IL-6)
biological activities of, **3**:342
identity with interferon β2, **2**:465
murine, amino acid sequence, **5**:180–81, 181f
receptor for, gp130, **5**:242
role in signal transduction, **2**:470–71
Interleukin-7 (IL-7), biological activities of, **3**:342
Interleukin-8 (IL-8), biological activities of, **3**:342–43
Interleukin-9 (IL-9)
biological activities of, **3**:343
sequencing, **5**:183
Interleukin-10 (IL-10), biological activities of, **3**:343
Interleukin-11 (IL-11), biological activities of, **3**:343
Interleukin-12 (IL-12)
biological activities of, **3**:343
role in production of Th1 cells, **2**:431f
Interleukin-13 (IL-13), biological activities of, **3**:343
Interleukin-14 (IL-14), biological activities of, **3**:343
Interleukin-15 (IL-15), biological activities of, **3**:344
INTERLEUKINS, **1**:442; **2**:469; **3:341–46**
activation of keratinocytes by, **2**:269–70
characterization of, **3**:336t
defined, **1**:441; **2**:463; **3**:331
immunological properties of, and vaccine development, **2**:431
as mediators in endotoxicity, **3**:471
nomenclature of, **1**:443t
Intermediate density lipoproteins (IDL), **3**:474
defined, **3**:404
Intermediate filaments (IFs), **4**:143
defined, **4**:112
Intermolecular interactions, formation of protein aggregates by, **5**:78f
Internal guide sequence (IGS), **4**:230
Internalin, **1**:106
Internal loops
defined, **5**:360
RNA, **5**:362, 377
Internal ribosome entry site (IRES), picornaviruses, **5**:17–19
International Commission for Radiological Protection (ICRP), **5**:211
International System for Cytogenetic Nomenclature (ISCN), human chromosome banding pattern designations, **2**:379
International System of Units, becquerel (Bq) defined in, **5**:211
International Union of Biochemistry, Commission on Biochemical Nomenclature, **3**:366
International Union of Pure and Applied Chemistry, guidelines for peptide nomenclature, **4**:363–64
Internucleoside linkage, defined, **4**:269
Internucleotide linkage, defined, **4**:269
Interphase, chromosomal organization of cell nuclei at, **1**:371; **6**:256
Interphase DNA, target in FISH, **2**:379, 389; **3**:119
Interprocess communication (IPC), software architecture incorporating, **5**:397
Intersegment transfer, DNA, **5**:281
Interspecific backcross, **3**:514
Interspersed elements, LINE and SINE, **5**:430
Interspersed repeat sequence (IRS) DNA amplification, **6**:254
Interstellar dust clouds, defined, **4**:289
Interstitial 18q deletion syndrome, **3**:212
Interstitial compartment, testicular, **3**:501
Interstitium, immunosuppression in, role of activin in, **3**:510

Intestinal calcium-binding proteins (ICaBP), cell cycle control by, **1**:242
Intracellular adhesion molecule 1 (ICAM-1), **1**:451, 453
 receptor for rhinoviruses, **1**:456; **6**:232
Intracellular membrane traffic, **4**:45
Intracellular phospholipase (iPLA$_2$), calcium-ion independent, **4**:423–24
INTRACELLULAR PROTEIN DEGRADATION, MECHANISMS OF REGULATED, **3:346–54**
Intraindividual factors
 affecting xenobiotic metabolism, **6**:260–61, 260t
 defined, **6**:259
Intramuscular injection, of plasmid DNA, **2**:435–36, 435f
Intrastrand deletion, **6**:4f, 5
Intrinsic factor, interaction with vitamin B$_{12}$, **6**:250–51
Intron, group, modeling the structure of, **5**:385, 387f
Intron-encoded endonuclease, **5**:300
 defined, **5**:296
Intron homing, **1**:351
Introns, **2**:371–73; **3**:512; **5**:356–57
 of *Caenorhabditis elegans* genes, **1**:229
 catalytic, pseudoknots of, **5**:362
 of *Chlamydomonas reinhardtii*, **1**:348, 350–51, 353
 defined, **1**:347; **2**:278, 364; **3**:28, 227; **4**:129, 226; **5**:338, 352, 372
 of eukaryotes, **2**:393, 393f
 and gene expression, **6**:102
 intron-exon distribution in the *FMR-1* gene, **2**:325f
 lack of, in IFN-α and INF-β genes, **1**:444
 requirement for, in expression of transgenes, **6**:105
 of self-splicing RNA, **5**:338–39, 339f
 simple DNA of, **5**:431
 strings of variable length describing, **6**:46–47
Invasins, bacterial surface proteins, **1**:106
Invasion plasmid antigens (*ipa*), **1**:105, 106
Invasiveness, of tumors, **1**:260–61
Inversion
 chromosome, in *Drosophila*, **2**:127–28
 DNA, site-specific, **2**:404
 mutation by, **2**:418; **3**:32
  and hemophilia A, **3**:32
Inverted region
 electron transfer reactions, **2**:195
 Marcus, in photosynthesis, **2**:197
Inverted repeats, chloroplast, **4**:488–89
In vitro activity
 DNA packaging, **6**:205–6
 of immunotoxins, **3**:299
 of oligonucleotides, antisense and triplex-forming, **1**:74–75
 transcription by RNA polymerase, **4**:228f
  defined, **4**:226
In vivo activity, of oligonucleotides, antisense and triplex-forming, **1**:75
Involucrin, **2**:267
Iodine, ^{125}I radiation hazard, **5**:212–14
Ion channel-coupled receptors, **5**:230–33, 233t
 for neurotransmitters, **5**:237–38
Ion channel receptors, **2**:208, 209
Ion channels
 annexin V role in formation of, **1**:61
 defined, **2**:130
 in heart failure, **1**:290, 292
 ligand-gated, **4**:30
 synaptic, **5**:514
 voltage-gated, **4**:30
Ion equilibrium, and rate constant, **2**:177–78
Ion exchange chromatography (IEC), **3**:79–81
Ionic composition, body fluids in humans, **1**:127, 127t
Ionic strength, structural effect of, virion fd, **5**:108, 109f
Ionization, process for achieving, for mass spectrometry, **4**:13–17
Ionization monitors, for radiation detection, **5**:218
Ionizing radiation
 defined, **3**:354
 models for effects of, **3**:357–58
IONIZING RADIATION DAMAGE TO DNA, **2**:41; **3:354–65**
Ionophore
 defined, **1**:313
 gramicidin A, **1**:337
Ion pump
 bacteriorhodopsin, **6**:131
 defined, **1**:108
 electron-flow-driven, **6**:131
 reversible, **1**:308
Ions
 flow of, and rate constant, **2**:178
 metal, radii of, **1**:127, 128t
Ion-selective electrodes (ISEs), **1**:186
Ion-selective field effect transistors (ISFETs), **1**:186–87, 187f
Ion-translocating antibiotics, **6**:128
Ion trap mass spectrometry, **4**:26–27
Iron
 biocatalyst, **4**:61–62, 61t
 chelation of, by *Escherichia coli* K-12, **5**:11
 in cytochrome P450, **1**:435
 ferrous and ferric, in hemoglobin, **3**:5
 heme, in enzymes, **4**:61–62, 62f
 and hydroxyl radical generation, **2**:334
 in nitrogenases, of methanogens, **4**:77–78
 in tryptophan hydroxylase, **1**:211
 in tyrosine hydroxylase, **1**:209
Iron-responsive element binding protein (IRE-BP), **6**:62
Iron sulfide, electron transfer and clusters of, **2**:194
Iron-sulfur clusters
 in enzymes, **2**:233
 of hyperthermophiles, **2**:244
 in photosynthetic energy transduction, **4**:444–45
Irreversible reactions, inhibition in enzyme mechanisms, **2**:235, 235f
Irving-Williams stability sequence, **1**:129
Ischemic heart disease, **1**:292
 inverse correlation with α-tocopherol, **6**:238
Islet cell antibody (ICA), screening the human islet cDNA expression library for, **2**:36
Isochores
 DNA, **5**:429
  defined, **2**:360; **5**:428
  human genome, **2**:363
 organization of, **2**:361, 361f
Isochromes, 18p tetrasomy syndrome and 18q syndrome, **3**:212
Isocitrate lyase, circular dichroism spectra, **1**:391–92, 392f
Isocratic elution
 in affinity chromatography, **5**:245, 246f
 in high performance liquid chromatography, **3**:77
Isoelectric focusing (IEF)
 first-dimension, technique, **2**:353, 354
 for protein purity determination, **5**:123–24
 for separation of isoenzymes, **3**:368
 for separation of plasma lipoproteins, **3**:475, 476
Isoelectric point (pI), defined, **2**:353
ISOENZYMES, **3:366–76**
Isoforms
 of genes, *NF2*, **4**:184f
 of proteins
  defined, **4**:129
  muscle-specific, **4**:130
  myosin heavy chain, **4**:134
  myosin light chain, tissue distribution of, **4**:135
  tropomyosins, **4**:138
  troponins, **4**:140–142
Isolated congenital ACTH deficiency, **2**:214
Isolation, of plasmid DNA, **5**:2
Isomerase, protein disulfide (PDI), **5**:199

Isomerization, cis-trans, in proline-containing peptides, **5**:158
Isomers
  defined, **1**:332
  of 4-coumarate:CoA ligase, **4**:480
Isomorphous replacement, for protein structure determination, **5**:114; **6**:279
Isoniazid
  peripheral neuropathy associated with, genetic susceptibility to, **1**:199
  polymorphism in metabolism of, **4**:404
  resistance to, gene mutation in, **4**:160
  for treating tuberculosis, **4**:157
Isoprenoid chains, of glycerol diether lipids, **3**:429
Isoprenoid pathway, vitamins synthesized in, **3**:461
Isopropylthiogalactoside (IPTG)-inducible promoter, **3**:23
Isoschizomers, **5**:301
  defined, **5**:296
Isothiopropylgalactoside, for regulating transcription, **2**:273
Isotopes, for nuclear magnetic resonance measurements, **4**:218
Isotopomers, fd virion, normal and deuterated Raman spectra of, **5**:105–8, 108f
Isotype, defined, **3**:275
Isovalericacidemia, gene for, mapping to chromosome 15, **3**:164
Isozymes, PI-PLC, classification of, **4**:427t
Isua Supercrustal System, Greenland, sedimentary rock of, **4**:289
Iterative cycling, in amplification, **4**:96
Ivm/ifv, defined, **6**:99

Jablonski diagram, for fluorescence, **2**:300–301, 301f
Jacobsen syndrome, **2**:318
  defined, **2**:313
JAK-STAT pathway, activation of, **5**:242
Janus kinase, **1**:446
Jasmonic acid, and plant resistance to pathogens, **4**:483, 493
J-coupling
  defined, **4**:218
  in NMR analyses, **4**:219
J gene, **3**:281
Jumping genes, **2**:216
  mutation by insertion of, **4**:150
Jumping libraries, **2**:74–76, 74f
  Not-1, **2**:74
Junctions
  four-way, recognition by SRY protein and HMG1, **5**:440
  multibranch, **5**:360, 362
  See also Adherens-type junctions; Gap junctions; Holliday junctions
Juvenile hormone, **4**:189–90, 190f
Juxtaglomerular apparatus (JGA)
  defined, **5**:264
  synthesis of renin in, **5**:265–66

Kallikrein genes, human, **3**:384
KALLIKREIN-KININOGEN-KININ SYSTEM, **3**:**377–84**; **5**:267–68
Kallikreins, **3**:377, 379, 382–84
  tissue, **5**:267
Kallistatin (KST), gene for, mapping to chromosome 14, **3**:155
Kallman's syndrome, **2**:292
Kanamycin, resistance to, as a gene marker, **4**:163; **6**:54
Kaposi sarcomas, **1**:251
  interferon-α therapy for, **3**:330
Karplus relation, for *J*-coupling, **4**:219, 219f
Karyotype
  bivariate flow, **2**:388f
  defined, **1**:370
  Drosophila, **2**:122, 122f
  visible chromosome bands of, **1**:371
  XY, role of P450s in expression of male characteristics associated with, **1**:437
Karyotyping, pulsed field gel electrophoresis (PFGE) for, **4**:307–8
Kasugamycin, **4**:396–97, 398f
Kearns-Sayre syndrome (KSS), **1**:10; **4**:82
  mtDNA deletion in, **4**:90
Kennedy's disease, **2**:328; **3**:242
Keratinocyte growth factor (KGF), in prostate cancer cells, **2**:469
Keratinocytes
  activated
    defined, **2**:263
    markers of, **2**:266
  autoantibodies against products of, **1**:86
Keratinopathy
  defined, **2**:263
  inherited, **2**:270
Keratins
  acidic and basic, **4**:143
  defined, **2**:263
  as markers of epithelial differentiation, **2**:265–66
Ketal, defined, **5**:485
3-Ketothiolase, **3**:434f
Ketotifen, glucuronidation of, **1**:198, 198f
Kexin, **4**:344–45
k-Gram, defined, **6**:38
Kidd blood group antigen, gene for, mapping to chromosome 18, **3**:207
Killer T lymphocytes, generation in response to MHC type 1, **2**:431
Killing yield, for ionizing radiation, **3**:355
Kinase marker, thymidine, **4**:475
Kinases
  cyclin-dependent, **3**:507
  defined, **6**:138
  genes coding for, in *Drosophila*, **2**:137
  rhodopsin, **5**:324
Kinesin-related proteins (KRPs), directional movement of, **4**:118
Kinesins, **4**:113–14, 114t
  characterization of, **4**:117–18, 117f
  defined, **4**:112
  roles of, **4**:120
  superfamily of, and flagellar apparatus of *Chlamydomonas*, **1**:358–59
Kinetic analysis, **5**:246, 248, 248f
Kinetic effects, interfacial, on phospholipases, **4**:416–18
Kinetic energy (KE), defined, **1**:304
Kinetics
  of binding by hemoglobin, **3**:8–9
  constants for assembly of subunits in oxyhemoglobin, **3**:12t
  of cyclobutane dimer repair, **6**:158–59, 159f
  enzyme, **2**:227–28
    effects of organic solvents on, **2**:255–56
    immobilized lipases, **2**:251
    tyrosine hydroxylase, **1**:212–13
  from filter binding information, **4**:246
  of growth, *Taxus cuspidata* cell suspension, **6**:16
  of hybridization, **4**:240–41, 241f
  of inhibition of multienzyme pathways, **1**:263–64
  isoenzyme, **3**:369, 374
  ligand binding, genetically engineered hemoglobin, **3**:24–25
  of metal ion complexation, **1**:133–34
  of nucleic acid-protein interactions, **4**:247
  of oxygen binding, **3**:24–25
  of protein degradation, **3**:347
  of protein folding, **5**:153–54, 160, 160f
    model for, **5**:158f
  of protein refolding, **5**:155
  of radical reactions, **3**:357–58
  of radioactive decay, **5**:210–11, 210f
  rate-limiting step in protein folding, **5**:155
  rate of DNA repair in human cells, **2**:45–46
  of reassociation, *Arabidopsis* genome, **1**:76–77
  of RNA secondary structure assembly, **5**:363
Kinetochore, binding of motor proteins to, **4**:118
Kinetoplastida, mtDNA of, **4**:79
Kininases, **3**:377; **5**:268
Kininogenases, genes encoding, regulated expression of, **3**:384

Kininogens, **4:**354; **5:**268
genes
processing of, **3:**383f
structure of, **3:**382f
high molecular weight, **3:**380, 380f
structure and function, **3:**380
substrates for, **3:**377
Kinins, **3:**377
chemical mediator for pain, **3:**379
effects on nonvascular smooth muscle, **3:**378
receptors for, **3:**384
Klebsiella pneumoniae, encapsulation to evade immune response, **1:**105
Kleiber equation, **4:**260
Klenow fragment, of DNA polymerase, **2:**383
compared to HIV-1 reverse transcriptase, **1:**30
random priming, **2:**382
KNF model (Koshland-Nemethy-Filmer model), of cooperative binding, **3:**9, 9f
Knockout mice, **3:**518; **6:**64
apo E deficient, **4:**505
for interferon and interferon receptor gene deficiencies, **3:**327–28, 344
for neurofibromin study, **4:**180
Knudson hypothesis
application of neurofibromatosis 2, **4:**183
defined, **4:**181
Koji processes, **2:**338, 338t
Kosambi mapping function, **3:**113
Kpn repeat, **3:**217–18
defined, **3:**215
Krabbe disease, gene for, mapping to chromosome 14, **3:**158
Krebs cycle, **2:**233
aconitase in, **4:**61–62
Kringle domains
apolipoprotein(a), **3:**472; **4:**504
defined, **3:**404
hepatocyte growth factor, **2:**469
Kringles
of apo(a), **1:**286; **3:**415
defined, **3:**404, 472
Kruppel family of zinc finger proteins, **3:**151
k-k-tuple, defined, **5:**428; **6:**38
Kugelberg-Welander disease, **4:**106

Label
defined, **3:**385
for a hybridization probe, defined, **6:**253
in restriction landmark genomic scanning, **5:**307
Label-free optical sensors, **1:**190
LABELING, BIOPHYSICAL, **3:385–90**
Labial (lab) class of genes, homeodomain of, **3:**67
LacI, in an Achilles' cleavage protocol, **1:**4
α-Lactalbumin, **6:**103
circular dichroism analysis, molten globule form, **1:**391
circular dichroism spectra of, **1:**393f
in the milk of transgenic mice, **6:**105–06
molten globule intermediate in folding of, **5:**77–78
β-Lactamase, unfolding-refolding of, **5:**77, 85
Lactams
δ- and γ, constraints on, **4:**377
formation of, in peptides, **4:**376–77, 376f
β-Lactams
antibiotic, **2:**156–57
synthesis of drugs based on, **2:**163–64, 164f
transformation of producers of, commercial applications, **2:**345–46
Lactate dehydrogenase (LD), **3:**367, 367f
unfolding-refolding of, **5:**79–80, 80f
Lactation
defined, **6:**99
requirement for α-lactalbumin, **6:**103
Lactobacillic acid, synthesis by bacteria, **3:**432
β-Lactoglobulin (β-LG), **6:**103
production by transgenic animals, **6:**105
Lactones, cytotoxic, **5:**492–93, 492f
Lactose, amperometric electrodes for determination of, **1:**185–86
Lactotetraosylceramide, **1:**280f, 282, 282f
LacZ gene, defined, **4:**167
Ladder sequencing, of proteins, **5:**175
Lagging strand, in DNA synthesis, **5:**458–59
Lambda-based vectors, **2:**69f
Lambda phage, for cloning antigen-binding fragments Fab and Fv, **1:**415–17
Lambda replicative intermediate, **4:**317–18, 318f
Laminin A chain (LAMA), gene for, mapping to chromosome 18, **3:**207
Laminins, **1:**453; **2:**283
Landmark
defined, **5:**305
*Not*I, direct labeling of, **5:**308–9
*Pac*I, direct labeling of, **5:**309–10
Landmark cleavage, **5:**307, 308
Landscapes, and the origins of order, **4:**94
Lane finding, in HUGE sequencing of DNA, **2:**101
Langmuir-Blodgett (LB) film
cadmium sulfide formation in the hydrophilic interlayer of, **3:**313f
deposition process, **1:**151–52, 151f; **3:**311–12, 311f
Langmuir-Blodgett technique
defined, **1:**148
for reconstituting photosynthetic reaction centers, **1:**151
Langmuir isotherm, **1:**180
Langmuir monolayers, **3:**311–12
Large multifunctional protease (LMP) genes, **2:**31
Large unilamellar vesicles (LUVs), **3:**416–17
Laron dwarfism, **2:**214
Laser, microbeam, for plant cell transformation, **4:**465
Laser-assisted time-of-flight mass spectrometry, **5:**395
Laser desorption, of DNA fragments, for mass spectrometry, **4:**3
Laser flow cytometry, **4:**486
Laser microbeam method, for chromosome microdissection, **1:**376
Latency
defined, **1:**19
of herpesviruses, **6:**172
and negative transcription control, **1:**26
from vaccines, **6:**173
in virus replication, **6:**210
Laue method, **6:**278
LDL-receptor-like protein (LRP), **3:**405
Lead
for bioassay, defined, **2:**155
compounds in medicinal chemistry, **4:**28–29, 29f, 32f, 100
immunosuppression by, **3:**298
for synthetic targets, drug synthesis, **2:**155–60
Leader, immunoglobulin peptide, **3:**281
Leading strand, in DNA synthesis, **5:**458
Learning
defective, in mutant *Drosophila* with modified courting behavior, **2:**137
defined, **4:**54
Least-squares fit, Raman sulfhydryl band, **5:**105f
Leber's hereditary optic neuropathy (LHON), **4:**82
from mtDNA base substitutions, **4:**89
Lec-cam. *See* Selectins
Lecithin-cholesterol acyltransferase (LCAT), **3:**408, 457, 474
Lecithin retinol acyl transferase (LRAT), retinyl ester biosynthesis by, **5:**325, 326f
Lectins, **2:**455
binding specificities of, immobilized columns, **2:**450t
for carbohydrate analysis, **1:**273t
carbohydrate analysis using, **1:**272
C-type and S-type, **1:**296

defined, **1**:266, 293; **2**:446
domain in P-selectins, **1**:294f
Legal considerations, in animal patenting, **6**:75–77
Legionella pneumophilia
inhibition by interferon-γ, **3**:326
integrin receptor use by, **1**:106
Leishmania genome project, **4**:311
Leishmaniasis, **4**:307
Length of DNA, and duplex stability, **2**:12
Lennard-Jones forces, between stacked base pairs, **2**:8
Lenses, intraocular, from poly(methyl methacrylate), **5**:40–41
Lentinan, antitumor activity of, **2**:344–45
Lentivirus
HIV, **1**:19
HIV-1, **1**:34
Tat and Rev regulatory proteins of, **1**:35f
Leprosy, **4**:157
Lesion
versus mutation, **4**:149
noncoding, **4**:152
in transgenic medaka, **6**:95–96
Lethal infantile respiratory failure, nuclear mutation in, **4**:90
Lethal mitochondrial myopathy, **4**:83
Lethal selection, environmental, defined, **2**:407
Lethal zygosis, **5**:502, 506
Leucine zipper
basic (b-ZIP), 5:130
defined, **5**:128
in plants, **4**:468, 492
of dimerization domains, **4**:492
family of, transcription factor, **3**:506
interleukin involvement in activation of, **3**:339
trans-acting factors, **4**:467, 467t
Leukemia
acute lymphoid, gene rearrangements in, **2**:64, 64t
acute megakaryoblastic, associated with Down's syndrome, **2**:119
acute promyelocytic, translocation in, **2**:65
acute T-cell, translocation in, **3**:159
chronic myelogenous, **1**:251
gene rearrangements in, **2**:64, 64t; **4**:286
hairy cell, cytokine therapy for, **1**:450
resistance to glucocorticoid therapy, **2**:215
Leukemia inhibitory factor (LIF), **1**:442; **2**:464; **5**:241
defined, **2**:285
female reproductive system expression of, **2**:292
Leukemia virus, murine, encapsidation signals, **6**:222f
Leukocyte adhesion deficiency (LAD), **1**:459
Leukocytes
blood, photochemical attack on, **6**:210
cytokine receptors of, **1**:445
defined, **3**:295
role of selectins in movement of, **1**:297f
Leukocyte tyrosine kinase (LTK), mapping to chromosome 15, **3**:167
Leukopenia, in interferon-α therapy, **3**:330
Leukopyrokinin (LeuPK), **4**:195
Leupeptin, accumulation of lipofuscin associated with, **1**:49
Levonorgestrel, controlled release from poly(ortho-ester) devices, **5**:36
Lewis antigens, of erythrocytes, **1**:282f
Lewy bodies, in Alzheimer's disease, **1**:50
Leydig cells
defined, **3**:500
development of, **3**:505
transforming growth factor-β regulation of testosterone in, **3**:509
Libraries
Arabidopsis genome, **1**:83
chimeras of related genes from sexual PCR, **5**:456f
chromosome 15, **3**:172–73
encoded, **4**:97
linking, **2**:74–76, 76f; **3**:219, 220
naive antibody, **1**:423–24
nonpeptide synthetic, **5**:518, 518f, 522
phage display, **4**:96–97
in vitro, for antibody identification, **1**:64–65
See also Combinatorial library; Peptide libraries
Library arraying
automatic, **5**:392, 393f
chromosome 9, **3**:101
Life cycle
Chlamydomonas, **1**:349f
of double-stranded DNA bacteriophages, **6**:202–5
HIV-1, **1**:25–26, 34–35, 36f; **6**:196f
sites for therapeutic intervention, **1**:37f
HIV, regulation of, **1**:19
plant, and size of genome, **4**:486
poliovirus, **5**:16f
RNA tumor virus, **5**:358f
streptomycetes, **5**:501
vaccinia virus, **6**:195–98, 195f
yeast, **6**:304–5, 305f
Life span
correlation with DNA damage and repair, **2**:79–80
and mechanism of aging, **1**:10
and repair of UV-damage to DNA, **1**:6–7
Lifetime-based fluorescence sensing, **2**:304–5, 305f
Li-Fraumeni cancer syndrome, **2**:428; **6**:142
defined, **6**:138
p53 mutations in, **1**:412; **4**:329; **6**:141
telomerase in, **6**:35
TP53 mutations in, **2**:423
Ligand-binding pathways, hemoglobin, **3**:11–12
Ligand-binding rate, hemoglobin, genetically engineered, **3**:24–25
Ligand-gated ion channels, **4**:30; **5**:514
acetylcholine receptor, **6**:128–29
calcium, **2**:210
Ligand-macromolecule interactions, characterization by NMR, **4**:224
Ligands
binding equilibrium, **2**:176
and circular dichroism spectra, **1**:394
c-*kit*, **2**:465
defined, **1**:**126**; **4**:28, 90; **5**:159; **6**:59
exchange of, for metal ion complexation, **1**:133–34
heme, **3**:5–9
discrimination between, **3**:25
immobilized, for analyzing molecular recognition, **5**:245
selection of, and combinatorial libraries, **4**:96–100
vitamin, interaction with receptors, **6**:235
Ligase
DNA
biology of, **3**:392
in Bloom's syndrome, **3**:160
defined, **3**:391, 400
mechanism of action, **3**:302
nick-sealing activity of, **3**:303
in plasmid replication, **5**:3
thermostable, **3**:302, 303, 401
RNA, **3**:393
T4, **3**:394f, 401
for preparing misacylated tRNAs, **5**:142
tRNA
chiral specificity of, **1**:336
reactions not involving the chiral center of substrates, **1**:342–43
Ligase chain reaction (LCR), **2**:65, 421; **3**:396, 396f; **4**:327
gap sites for, **3**:305t
infectious disease testing by, **3**:302–9
mechanism of, **3**:303, 303f
principles of, **3**:302–5
LIGASE-MEDIATED GENE DETECTION, **3**:**391–99**
Ligation
allele-specific, **2**:423

Ligation (*cont.*)
blunt-end, **3:**304
calcium ion, **1:**239
chemical, of nucleic acids, **3:**393, 393f
for cloning experiments, **5:**505
by HIV-1 integrase, **1:**33
of inserts to yeast artificial chromosome vectors, **6:**296–97
Ligation amplification, **3:**400–401
defined, **3:**400
reaction for, **3:**402f
LIGATION ASSAYS, **3:400–403**
defined, **3:**400
Ligation-mediated polymerase chain reaction (LM-PCR)
for analyzing ultraviolet radiation damage to DNA, **6:**156
for DNA-protein interaction assays, **4:**471
Light, and plant growth and development, **4:**493
Light-addressable potentiometry, immunoassay system using, **1:**142
Light-directed, spatially addressable parallel peptide synthesis, **4:**392
Light-directed synthesis
of DNA chips, **1:**157
of peptides, on chips, **5:**519
Light emission, measurement of, **1:**326
Light-harvesting antenna complex (LHCII), **4:**442, 442t
Light reactions, of photosynthesis, **1:**310
Light response element (LRE), **2:**221
Light scattering, for detecting protein aggregates, **5:**81
Lignin
biodegradation by white rot fungi, **1:**122
synthesis during pathogen attack, **4:**480
Ligninases, role in plant pathology, **4:**482
LIHs (LIs of *Homo sapiens*), **6:**134
Limb girdle muscular dystrophy type 2, **3:**163–64
LIM class, homeobox genes, **3:**72, 73f
LIM domain, structure of, **6:**334
Lincosamides, antibiotic, resistance induced by plasmids, **5:**10
Line, germ, defined, **6:**99
LINE (long interspersed repeat) element, **1:**380; **3:**244–45, 245f, 512; **6:**134–35, 135f
Alu repeats, **2:**73
of chromosome G bands, **1:**371
defined, **3:**215, **6:**134
in hemophilia A, **3:**31
for hybridization probes, **2:**380–81
Kpn, **3:**218
Linear assembly, PCR, **5:**450–51, 450f
Linear DNA, for transgenic animal modeling, **6:**67
Linear structure, of varicella-zoster virus and cytomegalovirus glycoproteins, **6:**176f
Linguistic theory, and code deciphering, **6:**45–47
Linkage, **3:**233
defined, **1:**221, 425; **2:**62; **3:**84, 222, 377
of genes on a chromosome, **5:**254, 254f
Lk, characteristic of supercoils, **2:**116–17
and QTL mapping, **2:**152
Linkage analysis, **3:**112–14, 224–25, 235, 237
in breast cancer, **1:**222–23
for cancer diagnosis, **2:**428
in colon cancer, **1:**411
defined, **1:**407; **2:**420; **3:**126
for genetic testing, **2:**421, 429
glycosidic, **1:**275–76, 276f
in HNPCC kindreds, **1:**410
for neoplastic disease diagnosis, **2:**64
in neurofibromatosis 1, **4:**179
Linkage disequilibrium
analysis using, **3:**186
defined, **4:**173
in Huntington's disease locus identification, **3:**248–49
Linkage group
conserved, mouse and human, **4:**126
defined, **1:**347
of keratins, **2:**265–66
nuclear, *Chlamydomonas reinhardtii*, **1:**348
uni, *Chlamydomonas*, **1:**359
Linkage maps
for chromosome 11, male and female, **3:**134f
constructing, **3:**235–36
for chromosome 16, **3:**175–76
PCR-based, for chromosome 16, **3:**177f
LINKAGE program, **3:**175, 237
Linkage testing stocks, experimental animals, **3:**514
N-Linked sugar chains, **2:**458
Linker
actin-integrin, **1:**458
DNA, **1:**366, 369, 370
molecular, in immuno-PCR, **3:**289
polypeptide, in immunoglobulins, **3:**277
Linker adapter, with PCR amplification, **1:**379–80
Linker histone
defined, **1:**360
and steric hindrance on nucleosomes, **1:**369
Linking clone, defined, **3:**215
Linking libraries, **2:**74–76, 76f; **3:**219, 220
γ-Linoleic acid (GLA)
synthesis by molds, **3:**439
production from fungi, **3:**438–39, 439f
Lipase
lipoprotein, **3:**404–6, 472; **4:**496
in transgenic mice, **4:**505–6
for interesterification of fats, **2:**250
modification for solubilization in organic media, **2:**251
LIPED program, **3:**237
Lipid A
and cell viability, **3:**464–65, 465f
defined, **3:**462
LIPID AND LIPOPROTEIN METABOLISM, **3:403–15**
LIPID-BASED GENE VECTORS, **3:415–26**
Lipid bilayer membranes, **1:**193
formation by fatty acids, **3:**455
model for electroporation, **2:**199
Lipid bilayers
defined, **6:**183
repulsion between, **4:**46, 46f
virus, **6:**184
Lipid binding domain, calcium ion, **4:**423
Lipid management, in liver, **4:**434–35, 505–6
Lipid metabolism, role of pyridoxal 5′-phosphate in, **1:**402
Lipid peroxidation, **2:**331
LIPIDS
chemical classification of, **3:**443, 443t
dietary, **4:**265–66
of enveloped viruses, **6:**188
fusogenic, **4:**47–48
incorporation of xenobiotic acids into, **6:**273
interactions with proteins, in foods, **2:**307, 310
MICROBIAL, **3:426–42**
and the origins of life, **4:**293–94
photooxidation of, **6:**208
STRUCTURE AND BIOCHEMISTRY OF, **3:442–62**
transfer protein, role in invagination and budding at membranes, **4:**51–52
type II, **4:**42
Lipid-soluble vitamins, **6:**235
Lipid tubules
formation in chiral, bilayer-forming lipids, **3:**315
self-assembly of, **1:**153
Lipocalin superfamily, apoD associated with, **3:**413
Lipocortin (annexin I), structure, **1:**60f
Lipofuscin (age pigment), association with amyloid precursor protein, **1:**49
Lipoglycans, defined, **3:**426
Lipopolylamine-nucleic acid condensation product, structure of, **3:**422f
LIPOPOLYSACCHARIDES, **3:**461, **462–71**

antigens of Gram-negative bacteria, **1:**279
hydroxy fatty acids of, **3:**433, 433f
macrophage, induction of interleukin-1 by, **3:**334–35
rough, **3:**466
S-type, **3:**469–70
LIPOPROTEIN ANALYSIS, **3:471–80**
Lipoprotein lipase (LPL), **3:**404–6, 472; **4:**496
in transgenic mice, **4:**505–6
Lipoproteins, **3:**404
analysis of subfractions, **3:**477–78
classes of, **1:**284t; **4:**497t
defined, **1:**283; **3:**442; **4:**495
low density, **1:**284
Lp(a), **3:**414–15, 472
metabolism of, **1:**284–87; **3:**473f; **4:**495–96
rabbit model of, **6:**69
plasma, **4:**495–506, 497t
transport of lipid-soluble vitamins involving, **6:**235
See also High density lipoproteins; Low density lipoproteins
Liposome-mediated gene transfer, **2:**433
Liposomes, **3:**416–17, 416f
aggregation and fusion on, induced by sPLC, **4:**428
biodistribution of, **2:**167
defined, **2:**165; **3:**415
encapsulation of drugs in, **2:**166
tethered, **1:**154
Lipoxygenase, modification for solubilization in organic media, **2:**251
Lipoyl function, coenzyme, **1:**405
Liquid chromatography, **3:**77
Liquid crystals
defined, **1:**134; **3:**442
lipid, **3:**445
packing and dimensional order, **3:**447f
properties of, **3:**446t
peptide polymers, biomimetic, **1:**137
Liquid phase extraction, of steroids, **5:**473–74
Liquid state, of lipids, **3:**445, 447–49
Listeria monocytogenes, **1:**447
defense against phagocytosis, **1:**105
interaction with actin in a host cell, **1:**106
Lithium, acquired nephrogenic diabetes insipidus from, **2:**24
Lithium salts, for yeast transformation, **6:**310
Lithography, nanometer-scale, **1:**140
Live attenuated vaccines, safety of, **6:**173
Liver
cytosolic α-tocopherol binding protein in, **6:**238
hepatocyte growth in, fetal and adult, **3:**481
lipid management in, **1:**284
formation of phosphatidylcholine, **4:**434–35
transgenic model for, **4:**505–6
organ regeneration, **2:**469; **3:**482
synthesis in
of apoB, **3:**411–12
of bile acids, **3:**409–10
of factor IX, **3:**37
of plasma fibronectin, **2:**283
transfection of genes into cells of, **3:**424
uptake of HDL cholesterol esters in, **3:**409
LIVER CANCER, MOLECULAR BIOLOGY OF, **3:480–91**
Live vaccines
defined, **6:**172
recombinant, **6:**179–81
for varicella-zoster virus and varicella-zoster virus, **6:**180t
Local alignment
defined, **5:**420
dynamic programming to establish, **5:**421
methods for determining, **5:**424
Local assignments, multiple, of sequence alignments, **5:**424
Local compositional complexity, **6:**47
Locally multiply damaged sites (LMDS), from ionizing radiation, **3:**355, 356
Local-point fusion, **4:**44, 47, 47f
defined, **4:**42
Local regulation, defined, **2:**285
Location cloning, Huntington's disease identification using, **3:**248, 248f
Locus
chromosomal, **3:**217
defined, **3:**112
for Huntington's disease, **3:**248–49, 248f
defined, **2:**420; **3:**142, 160, 233, 366
Locus ceruleus, effects of opiates in, **2:**150
Locus-control elements, tissue-specific, **4:**501–2, 502f
Locus-control region
hemoglobin, **3:**13, 14f
human globin, and transgene expression, **6:**102
Locus heterogeneity, **3:**223
Locus order
estimation of, **3:**237
in genetic maps, **3:**236
Locus-specific probe (LSP)
defined, **6:**253
for gene mapping, **6:**257
Logarithm of the odds (LOD), **3:**113, 113t, 224, 237, 237f
defined, **3:**112
dependence on marker allele frequencies, **3:**239
linearity of ratios with complexity distributions, **6:**48, 49–50f
LINKAGE program for using in mapping, **3:**175
programs for computing, **3:**237
for radiation hybrid mapping, **3:**121
score for familial PrP mutation, **5:**71–72
for testing linkage, **1:**223
Log cell kill concept, **1:**258
Log P, defined, **1:**193
Lomustine, as an anti-cancer drug, **1:**254
Long interspersed element (LINE) repeats. *See* LINE
Long QT syndrome, mapping of, to chromosome 11, **3:**127
Long-range mapping, jumping and linking libraries for, **2:**77f
Long-range restriction maps, **3:**121–22
chromosome 16, **3:**183, 186
Long terminal repeat (LTR)
defined, **1:**19; **6:**193
HIV-1, **1:**35f
of Moloney murine leukemia virus (MoMLV), **6:**88
nonspecific inhibition of viral, **1:**26
Tal-3, of *Arabidopsis*, **4:**488
and virus life cycle, **1:**26
HIV-1, **6:**197
Long-term potentiation (LTP)
and cellular memory, **4:**57–58
in the hippocampus, **5:**233
Long-term sensitization, in *Aplysia*, **4:**56f
Loop modeling, by homology, **5:**164
ω-Loop structure, of adipokinetic hormone I, **4:**197
Loricrin, **2:**267–68
Losartan, structure of, **4:**32f
Loss of activity, neurofibromin, in mutation of the *NF1* gene, **6:**144
Loss-of-function mutations
in colorectal carcinomas, **1:**410–11
fish, **6:**86f
of the *lin-14* gene, in *Caenorhabditis elegans*, **1:**235
transgene, **6:**87, 98
Loss of heterozygosity (LOH)
in breast tumor formation, chromosome 13q, **6:**142
in colon tumor formation, **1:**411
defined, **1:**407; **3:**480; **5:**319
in hepatocellular carcinoma, **3:**487, 487t
for identifying tumor suppressor genes, **3:**225–26
in retinoblastoma, **5:**320
for studying tumor suppressor genes of chromosome 14, **3:**159
in tumor cells, **6:**139–40

Loss of heterozygosity (LOH) (*cont.*)
  tumor map using, chromosome 18, **3:**197
Low density lipoproteins (LDL), **3:**472–73
  analysis of, **3:**478
    separation of subfractions, **3:**478
  analysis of,n estimate from triglyceride and HDL concentrations, **3:**475
  apo B of, expression in transgenic mice, **4:**504
  and atherosclerosis, **1:**284, 288; **4:**495–96
  and coronary heart disease, **3:**479
  defined, **3:**404
  receptors for, **5:**204
  structure of, **3:**407f
Lowry assay, **5:**121–22
Luciferase, **1:**326
  defined, **1:**323
  gene for
    as a chemiluminescent reporter for transcriptional activity, **1:**329
    to detect drug resistance, **4:**164
    reporter gene in plant transformations, **4:**490
  import into peroxisomes, **5:**206
  incorporation of para-substituted phenylalanine analogues in, **5:**138
  molten globule form, **1:**391
  reaction at chiral center of the substrate, **1:**342
Luciferin, **1:**323, 330
  defined, **1:**323
Lumbricus terrestris, hemoglobin of, **3:**12
Lumen fusion, with dielectrophoresis, **2:**201
Luminescence, cellular, **1:**330–31
Luminescent labeling, **3:**387–88
Luminol
  in enhanced chemiluminescence assay, **1:**327
  oxidation of, producing chemiluminescence, **1:**323
Luminometer, defined, **1:**323
LUNG CANCER
  amplification of c-*myc* in, **4:**286
  MOLECULAR BIOLOGY OF, **3:491–97**
  ras mutation in, **2:**65
Lungs
  alveolar type II cells, calcium ion independent PLA_2 in, **4:**423–24
  stress on, in cystic fibrosis, **1:**425
  surfactant of, composition of, **4:**433
  in vivo gene therapy in, **3:**425
Luteinizing hormone (LH), **2:**215, 454
  on Leydig cells, **3:**505
β-Lyase, pyridoxal phosphate dependent, **1:**200
Lymphoblastoid cell line, defined, **2:**386
Lymphocyte homing, **2:**455
Lymphocytes
  cytoskeleton, **1:**455
  regulation of, roles of cytokines, **1:**446–47
Lymphocyte-specific factor, effect on viral replication, **1:**19
Lymphoid hyperplasia, induction by interleukin-2, **3:**341
Lymphokine-activated killer cells (LAK cells), **3:**340
Lymphokines, **1:**442; **2:**469
  defined, **1:**441; **2:**463
Lymphomas, **1:**250–51
  B-cell, translocation in, **1:**302; **2:**65
  follicular, **2:**65
    low grade, translocation in, **4:**286
  resistance to glucorticoid therapy, **2:**215
Lymphopoiesis, suppression of, by transforming growth factor-β, **2:**468
Lymphotoxin (LT), human, fusion protein with an antiganglioside antibody, **1:**67
Lyon hypothesis, defined, **1:**250
Lyonization, and expression of X-Y translocation, **5:**436–37
Lyophilized enzyme powder, defined, **2:**249
Lysine, biosynthesis of, in yeast, **6:**313
Lysine hydroxylase, posttranslational modification by, **5:**54
8-Lysine vasopressin, **2:**20
Lysis, **6:**204
  buffers for
    gentle cellular disruption, **5:**344–45
    harsh cellular disruption, **5:**345
Lysogeny, **6:**204, 210
lauroyl-1-Lysophosphatidylcholine, absorption enhancement by, **2:**167
Lysophospholipase
  $cPLA_2$ as, **4:**423
  defined, **4:**416
Lysophospholipid, **2:**211; **3:**457
  inhibition of biomembrane fusion by, **4:**48
Lysosomal protease inhibitors, accumulation of lipofuscin associated with, **1:**49
Lysosomal storage disorders, among Ashkenazi Jews, **5:**48
Lysosomal targeting, **5:**202–3
Lysosomes
  defined, **3:**347; **5:**194
  protein degradation in, **3:**347–48
Lysozyme
  binding of N-acetylglucosamine derivatives by, effect on circular dichroism spectrum, **1:**394
  matrix-assisted laser desorption ionization spectrum, **4:**4f
McArdles's disease, **4:**146
  mapping of, to chromosome 11, **3:**127
McCune-Albright syndrome
  genetic defect in, **2:**292
  mosaicism in, **2:**215
Machado-Joseph disease, gene for, mapping to chromosome 14, **3:**158
Machine vision, for automatic colony picking, **5:**392
Macroautophagy, **3:**347–48
α_2-Macroglobulin receptor (LDL-receptor-like protein), **3:**405
Macroinjection, for plant cell transformation, **4:**464
Macrolides, antibiotic, resistance induced by plasmids, **5:**10
Macromolecular assemblies, electron microscopy of, **2:**190
Macromolecular conjugates, defined, **6:**259
Macromolecular interaction analysis, **5:**247f
Macromolecular pesticides, **4:**394–95
Macromolecules, defined, **5:**159
Macrophages
  activation of
    by interferon-γ, **1:**447
    and tumor cytotoxicity, **1:**262
  in atherosclerotic lesions, **1:**287–88
  binding of oxidized LDL by, **3:**408
  CD44 molecule on, association with the actin cytoskeleton, **1:**456
  cytotoxic activity against tumor cells, **1:**257
  defined, **3:**295, 331
  role in wound healing and angiogenesis, **1:**448
Macula densa cells, defined, **5:**264
MADS box, trans-acting factor, **4:**467, 467t
MAG (myelin-associated glycoprotein), **1:**294f
Magnesium
  in Achilles' cleavage, **1:**4
  in aldehyde transfers facilitated by thiazole pyrophosphate, **1:**397
  and ATP reactions, **1:**399–400
  comparison with calcium ion, **1:**127–28
  in RecA-Achilles' cleavage, **1:**5
Magnetic induction of electric fields, defined, **2:**172
Magnetobacteria, **2:**172
Magnetosomes, defined, **2:**172
Maillard reaction, **2:**307, 309–10, 310f
Major facilitator superfamily, **6:**129
Major grooves, defined, **6:**327
MAJOR HISTOCOMPATIBILITY COMPLEX (MHC), **1:**261; **3:499–500**
  association with insulin-dependent diabetes mellitus, **2:**29–36

class I, defined, **3**:499
class II, **3**:500; **5**:509–10
associated with insulin-dependent diabetes mellitus, proposed mechanism for, **2**:35
defined, **3**:499; **5**:508
genes for, in autoimmune disease, **1**:85
defined, **2**:28
functions of, **2**:32f
human, on chromosome 6, **2**:30f
role in antigenicity, **2**:431; **3**:285; **6**:179
regulation by interferons, **3**:327
single-chain engineering of fragments, **3**:286–87
structure of, **2**:31; **3**:280–81, 280f
Major intrinsic protein (MIP), **6**:127, 127t, 129
Malaria, cerebral, implication of interferon-γ in, **3**:328
Malaria genome project, **4**:307–9
Malarial protease, structure by homology, example, **5**:164–66
Malate dehydrogenase
semisynthetic, **1**:143–44, 144f
targeting sequence of, **5**:205
Male pseudohermaphrodite, androgen receptor mutation of, **2**:214
MALE REPRODUCTIVE SYSTEM, MOLECULAR BIOLOGY OF, **3:500–510**
Males
gametocytes of, mutation expression in fragile X syndrome, **2**:329
transmission of premutation, fragile X syndrome, **2**:328–29
XX, human, **5**:443–44
Malignant hyperthermia
and calcium regulation, **1**:249
response to general anesthesia, **4**:405
Malignant lymphocytic leukemia, rabbit transgenic model for, **6**:69
Malignant melanoma
cutaneous, gene for, mapping to chromosome 1, **3**:93
gene for, mapping to chromosome 15, **3**:165
interferon-α therapy for, **3**:330
Malnutrition, and oxidative stress, **2**:334
Malonaldehyde, from DNA radiolysis, **3**:364–65
Mammalian cells, expression of proteins in, **2**:276
MAMMALIAN GENOME, **3:510–20**
Mammalian-wide interspersed repeats (MIRs), **3**:243
Mammals, phospholipases of, **4**:431
Mammary gland, and milk, **6**:102–6
Mandelate racemase, **1**:343
Manganese ion
binding to HIV-1 Rnase H, **1**:33
binding to HIV integrase, **1**:33
binding to RNase H, **1**:32
in peroxidases, **4**:65
in superoxide dismutase, **2**:333
Manic depression, **4**:200
Manning-Record equation, **2**:7
Mannose, in oligosaccharides, **2**:458
Mannose-6-phosphate, posttranslational production of, **5**:202–3, 203f
Map functions, relation between map distance and recombination fraction, **3**:235
MAP kinase (MAPK), defined, **5**:229–30
Mapping
of the *Arabidopsis* gene, **1**:81f
of *Caenorhabditis elegans* chromosomes, **4**:167
of chromosome 21, **2**:119–20, 120f
deletion, in the Y chromosome, **5**:437, 437f
of a DNA sequence, **5**:429–30
of *Drosophila melanogaster* chromosomes, **2**:124, 125–26
epitope, with synthetic peptides, **4**:368
of the *Escherichia coli* genome, **2**:169–72, 170f
gene, **2**:378
of susceptibility to insulin-dependent diabetes mellitus, **2**:28–29
genome, **2**:129; **3**:216–17
jumping and linking libraries for, **2**:74–76, 76f
Genome Project, **3**:520
of human CENP-B and CENP-C cDNA clones, **5**:413f
of human disease genes, **3**:221–27
of human DNA repair genes, **2**:47
mouse genome, **4**:122
peptide, **5**:179–80
physical, **3**:217
of the fragile X region, **2**:322f
of the mouse genome, **4**:122
of *Streptomyces coelicolor*, **5**:507
using CpG islands, **2**:74
of the *SRA1* gene, **5**:443
of *SRVX* (*DSS*) gene, **5**:442
of susceptibility to drug abuse, **2**:151
of susceptibility to insulin-dependent diabetes mellitus, transracial studies, **2**:31–35
transcription, of chromosome 16, **3**:186
MARCKS protein, **1**:459
Marcus relation, **2**:195
Marfan syndrome, **2**:426; **3**:162–63
gene for, mapping to chromosome 15, **3**:160
neonatal, **3**:162–63
Marker gene
in genetic manipulation of plant cells, **4**:454–55, 454t, 456f
for plant cell transformation, tomato, **6**:54
TK, for recombinant vaccine, **6**:166
visible, in *Drosophila*, **2**:124
Markers
for alcoholism risk, **2**:147
antibiotic resistance, **4**:161–62
for breast cancer, **1**:224
cell-surface, immunologic roles of, **3**:276
chromosome 15, **3**:172t
cotransformation, interaction with gene of interest, **1**:236
CpG islands, **2**:68
of differentiation, epidermal keratinocytes, **2**:267–68
D-loop of mitochondrial DNA, sequence polymorphism, **4**:339–40
DNA
in immuno-PCR, **3**:289–90
mouse genome, **4**:122–24
polymorphic, **3**:97
extracellular, expression by keratinocytes, **2**:269
for genetic mapping, **2**:387; **3**:217, 224–25; **4**:338
Caenorhabditis elegans, **1**:231
chromosome 16, **3**:178–81t
chromosome 9, **3**:99–01t
male, female, and sex-equal patterns in chromosome 9, **3**:102f
molecular, indicating independent origin of multiple cancers, **3**:496
oncogene changes as, **2**:62
for plant cell transformation, **6**:54
polymorphic, **3**:236
in neoplastic disease diagnosis, **2**:64
population-specific, **5**:44–49
primer sequences as, **3**:260
reference, chromosome 11, **3**:135t
selectable, **2**:273; **6**:301–2
antibiotics as, **5**:506
thymidine kinase as, **6**:293
selection of, for genetic analysis of populations, **2**:398–401
selection, kanamycin resistance as, **4**:469, 469f
supF, for constructing jumping and linking libraries, **2**:75f
susceptibility, validation of, **2**:260–61
types of, **2**:67
for yeast artificial chromosome library preparation, **6**:297
yeast vector, **6**:311
Markov analyses, **6**:43
Markov chains, **6**:39
Marmur-Schildkraut-Doty equation, **2**:5

Mass action effect, and rate of adenosine triphosphate synthesis, **1:**312
Mass chromatography, **5:**476–77
MASS SPECTROMETRY
for amino acid sequencing, **5:**184, 192–93
for analysis of biomolecules in solution, **4:**218
for analysis of steroid hormones, **5:**476–77
for carbohydrate analysis, **1:**272, 274–76
defined, **5:**173, 388
for DNA sequencing, **5:**394–95
HIGH SPEED DNA FRAGMENT SIZING, **4:1–12**
OF BIOMOLECULES, **4:12–28**
of peptides, **5:**180
for protein analysis, **5:**186
and sequence analysis of proteins, **5:**175
Mass spectrum, MALDI, negative ion and positive ion, **4:**4f
Mass transfer, limitation on, enzymes in organic media, **2:**251
Mass transport, balance with surface reactions, biosensors, **1:**180, 180f
Material adequacy, of a model, **6:**53
Maternal inheritance, of mtDNA, **4:**81
Mating types, yeast, α and a, **6:**304–5
Matrix-assisted laser desorption ionization (MALDI), **4:**3–6, 15–17, 17f, 16t
defined, **4:**1, 12
proinsulin spectrum, **4:**27f
for protein identification, **5:**126
for protein mass spectrometry, **5:**192
for protein measurements, **4:**19; **5:**175
Matrix attachment region (MAR), **6:**102
chromosomal, **1:**370–71
defined, **6:**27
nuclear, role in apoB expression, **3:**412
Matrix compounds, for MALDI analysis, **4:**8t
Matrix metalloproteases (MMPs), tissue remodeling by, **2:**286, 290
Matrix protein
defined, **6:**183
and replication of rhabdoviruses, **6:**190–91
utilization by enveloped viruses, **6:**188
Maturation, of peptide hormone precursors, **4:**341, 343–45, 343f
Maturation promoting factor (MPF), in spermatogenesis, **3:**507
Maturity onset diabetes of the young (MODY)
defined, **2:**28
genetics of, **2:**28, 38–39, 214
Maximum entropy relationship, **6:**48–49, 48f
Mean residue weight, in circular dichroism reporting, **1:**388
Mechanical work, conversion of energy to, by motor proteins, **4:**114
Mechanism
of carboxylation, by biotin-dependent enzymes, **1:**406f
of copy number variation, **6:**3–5
defined, **6:**1
of enzyme reactions
CoB_{12}-dependent, **1:**403, 405f
hydroxylases, **1:**208–9
of regulation, of tyrosine hydroxylase, **1:**215–17
Mechanochemical enzyme, defined, **4:**112
Medaka
small-sized fish model, **6:**85
transgenic, **6:**88, 88f
for β-actin gene promoter and rtIGF transgene, **6:**95–96
histological examination, **6:**97f
Median effect equation, in combination chemotherapy, **1:**266
Mediators, from endotoxin-stimulated macrophages, **3:**471
MEDICINAL CHEMISTRY, **4:28–42**
applications of collagen fibers, **5:**33
defined, **2:**155
Medium, for plant tissue culture, **6:**12
Medium chain acyl coenzyme A dehydrogenase (MCAD) deficiency, **2:**425
Medium reiteration frequency repeats (MERs), **3:**245–46, 512
Medroxyprogesterone acetate
antitumor activity of, **5:**488
for cancer therapy, **5:**486
Megabase, defined, **1:**370
Megestrol acetate, antitumor activity of, **5:**488
Meiosis
crossing over during, **3:**513
defined, **2:**118; **3:**233; **6:**28
genome analysis during, **6:**309–10
motor protein roles in, **4:**118
recombination during, **3:**237; **5:**252, 254
near centromeres and telomeres, **6:**9–10
recombination repair during, **2:**82
telomere behavior during, **6:**31
yeast, **6:**304
Meiotic mapping, **3:**114–15, 115f
Mel-14, defined, **2:**446
Melanin, limitation of UV damage by, **6:**157
Melanization and reddish colorization hormone (MRCH), **4:**195
Melanocortin receptors, gene for, mapping to chromosome 18, **3:**207, 208
Melanoma, **1:**257
familial susceptibility gene, chromosome 9, **3:**103
injection of, with liposomes containing the HLA-B7 gene, **2:**433
α-Melanotropin (α-MSH), peptide and mimetics, **4:**380
Melphalan, structure of, **1:**255f
Melting, oligonucleotide, thermodynamics of, **4:**238–39, 238f
Melting curve
DNA, **2:**2, 4f
defined, **2:**1
and polynucleotide denaturation, **4:**235–36, 236f
Melting temperature (T_m)
defined, **1:**72–73; **4:**235
for DNA, **2:**2
for assessing hybrid stability, **2:**383
sodium ion dependence of, **2:**8
Membrane, proton electrochemical potential difference across, **4:**448
Membrane-associated proteins, of respiratory syncytial virus, **5:**290
Membrane-bounded organelles, **4:**120
defined, **4:**112
translocation by kinesins, **4:**119, 119f
Membrane breakdown
defined, **2:**198
in an electric field, **2:**198–200
Membrane curvature, defined, **6:**183
Membrane elution method, **1:**97, 100, 102
MEMBRANE FUSION
defined, **1:**59; **4:**42
with dielectrophoresis, **2:**201
MOLECULAR MECHANISM OF, **4:42–53**
role of annexins in, **1:**59
study of, with electroporation, **2:**205
Membrane oxygenators
from poly(tetrafluoroethylene), **5:**39
from silicone rubber, **5:**37
Membrane pores, mimicking, **1:**140
Membrane proteins
circular dichroism analysis of, **1:**394–95
intrinsic (CF_0) and extrinsic (CF_1), **4:**448–49, 448f
lysosomal targeting of, by cytoplasmic tail sequences, **5:**203
retention of, in the endoplasmic reticulum, **5:**199, 201
topography of, **5:**199
Membrane receptors, **2:**208–9, 208f
classification of, **5:**231t
Membranes
anchoring of proteins to, **4:**439–40
binding to, myristoylation step in, **5:**55
biophysical labeling to study, **3:**389–90
chromophore, **4:**441
external face of, electropermeabilized red blood cell, **2:**200f

lipid bilayer, **1**:193
formation by fatty acids, **3**:455
model for electroporation, **2**:199
mitochondrial, **1**:308f
origins of, evolutionary, **4**:294
proton electrochemical potential difference across, **4**:448
Membrane systems
amplification mode, **2**:180
energy-transducing, **1**:307f
Memory
defined, **4**:54
signal transduction and, **5**:238
MEMORY AND LEARNING, MOLECULAR BASIS OF, **4:53–58**
Mendelian inconsistency, for detecting linkage map errors, **3**:237–38
MENDEL program, **3**:237
Mendel's laws, transmission of a *Pac*I RLGS spot, **5**:317f, 318
Meningioma
defined, **4**:181
gene for, mapping to chromosome 14, **3**:159
mutation of the *NF2* gene in, **4**:185
Mephenytoin polymorphism, **4**:402
Mercury
autoimmunity induced by, **1**:91–92
toxicity of, **1**:134
Merocyanines
540
derivatives of light-activated, **6**:215f
preactivated, effect on HIV-1, **6**:215f
structure of, **6**:214f
as photosensitizers, **6**:213–14, 214f
Merrifield technique, **4**:383
chemistry of, **4**:386–88, 386f
Meselson-Radding model of homologous recombination, **5**:259, 259f
Mesolimbic dopamine system, response to drugs, **2**:150–51
Mesophile, defined, **2**:240
Messenger ribonucleoprotein complexes (mRNPs), **6**:122
Messengers, hormone and neurotransmitter, **2**:207. *See also* First messengers; Second messengers
Metabolic engineering, **1**:174
Metabolic labeling, in gel electrophoresis of proteins, **2**:355
Metabolic pathways, **2**:229
Metabolism
bacterial, and biodegradation of waste, **1**:119–22
cellular, thermodynamics of, **1**:305
of cholesterol, **3**:409–10
defined, **4**:257; **6**:259
of drugs, **1**:193, 260
phase I, **4**:399, 400–3
phase II, **4**:399, 404–5
effects of neuropeptides on, **4**:195
enzyme control of, **1**:113
of food, **4**:257–58
glucose, **1**:161
inborn errors of, **1**:113
isoenzyme control of, **3**:373–74
lipoprotein, **1**:285t; **4**:495–96
in transgenic animals, **4**:502–6
in living systems, **1**:304
neuropeptide, **4**:197
phospholipid, **3**:410
role of biotins in, **6**:241–42
xenobiotic, **1**:193
Metabolites
secondary
amino acid and hydroxy acid, **4**:208–9, 208t
defined, **4**:393
in plant defense against pathogens, **4**:479–80
in vivo NMR determination of, **4**:225–26
Metabolizable energy (ME), **4**:258
Metal ions
binding of, **1**:132
by HIV-1 Rnase H, **1**:33
bioavailability of, **4**:59–60
cofactors for ribozyme cleavage reactions, **4**:233–34, 234f
complexation kinetics, **1**:133–34
and control of photosynthetic gene expression, **1**:357–58
divalent, requirement of HIV-1 integrase, **1**:33
hydrolysis of, **1**:129
METALLOENZYMES, **4:58–66**; **6**:60
defined, **6**:59
yeast AP endonuclease I, **6**:287
zinc, sPLC, **4**:428
Metalloprotease inhibitor domain, amyloid precursor protein, **1**:46
Metalloproteases, activity of, around tumors, **1**:260
Metallothionein
enhancer, in transgenic animals, **6**:70
gene for, **6**:60, 61, 61f
Metallothionein-like 3, gene for, mapping to chromosome 18, **3**:208
Metal-response-element-binding transcription factors (MTFs), **6**:61
Metal response elements (MRE), **6**:59, 61
defined, **6**:59
Metals
enzyme cofactors, **1**:126
essential, composition of the human adult, **4**:59t
nutritional requirement of microbes, **1**:119
toxic, **1**:127
Metamorphosis, insect, **4**:189
Metaphase
chromosome 18, high-resolution GTG-banded, **3**:193f
fragile site appearance at, **2**:315–16, 316f
GTG banding for chromosome microdissection, **1**:372–73
images in FACS, **2**:388f
images in FISH, **2**:380f
scaffold at, **1**:371
Metaphase chromosomes, defined, **1**:370
Metarhizium anisopliae, source of mycoinsecticides, **2**:343
Metarhodopsin II, **5**:324
Metasomatism, in mineralization, **4**:300
Metastasis
defined, **1**:253
and invasion, **1**:260–61
Metazoans, mitochondrial genome of, **4**:79
organization, **4**:79
Meteorites, carbonaceous, **4**:291
Methane monooxygenase, **4**:61–62
Methanesulfonate, antitumor activity of, **5**:497
Methanofuran, **4**:69
Methanogen coenzymes, **1**:406
Methanogenesis
by anaerobic bacteria, **1**:121
biochemistry of, **4**:69–73
METHANOGENS AND ARCHAEA, MOLECULAR BIOLOGY OF, **4:66–78**
Methanol, methanogenesis from, **4**:71–72, 72f
Methanotrophs, **1**:121
lipids of, **3**:429
β-Methasone, **5**:483
Methemoglobin, **3**:23
Methemoglobinemia, **1**:113
gene for, mapping to chromosome 18, **3**:214
Methemoglobin reductase, **1**:113
Methionine, role of vitamin B_{12} in synthesis of, **6**:251
Methionine aminopeptidase, pathways for protein modification by, **3**:351, 351f
Methionyl-tRNA synthetase (MetRS), zinc-binding domain of, **6**:336
Methotrexate (amethopterin), **1**:402
inhibition of thymidylate synthase by, **1**:254
structure of, **1**:255f
for synchronizing cell growth, **2**:379
3-Methyladenine glycolase repair enzyme, **1**:7
Methylamines, methanogenesis from, **4**:71–72
Methylases, **5**:297
cognate, **5**:301

Methylases (*cont.*)
of a cytosine residue, **5**:301–2
DNA
change in circular dichroism spectrum with binding of S-adenosylmethionine, **1**:394
circular dichroism spectra, **1**:396f
Methylation
in Achilles' cleavage reactions, conditions for, **1**:4
in carbohydrate analysis, **1**:272, 275f
of a CpG island, **2**:321–23
CpG island, in fragile X syndrome, **2**:323
defined, **1**:407
DNA, defined, **2**:439
enzymes sensitive to, **5**:317t
and intensity of RLGS spots, **5**:318, 318f
patterns of, **3**:517
in genomic imprinting, **2**:444, 444f; **3**:161–62
posttranslational modification, **5**:54
sites of adenine and cytosine, **5**:298–99f
in xenobiotic metabolism, **6**:268–69, 268f
N-Methylation
effect on conformational constraint, **4**:381–82
Methylation interference, for assaying DNA-protein complex formation, **4**:471
Methylcobalamin, **1**:403
Methylcoenzyme M methylreductase (MCR), **4**:76–77
Methylene blue, as a guanylyl cyclase inhibitor, **4**:206
Methylenetetrahydromethanopterin (MTD), F_{420}-dependent, gene encoding, in methanogens, **4**:77
Methyl groups, chiral, analysis of, **1**:345–46, 346f
O^6-Methylguanine-DNA methyltransferase (MGMT), repair of alkylation damage to DNA by, **6**:284–85
Methylhydroxylamine linkage, thymidine nucleoside dimers, **4**:279, 279f
N^G-Methyl-L-arginine (NMA), inhibition of NO synthase by, **4**:205f
Methyl mercury, toxicity of, **1**:134
Methylnitrosourea (MNU), DNA adducts induced by, and aging, **1**:7
β-Methylphenylalanine, in peptide side chains, **4**:379, 379f
Methylphosphonates, oligonucleotide analogues, **4**:283
Methyltransferases, **1**:199
activity of, in Achilles' cleavage, **1**:4
defined, **1**:1
direct reversal of DNA damage by, **2**:44
DNA, circular dichroism for characterizing, **1**:395
S-, methylation of sulfhydryl groups by, **1**:200
β-Methyltryptophan, in peptide side chains, **4**:379
β-Methyltyrosine, in peptide side chains, **4**:379
Mevalonate kinase, insertional activation of, by hepatitis B virus, **3**:489
Mevalonate pathway, **3**:427, 429
Mica, substrate for DNA imaging with STM, **5**:409
Mice
genetically engineered, **1**:64
homeobox gene expression, **3**:65f
long-sleep and short-sleep, responses to ethanol, **2**:148
SxI deletion, in XY sex reversal, **5**:446
transgenic, **1**:64
defined, **2**:440
See also Transgenic mice
Micelles
casein, **2**:308
defined, **3**:447
lipid, defined, **3**:442
protein, **2**:307
reversed, **2**:250–51
Michaelis constant, defined, **3**:366
Michaelis-Menten equation, **1**:173, 174, **2**:234; **3**:369
Michaelis-Menten kinetics, for pancreatic lipase, **2**:257
Microbeads, magnetic, for immuno-PCR, **3**:292
Microbial Genome Initiative, **4**:68
Microbore column chromatography, **5**:176–77, 177t
Microcloning, of chromosomal DNA, **1**:377, 379
Microdissection
of chromosomes, **6**:254, 256
defined, **1**:370
methods, **1**:373–77
for genomic library construction, **2**:73–74
preparation of chromosomes for, **1**:371–73
Microdomain, membrane, **4**:51
Microemulsions, water-in-oil, **3**:314–15
Microfabrication, for DNA manipulation and analysis, **5**:395
Microfilaments
ATPase activity of myosin in the presence of, **4**:118
defined, **4**:112
Microfossils, **4**:290f
$β_2$-Microglobulin, **3**:169, 500
regulator, gene localization to chromosome 15, **3**:168
Microheterogeneity, defined, **2**:458
Microinjection
DNA, for transgenic animal modeling, **6**:67–69
with electroporation, **2**:206
for gene targeting, **4**:102, 453, 453f
in fish, **6**:87–88, 87t
in livestock, **6**:100–101
for introducing YACs into cells, **6**:301–2
for plant cell transformation, **4**:464
Microorganisms
chiral structures in, **1**:334
DNA fingerprinting of, **2**:61
Microsatellite marker, **2**:387; **3**:175, 224–25
chromosome 16, **3**:184–85f
defined, **2**:386; **3**:222
DNA, **2**:400–401
mouse genome, **4**:124
representation of locus, **3**:224f
variable number of random repeats, **2**:67–68
Microsatellites, **3**:229, 241–42
disease-related, amplification of, **6**:5
DNA, **6**:1
genetic analysis of populations using, **2**:401f
human genome, **3**:143
sequences in, **6**:2
tandem arrays in, characterizing, **6**:8–10
Microscopy
in molecular paleontology, **4**:303
in nucleic acid-protein complex characterization, **4**:251
Microsequencing, **2**:357, 359f, 360
Microsomal triglyceride transfer protein (MTP), **3**:404
Microsomes
defined, **5**:328
protein processing in, **5**:335–36
steroid synthesis in, **5**:479
Microtargeting, for plant cell transformation, **4**:461
Microtiter plate wells, for immuno-PCR, **3**:290–91, 290f
Microtubule-associated proteins (MAPs), **1**:210
gene localization to chromosome 15, **3**:168
Microtubule organizing center (MTOC), **4**:119f
Microtubules, **4**:119, 119f
ATPase of kinesin and dynein in the presence of, **4**:118
defined, **4**:112
Microviscosity, membrane, measuring, **2**:299, 299f
Midazolam, **2**:160
Milk proteins, **2**:308

Miller-Dieker syndrome, **4**:424
unbalanced translocation of chromosome ends in, **6**:35
Miller spread, defined, **2**:182
Mimetic agents
defined, **1**:135
peptide, **4**:372
Mimicry
of DNA by PNA, **4**:273
of glycoproteins and glycolipids, by bacteria, **1**:279
Mineralization, **4**:300
defined, **4**:299
Mineralocorticoids
role of, **5**:482
steroid subgroup, **5**:479
synthesis of, **5**:480
Minerals
defined, **4**:257
nutritional requirements, **4**:266, 268, 268t
Miniantibodies, engineered, **3**:286
Minimal base change, for converting a codon, **5**:422
Minisatellites, **2**:67–68; **3**:229, 241–42
copy number variation, **6**:8–9, 8f
defined, **2**:48
DNA of, **6**:1
as a genetic marker, **2**:400, 400f
sequences in, **6**:2
See also Variable number of tandem repeats (VNTR)
Minisatellite variable repeat (MVR), for individual identification, **4**:339
Minor grooves, defined, **6**:327
Minoxidil, prodrug, sulfation of, **1**:198
Minus strand, DNA. *See* Sense strand
Misacylation, of tRNA, **5**:139
Mismatch
base loop, **5**:360
correction of, and genetic stability, **2**:411
defined, **5**:368
internal loop, **5**:362
nucleotide, in PCR, **4**:337
RNA, G-A, **5**:377, 377f
and RNA stability, **5**:371
Mismatch cleavage analysis, **2**:423
Mismatch primer, **4**:282
Mismatch repair (MMR), **5**:260; **6**:290, 290f
defined, **5**:252; **6**:283
as a form of nucleotide excision repair, **6**:285
and gene conversion, **5**:255–56
gene conversion resulting from, **5**:261
Mispairing
by base alkylation, **4**:152
slipped-strand, **4**:151, 152f
and thermal denaturation, DNA, **2**:12
Missense mutation
defined, **4**:149
factor VIII gene, in hemophilia, **3**:33–34t
in hemophilia B, **3**:38, 38f
inactivation of p53 resulting from, **1**:412
Missense variant
defined, **4**:198
intragenic complementation with, **6**:319
in phenylketonuria, prevalence of, **4**:409
in X-chromosomal recessive nephrogenic diabetes insipidus, **2**:25
Mitochondria
bridge to the endoplasmic reticulum (ER), **4**:438
calcium-transporting systems of, **1**:247–48
defined, **1**:315; **4**:79
electron transport system of, **1**:309f
enzymes for phospholipid biosynthesis in, **4**:433
gene expression in plants, **4**:466
hsp70 of, **1**:317f
membranes of, **1**:308f
origin of, **4**:83–84
ribosomes of, **5**:328–29
role in cellular energetics, **1**:307f
single-stranded DNA binding protein of, **5**:462, 465
transfer of genes with chloroplasts, **4**:84
transformation of, **6**:311
Mitochondrial carrier family (MCF), **6**:128, 130
MITOCHONDRIAL DNA, **2**:399–400; **4:79–84**
Chlamydomonas reinhardtii, **1**:353
uniparental transmission of, **1**:348
EVOLUTION OF HUMAN, **4:84–90**
of *Plasmodium falciparum*, **4**:308
Mitochondrial encephalomyopathy
lactic acidosis and strokelike symptom (MELAs), **4**:82
mapping to chromosome 18, **3**:214
Mitochondrial genome
Arabidopsis, **1**:83–84
Saccharomyces cerevisiae, nomenclature describing, **6**:307–8, 307t
Mitochondrial import, **5**:205–6, 205f
Mitochondrial signal peptide, circular dichroism analysis of, **1**:390
Mitogen-activated protein kinase (MAPK), **2**:210; **5**:241
Mitogenesis, role of epidermal growth factor receptor kinase in, **2**:468
Mitosis
chromosome movement during, **4**:113
defined, **4**:113; **6**:28
histone changes in, **3**:50
karyotype in, **2**:387
kinesin-related proteins in, **4**:120
motor protein roles in, **4**:118
Mitotic nucleus, organization of, **6**:31
Mitoxantrone
as an anticancer drug, **1**:254
structure of, **1**:255f
Mitral valve prolapse syndrome, gene for, mapping to chromosome 15, **3**:163
Mixed-function oxidase, **1**:436
defined, **1**:435
Mixed-probe arrays, **3**:254
M-line proteins, **4**:143
Mnemogenetics, *Drosophila*, **2**:132
Mobile carriers, **6**:129–30
transport across a membrane by, **6**:128
Mobile elements, **2**:216
Mobility, genes affecting, in *Drosophila*, **2**:142
Mobilization
defined, **5**:1
for transfer of nonconjugative plasmids, **5**:8–9
Models
fish, small, medium and large, **6**:85
mouse, for colon cancer, **1**:413
from X-ray diffraction data, **6**:279–80
zinc finger motifs, from extended X-ray absorption fine structure, **6**:328
Modification, oligonucleotide, defined, **4**:281
Modified phosphate backbone, defined, **4**:269
Modifier gene
defined, **2**:439
effect on epigenetic programming, **2**:445
Molds, agricultural, biotechnology of, **2**:343–44
Molecular assembly pathways, dissecting, in *Caenorhabditis elegans*, **1**:236–37
Molecular backcrossing, **5**:450
defined, **5**:447
with sexual PCR, **5**:457
Molecular chaperones, **5**:81
defined, **5**:73, 194
Molecular characterization, of high temperature enzymes, **2**:246–247, 246t
Molecular clock, **2**:393
defined, **2**:392
Molecular design, **4**:93
Molecular diagnosis, conditions identified by, **2**:422t
MOLECULAR DIVERSITY, THEORETICAL ASPECTS OF, **4:90–100**
Molecular drive, bias from, in genetic analysis of populations, **2**:394
Molecular dynamics (MD), **5**:166–67
for RNA structure determination, **5**:379
Molecular electronics, principles of, **1**:149–50

MOLECULAR GENETIC MEDICINE, **4:100–105**
Molecular genetics
of fragile sites, **2**:313–15
of fragile X syndrome, **2**:324f
of hepatocellular carcinoma, **3**:486–87
of phenylketonuria, **4**:407–15
Molecular lesions, in the human phenylalanine hydroxylase gene, **4**:409
Molecular markers, for mapping *Arabidopsis* genes, **1**:79–80
Molecular mechanics
modified, for RNA three-dimensional structure modeling, **5**:380
optimization using, for RNA structure determination, **5**:379, 383
Molecular mechanism, in imprinting, **2**:444–45
Molecular modeling
of membrane channels, **1**:140
of membrane fusion, **4**:42–43, 43f
Molecular orientation, antibody-mediated, **1**:151
Molecular probes, inserting in cells, with electroporation, **2**:206
Molecular replacement, **6**:279
for protein structure determination, **5**:114
Molecular systems, organized, **3**:311–16
Molecular wires, **1**:154
Mollusc shell, hierarchically ordered structures of, **1**:147
Molsidomine, **4**:206, 206f
Molten globule state, **5**:148, 150
binding of GroEL family of molecular chaperones to, **5**:88
folding from, kinetic block to, **5**:154
and inclusion body formation, **5**:84
as an intermediate in protein folding, **5**:155
protein, **1**:391
in protein refolding, **5**:77
Molting, behavioral components of, **4**:192
Molybdenum
in nitrogenase, **4**:65
of methanogens, **4**:77–78
in transcription of genes for FDH, **4**:77
Molybdoenzymes, metal-containing cofactors for, **4**:60
Molybdopterin, **2**:248
Monensin, effect on viral budding, **6**:188
Monoamine oxidase, **1**:195
flavin adenine dinucleotide covalent attachment to, **1**:400
Monobactams, aztreonam, **2**:157
Monoclonal antibodies (Mabs)
antiphosphorylcholine McPC603, **3**:278f, 283, 284f
versus autoantibodies, **5**:417
cloning of, **1**:417
defined, **1**:62; **2**:165; **5**:31; **6**:172
for disease treatment, **1**:420
drug targeting with, **5**:34
homogenous amino acid sequences of, **3**:276
for identifying variants of respiratory syncytial virus, **5**:288
linkage of drugs to, **2**:166, 167
MEL-14, **2**:455
mouse class IgG_1, **3**:290
for neutralizing varicella-zoster virus or cytomegalovirus, **6**:179
production of, **1**:172, 175
receptor-specific, purification with affinity chromatography, **6**:232
recombinant, from HuSCID mice, **1**:422
saturation of viral receptors with, **6**:229
specificity of binding to isoenzymes, **3**:369
for studying virus assembly, **6**:189
techniques for identify catalytic antibodies, **1**:68
against tumor cells, **1**:257
Z-DNA, binding to bovine papillomavirus (BPV-1) DNA, **2**:187f
Monoclonal autoantibody, specificity of, **1**:90
Monoclonality
defined, **1**:250
of tumors, **1**:251
Monod equation, **1**:263
for whole-cell bioreactor kinetics, **1**:174
Monogenic resistance, defined, **4**:476
Monokine, **1**:442
defined, **1**:441
Monomeric proteins, aggregate formation in refolding, **5**:74–78
Monooxygenase, **1**:121; **6**:261–62
deficiency of, in fish odor syndrome, **4**:405
flavin-containing, oxidation of heteroatoms by, **1**:195, 195f
Monosaccharides, analysis of, **1**:267
Monoubiquitinylation, defined, **3**:347
Morphine, **2**:156, 156f
responses to, variability in, **2**:153
structure of, **4**:29f
Morphogenesis, and cell adhesion, **1**:293
Morpholino nucleoside oligomers, **1**:71; **4**:273
Morphology
crystalline
additives affecting, **3**:311, 312–14
calcite and calcitic cocliths, **3**:310f
defined, **3**:309
enzyme, effects of solvents on, **2**:252
Mortality
due to alcohol or drug abuse, **2**:147
from malaria, **4**:307
Mosaicism
defined, **2**:420; **4**:173
in Down syndrome, **2**:119
in fibromatosis 1, **4**:175
in fragile X linked mental retardation, **2**:323
genetic analysis in the presence of, **2**:423–24
in McCune-Albright syndrome, **2**:215
in Prader-Willi syndrome, **3**:172
in transgenic fish, **6**:90, 93
in XY hermaphrodites, **5**:446
Mosaics
defined, **6**:99
for examining lethal mutants, in *Drosophila*, **2**:135–36
of introns and exons in globin genes, **3**:13, 15f
from microinjection experiments, **6**:67
Mosquitos, anautogenous, **4**:192
Mother, mitochondrial ancestral, **4**:87
Motif
defined, **5**:428, 516; **6**:38
peptide, for predicting protein sequence, **5**:432
in sequence alignment by pattern matching, **5**:425
Motion
molecular, and external fields, **2**:174–75
thermal, and electric and magnetic fields, **2**:174
MOTOR NEURON DISEASE, **4:105–12**
defined, **4**:105
Motor neurons, regulation of, **4**:110–11
MOTOR PROTEINS, **4:112–20**
defined, **4**:113
Mouse, physical map compared with genetic map, **4**:125, 125f
MOUSE GENOME, **4:120–28**
Sry gene, **5**:439
Sxy deletion, sex reversal in, **5**:445
T-locus mutants, sex reversal in, **5**:446
XX males, genes affecting, **5**:444
XYd sex reversal, **5**:445–46
Mouse Genome Project, **4**:122
Mouse mammary tumor virus (MMTV)
defined, **5**:508
infection by, **5**:511
pseudoknot-related frameshifts of, **5**:362
Moving fluids, motion of charges in, **2**:174
M phase, cell cycle, defined, **6**:138
M protein, antibodies against, in rheumatic heart disease, **1**:105
mRNA
for activin type II receptor subunits, **3**:510
archaeal, **4**:75
binding of antisense oligonucleotides to, **1**:70, 70f
of *Caenorhabditis elegans*, **1**:229
editing of apo B, **4**:497–500, 500f

eukaryotic, **6:**118
expression of, in fragile X syndrome, **2:**323
factor VIII, **3:**30
factor IX, **3:**36–37
ferritin, IRE-BP affinity for, **6:**62
formation of an initiation complex from, **6:**114
and hemophilia A, **3:**35
histone, **3:**51
mapping *fem-3* genes to, in *Caenorhabditis elegans*, **1:**235
modification and processing of, **5:**356–57
for N-CAMs, **1:**295
for p53 protein, **6:**141–42
poly(A)-tailed, **1:**203
prokaryotic, **6:**114–17
and protein synthesis, **5:**355–57
PrP, constitutive expression in brain of adult animals, **5:**59
psaA, chloroplast of *Chlamydomonas reinhardtii*, **1:**357
for quantifying expression of genes, degradative pathways, **1:**124
Rb, **3:**495
RB, **6:**140–41
of respiratory syncytial virus, **5:**287–88
stability of, and control of protein translation, **6:**121–22
structure of, and the genetic code, **6:**111–12, 111f
Tat-induced levels of, **1:**25
for Tat transactivation, **1:**21
transcription of, **2:**366
transferrin receptor, IRE-BP affinity for, **6:**62
transgene-encoded expression of, **6:**72
tyrosine hydroxylase, **1:**211, 216
viral, translation of, **6:**198
mRNA 5′ cap structure, defined, **5:**328
mRNA splicing, defined, **2:**279
msh class, of homeobox genes, **3:**72
mtDNA
human, **4:**85f
lineage, defined, **4:**84
region V mutation, **4:**86
type, defined, **4:**84
See also MITOCHONDRIAL DNA
Mucins, glycoprotein antigens and TACAs from, **1:**282–83
Mucosa, immune mechanisms of, and respiratory syncytial virus infection, **5:**292
Mucosa-associated lymphoid tissue (MALT) lymphomas, translocation between chromosomes 11 and 18 associated with, **3:**211
Multibranch junction, defined, **5:**360
Multibranch loop
defined, **5:**360
RNA, **5:**362
Multichain synthesis, of peptides, **4:**392–93
Multichannel communication, **5:**388
Multicolor, multiprobe FISH, **2:**384–85
Multidimensional NMR, **4:**220–23, 223f
Multienzyme complexes, defined, **4:**207
Multienzyme polypeptides, defined, **4:**207
Multifactorial disorders, **3:**230
Multigene families
Arabidopsis genome, **1:**77
hsp70, **1:**317f
Multilamellar vesicles (MLV), **3:**416
Multimeric repeats, satellite, **6:**9–10
Multipin synthetic peptide libraries, **5:**518
Multiple antigen peptide system (MAPS), **4:**368–69, 369f
Multiple cassette mutagenesis, combinatorial, **5:**452
Multiple cloning site, **4:**255
Multiple endocrine neoplasia type 2A (MEN2A), **2:**429
mapping of, to chromosome 10, **3:**115
Multiple forks, defined, **1:**95
Multiple intestinal neoplasia (Min), mouse strain, **1:**413
Multiple myeloma
interferon-α therapy for, **3:**330
translocation in, **2:**65
Multiple sclerosis (MS), interferon-β therapy for, **3:**330
Multiple sequence alignment, **5:**433, 433f
Multiple tumor associated chromosome region I (MTACRI), mapping to chromosome 11, **3:**127
Multiple tumor suppressor 1 (MTS1), gene in tumors, **4:**287
Multiplex approach, to physical mapping, **3:**220
Multipoint linkage analysis, **3:**113–114, 114f
Multivaccines, recombinant, **4:**163
Mössbauer labels, **3:**388–389, 389f
Müllerian inhibitory substance (MIS), **2:**468, **3:**502
in the female reproductive system, **2:**288–89
in the male reproductive system, **3:**509
Mummification, **4:**300–301, 301t
defined, **4:**299
Muramyl peptide, antitumor activity of, **5:**498
Murchison meteorite, nonpolar organic compounds extracted from, **4:**294
Murein, recycling of, in cell growth, **1:**99
Muscarinic receptors, **5:**237
MUSCLE
assembly of, in *Caenorhabditis elegans*, **1:**236
contraction of
mechanism, **4:**132, 132f
role of allatostatins in, **4:**198
role of calcium in, **1:**238
direct injection of genetic constructs into, **2:**433, 435
as food, **2:**308
L-type calcium channels of, **1:**243
mammalian, DNA damage in, **2:**79
MOLECULAR GENETICS OF HUMAN, **4:129–47**
smooth, effect of vitamin E on, **6:**238
vitamin B_6 in, **6:**243
Muscular dystrophy
Baker's, **4:**147
genetic basis of, **4:**101
LI insertion in, **3:**245
limb girdle, **3:**163–64
mapping to chromosome 15, **3:**163–64
oculopharyngeal, gene for, mapping to chromosome 14, **3:**158
See also Duchenne muscular dystrophy; Myotonic dystrophy
N-Mustard esters, antitumor activity of, **5:**494
Mutable system, defined, **3:**227
MUTAGENESIS, **4:148–55**
alanine-scanning, **3:**285
algorithms for, sexual PCR, **5:**453
chemical, for characterizing mycobacteria, **4:**163
defined, **4:**148; **6:**327
deletion, tyrosine hydroxylase studies, **1:**211
for genetic analysis in *Drosophila*, **2:**124–26
for identifying biodegradative genes, **1:**125
Insertional, **3:**518; **6:**67
T-DNA, **4:**491
random, to construct enzymes, **1:**144
site-directed, **3:**43
of bacteriorhodopsin, **1:**111, 139
p-element, **1:**12
of proteinase 3C, poliovirus, **5:**19–20
and studies of phospholipase A_2, **4:**421
systematic strategy for, **1:**236
targeted, **3:**519f
transposon, identifying genes for toluene oxygenases, **1:**124
trinucleotide, **4:**95–96
in vitro, for yeast genetic studies, **6:**314
Mutagens
aflatoxin Bs1, **3:**485
pattern of change induced by, **4:**155
defined, **4:**148
environmental, **2:**404
ethylmethanesulfonate, patterns of mutations induced by, **4:**155

Mutagens (*cont.*)
insertional, **4:**456
thermally induced, in foods, **2:**312
and transposon mobilization, *Drosophila*, **2:**123
ultraviolet light, patterns of mutations induced by, **4:**155
Mutants
alanine to threonine, at codon 713, APP gene, **1:**45
alanine to valine, at codon 713, APP gene, **1:**44–45
amber, phage, **1:**65, 65f
amyloid precursor protein gene, **1:**43–45
C7, of hemoglobin, **3:**25
of *Caenorhabditis elegans*, **1:**228–29
cloning, **1:**231–32
recovering genes for, **1:**230–32
conditional-lethal, **6:**201
defined, **6:**200
courtship-specific, **2:**132
dominant-negative, **6:**318
*Eco*RI, for DNA mapping with atomic force microscopy, **5:**405–6, 406f
fragile X, *FMR 1*, **2:**322, 325t
glutamic acid to glutamine, at codon 693, APP gene, **1:**44
HCHWA-D, **1:**43–45
of hemoglobin β-globin, genetically engineered, **3:**22–23
HIV-1, affecting response to AZT and ddI, **1:**31
in Huntington's disease, **3:**249f
identification of, with PCR, **4:**338
mapping to chromosome 18, **3:**206t
of metHb, **3:**19
mouse, mapping to chromosome18, **3:**205t
of mtDNA, **4:**89–90
N-terminal, hemoglobin, **3:**25–26
p53 gene, **3:**488–89, 488t
predisposing to early onset Alzheimer's disease, **1:**42
protein, degradation of, **3:**353–54
radiation-sensitive, as hosts for YAC libraries, **6:**297
ras, in lung cancers, **3:**493
Saccharomyces cerevisiae genome, **6:**307–8, 307t
sensory transduction, in *Drosophila*, **2:**132–33
Sxr, and sex reversed XX males, **5:**436
TAR loop, effect on Tat transactivation, **1:**21
from ultraviolet radiation, outcomes, **6:**163
Mutation, **2:**418–19; **5:**46
accumulation of, **1:**10
defined, **1:**5
and aging, **1:**5–14
chromosome, DNA variations due to, **3:**228–29
in cloned genes, recovery of, **1:**232–33
in the cystic fibrosis transmembrane conductance regulator gene, **1:**432–33, 433f
defined, **1:**113, 407; **2:**40, 78, 407, 415; **3:**227; **4:**121, 226, **4:**148
disorders caused by, **3:**229–30
cystic fibrosis, **2:**421
Gaucher disease, **2:**349
hemophilia B, **3:**37–39, 37t
retinoblastoma, **5:**319–20, 320f, 322
distribution in the genome, **4:**151
drug-resistant, **4:**164, **6:**234
in DNA, **3:**228–29
and neoplastic disease diagnosis, **2:**62
and effectiveness
of drugs for HIV-1, **1:**38
of Nivirapine, **1:**31
effects on genes, **3:**229f
of hemoglobin structure, **3:**16–17, 17t
in the high mobility group of *SRY*, **3:**501
in the human *PAH* gene, distribution of, **4:**409–10
in the human V_2 receptor gene associated with NDI, **2:**26t
identification of, methods for, **1:**114–115
identifying with genes, **1:**232–33
induced, **4:**151–152
defined, **4:**148
and natural selection, **4:**96
nature of, **2:**392–93
nonsense, in hemophilia A, **3:**31
nuclear, affecting chloroplast genes, **1:**356–57
in pharmacogenetic defects, **4:**400
point, in the gene encoding mitochondrial NADH dehydrogenase, **1:**46
resequencing arrays to detect, **3:**259
rate of, **2:**393
defined, **4:**173
for HIV-1, **6:**199
RNA structure analysis using, **5:**365–66
site-specific, in bovine PLA_2, **4:**422t
spontaneous
defined, **2:**402; **4:**148
generation of, **4:**151
mechanisms and effects of, **2:**403
rate of, **1:**12
stepwise, in microsatellite tandem repeats, **6:**10
and tandem repeats, **6:**6
temperature-sensitive, **6:**201
transition, defined, **4:**84
transversion, defined, **4:**84
See also Chromosomes; Genes
Mutational cloning, **5:**503
defined, **5:**501
Mutational heterogeneity, in hemophilia, **3:**41
Mutation analysis, **5:**48
in neurofibromatosis 2, **4:**183–86
Mutation yield, of ionizing radiation, **3:**355
MVR-PCR (minisatellite variant repeat-PCR), **2:**51
defined, **2:**48
MWC model, of allostery, **3:**9–10, 9f
Myasthenia gravis, **1:**86
animal model for, **1:**90–91
MycDB database, **4:**157
MYCOBACTERIA, **4:156–65**
lipids of, **3:**431–32
Mycobacteriophages, **4:**157–59, 159f
Mycobacterium leprae, **1:**447; **4:**156
genetic map of, **4:**158f
Mycobacterium tuberculosis, **1:**447; **4:**156
integrin receptor use by, **1:**106
Mycoinsecticides, **2:**343
Mycolic acids, bacterial synthesis of, **3:**433
Mycoprotein, production of, *Fusarium graminearum* fermentation, **2:**340
Mycosporine amino acids (MASSa), protection from ultraviolet radiation by, **6:**157
Myelin basic protein (MBP), gene for, mapping to chromosome 18, **3:**208
Myelin protein zero (MPZ), mapping of, to chromosome 1, **3:**91
Myeloma proteins, **3:**276
Myoblasts
defined, **4:**129
myoblast determination genes, *MyoD*, **4:**145–46
Myocardial infarction, **1:**284
and atherosclerotic lesions, **1:**288
diagnosis of, with lactate dehydrogenase isoenzymes, **3:**374
Myocardium, damage to, **1:**288–92
Myoclonus epilepsy with ragged-red fibers (MERRF), **4:**82
from mtDNA base substitutions, **4:**89
Myocytes, changes in, and heart failure, **1:**289
Myofibrils, **4:**129, 130–31, 131f
protein components of, **4:**133t
Myogenesis, **4:**145–46
MyoD gene in, **1:**234
Myogenin, **4:**145
Myoglobin (Mb), **3:**21, 21f
alignment of sequences, gray whale and killer whale, **5:**432f
analysis of, with electrospray ionization mass spectrometry, **4:**22, 22f
in carbon tetrachloride, FTIR spectrum, **2:**254f

in chloroethanol, structure of, **2**:253
matrix-assisted laser desorption ionization spectrum, **4**:4f
oxygen-binding curve, **2**:239f
Myomesin, **4**:143
Myopathies, **4**:146–47
Myosin head, **4**:133
actin-activated ATPase activity of, **4**:134
conformation change of MLC of, in smooth muscle contraction, **4**:144
Myosin heavy chain (MHC), **4**:115, 133
mutants of, in *Caenorhabditis elegans*, **1**:236
Myosin heavy chain gene, heart failure related to mutation of, **1**:289
Myosin light chain (MLC), **4**:115, 133
Myosin light chain kinase (MLCK), **1**:241
Myosins, **4**:113–14, 114t
characterization of, **4**:115–16
defined, **4**:113
directionality and actions of, **4**:120
filament, **4**:131–32
and gene switching in heart failure, **1**:290
size and structure of, **4**:115
structure and function, **4**:133–34; **5**:159–60
in thick myofilaments, **4**:130
VII, encoded by the shaker-I gene, **4**:126
Myotonic dystrophy, **2**:328, 420, 427–28; **3**:231–32, 242
parental imprint in, **2**:445
as a single-gene disorder, **3**:230
Myotropins, **4**:195
Myotube
defined, **4**:129
developmental changes in, **4**:143
Myoxin gene, mapping to chromosome 15, **3**:165
Myristoylation, posttranslational modification, **5**:54–55
Mytilus edulis, adhesive protein of, **5**:28

Nail-patella syndrome (NPSI), mapping to chromosome 9, **3**:107
Naive antibody libraries, **1**:423–24
Naked DNA injection, defined, **2**:430
Nano- (prefix), defined, **3**:309
Nanomaterials, **3**:312
Nanometer-scale structures, biomimetic fabrication and patterning for, **1**:140–41; **3**:316
Nanotubules, cyclic peptide self-assembly as, **1**:153
Nanowires, **1**:150
composites for, **1**:155, 155f
Naphthalene biodegradation pathway, **1**:124
Naphthalimides
effectiveness against HIV-1, **6**:216
photodynamic effects of, **6**:214–16, 216f
Naphthalocyanines, for sensitization of viruses to light, **6**:213
National Cancer Institute, **4**:99
National Center for Human Genome Research (NCHGR), **3**:238
National Collection of Yeast Cultures, addresses of, **6**:304
National Institutes of Health (NIH), **3**:143
National Neurofibromatosis Foundation, e-mail address, **4**:182
National Toxicology Program (NTP), **3**:299
Native state, fully folded proteins, **5**:150
Natriuretic peptides, receptors for, **5**:243
Natural fields, magnetic and electric, **2**:173
Natural killer (NK) cells
activation by interferons, **1**:447; **3**:327
defined, **3**:296
immunological destruction of tumor cells by, **1**:261–62
innate immunity mediated by, **3**:296
Natural polymers
biodegradable, **5**:32–34
defined, **5**:31
Natural products, as drug synthesis leads, **2**:155–57
Natural selection, **2**:393; **5**:46–47
defined, **5**:45
and mutation, **4**:96
and tandem repeats, threshold values for fitness, **6**:6
Na^+-K^+ ATPases, **5**:273
N-cadherin (NCAD), gene for, mapping to chromosome 18, **3**:208
Nearest-neighbor effects, on melting temperature and sodium ion counterion interactions, **2**:8–10
Nearest-neighbor model
for predicting RNA secondary structure, **5**:364
of respiration control, **1**:311–12
Nebulization, by electrospray, **4**:13
NEC class, homeobox genes, **3**:72
Necrosis, **1**:299
Necrotroph, defined, **4**:476
Needleman-Wunsch-Sellers algorithm, **5**:432–33
Negative feedback, in ColE1 replication, **5**:3
Negative staining
defined, **2**:182
of proteins, **2**:187
Neisseria gonorrhoea
defense against host immune response, **1**:105
lipopolysaccharides of, **1**:279
Neisseria meningitidis
antigens, **1**:278–79, 278f
lipopolysaccharides of, **1**:279
mimicry of host markers, **1**:105
polysaccharide antigens of, **1**:278
Nemaline myopathy (NEM1), mutation of the α-tropomyosin gene in, **3**:93
Nematoda, body plan of, **4**:167–69
NEMATODES, NEUROBIOLOGY AND DEVELOPMENT OF, **4:167–72**
NEM-sensitive fusion protein (NSF), **5**:201–2
N-end rule, protein degradation, ubiquitin-dependent, **3**:350–53
Neolactotetraosylceramide, **1**:280f
Neomycin
B, aptamer for, **4**:229–30, 229f
streptomycetes transformant resistant to, **5**:502
Neonatal Marfan syndrome, **3**:162–63
Neonatal severe hyperparathyroidism, **2**:215
Neoplasms
and associated tumor suppressor genes, **6**:139t, 146t
defined, **1**:250
origin of, **2**:62
Neoplastic transformation, defined, **1**:253
Neoprontosil, binding to carbonic anhydrase, **1**:394
Neovascularization, **1**:261
Nephroblastoma (Wilms' tumor), **6**:142–44, 144f
Nephrogenic diabetes insipidus (NDI)
cause of, **2**:20
defined, **2**:19
mutations associated with, **2**:24–27t
Nernst equation, **1**:186
resting membrane potential, **5**:513
Nerve growth factor (NGF), **2**:465, 466–68, 468f; **5**:240
Nerve regeneration, using silicone rubber tubing and a collagen polymer, **5**:37
Nerve terminal, defined, **5**:512
Nervous system
of nematodes, **4**:168
networks in, **4**:95
Nesch-Nyhan syndrome, **2**:427
Net energy (NE), from food, **4**:258
N-ethylmaleimide-sensitive factor (NSF), regulatory role in intracellular membrane traffic, **4**:50
Net protein utilization (NPU), **4**:264, 264t
Networks
defined, **2**:378
immune and nervous systems, **4**:95
neural
defined, **5**:428
for secondary protein structure prediction, **5**:162

Neural cell adhesion molecule (NCAM), **4:**56f
Neuralgia, postherpetic, **6:**172
Neural networks
  defined, **5:**428
  for secondary protein structure prediction, **5:**162
Neuraminic acid, defined, **1:**293
Neurite fasciculation, **1:**295
Neurobiology
  Drosophila, **2:**133–37
  nematode, **4:**169–71
Neuroblastoma
  gene for, mapping to chromosome 1, **3:**93
  human liver, suppression with a fusion protein, **1:**67
  n-*myc* amplification in, **4:**286
Neurochemical mutants, in *Drosophila*, **2:**135–37
Neurodegenerative diseases, **1:**302; **5:**516
  cytokine role in, **1:**450
  mitochondrial mutations in, **4:**83
  of motor neurons, **4:**105–12
  prion-related, **5:**57
  spontaneous and inherited, transgenic models, **5:**65, 65f
  xeroderma pigmentosa, **6:**162
Neuroendocrine peptides, **4:**341t
Neuroendocrine system, interleukin-1 role in, **3:**340
NEUROFIBROMATOSIS, **2:**429
  Alu insertion in, **3:**244; **6:**136
  genes for, types 1 and 2, **6:**144
  1, **4:172–80**
    complications of, **4:**174t
    genes for, mapping, **3:**226
    mutation rate of, **5:**46
    as a single-gene disorder, **3:**230
  2, **4:180–88**
Neurofibromin, **1:**412; **4:**173; **6:**144
  function of, **4:**177–78, 177f
Neurofilaments
  degeneration of, and motor neuron disease, **4:**110
  proteins of, **4:**143
Neurogenic muscle weakness, ataxia, and retinitis pigmentosa (NARP), **4:**82
  from mtDNA base substitutions, **4:**89
Neurohemal organ, defined, **4:**188
Neuromedin B (NMB), gene localization to chromosome 15, **3:**169–70
Neuronal development, and adult plasticity, **4:**58
Neuron-neuron adhesion, **1:**295
Neurons
  in *Caenorhabditis elegans*, **1:**227–28
  defined, **4:**54
  histone gene transcription in, **3:**52
  loss of, in Huntington's disease, **3:** 247
  N-type calcium channels in, **1:**243
  peptidergic, in nematodes, **4:**171
Neuroparsins
  antidiuretic peptide of insects, **4:**196
  biological activities of, **4:**198
Neuropathology, of Huntington's disease, **3:**247
NEUROPEPTIDES
  defined, **2:**130; **4:**167, 188
  INSECT, **4:188–98**
    nonribosomal peptide sequences of, **4:**191–92t
  myotropic, **4:**195
  Y, **5:**238
NEUROPSYCHIATRIC DISEASES, **4:198–200**
Neurosecretion
  concept of, **4:**189–93
  defined, **4:**188
Neurosteroids, biosynthesis of, **1:**437
Neurotransmitter receptor, **5:**234t, 237–38
  in the brain, **1:**75
Neurotransmitters
  characterization of, **5:**514–15, 515t
  defects in, Alzheimer's disease, **1:**50
  defined, **1:**207; **5:**230, 512
  Drosophila, defined, **2:**130
  motor neuron, **4:**106
  nitric oxide, **4:**205
Neurotrophic factors, **4:**106
  defined, **4:**105
  effects on motor neurons, **4:**110–11
Neurotrophins, **1:**301; **2:**466–68
  receptors for, **5:**240
Neutrality theory, defined, **4:**84
Neutralizing antibody, **6:**233
  analysis using, **6:**183
  defined, **6:**172
Neutral lipid, defined, **3:**426
Neutral percolating domains, of sequence space, **4:**99
Neutral selection, **2:**411–12, 411f
Neutron diffraction, study of hydrogen bonding with, **3:**267
Neutrophils
  attraction by interleukin-8 to sites of injury, **3:**342–43
  impaired, in leukocyte adhesion deficiency, **1:**459
  interaction of α-actinin and integrin β_2 in, **1:**455
Nevirapine, **1:**29
  binding to reverse transcriptase, **1:**30
  HIV-1 resistance to, and reverse transcriptase binding, **1:**30–31
  as a noncompetitive inhibitor of HIV reverse transcriptase, **2:**236
  structure, **1:**39f
Nevoid basal cell carcinoma syndrome (NBCCS), mapping to chromosome 9, **3:**104
Newton's equation of motion, applied to the motion of atoms in a protein, **5:**166
Nexin II, protease inhibitor, **1:**46
NF-κB
  activation of, **2:**368, 368f
  inhibition by antisense oligonucleotides, **1:**75
  and long terminal repeat, **1:**19
Niacin (nicotinic acid), structure and function of, **6:**249–50, 250f
Nickel, **4:**65
  in hydrogenases of hyperthermophiles, **2:**245
  stability of metal-ligand complexes, **1:**131, 131t
  in urease, **4:**59
Nick translation, **2:**382
Nicotinamide, structure and function of, **6:**250
Nicotinamide adenine dinucleotide (NADH), reduced form, transfer of electrons to cytochromes P450, **1:**436
Nicotinamide adenine dinucleotide (NAD)
  cofactor of ligases, **3:**392, 392f
  as an enzyme cofactor, **2:**232–33, 232f
Nicotinamide adenine dinucleotide phosphate (NADPH)
  reduced form, **4:**399; **6:**250
    pentosephosphate pathway production of, **1:**397–98
  transfer of electrons catalyzed by cytochrome P450, **1:**435
Nicotine, impaired metabolism of, **4:**405
Nicotinic acetylcholine receptors, **5:**231–32, 232f, 237–38
Nicotinic antagonists, effect on tyrosine hydroxylase activation, **1:**216–17
Nicotinic receptors, subunit genes, chromosome 15, **3:**165
Nicotinic stimulation, activation of tyrosine hydroxylase by, **1:**215
Niemann-Pick disease
  ceramide level in, **3:**457
  mapping of, to chromosome 11, **3:**127
  sphingomyelinase deficiency in, **2:**349
  sphingomyelin levels in, **3:**458
  type C, mapping to chromosome 18, **3:**199, 214
Nifedipine, binding to L-type calcium channels, **1:**243
Ninhydrin analysis, to monitor solid phase peptide synthesis, **4:**391
Nisin, **1:**337
Nitrate reductase, deficiency in tobacco plants, relationship with *TutI* retrotransposon, **4:**488

Nitrate reduction, by anaerobic bacteria, **1**:121; **4**:204, 204f
NITRIC OXIDE (NO)
activation of, role of glutathione, **6**:275
IN BIOCHEMISTRY AND DRUG DESIGN, **4**:**200–6**
chemical reactivity of, **4**:201, 201f
defined, **4**:200
free radical, synthesis of in vivo, **2**:331
kinetics of binding to hemoglobin, **3**:8–9
in long-term potentiation, **4**:57f
regulatory roles of, **4**:29
examples, **4**:204t
in the renal system, **5**:268–69
as a second messenger, **1**:239
Nitric oxide reductases, defined, **4**:200
Nitric oxide synthetases, **2**:211; **4**:202; **5**:268–69
defined, **4**:200
genes for, structure, **4**:203f
inhibitors of, **4**:205f
Nitrilase, bromoxynil-specific, **6**:57
Nitrite reductases, **4**:204
defined, **4**:200
Nitrogen
fixation of
by methanogens, **4**:77–78
plasmids' roles in, **5**:12–13
fixed, nutritional requirement of microbes for, **1**:119
Nitrogenase
of archaea, genes for, **4**:77–78
FeMo cofactor for, **4**:60, 60f
iron-sulfur clusters of, **2**:233
metals in cofactors for, **4**:59
Nitrogen dioxide, immunomodulation by, **3**:298
Nitrogen mustards, as anti-cancer drugs, **1**:254
Nitrogen regulatory protein C (NRPC), enhancer-binding protein, **2**:370
Nitrosamines
metabolism by CYP2A6, **4**:402
tobacco-specific, lung tumors induced by, **3**:493
Nitrosoureas
as anti-cancer drugs, **1**:254
antitumor activity of, **5**:494, 494f
Nitrous oxide, withdrawal from, genetic studies, **2**:148–49
Nitrovasodilators, **4**:206, 206f
defined, **4**:200
Nitroxide radicals (NRs), **3**:385–86
NK-1 class, of homeobox genes, **3**:72
NK-2 class, of homeobox genes, **3**:72
N-linked oligosaccharides, classification of, **2**:459f
N-linked sugar chains, **2**:451–53, 451f
NMR (nuclear magnetic resonance), defined, **1**:267
Nomenclature
of apoB species, **3**:411–12
of chromosome banding patterns, **2**:379
of cytochrome P450 superfamily members, **1**:193–94
of cytokines, **1**:442–44, 444t
of enzymes, **2**:224–25, 226t
endonucleases and methylases, **5**:300–1
restriction endonucleases or methylases, **5**:297
of fatty acids, **4**:266t
of fragile sites, **2**:313
of hemopoietic growth factors, 443t
of high-resolution chromosomal bands, **3**:191
of hydrocarbon chain packing, **3**:444t
of insect neuropeptides, **4**:192n
of interferons, **3**:318–19
of interleukins, **3**:332–33, 333t
of isoenzymes, **3**:368
of peptides, **4**:363–64
of RNA structure, **5**:369, 369f
of tRNA, **6**:112n
of unsaturated fatty acid, **3**:438
of vitamins, **6**:245
of the yeast genome, **6**:306–08, 306t
Nonaqueous environments
bioinorganic, **1**:130
enzymes in, **2**:249–58
Nonassociative learning, defined, **4**:54
Noncoded amino acids
analogues of, **5**:138t
defined, **5**:135
incorporating into proteins by suppressing nonsense codons, **5**:143f
Noncoding region (NCR)
3′, of picornaviruses, **5**:23
5′, of picornaviruses, **5**:17, 18f
Noncomplementation, nonallelic, **6**:319–20, 319t
Nonconjugative plasmids, transfer of, **5**:8–9
Noncontiguous patterns, in nucleotide sequences, **6**:47
Noncovalent interactions, bioorganic chemistry, **1**:161–65
Noncovalent protein complexes, analysis of, with electrospray ionization mass spectrometry, **4**:22–23, 22f
Non-cross-resistant agents, in combination chemotherapy, **1**:264–65
Nonenveloped virus, defined, **6**:193
Nongenomic effects, of hormones, **2**:212
Nonhematopoietic growth factors, list, **1**:443t
Non-Hodgkin's lymphoma, interferon-α therapy for, **3**:330
Nonhost interaction, **4**:476
Non-insulin-dependent diabetes mellitus (NIDDM), **2**:28
Nonionic, defined, **4**:269
Non-maturity-onset diabetes of the young, glucokinase polymorphism in, **2**:39
Non-Mendelian determinants, yeast genome, **6**:308, 310
Nonobese diabetic mouse
diabetes mellitus model, **2**:29, 36
susceptibility loci for diabetes mellitus, **4**:126
Nonobviousness (inventive step)
defined, **6**:74
as a requirement for patentability, **6**:76
Nonoverlapping frequency counts, **6**:40–41
Nonpeptide libraries, synthetic, **5**:518, 518f, 522
Nonprocessive transcription
in the absence of Tat, **1**:21
accumulation of TAR RNA in, **1**:26
defined, **1**:19
NONRIBOSOMAL PEPTIDES, **4**:**207–17**
defined, **4**:207
Nonsense mutations
defined, **4**:149; **5**:135
of factor VIII genes, **3**:32t
in hemophilia A, **3**:31
in phenylketonuria, prevalence of, **4**:409
in X-chromosomal nephrogenic diabetes insipidus, **2**:25
Non-small-cell lung cancer
defined, **3**:491
epidermal growth factor receptor (EGFR) expression in, **3**:494
Nonsteroidal anti-inflammatory agents, metabolism by CYP2C9, **4**:402
Nonsyntenic probe, **3**:119–20, 120f
Noradrenaline
inhibitor, for tryptophan hydroxylase, **1**:219
interaction with G proteins, **5**:238
lead for drug synthesis, **2**:157–58
Norgestrel, **2**:164, 164f
Normal transmitting males (NTM)
defined, **2**:319
for fragile X linked mental retardation, **2**:320
Northern blot analysis, **4**:326
defined, **5**:342
method, **5**:347, 347t
Northern blot hybridization, in transgenic animal modeling analysis, **6**:72
Not I jumping library, **2**:74
construction of, comparison of methods, **2**:76t
Nottingham Arabidopsis Stock Centre (NASC), **1**:84

Novelty, as a requirement for patentability, **6**:76
Novrad, **1**:340
N-terminal
  blocking, in protein sequencing, **5**:189–90
  modification, defined, **5**:185
Nuclear envelope
  association of telomeres with, **3**:243
  defined, **6**:28
Nuclear factor κB (NF-κB), **1**:19–20
  activation of, **2**:368, 368f
  inhibition by antisense oligonucleotides, **1**:75
  and long terminal repeat, **1**:19
Nuclear factor of activated T cells (NFAT), **1**:19–20
Nuclear gamma resonance (NGR), **3**:388–89
Nuclear genes, as genetic markers, **2**:398–99
Nuclear genome, plant, **4**:486–88
Nuclear hormone receptors, **6**:331–33
Nuclear import, **5**:206, 206f
NUCLEAR MAGNETIC RESONANCE (NMR) OF BIOMOLECULES IN SOLUTION, **4:218–26**
  for carbohydrate analysis, **1**:276–77
  carbon 13, for carbohydrate analysis, **1**:277
  catalytic triad analysis, of α-lytic protease, **2**:252
  comparison with X-ray diffraction, **6**:276
  for following DNA denaturation, **2**:6t
  for hydrogen bond studies, **3**:267
  one-dimensional, **4**:253
  for protein structure determination, **5**:126, 159
  proton, for carbohydrate analysis, **1**:276–77
  for RNA structure modeling data, **5**:378–79
  secondary structure determination, RNase H domain of HIV-1 reverse transcriptase, **1**:31–32
  zinc finger motif studies, **5**:132
  for zinc finger structure determination, **6**:328–29
Nuclear magnetic resonance force microscopy, **1**:158, 158f
Nuclear matrix, defined, **6**:28
Nuclear matrix protein (NMP-1), **3**:55
Nuclear membrane, calcium ion pump in, rat liver cells, **1**:248
Nuclear moments, **4**:218
Nuclear Overhauser effect (NOE), **4**:219, 220
  analysis of GroEL binding, **5**:88
  defined, **4**:218
  for nucleic acid-protein complex study, **4**:253
  in phospholipases in the presence of inhibitor, **4**:421
  restraint for TAR RNA, bound to arginine, **5**:384
  transfer, **4**:224
Nuclear Overhauser effect spectroscopy (NOESY), **4**:220
  defined, **4**:218
Nuclear pore complexes, **5**:206, 206f
Nuclear receptors, **2**:207–8, 207f
  binding of, and structure, **5**:133
  as transcription factors, **2**:212
Nuclear RNA, **5**:343
Nuclear runoff assay, **5**:350
  defined, **5**:342
Nuclear SSB, human, **5**:462–65
Nuclear transcription factors, activation by interleukin-1, **3**:339
Nuclear transformation, of *Chlamydomonas reinhardtii*, **1**:354
Nuclear transplantation, defined, **2**:439
Nuclease protection assay, **5**:347–48, 347t
  defined, **5**:342
Nucleases
  defined, **4**:281
  editing by, **2**:409
  Rad27, **6**:287
  resistance of phosphorothioate DNA to, **1**:166
Nuclease stability, **1**:71–72
  defined, **1**:69
Nucleation
  in crystallization, **3**:309–10
  in virus assembly and budding, **6**:184
Nucleic acid-amino acid conjugates, synthesis of semirigid PNA diamide linkages, **4**:276f
NUCLEIC ACID ENZYMES, EVOLUTIONARY ENGINEERING OF, **4:226–34**
Nucleic acid hybridization assay, defined, **1**:323
NUCLEIC ACID HYBRIDS, FORMATION AND STRUCTURE OF, **4:235–43**
NUCLEIC ACID-PROTEIN COMPLEXES, LABORATORY TECHNIQUES IN, **4:243–53**
Nucleic acids
  amplification of, for mutation detection, **2**:421
  bases in meteorites, **4**:292
  cellular uptake of liposome-associated, **3**:417f
  crystal structures, hydrogen bonds of, **3**:270–71, 270f
  electron microscopy of, **2**:182
  formation of, and origin of life, **4**:292–93
  hydration of, **3**:272
  labeling methods, **3**:390; **5**:214–18
  metal binding by, **1**:132
  peptide, **1**:71, 73f
  peptide or polyamine, binding to a restriction enzyme target, **4**:274–75, 275f
  photooxidation of, **6**:209
  quality of paleontological samples, **4**:302–3, 302t
  structures of, NMR determination, **4**:224
  study of, in molecular paleontology, **4**:303–5
  3′ and 5′ ends of strands, defined, **3**:391
  transferable, in plants, **4**:451
NUCLEIC ACID SEQUENCING TECHNIQUES, **4:254–57**
  nucleic acid hybridization, **3**:252
Nucleobase radicals
  anion, **3**:359, 359f
  cation, reactions of, **3**:358–59, 358f
Nucleobases
  ionizing radiation reaction products from, **3**:360–63
  loss of, spontaneous mutation involving, **4**:151
Nucleocapsid, **6**:328f
  defined, **6**:193
  proteins of, respiratory syncytial virus, **5**:290
  release into the host cell, **6**:196
Nucleocapsid cysteine-rich domain, zinc finger DNA-binding motif, **6**:334, 334f
Nucleophile, defined, **1**:193
Nucleophilic catalysis, **1**:169–70
Nucleoproteins
  defined, **6**:183
  "frozen" filaments of, **1**:4
  of viral genome cores, **6**:184
Nucleosides
  analogues to, for HIV treatment, **1**:14, 37–38f
  metal binding by, **1**:132
  prebiotic synthesis of, **4**:292–93, 293f, 294t
Nucleosome-like complexes, from ssDNA and *Eco*SSb tetramers, **5**:462
Nucleosomes, **1**:361
  assembly of, **1**:367f
  chromatin, **1**:86
    electron microscope image, **2**:188f
  core particle of, **1**:363–64, 363f
    defined, **1**:360
    internal structure of, **1**:364, 364f
  defined, **1**:360; **2**:364; **3**:49; **6**:28

electron spectroscopic imaging of, **2**:193
filaments of, in chromatin fiber, **1**:364–65, 365f
structure of, **1**:370; **2**:117–18; **3**:50, 50f
ESI study of, **2**:193
and transcription, **1**:369, **2**:369
in vitro assembly of, **1**:366
Nucleotide bases, unstacked, **5**:460–61
Nucleotide-binding folds (NBFs), **1**:429
and chloride channel control, **1**:431–32, 432f
Nucleotide excision repair (NER), **6**:287–90
defined, **6**:283, 285
model for, **2**:46f
replication protein A role in, **5**:464–65
of UV damage to DNA, **2**:45; **4**:152–53; **6**:158
Nucleotides
antitumor activity of, **5**:495
cyclic, as second messengers, **1**:239; **2**:209–10
interactions of, **1**:165–66
polymers of, RNA, **5**:373–74, 373f
prebiotic formation of, **4**:293
stains for, **1**:166
tautomers of, in DNA replication, **2**:403
Nucleotidylation, posttranslational, in bacteria, **5**:55
Nucleus
acetylation of lysine side chains in, **5**:52
calcium in, **1**:248
enzymes for phospholipid biosynthesis in, **4**:433
exchange of genes with the chloroplast, **4**:84
pyridoxal phosphate binding in, **6**:243
retinoic acid receptors of, **6**:236–37
See also Nuclear *entries*
Null mutant mice, for alcohol and drug abuse studies, **2**:151
Nurse cells, **2**:85
Nutrient transfer, defined, **2**:440
NUTRITION, **4:257–68**
defined, **4**:257

O^6-methylguanine-DNA methyltransferase, defined, **6**:283
O-antigen, defined, **3**:463
Obesity, gene mutations producing, **2**:38
Ochre termination signal, **6**:201
Octadecanoyl-*sn*-glycerol, **3**:456f
Octopine dehydrogenase, unfolding and refolding of, **5**:76
Oculocutaneous albinism type II, gene for, mapping to chromosome 15, **3**:163
Oculopharyngeal muscular dystrophy, gene for, mapping to chromosome 14, **3**:158
Odor responses, to enantiomers, **1**:339–40
Ohno's law, **5**:441
Oil chamber method, for chromosome microdissection, **1**:373–74, 374f
Okadaic acid, protein phosphatase inhibitor, **5**:172
Okazaki fragments, **2**:87; **5**:355, 464; **6**:18f, 31
defined, **5**:352; **6**:28
in plasmid replication, **5**:3
Oleaginicity, biochemistry of, **3**:436, 436f
Olefinic analogues, peptides, **4**:373
Oligodeoxyribonucleotides, **4**:282–83
synthesis of, **4**:282f
synthetic, pairing with chromosomal DNA, **1**:3
Oligogenic etiology, **3**:223
Oligomeric proteins, aggregation in refolding, **5**:78–80
Oligomers
defined, **4**:289
mixture of, analysis by MALDI, **4**:6, 6f
OLIGONUCLEOTIDE ANALOGUES, PHOSPHATE-FREE, **4:269–81**
Oligonucleotide assembly, with assembly PCR, **5**:451
Oligonucleotide-directed mutagenesis, **4**:282; **5**:453; **6**:314
Oligonucleotide hybrids, **4**:242, 242f
dissociation of, **4**:237, 237f, 242
Oligonucleotide ligation assay (OLA), **3**:393–95, 395f, 400, 401f
OLIGONUCLEOTIDES, **4:281–85**
allele-specific, **2**:421–422
antisense, **1**:70
developing drugs from, **4**:40
atomic force microscopy for direct mapping of, **5**:406
cyclic, **1**:73
defined, **3**:391; **4**:235, 254; **5**:401
dephospho, **1**:71, 73f
design of, for probing gene banks, **5**:503–4
patchy, **6**:50–51
phosphodiester, backbone-modified, **4**:40, 40f
for producing biomimetic peptides, **1**:137
reactions on solid supports, **4**:242, 242f
in a sequencing chip, **3**:252
synthesis of, **1**:73–74
synthetic, for altering yeast chromosomal genes, **6**:310–11
for triple-helix Achilles' cleavage, **1**:4
triple helix forming, **1**:70, 70f, 74
Oligonucleotide synthesis, **4**:92, 282f
solid phase, **1**:165–66, 165f
defined, **1**:159
Oligoribonucleotides, **4**:284
Oligosaccharides
alcohols of, defined, **2**:446
analysis of, **1**:267; **2**:450–51
core, of lipopolysaccharides, **3**:466–67
defined, **2**:458
fucosylated, interaction with selectins, **1**:297
of lipopolysaccharides, **3**:463
maturation in the cisternae of the Golgi complex, **5**:202
Nod, **5**:13
structure of, **2**:458
Oligosaccharins, **4**:492
in plant defense against insect attack, **4**:493
O-linked sugar chains, **2**:452–53, 453f
Omnipotent suppressor, **6**:320
On-bead binding assay, **5**:520–21, 521f
Onchocerciasis, **4**:307
Oncofetal antigens, **1**:282
ONCOGENES, **1**:253; **2**:215–16; **4:285–88**
association with hepatocellular carcinoma, **3**:486–89
bcl-2, **1**:302; **4**:106
c-Ki-*ras*, mutation in colon cancer, **1**:251
classification by function of product, **4**:286–87
c-ski, in transgenic cattle and swine, **6**:70
defined, **1**:407; **2**:62, 207; **3**:227, 491; **4**:285; **6**:139
expression of, effect of interleukin-2 on, **3**:339
K-*ras*, mutation in colon cancer, **4**:329
mapping to chromosome 14, **3**:158–59
mutations in cancer, **4**:103
mutations in lung cancer, **3**:492t
myc, in lung cancer, **3**:492–93
N-*myc*, in woodchuck HBV infection, **3**:488
ras
activation in colon cancer, **1**:412
activation in lung cancer, **3**:493
viral, nonreceptor tyrosine kinases as products of, **5**:170
Oncogenesis, TCL class of genes implicated in, **3**:67–68
Oncoproteins
cytoplasmic/membrane, in vitro activity of antisense or triplex-forming oligonucleotides, **1**:74t
nuclear, in vitro activity of antisense or triplex-forming oligonucleotides, **1**:74t
One-dimensional chain, first-level structure of DNA, **2**:103
One gene-one enzyme concept, defined, **1**:113

Ontogeny
of globin types, **3**:16–17
metabolic patterns in, **3**:370
Oogenesis
histone mRNA transcripts in, **3**:52
hormonal control of, **4**:189
in mosquitos, **4**:192, 193f
Oomycin A, pesticide from *P. fluoroescens*, **4**:398–99
Opal termination signal, **6**:201
Open probability, of a channel, defined, **1**:425
Open reading frame (ORF)
chloroplast, in *Chlamydomonas reinhardtii*, **1**:349, 351
coding regions, **5**:431
cytomegalovirus DNA genome, **6**:174–75
defined, **1**:347
Escherichia coli chromosome, **2**:170, 171f
HIV-1, **1**:35, 35f
LI gene, **3**:245
mitochondrial DNA, **4**:79–80
sequencing information from, **3**:517
SRY gene, **5**:438–39, 438f
in a transcription unit, **2**:365
varicella-zoster virus DNA genome, **6**:175
Open sandwich domain, HLA proteins, **3**:281
Operant conditioning
versus classical conditioning, **4**:53–54
defined, **4**:54
Operator, repressor protein, **2**:95
Operon hypothesis, **5**:275
Operons
bacterial, comparison with introns of *Caenorhabditis elegans*, **1**:229
containing internalin genes, **1**:106
fusion of, **2**:404
hemoglobin, **3**:23
transcribing to RNA, **2**:86
tryptophan, attenuation of expression, **2**:372f
for virulence genes, **1**:107
Operons, specific
ars, **6**:132–33
lac, **1**:1–2
of *E. coli*, transcriptional repressor in, **2**:370–71, 370f
repression system in, **5**:276f
lux, bioluminescence genes of, **1**:122–23
trp, **2**:371
Ophthalmoplegia, chronic progressive external, **1**:10
Opiates, withdrawal from, neurobiological changes in, **2**:150
Opines, genes for, on T-DNA, **5**:12
Opioids, **2**:155–156
endogenous, **2**:150; **4**:355
peptide and mimetics of peptide, **4**:380–81
Opsin, **1**:111
Optical activity, defined, **1**:332
Optical biosensors, **1**:187–90
Optical fibers, in competitive binding assays, **1**:188, 188f
Optical purity, defined, **1**:52
Optical sensors, label-free, **1**:190
Optimization
for hybridization, in situ hybridization, **2**:383
for labeling in situ hybridization, **2**:382–83
of ligase chain reaction amplification, **3**:306
of ligase chain reaction amplification conditions, **3**:306
in RNA structure determination, **5**:379
Optoelectronic devices, biomolecular, **1**:150–52
Optomotor activity/optomotoring
in mutant *Drosophila*, **2**:137–38
related song deviance, **2**:143
Oral contraceptives, synthetic steroids in, **5**:483–84
Organelle ribosomes, preparation of, **5**:329–30
Organelles, membrane, **4**:120
Organic reagents, denaturation of DNA by, **2**:11t
Organismal aging, **1**:302
Organogenesis, during embryological development, **2**:468
Organophosphate insecticides, hydrolysis of, by paraoxonase, **4**:403
Organ regeneration, poly(L-lactic acid) scaffold for, **5**:35
Orientation, of biomolecules, in electronic systems, **1**:150
Origin of replication (ori), **2**:272–73, 272f; **5**:355
defined, **1**:95; **2**:86
DNA, **2**:86
isolation and structure of, **2**:90
in mammalian mtDNA, **4**:81
plasmid, **5**:2–3, 3f
types of, **2**:90f
yeast, **6**:291
ORIGINS OF LIFE, MOLECULAR BASIS OF, **4**:92–93, 288–97
Origins of order, and landscapes, **4**:94
Ornithine decarboxylase (ODCase), degradation of, **3**:353
Orphan genes, homeobox, **3**:72, 74
Orthologous genes, defined, **6**:134
Osler-Weber-Rendu syndrome, mapping to chromosome 9, **3**:107
Osmolarity, and cell fusion efficiency, **2**:204
Osmotic treatment, in plant cell transformation, **4**:461–62
Osmotin, **4**:480
O-specific antibodies, **3**:467
O-specific chain, **3**:463, 467–68, 468f
Osteoblasts, calcium-binding proteins secreted by, **1**:239
Osteocalcin, **1**:239
Osteodystrophy 2, Albright hereditary, deletion in chromosome 15, **3**:164
Osteogenesis imperfecta (OI), **2**:426
Osteolysis, familial expansile, gene for, mapping to chromosome 18, **3**:213–14
Osteomas, relative risk of, in heterozygotes for *RB*1 mutation, **5**:320
Osteonectin, **1**:239
Osteopetrosis, in null mutation of the *c-src* gene, **4**:103
Osteopontin, **1**:239
Osteoporosis, relationship with vitamin D receptors, **6**:237
Osteosarcoma, and *Rb* genes, **1**:251
Outcrossing, **2**:83–84
Outer core, of lipopolysaccharides, **3**:466
Outwardly rectifying chloride channel (ORCC), **1**:430
Ovarian adenocarcinoma, mapping to chromosome 18, **3**:211
Ovarian cancer, gene predisposing to, **3**:223
Ovary
gonadotropin receptors of, **2**:287
processes under local control, **2**:288t
remodeling in, **2**:290
Overdominance, maintaining genetic variation through, **2**:393
Overexpression, suppression by, **6**:320
Overlapping frequency counts, **6**:40–41
Overlapping libraries, **3**:217
clone, **3**:220
Oxazinones, diphenyl, **1**:54
Oxidants, endogenous, DNA damage from, **2**:42–43
Oxidases
D-amino acid, reaction at chiral center of the substrate, **1**:342
defined, **4**:59
pyridoxamine 5′-phosphate, **1**:401
terminal, in electron transport, **1**:306
Oxidation
β-, of fatty acids, **3**:454, 455f
in bioluminescence and chemiluminescence, **1**:323
of carbon, in xenobiotic metabolism, **6**:262–64
of guanine, transversion generated by, **4**:151
of low density lipoprotein, **1**:288; **3**:479

NAD-dependent, of aldehydes, **1**:196
of xenobiotics, **6**:266–67
Oxidative damage
defined, **2**:330
DNA,endogenous, **2**:79
Oxidative phosphorylation, **1**:306
comparison with photophosphorylation, **1**:311
defined, **1**:303–4
interference with, in the presence of environmental stressors, **2**:218
mitochondrial genome's role in, **4**:82
regulation of, **1**:308
uncoupled, **1**:314f
uncoupling respiration from, **1**:313
Oxidative stress
defined, **2**:216, 330
genes protecting against, in aging, **1**:12
genomic responses to, **2**:219–20
and human disease, **2**:335, 335f
and repair systems, **2**:334, 334t
Oxidoreductase, **6**:266
defined, **2**:240
ferredoxin-dependent, **2**:244
NAD or NADP dependent, **1**:400–1
Oxyanion hole, of serum proteases, **2**:230, 230f, 233
Oxygen, **1**:132–33
binding by hemoglobin, **3**:10–11, 11f
in blood and blood substitutes, **3**:19f
inhibition of tyrosine hydroxylase by, **1**:213
reactive species of, formed by radiation exposure, **6**:209
substrate for tryptophan hydroxylase, **1**:219
supplying to protein production systems, **2**:278
Oxygenase
biodegradation of toluene using, **1**:124
defined, **4**:59
mixed-function system, metabolism of aflatoxins in, **3**:485
Oxygenators, silicone rubber construction, **5**:37
Oxygen equilibrium curve, hemoglobin, **3**:8f
genetically engineered, **3**:24
Oxygen paradox, **2**:219–20
Oxygen radicals, mediators in endotoxin reactions, **3**:471
Oxygen radical theory, of aging, **1**:302
Oxyhemoglobin, equilibrium and kinetic constants for assembly of subunits, **3**:12t
Oxytocin, **2**:291
conformation of peptide and mimetics, **4**:379–80
defined, **2**:285
structure of, **2**:20, 20f; **4**:364
Ozone layer, destruction of
and biological damage from ultraviolet radiation, **6**:163
potential DNA damage resulting from, **2**:41

P1-derived artificial chromosomes (PAC), **2**:129, **3**:220
P_1 transgenic fish
defined, **6**:84
inheritance of transgenes, **6**:90
medaka, **6**:95–96, 96f
P450 variants
$P450_{arom}$, and estrogen biosynthesis, **5**:469
P_{cam}, genes encoding oxygenases on a plasmid, **1**:125
reactions catalyzed by, **5**:469
See also Cytochrome P450
Pacemaker
circadian, **2**:221
lead insulation from polyurethane, **5**:38
Packaging
defined, **6**:217
DNA pathway, **6**:205–06
Packing
defined, **3**:76
density of, in proteins, **5**:150
for high performance size exclusion chromatography, **3**:79
for hydrophobic interaction chromatography, **3**:81
in ion exchange chromatography, **3**:79–80, 79t
for reverse phase chromatography, **3**:82
Pacmids, for physical mapping of *Drosophila*, **2**:129
PADGEM (P-selectin), **1**:296
Padlock probes, ligation-mediated gene detection, **3**:397, 398f
PAGE, 2D- (two-dimensional polyacrylamide gel electrophoresis), **2**:353, 353f
Paired (prd) class, homeobox genes, **3**:68, 69f
Paired helical filaments (PHF), **1**:50
Paired-like (prd-like) class, homeobox genes, **3**:68
Pairing, DNA, **5**:225–26, 225f
Pairing energy, with an anticodon, **5**:429
PALEONTOLOGY
defined, **4**:299
MOLECULAR, **4:299–06**
Palindrome
consensus regulatory sequence, heat shock regulatory element, **2**:218
defined, **5**:296
Palmitoylation, posttranslational modification, **5**:54
Pancreas, involvement of, in cystic fibrosis, **1**:425
Pancreatic beta cell, **2**:39f
Pancreatic trypsin inhibitor (PTI), **2**:236
Pancreatitis, acute, protease inhibitor therapy for, **3**:379
Panning, **1**:416–17, 416f
combinatorial phage display library, **1**:419f
defined, **1**:415
for selection from antibody phage libraries, **1**:418–20
Pantetheine-binding sequences, of peptide synthetases, **4**:215–16, 216f
Pantothenic acid, **6**:251–52, 251f
Pantropic retrovirus, defective, **6**:88
Paper chromatography, for steroid purification, **5**:475
Papillary carcinoma, of the thyroid, **1**:410
Papilloma virus (BPV)
bovine, electron microscope image, **2**:192f
DNA-binding domain of the transcriptional activator, **5**:130
E6 protein of, **3**:170
human, **1**:412
binding to p53, **3**:352
inactivation of pRB by, **5**:321
Papua-New Guinea (PNG) origin, calibrating mtDNA mutations in terms of, **4**:88–89, 89f
Paracrine, defined, **2**:463
Parahormones, **2**:211
Parallel-architecture computers, **5**:433
Paralogies, chromosome, **3**:84
Paramagnetic interactions, in NMR structure determination, **4**:224–25, 226f
Paramecium aurelia, mtDNA of, **4**:79
Paramyxoviruses, **6**:167
Paraoxonase, **4**:403
Paraquat, metabolism of, in the lung, **2**:335
Parasexual cycle, fungal, for manipulating to improve commercial strains, **2**:345
PARASITE GENOME PROJECTS, **4:306–12**
Parasites
infection by, cytokine role in immune response to, **1**:447
inhibition of growth of, by interferon-induced reactions, **3**:326
Parathyroid hormone (PTH), **1**:238; **2**:215
mutation in the gene for, **2**:213
regulation of P450-dependent vitamin D_3 metabolism, **1**:439
Paratope
antibody, **3**:283
defined, **3**:275

Parental imprinting, **2:**445
  defined, **2:**440
Parenteral, defined, **2:**165
Parkinson's disease
  and metabolism of environmental toxins, **4:**402
  mitochondrial mutations associated with, **4:**83
Parthenogenesis, defined, **2:**440
Parthenogenetic embryos, development of, **2:**440, 441f
PARTIAL DENATURATION MAPPING, **4:312–23**
"Partial filling in", for constructing a genomic library, **2:**71, 71f, 75f
Particle bombardment
  defined, **4:**458
  DNA, for genetic vaccination, **2:**433, 434–35f
  for plant cell transformation, **4:**459–62
Particle inflow gun, for plant cell transformation, **4:**461
Partitioning
  defined, **2:**271
  of plasmids, **5:**4–5
Partition systems (par), for maintaining copy number, **2:**273
Parvalbumin, **1:**242
Parvoviruses, DNA and RNA of, **5:**357, 357t
Patch-clamp analysis, of membrane fusion, **4:**44
Patchy complexity, **6:**50–51, 51f
Patent
  alternatives to, **6:**81
  defined, **6:**74
  requirements for obtaining, **6:**76–77
Patent and Trademark Office (PTO), U.S., **6:**74
Paternal inheritance, of mtDNA, **4:**81
Pathogenesis
  defined, **4:**173
  of Down's syndrome, **2:**120
  of Huntington's disease, **3:**251
  role of cytokines in, **1:**444–45, 449–50
Pathogenesis-related (PR) proteins, **4:**480
Pathogenicity, defined, **1:**103
Pathogens
  and cytokine production, **1:**444
  identification of, using PCR, **4:**339
  infectious, prions, **5:**57
  plant, biochemical weapons of, **4:**481–82
PATHOLOGY
  kallikrein-kininogen-kinin system in, **3:**379–80, 379t
  MOLECULAR, **4:323–33**
Pathophysiology
  of heart failure, **3:**1
  of neurofibromatosis 1, **4:**178–80
  of regulatory peptides, **4:**355–57
  of sporadic motor neuron disease, **4:**107–08
Pathways, of gene action, **2:**418, 418f
Pattern, defined, **6:**38
Pattern matching, for sequence alignment of proteins, **5:**425
Patterson map, **5:**112
Paxillin
  focal adhesion protein, **1:**452, 454
  phosphorylation of, **1:**458
    in embryonic development, **1:**459
PBC class, homeobox genes, **3:**74, 75f
PCR TECHNOLOGY, **4:333–40**. *See also* Polymerase chain reaction
"Pearl chain" formation, in dielectrophoresis, **2:**200–01
  optimal conditions for, **2:**204
Pearson's syndrome, **4:**82–83
  from mtDNA base substitutions, **4:**90
Pectinases, role in plant pathology, **4:**482
Pectinmethylesterase (PME), in tomatoes, **6:**58
Pedigree analysis, **4:**124
P element, *Drosophila*, **2:**123
Pemphigus foliaceus, **3:**214
Pemphigus vulgaris, **3:**214
Penetrance
  defined, **1:**221, 407; **2:**420; **3:**246; **4:**173, 323
  of disease genes, **3:**223
  of fragile X syndrome, **4:**331
  in Huntington disease, **2:**427
  incomplete, allowing for genotype errors through, **3:**238, 238t
  of neurofibromatosis 1, **4:**175
Penicillins, **2:**156–57
  fungal biotechnology using, **2:**340
Penicillium camembertii, **2:**338
Penicillium chrysogenum, **2:**340
Penicillium roquefortii, **2:**338
Pennsylvania vs. *Pestinikis*, acceptance of DNA fingerprinting in court, **4:**339
Pentose phosphate pathway, **1:**397–98
Pepstatin, **4:**34, 34f
Peptidases
  clinically important, **4:**356–57, 357t
  inactivation of peptides by, **4:**354
  processing of glutathione conjugates by, **1:**200
Peptide analogue, defined, **4:**371
Peptide backbone, modifying amides of, **4:**372–74, 374f
Peptide bond, **5:**117; **6:**111f
  defined, **4:**361
Peptide conformations, defined, **4:**371
PEPTIDE HORMONES
  BIOSYNTHESIS, PROCESSING, AND SECRETION OF, **4:340–47**
  cholecystokinin, **4:**381–82
  circulating, **4:**341t
  in insect development, **4:**190, 190f
  structure determinations, historic summary, **4:**348t
Peptide libraries
  combinatorial, **4:**91
  comparison of methods, biological applications, **5:**524t
  synthetic, **4:**96, 370, 392; **5:**516–24, 524t
Peptide mass fingerprinting, defined, **5:**173
Peptide mimetic, defined, **4:**371. *See also* Peptidomimetics
Peptide polymers, biomimetic synthesis, **1:**137
Peptide receptors, nonpeptides interacting with, **4:**35
Peptide-resin linkage, solid phase peptide synthesis, **4:**389–90
Peptidergic neurons, in nematodes, **4:**171
PEPTIDES, **4:347–61**
  biomimetic, **4:**33–35
    and recombinant DNA technology, **1:**137, 138f
  bonding in, **4:**352, 352f
  bulk and conformation of side chains, **4:**374
  defined, **1:**148; **4:**361, 383; **5:**516
  diabetogenic, **2:**36
  direct synthesis of, **1:**172–73
  metal binding by, **1:**132
  natriuretic, receptors for, **5:**243
  nomenclature of, **4:**363–64
  opioid, **4:**341t
  pleiotropic, neuropeptides, **4:**194–95
  products from mRNA with a UAG termination codon, **5:**146f
  resolution of, using reversed phase chromatography, **3:**83
  secondary structure of, **1:**389–91
  structure of, **4:**371–72
  synthesis of mixtures, **4:**392
  synthesis procedures, **4:**364–66
  SYNTHETIC, **4:**359, **361–71**
    mimicking linear epitopes of native proteins, **6:**179
  vibrational analysis of, **5:**97
PEPTIDES AND MIMICS, DESIGN OF CONFORMATIONALLY CONSTRAINED, **4:371–82**
PEPTIDE SYNTHESIS, **4:**353
  SOLID PHASE, **1:**171, 171f; **4:383–93**
Peptide synthetases, **4:**208–9, 209–12, 212t
  amino acid activating domain of, **4:**214–17
  domains of, **4:**212–13, 213f
Peptide topography, defined, **4:**371
Peptidoglycan
  defined, **1:**95
  synthesis of, **1:**98–99, 99f

Peptidomimetics, **4:**33–35, 35f, 360f
defined, **4:**347
polymers, oligocarbamate, **4:**97
synthesis of, **4:**370
Peptidyl α-amidating enzyme (PAM), **4:**345
Peptidylglycine-α-amidating monooxygenase, **4:**197
Peptidylprolyl *cis-trans*-isomerase, **5:**82, 92
Period gene, **2:**221
Peripheral blood lymphocytes (PBLs), isolation of antibodies from, **1:**64, 416
Periplasmic space, defined, **1:**315
Periseptal annulus, **1:**100
Peritubular myoid cell, defined, **3:**501
Permanganate ion, in vivo footprinting with, **4:**249
Permatin. *See* Zeamatin
Permeases, bacterial, **6:**128, 130
Permeation enhancers, **2:**166–67
defined, **2:**165
Permineralization, **4:**300
Peroxidases, **6:**266–67
chemiluminescence assay for, **1:**329f
defined, **4:**59
detection of label, chemiluminescent reaction for, **1:**327
oxidation of lignin by, **1:**122
Peroxisomal import, **5:**206
Peroxisomes
enzymes for phospholipid biosynthesis in, **4:**433
β-oxidation in, **3:**454
syndromes associated with abnormal, **1:**115
Peroxyl radical
defined, **3:**354
formation by oxygen addition to DNA radicals, **3:**357
Perturbation calculations, free energy, for analysis of drug binding, **5:**167
Pertussis toxin (Ptx), **1:**104
inhibition or receptor-mediated reactions by, **5:**235
Peste des petits ruminants, vaccines for, **6:**168–69
PESTICIDAL COMPOUNDS FROM PROKARYOTES, **4:393–99**
Pesticides
defined, **4:**393
engineering plants to produce, **6:**56–57
secondary metabolites as, **4:**395–99
Petrochemicals, starting materials for drug production, **2:**161–63
Peutz-Jaegher syndrome, **1:**410
P gene, hepatitis B virus, **3:**483
pH
defined, **1:**126
and DNA denaturation or degradation, **2:**5; **4:**312
and DNA stability, **2:**5
and hormone action, **2:**211
liposomes sensitive to, **3:**418–19
and protein unfolding, **5:**151
for proton transfer reactions, bacteriorhodopsin, **1:**110
and stability of metal-ligand complexes, **1:**130–31, 131t
Phage
antibody isolation using, **1:**64
for antibody library construction, **1:**416; **3:**287–88
in archaea, **4:**74
assembly of, analyzing, **1:**236
DNA of, preparation for cloning experiments, **5:**505
for genomic library construction, **2:**68
immune system mimicking in, **3:**287f
M13 filamentous, antibody isolation using, **1:**417
Mμ, electron micrograph, **4:**322f
peptide libraries based on, **4:**353
Spi-, **2:**66
See also Bacteriophage
Phage display
defined, **1:**415
systems for, **4:**91
Phage display library, **4:**96–97
Phagemid systems, pComb3, **1:**419f
Phagocytosis
cellular luminescence for study of, **1:**330–31
pathway for, **5:**204
Phagolysosome, **1:**105
Phanerocheate spp., biodegradation of lignin by, **1:**122
Pharmaceutical production, from transgenic animals, patentable, **6:**77
PHARMACOGENETICS, **4:399–405**
Pharmacokinetics
defined, **1:**253
targeting drugs by manipulation of, **1:**253, 258–60; **2:**166
Pharmacophore, **4:**355
Phaseolotoxin, toxicity to plant, **4:**482
Phase
changes among, for lipids, **3:**447, 447f
defined, **3:**233
determination, in X-ray crystallography, **6:**278–79
Phase transition, and cooperativity, **2:**180
Phasmids, **2:**68
Phenazine, production by fluorescent pseudomonads, **4:**397–98, 398f
Phenobarbital, effect of, on phenytoin metabolism, **1:**202
Phenolic resins, peroxidase-catalyzed, **2:**250
Phenotype
correlation with genotype, in neurofibromatosis, **4:**178–79
defined, **1:**113, 227, 407; **2:**118, 146, 347, 440; **3:**227, 233; **4:**148, 173
Down's syndrome, **2:**118–19
and genotype, for cytochrome P450 alleles involved in debrisoquine metabolism, **4:**402f
heterogeneity of, in phenylketonuria, **4:**410–11
milchlos, of transgenic mice expressing whey acid protein, **6:**105
probability of occurrence of, **3:**237
Phenoxazines, reaction with NADH, **1:**183
Phenylalanine
deletion of triplet coding for, in cystic fibrosis, **1:**432–33
inhibitor, for tryptophan hydroxylase, **1:**219
as a precursor, taxol production, **6:**17
topographical features, in a peptide, **4:**371–72, 372f
Phenylalanine ammonium lyase (PAL)
in phenylpropanoid metabolism, **4:**479, 479f
reaction at chiral center of the substrate, **1:**342
Phenylalanine hydroxylase (PAH), **1:**207; **4:**406f
amino acid sequence, **1:**209–10, 210f
deficiency of, **4:**406
defined, **4:**405
gene for, structure and location of the human gene, **4:**407
Phenylalanine residue, **4:**372, 372f
Phenylalkylamines, binding to L-type calcium channels, **1:**243
Phenylethanolamine-*N*-methyltransferase (PNMT), **1:**207
Phenylglycine, *p*-hydroxy-D-, optically active, **1:**58
Phenylisothiocyanate (PITC), for protein degradation, **5:**174
PHENYLKETONURIA, **1:**116
defined, **4:**405
in Europe, **5:**48
founder effect, **5:**47
genetic defect in, **1:**114
MOLECULAR GENETICS OF, **4:405–15**
as a single-gene disorder, **3:**230
Phenylpropanoid metabolism, in plant defense against pathogens, **4:**479–80, 479f
Pheochromocytoma, **6:**144–45
tyrosine hydroxylase of, **1:**209, 211–12f
Pheophytin, as an electron acceptor, **4:**442
Pheromone biosynthesis activating neuropeptide (PBAN), **4:**195, 196–97

Pheromone biosynthesis activating neuropeptide (PBAN) (*cont.*)
  bioassay using, **4**:194
Pheromones
  defined, **2**:130
  Drosophila, genes affecting, **2**:144–45
  insect, chiral, **1**:340
  and plasmid conjugation, **5**:8–9
Philadelphia chromosome, **1**:251
  defined, **1**:250
  test for, and chronic myelogenous leukemia, **4**:328–29
  translocation creating, **4**:286
    chromosome 9, **2**:64, 64t; **3**:98
Phorbol esters
  activation of tyrosine hydroxylase by, **1**:215
  protein kinase C activation by, **5**:172
Phosphatase
  cytokine/*JAK/STAT* regulation by, **2**:471
  inhibition of, **5**:172
  nucleotide, activity of motor proteins as, **4**:114
Phosphate backbone, modified, defined, **4**:269
Phosphate bonds, modified oligonucleotide analogues with, **4**:269–70, 270f
Phosphate-free backbone, defined, **4**:269
Phosphates
  inorganic, concentration of, and respiratory control, **1**:312
  modifying
    in oligodeoxyribonucleotides, **4**:283
    in oligoribonucleotides, **4**:284
  nutritional requirement of microbes, **1**:119
Phosphatidic acid-phospholipase C, **4**:430
Phosphatidylcholine, **3**:416
  defined, **4**:432
  regulation of biosynthesis of, **4**:436–37, 437f
Phosphatidylethanolamine (PE)
  inverted hexagonal phase formation by, **4**:48
  for pH sensitive liposomes, **3**:418–19
Phosphatidylethanolamine methyltransferase, **4**:434–35
Phosphatidylglycerol (PG)
  synthesis in mitochondria and microsomes, **4**:435
  and transfection efficiency of liposomes, **3**:417
Phosphatidylglycerolphosphate synthetase, of *E. coli*, **4**:435
Phosphatidylinositol diphosphate (PIP_2), **1**:246
  hydrolysis of, **3**:46
  preferences of cPi-PLC isoforms for, **4**:427–28
  structure of, **4**:435f
Phosphatidylinositol-mediated signaling pathways, **4**:424
Phosphatidylinositol-phospholipase C enzymes (PI-PLC), secretory, **4**:424–26, 426f
Phosphatidylinositols (PIs)
  biosynthesis and phosphorylation of, **4**:435–36
  defined, **4**:416
  derivatives as second messengers, **1**:239
  linkage to phospholipase C, **4**:424
Phosphatidylserine (PS)
  membrane fusion role of, **4**:48
  and transfection efficiency of liposomes, **3**:417
  transport to the mitochondria, **4**:438
Phosphatidylserine decarboxylase (PSD), **4**:435
Phosphatidylserine synthetase (PSS), **4**:435
Phosphinothricin, active moiety of bialaphos, **4**:395, 396f
Phosphlipids, and tyrosine hydroxylase regulation, **1**:214
Phosphocreatine, muscle contraction and, **4**:145
Phosphodiester backbone
  and DNA stability, **2**:1
  modification of, antisense oligonucleotides, **1**:70–71
Phosphodiester linkage, DNA, **1**:160
  formation by ligases, **3**:391, 392–93, 392f
Phosphoenol pyruvate carboxykinase (PEPCK), effect of biotin on, **6**:243
Phosphoethanolamine cytidylyltransferase, **4**:434
Phosphofructokinase (PFK)
  deficiency of, in Tarui's disease, **4**:146
  and the glycolytic pathway, **2**:235
Phosphogluconate dehydrogenase-like 1 (PGDL1), gene for, mapping to chromosomes 1 and 18, **3**:208
Phosphoglycerate kinase
  aggregates of, detecting, **5**:80, 80f
  unfolding and refolding of, **5**:75, 75f, 76f
Phosphoglycoprotein, renal Na^+/H^+1 exchanger as, **5**:272
Phospholamban, **3**:2
  in heart failure, **1**:292
  role in the SR-Ca^{2+} pump, **1**:247
PHOSPHOLIPASES, **4:415–32**
  A, outer membrane, activation by a lysis protein, **5**:12
  A_1, **4**:424
  C, **3**:46, 47f; **4**:51, 424–30
    control by G protein, **3**:45
    gene for, in *Drosophila*, **2**:133
    stimulation by the vasopressin V_1 receptor, **2**:21
  D, **4**:430–31, 440
  defined, **4**:416
  γ, in the immune response, **3**:279
  PLA_2, **4**:420–24, 424f
PHOSPHOLIPIDS, **3**:427–28; **4:432–40**
  bacterial, **3**:432f
  catabolism, in Alzheimer's disease, **1**:50
  colorimetric assay for measuring, **3**:475
  defined, **1**:59; **2**:307; **3**:426
  glucosamine-based, **3**:464
  membrane, binding of talin to, **1**:456
  metabolism of, **3**:410
  plasma lipoprotein, **4**:495
  second messengers arising from, **2**:211
  structure of, **3**:457–60, 457f; **4**:433, 433f
Phospholipid transfer protein (PLTP), **3**:409
Phosphoprotein, of respiratory syncytial virus, **5**:290
Phosphoprotein phosphatases
  classification and properties of, **5**:169, 171
  defined, **5**:169
Phosphoramidate internucleotidic linkages, **4**:283
Phosphoramidite method, in synthesis of oligonucleotides, **1**:73; **4**:284
Phosphoramidites, defined, **4**:281
5-Phosphoribosyl-1α-pyrophosphate (PRPP), **1**:400
Phosphoric acid, condensation products of, **1**:160–61, 161f
Phosphorothioate-containing DNA, **1**:166
  defined, **1**:159
Phosphorothioate oligonucleotides, **4**:283, 284
Phosphorus, S32P radiation hazard, **5**:212–14
Phosphorylase, b, peptide mapping, **5**:182f
Phosphorylation
  activation of tyrosine hydroxylase by, **1**:213–15
  by adenosine 5′-triphosphate, **1**:160
  cascade in, **2**:238f; **5**:171–72
  catalysis by protein kinases, **5**:55
  control of transcription factors, **2**:212
  defined, **1**:59
  of initiation factor eIF-2, **6**:124
  of peptide hormones, **4**:343
  protein, **3**:46; **5**:50
    modulation of, **5**:172–73
    multisite, **5**:171–72
  regulation by hormones, **2**:212
  replication protein A, **5**:463–64
  and steroid activation of genes, **5**:472
  in xenobiotic metabolism, **6**:275

Phosphoryl intermediate ATPases (P-type ATPases), **6:**131–32
Phosphotransferases, **6:**275
Phosphotransferase system (PTS), **4:**425f; **6:**128
  bacterial, proteins of, **6:**130
Photinus pyralis, bioluminescence of, **1:**323
Photochemical damage, protection against, **6:**209–10
Photochemistry, light sources for, **6:**210–11, 211t
Photochrome-triplet labeling, **3:**388
Photochromic system, **3:**388
  defined, **1:**148
Photocycle, bacteriorhodopsin, **1:**110
Photodamage, **4:**449
Photodependent repair, **6:**157–58
Photodynamic action, **6:**207–8
Photodynamic damage, defined, **6:**207
Photodynamic protein, defined, **1:**135
Photoenhanced excision repair, **6:**158
Photoenzymatic repair (PER), **6:**157–58, 157t
  defined, **6:**149
Photofootprinting, **4:**249
Photography, for recording FISH images, **2:**385
Photohydrate, defined, **6:**149
Photoionization, **4:**1
Photoisomerization, **1:**150
  defined, **1:**148
  of rhodopsin, **5:**323f, 324
  signal generated by, bacteriorhodopsin, **1:**150
Photolithography, in DNA analysis, **5:**395
Photolyase
  DNA, **6:**209
    defined, **6:**283
  involvement in DNA repair, **2:**44, 44f
Photolysis, of CoB_{12}, **1:**403
Photon scanning-tunneling microscope (PSTM), **5:**410
Photophosphorylation, **1:**306
  ATP synthetase, **4:**448–49
  comparison with oxidative phosphorylation, **1:**311
  defined, **1:**304; **4:**440
Photophysical properties, of bacteriorhodopsin, **1:**138
Photoproduct, defined, **6:**149
Photoprotection, **6:**157
Photoprotein, defined, **1:**323
Photoreactivation
  reducing DNA damage through, **2:**44, 44f
  in repair, **4:**153
Photoresponses, genomic, **2:**221
Photosensitizers
  defined, **6:**207
  for destroying viruses, **6:**211–16
Photosynthesis, **4:**493
  in *Chlamydomonas*, **1:**347
  electron transfer in, **2:**194, 197
  and free energy, **1:**306
  photosystem I reaction center, **1:**149, 149f
PHOTOSYNTHETIC ENERGY TRANSDUCTION, **4:440–50**
Photosynthetic organisms
  adenosine triphosphate synthesis in, **1:**310–11
  in *Chlamydomonas reinhardtii*, **1:**354–55, 356f
Photosystem
  chlorophylls of, **4:**442–43
  defined, **1:**148; **4:**440
  for energy capture, **1:**310
  I
    electron transport in, **4:**447–48
    model of, **4:**445f
    proteins of, and corresponding genes, **4:**445t
    structure of, **4:**444
  II
    photodamage to reaction centers, **4:**449
    proteins of, and corresponding genes, **4:**445t
    reaction centers of, **4:**443–44, 444f
Phototransduction, *no receptor-potential*, in *Drosophila*, **2:**133; **3:**143
Phthalocyanines, **6:**209
  as photosensitizers, **6:**213
Phycobilins, **4:**441–42
Phylogenetic conservation, of *ZFY* and *SRY* genes, **5:**440t
Phylogenetic trees, **4:**67f
  constructing with mtDNA data, **4:**88
  for DNA studies, ancient material, **4:**304–5, 304f
Phylogeny, defined, **4:**84, 226
Physarum, intron homing in, **1:**351
Physical mapping
  of *Caenorhabditis elegans*, **4:**167, 169
  chromosome 1, **3:**95–97, 97f
  chromosome 9, **3:**101, 108–10t
  chromosome 10, **3:**118–23
  chromosome 11, **3:**135–40
  chromosome 14, **3:**144, 149–51
  chromosome 15, **3:**170–74
  chromosome 16, **3:**181–86
  chromosome 18, **3:**197–98, 202–3f
  defined, **3:**97, 112, 160, 191; **4:**121, 173
  of distance between markers, **3:**219
  of *Drosophila*, **2:**128–29
  of the fragile X region, **2:**320–21
  versus genetic mapping, **3:**123–24
  genome
    defined, **6:**291
    integration of, **6:**301
    mammalian, **3:**515
    mouse, **4:**127–28
  translocation breakpoints, neurofibromatosis gene, **4:**176
Physical properties, of lipids, **3:**443–49
Physicochemical bases, of amino acid changes, **5:**422
Physicochemical method of gene transfer, defined, **2:**430
Physiologically based pharmacokinetic (PB-PK) models, **1:**259–60, 259f
Physiology, of interleukins, **3:**344
Phytase, commercial production of from filamentous fungi, **2:**342
Phytoalexins, **4:**478, 479
  accumulation in fungus-resistant bean cultivars, **4:**483
  defined, **4:**476
Phytochrome
  defined, **4:**485
  light responses regulated by, **4:**493–94
Phytophthora palmivora, culture for herbicidal properties, **2:**344
π-bond cooperativity, in hydrogen bonding, **3:**266, 266f
Pick's disease, **5:**516
Picornaviruses, **6:**224, 224f
  defenses of, against antiviral reactions, **3:**325
  genera and representative species, **5:**14t
  poliovirus, **5:**13
  receptors for, and virus attachment interference, **6:**231t
  structural strategy for evading immune surveillance, **6:**229f
  translational enhancers of, **2:**377
Piezoelectric devices, in biosensors, **1:**191
Piezoelectric material, crystalline, defined, **5:**401
Pigmentation, protection from ultraviolet radiation by, **6:**157
Pigment-proteins
  defined, **4:**440
  transfer of light energy via, **4:**441–42
Pigments
  body, in mutant *Drosophila*, **2:**137
  eye, in mutant *Drosophila*, **2:**134
  red, identifying *ADE1* and *ADE2* yeast genes, **6:**313
Pigs, transgenic, as models for disease, **6:**70–71
Pilbara Block, Australia, microfossil source, **4:**289
Pilin
  and bacterial adherence, **1:**104
  genes for, and bacterial virulence, **1:**105
Pilot study, for genetic analysis of populations, **2:**394

Pilus
  defined, **5:**1
  F plasmid, **5:**5–6, 7f
Pineal gland, tryptophan hydroxylase activity in, **1:**218
Pinene cyclases, **1:**338
Ping-pong kinetics, **1:**183
  in amperometric biosensors, redox reactions, **1:**182
Ping-pong type of reaction, of tyrosine hydroxylase, **1:**208
Pinocytosis, **5:**204
  low-density lipoprotein transfer to cells by, **3:**473
"Pin" technology for peptide synthesis, **4:**392
Piperazine-2-ones, peptide derivatives, **4:**377
Pisatin dimethylase (PDA), **4:**481–82
Pituitary
  anterior, endocrine cells of, **2:**287
  endocrine cells of, **2:**291
Pituitary adenylate cyclase activating polypeptide (*PACAP*), gene for, mapping to chromosome 18, **3:**208
Pituitary gonadotropin, therapy using, and transmission of prions, **5:**70
Pituitary polypeptide 7B2, localization to chromosome 15, **3:**167
PKA (cAMP-dependent protein kinase)
  defined, **1:**207
  phosphorylation of tyrosine hydroxylase, **1:**210, 210t, 214, 214t
PKC (Ca^{2+}/phospholipid-dependent protein kinase)
  defined, **1:**207
  phosphorylation of tyrosine hydroxylase, **1:**210
  and regulation of tyrosine hydroxylase, **1:**215
Placeopecten megellanicus, mtDNA of, **4:**79
Planar chirality, **1:**335
Plant cell, walls of, **4:**478f
PLANT CELLS, GENETIC MANIPULATION OF, **4:450–58**
PLANT CELL TRANSFORMATION
  defined, **4:**458
  PHYSICAL METHODS FOR, **4:458–65**
Plant galls, **2:**221
PLANT GENE EXPRESSION REGULATION, **4:465–75**
PLANT PATHOLOGY, MOLECULAR, **4:475–85**
PLANTS
  chiral structures in, **1:**334
    growth regulators, **1:**340
  DNA fingerprints of, **2:**61
  extraction of DNA for fingerprinting, **2:**53
  extraction of DNA from, **2:**54f
  homeobox genes of, **3:**75–76
  mitochondrial DNA of, **4:**79, 80, 489
  MOLECULAR BIOLOGY OF, **4:485–95**
  phospholipases of, **4:**430
  role of calcium in, **1:**248–49
  tissue culture, defined, **4:**458
  transformation systems, **4:**489–90
Plant Variety Protection Act of 1970, **6:**78–79
Plaque-forming units (pfu), assay for number of binding sites, **6:**225
Plaques
  in Alzheimer's disease, dynamic nature of, **1:**47
  atherosclerotic, cytokine role in development of, **1:**450
  cytokines, and development of, **1:**450
  desmosomal, **1:**458
Plasma
  formation of HDL particles in, **3:**409
  lipoproteins of, **3:**472t
  separation of lipoproteins of, **3:**474–75
Plasma desorption mass spectrometry (PD-MS), **4:**13
PLASMA LIPOPROTEINS, **4:495–506**
Plasmalogen, defined, **1:**113
Plasma membrane
  activation of receptors of, **2:**217
  calcium transport systems of, **1:**243–45
  carbohydrates of, **1:**296
  cell surface molecules, **1:**294f
  as a condenser for storing electric energy, **2:**174
  receptors for viruses, **6:**232, 232f
Plasmanyl compounds, **4:**433
Plasmapheresis, with poly(vinyl alcohol), **5:**43
Plasmid DNA, preparation of, for cloning experiments, **5:**504
PLASMIDS, **5:1–13**
  in bioprocess engineering, **1:**172
  circular, yeast, **6:**305
  for constructing bioluminescent bacteria, **1:**122
  DNA of, for varicella-zoster virus and cytomegalovirus vaccines, **6:**181–82
  FLP recombinase, expression in *Drosophila*, **2:**124
  genes encoding oxygenases on, **1:**125
  for gene targeting, **5:**196
  inhibition involving PNAs, **4:**274
  insertion of a gene into, **6:**202, 202f
  linear, transformation by crossing over, **5:**257, 257f
  of mycobacteria, **4:**157–59
  for mycobacteria gene cloning, **4:**161, 163
  pAL5000, **4:**161
    restriction map, **4:**162f
  replication of, **1:**95, 99
    intermediate in, electron microscope image, **2:**185f
  Ri, **4:**451
  rol-6, of *Caenorhabditis elegans*, **1:**234
  stability of, defined, **2:**271
  structures of, circular, **4:**74
  synthesis of, with assembly PCR, **5:**451–52, 451f
  Ti, **1:**80; **4:**451, 465
    transformation vectors constructed from, **4:**490
  yeast vector, **6:**314–17, 322
Plasmid shuffle
  defined, **6:**302
  for yeast genetic studies, **6:**317
Plasmid substrates, recombination using, **5:**256–58
Plasmid vectors
  Escherichia coli
    components of, **2:**272–75
    in recombinant fusion proteins, **6:**176–77
  in FISH, **2:**381
  for gene targeting, **4:**101
  in yeast analyses, two-hybrid systems, **6:**322
  for yeast artificial chromosomes, **6:**292f
Plasminogen
  homology with apo(a) protein, **1:**286
  kringles of, **3:**415
Plasminogen activator, tissue-type, engineering thrombolytic agent from, **1:**67
Plasminogen activator inhibitor
  type 1 (PAI-1), folding and unfolding of, **5:**154, 154f
  type 2, gene for, mapping to chromosome 18, **3:**208
Plasmodium falciparum
  genome of, **4:**307
  resistance to, in sickle cell heterozygotes, **5:**47
Plasmon, surface resonance, **1:**190
Plasticity, cellular, in learning and memory, **4:**57–58
Plastic pin technology, for selection, **4:**98
Plastids, **4:**466
  chloroplasts and amyloplasts, **4:**488
Plastocyanin
  catalysis of electron transfer by, **1:**357–58
  in photosystem I, **4:**447
Plastoquinol, electron transfer to plastocyanin through, **4:**449
Plastoquinone, in photosystem I, **4:**447–48

Platelet-activating factor (PAF), **4:**439
defined, **4:**416
inactivation by PLA_2-type activity, **4:**424
mediator in endotoxin reactions, **3:**471
Platelet-derived growth factors (PDGF), **1:**448; **2:**468–69; **5:**239, 269–70
induction of, by interleukin-1, **3:**340
Platelets
cytokines in, **1:**445
cytoskeletal-membrane interactions in, **1:**451, 453, 456
talin with actin, **1:**456
irradiation of, to destroy viruses, **6:**210
serotonin uptake in Down syndrome, **2:**120–21
tropomyosins from, **4:**138
Platinum complexes
activity against quiescent cells, **1:**257
as anti-cancer drugs, **1:**254
antitumor activity of, **5:**495–496, 496f
Pleckstrin homology (PH) domain, cPI-PLC isoforms, **4:**427
Pleiotropy
antagonistic, **1:**10
defined, **1:**5
of cytokines, **1:**442
effects of heat shock genes, **2:**218
in Marfan syndrome, **3:**162–63
in neurofibromatosis 1, **4:**175
neuropeptide, **4:**197–98
Plexiform neurofibroma, in neurofibromatosis 1, **4:**175
Ploidy, defined, **1:**76
Pluripotent stem cells, **1:**287, 447–48
leukocytes arising from, **3:**296
PNA (polyamide nucleic acid), **1:**166, 166f
Point-accepted mutation (PAM) matrix, **5:**422, 432
evolutionary distance in terms of, **5:**433t
Point mutagenesis, PCR assembly application, **5:**452
Point mutations, **2:**415, 418; **4:**149
in α_1-antitrypsin deficiency, **3:**394–95
APC genes, **1:**410
in assembled V genes, **3:**283
carcinogen-induced, **4:**152–53, 154f
in CJD, **5:**72
Cu^{2+}/Zn^{2+}-dependent superoxide dismutase gene, **4:**106
defined, **1:**407; **2:**392–93; **6:**138
in familial hypertrophic cardiomyopathy, **3:**169
FMR-1 gene, in patients without CGG repeat expansion, **2:**326
in Gaucher disease, **2:**348
type 3, **2:**351
in hemoglobins, **3:**18t
identification of, **5:**302
in Lesch-Nyhan syndrome, **2:**427
in a male without an estrogen receptor, **2:**292
in Marfan syndrome, **3:**162
in MCAD, **2:**425
and mispair thermodynamics, DNA, **2:**16–17
mitochondrial, **4:**82
oncogene activation from, K-*ras*, **3:**493
oncogenic transformation caused by, **4:**286
PrP genes, **5:**58
in sickle cell anemia, **4:**105
SOD I gene, in familial ALS, **4:**109
in Wilms' tumors, **6:**142–43
WT1 gene, **6:**144
Polarity, defined, **6:**109
Polarization
defined, **2:**294
fluorescence, **2:**299
defined, **2:**294
Polar lipids, **3:**427–28
defined, **3:**426
Polar solvents, Stokes shifts in, **2:**298
Poles, formation of, division cycle, **1:**100
Pol gene, defined, **1:**27
POLIOVIRUS
electron microscope image, **2:**192f
infection in transgenic mice, **6:**223, **6:**225
MOLECULAR BIOLOGY OF, **5:13–25**
receptor for, **6:**232, 232f
in transgenic mice, **6:**229–30
RNA of, **5:**359
Pollen tube, for plant cell transformation, **4:**464
Pol protein, HIV-1, **1:**41
Polyacrylamide
defined, **2:**96
gradient gel for separation of plasma lipoproteins, **3:**475
hydrogel, **5:**43–44
Polyacrylamide gel electrophoresis (PAGE)
protein purification using, **5:**178
strategies for obtaining sequence information from, **5:**181f
2D-, **2:**353, 353f
for DNA analysis with RLGS, **5:**310–16, 310f
Polyadenylation
alternative, **2:**376f
recognition motif, in the nuclear genome of *Chlamydomonas reinhardtii*, **1:**348
Poly-ADP-ribosylation
in the nucleus, **5:**53
protein, reversibility of, **5:**50–51
Polyamide
defined, **4:**269
polar, support for solid phase peptide synthesis, **4:**388–89
Polyamide backbone, in DNA analogues, **1:**166
Polyamide linkages, in modified oligonucleotides, **4:**273–76, 276f
Polyamide nucleic acid, synthesis of a polyamide oligomer, **4:**274f
Polyamines, in transcription and translation, **2:**211
Poly-β-hydroxyalkanoate (PHA), bacterial and commercial production of, **3:**433–34
Polychlorinated biphenyls (PCBs), reductive dehalogenation of, **1:**121–22
Polycyclic aromatic hydrocarbons (PAHs)
effect on macrophages, **3:**298
oxidation by cytochromeP450, to epoxides, **1:**202
Polycystic kidney disease (PKD)
in adults, **3:**232, 232f
defined, **3:**174
overexpression of epidermal growth factor in, **5:**272
Polydipsia, in diabetes insipidus, **2:**23
Polyene spacer, between electron donor and acceptor molecules, **1:**154, 155f
Polyesters, for drug delivery, **2:**167
Polyestradiol phosphate, antitumor activity of, **5:**488–89
Polyethylene glycol (PEG)
enzyme modification with, for solubilization in organic media, **2:**251
for plant cell transformation using protoplasts, **4:**462
for solid support, peptide synthesis, **4:**388–89
stimulation of fusion by, **5:**505
Polygalacturonase (PG), altering to produce tomatoes that resist softening, **6:**57–58
Polygenic resistance, defined, **4:**476
Polyglutamic acid, and tyrosine hydroxylase regulation, **1:**214
Polyhydroxyalkanoates (PHA), bacterial, **3:**433–34
Polyketides
antibiotic, altering the backbone structures of, **5:**506–07
screening for pharmacologically active compounds, **4:**99
Polylinker, defined, **2:**66
Polymer, defined, **1:**135; **5:**26
Polymerase
active site, HIV-1 reverse transcriptase, **1:**29
defined, **4:**281
DNA, **2:**88–89; **5:**354
of prokaryotes and eukaryotes, **2:**88t

Polymerase (*cont.*)
and telomere length, **3**:242
thermostable, **3**:302
DNA I of *E. coli*, **2**:90
errors involving, as point mutations, **4**:151
γ, in mitochondrial replication, **4**:82
poly(ADR) synthetase, **1**:401
replication errors, modeling, **2**:412–14
of respiratory syncytial virus, **5**:290
RNA, **5**:359
of vaccinia virus, **6**:197
Taq, **4**:334–35
defined, **4**:333
See also DNA polymerase; RNA polymerase
Polymerase α-primase, interaction with RPA and large T antigen, **5**:464
Polymerase chain reaction (PCR), **1**:124; **2**:55f, 398f; **3**:224–25, 239, 513; **5**:342
amplification
of microdissected chromosomal DNA, **1**:380–81
of high complexity probes, **6**:254
for analysis
of foreign DNA in plant cells, **4**:455–56
of microdissected chromosomal DNA, **1**:377
of risk for thrombosis, **4**:332
for assay of genetic markers, **3**:99
in bioprocess engineering, **1**:172
for characterization of ancient nucleic acids, **4**:305–06
chromosome 16 markers based on, **3**:176
in constructing antibody libraries, **1**:416, 416f
cycles of denaturation and synthesis in, **2**:18
defined, **1**:370; **2**:48, 62, 386; **3**:112, 215, 400; **4**:101, 226, 235, 323; **5**:338, 342, 388; **6**:84
for detecting fragile X syndrome, **4**:331
for detecting mycobacteria, **4**:163
for determining mtDNA sequence, **4**:85
for determining telomerase activity, **6**:35
extraction of DNA for amplification with, **2**:53
in gene detection, **3**:391
for genetic analysis, **2**:57
in genetic analysis
of alcohol and drug susceptibility, **2**:151
of nephrogenic diabetes insipidus, **2**:25
of populations, **2**:396
in genetic testing, **2**:421
in genome research
automation of, **5**:394
restriction landmark genome scanning, **5**:305
identification of microsatellite markers with, **2**:68, 387
for infectious disease assays, **4**:329
ligase mediated, **3**:401
in ligation assay, **3**:400
and molecular pathology, **4**:325–26
multiplex, for muscular dystrophy testing, **2**:426
for probe labeling, **2**:382
reverse dot-blot, for identification of Vβ families, **5**:510
RNA, **5**:351
for screening embryonic-stem cells, **6**:68
for telomerase assay, **6**:23
Polymerase domain, HIV-1 reverse transcriptase, **1**:29–30
Polymeric immunoglobulin receptor (poly IgA-R), transcytosis of, **5**:204–5
Polymerization, electrochemical, for processing nanowires, **1**:155, 156f
Polymer reactions, RNA world theory, **4**:296–97
POLYMERS
biodegradable, from transgenic plants, **4**:456
FOR BIOLOGICAL SYSTEMS, **5:31–44**
for coating drugs, **2**:166, 167
defined, **2**:165
for field effect transistors, **1**:154
GENETIC ENGINEERING OF, **5:26–31**
natural, **5**:27
nondegradable, **5**:36–44
structures and properties, **5**:26
Polymorphic DNA marker, defined, **3**:97
Polymorphic marker map
for chromosome 18, **3**:195
for chromosome 9, **3**:103–4f
Polymorphic triple repeat, diseases associated with, **3**:231–32
Polymorphism
biallelic, detection of, **3**:400
in chromosome 14, **3**:145–48t
in crystalline structure, **3**:309
debrisoquine, **4**:400–402
and deficiency mutations, **2**:349
defined, **2**:392; **3**:84, 142; **5**:45
DNA, **3**:230f
defined, **1**:42
in the human *PAH* gene, **4**:408–9
lipid, **4**:47–48, 47f
Msp I, and apolipoprotein levels in serum, **3**:411
sequence, detecting, **4**:339–40
size, of apo(a) isoforms, **3**:415
Polymorphism information content (PIC), **3**:236
Polymorphism link-up, **3**:219
defined, **3**:215
Polynuclear aromatic hydrocarbons (PAHs), bacterial metabolism of, **1**:121
Polynuclear aromatic hydrocarbons (PAH), fluorescence of, **2**:294
Polynucleotide amino acid (PNA)
oligo-T, **4**:273
synthesis of diamide linkages, **4**:276f
Polynucleotide duplexes, stability of, **4**:235–36
Polynucleotides
antitumor activity of, **5**:495
hybridization with, **4**:239–42, 239f
prebiotic formation of, **4**:295–96, 295f
RNA chain structure, **5**:373–74, 373f
Polynucleotidyl transferases, HIV reverse transcriptase, **1**:33
Polypeptide backbone, **4**:361
Polypeptide chains
of immunoglobulins, **3**:276–77, 277f
gene rearrangement events controlling, **3**:281f
termination signals, **6**:201
Ti, structure of, **3**:279–80, 279f
Polypeptide fold, RNase H domain, **1**:31–32
Polypeptide hormones
defined, **4**:340
receptors for, **5**:478
Polypeptide linker, **3**:286
Polypeptides
calcium ion pump of plasma membranes, **1**:244
defined, **5**:115
detection techniques, **2**:355–56
formation in aqueous solution, **4**:296f
insulinlike growth factor II, **5**:271
motor protein, characteristic of, **4**:114–15
prebiotic formation of, **4**:295
Polyphosphoinositide pathway, **2**:211, 211f
Polyprotein, defined, **5**:13
Polyprotein precursor
defined, **1**:34
Gag-Pol, **1**:34
picornavirus, **5**:19–21
Polypyrimidine tract-binding protein (PTB), binding to poliovirus internal ribosome entry site, **5**:19
Polyribosome (polysome)
defined, **5**:328
preparation of, **5**:330
Polysaccharides
analysis of, **1**:267, 269
biodegradable, **5**:33–34

lipopolysaccharide, **3**:463, 465–68
O-, antigen, **1**:279
Polysialic acid, in N-CAM, **1**:295
Polysome display method, **4**:92
Polystyrene gels, for solid supports, peptide synthesis, **4**:390
Polytene chromosomes
defined, **2**:121
Drosophila melanogaster, **2**:123
genetic analysis of, **2**:126–28, 128f
Polytene grids, of yeast artificial chromosomes, **1**:232
Polyubiquitinylation, defined, **3**:347
Polyunsaturated fatty acids (PUFA)
bacterial synthesis of, **3**:431
defined, **2**:330; **3**:426
fungal synthesis of, **3**:431
of low density lipoprotein phospholipids, oxidation of, **3**:479
Polyurethanes, biomedical applications, **5**:38–39
Polyuria, in diabetes insipidus, **2**:23
Polyvinylpyrrolidone (PVP), for binding phenols, in plant tissue culture, **6**:12
Poly(2-hydroxyethyl methacrylate), **5**:41–42
Poly(α-esters), **5**:35–36
Poly(L-glutamic acid) (PLGA), for artificial protein preparation, **5**:29
Poly(amino acids), **5**:34
Poly(anhydrides), **5**:36
Poly(A), defined, **4**:129
Poly(A) site, **5**:431
Poly(A)-tailed mRNA
defined, **1**:203; **5**:342
minus, **5**:344
plus, **5**:343–44
Poly(dimethyl siloxane), biomedical applications of, **5**:36–38
Poly(ethylene oxide), medical applications of, **5**:43
Poly(ethylene), medical applications of, **5**:39–40
Poly(glycolic acid), **5**:35
Poly(lactic acid), **5**:35–36
Poly(L-lactic acid), **5**:35–36
Poly(methyl methacrylate), medical applications of, **5**:40–41
Poly(*N*-vinyl pyrrolidone), medical applications of, **5**:43
Poly(ortho-esters), **5**:36
Poly(tetrafluoroethylene), medical applications of, **5**:39
Poly(vinyl alcohol), medical applications of, **5**:42–43
Poly(vinyl chloride), medical applications of, **5**:40
Ponticulin, actin attachment to the plasma membrane via, **1**:453
Population bottleneck, change in heterozygosity, **2**:394f
Population genetics
in Gaucher disease, **2**:349
of phenylketonuria, **4**:411–12
population-based screening programs, moratorium on, **4**:332
Population processes, in tandemly repeated noncoding DNA, **6**:10
Population size, and repeat array increase in size, **6**:9–10
POPULATION-SPECIFIC GENETIC MARKERS AND DISEASE, **5:44–49**
Pores
fusion, **4**:52–53, 52f
formation of, **4**:46–47, 47f
membrane, mimicking, **1**:140
model of transport through a membrane, **6**:128
nuclear complexes, **5**:206, 206f
Porins, **6**:128
membrane, **1**:394–95
Porphyrin, **6**:209
defined, **1**:148
derivatives of, for photosensitization, **6**:211–13
toxic response to cytochrome P450 inducers associated with, **4**:405
Positional cloning, **2**:387, 420; **3**:112, 135, 216, 222, 515–17; **4**:127–28
of the cystic fibrosis gene, **1**:428f
of the *RB1* gene, **2**:429
of the fragile X syndrome gene, **2**:320–21
of the cystic fibrosis gene, **1**:426
Positional effects
and incorporation of genes in transgenic modeling, **6**:67, 102
telomeric, **6**:31
Positive negative selection (PNS), **4**:102, 102f
Positive-strand enveloped RNA viruses, **6**:221–22
release for translation, **6**:224
Positive transcription elongation factor (P-TEF), **1**:25
POST- AND CO- TRANSLATIONAL MODIFICATIONS OF PROTEINS, ENZYME CATALYZED, **5:49–56**
Postgastrectomy dumping syndrome (PDS), **3**:379
Postglomerular resistance, and nitric oxide, **5**:269
Posthybridization treatment, in FISH, **2**:383–84
Postionization, of DNA after desorption, MALDI, **4**:9–10
Postreplicative recombinational repair, **4**:153
Postsource decay (PSD)
defined, **4**:12
of peptide ions in mass spectrometry, **5**:193
utilizing in tandem mass spectrometry, **4**:18
Postsynaptic cell, defined, **4**:54
Postsynaptic events, in long-term potentiation, **4**:57f
Posttranscriptional control
in *Caenorhabditis elegans*, **1**:234–35
defined, **3**:49
Posttranslational control
of histone proteins, **3**:50
of the switch from early mRNA expression to late mRNA expression, **6**:197–98
Posttranslational event, defined, **1**:42
Posttranslational modification, **5**:50–51, 50t
active peptides from, **4**:353–54
of amino acids, **5**:116
defined, **4**:340; **5**:49
of heterologous proteins produced by filamentous fungi, **2**:342
identifying, with mass spectrometry, **4**:19–21, 20t
in mycobacteria, **4**:160
of nitric oxide synthetases, **4**:202–3
of peptide hormone precursors, **4**:342–45
proline hydroxylation, 5;54
protein phosphorylation, **5**:169
of virus glycoproteins, **6**:187–88
Posttranslational processes
conformational change in prion protein, **5**:57
defined, **5**:194
and expression of heterologous proteins in yeast, **6**:324
Posttranslational tags, peptide, **4**:353
Potassium
activation of tyrosine hydroxylase by, **1**:215
nutritional requirement of microbes, **1**:119
Potassium ion channels
atrial, **3**:46
cardiac, **3**:2
membrane, **5**:513
Drosophila, **2**:133
Potential diversity, in sexual PCR, **5**:453–55, 455f
Potential energy (PE), defined, **1**:304
Potential energy well, electron transfer, **2**:194–95, 195f
Potentiometric, defined, **1**:177
Potentiometric biosensors, **1**:179

Potentiometric enzyme electrodes, **1**:186–87
POU class, homeobox genes, **3**:68–70, 71f
  defined, **3**:56
Power stroke, in muscle contraction, **4**:145
Poxviruses
  avian, for vaccines, **6**:179
  gene transfer mediated by, **2**:437
Prader-Willi syndrome (PWS)
  clinical expression of, **3**:160–62
  genomic imprinting in, **2**:424, 443, 445; **3**:160
  imprinting in, **3**:517
  physical map of the chromosome 15 region involved in, **3**:161f
  small nuclear ribonucleoprotein polypeptide N in, **3**:168
Pragmatic inference, **6**:39–40, 40f
Prebiotic chemical processes, **4**:291
Precipitation
  for analysis of high density lipoprotein subfractions, **3**:477
  of plasma lipoproteins, **3**:475
Precision, of DNA fingerprint analysis, **2**:58
Preconjugation format, **3**:292
  for biotinylated marker DNA, **3**:292
  for PCR, **3**:293f
Precursor, p66/p66, of HIV reverse transcriptase, **1**:32–33
Prediction
  of function
    DNA sequence, **5**:431
    from sequence analysis, **5**:429
  from immunotoxicity evaluations, **3**:301, 301t
  of mRNA sequence function, **5**:431–32
  of protein sequence function, **5**:432
  of RNA secondary structure, **5**:363–64
  of RNA three-dimensional structure, **5**:381–87
  of tRNA secondary structure, **5**:373
Prednimustine, antitumor activity of, **5**:488
Prednisolone, **5**:483
  antitumor activity of, **5**:488–89
Prednisone, **5**:483
  antitumor activity of, **5**:488–89
  for cancer therapy, **5**:486
Pregnancy
  cytokines involved in, **1**:444t
  energy requirements during, **4**:261
Premutation, **6**:5
  defined, **2**:319
  in fragile X syndrome, **2**:427
Prenatal diagnosis, of Gaucher disease, **2**:351
Prenylation, posttranslational modification by, and oncogene function, **5**:55–56
Presegregation Methocell method, **1**:100
Presequences, protein folding role of, **5**:91–92
Presynaptic cell, defined, **4**:54
Presynaptic facilitation, **4**:55f
Prevention, of cancer, **5**:499–500, 500f
Price, and patent protection, **6**:78, 79–80
Primary CAMs, **1**:292, 295
Primary liver cancers (PLCs), **3**:481
Primary messengers, **1**:237
  recognition of, by cell surface receptors, **1**:239
Primary structure
  of mtSSB, **5**:465
  of peptides, **4**:350, 351t, 352, 352f
    defined, **4**:347
  of proteins, **2**:307
    defined, **5**:147
    *Eco*SSB, **5**:461
    Raman spectrum analysis of, **5**:103–4
  of RecA protein, **5**:222, 222f
  of RNA, **5**:360, 374
  substrate-directed posttranslational modification of, **5**:52
Primase
  association with helicase, in DNA replication, **2**:89, 89t
  DnaG, **2**:88
  in initiation of DNA replication, **2**:91
  in plasmid replication, **5**:3
Primate genome, alpha satellites in, **3**:241
Primers
  Alu-specific, **2**:73f
  binding site, human telomerase, **6**:20f
  defined, **4**:254, 281, 333; **5**:342
  double-nested, **4**:336
  mismatch, **4**:282
  nonspecfic, characteristics of, **1**:380
  for nucleic acid sequencing, **4**:255, 334
  oligonucleotide, for DNA synthesis, **2**:58
  for picornavirus replication, **5**:23
  replication, for Col E1, **5**:2–3
  for telomerase, **6**:19
Primordial germ cells, for producing transgenic animals, **6**:68
Primosome, **5**:355
Primosome proteins, **2**:87–88
Prion protein (PrP), **5**:57–58
  accumulation of, in transmissible dementias, **4**:199
PRIONS, **5:57–73**
  defined, **4**:198; **5**:57
  infection by, in use of transgenic animals to produce drugs, **6**:107–8
  yeast, **6**:308, 308t
Priority sequence, for determining *R* and *S* configurations, **1**:335
Probabilistic search, for RNA three-dimensional structure modeling, **5**:380–81
Probe, nonsyntenic, **3**:119–20, 120f
Probe complexity, defined, **6**:253
Probe detection, defined, **2**:378
Probe down, **3**:255
  for diagnostic sequencing, **3**:259
  experimental implementation, **3**:256–57
Probe microscopy, **5**:401–02
  for DNA imaging, **5**:407–09
  sequencing by direct imaging with, **5**:409–10, 409f
Probes
  autoantibodies as, **1**:87–90
  chromosome-specific, **6**:254
  defined, **1**:113; **2**:378; **3**:385; **6**:253
  for disease detection, **3**:306
  DNA
    for detecting mutations, **2**:421
    for detecting unique sequences, **2**:58
    for mycobacteria, **4**:163
  fluorescent, **2**:294
  gene, **1**:124
  gene, defined, **1**:118
  heterologous, for screen libraries, **4**:472
  hybridization, **2**:389; **3**:218–19
    for FISH, **2**:380–83, 383t
    nick translated, **4**:242
  for identifying RNA secondary structure, **5**:364–65
  ligase chain reaction, **3**:303
  MER elements as, **3**:246
  nitroxide radicals as, **3**:386
  nitroxide spin, quenching of triplet labels by, **3**:388
  nonsyntenic, **3**:119–20, 120f
  nuclear magnetic resonance, **3**:389
  nucleic acid, for assays, **5**:348
  for nucleic acid structures, **1**:166
  oligomeric, establishing favorable conditions for binding, **2**:12
  padlock, for gene detection, **3**:397
  preparing, for ligase chain reaction, **3**:302
  in restriction length fragment polymorphism analysis, **5**:302
  for screening cloned plant regulatory genes, **4**:472
  "sticky end" interaction, **3**:305
  for whole chromosome fluorescence staining, **6**:253–55
  yeast artificial chromosomes as, **6**:300
Probe up, **3**:255
  experimental implementation, **3**:256
Procapsid, viral, defined, **6**:200
Processive catalysis
  defined, **4**:416
  by phospholipases, **4**:417–18
Processive transcription, defined, **1**:19
Processivity factor, Tat as, **1**:24–25
Prochirality
  defined, **1**:52, 332

and enzyme specificity, **1**:343–46, 343f
Prodrugs, **1**:196
defined, **1**:193
minoxidil, sulfation of, **1**:198
Productivity
of antibiotic preparation, **5**:507
of domestic animal transgenic models, **6**:69–70
Product specificity, defined, **6**:259
Profile, sequence alignment, defined, **5**:420
Progeria, due to mutation, **1**:302
Progeroid syndromes, **1**:7–8
Progesterone, **2**:159
effect on metastatic endometrial carcinoma, **5**:490
role of, **5**:482
steroid subgroup, **5**:479
structure of, **2**:160f
synthesis of, **5**:480
Progesterone receptor (PR), **2**:290
Progestins, **2**:159
role of, in insulinlike growth factor receptor gene expression, **5**:270
synthetic, **5**:483
treatment for endometrial cancer, **1**:256
Prognosis
in neoplastic disease, **2**:65
in scleroderma, correlation with clinical presentation, **5**:418–19
Programmed cell death, **1**:449
defined, **1**:299
See also Apoptosis
Progressive enzyme, defined, **6**:28
Prohormone convertase, defined, **4**:340
Proinsulin
chromatographic and electrospray ionization analysis of, **4**:23–25, 25f
fragment of, LC/ESI-MS/MS spectrum, **4**:26f
spectrum of, matrix-assisted laser desorption ionization (MALDI), **4**:27f
Prokaryotes
defined, **4**:67
DNA replication in, **2**:87f
gene regulation in, **2**:369–71, 369f
lack of identified motor proteins in, **4**:114
mRNA of, **6**:111, 111f
nucleotide excision repair in, **2**:45
and oxygen, **1**:133
primer synthesis in, **2**:89
RNA polymerase of, **2**:91
transcriptional regulation in, **2**:95
transcription units, **2**:93–94f
Prokaryotic cells, energetics of, **1**:304
Prokaryotic model, of RNA-directed elongation regulation, **1**:25
Proliferating cell nuclear antigen (PCNA), **1**:90
Proline
in collagen, **5**:28
and protein folding, **5**:31
Proline hydroxylation, posttranslational, **5**:54
Proline racemase, **2**:231f
Prolyl *cis-trans*-isomerase, **5**:199
Promoter, **2**:273; **5**:128
alternative use of, collagen genes, **2**:280
casein, heterologous proteins from, **6**:103
from cauliflower mosaic virus, **6**:54–55
defined, **2**:86, 271, 364; **3**:227; **4**:129, 226; **6**:53
epigenetic modification of, **2**:369
eukaryotic, structure of, **2**:366
for eukaryotic transcription, **2**:90, 94–95
external and intragenic, **2**:95
functional DNA, **3**:511–12
GAL1, yeast, **6**:313–14
gonadotropin-releasing hormone, **2**:286
inducible, in expression systems, **3**:22
of keratin genes, **2**:266
lac, control of the gene for T7 RNA polyermase by, **2**:275
light-dependent gene activation by, **4**:469f
modeling of, **2**:414
mutation of, in hemophilia, **3**:40
mycobacterial, **4**:162f
for prokaryotic transcription, **2**:90
structure of, **4**:468–69, 468f
tapetum-specific, **4**:466
tissue-specific, in transgenic tobacco, **4**:473f
transcriptional, **3**:282
of archaea, **4**:74–75
as a vector for amplification, **6**:293
for vertebrate mtDNA, **4**:80–81
for transcription of the apoA-II gene, **3**:411
tumor, okadaic acid, **5**:172
Promoter element, insertion into cells, **1**:234
Promoter/enhancer module, defined, **1**:19
Promoter region
CpG islands in, **2**:322–23
defined, **1**:283; **2**:216
Sp1 module, **1**:20, 20f
Promoter regulatory elements
cell-cycle-dependent histone gene expression, **3**:53–55, 54f
defined, **3**:49
proximal, for apolipoproteins, **4**:500–501
Promoter sequences, **5**:356
long terminal repeat, HIV, **1**:20, 20f
for mRNA synthesis, vaccinia virus, **6**:197
Pronuclear injection, **3**:518
Proofreading polymerases, point mutagenesis with, **5**:452
Propane oxidizers, bacterial, **1**:121
Propeller-twist, in DNA, **2**:110, 110f
Propeptides, protein folding role of, **5**:91–92
Prophage, P1, partitioning in replication of, **5**:4–5
Prophase
chromosome bands in, **1**:371
preparation of chromosomes at, for FISH, **2**:379
Propidium iodide (PI), use in FISH, **2**:379
Propionic anhydride, defined, **5**:485
Propranolol
androgenic blocker, **4**:31
structure of, **4**:32f
Propylene, isotactic form of, **5**:26
PROSITE database, **5**:432
Prospective studies, cohort, in molecular epidemiology, **2**:262
prospero (pros) class, homeobox genes, **3**:74–75, 75f
Prostaglandin synthesis
in Down syndrome, **2**:120–21
enhancement by phospholipase C activity, **4**:428
Prostate cancer
gene predisposing to, **3**:223
keratinocyte growth factor (KGF) in cells of, **2**:469
KAI1 gene implicated in, **6**:146
response to flutamide, **5**:484
Prostate-specific antigen (PSA), gene encoding, **3**:384
Prostheses, vascular, poly(tetrafluoroethylene) for, **5**:39
PROTEAN program, for determining three-dimensional structure of RNA, **5**:381
Protease
angiotensin-converting enzyme, **1**:167–68
aspartic, **1**:14
aspartyl, renin as, **5**:265–66
cysteine, lysosomal, **3**:347–48
defined, **1**:34; **2**:240, 271
effect of inhibitors on tumors, **3**:379
HIV, **1**:14
crystal structure, **1**:41f
inhibition of, **1**:37f
inhibitors of, **1**:14, 38–40, 38f; **4**:34, 34f, 37–38
role of, **1**:34
structural and biological properties, **1**:14–15
structure of, **1**:27–28, 28f; **4**:36–37, 36f
structure change with inhibitor binding, **1**:15

Protease (*cont.*)
synthesis of, **4:**370–71
target for rational design of drugs, **1:**14, 14f
tertiary structure, **1:**27–28
inhibitors of
Kunitz type, **1:**46
locus on chromosome 14, **3:**151
K, transgenic integration analysis, **6:**71–72
malarial
structure by homology, example, **5:**164–66
mapping to chromosome 14, **3:**155
nexin-2 inhibitor, **1:**46
polyprotein processing by, **1:**32–33
serum, inhibitor of, in AAT deficiency, **2:**424
viral, 2A, **5:**21
3C, **5:**19–20
3CD, **5:**20–21
virus-specific, of retroviruses, **6:**191–92
Protease cascade
kallikrein-kininogen-kinin, **3:**377–78, 378f
poliovirus, **5:**19–20, 20f
Proteasomes, **1:**446; **4:**494
cytoplasmic, **3:**348–49, 348f
of methanogens, genes encoding, **4:**78
Protecting groups, in peptide synthesis, **4:**365, 365f, 367t, 383–84
Protein, food
digestibility of, **4:**263–64
functional interactions in, **2:**311–12
quality measures, **4:**263
Protein activity, in heart failure, **1:**291t
PROTEIN AGGREGATION, **5:73–93**
PROTEIN ANALYSIS
by mass spectrometry, **4:**19–21
BY RAMAN SPECTROSCOPY, **5:93–110**
BY X-RAY CRYSTALLOGRAPHY, **5:110–15**
PROTEIN ARCHITECTURE AND ANALYSIS, **5:115–27**
Proteinase inhibitor, production of, **4:**492–93
Proteinase inhibitor genes, **4:**480
Proteinases
K, for DNA fingerprint analysis, **2:**49
zinc, MMPs, **2:**290
Protein association, measuring, **2:**299f
Protein-binding aptamers, **4:**228–29
Protein C, activated, and risk of thrombosis, **4:**332
Protein databank, defined, **2:**169
PROTEIN DESIGNS, FOR THE SPECIFIC RECOGNITION OF DNA, **5:127–35**
Protein disulfide isomerase (PDI), **3:**353; **5:**199
formation of disulfide bonds in secretory proteins, **5:**92
and structure of secretory proteins, **5:**81–82
Protein-DNA binding, optical sensors for studying, **1:**190
Protein-DNA interactions, **5:**129
glucocorticoid and estrogen receptor DNA-binding domains, **6:**332
Protein-DNA recognition, **5:**301–2
Protein domain, defined, **5:**128
Protein-drug interactions, mass spectrometry for analyzing, **4:**21–22
Protein efficiency ratio (PER), in foods, **2:**309
Protein engineering
of enzymes for organic biocatalysis, **2:**251
in immunology, **3:**285–88, 288f
PROTEIN EXPRESSION BY EXPANSION OF THE GENETIC CODE, **5:135–47**
PROTEIN FOLDING, **5:147–58**
analysis of
with circular dichroism, **1:**391
with electrospray mass spectrometry, **4:**23
defined, **1:**315; **5:**73
effect of noncoding amino acid substitution on, **5:**138
prevention of, by hsp70s, **1:**316
pro region sequences as intramolecular chaperones in, **1:**322
role of chaperonins in, **1:**319–20, 320f
rules of, **5:**81, 92–93
in Tat transactivation, **1:**23
and tertiary structure, **5:**120
Protein Identification Resource, **2:**170
Protein kinase
A (cAMP-dependent), **3:**46
activation by growth factors, **1:**446
Ca^{2+}/calmodulin-dependent, **1:**207; **6:**124
Ca^{2+}/phospholipid-dependent, **1:**207
Ca^{2+}-dependent, in plants, **4:**480
calmodulin-activated, autophosphorylation of, **5:**51
cAMP-dependent, **1:**207; **3:**339
phosphorylation of phosphoprotein phosphatase by, **5:**172
classification and properties of, **3:**46–48; **5:**170
defined, **3:**42; **4:**432; **5:**169
tyrosine and serine, in immune responses, **3:**279
Protein kinase A (PKA)
in cystic fibrosis, **1:**429, 429f
and spermatogenesis, **3:**506
Protein kinase C (PKC), **1:**240, 455; **4:**55f, 439
activation of
by diacylglycerol, **4:**424
by phorbol esters, **5:**172
calcium-ion-dependent serine-threonine, **1:**459
cysteine-rich domain of, **6:**336
effects on annexin I, **1:**61
in idiopathic cardiomyopathy, **1:**289
and interleukin-2 gene induction, **3:**335
in mice bred for ethanol sensitivity, **2:**148
phosphorylation of $cPLA_2$ by, **4:**423
role in learning and memory, **4:**57f
Protein kinase-coupled receptors (PKR), **5:**238–42
Protein kinase domain, autoinhibition, **5:**51
Protein kinase G (PKG), **2:**210
Protein kinase R (PKR)
inhibition of protein synthesis by, **3:**325
shutoff of cellular syntheses after activation of, **6:**199
tumor suppression by, **3:**326
Protein labeling, with radioisotopes, **5:**218
Protein-ligand complexes, mass spectrometry for analysis of, **4:**19–21
Protein ligase, ubiquitin, **3:**349
Protein micelles, defined, **2:**307
PROTEIN MODELING, **5:158–69**
Protein-nucleic acid interactions, circular dichroism for observing, **1:**395
PROTEIN PHOSPHORYLATION, **5:169–73**
Protein processing
by picornaviruses, **5:**19–21
by poliovirus, **5:**20f
Protein-protein assemblies, optical sensors for studying, **1:**190
Protein-protein interactions
cascade of, and heterotrimeric G proteins, **3:**43
in foods, **2:**309
hirudin-thrombin, **5:**56
involving the LIM class of homeobox genes, **3:**72
involving T4gp32, **5:**461
phage display systems for studying, **4:**91–92
and regulation of promoter activity, **2:**367
regulation of transcription by RNA polymerase, **4:**467
in virus release from cells, **6:**184
Proteins
accessory, of viruses, **6:**188t
N-α-acetylated, ubiquitin-dependent degradation of, **3:**352

activator
for CaM PK II, **1:**215
in enzyme-directed posttranslational modification, **5:**51
adhesive, **5:**28
analysis of mixtures, mass spectrometry adaptations for, **4:**23–26
anchoring to membranes, **4:**439–40
artificial, repeating units of, **5:**29t
bioorganic chemistry of, **1:**166–71
channel-type, **6:**128–129
characterization of, by molecular mass determination, **4:**19–21
for chromosomal DNA replication, classes of, **2:**87–88
coding for, **5:**431
coiled-coil, **5:**29
control of translation from RNA, **6:**122
covalent modification by lipids, **3:**461–62
crystal structure of, and hydrogen bonding, **3:**269–70
denaturation of, **2:**311f
DNA-binding, assay of sequence-specific, **4:**470–71
EF-hand, **1:**237
electron microscopy of, **2:**185, 187, 190
export of, **2:**274
expression in transgenic animals, **6:**72–73
fibrous, **5:**27–29
as foods, **4:**262–65
heterologous, expression in the milk of farm animals, **6:**106–7, 106t
high mobility group (HMG1 and HMG2), **5:**440–41
hydration of, **3:**272
hyperthermostable, **2:**247–49
interactions of, **2:**309–11
in foods, **2:**311–12
intracellular calcium-binding, **1:**240–42
intracellular degradation of, mechanisms, **3:**346–54
membrane, regulating fusion, **4:**48–49, 49f
metal binding by, **1:**132
metalloregulatory, **6:**60
milk, **6:**102–3, 103t
M-line, **4:**143
in molecular paleontology, **4:**303–5
phospholipid transfer, **4:**438
photodynamic, defined, **1:**135
presynaptic membrane, **5:**514
production of, by filamentous fungi, **2:**341–43, 341t
purification of, with ion exchange chromatography, **3:**80
purity of, measuring, **5:**123–24
regulatory, **4:**138–40
spin labeling analyses, **3:**389
stability of, monitoring with circular dichroism, **1:**393–94
study of, in molecular paleontology, **4:**303–5
synaptic-vesicle associated, **5:**514
telomerase component, **6:**32–33
telomere-binding, **6:**29
thaumatin-like, **4:**480
translation of, in spermiogenesis, **3:**504
transmembrane
as channels, **4:**54
exchanger, **1:**237
pump, **1:**237
transport, **6:**125–33
viral
components of assembly and budding, **6:**186–87, 187t
polymers of, **5:**28–29
viral-encoded, membrane-anchored, **6:**184
See also Adaptor proteins; Membrane proteins
Proteins, specific
85K, **4:**142
apoproteins, apoB-100, apoAI, apoE, C, **3:**406
bacteriorhodopsin, **1:**108–9, 137
biotin-binding, **6:**240–41
Bt crystal, toxins, **4:**394
Bt toxin, **6:**56–57
C, **4:**142
defined, **2:**446
expression in milk of transgenic mice, **6:**105
expression in the milk of transgenic pigs and sheep, **6:**106
cap-binding, **6:**118
collagen, **1:**146
E6, gene mapping to chromosome 15, **3:**170
*Eco*SSB, single-stranded DNA-binding protein, **5:**461–62, 462f
gag, L-A yeast virus, **6:**220
gal4, **6:**333
GCN4, **1:**3
gp130, **5:**242
gp32, **5:**460–61
harpin, produced in fire-blight of pears, **4:**479
histone, **1:**361
IHF, **1:**3, 3f
interferon, **3:**319–20
interferon-induced, **3:**323–25, 324f, 325t
interleukin, **3:**334
keratin, **2:**265–66
LacR, intersegment transfer in, **5:**281–82
LexA, **2:**95
Lex A, autocatalytic cleavage of, **5:**222
MetJ, **5:**277
M-line, **4:**143
p21, mutation in lung cancer, **3:**493–94
p220, mRNA cap binding role of, **5:**25
p24 *gag*, **1:**26
p53
detection of damage to DNA, **1:**263
multiubiquitination of, **3:**352
p68, **1:**21
p78, poly(A)-binding protein, **6:**122
pRB (retinoblastoma), hypophosphorylated, **5:**320–22, 322f
protein A, defined, **3:**288
rab, vesicle docking and fusion regulation by, **4:**50
RB, p110*SRB*, **6:**140, 140f
RecA, **1:**3–4, 3f; **5:**220–29
RecE, F, G, J, Q, and T, **5:**228
RepA, in plasmid replication, **5:**4
RuvC, **5:**228–29
schwannomin, structure of, **4:**186
Sp1, transcriptional activator, **1:**20, 20f
SRY, **5:**440–41
TAR-binding, **1:**21
Tat, viral transcription activator for HIV-1, **6:**197
Tat and Rev, **1:**26
TFIIH complex, transcription factors, **6:**288–90, 290f
3A and 3B, viral, **5:**23
3CD, viral, **5:**22
α-tocopherol binding protein, **6:**238
Tox R, **1:**107
2B and 2C, viral, **5:**22
Proteins, structural
design of artificial, **5:**29–31
expression of, **5:**27–29
of muscle, **4:**142–43
scaffolding, defined, **6:**200
PROTEINS AND PEPTIDES, ISOLATION FOR SEQUENCE ANALYSIS, **5:173–85**
Protein sequence, MATα2, degradation pathways for, **3:**352–53
PROTEIN SEQUENCING TECHNIQUES, **5:185–93**
Protein serine/threonine kinases, **5:**170, 233
defined, **5:**230
Protein structure, **2:**307–9
correlation with function, in *Caenorhabditis elegans*, **1:**235–36
secondary, **1:**389–91
tertiary, from circular dichroism data, **1:**391
Protein synthesis
analysis of, via trichloroacetic acid precipitation, **5:**333t
cell-free systems for, **5:**332–36
defined, **5:**328

Protein synthesis (*cont.*)
elongation factor involved in aging, **5:**336
general requirements for assays of, **5:**331t
regulation of, by translational repressors or enhancers, **2:**376–77
Protein synthesis factors
ancillary to ribosomes, **5:**330–32
preparation of, from cells, **5:**330
PROTEIN TARGETING, **5:194–207**, 335–36
Protein translocation, **5:**196f
defined, **1:**315; **5:**194
Protein tyrosine kinases, defined, **5:**230
Protein tyrosine phosphatase, **3:**208, **5:**171
Proteoglycans, **2:**282, 283, 453
analysis of, **1:**269
sugar chains of, **2:**448
transforming growth factor-β receptor type III, **2:**468
Proteolipid protein gene, mouse, **4:**126
Proteolysis
cleavage of amyloid precursor protein, **1:**48, 48f
defined, **3:**347; **6:**327
in homologous protein production by filamentous fungi, **2:**343
interference with protein sequencing by, **5:**190
Proteolytic processing, defined, **5:**13
Prothoracicotropic hormone (PTTH), **4:**190
Protochlorophyllide reduction, light-independent, in *Chlamydomonas reinhardtii*, **1:**349
Proton electrochemical potential difference, across a membrane, **4:**448
Protonmotive force
bacterial rhodopsin, **1:**109
defined, **1:**108
and pH difference across a membrane, **1:**307–8
Proton pump
ATP-dependent, **5:**203
bacteriorhodopsin, **1:**149, 149f
electrogenic, bacteriorhodopsin, **1:**109
light-driven, **1:**110
photosynthetic energy transduction, **4:**445
Proton release domain (PRD), bacteriorhodopsin, **1:**110f, 112, 112f
Proton transfer
in bacteriorhodopsin, **1:**110–11
along hydrogen bonds, **3:**266–67
Proton uptake domain (PUD), bacteriorhodopsin, **1:**110f, 112, 112f
Proto-oncogenes, **4:**329; **6:**138
c-*met*, **5:**272
cell surface receptor, **3:**482
c-*myc* and c-*fos*, platelet-derived growth factor promotion of, **5:**270
c-*onc*, **4:**285
defined, **2:**62, 215; **3:**1, 480; **4:**323
growth regulation by, **1:**412
in hepatocellular carcinoma, **3:**487–88
k-*ras*, **2:**63
mapping to chromosome 14, **3:**158
Mas, and parental imprinting, **2:**443
in myocardial hypertrophy, **3:**1–2
N-*myc*, amplification of, accompanying retinoblastoma, **5:**320
p170, overexpression of, accompanying retinoblastoma, **5:**320
rel family, binding by NF-κB, **1:**19
translocations associated with, **2:**64t
Protoplasts
defined, **4:**458; **5:**501
for genetic transformation, **4:**452
gene transfer to, **4:**453f
plant cell transformation methods using, **4:**462–63
preparing for cloning experiments, **5:**505
for transformation of tomato plants, **6:**56
Protoporphyria, **3:**214–15
Prototroph, defined, **5:**135
Protozoal diseases, genetic immunization against, **2:**437–38
Protozoans, mitochondrial DNA of, **4:**79
Proviruses
defined, **6:**193
endogenous, human genome, **3:**246
for gene transfer, **2:**436
PrP, ablation of the gene for, and prion-resistance, **5:**64
PrP^C, defined, **5:**57
PrP^{Sc}
characteristics of, **5:**58
defined, **5:**57
formation of, **5:**59–65, 65f
PRX superclass of genes, **3:**68
P-selectin, **1:**294f
Pseudoatoms, for RNA modeling simplification, **5:**381
Pseudoautosomal regions
defined, **4:**198
Y chromosome, association with schizophrenia, **4:**200
Pseudogenes
CYB5P1, mapping to the X chromosome, **3:**201
defined, **2:**347
globin cluster, α-type , **3:**15
human repetitive elements as, **3:**240f
LINEs as, **6:**135
Pseudohermaphroditism, lack of estrogen production in, **2:**292
Pseudohypoparathyroidism type IA, **2:**214
Pseudoknot, **5:**361, 362–63
defined, **4:**226; **5:**360
mRNA secondary structure, **6:**122
RNA, aptamer binding to HIV-1 reverse transcriptase, **4:**228, 228f
Pseudomonas, defined, **4:**393
Pseudomonas aeruginosa
attack on antibodies, **1:**105
regulation of the virulence genes of, **1:**107–8
regulation of virulence genes of, **1:**107–8
Pseudomonas savastonoi, plasmid genes of, **5:**12
Pseudomurein, in Archaea cell walls, **4:**68
Pseudopeptides, **4:**354
Psi packaging signal, for RNA packaging, **6:**334
Psoralen, as a probe for determining RNA structure, **5:**365
Psoralen dyes, as photosensitizers, **6:**209, 216
Psychoactive drug, defined, **2:**146
Psychoses, functional, **4:**199–200
Psychosine, **3:**457
Pterin coenzymes, **1:**402
Public health, **2:**262
Public opinion
about animal biotechnology and patents, **6:**82
about transgenic livestock, **6:**107–8
Pulsed field gel electrophoresis (PFGE), **2:**74; **3:**216, 515
of chromosome 11, **3:**135–37, 137f
for copy number analysis, transgenic sheep, **6:**107
defined, **3:**126, 191, 215
for karyotyping, **4:**307–8
for mapping the *M. tuberculosis* genome, **4:**157
for physical mapping of chromosomes, **3:**121–22
Pulse radiolysis, defined, **3:**354–55
Pulse width, and efficiency of electroporation, **2:**203
Pump
calcium ion, **1:**243–45
sarcoplasmic reticulum, **1:**247
transmembrane protein, defined, **1:**237
Punctuations, defined, for a *k*-gram alphabet, **6:**37
Purification
of peptides, in solid phase synthesis, **4:**388
of RNA, goals of procedures, **5:**344
Purine base, products of oxidation and reduction of, **3:**362f
Purine base radical cations, **3:**358–59
Purines
binding sites for oligonucleotides, **1:**3

products from, after exposure to ionizing radiation, **3**:361f
Purple bacteria
mitochondrial rRNA of, **4**:83–84
reaction centers of, **4**:442
Pyridine nucleotide coenzymes, **1**:400–401
Pyridinones, reverse transcriptase inhibitors, **1**:39f
Pyridoxal 5′-phosphate (PLP), **1**:401–2, 402f
effect of, in blood substitutes, **3**:20
Pyridoxal kinase, **1**:401
Pyridoxal phosphate binding proteins, **6**:243
Pyridoxamine 5′-phosphate (PMP), **1**:401–2
Pyrimidine bases, nucleic acid, effects of irradiation on, **6**:209
Pyrimidine dimer-DNA glycosylase, **6**:285
Pyrimidine dimers, **2**:42f
removal from transcriptionally active genes, **2**:82
Pyrimidine free radicals, **3**:358–59, 361f
Pyrococcus furiosus
aldehyde oxidoreductase of, **2**:245
hydrogenase of, **2**:245, 246
protease of, **2**:244
sucrose α-glucohydrolase of, **2**:244
Pyrococcus woesei, amylase of, **2**:244
Pyrodictium brockii, hydrogenase of, **2**:245
Pyroglutamylation, of peptide hormones, **4**:346
Pyrroles, in peptide design, **4**:377
Pyrroliquinoline quinone (PQQ), coenzyme, **1**:406
Pyrrolnitrin, control of damping-off in cotton with, **4**:398
Pyrrolylalkanethiol, conducting layer on a gold surface, **1**:154, 154f
Pyruvate decarboxylase, **1**:343
Pyruvate kinase-3, gene localization to chromosome 15, **3**:169
Pyruvate oxidoreductase (POR), **2**:244
Pyruvoyl N-terminus, coenzyme, **1**:405

Q bands, chromosomal, **1**:371
Qβ amplification, ligation-initiated, **3**:397, 397f
Q cycle, **4**:449
electron transfer mechanism, **4**:447, 447f
Quadruplex structures, of G-strand DNAs, **6**:30–31
Quadrupole ion trap mass spectrometer, **4**:27
defined, **4**:12
Quadrupole mass analyzers, **4**:15
Quadrupole mass spectrometry, collision-induced dissociation in, **4**:17–18
Quality assurance, in DNA fingerprint analysis, **2**:51–52
Quantitative measurements, with matrix-assisted laser desorption ionization mass spectrometry (MALDI), **4**:5–6, 5f
Quantitative structure-activity relationships (QSARs), **4**:35–36, 36f
Quantitative trait, defined, **2**:146
Quantitative trait loci (QTL) mapping
in alcohol and drug sensitivity studies, **2**:152–54
defined, **2**:146
for drug addition studies, **2**:151
Quantum, defined, **1**:148
Quantum size effects, of inorganic materials, **3**:309
Quantum-size semiconductor particles
growing in phospholipid vesicles, **3**:314
growing in yeast cultures, **3**:316
Quantum yield, of a fluorophore, **2**:300–301
Quartz crystal microbalance (QCMB), **1**:192
Quasi-irreversibility, in protein unfolding-refolding, **5**:79
Quasi-vitamins, **6**:252
Quaternary structure
of hemoglobin, and ligand binding, **3**:10–11, 11f
of proteins, **2**:308; **5**:120–21, 120f
Raman spectrum analysis of, **5**:105
symmetry of, **5**:112
RNA, **5**:360
Quenching, **2**:298–99, 298f; **3**:388
collisional, **2**:301, 301f
defined, **5**:209
for protection against photochemical damage, **6**:209–10
reduction of scintillation efficiency by, **5**:219
Quinacrine staining, of chromosomes, **1**:371
Quinine
fluorescence of, **2**:295
structure of, **4**:29f
Quinones (Q), electron transport by, **1**:399, 400; **4**:445
Quinoproteins, **1**:406
Quorum-sensing, **1**:107
in expression of virulence genes, **1**:108

Rabbits
as models for studying cancers, **6**:69
reticulocyte lysate from, **5**:334, 337
use in antisense technology, **5**:336
Race, and occurrence of cystic fibrosis, **1**:425
Racemases, **1**:343
Racemic mixtures, **1**:336
defined, **1**:52, 332
Racemization, defined, **1**:332
Race-specific resistance, in plants, **4**:476
Radiation
effect on DNA, **2**:40
effect on life span, **1**:6
effect on risk of osteogenic sarcomas, in *RB1* mutation, **5**:320
protection from, **5**:211–14
isotopes used in molecular biology, **5**:213t
See also Ultraviolet radiation
Radiation hybrids
cell lines for cloning markers specifically on chromosomes, **2**:73
defined, **3**:126
fusion, for physical mapping of chromosome 10, **3**:122–23
linkage maps based on, **3**:120–21, 137, 139f, 140
chromosome 16, **3**:182–83
Radical
carbon-centered, addition to double bonds of nucleobases, **3**:365
defined, **2**:330
reactions of, after ionizing radiation exposure, **3**:357–60
Radical ion, defined, **3**:355
Radioactive decay, **5**:209–10
Radioactive solutions, handling in the laboratory, **5**:214
Radioactivity, defined, **5**:209
Radio frequency (RF) pulse, for transfection with reporter genes, **2**:202
Radioimmunoassay (RIA)
for comparing extinct and living organisms, **4**:300
of steroid hormones, **5**:476
Radioiodination, chloramine T procedure for labeling proteins, **5**:218
Radioisotope probe hybridization, for DNA fingerprinting, **2**:57
RADIOISOTOPES
defined, **5**:209
IN MOLECULAR BIOLOGY, **5**:**209–20**
for labeling proteins, **2**:356
molecules labeled with for electron microscopy, **2**:193
Radionuclides
choice of, for experimental use, **5**:214
comparison of utility and safety, **5**:215t
defined, **5**:209
Ramachandran plots, **5**:148
for evaluating protein models from X-ray diffraction data, **6**:280
Raman effect, defined, **5**:93
Raman microscopy, for hydrogen bond studies, **3**:267
Raman spectroscopy, for following DNA denaturation, **2**:6t

Raman spectrum, for thioredoxin, **5**:102–3, 103f
Rana catesbeiana, hemoglobin of, **3**:12
Random amplification of polymorphic DNA (RAPD), for mapping the *Arabidopsis* gene, **1**:80
Random coils
  conditions for formation from unfolded proteins, **5**:148
  of denatured DNA, **2**:1–2, 3f
  protein, **2**:307; **5**:119
Randomization, for macromolecular selection experiments, **4**:95–96
Randomized oligonucleotide selection, **4**:227
Randomized pool
  defined, **4**:226
  nucleic acid, generation of, **4**:227, 227f
Randomly primed cDNA library, defined, **1**:203
Random mutagenesis, to construct enzymes, **1**:144
Random peptide libraries, **5**:516–17
Random priming, **2**:382
RAPD (random amplified polymorphic DNA)
  defined, **2**:48
  for DNA fingerprinting, **2**:58
  for genetic analysis of populations, **2**:397
Raphillin, **4**:50
RARE (RecA-assisted restriction endonuclease) cleavage, **3**:219, 219f
  defined, **3**:215–16
RARE endonuclease restriction enzymes, for mapping, chromosome 16, **3**:183
Ras-MAPK cascade, activation of, **5**:241
Rate constants
  electric and magnetic field-induced changes, **2**:174, 177–78
  electron transfer, **2**:194
    distance dependence, **2**:196–97, 196f
  for metal ion complexation by ligand exchange, **1**:133–34
  photocycle reactions, bacteriorhodopsin, **1**:110
  and time for protein folding, **5**:153–54
Rational drug design, **4**:31–32
  antisense oligonucleotides for, **1**:69–70
  defined, **1**:14
  from life cycle information, **1**:34–35
  from structures of protease-inhibitor complexes, **1**:38
Rats, as models for hypertension, **6**:69
Raynaud's phenomenon, **5**:418, 418t
R bands, chromosomal, **1**:371
Reaction center (RC)
  defined, **1**:148; **4**:440
  electron transfer, antennae pigment matrices for, **4**:441, 441f
  photochemistry of, **4**:442–44, 444f
  photosynthetic, **1**:139, 149, 149f
    reconstructing, **1**:151
Reaction coordinate diagram, **2**:227–28
Reaction cycle, for P450, general, **1**:437f
Reaction intermediates, chiral specificity of, **1**:341, 341f
Reaction mechanisms, back-side displacement, glucuronidation, **1**:198
Reaction media, solid-phase peptide synthesis, **4**:390
Reaction quotient, **2**:225, 227
Reactive nitrogen intermediates (RNI), antimicrobial activity of, **3**:326
Reactive oxygen intermediates, antimicrobial reactions of, **3**:326
Reactive species
  oxygen, defined, **2**:216, 330
  produced by ionizing radiation, **3**:355–56
Reading frame
  change in, by alternative use of promoter, **2**:280
  defined, **6**:109
  See also Open reading frame
Reading head
  estrogen receptor complex, **6**:332
  α-helix, zinc-binding domains, **6**:336
Readout, direct and indirect, defined, **5**:275
Rearrangements
  Arabidopsis genome, **1**:84
  chromosomal, **3**:226; **4**:150, 150f
    analysis of, **6**:255–56
Reassociation, **4**:235
  kinetic studies of the *Arabidopsis* genome, **1**:76–77
RecA PROTEIN
  Achilles' cleavage protocol using, **1**:5
  gene for, localization to chromosome 15, **3**:169
  mediation of homologous recombination by, **5**:406
  specific, **2**:95
  strand exchange mediated by, **5**:255
  STRUCTURE AND FUNCTION OF, **4:220–29**
Receptor-associated protein (RAP), **3**:405
RECEPTOR BIOCHEMISTRY, **5:229–43**
Receptor-ligand interactions, optical sensors for studying, **1**:190
Receptors, **2**:207–9, 209f
  adrenergic, in heart failure, **1**:290
  angiotensin II, **5**:266–67
  bradykinin, **5**:268
  classification of, **5**:230
  defined, **2**:155, 285, 463; **4**:371; **5**:243; **6**:235
  drug, and drug design, **2**:160–61
  for enveloped viruses, **6**:226–28t
  fibronectinlike, **2**:208
  glutamate, **5**:232–33
  G protein, **3**:44–45
  G-protein-linked, **3**:43; **5**:233–38
  for growth factors, **2**:464, 467f
    genes mapping to chromosome 14, **3**:155
    hematopoietic, **2**:469
  in hormone-resistant tumors, **2**:215
  inositol 1,4,5-phosphate, **1**:246
  interferon, signaling by, **3**:322–23
  interleukin, **1**:445; **3**:335–40
  interleukin-2, **2**:464; **3**:301
  Ion channel, **2**:208, 209
  ion channel-coupled, **5**:230–33, 233t
  kinin, **3**:384
  lipoprotein, **1**:285t
  low density lipoprotein, **1**:284, 286
  membrane surface, interaction with kinins, **3**:378
  nicotinic acetylcholine, **5**:231–32
  opiate, in the brain, **4**:355
  opioid, μ, δ, and κ, **4**:380–81
  pharmacological, **1**:142–43; **4**:29–31
  rhodopsin as, **5**:324
  role in endocrinopathies, **2**:214
  serine protease inhibitor, genes mapping to chromosome 14, **3**:151
  steroid hormone, **5**:482–83
    and hsp90, **1**:321
  for steroid hormones, **5**:471–72
  surrogate, and uncoating of viruses, **6**:225
  vitamin D, **6**:237
Receptors, cell
  and cell adhesion, **1**:175
  cell surface, interferon-specific, **3**:322
  for cytokines, **1**:442
  defined, **4**:28
  host cell, for bacterial adherence, **1**:104
  membrane, **1**:239
  for site-specific drug delivery, **2**:166
  T-cell, **3**:279
  for viruses, **6**:195
Receptor tyrosine kinases (RTKs), **2**:208, 208f
Recessive allele, defined, **2**:420; **3**:142
Recessive genetic expression
  in cystic fibrosis, **1**:425
  defined, **1**:407
Recessive mutation, lethal, mapping of, **2**:126, 126f
Recessive trait, defined, **3**:227
Reciprocal translocation
  defined, **2**:386
  use in mapping with FISH, **2**:389
Recognition
  chiral, **1**:340
  defined, **4**:281

RECOGNITION AND IMMOBILIZATION, MOLECULAR, **5:243–51**
Recognition element, β-ribbon, **5:**131
Recognition helix, **5:**281f
  DNA, zinc-finger binding domain, **5:**133
  of a helix-turn-helix, **5:**129
  homeodomain, **3:**58
  zinc-finger binding domain, **5:**134f
Recognition motifs
  for blood clotting factor Xa, **3:**22
  for protein kinases, **5:**170
Recognition sequences
  for Achilles' cleavage, **1:**2–3
  defined, **1:**1
  for DNA cleavage, **5:**299
  *Fok*I, specific cleavage of DNA using, **5:**303
Recognition sites
  for CaM PK II, tryptophan hydroxylase, **1:**220
  for PKA, tryptophan hydroxylase, **1:**220
Recombinant DNA
  defined, **1:**172
  libraries, generation from microdissected chromosomal fragments, **1:**382–83
  viral proteins associated with assembly and budding identified by, **6:**189, 189t
Recombinant inbred (RI) lines, **3:**514–15
  for alcohol and drug sensitivity studies, **2:**152–54
Recombinant PCR, misincorporation in, **4:**337
Recombinant peptides
  defined, **4:**347–48
  preclinical leads, **4:**357–58, 358f
Recombinant products
  defined, **6:**165
  vaccinia virus vaccines, **6:**167–70
Recombinant protein
  application to scleroderma diagnosis, **5:**417–19, 419t
  defined, **5:**410
Recombinants
  genetic distance measurements using, **3:**234–35, 235f
  linkage analysis using, **3:**112–14, 114f
  percentage of, in a genomic library, **2:**70
Recombinant vaccines, **6:**166–67
  live, against respiratory syncytial virus, **5:**295
  safety and efficacy of, **6:**171
Recombinase, site-specific, plasmid replication, **5:**5
RECOMBINATION
  defined, **2:**407, 415; **3:**233, 377
  DNA, for searching molecular landscapes, **4:**95–96
  DNA rearrangement in, **2:**404
  Drosophila melanogaster chromosome studies with, **2:**126
  frequency of, in women and in men, **3:**236
  in Gaucher disease, **2:**348t
  geminate, defined, **3:**3
  genetic, modeling, **2:**413–14
  hot spots and cold spots, **3:**513
  illegitimate, **3:**513; **4:**163
    defined, **3:**511
  in immunoglobulin assembly, **3:**282, 282f; **4:**95
  integrative, in yeasts, **6:**303
  intrachromatid homologous, **3:**32
  mammalian genome, **3:**513
  and minisatellite tandem arrays, **6:**8–9
  MOLECULAR BIOLOGY OF, **5:251–64**
  molecular steps in, **5:**260–61, 261f
  among mtDNA molecules, **4:**81
  natural and artificial selection due to, **2:**419, 419f
  nonreciprocal, **5:**254–55
  proteins involved in initiation of, **5:**227–29
  RPA role in, **5:**465
  with sexual PCR, **5:**453
  in sexual reproduction, **2:**83–84
  telomere-promoted, **6:**31
Recombinational repair, **2:**82–83, 83f
  of DNA, **2:**45
Recombination counts, **3:**236t
Recombination frequency
  defined, **3:**112
  differences between male and female, **3:**114
  male and female, **3:**115–16
Recombination interference, *Arabidopsis*, **1:**79
Recombination intermediates
  electron micrographs, **4:**320f
  in partial denaturation mapping, **4:**318–21, 321f
Recombination protein A (RecA). *See* RECA PROTEIN
Recoverin, **1:**242
  inhibition of rhodopsin phosphorylation by, **5:**234
Recursive evolution, **5:**453, 455, 455f
Recycling schemes, amperometric electrodes utilizing, **1:**185–86
Red algae, phycobilisomes of, **4:**442
Red blood cells, **2:**471
  channel-forming integral membrane protein in, **5:**273
Redox-active molecules, monolayer on electrodes, **1:**154, 154f
Redox catalysis, **2:**232, 232f
  by protein disulfide isomerase, **5:**92
Redox centers
  modeling, **2:**194
  nitroxide radicals, **3:**386
Redox cycling toxins, **2:**335
Redox enzyme
  defined, **1:**148
  flavocoenzyme functions, **1:**400, 400f
Redox potentials, relative, and free energy, **1:**309–10
Redox reactions, biosensors using, **1:**182–83
Reducing equivalent
  in electron transfer, **1:**308–9
  rate of delivery of, as a respiratory control, **1:**312
Reductase
  α-aminoadipate, gene for, **6:**313
  cytochrome P450, **1:**194
  5α-, inhibitors of, **5:**492–93, 492f
  HMG CoA, control of cholesterol levels, **1:**287
  NADPH cytochrome P450, **1:**439
Reduction
  NADH-dependent, of carbonyl compounds, **1:**195–96, 196f
  in xenobiotic metabolism, **6:**267
Reductive dehalogenation, **1:**121–22
  defined, **1:**118
Redundancy
  cellular
    compensation for DNA damage through, **2:**80
    repair of DNA damage through, **2:**81
  of cytokines, **1:**442
  defined, **6:**43
  functional, of amino acids, and structural comparisons, **5:**426–27
  of nucleotide sequences, **6:**45t
Reference pedigrees, Centre d'Etude du Polymorphisme Humain (CEPH), **3:**143
Reflectron, for resolution improvement in time-of-flight mass spectrometry, **4:**5, 18
Reflex, sensitization of, **4:**54
Refolding of proteins, **5:**154–58
  aggregation in, **5:**74–81
  inclusion body, **5:**84–86, 86f
Refractive index, changes in, SPR analysis, **1:**190
Regan isoenzyme, in tumors, **3:**375
Regeneration
  plant, **6:**54, 56
  streptomycetes protoplasts, **5:**506
Regioselectivity
  in cytochrome P450 catalyzed hydroxylation, **6:**262–63
  substrate, **6:**260

Regular expression
  defined, **5:**420
  in pattern matching, **5:**425
Regulated secretory pathway, defined, **4:**340
Regulation
  of antioxidant defenses, **2:**334t
  of apolipoproteins, **4:**500–502
  of bacterial virulence, **1:**106–8
  of biomembrane fusion, **4:**44f, 49f
  of cellular gene expression, by interferons, **3:**322–23
  of cytochrome P450s, **1:**439–40
  of cytokine production, **1:**444–45
  of cytoskeleton-membrane interactions, **1:**458–59
  of DNA repair, genetic, **6:**161–63
  of enzyme reactions, **2:**224, 234–37
  of exocytosis, **4:**42
    and specificity of biomembrane fusion, **4:**45
  of gene expression, **2:**418; **5:**128–29
    nitric oxide synthetases, **4:**202–3
    in the pituitary gland, **2:**287, 287t
    role of steroid hormone receptors, **5:**483
    trace element role in, **6:**59
  global network for, *Streptomyces* metabolites, **5:**508
  of glycogenolysis by epinephrine, **5:**171–72
  of hormones, defective, **2:**214
  of immunoglobulin synthesis by interleukins, **3:**341–42
  of intracellular protein degradation, **3:**346–54
  metabolic, by isoenzymes, **3:**373–74
  of muscle contraction, proteins for, **4:**143–45
  of neuropeptides, **4:**197
  by nitric oxide, examples, **4:**204t
  of phosphotidylcholine biosynthesis, **4:**436–37
  of photosynthetic energy transduction, **4:**449–50
  of posttranslational modifications, **5:**51–52
  of promoter activity, **2:**366–67
  of protein synthesis, effect of biotin on, **6:**242–43
  of the reproductive system, female, **2:**286
  of transcription, **2:**95, 367–68, 367f
    by trace element micronutrients, **6:**60–61
  of transcription termination, polymerase pausing, **2:**371
  of tryptophan hydroxylase, **1:**219–20
Regulatory cascade, induced by cholera toxin, **1:**107
Regulatory domain, calcium ion pump, **1:**244–45
Regulatory peptides
  pathophysiology of, **4:**356–57, 357t
  physiological actions of, **4:**349t
Regulatory proteins, **4:**138–40
  GTP-binding, **5:**229
  nuclear, binding sites in the agpB gene promoter region, **3:**412
Regulons, **2:**220; **5:**275
Relatedness table, defined, **5:**420
Relative abundance, defined, **5:**342–43
Relative biological effectiveness, of ions, versus γ-rays, **3:**355
Relative entropy, *k*-gram, **6:**43
Relaxation shift, in luminescent labeling, **3:**387
Release factor (RF)
  defined, **5:**136, 328
  Escherichia coli, **5:**146–47
  eukaryotic, **6:**121, 121t
Reliability
  of a biological marker, **2:**261
  in immunotoxicology testing, **3:**300
Reliability index (*R* factor), for models based on X-ray diffraction, **6:**280
Remodeling, tissue
  by absence of a growth factor, **2:**464
  enzymes for, **2:**285, 290
Remyelination, by fibroblast growth factor, **2:**469
Renal cell carcinoma, **6:**144–45
  association with von Hippel-Lindau disease, **2:**429
  biological response modifiers for treating, **1:**257
  cytokine therapy for, **1:**450
  inhibition of, by IL-2, **2:**438
  interferon-α therapy for, **3:**330
  tumor suppressor genes for, mapping to chromosome 14, **3:**159
RENAL SYSTEM, **5:264–74**
Renaturation
  DNA, second-order rate constant for, **2:**2
  and hybridization, **4:**235
Renaturation curve, polynucleotide, **4:**240
Renin, **5:**265–66
  inhibitor of, **4:**34, 34f
  manufacture of, in yeast, **6:**324
Renin-angiotensin system (RAS), **4:**353–54; **5:**265–67
Reorganization energy
  defined, **2:**194
  in electron transfer, **2:**195
Reoviruses, defenses of, against antiviral reactions, **3:**325
Repair
  chemical, of DNA, **3:**357
  of DNA, biological indications of, **2:**43–44
  function of RecA protein, **4:**153; **5:**221
  prevention of mutagenesis through, **4:**153, 155
  recombinational, **2:**82–83
    of DNA, **2:**45
  relationship with recombination, **5:**261
  of UV damage to DNA, **6:**156
Repairadox, **6:**160
Repair chain reaction, **3:**396
Repairosome, DNA helicases of, **6:**288, 288f
Repeating units
  O-specific chain, **3:**463
  in prokaryotic replicons, as origins of replication, **2:**90
Repeat polymorphism, **3:**231f
  in fragile X syndrome, **2:**322
  in phenylketonuria, **4:**406
Repetition vector
  defined, **6:**47
  strings of symbols for determining, **6:**50
Repetitive elements
  amplification of
    CGG, in fragile X syndrome, **2:**323–25, 325f
    in inherited disorders, **2:**322; **3:**231–32, 232f
    detection procedure, **3:**397
    mechanism of expansion, in fragile X syndrome, **2:**329
  defined, **3:**240
  LI family of, **3:**245, 245f
  in mycobacteria, **4:**159
  See also LINE (long interspersed repeat) element; Short tandem repeats (STR); SINE (short interspersed repeat element)
Repetitive sequences
  of α-actinin, **4:**142
  analysis of, with a padlock probe, **3:**398f
  Arabidopsis genome, **1:**77
  defined, **3:**511
  dinucleotide, instability in HNPCC, **6:**146
  in DNA, of base pairs, **2:**114–15
  LI (*Kpn*) I and *Alu* types, **5:**430
  as probes, **2:**381
Replacement, cell, coping with DNA damage through, **2:**80
Replica casting, defined, **2:**182
Replicases
  Qβ, **3:**397; **4:**92
  and RNA viruses, **5:**359
Replication
  bidirectional, **4:**317
    defined, **1:**95
  DNA, **2:**416–17, 417f
    of bacteriophages, **4:**316–18

plasmid ColE1, **5**:2–4, 4f
retrotransposition model of, **6**:134
unidirectional, of mtSSB, **5**:465
Replication fork
defined, **2**:86
DNA, **1**:366; **2**:86–87, 87f; **5**:355
Replication frequency, **2**:272
Replication protein A (RPA), **5**:462–65
defined, **5**:458
Replication slippage, **4**:151
DNA turnover, **6**:3, 3f
Replicative capacity, and telomere length, **6**:34
Replicon, **2**:86
Replisomes, **2**:88, 88f; **5**:458
Reporter chromophore, for optical sensors, **1**:187
Reporter function, transgene, **6**:87
Reporter genes
bioluminescence and chemiluminescence assays using, **1**:329–30
defined, **3**:416
fish, **6**:86f
lacZ
for studying *Caenorhabditis elegans*, **4**:169
yeast, **6**:314
for mapping gene function in vertebrate development, **6**:98
in plant transformation studies, **4**:490
for screening recombinant vaccinia viruses, **6**:166–67
for study of transgenic animals, **6**:73
transcription of, **1**:234
transfection with, waveforms used for, **2**:202
for tyrosine hydroxylase transcription monitoring, **2**:151
for in vivo study of repressor-operator interactions, **5**:283
Reporter groups, protein folding pathway studies using, **5**:136
Reporter molecules, autoantibodies, **1**:92
Representivity of a genomic library, **2**:69–70
Repressor-operator complexes, **5**:277–79
REPRESSOR-OPERATOR RECOGNITION, **5:274–85**
Repressors
for control of a lactose promoter, **2**:273
MetJ, **5**:130, 131f, 277
indirect readout interaction of, **5**:279
regulation of protein synthesis by translational, **2**:376–77
transcriptional, **2**:370–71; **5**:275
control by, **2**:273
trp, activation by L-tryptophan binding, **5**:278f
TrpR, **5**:277
in vitro studies of, **5**:282–83
Reproductive system
female, molecular biology of, **2**:285–93
male, molecular biology of, **3**:500–10
Repulsion, cell-cell, **1**:297–99
Resequencing arrays, to detect mutations, **3**:259
Reserpine
delayed response to, by tyrosine hydroxylase, **1**:216
structure of, **4**:29f
Residue, defined, **4**:361
Resilin, **1**:146
Resin, defined, **4**:383
Resistance
to drugs, and reverse transcriptase binding, **1**:30–31
to pathogens
barriers in plants, **4**:478–80
plant metabolic defenses, **4**:479–80
Resolution
defined, **5**:252
of a Holliday junction, **5**:256, 256f
of radiation hybrid mapping, **3**:121
of X-ray diffraction measurements, **6**:277, 279–80
Resolvase, **5**:256, 261
Resonance assignment, NMR, **4**:223
Resonance-assisted hydrogen bonding, **3**:264, 266, 266f
Resonance-enhanced multiphoton ionization (REMI), **4**:1
Resonance ionization spectroscopy (RIS), **4**:1
Resonance Raman effect, defined, **5**:94
Respiration, **1**:308
cellular, control of, **1**:311–13
electron transfer in, **2**:194
Respiratory syncytial virus (RSV)
recombinant Fabs against, **1**:421
treatment with antibodies, animal model, **1**:420
RESPIRATORY SYNCYTIAL VIRUS VACCINE, APPROACHES TO, **5:286–96**
Response element (RE)
binding to, by HIV-1 Rev protein, **4**:228–29, 229f
defined, **5**:128; **6**:235
hormone receptors binding to, **5**:133
jasmonate, **4**:493
retinoic acid, in promoters of keratin genes, **2**:266
retinoid, **6**:236–37
Restrained molecular dynamics, for refining X-ray diffraction-based models, **6**:280
Restriction endonuclease (RE)
defined, **1**:1; **3**:227; **4**:129, 323; **5**:296, 304, 401
*Eco*DRI type, sequence specificity of mutations of, **5**:304
*Eco*RV, salt bridges of, **1**:163
effect on DNA acquisition, **2**:405
in Southern blot analysis, **4**:324–25
RESTRICTION ENDONUCLEASES AND DNA MODIFICATION METHYLTRANSFERASES, FOR MANIPULATION OF DNA, **5:296–304**
Restriction enzymes
B, for restriction landmark genome scanning, **5**:309, 310
defined, **2**:66; **5**:304
direct atomic force microscopy mapping with, **5**:405–6
Restriction fragment length polymorphisms, **3**:513
apoB, **3**:412
for mapping the nuclear genome, *Chlamydomonas reinhardtii*, **1**:348
Restriction fragment length polymorphism (RFLP), **2**:51, 67, 67f, 395; **3**:224, 229, 231f
analysis using, forensic applications, **5**:302
Arabidopsis, **1**:80, **4**:491
defined, **4**:323
for genetic mapping, **1**:231–32
for linkage analysis, **2**:421; **3**:99, 143
linked to the *uni* group, **1**:359–60
mapping with, **3**:516f
for nuclear gene variation study, **2**:399
1-myc, for assessing the prognosis in lung cancer, **4**:329
for population analysis, **2**:396f
protocol, **2**:49f
sites in the *PAH* gene, **4**:408, 408f
Europe, **4**:411–12, 411f
Restriction fragments, **3**:217
human chromosome, **3**:218f
RESTRICTION LANDMARK GENOMIC SCANNING METHOD, **5:304–19**
Restriction landmark genomic scanning profiles, **5**:316, 316f
Restriction landmarks
defined, **5**:304
direct labeling technique, **5**:308–10
Restriction maps, of pAL5000, **4**:162f
Restriction-modification systems, defined, **5**:296
Retention, of proteins, in the Golgi complex, **5**:202
Retention sequences, **5**:196, 199, 201
Reticular systems, intracellular, calcium in, **1**:238
Reticulocytes, lysate from rabbit cells, **5**:334

Reticulum, calcium ion transport systems of, **1**:245–47
Retina
axons of, **1**:297
bacteriorhodopsin of, **1**:109, 109f
membrane of, docosahexaenoate in, **4**:433
Retinal
binding to rhodopsin, **5**:234
11-*cis*-, regeneration of, **5**:324–25
trans-, defined, **1**:148
Retinaldehyde-binding protein, cellular, gene localization to chromosome 15, **3**:170
Retinitis pigmentosa, autosomal dominant mutation in, **2**:27
RETINOBLASTOMA, **1**:251; **2**:420, 429; **3**:223
diagnosis of, using PCR, **2**:64
gene for, **3**:495; **4**:287, 329; **6**:140–41, 140f
mapping to chromosome 11, **3**:127
MOLECULAR GENETICS OF, **5:319–22**
tumor suppressor gene in, **3**:489; **6**:139
Retinoblastoma protein, testicular, **3**:507
Retinoic acid binding protein, cellular, gene localization to chromosome 15, **3**:169
Retinoic acid receptors
nuclear, **6**:236–37
in Sertoli cells and germ cells, **3**:506
viral integration in the gene coding, hepatocellular carcinoma, **3**:489
Retinoic acid responsive elements (RAREs), in promoters of keratin genes, **2**:266
Retinoic acids, isomerization of, **5**:325–27
RETINOIDS, **5:322–27**
defined, **5**:322
effects on epithelia, **2**:268–69
X receptor, **6**:236–37
Retinol-binding protein (RBP), **6**:236
relationship with apoD, **3**:413
Retinoma, **5**:320
defined, **5**:319
Retro-inverso peptide analogues, **4**:373
Retroposition, transcription of *Alu* RNA in, **3**:244
Retroposons, **3**:243–45
Alu sequences introduced into the human genome by, **3**:242
defined, **3**:216, 240
Retropseudogenes
of plants, **4**:488
of protein-coding mRNAs, **3**:246
Retrotransposons, **3**:512
defined, **3**:240; **4**:485; **6**:134
non-LTR, **3**:245
plant, **4**:487–88
yeast genome, **6**:305–6
Retrovir, **1**:37
Retroviral integration, HIV-1, **1**:33
Retroviruses
assembly and budding of, **6**:191–92, 191f
cloned genome of, modifying, **6**:200
defined, **1**:27, 34; **5**:352
endogenous, **6**:136–37
defined, **6**:133
as enveloped polyhedral viruses, **6**:222
for genetic immunization, **2**:436
gene transfer mediated by, **2**:436
information transfer in, **5**:359
for introduction of transgenes, **6**:67–68, 88, 100–101
life cycle of, **5**:358f
mouse mammary tumor virus, **5**:508, 511
packaging of, **6**:221–22
recombinant, for gene therapy of phenylketonuria, **4**:413
reverse transcriptase of, **4**:326
zinc-binding motifs of, **6**:334
Rev-binding element (RBE), **5**:385–86
secondary structure of, **5**:387f
Reversal, direct, in DNA repair, **2**:44
Reversed micelles, enzyme, **2**:250–51
Reversed-phase chromatography (RPC), **3**:82–83, 82f
Reversed-phase high performance liquid chromatography
defined, **5**:173
for peptide mapping, **5**:180, 182f
for protein purification, **5**:174, 175–76
Reverse genetics, **2**:420
with *Caenorhabditis elegans*, **4**:169
defined, **2**:130; **4**:173
for *Drosophila* genetic analysis, **2**:125
with mycobacteria, **4**:163
with streptomycetes, **5**:503–4
Reverse phase evaporation (REV), **3**:417
Reverse transcriptase, **5**:352
with AG-LCR, to detect RNA, **3**:302
defined, **1**:34; **5**:338
HIV, photodynamic inactivation of, **6**:214f
HIV-1, **1**:28–31; **6**:196
crystal structure, **1**:40f
inhibition of, **1**:37–38; **4**:228, 228f
role of, **1**:34
three dimensional structure, **1**:38
inhibition of, HIV treatment, **1**:14, 18
of retrotransposons, **4**:487–8
telomere specific, **3**:245; **6**:31
Reverse transcriptase inhibitors, **1**:14, 18
nonnucleoside, **1**:38
Reverse transcription, defined, **4**:226
Reverse transcription-polymerase chain reaction (RT-PCR), **4**:326–27
for identifying transgenes, **6**:91, 91f
Reverse transport of cholesterol, **3**:473–74, 473f
role of apoD in, **3**:413
Reversible reactions, in posttranslational modification, **5**:50
Rev response element (RRE), HIV-1, **6**:197–98
Reynaud's phenomenon, association with antibody to CENP-B, **5**:417
Rhabdomyosarcoma
genomic imprinting associated with, **2**:445; **3**:127
mapping of, to chromosome 11, **3**:127
Rhabdoviruses, assembly and budding of, **6**:190–91
Rhamnolipid biosynthesis, transcriptional activator for, **1**:108
Rhegnylogic organization, peptide, **4**:372, 372f
Rheumatic diseases, **1**:85
heart, M protein antibodies in, **1**:105
Rheumatoid arthritis, uric acid degradation products in, **2**:336
Rhinovirus
human, (HRV), structural strategy for evading immune surveillance, **6**:229f
x-ray crystallographic structural analysis, **5**:16
Rhizobium, plasmids of, **5**:12–13
Rhizosphere-colonizing bacteria, pesticides from, **4**:397–99
Rhodamine, label for FISH, **2**:382
Rhodanese, unfolding of, reversible, **5**:75–76, 77f
Rhodobacter sphaeroides, structure of the reaction center of, **4**:442–43, 443f
Rhodococcus fasciens, plasmid of, **5**:12
Rhodopsin
interaction with Gα_1 subunit of protein G, **3**:45, 45f
mutation in the gene for, **2**:27f
Rhodopsin kinase, **5**:234
Rho factor, role in transcription, **2**:371
β-Ribbon motifs, **5**:130–31, 131f
Riboflavin, structure and function of, **6**:249, 249f
Ribonuclease
defined, **5**:343
for determining RNA secondary structure, **5**:365, 365t
inhibition of, during isolation of RNA, **5**:345–46
2–5A system, antiviral activity of, **6**:199
Ribonuclease A (RNase A)
chirality of components, **1**:340–41
phosphodiester cleavage by, **1**:169–70, 169f

Ribonuclease H (RNase H)
cleavage of RNA, DNA-RNA hybrids, **1**:74
defined, **1**:69
H1, and sealing of Okazaki fragments, **2**:90
HIV, activity of, **1**:31
HIV-1, **1**:29–30, 30f
drugs interfering with, **1**:41
Ribonuclease P, **5**:340–41, 340f
secondary structure of, **5**:367f
case study, **5**:366
structure of, **5**:341, 341f
Ribonucleoprotein
catalysis by, **5**:340
defined, **5**:338
signal sequences targeted to the endoplasmic reticulum by, **5**:197
3CD, **5**:22
Ribonucleoside, vanadyl complex, **5**:346
Ribosomal DNA (rDNA), repetitive clusters, *Arabidopsis* genome, **1**:77
Ribosomal misreading, due to omnipotent suppressor, **6**:320
Ribosomal protein L35A, mapping to chromosome 18, **3**:208
Ribosomal RNA (rRNA)
comparisons of, and mtDNA origin, **4**:83–84
defined, **5**:372
preparation from ribosomes, **5**:332
transcription of, **2**:91, 365–66
yeast, **6**:305–6
Ribosome-binding site, **2**:274
defined, **2**:271
RIBOSOME PREPARATIONS AND PROTEIN SYNTHESIS TECHNIQUES, **5:327–37**
Ribosomes
biosynthesis of peptides on, **4**:353
chloroplast, *Chlamydomonas reinhardtii*, **1**:353
cold sensitive hsp70s of, **1**:317f
cytoplasmic, **6**:122f
defined, **5**:328
engineered biosynthesis, **1**:136–137
and fidelity of protein synthesis, **6**:124
membrane-bound, and synthesis of polypeptide hormones, **4**:342
organelle, preparation of, **5**:329–330
properties of, **6**:115f
structure of, **5**:328–32
studies of, **5**:332
translation in, **5**:355
archaeal, **4**:75–76
RIBOZYMES, **5:338–42**, 357
altering, **4**:232–33, 233f
defined, **4**:91, 281; **5**:338, 352
hepatitis delta, pseudoknots of, **5**:362
RNA ligase, **3**:393; **4**:230–31, 230f
in the “RNA world”, **4**:296
secondary structure, rotamase, **4**:232f
selection using polynucleotide kinase activity, **4**:231–32, 231f
Ribulose bisphosphate carboxylase (Rubisco), **5**:86, 86f
reconstitution of unfolded, **5**:87, 87f
Ricin, administration with monoclonal antibody linkage, **2**:167
Ricinoleic acid, **3**:437
synthesis by molds, **3**:440
Rickets, vitamin-D-resistant, **1**:249
Rieske Fe-S centers, **4**:444–45, 447, 447t
Rifampicin
resistance to, **4**:161
as an RNA polymerase inhibitor, **4**:466
sensitivity of chloroplast DNA-dependent RNA polymerase to, **1**:350
for treating tuberculosis, **4**:157
Riley-Day syndrome, mapping to chromosome 9, **3**:107
Rinderpest, vaccinia virus recombinant vaccine for, **6**:167–68
Ring chromosome
chromosome 15, **3**:170–71
chromosome 18, **3**:212
RING finger, **6**:334–35, 335f
Ripening of fruit, altering with antiethylene genes, **6**:58
Risk factors, for heart failure, **3**:2
R loops, bacteriophage, **4**:322
R mutant lipopolysaccharides, **3**:468–70, 470f
RNA
analysis of
methods, **5**:346–51, 351t
with reverse transcriptase-polymerase chain reaction, **4**:326–27
antisense
in gene expression, **1**:70
in plasmid replication, **5**:4
binding of
by Tat, **1**:19
by TFIIIA, **6**:331
catalysis by, **2**:224; **5**:338
as a catalyst inhibitor, in gene expression, **1**:173
catalytic molecules, in vitro selection of, **4**:230–32
circular single-stranded, of yeast, **6**:305–6
coding for, **5**:431
nuclear, **5**:343
poliovirus, cap-independent translation of, **5**:17–19
posttranscriptional processing of, in *Caenorhabditis elegans*, **1**:235
processing of, defined, **5**:338
self-cleaving, **5**:339–40, 340f
defined, **5**:338
self-splicing, **5**:338–39, 339f
sequencing of, **4**:256–57
7SL, **3**:244
16S, predicting the structure of, **5**:382–83
splicing of
defined, **2**:364; **5**:338
differential, **1**:42
and regulation of gene expression, **2**:371–73
in spermatogenesis, **3**:503, 503t
strand of, **5**:354f
structure of, **5**:338
subpopulations of, **5**:343–44
synthesis of, during spermatogenesis, **3**:503–04
synthesis cycle, **2**:92f
telomerase, **6**:21t, 32
cloning of, **6**:20
virus
genes encoded by, **4**:285
packaging model for yeast virus L-A, **6**:220f
transcription of, **6**:196
in vitro selection of, **5**:341–42, 341f
RNA editing
defined, **4**:495
nuclear, of plasma lipoproteins, **4**:497
RNA editing protein (REPR), **4**:499–500
RNA ligase, T4, **5**:142
RNA METHODOLOGIES, **5:342–51**
RNA polymerase, **2**:417
archaeal, **4**:74–75
bacteriophage T7, **2**:274
effect of Tat on, **1**:25
eukaryotic, **2**:91–93, 91t, 365–66
types of, **2**:92t
nucleotidyl transferase, zinc requirement of, **6**:60
RNA-dependent
defined, **5**:13
telomerase as, **6**:19
$3D^{pol}$, **5**:22
viral $3D^{pol}$, **5**:22
role in transcription, **1**:369
of Sindbis virus, degradation of, **3**:351
virus-specific, **5**:358
vaccinia virus, **6**:197
yeast, as an independent replicon, **6**:306
RNA processing, and regulation of gene expression, **2**:373
RNA-processing factors, mutant *Drosophila* studies, **2**:142–43
RNA-protein interactions, in the 5′ noncoding region, **5**:19
RNA REPLICATION, **5:352–60**
picornavirus, **5**:22–25, 25f
RNA-responsive element binding protein, defined, **6**:59

RNA SECONDARY STRUCTURES, **5:360–68**
defined, **2:**271
RNases. *See* Ribonuclease *entries*
RNasin, inhibition of ribonuclease by, **5:**346
RNA splicing factors, defined, **2:**130
RNA STRUCTURE, NONHELICAL, **5:368–71**
RNA THREE-DIMENSIONAL STRUCTURES, COMPUTER MODELING OF, **5:371–88**
"RNA world" postulate, **4:**296
rNTP (ribonucleoside triphosphate), defined, **6:**17
Robertsonian fusion, in Down syndrome, **2:**119
Robotics, defined, **5:**389
ROBOTICS AND AUTOMATION IN GENOME RESEARCH, **5:388–99**
Robots
in stand-alone applications, **5:**391, 393f
thermal cycling integrated with, **5:**394
Rodent paradox, **6:**160
Roll, helix step
of AA(=TT) steps in DNA, **2:**115
to form DNA structure, **2:**112–13, 112f
defined, **2:**103, 107–9
Rolling-circle replication, **6:**4
RNA, **5:**340f
Root effect, **3:**8
in fish, **3:**27
Rotary motor complex, of bacterial flagella, **4:**114–15
Rotational diffusion
of luminescent labels, **3:**387, 387f
of nitroxide labels, **3:**385–86
Rotational positioning, of histone octamer, of nucleosomes, **1:**368
Rotations, in DNA structure, **2:**107
Rotaviruses
assembly and budding of, **6:**191
protein VP7, aggregation in the endoplasmic reticulum, **5:**201
Rough lipopolysaccharides, **3:**466
Rous sarcoma virus (RSV)
association of tyrosine kinase and hsp90 in, **1:**321
long terminal repeat of, transfer to Carp and catfish, **6:**92
proteases encoded by, **1:**14
Royalties
from breeders of patented animals, **6:**80
defined, **6:**74
R state (relaxed state, cooperativity model)
defined, **3:**20
hemoglobin, **3:**21
Ru 486, **5:**484
for cancer therapy, **5:**490
Rubella, recombinant Fabs against, **1:**421
Rubenstein-Tabi syndrome, **3:**186
localization to chromosome 16, **3:**190
Rubisco-binding protein, **5:**86–87
Rubredoxin, of hyperthermophiles, **2:**245
Rubredoxin knuckle, in zinc-binding folds, **6:**336
Russell-Silver syndrome, **3:**164
Ruthenium-bipyridyl complex, in semisynthetic cytochrome *c*, **1:**157, 157f
Ryanodine receptor, **3:**2
type 3, gene localization to chromosome 15, **3:**168

S03, S45, S100 supernatants, defined, **5:**328
S100 proteins, family of, **1:**242
Saccharomyces cerevisiae
Achilles' cleavage of genome, **1:**2, 3
cell-free translation systems from, **5:**335
defined, **6:**291
DNA repair in, **6:**283
DNA transformation in, **2:**276; **5:**256, 256f
homeobox genes of, **3:**75
hsp70 of, **1:**317f
lipid synthesis in, **3:**435
as a model eukaryote, **6:**303
production of recombinant hemoglobin in, **3:**23
2μm plasmid of, **5:**5, 6f
Safety considerations
in animal systems for testing vaccines, **6:**182–83
in handling radioactive iodine, **5:**218
in handling radioactive substances, **5:**211–14, 211t
in HIV vaccine development, **6:**170
in live attenuated vaccine use, **6:**173
in live recombinant vaccine use, **6:**169
in using hyperimmune serum clinically, **1:**415
in vaccine for rinderpest, genetically engineered, **6:**168
Salicylic acid, and plant response to pathogens, **4:**480
Saliva, extraction of DNA from, **2:**53
Salmonella
binding to a host cell, **1:**106
plasmids of, **5:**11
Salmonella enterica, lipopolysaccharide of, **3:**463
Salmonella typhimurium, defense against phagocytosis, **1:**105
Salt bridges, **1:**162–63
stabilizing proteins, **2:**309, 309f
Sample loading, horizontal ultrathin gel electrophoresis, **2:**99
Sample preparation
collection criteria, for genetic analysis of populations, **2:**394–95
electron microscopy, **2:**182, 183f
Raman spectroscopy, **5:**101
Sandhoff's disease, isoenzyme deficiency in, **3:**375
Sandwich assay format, immuno-PCR analysis, **3:**291–92, 292f
Sandwich immunoassay, **1:**189–90, 189f
double-tag, **3:**294
Sandwich structure, of partial charge for base pairs in DNA, **2:**110
Sanger sequencing method, **4:**282
for DNA fingerprint analysis, **2:**58
Saposin, cofactor for glucocerebrosidase, **2:**350
SAR (structure-activity relationship)
for cephalosporins, **2:**164
defined, **2:**155
for drug candidates, **4:**99
and drug design, **2:**158
Sarcolemma, **4:**129, 130
Sarcomas, **1:**412
neurofibrosarcomas, *p53* mutants in, **6:**144
Sarcomere, **4:**130
defined, **3:**1
Sarcoplasm, **4:**129, 131
Sarcoplasmic calcium-binding protein (SCP), **1:**241
Sarcoplasmic reticulum (SR), **4:**129, 131
calcium ion (Ca^{2+}) of, and muscle contraction, **4:**144
calcium release channel of, **1:**246–47
role in calcium metabolism, **1:**292
Sarcosome, **4:**129
Satellite DNA, **2:**123; **6:**1
α-, **3:**241
Drosophila, **2:**122
tandem repeats, **1:**371
Satellites
Bkm, **5:**435, 435f, 436, 436f
human repetitive elements, **3:**241, 241f
mammalian genome, **3:**511
tandem arrays of, **6:**9–10
Saturation
in enzyme kinetics, **2:**233
mathematical treatment of, Michaelis-Menten equation, **2:**234
Sauerbrey equation, for frequency shift as a function of mass loading, **1:**192
Scaffold-associated regions (SAR), chromosomal, **1:**370–71
Scaffolding approach, for nonpeptide library assembly, **5:**522
Scandanavian wood lemming, X-linked sex-determining genes of, **5:**441
Scanning, DNA, restriction landmark procedure, **5:**306–307, 306f
Scanning model, for initiation complex formation, **6:**118

SCANNING PROBE MICROSCOPY IN SEQUENCING OF DNA, **5:401–10**
Scanning-tunneling microscopy (STM), **1:**152
for DNA imaging, **5:**407–9, 409f
picosecond resolution in, **1:**158, 158f
techniques, **5:**402–3, 402f
Scan rate, for circular dichroism determinations, **1:**387
Scapharaca inaequivalvis, hemoglobin of, **3:**13
Scatchard equation, **1:**263
Scatter factor (HGF/SF), liver, **3:**482
Scavengers, free radical, **2:**333
Schemata (blocks of genetic code), defined, **2:**413–14
Schiff base (SB), bacteriorhodopsin, **1:**109, 109f, 111–12
Schistocerca gregaria, diplotene nucleus, **5:**253, 253f
Schistosoma genome project, **4:**310–11
Schistosomiasis, **4:**307
Schizophrenia, genetic factors in, **4:**199–200
mapping to chromosome 11, **3:**127
Schwannomas
defined, **4:**181
mutation of the *NF2* gene in, **4:**183, 185
Schwannomin (merlin), **4:**181, 185, 186–88
Scintillation counters, **5:**218–19
SCLERODERMA, **1:**87
defined, **5:**410
DIAGNOSIS WITH RECOMBINANT PROTEIN, **5:410–19**
Scleroglucan, antitumor activity of, **2:**344–45
Scope
defining, as a requirement for patentability, **6:**76
of a patent, defined, **6:**74
Scrapie, defined, **5:**57
Scrapie prions
diversity of, **5:**66
inhibition of synthesis of, genetic control, **5:**68t
Screening
assays for, defined, **2:**155
defined, **5:**243
for phenylketonuria, **4:**406–7
for recognition molecules, **5:**249
two-hybrid system, for detecting protein-protein interactions, **6:**303
for unrearranged clones, in yeast artificial chromosome preparation, **6:**298
Screening, drug identification, **1:**34–35; **2:**160
mechanism-based, **1:**38, 38f
penicillin inhibitor to HIV protease, **1:**16
Scurvy, vitamin C deficiency, **6:**244
SDS-PAGE, defined, **5:**173. *See also* Sodium dodecyl sulfate polyacrylamide gel electrophoresis
Second activation signals, in immune reactions, **3:**276
Secondary CAMs, **1:**292
N-CAM, **1:**295
Secondary metabolites
defined, **4:**393; **5:**501
as pesticides, **4:**395–99
Secondary structure
defined, **5:**368
G-protein-coupled receptors, **3:**43, 43f
of a hybridization probe, **3:**257
NMR analysis of, **4:**223
of polypeptides, **5:**117
of ribozymes, **5:**338
of ribozyme-substrate complex, **5:**340, 340f
of RNA, **5:**360–68, 374–77
RNase H domain, **1:**31–32
three dimensional structure of RecA protein, **5:**222–223, 223f
Secondary structure, peptide, **4:**350
defined, **4:**348
designed, **4:**377–378
synthetic, **4:**363
effect of size on, **4:**361
Secondary structure, protein, **2:**307–8, 308f, 309f; **5:**117–19
analysis of predictions, **5:**167–68
combinatorial arrangements, **5:**163f
defined, **5:**147, 159
predicting, **5:**161–62, 432
PrP, **5:**60–61
Raman spectrum analysis of, **5:**104
Secondary structure analysis
circular dichroism used in, **1:**385, 389–91, 389f
defined, **1:**384
Second law of thermodynamics, **1:**305
Second messengers, **2:**209–12
calcium, **1:**237
ceramide as, **3:**458
defined, **1:**237; **2:**285; **3:**42; **4:**347, 416, 432; **5:**169, 230
eclosion hormone stimulation of production of, **4:**195–96
generation by interleukin-1, **3:**339
generation by regulatory phospholipases, **4:**438–39
inositol 1,4,5-phosphate, **1:**246; **3:**44
interaction of
with G proteins, **3:**45–46, 45f; **5:**233
with protein kinases, **5:**170
lipids as, **3:**460
in long-term sensitization, **4:**56
as oncogenes, **2:**215
origination of concept, **1:**238
peptide, **4:**355
phospholipases, and generation of, **4:**430
in postsynaptic cells, **4:**54
signal transduction by, **1:**239–40; **4:**55f
Secretase
α-
cleavage of amyloid precursor protein by, **1:**47
pathway involving, cleavage of amyloid precursor protein, **1:**48–49
β-, APP cleavage by, **1:**49
γ-, accumulation of β-peptide associated with, **1:**49
Secretion
constitutive and regulated, **5:**203–4
of peptide hormones, **4:**346–47, 346f
by streptomycetes, **5:**507
Secretory glycoproteins, defined, **2:**458
Secretory granules, **4:**346
targeting polypeptide hormones to, **4:**346
Secretory pathway, for peptide hormones, **4:**341–42, 342f, 346
Secretory phosphatidylinositol-phospholipase C enzymes (sPI-PLC), **4:**424–26, 426f
Secretory phospholipase A_2 (sPLA_2), **4:**420–22
Secretory phospholipase C (sPLC), **4:**428
Secretory proteins
mycobacterial, **4:**159–60
transport of, **3:**353
Sedimentation rate
for classification of ribosomes, **5:**328–29
for following DNA denaturation, **2:**6t
Segment matches, probability of, in sequence alignment, **5:**423, 423f
Segregation
in cell replication, **1:**99–100
defined, **1:**95
genetic, defined, **1:**221
and heteroplasmy, **4:**81
in plasmid replication, **5:**5
Segregational stability, **2:**271
Segregation analysis, **1:**221–22
Seizure susceptibility, in drug withdrawal, animal models, **2:**148–49
Selected line, defined, **2:**146
Selectide process, synthetic peptide library, **5:**520–22, 522f
Selectins
E-, **1:**296
family of cell adhesion molecules, **1:**296–97
L-, **1:**296, 451, 456
P-, **1:**294f

Selection
from antibody phage libraries, **1**:418–20
cell line, in culture of yew cells, **6**:17
genetic, in aging studies, **1**:13
nonlethal environmental, defined, **2**:407
randomized, of oligonucleotides, **4**:95–96
strategies of, **4**:98
Selective induction of enzymes, **1**:194
Selectivity
adhesive, **1**:293
defined, **6**:259
of drugs
biochemical basis, **1**:263–64
and cytokinetic parameters, **1**:257–58
factors determining, **1**:253–54
of enzymes, control by solvents, **2**:256
of metabolic processes, **6**:260
of opioid ligands, **4**:381
of phosphoprotein phosphatases, **5**:171
of restriction endonucleases and methylases, **5**:297
Selenium
in glutathione peroxidase, **2**:333, 333f
protection against photochemical damage by, **6**:210
Self-administration studies, animal models for drug use, **2**:149
Self-assembly
defined, **1**:148
and nanoengineering, **1**:152–54
nanometer-scale systems, **1**:141
Self-cleaving RNA, defined, **5**:338
Selfish DNA, **6**:1, 5
defined, **6**:1
Self-organizing systems, mimicking nanometer-scale structural arrays, **1**:140
Self-overlap capacity, **6**:41
Self recognition, transplantation antigens responsible for, **3**:499–500
Self-replicating systems, in vivo and in vitro, **4**:92
Sellers method, for local alignment assessment, **5**:424
Semaphorins, collapsin, **1**:299
Semiconductor diodes, light emitted by, **6**:211
Semiconductors
binary, **3**:309
oriented growth of crystals, **3**:312
Semifusion
defined, **4**:42
of lipid bilayers, **4**:46
Seminiferous epithelium, **3**:501
expression of cyclins A,B, and D in, **3**:507
Seminiferous tubule, epithelium of, **3**:502f
Semustine, as an anti-cancer drug, **1**:254
Senescence
defined, **1**:299; **6**:28
and telomerase, **6**:34–35
Sense strands, **5**:355–56
defined, **5**:352
Sensitivity
of anticentromere antibody detection of scleroderma antigens, **5**:411
of autoradiography and fluorography, **5**:220t
of bioluminescence and chemiluminescence assays, **1**:323
of immuno-PCR, **3**:291, 293–94, 293f
of ligase chain reaction, **3**:302
of microbore column chromatography, **5**:177f
of RT-PCR, **6**:72
Sensitization
in *Aplysia*, **4**:54, 55f
as nonassociative learning, defined, **4**:54
Sensor development, from biomolecular systems, **1**:141
Sensory processes
association with electric and magnetic fields, **2**:172–73
and electric and magnetic fields, **2**:172
Sensory transduction
defined, **2**:130–31
mutants involving, *Drosophila*, **2**:132–33
Separation, defined, **5**:243
Sepsis, role of interleukin-1 in, **3**:345
Septic shock, lipopolysaccharide as a cause of, **3**:470
Sequence
amino acid
defined, **5**:420
in human interferons, **3**:319f
and protein structure, **5**:160–61
for histone proteins, chicken erythrocyte, **1**:362f
homeodomain, **3**:56–57, 61–63f
nucleotide
defined, **5**:420
in DNA packaging by bacteriophages, **6**:205
satellite, **6**:2–3
Sequence addition, by telomerase, **6**:33–34, 33f
Sequence alignment
from dot plots, **5**:432
in homology modeling, **5**:164
multiple, **5**:425
SEQUENCE ALIGNMENT OF PROTEINS AND NUCLEIC ACIDS, **5:419–28**
SEQUENCE ANALYSIS, **5:428–34**
in biopharmaceutical product development, **5**:174f
comparative, for RNA structure determination, **5**:378
C-terminal, defined, **5**:173
by hybridization, **1**:157
using coincidence indices, **6**:44–7
Sequence checking, **3**:258–59
Sequence comparison, **5**:429
Sequence database searching, **5**:433
Sequence dictionaries, **5**:434
Sequence divergence, DNA, **2**:12
Sequence divergence estimation, measuring in repetitive families, **2**:19
Sequence homogenization, in tandem arrays, **6**:7
Sequence indexed clusters, **5**:434
Sequence pattern
autoantigens, scleroderma, **5**:416
defined, **6**:38
glycoprotein genes of varicella-zoster virus and cytomegalovirus, **6**:175–76
Sequence polymorphisms, typing of, **2**:57–58
Sequences, insertion, defined, **4**:156
Sequence space, and shape space, **4**:99
Sequence-specific oligonucleotide (SSO), **4**:338
Sequence-specific recognition, **5**:129
Sequence-tagged restriction site (STAR), **3**:220
defined, **3**:216
Sequence-tagged sites (STSs)
chromosomal, **1**:4, 380, 383; **3**:122, 520; **4**:338
defined, **2**:66; **3**:174, 191, 216; **4**:306
for mapping, **3**:220; **6**:301
chromosome 18, **3**:200f
Plasmodium falciparum, **4**:308
for mapping chromosome 14, **3**:151
mouse genome, **4**:124, 124–25, 125f
Sequence threading, in structure-based alignment, **5**:426
Sequencing
comparative, **4**:128
DNA, **5**:406–7
chemical and dideoxy, **3**:517
defined, **5**:401
direct, **3**:226; **4**:128
of a genome, **4**:128
by hybridization, **3**:253f
defined, **3**:252
hypothetical project, cost and time analysis, **5**:389–390
large-scale, **3**:520
of proinsulin, LC/ESI/MS for, **4**:24–26, 24f
RNA, **3**:517
unambiguous, from two overlapping sequences, **3**:255f

Sequencing chip, **3**:256f
defined, **3**:252
hybridization of cloned genes to, **3**:258–59
Sequencing gel, **4**:254, 254f
for analyzing ultraviolet radiation damage to DNA, **6**:156
SequeNet Entrapment Procedure, protein preparation for sequencing, **5**:188, 188t
Sequinavir, HIV protease inhibitors, clinical trials, **1**:18
Serine, β-lactone derivatives of, **1**:54
Serine kinases, phosphorylation by, **5**:55
Serine protease inhibitor (serpin) gene family, *PA12*, **3**:208
Serine proteases, **2**:229–30, 233
defined, **3**:28
genes for, mapping to chromosome 14, **3**:155
kallikreins, **3**:382
kininogen substrates for, **3**:380
spin-labeling evaluation using, **3**:389
subtilisin, effects of organic solvents on, **2**:254, 254f
Serine/threonine kinases, **2**:211t
association with cyclin B, in spermatogenesis, **3**:507
association with interferon-γ receptor, **3**:323
of cDNA encoding CDPK, **1**:249
in transforming growth factor-β receptors, **5**:242
Serotonin
biosynthesis of, **1**:207, 208f
gene controlling in *Drosophila*, **2**:136–37
in the pineal gland, **1**:218
transgenic mice deficient in 1B receptors for, **2**:151
as a transmitter, **4**:54, 55f
Serpentine receptors, **2**:208–9, 208f
activating mutations in, **2**:215
Serpins, kinetic factors determining the structure of, **5**:154
Sertoli cell, defined, **3**:501
Serum cholinesterase, mutant alleles in deficiency of, **4**:400, 403
Seven-transmembrane helical receptors, **2**:208–209; **5**:324
Severe combined immunodeficiency, human, correction of adenosine deaminase deficiency in, **6**:200
Severe combined immunodeficiency syndrome (SCID)
gene therapy for, **4**:40–41
mice
genetic mechanism, **6**:136
for immunotoxicology testing, **3**:300–301
transplanting human cells to, **3**:301
X-chromosome-related, interleukin-2 receptors in, **3**:344
Severity, defined, **2**:420
Severity score, Gaucher disease, **2**:349, 349f
Sex
and crossover frequency, **3**:142
defined, **2**:78
DNA repair theory of, **2**:83–85
Sex chromosomes, **3**:511
in fragile X syndrome, **2**:320, 320f
SEX DETERMINATION, 5:434–47
defined, **5**:435
in *Drosophila*
mutant studies, **2**:138, 141–42, 141f
Sex-lethal, *fruitless*, *transformer* mutants, **2**:142
male-determining factor, **3**:518–20
Sex-determining genes
autosomal, **5**:443
X-linked, **5**:441–442
Sex-determining region Y (SRY), **3**:501
Sex-determining region Y (SRY) gene, **5**:438–41, 438f
Sex-hormone binding globulin (SHBG), **5**:481
Sex-linked trait, defined, **3**:142
Sex peptide, in *Drosophila*, **2**:143
Sex reversal
defined, **5**:435
XX, **5**:443–45
XY, **5**:445–46
Sex-reversed phenotypes, mutations in, **3**:501
Sexual PCR, **5**:448–50, 450f
applications, **5**:455–57
defined, **5**:448
theory of, **5**:453–55, 455f
SEXUAL PCR AND ASSEMBLY PCR, 5:447–57
Sexual reproduction
benefits of, **2**:84–85
costs of, **2**:84
frequency of, **2**:84
SH2 sites, defined, **5**:230
Shannon entropy, and complexity measures, **6**:47
Shape space, and sequence space, **4**:99
Sherman paradox, **2**:325
Shielding, in work with radioisotopes, **5**:213–14
Shigella, plasmids of, **5**:11
Shigella flexneri, defense against phagocytosis, **1**:105
Shigella spp., invasion plasmid antigens of, **1**:106
Shikimate pathway, chiral specificity in, **1**:341
Shine-Delgarno mechanism, **6**:116
Shine-Delgarno sequences, **1**:353
Shingles, from reactivated varicella-zoster virus, **6**:172
Shock
hypotensive, role of interleukin-1 in, **3**:345
role of the kallikrein-kininogen-kinin system in, **3**:379–80
septic, **3**:470
See also Toxic shock syndrome
Short interspersed repeat element (SINE). *See* SINE
Short tandem repeats (STR), **2**:51
characteristics of, **2**:59t
defined, **2**:420
for genetic testing, **2**:421; **4**:339
as hot spots, **4**:151
on the phenylketonuria gene, **4**:408
Shotgun sequencing, for long-range genome mapping, **2**:78
Shuttle vectors
defined, **6**:302
plasmid, for yeast studies, **6**:311–12
viruses as, **2**:436
Sialic acid
defined, **1**:293
of N-CAM, **1**:294
Sickle cell anemia, **1**:113–14; **2**:418, 424–25; **3**:19
diagnosis of, **3**:230, 230f
point mutation in, **4**:105
point mutation analysis in, **2**:423; **4**:332
prevalence of, **5**:45
as a single-gene disorder, **3**:230
Side chains
amino acid, structure of, **5**:118f
peptide
constraint on conformations, **4**:374, 374f, 378–79, 379f
identification with Raman spectroscopy, **5**:104
protecting during synthesis, **4**:383, 387, 388
placement by homology modeling, **5**:164
protein, identification with Raman spectroscopy, **5**:97–101
Sievert (Sv), unit of equivalent dose of radiation, **5**:211
Sigma factors
in prokaryotic transcription, **2**:369
RNA polymerase, **4**:161
Sigma replicative intermediate, **4**:317–18, 317f
Signal detection, in FISH, **2**:378f, 384–85
Signaling pathways
intracellular, interleukin receptors in, **3**:339–40
phosphatidylinositol-mediated, **4**:424
Signal peptidase, microsome, **5**:335–36

Signal peptide
defined, **4**:340
protein folding role of, **5**:81
Signal recognition particle (SRP), **4**:342; **5**:197–98, 198f
defined, **5**:194
Signal recognition particle pathway, insertion of P450s into the endoplasmic reticulum by, **1**:439
Signal sequence
defined, **5**:194, 428
eukaryotic gene structure, **5**:431
for glycoprotein translocation out of the endoplasmic reticulum, **6**:188
Signal/signaling
by cytokines and hormones, **5**:238
defined, **2**:378
generation with receptor proteins, **1**:142–43
intracellular, pathways for, **5**:242
by the *JAK/STAT* system, **2**:465
by metal-binding proteins, **6**:60
in plant defense against pathogens, **4**:480–81
for protein degradation, ubiquitin-dependent, **3**:350–53
synaptic, **5**:512–14
targeting, **5**:195, 195t
transmembrane, and integrin interaction with extracellular matrix, **1**:459
via surface receptors, **1**:452
Signal transducers and activators of transcription (STATs), **1**:446
activation by cytokines, **5**:242
Signal transducing factor, calcium ion, **1**:239
Signal transduction, **5**:514
in *Caenorhabditis elegans*, **4**:171
control of
by phosphorylation, **5**:169, 172
by vasopressin, **2**:20, 21–23
cyclophilin and FK506-binding protein involvement in, **5**:92
defined, **2**:216, 285, 463
disruption of, and colon cancer, **1**:412
in heart failure, **1**:289–90
by heterotrimeric G proteins, **3**:41–48
by interferon systems, **3**:324f
by interleukins, **2**:470–71; **3**:337–38
involving epidermal growth factor, **2**:269
by peptide hormones, **4**:355
phytochrome, mutations in the pathway for, **4**:494
in plants, **4**:492–93
principles of, **1**:239–40; **4**:95
protein kinase used in, **5**:171
role of phospholipases in, **4**:431–32, 431f
schwannomin's role in, **4**:186
second messenger systems, G proteins in, **2**:287
sphingolipids involved in, **3**:458
study of, using electroporation, **2**:206
Signal transduction pathways
calcium, of plants, **1**:249
p21Ras, neurofibromin in, **4**:177–78, 178f
peptides and receptors in, **4**:356t
Signal visualization, in FISH, **2**:385–86
Silicon carbide whiskers, for plant cell transformation, **4**:464
Silicone elastomers, medical, **5**:36–37
Silks, genetic engineering of, **5**:27
Silver staining, for protein detection, **2**:355–56, 356t
Simian immunodeficiency virus (SIV)
activation domain of Tat, **1**:21
experimental vaccination for, **6**:169–70
infection with, and response to rinderpest vaccination, **6**:168
proteases encoded by, **1**:14
response to light in the presence of merocyanine 540, **6**:215t
Simian sarcoma associated virus-1, mapping to chromosome 18, **3**:208–9
Simian virus SV40, **1**:412
chromatin of, electron microscope image, **2**:189f
electron microscopy of, **2**:183f
Simple disease, defined, **3**:222
Simple sequence length polymorphisms (SSLPs), for mapping
the *Arabidopsis* gene, **1**:80
of microsatellites, mouse, **4**:124
Simple sequence repeat polymorphisms (SSRPs), **3**:143
Simple sequence repeats (SSRs), **3**:241–42, 513
mapping with, **3**:516f
Simple sequences, **5**:430–31
Simple tandem repeat polymorphisms (STRPs), **3**:143
Simulated distributions, of sequence alignment, **5**:423–24
Sinapinic acid, matrix for MALDI, **4**:16
Sindbis virus, packaging of, **6**:220–21, 221f
SINE (short interspersed repeat element), **1**:380; **3**:243–44, 512
Alu, **2**:73, **3**:217–18; **6**:133, 135–36, 135f
in chromosome R bands, **1**:371
defined, **3**:216
for hybridization probes, **2**:380–81
Single-cell oils
synthesis by molds, **3**:438–40
yeast, **3**:435–36
Single-chain fragment (scFv), antibody-binding site, **1**:63f
Single-chain fusion proteins, for tumor treatment, **1**:68
Single-chain technology, for engineering proteins, **3**:286–87, 286f
Single-gene disorders, **3**:222, 229–30
Single-residue hypotheses, for susceptibility to insulin-dependent diabetes mellitus, **2**:34
Single sequences, analysis of, **5**:429–32
Single strand, DNA, efficiency of repair of, **2**:82
Single-strand conformational polymorphism (SSCP)
analysis of mutations with, **5**:45
analysis of point mutations with, **5**:302
defined, **2**:420; **4**:173
for mutation identification, **2**:396–97, **2**:423; **3**:226
Single-stranded DNA (ssDNA)
defined, **5**:458
from polymerase chain reactions, **4**:251
SINGLE-STRANDED DNA-BINDING PROTEINS, **2**:89–90; **5:457–66**
defined, **5**:458
Single-stranded RNA (ssRNA)
in the genome of alphaviruses, **6**:189–90
structure of, **6**:218
Singlet oxygen
cellular, damage from, **2**:42
defined, **6**:207
Sister chromatids, unequal exchange between, and tandem arrays, **6**:8–9
Site-directed mutagenesis, **1**:116
defined, **5**:136
to determine RNA secondary structure, **5**:365–66
to modify enzymes, **1**:142
for nucleic acid-protein complex characterization, **4**:252
to produce biomimetic enzymes, **1**:143–44
for RNA structure determination, **5**:378
for studying the calcium ion pump reaction cycle, **1**:247
of synthetase, to produce novel proteins, **5**:137–38
Site-specific DNA inversion, **2**:404
Site-specific drug delivery, **2**:166
transfection with liposomes, into an artery, **3**:424–25
Site-specific free radical damage, **2**:332–33
Site-specific mutagenesis
of bacteriorhodopsin, **1**:111
of the consensus sequence for luciferase import by peroxisomes, **5**:206
in engineered ribosomal biosynthesis, **1**:136–37
helix swap experiment, **5**:277

for identifying regions for binding extracellular ligands, **3**:43
for probing structure-function relationships, **1**:433
for studying alphavirus, **6**:190
Site-specific recognition, and formation of vesicular aggregates, **1**:154
Site-specific recombination, **2**:413
Situs inversus, **1**:333
Size exclusion chromatography (SEC), **3**:78–79
Sjøgren-Larsson syndrome, fatty acid metabolism in, **3**:452
Sjøgren's syndrome, autoantibody responses in, **1**:86
Skeletal muscle, **4**:130–32
fast-fiber, alkali light chains of, **4**:135
organization of, **4**:131f
slow fiber, **4**:135
Skin cancer, from UV damage, **6**:163
Skin regeneration, using a silicone rubber membrane covering, **5**:37
Slide, to form DNA structure, **2**:107–09
defined, **2**:103
Slippage mechanism, DNA, mutation due to, **2**:392
Slot-blot analysis, **1**:124
defined, **5**:342
in transgenic animal modeling, **6**:72
Slovak Collection of Yeasts, addresses of, **6**:304
Small-cell lung cancer
bombesin production by, **3**:494
defined, **3**:491
Small hydrophobic (SH) protein, of respiratory syncytial virus, **5**:290
Small molecule libraries, **5**:518, 518f
nonpeptide, **5**:522
Small nuclear ribonucleoprotein (smRNP), spliceosome, **1**:86
Small nuclear ribonucleoprotein polypeptide N (SNRPN), gene for, mapping to chromosome 15, **3**:168
Small-peptide-related genes, in *Drosophila*, **2**:143–44
Smallpox, development of vaccination to prevent, **6**:165
Smith-Lemli-Opitz syndrome, **1**:116
Smith-Waterman method, **5**:433
for local alignment, **5**:434
for local alignment assessment, **5**:424
Smooth colony morphology, bacterial, **3**:466
Smooth muscles, **4**:130
contraction of, **4**:144
nonvascular, effects of kinins on, **3**:378
organization of, **4**:132
Snap-25, **5**:514
Snap-back structures, in picornavirus replication, **5**:23
SNAP receptors (SNAREs), regulatory role in intracellular membrane traffic, **4**:50
SNAPs (soluble NSF attachment proteins), **4**:346
Soaps, fatty acids at high pH, **3**:455
Sodium channel
gene for, in *Drosophila*, **2**:143
membrane, **5**:513
mutants affecting in *Drosophila*, **2**:133
Sodium dodecyl sulfate polyacrylamide gel electrophoresis (SDS-PAGE)
for apolipoprotein determination, **3**:475, 476
comparison with mass spectrometry, **4**:19
procedure for, **5**:123–24
for protein/peptide purification, **5**:178, 188–89
for protein purification, **5**:174
two-dimensional, **2**:353, 354–55
Sodium ion/calcium ion exchanger, **1**:245
Sodium/neurotransmitter symporters (SNF), **6**:129
Software, architecture of, automated genome research, **5**:397
Solenhofen deposits, **4**:300
Solid phase chemistry, for synthesis of oligonucleotides, **1**:73
Solid phase extraction, of steroids, **5**:474
Solid phase library, synthetic peptide, **5**:518
Solid phase peptide synthesis (SPPS), **1**:171, 171f; **4**:361, 365, 365f
defined, **4**:383
steps in, **4**:384, 384f
Solid phase technique
for sequencing of peptides, **5**:187
supports for, synthesis of peptides, **4**:388–89, 390
Solid state, lipid packing in, **3**:445
Solid substrate fermentation (SSF), **2**:338, 339f
defined, **2**:337
Solid tumors
diagnosis of, **2**:65–66
gene for, mapping to chromosome 1, **3**:95
Soluble guanylyl cyclase, defined, **4**:200
Soluble ions, association with proteins, in food, **2**:310–11
Soluble NSF attachment proteins (SNAPs), regulatory role in intracellular membrane traffic, **4**:50
Soluble proteins
lysosomal targeting of, **5**:202–3
retrieval of, in the endoplasmic reticulum, **5**:199
Solution phase assay, **5**:521, 521f
Solvent engineering, for enzymes in nonaqueous media, **2**:256–57
Solvents, effects of
on DNA denaturation, **2**:10–12
on organic biocatalysis, **2**:252
Somaclonal variation, **2**:221
Somatic cell hybrid
for chromosome 15, **3**:173, 173t
defined, **2**:386; **3**:191; **6**:28
for mapping chromosome 18, **3**:197–98f
for mapping microsatellite markers, **2**:387
Somatic cells
damage to DNA of, **2**:79
in gene therapy, **3**:231
mortality of, **6**:23–24
attributed to lack of telomerase, **6**:18
Somatic mutation theory
in aging, **1**:6
defined, **1**:5
Somatic recombination, **3**:513
Somatic self, **3**:275
Somatomedin, **2**:466
Somatostatins, analogues of, **4**:381, 381f
Song genes, in *Drosophila*
related changes in mutants, **2**:134, 135f, 140f
related to a visual behavior gene, **2**:142–43
related to clock genes, **2**:138
Sophorolipids, synthesis by *Candida bombicola*, **3**:437
Sorbitol dehydrogenase (SORD), **3**:168
SOS box, **5**:12
SOS regulatory system
in bacteria, **5**:226–27
defined, **5**:221
SOS repair, of damaged DNA, **4**:149
error-prone, **4**:155
Source collision-induced dissociation, **4**:18
Source gene, defined, **3**:240
Southern blot analysis
defined, **4**:323
for DNA quality evaluation, **2**:49
FMR-1 gene, **2**:326, 326f
of foreign DNA in plant cells, **4**:455, 456–57f
and molecular pathology, **4**:324–25
for mutation detection, **2**:421
in fragile X syndrome, **4**:331
for physical mapping, **3**:118
of ultraviolet radiation damage to DNA, **6**:156
Southern blot hybridization, and landmark identification, **5**:305
Space group, crystals, **5**:112
Spatial light modulators (SLMs), **1**:150–51
voltage-controlled, **1**:151, 151f
Spatially addressable format, synthetic peptide libraries, **5**:518

Species, selection of animal models for immunotoxicity evaluation, **3**:299
Species barriers, for transmission of prion diseases, **5**:62, 64
Species integrity, animal right to, **6**:82–83
Species variation
  in hemoglobins, **3**:12–13
  and immunotoxicity testing, **3**:300
Specific activity, radioactivity per unit mass, **5**:209, 211
Specificity
  of the ABC-type transport system, **6**:130
  of Achilles' cleavage, **1**:2, 2f
  of Achilles' cleavage reactions, experimental considerations, **1**:4
  of amplification, **4**:339
  of antibody binding, expression of, **1**:64–65
  of autoantibodies, in clinical diagnosis, **1**:86–87, 86t
  of binding bacterial cells, **1**:104
  of biological reactions, mimicking, **1**:141–45
  of cadherin binding, **1**:295
  of cellular receptors for viruses, **6**:225–31
  defined, **6**:259
  of enzyme reactions, **6**:260
  of enzymes, **2**:223, 223f
    and chirality, **1**:341
    and solvent characteristics, **2**:256
  of genetic switching on and off, **6**:327
  geographic, of mtDNA patterns, **4**:86
  of G protein interaction with receptors, **5**:237
  of hormones in the female reproductive system, **2**:286
  of immune phenomena, **3**:275, 276, 283–85, 285f, 296
  of immunoassays for steroid hormones, **5**:476
  of isoenzymes, **3**:374
  of ligase chain reaction analysis, **3**:303
  of ligase reactions, **3**:303–4
  of membrane fusion, **4**:45
  mutational, **4**:155
  of PCR amplifications, **4**:325, 336
  of protein kinases, **3**:46
  of receptor interactions, G protein cycle, **3**:44–45, 45f
  of repressor-operator interactions, **5**:279
  of restriction enzyme sequence recognition, **5**:305–6
  of RNA reactions with proteins, **4**:284
  of synthetic oligonucleotides, **1**:70
  of ubiquitin protein ligase, **3**:349
  See also Site-specific mutagenesis
Spectra
  one-dimensional NMR, **4**:218–19, 219f
  two-dimensional NMR, **4**:220–21, 221f
Spectral bandwidth, in circular dichroism determinations, **1**:387
Spectral intensity, Raman, **5**:96
Spectrin, **1**:452
Spectroscopic sensors, **1**:187
Spectroscopy, for studying nucleic acid-protein interaction, **4**:244
Sperm
  binding to eggs, **2**:292
  protocol for separating sperm and nonsperm cells, **2**:50f
Spermatid, defined, **3**:501
Spermatocyte, defined, **3**:501
Spermatogenesis, **3**:501
  histone transcription in, **3**:53
  macromolecular synthesis during, **3**:503–4
Spermatogonium, defined, **3**:501
Spermatozoon, defined, **3**:501
Sperm-containing stains, differential extraction of DNA from, **2**:52–53
S phase, histone protein synthesis in, **3**:51
Sphenoid wing dysplasia, in neurofibromatosis 1, **4**:175
Spheroplasts
  double transformants in yeast artificial chromosomes, **6**:298
  fusion of, for constructing transgenic cell lines, **6**:301–2
  for transformation, yeast, **6**:310–11
  for transformation of DNA into yeast cells, **6**:297
Sphingolipids, **3**:457
  metabolism of, **3**:458
  occurrences of, **3**:460
Sphingolipid signaling, **4**:430
Sphingomyelin
  biosynthesis of, **3**:459f
  and hormone mediation, **2**:211
Sphingomyelinase deficiency, in Niemann-Pick disease, **2**:349
Spin, defined, **3**:385
Spinal and bulbar muscular atrophy (SBMA), **2**:328
Spinal muscular atrophy (SMA), **3**:223; **4**:105
  motor neuron degeneration in, **4**:106
  X-linked, gene defect in, **4**:106–7
Spin-glass theories, **4**:94
Spin labeling, **3**:385–86, 386f
Spinocerebellar ataxia (SCA), **3**:242
  type 1, **2**:328
  type 3, gene for, mapping to chromosome 14, **3**:158
Spin oxymetry, **3**:386
Spirolactam analogues, in peptide design, **4**:377
Spiro-(bicyclic thiazolidine) analogues, in peptide design, **4**:377
Splanchnic vasoconstriction, role of vasopressin in, **2**:27
Spliceosome, **5**:340
  assembly pathway, **2**:373f
  RNA splice site choice during assembly of, **2**:371–72
Splicing
  abnormal, in hemophilia, **3**:31, 32, 38, 38t
  alternative
    of genes, **3**:512
    of mRNA in collagen genes, **2**:280
    of mRNA in elastin genes, **2**:282
    of mRNA in fibronectin genes, **2**:283
    of RNA, **1**:407; **2**:279
  APC gene, **1**:410
  in the chloroplast of *Chlamydomonas reinhardtii*, **1**:353
  defined, **1**:347; **4**:129
  gene, high and low molecular weight kininogens from, **3**:377
  process of, **5**:357
Split synthesis, of peptides, **5**:517, 520
Spodoperta frugiperda, host, **2**:277
Sporadic disease, defined, **4**:173
Sporulation, yeast, **6**:304
Spot intensity, reflecting copy number, of restriction landmarks, **5**:316, 317, 317f
SPOTs membrane synthetic peptide libraries, **5**:518–19
"Spots" technology for peptide synthesis, **4**:392
Squalene, comparisons of structure with beta-carotene, **3**:451f
Squalene synthetase, **3**:409
Squamous cell carcinoma of the head and neck (HNSCC), **3**:211
Squamous epithelial tumors, interferon-α therapy for, **3**:330
Stability
  of acyl derivatives of carboxylic acids, **1**:160
  defined, **4**:281
  of DNA, dependence on base sequence, **2**:12–14, 14f
  enzyme, control by organic solvents, **2**:256–57
  of hybrids, **4**:235–39
  metal ion, dependence on ligand, **1**:128–29
  of oligonucleotide duplexes, **4**:237–39, 237f
  of proteins, estimating from circular dichroism data, **1**:393–94
Stability ruler, metallic ions, and toxicity measures, **1**:129, 129t
Stabilization, of antigen-antibody complexes, hydrophobic effect on, **3**:283

Stacked pairs, in duplex DNA, forces between, **2:**1
Stacking forces, defined, DNA, **2:**1
Stacking interactions
of DNA
mispairing, **2:**17
preferential, **2:**111
of RNA, **5:**369–70
in sequencing chips, **3:**257
Stacking rule, for bulge loops, RNA, **5:**376–77
Standard reduction potential (*E*), **1:**309–10
Standard strains, yeast, **6:**304
Staphylococcal nuclease, in analytical affinity chromatography, **5:**245
Staphylococcus aureus, **3:**289
α-hemolysin of, **1:**140
"Star" activity
defined, **5:**296
of restriction endonucleases, **5:**300
Starch, **5:**33–34
Static model, protein structure versus dynamic model, **5:**161, 161f
Steady state
enzyme reaction, **2:**234
nonequilibrium, **1:**305
Steady state RNA, defined, **5:**343
Stefin gene, **3:**377, 381
Stem cell factor, **2:**465, 469; **3:**508–9
Stem cells
defined, **2:**263, 463
keratinocyte, **2:**263–264, 264f
pluripotent, **1:**447–448
leukocytes arising from, **3:**296
See also Embryonic stem (ES) cell; Pluripotent stem cells
Step, of consecutive base pairs, DNA, **2:**108f
Step gradient, in high performance liquid chromatography, **3:**77
Step number *n*, and roll angle, in DNA, **2:**113, 113f
Stereochemistry, of medicinal compounds, **4:**38–40, 40f
Stereoisomers, defined, **1:**332
Stereoselectivity, substrate, **6:**260
Stereospecificity, enzyme, **1:**340–41
Sterility
genetic basis of, associated with cystic fibrosis, **1:**432
in mutant *Drosophila*, and related behavior, **2:**141–42
Sterilization, in plant tissue culture, **6:**11
Stern-Volmer equation, **3:**388
Steroid action, defined, **5:**466
STEROID ANALYSIS, **5:466–78**
Steroid biosynthesis and metabolism pathways, defined, **5:**466
Steroid hormone receptors, superfamily of, **5:**463
Steroid hormone regulatory element, in the gene for apoD, **3:**413
Steroid hormones
biosynthesis of
with fungi, **2:**344
role of P450, **1:**436–37
sex, for treating breast and prostate cancers, **2:**215–16
synthesis of, **2:**164–65, 165f
STEROID HORMONES AND RECEPTORS, **5:478–85**
Steroidogenic enzymes
defined, **2:**285
expression of, **2:**289–90
Steroidogenic factor 1 (SF-1), induction of Müllerian inhibiting substance by, **3:**502
Steroid receptors
hsp90 binding to, **1:**321
interaction with hormone-responsive elements, **2:**290
Steroid responsive element (SRE), **5:**472
STEROIDS, **3:**460–61
ANTITUMOR, **5:485–500**
plasma-binding proteins for, **5:**481
Steroid sulfatase, gene encoding, X chromosome, **3:**503
Sterol response elements (SREs), **3:**404
and cholesterol metabolism, **1:**287
Sterols
antitumor activity of, **5:**496–97
synthesis by molds, **3:**440
"sticky end" interaction, asymmetric gap ligase chain reaction, **3:**305
Stiff skin syndrome, **3:**164
Stigmasterol, starting material for steroid production, **2:**164–65, 165f
Stilbenes, as drugs, **2:**158–59
Stirred tank reactor (STR), **2:**339–40
Stochastic process, defined, **4:**173
Stoichiometry, affinity biosensor techniques for determining, **5:**249
Stokes shifts, **2:**298, 300
and emission spectra, **2:**298
Stop codons
defined, **3:**227
in hemophilia A, **3:**35
release factors recognizing, in *E. coli*, **5:**146–47
suppression of recognition of, experimental, **5:**143–47
Stowaway transposable elements, **4:**487
STR. *See* Short tandem repeat
Strains, prion, molecular basis of, **5:**66–67
β-Strand, defined, **5:**159
Strand exchange
defined, **5:**252
DNA, role of RecA protein in, **5:**226
Strand separation temperature, defined, **4:**235
Stratum corneum, filaggrin in, **2:**268
Strecker synthesis, asymmetric, synthetic amino acids, **1:**56–57
Strength, hydrogen bond interactions
of donors and acceptors, **3:**262–63
moderate and weak, **3:**263–65, 263t
Streptavidin
biotin affinity for, **6:**240
defined, **3:**288
in oligonucleotide ligation assay, **3:**400
Streptavidin-protein A chimera, **3:**289
Streptococcus, group B, capsular polysaccharides, **1:**279, 279t
Streptococcus pneumoniae, vaccination against, **1:**279
Streptococcus pyogenes
fibronectin, **1:**105
immune system responses to, **1:**105
mimicry of host markers, **1:**105
Streptogramin B, resistance induced by plasmids, **5:**10
Streptomyces, defined, **4:**393
Streptomyces avermitilis, **4:**396
Streptomyces avidinii, **3:**289
Streptomyces griseochromogenes, **4:**396
Streptomyces hygroscopicus, **4:**395
Streptomyces kasugaspinus, **4:**396
Streptomyces viridochromogenes, **4:**395
STREPTOMYCETES, MOLECULAR BIOLOGY OF, **5:501–8**
Streptomycin, **1:**338
resistance to, gene involved in, **4:**161
Streptozotocin, as an anti-cancer drug, **1:**254
Stress
environmental change as, **2:**216
response to
and hsp70s, **1:**319
and synthesis of heat shock proteins, **5:**336–37
and tyrosine hydroxylase regulation, **1:**217
Stress fibers, actin filaments in focal adhesion, **1:**454
Stress resistance, correlation with longevity, genetic studies, **1:**13
Striated muscles, **4:**130
Stringency, of a hybridization scheme, **2:**17
String length test, **6:**39, 45–47, 47t
String repetition test, **6:**39, 45
Structural alteration, in activation of oncogenes, **4:**286
Structural proteins, defined, **5:**26
Structural refinement, X-ray crystallography, **5:**114–15
Structural reporter groups, **1:**276–77
Structural stability, defined, **2:**271
Structure
of acetylcholine receptors, **6:**128–29

Structure (*cont.*)
of bacteriophages, DNA double-stranded, **6:**202–4
of chaperonin GroEL, **5:**88, 89f
common motifs in transport proteins, **6:**127–28, 127t
of δ-endotoxins, **4:**394
dependence on metallic trace elements, **6:**59, 60
determination of, X-ray diffraction, **6:**277–79, 277f
diphenyl oxazinones, **1:**53
DNA, **5:**352–55, 355f
and denaturability, **2:**12
enzyme, integrity in organic solvents, **2:**252–55
of eukaryotic genes, **5:**431
of HIV-1, **6:**194f
of homeodomains, **3:**57–58, 58f
imidazolidinones, **1:**53
ligation reactions for analysis of, **3:**397, 399f
methods for obtaining information about, RNA, **5:**378–79
mRNA
and control of protein synthesis, **6:**122–23
control of translation to protein by, **6:**122–23, 123f
of Na^+/K^+ ATPase, **6:**131–32
of peptides, **4:**350–52, 350f
physical, of respiratory syncytial virus, **5:**286–87
of prion protein, PrP^C and PrP^{Sc}, **5:**61, 61f
of the prion protein gene, **5:**59, 59f
of proteins, and function-inhibitor design, **5:**159–60
of psoralen, **6:**216f
of replication protein A, **5:**463
of repressor proteins, **5:**275–77
of ribosomes, **6:**114
RNA, **5:**352–55, 355f, 373–78; **6:**110f
primary, secondary and tertiary, **5:**361f
types of, **5:**360
and sequence alignment, **5:**426–27
steroid hormones, **5:**467f
of telomeric DNA, **6:**29
of vaccinia virus, **6:**194f
of varicella-zoster virus and cytomegalovirus, **6:**173–79, 179f
of viruses, and susceptibility to photochemical damage, **6:**210
Structure-based drug design, **4:**36–38, 38f
Structure-biased alignment, **5:**425–26
Structure factor
defined, **5:**110
equation for, **5:**112
STSs. *See* Sequence-tagged sites
Subcell, hydrocarbon packing in liquids, defined, **3:**442
Subcellular mechanism of action, of steroid hormones, **5:**471–72
Subcellular organelle, defined, **5:**194
Submerged liquid fermentation (SLF), **2:**338–40, 340f
defined, **2:**337
Subphase
defined, for a surfactant monolayer film, **3:**309
of a surfactant monolayer film, **3:**311–12
Substitution, mutation by, **4:**150f
Substitution bias, in structure-based alignment, **5:**426
Substitution reactions, for metal ion complexation, **1:**133
Substrate
availability of, in enzyme-directed post-translational modification, **5:**51
defined, **2:**223; **5:**159
effect of aggregation on rate of reaction, for phospholipases, **4:**417
for enzymatic ligation, **3:**393
interactions of solvents with, **2:**252
for methanogenesis, **4:**68–69, 68t, 69–72
for phosphlipases, detergent mixed micelles, **4:**418
for T4gp32 binding, **5:**460
Substrate adhesion molecules (SAMs), **1:**293
Substrate-directed influence, defined, **5:**49
Substrate specificity, **6:**260
defined, **6:**259
Subtelomeric repeats, **3:**243
Subtilisin
FTIR analysis of, in nonaqueous solvents, **2:**253
nonaqueous biocatalysis by, **2:**251
kinetics of, **2:**256
reaction not involving the chiral center of substrates, **1:**343
superfamily of, **4:**344–45, 344f
Subtractive hybridization, for cloning CpG islands from defined regions of chromosomes, **2:**76
Subunits
polypeptide, defined, **5:**115
protein
defined, **1:**27; **2:**223
and quaternary structure, **5:**120–21
Subunit vaccines
defined, **6:**172
development of, **6:**176–79, 176f
protein, for varicella-zoster virus and cytomegalovirus, **6:**177t
Succinate dehydrogenase
flavin adenine dinucleotide attachment to, **1:**400
reaction at chiral center of the substrate, **1:**342
Sugar chains, **2:**447–48, 447f
Sugar-phosphate chains
covalent connections of, **2:**110
DNA, **2:**105–6
Sugars
DNA, products of reactions induced by ionizing radiation, **3:**363–65
modifying
in oligodeoxyribonucleotides, **4:**283
in oligoribonucleotides, **4:**284
puckering modes, RNA, **5:**374, 374f
Suicide enzymes, repair of alkylation damage to DNA by, **6:**284–85
Sulfamate linkage, oligonucleotide analogues, **4:**277
Sulfates
reduction of, by anaerobic bacteria, **1:**121
of steroids, **5:**471
Sulfatides, **3:**458
Sulfation
of peptide hormones, **4:**343, 353
posttranslational modification by, **5:**56
Sulfhydrogenase, **2:**246
Sulfhydryl groups, and metal ion toxicity, **1:**129
Sulfide dehydrogenase, **2:**245
Sulfoconjugation, in xenobiotic metabolism, **6:**269, 269f
Sulfolipid, defined, **3:**426
Sulfonamide linkages, oligonucleotide analogues, **4:**276–77, 277f
Sulfonamides, **2:**160
Sulfotransferases, **1:**198; **6:**269
Sulfur-based linkages, in phosphate-free oligonucleotide analogues, **4:**276–77
Sulfur-dependent organisms, **2:**241
enzymes from, **2:**243t
Sunlight, effect on human cells in culture, **2:**41
SUPERANTIGENS, **5:508–12**
Supercoiling
of DNA, defined, **2:**103
of long DNA, **2:**116–18
Supercritical fluid
defined, **2:**249
freeze-dried enzymes in, **2:**252
Superfamily
defined, **1:**435; **3:**511
kinesin, **4:**117–18, 117f
myosin, **4:**115
P450, **1:**435, 438
Superhelix, DNA, **2:**114–15
Superoxide anion
cellular, damage from, **2:**42
oxidative damage from, in cells, **2:**80

Superoxide dismutase (SOD), **1:**133; **2:**220, 333; **4:**63; **6:**210
involvement in amyotrophic lateral sclerosis, **4:**108–11
manganese, **2:**220
metals involved in action of, copper and zinc, **4:**64f
and mycobacterial virulence, **4:**160
and necrosis of infected plant tissue, **4:**480–81
point mutation of the gene for, in familial amyotrophic lateral dystrophy, **4:**106
SOD1, in familial amyotrophic lateral sclerosis, **2:**425
Superoxide radical
anions, produced in photosensitization reactions, **6:**208
chemical reactivity of, **2:**331
Supersaturation, defined, **3:**309
Support film, defined, **2:**182
Suppression, defined, **5:**136
Suppression efficiency, defined, **5:**136
Suppressor
fungal-derived, role in plant pathology, **4:**483
informational and metabolic, **6:**320
omnipotent, **6:**320
Suppressor-mutator (*Spm*) autonomous elements, **4:**487
Suppressor tRNA, **6:**320
aminoacyl, modification summary, **5:**142
defined, **5:**136
Suprachromosomal subfamily, of alpha satellites, **3:**241
Supramolecular switch, **1:**156
Surface acoustic wave (SAW) device, **1:**192
Surface behavior, of lipids, **3:**448–49
Surface dilution
defined, **4:**416
and kinetics of phospholipase reactions, **4:**417
Surface exclusion, **5:**6
Surface mass loading, in acoustic wave biosensors, **1:**191
Surface plasmon resonance (SPR), **1:**190, 190f; **5:**245, 283
Surfactants
organized array of, **3:**311–12, 311f, 315
synthesis by molds, **3:**440
synthesis by yeasts, **3:**437
Surfactin, **3:**434
Surprisals, **6:**48–49
relationship with complexity, **6:**50f
slope of, versus complexity, **6:**52f
Surrogate markers, for disease progression, HIV-1, **1:**41
Swarming, and turbulence, **4:**94
Sweat gland, involvement of, in cystic fibrosis, **1:**426
Switching, by G protein interactions, **2:**209
Swyer syndrome, **5:**445
Sychnologic organization, peptide, **4:**372, 372f
Symmetric inhibitors, C_2-, of HIV-1 protease, **1:**38–39f, 40
Symmetry, bilateral, in animals, **1:**333–34
Symport, **6:**127
defined, **6:**126
by solute/Na^+ proteins, **6:**129
SYNAPSES, **5:512–16**
defined, **4:**54
exocytosis of, **4:**45
Synapsins, **5:**514
Synaptic cleft, **5:**512
Synaptic function
excitatory transmissions, glutamate receptor role in, **5:**233
modulation of, **5:**515
Synaptic transmission, as regulated secretion, **5:**203–4
Synaptobrevin, **4:**48–50, 346, 346f; **5:**514
Synaptosomal-associated protein receptors (SNAREs), **5:**201–2
Synaptotagmin, **4:**48–50
Synchrotron radiation, for diffraction measurements of crystals with large unit cells, **6:**278
Syneresis, defined, **2:**307
Synergism, defined, **1:**253
Synexin (annexin VII)
promotion of intermembrane contacts, hypothetical, **1:**61f
structure, **1:**60f
Synovial sarcoma
fusion gene associated with, **3:**211
translocation gene associated with, mapping to chromosome 18, **3:**209
Syntaxins, **4:**48–50, 346, 346f; **5:**514
Synteny, **3:**112–14, 511
conserved, defined, **4:**180
defined, **3:**84, 160
demonstration of, with FISH and WCP, **6:**256
human chromosome 15 with mouse chromosomes 2, 7, and 13, **3:**170
mouse chromosome 18 with human chromosomes 1 10, and 18, **3:**198–99
Synthesis
biomimetic, advantages of, **1:**137
of peptide hormones, **4:**341–46
of polymers, **5:**27
Synthetase
adenosine triphosphate, **1:**308
ATP, **6:**132
binding of, effect on tRNA, **5:**371, **6:**122
biotin-holocarboxylase, **6:**240–41
flavin adenine dinucleotide, **1:**399–400
genetically modified, **5:**137–38
HMG CoA, control of cholesterol levels, **1:**287
interactions with tRNA, **6:**112
misacylation of tRNA by, **5:**136–38
phosphoribosylaminoimidazole-succinocarbozamide, gene for, **6:**313
polyhydroxybutyrate, **3:**434f
poly(ADPR), **1:**401
squalene, **3:**409
2–5A, interferon-inducible antiviral protein, **3:**325–26
Synthetic combinatorial libraries, convergent versus parallel methods, **5:**522f
Synthetic enhancement, by two mutant genes, **6:**320–21
Synthetic lethality, **6:**320–21, 321f
Synthetic peptide, PrP, amyloid formation by, **5:**61
SYNTHETIC PEPTIDE AND NON-PEPTIDE COMBINATORIAL LIBRARIES, **5:516–24**
Synthetic peptides, defined, **4:**348
Synthetic polymers
biodegradable, **5:**34–36
defined, **5:**31
Synthetic steroids, **5:**483–84
Synthetic unrandomization of randomized fragments (SURF), **4:**98
System, automation, attributes of, **5:**390
Systematic evolution of ligands by exponential enrichment (SELEX), **4:**227
Systemic acquired resistance (SAR)
defined, **4:**476
plant, **4:**481
Systemic lupus erythematosus (SLE)
animal model for, **1:**92
anti-Smith antigen response in, **1:**86
autoantibodies in, **1:**86; **5:**415
interferons in, **3:**328
Systemin, in plant response to wounding, **4:**492, 493
System integration
defined, **5:**389
design for, **5:**395–98
Szilard, Leo, on radiation exposure and senescence, **1:**6

T_m, defined, DNA, **2:**1
T4gp32, binding characteristics and substrate preferences of, **5:**460
Tabtoxin, toxicity to plants, **4:**482
Tabtoxinine-β-lactam (TBL), **4:**395–96, 396f

Tacrine (tetrahydroaminoacridine, THA), in Alzheimer's disease, **1**:50
Talin
 in focal adhesions, **1**:452, 453
 function of, in vivo, **1**:455–56
Tamoxifen
 breast cancer response to, **1**:256; **5**:484, 484f
 structure of, **2**:159f
Tandem arrays
 dynamics of, **6**:6–7
 persistence time of, **6**:7, 7f
 properties of, **6**:2–3
TANDEMLY REPEATED NONCODING DNA SEQUENCES, **6:1–10**
 defined, **6**:1
Tandem mass spectrometry (MS/MS), **4**:17–19, 19f
 defined, **4**:12
 for protein identification, **5**:126
Tandem repeats, **2**:50–51; **3**:229, 240, 241–42
 in collagen fibers, **5**:28
 defined, **2**:48, 420
 in rodent kallikrein gene families, **3**:383
 in telomeres, **6**:28–29, 28t
T antigen, large, interaction of RPA and polymerase α-primase with, **5**:464
Tapetal cells, destruction of, by barnase, **4**:473
TAR binding protein 1 (TRP-1 or TRP-185), **1**:21, 22f
Target, hybridization, preparation of, **2**:383
Target-based drug design, **2**:160–61
Target cell, hormone, **5**:478
Targeted insertion, for genetic immunization, **2**:436–37
Targeted mutagenesis, **3**:518, 519f
 defined, **3**:511
Targeted transformation, **5**:257
Targeting
 antibiotic, modification by plasmids, **5**:10
 infectious disease testing with LCR, **3**:306
 polynucleotide, defined, **3**:302
Targeting patch, defined, **5**:194
Targeting pathways, **5**:195–96, 196f
Target sequence, **5**:195t
 defined, **2**:378; **5**:194
Target sites, substrate-directed posttranslational modification of, **5**:51–52
TAR hairpin structure, **4**:224
TAR RNA-arginine complex, NMR definition of, **4**:224
Tartaric acid, enantiomers of, **1**:338
Tarui's disease, **4**:146
Taste, of enantiomers, **1**:339
Tat
 defined, **1**:19
 HIV-1
  inhibitors of, **1**:38, 39f
  model for, **1**:23f
  role of, **1**:34
  transactivator, **1**:21–25
 transcription in the absence of, **1**:21
TATA-binding protein (TBP)
 homology with COP1 protein, **4**:494
 and poliovirus 3C proteinase, **5**:25
 in RNA polymerase transcription, **2**:366; **4**:491–92
 three-dimensional structure of, **4**:467
TATA box, **1**:25; **4**:468–69, 468f; **5**:131
 Arabidopsis genome, **1**:78
 counterpart in nuclear genes of *Chlamydomonas reinhardtii*, **1**:348
 defined, **2**:364; **4**:129
 promoter region, HIV-1 LTR, **1**:20, 20f
 promotor sequence
  eukaryotic, **2**:95
  myosin light chain, **4**:135, 135f
Tat-binding protein-1 (TBP-1), **1**:23
Tau protein
 association with neurofibrillary tangles, **1**:50
 in tangles associated with Alzheimer's disease, **1**:43
Taxanes
 as anticancer drugs, **1**:255–56
 defined, **6**:10
 standards for quantification and identification of, **6**:15
Taxol, **4**:29
 defined, **6**:11
 production of, **1**:174
  P450-dependent pathway, **1**:437
 structure of, **4**:30f
TAXOL AND TAXANE PRODUCTION BY CELL CULTURE, **6:10–17**
Taxus, defined, **6**:11
Tay-Sachs disease, **1**:115–16; **2**:425; **3**:375
 accumulation of cerebrosides in, **3**:460
 among Ashkenazi Jews, founder effect and heterozygote advantage, **5**:48
 chromosome 15 involvement in, **3**:164–65
 among French Canadians, independent origin of, **5**:49
 hexosaminidase A deficiency in, **2**:349
 prevalence of, **5**:45
 as a single-gene disorder, **3**:230
"T-bag" technology, for peptide synthesis, **4**:392
T-Bu, defined, **5**:485
T-cell lymphomas
 cutaneous, interferon-α therapy for, **3**:330
 and interleukin-9 expression, **3**:343
T-cell receptors, **3**:279–80
 antigenicity of, **3**:285
 binding site on superantigen, **5**:510–11
 complex of, activation of lymphocytes using, **1**:455
 defined, **5**:508
 genes coding for, **3**:281–82, 282f
  chromosome 14, **3**:159
 interaction with superantigen, **5**:510–11, 510f
 single-chain engineering of fragments, **3**:286–87
 specificity of, **3**:283–85, 285f
T cells
 activation of, and cytokine production, **1**:445
 apoptosis of, **1**:302
 cellular attack by, **3**:296
 cytotoxic, activation of, **1**:447
 and immunity to respiratory syncytial virus, **5**:292–93
TCL class, homeobox genes, **3**:67–68
T-DNA
 fragment from *Agrobacterium tumefaciens* plasmid, **5**:11–12
 tagging with, for cloning *Arabidopsis* genes, **1**:80
T-DNA insertional mutagenesis, **4**:471–72; **5**:9
Technology, of fungal processing for agricultural applications, **2**:343
Telephone coil model of DNA, **2**:115
TELOMERASE, **3**:242; **6:17–27**, 31–34
 assays for, **6**:21–23, 23t
 defined, **6**:18
Telomerase inhibitors, for cancer treatment, **6**:35
Telomerase primer, defined, **6**:18
Telomere end factor, from *Xenopus* (XTEF), **6**:29
Telomere repeat amplification protocol (TRAP), **6**:23, 35
TELOMERES
 Arabidopsis genome, **1**:77–78
 defined, **6**:18
 invariant repeats of, *Chlamydomonas reinhardtii*, **1**:348
 MAMMALIAN, **6:27–37**
 position effects, **6**:31
 preserving end replication, **6**:18–19, 18f
 structure of, **1**:371
 tandem repeats of, **3**:242–43
Tempeh, production of, **2**:338
Temperature
 Arrhenius-type dependence on in electron transfer, **2**:195
 and DNA degradation, **2**:2
 and DNA denaturation, **2**:5
 electron transfer independence of, **2**:194
 food processing, **2**:312
 high
  effect on mutant *Drosophila*, **2**:133

enzymes active at, **2:**240–41
sensitivity of alleles to, **2:**125
sodium channel function and sensitivity to, in *Drosophila*, **2:**143
and inclusion body formation, **5:**83
and incorporation of noncoding amino acids into proteins, **5:**137
normal human, and DNA damage, **2:**42
and transcription regulation, **2:**273
Temperature gradient gel electrophoresis (TGGE)
for following DNA denaturation, **2:**4
effect of mispair stacking in, **2:**17, 17f
Template
defined, **4:**254
generation of, nucleic acid sequencing, **4:**254–55
Template-assembled synthetic protein (TASP), **4:**368–69, 369f
Template-directed synthesis
defined, **4:**289
RNA world hypothesis, **4:**296–97
Template domain, **6:**19
of telomerase, **6:**21f
Template strand, defined, **6:**109
Tenascin, **2:**283
Tendon, structure, **1:**146f
Tensegrity model, cytoskeleton, **1:**457
Tensin, focal adhesion protein, **1:**454
Teratocarcinoma cells, **6:**134–35
Terbutalin, sulfation of, **1:**198, 199f
Terminal inverted repeat (TIR), in maize transposons, **4:**486–87
Terminal modification, of peptide hormones, **4:**345–46
Terminal restriction fragment (TRF), **6:**19, 19f
defined, **6:**18, 28
mouse, length of, **6:**29
Termination
codons for, **6:**121
modifying with noncoding amino acids, **5:**139
splicing into mRNA, **2:**374f
UAG, **5:**145
in polypeptide synthesis, defined, **6:**109
in RNA replication, **2:**365, 371
Termination factors
defined, **5:**136, 328
Escherichia coli, **5:**146–47
eukaryotic, **6:**121, 121t, 121, 121t
prokaryotic, **6:**120t
See also Release factorsTerminator, transcriptional, **2:**91
of archaea, **4:**74–75
Terminus, replication, defined, **1:**95
Terminus binding protein (Tus), **2:**87
Ternary complex enzymes, in amperometric biosensors, **1:**182–83
Terpenes, chirality of, **1:**338
Terpenoid lipids, **3:**427
defined, **3:**426
synthesis by yeasts, **3:**437
Tertiary structure
carbonic anhydrase, **2:**308f
defined, **5:**368
determining with NOE, **4:**219
peptides, **4:**350, 363
protein, **2:**308, 309f; **5:**119–20, 120f
from circular dichroism data, **1:**384, 385–86, 391
defined, **5:**147, 159
evaluation of predicted forms, **5:**168–69
predicting, **5:**162
Raman spectrum analysis of, **5:**104
screening of proposed forms, **5:**168
RNA, **5:**360, 377–78, 377f
Testicular feminization syndrome, **5:**435f
from receptor defects, **2:**214
Testis
architecture of, **3:**501
development of, effect of follicle-stimulating hormone, **3:**504–5
genetic control of determination of, **3:**501–3
growth factor and receptor expression in, **3:**508f
Testis-determining factor (TDF), **3:**511
defined, **5:**435
search for, **3:**518–20; **5:**435–41
α^1-Testololactone, for cancer therapy, **5:**490
Testosterone
embryonic development of males determined by, **5:**435
and spermatogenesis, **3:**505
Testosterone propionate, cancer therapy using, **5:**490
Testotoxicosis, **2:**292
Tetracycline, **4:**99
resistance to, mechanism of plasmid-induced, **5:**9–10
Tetrad, in meiosis
defined, **6:**303
using for genome analysis, **6:**309–10, 309f
Tetrahydrobiopterin (BH_4)
cofactor for tryptophan hydroxylase, **1:**218
defined, **1:**207
Tetrahydrofolate (THF) coenzymes, **1:**402
1,2,3,4-Tetrahydroisoquinoline-2-carboxylic acid (Tic), in peptides, **4:**378–79, 379f
Tetrahydromethanopterin (H_4MPT), **4:**69
Tetrahydropterin (BH_4), specific requirement for, tyrosine and tryptophan hydroxylases, **1:**208
Tetrahydropteroylglutamates, **1:**402
structures of, **1:**403f
Tetrahymena ribozyme, selection of variants, **4:**233, 233f
Tetrahymena thermophila, group intron of, secondary structure, **5:**386
Tetraloops, **5:**362
defined, **5:**360
Tetramethyl rhodamine, label for FISH, **2:**384, 384t
Tetrazole analogues, peptides, **4:**373
Texas red, label for FISH, **2:**380f, 384, 384t
Thalassemia, **6:**136
as a single-gene disorder, **3:**230
Thalassemia
α-, **2:**425
β-, **3:**223
Southern blot analysis to detect, **4:**332
α-Thalassemia/mental retardation syndrome (ATR-16), **6:**35
Thalidomide, **1:**340
Thaumatin-like proteins, **4:**480
T-helper cells (Th)
defined, **5:**286
and immunity to respiratory syncytial virus, **5:**292–293
Theophylline, aptamer for, **4:**229, 229f
THEORETICAL MOLECULAR BIOLOGY, **6:37–53**
Therapeutic, defined, **5:**244
Therapy
dietary, for PKU, **4:**406–7
with engineered immunotoxins, **3:**286
in Gaucher disease, **2:**351
gene, **4:**101
genetic, **3:**231
in heart failure, **3:**2
hematopoietic growth factors used in, **2:**469
with interferons, side effects, **3:**329t
for manic-depressive illness, and nephrogenic diabetes insipidus, **2:**24
and molecular information about cancers, **1:**252
with peptides and peptidomimetics, **4:**348, 350, 357–60, 360t
retinoic acid, in acute promyelocytic leukemia, **2:**65
strategies involving cytokines, **1:**450
Thermal cycling, automated, **5:**394
Thermal denaturation, of monomeric proteins, **5:**74
Thermal processing, of proteins, **2:**312
Thermal stress, genomic response to, **2:**217–19
Thermococcus litoralis
alcohol dehydrogenase of, **2:**245

Thermococcus litoralis (cont.)
formaldehyde oxidoreductase of, **2**:244
Thermocycler, defined, **3**:302
Thermodynamics
analysis of denaturation, **2**:12–16
of antibody-antigen reactions, **3**:283
of biological energy transfer, **1**:304–6
of enzyme reactions, **2**:225–27
of hybridization, oligonucleotides and polynucleotides, **4**:237–39, 239f
of mispairing and point mutations, DNA, **2**:16–17
of sequence-structure relationships in proteins, **5**:160
Thermogenesis, in brown adipose tissue, **1**:313
Thermolysin, synthesis of aspartame catalyzed by, **2**:250
Thermophiles
lipids of, **3**:429
sulfur-dependent, **4**:69
Thermostability, of enzymes in organic media, **2**:255
Thermostable DNA ligase
for analytic reactions, **3**:392
defined, **3**:302
Thermostable DNA polymerase, defined, **3**:302
Thermotoga maritima
ferredoxin of, **2**:244
hydrogenase of, **2**:245
Thermotolerance, and survival of biological systems, **2**:218
Thermus aquaticus, source of *Taq* polymerase, **4**:333
Theta replicative intermediate, **4**:317f
Thiamine
beri-beri from deficiency of, historic record, **6**:244
structure and function of, **6**:248, 248f
Thiamine pyrophosphate, **1**:397–98
Thiaminokinase, **1**:397
Thiazine pyrophosphate, function of, in α-keto acid decarboxylation, **1**:399f
Thiazole pyrophosphate (TPP), formation from thiamine, **1**:397–98
Thiazolidines, bicyclic, in peptide ANALOGUES, **4**:377, 377f
Thick filament, myosin, **4**:133
Thin-layer chromatography (TLC)
for carbohydrate analysis, **1**:271
for steroid purification, **5**:475
Thioesterase, and release of enzyme-bound peptides, **4**:217
Thioesters, **1**:160
acyl-coenzyme A, as intermediates in xenobiotic metabolism, **6**:272–73, 272f
Thioether linkages, in oligonucleotide analogues, **4**:276, 277f
Thioformacetal linkages, in deoxyoligonucleotide analogues, **4**:278–79, 279f
Thionins, **4**:480
Thiopurine *S*-methyltransferase, deficiency of, **4**:405
Thioredoxin, Raman spectra, cysteine pK_a values from, **5**:104–05, 104f
Thioredoxin protein disulfide oxidoreductase system, **4**:449
Thiostatin (T-kininogen), **3**:380, 380f, 381–82
Thiostrepton, sensitivity of streptomycetes to, **5**:502
Third-codon positions, human genome fragments, **2**:363
analysis of, **2**:362f
Three-center hydrogen bond, **3**:265, 265t
in α helices, **3**:269–70
defined, **3**:261
geometry of, **3**:268
and water movement in biological systems, **3**:273
Three-point attachment hypothesis, chiral recognition, **1**:344–45, 345f
Threonine, derivatives of, **1**:54
Threshold potential, for electroporation, **2**:198–99
Thrombin
apoB cutting by, **3**:411–12
inhibition of by an aptamer, **4**:228, 228f
interactions with inhibitors, electrospray mass spectra, **4**:21–22, 21f
peptidomimetics of, for preventing thrombus formation, **4**:357
Thrombolysis
genetically engineered fusion protein for, **1**:67
inhibition of, by apo(a), **3**:414
Thrombopoietin (tpo), for platelet recovery in thrombocytopenia, **2**:469
Thrombosis, risk for, and resistance to activated protein C, **4**:332
Thrombospondin
as an adherence molecule, **1**:453
binding to the D3 domain of kininogen, **3**:380
in the extracellular matrix, **2**:284
gene for, mapping to chromosome 15, **3**:168–69
Thylakoid lumen, **4**:446f
Thylakoid membrane, **4**:444
of the *Chlamydomonas* photosynthetic system, **1**:354–55, 355f
proton electrochemical potential difference across, **4**:449
Thylakoids
defined, **4**:440
in oxygen-evolving organisms, **4**:441
regulation of energy distribution between photosystems I and II, **4**:449
Thymidine
interaction with psoralen, **6**:209
for synchronizing cell growth, **2**:379
Thymidine kinase gene, locus for preparing recombinant vaccines, **6**:166, 180
Thymidine kinase marker, **4**:475; **6**:293
Thymidylate synthetase
gene for, mapping to chromosome 18, **3**:209
inhibitors of, **1**:254
Thymine (T), **2**:415
in Achilles' cleavage oligonucleotide, **1**:4
photooxidation of, **6**:209
Thymine base, oxidation, reduction and dimerization products, **3**:362f
Thymine glycol, **2**:81
Thyroid
calcium sensors in, **1**:248
papillary carcinoma of, **1**:410
Thyroid disease, and Alzheimer's disease, **1**:42
Thyroid hormones
effect on Lp(a) levels, **3**:414–15
and hypocholesterolemia, **3**:410
and LDL receptor expression, **3**:408
nuclear receptors for, **2**:208
in spermatogenesis, **3**:504–5
stimulation of production of ApoA-I, **3**:410
Thyroiditis, autoimmune, **1**:85
Thyroid-stimulating hormone, **2**:454
Thyrotropin-releasing hormone (TRH) receptor, **2**:287
Tiazofurin, inosine monophosphate dehydrogenase inhibition by, **1**:252
Tibial pseudoarthrosis, in neurofibromatosis 1, **4**:175
TIBO, reverse transcriptase inhibitor, structure, **1**:39f
Tight coupling/linkage
of histone gene expression and DNA replication, **3**:51
of tropomyosin I and familial hypertrophic cardiomyopathy genes, **3**:169
Tilapia, medium-sized fish model, **6**:85
Time constant, for circular dichroism determinations, **1**:387
Time domain (TD) measurement, of fluorescence, **2**:301–2, 302f
Time line, for DNA studies, ancient material, **4**:304f
Time-of-flight mass spectrometer, **4**:2–3, 2f
defined, **4**:1, 12
for protein sequencing and analysis, **5**:192
use with MALDI-MS, **4**:16, 16f

Time-resolved fluorescence spectroscopy, **2:**300–303
anisotropy decay in, **2:**303, 303f
decay in biological molecules, **2:**302–3, 303f
Time-resolved spectroscopy, defined, **1:**108
Time-resolved X-ray diffraction studies, **6:**276
rate of data collection in, **6:**278
Timolol, synthesis of, **2:**163, 163f
Ti plasmids, **4:**451
defined, **4:**465
transformation vectors constructed from, **4:**490
Tissue culture, genomic changes induced by, **2:**221
Tissue distribution, of cystic fibrosis transmembrane conductance regulator expression, **1:**434
Tissue plasminogen activator, production from recombinant organisms, **1:**174
Tissue remodeling
by absence of a growth factor, **2:**464
enzymes involved in, **2:**290
defined, **2:**285
Tissue-specific locus control element, defined, **4:**495
Tissue-specific transcription, **2:**369
Tissue tropism
defined, **6:**193
of viruses, **6:**195
Titin, in muscle, **4:**143
Titration curve, showing denaturation of DNA, **2:**5
TLC (thin-layer chromatography), defined, **1:**267
T lymphocytes
in atherosclerotic lesions, **1:**287–88
$CD4^+$, antigen and superantigen recognition by, **5:**508
production of interleukins by, **3:**337t
$CD8^+$, production of interleukins by, **3:**337t
cytotoxic responses of, elicited by live recombinant vaccines, **6:**179
defined, **3:**296, 331–32
helper $CD4^+$, in HIV infection, **5:**511
immunological destruction of tumor cells by, **1:**261
peripheral, effect of activin on, **3:**510
production of interleukins, **3:**335
Tobacco, *TntI* retrotransposon of, **4:**488
Tobacco mosaic virus (TMV), **5:**359; **6:**57, 217–18, 217f
Tocopherols
α, and ischemic heart disease, **6:**238
free radical scavenger, **2:**333
Tolbutamide, metabolism by CYP2C9, **4:**402
Tolerance, defined, **2:**146
Toluene, biodegradation of, **1:**124
Toluidinylnaphthalene sulfonate (TNS), fluorescence probe, **2:**298
TOMATOES, GENE ALTERATIONS OF, **6:53–58**
Tomato mosaic virus (ToMV), **6:**57
Topliss decision tree, **4:**36
Topographical constraints, on peptide mimetics, **4:**378–79, 379f
Topoisomerase, **3:**392
anticancer drugs from, **1:**254
DNA, **2:**90
I
identification with scleroderma, **5:**414, 417
relaxation of supercoiling controlled by, **5:**1
II, chromatid, **1:**370–71
IV, DNA, **2:**87
Topology
control of, and structure of polymers, **5:**26
of the cystic fibrosis transmembrane conductance regulator, **1:**429–30, 430f
Total genomic DNA, preparation of, for cloning, **5:**504
Totipotency, defined, **4:**450
Touch cells, genes affecting development of, *Caenorhabditis elegans*, **1:**233–34
Touch response, in plants, role of calcium in, **1:**248–49
Tourist transposable elements, **4:**487
Toxication, defined, **6:**260
Toxicity
of acetaminophen, **1:**200, 202
of amyloid precursor protein, **1:**49–50
of antisense oligonucleotides, **1:**75
of interleukins, **3:**333
nonoverlapping, in combination chemotherapy, **1:**264
of nucleoside analogues, **1:**38
of oxygen, **1:**133
of radioisotopes, **5:**211
testing for, **1:**331
of vitamins, overdose, **6:**252
and xenobiotic metabolism, **1:**202
Toxic shock syndrome, **5:**511
binding of toxin in, **5:**509–10
mediated by TNF-α, **1:**449
toxin in, X-ray crystallographic studies, **6:**280, 280f
Toxic substances, degradation by fungi, **2:**345
Toxin genes, for studying *Drosophila* behavior, **2:**144
Toxins
bacterial, interaction with ADP-ribosylation factor protein, **5:**51
botulinum, **1:**282
cholera, **1:**282
constitutive activation of adenylate cyclase by, **5:**52
G-protein α subunit modification by, **5:**201
defined, **2:**165
diphtheria, ADP-ribosylation by, **5:**52
implication in sporadic motor neuron disease, **4:**108
to insects, incorporating in the tomato genome, **6:**56–57
lytic pore-forming, **1:**140
mimicking Huntington's disease neuropathology, **3:**251
oxidative stress imposed by, **2:**334–35
in plant diseases, **4:**482
of maize, **4:**489
plasmids coding for, **5:**1
protection against, **6:**165
secretion by bacterial pathogens, **1:**106
tetanus, **1:**282
TRACE ELEMENT MICRONUTRIENTS, **6:58–63**
defined, **6:**59
Track-end effect, of damage by low-energy electrons, **3:**355
Trade secrets, legally protected, **6:**81
Tramtrak, **6:**328f
zinc finger motif of, **6:**329
Trans-acting factor
cloning genes that encode, **4:**471
defined, **4:**465
effect on transcription of transgenes, **6:**102
of the *lin-14* gene, in *Caenorhabditis elegans*, **1:**235
overexpression of, in plants, **4:**475
plant, types of, **4:**467, 467t
types of, **4:**468–69
metal binding/sensing proteins, **6:**61, 61f
Trans-acting proteins, **1:**233
Transactivation, defined, **1:**34, 38
Transactivation functions (TAF), of hormone receptors, **2:**290
Transactivation response (TAR) element
bound to arginine, structure of, **5:**384, 384f
defined, **1:**19
effects on translation, **1:**25
HIV-1 LTR, **1:**20–21
HIV-1 RNA sequence, **1:**21
predicting the structure of, **5:**383–85, 383f
RNA-arginine complex, internucleotide distances, **4:**224

Transacylase, $cPLA_2$ as, **4:**423
Transcortin, gene for, mapping to chromosome 14, **3:**155
Transcriptase, reverse, crystalline, **1:**29
Transcription, **2:**417
- in archaea, **4:**74–75
- blocking of, by ultraviolet radiation-induced DNA damage, **2:**79
- defined, **6:**235
- eukaryotic, apparatus for, **2:**91–93
- factors mediating hormone action, **2:**212, 214
- of genes, repression by histone proteins, **1:**369
- in heterologous protein secretion, **2:**342
- histone, **3:**54f
- HIV, in the absence of Tat, **1:**21
- HIV-1, **6:**197–98
- inactivation by triplex-forming oligonucleotides, **1:**74
- in metazoa, **4:**80–81
- mutations affecting, **3:**37–38
- and nucleosomes, **1:**369; **2:**118
- for origin recognition and duplex opening, DNA replication, **2:**91
- process of, **3:**505–6; **5:**356f
- processive and nonprocessive, **1:**19
- rate of
  - in the *Chlamydomonas reinhardtii* chloroplast, **1:**350
  - for transgenes, **6:**71
- regulation by trace elements, **6:**61–63, 61f
- and regulation of gene expression, **1:**233–35
- RNA, in the immune response, **3:**279–80
- by RNA polymerase, involvement of GNER proteins in, **6:**288–90, 290f
- during spermatogenesis, **3:**505–6
- vaccinia virus, **6:**197
- in vitro, RNA probes generated by, **2:**382

Transcription activation domain (TAD), **2:**207–8
Transcription activator, **5:**130
Transcription activator proteins, acidic stretches of, **5:**463
Transcriptional control
- defined, **3:**49
- operon hypothesis of, **5:**275
- of peptide hormone production, **4:**342
- of plant genes, **4:**491–92
- regulator *oxyR*, **2:**220
- by trace elements, **6:**60–61, 61t

Transcription factor, **1:**257
- activating, **3:**54
- in archaea, **4:**75
- ATF/AP-1 family of, **3:**54
- D, for polymerase II (TFIID), defined, **2:**364
- defined, **2:**86; **4:**485–6, 492; **5:**128, 319; **6:**138, 235, 327
- DNA-binding, **5:**129
- eukaryote, for two-hybrid analyses in yeast, **6:**321–22, 322f
- in eukaryotes, **5:**128
- of the female reproductive system, **2:**290–91
- GAL4, yeast, **6:**333, 333f
- GATA-1, **6:**328f, 332, 332f
- GCN4, analyzing with circular dichroism, **1:**395
- genes for, mapping to chromosome 14, **3:**151
- in histone gene expression, **3:**53–54, 54f
- interaction of replication protein A with, **5:**464
- ligand-activated, nuclear hormone receptor superfamily of, **6:**331–33
- mutants involving mating behavior in *Drosophila*, **2:**137–42
- MyoD, **1:**234
- oncogenes as, **2:**215
- plant nuclear, **4:**467
- prd homeodomain genes, **3:**68
- second messengers in activation of, **2:**212
- TFIIS, **6:**334
- TFIIIA, **6:**327–28
  - and structure of the linker of a zinc finger motif, **6:**328
- UAS-binding, **2:**367, 367t

Transcription map, chromosome 16, **3:**186
Transcription rate assays, RNA, **5:**348, 350
Transcription-repair coupling factor (TRCF), **2:**45
Transcription signal, **2:**273–74, 273f
Transcription-translation systems, cell-free, **5:**336–37
Transcription unit, **2:**94f
- defining, **2:**365
- eukaryotic, **2:**94–95

Transcripts, identifying, human disease mapping, **3:**226
Transcytosis, protein targeting in, **5:**204–5
Transducer, defined, **1:**177
Transduction
- bacteriophage-mediated, **2:**402
- defined, **4:**285
- mechanical force, via integrins, **1:**456–57
- pathways for, effects of defects in, **2:**214
- photosynthetic energy, **4:**440–50
- of plasmids, **5:**5
- by steroid receptors, chaperone role in, **5:**91
- virus-mediated, **2:**405
- visual, **5:**324, 324

Transduction domain, calcium ion pump, **1:**244
Trans factors, and regulation of gene expression, **1:**233–34
Transfection
- cationic lipid systems for, comparison of efficiencies, **3:**422f
- defined, **3:**416; **4:**101; **6:**28
- delivering neutral and negatively charged liposomes, **3:**417–18
- dioctadecyl-amidoglycylspermine (DOGS) for, **3:**421
- for protein targeting, **5:**196
- studies of cytoskeleton-plasma membrane interactions, **1:**452
- techniques for, in vivo, **3:**423–25
- *See also* Vectors

Transfer, of genes, methodology, **6:**87–88
Transferases
- defined, **4:**399
- direct reversal of DNA damage by, **2:**44
- glycosyl, **2:**459
- methylases as, **5:**297
- oligosaccharyl, **2:**459
- in xenobiotic metabolism, **6:**267–68

Transfer protein, for lipids, role in membranes, **4:**51–52
Transferrin, receptors for, **5:**204; **6:**122
Transferrin fusion proteins, targeting to the brain, **1:**68
Transform, defined, **6:**53
Transformant
- defined, **5:**501
- selection of, by lethal zygosis, **5:**506

Transformation
- of cells, adhesion changes, **1:**458
- defined, **1:**253; **4:**285, 450; **5:**252; **6:**138
- DNA, **2:**402, 405
- efficiency of, in cloning hosts, **5:**503
- of fungi, commercial applications, **2:**345
- introducing DNA into protoplasts, **5:**505–6
- *Lycopersicon esculentum*, **6:**54
- of plasmids, **5:**5
- of yeast, **6:**310–11
  - using a plasmid vector, **5:**256–57f; **6:**303

Transformation scheme for higher plants, **4:**469f
Transformation vectors
- *Drosophila* gene, **2:**123
- Ti plasmid, **6:**55

Transforming growth factors (TGF)
- α
  - production by hepatocytes, **3:**481–82, 482t

production by lung cancer cell lines, **3:**495
relationships with epidermal growth factor, **5:**271
transcription of keratin genes induced by, **2:**266
β
family of, and the extracellular matrix, **2:**468
inhibition of hepatocytes by, **3:**482
receptors for, **5:**242
receptor superfamily, **2:**289
role in ascorbic acid transport, **6:**239
superfamily of, **3:**509–10
defined, **2:**285
embryonic development influenced by, **2:**287–88
Transgenes
in *Caenorhabditis elegans*, **1:**234
c*myc*, in rabbits, **6:**69
defined, **6:**84
gain-of-function, **6:**86–87, 86f, 98
identification of, in successful transfer, **6:**88–89
loss-of-function, **6:**87, 98
overexpression of, wtPrP, **5:**64
presentation to the immune system, **2:**431
reporter function, **6:**87
Sry, induction of testicular differentiation in the mouse, **5:**439
Transgenic, defined, **1:**172
TRANSGENIC ANIMAL MODELING, **6:63–74**
of Alzheimer's disease, **1:**50–52
in tumor suppressor gene studies, **6:**146
Transgenic Animal Patent Reform Act, **6:**77
TRANSGENIC ANIMAL PATENTS, **6:74–84**
Transgenic animals
defined, **3:**480; **6:**64
production of, **6:**66–69, 100–101, 101t
by electroporation, **2:**204–5
as production systems, **2:**277
renin gene in, **5:**265
TRANSGENIC FISH, **6:84–99**
defined, **6:**84
Transgenic founder, defined, **6:**64
Transgenic line
constructing, **6:**301–2
using yeast artificial chromosomes, **6:**293
defined, **6:**64
TRANSGENIC LIVESTOCK AS BIOREACTORS, **6:99–108**
Transgenic mice
for aging studies, **1:**12
for alcohol and drug abuse studies, **2:**151
apolipoproteins expressed in, **4:**503–5, 505f
defined, **2:**440; **3:**511; **4:**495; **5:**57
for Down syndrome studies, **2:**121
from embryonic stem cell transfer, **6:**65
for hepatocellular carcinoma studies, **3:**484
for mammalian genome studies, **3:**517–18
for poliovirus disease studies, **6:**223, 225
prion titers in brains after RML scrapie inoculation, **5:**65t
for producing human antibody molecules, **1:**64
proteins expressed in the milk of, **6:**104t
for *RB1* mutation studies, **5:**322
for scrapie studies, **5:**63f
for superoxide dismutase and motor neuron disease studies, **4:**109
Transgenic organisms
as bioreactors, **1:**175
defined, **1:**283; **6:**99
Drosophila, for studying heat shock protein gene and learning, **2:**137
for lipoproteins studies, **1:**286–87
Transgenic plants
analysis of materials from, **4:**454–55
defined, **4:**450, 466
for producing antibody molecules, **1:**64
as production systems, **2:**277
selection of, kanamycin resistance marker, **4:**455f
Transglutaminases, epidermal, **2:**268
Transistors
all-polymer, **1:**155f
fabrication using the atomic force microscopy, **1:**158
and nanowires, **1:**154–56
Transition metal ions
in redox catalysis, **2:**232
role in hydroxyl radical formation, **2:**332–33
Transition mutation, defined, **4:**84
Transitions
from point mutations, **4:**149, 149f
in protein folding, **5:**150–53
Transition state
analogue,for preparation of catalytic antibodies, **2:**239–40
defined, **2:**223
in enzymatic catalysis, effect of organic media, **2:**257
high energy, stabilization of, **1:**145
for protein folding, **5:**155–56
stabilization of, in enzymatic catalysis, **2:**227, 229–30
structure of, and a phosphonate-containing analogue, **1:**170, 170f
Transition-state analogues
and corresponding catalytic antibodies, **1:**170
defined, **1:**159
HIV-1 protease inhibitors, **1:**38, 39f
Transketolase, triazole pyrophosphate dependent, **1:**397–98
Translation
by archaea, **4:**75–76
cap-dependent
shutoff by host cell, **5:**20f
shutoff by poliovirus, **5:**25
cap-independent, by picornaviruses, **5:**17
defined, **2:**271, 364; **3:**366; **5:**328; **6:**109
in DNA structure, **2:**107
inactivation of, by antisense oligonucleotides, **1:**74
mRNA to protein, **2:**418; **5:**355; **6:**116f
control by translation factor modification, **6:**124
mutations affecting, in factor the IX gene, **3:**38–39, 39f
regulation of, trace element role in, **6:**61–63
regulation of gene expression at the level of, **2:**374
ribosome cycle in, **6:**114–21
in spermatogenesis, **3:**506–7
in vitro, for RNA analysis, **5:**351
Translational positioning, of histone octamers in nucleosomes, **1:**368
Translational regulation
in *Caenorhabditis elegans*, **1:**235
of peptide hormone production, **4:**342
Translational repressors, regulation of protein synthesis by, **2:**376–77
TRANSLATION OF RNA TO PROTEIN, **6:108–25**
Translation signal, **2:**273–74, 273f
Translocation, **2:**418
associated with B-cell lymphoma, **3:**209–10
in the B-cell lymphoma 2 (*BCL2*) gene, **3:**201
chromosome, defined, **6:**138
defined, **2:**62; **3:**191, 480; **5:**328; **6:**109
detection of, with FISH, **1:**382
and WCPs, **6:**255, 255f
in DNA replication, **6:**19
in follicular lymphomas, **1:**302
group, **6:**125, 126
identification with PCR, **4:**339
into the endoplasmic reticulum, **5:**197
ion-translocating ATPases, **6:**131–32
oncogene activation resulting from, **4:**286
in promyelocytic leukemias, **4:**176

Translocation (*cont.*)
reciprocal, in hepatitis B virus integration, **3:**207
in translation of RNA to protein, **6:**121
Translocation, chromosome
aberrant, **2:**128
balanced, 17 and 22 in neurofibromatosis 1, **4:**175, 175f
BCR/*abl* gene rearrangement in, **4:**328–29
characteristic of follicular lymphomas, **4:**329
crossing heterozygotes for, **2:**440–41
defined, **3:**160; **6:**138
in Down syndrome, **2:**119–21
flow dot-blot analysis of, **2:**390t
leading to malignancies, **3:**159
in neurofibromatosis 1, **4:**176
Philadelphia chromosome creation by, **2:**64–65, 64t
reciprocal, **2:**65; **3:**512
defined, **2:**386
and telomere length, **3:**242
unbalanced, of chromosome ends, **6:**35
Y-X, in XX males, **5:**436–38, 438f, 443–44
Y;22, **5:**437
Translocation, protein, **5:**335–36
defined, **1:**315
by hsp60, **1:**321f
by hsp70, **1:**317, 321f
into the mitochondrial matrix, **1:**316
model, **1:**321f
by sec61p proteins of the endoplasmic reticulum, **5:**198
Transmembrane electron transfer, conditions for, **2:**197
Transmembrane glycoproteins, defined, **6:**184
Transmembrane helices, defined, **2:**285
Transmembrane proteins, p75, neurotrophin receptor, **5:**240
Transmembrane receptor kinases, **5:**170
Transmembrane signaling
by G proteins, **3:**43–44, 44f
integrins' role in, **1:**452
by vasopressin, **2:**23f
Transmission, of plasmids, **5:**5–9
Transmission electron microscopy (TEM), **2:**182
of denatured DNA, **2:**8
Transmitters. *See* Neurotransmitters
Transplantation
of bone marrow, in Gaucher disease, **2:**351
of fetal brain tissue, in Alzheimer's disease, **1:**50
nuclear, defined, **2:**439
Transplastomic plants, defined, **4:**466
Transport
of α-tocopherol, **6:**238
biotinidase as a biotin carrier protein, **6:**241–42
cell, **1:**175
endogenous, of lipids, **3:**406–8, 406f
exogenous, of lipids, **3:**404–6, 406f
facilitation by permeases, **6:**128
of lipids, **1:**284; **3:**404
mass, in biosensors, **1:**178–80
mitochondrial, of calcium ion, **1:**247–48
of phospholipids, **4:**438
process of, **6:**126, 126t
of proteins, in vesicles, **5:**201–2, 201f
proton, in the photocycle of bacteriorhodopsin, **1:**110–12
reticular system, of calcium ion, **1:**245–47
reverse, of cholesterol, **1:**284; **3:**408–9
of secretory glycoproteins, intracellular, **2:**461
of secretory proteins, **3:**353
of steroid hormones, **5:**481
transmembrane, of calcium ion, **1:**243–45
tubular, **5:**272–74
of vitamin C, **6:**239–40, 252
of vitamins, **6:**236
Transporter
ATP-binding transmembrane, *w* gene encoding, **2:**134–35
defined, **6:**126
Transporters important in antigen presentation (TAP), **1:**446
TRANSPORT PROTEINS, **6:125–33**
Transposable elements
activation of, in response to trauma, **2:**220–21
in *Arabidopsis*, **1:**78
defined, **2:**216
in *Drosophila*, **2:**122, 123–24
inteins, **4:**159
mutations induced by, in *Drosophila*, **2:**124
in mycobacteria, **4:**159
in plants, **4:**486–87
tagging genes with, **1:**232
tourist, **4:**487
Transposase, **4:**487
Transposition
chromosomal, in *Drosophila*, **2:**128
defined, **2:**402
of L1H, **6:**134
of mobile genetic elements, **2:**405
replicative, **5:**7
Transposonlike human element 1 (THE1), **3:**246, **6:**137
TRANSPOSONS
of *Caenorhabditis elegans*, **4:**169
composite, **2:**405
conjugative, **5:**9
defined, **1:**76; **3:**240; **4:**156, 167, 486
of *Drosophila*, defined, **2:**131
IN THE HUMAN GENOME, **6:133–38**
for isolating plant regulatory genes, **4:**471
mutagenesis of, **4:**163
for identifying genes for toluene oxygenases, **1:**124
mutation by insertion of, **4:**150
mycobacterial, **4:**160t
recombination involving, modeling of, **2:**414, 414f
of R plasmids, **5:**9
Tn160, **4:**159
Transposon tagging
Arabidopsis gene, **1:**82
for cloning *Arabidopsis* genes, **1:**80; **4:**491
Trans-splicing
for forming mRNA, **2:**373
and maturation of chloroplast *psaA* mRNA of *Chlamydomonas reinhardtii*, **1:**356–57, 357f
Transthyretin (TTR), gene for, mapping to chromosome 18, **3:**209
Transverse tubules (T-tubules)
signaling through, for muscle contraction, **4:**144
system of striated muscles, **1:**246
Transversion mutation
defined, **4:**84
from oxidation of guanine, **4:**151
point, **4:**149, 149f
β-Trefoil structure, of interleukin-1β, **3:**334
Treponema pallidum, transmission of, **1:**103
Triacylglycerols
biosynthesis of, **3:**456
defined, **3:**427
Triazolam, **2:**160
manufacture of, **2:**161–62, 162f
Tricarboxylic acid cycle (TCA), **2:**233
essential function of CoB_{12} in, **1:**402
generation of ATP through, **1:**307n
role in the electron transport system, **1:**308–9
sensitivity to feedback, **1:**312
stereochemical components of, **1:**344
Trichloroethylene (TCE)
bacterial metabolism of, **1:**120f
biodegradation of, **1:**121
dehalogenation of, **1:**125
oxidation by toluene oxygenases, **1:**124
Trichothiodystrophy (TTD)
DNA excision repair disorder in, **6:**161
excision repair defects in, **2:**47
Tridecapeptide, synthetic, COSY spectrum, **4:**222f

Trigger, for membrane fusion, **4:**44–45
Triglycerides
  metabolism of, **3:**410; **4:**265–66, 495
  plasma levels of, and coronary heart disease, **3:**479
Trigonal carbon atom, distinguishing faces of, **1:**344
Trimannosyl core, of N-linked sugar chains of glycoproteins, **2:**451–53
Trinucleotide mutagenesis, **4:**95–96
Trinucleotide repeat, defined, **2:**313
Trinucleotide repeat conditions
  exhibiting expansion, **2:**328, 328t; **3:**242
  Machado-Joseph disease, **3:**158
  in myotonic dystrophy, **2:**427–28
Tripeptides, NMR analysis of, **4:**219–20, 219f
Triple helix
  Achilles' cleavage reaction, **1:**4
  collagen gene structure for, **2:**280
Triple quadrupole mass spectrometer, **4:**17–18
Triplet labeling, **3:**388
Triplex, defined, **5:**221
Trisomic lines, *Arabidopsis*, **1:**77
Trisomy
  defined, **2:**118
  16
    human, spontaneous abortion in, **3:**186
    in mice, **2:**121
  18, human, **3:**212–13
  18q$^-$ syndrome, **3:**213
  See also Down syndrome
tRNA
  abundance of, and protein synthesis, **6:**124
  aminoacyl-, **6:**120–21
  anticodon hairpin loops, **5:**370–71
  charged, defined, **5:**135
  cleavage from HIV RNA strand, **1:**31
  in engineered biosynthesis, **1:**136–37
  misacylated
    with a photoactivable amino acid, **5:**142f
    synthesis of, **5:**141f
  modification of aminoacylated, **5:**138–39
  phenylalanyl, structure of, **6:**113f
  predicting the structure of, **5:**381–82, 382f
  and protein synthesis, **5:**355–57
  structure of, **5:**354f
  suppressor, acylated, **5:**139
  transcription of, **2:**366
    in archaea, **4:**75
  translation by, **6:**112–14, 112f
  yeast, structure of, **5:**375f
  secondary, circular representation, **5:**376, 376f
tRNA ligases
  chiral specificity of, **1:**336
  reactions not involving the chiral center of substrates, **1:**342–43
Trophic factors, **1:**301
Trophoblast, development of, in monoparental diploid embryos, **2:**440
Trophoblast cells, interferon-τ production by, **3:**322
Trophoblast specific element binding protein (TSEB), **2:**291
Tropism, viral
  and cellular receptors, **6:**225
  defined, **6:**223
  receptors as determinants of, **6:**232–33
Tropocollagen, **5:**32–13
Tropoelastin, **2:**282, 282f
  natural, and synthetic analogues, **5:**28
Tropomyosin (TM)
  gene localization to chromosome 15, **3:**169
  inhibition of myosin ATPase activity by, **4:**144
  in thin myofilaments, **4:**130, 138–40, 140f
Troponin (TN)
  calcium ion binding sites of, **1:**241
  inhibition of myosin APTase activity by, **4:**144
  in thin myofilaments, **4:**130
Troponin-C (Tn-C), **4:**140
  calcium-binding sites of, **1:**240
  function in muscles, **1:**242
Troponin-I (Tn-I), inhibition of myosin ATPase by, **4:**140
Troponin-T (Tn-T), gene for
  alternative splicing by, **2:**373, 375
  and protein, **4:**140–42, 142f
Trypanosoma brucei genome project, **4:**311
Trypanosoma cruzi genome project, **4:**311
Trypanosomiasis, **4:**307
Trypsin, inhibition of, reversible and irreversible, **2:**236
Trypsin-modulating oostatic factor (TMOF), **4:**192
Tryptophan
  conformation-sensitive Raman bands, **5:**101t
  fluorescence of, **2:**294
  Raman bands, **5:**100–101
  regulation of production, in prokaryotes, **2:**371
Tryptophanase, quasi-irreversible unfolding-refolding of, **5:**79
Tryptophan dioxygenase, reaction not involving the chiral center of substrates, **1:**342
Tryptophan hydroxylase (TPH), **1:**211, 217–20
  amino acid sequence, **1:**209–10, 210f
  reactions involving, **1:**207–09
Tryptophan synthetase, refolding of, **5:**79, 79f
Tsc2, defined, **3:**174
TsOH (*p*-toluenesulfonic acid), defined, **5:**485
T state (tense state, cooperativity model)
  defined, **3:**20
  hemoglobin, **3:**21
    in animals, **3:**27
Tuberculosis, **4:**156–57
Tuberculostearic acid, **3:**431–32
Tuberous sclerosis (TSC), localization of
  to chromosome 9, **3:**107, 188
  to chromosome 16, **3:**188
Tubular transport systems, **5:**272–74
Tubulin-binding drugs
  activity of, around tumors, **1:**260
  anticancer, **1:**254–56
Tubulins, in *Chlamydomonas reinhardtii* flagellae, **1:**359
Tubuloglomerular feedback mechanism, role of nitric oxide in, **5:**269
Tumor-associated carbohydrate antigens (TACA), **1:**281f
  defined, **1:**278
Tumor cell, defined, **3:**317
Tumor growth factor-β, inflammation limitation by, **1:**447
Tumorigenesis
  in neurofibromatosis 1, **4:**179
  in neurofibromatosis 2, **4:**184f
  telomerase activity in, **6:**35
Tumor models, development of anticancer agents, **5:**490–98
Tumor necrosis factor (TNF), **1:**441
  defined, **3:**463
  fusion protein with antitransferrin receptor antibody, **1:**67
  α, mediator in endotoxin reactions, **3:**471
  nomenclature, **1:**444t
  α, production by macrophages, **1:**445
Tumor progression models, for neoplastic disease diagnosis, **2:**63
Tumor rejection antigen, gene for, chromosome 15, **3:**165
Tumors
  blood vessel formation in, role of fibroblast growth factors, **2:**469
  chimeric antibodies targeting, **1:**66
  defined, **1:**250
  DNA-liposome complexes for stimulating immunogenic responses again, **3:**424
  effects on
    of antisense oligonucleotides, **1:**75

Tumors (*cont.*)
of hepatocyte growth factor, **2**:469
of photodynamic photosensitizers, **6**:212
embryonal, genomic imprinting associated with, **2**:445
endocrine, peptide hormone precursor expression in, **4**:341
epidermal growth factor receptor involvement in, **2**:468
granulosa cell of the ovary, inhibin as a marker for, **2**:293
innervation of, **2**:466
resistance to therapy, **2**:215–16
retrovirus induction of, **4**:285
telomere dynamics and telomerase activity in, **6**:25t
TUMOR SUPPRESSOR GENES, **6:138–47**
as antioncogenes, **4**:287
APC, *p53*, *DCC*, **1**:411
association with chromosomal loss, **2**:66
and colon carcinomas, **1**:411
defined, **1**:407; **2**:62; **3**:227, 480, 491; **4**:173, 285; **6**:138
identifying with loss of heterozygosity regions, **3**:225–26
inactivation of, **2**:63
inhibin α, **2**:293
MCC, **6**:138
mutation in lung cancer, **3**:492t
NF1, **6**:138
NF2, **4**:183
p53, **5**:320; **6**:138
and liver cancer, **3**:488–89
and lung cancer, **3**:495–96
products of, and antiviral reactions, **3**:326
RB, defined, **6**:138
TP53, **3**:223–24
WT1, defined, **6**:138
Tumor viruses, information flow in, **5**:352
Tungsten, in enzymes of hyperthermophiles, **2**:244, 248
Tunicamycin, effect on viral budding, **6**:188
Tunneling
defined, **1**:148; **5**:401
in electron transfer, **2**:194
Tuple, **5**:424–25
defined, **5**:420
k-, method for determining protein coding, **5**:431
Turbulence, and swarming, **4**:94
Turcot syndrome, **1**:408–9, 410
Turner syndrome, **5**:445
Turnover, DNA, **6**:6–7
defined, **2**:391
process of, and copy number change, **6**:3–5
Twisted helix, **2**:106, 106f
Twist *Tw*
in DNA structure, **2**:107–9, 115, 116–17
defined, **2**:103
Two-center hydrogen bond, **3**:262t, 265
defined, **3**:261
geometry of, **3**:268
Two-dimensional gel electrophoresis, defined, **5**:173
Two-fold symmetry
C_2, peptidic inhibitors, HIV protease, **1**:15–16
HIV protease, **1**:28
myosin light chain kinase, **1**:241
rotational, HIV-1 protease, **1**:38, 40
Two-hit model, in retinoblastoma, **6**:139
Two-hybrid systems
defined, **6**:303
for detecting protein-protein interactions, **6**:303
for yeast genome analyses, **6**:321–22, 322f
Type II lipids, defined, **4**:42
Type II restriction-modification systems, defined, **5**:296
Tyrocidine, structure of, **4**:210f
Tyrocidine synthetase, **4**:209
Tyrosinase, streptomycetes transformant containing, **5**:502
Tyrosine, phosphorylation in normal cells, **1**:459
Tyrosine hydroxylase (TH), **1**:207, 209–17
amino acid sequence, **1**:209–10, 210f
reactions involving, **1**:207–09
Tyrosine kinases (TKs)
association of hsp90 with, **1**:321
cytoplasmic, gene localization to chromosome 15, **3**:170
epidermal growth factor receptor, **2**:464
and interferon receptor signaling, **3**:322–23
intracellular, **2**:465
leukocyte, mapping to chromosome 15, **3**:167
ligand-stimulated, **5**:241
phosphorylation by, **5**:55
platelet-derived growth factor receptor, **5**:270
pp60SV-*src*, **1**:458
protein, defined, **5**:230
receptor, **2**:208, 208f, **4**:30
in retrovirally induced neoplasia, **3**:494
role in cytoskeleton-membrane interaction, **1**:454
stimulation by interleukin receptor binding, **1**:445; **3**:339
3 (TYRO3), gene for, on chromosome 15, **3**:165
Tyrosylprotein sulfotransferase, **4**:343
Ubiquinones, **1**:405–6
controlled electron transfer in vivo, **1**:309
Ubiquitin
circular dichroism analysis of, **1**:390
dependence of acetylated protein degradation on, **3**:352
plant, **4**:494
reaction with proteins, **3**:349–50, 349f
Ubiquitin protein ligase (E3), selection of proteins for multiubiquitination, **3**:350
UDP-glucuronosyltransferase, deficiency of, **4**:405
Ultracentrifugation, for analysis of high density lipoprotein subfractions, **3**:477
Ultrathin gel electrophoresis (HUGE), horizontal, **2**:96
Ultraviolet absorption spectrum, bovine serum albumin, **5**:122f
Ultraviolet-induced damage, in progerias, **1**:8
Ultraviolet radiation (UVR)
biological responses to exposure to, **6**:150f
cross-linking by
as a probe for determining RNA structure, **5**:365
"zero length", **4**:250
defined, **6**:149
dimerization caused by, **3**:356
footprinting with, **4**:249
mechanisms for tolerance to exposure, **6**:156–58
site-specific cleavage of dyneins and myosins by, **4**:118
ULTRAVIOLET RADIATION DAMAGE TO DNA, **2**:40, 41; **6:149–64**
formation of cyclobutane pyrimidine dimers, **6**:284
genes affecting cell sensitivity to, *ruvA* and *ruvB*, **5**:228
Ultraviolet resonance Raman (UVRR) spectroscopy, **5**:94
amide II band, **5**:97
spectra of virion fd, **5**:108–09, 109f
Uncoating
of poliovirus, in a host endolysosome, **5**:17
of vaccinia virus, **6**:196
of viruses, **6**:225
virus genome, defined, **6**:223
Uncoupling protein (UCP), of adipocytes, **1**:313
Underwinding, of long DNA, **2**:116, 118
Unequal crossing over
defined, **3**:377
and molecular drive, **2**:394

Unequal exchange
and DNA turnover, **6:**6–7
between homologous chromosomes of sister chromatids, **6:**6
and replication slippage, **6:**4, 4f
Unfolded state (U), proteins, **5:**148
Unfolding, of proteins, **5:**154–58, 154f
Unified theory, of aging and sexual reproduction, **2:**78
Unilamellar vesicles, defined, **4:**416
Uniparental disomy, **2:**440–41
defined, **2:**420, 440
effect on genetic analysis, **2:**424
Uniport, **6:**127
defined, **6:**126
Unit cell
defined, **3:**442
volume of, **5:**112
Units
of DNA transcription, **2:**93–95, 95f
of energy, **4:**258
of genetic distance, **3:**217
of molecular weight, **4:**129
of radioactivity, **5:**211, 211t
of RNA transcription, **2:**365
Unity of cell biology, **1:**102
Universal ancestor, defined, **2:**240
Universal epitope, **5:**415–17
Universal primers, defined, **1:**380
Unpaired electrons, free radical, **2:**330
Unsaturated fatty acids, defined, **3:**427
Upstream activating sequences (UAS), **2:**366–68, 366f
in eukaryotes, **2:**95
Upstream stimulatory factor (USF), **1:**20f
Uracil, biosynthesis of, in yeast, **6:**313
Urea, enzyme sensors for measuring, **1:**186
Urease, nickel in, **4:**59
Urechis caupo, hemoglobin of, **3:**13
Uric acid, and oxidative damage, **2:**336
Uridine-rich small nuclear ribonucleoprotein particles (U snRNPs), defined, **2:**364
Urogastrone (epidermal growth factor), **2:**466
Urokinase, secretion in milk of a transgenic mouse, **6:**103
Usher syndrome, gene for, **4:**126
mapping to chromosome 1, **3:**93, 95
mapping to chromosome 11, **3:**127
type 1, mapping to chromosomes 1, 11 and 14, **3:**158
U snRNP, defined, **2:**364
Utah Genome Center Genetic Marker and Mapping Group, genetic map for chromosome 14, **3:**144
Uterus, remodeling process under local control, **2:**290, 290t
Utility, as a requirement for patentability, **6:**76
U-turn, RNA structural motif, **5:**370, 377
Uvomorulin, **1:**239, 295

V_{max}
defined, **2:**223
for enzymatic catalysts, **2:**233–34, 234f
Vaccenic acid, synthesis by bacteria, **3:**431
Vaccination
with polysaccharide antigen, **1:**279
against respiratory syncytial virus, trials, **5:**293–94
VACCINE BIOTECHNOLOGY, **6:165–71**
vaccina virus-encoded proteins for immunization, **6:**200
Vaccines
inactivated, **6:**165
live attenuated, against respiratory syncytial virus, **5:**295–96
monreplicating subunit, against respiratory syncytial virus, **5:**294–95
subunit
defined, **6:**172
development of, **6:**176–79, 176f
protein, for varicella-zoster virus and cytomegalovirus, **6:**177t
Vaccinia growth factor (VGF), **6:**199
Vaccinia virus
defenses of, against antiviral reactions, **3:**325
defined, **6:**165
as a recombinant vector, **6:**166, 179, 180–81, 180f
replication of, **6:**198
smallpox vaccination with, **6:**199
uncoating of, **6:**196
for vaccine vector system use, **2:**275, 277, 437
Validation, of protein models, **5:**164
Validity, of molecular biological markers, **2:**260
Valine, at codon 717, APP gene, **1:**44
Vanadium
in metalloenzymes, **4:**65
in nitrogenases, of methanogens, **4:**77–78
Vanadyl ribonucleoside complexes, RNA purification using, **5:**346
van der Waals forces
defined, **2:**307
quaternary protein structure attributed to, **2:**308
and stability of folded proteins, **5:**151
Van der Woude syndrome (VWS), gene for, mapping to chromosome 1, **3:**95
van't Hoff plot, equilibrium between native and unfolded proteins, **5:**156f
Variable domains
antibody-binding site, **1:**63f
of immunoglobulin light and heavy chains, **3:**278
Variable expressivity
defined, **4:**173
of neurofibromatosis 1, **4:**174–75
Variable number of tandem repeats (VNTR), **2:**67–68, 67f, **3:**224
analysis using, **2:**51
autoradiogram, paternity analysis, **2:**60f
characteristics of forensic DNA typing probes, **2:**59t
defined, **2:**48, 420; **3:**142
for genetic analysis of populations, **2:**397
for genetic marker assay, **3:**99
for genetic testing, **2:**421
hybridization solution for probing, DNA fingerprinting, **2:**57
IDDM-2, insulin gene, **2:**37
of the phenylketonuria gene, **4:**408
for polymerase chain reaction analysis, **4:**337
Variation generator, defined, **2:**402
Varicella-zoster virus (VZV)
DNA genome of, **6:**174f
recombinant Fabs against, **1:**421
VARICELLA-ZOSTER VIRUS AND CYTOMEGALOVIRUS VACCINES, **6:172–83**
Variegation, position effect, **2:**125
Variola virus, **6:**193
Vascular endothelial growth factor (VEGF), **2:**469
Vasoactive intestinal peptide (VIP), **5:**238
Vasoactive systems, **5:**264–69
Vasopressin receptors, **2:**21–23
defined, **2:**19
regulation of water permeability by, **5:**274
V_2, structure of, **2:**22f
Vasopressins, **2:**20–21
defined, **2:**19
effect on calcium ion in cells, **1:**240
regulation of urine production by, **5:**273
structure of, **2:**20–21, 20f
Vaterite crystals, formation from supersaturated calcium carbonate, **3:**312
VEC-DIC microscopy, defined, **4:**113
Vector DNA, dephosphorylation of, for cloning experiments, **5:**505
Vectors
adenoviral, for transfection, **4:**101
baculovirus expression, **6:**170–71
binary, **6:**55, 55f
centromere plasmid vectors,YCp, **6:**314
circular, yeast artificial chromosome, **6:**322–23
for cloning, **4:**161, **6:**314

Vectors (*cont.*)
defined, **6:**165
delivery of genes by, **4:**163
in phenylketonuria, **4:**413
lambda phage, **2:**67
lipid-based, **3:**417–23
pComb3, **1:**417
for recombinant vaccines, **6:**166
replacement, in homologous recombination, **4:**102, 102f
retroviral, for transfection, **3:**494; **4:**101; **6:**88
shuttle vector systems, **1:**9; **6:**311–12
for studying age-dependent mutations, **1:**12
transformation
Drosophila gene, **2:**123
Ti plasmid, **6:**55
yeast, **6:**311–12, 311f
yeast artificial chromosome, **6:**291–93
See also Plasmid vectors
Velocity, maximum, of an enzymatic reaction, **2:**223, 233–34
Venous malformation (VMCM), mapping to chromosome 9, **3:**107–8, 111
Verapamil
binding to L-type calcium channels, **1:**243
effect on viral budding, **6:**188
stereoselectivity of, **4:**39–40, 40t
Veratridine
activation of tyrosine hydroxylase by, **1:**215
and tyrosine hydroxylase regulation, **1:**217
Vertebrates, substitution rate for mtDNA in, **4:**82
Verticillium lecanii, source of mycoinsecticides, **2:**343
Very large scale immobilized polymer synthesis (VLSIPS), **5:**519
Very low density lipoprotein-low density lipoprotein pathway, **4:**499f
Very low density lipoproteins (VLDL), **3:**472
apoC-III of, expression in transgenic mice, **4:**504–5
apolipoproteins associated with, **4:**496
assembly of, **3:**406
and atherosclerosis, **1:**284
β, clearance impairment and atherosclerosis, **4:**505
defined, **3:**404
Vesicle-associated membrane protein (VAMP), **4:**346, 346f
Vesicle-membrane fusion, intracellular, **4:**48–50
Vesicles
aggregation of, induced by specific recognition, **1:**154, 154f
phospholipid, for synthesis of inorganic solids, **3:**314f
phospholipid transport by, **4:**438
preparation of synthetic unilamellar, **3:**314
Vesicular stomatitis virus (VSV)
assembly and budding of, **6:**186f
G protein of, in an envelope with transgenes, **6:**88
inactivation by light exposure, **6:**209–10
matrix protein of, **6:**188
photosensitization of, with phthalocyanines, **6:**213
producing a pantropic defective retroviral vector, **6:**89f
vaccinia virus recombinant vaccine for, **6:**169
Veterinary applications, DNA fingerprint analysis, **2:**61
V gene, **3:**281
Vibrational analysis, Raman, of proteins, **5:**97–101
Vibrational normal modes
defined, **5:**94
group, and Raman frequencies, **5:**95–96
for *trans-N*-methylacetamide, **5:**98f
trans peptide group, **5:**99t
Vibrio cholerae
binding of, to GM_1, **1:**282
transmission of, **1:**103
Vibrio fischeri, regulation of the virulence genes of, **1:**107–8
Videomicroscope method
for chromosome microdissection, **1:**374–75, 375f
collection of chromosome fragments, **1:**376–77
Vimentin, **4:**143
attachment sites for, cytoskeleton, **1:**458
Vinca alkaloids, effect on tubules, **1:**254–55
Vincristine, structure of, **1:**255f
Vinculin
binding domain for, **1:**453
focal adhesion protein, **1:**453
phosphorylation of, **1:**458
Vinestramide, antitumor activity of, **5:**497
Viral attachment protein, **6:**195
defined, **6:**193
Viral capsid, difference Raman spectra of, **5:**106f
VIRAL ENVELOPE ASSEMBLY AND BUDDING, **6:183–92**
Viral integration, defined, **3:**480
Viral oncogenes (v-*onc*), **4:**285
Viral oncoproteins, defined, **5:**319
Viral vectors, defined, **2:**430
Virion, **6:**193–95, 195f
assembly and release of, **6:**198
defined, **6:**193, 207
entry into a host cell, **6:**224, 224f
HIV-1, **6:**196
assembly of, **6:**198
Virion attachment protein (VAP), **6:**224, 224f
defined, **6:**223
Virion attachment site (VAS), **6:**224
defined, **6:**223
structure, and accessibility to antibody molecules, **6:**225
VIROLOGY, MOLECULAR, **6:192–200**
Virosomes, **3:**418
Virulence
bacterial, **1:**107t
defined, **1:**103
Virulence genes, mycobacterial, **4:**160
Viruses
animal
DNA packaging by, **6:**206
types of, **6:**193–95, 193t
cancer-causing, **1:**250
cellular compartment for assembly and budding of, **6:**187f
chemical inactivation of, for vaccines, **6:**176
classification based on nucleic acid in the genome, **5:**357t
defenses against biological antiviral reactions, **3:**325, 326t
defined, **2:**271; **6:**217
degenerate, of fungi, **6:**220
diseases caused by, genetic immunization against, **2:**437
DNA of, **5:**357–59
enveloped
fusion with a target membrane, **4:**45
response to photosensitizers, **6:**210
fd, Raman spectroscopic analysis of structure, **5:**105–9
fibrous proteins of, **5:**28–29
infection by, cytokine role in immune response to, **1:**447
insect, expression systems for the production of proteins, **2:**277
mRNA encoding in development of, **5:**337
oncogenic, **2:**215
paramyxoviruses, **6:**167
release from cells, **6:**184
RNA, **5:**357–59
double-stranded, of yeasts, **6:**305
icosahedral positive-strand, **6:**218
respiratory syncytial virus, **5:**286
tumor, information flow in, **5:**352
virus-membrane fusion, **4:**52–53
in vitro activity of antisense or triplex-forming oligonucleotides, **1:**74t
yellow fever, assembly and budding of, **6:**190
See also Influenza virus

VIRUSES, DNA PACKAGING OF, **6:200–207**
VIRUSES, PHOTODYNAMIC INACTIVATION OF, **6:207–16**
VIRUSES, RNA PACKAGING OF, **6:217–23**
Virus host-cell interactions, **6:**198–99
VIRUS-RECEPTOR INTERACTIONS, **6:223–34**
Virus-responsive elements, of interferon-α, **3:**321
Viscometry, for following DNA denaturation `, **2:**6t
Visual cycle
  defined, **5:**322, 325
  role of vitamin A in, **6:**246
Visual system
  ganglia of, defined, **2:**131
  mutation of, in *Drosophila*, **2:**137–38
  recoverin in, **1:**242
Visual transduction, defined, **5:**322
Visual transduction pathway, **3:**46, 47f
Vitamer
  ascorbic acid, **6:**239
  defined, **6:**235
Vitamin D-dependent calcium-binding proteins, **1:**242
VITAMIN RECEPTORS, **6:235–44**
  vitamin D, mutations leading to rickets, **1:**249
VITAMINS
  A
    receptors for, **6:**236–37
    regulation of, **1:**437
    and spermatogonia proliferation, **3:**506
    structure and function of, **6:**245–46
    in the visual cycle, **5:**324
  B_6, **1:**401–02; **6:**243–44
    structure and function of, **6:**249, 249f
  B_{12}
    cobalt in, **4:**60
    structure and function of, **6:**250–51
  B complex, **2:**233
    coenzymatic forms and functions, **1:**397, 398t
  C
    scurvy from deficiency of, historic record, **6:**244
    structure and function of, **6:**252, 252f
  D
    receptors for, **6:**237
    role in calcium homeostasis, **1:**238, 238f
    structure and function of, **6:**246–47, 246f
  defined, **1:**397; **4:**257; **6:**235
  E, **6:**238
    and coronary heart disease, **3:**479
    effect on life span, **2:**80
    structure and function of, **6:**247, 247f
  fat-soluble, **6:**245–48, 245t
  functions and requirements for, **4:**267–68t
  K, structure and function of, **6:**247–48, 247f
  as micronutrients, **6:**59
  nutritional requirements, **4:**266
  STRUCTURE AND FUNCTION OF, **6:244–52**
  water-soluble, **6:**248–52, 248t
Vitellogenin, production of, regulation in the mosquito, **4:**192, 193f
Vitronectin, as an adherence molecule, **1:**453
VNTR. *See* Variable number of tandem repeats
Voltage-gated ion channels, **4:**30; **5:**514; **6:**127, 127t, 129
  calcium ion
    antagonists active at, **4:**36, 36f
    in exocytosis, **4:**44–45
    protein of, **1:**312
von Hippel-Lindau (VHL) syndrome, **2:**429; **6:**144–45
von Recklinghousen neurofibromatosis, **2:**429; **6:**138, 144
von Willebrand factor, **1:**453
  binding of factor VIII to, **3:**30
  release by desmopressin, **2:**22
von Willebrand's disease, in Finland, founder effect, **5:**47
VP-16 transcriptional initiator, **1:**25

Waardenburg syndrome, **3:**223, 226
  Pax-3 defect in, **3:**68
WAGR syndrome (Wilms' tumor, aniridia, genitourinary abnormalities, mental retardation), **6:**142–143
  mapping to chromosome 11, **3:**126–27
Walk-along theory of contraction, **4:**144–45, 144f
Warfarin, **6:**248
  metabolism by CYP2C9, **4:**402
Waste
  organic, examples, **1:**119t
  from recombinant protein production, **2:**278
Water
  bound, bacteriorhodopsin, **1:**111, 112
  desolvation energy in electrostatic interactions, **1:**162–63
  as an electron donor, in energy production by eukaryotic autotrophs, **1:**310
  "essential", of enzymes suspended in organic solvents, **2:**252, 257
  in foods, **4:**261–62, 262f
  lipid interaction with, **3:**449
  oxidation by photosystem II, **4:**445–47
Water channel in the collecting duct (WCH-CD), **5:**274
Water channels
  and activation of the vasopressin V_2 receptor, **2:**21
  kidney, **5:**273–74
Water molecule
  flap region, HIV protease, **1:**28
  hydrogen bonding in, **3:**261, 261f, 271–72, 271f
Water signal, suppression of, NMR, **4:**221, 223
Water-soluble vitamins, **6:**235, 248–52, 248t
Watson-Crick base pairs, **1:**70–71, 71f; **5:**353–54
  relative stabilities of, **2:**18t
  reverse, G-C, **5:**377, 377f
  RNA secondary structure, **5:**374–76, 376f
Watson-Crick-Chargaff base pairs, **6:**39
Watson-Crick hydrogen bonds, in duplex DNA, **2:**1
Waveforms, for electroporation, **2:**202, 204
Waxes, **3:**455–56
Wax esters, **3:**452
  synthesis by bacteria, **3:**433–34
Weak-field approximation, **2:**175–76
Weeds, control of, using fungi, **2:**344
Weight array, **5:**431
  defined, **5:**428
Weight matrices
  defined, **5:**428
  search for signal sequences using, **5:**431
Wellcome Trust, malaria genome project of, **4:**308
Werdnig-Hoffmann disease, **4:**106
Werner's syndrome, **1:**8
  chromosome breakage in, **1:**413
  mutation in, **1:**302
  premature aging in, **2:**80; **6:**34
Western blotting
  defined, **3:**472
  for protein identification, **5:**124–25, 125f
Western electroblotting, defined, **2:**353
Western immunoblotting, **2:**357, 358f
Wheat germ, extracts for protein synthesis experiments, **5:**334–35
Whey acidic protein (WAP), **6:**103
  expression in transgenic mammary cell lines, **6:**105
WHOLE-CHROMOSOME COMPLEMENTARY PROBE FLUORESCENCE STAINING, **6:253–57**
Whole-chromosome probe (WCP)
  defined, **6:**253
  generation of, **6:**254–55, 254t
Whole genome molecular analysis, **3:**226

Wilms' tumor, **2:**429
defined, **6:**138
gene for, **5:**320; **6:**142–44, 144f
mapping to chromosome 11, **3:**126
imprinting of the gene for, **3:**127
Wishart neurofibromatosis, **4:**181
Withdrawal, defined, **2:**146
Wolf-Hirschhorn syndrome, unbalanced translocation of chromosome ends in, **6:**35
Words, classification code, determining, **6:**39
Work
cellular, energy for, **1:**304–5
defined, **1:**304
World Health Organization (WHO), genome projects, **4:**309–10
World Intellectual Property Organization (WIPO), **6:**75
Wound dressing
polyurethane, **5:**39
poly(2-hydroxyethyl methacrylate), **5:**42
Wound healing
cytokines in, **1:**448
fibronectins in, **2:**283
Writhe *Wr*, **2:**117
in the fifth-level structure of DNA, **2:**104

Xanthine oxidase
hydrogen peroxide production from the actions of, **2:**331–32
molybdenum in, **4:**65
Xanthomas, in apoE deficiency, **3:**414
X antigen, of hepatitis B virus, **3:**489
Xenobiotic interaction, defined, **3:**296
XENOBIOTIC METABOLISM, **6:259–75**
enzymes of, **4:**399–400
P450-dependent, **1:**437
Xenobiotics
as cofactors, for peroxidases, **6:**266–67
defined, **1:**193; **5:**343; **6:**260
exposure to, molecular epidemiology of, **2:**261
Xenogeneic extrapolation, of immunotoxicity evaluations, **3:**300
Xenon arc lamp, **6:**211
Xenopus laevis
defined, **6:**327
ribosomal DNA of, electron microscopy, **2:**183f
Xeroderma pigmentosum (XP), **1:**8
defined, **6:**149
excision repair defects in, **2:**46, 47; **3:**164; **6:**161, 162
neuropeptide, **5:**464–65
Xeroderma pigmentosum complementation group F (XPF), chromosome 15 involvement in, **3:**164
X gene, of hepatitis B virus, **3:**483
X inactivation, **3:**517
defined, **3:**511
X-linked characteristics, defined, **3:**28
X-linked conditions
defined, **2:**420
Duchenne muscular dystrophy, **2:**426
recessive, **2:**426–27
X-ray crystallography
antibody-binding site structure determined by, **3:**283
defined, **1:**27
of HIV protease, **1:**15–16
nucleic acid-protein interactions studied with, **4:**252–53
protein structure determination by, **5:**113f, 126, 159
structures determining antigenicity, **3:**285
X-RAY DIFFRACTION
by a crystal, **5:**112
studies of hydrogen bonding, **3:**267
defined, **3:**442; **5:**110
OF BIOMOLECULES, **6:275–82**
structures from
of annexin V, **1:**60, 60f
of DNA, **2:**103–4
of DPPC phase transitions, **3:**448, 448f
of poliovirus, **5:**16
three dimensional structure of unliganded HIV-1 reverse transcriptase from, **1:**38
X-ray scattering
by electrons, **5:**111, 111f
labels for, **3:**389
XX males, **5:**443
XY females, **5:**445–46
XY gonadal dysgenesis, **5:**441, 441f, 445
XY hermaphrodites, **5:**446
XY reversal, Xp duplication in, **5:**442

Yamaguchi sarcoma virus proto-oncogene 1 (*YES1*), **3:**209
YEAST
DNA REPAIR IN, **6:283–91**
eukaryotic model, **6:**283
host cell, **2:**275, 276
lipids of, **3:**434–37, 437t
mutant, for protein targeting studies, **5:**196
strains of, **6:**304
$tRNA^{Phe}$, structure of, **5:**375f, 376f
viruses of, L-A dsRNA virus, **6:**220
YEAST ARTIFICIAL CHROMOSOME (YAC), **4:**127
clones for gene mapping, **1:**83; **2:**73, 320, 321f; **3:**122, 515
mouse, **4:**125f
Plasmodium falciparum, **4:**308
CONSTRUCTION AND APPLICATIONS, **6:291–302**
contig map, chromosome 16, **3:**183
defined, **1:**370; **2:**386; **3:**142, 174, 191, 216; **4:**306; **6:**291
for eukaryotic genome characterization, **6:**322–24, 324f
fragments cloned into, **3:**220
framework map, chromosome 18, **3:**197–98
genomic library construction with, for *Caenorhabditis elegans*, **1:**230
with human amyloid precursor protein sequences, **1:**51
human chromosome-18-specific, **3:**199f
libraries
for *Arabidopsis* DNA, **4:**491
for chromosome Y and chromosome 21, **3:**149
for physical mapping of *Drosophila*, **2:**128
probe made from, **2:**389
YEAST GENETICS, **6:302–25**
Yeast Genetic Stock Culture Center, addresses of, **6:**304
Yellow fever virus, assembly and budding of, **6:**190
Yersinia pestis, transmission of, **1:**103
Yersinia pseudotuberculosis, integrin receptor use by, **1:**106
Yield, from *Taxus* cultures, **6:**15–17, 15f
Yops proteins, **1:**106
Yule distribution/equation
fit to ranked hexanucleotide frequencies, **6:**43t
for hexanucleotide frequencies, **6:**44f
ranked oligonucleotide frequencies, **6:**41

Z disk (muscle), **4:**130, 132
85K protein of, **4:**142
Zeamatin, **4:**480
Zea mays, transposable elements of, **4:**486–87
Zebrafish, small-sized fish model, **6:**85
Zellweger syndrome, **1:**116; **5:**206
peroxisome abnormalities in, **1:**115
ZF class, homeobox genes, **3:**70, 72
Zidovudine, **1:**37
Ziegler-Natta catalyst systems, for isotactic polyproplyene formation, **5:**26
Zinc
in enzymes, **4:**64–65, 65f
MMPs, **2:**290
in flavokinase, **1:**399
in hippocampal mossy fibers, relationship with seizures, **2:**149
in pyridoxal kinase, **1:**401
sites in phospholipase C, **4:**429f

stability of metal-ligand complexes, **1:**131, 131t
in superoxide dismutase, **2:**334
Zinc-binding motifs
COP1 proteins, **4:**494
HIV integrase protein, **1:**33
retrovirus, **6:**334
Zinc finger
of basonuclin, **2:**266
genes containing, mapping to chromosome 14, **3:**151
in receptors, defined, **5:**478
in steroid receptors, **2:**290; **5:**471
trans-acting factors, **4:** 467, 467t
ZINC FINGER DNA-BINDING MOTIFS, **6:327–36**
Zinc finger domains, **5:**128
LIM motif in, **3:**72
of the Wilms' tumor gene, **6:**143, 143f
Zinc finger motif
classical, **6:**327–31, 331f
defined, **6:**59
DNA-binding, nucleocapsid cysteine-rich domain, **6:**334, 334f
regulatory functions of, **6:**60
of replication protein A, **5:**463
Zinc finger protein 24 (ZNF24), gene for, mapping to chromosome 18, **3:**209
Zinc finger proteins
first class of, **5:**131–32, 132f
TFIIIA, **5:**135
gene encoding *ZNF127*, mouse, **2:**443
Kruppel family of, **3:**151
second class of, **5:**132–33
third class of, **5:**133
Zinc-finger Y gene (*ZFY*), **3:**506; **5:**436–38
Zipf distribution, **6:**41
Zona pellucida (ZP), glycoproteins of, **2:**455–456
Zona pellucida proteins, **2:**292
defined, **2:**285
Zonular pulverent cataract (CAEI), gene for, mapping to chromosome 1, **3:**95
Zoo blot approach, **3:**226
Z scheme of photophosphorylation, **1:**311f
Zwitterionic molecule, defined, **1:**52
Zygospore, *Chlamydomonas reinhardtii*, **1:**348
Zymolyase, for dissolving cell walls, **6:**310
Zyxin, binding domain for, **1:**454
comparisons of, and mtDNA origin, **4:**83–84
defined, **5:**372
preparation from ribosomes, **5:**332
transcription of, **2:**91, 365–66